# Foodborne Pathogens
## Microbiology and Molecular Biology

# Foodborne Pathogens
## Microbiology and
## Molecular Biology

Edited by Pina M. Fratamico, Arun K. Bhunia, and James L. Smith

Caister Academic Press

Copyright © 2005

Caister Academic Press
32 Hewitts Lane
Wymondham
Norfolk NR18 0JA
UK

www.caister.com

British Library Cataloguing-in-Publication Data
A catalogue record for this book is available from the British Library
ISBN: 1-904455-00-X

Printed and bound in Great Britain

# Contents

# Contributors

**George P. Anderson**
Naval Research Laboratory
Center for Bio/Molecular Science and Engineering
Code 6900
Washington, DC
USA

Email: ganderson@cbmse.nrl.navy.mil

**Helene L. Andrews**
Department of Medical Microbiology and Immunology
Texas A & M University
407 Reynolds Medical Building
College Station, TX
USA

**Andreas J. Bäumler**
Department of Medical Microbiology and Immunology
University of California at Davis
1 Shields Avenue
3146 Tupper Hall
Davis, CA
USA

Email: ajbaumler@ucdavis.edu

**Darrell O. Bayles**
United States Department of Agriculture
Agricultural Research Service
Eastern Regional Research Center
600 East Mermaid Lane
Wyndmoor, PA
USA

**Arun K. Bhunia**
Department of Food Science
Purdue University
745 Agriculture Mall Drive
West Lafayette, IN
USA

Email: bhunia@purdue.edu

**Clifford Clark**
National Laboratory for Enteric Pathogens
National Microbiology Laboratory
Health Canada
Canadian Science Centre
1015 Arlington Street
Winnipeg, Manitoba
Canada

**Nathalie Corneau MSc**
Bureau of Microbial Hazards
Food Directorate, Health Canada
Address Locator 2204A2
Sir Frederick G. Banting Building
Ottawa, Ontario
Canada

**Maribeth A. Cousin**
Purdue University
1160 Food Science Building
West Lafayette, IN
USA

Email: cousin@purdue.edu

**Angelo DePaola**
Food and Drug Administration
Gulf Coast Seafood Laboratory
PO Box 158
Dauphin Island, AL
USA

**Jeffrey M. Farber**
Health Canada
Tunney's Pasture, Banting Research Center
Postal Locator 2203G3
Ottawa, Ontario
Canada

**Pina M. Fratamico**
United States Department of Agriculture
Agricultural Research Service
Eastern Regional Research Center
600 East Mermaid Lane
Wyndmoor, PA
USA

Email: pfratamico@errc.ars.usda.gov

**Per Einar Granum**
Department of Food Safety and Infection Biology
Norwegian School of Veterinary Science
PO Box 8146 Dep
0033 Oslo
Norway

Email: per.e.granum@veths.no

**Patricia Guerry**
Enteric Diseases Department
Naval Medical Research Center
503 Robert Grant Avenue
Silver Spring, MD
USA

**Eric A. Johnson**
Food Research Institute
University of Wisconsin
1925 Willow Drive
Madison, WI
USA

**Vijay K. Juneja**
United States Department of Agriculture
Agricultural Research Service
Eastern Regional Research Center
600 East Mermaid Lane
Wyndmoor, PA
USA

**Keith A. Lampel**
United States Food and Drug Administration
HFS-025
8301 Muirkirk Road
Laurel, MD
USA

Email: keith.lampel@cfsan.fda.gov

**Peter Leopold**
President
BioAnalyte Inc.
264 Eastern Promenade
Portland, ME
USA

**Irving Nachamkin**
University of Pennsylvania
Department of Pathology and Laboratory Medicine
3400 Spruce Street
Philadelphia, PA
USA

Email: nachamki@mail.med.upenn.edu

**Truls Nesbakken**
The Norwegian School of Veterinary Science
Department of Food Safety and Infection Biology
PO Box 8146 Dep
0033 Oslo
Norway

Email: Truls.Nesbakken@veths.no

**Mitsuaki Nishibuchi**
Center for Southeast Asian Studies
Kyoto University
46 Shimoadachi-cho
Yoshida
Sakyo-ku
Kyoto
Japan

Email: nisibuti@cseas.kyoto-u.ac.jp

**John S. Novak**
Research Scientist
American Air Liquide
5230 South East Avenue
Countryside, IL
USA

Email: john.novak@airliquide.com

**James D. Oliver**
University of North Carolina at Charlotte
Department of Biology
9201 University City Boulevard
Charlotte, NC
USA

Email: jdoliver@email.uncc.edu

**Ynes R. Ortega**
Center for Food Safety and Quality Enhancement
Department of Food Science and Technology
University of Georgia
1109 Experiment Sreett
Griffin, GA
USA

Email: ortega@uga.edu

**Franco Pagotto**
Bureau of Microbial Hazards
Food Directorate, Health Canada
Address Locator 2204A2
Sir Frederick G. Banting Building
Ottawa, Ontario
Canada

Email: franco_pagotto@hc-sc.gc.ca

**George C. Paoli**
United States Department of Agriculture
Agricultural Research Service
Eastern Regional Research Center
600 East Mermaid Lane
Wyndmoor, PA
USA

Email: gpaoli@errc.ars.usda.gov

**Michael W. Peck**
Institute of Food Research
Norwich Research Park
Norwich
UK

**James J. Pestka**
Department of Food Science and Human Nutrition
Michigan State University
East Lansing, MI
USA

**Gary P. Richards**
United States Department of Agriculture
Agricultural Research Service
Delaware State University
James W. W. Baker Center
Dover, DE
USA

Email: grichard@desu.edu

**Ronald T. Riley**
Toxicology and Mycotoxin Research Unit
United States Department of Agriculture
Agricultural Research Service
PO Box 5677
Athens, GA
USA

**Robin J. Rowbury**
1 Church View
Shaston Road
Stourpaine
Blandford Forum, Dorset
UK

Email: robinrowbury@microbiol.freeserve.co.uk

**Chris Scherf**
National HIV and Retrovirology Laboratories
Public Health Agency of Canada
Address Locator 0603A1
Tunney's Pasture
Ottawa, Ontario
Canada

**James L. Smith**
United States Department of Agriculture
Agricultural Research Service
Eastern Regional Research Center
600 East Mermaid Lane
Wyndmoor, PA
USA

Email: jsmith@arserrc.gov

**Jeremy Sobel**
Foodborne and Diarrheal Diseases Branch
Centers for Disease Control and Prevention
1600 Clifton Road
Atlanta, GA
USA

Email: qzs3@cdc.gov

**George C. Stewart**
McKee Professor of Microbial Pathogenesis
Department of Veterinary Pathobiology
Life Sciences Center
1201 Rollins Road
University of Missouri
Columbia, MO
USA

Email: stewartgc@missouri.edu

**Chris Rowe Taitt**
Naval Research Laboratory
Center for Bio/Molecular Science and Engineering
Code 6900
Washington, DC
USA

**Mark L. Tamplin**
United States Department of Agriculture
Agricultural Research Service
Eastern Regional Research Center
600 East Mermaid Lane
Wyndmoor, PA
USA

Email: mtamplin@errc.ars.usda.gov

**Jennifer L. Wampler**
Department of Food Science
Purdue University
745 Agriculture Mall Drive
West Lafayette, IN
USA

# Molecular Approaches for Detection, Identification, and Analysis of Foodborne Pathogens

Pina M. Fratamico and Darrell O. Bayles

## Abstract

Traditional microbiological methods for testing foods for the presence of pathogens rely on growth in culture media, followed by isolation, and biochemical and serological identification. Traditional methods are laborious and time consuming, requiring a few days to a week or longer to complete. Rapid detection of pathogens in food is essential for ensuring the safety of food for consumers, and in the past 25 years, advances in biotechnology have resulted in the development of rapid methods that reduce the analysis time. Two major categories of rapid methods include immunologic or antibody-based assays and genetic-based assays such as the polymerase chain reaction. Next generation assays under development include biosensors and DNA chips that potentially have the capability for near real-time and on-line monitoring for multiple pathogens in foods. In addition to the identification and classification of microorganisms associated with foods or the food chain, global analysis methods are becoming increasingly available for analyzing these microorganisms and their environments in ways that will lead to a more complete understanding of how these organisms respond to their environments. The use of global analyses offers the opportunity to identify cellular genes, proteins, metabolites, and their interconnected networks without the need for a complete understanding of the organism prior to testing. These types of analyses have the potential to improve public health by providing the information needed to determine what causes particular organisms to become pathogenic and what causes organisms to persist in environments related to foods. This information will also allow more specific detection tests to be developed, will provide the information needed to enhance current interventions, and may possibly lead to the design and validation of new interventions.

## Introduction

Microbiological analysis is important to determine the safety and quality of food. For many years, detection and identification of microorganisms in foods, animal feces, and environmental samples have relied on cultural techniques regarded as the "gold standard." Conventional methods are labor intensive, time consuming, and costly, and advances in these methods have been limited to the development of instruments such as the Stomacher or Pulsifier for sample processing, improved liquid and selective/differential agar media, instruments for plating and counting bacteria, and identification test kits (de Boer and Beumer, 1999). More recently, advances in biotechnology have led to the development of "rapid methods" that minimize manipulation, provide results in less time, and reduce cost. Rapid methods generally include immuno-based and DNA-based assays. Immunological or antibody-based assays include enzyme linked-immunosorbent assays (ELISA) and immunochromatographic or "dipstick" assays. Genetic methods include the polymerase chain reaction (PCR), DNA hybridization, and DNA microarrays, also known as GeneChips.

There is a zero tolerance policy for *Listeria monocytogenes* in ready-to-eat products and for *Escherichia coli* O157:H7 in non-intact fresh beef products; thus, appropriate methods must have the ability to detect one colony-forming unit in the sample being analyzed after enrichment culturing. Regardless of the technology employed, food analysis remains a challenging task. Problems that complicate pathogen detection include: (1) non-uniform distribution of pathogens in the food, thus the sample analyzed may not be representative of the entire lot; (2) low level of the target pathogen compared to that of the indigenous microbiota, which may be present at levels as high as $10^8$ CFU/g in raw products; (3) heterogeneity of food matrices and food components interfering with growth or detection of the target organism; and (4) inability to recover injured target organisms using selective enrichment media.

One of the exciting developments in food microbiology has been the availability and application of molecular analyses that have allowed scientists to address microbial food safety questions beyond merely determining whether particular pathogens are in a food. Such global analyses are allowing scientists to ask deeper questions regarding foodborne pathogens and are currently leading the way to ascertaining the genes, proteins, networks, and cellular mechanisms that determine the persistence of strains in foods and other environments, determine why certain strains are more commonly isolated from foods, and determine why certain strains are more pathogenic.

Such molecular tools are also making it possible to more fully determine the microflora present in foods along with pathogens, and to assess the effect that the food microbiota has on the death, survival, and pathogenicity of foodborne pathogens. As the application of molecular analyses improves our understanding of the responses of pathogens to foods and food environments, we anticipate that the information will lead to the development of more specific detection tests, to the enhancement of current interventions, and to the development of new interventions.

## Culture-based methods

Detection of pathogens by conventional culture-based methods involves enrichment of target pathogens in a representative food sample using specific liquid growth media, followed by plating of culture-enriched samples onto selective/differential agar/s and confirmation of the pure culture isolate using a series of morphological, biochemical, serological, and other tests. Cultural methods are costly and labor intensive and yield results generally after five days or more of culture and confirmation steps. An effective culture-based method should suppress the growth of competitive microorganisms to allow easy and reliable detection of the target species.

As an example, the method used by the Food and Drug Administration, Center for Food Safety and Applied Nutrition for enrichment, detection, and isolation of *E. coli* O157:H7 from foods involves enrichment in EHEC enrichment broth (EEB) containing cefixime cefsulodin and vancomycin followed by plating onto sorbitol MacConkey agar containing potassium tellurite and cefixime (CT-SMAC). The food sample (25 g) is added to 225 mL of EEB and briefly pummeled in a Stomacher. After incubation for 24 h at 37°C with aeration, the culture-enriched samples are diluted and plated onto CT-SMAC agar. Presumptive positive colonies are restreaked onto tryptic soy agar containing yeast extract, and colonies are then subjected to a number of biochemical and serological tests and tested for the presence of Shiga toxin genes by the PCR. The method used by the USDA Food Safety and Inspection Service described in the Microbiology Laboratory Guidebook (FSIS, 2002) for detection, isolation, and identification of *Escherichia coli* O157:H7 and O157:NM (non-motile) from meat products is similar to that used by the FDA; however, different enrichment and plating media are used, and culture-enriched samples are tested using a rapid screening test and subjected to immunomagnetic separation using magnetic beads coated with antibodies against *E. coli* O157:H7 to capture and concentrate the organism prior to plating onto a selective agar medium. Presumptive positive colonies are then confirmed biochemically and serologically and are tested for the production of Shiga toxins or the presence of Shiga toxin genes. Although, conventional culture methods are effective and are commonly used in food testing, they are not adequate for making rapid assessments on the microbiological safety of foods.

Enumeration of microorganisms involves making ten-fold serial dilutions of samples to obtain a population countable on appropriate agar media. The aerobic plate count is one application of the colony count method. A common method used to estimate the number of microorganisms in a sample is the most probable number (MPN) technique, in which results are reported as positive or negative in one or more dilutions of the sample. As samples are diluted, some will contain no target microorganisms, and the most probable number of target microorganisms per gram or milliliter of sample is computed based on combinations of positive and negative tubes for any numbers of dilutions used. The MPN method is used for coliform testing of water and for general testing foods for specific microorganisms.

## Injured bacteria

Intrinsic factors within the food and/or food processing or treatment procedures, designed to inactivate or control the growth of pathogens, such as heating, freezing, or exposure to acid or sanitizing compounds, can cause stress/sublethal injury in the bacteria, thus target pathogens may easily be outgrown or out competed during enrichment by the background microbiota. When consumed with food, stressed/injured pathogens may go on to repair, grow, and regain pathogenicity; therefore, it is critical that methods be evaluated to determine their ability to detect severely stressed target organisms. Use of a non-selective enrichment step to allow bacteria to recover from injury, followed by selective enrichment has proven to be effective; however, this may increase the cost and time required for detection and isolation (Suslow *et al.*, 2002). Alternatively, one enrichment medium may have the capability to facilitate the recovery of injured bacteria and the rapid growth of the target bacteria. A novel broth method, called SPRINT *Salmonella*, based on the timed release of a selective agent formulation making the non-selective pre-enrichment broth into a selective medium, improved the rate of detection of low numbers of heat-injured *S.* Typhimurium in ice-cream and skimmed milk powder (Baylis *et al.*, 2000). Selective medium components, including phenylethanol, acriflavin, acriflavin-polymixin, and sodium chloride inhibited recovery of sublethally stressed *L. monocytogenes* (Ryser and Donnelly, 2001). *Listeria* repair broth was consistently superior to University of Vermont modified *Listeria* enrichment broth for recovery of sodium nitrite-injured *L. monocytogenes* from frankfurters (Ngutter and Donnelly, 2003). Another study showed that enrichment for 24 hours was necessary for detection of cold-stressed (exposure to 4°C and −20°C) enterohemorrhagic *E. coli* in ground beef, since cold stress likely lengthened the bacterial lag phase (Uyttendaele *et al.*, 1998).

## Immunologic-based methods

Immunological methods rely on the specific binding of a bacterial antigen to either monoclonal or polyclonal antibodies.

Different types of immunoassays have been described for detection of specific foodborne pathogens and bacterial toxins. These include enzyme-linked immunosorbent assays (ELISA) and immunochromatographic assays, which are commonly used formats, and immunofluorescence assays. A popular assay format is the "sandwich" ELISA, which involves immobilization of antibody onto a solid phase, such as wells of a microtiter plate, capture of the target microbial cells or toxin from culture-enriched samples, rinsing, addition of a second antibody conjugated to an enzyme, followed by addition of enzyme substrate and colorimetric detection of the product of the reaction. Imunochromatographic assays, also called lateral flow or strip tests, are rapid and easy to use. The sample is applied to a well in the device and is wicked through a chromatographic matrix to an area containing antibodies conjugated to a precipitable material such as colloidal gold. If the target antigen is present, the antigen-antibody complex flows to an area containing a second antibody. The antibody-antigen-antibody complex results in a colored immunoprecipate. A very useful technique is immunomagnetic separation, in which antibody-coated magnetic particles are used to capture the target organisms. A magnet applied to the side of the sample tube captures the beads, and unbound material, including food components and non-target bacteria is removed. The target bacteria bound to the beads can then be detected by plating on agar media, the PCR, microscopy, or electrochemiluminescnece assays. Other types of immunoassays have also been described (Entis *et al.*, 2001). Problems associated with immunologic assays include cross-reactivity of the antibody with similar antigens found in non-target bacteria and a detection limit ranging from $5 \times 10^4$ to greater than $10^6$ CFU/mL. False-negative results will be obtained if target pathogen levels are in concentrations below the limit of detection.

## Nucleic acid-based methods

### DNA hybridization
DNA hybridization or gene probe assays consist of detection of DNA or RNA targets using complementary labeled nucleic acid probes. One format is colony hybridization, involving transfer of colonies from the surface of an agar medium onto a solid support, lysis of the cells, denaturation of the DNA, and linkage to the support, followed by hybridization with a labeled probe. Commercially available kits from GeneTrak use a dipstick, which have bound a capture probe targeting specific regions of rRNA genes of the target microorganism. GeneTrak kits are available for detection of *Salmonella*, *Listeria*, *Campylobacter*, and *E. coli* O157:H7 (Fung, 2002). Hybridization formats have also been developed in which a hybridization product is formed in solution and measured directly in the reaction tube. These systems utilize fluorescent dyes selective for double-stranded DNA or fluorescence resonance energy transfer (FRET) probes (see below).

## The polymerase chain reaction
The PCR is a powerful technique that has revolutionized molecular biology research and has application in the diagnosis of microbial infections and genetic diseases, as well as in detection of pathogens in food, fecal, and environmental samples. The PCR is an *in vitro* method that employs a DNA polymerase enzyme and olignonucleotide primers to amplify a specific region of DNA. The choice of the DNA region to be amplified determines the specificity of detection. Suitable targets for detecting foodborne pathogens include ribosomal RNA genes and protein (enzymes, outer membrane proteins, heat shock proteins, toxins, and other virulence factors) genes. Assays based on the PCR are now accepted methods for rapidly confirming the presence or absence of specific pathogens in foods, and PCR-based kits are commercially available for detection of specific foodborne pathogens.

Typing of *E. coli* isolates is traditionally performed by serotyping, which relies on agglutination reactions using antisera raised against the ca. 179 O (somatic) and 53 H (flagellar) serogroup antigens. Serotyping, however, is generally only performed in specialized laboratories, the procedure is labor intensive and may require several days to complete, and cross-reactivity of antisera with several *E. coli* serogroups frequently occurs. Determination of the sequences of the O antigen genes, clustered between the *galF* and *gnd* genes on the *E. coli* chromosome permits identification of unique genes/sequences that can be used to design serogroup-specific PCR assays for detection, as well as typing of *E. coli* as an alternative to serotyping. Several O antigen gene clusters have been sequenced, including, those of serogroups O26, O55, O91, O104, O111, O113, and O157, and serogroup-specific PCR assays based on genes in the respective O antigen clusters have been developed (Wang *et al.*, 2001; DebRoy *et al.*, 2004). As an example, the O antigen gene cluster of an *E. coli* serogroup O121 strain was sequenced, and PCR assays using primers based on the *wzx* (O antigen flippase) and *wzy* (O antigen polymerase) genes in the cluster were designed and used to detect *E. coli* O121 strains in swine (Fratamico *et al.*, 2003). In addition, multiplex PCR assays targeting O antigen genes and virulence genes can be employed to detect and identify specific pathogenic *E. coli* strains (Wang *et al.*, 2002).

The PCR and other rapid methods are often inhibited by food matrix components, thus there is a need to separate and concentrate bacteria or to purify bacterial DNA from food samples or enrichments prior to detection. When performing the PCR, sample preparation is necessary to remove inhibitory substances such as complex polysaccharides, fats, and proteins that may reduce amplification efficiency, to increase the concentration of the target organism/DNA, and to produce a homogenous sample ensuring reproducibility of the assay. Some of the methods that have been used for this purpose include aqueous polymer two-phase systems, immunomagnetic separation, centrifugation, or filtration (Benoit and Donahue, 2003). Furthermore, many DNA extraction kits and reagents such as PrepMan Ultra (Applied Biosystems) and the UltraClean Fecal DNA Kit (Mo Bio Laboratories, Inc., Carlsbad, CA) are commercially available.

Various techniques have been developed to improve the performance of the PCR. For example, non-specific amplification can be reduced using hot-start PCR, in which enzyme activity is released immediately before the first primer binding step. Hot start PCR is commonly performed using a DNA polymerase that is supplied in an inactive form, requiring a few minutes of heating at 94 to 96°C for activation. Nested PCR, which involves the use of two pairs of PCR primers for a single locus, can overcome problems associated with low numbers of target organisms in the presence of high levels of non-target bacteria and PCR inhibitors. Nested PCR involves a first round of amplification for 20 to 30 cycles followed by a second run using the product of the first PCR. If the wrong locus were amplified using the first set of primers, the probability is very low that the wrong locus would also be amplified a second time by a second PCR using the nested primer pair. PCR enhancers, such as formamide, DMSO, glycerol, betaine, and others, lower the template melting temperature, aiding primer annealing and product extension of the template through regions of secondary structure. In addition, the magnesium concentration in the reaction can alter the yield by affecting DNA polymerase activity, and it can alter specificity by affecting primer annealing. Therefore, the optimal magnesium concentration needs to be determined for each primer pair and template DNA combination.

*Real-time PCR*

Conventional PCR generally involves four steps: (1) nucleic acid extraction, (2) DNA amplification, (3) product detection and sizing by agarose gel electrophoresis, and (4) amplicon confirmation, for example, by hybridization with a DNA probe. Combining the PCR with a hybridization step enhances assay sensitivity and specificity. A disadvantage of traditional PCR is that post-PCR sample handling may result in carryover contamination of the amplicon to future PCR assays, potentially causing false positive results.

An important advance in PCR technology is the development of real-time assays, which combine target amplification and detection in one step. Detection of the amplicon is monitored as it occurs, i.e., in a "real-time" mode. Applications of real-time PCR include gene expression analysis (Mayer *et al.*, 2003), single nucleotide polymorphism (SNP) typing (Mhlanga and Malmberg, 2001), and pathogen detection (Vishnubhatla *et al*,. 2000; Fortin *et al.*, 2001; Schaad and Frederick, 2002; Sharma, 2002). Advantages of real-time compared to conventional PCR include: (1) closed-tube format that lowers the potential for carryover contamination and false positive results; (2) shorter analytical turnaround time with higher sensitivity and precision; (2) no need for post-PCR processing steps to detect the product; and (3) a larger dynamic range for quantitative real-time PCR assays compared to traditional PCR (6–7 logs versus 2–3 logs, respectively).

Real-time PCR systems rely upon detection and quantitation of signal generated from a fluorescent reporter, whose signal increases in direct proportion to the amount of PCR product in a reaction. The cycle threshold ($C_t$) is defined as the first cycle in which there is a significant increase in fluorescence above a specified threshold, and plotting fluorescence against cycle number yields a curve that represents the accumulation of PCR product over the duration of the PCR reaction. A standard curve can be generated from $C_t$ values for a series of reactions using dilutions containing known quantities of target DNA. The amount of target in unknown samples can be derived by measuring the $C_t$ and using the standard curve to determine starting copy number. Alternatively, quantitation can be "relative," where the $C_t$ values are compared to those of an internal standard, which is usually a housekeeping gene.

A number of fluorescence systems have been employed for real-time PCR, and among the various chemistries available, SYBR Green I is the most economical and convenient to use. SYBR Green I is a thermostable intercalating dye that binds to double-stranded DNA, and emits a fluorescent signal that increases proportionately with the amount of PCR product. After the PCR, an additional time-temperature program is used to generate a melting curve. A drop in fluorescence is observed at the point where the double-stranded DNA melts and the dye dissociates. Melt curves can distinguish between specific and non-specific products based on the melting temperature ($T_m$) of the amplicons, since each product has a unique melting temperatures ($T_m$) due to differences in sequence and length. Hydrolysis or TaqMan probes and hybridization probes like molecular beacons are oligonucleotides that are labeled with a reporter dye and a quencher at opposite ends of the sequence. TaqMan probes bind to an internal region of the PCR product. When intact, the fluorescence of the reporter dye is suppressed by the quencher. The probe hybridizes to a complementary sequence on the template between the two primer binding sites, and during replication of the template, the polymerase exonuclease activity cleaves the probe, which separates the reporter dye from the quencher. The fluorescence increases in proportion to the quantity of amplicons generated. TaqMan-based real-time PCR assays have been used for detection of foodborne pathogens (Vishnubhatla *et al.*, 2000; Sharma, 2002; Cheung *et al.*, 2004). Molecular beacons are stem-and-loop shaped probes, consisting of a sequence-specific portion (loop, 20–24 nucleotides) and complementary stem sequences (4–6 nucleotides) that hold the probe in a hairpin configuration and are end labeled with the reporter and quencher dyes. The molecular beacon binds to a complementary sequence in the amplicon during the PCR, and the conformational change that occurs spatially separates the quencher from the reporter dye, resulting in detectable fluorescence (Vet *et al.*, 1999; Chen *et al.*, 2000; Fortin *et al.*, 2001). Another hybridization probe detection system that relies on the fluorescence resonance energy transfer (FRET) principle was initially developed for use with the LightCyler instrument. This system employs two probes, one labeled with a donor fluorochrome (fluorescein) at the 3′ end and the other labeled with an acceptor dye at the 5′ end. The probes are designed to hybridize to the template during the annealing phase of the PCR in a head-to-tail arrangement so that they are distanced by one to five bases. When both are hybridized, the energy emitted from the donor excites the acceptor dye, which then emits red fluorescent light at a longer wavelength that can be measured (Palladino et al., 2003). Scorpion probes consist of a PCR primer; a PCR stopper to prevent undesirable PCR read-through of the probe by Taq DNA polymerase, a specific probe sequence, and a fluorescence detection system consisting of reporter and quencher dyes. In the unhybridized state, the Scorpion probe maintains a stem-loop configuration. The 3′ portion of the stem contains a sequence linked to the 5′ end of a specific primer via a non-amplifiable monomer that is complementary to a region in the PCR product. The probe binds to its target after extension of the Scorpion primer, producing an increase in fluorescence due to separation of the reporter from the quencher dyes. LUX (light upon extension) primers are designed to be self quenched until they are incorporated into the PCR product where they generate fluorescence due to a change in the secondary structure. Additional real-time PCR fluorescence systems include Amplifluors (Rodríguez-Lázaro et al., 2004a) and MGB Eclipse probes (Afonina et al., 2002).

Instruments currently available for performing real-time PCR include the LightCycler (Roche Diagnostics Corp.), the RAPID and RAZOR (Idaho Technologies), a hand-portable instrument that was co-developed with the Department of Defense, the iCycler iQ (Bio-Rad), ABI Prism 7000, 7300, 7500, and 7900HT (Applied Biosystems), the Quantica (Techne), the MX4000 (Stratagene), the Rotor Gene (Corbett Research), DNA Engine cyclers (MJ Research), and the Smart Cycler systems (Cepheid). The Smart Cycler is unique in that each processing block contains 16 independently controlled, programmable I-CORE (intelligent cooling/heating optical reaction) modules, such that 16 different PCR protocols can by run simultaneously, facilitating optimization of PCR assays. Furthermore, up to six processing blocks can be linked together allowing simultaneous analysis of 96 discrete samples. Automated nucleic acid extraction instruments called the MagNa Pure LC and the ABI Prism 6700 or 6100, can be coupled with the Light Cycler and ABI real-time PCR instruments, respectively.

*Conventional and real-time reverse transcriptase PCR*
Real-time reverse transcriptase PCR (RT-PCR) allows amplification of a few target RNA molecules (mRNA or total RNA) using the enzyme reverse transcriptase to reverse transcribe RNA into cDNA, which is subsequently amplified. In real-time assays, fluorescent dyes or probes are incorporated into the product as the cDNA is copied. A one-step Taq-Man-based real-time RT-PCR assay developed for detection and quantitation of the Barley yellow dwarf virus in infected plants and in viruliferous aphids was 10–1000 times more sensitive than standard RT-PCR and ELISA assays. (Fabre et

al., 2003). DNA persists after cell death or viral inactivation; therefore, a disadvantage of traditional PCR techniques is that DNA from both viable and non-viable cells is detected. When testing for bacterial pathogens, this problem is generally overcome by the inclusion of an enrichment step that dilutes out non-viable cells and allows growth of viable cells. A comparison of results of PCR assays performed on samples prior to and after enrichment would determine the presence of viable cells. Cell viability could be assessed directly through detection of specific mRNAs since these molecules have a short half-life and would only be found in viable cells. Using an RT-PCR assay after a 2-hour enrichment, Klein and Juneja (1997) were able to detect a L. monocytogenes iap gene-specific product in cooked ground beef samples initially inoculated with 3 CFU/g. Amplification of three L. monocytogenes mRNAs, iap, hly, and prfA, was compared, and the RT-PCR targeting the iap message showed higher sensitivity, indicating a higher level of the iap mRNA within the cell. Other investigators were able to detect 1 CFU of viable Shiga toxin-producing E. coli in 25 g of cooked ground beef following a 12-hour enrichment using an RT-PCR assay targeting $stx_2$ mRNA (McIngvale et al., 2002). Morin et al. (2004) used a multiplex reverse transcriptase PCR assay to detect E. coli O157:H7, Vibrio cholerae O1, and Salmonella Typhi. The sensitivity of the assay was 30 bacteria, and there was no interference from non-target bacteria.

*Conventional and real-time multiplex PCR*
In multiplex PCR assays, two or more primer pairs are used in a single reaction enabling simultaneous amplification of multiple targets, resulting in savings in time, labor, and cost. However, compared to PCR assays in which one sequence is amplified, multiplex assays can be tedious and time consuming to establish since more extensive optimization of reaction parameters is required to avoid the formation of spurious amplification products and uneven amplification of target sequences. A multiplex PCR assay, targeting the $fliC_{h7}$, $stx_1$, $stx_2$, eaeA, and $hly_{933}$ genes was used to detect E. coli O157:H7 in bovine feces, ground beef, and other types of food samples inoculated with 1 CFU/g and subjected to enrichment for 24 h (Fratamico et al., 2000). Moreover, a multiplex PCR assay targeting conserved sequences of the $stx_1$ and $stx_2$ genes and the eae and $hly_{933}$ genes of E. coli O157:H7, and the invA gene of Salmonella spp. was developed for simultaneous detection of E. coli O157:H7 and Salmonella spp. in ground beef, apple cider, bovine feces, and beef carcass wash water samples inoculated with both pathogens, subjected to enrichment in mEC broth containing novobiocin or buffered peptone water, followed by DNA extraction or IMS prior to the PCR (Fratamico and Strobaugh, 1998).

In multiplex real-time PCR assays, probes that bind to their respective PCR products are labeled with different fluorophores that have unique emission spectra. Josefsen et al. (2004) compared real-time PCR using the ABI-PRISM 7700 and the Rotor-Gene 3000 instruments for detection of thermotolerant Campylobacter species, C. jejuni, C. coli, and C. lari, in naturally contaminated chicken rinse samples. There was good agreement between PCR assays performed in both instruments and with an International Standard Organization (ISO)-based culture method. Bellin *et al.* (2001) developed a real-time multiplex PCR assay targeting the Shiga toxin 1 and Shiga toxin 2 genes in enterohemorrhagic *E. coli* using FRET hybridization probes and the Light Cycler instrument. The assay was carried out in a single reaction capillary and could be performed in 45 minutes. A melting curve analysis could discriminate between *stx2* and *stx2e*, which is the gene variant associated with pig edema disease.

*PCR on a chip*
Procedures involving on-chip PCR amplification and detection of products have been described (Kricka, 1998; Kricka, 2001; Nickisch-Rosenegk *et al.*, 2005). The PCR described by Nickisch-Rosenegk and coworkers (2005) occurred with primers immobilized on the chip and the products were visualized through the use of intercalating dyes or by confocal microarray scanning or fluorescence microscopy using Cy5-dye fluorescence of the modified free primer. Moreover, microchip devices are being developed that integrate cell lysis, multiplex PCR amplification, electrophoretic separation of PCR products, and product detection (Cheng et. al., 1996; Woolley *et al.*, 1996; Waters *et al.*, 1998). Microminiaturization of nucleic acid-based techniques will have a significant impact on diagnostic and food testing, potentially enabling the assays to be performed by relatively unskilled operators in non-laboratory settings.

## Alternative nucleic acid amplification techniques
A number of other nucleic acid amplification techniques have been described, including isothermal amplification methods known as nucleic acid sequence-based amplification (NASBA), strand displacement amplification (SDA), and the ligase chain reaction (LCR) (Fratamico, 2001). The isothermal SDA technique provides about $10^8$-fold amplification of the target after 2 hours at ca. 40°C. A restriction enzyme is used to nick a hemimodified recognition site, and a DNA polymerase initiates synthesis of a new DNA strand at the nick, while displacing the existing strand. The modified target sequences are amplified exponentially by repeated nicking, strand displacement, and priming of displaced strands. A multiplex SDA procedure was used to amplify DNA from *Mycobacterium* species, and the amplified target molecules were detected by DNA polymerase-catalyzed extension of $^{32}$P-labeled probes hybridized to the different targets (Walker *et al.*, 1994). Westin *et al.* (2000) developed a microchip array with primers anchored to the surface which were used to perform multiplex SDA-based amplification and detection *in situ* on the chip. The NASBA technique involves isothermal amplification of an RNA target with the coordination of three enzymes, a reverse transcriptase, RNaseH, and T7 polymerase (Cook, 2003). Since NASBA is an isothermal technique, RNA is amplified below

the melting temperature of DNA; therefore, there is no interference by DNA. NASBA has been employed for detection of *Campylobacter* spp., *L. monocytogenes*, *Salmonella* spp., *Cryptosporidium parvum*, and foodborne viruses using 16S rRNA and mRNA as target molecules (Cook, 2003). The technique was used to detect viable *S. enterica* using the NucliSens kit, and a NASBA electrochemiluminescnce assay was used to detect the organism in fresh meats, poultry, and other foods, with both methods targeting mRNA transcribed from the *Salmonella dnaK* gene (Simpkins *et al.*, 2000; D'Souza and Jaykus, 2003). Rodríguez-Lázaro *et al.* (2004b) developed a molecular beacon-based real-time NASBA targeting the *dnaA* gene for detection of *Mycobacterium avium* subsp. *paratuberculosis* in water and milk. The assay had a sensitivity of 150–200 cells per reaction, and was consistently able to detect $10^3$ target bacteria in artificially contaminated drinking water. Another isothermal amplification technique, termed ramification amplification (RAM), utilizes a circular probe (C-probe) formed through ligation of its ends after hybridization to its target (Zhang *et al.*, 2001). The C-probe can be amplified by rolling circle amplification generating a multimeric single-stranded DNA, which serves as a template for multiple reverse primers to hybridize, extend, and displace downstream DNA. This results in a large branching or ramified DNA complex that can be further amplified by RAM through primer extension and downstream DNA displacement.

## DNA microarrays
DNA microarrays are the latest development in pathogen detection technology and represent a potential solution to the challenge of multiplexed pathogen detection. DNA arrays employed for pathogen detection consist of glass slides or nylon membranes onto which PCR products of target-specific sequences or olignonucleotides are bound. DNA samples for analysis are chemically labeled with fluorescent dyes or are labeled during PCR amplification and are then hybridized with their complementary sequences on the chip. Following hybridization and washing steps, the arrays are examined using a high-resolution scanner. Microarrays can be used to detect PCR products by hybridization of labeled amplicons to an array composed of pathogen-specific probes. Sergeev *et al.* (2004) used degenerate primers to amplify variable regions of *Staphylococcus aureus* enterotoxin genes and hybridized chemically labeled ssDNA (derived from PCR reactions specific for enterotoxin genes) to microarrays onto which probes specific for each of the different enterotoxin genes were bound. Call *et al.* (2001) were able to detect 55 CFU of *E. coli* O157:H7 per milliliter of chicken rinsate without an enrichment step using immunomagnetic capture followed by PCR amplification and hybridization of the products onto a microarray containing oligonucleotide probes (25- to 30-mer) complementary for four virulence genes of *E. coli* O157:H7. The microarray assay was 32-fold more sensitive than gel electrophoresis for PCR product detection. Chizhikov *et al.* (2001) detected microbial virulence factors (*eaeA*, *slt-I*, *slt-II*, *fliC*, *rfbE*, and *ipaH*)

of multiple pathogens by hybridizing Cy5-labeled fluorescent PCR products to the gene-specific oligonucleotides spotted on microarrays. A similar approach was used by Keramas et al. (2004) to detect *Campylobacter* spp. directly from fecal cloacal swabs targeting the 16S rRNA, the 16S–23S rRNA intergenic region, and specific *Campylobacter* genes. Vora et al. (2004) investigated the efficacy of different DNA amplification strategies, including random primed, isothermal Klenow fragment-based, Φ29 DNA polymerase-based, and multiplex PCR for use in oligonucleotide microarray applications. It was found that a method using random PCR primers and a 70-mer oligonucleotide microarray was unbiased and sensitive and retained specificity via hybridization to the probes on the microarrays. Moreover, microarrays can be used for direct detection of labeled DNA or RNA without PCR amplification and also for genotyping, "fingerprinting," or speciating bacterial isolates by analyzing hybridization of labeled DNA to oligonucleotides covering all known sequence variants of the target. A number of applications for pathogen detection and characterization with microarrays are described in a review by Call et al. (2003).

## Testing food for the presence of viruses and parasites

Although foodborne viruses cause a sizable number of foodborne illnesses each year worldwide, no established or validated methods yet exist, except possibly for shellfish, for reliably detecting viruses in food items (Koopmans et al., 2002). Furthermore, culture systems for specific viruses are lacking. In the United States, molecular detection techniques, including PCR-based assays, are being implemented in state public health laboratories under the guidance of the Centers for Disease Control and Prevention. A multiplex NASBA method has recently been reported for simutaneous detection of hepatitis A and genogroup I and II noroviruses in ready-to-eat foods (Jean et al., 2004). A drawback in control programs for parasites such as *Trichinella, Toxoplasma, Taenia, Cyclospora, Cryptosporidium*, and *Giardia* has been the absence of rapid, accurate, and sensitive diagnostic tests for these and other parasites. For the most part, current techniques for detecting foodborne parasites are labor-intensive and tedious; however, efforts are being directed to refine techniques and to develop additional rapid methods.

## Molecular analysis of foodborne pathogens

There are an increasing number of molecular analysis methods that are available for characterizing microorganisms in their environments. We have chosen to discuss the available methods beginning with techniques that are rooted in genetics and finishing with methods that are based on the protein output of the cells. The genetic-based methods are further subdivided into methods primarily focusing on genomic DNA analyses,

mRNA expression analyses, and computational (*in silico*) analyses.

## DNA-based analyses

There are a variety of methods available for examining how the DNA of one organism differs from the DNA of another organism. Defining such differences provides information that can lead to the identification of the genes that contribute unique characteristics to the organism being studied. The DNA-based methods include techniques such as comparative genomic hybridization (Fukiya et al., 2004), subtractive hybridization (Wu and Muriana, 1995; Wu and Muriana, 1999; Pradel et al., 2002; Sorsa et al., 2004), representational difference analysis (RDA) (Allen et al., 2001; Allen et al., 2003), some community-based analyses (Rudi et al., 2002a; Rudi et al., 2002b; Filion et al., 2004; Bartosch et al., 2004), and gradient gel electrophoresis using either chemical or temperature gradients to create a denaturing gradient.

Comparative genomic hybridization takes advantage of the information available from whole genome sequences by hybridizing fluorescently labeled DNA from one organism's genes to a DNA microarray that contains the genes of another organism. This type of comparison detects which of the genes represented in the strain-specific genetic elements imprinted on the DNA array are found in each genome of the organisms being tested. One such study has recently been used to define the genomic diversity present in pathogenic strains of *E. coli* and *Shigella* (Fukiya et al., 2004). In that study, randomly amplified elements representing the genes from each of 22 pathogenic strains of *E. coli* and *Shigella* were hybridized to DNA arrays containing the open reading frames (ORFs) of *E. coli* K-12 strain W3110. Using these comparisons, the authors were able to determine that 1424 of the K-12 strain W3110 ORFs were absent in at least one of the pathogenic strains. Each pathogenic strain lacked between 221 and 724 of the W3110 ORFs, and 96 W3110 ORFs were not found in any of the 22 pathogenic strains tested.

Another group of techniques that compare the DNA sequences of whole genomes are the subtractive hybridization techniques. Subtractive hybridizations do not rely on DNA microarrays and do not identify the genes common between two strains; rather, subtractive hybridizations preferentially isolate unique genetic differences between two organisms. The genetic elements isolated using subtractive procedures can be cloned and sequenced and the sequence information can be used to conduct searches of molecular biology databases. While there are many variations of subtractive hybridization techniques, the core methodology is relatively straightforward. The DNA from the strain of interest (tester DNA) and the DNA from the control strain (driver DNA) are isolated. The tester DNA is digested with a restriction endonuclease while the driver DNA is typically fragmented using some type of mechanical shearing. The tester DNA is modified, typically by adding an oligonucleotide adapter, so that the desired fragments can be amplified in later steps using the PCR and primers that are specific for the adapter sequence. The tester DNA is mixed with an excess of driver DNA and the mixed sample is denatured and allowed to reanneal. Hybridized sequences that are common to both the tester and driver are either removed using a capture technique (when the driver DNA has been biotinylated) or are not amplified (using amplification strategies that only amplify the tester DNA that contains the added primer specific sequence.) Multiple rounds of subtraction are often used to remove the sequences common to both the tester and driver DNA. At the end of the procedure, the sequences that remain are largely genetic elements unique to the tester strain. These sequences of interest are amplified by PCR for further cloning and sequencing. While many organisms have been studied using subtractive hybridization methods, experiments conducted with *L. monocytogenes* and *E. coli* serve as good examples (Wu and Muriana, 1995; Wu and Muriana, 1999; Pradel *et al.*, 2002; Sorsa *et al.*, 2004). Subtractive hybridization has been used to clone and sequence *L. monocytogenes* specific genes not found in *L. innocua* or *L. ivanovii* (Wu and Muriana, 1999), to identify virulence loci in uropathogenic *E. coli* (Sorsa *et al.*, 2004), and to identify loci unique to Shiga toxin-producing *E. coli* O91:H21 (Pradel *et al.*, 2002). Representational difference analysis (RDA) is a technique closely related to subtractive hybridization. RDA has been successfully used to isolate unique DNA sequences from a variety of pathogens (see Allen *et al.*, 2003, for a list). The primary modification of RDA compared to subtractive hybridization is that a round of PCR amplification of DNA sequences unique to the tester organism is performed after each round of subtraction. This modification has the effect of substantially enriching the unique tester DNA sequences. In one instance, RDA was used to identify *E. coli* O157:H7 specific DNA sequences (Allen *et al.*, 2001).

There is considerable interest in gaining a better understanding of the community of microorganisms that are associated with foods since food microflora will have an effect on the survival and pathogenicity of human pathogens associated with foods. Profiling the communities of microorganisms associated with foods can assist in predicting the microbial diversity of different foods, and can be used to determine how food processing and storage influences pathogens along with other microorganisms in the community. Most of the community analysis techniques employ primers that anneal to 16S DNA sequences specific to particular organisms or groups of organisms and use PCR to amplify a region of 16S rDNA. Typically the amplified sequences are subcloned into a plasmid vector for further analysis. The subcloned 16S rDNA sequences can be analyzed by a variety of gel-based methods that include the length of the PCR product (length heterogeneity PCR or LH-PCR), restriction fragment length polymorphisms (RFLP), denaturing gradient gel electrophoresis (DGGE), temperature gradient gel electrophoresis (TGGE), and temporal temperature gradient electrophoresis (TTGE). There are a large number of recent reviews on the use of community analysis techniques. There are also some current re-

views that are more oriented to profiling communities in foods and other communities of interest to food-safety researchers (McCartney, 2002; Rudi *et al.*, 2002b; Maukonen *et al.*, 2003; Ercolini, 2004; Meays *et al.*, 2004). Community analysis methods have been used to track the changes in the microbiological populations of milk as a result of refrigeration (Lafarge *et al.*, 2004), to detect and quantify the microbiological populations in packaged vegetable salads (Rudi *et al.*, 2002a), and have been used to identify lactic acid bacteria in a food processing plant and in meats (Takahashi *et al.*, 2004). Recently, the term metagenomics has been used to describe community analyses that are performed on uncultured or presently unculturable microorganisms (Handelsman, 2004). This may prove a promising approach to further sample and define the organisms relevant to food environments.

## mRNA expression analyses

Expression analyses comprise a variety of methods used to identify genes that are being expressed in an organism under a defined set of conditions. Many of the mRNA expression analysis methods are similar to methods previously described in the "DNA-based analyses" section above, but instead of using genomic DNA, the expression analyses rely on the isolation of transcribed mRNAs with analyses being performed using cDNA generated from the mRNA. For instance, expression analyses that utilize high-density DNA microarrays operate similar to comparative genomic DNA analyses that are done with DNA microarrays except that the expression analyses use labeled cDNAs produced from the mRNAs captured under the condition of interest instead of labeled genomic DNA. The labeled cDNA is used to probe the DNA microarray, which consists either of gene-specific PCR products or oligonucleotides affixed to the slide coordinates. There are a number of excellent reviews available on the application of DNA microarrays for analysis of gene expression. Recently, there have begun to be reviews (Schena *et al.*, 1998; Wells and Bennik, 2003; Liu-Stratton *et al.*, 2004) and microarray based experiments (Vasil, 2003; Arous *et al.*, 2004) that more specifically address the interests of food safety and the responses of foodborne pathogens.

Two analysis methods that do not rely on DNA microarrays, but have been successfully used to identify genes that are differentially expressed, either increasing or decreasing in expression, are differential display (Gery and Lavi, 1997; Chia *et al.*, 2001) and selective capture of transcribed sequences (Hou *et al.*, 2002; Liu *et al.*, 2002). These methods have the advantage of being able to identify differences without the need for comprehensive or for that matter any information about the DNA sequence of genes in the genomes. Another method for identifying expressed genes relies on the amplification of transcribed sequences through the use of random primers that amplify portions of the cDNA generated from reverse transcription of the mRNA (Frias-Lopez *et al.*, 2004).

*In vitro* expression technology (IVET) is a technique that has allowed a number of genes to be identified based on the expression of pathogen genes inside a host cell (Wang *et al.*, 1996; Valdivia and Falkow, 1996; Fernández *et al.*, 2004). The IVET methods generally utilize a reporter gene, like green fluorescent protein (GFP), fused to genes from the organism under study. These hybrid genes are put back into the pathogen and the pathogen is allowed to infect eukaryotic host cells. Using flow cytometry and repeated rounds of differential selection, pathogen cells that only express the reporter gene while inside a eukaryotic host are selected and collected. Gene trap methods that do not use flow cytometry have also been used to identify genes essential for pathogenesis (Cecconi and Meyer, 2000). Whole-genome bioluminescence (Van Dyk *et al.*, 2001) is a reporter-based methodology that has been successfully used to identify expressed genes in a genome-wide profiling format. It may be possible to extend these methods to food safety and thus identify genes that are essential for pathogen survival in foods or under conditions commonly encountered in the processing and storage of foods.

The availability of completely sequenced pathogen genomes has made possible the analysis of these genomes solely using computational methods (in silico) to perform the analyses (Whittam and Bumbaugh, 2002). Comparisons can be made between strains of a particular species (Nelson *et al.*, 2004) and between different genus or species (Glaser *et al.*, 2001; Fukiya *et al.*, 2004). These methods of analysis are helping to identify genes that are potential targets for interventions, genes that may play are role in pathogenesis, and genes that are responsible for some specific survival and pathogenicity characteristics. As more sequences for food pathogens become available, the use of in silico techniques to determine important core genetic similarities and differences in these pathogens will provide information critical for improved detection and intervention of these pathogens and will lead to treatments that inhibit the survival and disease causing aspects of these organisms.

## Protein-based analyses

The global genetic-based analyses are able to provide information regarding which genes an organism contains or which genes are expressed under some set of conditions; however, examining the post-translational protein output of an organism allows the researcher to query the ultimate outcome of an organism's genetic and regulatory activities. Combining the information obtained in both genetic and protein-based analyses provides a more comprehensive view of the cellular activities being carried out by an organism. Global analysis of cellular protein output generally involves the use of techniques that fall within either the category of proteomics or protein arrays. There are a number of recent reviews discussing the various methods and applications of proteomics for analysis of microorganisms (Jungblut, 2001; Cordwell *et al.*, 2001; Graves and Haystead, 2002; Cash, 2003; Hecker and Volker, 2004; Cordwell, 2004; Walduck *et al.*, 2004). There are some examples of foodborne pathogen analysis via proteomics. One group of examples is the use of proteomics to begin defining the protein output of *L. monocytogenes* (Trémoulet *et al.*, 2002;

Helloin *et al.*, 2003) and *C. jejuni* (Dykes *et al.*, 2003) in biofilms.

Protein arrays, also known as biochips, represent a group of rapidly developing technologies that have potential application in the analysis of foodborne pathogens. Protein arrays, which can be used to analyze cellular proteins based on interactions with proteins in an array affixed to a platform, are roughly the protein equivalent of the DNA microarrays. In addition to protein-protein interactions, cellular proteins can also be profiled for their interaction with a variety of other compounds in a protein capture microarray (Templin *et al.*, 2003). The detection methods appropriate for identifying bound proteins are dependent upon the nature of the protein interaction and profiling technique used. Methods for detection or identification of bound proteins includes: surface enhanced laser desorption ionization time of flight mass spectrometry (SELDI TOF MS); matrix assisted laser desorption ionization TOF MS (MALDI TOF MS); surface plasmon resonance (SPR); radiolabeling; chemiluminescence; fluorescence; and others. A distinct advantage of protein arrays is that they provide direct information on which proteins are interacting with the proteins or other compounds contained in the array, something that DNA arrays cannot do. While most of the early work with protein arrays has focused on eukaryotic systems, there are reviews available that are instructive in understanding how protein arrays are created, how they function, and how they can be used to address questions of biological significance (Marko-Varga *et al.*, 2003; Schweitzer *et al.*, 2003; Templin *et al.*, 2003; Cahill and Nordhoff, 2003; Panicker *et al.*, 2004).

## Conclusions and additional information on microbiological analysis of foods

Further information on detection and sample preparation techniques and commercially available kits for microbiological analysis of foods can be found in the Bacteriological Analytical Manual (BAM), available on-line at http://www.cfsan.fda.gov/~ebam/bam-toc.html, the Microbiology Laboratory Guidebook (FSIS, 1998), in the Compendium of Methods for the Microbiological Examination of Foods (Downes and Ito, 2001), and in several book chapters and review articles by Notermans *et al.* (1997), Barbour and Tice (1997), Feng (2001), Brown *et al.* (2000), and Stevens and Jaykus (2004). In addition, a review by Maukonen *et al.* (2003) and a book edited by McMeekin (2003) cover the role of microbiological testing in managing microbial food safety in food and industrial environments, sampling techniques, and types of testing methods currently available.

Although reliable, traditional testing methods involving culture and confirmation of isolates are labor intensive and time consuming. In recent years, advances in biotechnology have led to the development of rapid methods, including antibody-and nucleic acid-based techniques, that expedite the detection process and that are relatively sensitive and reliable. Nucleic acid-based assays, including the PCR will likely continue to revolutionize the ability of the food industry and regulatory agencies to improve the safety of foods. Although enrichment may still be necessary prior to detection, the PCR can shorten the time needed to identify pathogens in foods and environmental samples. Compared to conventional PCR, real-time PCR systems reduce overall assay time by eliminating post-PCR processing steps and also permit quantification of template DNA. Technology continues to advance at a rapid pace, and the next generation assays under development include multiplex real-time PCR, on-chip PCR, and DNA macro or microarrays, which have the potential for near real-time and on-line monitoring of multiple pathogens in animals, food, and food processing environments. Challenging problems that remain for food testing include: ensuring (1) recovery of injured target organisms; (2) capture of low levels of pathogens that are generally non-homogenously distributed in food; and (3) separation of target pathogens interfering components in the food matrix and concentration prior to detection using nucleic acid-based methods or other types of assays. Research is needed in the development of testing procedures combining sensitive enrichment with immunological or genetic-based techniques to increase reliability. For example, enrichment procedures should allow the recovery of injured target pathogens, while suppressing the growth of the background microbiota, and antibodies may be used for capture and concentration of target organisms followed by nucleic acid-based methods for detection. For nucleic acid-based technologies to be adopted for routine analysis in food and environmental testing laboratories, simpler and more user-friendly systems are needed for high-throughput automated sample processing and detection. The future of food microbiology lies in the development and integration of molecular methods that can be automated into high-throughput testing regimens for food and environmental samples.

## References

Afonina, I.A., Reed, M.W., Lusby, E., Shishkina, I.G., and Belousov, Y.S. 2002 Minor groove binder-conjugated DNA probes for quantitative DNA detection by hybridization-triggered fluorescence. Biotechniques 32: 940–949.

Allen, N.L., Hilton, A.C., Betts, R., and Penn, C.W. 2001. Use of representational difference analysis to identify *Escherichia coli* O157-specific DNA sequences. FEMS Microbiol. Lett. 197: 195–201.

Allen, N.L., Penn, C.W., and Hilton, A.C. 2003. Representational difference analysis: critical appraisal and method development for the identification of unique DNA sequences from prokaryotes. J. Microbiol. Methods 55: 73–81.

Arous, S., Buchrieser, C., Folio, P., Glaser, P., Namane, A., Hebraud, M., and Hechard, Y. 2004. Global analysis of gene expression in an *rpoN* mutant of *Listeria monocytogenes*. Microbiology 150: 1581–1590.

Barbour, W.M. and Tice, G. 1997. Genetic and immunologic techniques for detecting foodborne pathogens and toxins. In: Food Microbiology Fundamentals and Frontiers. M.P. Doyle, L.R. Beuchat, and T.J. Montville, ed. Washington, D C, ASM Press, pp. 710–739.

Bartosch, S., Fite, A., Macfarlane, G.T., and McMurdo, M.E. 2004. Characterization of bacterial communities in feces from healthy elderly volunteers and hospitalized elderly patients by using real-time PCR and effects of antibiotic treatment on the fecal microbiota. Appl. Environ. Microbiol. 70: 3575–3581.

Baylis, C.L., MacPhee, S., and Bett, R.P. 2000. Comparison of methods for the recovery and detection of low levels of injured *Salmonella* in ice-cream and milk powder. Lett. Appl. Microbiol. 30: 320–324.

Bellin, T., Pulz, M., Matussek, A., Hempen, H., and Gunzer, F. 2001. Rapid detection of enterohemorrhagic *Escherichia coli* by real-time PCR with fluorescent hybridization probes. J. Clin. Microbiol. 39: 370–374.

Benoit, P.W. and Donahue, D.W. 2003. Methods for rapid separation and concentration of bacteria in food that bypass time-consuming cultural enrichment. J. Food Prot. 66; 1935–1948.

Brown, M.H., Gill, C.O., Hollingsworth, J., Nickelson, R. 2nd, Seward, S., Sheridan J.J., Stevenson, T., Sumner, J.L., Theno, D.M., Usborne, W.R., and Zink, D. 2000. The role of microbiological testing in systems for assuring the safety of beef. Int. J. Food Microbiol. 62: 7–16.

Cahill, D.J. and Nordhoff, E. 2003. Protein arrays and their role in proteomics. Adv. Biochem. Eng Biotechnol. 83: 177–187.

Call, D.R., Brockman, F.J., and Chandler, D.P. 2001. Detecting and genotyping *Escherichia coli* O157:H7 using multiplexed PCR and nucleic acid microarrays. Int. J. Food Microbiol. 67: 71–80.

Call, D.R., Borucki, M.K., and Loge, F.J. 2003. Detection of bacterial pathogens in environmental samples using DNA microarrays. J. Microbiol. Meth. 53: 235–243.

Cash, P. 2003. Proteomics of bacterial pathogens. Adv. Biochem. Eng Biotechnol. 83: 93–115.

Cecconi, F., and Meyer, B.I. 2000. Gene trap: a way to identify novel genes and unravel their biological function. FEBS Lett. 480: 63–71.

Chen, W., Martinez, G., and Mulchandani, A. 2000. Molecular beacons: a real-time polymerase chain reaction assay for detecting *Salmonella*. Anal. Biochem. 280: 166–172.

Cheng, J., Shoffner, M.A., Hvichia, G.E., Kricka, L.J., and Wilding, P. 1996. Chip PCR. II. Investigation of different PCR amplification systems in microfabricated silicon-glass chips. Nucl. Acids Res. 24: 380–385.

Cheung, P.W., Chan, C.W., Wong, W., Cheung, T.L., and Kam, K.M. 2004. Evaluation of two real-time polymerase chain reaction pathogen detection kits for *Salmonella* spp. in food. Lett. Appl. Microbiol. 39: 509–15.

Chia, J.S., Lee, Y.Y., Huang, P.T., and Chen, J.Y. 2001. Identification of stress responsive genes in *Streptococcus mutans* by differential display reverse transcription PCR. Infect. Immun. 69: 2493–2501.

Chizhikov, V., Rasooly, A., Chumakov, K., and Levy, D.D. 2001. Microarray analysis of microbioal virulence factors. Appl. Environ. Microbiol. 67: 3258–3263.

Cook, N. 2003. The use of NASBA for the detection of microbial pathogens in food and environmental samples. J. Microbiol. Meth. 53: 165–174.

Cordwell, S.J. 2004. Exploring and exploiting bacterial proteomes. Methods Mol. Biol. 266: 115–135.

Cordwell, S.J., Nouwens, A.S., and Walsh, B.J. 2001. Comparative proteomics of bacterial pathogens. Proteomics 1: 461–472.

de Boer, E. and R. R. Beumer. 1999. Methodology for detecting and typing of foodborne microorganisms. Int. J. Food Microbiol. 50: 119–130.

DebRoy, C., Roberts, E., Kundrat, J., Davis, M.A., Briggs, C.E., and Fratamico, P.M. 2004. Detection of *Escherichia coli* serogroups O26 and O113 by PCR amplification of the *wzx* and *wzy* genes. Appl. Environ. Microbiol. 70: 1830–1832.

D'Souza, D.H. and Jaykus, L.-A. 2003. Nucleic acid sequence based amplification for the rapid and sensitive detection of *Salmonella enterica* from foods. J. Appl. Microbiol. 95: 1343–1350.

Downes, F P and Ito, K. 2001. Compendium of Methods for the Microbiological Examination of Foods, 4th edition, American Public Health Association, Washington, D.C.

Dykes, G.A., Sampathkumar, B., and Korber, D.R. 2003. Planktonic or biofilm growth affects survival, hydrophobicity and protein expression patterns of a pathogenic *Campylobacter jejuni* strain. Int. J. Food Microbiol. 89: 1–10.

Entis, P., Fung, D.Y.C. Griffiths, M.W., McIntyre, L., Russell, S., Sharpe, A.N., and Tortorello, M.L. 2001. Rapid methods for detection, identification, and enumeration. In: Compendium of Methods for the Microbiological Examination of Foods, 4th edition, American Public Health Association, Washington, D.C. p. 89–126.

Ercolini, D. 2004. PCR-DGGE fingerprinting: novel strategies for detection of microbes in food. J. Microbiol. Methods 56: 297–314.

Fabre, F., Kervarrec, C., Mieuzet, L., Riault, G., Vialatte, A., and Jacquot, E. 2003. Improvement of *Barley yellow dwarf virus*-PAV detection in single aphids using a fluorescent real time RT-PCR. J. Virol. Methods 110: 51–60.

Feng, P. 2001. Development and impact of rapid methods for detection of foodborne pathogens. In: Food Microbiology Fundamentals and Frontiers. M.P., Doyle, L.R. Beuchat, and T.J. Montville, 2nd ed. Washington, DC, ASM Press, pp. 775–796.

Fernández, L., Márquez, I., and Guijarro, J.A. 2004. Identification of specific *in vivo*-induced (ivi) genes in *Yersinia ruckeri* and analysis of ruckerbactin, a catecholate siderophore iron acquisition system. Appl. Environ. Microbiol. 70: 5199–5207.

Filion, M., Hamelin, R.C., Bernier, L., and St Arnaud, M. 2004. Molecular profiling of rhizosphere microbial communities associated with healthy and diseased black spruce (*Picea mariana*) seedlings grown in a nursery. Appl. Environ. Microbiol. 70: 3541–3551.

Fortin, N.Y., Mulchandani, A., and Chen, W. 2001. Use of real-time polymerase chain reaction and molecular beacons for the detection of *Escherichia coli* O157:H7. Anal. Biochem. 289: 281–288.

Fratamico, P.M. and Strobaugh, T.P. 1998. Simultaneous detection of *Salmonella* spp. and *Escherichia coli* O157:H7 by multiplex PCR. J. Indust. Microbiol. Biotechnol. 21: 92–98.

Fratamico, P.M. 2001. Applications of the polymerase chain reaction (PCR) for detection, identification, and typing of foodborne microorganisms. In: Microbial Food Contamination, C.Wilson ed. CRC Press, Boca Raton, FL, pp. 95–115

Fratamico, P.M., Bagi, L.K., and Pepe, T. 2000. A multiplex polymerase chain reaction assay for rapid detection and identification of *Escherichia coli* O157:H7 in foods and bovine feces. J. Food Prot. 63: 1032–1037.

Fratamico, P.M., Briggs, C.E., Needle, D., Chen, C.-Y., and DebRoy, C. 2003. Sequence of the *Escherichia coli* O121 O-antigen gene cluster and detection of enterohemorrhagic *E. coli* O121 by PCR amplification of the *wzx* and *wzy* genes. J. Clin. Microbiol. 41: 3379–3383.

Frias-Lopez, J., Bonheyo, G.T., and Fouke, B.W. 2004. Identification of differential gene expression in bacteria associated with coral black band disease by using RNA-arbitrarily primed PCR. Appl. Environ. Microbiol. 70: 3687–3694.

FSIS (Food Safety and Inspection Service). 1998. Microbiology Laboratory Guidebook, 3rd Edition, USDA, FSIS, Office of Public Health and Safety, Microbiology Division.

FSIS (Food Safety and Inspection Service). 2002. Detection, isolation, and identification of *Escherichia coli* O157:H7 and O157:NM (non-motile) from meat products. In: Microbiology Laboratory Guidebook, revision MLG 5.03, chapter 5, 3rd Edition, USDA, FSIS, Office of Public Health and Safety, Microbiology Division.

Fukiya, S., Mizoguchi, H., Tobe, T., and Mori, H. 2004. Extensive genomic diversity in pathogenic *Escherichia coli* and *Shigella* strains revealed by comparative genomic hybridization microarray. J. Bacteriol. 186: 3911–3921.

Fung, D.Y.C. 2002. Rapid methods and automation in microbiology. Compr. Rev. Food Sci. Food Saf. 1: 3–22.

Gery, S., and Lavi, S. 1997. Purification and cloning of differential display products. BioTechniques 23: 198–202.

Glaser, P., Frangeul, L., Buchrieser, C., Rusniok, C., Amend, A., Baquero, F., Berche, P., Bloecker, H., Brandt, P., Chakraborty, T., Charbit, A., Chetouani, F., Couve, E., de Daruvar, A., Dehoux, P., Domann, E., Dominguez-Bernal, G., Duchaud, E., Durant, L., Dussurget, O., Entian, K.D., Fsihi, H., Portillo, F.G., Garrido, P., Gautier, L., Goebel, W., Gomez-Lopez, N., Hain, T., Hauf, J., Jackson, D., Jones, L.M., Kaerst, U., Kreft, J., Kuhn, M., Kunst, F., Kurapkat, G., Madueno, E., Maitournam, A., Vicente, J.M., Ng, E., Nedjari, H., Nordsiek, G., Novella, S., de Pablos, B., Perez-Diaz, J.C., Purcell, R., Remmel, B., Rose, M., Schlueter, T., Simoes, N., Tierrez, A., Vázquez-Boland, J.A., Voss, H., Wehland, J., and Cossart, P. 2001. Comparative genomics of *Listeria* species. Science 294: 849–852.

Graves, P.R. and Haystead, T.A.J. 2002. Molecular Biologist's Guide to Proteomics. Micro. Mol. Biol. Rev. 66: 39–63.

Handelsman, J. 2004. Metagenomics: Application of genomics to uncultured microorganisms. Microbiol. Mol. Biol. Rev. 68: 669–685.

Hecker, M. and Volker, U. 2004. Towards a comprehensive understanding of *Bacillus subtilis* cell physiology by physiological proteomics. Proteomics. 4: 3727–3750.

Helloin, E., Jansch, L., and Phan-Thanh, L. 2003. Carbon starvation survival of *Listeria monocytogenes* in planktonic state and in biofilm: A proteomic study. Proteomics 3: 2052–2064.

Hou, J.Y., Graham, J.E., and Clark-Curtiss, J.E. 2002. Mycobacterium avium genes expressed during growth in human macrophages detected by selective capture of transcribed sequences (SCOTS). Infect. Immun. 70: 3714–3726.

Jean, J., D'Souza, D.H., and Jaykus, L.A. 2004. Multiplex nucleic acid sequence-based amplification for simultaneous detection of several enteric viruses in model ready-to-eat foods. Appl. Environ. Microbiol. 70: 6603–6610.

Josefsen, M.H., Jacobsen, N.R., and J. Hoorfar. 2004. Enrichment followed by quantitative PCR both for rapid detection and as a tool for quantitative risk assessment of foodborne thermotolerant campylobacters. Appl. Environ. Microbiol. 70: 3588–3592.

Jungblut, P.R. 2001. Proteome analysis of bacterial pathogens. Microbes Infect. 3: 831–840.

Klein, P.G. and Juneja, V.K. 1997. Sensitive detection of viable *Listeria monocytogenes* by reverse transcriptase-PCR. Appl. Environ. Microbiol. 63: 4441–4448.

Keramas, G., Bang, D.D., Lund, M., Madsen, M., Bunkenborg, H., Telleman, P., and Christensen, C.B.V. 2004. Use of culture, PCR analysis, and DNA microarrays for detection of *Campylobacter jejuni* and *Campylobacter coli* from chicken feces. J. Clin. Microbiol. 42: 3985–3991.

Koopmans, M., von Bonsdorff, C.H., Vinjé, j., de Medici, D., and Monroe, S. 2002. Foodborne viruses. FEMS Microbiol. Lett. 26: 187–205.

Kricka, L.J. 1998. Miniaturization of analytical systems. Clin. Chem. 44: 2008–2014.

Kricka, L.J. 2001. Microchips, microarrays, biochips and nanochips: personal laboratories for the 21st century. Clin. Chim. Acta 307: 219–223.

Lafarge, V., Ogier, J.C., Girard, V., Maladen, V., Leveau, J.Y., Gruss, A., and Delacroix-Buchet, A. 2004. Raw cow milk bacterial population shifts attributable to refrigeration. Appl. Environ. Microbiol. 70: 5644–5650.

Liu, S., Graham, J.E., Bigelow, L., Morse, P.D., and Wilkinson, B.J. 2002. Identification of *Listeria monocytogenes* genes expressed in response to growth at low temperature. Appl. Environ. Microbiol. 68: 1697–1705.

Liu-Stratton, Y., Roy, S., and Sen, C.K. 2004. DNA microarray technology in nutraceutical and food safety. Toxicol. Lett. 150: 29–42.

McCartney, A.L. 2002. Application of molecular biological methods for studying probiotics and the gut flora. Br. J. Nutr. 88 Suppl 1: S29-S37.

McIngvale, S.C., Elhanafi, D., and Drake, M.A. 2002. Optimization of reverse transcriptase PCR to detect viable Shiga-toxin-producing *Escherichia coli*. Appl. Environ. Microbiol. 68: 799–806.

McMeekin, T. A. 2003. Detecting Pathogens in Food. CRC Press, New York.

Marko-Varga, G., Nilsson, J., and Laurell, T. 2003. New directions of miniaturization within the proteomics research area. Electrophoresis 24: 3521–3532.

Maukonen, J., Mättö, Wirtanen, G., Raaska, L., Mattila-Sandholm, T., and Saarela, M. 2003. Methodologies for the characterization of microbes in industrial environments: a review. J. Indust. Microbiol. Biotechnol. 30: 327–356.

Mayer, Z., Färber, P., and Geisen, R. 2003. Monitoring the production of alflatoxin B1 in wheat by measuring the concentration of *nor-1* mRNA. Appl. Environ. Microbiol. 69: 1154–1158.

Meays, C.L., Broersma, K., Nordin, R., and Mazumder, A. 2004. Source tracking fecal bacteria in water: a critical review of current methods. J. Environ. Man. 73: 71–79.

Mhlanga, M.M. and Malmberg, L. 2001. Using molecular beacons to detect single-nucleotide polymorphisms with real-time PCR. Methods 25: 463–471.

Morin, N.J., Gong, Z., and Li, X.F. 2004. Reverse transcription-multiplex PCR assay for simultaneous detection of *Escherichia coli* O157: H7, *Vibrio cholerae* O1, and *Salmonella* Typhi. Clin. Chem. 50: 2037–2044.

Nelson, K.E., Fouts, D.E., Mongodin, E.F., Ravel, J., DeBoy, R.T., Kolonay, J.F., Rasko, D.A., Angiuoli, S.V., Gill, S.R., Paulsen, I.T., Peterson, J., White, O., Nelson, W.C., Nierman, W., Beanan, M.J., Brinkac, L.M., Daugherty, S.C., Dodson, R.J., Durkin, A.S., Madupu, R., Haft, D.H., Selengut, J., Van Aken, S., Khouri, H., Fedorova, N., Forberger, H., Tran, B., Kathariou, S., Wonderling, L.D., Uhlich, G.A., Bayles, D.O., Luchansky, J.B., and Fraser, C.M. 2004. Whole genome comparisons of serotype 4b and 1/2a strains of the foodborne pathogen *Listeria monocytogenes* reveal new insights into the core genome components of this species. Nucleic Acids Res. 32: 2386–2395.

Ngutter C. and Donnelly C. 2003. Nitrite-induced injury of *Listeria monocytogenes* and the effect of selective versus nonselective recovery procedures on its isolation from frankfurters. J. Food Prot. 66: 2252–2257.

Nickisch-Rosenegk, M., Marschan, X., Andresen, D., Abraham, A., Heise, C., and Bier, F.F. 2005. On-chip PCR amplification of very long templates using immobilized primers on glassy surfaces. Biosens. Bioelectron. 20: 1491–1498.

Notermans, S., Beumer, R., and Rombouts, F. 1997. Detecting food-borne pathogens and their toxins – conventional versus rapid and automated methods. In: Food Microbiology Fundamentals and Frontiers, Doyle, M P, Beuchat, L R, and Montville, T J, ed. ASM Press, Washington, D.C. p. 697–709.

Palladino, S., Kay, I.D., Costa, A.M., Lambert, E.J., and Flexman, J.P. 2003. Real-time PCR for the rapid detection of *vanA* and *vanB* genes. Diag. Microbiol. Infect. Dis. 45: 81–84.

Panicker, R.C., Huang, X., and Yao, S.Q. 2004. Recent advances in peptide-based microarray technologies. Comb. Chem. High Throughput Screen 7: 547–556.

Pradel, N., Leroy-Setrin, S., Joly, B., and Livrelli, V. 2002. Genomic subtraction to identify and characterize sequences of Shiga toxin-producing *Escherichia coli* O91:H21. Appl. Environ. Microbiol. 68: 2316–2325.

Rodríguez-Lázaro, D., Hernández, M., Scortti, M., Esteve, T., Vázquez-Boland, J.A., and Pla, M. 2004a. Quantitative detection of *Listeria monocytogenes* and *Listeria innocua* by real-time PCR: assessment of *hly*, *iap*, and *lin02483* targets and AmpliFluor technology. Appl. Environ. Microbiol. 70: 1366–1377.

Rodríguez-Lázaro, D., Lloyd, J., Herrewegh, A., Ikonomopoulos, J., D'Agostino, M., Pla, M., and Cook, N. 2004b. A molecular beacon-based real-time NASBA assay for detection of *Mycobacterium avium* subsp. *paratuberculosis* in water and milk. FEMS Microbiol. Lett. 237: 119–126.

Rudi, K., Flateland, S.L., Hanssen, J.F., Bengtsson, G., and Nissen, H. 2002a. Development and evaluation of a 16S ribosomal DNA array-based approach for describing complex microbial communities in ready-to-eat vegetable salads packed in a modified atmosphere. Appl. Environ. Microbiol. 68: 1146–1156.

Rudi, K., Nogva, H.K., Moen, B., Nissen, H., Bredholt, S., Møretrø, T., Naterstad, K., and Holck, A. 2002b. Development and application of new nucleic acid-based technologies for microbial community analyses in foods. Int. J. Food Microbiol. 78: 171–180.

Ryser, E.T. and Donnelly, C.W. 2001. *Listeria*. In: Compendium of Methods for the Microbiological Examination of Foods. 4th edition, American Public Health Association, Washington, D.C.

Schaad, N.W. and Frederick, R.D. 2002. Real-time PCR and its application for rapid plant disease diagnostics. Can. J. Plant Pathol. 24: 250–258.

Schena, M., Heller, R.A., Theriault, T.P., Konrad, K., Lachenmeier, E., and Davis, R.W. 1998. Microarrays: biotechnology's discovery platform for functional genomics. Trends Biotechnol. 16: 301–306.

Schweitzer, B., Predki, P., and Snyder, M. 2003. Microarrays to characterize protein interactions on a whole-proteome scale. Proteomics 3: 2190–2199.

Sergeev, N., Volokhov, D., Chizhikov, V., and Rasooly, A. 2004. Simultaneous analysis of multiple staphylococcal enterotoxin genes by an oligonucleotide microarray assay. J. Clin. Microbiol. 42: 2134–2143.

Sharma, V.K. 2002. Detection and quantitation of enterohemorrhagic *Escherichia coli* O157, O111, and O26 in beef and bovine feces by real-time polymerase chain reaction. J. Food Prot. 65: 1371–1380.

Simpkins, S.A., Chan, A.B., Hays, J., Pöpping, B., and Cook, N. 2000. A RNA transcription-based amplification technique (NASBA) for the detection of viable *Salmonella enterica*. Lett. Appl. Microbiol. 20: 75–79.

Sorsa, L.J., Dufke, S., and Schubert, S. 2004. Identification of novel virulence-associated loci in uropathogenic *Escherichia coli* by suppression subtractive hybridization. FEMS Microbiol. Lett. 230: 203–208.

Stevens, K.A. and Jaykus, L.A. 2004. Bacterial separation and concentration from sample matrices: a review. Crit. Rev. Microbiol. 30: 7–24.

Suslow, T.V., Wu, J., Fett, W. F., and Harris, L.J. 2002. Detection and elimination of *Salmonella* Mbandaka from naturally contaminated alfalfa seeds by treatment with heat or calcium hypochlorite. J. Food Prot. 65: 452–458.

Takahashi, H., Kimura, B., Yoshikawa, M., Gotou, S., Watanabe, I., and Fujii, T. 2004. Direct detection and identification of lactic acid bacteria in a food processing plant and in meat products using denaturing gradient gel electrophoresis. J. Food Prot. 67: 2515–2520.

Templin, M.F., Stoll, D., Schwenk, J.M., Potz, O., Kramer, S., and Joos, T.O. 2003. Protein microarrays: promising tools for proteomic research. Proteomics 3: 2155–2166.

Trémoulet, F., Duché, O., Namane, A., Martinie, B., The European *Listeria* Genome Consortium, and Labadie, J.C. 2002. Comparison of protein patterns of *Listeria monocytogenes* grown in biofilm or in planktonic mode by proteomic analysis. FEMS Microbiol. Lett. 210: 25–31.

Uyttendaele, M., Grangette, C., Rogerie, F., Pasteau, S., Debevere, J., and Lange, M. 1998. Influence of cold stress on the preliminary enrichment time needed for detection of enterohemorrhagic *Escherichia coli* in ground beef by PCR. Appl. Environ. Microbiol. 64: 1640–1643.

Valdivia, R.H., and Falkow, S. 1996. Bacterial genetics by flow cytometry: rapid isolation of *Salmonella typhimurium* acid-inducible promoters by differential fluorescence induction. Mol. Microbiol. 22: 367–378.

Van Dyk, T.K., DeRose, E.J., and Gonye, G.E. 2001. LuxArray, a high-density, genomewide transcription analysis of *Escherichia coli* using bioluminescent reporter strains. J. Bacteriol. 183: 5496–5505.

Vasil, M.L. 2003. DNA microarrays in analysis of quorum sensing: strengths and limitations. J. Bacteriol. 185: 2061–2065.

Vet, J.A.M., Majithia, A.R., Marras, S.A.E., Tyagi, S., Dube, S., Poiesz, B.J., and Kramer F.R. 1999. Multiplex detection of four pathogenic retroviruses using molecular beacons. Proc. Natl. Acad. Sci. USA 96: 6394–6399.

Vishnubhatla, A., Fung, D.Y.C., Oberst, R.D., Hays, M.P., Nagaraja, T.G., and Flood, S.J.A. 2000. Rapid 5′ nuclease (TaqMan) assay for detection of virulent strains of *Yersinia enterocolitica*. Appl. Environ. Microbiol. 66: 4131–4135.

Vora, G.J., Meador, C.E., Stenger, D.A., and Andreadis, J.D. 2004. Nucleic acid amplification strategies for DNA microarray-based pathogen detection. Appl. Environ. Microbiol. 70: 3047–3054.

Walduck, A., Rudel, T., and Meyer, T.F. 2004. Proteomic and gene profiling approaches to study host responses to bacterial infection. Curr. Opin. Microbiol 7: 33–38

Walker, G.T., Nadeau, J.G., Spears, P.A., Schram, J.L., Nycz, C.M., and Shank, D.D. 1994. Multiplex strand displacement amplification (SDA) and detection of DNA sequences from *Mycobacterium tuberculosis* and other mycobacteria. Nucleic Acids Res. 22: 2670–2677.

Wang, G., Clark, C.G., and Rodgers, F.G. 2002. Detection in *Escherichia coli* of the genes encoding major virulence factors, the genes defining the O157:H7 serotype, and components of the type 2 Shiga toxin family by multiplex PCR. J. Clin. Microbiol. 40: 3613–3619.

Wang, J., Lory, S., Ramphai, R., and Jin, S. 1996. Isolation and characterization of *Pseudomonas aeruginosa* genes inducible by respiratroy mucus derived from cystic fibrosis patients. Mol. Microbiol. 22: 1005–1012.

Wang, L., Brigg, C.E., Rothemund, D., Fratamico, P., Luchansky, J.B., and Reeves, P.R. 2001. Sequence of the *E. coli* O104 antigen gene cluster and identification of O104 specific genes. Gene 270: 231–236.

Waters, L.C., Jacobson, S.C., Kroutchinina, N., Khandurina, J., Foote, R.S., and Ramsey, J.M. 1998. Multiple sample PCR amplification and electrophoretic analysis on a microchip. Anal. Chem. 70: 158:162.

Wells, J.M., and Bennik, M.H.J. 2003. Genomics of foodborne bacterial pathogens. Nutr. Res. Rev. 16: 21–35.

Westin, L., Xu, X., Miller, C., Wang, L., Edman, C.F., and Nerenberg, M. 2000. Anchored multiplex amplification on a microelectronic chip array. Nature Biotechnol. 18:199–204.

Whittam, T.S., and Bumbaugh, A.C. 2002. Inferences from whole-genome sequences of bacterial pathogens. Curr. Opin. Genet. Dev. 12: 719–725.

Woolley, A.T., Hadley, D., Landre, P, deMello, A.J., Mathies, R.A., and Northrup, M.A. 1996. Functional integration of PCR amplification and capillary electrophoresis in a microfabricated DNA analysis device. Anal. Chem. 68: 4081–4086.

Wu, F.M., and Muriana, P.M. 1995. Genomic subtraction in combination with PCR for enrichment of *Listeria monocytogenes*-specific sequences. Int. J. Food Microbiol. 27: 161–174.

Wu, F.M., and Muriana, P.M. 1999. Cloning, sequencing, and characterization of genomic subtracted sequences from *Listeria monocytogenes*. Appl. Environ. Microbiol. 65: 5427–5430.

Zhang, D.Y., Brandwein, M., Hsuih, T., and Li, H.B. 2001. Ramification amplification: a novel isothermal DNA amplification method. Mol. Diagn. 6: 141–150.

# Animal and Cell Culture Models for Foodborne Bacterial Pathogens

2

Arun K. Bhunia and Jennifer L. Wampler

## Abstract

In the field of medical microbiology, pathogens that transmit through food are not well studied. The importance of foodborne pathogens was not realized until the 1980s when several outbreaks caused by *Listeria monocytogenes* and *Escherichia coli* O157:H7 resulted in a large number of fatalities. Since then, in-depth molecular studies have been undertaken to understand the mechanism of pathogenesis and immunity using animal and cell culture models. Several models already exist and efforts continue to find suitable ones that address unique pathogenic feature or ultimately could be used for identification and detection purposes. A specific model may be sensitive to a pathogen or a group of pathogens but may be unresponsive to others. Some pathogens are host-specific and the degree of virulence response in one model may be different from that of another. In view of this, a general description of animal and cell culture models is presented. In addition, how animal and cell culture models facilitated our understanding of selected foodborne pathogens such as *Listeria monocytogenes*, *Salmonella* species, *Campylobacter* species, diarrheagenic *Escherichia coli*, *Bacillus cereus*, *Clostridium perfringens*, and *Shigella* species are described.

## Animal models

Humans would be the ideal model to study human diseases; however, safety and ethical concerns limit or often forbid human use. Animal models have been used widely and data are converted to approximate the human form of the disease. Attenuated or mildly infective or nonpathogenic strains, however, have been used in human trials. The single most important criterion for selecting an animal model is that the animal must be sensitive to the infection. Furthermore, the animal model should also manifest clinical syndromes equivalent or comparable to that of a human form of infection. It should also provide data on the role of a specific virulence factor (using a recombinant strain or a mutant strain), be suitable for studying immune response, or provide definitive data on prophylactic treatments for vaccines or probiotics. In the animal model, route of infection, tissue distribution, and degree of virulence should be equivalent to human infection. Commonly used animals are rodents including mice, rats, rabbits and guinea-pigs. Unconventional models such as ferrets, gerbil, silkworm larvae (*Bombyx mori*) (Kaito *et al.*, 2002), and *Xenopus* (frog) oocyte have also been tested. Primates such as monkey and baboons serve as very good models due to their relative genetic homogeneity with humans; housing, maintenance, variability in immune status, and ethical considerations, however, limit their routine use.

A single animal model may not reproduce all clinical signs for a disease; hence, multiple models had been employed to study each form of disease (see descriptions for specific pathogens). Furthermore, new models or transgenic animal models have been developed or tested to reproduce a specific clinical signature for a pathogen. No single animal or a single species can justifiably be sufficient to replace a human subject.

Rodents such as mice, rats, and guinea-pigs are commonly used and have several advantages. They are often inbred thus are genetically homogeneous and will give reproducible results. They breed rapidly and large numbers can be used to design a statistically significant experimental method. They are small and require less housing spaces. However, there are some limitations, which may hinder their use; they may not be sensitive to some human pathogens. Therefore, neonatal or immunocompromised animals are used. Examples of immunocompromised animals are; (i) irradiated mice – in which immune cells and stem cells are killed after exposure to irradiation, (ii) nude mice (no functional T cells), and (iii) severe combined immunodeficient (SCID) mice, which lack functional T and B cells.

Differences in physiology and behavior may also influence the pathogenesis result. They are often coprophagy thus reingestion of same microorganism through fecal materials may affect the outcome of the experiment. Population of resident indigenous microflora in animals may be different from that of humans, thus their response may differ regarding a given pathogen.

### Measurement of pathogenesis

Pathogenicity is often measured by determining the lethal dose (LD) or infective dose (ID) of the test organism. However, infectivity or lethality may vary depending on the route of administration used, i.e., oral, intragastric (ig), intravenous (iv), intraperitoneal (ip), intramascular (im), subcutaneous (sc) or intradermal (id). Generally, to assess infective dose for a given pathogen, natural route of infection are used, i.e., for foodborne pathogens, oral or ig route are used, while to determine lethal dose, alternative routes are used. The pathogenesis measurements are expressed as $ID_{50}$ or $LD_{50}$. $ID_{50}$ is defined as the number of bacteria required to infect 50% of animals, whereas $LD_{50}$ is defined as the number of bacteria required to kill 50% of animals (Reed and Muench, 1938). Infectious dose or lethal dose data provide crude measure of infection that includes cumulative effects of colonization, invasion, spread or tissue damage. Thus, $ID_{50}$ or $LD_{50}$ values could not be compared for a given pathogen because these provide relative measure of virulence and cannot compare two different diseases.

Alternatively, pathogenesis has been measured by determining bacterial cell counts in infected organs or tissues following oral, intravenous or intraperitoneal administration. Differences in counts for strains in target organs and tissues are indicative of their virulence traits. This type of experiment is very informative and will provide data on the specific action of virulence traits for adhesion, invasion, cell-to-cell movement or overall spread and translocation to distant organs.

## Organ culture

Animal organs can be used to investigate certain bacterium–host interactions. Organs are genetically intact, multiple cell types are represented, and cells retain their original shape and configurations. The most commonly used organ for foodborne pathogen is the ligated ileal loop (LIL) assay (Spira *et al.*, 1981) and is used for diarrheagenic microorganisms described later. Both rabbits and rats have been used for this test. After animals are anaesthetized, the abdomens are opened and loops are created in the intestine 5 cm apart with no more than four loops per animal. Bacteria or toxin suspension are injected into the loops, the intestine is placed back inside, and the incision is closed. Depending on the specific experimental set up, the abdomen will be opened again at a specified time (from a few minutes to several hours) and the intestine is examined for acute inflammation or fluid accumulation (ballooning) to determine the toxic effect of test materials.

In the embryonated egg assay, 12- to 14-day-old embryonated hen's eggs are injected aseptically with test organisms in the chorioallantoic membrane (CAM) cavity and incubated for 3–4 days. Death of embryos or characteristic cytopathic effects is indicative of the infective nature of the test organisms. This model is used for certain bacterial species and viruses.

## Cultured cell lines

Cells originated from human or animals are widely used in many fields of biological research including food microbiology. It is an attractive and popular model to study mechanism of pathogenesis in cellular and molecular levels. Once the detailed mechanism is established, an animal model could be used to further validate *in vitro* findings.

Primary cells will provide the best results since cells maintain their original cellular characteristics with expression of various surface molecules. However, it is rather tedious to harvest and purify a cell type for each experiment. Another disadvantage is the short-lived nature of these cells. To overcome such problems, secondary cells are created and many of those are readily available from commercial sources. Secondary cells can survive indefinitely as long as the source of nutrient and proper growth environment is maintained. Often these cells are referred to as immortal cells and originated from cancer cells. Sometimes primary cells could be transformed by genes from oncogenic viruses to create secondary cell lines. For example, simian virus (SV40), papillomavirus, adenovirus, human T-cell leukemia virus or Epstein–Barr virus has been used to transform cells. Hybridoma cells created by fusing primary cells with myeloma cells have also been used. These cells maintain dual characteristics of both partner cells and are an elegant model for immunological research.

Cell culture models are ideal for studying host–bacterium interaction in a simple and controlled environment. Cells multiply rapidly and maintenance is relatively cheaper than animals. Also this system can be used with radioactive materials, toxins, or transfected with foreign DNA to study genetic regulation of a disease. However, there are some limitations for use with cultured cell lines. A majority of cell lines derived from tumor cells may thus result in genetic aberrations, which are not genetically identical to the parental cells. Furthermore, continuous growth and subculture of these cells may spawn mutations or genetic rearrangements in the chromosome. Consequently, cells may lose original traits such as tissue specific receptors for bacterial adhesins or express alternative receptors to promote increased adhesion or invasion. Normal cells are polarized in their natural environment of the body, i.e., different parts of the cells are exposed to different environments (lumen, adjacent cells, underlying blood vessels and tissues). Polarized cells contain different sets of proteins important for their functions. Disruption of cells will expose some surface proteins that are normally hidden. Cultured cell lines can be made polarized by growing them in the presence of hormones. In spite of this, fundamentally immortalized cells are not equal to the cells in the body.

Host tissue consists of multiple cell types such as epithelial, fibroblast, muscle cells and neurons whereas the cultured cells consisted of only one cell type. During the infection process multiple host cells, mucus or other secretory components (i.e. sIgA, lactoferrin, etc.) interact with a pathogen. In the cultured cell line that interaction cannot be duplicated.

## Measurement of virulence

### Cytopathogenicity assay

An assay of pathogen or toxin interactions with eukaryotic cells is often referred to as a cytotoxicity or cytopathogenicity assay. This assay could be used to confirm the pathogenic or virulence potential of a microorganism after being isolated from a sample. This assay also could provide insight of molecular mechanism of pathogenesis. Cytopathogenicity can be measured by (i) microscopy, (ii) trypan blue exclusion test, (iii) alkaline phosphatase (ALP) assay, (iv) lactate dehydrogenase (LDH) assay and (v) MTT (3 [4,5-dimethyl thiazolyl-2]-2, 5-diphenyltetrazolium bromide) assay.

*Microscopy*
Degree of cell damage can be assessed under a phase-contrast light microscope. The typical cytopathic effect (CPE) consists of cell detachment, flocculation, rounding or elongation, cell lysis, cytoplasmic granulation, and cell death (Figure 2.1). Highest dilutions of the toxin or pathogen showing CPE are considered as the cytotoxic titer for that agent. Furthermore, the nature of cell death (necrosis vs. apoptosis) can be detected by fluorescence microscopy after staining cells with acridine orange, ethidium bromide, or propidium iodide. Scanning electron microscopy can be used to examine membrane damage, pore formations, and apoptotic bodies in eukaryotic cells (Figure 2.1).

*Trypan blue exclusion test*
Eukaryotic cell suspensions are mixed with equal volumes of trypan blue (0.4%) solution and placed in a hemacytometer to count viable or dead cells under a microscope. Viable cells with intact membrane will exclude dye diffusion, therefore, will appear as bright translucent while dead cells with damaged membrane will allow dye penetration and will appear blue.

*Alkaline phosphatase assay*
It is present in various tissues such as kidney tubule, bone (osteoblasts), liver and placenta. It is also present in selected mammalian cells like lymphocytes, fibroblasts and certain intestinal epithelium cells and has been used as a marker for cell cytotoxicity. Alkaline phosphatase (ALP) is a glycosyl-phosphatidylinositol (GPI)- anchored membrane protein and could be released from the damaged membrane upon exposure to the membrane active toxins. Isozyme pattern of ALP is variable and may range from 100 to 190 kDa thus membrane pores have to be substantially large to allow the excretion of these large molecules. *p*-Nitrophenyl phosphate (PNPP) or methyl umbellypherol phosphate (MUP) is used as substrate for ALP, which could produce colored or fluorescence end products, respectively and are detected by spectrometry (Bhunia and Westbrook, 1998; Shroyer and Bhunia, 2003). The drawback of this system is that it is present in low concentrations in various epithelial cells, which are commonly used for cytotoxicity assays. In contrasts, cells of B- or T-lymphocyte origin carry abundant quantities of ALP and are used as a marker for cell activation and cytotoxicity (Souvannavong *et al.*, 1995; Menon *et al.*, 2003).

*Lactate dehydrogenase assay*
A majority of cells carry lactate dehydrogenase (LDH) and is widely used as a marker for cell cytotoxicity. It is a low molecular weight enzyme (35 kDa) and may exist in tetrameric form. LDH converts lactate to pyruvate and release hydrogen ions that reduces $NAD^+$ to $NADH/H^+$, which catalyzes transfer of H ion to tetrazolium salt converting it into red formazan and is measured by a spectrophotometer at 490/655 nm. Because of its small size, LDH can be easily released from the cells with minor damage or destabilization in the cell membrane (Bhunia and Westbrook, 1998; Roberts *et al.*, 2001).

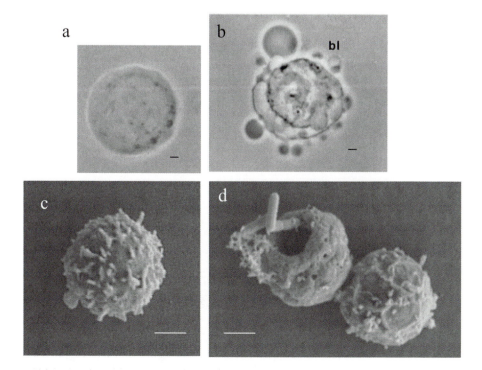

**Figure 2.1** Light microscopic (a and b) and scanning electron microscopic (c and d) photographs of mouse hybridoma B cell (Ped-2E9) and primary B cells infected with *Listeria monocytogenes*. Cells in panels a and c are healthy while those in panels b and d are infected. Blebs (b), pores (d) and destruction of surface molecules (d) are evident due to infection. Bar = 2 micron. [Reproduced with permission from Bhunia and Feng, (1999) and Menon *et al*. (2003).]

*MTT assay*

Eukaryotic cell viability or proliferation could be assayed by metabolic staining method. The yellow tetrazolium MTT (3-[4,5-dimethyl thiazolyl-2]-2,5-diphenyltetrazolium bromide) is a water-soluble compound, which is reduced by metabolically active cells. MTT is converted to water-insoluble purple formazan by the action of dehydrogenase enzymes that generates NADH and NADPH. Purple formazan could be quantified by a spectrophotometer at 570 nm. This assay has been used to measure enterotoxin action of several foodborne pathogens on eukaryotic cells (Finlay *et al.*, 1999; Fletcher and Logan, 1999).

*Adhesion assay*

Pathogenesis is most often initiated by a successful attachment of microorganisms to a target cell. This could be modeled by using a cell line *in vitro*. Cell monolayers are inoculated with test organisms for a brief period (30 min to 2 h) and unbound organisms are washed away with buffer. Bacterial attachment could be examined by specific immunostaining, fluorescence staining or by Giemsa staining of cell monolayers (Figure 2.2). Attachment could also be analyzed by lysing the cell layers with mild detergent (Triton-X or SDS) and subsequently plating on the nutrient agar plates for enumeration of bacteria. Differential fluorescence staining could also be used to distinguish surface attached bacteria from intracellular bacteria (Drevets and Campbell, 1991).

*Invasion assay*

Antibiotics are used to selectively kill extracellular bacteria without affecting intracellular ones. In this assay, cell monolayers are infected with test bacteria for 1–2 h to allow bacterial entry in the cell. Antibiotic (generally gentamicin) is added to kill extracellular bacteria. Cell monolayers are washed and treated with detergent or cold buffer solution to release bacterial cells and then plated to enumerate intracellular bacteria (Gaillard *et al.*, 1987).

*Plaque assay*

The plaque assay has been used to determine the cell-to-cell movement of certain intracellular bacterial pathogens like *Shigella* and *Listeria* species (Sun *et al.*, 1990). Washed fibroblast or epithelial cell monolayers are inoculated with the test organisms for 1–2 h and then washed to remove unbound bacteria and treated with gentamicin (10 µg/mL) for 1.5 h. After thorough washing, cell monolayers are overlaid with tempered agarose (0.7%) containing gentamicin (10 µg/mL) and incubated at 37°C. Agarose layer is stained after about 72 h with neutral red (0.1%) and the plaques (clear zone) are visualized. Centripetal movement of the pathogen will result in massive cell

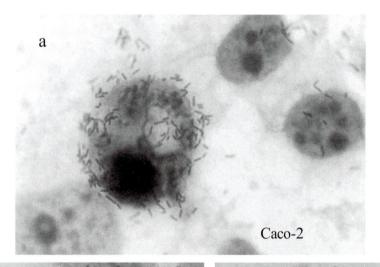

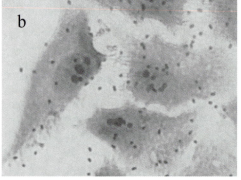

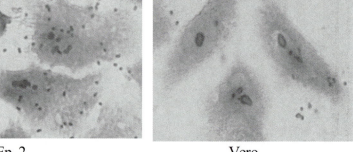

a

Caco-2

b

c

HEp-2

Vero

**Figure 2.2** Giemsa staining analyses of *Listeria monocytogenes* attachment to Caco-2 (a) and *Escherichia coli* O157:H7 attachment to HEp-2 (b) and Vero (c) cells (photographs courtesy of Yadilka Maldonado and Arun Bhunia).

death causing a clearing zone in the cell monolayers to form a plaque. Sizes and numbers of plaque are indicative of their virulence or infectivity potential.

*Cell death analysis*

Pathogen interaction often results in cellular death, which could be either necrotic or apoptotic (Squier and Cohen, 2001). Necrosis, or nonspecific cell death, usually leads to cell swelling with a loss of membrane integrity and initiation of a host inflammatory response. In contrast, apoptosis is a tightly regulated method for removal of damaged or unneeded "self" cellular debris, especially during growth, development, and cell homeostasis. Apoptosis is characterized by cell shrinkage, chromatin condensation and margination, DNA fragmentation and membrane blebbing. Many cytokines and cellular proteins regulate cells death via two main pathways: (i) By activating tumor necrosis factor receptor (TNFR) and (ii) by activating CD95 receptor (also known as Fas). These receptors bind effector molecules and set an intercellular apoptotic cascade in motion by recruiting, binding, and activating Caspase (Cas) proteins or down-regulating apoptosis by interacting with inhibitor proteins, such as proteins in the Bcl-2 family (Weinrauch and Zychlinsky, 1999).

Apoptosis is usually regarded as a non-inflammatory response that pinpoints specific damaged cells without full-scale tissue damage. However, in some cases of bacteria induced apoptosis, this process is often pro-inflammatory. Consequently, speculation surrounds the question of whether host cells initiate apoptosis in an attempt to reduce widespread infection or if bacterial cells themselves may trigger apoptosis through activation of the apoptotic protein cascade as an infection strategy (Zychlinsky and Sansonetti, 1997; Gao and Kwaik, 2000; Monack and Falkow, 2000).

Apoptosis or necrosis could be differentiated using a DNA fragmentation assay, by flow cytometry analysis of annexin–V binding to outer leaflet of the apoptotic cells membrane, or through a specific cell staining method such as acridine orange, ethidium bromide, or propidium iodide staining which binds to DNA, etc.

## Pathogenicity testing of selected foodborne pathogens

*Listeria monocytogenes*

*Animal models*

*Listeria monocytogenes* is a facultative intracellular pathogen and causes listeriosis characterized by febrile neurological disorders in immune compromised individuals. It also causes rare but unusual gastroenteritis in healthy persons and abortion and stillbirth in pregnant women (Vazquez-Boland *et al.*, 2001). Different animal models have been used depending on the pathogenic events that required study (Lecuit and Cossart, 2002). No single model is capable of reproducing mul-

tiple pathophysiological events that are evoked by *L. monocytogenes*. The mouse is the most widely used model to study pathogenicity, invasiveness, and uteroplacental translocation given the similarity to the human immune system (Buer and Balling, 2003). Species of mice used influence the outcome of the disease. For example, A/J mice manifest listeriosis upon intragastric administration while C57BL/6 mice do not (Czuprynski *et al.*, 2003). Swiss white, Balb/C and ICR white mice have been used routinely for pathogenicity testing (Stelma *et al.*, 1987; Conner *et al.*, 1989; Jaradat and Bhunia, 2003; Liu, 2004). Immune status of the animal is also critical. Since *L. monocytogenes* is a facultative intracellular pathogen and mostly causes infection in immunocompromised humans, it is logical to test pathogenicity using immune-suppressed animals. Immunosuppressive drugs such as pyrimethanine, cyclosporin A, carrageenan or hydrocortisone have been used in animals to simulate immunesuppressed conditions in humans. However there is no animal model to study *L. monocytogenes*-induced gastroenteritis.

The natural route of infection is oral. However, for lethal dose analysis, ip route has been used. Intraperitoneal route increases the infectivity of a strain than the oral or intragastric route. $LD_{50}$ for adult healthy mice in ip route is about $10^5$–$10^7$ CFU (Stelma *et al.*, 1987) while greater than $10^9$ CFU for ig route. In immunesuppressed mice, the $LD_{50}$ could be less than $10^2$ CFU (Stelma *et al.*, 1987). Hypovirulent or avirulent strains of *L. monocytogenes* had $LD_{50}$ values $10^8$–$10^9$ CFU (Roche *et al.*, 2003). Pregnancy does not affect pathogenicity but prolongs the infection and organisms could be isolated from the fetus (Lammerding *et al.*, 1992).

Routes of administration also affect infectivity. Oral or intragastric routes are the preferred routes to study bacterial colonization, crossing of intestinal and feto-placental barrier (Stephens *et al.*, 1991; Lammerding *et al.*, 1992). Intranasal administration by means of aerosol inoculation appeared to be more infective than the intragastric route of infection (Bracegirdle *et al.*, 1994) but much lower than the intravenous route (Mizuki *et al.*, 2002). Aerosol infection, though rare, but may occur in certain situations. $ID_{50}$ values for aerosol inoculation was $10^5$ CFU and death occurred in 4 days while in comparison, intragastric inoculation needed $10^9$–$10^{10}$ CFU to cause death in 7 days (Bracegirdle *et al.*, 1994). Mice immunized by intranasal route display higher resistance to infection than when inoculated by iv route (Mizuki *et al.*, 2002). Mice repeatedly fed with *Listeria* can cause central nervous system (CNS) infection (Altimira *et al.*, 1999). Variation in infectivity among *L. monocytogenes* strains has been reported when animals are challenged orally or by ig route (Barbour *et al.*, 2001; Jaradat and Bhunia, 2003; Jaradat *et al.*, 2003; Roche *et al.*, 2003). It has been documented that *L. monocytogenes* can enter through Peyer's patch-dependent or -independent routes during the intestinal phase of infection. This has been confirmed by using a rat ligated ileal loop (Pron *et al.*, 1998) and by using intrarectal administration of mice (Nishikawa *et al.*, 1998).

In some cases, it may be difficult to reproduce *L. mono-cytogenes* infections in mice following oral challenge. A transgenic mouse model has been developed expressing a specific receptor for internalin A, an invasion protein required for invasion of epithelial cells (Gaillard *et al.*, 1991). Internalin binds to E-cadherin, whose sequence is different in mouse. Human E-cadherin gene (hEcad) has been cloned in mouse to facilitate increased binding and translocation of *L. monocytogenes* to spleen and liver during intestinal phase of infection (Lecuit *et al.*, 2001). Internalin specific E-cad has been found in rat and guinea-pig; thus these models could be used as an alternative to study intestinal phase of listeriosis (Lecuit *et al.*, 2001).

In a rat model, both pregnant and juvenile rats have been used. Pregnancy did not affect $ID_{50}$ values although animals displayed abnormal reproductive outcomes (Schlech, 1993). *L. monocytogenes* colonization took place as early as 6 h at an infective dose of $10^6$ CFU. Unlike the mouse model, immunosuppression did not affect the $ID_{50}$ values in the rat model but prolonged the infection. Gastric acidity, however, affected infection and invasiveness. When stomach acidity was experimentally raised by cimetidine (raised stomach pH to 4.5), the infection rate was increased and significantly lowered the infective dose (Schlech, 1993).

A guinea-pig model has been introduced to simulate the gastrointestinal form of listeriosis (Cossart *et al.*, 2003). This model displayed a dose-dependent lethality and is recommended for orally acquired listeriosis (Cossart *et al.*, 2003).

The monkey (*Macaca fasicularis*) has been used to determine the infective dose for healthy individuals. Only a dose of $10^9$ cells showed septicemia, irritability, loss of appetite and occasional diarrhea and the fecal shedding continued for approximately 21 days (Farber *et al.*, 1991).

There are no specific animal models to study central nervous system disorders. However, a gerbil model has been used to reproduce rhombencephalitis (Blanot *et al.*, 1997). Animals were inoculated in the middle ear with a low infective dose of *L. monocytogenes* causing otitis media, bacteremia, ataxia, and circling syndrome. Histological lesions on the brainstem showed characteristic invasion of bacteria with necrotic abscesses.

Experimental inoculation of the chick embryo via chorioallantoic membrane (CAM) also provides pathogenicity data for *L. monocytogenes* in 72 h (Terplan and Steinmeyer, 1989). This model is comparable to the mouse pathogenicity testing model (Notermans *et al.*, 1991) and capable of differentiating highly virulent strains from slightly attenuated or weakly virulent strains (Olier *et al.*, 2002).

*Cell culture model.* In a murine model of infection, there is extensive neutrophil aggregation in the liver and spleen (Merrick *et al.*, 1997; Conlan, 1999) and lymphocyte death are hallmarks of the early pro-inflammatory response to *L. monocytogenes* infection. During infection bacteria also adhere, invade, and multiply inside the host cells. Cultured cell lines derived from target host tissues and organs could be used to investigate spe-

cific events including cytotoxicity (Table 2.1). Models of cell toxicity include neurons (Dons *et al.*, 1999; Jin *et al.*, 2001) macrophages (Dechastellier and Berche, 1994; Dallas *et al.*, 1996; Barsig and Kaufmann, 1997) fibroblasts (Marquis *et al.*, 1993) epithelial cells (Gaillard *et al.*, 1987) and B cells (Bhunia *et al.*, 1994; Bhunia *et al.*, 1995; Bhunia and Westbrook, 1998). Both *in vitro* and *in vivo* studies have identified either necrosis or apoptosis as the cytotoxic mechanism in many cell types. *L. monocytogenes* causes necrosis of murine bone-marrow derived macrophages (Barsig and Kaufmann, 1997) and apoptosis in primary murine hepatocytes (Rogers *et al.*, 1996; Mueller *et al.*, 2002) enterocyte-like Caco-2 cells (Valenti *et al.*, 1999) murine hybridoma Ped-2E9 B cells (Bhunia and Feng, 1999), human B lymphoma cells (Menon *et al.*, 2003) and primary murine B cells (Menon *et al.*, 2003). *L. monocytogenes*-infected mice displayed severe dosage-dependent splenic lesions distinguished by infiltration of infected monocytes and apoptotic depletion of both B- and T-lymphocytes within 48 h (Merrick *et al.*, 1997). *In vivo* infection also caused substantial apoptotic death of murine hepatocytes (Rogers *et al.*, 1996).

The mechanisms governing *L. monocytogenes*-mediated apoptosis are less well understood. Both listeriolysin (LLO)-positive *L. monocytogenes* strains and purified LLO triggered apoptosis in murine dendritic cells (Guzman *et al.*, 1996). However, activated dendritic cells could process and present bacterial antigens (Paschen *et al.*, 2000). Furthermore, pro-apoptotic molecules that are essential for activation of cell's apoptotic machinery are currently under investigation. Although Cas-11 is a protease involved in activating the apoptotic protein cascade, Cas-11-negative and wild type mice appeared equally affected by *L. monocytogenes* infection, each presenting normal inflammatory responses and levels of apoptotic cells (Mueller *et al.*, 2002). Conversely, T cells harvested from murine spleens displayed apoptotic death receptor Fas-dependent apoptotic cell death following infection (Mukasa *et al.*, 2002).

To study specific pathogenic events such as adhesion, invasion and cell-to cell spread during infection numerous cell types has been used. Caco-2, HEp-2, HepG2, HUVEC, 3T3, etc., have been used to study adhesion (Bubert *et al.*, 1992; Pandiripally *et al.*, 1999; Santiago *et al.*, 1999; Jaradat and Bhunia, 2002; Jaradat *et al.*, 2003; Wampler *et al.*, 2004), invasion (Gaillard *et al.*, 1991; Gaillard and Finlay, 1996; Braun *et al.*, 1997; Parida *et al.*, 1998), and cell-to-cell spread. The macrophage like cell line J774 has been used to study *L. monocytogenes* survival in the phagosome and cell-to-cell spread that are controlled by hemolysin, phopholipase C and Actin polymerization proteins (Camilli *et al.*, 1993; Moors *et al.*, 1999; Gedde *et al.*, 2000; Vazquez-Boland *et al.*, 2001).

## Salmonella

*Animal model.* *Salmonella* serovars including *Salmonella enterica* serovar *Typhimurium*, *Salmonella enterica* serovar Enteritidis or *Salmonella enterica* serovar Typhi cause bacteremia, enterocolitis and typhoid fever in humans. For foodborne salmonello-

**Table 2.1** Cell lines used for foodborne bacteria

| Bacteria | Cell line | Cell type | Source | Cytopathic effects |
|---|---|---|---|---|
| Salmonella | CHO | Epithelial | Chinese hamster ovary | Elongation, detachment |
| | Vero | Fibroblast | Monkey kidney | Lysis, protein synthesis inhibition |
| | HEp-2 | Epithelial | Human laryngeal | Invasion |
| | JY | B-cell | Mouse | Invasion |
| | H9 | T-cell | Mouse | Invasion |
| | Henle-407 | Epithelial | Human jejunal | Intracellular growth |
| | J774 | Macrophage | Mouse | Intracellular growth |
| | HeLa | Epithelial | Human cervix | Toxicity, actin polymerization |
| | RAW264.7 | Macrophage | Mouse | Intracellular growth |
| | HT-29 | Epithelial | Human colon | Apoptosis |
| Diarrheagenic E. coli | Vero | Fibroblast | Monkey kidney | Lysis, protein synthesis inhibition |
| | CHO | Epithelial | Chinese hamster ovary | Lysis, toxicity |
| | Henle-407 | Epithelial | Human jejunal | Adhesion |
| | HEp-2 | Epithelial | Human laryngeal | Adhesion, toxicity, vaculation |
| | MAC-T | Epithelial | Bovine mammary gland | Adhesion, invasion |
| | MDBK | Epithelial | Madine-Darby Canine kidney | Invasion |
| | HeLa | Epithelial | Human cervix | Toxicity, apoptosis |
| | T84 | Epithelial | Human colon | Apoptosis |
| | Y1 | Epithelial | Human adrenal cell line | Rounding, detachment, cAMP |
| | W138 | Fibroblast | Human lung | Toxicity |
| | J774 | Macrophage | Mouse | Apoptosis |
| | Ramos | Lymphocyte | Human | Apoptosis |
| | HT-29 | Epithelial | Human colon | Apoptosis |
| Shigella | HeLa | Epithelial | Human cervix | Cell death, protein synthesis inhibition |
| | Vero | Fibroblast | Monkey kidney | Lysis, protein synthesis inhibition |
| | 3T3 | Fibroblast | Mouse | Invasion, actin polymerization |
| | U937 | Monocyte | Human | Apoptosis |
| | Mφ | Macrophage | Mouse | Invasion, apoptosis |
| Campylobacter | HeLa | Epithelial | Human cervix | Distended cells (CDT effect) |
| | Vero | Fibroblast | Monkey kidney | Distended cells (CDT effect) |
| | CHO | Epithelial | Chinese hamster ovary | Distended cells (CDT effect) |
| | AZ-521 | Epithelial | Human stomach | Vacuolation |
| | Ped-2E9 | B-cells | Mouse | Toxicity |
| Clostridium perfringens | Vero | Fibroblast | Monkey kidney | Cytotoxicity |
| | Caco-2 | Epithelial | Human colon | Cytotoxicity |
| | HUVEC | Endothelial | Human umbalical vein endothelial cell | Cytotoxicity, increased membrane permeability |
| | MDBK | Epithelial | Madine–Darby canine kidney | Cytotoxicity |
| Bacillus cereus | HEp-2 | Epithelial | Human laryngeal | Used for emetic toxin: Vacuolation, cytostatic action |
| | Henle-407 | Epithelial | Human jejunml | Vacuolation |
| | McCoy cell | Fibroblast | Mouse | Used for diarrheagenic toxin: detachment of cell monolayers |
| | CHO | Epithelial | Chinese hamster ovary | Used for both emetic and diarrheagenic toxin activity |
| Listeria monocytogenes | Caco-2 | Epithelial | Human colon | Adhesion, invasion, apoptosis |
| | CHO | Epithelial | Chinese hamster ovary | Detachment, lysis |
| | Henle-407 | Epithelial | Human jejunum | Intracellular growth, death |
| | Vero | Fibroblast | Monkey kidney | Toxicity |
| | HEp-2 | Epithelial | Human larynx | Invasion |
| | J774 | Macrophage | Mouse | Intracellular growth |
| | RAW | Macrophage | Mouse | Intracellular growth |
| | CB1 | Dendritic cell | Mouse | Apoptosis, intracellular growth |
| | HeLa | Epithelial | Human cervix | Toxicity |
| | HUVEC | Endothelial | Human umbilical vein endothelial cell | Intracellular growth |
| | Hep-G2 | Epithelial | Human liver | Intracellular growth, apoptosis |
| | 3T3 | Fibroblast | Mouse | Invasion, plaque formation, lysis |
| | L-M | Fibroblast | Mouse | Invasion, plaque formation, lysis |
| | NS1 | Myeloma | Mouse | Lysis, toxicity |
| | Ped-2E9 | B-cell | Mouse hybridoma | Lysis, apoptosis |
| | RI-37 | B-cell | Human–mouse hybridoma | Lysis, apoptosis |
| | Ramos RA-1 | B-cell | Human | Lysis, apoptosis |

sis, enterocolitis is the most significant clinical feature. During infection, *Salmonella* elicits inflammation and neutrophil infiltration resulting in necrosis, increased vascular permeability and fluid loss. To study enterocolitis, various animal models have been evaluated; however, the mouse is most widely used. *Salmonella* infection (except *Salmonella enterica* serovar *Typhi*) causes mostly self-limiting gastroenteritis (enterocolitis) and occasionally systemic infections after oral inoculation. The organism could be found in Peyer's patches, mesenteric lymph nodes, liver, and spleen (Collins and Carter, 1978). $LD_{50}$ values in mice have been tested by oral administration and depending on the strain, the $LD_{50}$ values may vary from $10^3$–$10^7$ CFU (Peluffo *et al.*, 1981). $LD_{50}$ value for a *Salmonella* Enteritidis in germ-free mice was only 10 CFU and mice died within 5–8 days (Collins and Carter, 1978). In the ip route, *Salmonella* is extremely pathogenic and $LD_{50}$ may be less than 20 CFU (Sukupolvi *et al.*, 1997). Mice are also routinely used to study immune response to *Salmonella* antigens, recombinant bacteria, or probiotics for the development of suitable vaccines (Michetti *et al.*, 1992; Zhang *et al.*, 1999; Gill *et al.*, 2001; Henriksson and Conway, 2001; Fierer *et al.*, 2002; Matsui *et al.*, 2003). The mouse paw edema test (POT) has been developed to determine the enterotoxic effect of *Salmonella* lysate (Harne *et al.*, 1990). The percent thickness of paw changes dramatically within 48 h post injection and the results are comparable to the standard rabbit ileal loop assay.

A rat model has been used to determine *Salmonella* Typhimurium or *Salmonella* Enteritidis infection (Naughton *et al.*, 1996) or the dose–response relationship of *Salmonella* Enteritidis (Havelaar *et al.*, 2001). Peroral administration of $10^8$ showed illness, and the colonization occurred in ileum within 2–4 h and high numbers of bacteria were recovered from spleen, liver, and the mesenteric lymph node.

Likewise in the rabbit model, *Salmonella* caused diarrheal disease and the organisms could spread to blood, liver and spleen (Hanes *et al.*, 2001).

Owing to similarity between the human and pig gastrointestinal tract, a porcine model was tested to mimic *Salmonalla* Typhi infection; the disease, however, could not be reproduced in this animal (Metcalf *et al.*, 2000).

A neonatal calf model has been used where *Salmonella* Typhimurium caused localized infection resulting in diarrhea, fever, anorexia and dehydration (Tsolis *et al.*, 1999). The pathological lesions were mostly concentrated in the intestinal tract without systemic spread, thus the neonatal calf was thought to be the most attractive model to study infection (Zhang *et al.*, 2003).

*Cell culture model.* During the intestinal phase of infection, *Salmonella* induces inflammation leading to massive infiltration of neutrophils. *Salmonella* is a facultative intracellular organism with growth and multiplication inside epithelial or macrophages leading to cell death (Darwin and Miller, 1999). In a cultured cell model, *Salmonella* induced apoptosis in macrophage J774A.1 (Chen *et al.*, 1996a; Lindgren *et al.*, 1996;

Schwan *et al.*, 2000), human and murine osteoblasts (Alexander *et al.*, 2001), human-derived dendritic cells (Dreher *et al.*, 2002), human keratinocyte cells (Nuzzo *et al.*, 2000), and bovine macrophages (Monack *et al.*, 1996; Lundberg *et al.*, 1999; Santos *et al.*, 2001). During infection, apoptosis has been seen in murine Peyer's patches (Monack *et al.*, 2000; Cerquetti *et al.*, 2002), and in murine liver neutrophils and macrophages (Richter-Dahlfors *et al.*, 1997).

It has been shown that wild-type *Salmonella* Typhimurium causes apoptosis in human blood monocyte-derived macrophages whereas a noninvasive mutant did not (Zhou *et al.*, 2000). In virulent strains, a type III secretion system injects *Salmonella* invasion protein (SipB) into the host cytoplasm, which binds and activates Cas-1 (Hersh *et al.*, 1999). The mechanism of *Salmonella*-induced apoptosis thus appears analogous to that of *Shigella*. Cas-1 negative mice resist macrophage infection (Hersh *et al.*, 1999) and have an oral *Salmonella* $LD_{50}$ 1000-fold higher than that of wild type mice (Monack *et al.*, 2000). However, an alternative long-term Cas-1 independent apoptotic mechanism has also been suggested where Cas-1 knockout macrophages die in 4–6 hours post infection through activation of a Cas-2 pathway (Jesenberger *et al.*, 2000). Furthermore, Cas-3 activation preceded apoptosis in cultured HT-29 epithelial cells (Paesold *et al.*, 2002) and pre-treating macrophage with a Cas-3 inhibitor decreased apoptosis by 40% (Zhou *et al.*, 2000).

*Campylobacter*
*Animal model.* *Campylobacter* invades colonic epithelial cells causing inflammation and diarrhea. The infection can lead to extraintestinal sequelae and Guillain–Barré syndrome that can be life threatening (Wassenaar and Blaser, 1999). Several animal models have been tested to study *Campylobacter jejuni* or *C. coli* colonization and gastroenteritis (Newell, 2001). The primate (monkey) manifests clinical signs with vomiting and diarrhea with occasional blood in the stool mimicking human campylobacteriosis (Russell *et al.*, 1989). However, it is not a convenient model because of the ethical issues for use, requirement for trained personal to handle the animals, and cost.

A ferret model (3–6 weeks of age) has been proposed, which exhibited a mucoid stool with occult blood following a challenge with *Campylobacter* (Fox *et al.*, 1987). The clinical signs consisted of anorexia, dehydration and bacteremia and were often self-limiting. Drawbacks of this model include the cost and availability of *Campylobacter*-free ferrets.

In the piglet model, *Campylobacter* colonize but do not produce disease. If the piglets are deprived of colostrum, *Campylobacter* could induce diarrhea with most of the histopathological lesions seen in the large intestine (Babakhani *et al.*, 1993).

In the mouse model, immunocompromised (SCID) animals have shown possible *Campylobacter* colonization following oral administration, but results were not reproducible. However, in intranasal administration route, the infection was systemic (found in liver and spleen) causing death within 6 days without gastrointestinal symptoms. Later, a specific role for cytolethal distending toxin (CDT) in translocation of *C. jejuni* to blood, liver and spleen was demonstrated in SCID mice (Purdy *et al.*, 2000). In a suckling mouse model, *C. jejuni* caused profuse watery diarrhea and massive tissue damage in the descending colon (Okuda *et al.*, 1997).

In rabbits, the organism did not show disease but exhibited transient colonization. However, in sealed intestinal segment that physically block the normal peristalsis of rabbit intestine in RITARD (removable intestinal tie adult rabbit diarrhea) model, *Campylobacter* induced mucoid bloody diarrhea (Caldwell *et al.*, 1983). This model is also useful for studying long-term pathological effects. In the ligated ileal loop assay, *Campylobacter* inoculation caused acute inflammation and fluid accumulation in the loop due to increased cAMP, prostaglandins, and leukotriens (Everest *et al.*, 1993).

In a 1-day-old chick model, *Campylobacter* preferentially colonizes the cecum and often disseminates into liver and spleen (Sanyal *et al.*, 1984; Stern *et al.*, 1988; Wassenaar *et al.*, 1993). The chicken model has been used to study systemic and mucosal immune response and to determine the efficacy of certain vaccines against the infection.

*Cell culture model. Campylobacter* spp. cause enteritis and induce inflammatory responses in the gut epithelial cells. Numerous cell lines have been examined for suitability to detect cytotoxins and inflammatory responses. Heat-labile toxin production by *Campylobacter* spp. was cytolethal to CHO, Vero, and HeLa, but showed no effect on the Y-1 adrenal cell line (Johnson and Lior, 1988). In a chicken lymphocyte cell line, activity of heat-labile toxin from *C. jejuni* was demonstrated by assaying $^{51}$Cr release (Lam, 1993). Later it was shown that outer membrane protein extracts from *C. jejuni* caused apoptosis in chicken lymphocytes (Zhu *et al.*, 1999). A CHO cell assay was later used to demonstrate the influence of animal serum on cytotoxic activity of the toxin (Misawa *et al.*, 1996). Calf or adult bovine serum appeared to exert highest cytotoxicity action with the toxin. The major toxin produced by *Campylobacter* is cytolethal distending toxin, which causes distention, detachment and cell cycle arrest in target cells such as HeLa and CHO cell lines (Hanel *et al.*, 1998; Purdy *et al.*, 2000; Nadeau *et al.*, 2003). CDT nuclease activity may cause limited damage to the DNA leading to cell cycle arrest (Lara-Tejero and Galan, 2002). Using HeLa and Vero cell lines, Coote and Arain (1996) developed the MTT assay for *Campylobacter* cytotoxins. *C. upsaliansis* produces CDT and shows cytotoxicity (distension and cell cycle arrest) and apoptosis in HeLa and T-lymphocytes (Mooney *et al.*, 2001).

Adhesion and invasion properties of *Campylobacter* were investigated in HEp-2 (Konkel and Joens, 1989) and Int-407 cell lines (Nadeau *et al.*, 2003). Using these models, it was shown that clinical isolates are more invasive than the nonclinical isolates (Konkel and Joens, 1989) and human isolates are more invasive than chicken isolates (Nadeau *et al.*, 2003). The proinflammatory response of *Campylobacter* was evaluated on a human monocyte cell line, THP-1, that caused increased release of cytokines (Jones *et al.*, 2003).

## Foodborne diarrheagenic *E. coli*

Among the foodborne diarrheagenic *E. coli* (FDEC), enterohemorrhagic (EHEC), enterotoxigenic (ETEC) and enteropathogenic (EPEC) *E. coli* are important food- or water-borne pathogens of interest in humans. Human volunteers have been used to study diarrheagenic potential of select toxins.

## Animal models

*Enterohemorrhagic E. coli (EHEC)*
Gnotobiotic piglet model is the most widely used model for EHEC and oral inoculation caused anorexia, lethargy and watery diarrhea (Francis *et al.*, 1986). Histopathological study revealed damage in the microvilli showing irregular shaped epithelial cells that were rounded or detached. Involvement of *eaeA* in generation of diarrhea and neurological disorder was also studied in a piglet model (McKee *et al.*, 1995; Tzipori *et al.*, 1995). Neurological disorder is more severe in suckling piglets than in the colostrum-deprived neonatal piglets (Dean-Nystrom *et al.*, 2000). Oral challenge of piglets with StxII producing O157:H7 bacterium also caused brain lesion (edema disease) within 2–8 days (Francis *et al.*, 1989), whereas Stx-negative strains caused only attachment effacement lesions (Dean-Nystrom *et al.*, 2000). Using this model, it was later established that only StxII positive strain is capable of developing extraintestinal complications such as thrombotic microangiopathy in the kidney similar to HUS in humans (Gunzer *et al.*, 2002).

Later in a baboon model, it was confirmed that StxII, not StxI, is responsible for development of HUS (Siegler *et al.*, 2003). In this study, purified toxin preparations were injected intravenously and animals receiving StxII developed progressive thrombocytopenia, hemolytic anaemia and glomerular thrombotic microangiopathy in the kidney. StxI-injected animals exhibited no clinical signs of HUS.

Diarrheagenic properties of *E. coli* O157:H7 strains could also be induced in an infant rabbit model (Pai *et al.*, 1986; Fratamico *et al.*, 1993). Animals fed orally with ~$5 \times 10^7$ cells using an animal feeding needle developed diarrhea in 24 h. Necropsy samples revealed bacterial attachment, mucosal damage, and sloughing of cells mostly associated with the cecum and colon.

Subcutaneous injection of three-week old mice with $10^8$ O157:H7 cells induced systemic infection equivalent to HUS. Animals developed diarrhea within 24 h and the disease progressed to respiratory distress, paralysis of hind legs, and death (Lai *et al.*, 1991). Histological lesions were observed in the distal part of the small intestine, kidney, liver, spleen, lymph node, and brain. Similar results were observed in germ-free mice in which ruffled fur, hind leg weakness, and death occurred within 5 days (Taguchi *et al.*, 2002). Injection of radio ($^{125}$I)-labeled StxI and StxII to mice and whole-body autoradiography revealed StxI distribution in lungs, nasal turbinates, bone marrow and the kidney, while StxII were accumulated primarily in the kidney with much higher concentrations than StxI (Rutjes *et al.*, 2002). The serum half-life for StxI was determined to be about 2.7 min whereas StxII is 3.9 min.

A ferret model has been tested for production of colitis and extraintestinal form of the disease, i.e., generation of HUS (Woods *et al.*, 2002). *Campylobacter*-free ferrets at age of 6 weeks were challenged orally with Stx-positive *E. coli* (STEC) strains including O157:H7, which caused bloody diarrhea and hematuria, and also revealed histological damage to glomeruli and thrombocytopenia equivalent to HUS. However, only 23% of animals developed HUS.

*Enterotoxigenic E. coli (ETEC)*
ETEC produces heat-labile (LT) or heat-stable (ST) toxins and is responsible for watery diarrhea. Suckling mice have been used to determine the toxic effects (Giannella, 1976). Mice are injected with toxin preparations via percutaneous intragastric route and sacrificed after 3 days. The intestine is removed and the ratio of gut to carcass weight is calculated to determine toxicity. In a rat ligated loop assay toxic effects of *E. coli* STb were examined with fluid accumulation exhibited within 3 h (Cohen *et al.*, 1989).

ETEC action was also examined in a mouse model using intranasal inoculation (Byrd *et al.*, 2003). Strains expressing both LT and ST at high dose (~$10^9$ CFU) caused illness and death while at low dose (~$10^8$ CFU), the organisms were cleared within 2 weeks and the mice showed strong immune response against the pathogen. This suggests that an intranasal route could be used for pathogenicity analysis and vaccine trials.

*Enteropathogenic Escherichia coli (EPEC)*
EPEC causes diarrhea in infants under 2 years of age and the disease is characterized by formation of attaching and effacing (A/E) lesion in microvilli. In infant rabbit model, orally gavaged EPEC with $10^6$ cells showed diffused to patchy adhesion on epithelial cells in Peyer's patch (von Moll and Cantey, 1997) and showed typical A/E lesion within 24 h (Heczko *et al.*, 2001). In the ileal loop model, EPEC showed adherence and typical A/E lesion leading to pedestal formation within 24 h (Batt *et al.*, 1987).

## Cell culture model
The Vero cell is the most predominantly used model to evaluate cytotoxicity (Konowalchuk *et al.*, 1977). These cells are highly sensitive to verotoxins or cytotoxins from *Shigella* species, *E. coli*, *Campylobacter* spp. and *Clostridium difficile*. Cytopathic effects are hallmarked by detachment, flocculation, rounding and cytoplasmic granulation, which are scored visually after examining the cell monolayers under a microscope (Konowalchuk *et al.*, 1977; Janda *et al.*, 1991). Using this cell line, a colorimetric cytotoxicity assay was developed that measures the release of LDH from the affected cells (Roberts *et al.*, 2001; Maldonado

*et al.*, 2005). Many other cell lines have been used to study specific interaction of the pathogen or their virulence factors. In the HEp-2 cells, Shiga-toxigenic strains cause vacuolation (Roberts *et al.*, 2001). In the bovine epithelial cell line, MAC-T, adhesion and invasion property of Shiga-toxigenic strains were investigated (Matthews *et al.*, 1997). Pathogenic *E. coli* strain BH-5 induced apoptosis in human monocyte U937 or murine macrophage J774 cell lines (Stravodimos *et al.*, 1999). Diarrheagenic *E. coli*, except the ETEC strain, has been shown to induce apoptosis in human primary monocytes or the murine J774 cell line (Lai *et al.*, 1999). Later, it was shown that purified hemolysin (ClyA) or ClyA-expressing *E. coli* strains are responsible for such action (Lai *et al.*, 2000).

EHEC and EPEC bind to phosphatidylethanolamine (PE), induce apoptosis in HEp-2 epithelial cells which disrupts membrane, expose more PE, and allow increased binding of bacteria (Foster *et al.*, 2000). EPEC manifested mixed features of apoptosis and necrosis in human cervical cancer, HeLa cells, and human colon carcinoma T84 cells (Crane *et al.*, 1999). EPEC-induced cell death is not beneficial to the bacterium, since it is an adherent, non-invasive organism and death of intestinal epithelial cells will only lead to the loss of adhesion sites (Crane *et al.*, 1999)

## *Shigella* species

### *Animal model*
All four species of the genus *Shigella*, i.e., *S. flexneri*, *S. dysenteriae*, *S. boydii* and *S. sonnei* are known to cause human infections resulting in bloody mucoid diarrhea. Contaminated food or water or humans are potential sources of this organism. Intracellular propagation and production of Shiga toxin are two major virulence events that contribute to the pathogenesis. Finding an appropriate animal model to study this organism is quite challenging. Rabbits are generally non-responsive to infection, yet rabbits were used in several studies (Butler *et al.*, 1985). In most studies rabbits were starved or pretreated with a battery of antibiotics or antacids prior to challenge to make the animals susceptible to the infection. Animals were lethargic and developed diarrhea 20 h after oral inoculation (Etheridge *et al.*, 1996). Histopathologically, severe diffused necrotic lesions were observed in the ileum. Ligated intestinal loops from those pretreated animals was also used to study the effect of the bacteria (Rabbani *et al.*, 1995), Shiga toxin (Sjogren *et al.*, 1994) or enterotoxins (Fasano *et al.*, 1997).

In a guinea-pig model, *Shigella* was inoculated into the cornea to produce keratoconjunctivitis. The efficacy of a vaccine strain was evaluated by recovery of the animal from conjunctivitis (Hartman *et al.*, 1991). A pig model was used to study *Shigella* infection; however, the pigs did not show any response (Maurelli *et al.*, 1998).

### *Cell culture model*
Cytotoxicity resulting from *Shigella* infection has been established both *in vivo* and *in vitro* in a variety of cell types. Vero

cells are used to assay Shiga toxins. *Shigella* typically induces apoptosis in macrophages (Zychlinsky *et al.*, 1992; Guichon and Zychlinsky, 1997; Runyen-Janecky and Payne, 2002) which is pivotal in the initial entry and spread of infection in the host. *Shigella flexneri* infection also caused *in vivo* apoptosis of B and T cells in lymphoid follicles of rabbit ileal loops (Zychlinsky *et al.*, 1996) and induced apoptosis in human rectal tissues in patients infected with *Shigella* (Islam *et al.*, 1997; Raqib *et al.*, 2002). Although monocytes from healthy human donors have been shown to kill *S. flexneri* instead of allowing escape, these cells still die by bacterially mediated apoptosis (Hathaway *et al.*, 2002). Furthermore, the induction of apoptosis in the human monoblastic cell line U937 resulting from *Shigella* infection seemed to be based on the extent of host cell differentiation (Nonaka *et al.*, 1999).

The mechanism of *Shigella*-induced apoptosis relies on injection of the invasion plasmid antigen (IpaB) into host cell by a Type III secretion system. IpaB binds to and activates Cas-1 (Zychlinsky *et al.*, 1992; Zychlinsky *et al.*, 1994b; Chen *et al.*, 1996b; Hilbi *et al.*, 1998) a protease that triggers apoptosis by activating the release of pro-inflammatory cytokine interleukin-1 (IL-1) as shown in both macrophage (Zychlinsky *et al.*, 1994a; Hilbi *et al.*, 1997) and dendritic cells (Edgeworth *et al.*, 2002). Moreover, macrophages from Cas-1 knockout mice are not susceptible to *Shigella*-mediated apoptosis, although macrophages from Cas-3 and Cas-11 knockout mice are still killed (Hilbi *et al.*, 1998). In humans, acute shigellosis is also accompanied by increased production of Fas/Fas Ligand, Cas-1, and Cas-3, which are involved in programmed cell death (Raqib *et al.*, 2002).

## Shiga toxin

Both EHEC and *Shigella* spp. produce Shiga toxins (Stx) that bind to the globotriaosylceramide receptor (Gb3) expressed on the host cell surface (Waddell *et al.*, 1988). Cells that express Gb3, such as human renal endothelial cells and Vero cells have proven very susceptible to Stx (Boyd and Lingwood, 1989; Williams *et al.*, 1999). Stx induces apoptosis in dose dependent manner in Hep-2 cells, which express Gb3 (Ching *et al.*, 2002). Stx1 induces apoptosis in Gb3-expressing Hep-2 and Caco-2 cells, but not in non-Gb3-expressing T84 cells (Jones *et al.*, 2000). Pre-treating cells *in vitro* with TNF-α increases cytotoxicity and cell sensitivity to Stx by inducing Gb3 receptor expression (Kaye *et al.*, 1993; van de Kar *et al.*, 1995). Upregulation of Gb3 with TNF-α has been demonstrated in human cerebral endothelial cells usually insensitive to Stx (Ramegowda *et al.*, 1999; Eisenhauer *et al.*, 2001) and cytotoxicity by apoptosis in TNF-α -treated human umbilical vein endothelial cells (HUVEC) (Pijpers *et al.*, 2001) and human cerebral endothelial cells (Ergonul *et al.*, 2003). However, TNF-α shown to be a very weak inducer of toxicity in murine kidney cells *in vivo* (Wolski *et al.*, 2002).

Binding of Stx also causes apoptosis in Vero cells (Inward *et al.*, 1995), astrocytoma cells (Arab *et al.*, 1998), primary human kidney epithelial cells (Kiyokawa *et al.*, 1998), pulmo-

nary epithelium-derived cells (Uchida *et al.*, 1999), primary HUVEC (Yoshida *et al.*, 1999), human-derived amniotic cells (Yoshimura *et al.*, 2002), and Ramos Burkitt's B lymphoma cells (Mangeney *et al.*, 1991; Mangeney *et al.*, 1993; Marcato *et al.*, 2002). Biopsies of tissues from humans affected by hemolytic–uremic syndrome (HUS) (Karpman *et al.*, 1998; Kaneko *et al.*, 2001) and kidneys from mice experimentally infected with StxII-positive *E. coli* O157:H7 (Karpman *et al.*, 1998) have displayed signs of *in vivo* apoptosis.

Stx inhibits host protein synthesis (Endo *et al.*, 1988; Lingwood, 1996); however, an apoptosis mechanism may be involved independently of protein synthesis inhibition. As an initial step in pathogenesis, cell death by apoptosis could allow more toxin to enter the host through damaged tissue (O'Loughlin and Robins-Browne, 2001; Ching *et al.*, 2002). In several instances, pro-apoptotic proteins have been associated with Stx cytotoxicity. StxI and II activated Cas-3 in apoptotic cells whereas a specific Cas-3 inhibitor decreased numbers of apoptotic cells (Kojio *et al.*, 2000). Stx1 also activated Cas-3 and enhanced pro-apoptotic Bax expression which thereby induced apoptosis (Jones *et al.*, 2000) in human cells. Mitochondrial Bcl-2 associated with StxII to promote apoptosis in human HepG2 cells (Suzuki *et al.*, 2000). In Burkitt's lymphoma cells, Stx-1 activates Cas-3, -7, and -8 to initiate apoptosis (Kiyokawa *et al.*, 2001). Finally, *in vitro* cleavage of Poly (ADP-ribose) Polymerase (PARP), which acts as indicator of apoptosis, was identified in Stx-induced apoptotic cells (Ching *et al.*, 2002).

### *Bacillus cereus*

*Animal model.* *Bacillus cereus* produces enterotoxins, hemolysins, phospholipase C and multiple enzymes. Of these the emetic toxin and the diarrheagenic enterotoxin are important in gastrointestinal disorder (Beecher *et al.*, 1995; Kotiranta *et al.*, 2000). Generally primates are used to determine emetic toxicity. However, a new animal model using *Suncus murinus* (the house shrew) has been developed (Agata *et al.*, 1995). In this animal model, oral or ip administration with the emetic toxin (cereulide) caused vomiting while a receptor ($5-HT_3$) antagonist (ondansetron hydrochloride) abolished vomiting (Agata *et al.*, 1995).

*Cell culture model.* In the cell culture model, the emetic toxin produced vacuoles in HEp-2 and Int-407 cells (Szabo *et al.*, 1991); however, HEp-2 cells proved to be the better model for cytotoxicity assay. Later HEp-2 cell model was used to determine the adhesion, invasion and cytotoxicity of veterinary isolates of *Bacillus* species (Rowan *et al.*, 2003). The toxin also inhibits proliferation of HEp-2 (Mikami *et al.*, 1994). In the quest for better assay method for the emetic toxin, a new approach was taken where boar spermatozoa was exposed to the purified toxin. The spermatozoa exhibited swelling of the mitochondria and a loss in motility. Apparently, the emetic toxin caused paralysis of the spermatozoa by damaging mitochondria and the oxidative phosphorylation process (Andersson *et al.*, 1998). A HEp-2 cell-based colorimetric MTT assay has been developed for emetic toxin that utilizes metabolic staining of cells (Finlay *et al.*, 1999). HEp-2

The diarrheagenic toxin is a heat-labile toxin and the toxicity is most commonly determined by rabbit ileal loop assay. The rat ileal loop has been used but found to be less sensitive than rabbits (Ting and Banwart, 1985). In the cell culture assay, enterotoxin activity of *B. cereus* was determined by the McCoy cell cytotoxicity assay, in which the toxin causes progressive destruction of cell monolayers (Jackson, 1993). In a modified McCoy cell-based assay the cytopathic effects were determined by visual examination and staining (Fletcher and Logan, 1999).

For detection of both emetic and diarrheagenic toxin, the CHO cell line has been used and cytotoxicity was examined by MTT assay (Beattie and Williams, 1999; Hsieh *et al.*, 1999; Pedersen *et al.*, 2002).

### Clostridium perfringens

*Animal model.* *C. perfringens* produces 14 different toxins and can cause varieties of diseases in humans, animals and birds (Rood, 1998). As a foodborne pathogen, it is responsible for diarrhea and vomiting due to the action CPE (*Clostridium perfringens* enterotoxin) in humans. In severe cases, it causes enteritis necroticans or "pigbel" in humans (Brynestad and Granum, 2002). The cytotoxic effect of CPE starts with binding of the toxin to the receptor (50 kDa) on the sensitive cell membrane. This small complex then interacts with a 70 kDa protein on the membrane and form a larger complex, which alters the membrane permeability to cause fluid and ion ($Na^+$, $Cl^-$) losses. CPE also modulates the architecture of epithelial tight junctions (TJ) by forming a complex with TJ protein occludin (Singh *et al.*, 2001).

Necrotic enteritis is primarily caused by strains of *C. perfringens* type C in the piglet, chicken, calf, lamb, and goat. It produces β-toxin (34 kDa), a necrotizing toxin that acts on autonomic nervous system causing arterial constriction leading to mucosal necrosis. The toxin also forms multimeric transmembrane pores and facilitates the release of cellular-arachidonic acid and inositol (Shatursky *et al.*, 2000; Steinthorsdottir *et al.*, 2000). *C. perfringens* type C also produces α–toxin which has phospholipase/sphingomyelinase C activity, also responsible for necrotizing effects and shock-like syndrome (Songer, 1996; Tweten, 2001).

The guinea-pig model is used to study necrotic enteritis. Administration of *Clostridium welchii* Type C culture in guinea-pigs produced necrotic lesions for which the microscopic and macroscopic features were similar to "pigbel" in humans (Lawrence and Cooke, 1980). *C. perfringens* enterotoxin and α-toxin along with *E. coli* STa toxin, staphylococcal enterotoxin B, and *C. difficile* enterotoxin A and B cause sudden infant death syndrome (SIDS). This disease could be reproduced in a rabbit model showing reduced heart rate and blood pressure and onset of sudden death (Siarakas *et al.*, 1995). These clinical features are attributed to the increased level of catecholamine

causing a sharp rise in adrenaline and noradrenaline levels in rabbits (Siarakas *et al.*, 1997).

*Cell culture models.* CPE action has been studied using Vero (Granum and Richardson, 1991), Caco-2 (Singh *et al.*, 2001) and MDDK (Madin-Darby Canine Kidney) (Miyata *et al.*, 2002) cell models. Caco-2 is the most appropriate model since it is an enterocyte-like cell, representing cells of intestinal origin. Loss of membrane integrity due to CPE action has been assayed by the release of $^{86}$Rb (Singh *et al.*, 2001). Toxic effects of β-toxin have been studied using human umbilical vein endothelial cells (HUVEC), in which toxin affects membrane permeability (Steinthorsdottir *et al.*, 2000), resulting in loss of essential cellular materials.

## Conclusions and future directions

Cell culture and animal models are often used to mimic or emulate development of disease in the human host. In recent years, both models have facilitated our understanding of disease mechanisms for foodborne pathogens. Because of emerging new pathogens or reemergence of old pathogens, new and sensitive models need to be developed, not only to investigate molecular mechanisms of pathogenesis, but also to use for pathogenicity testing assays. The techniques and tools presented in this review represent a foundation for future investigations and it is anticipated these would tremendously aid in understanding foodborne pathogens beyond traditional food microbiological techniques.

## References

Agata, N., Ohta, M., Mori, M., and Isobe, M. 1995. A novel dodecadepsipeptide, cereulide, is an emetic toxin of *Bacillus cereus*. FEMS Microbiol. Lett. 129: 17–19.

Alexander, E.H., Bento, J.L., Hughes, F.M., Marriott, I., Hudson, M.C., and Bost, K.L. 2001. *Staphylococcus aureus* and *Salmonella enterica* serovar Dublin induce tumor necrosis factor-related apoptosis-inducing ligand expression by normal mouse and human osteoblasts. Infect. Immun. 69: 1581–1586.

Altimira, J., Prats, N., Lopez, S., Domingo, M., Briones, V., Dominguez, L., and Marco, A. 1999. Repeated oral dosing with *Listeria monocytogenes* in mice as a model of central nervous system listeriosis in man. J. Comp. Pathol. 121: 117–125.

Andersson, M.A., Mikkola, R., Helin, J., Andersson, M.C., and Salkinoja-Salonen, M. 1998. A novel sensitive bioassay for detection of *Bacillus cereus* emetic toxin and related depsipeptide ionophores. Appl. Environ. Microbiol. 64: 1338–1343.

Arab, S., Murakami, M., Dirks, P., Boyd, B., Hubbard, S.L., Lingwood, C.A., and Rutka, J.T. 1998. Verotoxins inhibit the growth of and induce apoptosis in human astrocytoma cells. J. Neuro-Oncol. 40: 137–150.

Babakhani, F.K., Bradley, G.A., and Joens, L.A. 1993. Newborn piglet model for capylobacteriosis. Infect. Immun. 61: 3466–3475.

Barbour, A.H., Rampling, A., and Hormaeche, C.E. 2001. Variation in the infectivity of *Listeria monocytogenes* isolates following intragastric inoculation of mice. Infect. Immun. 69: 4657 4660.

Barsig, J., and Kaufmann, S.H.E. 1997. The mechanism of cell death in *Listeria monocytogenes* infected murine macrophages is distinct from apoptosis. Infect. Immun. 65: 4075–4081.

Batt, R.M., Hart, C.A., McLean, L., and Saunders, J.R. 1987. Organ culture of rabbit ileum as a model for the investigation of the mechanism of intestinal damage by enteropathogenic *Escherichia coli*. Gut 28: 1283–1290.

Beattie, S.H., and Williams, A.G. 1999. Detection of toxigenic strains of *Bacillus cereus* and other *Bacillus* spp. with an improved cytotoxicity assay. Lett. Appl. Microbiol. 28: 221–225.

Beecher, D.J., Schoeni, J.L., and Wong, A.C.L. 1995. Enterotoxic activity of hemolysin Bl from *Bacillus cereus*. Infect. Immun. 63: 4423–4428.

Bhunia, A.K., Steele, P.J., Westbrook, D.G., Bly, L.A., Maloney, T.P., and Johnson, M.G. 1994. A six-hour *in vitro* virulence assay for *Listeria monocytogenes* using myeloma and hybridoma cells from murine and human sources. Microb. Pathog. 16: 99–110.

Bhunia, A.K., Westbrook, D.G., Story, R., and Johnson, M.G. 1995. Frozen stored murine hybridoma cells can be used to determine the virulence of *Listeria monocytogenes*. J. Clin. Microbiol. 33: 3349–3351.

Bhunia, A.K., and Westbrook, D.G. 1998. Alkaline phosphatase release assay to determine cytotoxicity for *Listeria* species. Lett. Appl. Microbiol. 26: 305–310.

Bhunia, A.K., and Feng, X. 1999. Examination of cytopathic effect and apoptosis in *Listeria monocytogenes*-infected hybridoma B-lymphocyte (Ped-2E9) line *in vitro*. J. Microbiol. Biotechnol. 9: 398–403.

Blanot, S., Joly, M.M., Vilde, F., Jaubert, F., Clement, O., Frija, G., and Berche, P. 1997. A gerbil model for rhombencephalitis due to *Listeria monocytogenes*. Microb. Pathog. 23: 39 48.

Boyd, B., and Lingwood, C. 1989. Verotoxin receptor glycolipid in human renal tissue. Nephron 51: 207–210.

Bracegirdle, P., West, A.A., Lever, M.S., Fitzgeorge, R.B., and Baskerville, A. 1994. A comparison of aerosol and intragastric routes of infection with *Listeria* spp. Epidemiol. Infect. 112: 69–79.

Braun, L., Dramsi, S., Dehoux, P., Bierne, H., Lindahl, G., and Cossart, P. 1997. InIB: An invasion protein of *Listeria monocytogenes* with a novel type of surface association. Mol. Microbiol. 25: 285–294.

Brynestad, S., and Granum, P.E. 2002. *Clostridium perfringens* and foodborne infections. Int. J. Food Microbiol. 74: 195–202.

Bubert, A., Kuhn, M., Goebel, W., and Kohler, S. 1992. Structural and functional properties of the p60 proteins from different *Listeria* species. J. Bacteriol. 174: 8166–8171.

Buer, J., and Balling, R. 2003. Mice, microbes, and models of infection. Nat. Rev. Genet. 4: 195 205.

Butler, T., Rahman, H., Almahmud, K.A., Islam, M., Bardhan, P., Kabir, I., and Rahman, M.M. 1985. An animal model of Hemolytic Uremic Syndrome in shigellosis: Lipopolysaccharides of *Shigella dysenteriae* I and *Shigella flexneri* produce leukocyte-mediated renal cortical necrosis in rabbits. Brit. J. Exp. Pathol. 66: 7–15.

Byrd, W., Mog, S.R., and Cassels, F.J. 2003. Pathogenicity and immune response measured in mice following intranasal challenge with enterotoxigenic *Escherichia coli* strains H10407 and B7A. Infect. Immun. 71: 13–21.

Caldwell, M.B., Walker, R.I., Stewart, S.D., and Rogers, J.E. 1983. Simple adult rabbit model for *Campylobacter jejuni* enteritis. Infect. Immun. 42: 1176–1182.

Camilli, A., Tilney, L.G., and Portnoy, D.A. 1993. Dual roles of *plcA* in *Listeria monocytogenes* pathogenesis. Mol. Microbiol. 8: 143–157.

Cerquetti, M.C., Goren, N.B., Ropolo, A.J., Grasso, D., Giacomodonato, M.N., and Vaccaro, M.I. 2002. Nitric oxide and apoptosis induced in Peyer's patches by attenuated strains of *Salmonella enterica* serovar Enteritidis. Infect. Immun. 70: 964–969.

Chen, L.M., Kaniga, K., and Galan, J.E. 1996a. *Salmonella* spp. are cytotoxic for cultured macrophages. Mol. Microbiol. 21: 1101–1115.

Chen, Y.J., Smith, M.R., Thirumalai, K., and Zychlinsky, A. 1996b. A bacterial invasin induces macrophage apoptosis by binding directly to ICE. EMBO J. 15: 3853–3860.

Ching, J.C.Y., Jones, N.L., Ceponis, P.J.M., Karmali, M.A., and Sherman, P.M. 2002. *Escherichia coli* Shiga-like toxins induce apoptosis and cleavage of poly(ADP-ribose) polymerase via *in vitro* activation of caspases. Infect. Immun. 70: 4669–4677.

Cohen, M.B., Thompson, M.R., and Giannella, R.A. 1989. Differences in jejunal and ileal response to *Escherichia coli* enterotoxin: Possible mechanisms. Am. J. Physiol. 257: G118 G123.

Collins, F.M., and Carter, P.B. 1978. Growth of *Salmonellae* in orally infected germfree mice. Infec. Immun. 21: 41–7.

Conlan, J.W. 1999. Early host-pathogen interactions in the liver and spleen during systemic murine listeriosis: an overview. Immunobiology 201: 178–187.

Conner, D.E., Scott, V.N., Sumner, S.S., and Bernard, D.T. 1989. Pathogenicity of foodborne, environmental and clinical isolates of *Listeria monocytogenes* in mice. J. Food Sci. 54: 1553–1556.

Coote, J.G., and Arain, T. 1996. A rapid, colourimetric assay for cytotoxin activity in *Campylobacter jejuni*. FEMS Immunol. Med. Microbiol. 13: 65–70.

Cossart, P., Pizarro-Cerda, J., and Lecuit, M. 2003. Invasion of mammalian cells by *Listeria monocytogenes*: functional mimicry to subvert cellular functions. Trends Cell Biol. 13: 23 31.

Crane, J.K., Majumdar, S., and Pickhardt, D.F. 1999. Host cell death due to enteropathogenic *Escherichia coli* has features of apoptosis. Infect. Immun. 67: 2575–2584.

Czuprynski, C.J., Faith, N.G., and Steinberg, H. 2003. A/J mice are susceptible and C57BL/6 mice are resistant to *Listeria monocytogenes* infection by intragastric inoculation. Infect. Immun. 71: 682–689.

Dallas, H.L., Thomas, D.P., and Hitchins, A.D. 1996. Virulence of *Listeria monocytogenes*, *Listeria seeligeri*, and *Listeria innocua* assayed with *in vitro* murine macrophagocytosis. J. Food Prot. 59: 24–27.

Darwin, K.H., and Miller, V.L. 1999. Molecular basis of the interaction of *Salmonella* with the intestinal mucosa. Clin. Microbiol. Rev. 12: 405–428.

Dean-Nystrom, E.A., Pohlenz, J.F.L., Moon, H.W., and O'Brien, A.D. 2000. *Escherichia coli* O157:H7 causes more-severe systemic disease in suckling piglets than in colostrum-deprived neonatal piglets. Infect. Immun. 68: 2356–2358.

Dechastellier, C., and Berche, P. 1994. Fate of *Listeria monocytogenes* in murine macrophages – Evidence for simultaneous killing and survival of intracellular bacteria. Infect. Immun. 62: 543–553.

Dons, L., Weclewicz, K., Jin, Y.X., Bindseil, E., Olsen, J.E., and Kristensson, K. 1999. Rat dorsal root ganglia neurons as a model for *Listeria monocytogenes* infections in culture. Med. Microbiol. Immunol. 188: 15–21.

Dreher, D., Kok, M., Obregon, C., Kiama, S.G., Gehr, P., and Nicod, L.P. 2002. *Salmonella* virulence factor SipB induces activation and release of IL-18 in human dendritic cells. J. Leukoc. Biol. 72: 743–751.

Drevets, D.A., and Campbell, P.A. 1991. Macrophage phagocytosis – Use of fluorescence microscopy to distinguish between extracellular and intracellular bacteria. J. Immunol. Methods 142: 31–38.

Edgeworth, J.D., Spencer, J., Phalipon, A., Griffin, G.E., and Sansonetti, P.J. 2002. Cytotoxicity and interleukin-1 beta processing following *Shigella flexneri* infection of human monocyte-derived dendritic cells. Eur. J. Immunol. 32: 1464–1471.

Eisenhauer, P.B., Chaturvedi, P., Fine, R.E., Ritchie, A.J., Pober, J.S., Cleary, T.G., and Newburg, D.S. 2001. Tumor necrosis factor alpha increases human cerebral endothelial cell Gb(3) and sensitivity to Shiga toxin. Infect. Immun. 69: 1889–1894.

Endo, Y., Tsurugi, K., Yutsudo, T., Takeda, Y., Ogasawara, T., and Igarashi, K. 1988. Site of action of a vero toxin (Vt2) from *Escherichia coli* O157:H7 and of Shiga toxin on eukaryotic ribosomes: Rna N-glycosidase activity of the toxins. Eur. J. Biochem. 171: 45 50.

Ergonul, Z., Hughes, A.K., and Kohan, D.E. 2003. Induction of apoptosis of human brain microvascular endothelial cells by Shiga toxin 1. J. Infect. Dis. 187: 154–158.

Etheridge, M.E., Hoque, A., and Sack, D.A. 1996. Pathologic study of a rabbit model for shigellosis. Lab. Anim. Sci. 46: 61–66.

Everest, P.H., Goossens, H., Sibbons, P., Lloyd, D.R., Knutton, S., Leece, R., Ketley, J.M., and Williams, P.H. 1993. Pathological changes in the rabbit ileal loop model caused by *Campylobacter jejuni* from human colitis. J. Med. Microbiol. 38: 316–321.

Farber, J.M., Coates, E.D.F., Beausoleil, N., and Fournier, J. 1991. Feeding trials of *Listeria monocytogenes* with a nonhuman primate model. J. Clin. Microbiol. 29: 2606–2608.

Fasano, A., Noriega, F.R., Liao, F.M., Wang, W., and Levine, M.M. 1997. Effect of *Shigella* enterotoxin 1 (ShET1) on rabbit intestine *in vitro* and *in vivo*. Gut 40: 505–511.

Fierer, J., Swancutt, M.A., Heumann, D., and Golenbock, D. 2002. The role of lipopolysaccharide binding protein in resistance to *Salmonella* infections in mice. J. Immunol. 168: 6396–6403.

Finlay, W.J.J., Logan, N.A., and Sutherland, A.D. 1999. Semiautomated metabolic staining assay for *Bacillus cereus* emetic toxin. Appl. Environ. Microbiol. 65: 1811–1812.

Fletcher, P., and Logan, N.A. 1999. Improved cytotoxicity assay for *Bacillus cereus* diarrhoeal enterotoxin. Lett. Appl. Microbiol. 28: 394–400.

Foster, D.B., Abul-Milh, M., Huesca, M., and Lingwood, C.A. 2000. Enterohemorrhagic *Escherichia coli* induces apoptosis which augments bacterial binding and phosphatidylethanolamine exposure on the plasma membrane outer leaflet. Infect. Immun. 68: 3108–3115.

Fox, J.G., Ackerman, J.I., Taylor, N., Claps, M., and Murphy, J.C. 1987. *Campylobacter jejuni* infection in the ferret – An animal model of human campylobacteriosis. Am. J. Vet. Res. 48: 85–90.

Francis, D.H., Collins, J.E., and Duimstra, J.R. 1986. Infection of gnotobiotic pigs with an *Escherichia coli* O157:H7 strain associated with an outbreak of hemorrhagic colitis. Infect. Immun. 51: 953–956.

Francis, D.H., Moxley, R.A., and Andraos, C.Y. 1989. Edema disease-like brain lesions in gnotobiotic piglets infected with *Escherichia coli* serotype O157:H7. Infect. Immun. 57: 1339–1342.

Fratamico, P.M., Buchanan, R.L., and Cooke, P.H. 1993. Virulence of an *Escherichia coli* O157-H7 sorbitol-positive mutant. Appl. Environ. Microbiol. 59: 4245–4252.

Gaillard, J.L., Berche, P., Mounier, J., Richard, S., and Sansonetti, P. 1987. *In vitro* model of penetration and intracellular growth of *Listeria monocytogenes* in the human enterocyte-like cell line Caco-2. Infect. Immun. 55: 2822–2829.

Gaillard, J.L., Berche, P., Frehel, C., Gouin, E., and Cossart, P. 1991. Entry of *L. monocytogenes* into cells is mediated by internalin, a repeat protein reminiscent of surface-antigens from Gram-positive cocci. Cell 65: 1127–1141.

Gaillard, J.L., and Finlay, B.B. 1996. Effect of cell polarization and differentiation on entry of *Listeria monocytogenes* into the enterocyte-like Caco-2 cell line. Infect. Immun. 64: 1299 1308.

Gao, L.Y., and Kwaik, Y.A. 2000. The modulation of host cell apoptosis by intracellular bacterial pathogens. Trends Microbiol. 8: 306–313.

Gedde, M.M., Higgins, D.E., Tilney, L.G., and Portnoy, D.A. 2000. Role of listeriolysin O in cell-to-cell spread of *Listeria monocytogenes*. Infect. Immun. 68: 999–1003.

Giannella, R.A. 1976. Suckling mouse model for detection of heat-stable *Escherichia coli* enterotoxin: characteristics of the model. Infect. Immun. 14: 95–9.

Gill, H.S., Shu, Q., Lin, H., Rutherfurd, K.J., and Cross, M.L. 2001. Protection against translocating *Salmonella typhimurium* infection in mice by feeding the immuno-enhancing probiotic *Lactobacillus rhamnosus* strain HN001. Med. Microbiol. Immunol. 190: 97–104.

Granum, P.E., and Richardson, M. 1991. Chymotrypsin treatment increases the activity of *Clostridium perfringens* enterotoxin. Toxicon 29: 898–900.

Guichon, A., and Zychlinsky, A. 1997. Clinical isolates of *Shigella* species induce apoptosis in macrophages. J. Infect. Dis. 175: 470–473.

Gunzer, F., Hennig-Pauka, I., Waldmann, K.H., Sandhoff, R., Grone, H.J., Kreipe, H.H., Matussek, A., and Mengel, M. 2002. Gnotobiotic piglets develop thrombotic microangiopathy after oral infection with enterohemorrhagic *Escherichia coli*. Am. J. Clin. Pathol. 118: 364–375.

Guzman, C.A., Domann, E., Rohde, M., Bruder, D., Darji, A., Weiss, S., Wehland, J., Chakraborty, T., and Timmis, K.N. 1996. Apoptosis of mouse dendritic cells is triggered by listeriolysin, the major virulence determinant of *Listeria monocytogenes*. Mol. Microbiol. 20: 119–126.

Hanel, I., Schulze, F., Hotzel, H., and Schubert, E. 1998. Detection and characterization of two cytotoxins produced by *Campylobacter jejuni* strains. Zent.bl. Bakteriol.-Int. J. Med. Microbiol. Virol. Parasitol. Infect. Dis. 288: 131–143.

Hanes, D.E., Robl, M.G., Schneider, C.M., and Burr, D.H. 2001. New Zealand white rabbit as a nonsurgical experimental model for *Salmonella enterica* gastroenteritis. Infect. Immun. 69: 6523–6526.

Harne, S.D., Sharma, V.D., and Rahman, H. 1990. Paw oedema test for detection of *Salmonella* enterotoxin – Modification and standardization. Indian J. Exp. Biol. 28: 1141–1144.

Hartman, A.B., Powell, C.J., Schultz, C.L., Oaks, E.V., and Eckels, K.H. 1991. Small animal model to measure efficacy and immunogenicity of *Shigella* vaccine strains. Infect. Immun. 59: 4075–4083.

Hathaway, L.J., Griffin, G.E., Sansonetti, P.J., and Edgeworth, J.D. 2002. Human monocytes kill *Shigella flexneri* but then die by apoptosis associated with suppression of proinflammatory cytokine production. Infect. Immun. 70: 3833–3842.

Havelaar, A.H., Garssen, J., Takumi, K., Koedam, M.A., Dufrenne, J.B., van Leusden, F.M., de la Fonteyne, L., Bousema, J.T., and Vos, J.G. 2001. A rat model for dose-response relationships of *Salmonella enteritidis* infection. J. Appl. Microbiol. 91: 442–452.

Heczko, U., Carthy, C.M., O'Brien, B.A., and Finlay, B.B. 2001. Decreased apoptosis in the ileum and ileal Peyer's patches: A feature after infection with rabbit enteropathogenic *Escherichia coli* O103. Infect. Immun. 69: 4580–4589.

Henriksson, A., and Conway, P.L. 2001. Isolation of human faecal *Bifidobacteria* which reduce signs of *Salmonella* infection when orogastrically dosed to mice. J. Appl. Microbiol. 90: 223–228.

Hersh, D., Monack, D.M., Smith, M.R., Ghori, N., Falkow, S., and Zychlinsky, A. 1999. The *Salmonella* invasin SipB induces macrophage apoptosis by binding to caspase-1. Proc. Natl. Acad. Sci. USA 96: 2396–2401.

Hilbi, H., Chen, Y.J., Thirumalai, K., and Zychlinsky, A. 1997. The interleukin 1 beta-converting enzyme, caspase 1, is activated during *Shigella flexneri*-induced apoptosis in human monocyte-derived macrophages. Infect. Immun. 65: 5165–5170.

Hilbi, H., Moss, J.E., Hersh, D., Chen, Y.J., Arondel, J., Banerjee, S., Flavell, R.A., Yuan, J.Y., Sansonetti, P.J., and Zychlinsky, A. 1998. *Shigella*-induced apoptosis is dependent on Caspase-1 which binds to IpaB. J. Biol. Chem. 273: 32895–32900.

Hsieh, Y.M., Sheu, S.J., Chen, Y.L., and Tsen, H.Y. 1999. Enterotoxigenic profiles and polymerase chain reaction detection of *Bacillus cereus* group cells and B. cereus strains from foods and foodborne outbreaks. J. Appl. Microbiol. 87: 481–490.

Inward, C.D., Williams, J., Chant, I., Crocker, J., Milford, D.V., Rose, P.E., and Taylor, C.M. 1995. Verocytotoxin-1 induces apoptosis in Vero cells. J. Infect. 30: 213–218.

Islam, D., Veress, B., Bardhan, P.K., Lindberg, A.A., and Christensson, B. 1997. In situ characterization of inflammatory responses in the rectal mucosae of patients with shigellosis. Infect. Immun. 65: 739–749.

Jackson, S.G. 1993. Rapid screening test for enterotoxin-producing *Bacillus cereus*. J. Clin. Microbiol. 31: 972–974.

Janda, J.M., Desmond, E.P., and Abbott, S.L. 1991. Animal and cell culture systems. In: Manual of Clinical Microbiology. A. Balows, W.J. Hausler, K.L. Herrmann, H.D. Isenberg, and H.J. Shadomy, ed. ASM Press, Washington, D.C. 137–146.

Jaradat, Z.W., and Bhunia, A.K. 2002. Glucose and nutrient concentrations affect the expression of a 104-kilodalton *Listeria* adhesion protein in *Listeria monocytogenes*. Appl. Environ. Microbiol. 68: 4876–4883.

Jaradat, Z.W., and Bhunia, A.K. 2003. Adhesion, invasion, and translocation characteristics of *Listeria monocytogenes* serotypes in Caco-2 cell and mouse models. Appl. Environ. Microbiol. 69: 5736–5736.

Jaradat, Z.W., Wampler, J.L., and Bhunia, A.K. 2003. A *Listeria* adhesion protein-deficient *Listeria monocytogenes* strain shows reduced adhesion primarily to intestinal cell lines. Med. Microbiol. Immunol. 192: 85–91.

Jesenberger, V., Procyk, K.J., Yuan, J.Y., Reipert, S., and Baccarini, M. 2000. *Salmonella* induced caspase-2 activation in macrophages: A novel mechanism in pathogen mediated apoptosis. J. Exp. Med. 192: 1035–1045.

Jin, Y.X., Dons, L., Kristensson, K., and Rottenberg, M.E. 2001. Neural route of cerebral *Listeria monocytogenes* murine infection: Role of immune response mechanisms in controlling bacterial neuroinvasion. Infect. Immun. 69: 1093–1100.

Johnson, W.M., and Lior, H. 1988. A new heat-labile cytolethal distending toxin (Cldt) produced by *Campylobacter* spp. Microb. Pathog. 4: 115–126.

Jones, M.A., Totemeyer, S., Maskell, D.J., Bryant, C.E., and Barrow, P.A. 2003. Induction of proinflammatory responses in the human monocytic cell line THP-1 by *Campylobacter jejuni*. Infect. Immun. 71: 2626–2633.

Jones, N.L., Islur, A., Haq, R., Mascarenhas, M., Karmali, M.A., Perdue, M.H., Zanke, B.W., and Sherman, P.M. 2000. *Escherichia coli* Shiga toxins induce apoptosis in epithelial cells that is regulated by the Bcl-2 family. Am. J. Physiol.-Gastroint. Liver Physiol. 278: G811 G819.

Kaito, C., Akimitsu, N., Watanabe, H., and Sekimizu, K. 2002. Silkworm larvae as an animal model of bacterial infection pathogenic to humans. Microb. Pathog. 32: 183–190.

Kaneko, K., Kiyokawa, N., Ohtomo, Y., Nagaoka, R., Yamashiro, Y., Taguchi, T., Mori, T., Fujimoto, J., and Takeda, T. 2001. Apoptosis of renal tubular cells in Shiga-toxin-mediated hemolytic uremic syndrome. Nephron 87: 182–185.

Karpman, D., Hakansson, A., Perez, M.T.R., Isaksson, C., Carlemalm, E., Caprioli, A., and Svanborg, C. 1998. Apoptosis of renal cortical cells in the hemolytic-uremic syndrome: *In vivo* and *in vitro* studies. Infect. Immun. 66: 636–644.

Kaye, S.A., Louise, C.B., Boyd, B., Lingwood, C.A., and Obrig, T.G. 1993. Shiga toxin-associated Hemolytic Uremic Syndrome – Interleukin-1-Beta enhancement of Shiga toxin cytotoxicity toward human vascular endothelial-cells *in vitro*. Infect. Immun. 61: 3886 3891.

Kiyokawa, N., Taguchi, T., Mori, T., Uchida, H., Sato, N., Takeda, T., and Fujimoto, J. 1998. Induction of apoptosis in normal human renal tubular epithelial cells by *Escherichia coli* Shiga toxins 1 and 2. J. Infect. Dis. 178: 178–184.

Kiyokawa, N., Mori, T., Taguchi, T., Saito, M., Mimori, K., Suzuki, T., Sekino, T., Sato, N., Nakajima, H., Katagiri, Y.U., Takeda, T., and Fujimoto, J. 2001. Activation of the caspase cascade during Stx1-induced apoptosis in Burkitt's lymphoma cells. J. Cell. Biochem. 81: 128–142.

Kojio, S., Zhang, H.M., Ohmura, M., Gondaira, F., Kobayashi, N., and Yamamoto, T. 2000. Caspase-3 activation and apoptosis induction coupled with the retrograde transport of Shiga toxin: inhibition by brefeldin A. FEMS Immunol. Med. Microbiol. 29: 275–281.

Konkel, M.E., and Joens, L.A. 1989. Adhesion to and invasion of Hep-2 cells by *Campylobacter* spp. Infect. Immun. 57: 2984–2990.

Konowalchuk, J., Speirs, J.I., and Stavric, S. 1977. Vero response to a cytotoxin of *Escherichia coli*. 18: 775–779.

Kotiranta, A., Lounatmaa, K., and Haapasalo, M. 2000. Epidemiology and pathogenesis of *Bacillus cereus* infections. Microbes Infect. 2: 189–198.

Lai, X.H., Liu, B.Y., and Xu, J.G. 1991. Experimental infection of specific pathogen-free mice with enterohemorrhagic *Escherichia coli* O157: H7. Microbiol. Immunol. 35: 515–524.

Lai, X.H., Xu, J.G., Melgar, S., and Uhlin, B.E. 1999. An apoptotic response by J774 macrophage cells is common upon infection with diarrheagenic *Escherichia coli*. FEMS Microbiol. Lett. 172: 29–34.

Lai, X.H., Arencibia, I., Johansson, A., Wai, S.N., Oscarsson, J., Kalfas, S., Sundqvist, K.G., Mizunoe, Y., Sjostedt, A., and Uhlin, B.E. 2000. Cytocidal and apoptotic effects of the ClyA protein from *Escherichia coli* on primary and cultured monocytes and macrophages. Infect. Immun. 68: 4363–4367.

Lam, K.M. 1993. A cytotoxicity test for the detection of *Campylobacter jejuni* toxin. Vet. Microbiol. 35: 133–139.

Lammerding, A.M., Glass, K.A., Gendronfitzpatrick, A., and Doyle, M.P. 1992. Determination of virulence of different strains of *Listeria monocytogenes* and *Listeria innocua* by oral inoculation of pregnant mice. Appl. Environ. Microbiol. 58: 3991–4000.

Lara-Tejero, M., and Galan, J.E. 2002. Cytolethal distending toxin: limited damage as a strategy to modulate cellular functions. Trends Microbiol. 10: 147–152.

Lawrence, G., and Cooke, R. 1980. Experimental pigbel: the production and pathology of necrotizing enteritis due to *Clostridium welchii* type C in the guinea-pig. Br. J. Exp. Pathol. 61: 261–71.

Lecuit, M., Vandormael-Pournin, S., Lefort, J., Huerre, M., Gounon, P., Dupuy, C., Babinet, C., and Cossart, P. 2001. A transgenic model for listeriosis: Role of internalin in crossing the intestinal barrier. Science 292: 1722–1725.

Lecuit, M., and Cossart, P. 2002. Genetically modified-animal models for human infections: the *Listeria* paradigm. Trends Mol. Med. 8: 537–542.

Lindgren, S.W., Stojiljkovic, I., and Heffron, F. 1996. Macrophage killing is an essential virulence mechanism of *Salmonella typhimurium*. Proc. Natl. Acad. Sci. USA 93: 4197–4201.

Lingwood, C.A. 1996. Role of verotoxin receptors in pathogenesis. Trends Microbiol. 4: 147–153.

Liu, D. 2004. *Listeria monocytogenes*: comparative interpretation of mouse virlence assay. FEMS Microbiol. Lett. 233: 159–164.

Lundberg, U., Vinatzer, U., Berdnik, D., von Gabain, A., and Baccarini, M. 1999. Growth phase-regulated induction of *Salmonella*-induced macrophage apoptosis correlates with transient expression of SPI-1 genes. J. Bacteriol. 181: 3433–3437.

McKee, M.L., Meltoncelsa, A.R., Moxley, R.A., Francis, D.H., and Obrien, A.D. 1995. Enterohemorrhagic *Escherichia coli* O157-H7 requires intimin to colonize the gnotobiotic pig intestine and to adhere to Hep-2 cells. Infect. Immun. 63: 3739–3744.

Maldonado, Y., Fiser, J.C., Nakatsu, C.H., and Bhunia, A.K. 2005. Cytotoxicity potential and genotypic characterization of *Escherichia coli* isolates from environmental and food sources. Appl. Environ. Microbiol. 71: 1890–1898.

Mangeney, M., Richard, Y., Coulaud, D., Tursz, T., and Wiels, J. 1991. Cd77 – An antigen of germinal center B-cells entering apoptosis. Eur. J. Immunol. 21: 1131–1140.

Mangeney, M., Lingwood, C.A., Taga, S., Caillou, B., Tursz, T., and Wiels, J. 1993. Apoptosis induced in Burkitts lymphoma cells via Gb3/Cd77, a glycolipid antigen. Cancer Res. 53: 5314–5319.

Marcato, P., Mulvey, G., and Armstrong, G.D. 2002. Cloned Shiga toxin 2 B subunit induces apoptosis in ramos Burkitt's lymphoma B cells. Infect. Immun. 70: 1279–1286.

Marquis, H., Bouwer, H.G.A., Hinrichs, D.J., and Portnoy, D.A. 1993. Intracytoplasmic growth and virulence of *Listeria monocytogenes* auxotrophic mutants. Infect. Immun. 61: 3756–3760.

Matsui, H., Suzuki, M., Isshiki, Y., Kodama, C., Eguchi, M., Kikuchi, Y., Motokawa, K., Takaya, A., Tomoyasu, T., and Yamamoto, T. 2003. Oral immunization with ATP-dependent protease-deficient mutants protects mice against subsequent oral challenge with virulent *Salmonella enterica* serovar Typhimurium. Infect. Immun. 71: 30–39.

Matthews, K.R., Murdough, P.A., and Bramley, A.J. 1997. Invasion of bovine epithelial cells by verocytotoxin-producing *Escherichia coli* O157:H7. J. Appl. Microbiol. 82: 197–203.

Maurelli, A.T., Routh, P.R., Dillman, R.C., Ficken, M.D., Weinstock, D.M., Almond, G.W., and Orndorff, P.E. 1998. *Shigella* infection as observed in the experimentally inoculated domestic pig, Sus scrofa domestica. Microb. Pathog. 25: 189–196.

Menon, A., Shroyer, M.L., Wampler, J.L., Chawan, C.B., and Bhunia, A.K. 2003. *In vitro* study of *Listeria monocytogenes* infection to murine primary and human transformed B cells. Comp. Immunol. Microbiol. Infect. Dis. 26: 157–174.

Merrick, J.C., Edelson, B.T., Bhardwaj, V., Swanson, P.E., and Unanue, E.R. 1997. Lymphocyte apoptosis during early phase of *Listeria* infection in mice. Am. J. Pathol. 151: 785–792.

Metcalf, E.S., Almond, G.W., Routh, P.A., Horton, J.R., Dillman, R.C., and Orndorff, P.E. 2000. Experimental *Salmonella* typhi infection in the domestic pig, Sus scrofa domestica. Microb. Pathog. 29: 121–126.

Michetti, P., Mahan, M.J., Slauch, J.M., Mekalanos, J.J., and Neutra, M.R. 1992. Monoclonal secretory immunoglobulin A protects mice against oral challenge with the invasive pathogen *Salmonella typhimurium*. Infect. Immun. 60: 1786–1792.

Mikami, T., Horikawa, T., Murakami, T., Matsumoto, T., Yamakawa, A., Murayama, S., Katagiri, S., Shinagawa, K., and Suzuki, M. 1994. An improved method for detecting cytostatic toxin (Emetic Toxin) of *Bacillus cereus* and its application to food samples. FEMS Microbiol. Lett. 119: 53–57.

Misawa, N., Ohnishi, T., Itoh, K., and Takahashi, E. 1996. Detection of serum-dependent cytotoxic activity of *Campylobacter jejuni* and its characteristics. J. Vet. Med. Sci. 58: 91–96.

Miyata, S., Minami, J., Tamai, E., Matsushita, O., Shimamoto, S., and Okabe, A. 2002. *Clostridium perfringens* epsilon-toxin forms a heptameric pore within the detergent-insoluble microdomains of Madin-Darby canine kidney cells and rat synaptosomes. J. Biol. Chem. 277: 39463–39468.

Mizuki, M., Nakane, A., Sekikawa, K., Tagawa, Y., and Iwakura, Y. 2002. Comparison of host resistance to primary and secondary *Listeria monocytogenes* infections in mice by intranasal and intravenous routes. Infect. Immun. 70: 4805–4811.

Monack, D.M., Raupach, B., Hromockyj, A.E., and Falkow, S. 1996. *Salmonella typhimurium* invasion induces apoptosis in infected macrophages. Proc. Natl. Acad. Sci. USA 93: 9833–9838.

Monack, D.M., and Falkow, S. 2000. Apoptosis as a common bacterial virulence strategy. Int. J. Med. Microbiol. 290: 7–13.

Monack, D.M., Hersh, D., Ghori, N., Bouley, D., Zychlinsky, A., and Falkow, S. 2000. *Salmonella* exploits caspase-1 to colonize Peyer's patches in a murine typhoid model. J. Exp. Med. 192: 249–258.

Mooney, A., Clyne, M., Curran, T., Doherty, D., Kilmartin, B., and Bourke, B. 2001. *Campylobacter upsaliensis* exerts a cytolethal distending toxin effect on HeLa cells and T lymphocytes. Microbiology-(UK) 147: 735–743.

Moors, M.A., Levitt, B., Youngman, P., and Portnoy, D.A. 1999. Expression of listeriolysin O and ActA by intracellular and extracellular *Listeria monocytogenes*. Infect. Immun. 67: 131–139.

Mueller, N.J., Wilkinson, R.A., and Fishman, J.A. 2002. *Listeria monocytogenes* infection in caspase-11-deficient mice. Infect. Immun. 70: 2657–2664.

Mukasa, A., Lahn, M., Fleming, S., Freiberg, B., Pflum, E., Vollmer, M., Kupfer, A., O'Brien, R., and Born, W. 2002. Extensive and preferential Fas/Fas ligand-dependent death of gamma delta T cells following infection with *Listeria monocytogenes*. Scand. J. Immunol. 56: 233–247.

Nadeau, E., Messier, S., and Quessy, S. 2003. Comparison of *Campylobacter* isolates from poultry and humans: Association between *in vitro* virulence properties, biotypes, and pulsed-field gel electrophoresis clusters. Appl. Environ. Microbiol. 69: 6316–6320.

Naughton, P.J., Grant, G., Spencer, R.J., and Bardocz, S. 1996. A rat model of infection by *Salmonella typhimurium* or *Salmonella enteritidis*. J. Appl. Bacteriol. 81: 651–656.

Newell, D.G. 2001. Animal models of *Campylobacter jejuni* colonization and disease and the lessons to be learned from similar *Helicobacter pylori* models. J. Appl. Microbiol. 90: 57S-67S.

Nishikawa, S., Hirasue, M., Miura, T., Yamada, K., Sasaki, S., and Nakane, A. 1998. Systemic dissemination by intrarectal infection with *Listeria monocytogenes* in mice. Microbiol. Immunol. 42: 325–327.

Nonaka, T., Kuwae, A., Sasakawa, C., and Imajoh-Ohmi, S. 1999. *Shigella flexneri* YSH6000 induces two types of cell death, apoptosis and oncosis, in the differentiated human monoblastic cell line U937. FEMS Microbiol. Lett. 174: 89–95.

Notermans, S., Dufrenne, J., Chakraborty, T., Steinmeyer, S., and Terplan, G. 1991. The chick embryo test agrees with the mouse bioassay for assessment of the pathogenicity of *Listeria* species. Lett. Appl. Microbiol. 13: 161–164.

Nuzzo, I., Sanges, M.R., Folgore, A., and Carratelli, C.R. 2000. Apoptosis of human keratinocytes after bacterial invasion. FEMS Immunol. Med. Microbiol. 27: 235–240.

Okuda, J., Fukumoto, M., Takeda, Y., and Nishibuchi, M. 1997. Examination of diarrheagenicity of cytolethal distending toxin: Suckling mouse response to the products of the cdtABC genes of *Shigella dysenteriae*. Infect. Immun. 65: 428–433.

Olier, M., Pierre, F., Lemaitre, J.P., Divies, C., Rousset, A., and Guzzo, J. 2002. Assessment of the pathogenic potential of two *Listeria monocytogenes* human faecal carriage isolates. Microbiology-(UK) 148: 1855–1862.

O'Loughlin, E.V., and Robins-Browne, R.M. 2001. Effect of Shiga toxin and Shiga-like toxins on eukaryotic cells. Microbes Infect. 3: 493–507.

Paesold, G., Guiney, D.G., Eckmann, L., and Kagnoff, M.F. 2002. Genes in the *Salmonella* pathogenicity island 2 and the *Salmonella* virulence plasmid are essential for *Salmonella*-induced apoptosis in intestinal epithelial cells. Cell Microbiol. 4: 771–781.

Pai, C.H., Kelly, J.K., and Meyers, G.L. 1986. Experimental infection of infant rabbits with verotoxin-producing *Escherichia coli*. Infect. Immun. 51: 16–23.

Pandiripally, V.K., Westbrook, D.G., Sunki, G.R., and Bhunia, A.K. 1999. Surface protein p104 is involved in adhesion of *Listeria monocytogenes* to human intestinal cell line, Caco-2. J. Med. Microbiol. 48: 117–124.

Parida, S.K., Domann, E., Rohde, M., Muller, S., Darji, A., Hain, T., Wehland, J., and Chakraborty, T. 1998. Internalin B is essential for adhesion and mediates the invasion of *Listeria monocytogenes* into human endothelial cells. Mol. Microbiol. 28: 81–93.

Paschen, A., Dittmar, K.E.J., Grenningloh, R., Rohde, M., Schadendorf, D., Domann, E., Chakraborty, T., and Weiss, S. 2000. Human dendritic cells infected by *Listeria monocytogenes*: induction of maturation, requirements for phagolysosomal escape and antigen presentation capacity. Eur. J. Immunol. 30: 3447–3456.

Pedersen, P.B., Bjornvad, M.E., Rasmussen, M.D., and Petersen, J.N. 2002. Cytotoxic potential of industrial strains of *Bacillus* spp. Regul. Toxicol. Pharmacol. 36: 155–161.

Peluffo, C.A., Irino, K., and Demello, S. 1981. Virulence in mice of epidemic strains of *Salmonella typhimurium* isolated from children. J. Infect. Dis. 143: 465–469.

Pijpers, A., Van Setten, P.A., Van Den Heuvel, L., Assmann, K.J.M., Dijkman, H., Pennings, A.H.M., Monnens, L.A.H., and Van Hinsbergh, V.W.M. 2001. Verocytotoxin-induced apoptosis of human microvascular endothelial cells. J. Am. Soc. Nephrol. 12: 767–778.

Pron, B., Boumaila, C., Jaubert, F., Sarnacki, S., Monnet, J.P., Berche, P., and Gaillard, J.L. 1998. Comprehensive study of the intestinal stage of listeriosis in a rat ligated ileal loop system. Infect. Immun. 66: 747–755.

Purdy, D., Buswell, C.M., Hodgson, A.E., McAlpine, K., Henderson, I., and Leach, S.A. 2000. Characterisation of cytolethal distending toxin (CDT) mutants of *Campylobacter jejuni*. J. Med. Microbiol. 49: 473–479.

Rabbani, G.H., Albert, M.J., Rahman, H., Islam, M., Mahalanabis, D., Kabir, I., Alam, K., and Ansaruzzaman, M. 1995. Development of an improved animal model of Shigellosis in the adult rabbit by colonic infection with *Shigella flexneri* 2a. Infect. Immun. 63: 4350–4357.

Ramegowda, B., Samuel, J.E., and Tesh, V.L. 1999. Interaction of Shiga toxins with human brain microvascular endothelial cells: Cytokines as sensitizing agents. J. Infect. Dis. 180: 1205–1213.

Raqib, R., Ekberg, C., Sharkar, P., Bardhan, P.K., Zychlinsky, A., Sansonetti, P.J., and Andersson, J. 2002. Apoptosis in acute shigellosis is associated with increased production of Fas/Fas ligand, perforin, caspase-1, and caspase-3 but reduced production of Bcl-2 and interleukin-2. Infect. Immun. 70: 3199–3207.

Reed, L.J., and Muench, H. 1938. A simple method of estimating fifty per cent endpoints. Am. J. Hyg. 27: 493–497.

Richter-Dahlfors, A., Buchan, A.M.J., and Finlay, B.B. 1997. Murine salmonellosis studied by confocal microscopy: *Salmonella typhimurium* resides intracellularly inside macrophages and exerts a cytotoxic effect on phagocytes *in vivo*. J. Exp. Med. 186: 569–580.

Roberts, P.H., Davis, K.C., Garstka, W.R., and Bhunia, A.K. 2001. Lactate dehydrogenase release assay from Vero cells to distinguish verotoxin producing *Escherichia coli* from non-verotoxin producing strains. J. Microbiol. Methods 43: 171–181.

Roche, S.M., Gracieux, P., Albert, I., Gouali, M., Jacquet, C., Martin, P.M.V., and Velge, P. 2003. Experimental validation of low virulence in field strains of *Listeria monocytogenes*. Infect. Immun. 71: 3429–3436.

Rogers, H.W., Callery, M.P., Deck, B., and Unanue, E.R. 1996. *Listeria monocytogenes* induces apoptosis of infected hepatocytes. J. Immunol. 156: 679–684.

Rood, J.I. 1998. Virulence genes of *Clostridium perfringens*. Annu. Rev. Microbiol. 52: 333–360.

Rowan, N.J., Caldow, G., Gemmell, C.G., and Hunter, I.S. 2003. Production of diarrheal enterotoxins and other potential virulence factors by veterinary isolates of *Bacillus* species associated with non-gastrointestinal infections. Appl. Environ. Microbiol. 69: 2372–2376.

Runyen-Janecky, L.J., and Payne, S.M. 2002. Identification of chromosomal *Shigella flexneri* genes induced by the eukaryotic intracellular environment. Infect. Immun. 70: 4379–4388.

Russell, R.G., Blaser, M.J., Sarmiento, J.I., and Fox, J. 1989. Experimental *Campylobacter jejuni* Infection in *Macaca nemestrina*. Infect. Immun. 57: 1438–1444.

Rutjes, N.W.P., Binnington, B.A., Smith, C.R., Maloney, M.D., and Lingwood, C.A. 2002. Differential tissue targeting and pathogenesis of verotoxins 1 and 2 in the mouse animal model. Kidney Int. 62: 832–845.

Santiago, N.I., Zipf, A., and Bhunia, A.K. 1999. Influence of temperature and growth phase on expression of a 104-kilodalton *Listeria* adhesion protein in *Listeria monocytogenes*. Appl. Environ. Microbiol. 65: 2765–2769.

Santos, R.L., Tsolis, R.M., Baumler, A.J., Smith, R., and Adams, L.G. 2001. *Salmonella enterica* serovar *Typhimurium* induces cell death in bovine monocyte-derived macrophages by early sipB-dependent and delayed sipB-independent mechanisms. Infect. Immun. 69: 2293–2301.

Sanyal, S.C., Islam, K.M.N., Neogy, P.K.B., Islam, M., Speelman, P., and Huq, M.I. 1984. *Campylobacter jejuni* diarrhea model in infant chickens. Infect. Immun. 43: 931–936.

Schlech, W.F. 1993. An animal model of foodborne *Listeria monocytogenes* virulence – Effect of alterations in local and systemic immunity on invasive infection. Clin. Invest. Med.-Med. Clin. Exp. 16: 219–225.

Schwan, W.R., Huang, X.Z., Hu, L., and Kopecko, D.J. 2000. Differential bacterial survival, replication, and apoptosis-inducing ability of *Salmonella* serovars within human and murine macrophages. Infect. Immun. 68: 1005–1013.

Shatursky, O., Bayles, R., Rogers, M., Jost, B.H., Songer, J.G., and Tweten, R.K. 2000. *Clostridium perfringens* beta-toxin forms potential-dependent, cation-selective channels in lipid bilayers. Infect. Immun. 68: 5546–5551.

Shroyer, M.L., and Bhunia, A.K. 2003. Development of a rapid 1-h fluorescence-based cytotoxicity assay for *Listeria* species. J. Microbiol. Meth. 55: 35–40.

Siarakas, S., Damas, E., and Murrell, W.G. 1995. Is cardiorespiratory failure induced by bacterial toxins the cause of sudden infant death syndrome? Studies with an animal model (the rabbit). Toxicon 33: 635–49.

Siarakas, S., Damas, E., and Murrell, W.G. 1997. The effect of enteric bacterial toxins on the catecholamine levels of the rabbit. Pathology 29: 278–85.

Siegler, R.L., Obrig, T.G., Pysher, T.J., Tesh, V.L., Denkers, N.D., and Taylor, F.B. 2003. Response to Shiga toxin 1 and 2 in a baboon model of hemolytic uremic syndrome. Pediatr. Nephrol. 18: 92–96.

Singh, U., Mitic, L.L., Wieckowski, E.U., Anderson, J.M., and McClane, B.A. 2001. Comparative biochemical and immunocytochemical studies reveal differences in the effects of *Clostridium perfringens* enterotoxin on polarized Caco-2 cells versus Vero cells. J. Biol. Chem. 276: 33402–33412.

Sjogren, R., Neill, R., Rachmilewitz, D., Fritz, D., Newland, J., Sharpnack, D., Colleton, C., Fondacaro, J., Gemski, P., and Boedeker, E. 1994. Role of Shiga-like toxin 1 in bacterial enteritis: Comparison between

isogenic *Escherichia coli* strains induced in rabbits. Gastroenterology 106: 306–317.

Songer, J.G. 1996. Clostridial enteric diseases of domestic animals. Clin. Microbiol. Rev. 9: 216–234.

Souvannavong, V., Lemaire, C., Denay, D., Brown, S., and Adam, A. 1995. Expression of alkaline phosphatase by a B cell hybridoma and its modulation during cell growth and apoptosis. Immunol. Lett. 47: 163–170.

Spira, W.M., Sack, R.B., and Froehlich, J.L. 1981. Simple adult rabbit model for *Vibrio cholerae* and enterotoxigenic *Escherichia coli* diarrhea. Infect. Immun. 32: 739–747.

Squier, M.K.T., and Cohen, J.J. 2001. Standard quantitative assays for apoptosis. Mol. Biotechnol. 19: 305–312.

Steinthorsdottir, V., Halldorsson, H., and Andresson, O.S. 2000. *Clostridium perfringens* beta-toxin forms multimeric transmembrane pores in human endothelial cells. Microb. Pathog. 28: 45–50.

Stelma, G.N., Reyes, A.L., Peeler, J.T., Francis, D.W., Hunt, J.M., Spaulding, P.L., Johnson, C.H., and Lovett, J. 1987. Pathogenicity test for *Listeria monocytogenes* using immunocompromised mice. J. Clin. Microbiol. 25: 2085–2089.

Stephens, J.C., Roberts, I.S., Jones, D., and Andrew, P.W. 1991. Effect of growth temperature on virulence of strains of *Listeria monocytogenes* in the mouse – Evidence for a dose dependence. J. Appl. Bacteriol. 70: 239–244.

Stern, N.J., Bailey, J.S., Blankenship, L.C., Cox, N.A., and McHan, F. 1988. Colonization characteristics of *Campylobacter jejuni* in chick ceca. Avian Dis. 32: 330–334.

Stravodimos, K.G., Singhal, P.C., Sharma, S., Reddy, K., and Smith, A.D. 1999. *Escherichia coli* promotes macrophage apoptosis. J. Endourol. 13: 273–277.

Sukupolvi, S., Edelstein, A., Rhen, M., Normark, S.J., and Pfeifer, J.D. 1997. Development of a murine model of chronic *Salmonella* infection. Infect. Immun. 65: 838–842.

Sun, A.N., Camilli, A., and Portnoy, D.A. 1990. Isolation of *Listeria monocytogenes* small plaque mutants defective for intracellular growth and cell-to-cell spread. Infect. Immun. 58: 3770–3778.

Suzuki, A., Doi, H., Matsuzawa, F., Aikawa, S., Takiguchi, K., Kawano, H., Hayashida, M., and Ohno, S. 2000. Bcl-2 antiapoptotic protein mediates verotoxin II-induced cell death: possible association between Bcl-2 and tissue failure by *E. coli* O157:H7. Genes Dev. 14: 1734–1740.

Szabo, R.A., Speirs, J.I., and Akhtar, M. 1991. Cell culture detection and conditions for production of a *Bacillus cereus* heat-stable toxin. J. Food Prot. 54: 272–276.

Taguchi, H., Takahashi, M., Yamaguchi, H., Osaki, T., Komatsu, A., Fujioka, Y., and Kamiya, S. 2002. Experimental infection of germ-free mice with hyper-toxigenic enterohaemorrhagic *Escherichia coli* O157:H7, strain 6. J. Med. Microbiol. 51: 336–343.

Terplan, G., and Steinmeyer, S. 1989. Investigations on the pathogenicity of *Listeria* spp. by experimental infection of the chick embryo. Int. J. Food Microbiol. 8: 277–280.

Ting, W.T., and Banwart, G.J. 1985. Detection of *Bacillus cereus* diarrheagenic toxin using a rat ligated intestinal loop assay. J. Food Safety 7: 57–63.

Tsolis, R.M., Adams, L.G., Ficht, T.A., and Baumler, A.J. 1999. Contribution of *Salmonella typhimurium* virulence factors to diarrheal disease in calves. Infect. Immun. 67: 4879–4885.

Tweten, R.K. 2001. *Clostridium perfringens* beta toxin and *Clostridium septicum* alpha toxin: their mechanisms and possible role in pathogenesis. Vet. Microbiol. 82: 1–9.

Tzipori, S., Gunzer, F., Donnenberg, M.S., Demontigny, L., Kaper, J.B., and Donohuerolfe, A. 1995. The role of the eaeA gene in diarrhea and neurological complications in a gnotobiotic piglet model of enterohemorrhagic *Escherichia coli* infection. Infect. Immun. 63: 3621–3627.

Uchida, H., Kiyokawa, N., Taguchi, T., Horie, H., Fujimoto, J., and Takeda, T. 1999. Shiga toxins induce apoptosis in pulmonary epithelium-derived cells. J. Infect. Dis. 180: 1902–1911.

Valenti, P., Greco, R., Pitari, G., Rossi, P., Ajello, M., Melino, G., and Antonini, G. 1999. Apoptosis of Caco-2 intestinal cells invaded by

*Listeria monocytogenes*: Protective effect of lactoferrin. Exp. Cell Res. 250: 197–202.

van de Kar, N.C.A.J., Kooistra, T., Vermeer, M., Lesslauer, W., Monnens, L.A.H., and Vanhinsbergh, V.W.M. 1995. Tumor necrosis factor alpha induces endothelial galactosyl transferase activity and verocytotoxin receptors: Role of specific tumor necrosis factor receptors and protein kinase C. Blood 85: 734–743.

Vazquez-Boland, J.A., Kuhn, M., Berche, P., Chakraborty, T., Dominguez-Bernal, G., Goebel, W., Gonzalez-Zorn, B., Wehland, J., and Kreft, J. 2001. *Listeria* pathogenesis and molecular virulence determinants. Clin. Microbiol. Rev. 14: 584–640.

von Moll, L.K., and Cantey, J.R. 1997. Peyer's patch adherence of enteropathogenic *Escherichia coli* strains in rabbits. Infect. Immun. 65: 3788–3793.

Waddell, T., Head, S., Petric, M., Cohen, A., and Lingwood, C. 1988. Globotriosyl ceramide Is specifically recognized by the *Escherichia coli* verocytotoxin-2. Biochem. Biophys. Res. Commun. 152: 674–679.

Wampler, J.L., Kim, K.-P., Jaradat, Z.W., and Bhunia, A.K. 2004. Heat shock protein acts as a receptor for the *Listeria* adhesion protein in Caco-2 cells. Infect. Immun. 72: 931–936.

Wassenaar, T.M., Vanderzeijst, B.A.M., Ayling, R., and Newell, D.G. 1993. Colonization of chicks by motility mutants of *Campylobacter jejuni* demonstrates the importance of flagellin A expression. J. Gen. Microbiol. 139: 1171–1175.

Wassenaar, T.M., and Blaser, M.J. 1999. Pathophysiology of *Campylobacter jejuni* infections of humans. Microbes Infect. 1: 1023–1033.

Weinrauch, Y., and Zychlinsky, A. 1999. The induction of apoptosis by bacterial pathogens. Annu. Rev. Microbiol. 53: 155–187.

Williams, J.M., Boyd, B., Nutikka, A., Lingwood, C.A., Foster, D.E.B., Milford, D.V., and Taylor, C.M. 1999. A comparison of the effects of verocytotoxin-1 on primary human renal cell cultures. Toxicol. Lett. 105: 47–57.

Wolski, V.M., Soltyk, A.M., and Brunton, J.L. 2002. Tumour necrosis factor alpha is not an essential component of verotoxin 1-induced toxicity in mice. Microb. Pathog. 32: 263–271.

Woods, J.B., Schmitt, C.K., Darnell, S.C., Meysick, K.C., and O'Brien, A.D. 2002. Ferrets as a model system for renal disease secondary to intestinal infection with *Escherichia coli* O157:H7 and other Shiga toxin-producing *E. coli*. J. Infect. Dis. 185: 550–554.

Yoshida, T., Fukada, M., Koide, N., Ikeda, H., Sugiyama, T., Kato, Y., Ishikawa, N., and Yokochi, T. 1999. Primary cultures of human endothelial cells are susceptible to low doses of Shiga toxins and undergo apoptosis. J. Infect. Dis. 180: 2048–2052.

Yoshimura, K., Tanimoto, A., Abe, T., Ogawa, M., Yutsudo, T., Kashimura, M., and Yoshida, S. 2002. Shiga toxin 1 and 2 induce apoptosis in the amniotic cell line WISH. J. Soc. Gynecol. Invest. 9: 22–26.

Zhang, S.P., Adams, L.G., Nunes, J., Khare, S., Tsolis, R.M., and Baumler, A.J. 2003. Secreted effector proteins of *Salmonella enterica* serotype *Typhimurium* elicit host-specific chemokine profiles in animal models of typhoid fever and enterocolitis. Infect. Immun. 71: 4795–4803.

Zhang, X., Kelly, S.M., Bollen, W., and Curtiss, R. 1999. Protection and immune responses induced by attenuated *Salmonella typhimurium* UK-1 strains. Microb. Pathog. 26: 121–130.

Zhou, X., Mantis, N., Zhang, X.R., Potoka, D.A., Watkins, S.C., and Ford, H.R. 2000. *Salmonella typhimurium* induces apoptosis in human monocyte-derived macrophages. Microbiol. Immunol. 44: 987–995.

Zhu, J.T., Meinersmann, R.J., Hiett, K.L., and Evans, D.L. 1999. Apoptotic effect of outer-membrane proteins from *Campylobacter jejuni* on chicken lymphocytes. Curr. Microbiol. 38: 244–249.

Zychlinsky, A., Prevost, M.C., and Sansonetti, P.J. 1992. *Shigella flexneri* induces apoptosis in infected macrophages. Nature 358: 167–169.

Zychlinsky, A., Fitting, C., Cavaillon, J.M., and Sansonetti, P.J. 1994a. Interleukin-1 is released by murine macrophages during apoptosis induced by *Shigella flexneri*. J. Clin. Invest. 94: 1328–1332.

Zychlinsky, A., Kenny, B., Menard, R., Prevost, M.C., Holland, I.B., and Sansonetti, P.J. 1994b. IpaB mediates macrophage apoptosis induced by *Shigella flexneri*. Mol. Microbiol. 11: 619–627.

Zychlinsky, A., Thirumalai, K., Arondel, J., Cantey, J.R., Aliprantis, A.O., and Sansonetti, P.J. 1996. *In vivo* apoptosis in *Shigella flexneri* infections. Infect. Immun. 64: 5357–5365.

Zychlinsky, A., and Sansonetti, P.J. 1997. Apoptosis as a proinflammatory event: What can we learn from bacteria-induced cell death? Trends Microbiol. 5: 201–204.

# Biosensor-based Detection of Foodborne Pathogens

3

George P. Anderson and Chris Rowe Taitt

## Abstract

The sensitive, rapid, and specific detection of microorganisms and toxins that taint the food supply has become increasingly important as large-scale manufacture with wide distribution can threaten large populations when a contamination occurs. Biosensors have been seen as a means to provide a higher level of surveillance in a more automated and rapid manner. Only in the last few years have biosensors matured to the point where they can begin to meet these demanding applications. In this review we will endeavor to touch upon several biosensor methodologies and give the reader some insight as to what technologies are available and how they might be applied for their particular detection needs. There are, however, numerous detection technologies, with more being developed constantly. Regrettably therefore, not all methods will be covered, but we do hope to present a representative cross-section of techniques.

## Introduction

The development of biosensors for detection of foodborne pathogens has been in many cases a directed effort to meet a particular need in this arena (BioControl Systems' Lightning for ATP-detection), while in other cases useful technology was first developed for other applications such as biological warfare defense (Research International's RAPTOR) or as laboratory assay instrumentation (BioVeris/IGEN's ORIGEN or Luminex's Luminex[100]). Whatever the overt motivating factor in the development of the detection methodology, there are two underlying factors in common: the desire to obtain a faster answer and the preference to automate the procedure.

The bulk of the detection methodologies currently in use are not biosensor based. However they do represent extremely sensitive techniques, which biosensors will not easily displace. Cell culture has been a mainstay of microbiology from the beginning, and while time consuming, is still recognized by the US Food and Drug Administration (FDA), the Center for Food Safety and Applied Nutrition (CFSAN), the Food Safety Inspection Service (FSIS), and the US Department of Agriculture (USDA) as the gold standard for determination

of bacterial contamination. Typically, these analyses involve a non-selective "pre-enrichment", followed by one or more selective enrichments, with the extent of contamination determined by most probable number calculations. Biochemical and immunological analyses are also generally performed on strains isolated from foodstuffs; these may include tests for cytochrome oxidase activity, production of $H_2S$, fermentation of various energy sources, gram stain, and somatic and flagellar antigen agglutination tests, among others. Although a number of these tests are available commercially in user-friendly form (e.g., VITEK, RapID), the requirement for multiple dilutions and culture steps precludes rapid, on-site quantification and identification of bacterial contaminants. Additional laboratory-based detection methodologies include tests for DNA identification, and standard manual immunoassays (i.e., ELISAs). It is primarily, but not entirely, the automation of these DNA analyses and immunoassays that biosensors entail.

Biosensors are detection devices that use biological molecules to recognize and quantify analytes of interest. These recognition elements may be enzymes, antibodies, receptors, nucleic acids, and either combinatorial or naturally occurring peptides and oligosaccharides. Chosen on the basis of their affinity and specificity, these biomolecules transduce a recognition event (binding or catalysis) into a measurable optical, chemical, or electrical/electronic signal. These devices must also be capable of continuously monitoring analyte concentrations, providing quantitative or semi-quantitative information (Thevenot et al., 2001); this latter requirement distinguishes biosensor technology from those of single-use bioprobes and techniques such as ELISA. Most of what we currently refer to as biosensors fall short of being able to continuously monitor analyte concentrations in a manner similar to monitoring pH, but clearly this is the direction in which the biosensor field is striving to move.

We have divided this chapter into several sections based on the method of signal transduction. We start with discussions of several methodologies, including ATP detection, PCR and nucleic acid hybridization, and enzyme-linked immunosorbent assays (ELISAs). These methodologies are single-use techniques and are therefore not considered to be biosensor technologies. However, we are including these methods in this chapter, as they form the basis of many biosensors or have the potential for transition into biosensor platforms. We then continue with description of biosensors used for pathogen and toxin detection. These biosensors have been grouped according to the mechanism of signal transduction: first optical methods, then electrochemical methods, and finally miscellaneous methods. It should be noted that there is sufficient work in each of these categories to write entire books, so this chapter will be by necessity an overview, and the reader is directed to additional sources if they desire more in depth information. Indeed, there have been several excellent recent reviews that we would like to acknowledge and to direct the readers to as well: Hobson *et al.* (1996), Paddle (1996), McMeekin *et al.* (1997), Ivnitski *et al.* (1999), Vo-Dinh and Cullum (2000), Luppa *et al.* (2001), Hall (2002), Mello and Kuboto (2002), Patel (2002), Baeumner (2003), Dickert *et al.* (2003), and Leonard *et al.* (2003).

## ATP detection

ATP bioluminescence techniques are increasingly used for measuring the efficiency of cleaning surfaces and utensils. Typically, a swab sample is combined with a mixture of the enzyme luciferase and its substrate, luciferin. Reaction of luciferase with luciferin is dependent on the presence of adenosine triphosphate (ATP), which is found in all living, but not dead cells. Luciferin combines with ATP to form luciferyl adenylate and pyrophosphate ($PP_i$) on the surface of the luciferase enzyme:

$$\text{luciferin} + \text{ATP} \rightarrow \text{luciferyl adenylate} + PP_i \qquad (3.1)$$

The luciferyl adenylate then reacts with oxygen to form oxyluciferin and adenosine monophosphate (AMP), with the production of light:

$$\text{luciferyl adenylate} + O_2 \rightarrow \text{oxyluciferin}$$
$$+ \text{AMP} + \text{light} \qquad (3.2)$$

This enzymatic process, originally described in fireflies, measures the ATP extracted from the bacteria by the amount of light produced. Commercial reagents and luminometers equipped with photon multiplier tubes (PMTs) are now available, some in hand-held formats: Lightning (BioControl Systems), Uni-Lite and Clean-Trace (Biotrace), Advance and Spot Check (Celsis), and Pocket Swab (Charm). MicroStar™, produced by Millipore, uses a CCD camera to detect ATP from multiple spots and has demonstrated a limit of detection (LOD) of between 1 and 200 colony-forming units (CFU) per spot. While these commercial units can be an excellent aid in nonspecific detection of bacterial contamination, they may have limitations. Carrick *et al.* (2001) described significant errors, including false positives and false negatives, resulting from different sampling techniques. However, the high sensitivity that luminescence provides continues to attract adherents, and

significant research continues in this methodology (Deininger and Lee, 2001). Hattori *et al.* (2003) used a mutant luciferase, resistant to benzalkonium chloride, used to extract ATP from samples, to detect $1.8 \times 10^{-18}$ moles of ATP, a 500-fold improvement over some commercial systems. This concentration of ATP corresponded to 93 CFU/mL *Escherichia coli*, 170 CFU/mL *Pseudomonas aeruginosa*, 68 CFU/mL *Staphylococcus aureus*, and 7.7 CFU/mL of *Bacillus subtilis*. Sakakibara *et al.* (2003) have recently used an enzymatic cycling assay for ATP and AMP to amplify the bioluminescent signal. Using firefly luciferase and pyruvate orthophosphate dikinase (PPDK), AMP and pyrophosphate produced from ATP by firefly luciferase (eqn 3.1) were converted back into ATP by PPDK, with ATP detection limits at the level of single bacterial cells ($3.1 \times 10^{-19}$ moles/spot) in a 5 min assay.

In general, ATP assays are nonspecific, providing no information as to the identity of the ATP source. However, ATP detection has been combined with other detection and identification techniques by a number of workers. Takahashi *et al.* (2000) described the conversion of a fluorescent microscope to work in a manner similar to the MicroStar, but for the detection of fluoro-immuno stained cells, as well as detection of ATP. Seaver et al. (2001) developed a similar type of instrument to detect nuclear stained bacterial cells in blood platelets. Vermicon markets gene probes, small pieces of DNA labeled with a fluorescent dye, for bacterial detection and identification by fluorescent microscopy. Others have looked at combining bioluminescence with immuno-separation techniques to achieve the desired specificity. Tu *et al.* (2000) detected *E. coli* using immunomagnetic beads to capture the desired bacteria. Squirrel *et al.* (2002) combined immunomagnetic separation with targeted cell lysis using bacteriophage to detect as few as 100 cell/mL *E. coli*. In another method, Stender *et al.* (2001) combined ATP-dependent bioluminescence with a chemiluminescent *in situ* hybridization (CISH) method using peroxidase-labeled peptide nucleic acid (PNA) probes targeting species-specific rRNA sequences to provide identification of specific microorganisms on the MicroStar system.

## PCR/nucleic acid hybridization

In contrast to detection of ATP by luminescence, nucleic acid amplification and hybridization do provide identification of the source of contamination and have recently become more accepted as standard methods for testing of foods. The general principles and procedures of the polymerase chain reaction (PCR) have been well established and the reader is referred to Chapter 1 in this volume and elsewhere for more detailed descriptions of this technology (Sambrook *et al.*, 1989; Ausubel *et al.*, 1994). In general, PCR involves the repetitive cycling of the following steps: "melting" of duplex DNA to reveal sequences of interest, hybridization of oligonucleotide primers to specific (complementary) locations on the target DNA, and creation of new strands of DNA complementary to the target DNA. Amplification of DNA sequences through PCR has been the mainstay of many techniques for detection of pathogens, with PCR products detected electrophoretically, by Southern Blot, or by hybridization to probes immobilized in microarrays; a modification of standard PCR techniques, real-time PCR (RT-PCR), will be discussed later, as this is considered a biosensor technique. A number of direct PCR-based detection methods have become commercially available (e.g., BAX, Probelia).

While PCR (and RT-PCR, discussed later) holds great promise for rapid analysis of foods, LODs for PCR are in the range of $10^4$ CFU/g food. These high detection limits are primarily due to small reaction volumes, amplification of non-specific sequences at high cycle numbers, and the presence of inhibitors (e.g., proteases, phenolics) in many foodstuffs. Even though PCR is typically coupled to an enrichment step, the time savings (over standard methods) and high sensitivity obtained have made PCR a highly effective method for determination of bacterial contamination.

## Immunoassays

Cell culture, ATP analyses, and DNA-based detection methodologies, however, fail to account for the presence of non-culturable, non-viable microbial agents of disease and toxins. While many proteinaceous toxins, such as ricin and botulinum toxin, are inactivated by standard food sterilization methods, staphylococcal enterotoxins (Bergdoll, 1983), tetrodotoxin, saxitoxin, and mycotoxins are notoriously heat-stable and remain toxic even in processed or cooked foods. In addition to traditional laboratory methods for detection of non-protein toxins (e.g., gas chromatography, gas chromatography/mass spectroscopy, thin layer chromatography, high performance liquid chromatography, and bioassays), antibody-based techniques are fast gaining acceptance as methods of choice for detection and identification of foodborne contaminants, either as immunoaffinity purification steps (Maragos et al., 1997; Moller and Gustavsson, 2000), or as the core of the detection method (enzyme-linked immunosorbent assays, radioimmunoassay, dye-binding; Hokama et al., 1987, 1997, 1998; Yang et al., 1987). While radioimmunoassays and dye binding assays are considered tried and true methods, enzyme-linked immunosorbent assays (ELISAs) have become more widespread, in large part due to their adaptability for detection and quantification of a wide variety of analytes. For detection purposes, two ELISA formats are generally used: sandwich and competitive (Figure 3.1). Sandwich assays, most commonly used for detection of larger molecules (e.g., protein toxins, bacteria, and viruses), use an immobilized "capture" antibody to recognize and capture analyte from the sample. Addition of a second, "tracer" antibody forms a capture antibody-antigen-tracer antibody sandwich. The tracer antibody can be either directly coupled to an enzyme such as horseradish peroxidase or can be detected indirectly by use of an HRP-conjugated secondary antibody (e.g., anti-species antibody). The number of sandwiches formed is then quantified by an enzymatic reaction producing a colored or fluorescent product.

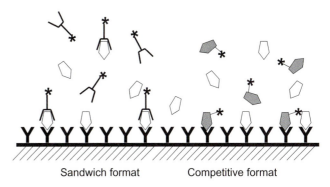

**Figure 3.1** Schematic of sandwich (left) and competitive (right) immunoassay formats.

Competitive ELISAs are most often used for detection of analytes too small to possess two epitopes or binding sites, as required in sandwich assays; mycotoxins are among those food-related analytes for which competitive assays have been developed (AOAC, 1995). Competitive assays are generally of two types. In one format, a labeled analog of the toxin is added to the sample; the spiked sample (containing labeled toxin and unlabeled, endogenous toxin) is then incubated with an immobilized capture antibody. Competition between the labeled and unlabeled toxins for binding to the immobilized antibody is measured as a decrease in signal over control levels (no endogenous toxin in the sample). The second format of competitive assay measures the competition between analyte in the sample and an immobilized analog for binding to tracer antibodies added to the sample. Again, the signal will decrease as more analyte in the sample prevents the tracer antibody from binding to immobilized analog. Generally competitive assays are less sensitive, since the challenge is to differentiate a small decrease in a large signal.

## Biosensors based on optical transduction

Optical biosensors have the advantage of rapid sensing, where the speed of analyte detection is most often limited by rate of the biomolecular recognition event, rather than by the speed of optical transduction. These biosensors measure changes in refractive index, fluorescence emission or quenching, chemiluminescence, and fluorescence energy transfer. As such, these sensors are relatively immune to sources of interference that may plague sensors with electrochemical transducers: voltage surges, electromagnetic radiation, harmonic induction, corrosion of transducing elements, and radio frequency interference.

Several types of optical biosensors are based on the principle of total internal reflectance and production of the evanescent wave. When light is launched into a waveguide that is placed into a dielectric medium of lower refractive index ($n_{waveguide} > n_{medium}$), the light may be partially reflected back into the waveguide and partially refracted into the medium. Under conditions where θ, the angle of incidence of the light entering the waveguide, is greater than the critical angle, $\theta_c$:

$$\theta_c = \sin^{-1}(n_{medium}/n_{waveguide}) \qquad (3.3)$$

all of the light is reflected within the waveguide and none is refracted. This phenomenon is known as total internal reflection.

Under these conditions, an electromagnetic component of the light, the evanescent wave, is produced and extends out from the surface of the waveguide into the medium. The strength of the evanescent field decays exponentially with distance from the surface. For multimode waveguides, the penetration depth ($d_p$), the distance at which the power in the evanescent field decays to $1/e$ of its original value at the waveguide surface, depends on the difference in refractive indices and both the wavelength and angle of incidence of the light within the waveguide:

$$d_p = \lambda/(4\pi[(n_{waveguide})^2 \sin\theta - (n_{medium})^2]^{1/2}) \qquad (3.4)$$

The surface selectivity of the evanescent wave has been exploited by a number of biosensors discussed hereafter: interferometers, resonant mirror biosensors, surface plasmon resonance biosensors, and fiber optic and planar array fluorescence biosensors.

## Interferometric biosensors

Interferometers measure the effect that changes in the refractive index of the medium have upon the phase of the light propagated within the waveguide. The phase (φ) is dependent on the pathlength ($L$), the refractive index ($n$) within the evanescent field, and the wavelength (λ):

$$\phi = 2\pi Ln/\lambda \qquad (3.5)$$

Thus, if a biomolecular or chemical interaction takes place at the surface (within the evanescent field), the resultant refractive index change will cause a change in the phase of the propagating light. The light's phase and any changes thereof are detected by the use of a reference beam. The reference beam is placed adjacent to the sensing beam but does not encounter the molecular interaction event and hence, does not reflect the change in refractive index. Lightwaves from the reference beam and the sensing beam are combined to create an interference pattern of alternating light and dark bands. Any changes in refractive index will produce a concomitant change in this interference pattern.

Primarily still a laboratory instrument for sample analysis under controlled conditions (i.e., pristine buffer), an interferometric biosensor assay for *Salmonella typhimurium* has been developed by Campbell and coworkers for use in chicken carcass rinse (Seo *et al.*, 1999). Carcass rinse samples inoculated with 20 CFU/mL generated positive signals in the in-

terferometric assay after a 12-hour non-selective enrichment. Specificity of the system was demonstrated by the lack of interference in these assays by the presence of a 10-fold excess of non-relevant bacteria. A similar system for detection of *Salmonella*, the Hartman interferometer (Schneider *et al.*, 1997), uses linearly polarized light with a planar waveguide format, thereby avoiding the problems associated with circular polarization and channel waveguides.

## Reflectometric interference spectroscopic biosensors

An interesting variation of these interferometric sensors is a sensor based on reflectometric interference spectroscopy (RIfS). Spearheaded by Gauglitz's group at University of Tuebingen, this technique uses the reflection of white light at a thin waveguide placed on a glass substrate to produce an interference pattern over a wide spectrum. A third (sensing) layer is formed by recognition molecules (e.g., antibodies) immobilized on the surface of the thin waveguide layer and light reflected off this layer is also measured. A change in optical thickness of this third layer occurs upon a binding event (e.g., binding of analyte to antibodies). The resulting shift of the interference pattern is then detected by a diode array. This method has been used for detection of antigen-antibody binding (Piehler *et al.*, 1996) and DNA hybridization (Jung *et al.*, 2001; Kroger *et al.*, 2002). Recently, Gauglitz's group has expanded the capabilities of this technology to include lateral resolution of different loci on their sensing surface. They were able to distinguish between 30 similar compounds using a microarray of immobilized recognition species (Willard *et al.*, 2003). While this method has not yet been applied to detection of food-related analytes, with further development, this methodology holds great promise for detection of both small molecules, such as mycotoxins, and larger analytes such as protein toxins, viruses, and bacteria.

## Resonant mirror biosensors

Resonant mirror (RM) biosensors also utilize the evanescent wave to detect changes in refractive index or mass in close proximity to the sensor surface. In RM sensors, the evanescent wave is produced by a resonant structure composed of a high-index resonant layer and a low-index resonant coupling layer (Figure 3.2). While the latter layer couples incident light from an optical prism into the Resonant Structure, the high refractive index layer acts as both a waveguiding structure and the sensing surface; it is from this high index layer that the evanescent field extends into the bulk medium or sample. Coupling of light into and out of the resonant structure is highly dependent on the angle of the incident light. Light coupled into the resonant structure undergoes a 90° phase change and is detected (after out-coupling) after a polarizer rejects all other (non-coupled) light. The resonant angle is sensitive to changes in the refractive index within the evanescent wave. Thus, like interferometric sensors, this instrument can also be used to trace minute changes in refractive index at the surface

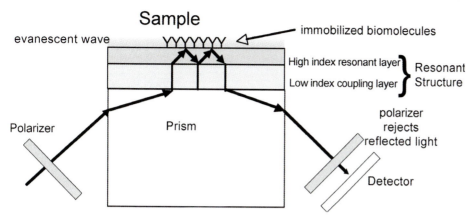

**Figure 3.2** Schematic of resonant mirror biosensor.

brought about by biomolecular binding events. One RM sensor has been described for detection of *S. aureus* in buffer and milk (Watts *et al.*, 1994). While a limit of detection (LOD) of $6 \times 10^6$ cells/mL was observed in direct binding assays, a 1000-fold improvement was observed in sandwich assays that incorporated colloidal gold particles to amplify the refractive index change. Using this colloidal gold-based sandwich assay, $4 \times 10^3$ cells/mL could be detected in spiked whole milk. This sensor has since been commercialized by Affinity Sensor (formerly Fisons Applied Sensor Technology) and is available as IAsys™.

## Surface plasmon resonance

Surface plasmon resonance (SPR) biosensors are very similar to the RM sensors described above in that light coupled into an intermediate layer forms the basis for detection of surface-specific events via changes in refractive index. While RM sensors use the resonant structure as this intermediate layer (Figure 3.2), SPR sensors use a thin (50 nm) metal film (Figure 3.3). In SPR-based detection, light launched through a high refractive index prism or an optical waveguide, under conditions of total internal reflection, produces an evanescent wave that extends into the metal layer. This evanescent wave excites surface electrons (plasmons) to produce a surface plasma wave

(SPW) that projects into the medium. Coupling of light from the prism/waveguide into the metal layer and excitation of surface plasmons can be measured as a *decrease* in the reflectivity of light from the metal-prism or metal-waveguide interface at the specific coupling conditions (coupling angle, wavelength, polarization). Any change in the optical properties in the bulk medium will modify the characteristics of the SPW, and hence, also the characteristics of the light interacting with the metal layer. These latter changes can be measured as a change in one of the characteristics of the light wave interacting with the SPW: wavelength, intensity, phase, or angle of incidence. Based on which characteristic of the light is measured, SPR sensors can be classified into angular-, wavelength-, intensity-, or phase-modulated systems. Angular-modulated systems (schematic shown in Figure 3.3) work at a single fixed wavelength and use a photodiode to monitor changes in the angle of incident light where coupling occurs; BIAcore™ and Spreeta™ systems are examples of commercial angular-modulated systems. Wavelength-modulated systems, on the other hand, utilize a constant angle of polychromatic light while measuring changes in the spectrum of the light reflected from the metal-prism or metal-waveguide interface. Phase- and intensity-modulated systems, less commonly used, keep both the wavelength and angle of incident light constant and measure changes in the

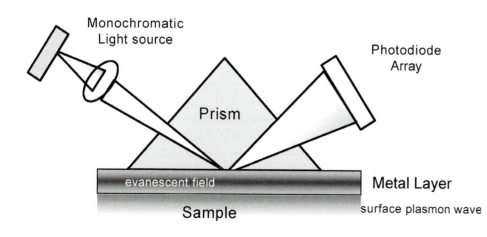

**Figure 3.3** Schematic of angular-modulated SPR sensor.

**Table 3.1** SPR-based detection of foodborne pathogens and toxins

| Analyte | Detection limit | Matrix | Reference |
|---|---|---|---|
| *Salmonella* sp., direct | $10^7$ CFU/mL | Buffer, interfering bacteria, carcass swabs | Bokken *et al.* (2003) |
| *S. enteritidis*, direct | $10^6$ CFU/mL | Buffer | Koubova *et al.* (2001) |
| *Listeria monocytogenes*, direct | $10^6$ CFU/mL | Buffer | Koubova *et al.* (2001) |
| Fumonisin, direct | 50 ng/mL | Buffer | Mullett *et al.* (1998) |
| Deoxynivalenol, competitive | 2.5 ng/mL | Buffer, wheat | Tudos *et al.* (2003) |
| *E. coli* heat-labile enterotoxin, ganglioside capture | 6 µg/mL | Buffer | Spangler *et al.* (2001) |
| SEA, sandwich | 10–100 ng/mL | Buffer, hot dogs, potato salad, milk, mushrooms | Rasooly and Rasooly (1999) |
| SEB, sandwich | | | |
|   SEB | 10 ng/mL | Buffer, milk, meat | Rasooly (2001) |
|   Direct | 5 ng/mL | Buffer, milk | Homola *et al.* (2002a) |
|   Sandwich | 0.5 ng/mL | Buffer, milk | Homola *et al.* (2002a) |
| SEB | | | |
|   Direct | 10 ng/mL | Buffer, milk, urine, seawater | Naimushin *et al.* (2002) |
|   Sandwich | 0.5 ng/mL | Buffer, milk, urine, seawater | Naimushin *et al.* (2002) |
| Two-step sandwich | 3 pg/mL | Buffer, milk, urine, seawater | Naimushin *et al.* (2002) |
| SEB, SPR–MALDI–ToF | 1 ng/mL | Buffer, milk, mushrooms | Nedelkov *et al.* (2000), Nedelkov and Nelson (2003a) |

phase or intensity of light when coupling occurs. See Homola *et al.* (2002b) for an excellent review of the physics and optics of SPR sensors.

SPR-based detection of foodborne pathogens and toxins has seen wide use (Table 3.1); however, one limitation of SPR-based detection is the potential for alterations in refractive index at the sensing surface due to non-homogeneous (complex) sample matrices and/or nonspecific binding of interfering species. Several groups have performed direct binding assays in the presence of non-relevant analytes or complex matrices with little or no observed signal from these non-specific components (Koubova *et al.*, 2001; Rasooly 2001; Homola *et al.*, 2002a; Bokken *et al.*, 2003). However, others have chosen to use sandwich format assays to circumvent this potential problem. Addition of one or more washing steps, in addition to incubation with a second ("tracer") antibody, resulted in lower levels of nonspecific binding or other matrix effects. Sandwich-format assays have been developed for detection of staphylococcal enterotoxins A and B (SEA, SEB) in milk, mushrooms, hot dogs, meat, and potato salad (Rasooly and Rasooly, 1999; Rasooly, 2001; Homola *et al.*, 2002a).

In addition to decreasing any effects from sample matrices, use of a second antibody in sandwich assays has the additional advantage of increasing the mass associated with surface-bound complexes. This additional mass increases the net change in refractive index accordingly. Homola and coworkers (2002a) observed a 10-fold improvement in detection limit when comparing a sandwich assay with an analogous direct binding assay for SEB. Naimushin *et al.* (2002) took this idea a step further and improved detection limits by > 1000-fold through use of *two* amplification steps. However, while this latter group was able to detect pg/mL concentrations of SEB in their samples, assay times were lengthened considerably over the direct assays due to the additional steps. Although antibodies are still the primary recognition species in most SPR-based schemes to detect food-related pathogens and toxins, several systems have been described that use receptors for toxin detection (Spangler *et al.*, 2001), and oligonucleotides or peptide-nucleic acid probes to detect PCR products from pathogens (Kai *et al.*, 2000; Pollard-Knight *et al.*, 1990).

Tracer antibodies are also frequently necessary for SPR-based detection of small analytes, where direct binding of analyte to immobilized antibodies does not produce a measurable response. Thus, assays for many smaller analytes take the form of competitive, or binding inhibition, assays. A competitive assay for the mycotoxin deoxynivalenol (DON) was recently described (Tudos *et al.*, 2003). Samples containing DON were mixed with anti-DON antibodies and then incubated with a DON/casein conjugate-coated SPR sensor; by comparing the loss in signals against negative controls (no DON in the sample), a limit of detection of 2.5 ng/mL was obtained.

Nedelkov and Nelson (Nedelkov *et al.*, 2000; Nedelkov and Nelson, 2003) have recently described use of an integrated SPR–MALDI–MS system; after SPR analysis of SEB-spiked mushroom samples, they were able to distinguish SEB from other proteins using subsequent analysis of the SPR-captured species by MALDI–TOF mass spectrometry. They further describe the potential of this system for identifying multiple *related* toxins using capture antibodies of broader specificities, with MALDI–TOF–MS providing the specificity required for identification of individual species (Nedelkov and Nelson 2003).

One recent trend in SPR biosensors is the development of additional sensing channels for background subtraction of complex samples (Homola *et al.*, 2001; Naimushin *et al.*, 2002) and multiplexing of assays (www.biacore.com; Berger *et al.*, 1998; Homola *et al.*, 1999; Lu *et al.*, 2001; Nedelkov and Nelson 2003). Homola and coworkers have also developed several methods to multiplex assays on a single chip. These methods include use of a beveled prism coupler to serially couple light at different areas of the sensing element (Homola *et al.*, 2001) and use of a high refractive index overlayer to shift the coupling wavelength in a portion of the sensing chip (Homola *et al.*, 1999). These methods enable the user to distinguish between multiple binding binding events on a *single* chip in a single channel, paving the way for incorporation of multi-analyte chips into future SPR systems.

## Fluorescence-based evanescent wave biosensors

In contrast to the "label-free" methods described above, evanescent wave fluorescence fiber optic and array biosensors require the presence of a fluorescent species for detection of analyte. The evanescent wave is therefore used for excitation of surface-bound fluorophores, rather than for interrogation of the refractive index of the medium. Upon evanescent excitation of surface-bound fluorophores, fluorescence is either coupled back into the waveguide to be quantified (fiber optic sensors) or imaged using a charge-coupled device (Wadkins *et al.*, 1998) or a PMT (Schuderer *et al.*, 2000). Owing to their requirement for a fluorescent tag, these fluorescence-based systems are generally more immune to interference from fluctuations in temperature and non-homogeneous sample matrices; these influences may plague systems measuring changes in refractive index that are not appropriately controlled. However, these fluorescence-based systems do require additional reagents and time.

Since the early 1990s, a large number of papers have described new biochemical assays targeted towards detection of multiple analytes in complex samples. Evanescent fiber optic biosensors have evolved from single-fiber, manually operated system requiring an expert user (Hirschfeld and Block, 1984; Golden *et al.*, 1992) to multiplexed, non-automated systems (Golden *et al.*, 1997) to single- and multi-fiber systems that possess fully automated fluidics delivery and data analysis systems (Oroszlan *et al.*, 1993; Smith *et al.*, 1999; King *et al.*, 2000; Anderson *et al.*, 2000, 2001; Jung *et al.*, 2003); two of these automated instruments have been commercialized: RAPTOR (Research International) and Endotect™ (ThreeFold Sensors). Evanescent planar array sensors have been developed more recently, taking advantage of the breakthroughs in the areas of genomics and proteomics; these sensors use multiple recognition species immobilized in discrete locations on planar waveguides for detection of multiple analytes. While many fiber optic and planar array immunoassays have been developed for food-related pathogens and toxins

**Table 3.2** Evanescent wave fluorescence biosensor assays for food-related pathogens and toxins

| Analyte | Detection limit | Food matrix | Reference |
| --- | --- | --- | --- |
| **Fiberoptic biosensors** | | | |
| SEB | 100 ng/mL | Ham | Tempelman *et al.* (1996) |
| E. coli O157:H7 | 500 CFU/mL | Ground beef | DeMarco *et al.* (1999) |
| | 300 to $3 \times 10^4$ CFU/mL | Apple cider, other bacteria | DeMarco *et al.* (2001) |
| Botulinum toxin A | 200 ng/mL | | Kumar *et al.* (1994) |
| | 5 ng/mL | | Ogert *et al.* (1992) |
| *L. monocytogenes* | $4 \times 10^8$ CFU/mL | | Tims *et al.* (2001) |
| *S. typhimurium* | $10^5$ CFU/mL | | Zhou *et al.* (1998) |
| Fumonisin | 3.2 mg/g | Wheat | Maragos and Thompson (1999) |
| *L. monocytogenes* (PCR products) | | | Strachan and Gray (1995) |
| Aflatoxin B1 | 2 ng/mL | Wheat | Maragos and Thompson (1999) |
| *S. aureus* | 1 ng/mL protein A | | Chang *et al.* (1996) |
| **Planar array biosensors** | | | |
| *L. monocytogenes* | $2 \times 10^4$ CFU/mL | | Taitt *et al.* (2003 |
| SEB | 0.5 ng/mL | Milk, ham, ground beef, egg | Shriver-Lake *et al.* (2003) |
| S. typhimurium | $10^4$–$10^6$ CFU/mL | Carcass rinse, cantaloupe, sprouts, sausage, egg | Taitt *et al.* (2004) |
| *Campylobacter jejuni* | $3 \times 10^3$ CFU/mL | Carcass rinse, turkey | |
| *Shigella dysenteriae* | $5 \times 10^4$ to $8 \times 10^5$ CFU/mL | Carcass rinse, ground turkey, milk, lettuce | Sapsford *et al.* (2005) |
| Fumonisin B1 | 250 ng/mL | | Ligler *et al.* (2003) |

(Table 3.2), to date, only a limited number of studies demonstrating detection of these analytes in appropriate foodstuffs have been published.

Key disadvantages of these evanescent wave fluorescence-based systems are the lack of power within the evanescent wave available for excitation of surface-bound fluorophores and, for fiberoptic biosensors, the poor coupling efficiency of fluorescence back into the waveguides. Novartis' Evanescent Resonator chip is a glass chip with corrugated surface coated with metal oxide (Neuschafer et al., 2003). The surface architecture is such that surface confinement of energy creates a larger evanescent field, leading to enhanced fluorescence signal from surface-bound molecules and a significant improvement in detection. Poor coupling efficiency has been at least partially remedied through use of fiber bundles (Golden et al., 1997) or through tapering of the optical fibers and the use of specialized lenses to collimate fluorescence returning to the detector (Jung et al., 2003).

A recent paper describes an *assay-based* method to improve signals in evanescent fiber optic biosensors. Zhou et al. (1998) described a method for increasing the local concentration of *Salmonella* in a fiber optic biosensor by an acoustic standing wave. The authors demonstrated that ultrasonic treatment concentrated the cells into parallel layers, forcing the cells (and cells bound to polystyrene microspheres) to move to the axis of the test cell. Fluorescent signals were an order of magnitude higher in tests utilizing the acoustic standing-wave than those without ultrasonic treatment.

## Optrode sensors

Fluorescence-based fiber optic biosensors may utilize a second configuration – the optrode configuration. While optrode sensors also rely on total internal reflection for signal propagation and guiding, an evanescent field is not produced. Rather, light shining out the end of the fiber is used to excite fluorescence and the same end surface or surrounding fibers collect the emitted light. Thus, sensors based on the optrode configuration utilize recognition elements immobilized at the distal end of the fiber, with the optical fiber serving only as a light pipe, transporting light to and from the sensing region. Although no reports describing use of bio-optrodes for detection of foodborne pathogens are available, this technology is included here due to its potential for use in future sensors.

A key advantage of these sensors is the relative ease of multiplexing. Walt's group at Tufts leads this field and has explored a number of different methods for detecting multiple species using optrode-based sensing. The simplest approach, creation of optical fiber bundles, was first used to detect eight different fluorescently labeled DNA sequences (Ferguson et al., 1996). This idea was developed further by taking advantage of large number of individual sensing elements present in optical imaging fibers; imaging fibers consist of 5000 to 50,000 optical fibers bundled together, with each fiber able to act as an individual light pipe. Microwells were etched into the ends of all fibers in the imaging fiber and were subsequently filled with DNA-conjugated microspheres (Walt, 2000). The diameter and number of microspheres was such that only a single microsphere occupied the end of each individual fiber. After random distribution of the microspheres, the location and identity of each microsphere was determined by the ratio of different fluorophores incorporated within each sphere. Hybridization and identification of up to 25 different fluorescently labeled DNA species was demonstrated (Ferguson et al., 2000). The power of this technique lies in the ability to multiplex, but also in the potential for hundreds of replicate measurements; the inherent redundancy increased the signal-to-noise ratio such that limits of detection were in the range of $10^{-21}$ mol when only three species of DNA were tested (Epstein et al., 2002)

In contrast to the reagentless optical methods (SPR, RM, interferometers), the majority of optrode biosensors require the presence of a fluorescent molecule and washing steps to distinguish signals caused by biomolecular recognition from the background. While the experiments described above utilized fluorescently labeled DNA species (Ferguson et al., 2000; Walt 2000; Epstein et al., 2002), the potential for reagentless sensing has also been demonstrated. Steemers and coworkers (2000) used molecular beacons (described later in this chapter) for detection of three unlabeled DNA species using optical imaging fibers. Grant and Glass (1999) developed a sensor for D-dimer, a marker for thrombosis, using a fluorescein-labeled anti-D-dimer antibody for recognition. When D-dimer bound to the immobilized capture antibody, fluorescence was quenched; by comparison to standard curves, D-dimer could be detected in whole blood at physiologically relevant concentrations. Hanbury and coworkers (1997) used a similar scheme to detect and quantify myoglobin, a marker of myocardial infarction. Antibody labeled with Cascade Blue was immobilized on the optrode sensor; when myoglobin bound, fluorescence resonance energy transfer between the Cascade Blue and the myoglobin heme group effectively quenched the antibody's fluorescence. Physiologically relevant concentrations of myoglobin were detected and no significant interference was observed in the presence of hemoglobin.

Limitations of this technology include the small surface area available for immobilization of recognition molecules, the tendency of these sensors to foul in complex samples, and sensitivity to ambient light and interference from components of the sample matrix. Reports describing use of these sensors for analysis of complex or real-world samples are limited to those involving detection of relatively small analytes (Marazuela et al., 1997; Barker et al., 1998; Grant and Glass, 1999; Vo-Dinh et al., 2000); detection of bacterial or viral analytes using bio-optrodes has not been demonstrated. However, provided that sources of interference from sample matrices and environmental sources can be controlled (presumably through the use of molecular beacons and similar technologies), these sensors have tremendous potential for future use in food safety monitoring. The degree of multiplexing in these systems is so great, especially when using imaging fibers, that losses in sensitivity due to biochemical limitations may be minimized when statistically analyzing data from thousands of replicate fibers.

## Nucleic acid sensors/real-time PCR

Advances in nucleic acid analysis during the mid-late 1980s and early 1990s led to the development of real-time PCR (RT-PCR), a technique whereby the kinetics of PCR, described earlier, can be followed optically (Higuchi et al., 1992, 1993). RT-PCR, not to be confused with reverse-transcriptase PCR (also given the same acronym), involves incorporation of fluorophores within the polymerase reaction, such that formation of PCR products can be tracked in real time. One method for RT-PCR uses a dye such as SYBR Green I; this dye binds specifically to double-stranded products but not to single-stranded templates and probes.

Another method of detecting and quantifying PCR products, termed "TaqMan" technology, involves use of the endogenous 5′ to 3′ nuclease activity of a DNA polymerase isolated from *Thermus aquaticus* (Taq DNA polymerase). Oligonucleotide probes (distinct from the primers used) used in TaqMan reactions are labeled with both a 5′-terminal fluorophore and a 3′-terminal quencher molecule such that, when intact, fluorescence from the probe is quenched. Annealing of the quenched probe to the target DNA occurs simultaneously with that of the forward and reverse primers. However, during the extension step of PCR, the 5′ end of the labeled probe is displaced by Taq DNA polymerase and is subsequently cleaved from the probe. Once released from the probe, the reporter dye is no longer quenched and the emitted fluorescence can be tracked over time.

An alternative to SYBR Green and TaqMan technology for direct detection of PCR products is the use of molecular beacons. Like TaqMan probes, molecular beacons contain 5′ and 3′ terminal fluorophores and quenchers; however, molecular beacons adopt a hairpin structure while free in solution, bringing the fluorescent dye and quencher in close proximity. When a molecular beacon hybridizes to a target, the fluorescent dye and quencher are separated and the fluorophore can then emit fluorescence. Thus, in contrast to the TaqMan probes, molecular beacons are fluorescent upon interaction with the target sequence and are not degraded during the amplification process. Both TaqMan and molecular beacon technologies allow detection of multiple DNA species (multiplexing) by the use of different reporter fluorophores on different probes or beacons.

A large variety of instruments are commercially available for use in both PCR and RT-PCR, the most well known of which are the LightCycler (Roche), SmartCycler II (Cepheid), R.A.P.I.D. (Idaho Technologies), and ABI Prism (Applied Biosystems). All commercially available models are able to perform PCR simultaneously on multiple samples and, with the exception of a few low-cost models, most are able to perform multiplexed assays with different fluorophores. However, in spite of advances in the instrumentation for RT-PCR, the rate-limiting step is typically a non-selective enrichment step to both increase the number of targets present and effectively dilute out any inhibitors present in the food matrix. Furthermore, the low detection limits quoted in many papers (in the range of 0.1 to 100 CFU/mL or CFU/g) refer to the concentration of inoculum used in spiked foodstuffs, rather than the actual concentration of bacteria used after the requisite non-selective enrichment step. A PCR system able to detect *Salmonella* and *Campylobacter* in carcass rinse *without pre-enrichment* has recently been described (Hong et al., 2003). This methodology, termed PCR-ELISA (Lazar, 1995), incorporated tagged nucleotides into the PCR amplification reaction. The resultant tagged PCR product was detected with an enzyme-antibody conjugate that recognized the tag. Inclusion of the ELISA step into the PCR protocol increased the sensitivity by 100 to 1000-fold. While the *Campylobacter* PCR-ELISA proved to be sensitive to concentrations of 5 CFU/mL in carcass rinse, the PCR-ELISA for *Salmonella* had a significantly higher detection limit, as well as relatively high rates of false positives and false negatives (Hong et al., 2003).

## Flow cytometry/flow cytometric biosensors

Flow cytometry involves the optical characterization of single cells as they pass at high speed (50,000 cells/second) through a laser beam. Flow cytometers use a fluidics system to precisely deliver the cells to the intersection of the laser beam and a light-gathering lens by hydrodynamic focusing; a single stream of cells is injected and confined within an outer stream at greater pressure, thereby allowing the cells to be individually optically interrogated (Figure 3.4). The laser light source is used to both characterize parameters of light scatter and to excite fluorescent molecules used to label the cell. Cells are characterized individually by their physical size and shape, as well

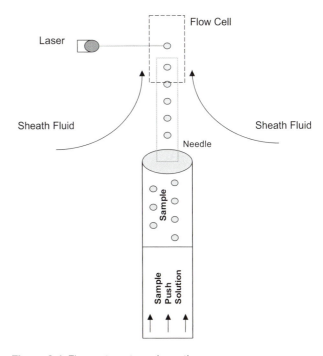

**Figure 3.4** Flow cytometer schematic.

for the presence and quantity of biological receptors or binding epitopes. The physical profile of a cell can be observed by combining forward light scatter and orthogonal or side light scatter. Interruption of the laser beam by the cell and measurement of the light that passes around the cell define the forward scatter. This measurement is an indication of the cell's unique refractive index which depends on a cell's size, organelles, water and molecular contents. Cellular side scatter is the light that is reflected 90° to the laser beam and is an indication of cytoplasmic density or cell surface granularity.

For use as a biosensor, target cells are incubated with fluorescently labeled antibodies and the number of fluorescently labeled cells determined by subsequent flow cytometry. High end instruments have several laser light sources that allow several different fluorophores to be excited, making discrimination of numerous cell types and their level of antigen expression possible. Multiplexing of different fluorophores significantly increases the amount of information available from each sample analysis. For example, while two-color assays generally provide four analytical areas, three-color assays can discriminate between twelve subsets and four-color assays between twenty-four quantifiable subsets.

Flow cytometers can also be used to sort cells based on all of the measured properties of fluorescence and light scatter. If the system is capable of sorting, after the measurement process, a high frequency vibration is applied and causes the flow stream to break up into droplets. The droplets can then be charged and charged deflection plates (+/-) redirect the drops to the left or right of the flow stream into a collection vessel. The cells collected are essentially unharmed and can then be analyzed at a highly purified level. This capability has not yet been fully exploited for pathogen detection applications.

Small portable flow cytometers are being tested as fieldable sensor systems. Stopa (2000) used a Coulter Epics XL flow cytometer to detect *B. anthracis* spores within 5 minutes at only $10^3$ CFU/mL. In other work, Stopa and Mastromanolis (2001) evaluated the use of flow cytometry to detect *E. coli* from well water, after labeling with various DNA stains. YORPO-1 and YOYO-1 gave detection limits of $10^4$ and $10^5$ CFU/mL respectively, indicating that this method would require a prefiltration step for most applications. Flow cytometry has also been used to measure *E. coli* in beef ruminal fluid or feces (Hussein *et al.*, 2002). After dilution and filtration, samples were spiked with *E. coli*, labeled with a FITC-conjugated antibody and analyzed by flow cytometry; LODs of $10^4$ cells/mL were obtained. Rattanasomboon *et al.* (1999) used flow cytometry to quantify *Brochothrix thermosphacta*, a common meat spoilage bacterium, at levels as low as $10^5$ cells/mL. Other examples where flow cytometry was utilized to identify and characterize bacteria from complex samples include Sgorbati *et al.* (1996) and Laplace-Builhe *et al.* (1993). From these examples, it can be seen the flow cytometry offers many attractive features for foodborne pathogen detection. A key advantage of flow cytometry is that the cells of interest are directly counted. Furthermore, what was once a very expensive and complicated instrument has moved into the realm of fielded biosensors, as there are a number of commercial systems now available. The RBD 3000 made by Advanced Analytical is marketed for near real-time monitoring during fermentation processes. They have also tested flow cytometry for the enumeration of water samples for *E. coli* and *Enterococcus*.

Another commercial flow cytometry-based instrument, Luminex, has adapted the flow cytometric instrumentation for use with dual labeled microspheres in multiplexed immunoassays. Using microspheres incorporating 10 different concentrations of two dyes, Luminex[100] can uniquely identify 100 different bead sets based on their flow cytometric profiles. Sandwich immunoassays can be developed using multiple sets of beads, with a different capture antibody immobilized on each bead set. After the beads are mixed with a sample and one or multiple fluorescent tracer antibodies, the resultant immunocomplexes are then analyzed by Luminex. Luminex identifies each bead type and the amount of tracer antibody bound, thereby processing a sample for up to 100 targets simultaneously. Ye *et al.* (2001) used this technology to identify bacterial 16S rDNA, which permitted multiplex analysis of 17 bacterial species representing a broad range of gram-negative and gram positive bacteria. McBride *et al.* (2003) used this technology to sensitively identify the four biological warfare simulants, *B. globigii*, *Erwinia herbicola*, bacteriophage MS2, and the protein ovalbumin; these simulants were used to safely test the instrument's capabilities for the range of threats: viruses, protein toxins, bacterial spores, and vegetative cells. In a study of bacterial pathogens implicated in foodborne illnesses (*E. coli*, *Salmonella*, *L. monocytogenes* and *C. jejuni*), Dunbar *et al.* (2003) used a Luminex instrument to detect PCR-amplified variable regions of bacterial 23S ribosomal DNA, with a detection sensitivity of $10^3$ to $10^5$ genome copies. Using Luminex-based immunoassays for the same organisms, limits of detection were in range of 1000 organisms/mL or lower. These studies exemplify the utility and flexibility of flow cytometry for sensitive target detection and identification.

## Electrochemical methods

### Potentiometric and amperometric sensors

Electrochemical sensors can be classified as either potentiometric or amperometric. Potentiometric sensors measure the change in potential that occurs when a bioactive material, usually an enzyme, generates or consumes an electroactive species, which is detected by an ion selective electrode. Amperometric sensors directly measure the resulting change in current. These sensors appear to have significant advantages over many competing technologies. They can easily operate in turbid or complex matrices that might confound optical methods, and even more importantly, the instrumentation is inexpensive and well suited to miniaturization.

While a number of research efforts continue in this area (see Shah and Wilkins (2003) for a recent review), the commercial instrument that has made the largest impact in the market is the Threshold™ from Molecular Devices. Although this light-addressable potentiometric sensor (LAPS) instrument currently requires manual operation, it can be adapted to perform 96 assays simultaneously. The system achieves good sensitivity through a preliminary step that involves formation of an immunocomplex in solution; this complex includes streptavidin and a secondary antibody conjugated with urease. Solution-phase interactions in this step overcome diffusion limitations often encountered with binding events occurring on surfaces. This complex is then concentrated on a biotinylated filter and, upon addition of urea, causes a rise in pH and a concomitant change in potential. The instrument transduces this change in potential between the surface of the insulator and the field effect transistor to voltage per time differential by monitoring the change in photocurrent produced by a modulated light emitting diode.

The LAPS system has been demonstrated to be a very capable immunosensor (Dill *et al.*, 1997; Lee *et al.*, 2000). Of particular interest, Dill *et al.* (1999) used LAPS to detect *Salmonella* as low as 119 CFU, and showed recoveries of up to 90 percent from spiked chicken carcass washings. *E. coli* O157: H7 has also been detected at levels as low as $2.5 \times 10^4$ cells/mL in buffer (Gehring *et al.*, 1998). Work still continues with custom built instruments, on which impressive results (10 cell/mL in 1.5 h) for the detection of *E. coli* have been reported (Ercole *et al.*, 2002, 2003).

Work on amperometric sensors has been equally promising. Brewster and Mazenko (1998) used a filter capture membrane in combination with an amperometric sensor to detect *E. coli* at concentrations as low as $5 \times 10^3$ cells/mL. Using immunomagnetic beads to concentrate *Salmonella* from buffer, they could detect as few as $8 \times 10^3$ cells/mL (Gehring *et al.*, 1996). Abdel-Hamid *et al.* (1999) reported detecting *E. coli* at concentrations as low as 100 cell/mL in another amperometric immunofiltration system of their own design. Peng *et al.* (2000) developed an amperometric biosensor for *E. coli* heat-labile enterotoxin on a sol-gel thin-film electrode; the detection limit of this system was 36 nmol/L.

## Electrochemiluminescence

The methodology of electrochemiluminescence (ECL), like chemiluminescence (CL), is extremely sensitive, exhibiting a very high signal-to-noise ratio. Unlike CL, however, ECL is controlled by the voltage potential present at the surface of an electrode; the voltage potential powers the luminescent redox reaction, typically of a ruthenium (II) trisbipyridal chelate coupled with tripropyl amine. This process has been developed into a commercial sensor platform, ORIGEN by BioVeris Corp. (formerly IGEN Corp.). For use as a biosensor, the anode is magnetized and immunomagnetic microspheres (IM) are used as the assay surface. In solution, a sandwich is formed between the capture antibody on the microsphere, the antigen (i.e., bacterial target), and a $Ru(bpy)_3^{2+}$ labeled antibody. After a 30- to 40-minute incubation time, the microspheres

are drawn to the electrode and assayed for the presence of the $Ru(bpy)_3^{2+}$-labeled antibody. $Ru(bpy)_3^{2+}$ label can be repeated raised to an excited state by the electrode and the emitted photons detected.

The ORIGEN ECL-IM biosensor was used by Yu and Bruno (1996) to detect *E. coli* and *Salmonella* in a wide variety of foods and environmental water samples. A detection limit of less than 100 cells/mL was reported for *E. coli* and less than 1000 cells/mL for *Salmonella*. More recently, Crawford *et al.* (2000) successfully applied this sensor to detect *E. coli* in ground beef at levels of 0.5 CFU/g following an 18-hour enrichment period at 37°C. Kuczynska *et al.* (2003) also used ORIGEN to detect as few as five *Crytosporidium parvum* oocysts in water samples. However, they observed significant background problems from nonspecific binding of organic components in environmental samples to beads, which often blocked the assay.

Work also continues with other immunomagnetic chemiluminescence sensors. Ye *et al.* (2002) have detected low levels (180 cells/mL) of *E. coli* using a magnetic bead capture surface and horseradish peroxidase-(HRP) labeled antibody to generate the signal upon addition of luminol and hydroden peroxide to the reaction cell. Yacoub-George *et al.* (2002) used this detection format to develop a multi-channel biosensor. This instrument could test a sample for three targets in less than 30 minutes, and had a detection limit for *E. coli* of $10^5$ CFU/mL. Using this same format, Liu *et al.* (2003) were also able to detect *E. coli* (< 1000 cells/mL) in ground beef, chicken carcass rinse, and lettuce samples without enrichment. These efforts point to the great potential of luminescence measurements in the development of biosensors that meet the critical detection limits required for food pathogen detection.

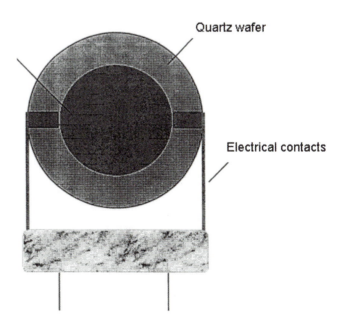

**Quartz wafer**

**Electrical contacts**

**Figure 3.5** Schematic of quartz crystal microbalance.

## Piezoelectric devices

Piezoelectric sensors create an electrical charge upon being mechanically stressed. The quartz crystal microbalance (QCM) is the most common piezoelectric device that has been applied for pathogen detection. The QCM consists of a thin quartz disk upon which is plated one or several electrodes (Figure 3.5). When an oscillating electric field is applied across the device, an acoustic wave propagates through the crystal. This acoustic wave has minimal impedance when the thickness of the device is a multiple of a half-wavelength of the acoustic wave. Deposition of material onto this surface decreases the frequency in proportion to the mass deposited, thereby, allowing binding events to be monitored.

The use of QCM as the basis of immunosensors was recently reviewed (Suleimann and Guilbault, 1994; O'Sullivan and Guilbault, 1999; O'Sullivan et al., 1999; Su et al., 2001). Antibodies immobilized onto the sensor surface capture and hold large biomolecules or bacteria to the surface, thereby inducing a measurable change in the device's resonant frequency. This permits direct, label-free measurement of the binding event and real-time observation of the association kinetics. Thus, unless nonspecific adsorption from complex matrices is present, sensitive and rapid measurements can be acquired.

Quartz crystal microbalances have been utilized for a wide range of sensing applications, including detection of pathogenic bacteria (Table 3.3), viruses (Eun et al, 2002; Uttenthaler et al., 2001), enterotoxins (Harteveld et al., 1997; Spangler and Tyler, 1999; Lin and Tsai, 2003) and PCR products (Mo et al., 2002; Mannelli et al., 2003). Thus, it is clear that QCM can be utilized to detect relatively large biological targets.

Small molecules, on the other hand, do not generate large changes in mass when directly bound to the QCM. These molecules must therefore be detected using either a sandwich assay format (with an additional antibody or ligand adding mass to the captured analyte; Saha et al., 2002) or a competitive assay format (Yokoyama et al., 1995). Liu et al. (1999)

developed a competitive assay for polycyclic aromatic hydrocarbons using QCM. They immobilized an analog of the antigen, benzo[a]pyrene-BSA conjugate, onto the surface. Upon binding of a monoclonal antibody to the immobilized conjugate, a decrease in the frequency was observed. When samples containing benzo[a]pyrene were injected, the release of antibody from the QCM surface to bind free benzo[a]pyrene was measured as an increase in resonant frequency. Nanomolar levels (10 nmol/L) of benzo[a]pyrene could be detected in this manner.

Recent notable applications of QCMs include the work of Su et al. (2001), who developed a screening test for S. enteritidis infection in chickens and eggs. By immobilizing recombinant S. enteritidis proteins onto the QCM, the investigators were able to detect the cognate antibodies from chicken serum or egg white, indicating exposure of the birds to this pathogen. For this application, the sensitivity and specificity of the sensors were 100% and 92.9%, respectively. Another example utilized an antigen displacement assay for detection of P. aeruginosa in milk and dairy samples (Bovenizer et al., 1998). A linear response from $2 \times 10^6$ cell/mL to $1 \times 10^8$ cell/mL and a limit of detection of 100,000 cells were observed. Minunni et al. (1996) developed a similar assay for Listeria. Their 15 minute assay yielded a calibration curve with a response range from $2.5 \times 10^5$ to $2.5 \times 10^7$ cells/crystal.

More recently, investigators have examined different methods to amplify the signal produced per binding event as a means to improve sensitivity. One amplification method involves use of microspheres or other microparticles to increase the weight deposited. Kim et al. (2003) used antibody-coated paramagnetic microspheres to both enrich samples for Salmonella, as well as to increase the mass deposited on the QCM. The observed LOD was decreased to about $10^3$ cells/mL.

In the near term, QCMs are making their biggest impact as air space sensors, functioning as an electronic nose. Ali et al. (2003) have recently described the use of a six-element array QCM to analyze volatiles in the headspace of milk. Analysis of milk contaminated with either P. fragi or E. coli showed significant changes in mass in the presence of samples fermented by E. coli. This application of QCM, as an equivalent to surface acoustic wave sensors (Leonard et al., 2003), looked at the variable adsorption of volatile bacterial fermentation products to various coatings applied to the sensor surface. This type of sensing is somewhat simpler than those described above, in that no liquid handling is required, and the sensors surfaces are usually reversible and long lasting. The primary limitations are, of course, the lack of sensitivity and specificity.

Commercial QCMs are available from Elchema (Potsdam, NY) and QCM Research (Laguna Beach, CA). While these systems are able detect changes of approximately 1 ng/cm$^2$, they are primarily research tools and not suitable for routine testing. In order for QCM to move from the laboratory to field use, a number of improvements need to be realized. Perhaps most importantly, the limits of detection will require further improvements. Beyond that, a number of practical issues also

**Table 3.3** Detection of pathogenic bacteria using QCM

| Organism | Reference |
| --- | --- |
| B. cereus | Vaughan et al. (2003) |
| E. coli | Plomer et al. (1992) |
| | Pyun et al. (1998) |
| | Spangler and Tyler (1999) |
| L. monocytogenes | Minunni et al. (1996) |
| Salmonella sp. | Babacan et al. (2002) |
| | Wong et al. (2002) |
| | Kim et al. (2003) |
| Shigella | Konig and Gratzel (1993) |
| S. aureus | Bao et al. (1998) |
| Yersinia | Konig and Gratzel (1993) |
| Vibrio cholerae | Carter et al. (1995) |
| Mycobacterium tuberculosis | He et al. (2003) |

must be addressed: reproducible immobilization of the biological components, improved resistance to non-specific binding, and consistent reusability. If these issues can be solved, a QCM might be able to deliver the same functionality as SPR sensors, but at considerably lower cost.

## Other detection techniques

### Bead array counter (BARC)

A unique and promising method has been described by Colton's group, who have invented a Bead Array Counter (BARC) biosensor (Baselt et al., 1998; Edelstein et al., 2000). This system utilized giant magnetoresistive (GMR) materials, thin film metal multilayers whose resistance is modified with changes in magnetic field. As current was passed through the metal layers, localized magnetic fields were sensed. Recognition elements immobilized onto a GMR substrate were incubated with biotinylated sample. After a wash step, avidin-modified magnetic beads were added. Formation of avidin-biotin bridges effectively bound the magnetic beads to the appropriate loci on the sensor array and localized magnetic fields were determined (Edelstein et al., 2000). A key to this system's sensitivity is that a magnetic field can be applied to remove any beads not specifically bound to the surface. This is a very promising technology which is still undergoing development, but might one day make a significant impact in the field of rapid pathogen detection.

### Membrane-based biosensors

Cornell's group at Ambri (Chatswood, NSW, Australia) has pioneered a novel biosensor that measures changes in conductance/impedance through a molecular membrane. Known as the Ion Channel Switch (ICS™), the sensor is comprised of a lipid bilayer in which is integrated the ion channel, gramicidin. Gramicidin occurs as two monomeric subunits, with each monomer localized within either the inner or outer lipid leaflet; upon alignment of two monomers, a functional transmembrane pore is formed, allowing ions to flow between the two sides of the membrane.

The membrane, with its integral monomers, is separated from a gold conducting surface by a polar spacer that provides a reservoir for ions. Membrane stability is enhanced by tethering of the inner leaflet to the gold surface and is further increased by incorporation of archaebacterial lipids that possess hydrocarbon chains spanning the entire membrane. Monomers of gramicidin within the inner leaflet are also tethered to the gold layer by alkane thiols. Thus, while the outer leaflet of the membrane is essentially a two-dimensional liquid crystal, with relatively large mobility of membrane components, the inner leaflet is essentially locked in place.

For biosensor purposes, gramicidin monomers and archaelipids in the outer leaflet are conjugated with antibodies against the analyte of interest. In the presence of analyte, an immunocomplex is formed, with the analyte forming a bridge

between the antibody-labeled archaelipid and the antibody-labeled gramicidin monomer; formation of this immunocomplex shifts the conformation of gramicidin from conducting dimer-pores to non-conducting monomers (Figure 3.6). This change can be measured as a decrease in current applied across the membrane. This sensor was used to detect the 16S rRNA of *L. monocytogenes* (Wright and Harding, 2000), *E. coli* (Cornell, 2002), and SEB (Cornell, 2002). While the ICS has not been used for analysis of foods for contaminants, the lack of interference observed in the presence of whole blood (Cornell, 2002) points to potential use for this sensor in monitoring for foodborne contaminants.

### Toxin guard™

Toxin Alert Inc. (Mississauga, Ontario, Canada) is marketing a product for food wrapping that includes an integral immunoassay to detect for the presence of food spoilage bacteria. It functions by coating the inside surface of standard polyethylene food wrap with a pattern of anti-pathogen antibodies. Below that is a nutrient gel that contains dye-labeled anti-pathogen antibodies. As bacteria migrate through the gel, they are coated by antibodies, and eventually captured by the antibodies affixed to surface, creating a visible warning. The primary limiting factor of this system is that its poor sensitivity could create a false sense of security. Furthermore, given its current limited sensitivity, the sense of smell of most consumers would likely equal or surpass the responsiveness of this sensor.

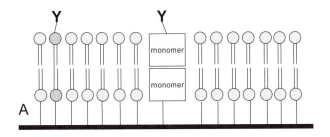

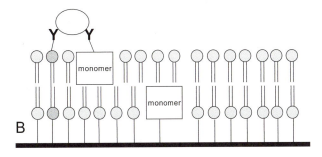

**Figure 3.6** Schematic of membrane-based biosensor. (A) In the absence of analyte, gramicidin monomers align, producing a functional ion channel. (B) After addition of sample, analyte forms a bridge between tethered archaelipid-antibody conjugate and gramicidin-antibody conjugate in the outer leaflet. Formation of this "sandwich" disrupts alignment of gramicidin monomers, resulting in decreased conductance.

## Electronic noses

Most electronic noses are arrays of sensors that operate in the gas phase and monitor how various organic components adsorb to the various coatings on the sensors. Some electronic noses are surface acoustic wave devices, which also function effectively as chemical agent detectors, or an electronic nose could be an array of conductivity sensors, or in fact any of several types capable of responding to volatile vapors. By training the sensor to recognize the pattern produced by differences in how the volatile materials adsorb to the various coatings, it can then recognize the bacteria by their fermentation byproducts. The factors which have limited its application in the food industry are ones of sensitivity and specificity. This is typified by the recent work of Canhoto and Magan (2003), who used an array of 14 conductivity sensors to detect for the presence of bacteria (~100 CFU/mL) 24 hours after spiking into the water sample. Keshri et al. (2002) found similar time requirements when detecting mold growth on bread. Thus, while the simplicity of the method is attractive, the serious limitations of sensitivity and specificity will need to be overcome before it will have applicability for food pathogen detection.

## Summary

We have described a number of biosensor technologies currently used or demonstrating potential, for detection of foodborne pathogens and toxins. While microbiological culture and ELISA remain the gold standards for detection of these contaminants, biosensors offer a viable alternative as routine, use-friendly, fieldable instruments designed for rapid and sensitive measurements. Although nucleic acid-based detection methods may be more specific and sensitive than antibody-based techniques, the latter are faster, generally more robust, often require fewer sample preparation steps, and detect not only pathogenic organisms, but also toxins produced that may survive pasteurization or other decontaminating processes.

Currently, biosensors have not yet reached the stage where they can be used for on-site testing of unprocessed food samples, with relevant levels of pathogens or toxins detectable in real-time. However, there is a great deal of continuing progress in the fields of microelectronics, micro- and nano-fluidics, optics, alternative recognition molecules, and signal generation schemes that we do not have sufficient space to discuss in this chapter. We believe that, with incorporation of many of these ever-developing technologies, biosensors may soon see the same commercial success and widespread usage in food testing as glucose test kits and home pregnancy tests have attained in their respective markets.

## References

Abdel-Hamid, I., Dmitri, I., Plamen, A., and Wilkins, E. 1999. Flow-through immunofiltration assay system for rapid detection of *E. coli* O157:H7. Biosens. Bioelectron. 14: 309–316.

Ali, Z., O'Hare, W.T., and Theaker, B.J. 2003. Detection of bacterial contaminated milk by means of a quartz crystal microbalance based electronic nose. J. Ther. Anal. Cal. 71: 155-161.

Anderson, G.P., King, K.D., Gaffney, K.L., and Johnson, L.H. 2000. Multianalyte interrogation using the fiber optic biosensor. Biosens. Bioelectron. 14: 771–778.

Anderson, G.P., Rowe-Taitt, C.A., and Ligler, F.S. 2001. RAPTOR: A portable, automated biosensor. Proc. 1st Conf. Point Detect. Chem. Biol. Defense. Oct. 2000, Williamsburg, VA.

AOAC. 1995. Official Methods of Analysis of AOAC International, Volume II, Chapter 49. P. Cunniff, ed. AOAC International, Arlington, VA.

Ausubel, F.M., Brent, R., Kingston, R.E., Moore, D.D., Seidman, J.G., Smith, J.A., and Struhl, K. 1994. Current Protocols in Molecular Biology. Wiley & Sons.

Babacan, S., Pivarnik, P., Letcher, S., and Rand, A. 2002. Piezoelectric flow injection analysis biosensor for the detection of *Salmonella typhimurium*. J. Food Sci. 67: 314–320.

Baeumner, A.J. 2003. Biosensors for environmental pollutants and food contaminants. Anal. Bioanal. Chem. 377: 434–445.

Bao, L.L., Tan, H.W., Duan, Q., Su, X.L., and Wei, W.Z. 1998. A rapid method for determination of *Staphylococcus aureus* based on milk coagulation by using a series piezoelectric quartz crystal sensor. Anal. Chim. Acta 369: 139–145.

Barker, S.L.R., Kopelman, R., Meyer, T.E., and Cusanovich, M.A. 1998. Fiber-optic nitric oxide-selective biosensors and nanosensors. Anal. Chem. 70: 971–976.

Baselt, D.R., Lee, G.U., Natesan, M., Metzger, S.W., Sheehan, P.E., and Colton, R.J. 1998. A biosensor based on magnetoresistance technology. Biosens. Bioelectron. 13: 731–739.

Bergdoll, M.S. 1983. Enterotoxins. In: Staphylococci and Staphylococcal Infections. C. Adlam and C.S.F. Easmon, eds. Academic Press, Ltd., London, pp. 559–598.

Berger, C.E.H., Beumer, T.A.M., Kooyman, R.P.H., and Greve, J. 1998. Surface plasmon resonance multisensing. Anal. Chem. 70: 703–706.

Bokken, G.C., Corbee, R.J., van Knapen, F., and Bergwerff, A.A. 2003. Immunochemical detection of *Salmonella* group B, D, and E using an optical surface plasmon resonance biosensor. FEMS Microbiol. Lett. 222: 75–82.

Bovenizer, J.S., Jacobs, M.B., O'Sullivan, C.K., Guilbault, G.G. 1998. The detection of *Pseudomonas aeruginosa* using the quartz crystal microbalance. Anal. Lett. 31: 1287–1295.

Brewster, J.D., and Mazenko, R.S. 1998. Filtration capture and immuno-electrochemical detection for rapid assay of *Escherichia coli* O157:H7. J. Immunol. Meth. 211: 1–8.

Canhoto, O.F., and Magan, N. 2003. Potential for detection of microorganisms and heavy metals in potable water using electronic nose technology. Biosens. Bioelectron. 18: 751–754.

Carrick, K., Barney, M., Navarro, A., and Ryder, D. 2001. The comparison of four bioluminometers and their swab kits for instant hygiene monitoring and detection of microorganisms in the brewery. J. Inst. Brewing 107: 31–37.

Carter, R.M., Mekalanos, J.J., Jacobs, M.B., Lubrano, G.J., and Guilbault, G.G. 1995. Quartz-crystal microbalance detection of *Vibrio cholerae* O139 serotype. J. Immunol. Meth. 187: 121–125.

Chang, Y.H., Chang, T.C., Kao, E.F., and Chou, C. 1996. Detection of protein A produced by *Staphylococcus aureus* with a fiber-optic based biosensor. Biosci. Biotechnol. Biochem. 60: 1571–1574.

Cornell, B.A. 2002. Membrane-based biosensors. In: Biosensors: Present and Future. F.S. Ligler, and C.R. Taitt, eds. Amsterdam: Elsevier, pp. 457–495.

Crawford, C.G., Wijey, C., Fratamico, P., Tu, S.I., and Brewster, J. 2000. Immunomagnetic-electrochemiluminescent detection of *E. coli* O157:H7 in ground beef. J. Rapid Meth. Auto. Micro. 8: 249–264.

Deininger, R.A., and Lee, J. 2001. Rapid determination of bacteria in drinking water using an ATP assay. Field Anal. Chem. Tech. 5: 185–189.

DeMarco, D.R., Saaski, E.W., McCrae, D.A., and Lim, D.V. 1999. Rapid detection of *Escherichia coli* O157:H7 in ground beef using a fiber optic biosensor. J. Food Prot. 62: 711–716.

DeMarco, D.R., and Lim, D.V. 2001. Direct detection of *Escherichia coli* O157:H7 in unpasteurized apple juice with an evanescent wave biosensor. J. Rap. Meth. Automat. Microbiol. 9: 241–257.

Dickert, F.L., Lieberzeit, P., and Hayden, O. 2003. Sensor strategies for microorganism detection- from physical principles to imprinting procedures. Anal. Bioanal. Chem. 377: 540–549.

Dill, K., Song, J.H., Blomdahl, J.A., and Olson, J.D. 1997. Rapid, sensitive and specific detection of whole cells and spores using the light-addressable potentiometric sensor. J. Biochem. Biophys. Meth. 34: 161–166.

Dill, K., Stanker, L.H., and Young, C.R. 1999. Detection of *Salmonella* in poultry using a silicon chip-based biosensor. J. Biochem. Biophys. Meth. 41: 61–67.

Dunbar, S.A., Vander Zee, C.A., Oliver, K.G., Karem, K.L., and Jacobson, J.W. 2003. Quantitative, multiplexed detection of bacterial pathogens: DNA and protein applications of the Luminex LabMAP (TM) system. J. Microbial. Meth. 53: 245–252.

Edelstein, R.L., Tamanaha, C.R., Sheehan, P.E., Miller, M.M., Baselt, D.R., Whitman, L.J., and Colton, R.J. 2000. The BARC biosensor applied to the detection of biological warfare agents. Biosens. Bioelectron. 14: 805–813.

Epstein, J.R., Lee, M., and Walt, D.R. 2002. High-density fiber-optic genosensor microsphere array capable of zeptomole detection limits. Anal. Chem. 74: 1836–1840.

Ercole, C., Del Gallo, M., Pantalone, M., Santucci, S., Mosiello, L., Laconi, C., and Lepidi, A. 2002. A biosensor for *Escherichia coli* based on a potentiometric alternating biosensing (PAB) transducer. Sens. Actuat. B-Chem. 83: 48–52.

Ercole, C., Del Gallo, M., Mosiello, L., Baccella, S., and Lepidi, A. 2003. *Escherichia coli* detection in vegetable food by a potentiometric biosensor. Sens. Actuat. B-Chem. 91: 163-168.

Eun, A.J.C., Huang, L.Q., Chew, F.T., Li, S.F.Y., and Wong, S.M. 2002. Detection of two orchid viruses using quartz crystal microbalance-based DNA biosensors. Phytopath. 92: 654–658.

Ferguson, J.A., Boles, T.C., Adams, C.P., and Walt, D.R. 1996. A fiber-optic DNA biosensor microarray for the analysis of gene expression. Nat. Biotechnol. 14: 1681–1684.

Ferguson, J.A., Steemers, F.J., and Walt, D.R. 2000. High density fiber-optic DNA random microsphere array. Anal. Chem. 72: 5618–5624.

Gehring, A.G., Crawford, C.G., Mazenko, R.S., VanHouten, L.J., and Brewster, J.D. 1996. Enzyme-linked immunomagnetic electrochemical detection of *Salmonella typhimurium*. J. Immunol. Meth. 195: 15–25.

Gehring, A.G., Patterson, D.L., and Tu, S.I. 1998. Use of a light-addressable potentiometric sensor for the detection of *Escherichia coli* O157: H7. Anal. Biochem. 258: 293–298.

Golden, J.P., Shriver-Lake, L.C., Anderson, G.P., Thompson, R.B., and Ligler, F.S. 1992. Fluorometer and tapered fiber optic probes for sensing in the evanescent wave. Opt. Eng. 31: 1458–1462.

Golden, J.P., Saaski, E.W., ShriverLake, L.C., Anderson, G.P., and Ligler, F.S. 1997. Portable multichannel fiber optic biosensor for field detection. Opt. Eng. 36: 1008–1013.

Grant, S.A., and Glass, R.S. 1999. Sol-gel-based biosensor for use in stroke treatment. IEEE Trans. Biomed. Eng. 46: 1207–1211.

Hall, R.H. 2002. Biosensor technologies for detection microbiological foodborne hazards. Microbes Infect. 4: 425–432.

Hanbury, C.M., Miller, W.G., and Harris, R.B. 1997. Fiber-optic immunosensor for measurement of myoglobin. Clin. Chem. 43: 2128–36.

Harteveld, J.L.N., Nieuwenhuizen, M.S., and Wils, E.R.J. 1997. Detection of staphylococcal enterotoxin B employing a piezoelectric crystal immunosensor. Biosens. Bioelectron. 12: 661–667.

Hattori, N., Sakakibara, T., Kajiyama, N., Igarashi, T., Maeda, M., and Murakami, S. 2003. Enhanced microbial biomass assay using mutant luciferase resistant to benzalkonium chloride. Anal. Biochem. 319: 287–295.

He, F.J., Zhao, J.W., Zhang, L.D., and Su, X.N. 2003. A rapid method for determining *Mycobacterium tuberculosis* based on a bulk acoustic wave impedance biosensor. Talanta 59: 935–941.

Higuchi, R., Dollinger, G., Walsh, P.S., and Griffith, R. 1992. Simultaneous amplification and detection of specific DNA-sequences. Bio-Technol. 10: 413–417.

Higuchi, R., Fockler, C., Dollinger, G., and Watson, R. 1993. Kinetic PCR analysis – real-time monitoring of dna amplification reactions. Bio-Technol. 11: 1026–1030.

Hirschfeld, T.E., and Block, M.J. 1984. Fluorescent immunoassay employing optical fiber in a capillary tube. US Patent No 4,447,546.

Hobson, N.S., Tothill, I., and Turner, A.P.F. 1996. Microbial detection. Biosens. Bioelectron. 11: 455–477.

Hokama, Y., Shirai Lk., and Iwamoto, Lm. 1987. Assessment of a rapid enzyme-immunoassay stick test for the detection of ciguatoxin and related polyether toxins in fish-tissues. Biol. Bull. 172: 144–153.

Hokama, Y., Nishimura, K.L., Takenaka, W.E., and Ebesu, J.S.M. 1997. Latex Antibody Test (LAT) for detection of marine toxins in ciguateric fish. J. Nat. Toxins. 6: 35–50.

Hokama, Y., Takenaka, W.E., Nishimura, K.L., Ebesu, J.S.M., Bourke, R., and Sullivan, P.K. 1998. A simple membrane immunobead assay for detecting ciguatoxin and related polyethers from human ciguatera intoxication and natural reef fishes. J. AOAC Int. 81: 727-735.

Homola, J., Lu, H.B., and Yee, S.S. 1999. Dual-channel surface plasmon resonance sensor with spectral discrimination of sensing channels using a dielectric overlayer. Electron. Lett. 35: 1105–1106.

Homola, J., Lu, H.B., Nenninger, G.G., Dostálek, J., and Yee, S.S. 2001. A novel multichannel surface plasmon resonance biosensor. Sens. Actuat. B-Chem. 76: 403–410.

Homola, J., Dostalek, J., Chen, S., Rasooly, A., Jiang, S., and Yee, S.S. 2002a. Spectral surface plasmon resonance biosensor for detection of staphylococcal enterotoxin B in milk. Int. J. Food Microbiol. 75: 61–69.

Homola, J., Yee, S.S., and Myszka, D. 2002b. Surface plasmon resonance biosensors. In: Optical Biosensors: Present and Future. F.S. Ligler and C.A. Rowe Taitt, eds. Elsevier Science B.V., Amsterdam, Neth. p. 207–251.

Hong, Y., Berrang, M. E., Liu, T., Hofacre, C. L., Sanchez, S., Wang, L., and Maurer, J. J. 2003. Rapid detection of *Campylobacter coli*, *C. jejuni*, and *S. enterica* on poultry carcasses by using PCR-enzyme-linked immunosorbent assay. Appl. Environ. Microbiol. 69: 3492-3499.

Hussein, H.S., Thran, B.H., and Redelman, D. 2002. Detection of *Escherichia coli* O157:H7 in bovine rumen fluid and feces by flow cytometry. Food Control. 13: 387–391.

Ivnitski, D., Abdel-Hamid, I., Atanasov, P., and Wilkins, E. 1999. Biosensors for detection of pathogenic bacteria. Biosens. Bioelectron. 14: 599–624.

Jung, A., Stemmler, I., Brecht, A., and Gauglitz, G. 2001. Covalent strategy for immobilization of DNA-microspots suitable for microarrays with label-free and time-resolved optical detection of hybridization. Fresn. J. Anal. Chem. 371: 128–136

Jung, C.C., Saaski, E.W., McCrae, D.A., Lingerfelt, B.M., and Anderson G.P. 2003. RAPTOR: A fluoroimmunoassay-based fiber optic sensor for detection of biological threats. IEEE Sens. J. 3: 352–360.

Kai, E., Ikebujuro, K., Hoshina, S., Watanabe, H., and Karube, I. 2000. Detection of PCR products of *Escherichia coli* O157:H7 in human stool samples using surface plasmon resonance (SPR). FEMS Immunol. Med. Microbiol. 29: 283–288.

Keshri, G., Voysey, P., and Magan, N. 2002. Early detection of spoilage moulds in bread using volatile production patterns and quantitative enzyme assays. J. Appl. Microbiol. 92: 165-172.

Kim, G.H., Rand, A.G., and Letcher, S.V. 2003. Impedance characterization of a piezoelectric immunosensor part II: *Salmonella typhimurium* detection using magnetic enhancement. Biosens. Bioelectron. 18: 91–99.

King, K.D., Anderson, G.P., Bullock, K.E., Regina, M.J., Saaski, E.W., and Ligler, F.S. 1999. Detecting staphylococcal enterotoxin B using an automated fiber optic biosensor. Biosens. Bioelectron. 14: 163–170.

Konig, B., and Gratzel, M. 1993. Detection of viruses and bacteria with piezoelectric immunosensors. Anal. Lett. 26: 1567–1585.

Koubova, V., Brynda, E., Karasova, L., Skvor, J., Homola, J., Dostalek, J., Tobiska, P., and Rosicky, J. 2001. Detection of foodborne pathogens using surface plasmon resonance biosensors. Sens. Actuat. B-Chem. 74: 100–105.

Kroger, K., Jung, A., Reder, S., and Gauglitz, G. 2002. Versatile biosensor surface based on peptide nucleic acid with label free and total internal reflection fluorescence detection for quantification of endocrine disruptors. Anal. Chim. Acta 469: 37–48.

Kuczynska E., Boyer D.G., and Shelton D.R. 2003. Comparison of immunofluorescence assay and immunomagnetic electrochemiluminescence in detection of *Cryptosporidium parvum* oocsts in karst water samples. J. Microbiol. Meth. 53: 17–26.

Kumar, P., Colston, J.T., Chambers, J.P., Rael, E.D., and Valdes, J.J. 1994. Detection of botulinum toxin using an evanescent wave immunosensor. Biosens. Bioelectron. 9: 57–63.

Laplace-Builhe, C., Hahne, K., Hunger, W., Tirilly, Y., and Drocourt, J.L. 1993. Application of flow-cytometry to rapid microbial analysis in food and drinks industries. Biol.Cell. 78: 123-128.

Lazar, J.G. 1994. Advanced methods in PCR product detection. PCR-Methods and Applications. 4: S1–S14.

Leonard, P., Hearty, S., Brennan, J., Dunne, L., Quinn, J., Chakraborty, T., and O'Kennedy, R. 2003. Advances in biosensors for detection of pathogens in food and water. Enzyme Microb. Tech. 32: 3–13.

Lee, W.E., Thompson, H.G., Hall, J.G., and Bader, D.E. 2000. Rapid detection and identification of biological and chemical agents by immunoassay, gene probe assay and enzyme inhibition using a silicon-based biosensor. Biosens. Bioelectron. 14: 795–804.

Ligler, F.S., Taitt, C.R., Shriver-Lake, L.C., Sapsford, K.E., Shubin, Y., and Golden, J.P. 2003. Array biosensor for detection of toxins. Anal. Bioanal. Chem. 377: 469–477.

Lin, H.C., and Tsai, W.C. 2003. Piezoelectric crystal immunosensor for the detection of staphylococcal enterotoxin B. Biosens. Bioelectron. 18: 1479–1483.

Liu, M., Li, Q.X., and Rechnitz, G.A. 1999. Flow injection immunosensing of polycyclic aromatic hydrocarbons with a quartz crystal microbalance. Anal. Chim. Acta 387, 29–38.

Liu, Y.C., Ye, J.M., and Li, Y.B. 2003. Rapid detection of *Escherichia coli* O157:H7 inoculated in ground beef, chicken carcass, and lettuce samples with an immunomagnetic chemiluminescence fiber-optic biosensor. J. Food Prot. 66: 512–517.

Lu, H.B., Homola, J., Campbell, C.T., Nenninger, G.G., Yee, S.S., and Ratner, B.D. 2001. Protein contact printing for a surface plasmon resonance biosensor with on-chip referencing. Sens. Actuat. B-Chem. 74: 91–99.

Luppa, P.B., Sokoll, L.J., and Chan, D.W. 2001. Immunosensors – principles and applications to clinical chemistry. Clin. Chem. Acta 314: 1–26.

Mannelli, I., Minunni, M., Tombelli, S., and Mascini, M. 2003. Quartz crystal microbalance (QCM) affinity biosensor for genetically modified organisms (GMOs) detection. Biosens. Bioelectron. 18: 129–140.

Maragos, C.M., Bennett, G.A., and Richard, J.L. 1997. Affinity column clean-up for the analysis of fumonisins and their hydrolysis products in corn. Food Agric. Immunol. 9: 3-12.

Maragos, C.M., and Thompson, V.S. 1999. Fiber-optic immunosensors for mycotoxins. Nat. Toxins. 7: 371–376.

Marazuela, M.D., Cuesta, B., Moreno-Bondi, M.S., and Qeujido, A. 1997. Free cholesterol fiber-optic biosensor for serum samples with simplex optimization. Biosens. Bioelectron. 12: 233–240.

McBride, M.T., Gammon, S., Pitesky, M., O'Brien, T.W., Smith, T., Aldrich, J., Langlois, R.G., Colston, B., and Venkateswaran, K.S. 2003. Multiplexed liquid arrays for simultaneous detection of simulants of biological warfare agents. Anal. Chem. 75: 1924–1930.

McMeekin, T.A., Brown, J., Krist, K., Miles, D., Neumeyer, K., Nichols, D.S., Olley, J., Presser, K., Ratkowsky, D.A., Boss, T., Salter, M., and Soontranon, S. 1997. Quantitative microbiology: A basis for food safety. Emerg. Infect. Dis. 3: 541–549.

Mello, L.D., and Kuboto, L.T. 2002. Review of the use of biosensors as analytical tools in the food and drink industries. Food Chem. 77: 237–256.

Minunni, M., Mascini, M., Carter, R.M., Jacobs, M.B., Lubrano, G.J., and Guilbault, G.G. 1996. A quartz crystal microbalance displacement assay for *Listeria monocytogenes*. Anal. Chim. Acta. 325: 169–174.

Mo, X.T., Zhou, Y.P., Lei, H., and Deng, L. 2002. Microbalance-DNA probe method for the detection of specific bacteria in water. Enz. Microbial Tech. 30: 583–589.

Moller, T.E., and Gustavsson, H.F. 2001. Determination of fumonisins B1 and B2 in corn using immunoaffinity column clean-up and thin layer chromatography/densitometry. Food Addit. Contam. 17: 463–468.

Mullett, W., Lai, E.P.C., and Yeung, J.M. 1998. Immunoassay of fumonisins by a surface plasmon resonance biosensor. Anal. Biochem. 258: 161–167.

Naimushin, A.N., Soelberg, S.D., Nguyen, D.K., Dunlap, L., Bartholeomew, D., Elkind, J., Melendez, J., and Furlong, C.E. 2002. Detection of *Staphylococcus aureus* enterotoxin B at femtomolar levels with a miniature integrated two-channel surface plasmon resonance (SPR) sensor. Biosens. Bioelectron. 17: 573–584.

Nedelkov, D., Rasooly, A., and Nelson, R.W. 2000. Multitoxin biosensor-mass spectrometry analysis: A new approach for rapid, real-time, sensitive analysis of staphylococcal toxins in food. Int. J. Food Microbiol. 60: 1–13.

Nedelkov, D., and Nelson, R.W. 2003. Detection of staphylococcal enterotoxin B via biomolecular interaction analysis mass spectrometry. Appl. Environ. Microbiol. 69: 5212-5215.

Neuschafer, D., Budach, W., Wanke, C., and Chibout, S. 2003. Evanescent resonator chips: a universal platform with superior sensitivity for fluorescence-based microarrays. Biosens. Bioelectron. 18: 489–497.

Ogert, R.A., Brown, J.E., Singh, B.R., Shriver-Lake, L.C., and Ligler, F.S. 1992. Detection of *Clostridium botulinum* toxin A using a fiber optic-based biosensor. Anal. Biochem. 205: 306–312.

Oroszlan, P., Thommen, C., Wehrli, M., Duveneck, G., and Ehrat, M. 1993. Automated optical sensing system for biochemical assays: A challenge for ELISA? Anal. Meth. Instrum. 1: 43–51.

O'Sullivan, C.K., Vaughan, R., and Guilbault, G.G. 1999. Piezoelectric immunosensors – Theory and applications. Anal. Let. 32: 2353–2377.

O'Sullivan, C.K., and Guilbault, G.G. 1999. Commercial quartz crystal microbalances – theory and applications. Biosens. Bioelectron. 14: 663–670.

Paddle, B.M. 1996. Biosensors for chemical and biological agents of defense interest. Biosens. Bioelectron. 11: 1079–1113.

Patel, P.D. 2002. (Bio)sensors for measurement of analytes implicated in food safety: a review. Trends Anal. Chem. 21: 96–115.

Peng, T., Cheng, Q., and Stevens, R.C. 2000. Amperometric detection of *Escherichia coli* heat-labile enterotoxin by redox diacetylenic vesicles on a sol-gel thin-film electrode. Anal. Chem. 72: 1611–1617.

Piehler, J., Brecht, A., Geckeler, K.E., and Gauglitz, G. 1996. Surface modification for direct immunoprobes. Biosens. Bioelectron. 11: 579–590.

Plomer, M., Guilbault, G.G., and Hock, B. 1992. Development of a piezoelectric immunosensor for the detection of enterobacteria. Enzyme Microb. Tech. 14: 230–235.

Pollard-Knight, D., Hawkins, E., Yeung, D., Pashby, D.P., Simpson, M., McDougall, A., Buckle, P., and Charles, S.A. 1990. Immunoassays and nucleic acid detection with a biosensor based on surface plasmon resonance. Ann. Biol. Clin. (Paris) 48: 642–646.

Pyun, J.C., Beutel, H., Meyer, J.U., and Ruf, H.H. 1998. Development of a biosensor for *E. coli* based on a flexural plate wave (FPW) transducer. Biosens. Bioelectron. 13: 839–845.

Rasooly, L., and Rasooly, A. 1999. Real time biosensor analysis of staphylococcal enterotoxin A in food. Int. J. Food Microbiol. 49: 119–127.

Rasooly, A. 2001. Surface plasmon resonance analysis of staphylococcal enterotoxin B in food. J. Food Prot. 64: 37–43.

Rattanasomboon, N., Bellara, S.R., Harding, C.L., Fryer, P.J., Thomas, C.R., Al-Rubeai, M., and McFarlane, C.M. 1999. Growth and enumeration of the meat spoilage bacterium *Brochothrix thermosphacta*. Int. J. Food Microbiol. 51: 145–158.

Saha, S., Raje, M., and Suri, C.R. 2002. Sandwich microgravimetric immunoassay: sensitive and specific detection of low molecular weight analytes using piezoelectric quartz crystal. Biotech. Lett. 24: 711–716.

Sakakibara, T., Murakami, S., and Imai, K. 2003. Enumeration of bacterial cell numbers by amplified firefly bioluminescence without cultivation. Anal. Biochem. 312: 48–56.

Sambrook, J., Fritsch, E.F., and Maniatis, T. 1989. Molecular Cloning: A Laboratory Manual. Cold Spring Harbor Laboratory, Cold Spring Harbor, NY.

Sapsford, K.E., Rasooly, A., Taitt, C.R., and Ligler, F.S. 2005. Detection of *Campylobacter* and *Shigella* species in food samples using an array biosensor. Anal. Chem. 76: 433–440.

Schneider, B.H., Edwards, J.G., and Hartman, N.F. 1997. Hartman interferometer: Versatile integrated optic sensor for label-free, real-time quantification of nucleic acids, proteins, and pathogens. Clin. Chem. 43: 1757–1763.

Schuderer, J., Akkoyun, A., Brandenburg, A., Bilitewski, U., and Wagner, E. 2000. Development of a multichannel fluorescence affinity sensor system. Anal. Chem. 72: 3942-3948.

Seaver, M., Crookston, J.C., Roselle, D.C., and Wagner, S.J. 2001. First results using automated epifluorescence microscopy to detect *Escherichia coli* and *Staphylococcus epidermidis* in WBC-reduced platelet concentrates. Transfusion 41: 1351–1355.

Seo, K.H., Brackett, R.E., Hartman, N.F., and Campbell, D.P. 1999. Development of a rapid response biosensor for detection of *Salmonella typhimurium*. J. Food Prot. 62: 431–437.

Sgorbati, S., Barbesti, S., Citterio, S., Bestetti, G., and DeVecchi, R. 1996. Characterization of number, DNA content, viability and cell size of bacteria from natural environments using DAPI/PI dual staining and flow cytometry. Minerva Biotecnologica 8: 9–15.

Shah, J., and Wilkins, E. 2003. Electrochemical biosensors for detection of biological warfare agents. Electroanalysis 15: 157–167.

Shriver-Lake, L.C., Shubin, Y.S., and Ligler, F.S. 2003. Detection of staphylococcal enterotoxin B in spiked food samples. J. Food Prot. 66: 1851–1856.

Smith, R.H., Lemon, W.J., Erb, J.L., Erb-Downward, J.R., and Downward, J.G. 1999. Development of kinetic ligand binding assays using a fiber optic sensor. J. Clin. Chem. 45: 1683–1685.

Spangler, B.D., and Tyler, B.J. 1999. Capture agents for a quartz crystal microbalance-continuous flow biosensor: functionalized self-assembled monolayers on gold. Anal. Chim. Acta 399: 51–62.

Spangler, B.D., Wilkinson, E.A., Murphy, J.T., and Tyler, B.J. 2001. Comparison of the Spreeta surface plasmon resonance sensor and a quaretz crystal microbalance for detection of *Escherichia coli* heat-labile enterotoxin. Anal. Chim. Acta. 444: 149–161.

Squirrel, D.J., Price, R.L., and Murphy, M.J. 2002. Rapid and specific detection of bacteria using bioluminescence. Anal. Chim. Acta 457: 109–114.

Steemers, F.J., Ferguson, J.A., and Walt, D.R. 2000. Screening unlabeled DNA targets with randomly ordered fiber-optic gene arrays. Nature Biotechnol. 18: 91–94.

Stender, H., Sage, A., Oliveira, K., Broomer, A.J., Young, B., and Coull, J. 2001. Combination of ATP-bioluminescence and PNA probes allows rapid total counts and identification of specific microorganisms in mixed populations. J. Microbiol. Meth. 46: 69–75.

Stopa, P.J. 2000. The flow cytometry of *Bacillus anthracis* spores revisited. Cytometry 41: 237-244.

Stopa, P.J., and Mastromanolis, S.A. 2001. The use of blue-excitable nucleic-acid dyes for the detection of bacteria in well water using a simple field fluorometer and a flow cytometer. J. Microbiol. Meth. 45: 143–153.

Strachan N.J. and Gray, D.I. 1995. A rapid general method for the identification of PCR products using a fibre-optic biosensor and its application to the detection of *Listeria*. Lett. Appl. Microbiol. 21: 5–9.

Su, X.D., Low, S., Kwang, J., Chew, V.H.T., and Li, S.F.Y. 2001. Piezoelectric quartz crystal based veterinary diagnosis for *Salmonella*

*enteritidis* infection in chicken and egg. Sens. Actuat. B-Chem. 75: 29–35.

Suleimann A.A., and Guilbault, G.G. 1994. Recent developments in piezoelectric immunosensors. A review. Analyst 119: 2279–2282.

Taitt, C.R., Golden, J.P., Shubin, Y.S., Shriver-Lake, K.E., Sapsford, K.E., Rasooly, A., and Ligler, F.S. 2003. A portable array biosensor for detecting multiple analytes in complex samples. Microb. Ecol. 47: 175–185.

Taitt, C.R., Shubin, Y.S., Angel, R., and Ligler, F.S. 2004. Detection of *Salmonella typhimurium* using a rapid array-based immunosensor. Appl. Environ. Microbiol. 70: 152–158.

Takahashi, T., Nakakita, Y., Watari, J., and Shinotsuka, K. 2000. A new rapid technique for detection of microorganisms using bioluminescence and fluorescence microscope method. J. Biosci. Bioeng. 89: 509–513.

Tempelman, L.A., King, K.D., Anderson, G.P., and Ligler, F.S. 1996. Quantitating staphylococcal enterotoxin B in diverse media using a portable fiber-optic biosensor. Anal. Biochem. 233: 50–57.

Thevenot, D.R., Toth, K., Durst, R.A., and Wilson, G.S. 2001. Electrochemical biosensors: recommended definitions and classification. Biosens. Bioelectron. 16: 121–131.

Tims, T.B., Dickey, S.S., DeMarco, D.R., and Lim, D.V. 2001. Detection of low levels of *Listeria monocytogenes* within 20 hours using an evanescent wave biosensor. Am. Clin. Lab. 20: 28–29.

Tu, S.I., Patterson, D., Uknalis, J., and Irwin, P. 2000. Detection of *Escherichia coli* O157:H7 using immunomagnetic capture and luciferin-luciferase ATP measurement. Food Res. Int. 33: 375–380.

Tudos, A.J., Lucas-van den Bos, E.R., and Stigter, E.C. 2003. Rapid surface plasmon resonance-based inhibition assay of deoxynivalenol. J. Agric. Food Chem. 51: 5843–5848.

Uttenthaler, E., Schraml, M., Mandel, J., and Drost, S. 2001. Ultrasensitive quartz crystal microbalance sensors for detection of M13 phages in liquids. Biosens Bioelectron. 16: 735-743.

Vaughan, R.D., Carter, R.M., O'Sullivan, C.K., and Guilbault, G.G. 2003. A quartz crystal microbalance (QCM) sensor for the detection of *Bacillus cereus*. Anal. Lett. 36: 731–747.

Vo-Dinh, T., and Cullum, B. 2000. Biosensors and biochips: advances in biological and medical diagnostics. Fresenius J. Anal. Chem. 366: 540–551.

Vo-Dinh, T., Alarie, J.P., Cullum, B.M., and Griffin, G.D. 2000. Antibody based nanoprobe for measurement of a fluorescent analyte in a single cell. Nature Biotechnol. 18: 764–767.

Wadkins, R.M., Golden, J.P., Pritsiolas, L.M., and Ligler, F.S. 1998. Detection of multiple toxic agents using a planar array immunosensor. Biosens. Bioelectron. 13: 407–415.

Walt, D.R. 2000. Bead-based fiber-optic arrays. Science 287: 451–451

Watts, H.J., Lowe, C.R., and Pollard-Knight, D.V. 1994. Optical biosensor for monitoring microbial cells. Anal. Chem. 66: 2465–2470.

Willard, D., Proll, G., Reder, S., and Gauglitz, G. 2003. New and versatile optical-immunoassay instrumentation for water monitoring. Environ. Sci. Pollut. Res. 10: 188–191.

Wong, Y.Y., Ng, S.P., Ng, M.H., Si, S.H., Yao, S.Z., and Fung, Y.S. 2002. Immunosensor for the differentiation and detection of *Salmonella* species based on a quartz crystal microbalance. Biosens. Bioelectron. 17: 676–684.

Wright, L.S., and Harding, H. 2000. Detection of DNA via an ion channel switch biosensor. Anal. Biochem. 282: 70–76.

Yacoub-George, E., Meixner, L., Scheithauer, W., Koppi, A., Drost, S., Wolf, H., Danapel, C., and Feller, K.A. 2002. Chemiluminescence multichannel immunosensor for biodetection. Anal. Chim. Acta 457: 3–12.

Yang, G.D., Imagire, S.J., Yasaei, P., Ragelis, E.P., Park, D.L., Page, S.W., Carlson, R.E., and Guire, P.E. 1987. Radioimmunoassay of paralytic shellfish toxins in clams and mussels. Bull. Environ. Contam. Toxicol. 39: 264–271.

Ye, F., Li, M.S., Taylor, J.D., Nguyen, Q., Colton, H.M., Casey, W.M., Wagner, M., Weiner, M.P., and Chen, J.W. 2001. Fluorescent microsphere-based readout technology for multiplexed human single nucleotide polymorphism analysis and bacterial identification. Human Mutation 17: 305–316.

Ye, J., Liu, Y., and Li, Y. 2002. A chemiluminescence fiber-optic biosensor coupled with immunomagnetic separation for rapid detection of *E. coli* O15:H7. Trans. ASAE 45, 473-478.

Yokoyama, K., Ikebukuro, K., Tamiya, E., Karube, I., Ichiki, N., and Arikawa, Y. 1995. Highly sensitive quartz-crystal immunosensors for multisample detection of herbicides. Anal. Chim. Acta 304: 139–145.

Yu, H., and Bruno, J.G. 1996. Immunomagnetic-electrochemilumines-cent detection of *Escherichia coli* O157 and *Salmonella typhimurium* in foods and environmental water samples. Appl. Environ. Micro. 62: 587–592.

Zhou, C.H., Pivarnik, P., Rand, A.G., and Letcher, S.V. 1998. Acoustic standing-wave enhancement of a fiber-optic *Salmonella* biosensor. Biosens. Bioelectron. 13: 495–500.

# Molecular Typing and Differentiation of Foodborne Bacterial Pathogens

Franco Pagotto, Nathalie Corneau, Chris Scherf, Peter Leopold, Clifford Clark , and Jeffrey M. Farber

4

## Abstract

Molecular typing of foodborne pathogens is used to generate approximations of population variation, definition of specific clonal lineages, comparison of isolates of similar species from different geographical locations, and changes of types within the population over time. Thus, it can be used to confirm the identity of organisms responsible for sporadic cases or foodborne outbreaks, as well as facilitating trace-back investigations and food product recalls. Validation of typing methods should include tests for intra- and inter-laboratory reproducibility for sets of known and test strains to ensure that appropriate standardization of methodology can be achieved. Typeability, the proportion of the population of micro-organisms that can be typed, and discriminatory power, the ability to distinguish unrelated strains, are two important endpoints when developing a molecular characterization scheme. The utility of any typing or subtyping method depends in part on the size and extent of the database with which any new isolate can be compared. It has become widely accepted that no single typing or fingerprinting method will be completely accurate or informative, and that a combination of methods must often be used. The output from these methods, however, cannot be looked at in isolation, and should be combined with epidemiological data when investigating foodborne illnesses. The following chapter describes some of the current phenotypic (serology, phage typing), genotypic (PFGE, RAPD, ribotyping) and emerging (Maldi-Tof mass spectroscopy, AFLP, MLST, VNTR, Microarray technology) methods used to characterize bacterial foodborne pathogens.

## Introduction

Molecular typing or fingerprinting of bacteria may be used to achieve a variety of goals. Most generally, these include the estimation of the variability within populations, definition of specific clonal lineages (especially virulent clones), comparison of isolates of similar species from different geographical locations, and changes of types within the population over time. Data acquired from investigations of this sort can provide a useful baseline against which unusual events can be detected, such as the emergence of new pathogens, outbreaks, or bioterrorist attacks. At the most specific level, molecular typing/fingerprinting can be used to confirm the identity of organisms responsible for such public health threats in putative vehicles of infection, as well as facilitating trace-back investigations and product recalls. In all the types of investigations summarized above, the molecular type or fingerprint represents a surrogate marker for the total genetic content of the organism under study. The fidelity with which each typing scheme does so must often be determined empirically through multi-disciplinary studies that often include epidemiological investigations.

Molecular characterization is further complicated by the fact that each typing method may be more useful for one organism than another, depending on the purpose for which typing is performed. For instance, phage typing and pulsed-field gel electrophoresis (PFGE) together, or separately, both appear to be excellent methods for subtyping *Escherichia coli* O157:H7 isolates in an epidemiologically relevant manner. These techniques are not as useful for *Salmonella* Enteritidis, where ribotyping using a combination of the restriction enzymes *Sph*I and *Pst*I appears to result in types that correlate well with epidemiologic data. PFGE appears to be useful for defining outbreak strains of *Campylobacter jejuni*, but because of its high discriminatory power, may not be helpful for analyzing populations of this organism over larger geographical areas or longer time periods. Recent investigations suggest that analysis of the *Campylobacter* flagellar (*fla*) gene sequence or restriction fragment length polymorphism analysis of these sequences (fla-RFLP), alone or in conjunction with multi-locus sequence typing (MLST), may provide a much better estimate of relationships in larger *Campylobacter* populations. Many investigators working with typing and fingerprinting methods recommend the use of a combination of methods for the optimal characterization of most organisms.

Investigators that use molecular characterization technologies commonly evaluate them against several criteria. Discriminatory power measures the ability of the method to differentiate strains that are considered unrelated based on epidemiologic data. The epidemiologic relevance of typing methods is often evaluated by comparison of typing/fingerprinting results with the results of epidemiologic investigations as to whether isolates should be included in clusters or outbreaks. However, in many cases, it may not be possible to use either the epidemiologic or typing data as a reference standard for inclusion of strains in clusters or outbreaks, making the assessment of the epidemiologic relevance of a typing method somewhat empirical. Results from many typing methods correlate fairly well with the phylogenetic structure of the population analyzed. However, it is not clear whether all such typing methods are phylogenetically relevant, in that it would be difficult to use them to predict the phylogenetic placement of an organism, when used in isolation. Some methods, such as random amplified polymorphic DNA PCR (RAPD) analysis, may evaluate too few characters to be truly phylogenetically relevant (Tyler *et al.*, 1997).

Finally, pattern identity in some DNA typing/fingerprinting methods may not necessarily indicate that the DNA fragments from which the patterns are derived are identical. Mutations may occur within the genome. When using some PFGE protocols, not all fragments are resolved; indistinguishable banding patterns could theoretically result from organisms with somewhat different genetic backgrounds. The results of any typing/fingerprinting analyses must therefore be interpreted with some care.

## Evaluation of typing method results

Validation of typing methods should include tests for intra- and inter-laboratory reproducibility for sets of known and test strains to ensure that appropriate standardization of methodology can be achieved. Care must be taken in ensuring that isolates are not subcultured too often after receipt, since it is known that some strains have genetic elements that can make their phenotype or genotype unstable, and that patterns based on enzyme restriction of DNA can change after multiple subcultures. Typing and subtyping methods must also detect substantial variability in populations of apparently unrelated isolates, and should group isolates that are epidemiologically related.

Two other important criteria are used to characterize typing methods. Typeability describes the proportion of the population of micro-organisms that can be typed using a particular method. Ideally, all typing and subtyping methods would have a typeability of 100%. However, some methods, such as serotyping of the heat-labile antigens (known as Lior typing) of *Campylobacter*, may have considerably lower typeability. Discriminatory power refers to the ability to distinguish unrelated strains, and can be quantitated using Simpson's Index of Diversity (Hunter and Gaston, 1988). Methods with the highest discriminatory ability are often considered the best methods to use for subtyping bacterial populations. However, some methods may be too discriminatory, and may separate isolates or strains that would be grouped by other typing methods or by epidemiological data. In practice, it is often useful to empir-

ically determine the level of discrimination that is appropriate for the analysis to be undertaken.

When isolates give indistinguishable results or patterns in bacterial characterization and typing tests, the assumption often made is that they are clonal, i.e., that they are directly descended from a single progenitor and/or genetically related to a high degree (Dijkshoorn *et al.*, 2000). However, the prevalence and stability of certain types may make interpretation of these results somewhat problematic. As an example, in our experience, the same PFGE type of *E. coli* O157:H7 has been isolated from cattle and again, after a five year interval, from a human patient in a geographically distant region. Obviously a direct connection cannot be inferred from typing data alone. In fact, in most cases, typing information can only be interpreted in the context of epidemiologic information associated with the isolate. It is necessary to know the date and source of isolation, geographical location from which the isolate was obtained, and whether any implicated vehicle of infection could have accounted for the transmission of strains between infected animals or contaminated foods and humans. In addition, it is helpful to know if the type(s) or subtype(s) detected is rare or common, and if the incidence of isolation or of disease is unusual.

The utility of any typing or subtyping method depends in part on the size and extent of the database with which any new isolate can be compared. *Salmonella* serotypes represent a comprehensive database with global distribution and a long history, which can also be universally understood and interpreted. Many other typing/fingerprinting databases are smaller, more restricted in duration, or can be interpreted only by specialists in the field. Therefore, the utility of any typing method depends not only on the ability to obtain results using standardized methods with good inter-laboratory reproducibility (portability), but also on the number of distinct entries into the database. Another important factor influencing the utility of any typing method is the completeness of data entered into the database. PulseNet, for instance, encourages contributing laboratories to type as close to 100% of isolates of specific species arriving at the laboratory by PFGE, so that any deviation from the normal incidence of that type can be detected and investigated. Databases must be properly maintained to ensure the quality of data placed therein, and there must be an active program to ensure that type designations do not arise from minor variations in methodology or interpretation within or between laboratories. Collections of type strains are essential to support these database validation activities, so that isolates having new types can be compared against very closely related types in the same analysis. Type strains are also useful in the development of novel typing schemes and serve as excellent reference points to compare all methods against themselves.

The characteristics measured by different typing or fingerprinting methods may exhibit different rates of change in different organisms. Multi-locus enzyme electrophoresis (MLEE) has been used extensively to determine phylogenetic relationships among many bacteria. In most cases, groupings of organisms obtained by MLEE can be subdivided much more finely by other typing or fingerprinting methods. *E. coli* MLEE group B contains organisms with most of the *E. coli* virulence factors, and therefore represents a number of *E. coli* serotypes and pathotypes. MLEE types may change more slowly than ribotypes, PFGE types, types obtained using PCR for repetitive elements, or analysis of variable nucleotide tandem repeats in the genome.

Different typing and subtyping methods appear to group strains or isolates at different levels of biological variability, making each method more useful for some purposes than others. MLEE has been widely used to evaluate phylogenetic relationships among isolates from different species as well as within species. The method has been useful for establishing the phylogenetic relationships between *Salmonella enterica* subsp. *enterica* serotypes, defining large groupings of strains within *E. coli*, and establishing the genetic lineage leading to the emergence of *E. coli* O157:H7 from a particular O55:H7 clone (Boyd *et al.*, 1996; Feng *et al.*, 1998; Pupo *et al.*, 1997). However, this method would not be useful for subtyping *E. coli* O157:H7 strains. Ribotyping, on the other hand, is a phylogenetically valid method for subtyping several bacterial species, and appears to be useful for establishing differences and relationships on a global basis over extended periods of time for organisms such as *Vibrio cholerae*. Changes in *V. cholerae* ribotype patterns have been estimated to take place at a rate of one band every 6 years (Karaolis *et al.*, 1994). PFGE patterns of *E. coli* O157:H7 may change at a much faster rate, and some variation in patterns may be seen within the course of extended outbreaks. Rates of change of other typing and subtyping methods may change even more rapidly, and could tend to subtype strains into groups that have little epidemiological relevance, but which could be useful for traceback investigations. The choice of method to be used depends on an understanding of how the typing or subtyping method discriminates within the population under investigation and fits with the purpose of the investigation.

In practice, most investigators currently use several kinds of typing and fingerprinting information for characterizing foodborne pathogens. Isolation of suspect organisms is followed by the determination of species and genus by biochemical testing. For many enteric bacteria, serotyping is routinely done at the primary laboratory or in a first-line reference laboratory using commercially available antisera. Unusual or difficult isolates may be sent to a national reference laboratory for unambiguous serotype characterization. For *Salmonella enterica*, definition of the serotype allows a rapid, informal risk assessment of the potential public health risk associated with the isolate, and may give sufficient data to characterize outbreak strains. The value of serotyping is not as clear for *E. coli* other than *E. coli* O157:H7, since the virulence potential of the organism depends on the pathotype, or set of virulence

genes defining the pathogenic potential and probable clinical presentation resulting from infection with the organism.

For most foodborne pathogens, further subtyping is necessary to obtain information useful for epidemiologic investigations. Phage typing, where available, and PFGE have become powerful tools for subtyping many bacteria when used together or individually. Ribotyping is commonly used by some laboratories, and may provide additional information, while PCR methods based on repeated chromosomal consensus sequences or on random amplification of DNA have come into widespread use. Newer, sequence-based methods such as multi-locus sequence typing (MLST), analysis of variable number of tandem repeats (VNTR), and others, are gaining prominence as more data accumulate in support of their efficacy. It has become widely accepted that no single typing or fingerprinting method will be completely accurate or informative, and that a combination of methods must often be used. Which methods should be used often depends on the type of analysis undertaken.

## Phenotypic typing (non-chromosomal)

## Non-nucleic acid typing methods

*Serology*
Exposed surface components of bacterial cell surfaces interact with the immune systems of mammalian hosts, with the result that antibodies are made by the host to specific bacterial structures. Microorganisms of the same species may differ in the antigenic structure of surface molecules, resulting in different antibodies being made to different bacterial types. The strongest immune reactions are usually to proteins such as flagella or fimbriae, lipopolysaccharides (LPS) or capsular polysaccharide material. Reactivity to these structures forms the basis of heat-labile flagellar (H) typing, heat-stable LPS (O) typing, and capsular (K) typing. Serotyping schemes have been developed for almost all enteric pathogens of clinical interest. In assays for serotype specificity, antiserum with known reactivity prepared against standard strains with known antigenic determinants, is reacted with bacteria cells. In positive tests, the cross-linking of bacteria by antibodies results in visible agglutination that is absent when the antigen of interest is not present. In practice, for some serotyping schemes with large numbers of factors that may be serotype determinants, it is necessary to use reagents containing combinations of pooled antiserum to narrow the range of possibilities to a number that can be tested by individual, absorbed serum. It may be necessary to passage bacteria several times in motility media to select flagellated variants, while for strains of *Salmonella*, it is often necessary to select for flagellar phase variants to obtain the full antigenic formula. Serotyping can take time and require considerable expertise.

For *Salmonella enterica*, the description of the O:H serotype, often expressed as serotype names such as Typhi, Typhimurium, Enteritidis, etc., is a useful descriptor for both the epidemiological and clinical behavior of the organism. Unusual clusters of rare *Salmonella* serotypes are often sufficient to detect outbreaks of the organism and confirm the identity of the vehicle of infection. Even for *E. coli*, the serotype is often the best predictor of the pathotype of the organism and its clinical significance (Meng *et al.*, 2001), even though some serotypes may carry virulence gene sets that group them into different pathotypes. While *E. coli* O157:H7 is closely associated with severe bloody diarrhea and hemolytic uremic syndrome (HUS), members of serotype O26 or O111 may be enterohemorrhagic *E. coli* (EHEC) or enteropathogenic *E. coli* (EPEC), with quite different implications for the epidemiology and clinical significance of the organism. *Listeria monocytogenes* serotype 4b is much more closely associated with large outbreaks than other serotypes, making serotype determination extremely useful for this pathogen. Similarly, specific O:H serotype combinations of *Campylobacter* are closely associated with progression to Guillain–Barré syndrome, presumably because of molecular mimicry of the bacterial exopolysaccharide moiety.

Serotyping requires the development of large stocks of unabsorbed and absorbed antiserum to all the factors included in the serotyping scheme. A full set of reference strains is required for quality control and quality assurance purposes. Antibodies are usually generated in rabbits, requiring ethics approval and specially trained animal care staff. Testing may involve direct slide agglutination, enzyme-linked immunosorbent assay (ELISA), latex co-agglutination, and erythrocyte co-agglutination. The preparation of antigen and interpretation of test results require highly trained and experienced laboratory staff. While commercial antisera are available for the more common antigens for most bacterial serotypes, full typing schemes are extremely expensive to produce and maintain, and are usually done so only at national reference or other highly specialized laboratories.

Serotypes represent markers that can readily be understood by epidemiologists and laboratory scientists alike, in many cases backed by very extensive, long-term, and comprehensive databases. However, serotyping alone often does not provide the discriminatory power required for detecting outbreaks or following the spread of interesting or particularly virulent clones. Some strains may be particularly difficult to serotype, resulting in long turnaround times for results, and in some serotyping schemes (such as the heat-labile serotyping scheme for *Campylobacter*), there may be a significant number of non-typeable strains. At the moment, serotyping appears to be one of the necessary first steps in the hierarchy of typing and subtyping methods, facilitating the choice of subsequent methods used and organization of information obtained. However, with the increasing sophistication of DNA sequence-based typing and subtyping methods, there may come a time in the near future when serological assays will not be relied upon for initial investigations of bacterial populations.

## Phage typing

Phage typing is based on the ability of selected bacteriophages to selectively enter, infect, and lyse host bacterial cells. The development of phage typing schemes involves the isolation of phages that can produce differential infection (lysis phenotypes) on bacterial populations of interest, often by adaptation in different propagating strains (Felix and Callow, 1943; Schmeiger, 1999). Lysis depends on the presence of specific receptors for phage adherence, while carriage of lysogenic phages in bacterial strains may confer immunity to infection, due to repressors or superinfection exclusion systems (Schmeiger, 1999). Phage infection may also be abrogated by restriction of phage DNA after entry into cells. Reproducible lytic patterns may range from confluent lysis to the production of only a single plaque. This can make the interpretation of phage type complex. Experience with the phage typing schemes in use is a definite asset. Effective use of phage typing as a subtyping method therefore requires specially trained personnel. Phage types may change with the introduction of plasmids, other phages, or genes that change the receptor content of the bacterial cell wall. The phage type of a particular strain may also change with the loss of a lipopolysaccharide (Baggesen *et al.*, 1997) or the loss or acquisition of plasmids (Brown *et al.*, 1999). However, this does not necessarily detract from the utility of the method for epidemiologic analysis. It should be emphasized that phage typing is a phenotypic method that does not necessarily correlate absolutely with the phylogenetic relationships of a population, or with molecular subtyping methods. This can be an advantage, however, as the use of phage typing in conjunction with molecular methods such as PFGE, can actually increase the discriminatory power of the overall analysis.

Phage genotypes and phenotypes are stabilized by growth on a single propagating bacterial strain and the ability of each phage to cause the characteristic patterns of lysis on standard strains is tested routinely. To ensure reproducibility among laboratories, in practice, phages for each typing scheme are prepared at a single source laboratory and distributed world-wide. For these reasons, the use of phage typing has been limited to reference laboratories, despite the fact that for many bacterial species the method is rapid, has relatively high discriminatory power, and is inexpensive to perform.

In many reference laboratories, phage typing has become the preferred method for screening large numbers of sporadic isolates to detect clusters that may indicate outbreaks are occurring. When outbreaks occur, phage typing is an extremely useful method for initial screening of patients into the outbreak. Because of the rapidity with which results can be obtained, the information may be extremely useful to epidemiologists in developing hypotheses with which to investigate outbreaks. The choice of phage sets to be used and interpretation of phage lytic patterns depends on the unambiguous identification of at least the genus and species of bacteria to be tested, and usually requires knowledge of the serotype as well. Different sets of phages are used to investigate, for instance, *Salmonella enterica* serovars Typhimurium, Typhi, Enteritidis, and Oranienberg.

Recent evidence suggests that the Anderson phage typing system for *S.* Typhimurium was developed using propagating strains carrying prophages, and that the typing phages derived from propagation on these hosts may have arisen from host-specific modification or recombination events. It may not be possible to reproduce these typing phages once the original stocks are exhausted, marking an end to this typing system (Schmeiger, 1999). It is not known whether any of the other phage typing systems suffer from the same problem. A summary of some existing phage typing systems for foodborne bacterial pathogens is presented in Table 4.1.

*Fatty acid profiling by fatty acid methyl ester (FAME) analysis*

Fatty acid profiling of whole bacterial cells can be used for species identification of some organisms (Tang *et al.*, 1998), and may provide subtyping below the species level for a few bacteria. The method depends on the detection of characteristic fatty acid methyl esters by gas chromatography after chemical derivitization of the fatty acids to make them more volatile. Lipids from cell membranes are first saponified with sodium hydroxide to release fatty acids, which are then methylated by an acid-methanol method to make the resulting fatty acid methyl esters volatile. After extraction into an organic solvent and removal of contaminants with a NaOH wash, samples are analyzed using a gas chromatograph equipped with a flame ionization detector and a very accurate automated injection system for samples.

Both growth conditions and the analytical process must be carefully standardized for valid comparisons which are carried out using specific softwares against a library containing thousands of sample entries to be made. Instrumentation and software for performing these analyses are available from Microbial Identification, Inc. (also known as MIDI).

*Burkholderia pseudomallei*, the cause of melioidosis, has been shown to be accurately distinguished from its closely related but nonpathogenic *B. thailandensis* using FAME analysis (Inglis *et al.*, 2003a). Using *Enterococcus* spp., FAME characterization was compared to biotype profiling and ribotyping using strains of *Enterococcus* (Lang *et al.*, 2001). However, FAME was shown to be the least discriminatory of the methods. The method has been used for the epidemiological typing of *Campylobacter* species, but with a relatively low discriminatory index (Steele *et al.*, 1998). It can be a helpful method to confirm typing results for these organisms obtained using other techniques. However, FAME analysis was less useful for subgrouping Shiga-toxigenic *Escherichia coli* and *Salmonella enterica* serovar Enteritidis strains (Steele *et al.*, 1997). It is not currently in common use for typing of most enteric organisms.

*Two-dimensional gel electrophoresis*

Two-dimensional polyacrylamide gel electrophoresis (2-DE) is best known as a method for high resolution differentiation of large numbers of cellular proteins, and is often the first step in proteomics approaches to protein expression profiling (Anderson and Anderson, 1996; Graves and Haystead, 2002). The use of immobilized pH gradients has greatly improved the reproducibility of this technique, and recent advances have allowed at least partial automation of the method. More recently, 2-DE has been used as a molecular subtyping method for the resolution of *E. coli* variants that appeared indistinguishable by other methods, including PFGE (Yokoyama *et al.*, 2001). The method has also proved valuable for the identification and speciation of *Listeria* spp. (Gormon and Phan-Than, 1995).

Proteins for 2-DE may be obtained by solubilization of whole bacteria or by precipitating the protein component of bacterial supernatants or other cellular fractions. These protein fractions are analyzed in the first dimension by isoelectric focusing in tube gels or in immobilized pH gradients until equilibrium is reached. Analysis in the second dimension is by molecular mass and is accomplished by adding the first dimension gel to an SDS-PAGE slab or gradient gel. Proteins may be visualized by staining with Coomassie blue or a silver stain, or alternatively by incorporation of radioactive amino acids and autoradiography. Comparison of gels usually requires imaging of the gel and the use of software that is developed specifically for the analysis of 2-DE.

Though the usefulness of the method is perhaps only now becoming apparent, there are still few typing laboratories

**Table 4.1** Summary of existing phage typing schemes for foodborne bacteria[1]

| Organism | Laboratory where phages are propagated[2] |
|---|---|
| *E. coli* O157:H7 | NML |
| *Campylobacter* spp. | CPHL |
| *Listeria monocytogenes* | Martin Luther University, Halle, Germany; NLEP |
| *Salmonella* Enteritidis | CPHL |
| *Salmonella* Hadar | CPHL |
| *Salmonella* Heidelberg | NML |
| *Salmonella* Infantis | NML |
| *Salmonella* Newport | NML |
| *Salmonella* Oranienberg | NML |
| *Salmonella* Panama | NML |
| *Salmonella* Paratyphi A | CPHL |
| *Salmonella* Paratyphi B | CPHL |
| *Salmonella* Thompson | NML |
| *Salmonella* Typhi | CPHL |
| *Salmonella* Typhimurium | CPHL |
| *Salmonella* Virchow | CPHL |
| *Shigella sonnei* | NML |
| *Staphylococcus aureus* | CPHL |

[1]Rafiq Ahmed, Health Canada (personal communication).

[2]NML, Enteric Disease Division, Enteric and Bacteriology Program, National Microbiology Laboratory Health Canada, Winnipeg, Manitoba, Canada; CPHL, Central Public Health Laboratory, Colindale, UK.

that have the necessary equipment to perform this technique. Standardization of all experimental parameters, including bacterial culture conditions, is necessary to achieve reproducibility, and it may be necessary to run the same sample three times to ensure reproducibility. The results of Graves and Haystead (2002) suggest that 2-DE may find applicability in resolving differences between very closely related strains, when such differences may be anticipated from epidemiogical findings.

*MALDI–TOF mass spectroscopy*
Mass spectroscopy is an important tool in the detection and identification of proteins. Modern techniques such as MALDI (matrix-assisted laser desorption/ionization) and MALDI time-of-flight mass spectroscopy (also termed MALDI–TOF MS) allows for the generation of large molecular ions. These can then be analyzed using time-of-flight (TOF) mass spectroscopy. MALDI–TOF MS, developed in 1988 by Tanaka *et al.* and Karas *et al.*, involves an UV-light absorbing matrix and a biomolecule that is irradiated by a nanosecond laser pulse (Figure 4.1A). As the energy from the laser hits the matrix, biomolecules become ionized and can be accelerated within an electric field (Figure 4.1B). In MALDI–TOF MS, most of the energy generated by the laser is absorbed by the matrix. This helps to prevent fragmentation of the biomolecules of interest. The ionized biomolecules are accelerated in an electric field and enter the flight tube (Figure 4.1B). Separation is done under vacuum. Mass-to-charge ratios are used to separate the molecules which arrive at a detector at different times (t) [A and B are calibration parameters]. Each molecule then gives a distinct signal which can be measured. Proteins, peptides, oligosaccharides and even nucleic acids can be detected and characterized by this method.

The use of MALDI–TOF MS has allowed for the rapid determination, along with great accuracy and precision, of molecular weights of individual proteins within a complex mixture of macromolecules (Allmaier *et al.*, 1995; Kaufmann, 1995). In fact, preparations of whole bacterial organisms are possible, with distinct and reproducible mass spectra being generated (Cain *et al.*, 1994; Conway *et al.*, 2001; Holland *et al.*, 1996; Krishnamurthy and Ross, 1996; Wang *et al.*, 1998; Welham *et al.*, 1998). In practice, a suspension of whole organisms is washed, co-crystallized with a UV-absorbing organic acid matrix, and then the analyte-matrix preparation is irradiated with a UV laser, which ablates the matrix and volatizes and ionizes the analytes. The mixture of ionized analytes is accelerated through an electric field, and time-of-flight in a 1- to 2-meter-long vacuum tube is recorded. The mass of an analyte is proportional to the square of its flight time. The resulting

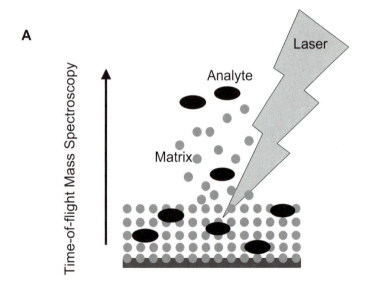

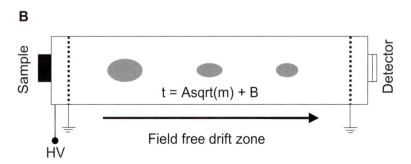

**Figure 4.1** Matrix-assisted laser desorption, time of flight (MALDI–TOF) mass spectroscopy. (A) MALDI depiction showing the biomolecules becoming ionized from the energy of the laser. (B) Ionized molecules are separated based on mass-to-charge ratios and their signals are detected.

complex spectra usually comprises over 35 peaks that represent proteins up to 50 kDa in mass and, depending on the analyses, greater than 50 kDa. The ability to directly study preparations of whole organisms permits a new approach to characterizing bacterial isolates at the level of proteome expression.

A recent study used MALDI-Tof to analyze isolates of *Enterobacteriaceae*, with particular emphasis on a well-characterized set of genotypically diverse *E. coli* isolates (Smole *et al.*, 2002). Specific digestion experiments established that the observed macromolecules represent primarily intracellular proteins. The patterns of mass spectrometric biomarkers observed for each isolate were evaluated by cluster analysis, which yielded a dendrogram that was essentially indistinguishable from that previously obtained using independent, extensively validated methods such as multilocus enzyme electrophoresis (MLEE) and ribotyping (Maslow *et al.*, 1995). One of the major advantages of this technology is its short measuring time (minutes), combined with relatively little sample requirements (picomole range). The development of MALDI-TOF mass spectrometry methods for the characterization of bacteria has been reviewed (Lay, 2001). While molecular biology has benefited from the development of DNA and protein microarrays, there is a need to advance protein analysis. The use of MALDI–TOF MS in the era of proteomics will have a tremendous impact.

### Hemagglutination

Hemagglutination of erythrocytes from different animal species has been commonly used as a method for the rapid differentiation of bacteria with different adherence characteristics. Erythrocytes obtained from humans, rabbits, guinea-pigs, rats, mice, sheep, chickens, and cattle differ in their content of surface receptors, especially glycoproteins and glycolipids. Hemagglutination assays are performed by mixing a standardized concentration of washed bacterial suspension with washed erythrocytes in either a slide agglutination test or microtitre hemagglutination assay (Qadri *et al.*, 1994). The amount of agglutination is read visually and quantitated by comparison with known controls. Specific sugars, oligosaccharide, or glycoprotein inhibitors can be used to inhibit the hemagglutination reaction and provide more information on the ligand-receptor interaction. This has been useful for elucidating the nature of different fimbrial colonization factors of enterotoxigenic *E. coli* strains (Evans *et al.*, 1980), but resulted in patterns too complex for useful subtyping of enteroaggregative *E. coli* strains (Qadri *et al.*, 1994). Though the method is rapid and simple to perform, it has largely been replaced by more specific methods that characterize the precise nature of the adhesin(s) present in a particular strain, when such methods are available.

### Multi-locus enzyme electrophoresis

Multi-locus enzyme electrophoresis (MLEE or MEE) assesses charge differences in housekeeping enzymes that represent alleles of each particular enzyme, a property scored as differences in the mobility of the enzyme. Such mobility changes, for example, can result from the mutation of a single amino

acid. Since genes encoding cytoplasmic housekeeping enzymes are not subject to diversifying selection, the rate of change of each gene is thought to be relatively constant, though the genes for different enzymes may change at somewhat different rates (Selander *et al.*, 1986).

In MLEE, bacteria are lysed without denaturation and the cytoplasmic enzymes are separated by electrophoresis on starch gels. Each enzyme is then visualized by developing the gel with the appropriate enzyme substrate and indicator dye. The distance of migration of each enzyme is determined and assigned a numerical designation corresponding to each allele (Figure 4.2). All enzymes migrating the same distance (therefore with identical alleles) are assumed to be the products of identical genes. Groups of strains with identical alleles for all enzymes tested are assigned to the same electrophoretic type (ET), while strains having different alleles are classed as different ETs. Genetic differences between strains with different ETs can be calculated by the number of allelic differences. When applied to bacterial populations, these differences can be presented as a dendrogram.

In rare cases, one or more of the assumptions underlying MLEE may not hold true. For instance, for some enzymes, an identical migration distance may not mean the enzymes are identical, as compensating mutations may be present. It is possible that, for some organisms, many enzymes may have null alleles, requiring that a large number of enzymes be tested to obtain phylogenetically valid results. This increases both the number of tests required and the time to completion. However, relatively large numbers of strains can be tested. For

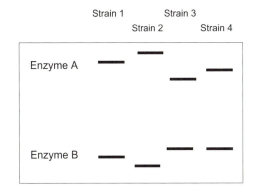

| | Test strain number | | | |
|---|---|---|---|---|
| Enzyme | 1 | 2 | 3 | 4 |
| A | 2 | 1 | 4 | 3 |
| B | 3 | 4 | 2 | 2 |

**Figure 4.2** Multi-locus enzyme electrophoresis. Housekeeping enzymes are separated based on their amino acid composition (size, upper panel) followed by staining with the appropriate substrate. The electrophoretic type is assigned based on the comparison with the mobility to a reference one. Any enzyme running at a different mobility is assigned a different number.

these reasons, MLEE is most often used to assess the population structure and phylogenetic relationships among food- and water-borne bacteria. The development of MLEE has provided both a strong theoretical basis and experimental validation for the development of multi-locus sequence typing methods (discussed elsewhere in this chapter) that are beginning to find application as rapid typing methods.

## Nucleic acid typing

*Plasmid profiling*
Of all the nucleic acid typing methods that currently exist, plasmid DNA profiling is perhaps the oldest genotypic method. It is the only nucleic acid method that analyses extra-chromosomal genetic elements. The method is relatively quick and easy to perform by most laboratories. Because of the nature of plasmids (small, usually supercoiled, circular), methods that take advantage of the biochemistry of plasmid biology to isolate them from the chromosome exist for all foodborne bacterial pathogens.

In plasmid profiling, plasmids are isolated from a bacterial isolate grown in pure culture. The plasmids are then separated electrophoretically (usually agarose gels are used) based on their size and number (Figure 4.3). It is important to note that plasmids that appear to be the same size on an agarose

gel may in fact be different in nucleotide sequences. As such, plasmid profiling is usually accompanied by restriction endonuclease digestion (see PCR-RFLP later in this chapter) prior to separation on a gel. While it appears that plasmid profiling is becoming less common in epidemiological investigations, it has been successfully used in several investigations (Eisgruber *et al.*, 1995; Kumao *et al.*, 2002; Sorum *et al.*, 1990).

## Genotypic (chromosomal) typing

### Amplification-based methods

*Polymerase chain reaction (gene specific detection)*
Both typing and subtyping may be carried out by PCR for specific genes or gene variants, though in most cases only typing in its crudest form is possible. Rather than go into detail here, a few examples will be given (Figure 4.4). The reader is referred to the extensive literature on the topic for further information.

*E. coli* pathotypes have very different disease manifestations and, for the most part, different sets of virulence genes (Table 4.2). Elucidation of the virulence gene sets within each isolate of importance is important for the designation of these pathotypes, though the demonstration of the serotype is often necessary for complete characterization (Karmali *et al.*, 2003). PCR, multiplex PCR, real-time PCR, and multiplex real-time PCR methods have been developed for the identification of specific genes and gene variants (Davis *et al.*, 2003; Fukushima *et al.*, 2003; Silviera *et al.*, 2001; Wang *et al.*, 2002). PCR tar-

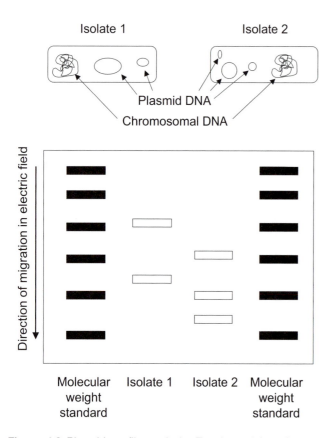

**Figure 4.3** Plasmid profile analysis. Two bacterial strains are shown, each having a different number of plasmids. Plasmids are isolated and separated on an agarose gel (lower panel) based on size and number.

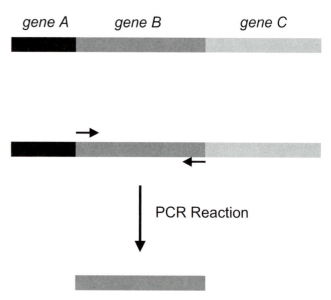

**Figure 4.4** Polymerase chain reaction. Gene specific primers are used to amplify a region of interest. In this example, the desired target is *gene B*. PCR amplification conditions are arranged so that the primers will only bind to the *gene B* complementary region and will allow the amplification of the region.

**Table 4.2** Partial list of foodborne bacteria with examples of some of their characteristic genes

| Organism | Pathotype/serovar | Gene profile | Additional useful information |
|---|---|---|---|
| *Campylobacter* | | *cdtA, B, C*; virulence plasmid genes (roles currently not well characterized) | Species, capsule and glycosylation genes, serotype |
| *Escherichia coli*[1] | EHEC (seropathotypeA), *Shigella dysenteriae* type I | *stx1 + stx2, stx2, eae, ehx, espP, katP* | Serotype |
| | EHEC (seropathotype B) | *stx1* or *stx2, eae, (ehx), (espP), (katP)* | Serotype, *ureC* |
| | EHEC (seropathotype C) | *(stx1), stx2, (eae), (ehx), (espP), (katP)* | Serotype, *ureC* |
| | STEC (seropathotype D) | *stx1, stx2,* or both; *(eae), (ehx), (espP), (katP)* | Serotype |
| | STEC (seropathotype E) | *stx1, stx2,* or both; *ehx, espP* | Serotype |
| | DAEC | *afaA, aida1, daaC, daaE,* | Diffuse adherence phenotype |
| | EPEC | *bfpA, eae* without *stx, espC, perA,* EAF region | Serotype, lack of virulence genes for other pathotypes |
| | ETEC | *est, elt* | Adhesin genes |
| | EIEC/*Shigella* | *invX, ipaC, ipaH,* ial | Serotype, plasmid genes, PAI genes |
| | EAggEC | *aggA, aggR, afa1* | Aggregative phenotype; EAST 1 toxin |
| | UPEC/ExPEC | *hlyA, cnf1* | Serotype, adhesin genes |
| | neonatal meningitis strains | *ibe, iucD,* K1 capsule genes, *neuC* | Serotype |
| *Listeria monocytogenes* | | *actA, inlA, inlB, hlyA* | Serotype |
| *Salmonella* | Typhimurium | *gogB, nanH, sodCI, sodCIII, sspH1, sspH2* | Serotype, presence and expression of *sopE* |
| | Typhi | *tviA, vexA* | Adhesin genes |
| | Enteritidis | *sefA, sefD* | Presence and expression of *sopE* |
| *Vibrio cholerae* | Pandemic strains | *ctx* | Serotype O1, 0139 |
| *Vibrio parahaemolyticus* | | *tdh, trh* | Serotype |

The EHEC and STEC seropathotypes refer to the VTEC seropathotypes defined by Karmali *et al.* (2003). Genes in brackets represent those that are present, but not found in most strains of a particular seropathotype. The designation *ehx* is used here for the EHEC hemolysin to avoid confusion with other hemolysins, especially the *E. coli* alpha-hemolysin (*hlyA*).

geting the *stx* genes provided useful subtyping information during the *E. coli* O157:H7 outbreak in Walkerton, Ontario. The *stx2* genotype that was characterized in the outbreak strain was a rare genotype in Canada, facilitating identification of the outbreak strain during screening of patient isolates. Note that a large, comprehensive database of *E. coli* O157:H7 *stx* genotypes was required for this information to be useful. Detection of the *sopE* gene in *Salmonella* serovars may allow the detection of particularly virulent strains within each serovar (Ehrbar *et al.*, 2002; Mirold *et al.*, 2001). For the most part, methods for gene detection using PCR do not subtype strains below the serotype level (Table 4.2).

*Non-specific amplification methods (RAPD)*
RAPD, also known as randomly amplified polymorphic DNA, is a typing technique based on the PCR reaction (Williams *et al.*, 1990). PCR in the context of RAPD analysis has two major differences. The first is that the primers used are very short (usually 9- or 10-mers). The sequences of the primers are random and, as such, no prior knowledge of target nucleic acid is required for RAPD analyses to be done. As the primers are short, they should be able to bind many genomic sites that

are complementary in sequence found throughout the bacterial genome. The second important aspect to RAPD analysis is the use of relatively low annealing temperatures. This takes advantage of the fact that a single primer with perhaps no or little known homology to a genome can anneal at many sites (even if there are mismatches). The RAPD reaction is set up to allow the amplification of fragments that are in the 200 to 2000 bp range (Figure 4.5).

RAPD analysis can be useful in outbreak investigations due to its rapid nature. Band differences amongst isolates can be attributed to deletion of priming sites, insertion of DNA elements (such as transposons) that move priming sites too far from each other, or perhaps insertions/deletions in the genome that change the size of the DNA fragment that could be amplified under the conditions used. An attractive feature of RAPD is the fact that it can be done from a bacterial isolate grown on an agar medium (plate) (Mazurier *et al.*, 1992). The biggest disadvantage of RAPD analyses stems from the random priming events, and as such, reproducibility issues often arise. The RAPD method has been used with many foodborne pathogens, an example of which can be seen in Table 4.3.

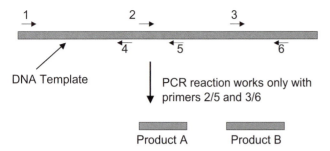

**Figure 4.5** Randomly amplified polymorphic DNA (RAPD). A nonspecific, short primer is used to increase its chances of binding the genome of a bacterial isolate. PCR amplification conditions are set so that amplicons of a desired size are obtained. In this example, the PCR conditions have been set to allow the amplification of the genomic sequences contained within primer binding sites 2 and 5, and 3 and 6.

## REP-PCR (VNTR, PATS, ERIC)

Repetitive DNA sequence elements of different types are universally found in bacterial genomes. Among the elements that have been used for PCR subtyping are repetitive extragenic palindromic (REP) elements (Figure 4.6), enterobacterial repetitive intergenic consensus (ERIC) sequences, *Salmonella* serotype Enteritidis repeat element (SERE), and BOX elements (Hulton *et al.*, 1991; Rajashekara *et al.*, 1998; Versalovic *et al.*, 1991; Versalovic *et al.*, 1994). REP and ERIC

elements are widespread in Gram-negative bacteria, and have been used extensively to subtype strains. A number of PCR primers have been constructed that are homologous to REP and ERIC sequences (Versalovic *et al.*, 1991). Single primers or combination of primers (e.g., REP1R-I, REP2-I, REP2-D; ERIC1R, ERIC2) can be used in PCR reactions to amplify sequences where the distance between elements is within the limits of polymerase extension (~5 kb). Because different strains will have elements located within different intergenic regions throughout the chromosome, each pattern produced is expected to be characteristic of the strain. The complexity of patterns can be adjusted by using different combinations of primers, and subtyping assays are normally developed after testing different primer combinations to determine which gives the appropriate discriminatory power.

Though ERIC PCR can be done under high stringency conditions, the data obtained from computer analysis of banding patterns are still not sufficiently reproducible for comparison of results at different times or for the development of continuous databases (Meacham *et al.*, 2003). There is considerable day-to-day and cycler-to-cycler variation in results (Johnson and Clabots, 2000; Johnson and O'Brian, 2000). This is a problem common to all PCR assays based on repetitive consensus sequence amplification (Tyler *et al.*, 1997). Despite this, repetitive element PCR has been used successful-

**Table 4.3** General comments regarding selected typing methods[1]

| Method | Typeability | Reproducibility | Discrimination | Interpretation[2] | Ease of use | Cost[3] |
|---|---|---|---|---|---|---|
| AFLP | All | Excellent | Excellent | Good | Difficult | 25.00 |
| Biotyping | All | Good | Very good | Easy | Easy | 15.00 |
| DNA microarrays | All | Good | Very good | Difficult | Difficult | 100.00[4] |
| FAME | All | Very good | Poor to good | Difficult | Difficult | 6.00 |
| Hemagglutination | Variable[5] | Good | Good | Good | Medium | 1.00 |
| MALDI–ToF | All | Very good | Excellent | Easy | Easy | 300.00[6] |
| MLEE | All | Good | Very good | Difficult | Medium | |
| MLST | All | Excellent | Very good | Easy | Easy | 250.00[7] |
| PFGE | Variable | Excellent | Excellent | Easy | Medium | 30.00 |
| Plasmid analysis | Variable | Fair to good | Good | Difficult | Medium | 5.00 |
| PCR-typing | All | Very good | Good | Easy | Easy | 5.00 |
| PCR-RFLP | Variable | Excellent | Good | Easy | Medium | 17.00 |
| Phage typing | Variable | Very good | Good | Easy | Medium | 1.00 |
| RAPD | All | Fair | Fair | Difficult | Easy | 15.00 |
| REP-PCR | All | Very good | Very good | Easy | Easy | 15.00 |
| Ribotyping | All | Excellent | Good | Easy | Easy[8] | 50.00 |
| Serology | Variable | Good | Fair | Good | Medium | 15.00 |
| Toxin/bacteriocin typing | Variable | Good | Fair | Good | Medium | |
| 2-D gels | All | Very good | Good | Difficult | Difficult | 35.00 |

[1]Adapted from Olive and Bean (1999), Farber (1996), Tyler and Farber (2003), and also based on the experience of the authors.
[2]Ease of data interpretation using a particular method.
[3]Approximate costs (Canadian currency) per isolate analyzed.
[4]Assumes home-made array platform based on PCR amplicons being used as features.
[5]Not all bacterial isolates may be typed using the method.
[6]Includes MALDI analysis.
[7]Includes seven-gene set, plus double-stranded sequencing of amplicons.
[8]Ribotyping done using an automated riboprinter.

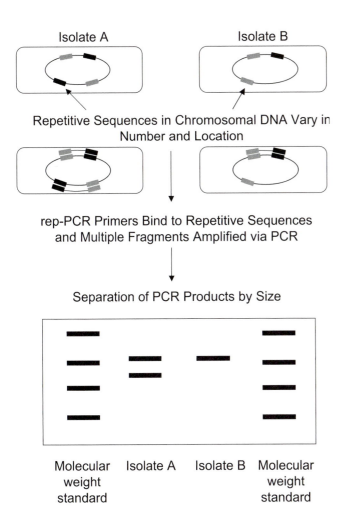

**Figure 4.6** Repetitive extragenic palindromic polymerase chain reaction (REP-PCR). Single primers (or combination of primers) are used to amplify regions of genomic DNA located between repetitive elements. The primers bind the repetitive elements (upper panel) and amplification products are usually separated on agarose gels (lower panel). Conditions are set so that non-related strains generate unique patterns.

ly for characterizing a hospital outbreak of *S*. Infantis caused by turkey dinners (Johnson *et al.*, 2001), for characterizing *V. cholerae* strains involved in a recent epidemic in Ukraine (Clark *et al.*, 1998), and for subtyping *V. parahemolyticus* strains causing an outbreak associated with oysters on Canada's west coast (Marshall *et al.*, 1999). Fingerprinting with ERIC primers was also found to allow typing of *Salmonella* to the serotype level (Van Lith and Aarts, 1994). However, ERIC PCR does not discriminate well among *S*. Enteritidis strains (Clark *et al.*, 2003b; Lopez-Molina *et al.*, 1998), supporting the concept that subtyping methods must be validated for each pathogen for which they are to be used. Profiles obtained from repetitive and arbitrary primer PCR may not match types based on other molecular typing or subtyping methods (Iriarte and Owen, 1996b), further complicating the interpretation of results and suggesting that PCR subtyping methods are best used in conjunction with other molecular typing protocols.

Polymorphic amplified typing sequences (PATS) analysis is a recently developed method for subtyping *E. coli* O157:

H7 (Kudva *et al.*, 2002). The method uses PCR primer sets complementary to chromosomal regions flanking *Xba*I restriction sites in *E. coli* chromosomes. This method therefore analyses the same sites as PFGE, but is not capable of assessing the presence of insertions or deletions near these sites. It is therefore currently less discriminatory than PFGE. Though the method is still under development, it does represent an approach that may have broad applicability.

Variable number of tandem repeat (VNTR) and multiple-locus VNTR analysis (MLVA) methods are novel subtyping methods that have been made possible by data gathered from genomic sequencing projects (Jansen *et al.*, 2002). These methods take advantage of DNA elements repeated in tandem within bacterial chromosomes, and are based on the observation that individual strains often carry the same elements with different copy numbers. Tandem repeats are found in genome sequences with software such as the tandem repeat finder program (http://tandem.biomath.mssm.edu/trf/trf.html). PCR primers capable of amplifying the region containing the tandem repeat unit are designed and used to amplify sequence from isolates of interest. The length of the DNA repeat unit is assessed on agarose gels or by separation of labeled amplicon using capillary electrophoresis. Results may be confirmed by DNA sequencing. This allows the determination of any mutations introduced into the region, as well as the number of repeat units. MLVA has recently been used to distinguish outbreak and sporadic *E. coli* O157:H7 isolates (Noller *et al.*, 2003a), to genotype *C. jejuni* (Schouls *et al.*, 2003), and to type *Salmonella* Typhi isolates (Liu *et al.*, 2003). In one study, VNTR of *Salmonella* Typhimurium DT104 strains provided improved discrimination over other methods such as *Xba*I pulsed-field gel electrophoresis, amplified fragment length polymorphism analysis, integron-cassette profiles and gene PCR of *intI1*, *qacE1*, *sulI1*, and *floR* (Lindstedt *et al.*, 2003). The high discriminatory power of VNTR and MLVA for many organisms may make them the methods of choice for trace-back investigations.

## Restriction endonuclease-based methods

### Pulsed-field gel electrophoresis

Pulsed-field gel electrophoresis (PFGE) has been the workhorse of molecular subtyping since it was found useful for the analysis of a large outbreak of *E. coli* O157:H7 in the western USA (Barrett *et al*, 1994), and it was subsequently selected as the method around which PulseNet would initially be developed (Swaminathan *et al.*, 2001). There are many PFGE methods for each organism studied, differing in the restriction enzymes, pulse conditions/switch times, run times, and size standards used. This is partly a reflection of the rapid development of PFGE methods independently in many laboratories around the world, and partly a result of the desire to use PFGE to achieve different ends. For some scientific applications, such as genome sizing or mapping, the most important parameter may be resolution of all bands produced (Chang and Taylor,

1990). Rapid turnaround time and reproducibility of results using a standardized method is often of higher priority for public health investigations (Ribot *et al.*, 2001).

PFGE methods are designed to restrict intact chromosomal DNA using rare cutting enzymes that optimally produce between 10 and 20 well-resolved bands between 10 to 800 kb that can be easily scored and compared to restriction patterns from other organisms, especially those that appear to have closely related patterns (Figure 4.7). Since most methods for DNA isolation result in shearing of the DNA, intact chromosomal DNA is obtained by embedding standard concentrations of bacteria in agarose plugs and digesting away all cellular components except DNA using enzymes (e.g., lysozyme and proteinase K) and detergent. Digestion products, detergent, and other contaminants are washed out of the plug. Restriction buffer and enzymes are then added to produce fragments of chromosomal DNA. The restriction enzymes used are usually chosen empirically taking into account the G + C content of the DNA under study, and most often recognize 8-base or 6-base sequences. Enzymes commonly used are *Xba*I, *Bln*I, *Spe*I, *Not*I, *Sma*I, and *Apa*I, among others.

Since the sizes of the DNA fragments produced are too large to move easily through pores of the agarose gel, PFGE methods depend on periodic inversion of the electric field to allow reorientation of strands trapped on agarose pieces

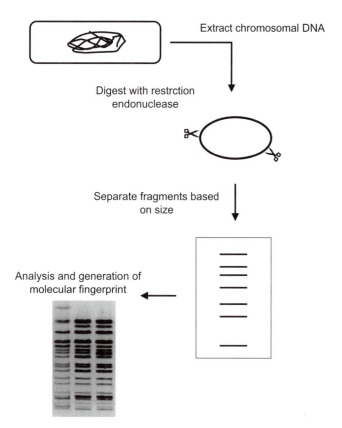

Extract chromosomal DNA

Digest with restrction endonuclease

Separate fragments based on size

Analysis and generation of molecular fingerprint

**Figure 4.7** Pulsed-field gel electrophoresis (PFGE). Chromosomal DNA is prepared in agarose plugs and digested *in situ*. Restriction endonucleases used are rare-cutters, generating 10–15 large bands that are then separated based on size.

(Swaminathan and Matar, 1993). Contour-clamped homogenous electric field (CHEF) electrophoresis, the most common method currently in use, uses a hexagonal array of electrodes to generate uniform electric fields at 120° to each other. Under these conditions, DNA fragments move in a straight line with no distortion. Careful control of input DNA (input bacterial colony-forming units), as well as standardization of other parameters, results in very thin, flat bands with superb resolution between bands (Ribot *et al.*, 2001).

After electrophoresis, gels are stained with ethidium bromide and visualized using UV light. Gel images are either captured using CCD camera systems or photographed and scanned. When stored as TIFF files, these gel images can be transferred electronically to other locations without analysis or imported into analytic software such as Molecular Analyst, Bionumerics, or similar programs for analysis. Many of the analytic programs available offer the option of sharing analyzed gels and associated organism or epidemiologic information with other users of the program. Each gel is normalized against a common standard pattern obtained from a single lane of the size standard. DNA band patterns are converted to densitometric scans by the program, and the centers of bands identified as the peak of density and confirmed visually against photographs of the original gel. Individual patterns can be compared against a library of patterns by construction of a dendrogram or by scanning for similar pattern with a high level of identity. Pattern similarities or differences are best confirmed by eye.

There is often confusion over the interpretation of PFGE pattern similarities. The most often quoted interpretive criteria are those of Tenover and colleagues (Tenover *et al.*, 1995), which were originally intended to be used by clinical microbiologists in hospital laboratories for the examination of small sets of isolates related to putative outbreaks of disease. These criteria assume epidemiologic relatedness of strains and assess the likelihood that strains are related and the epidemiologic interpretation given differences of 0 bands, 2–3 bands, 4–6 bands, and more than seven bands. Other criteria must be used for organisms with limited diversity, for larger number of isolates, or isolates collected over extended periods. Under many circumstances, a one band difference between two *E. coli* O157:H7 isolates is enough to classify them as unlikely to be part of the same outbreak (Barrett *et al.*, 1994), unless there is compelling epidemiologic evidence that they should be included. It must be noted that PFGE patterns of related strains have been observed to change during the course of an outbreak (Steinbruckner *et al.*, 2001), with repeated subculturing and/or prolonged storage (Iguchi *et al.*, 2002; On, 1998), or during passage through the intestine (Hänninen *et al.*, 1999). Furthermore, differences in the PFGE patterns of *E. coli* O157:H7 strains were found to be predominantly due to insertions or deletions of DNA in O-islands, not single-nucleotide polymorphisms (Kudva *et al.*, 2002).

PFGE may have the optimal typeability, resolution, and discriminatory power for subtyping *E. coli* O157:H7, especially in outbreak situations (Izumiya, et al., 1977; Johnson et al., 1995), but also for surveillance (Lee et al., 1996; Rice et al., 1999) and trace-back activities (Louie et al., 1999). PFGE was found to be superior to other typing methods for C. jejuni strain discrimination in an outbreak situation (Fitzgerald et al., 2001), but may not be appropriate for routine typing of this organism (Hedberg et al., 2001). Salmonella enterica serotype Enteritidis is a highly clonal organism that is not readily resolved using most molecular methods, including PFGE. In this case, a recently developed riboyping method using both PstI and SphI appears to provide data that is more epidemiologically relevant (Clark et al., 2003b; Landeras and Mendoza, 1998). The utility of PFGE must therefore be evaluated for each organism for which it is to be used.

### PCR-RLFP

PCR, followed by a restriction digest of the amplicon, has been shown to be a useful tool in the molecular characterization of foodborne pathogens (Arbeit, 1995). Restriction fragment length polymorhism (RFLP) analysis of individual genes has been used routinely for confirmation of PCR product identity, for subtyping gene variants, and for analysis of the variability of strains carrying specific genes. After PCR amplification, amplicons are cut using a restriction enzyme that produces fragments of known size based on the DNA sequence of the gene. Variants can arise when the DNA sequence of variant genes or intergenic regions is different from the reference sequence, adding or removing restriction enzyme cut sites (Figure 4.8). Rather than providing an exhaustive listing of PCR-RFLP methods, only a few examples are listed here.

PCR ribotyping has been used, targeting the 16S and 23S rRNA spacer region of bacterial pathogens (Kostman et al., 1995). While this method was somewhat useful for subtyping some serovars of Salmonella, others produced only a single PCR-ribotype (Lagatolla et al., 1996). Though promising for some organisms, this PCR-ribotyping has not been widely adopted for suptyping enteric bacteria. One reason for this may be that PCR-ribotyping is generally less discriminatory than conventional ribotyping (Severino et al., 1999). Sontakke and Farber (1995) have shown that polymorphisms existed in amplified DNA products to enable distinction of various strains of Listeria spp. and between serotypes of L. monocytogenes. Results were quicker and the method less challenging technically than PFGE.

PCR combined with RFLP analysis of individual rRNA genes has also been investigated. When applied to C. jejuni, this method had low discriminatory power (Iriarte and Owen, 1996a). The presence of multiple copies of rRNA genes within individual strains, some of which may exhibit sequence variation, further complicates the interpretation of this kind of analysis.

Flagellar gene restriction fragment length polymorphism (fla-RFLP), a type of PCR-RFLP method, appears to be particularly effective for subtyping C. jejuni and C. coli (Clark

et al., 2003b; Petersen and Newell, 2001; Petersen and On, 2000). Either the complete flaA gene or both the flaA and flaB genes are amplified, the product is restricted with a restriction enzyme (AluI, DdeI, EcoRI, HinfI, PstI), and the resulting fragments are separated on agarose or polyacrylamide gels and compared (Petersen and Newell, 2001). Campylobacter fla-RFLP does not correlate well with heat-labile (HL) serotyping (Burnens et al., 1995; Nachamkin et al., 1996). The known instability of the flagellin locus may contribute to the variability seen when this method is used (Harrington et al., 1997). More than one Campylobacter fla-RFLP method currently exists; standardization of methods would be necessary if fla-RFLP were to be used as a typing method by bacterial surveillance networks. Sequencing of the flagellin locus may provide an alternative method that is more portable and more conducive to the construction of large databases than fla-RFLP.

Though not strictly used for strain subtyping, PCR-RFLP of stx genes carried by Shiga-toxigenic E. coli (STEC) provides data useful in epidemiological investigations. Both Stx1 and Stx2 represent families of genes with related sequences, with sequence variation thought to arise mainly through recombi-

Sequence Specific Amplification via PCR Followed by Restriction Endonuclease Digestion of PCR Product

Restriction Endonuclease Site

Separation of fragments by gel electrophoresis. In the example below, isolate B has lost the restriction endonuclease site.

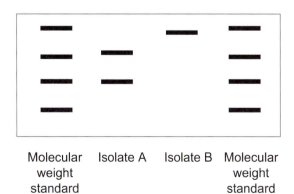

| Molecular weight standard | Isolate A | Isolate B | Molecular weight standard |

**Figure 4.8** Polymerase chain reaction followed by restriction fragment length polymorphism (PCR-RFLP). Genomic areas of interested are generated by using specific primers (see Figure 4.4). The resulting amplicon is then digested using a restriction endonuclease and the products are then separated on a gel.

nation of genes carried by lambdoid bacteriophages (Johansen *et al.*, 2001). Different genes appear to have different virulence properties, so that the determination of genotype is important for inferring the pathogenic potential of the strain. Various PCR-RFLP methods have been developed for detection and characterization of *stx* gene variants (Brett *et al.*, 2003; Eklund *et al.*, 2002; Gobius *et al.*, 2003; Osawa *et al.*, 2000; Tyler *et al.*, 1991; Zhang *et al.*, 2002).

This model for typing may be appropriate for the development of inexpensive PCR-RFLP methods for subtyping other organisms. For instance, *Salmonella* effector genes such as *sopB* and *sopE2* exhibit extensive sequence heterogeneity that may be useful for PCR-RFLP-based subtyping methods. Prager and colleagues (2000) used this sequence variability to develop a PCR-RFLP for the *sopB*, *sopD*, and *sopE1* genes. A limited analysis of the *sopB* PCR-RFLP suggested this method was useful for differentiating many *Salmonella* serotoypes. Though a PCR-RFLP was also developed for the *sopE1* gene (Prager *et al.*, 2003), this gene is present only in a small subset of *Salmonella* isolates, limiting this PCR-RFLP method as an epidemiologically useful tool. It would appear that the *sopE2* is more conserved in all *Salmonella* serovars (Mirold *et al.*, 2001), and therefore may provide a useful target for PCR-RFLP.

*AFLP*

Amplified fragment-length polymorphism (or AFLP) is a DNA molecular typing technique that detects restriction fragments of the genome using PCR amplification technology. Typically, the procedure is as follows (Figure 4.9A): the genome of a pathogen is digested using two restriction endonucleases (usually hexa and tetra-cutters), followed by ligation of double stranded adapters to the ends of the restricted fragments. Selective PCR amplification of a subset of the adapted restriction fragments follows and these amplified fragments are usually visualized on denaturing polyacrylamide gels either through autoradiographic or fluorescence methodologies.

The specific details of the AFLP procedure is outlined schematically in Figure 4.9B. In this example, complementary adaptors A and B, corresponding to restriction sites 1 and 2, respectively, are ligated to the restricted DNA. The new template is then subjected to pre-amplification. In the pre-amplification, primers 1 and 2, complementary to the adaptor sequences, have one additional base extending into the "unknown" region of DNA, in order to reduce the number of fragments to be amplified. In the example, primer 1 and primer 2 have the nucleotide G and A, respectively. Following preamplification, specific amplification occurs by using primers 3 and 4. These

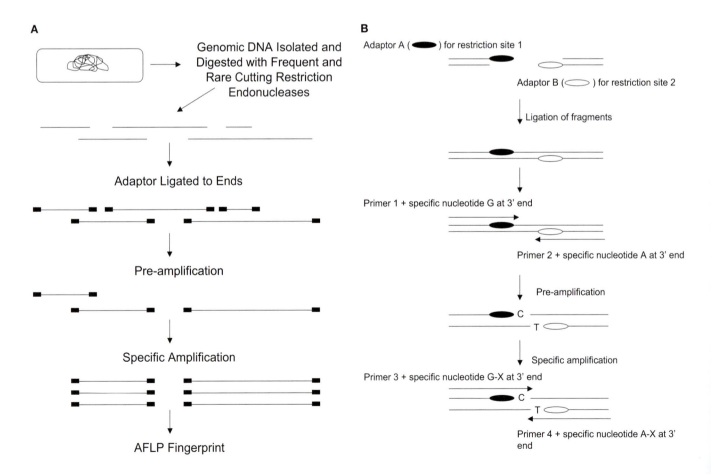

**Figure 4.9** Amplified fragment length polymorphism (AFLP). A. General schematic illustrating the AFLP procedure. B. Schematic depicting how the pre-amplification and the subsequent specific amplification of genomic fragments are done. No previous genomic DNA sequence information is required for the AFLP technique.

primers are specific for the preamplified products only, as they have the nucleotides G and A, respectively, in addition to a new nucleotide X (where X can be A, G, C, or T). Amplified products are visualized on gels. Alternatively, one of the primers (3 or 4) may be labeled with a fluorophore and the products visualized with an automated DNA sequencer.

In summary, AFLP technology uses selective amplification of a subset of the genomic restriction fragments. The sequence of the adapters (along with the restriction endonuclease site that is adjacent) acts as the primer binding site for amplification by PCR. Selective nucleotides are added to the extending restriction fragments at the 3′ end. As such, only this subset will be recognized by the selective amplification process. AFLP is a random amplification technique, however, its major difference when compared to RAPD, for example, is that it utilizes stringent PCR amplification conditions. The primers are usually 17–21 nucleotides in length and anneal perfectly to their complementary sequence target sites. The target sites are the adapter and restriction sites, as well as a small number of nucleotides found adjacent to the restriction recognition site.

When compared to RAPD or RFLP, AFLP provides equal or greatly enhanced performance in terms of reproducibility, resolution, and time efficiency. Probably the single greatest advantage of AFLP is its sensitivity to polymorphism detection at the total-genome level. With all of these assets, AFLP is becoming more popular as an investigative tool that gives insight into bacterial systematics and population genetics.

## Sequence-based typing

### Multi-locus sequence typing (MLST)

Multi-locus sequence typing, or MLST, is a nucleotide-based approach to molecular typing of bacterial pathogens. The technique is based on MLEE (see MLEE in this chapter). In the MLEE technique, bacterial isolates are characterized by the electrophoretic mobilities of 20 of their housekeeping enzymes (Selander *et al.*, 1994). Many pathogens have been successfully clustered by this method (Achtman, 1998) and it has provided valuable information that has greatly contributed to a better understanding of global epidemiology. However, the results obtained with this method are sometimes hard to compare between laboratories. As a solution to this problem, MLST was developed, to classify bacterial isolates directly from their nucleotide sequence (Maiden *et al.*, 1998). In the MLST method, internal fragments (approximately 450–500 bp) of approximately seven housekeeping genes are amplified and sequenced to determine the allelic profile of each isolate. For each housekeeping gene, the sequence differences found between all the isolates are each assigned a distinct allele (Figure 4.10). Each isolate is therefore unambiguously characterized by the combination of alleles for the seven housekeeping genes loci (Anon, 2001; Enright and Spratt, 1999; Maiden *et al.*, 1998). Since nucleotide sequences reveal all possible variation that might exist at a locus, MLST in comparison to MLEE allows a greater number of alleles per locus and thus a greater

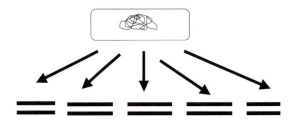

Amplify and Sequence Internal Regions of Housekeeping Genes. For Each Isolate, Assign Alleles for the Genes to Give Allelic Profile. Dendrograms are then Constructed.

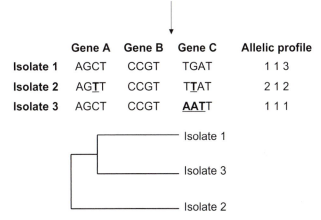

| | Gene A | Gene B | Gene C | Allelic profile |
|---|---|---|---|---|
| **Isolate 1** | AGCT | CCGT | TGAT | 1 1 3 |
| **Isolate 2** | AG**TT** | CCGT | T**T**AT | 2 1 2 |
| **Isolate 3** | AGCT | CCGT | **AAT**T | 1 1 1 |

**Figure 4.10** Multi-locus sequence typing (MLST). Genes of interest (in this example, housekeeping genes are used) and their DNA sequences are ascertained. Each gene is assigned a numerical allelic profile and compared to other strains. Strains showing differences (the number of differences does not matter) are given different allelic numbers. The allelic profile is a combination of several genes (middle panel). A dendrogram depicting the relationship of the strains is then generated using bioinformatic analyses (lower panel).

discriminatory potential between isolates than is provided with PFGE (Enright and Spratt, 1999). Given the fact that MLST is based on sequence data, this method is not only unambiguous, but also highly discriminatory. Other advantages include the ease of sequence data exchange between laboratories through the use of the Internet, making it a powerful tool for global epidemiology (Enright and Spratt, 1999; Maiden *et al.*, 1998). Housekeeping genes are not subject to unusual selective forces, and therefore diversification occurs slowly by the accumulation of neutral variations. As a result, they are conserved within a species (Slade, 1992). Rapid accumulation of variation within a clone makes it difficult to discern whether its descendants are derived from a common ancestor. Thus, studying long-term epidemiological questions using housekeeping gene data provides more reliable information about the relationship between isolates (Enright and Spratt, 1999). However, in short-term epidemiology, if the source of a sporadic case of, for example, listeriosis needs to be determined, more variation between isolates may be required to properly identify the outbreak strain. In the event that MLST itself is not sufficiently discriminatory in an outbreak setting, the inclusion of genetic (i.e., sequence) information from virulence genes may provide an appropriate level of discrimination. Surveying the variation seen in virulence genes may also allow a better understanding of the pathogenic potential of the organism.

MLST methods have been developed for several foodborne pathogens, including *C. jejuni* (Dingle *et al.*, 2001; Sails *et al.*, 2003), *Salmonella enterica* (Boyd *et al.*, 1996; Kotetishvili *et al*, 2002, Selander *et al.*, 1994), *E. coli* (Adiri *et al.*, 2003; Tarr *et al.*, 2002) and *L. monocytogenes* (Salcedo *et al.*, 2003; Revazishvili *et al.*, 2004; Zhang *et al.*, 2004), though these methods have rarely been used for subtyping. This approach is not useful for subtyping *E. coli* O157:H7 due to the clonality of this organism (Noller *et al.*, 2003b). An Internet based MLST database is available at www.mlst.net.

*Genomic sequencing and comparative genomics*
Genomics can be described as the investigations and understanding of complete genomes. It does not involve individual genes, but aims at treating the complete genome as a unit of investigation. Investigations into genomics must consider all the changes that may arise from an event that is occurring to the bacterium in question. Classical genomics revolves around the sequencing of genomes, followed by extensive bioinformatics analyses. The first free-living organism to be sequenced was *Haemophilus influenzae*, implicated in a range of diseases including otitis media and pneumonia (Fleischmann *et al.*, 1995). Since then, more than 100 microbial genomes have been published, and another 300 sequencing projects are currently under way (Nelson, 2003). The availability of a complete genome has allowed insight into gene transfer, virulence mechanism(s), effects of the environment on gene regulation and identification of metabolic pathways (Nelson, 2003). Perhaps the biggest advancement of microbial genome sequencing is that it allows for the attainment of genomes from unculturable organisms. Equally important, whole genome sequencing can shed insight into physiological phenomena seen in individual bacterial species, that are not definable by 16S rDNA analyses (Perna *et al.*, 2001).

Comparing gene (or protein) sequences in an attempt to ascertain functional (and evolutionary) significance has been used in the past. In the context of complete genome sequences, this application becomes more powerful. Comparative genomics, the analysis and comparison of genomes from different species, aims at understanding how species have evolved and at determining the function of regions of the genome (both coding and non-coding). By comparing and analysing a new genome with existing ones, much can be learned about genome products such as proteins without doing laboratory experiments to prove functionality.

Bioinformatics is a critical component of comparative genomics. Using computer programs, it is possible to align and compare multiple genomes and to examine regions that have (or may not have) similarity amongst them. Some of these sequence-similarity tools are accessible to the public over the Internet, including BLAST (available from the National Center for Biotechnology Information). Comparative genom-

ics has led to the development of functional genomics, which uses various methods to go beyond the direct sequence analyses of genomes. Functional genomics aims at using more experimental data (in contrast to raw genome sequence data). DNA microarray (see next section) plays an important role in addressing the post-genomics era, where genome sequences are becoming more available every day.

## Hybridization technologies

*DNA microarrays*
As the complexity of bacterial virulence has become better understood, it has become apparent that it is difficult to assess the full complement of virulence genes present in any individual strain. Early attempts to utilize hybridisation assays for demonstrating virulence gene sets only assessed a relatively few genes (Kuhnert *et al.*, 1997). In the last few years, DNA microarray technology has emerged as one of the most powerful tools in the biotechnology field. Microarrays (also known as biochips or gene chips) are a series of DNA molecules of known sequences (called features), fixed on a substrate (usually glass or nylon) at a defined location. Chemically modified microscope glass slides are the most commonly used support for high density DNA microarrays. The DNA molecules used to develop DNA arrays can be partial gene sequences generated by polymerase chain reaction (PCR), full-length cDNAs or oligonucleotides. The miniaturization behind this technique allows many thousand features to be deposited on a slide. As a result, DNA microarray technology is ideal to compare bacterial strains at a genomic level (genomotyping), do profiling expression studies on selected genes or even an entire genome at once. Applications for chip technology include gene discovery and expression, detection of mutations or polymorphisms, sequencing, as well as detection and molecular typing of pathogens (Call *et al.*, 2001; Call *et al.*, 2003; Chizhikov *et al.*, 2001; Dorrell *et al.*, 2001; Pollock, 2002).

In most applications, DNA arrays are used as a tool for studying differential gene expression using two fluorescent dyes (Cy3 and Cy5 are the most common), in which both experimental and control samples are hybridized to the same array (Figure 4.11). In the latter example, free nucleic acid samples whose identity and/or abundance are being detected, have been prepared from mRNA expressed in control and experimental cells. There are three main ways to prepare targets from RNA; 1) RNA can be labeled directly with the fluorophores Cy3 or Cy5; 2) a cDNA is produced with the Cy3 or Cy5 conjugated nucleotides being incorporated during the reverse transcriptase (RT) reaction; and 3) the dyes can be incorporated in a PCR assay (following the initial RT-reaction) to increase detection sensitivity. In other applications (genomotyping, polymorphism studies, etc.), targets can be made from chromosomal DNA or even from antibodies for protein arrays.

In microarray experiments, hybridization and washing are critical steps in generating high quality data. As glass is a non-

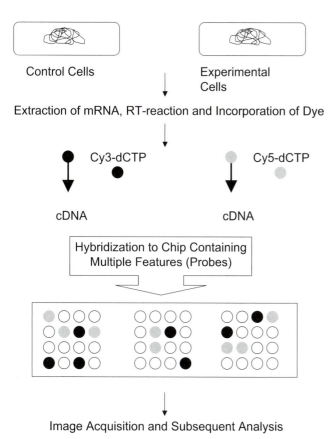

Control Cells          Experimental Cells

Extraction of mRNA, RT-reaction and Incorporation of Dye

Cy3-dCTP          Cy5-dCTP

cDNA          cDNA

Hybridization to Chip Containing Multiple Features (Probes)

Image Acquisition and Subsequent Analysis

**Figure 4.11** DNA microarray technology. In the example shown, identical cells are grown under different conditions (i.e., control and experimental). The messenger RNA (mRNA) is extracted from cells grown in both conditions, reverse transcribed, labeled with different fluorescent dyes (middle panel) and hybridzed to a platform containing features (DNA probes) that are specific for certain sequences (lower panel). Subsequent analyses using bioinformatics software(s) reveals those genes that are expressed in equal amounts under both conditions (white circles); genes that are expressed only in cells grown in experimental conditions (grey circles); and genes only expressed in control cells (black circles).

porous substrate, the hybridization volume can be kept to a minimum (8–100 μL), thus enhancing the kinetics of annealing targets to probes. The classical method of hybridization is usually done under a cover-slip. That is, the labeled RNA is pipetted on the printed slide, and a cover-slip is placed over the array(s). This setup allows for random and passive interactions between the probes and their target(s) to occur. Hybridization cassettes have been developed to maintain a saturated humidity, thereby avoiding evaporation of the targets. An alternative to this method is to use an automated hybridization station, where a specially designed cassette that covers the whole chip with the hybridization solution (~100 μL) is used. Automated hybridization stations have the advantage of automating all thermal and wash cycles for the hybridization of microarrays, thereby minimizing the handling of microarrays. This can substantially reduce human error. This approach is gaining in popularity since the circulation of fluid can be controlled to allow a better interaction between DNA molecules (i.e., probe and target). In order to optimize the hybridization process, strin-

gency can be modified by temperature variation, the use of different salt concentrations, or the addition of sodium dodecyl sulfate (SDS) in the washing solutions. The resulting fluorescence pattern is recorded by a two-color confocal scanner that generates light of the correct excitation wavelength (550 nm for Cy3; 650 nm for Cy5) for detection. The resulting image is then quantified and analysed.

The first eucaryotic chip was developed for *Saccharomyces cerevisiae* (Lashkari *et al.*, 1997). Sequences representing the entire genome (i.e., up to 2479 open reading frames) were spotted as a grid onto a single microscope slide. Following this publication, the use of microarray technology for the food industry was initiated. For example, some studies were done on the genetics of cereals such as rice, corn and soybean (McGonigle *et al.*, 2000; Yazaki *et al.*, 2000). Microarrays have also been developed for the comprehensive analysis of plant diseases (Schenk *et al.*, 2000), as well as major foodborne pathogens. For example, arrays for *L. monocytogenes*, *E. coli* O157:H7 and *C. jejuni* have been used successfully for detection and genomotyping (Call *et al.*, 2001; Call *et al.*, 2003; Dorell *et al.*, 2001). A virulence gene microarray has also been developed for the rapid identification of *E. coli* pathotypes (Bekal *et al.*, 2003) that may be present in foods.

Microarray technology is currently having its biggest impact on the understanding of gene function and expression. It is being used to complement and/or replace traditional, labor-intensive methods, such as Northern analysis. With the evolution of this technology, detections/typing and virulence factors analysis of foodborne pathogens will go a long way towards improving the safety of our food supply.

*Ribotyping (automated and traditional)*
Ribotyping is a method whereby nucleic acid probes are used to recognize ribosomal genes. In prokaryotic organisms, the ribosomal RNA consists of three major species: 5S, 16S, and 23S rRNA. The genes encoding the rRNAs are highly conserved and, for the most part, are present in *rrn* operons (continuous genes under the control of a common promoter). The *rrn* operon can be present in from 2 to 11 copies per genome, and its distribution can also vary from loci to loci. Ribotyping takes advantage of these two aspects.

Ribotyping examines RFLPs of rRNA genes within the chromosomes of bacteria (Grimont and Grimont, 1986; Stull *et al.*, 1988). Bacterial chromosomal DNA is purified, digested with one or more restriction endonucleases, separated on agarose gels, and then transferred to an appropriate membrane by Southern blotting. Immobilized chromosomal DNA fragments are probed with labeled DNA or RNA (Figure 4.12). The resulting banding pattern can be compared with profiles from other strains, allowing typing, subtyping, and phylogenetic analysis of bacterial strains and populations.

Many ribotyping methods have been developed for most or all of the common foodborne pathogens. Lack of standardization and common, centralized, standard databases and pattern nomenclatures is a major problem yet to be dealt with.

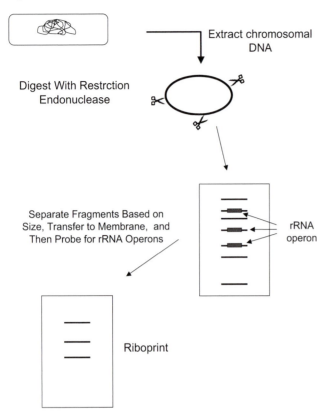

**Figure 4.12** Ribonucleic acid typing (ribotyping). Similar in certain aspects to PFGE (see Figure 4.7), the chromosome of a bacterial strain is digested using restriction endonucleases and separated based on size (automated ribotyping). The separated fragments are then transferred to a membrane and probed for the rRNA operons. As the number and location of the rRNA operons differ from one bacterial species to another (even within a bacterial genus), unique patterns are generated for non-similar isolates. Note: in manual ribotyping, the rRNA operons may be amplified by PCR followed by restriction endonuclease digestion and separation using gel electrophoresis.

Methods may differ in the restriction enzyme(s) and probes used, as well as the conditions for separation of DNA fragments on agarose gels. Probes can be the 16S or 23S rRNA genes, fragments amplified from within a conserved sequence of rRNA, or the entire 16S–5S–23S rRNA region; each probe produces data with different discriminatory power. Manual ribotyping is increasingly seen as slow, labor-intensive, and cumbersome, and also suffers from the difficulties in standardization common to all DNA fragment banding pattern methods. Despite this, the appropriateness of the discriminatory power of the method, its usefulness as an accessory typing or subtyping method, the existence of large ribotyping databases, and its value as a phylogenetically valid method all suggest that ribotyping will continue to be useful in the future.

Ribotyping has been used extensively for subtyping *C. jejuni* (Fitzgerald *et al.*, 1996; Jackson *et al.*, 1996; Wassenaar and Newell, 2000), *Salmonella* serotypes (Esteban *et al.*, 1993; Hilton and Penn, 1998;), *Shigella* (Coimbra *et al.*, 2001; Rolland *et al.*, 1998), *V. cholerae* (Karaolis *et al.*, 1994, Koblavi

*et al.*, 1990, and Popovic *et al.*, 1993) and *V. parahaemolyticus* (Marshall *et al.*, 1999). The method has been useful for discriminating among outbreak and non-outbreak strains, and between human and non-human strains of *Salmonella* (Baquar *et al.*, 1994; Landeras *et al.*, 1998) and other bacteria. The enzyme combinations used in ribotyping can be adjusted to achieve the required level of discrimination. Thus, ribotyping with a mixture of *Pst*I and *Sph*I has proved to be the most effective method for subtyping *Salmonella* Enteritidis (Clark *et al.*, 2003b; Landeras and Mendoza, 1998), a very clonal serovar of *Salmonella enterica*. As mentioned earlier, the selection of appropriate enzymes and conditions is critical for obtaining appropriate discrimination. While ribotyping was found not to be useful for subtyping *E. coli* O157:H7 in one study (Martin *et al.*, 1996), a second study on the same organism found that ribotyping had a discriminatory power similar to PFGE, and that a combination of the two methods provided optimal discrimination (Avery *et al.*, 2002).

Automated ribotyping uses the DuPont Riboprinter™ to perform bacterial cell lysis, restriction of DNA, gel electrophoresis, and analysis of results. Methods are standardized for all operators of the equipment and data can therefore be entered into a comprehensive database compiled from different sources. Run conditions may be different from those used in manual ribotyping. Though the interpretation of data obtained from automated ribotyping may be equivalent to that obtained by manual ribotyping, the data themselves may look quite different (Clark *et al.*, 2003a). In addition, the Riboprinter contains a program that groups closely related patterns. However, in our experience, the best discrimination is found when pattern differences are interpreted on the basis of single band differences. Automated ribotyping has proved useful for subtyping pathogenic *E. coli* strains (Clermont *et al.*, 2001), *L. monocytogenes* (Aarnisalo *et al.*, 2003; Inglis *et al.*, 2003b; Sauders *et al.*, 2003), and other foodborne bacteria. Because rDNA has some highly conserved regions across bacterial species, the use of a common, or universal probe (usually based from *E. coli*) is often used. The automated RiboPrinter® microbial characterization system (see www.qualicon.com) uses a 16S rRNA probe from *E. coli*. The system can analyse 32 strains within an 8 h working day, with complete results obtained in 16 h. Table 4.4 lists some examples of the various methods and their uses for characterization of foodborne bacterial pathogens.

## Epidemiology of foodborne diseases (databases and networks for the communication of molecular typing and subtyping results)

Though molecular typing and subtyping data are useful even when used only by a single laboratory, the value of these methods is greatly enhanced when data can be shared among a large number of laboratories. In the past, many methodologies for subtyping have been developed and have gained wide acceptance. However, results were not easily communicated among

**Table 4.4** Selected examples and applications of some molecular-based typing methods for foodborne pathogens

| Method | Organism | Purpose | Comments | Reference |
|---|---|---|---|---|
| AFLP | *Escherichia coli* | Evaluation of method | AFLP very good for large-scale screening due to speed and ease of performance over PFGE | Hahm *et al.* (2003) |
| DNA microarray | *Listeria moncytogenes* | Serotype/lineage-specific genome identification | Identification and screening for genetic markers from putative niches which are specific to certain serotypes | Call *et al.* (2003) |
| MALDI–ToF | *Escherichia coli* | Rapid diagnostic detection | Identification of unknown bacterial isolates based on similarities within protein biomarker databases | Conway *et al.* (2001) |
| MLST | *Campylobacter jejuni* | Evaluation of method | MLST less discriminatory than PFGE or Fla-SVR typing | Sails *et al.* (2003) |
| PFGE | *Shigella dysenteriae* | Investigation of intentional contamination of food | Outbreak linked to laboratory's stock culture, shown to contaminate pastries eaten by co-workers in a large medical center | Kolavic *et al.* (1997) |
| Plasmid analysis | *Salmonella enterica* serovar Oranienburg | Define limitations of chromosome-based and plasmid-based typing methods | Shown to be more discriminatory than PFGE or ribotyping | Kumao *et al.* (2002) |
| PCR-typing | *Listeria monocytogenes* | Identification/characterization of presumptive positive isolates from foods | PCR amplification of *hlyA* gene used to allow high throughput testing of unknown isolates | Blais *et al.* (2002) |
| PCR-RFLP | *Campylobacter jejuni* | Compare typeability, discriminatory ability, and inter-laboratory reproducibility of three flagellin PCR/RFLP (*fla* typing) methods | Inter-laboratory standardization of *fla* typing, using complete *flaA* gene followed by *Dde*I digestion | Harrington *et al.* (2003) |
| RAPD | *Listeria monocytogenes* | Compare food and clinical isolates | RAPD identified causative food item that resulted in listeriosis in patients | Farber *et al.* (2000) |
| REP-PCR | *Salmonella enterica* subsp. *Enterica* serovar Typhimurium | Novel VNTR method development | VNTR assay had greater resolution compared to PFGE, AFLP, and various integron-based PCR profiling | Lindstedt *et al.* (2003) |
| Ribotyping | *Clostridium perfringens* | Nursing home outbreak investigation | Ribotyping was more discriminatory than PFGE | Schalch *et al.* (2003) |

different groups due to the lack of a common nomenclature for designating subtype variants and the lack of common databases for providing a standard for comparison. Electronic data reporting from public health laboratories and the creation of comprehensive large-scale surveillance public health databases provides a solution to these problems, although the design and implementation of electronic data communication systems needs to be carefully considered (Bean and Martin, 2001).

In general, there appears to be two different kinds of solutions to the problem of creating a universal typing or subtyping system that is available for widespread use. These comprise a spectrum from the development of data repositories for specific data sets, for example GenBank for DNA sequence data, to the creation of laboratory networks for the communication of subtyping results obtained, using highly standardized methods and interpretive criteria such as PulseNet.

One of the common factors contributing to the success of these systems is the creation of centralized, comprehensive databases containing datasets that are as complete as possible. The Oxford *Campylobacter* multi-locus sequence type (MLST) and flagellar short variable region sequence database (*fla-svr*; http://mlst.zoo.ox.ac.uk) is an example of a very good curated database that can be accessed by users throughout the world, and provides a standard analytic framework for determining the status of isolates within the larger *Campylobacter* population. This includes a standardized nomenclature for describing subtypes and standard primer sets for obtaining DNA fragments for sequence analysis (Dingle *et al.*, 2001). As more methods are developed, the number of databases dealing with specific methods, organisms, or gene sequences can be expected to proliferate. These will greatly facilitate our understanding of the genetic variability among populations of microorganisms. However, this kind of system is passive and does not provide a mechanism for rapid exchange of subtyping data for the purposes of surveillance or outbreak detection.

Networks that have been implemented for these latter functions include Enter-net in Europe, FoodNet and PulseNet in the USA, and the W.H.O. Global Salm-Surv (Yang, 1998). Enter-net regularly collects summary data from participating European Union nations on *Salmonella* and Shiga-toxin *E. coli* (STEC) infections. Methods for subtyping of isolates or for the determination and characterization of virulence factors have not all been standardized (but see PulseNet, below). However, the equivalency of data generated using different methods has been investigated and is quite well understood. Enter-net is actively involved in suggesting there be standards for pathogen surveillance networks, so that real-time global surveillance can be attained. FoodNet utilizes active surveillance of sentinel sites to obtain a more precise estimate of the burden of enteric infectious disease.

PulseNet is probably best known for its optimization and use of standardized PFGE methods for the comparison of molecular subtyping results from laboratories across North America (Centers for Disease Control and Prevention, 1996). The fundamental innovation that has made PulseNet success-

ful is an agreement by participants to rapidly perform highly standardized testing on isolates arriving at participating laboratories, and to share information about unusual isolates or clusters with the entire network as soon as the analysis has been completed (Swaminathan *et al.*, 2001). This has led to very rapid communication and the fulfillment of the concept of real-time surveillance. Communication of raw data (TIFF files) and analyzed data containing normalized PFGE patterns (bundle files) was accomplished initially through a listserv that included all participants. More recently, information has also been posted on WebBoards that provides access only to participants and ensures security of data. In addition, the development of a national database of PFGE patterns and associated epidemiologic information allows more comprehensive analysis of data nation-wide, decreased time for laboratories with access to the database to find PFGE pattern matches, and implementation of algorithms designed to detect unusual events that could indicate outbreaks are occurring.

Within PulseNet, the analysis of laboratory and epidemiologic data is highly integrated, allowing continuing surveillance for case clusters and outbreaks against a background of sporadic cases. Confirmation of indistinguishable PFGE patterns in human patients and foods may facilitate trace-back investigations and product recalls. Such interventions can lead to earlier control of outbreaks with reduced human and economic costs (Elbasha *et al.*, 2000). In order to develop databases comprehensive enough to allow detection of outbreaks and to rule out pseudo-outbreaks, participating laboratories perform PFGE on as close to 100% of selected pathogens as possible. The inclusion of all state public health laboratories, as well as other federal laboratories, including the US Food and Drug Administration laboratories in the Food Safety and Inspection Service Laboratory (FDA-CFSAN), the Center for Veterinary Medicine, and the US Department of Agriculture's Food Safety and Inspection Service Laboratory (USDA-FSIS), as well as the CDC, ensures that subtyping data are available from human and non-human sources.

The PulseNet concept has been adopted in other jurisdictions, and implementation of networks using the same standardized methods, interpretive criteria, equipment, and analytic software are under development in Canada, Europe, and the Pacific Rim. In the near future other areas, such as some or all South American countries belonging to the Pan American Health Organization (PAHO) may develop associated molecular subtyping networks. The potential exists to create a global network for real-time exchange of laboratory subtyping data and associated epidemiologic information using a single set of standardized methods, common proficiency testing, and common QA/QC protocols. As alternative methods become available for molecular subtyping and the corresponding comprehensive databases of laboratory and epidemiologic information are developed, these networks will have the option of adopting methods that may provide faster turnaround time, greater ease of interpretation, or better portability between individual laboratories.

## Conclusions and future considerations

Genotypic typing methods have greatly increased our ability to discriminate between bacterial microorganisms and have become a very common and powerful tool in the area of molecular epidemiology. The food arena has greatly benefited from these techniques, as they have helped to identify potential food vehicles in investigations of outbreaks, as well as in tracing/tracking the dissemination of foodborne pathogens in food plants, and in strictly defining focal points of food contamination. Some of the large food companies routinely use molecular typing to help staff understand sources and spreading patterns of foodborne pathogens within the plant environment. Although genotypic methods will become increasingly popular, phenotypic methods will still remain important in diagnostic microbiology.

There are still a number of issues that need to be resolved in the area of molecular typing. One is the lack of a "gold standard" by which to judge a typing method, while another issue relates to standardization of laboratory protocols, and classification of strains using genotyping. The development of standardized reference material (DNA, bacterial strains) will become important in comparing current and novel methodologies against one another. It is also recognized that, in general, the use of more than one typing method can be of great benefit in terms of one's confidence in the results obtained. However, this must be balanced against the cost and time implications (Table 4.4).

What one can expect to see in the future is more and more reliance on genotypic typing methods that are amenable to automation. In particular, microarrays, notwithstanding some of the technical issues that currently exist, appear to be poised to hold great promise for the simultaneous identification, serotyping and genotyping of microorganisms. In addition, genotypic methods based on nucleotide sequence analysis such as MLST (or variants of this method), will become much more common, especially as we get better and faster at sequencing genes/whole genomes. As always, the ultimate goal is the development of an "ideal" molecular typing method; one that is easy to perform, cost-effective, relatively rapid, amenable to statistical analysis and automation, types all possible strains, is reproducible and has a balance between increased discriminatory power and applicability. Although such an "ideal" method does not currently exist, great strides have been made in the last 10 years in moving closer and closer to attaining this goal.

## References

Aarnisalo, K., Autio, T., Sjoberg, A.M., Lunden, J., Korkeala, H., and Suihko, M.L. 2003. Typing of *Listeria monocytogenes* isolates originating from the food processing industry with automated ribotyping and pulsed-field gel electrophoresis. J. Food Prot. 66: 249–255.

Achtman, M. 1998. A phylogenetic perspective on molecular epidemiology. In: Molecular Medical Microbiology. S. Sussman ed. Academic Press, London, pp. 485–510.

Adiri, R.S., Gophna, U., and Ron, E.Z. 2003. Multilocus sequence typing of *Escherichia coli* O78 strains. FEMS Microbiol. Lett. 222: 199–203.

Allmaier, G., Schaffer, C., Messner, P., Rapp, U., and Myer-Posner, F.J. 1995. Accurate determination of the molecular weight of the major surface layer protein isolated from *Closteridium thermosaccharolyticum* by time of flight mass spectrometry. J. Bacteriol. 177: 1402–1404.

Anderson, N.G., and Anderson, N.L. 1996. Twenty years of two-dimensional electrophoresis: Past, present, and future. Electrophoresis. 17: 443–453.

Anon. 2001. Multi-locus sequence typing. Building for pathogen research, University of Oxford, Oxford, UK

Arbeit, R.D. 1995. Laboratory procedures for the epidemiologic analysis of microorganisms. In: Manual of Clinical Microbiology. P.R. Murray, E.J. Baron, M.A. Pfaller, F.C. Tenover, and R.H. Yolken, ed. ASM Press, Washington, DC, pp. 190-194.

Avery, S.M., Liebana, E., Reid, C.A., Woodward, M.J., and Buncie, S. 2002. Combined use of two genetic fingerprinting methods, pulsed-field gel electrophoresis and ribotyping, for characterization of *Escherichia coli* O157:H7 isolates from food animals, retail meats, and cases of human disease. J. Clin. Microbiol. 40: 2806–2812.

Baggesen, D.L., Wegener, H.C., and Madsen, M. 1997. Correlation of conversion of *Salmonella enterica* serovar Enteritidis phage type 1, 4, or 6 to phage type 7 with loss of lipopolysaccharide. J. Clin. Microbiol. 35: 330–333.

Baquar, N., Burnens, A., and Stanley, J. 1994. Comparative evaluation of molecular typing of strains from a national epidemic due to *Salmonella brandenburg* by rRNA gene and IS200 probes and pulsed-field gel electrophoresis. J. Clin. Microbiol. 32: 1876–1880.

Barrett, T.J., Lior, H., Greene, J.H., Khakhria, R., Wells, J.G., Bell, B.P., Greene, K.D., Lewis, J., and Griffin, P.M. 1994. Laboratory investigation of a multi-state foodborne outbreak of *Escherichia coli* O157: H7 by using pulsed-field gel electrophoresis and phage typing. J. Clin. Microbiol. 32: 3013–3017.

Bean, N.H., and Martin, S.M. 2001. Implementing a network for electronic surveillance reporting from public health reference laboratories: an international perspective. Emerg. Infect. Dis. 7: 773–779.

Bekal, S., Brousseau, R., Masson, L., Prefontaine, G., Fairbrother, J., and Harel, J. 2003. Rapid identification of *Escherichia coli* pathotypes by virulence gene detection with DNA microarrays. J. Clin. Microbiol. 41: 2113–2125.

Blais, B.W., Phillippe, L., Pagotto, F., and Corneau, N. MFLP 78: Identification of presumptive positive *Listeria monocytogenes* from foods by the polymerase chain reaction (PCR). Compendium of Analytical Methods. (http://www.hc-sc.gc.ca/food-aliment/mh-dm/mhe-dme/compendium/volume_3/e_mflp7801.html).

Boyd, E.F., Wang, F.-S., Whittam, T.S., and Selander, R.K. 1996. Molecular genetic relationships of the *Salmonellae*. Appl. Environ. Microbiol. 62: 804–808.

Brett, K.N., Ramachandran, V., Hornitzky, M.A., Bettelheim, K.A., Walker, M.J., and Djordjevic, S.P. 2003. *stx1c* is the most common Shiga toxin 1 subtype among Shiga toxin-producing *Escherichia coli* isolates from sheep but not among isolates from cattle. J. Clin. Microbiol. 41: 926–936.

Brown, D.J., Baggesen, D.L., Platt, D.J., and Olsen, J.E. 1999. Phage type conversion in *Salmonella enterica* serotype Enteritidis caused by the introduction of a resistance plasmid of incompatibility group X (IncX). Epidemiol. Infect. 122: 19–22.

Burnens, A.P., Wagner, J., Lior, H., Nicolet, J., and Frey, J. 1995. Restriction fragment length polymorphisms amont the flagellar genes of the Lior heat-labile serogroup reference strains and field strains of *Campylobacter jejuni* and *C. coli*. Epidemiol. Infect. 114: 423–431.

Cain, T., Lubman, D., and Weber, J.W. 1994. Differentiation of bacteria using protein profiles from matrix assisted laser desorption/ionization time of flight mass spectrometry. Rapid Commun. Mass Spectrom. 8: 1026–1030.

Call, D.R., Brockman, F.J., and Chandler, D.P. 2001. Detecting and genotyping *Escherichia coli* O157:H7 using multiplexed PCR and nucleic acid microarrays. Int. J. Food Microbiol. 67: 71-80.

Call, D.R., Borucki, M.K., and Besser, T.E. 2003. Mixed-genome microarrays reveal multiple serotype and lineage-specific differences among strains of *Listeria monocytogenes*. J. Clin. Microbiol. 41: 632–639.

Centers for Disease Control and Prevention. 1996 (updated 2000). Standardized molecular subtyping of foodborne bacterial pathogens by pulsed-field gel electrophoresis: a manual. Atlanta: National Center for Infectious Diseases.

Chang, N., and Taylor, D.E. 1990. Use of pulsed-field agarose gel electrophoresis to size genomes of *Campylobacter* species and to construct a *Sal* I map of *Campylobacter jejuni*. J. Bacteriol. 172: 5211–5217.

Chizhikov, V., Rasooly, A., Chumakov, K., and Levy, D. 2001. Microarray analysis of microbial virulence factors. Appl. Environ. Microbiol. 67: 3258–3263.

Clark, C.G., Kravetz, A.N., Dendy, C. Wang, G., Tyler, K.D., and Johnson, W.M. 1998. Investigation of the 1994–5 Ukranian *Vibrio cholerae* epidemic using molecular methods. Epidemiol. Infect. 121: 15–19.

Clark, C.G., Price, L., Ahmed, R., Woodward, D.L., Melito, P.L., Rodgers, F.G., Jamieson, F., Ciebin, B., Li, A., and Ellis, A. 2003a. Characterization of water-borne outbreak-associated *Campylobacter jejuni*, Walkerton, Ontario. Emerg. Infect. Dis. 9: 1232–1241.

Clark, C.G., Kruk, T.M.A.C., Bryden, L., Hirvi, Y., Ahmed, R., and Rodgers, F.G. 2003b. Subtyping of *Salmonella enterica* serotype Enteritidis strains by manual and automated *Pst* I – *Sph* I ribotyping. J. Clin. Microbiol. 41: 27–33.

Clermont, O., Cordevant, C., Bonacorsi, S., Marecat, A., Lange, M., and Bingen, E. 2001. Automated ribotyping provides rapid phylogenetic subgroup affiliation of clinical extraintestinal pathogenic *Escherichia coli* strains. J. Clin. Microbiol. 39: 4549–4553.

Coimbra, R.S., Nicastro, G., Grimont, P.A., and Grimont, F. 2001. Computer identification of *Shigella* species by rRNA gene restriction patterns. Res. Microbiol. 152: 47–55.

Conway, G., Smole, S.C., Sarracino, D.A., Arbeit, R.D., and Leopold, P.E. 2001. Species identification of clinical isolates of *Escherichia coli* using matrix assisted laser desorption/ionization time of flight mass spectrometry. J. Mol. Microbiol. Biotechnol. 3: 103-112.

Davis, K.C., Nakatsu, C.H., Turco, R., Weagant, S.D., and Bhunia, A.K. 2003. Analysis of environmental *Escherichia coli* isolates for virulence genes using the TaqMan PCR system. J. Appl. Microbiol. 95: 612–620.

Dijkshoorn, L., Ursing, B.M., and Ursing, J.B. 2000. Strain, clone, and species: comments on three basic concepts of bacteriology. J. Med. Microbiol. 49: 397–401.

Dingle, K.E., Colles, F.M., Wareing, D.R.A., Ure, R., Fox, A.J., Bolton, F.E., Bootsma, H.J., Willems, R.J.L., Urwin, R., and Maiden, M.C.J. 2001. Multilocus sequence typing system for *Campylobacter jejuni*. J. Clin. Microbiol. 39: 14–23.

Dorrell, N., Mangan, J.A., Laing, K.G., Hinds, J., Linton, D., Al-Ghusein, H., Barrell, B.G., Parkhill, J., Stoker, N.G., Karlyshev, A.V., Butcher, P.D., and Wren, B.W. 2001. Whole genome comparison of *Campylobacter jejuni* human isolates using a low-cost microarray reveals extensive genetic diversity. Genome Res. 11: 1706–1715.

Ehrbar, K., Mirold, S., Friebel, A., Stender, S., and Hardt, W.-D. 2002. Characterization of effector proteins translocated via the SP1 type III secretion system of *Salmonella typhimurium*. Int. J. Med. Microbiol. 291: 479–485.

Eisgruber, H., Wiedmann, M., and Stolle, A. 1995. Use of plasmid profiling as a typing method for epidemiologically related *Clostridium perfringens* isolates from food poisoning cases and outbreaks. Lett. Appl. Microbiol. 20: 290–304.

Eklund, M., Leino, J., and Siitonen, A. 2002. Clinical *Escherichia coli* strains carrying *stx* genes; *stx* variants and *stx*-positive virulence profiles. J. Clin. Microbiol. 40: 4585–4593.

Elbasha, E.H., Fitzsimmons, T.D., and Meltzer, M.I. 2000. Costs and benefits of a subtype-specific surveillance system for identifying *Escherichia coli* O157:H7 outbreaks. Emerg. Infect. Dis. 6: 293–297.

Enright, M.C., and Spratt, B. 1999. Multilocus sequence typing. Trends in Microbiology 12: 482-487.

Esteban, E., Snipes, K., Hird, D., Kasten, R., and Kinde, H. 1993. Use of ribotyping for characterization of *Salmonella* serotypes. J. Clin. Microbiol. 31: 233–237.

Evans, D.J., Jr., Evans, D.G., Young, L.S., and Pitt, J. 1980. Hemagglutination typing of *Escherichia coli*: definition of seven hemagglutination types. J. Clin. Microbiol. 12: 235–242.

Farber, J.M. 1996. An introduction to the hows and whys of molecular typing. J. Food Prot. 59: 1091–1101.

Farber, J.M., Daley, E.M., MacKie, M.T., and Limerick, B. 2000. A small outbreak of listeriosis potentially linked to the consumption of imitation crab meat. Lett. Appl. Microbiol. 31: 100-104.

Felix, A., and Callow, B.R. 1943. Typing of paratyphoid B bacilli by means of Vi bacteriophage. Br. Med. J. 2: 127–130.

Feng, P., Lampel, K.A., Karch, H., and Whittam, T.S. 1998. Genotypic and phenotypic changes in the emergence of *Escherichia coli* O157:H7. J. Infect. Dis. 177: 1750–1753.

Fitzgerald, C., Owen, R.J., and Stanley, J. 1996. Comprehensive ribotyping scheme for heat-stable serotypes of *Campylobacter jejuni*. J. Clin. Microbiol. 34: 265–269.

Fitzgerald, C., Helsel, L.O., Nicholson, M.A., Olsen, S.J., Swerdlow, D.L., Flahart, R., Sexton, J., and Fields, P.I. 2001. Evaluation of methods for subtyping *Campylobacter jejuni* during an outbreak involving a food handler. J. Clin. Microbiol. 39: 2386–2390.

Fleischmann, R.D., Adams, M.D., White, O., Clayton, R.A., Kirkness, E.F., Kerlavage, A.R., Bult, C.J., Tomb, J.F., Dougherty, B.A., Merrick, J.M. *et al.* 1995. Whole-genome random sequencing and assembly of *Haemophilus influenzae* Rd. Science. 269: 496–512.

Fukushima, H., Tsunomori, Y., and R. Seki. 2003. Duplex real-time SYBR green PCR assays for detection of 17 species of food- or water-borne pathogens in stools. J. Clin. Microbiol. 41: 5134-5146.

Gobius, K.S., Higgs, G.M., and Desmarchelier, P.M. 2003. Presence of activable Shiga toxin genotype (*stx*(2d)) in Shiga toxigenic *Escherichia coli* from livestock sources. J. Clin. Microbiol. 41: 3777–3883.

Gormon, T., and Phan-Thanh, L. 1995. Identification and classification of *Listeria* by two-dimensional protein mapping. Res. Microbiol. 146: 143–154.

Graves, P.R., and Haystead, T.A.J. 2002. Molecular biologist's guide to proteomics. Microbiol. Molec. Biol. Rev. 66: 39–63.

Grimont, F., and Grimont, P.A.D. 1986. Ribosomal ribonucleic acid gene restriction patterns as potential taxonomic tools. Ann. Inst. Pasteur/Microbiol. (Paris) 137B: 165–175.

Hahm, B.K., Maldonado, Y., Schreiber, E., Bhunia, A.K., and Nakatsu, C.H. 2003. Subtyping of foodborne and environmental isolates of *Escherichia coli* by multiplex-PCR, rep-PCR, PFGE, ribotyping and AFLP. J. Microbiol. Methods. 53: 387–399.

Hänninen, M.-L., Hakkinen, M., and Rautelin, H.. 1999. Stability of related human and chicken *Campylobacter jejuni* genotypes after passage through chick intestine studied by pulsed-field gel electrophoresis. Appl. Environ. Microbiol. 65: 2272–2275.

Harrington, C.S., Thomson-Carter, F.M., and Carter, P.E. 1997. Evidence for recombination in the flagellin locus of *Campylobacter jejuni*: implications for the flagellin typing scheme. J. Clin. Microbiol. 35: 2386–2392.

Harrington, C.S., Moran, L., Ridley, A.M., Newell, D.G., and Madden, R.H. 2003. Inter-laboratory evaluation of three flagellin PCR/RFLP methods for typing *Campylobacter jejuni* and *C. coli*: the CAMPYNET experience. J. Appl. Microbiol. 95: 1321–33.

Hedberg, C.W., Smith, K.E., Besser, J.M., Boxrud, D.J., Hennessy, T.W., Bender, J.B., Anderson, F.A., and Osterholm, M.T. 2001. Limitations of pulsed-field gel electrophoresis for the routine surveillance of *Campylobacter* infections. J. Infect. Dis. 184: 242–243.

Hilton, A.C., and Penn, C.W. 1998. Comparison of ribotyping and arbitrarily primed PCR for molecular typing of *Salmonella enterica* and relationships between strains on the basis of these molecular markers. J. Appl. Microbiol. 85: 933–940.

Holland, R., Wilkes, J., Rafii, F., Sutherland, J.B., Persons, C.C., Voorhees, K.J., and Lay, J.O. Jr. 1996. Rapid identification of intact whole bacteria based on spectral patterns using matrix assisted laser desorption/ionization with time of flight mass spectrometry. Rapid Commun. Mass Spectrom. 10: 1227–1232.

Hulton, C.S.J., Higgins, C.F., and Sharp, P.M. 1991. ERIC sequences: a novel family of repetitive elements in the genomes of *Escherichia coli*,

*Salmonella typhimurium* and other enterobacteria. Mol. Microbiol. 5: 825–834.

Hunter, P.R., and Gaston, M.A. 1988. Numerical index of the discriminatory ability of typing systems: an application of Simpson's Index of Diversity. J. Clin. Microbiol. 26: 2465–2466.

Iguchi, A., R. Osawa, J. Kawano, A. Shimizu, J. Terajima, and H. Watanabe. 2002. Effects of subculturing and prolonged storage at room temperature of enterohemorrhagic *Escherichia coli* O157:H7 on pulsed-field electrophoresis profiles. J. Clin. Microbiol. 40: 3079–3081.

Inglis, T.J., Aravena-Roman, M., Ching, S., Croft, K., Wuthiekanun, V., and Mee, B.J. 2003a. Cellular fatty acid profile distinguishes *Burkholderia pseudomallei* from avirulent *Burkholderia thailandensis*. J. Clin. Microbiol. 41: 4812–4824.

Inglis, T.J., Clair, A., Sampson, J., O'Reilly, L., Vandenberg, S., Leighton, K., and Watson, A. 2003b. Real-time application of automated ribotyping and DNA macrorestriction analysis in the setting of a listeriosis outbreak. Epidemiol. Infect. 131: 637–645.

Iriarte, P., and Owen, R.J. 1996a. PCR-RFLP analysis of the large subunit (23S) ribosomal RNA genes of *Campylobacter jejuni*. Lett. Appl. Microbiol. 23: 163–166.

Iriarte, P., and Owen, R.J. 1996b. Repetitive and arbitrary primer DNA sequences in PCR-mediated fingerprinting of outbreak and sporadic isolates of *Campylobacter jejuni*. FEMS Immunol. Med. Microbiol. 15: 17–22.

Izumiya, H., Terajima, J., Wada, A., Inagaki, Y., Itoh, K.-I., Tamura, K., and Watanabe, H. 1997. Molecular typing of enterohemorrhagic *Escherichia coli* O157:H7 isolates in Japan by using pulsed-field gel electrophoresis. J. Clin. Microbiol. 35: 1675–1680.

Jackson, C.J., Fox, A.J., Wareing, D.R.A., Hutchinson, D.N., and Jones, D.M. 1996. The application of genotyping techniques to the epidemiological analysis of *Campylobacter jejuni*. Epidemiol. Infect. 117: 233–244.

Jansen, R., Van Embden, J.D.A., Gaastra, W., and Schouls, L.M. 2002. Identification of genes that are associated with DNA repeats in prokaryotes. Mol. Microbiol. 43: 1565–1576.

Johansen, B.K., Wasteson, Y., Granum, P.E., and Brynestad, S. 2001. Mosaic structure of Shiga-toxin-2-encoding phages isolated from *Escherichia coli* O157:H7 indicates frequent gene exchange between lambdoid phage genomes. Microbiol. 147: 1929–1936.

Johnson, J.M., Weagant, S.D., Jinneman, K.C., and Bryant, J.L. 1995. Use of pulsed-field gel electrophoresis for epidemiological study of *Escherichia coli* O157:H7 during a foodborne outbreak. Appl. Environ. Microbiol. 61: 2806–2808.

Johnson, J.R., and Clabots, C. 2000. Improved repetitive element-polymerase chain reaction fingerprinting of *Salmonella* with the use of extremely elevated annealing temperatures. Clin. Diagn. Lab. Immunol. 7: 258–264.

Johnson, J.R., and O'Brien, T.T. 2000. Improved repetitive element-polymerase chain reaction fingerprinting for resolving pathogenic and non-pathogenic phylogenetic groups within *Escherichia coli*. Clin. Diagn. Lab. Immunol. 7: 265–273.

Johnson, J.R., Clabots, C., Azur, M., Boxrud, D.J., Besser, J.M., and Thurn, J.R. 2001. Molecular analysis of a hospital cafeteria-associated Salmonellosis outbreak using modified repetitive element PCR fingerprinting. J. Clin. Microbiol. 39: 3452–3460.

Karaolis, D.K.R., Lan, R., and Reeves, P.R. 1994. Molecular evolution of the seventh-pandemic clone of *Vibrio cholerae* and its relationship to other pandemic and epidemic *V. cholerae* isolates. J. Bacteriol. 176: 6199–6206.

Karas, M., and Hillenkamp, F. 1988. Laser desorption ionization of proteins with molecular masses exceeding 10,000 daltons. Anal. Chem. 60: 2299–301.

Karmali, M.A., Mascarenhas, M., Shen, S., Ziebell, K., Johnson, S., Reid-Smith, R., Isaac-Renton, J., Clark, C., Rahn, K., and Kaper, J.B. 2003. Association of genomic O island 122 of *Escherichia coli* EDL 933 with verocytotoxin-producing *Escherichia coli* seropathotypes that are linkid to epidemic and/or serious disease. J. Clin. Microbiol. 41: 4930–4940.

Kaufmann, R. 1995. Matrix assisted laser desorption ionization (MALDI) mass spectrometry: a novel analytical tool in molecular biology and biotechnology. J. Biotechnol. 41: 155–175.

Koblavi, S., Grimont, F., and Grimont, P.A.D. 1990. Clonal diversity of *Vibrio cholerae* O1 evidenced by rRNA gene restriction patterns. Res. Microbiol. 141: 645–657.

Kolavic, S.A., Kimura, A., Simons, S.L., Slutsker, L., Barth, S., and Haley, C.E. 1997. An outbreak of *Shigella dysenteriae* type 2 among laboratory workers due to intentional food contamination. JAMA 278: 396–398.

Kostman, J.R., Alden, M.B., Mair, M., Edlind, T.D., LiPuma, J.J., and Stull, T.L. 1995. A universal approach to bacterial molecular epidemiology by polymerase chain reaction ribotyping. J. Infect. Dis. 171: 204–208.

Kotetishvili, M., Stine, O.C., Kreger, A., Morris, J.G., Jr., and Sulakvelidze, A. 2002. Multilocus sequence typing for characterization of clinical and environmental salmonella strains. J. Clin. Microbiol. 40: 1626–1635.

Krishnamurthy, T., and Ross, P. 1996. Rapid identification of bacteria by direct matrix assisted laser desorption/ionization mass spectrometry analysis of whole cells. Rapid Commun. Mass Spectrom. 10: 1992–1996.

Kudva, I.T., Evans, P.S., Perna, N.T., Barrett, T.J., DeCastro, G.J., Ausubel, F.M., Blattner, F.R., and Calderwood, S.B. 2002. Polymorphic amplified typing sequences provide a novel approach to *Escherichia coli* O157:H7 strains typing. J. Clin. Microbiol. 40: 1152–1159.

Kuhnert, P., J. Hacker, I. Mühldorfer, A.P. Burnens, J. Nicolet, and J. Frey. 1997. Detection system for *Escherichia coli*-specific virulence genes: absence of virulence determinants in B and C strains. Appl. Environ. Microbiol. 63: 703–709.

Kumao, T., W. Ba-Thein, and H. Hayashi. 2002. Molecular subtyping methods for detection of *Salmonella enterica* serovar Oranienburg outbreaks. J. Clin. Microbiol. 40: 2057–2061.

Lagatolla, C., Dolzani, L., Tonin, E., Lavenia, A., Di Michele, M., Tommasini, T., and Monti-Bragadin, C. 1996. PCR ribotyping for characterizing *Salmonella* isolates of different serotypes. J. Clin. Microbiol. 34: 2440–2443.

Landeras, E., and M.C. Mendoza. 1998. Evaluation of PCR-based methods performed with a mixture of *Pst* I and *Sph* I to differentiate strains of *Salmonella* serotype Enteritidis. J. Med. Microbiol. 47: 427–434.

Landeras, E., González-Hevia, M.A., and Mendoza, M.C. 1998. Molecular epidemiology of *Salmonella* serotype Enteritidis. Relationships between food, water, and pathogenic strains. Int. J. Food Microbiol. 43: 81–90.

Lang, M.M., Ingham, S.C., and Ingham, B.H. 2001. Differentiation of *Enterococcus* spp. by cell membrane fatty acid methyl ester profiling, biotyping and ribotyping. Lett. Appl. Microbiol. 33: 65–70.

Lashkari, D.A., DeRisi, J.L., MCCusker, J.H., Namath, A.F., Gentiles, C., Hwang, S.Y., Brown, P.O., and Davis, R.W. 1997. Yeast microarrays for genome wide parallel genetic and gene expression analysis. Proc. Natl. Acad. Sci. USA 94: 13057–13062.

Lay, J.O. Jr. 2001. MALDI-TOF mass spectrometry of bacteria. Mass Spectrom. Rev. 20: 172–94.

Lee, M.-S., Kaspar, C.W., Brosch, R., Shere, J., and Luchansky, J.B. 1996. Genomic analysis using pulsed-field gel electrophoresis of *Escherichia coli* O157:H7 isolated from dairy calves during the United States national dairy heifer evaluation project (1991–1992). Vet. Microbiol. 48: 223–230.

Lindstedt, B.-A., Heir, E., Gjernes, E., and Kapperud, G. 2003. DNA fingerprinting of *Salmonella enterica* subsp. *enterica* serovar *Typhimurium* with emphasis on phage type DT104 based on variable number of tandem repeat loci. J. Clin. Microbiol. 41: 1469–1479.

Liu, Y., Lee, M.-A., Ooi, E.-E., Mavis, Y., Tang, A.-L., and Quek, H.-H. 2003. Molecular typing of *Salmonella enterica* serovar Typhi isolates from various countries in Asia by a multiplex PCR assay on variable-number tandem repeats. J. Clin. Micriobiol. 41: 4388–4394.

López-Molina, N., Laconcha, I., Rementeria, A., Audicana, A., Perales, I., and Garaizar, J. 1998. Typing of *Salmonella enteritidis* of different phage types of PCR fingerprinting. J. Appl. Microbiol. 84: 877–882.

Louie, M., Read, S., Louie, L., Ziebell, K., Rahn, K., Borczyk, A., and Lior, H. 1999. Molecular typing methods to investigate transmission of *Escherichia coli* O157:H7 from cattle to humans. Epidemiol. Infect. 123: 17–24.

Maiden, M.C.J., Bygraves, J.A., Feil, E., Morelli, G., Russell, J.E., Urwin, R., Zhang, Q., Zhou, J., Zurth, K., Caugant, D.A., Feavers, I.M., Achtman, M., and Spratt, B.G. 1998. Multilocus sequence typing: a portable approach to the identification of clones within populations of pathogenic microorganisms. Proc. Nat. Acad. Sci. USA 95: 3140–3145.

Marshall, S., Clark, C.G., Wang, G., Mulvey, M., Kelly, M.T., and Johnson, W.M. 1999. Comparison of molecular methods for typing *Vibrio parahaemolyticus*. J. Clin. Microbiol. 37: 2473–2478.

Martin, I.E., Tyler, S.D., Tyler, K.D., Khakhria, R., and Johnson, W.M. 1996. Evaluation of ribotyping as epidemiologic tool for typing *Escherichia coli* serogroup O157 isolates. J. Clin. Microbiol. 34: 720–723.

Maslow, J.N., Whittam, T., Wilson, R.A., Mulligan, M.E., Adams, K.S., and Arbeit, R.D. 1995. Clonal relationship among bloodstream isolates of *Escherichia coli*. Infect. Immun. 63: 2409–2417.

Mazurier, S., van de Giessen, A., Heuvelman, K., and Wernars, K. 1992. RAPD analysis of *Campylobacter* isolates: DNA fingerprinting without the need to purify DNA. Lett. Appl. Microbiol. 14: 260–262.

McGonigle, B., Keeler, S.J., Lau, S.M., Koeppe, M.K., and O'Keefe, D.P. 2000. A genomics approach to the comprehensive analysis of the glutathione S-transferase gene family in soybean and maize. Plant Physiol. 124: 1105–1120.

Meacham, K.J., Zhang, L., Foxman, B., Bauer, R.J., and Marrs, C.F. 2003. Evaluation of genotyping large numbers of *Escherichia coli* isolates by enterobacterial repetitive intergenic consensus-PCR. J. Clin. Microbiol. 41: 5224–5226.

Meng, J., Doyle, M.P., Zaho, T., and Zhao, S. 2001. Enterohemorrhagic *Escherichia coli*. In: Food Microbiology: Fundamentals and Frontiers, 2nd edition. M.P. Doyle, L.R. Beuchat, and T.J. Montville eds. ASM Press, Washington, D.C, pp. 193–213.

Mirold, S., Ehrbar, K., Weissmüller, A., Prager, R., Tschäpe, H. Rüssman, H., and Hardt, W.-D. 2001. *Salmonella* host cell invasion emerged by acquisition of a mosaic of separate genetic elements, including *Salmonella* pathogenicity island 1 (SPI 1), SPI5, and *sopE2*. J. Bacteriol. 183: 2348–2358.

Nachamkin, I., Ung, H., and Patton, C.M. 1996. Analysis of HL and O serotypes of *Campylobacter* strains by the flagellin gene typing system. J. Clin. Microbiol. 34: 277–281.

Nelson, K.E. 2003. The future of microbial genomics. Env. Microbiol. 5: 1223–1225.

Noller, A.C., McEllistrem, M.C., Pacheco, A.G.F., Boxrud, D.J., and Harrison, L.H. 2003a. Multilocus variable-number tandem repeat analysis distinguishes outbreak and sporadic *Escherichia coli* O157: H7 isolates. J. Clin. Microbiol. 41: 5389–5397.

Noller, A.C., McEllistrem, M.C., Stine, O.C., Morris, J.G. Jr., Boxrud, D.J., Dixon, B., and Harrison, L.H. 2003b. Multilocus sequence typing reveals a lack of diversity among *Escherichia coli* O157:H7 isolates that are distinct by pulsed-field gel electrophoresis. J. Clin. Microbiol. 41: 675–679.

Olive, D.M., and Bean, P. 1999. Principles and applications of methods for DNA-based typing of microbial organisms. J. Clin. Microbiol. 37: 1161–1169.

On, S.L. 1998. *In vitro* genotypic variation of *Campylobacter coli* documented by pulsed-field gel electrophoretic DNA profiling: implications for epidemiological studies. FEMS Microbiol. Lett. 165: 379–385.

Osawa, R., Iyoda, S., Nakayama, S.I., Wada, A., Yamai, S., and Watanabe, H. 2000. Genotypic variations of Shiga toxin-converting phages from enterohaemorrhagic *Escherichia coli* O157:H7 isolates. J. Med. Microbiol. 49: 565–574.

Perna, N.T., Plunkett, G. III, Burland, V., Mau, B., Glasner, J.D., Rose, D.J., Mayhew, G.F., Evans, P.S., Gregor, J., Kirkpatrick, H.A., Posfai, G., Hackett, J., Klink, S., Boutin, A., Shao, Y., Miller, L., Grotbeck, E.J., Davis, N.W., Lim, A., Dimalanta, E.T., Potamousis, K.D.,

Apodaca, J., Anantharaman, T.S., Lin, J., Yen, G., Schwartz, D.C., Welch, R.A., and Blattner, F.R. 2001. Genome sequence of enterohaemorrhagic *Escherichia coli* O157:H7. Nature 409: 529–533.

Petersen, L., and On, S.L.W. 2000. Efficacy of flagellin genes typing for epidemiological studies of *Campylobacter jejuni* in poultry estimated by comparison with macrorestriction profiling. Lett. Appl. Microbiol. 31: 14–19.

Petersen, L. and Newell, D.G. 2001. The ability of *Fla*-typing schemes to discriminate between trains of *Campylobacter jejuni*. J. Appl. Microbiol. 91: 217–224.

Pollock, J.D. 2002. Gene expression profiling: methodological challenges, results, and prospects for addiction research. Chem. Phys. Lipids 121: 241–256.

Popovic, T., Bopp, C., Olsvik, Ø., and Wachsmuth, K. 1993. Epidemiologic application of a standardized ribotype scheme for *Vibrio cholerae* O1. J. Clin. Microbiol. 31: 2474–2482.

Prager, R., Mirold, S., Tietze, E., Strutz, U., Knueppel, B. Rabsch, W., Hardt, W.-D., and Tscähpe, H. 2000. Prevalence and polymorphism of genes encoding translocated effector proteins among clinical isolates of *Salmonella enterica*. Int. J. Med. Microbiol. 290: 605–617.

Prager, R., Rabsch, W., Streckel, W., Voigt, W., Tietze, E., and Tschäpe, H. 2003. Molecular properties of *Salmonella enterica* serotype Paratyphi B distinguish between its systemic and its enteric pathovars. J. Clin. Microbiol. 41: 4270–4278.

Pupo, G.M., Karaolis, D.K.R., Lan, R., and Reeves, P.R. 1997. Evolutionary relationships among pathogenic and non-pathogenic *Escherichia coli* strains inferred from multilocus enzyme electrophoresis and *mdh* sequence studies. Infect. Immun. 65: 2685–2692.

Qadri, F., Haque, A., Faruque, S.M., Bettelheim, K.A., Robins-Browne, R., and Albert, M.J. 1994. Hemagglutinating properties of enteroaggregative *Escherichia coli*. J. Clin. Microbiol. 32: 510-514.

Rajashekara, G., Koeuth, T., Nevile, S., Back, A., Nagaraja, K.V., Lupski, J.R., and Kapur, V. 1998. SERE, a widely dispersed bacterial repetitive DNA element. J. Med. Microbiol. 47: 489–497.

Revazishvili, T., Kotetishvili, M., Stine, O.C., Kreger, A.S., Morris, J.G. Jr., and Sulakvelidze, A. 2004. Comparative analysis of multilocus sequence typing and pulsed-field gel electrophoresis for characterizing *Listeria monocytogenes* strains isolated from environmental and clinical sources. J. Clin. Microbiol. 42: 276–85.

Ribot, E.M., Fitzgerald, C., Kubota, K., Swaminathan, B., and Barrett, T.J. 2001. Rapid pulsed-field gel electrophoresis protocol for subtyping of *Campylobacter jejuni*. J. Clin. Microbiol. 39: 1889–1894.

Rice, D.H., McMenamin, K.M., Pritchett, L.C., Hancock, D.D., and Besser, T.E. 1999. Genetic subtyping of *Escherichia coli* O157 isolates from 41 Pacific Northwest cattle farms. Epidemiol. Infect. 122: 479–484.

Rolland, K., Lambert-Zechovsky, N., Picard, B., and Denamur, E. 1998. *Shigella* and enteroinvasive *Escherichia coli* strains are derived from distinct ancestral strains of *E. coli*. Microbiol. 144: 2667–2672.

Sails, A.D., Swaminathan, B., and Fields, P.I. 2003. Utility of multilocus sequence typing as an epidemiological tool for investigation of outbreaks of gastroenteritis caused by *Campylobacter jejuni*. J. Clin. Microbiol. 41: 4733–4739.

Salcedo, C., Arreaza, L., Alcala, B., de la Fuente, L., and Vazquez, J.A. 2003. Development of a multilocus sequence typing method for analysis of *Listeria monocytogenes* clones. J. Clin. Microbiol. 41: 757–62.

Sauders, B.D., E.D. Fortes, D.L. Morse, N. Dumas, J.A. Kiehlbauch, Y. Schukken, J.R. Hibbs, and M. Wiedmann. 2003. Molecular subtyping to detect human listeriosis clusters. Emerg. Infect. Dis. 9: 672–680.

Schalch, B., L. Bader, H.P. Schau, R. Bergmann, A. Rometsch, G. Maydl, and S. Kessler. 2003. Molecular typing of *Clostridium perfringens* from a foodborne disease outbreak in a nursing home: ribotyping versus pulsed-field gel electrophoresis. J. Clin. Microbiol. 41: 892–895.

Schenk, P.M., K. Kazan, I. Wilson, J.P. Anderson, T. Richmond, S.C. Somerville, and J.M. Manners. 2000. Coordinated plant defense responses in Arabidopsis revealed by microarray analysis. Proc. Natl. Acad. Sci. USA 97: 11655–11660.

Schmeiger, H. 1999. Molecular survey of the *Salmonella* phage typing system of Anderson. J. Bacteriol. 181: 1630–1635.

Schouls, L.M., Reulen, S., Duim, B., Wagenaar, J.A., Willems, R.J.L., Dingle, K.E., Colles, F.M., and Van Embden, J.D.A. 2003. Comparative genotyping of *Campylobacter jejuni* by amplified fragment length polymorphism, multilocus sequence typing, and short repeat sequencing: strain diversity, host range, and recombination. J. Clin. Microbiol. 41: 15–26.

Selander, R.K, Caugant, D.A., Ochman, H., Musser, J.M., Gilmour, M.N., and Whittam, T.S. 1986. Methods of multilocus enzyme electrophoresis for bacterial population genetics and systematics. Appl. Environ. Microbiol. 51: 873–884.

Selander, R.K., Li, J., Boyd, E.F., Wang, F.-S., and Nelson, K. 1994. DNA sequence analysis of the genetic structure of populations of *Salmonella enterica* and *Escherichia coli*. In: Bacterial Diversity and Systematics. F.G. Priest, A. Ramos-Cormenzana, and B.J. Tindall, ed. Plenum Press, New York. p. 17–50.

Severino, P., Darini, A.L., and Magalhaes, V.D. 1999. The discriminatory power of ribo-PCR compared to conventional ribotyping for epidemiological purposes. APMIS 107: 1079–1084.

Silviera, W.D., Benetti, F., Lancellotti, M., Ferreira, A., Solferini, V.N., and Brocchi, M. 2001. Biological and genetic characteristics of uropathogenic *Escherichia coli* strains. Rev. Inst. Med. Trop. Sao Paulo 43: 303–310.

Slade, P.J. 1992. Monitoring *Listeria* in the food production environment III. Typing methodology. Food Res. Int. 25: 215–225.

Smole, S.C., King, L.A., Leopold, P.E., and Arbeit, R.D. 2002. Sample preparation of Gram-positive bacteria for identification by matrix assisted laser desorption/ionization time-of-flight. J. Microbiol. Methods 48: 107–115.

Sontakke, S., and Farber, J.M. 1995. The use of PCR ribotyping for typing strains of *Listeria* spp. Eur. J. Epidemiol. 11: 665–73.

Sorum, H., Bovre, K., Bukholm, G., Lassen, J., and Olsvik, O. 1990. A unique plasmid profile characterizing *Salmonella enteritidis* isolates from patients and employees in a hospital. APMIS 98: 25–29.

Steele, M., McNab, B., Fruhner, L., DeGrandis, S., Woodward, D., and Odumeru, J.A. 1998. Epidemiological typing of *Campylobacter* isolates from meat processing plants by pulsed-field gel electrophoresis, fatty acid profile typing, serotyping, and biotyping. Appl. Environ. Microbiol. 64: 2346–2349.

Steele, M., McNab, W.B., Read, S., Poppe, C., Harris, L., Lammerding, A.M., and Odumeru, J.A. 1997. Analysis of whole-cell fatty acid profiles of verotoxigenic *Escherichia coli* and *Salmonella enteritidis* with the Microbial Identification System. Appl. Environ. Microbiol. 63: 757–760.

Steinbruckner, B., Ruberg, F., and Kist, M. 2001. Bacterial genetic fingerprint: a reliable factor in the study of the epidemiology of human *Campylobacter* enteritis? J. Clin. Microbiol. 39: 4155-4159.

Stull, T., Lipuma, J., and Edlind, T. 1988. A broad-spectrum probe for molecular epidemiology of bacteria: ribosomal RNA. J. Infect. Dis. 157: 280–286.

Swaminathan, B., and Matar, G.M. 1993. Molecular typing methods. In: D.H.T. Persing, T.F. Smith, F.C. Tenover, and T.J. White, ed. Diagnostic Molecular Microbiology: Principles and applications. American Society for Microbiology, Washington, D.C. p. 26–50.

Swaminathan, B., Barrett, T.J., Hunter, S.B., Tauxe, R.V., and the CDC PulseNet Task Force. 2001. PulseNet: The molecular subtyping network for foodborne bacterial disease surveillance, United States. Emerg. Infect. Dis. 7: 382–389.

Tanaka, K., Waki, H., Ido, Y., Akita, S., Yoshida, Y., and Yohida, T. 1988. Protein and polymer analyses up to m/z 100 000 by laser ionization time-of flight mass spectrometry. Rapid Comm. Mass Spectrom. 2: 151–153.

Tang, Y.-W., Ellis, N.M., Hopkins, M.K., Smith, D.H., Dodge, D.E., and Persing, D.H. 1998. Comparison of phenotypic and genotypic techniques for identification of unusual aerobic pathogenic gram-negative bacilli. J. Clin. Microbiol. 36: 3674–3679.

Tarr, C.L., Large, T.M., Moeller, C.L., Lacher, D.W., Tarr, P.I., Acheson, D.W., and Whittam, T.S. 2002. Molecular characterization of a serotype O121:H19 clone, a distinct Shiga toxin-producing clone of pathogenic *Escherichia coli*. Infect. Immun. 70: 6853–6859.

Tenover, F.C., Arbeit, R.D., Goering, R.V., Mickelsen, P.A., Murray, B.E., Persing, D.H., and Swaminathan, B. 1995. Interpreting chromosomal DNA restriction patterns produced by pulsed-field gel electrophoresis: criteria for bacterial strain typing. J. Clin. Microbiol. 33: 2233-2239.

Tyler, K., and Farber, J.M. 2003. Traditional and automated rapid methods for species identification and typing. In: Rapid Microbiological Methods in the Pharmaceutical Industry. M.C. Easter, ed. CRC Press, London. p. 125–160.

Tyler, K.D., Wang, G., Tyler, S.D., and Johnson, W.M. 1997. Factors affecting reliability and reproducibility of amplification-based DNA fingerprinting of representative bacterial pathogens. J. Clin. Microbiol. 35: 339–346.

Tyler, S.D., Johnson, W.M., Lior, H., Wang, G., and Rozee, K.R. 1991. Identification of verotoxin type 2 variant B subunit genes in *Escherichia coli* by the polymerase chain reaction and restriction fragment length polymorphism analysis. J. Clin. Microbiol. 29: 1339–1343.

Van Lith, L.A.J.T., and Aarts, H.J.M. 1994. Polymerase chain reaction identification of *Salmonella* serotypes. Lett. Appl. Microbiol. 19: 273–276.

Versalovic, J., Koeuth, T., and Lupiski, J.R. 1991. Distribution of repetitive DNA sequences in eubacteria and application to fingerprinting of bacterial genomes. Nucl. Acids Res. 19: 6823-6831.

Versalovic, J.T., Schneid, M., de Bruijn, F.J., and Lupski, J.R. 1994. Genomic fingerprinting of bacteria using repetitive sequence-based polymerase chain reaction. Methods Mol. Cell. Biol. 5: 25–40.

Wang, G., Clark, C.G., and Rodgers, F.G. 2002. Detection in *Escherichia coli* of the genes encoding the major virulence factors, the genes defining the O157:H7 serotype, and components of the type 2 Shiga toxin family by multiplex PCR. J. Clin. Microbiol. 40: 3613–3619.

Wang, Z., Russon, L., Li, L., Roser, D., and Long, S. 1998. Investigation of spectral reproducibility in direct analysis of bacteria proteins by matrix assisted laser desorption/ionization time of flight mass spectrometry. Rapid Commun. Mass Spectrom. 12: 456–464.

Wassenaar, T.M., and Newell, D.G. 2000. Genotyping of *Campylobacter* spp. Appl. Environ. Microbiol. 66: 1–9.

Welham, K., Domin, M., Scannell, D., Cohen, E., and Ashton, D. 1998. The characterization of microorganisms by matrix assisted laser desorption/ionization time of flight mass spectronmetry. Rapid Commun. Mass Spectrom. 12: 176–180.

Williams, J.G., Kubelik, A.R., Livak, K.J., Rafalski, J.A., and Tingey, S.V. 1990. DNA polymorphisms amplified by arbitrary primers are useful as genetic markers. Nucleic Acids Res. 18: 6531–6535.

Yang, S. 1998. FoodNet and Enter-net: emerging surveillance programs for foodborne diseases. Emerg. Infect. Dis. 4: 457–458.

Yazaki, J., Kishimoto, N., Nakamura, K., Fujii, F., Shimbo, K., Otsuka, Y., Wu, J., Yamamoto, K., Sakata, K., Sasaki, T., Kikuchi, S. 2000. Embarking on rice functional genomics via cDNA microarray: use of 3' UTR probes for specific gene expression analysis. DNA Res. 7: 367–70.

Yokoyama, K., Iinuma, Y., Kawano, Y., Nakano, M., Kawagishi, M., Yamashino, T., Hasegawa, T., and Ohta, M. 2001. Resolution of *Escherichia coli* O157:H7 that contaminated radish sprouts in two outbreaks by two-dimensional gel electrophoresis. Curr. Microbiol. 43: 311–315.

Zhang, W., Bielaszewska, M., Kuczius, T., and Karch, H. 2002. Identification, characterization, and distribution of Shiga toxin 1 gene variant (*stx*(1c)) in *Escherichia coli* strains isolated from humans. J. Clin. Microbiol. 41: 1441–1446.

Zhang, W., Jayarao, B.M., and Knabel, S.J. 2004. Multi-virulence-locus sequence typing of *Listeria monocytogenes*. Appl. Environ. Microbiol. 70: 913–920.

# Stress Responses of Foodborne Pathogens, with Specific Reference to the Switching on of Such Responses

5

Robin J. Rowbury

## Abstract

Bacteria can be subjected to chemical, physical and biological stresses in the environment, in foods and food preparation/ production processes and in the body, and several such stresses influence survival in foods. Stress exposure can lead to damage at several cellular sites, but for most stresses, damage to DNA, to membranes and to enzymes is most significant for lethality. Ability to survive stress challenges not only depends on inherent tolerance, but also on the extent to which tolerance responses are induced. Many of these responses involve tolerance specifically to the inducing stress, but cross-tolerance responses occur, as does cross-sensitisation. Here, regulation of such processes is considered, with the conclusion that rapid induction of responses and early warning against impending stress exposures, depends on the functioning of extracellular sensing components (ESCs) and extracellular induction components (EICs). The pheromonal/alarmonal intercellular communication afforded by these agents may aid survival of contaminating organisms in foods, allowing such organisms to go on to cause disease on ingestion. A range of biochemical changes occur following switching on of responses, and the structure, properties and functioning of heat-shock proteins are emphasized here, as are those changes that lead to thermotolerance, acid tolerance and irradiation tolerance.

## Damaging and lethal stresses encountered in foods and in food preparation and production processes

The main aim of the present article is to consider factors that influence the ability of potentially pathogenic bacteria to resist or overcome stresses likely to be faced in foods or in food preparation or production processes, and, accordingly, all such stresses will be considered here. In addition, however, it will also be necessary to give an account of those stresses that may be faced by potential pathogens in water (Elliott and Colwell, 1985), and certain other locations like contaminated soils, since organisms from contaminated water may enter foods if such water is used in food production or preparation, or if food materials contaminated by, for example, sewage polluted water,

are employed. Also, food materials, e.g. such as vegetables, may be contaminated if they have been grown in fields which have been dressed with certain dried sewage sludges.

In food preparation and production processes, and in cooking itself, heat is the most important stress (Mackey and Derrick, 1982), and it should be noted that mild heating or heating at increasing temperatures not only enhances thermotolerance (Mackey and Derrick, 1986; Humphrey et al., 1993) but can also lead to other stress tolerances. Heat effects in the aquatic environment are also of significance, since polluting organisms in such natural waters may be induced to stress tolerances by such exposures, and if polluted waters are then used in food preparation or production, organisms ingested in foods may show such stress tolerances.

Cold stress can also be faced by organisms in foods, and such exposures can vary considerably (Mackey, 1984) from relatively mild exposure regimes, which can easily be survived, but which induce tolerance responses, to extreme exposures, for survival of which a range of cold-shock proteins (Wolffe, 1995; Weber and Marahiel, 2003) will be needed. Cold-shock also occurs in natural waters, and subjection of polluting organisms to such shocks may allow them to resist subsequent stresses in foods or, following ingestion, in the animal body.

Irradiation is another stress commonly used in food treatments, especially for preservation, and can also be used in sterilization treatments, e.g. of shellfish (Hicks and Rowbury, 1987) and, occasionally, of water for drinking; aside from inducing tolerance to irradiation itself (Walker, 1984), many other tolerances can be induced in organisms subjected to irradiation.

Acidity and alkalinity can also be used in food production processes (Humphrey, 1981), and in foods themselves, acidity, alkalinity (Humphrey et al., 1991), high salt concentrations and extremes of osmotic pressure can be faced, these agents or conditions, arising in the foods or being added to them, both to enhance flavour and for preservation. Where organisms have been exposed to osmotic stress in food materials (Csonka, 1989), this may influence their ability to resist cooking (since osmotic stress induces thermotolerance); they may then on ingestion, go on to cause disease. Acid and alkali stress exposures can also occur in natural waters (Rowbury et al., 1989), including sewage-polluted ones and in sewage itself, and organisms that entered food materials via the use of polluted waters or by the dressing of agricultural land with sewage sludge might be stress tolerant and, if ingested, able to resist host defense agents, and lead to disease. Exposures to weak acids at neutral pH (Kwon and Ricke, 1998) or at mildly acidic pH (Guilfoyle and Hirshfield, 1996) can also occur in both foods and natural waters. These can induce acid tolerance in contaminating or polluting organisms, under suitable conditions, and could influence subsequent survival of organisms, on later exposure to acidity.

Some inhibitory agents and conditions that do not occur in foods, or in food preparation or production processes, can nonetheless influence the ability of contaminating organisms to survive in response to potentially lethal stresses in foods. For example, organisms can be subjected to starvation in natural waters; since cross-tolerance responses can be induced by starvation (Jenkins et al., 1988, 1990; Matin, 1991; Hengge-Aronis, 1993), organisms starved in water may, if they later enter food materials, resist inhibitory agents or conditions there, e.g. high levels of salt, or high osmotic pressure or extreme acidity, and survive. Similarly, exposure of organisms in aquatic environments to metal ions can induce cross-tolerances. For example, organisms subjected to $Cu^{2+}$ become thermotolerant (Rowbury, 2002); if these organisms subsequently enter food materials, they may survive heating in food production processes or in cooking and, if later ingested might go on to cause disease. Also, organisms in natural waters may be exposed to oxidative components, e.g. as a result of outfalls from chemical plants. These can also induce tolerance (Demple and Halbrook, 1983) and cross-tolerance responses, and any polluting organisms exposed to oxidative components in water, and then entering foods or food materials, may survive lethal agents occurring there, or lethal procedures such as cooking.

## Sites of damage and lethality, and other effects produced by chemical, physical and biological stressing agents

### DNA

Numerous stresses act to damage the DNA, and at suitable stress levels the damaging effects can lead to bacterial death. For several chemical stressors, examination of the effects of the agents on free DNA or on cellular DNA, both chromosomal and plasmid forms, has established that the DNA strands can be broken on exposure to stresses (Woodcock and Grigg, 1972); SS and DS breaks can occur (e.g. with heat, Sedgwick and Bridges, 1972; Pauling and Beck, 1975; Kadota et al., 1978) and SS breaks probably occur at low stressor levels and DS breaks at higher levels. If mutants altered in DNA repair are tested, e.g. with low levels of heat, damage to the DNA occurs as for parental strains, but repair is virtually absent (Woodcock and Grigg, 1972). Similarly, strains of *Escherichia*

coli exposed to acidity (Raja et al., 1991) or alkalinity (Rowbury, 1994) showed SS breaks in chromosomal and plasmid DNAs, with loss of transformation ability, whilst free $R^+$ plasmid DNA, after challenge at low or high pH, lost its ability to transform organisms to antibiotic resistance. Studies with strains mutated in DNA repair activities strongly suggested that damage to DNA is a major, if not the prime, cause of death following exposure of organisms to acidity and alkalinity (Sinha, 1986; Goodson and Rowbury, 1990, 1991), as it is in those exposed to heat.

Other stress exposures, e.g. to hydrogen peroxide (Ananthaswamy and Eisenstark, 1977) also damage DNA, and such damage may again be the major cause of death, following these challenges; interestingly, it has been proposed that DNA damage by heat could result partially from the presence of $H_2O_2$ in media, the heat challenge allowing this oxidative component subsequently to penetrate into the cells and damage the DNA (Mackey and Seymour, 1987).

### Ribosomes

Low doses of many stresses lead to inhibition or cessation of protein synthesis and inducible enzyme formation, and analysis of these results suggests that ribosome function rather than transcription or translation defects are the primary cause of such inhibitions. Specifically, it has been proposed that mildly stressing heat interferes with ribosome function, by leading to a fall in charged tRNAs, resulting in empty A-sites and reduced ribosomal functioning (van Bogelen and Neidhardt, 1990). Greater heat stress can lead to ribosomal RNA breakdown (Allwood and Russell, 1968), although this is unlikely to be the major reason for the lethal effects of this stress at low doses. Mildly stressing cold, it is proposed, leads in contrast to accumulating charged tRNAs, with the blockage of A-sites (van Bogelen and Neidhardt, 1990); any inhibitory effect of cold on ribosome functioning is, however, probably secondary, with respect to reduced or abolished protein synthesis, to the reduced translation of mRNAs which occurs with mild cold stress.

### The Gram-negative outer membrane (OM)

The bacterial outer membrane forms a barrier to hydrophobic molecules and other agents (Nikaido and Vaara, 1985; Nikaido, 1994), but many challenges with chemical and physical stresses lead to defects in this OM barrier (Russell, 1984; Hancock, 1984; Mackey, 1983). For example, treatment of organisms with acidity or at elevated temperatures, can cause structural damage (Katsui et al., 1982) and can lead to permeabilisation of the OMs (Tsuchido et al., 1985, 1989), allowing, e.g. passage of molecules such as detergents, dyes and agents like nisin (Bozarias and Adams, 2001), to which OMs normally present an impenetrable barrier. Many agents which have these permeabilizing effects on the OM also lead to loss of lipopolysaccharide (LPS) from this envelope layer (Nikaido and Vaara, 1985), and leakage out into the medium of periplasmic components.

## The cytoplasmic membrane (CM)

*Damage or inhibition of specific components or CM processes*
Many CM components can be damaged by chemical and physical stressors, for example the $F_1 F_0$ ATPase and a range of permeases; accordingly, active transport and other processes dependent on these components, such as flagellar rotation and certain uptake processes, may fail to occur, when organisms are stressed. One stress which damages the CM is that caused by low pH plus weak acids; this certainly leads to growth inhibition and sometimes death, due to the collapse of $\Delta$pH, but other inhibitory effects on the CM have been proposed (Salmond et al., 1984).

*More general CM damage*
Many agents can act by interfering with the overall functioning of the CM. Detergents act in this way, as does the bacteriocin nisin; with both types of agent, the CMs of many Gram-positive bacteria are susceptible without other damage, but for most Gram-negatives, these agents will not act unless the OM has been permeabilized to them, e.g. by mild heat treatments (Boziaris and Adams, 2001)

## Envelope appendages, e.g. flagella and pili

Several stresses can affect the bacterial flagellum, for example, flagellar rotation depends on $\Delta$p, and, accordingly, any stresses which reduce or abolish this parameter will prevent motility. For example, acidity plus weak acids, which abolishes $\Delta$pH, will often have this effect, and starvation is likely, by affecting energy supply, to influence rotation. Also, relatively low levels of heat will interfere with the functioning of the $F_1 F_0$ ATPase (and, therefore, rotation) and may damage flagellar structure: temperature also affects flagellar synthesis. No doubt, extremes of stress will damage pili also, but increases in temperature, in the physiological range, actually induce sex pilus synthesis.

## The cell division process

Moderate levels of several stresses (heat, cold, low pH, high pH or low $a_w$) lead to cellular filamentation and this applies to *Salmonella enterica* serovar Typhimurium DT104, to *S. enterica* serovar Enteritidis PT4, to *Escherichia coli* O157: H7 and other *E. coli* strains (Mattick et al., 2003b). The temperatures and pH and $a_w$ values giving the effects varied somewhat from organism to organism (Table 5.1) but a wide range of isolates showed the effects. The stresses only inhibited cell division, with nuclear segregation going on as usual during challenge, and division restarted rapidly after removal of the stress (Mattick et al., 2000, 2003a). The latter finding is of major importance in food microbiology, since foodstuffs with virtually no viable organisms, as indicated by plate counts, could show large cell numbers shortly after transfer to non-stressing conditions, due to the above rapid septation.

## Cellular enzymes and other proteins: inactivation by stress

Stresses can inactivate essential enzymes or other essential proteins, and if such inactivation occurred at low stress levels, this could be the basis for the growth inhibitory or lethal effects of a stress. A major example of such a target is the homoserine trans-succinylase enzyme. *E. coli* fails to grow at 45°C in poor media (Ron and Davis, 1971), due to the inactivation of this enzyme, its loss of function leading to methionine starvation, and growth inhibition.

There is no doubt, although DNA is a major damage site, that destruction of proteins by stresses is also highly significant, at least under some conditions. One piece of evidence for this is the finding that loss of certain protein-repairing chaperones, e.g. the DegP or IbpA or IbpB proteins, leads to reduced tolerance to some stresses (e.g. Kitigawa et al., 2000).

**Table 5.1** Filamentation: percentage of organisms, after exposure of cultures to marginal stress, which are of greater than four times the typical cell length

| Strain | Source | % long cells (± SEM) after culture under conditions of | | | | |
| --- | --- | --- | --- | --- | --- | --- |
| | | Heat stress | Cold stress | High pH | Acidity | $a_w$ stress |
| *S. Typhimurium* DT104, strain 30 | Feces | 56 ± 12 | 83 ± 6 | NT | 30 ± 6 | 66 ± 6 |
| *S. Enteritidis* PT4, strain E | Human | 50 ± 6 | 72 ± 9 | 53 ± 10 | 78 ± 1 | 77 ± 2 |
| *S. Enteritidis* PT4, LA5 | Chicken | 86 ± 2 | 80 ± 10 | 58 ± 4 | 75 ± 4 | 61 ± 5 |
| *E. coli*, strain C | Sea water | 88 ± 6 | 68 ± 3 | 36 ± 12 | 40 ± 4 | 93 ± 7 |
| *E. coli*, strain D | Rice | 17 ± 7 | 50 ± 19 | 44 ± 0 | 100 ± 0 | 41 ± 2 |
| *E. coli* O157, strain E112586 | Bird | 83 ± 12 | 84 ± 3 | 34 ± 7 | 81 ± 5 | 91 ± 5 |
| *E. coli* O157, strain 30858 | Human | 73 ± 1.5 | 80 ± 4 | NT | 100 ± 0 | 90 ± 5 |
| *E. coli* O157, strain 510299 | Human | 28 ± 4 | 69 ± 9 | NT | 100 ± 0 | 96 ± 4 |
| *E. coli* O157, strain E100793 | Beef | 77 ± 3 | 89 ± 5 | 38 ± 7 | 100 ± 0 | 94 ± 6 |
| *E. coli* O157, atoxigenic strain | NCTC | 17 ± 2 | 47 ± 10 | 25 ± 3 | 21 ± 1 | 86 ± 4 |

Growth was with the following stress exposures: Heat, salmonellas, 44°C; *E. coli* O157, 46.5°C; other *E. coli* strains, 46°C. Cold, 8°C, all strains. Alkalinity, salmonellas, pH 9.5; *E. coli* strains, pH 9.4. Acidity, salmonellas, pH 4.4; *E. coli* O157 (except atoxigenic one), pH 4.1; other *E. coli* strains and atoxigenic O157 strain, pH 4.4. $a_w$, salmonellas, 0.95; *E. coli* strains, 0.96.

## Cellular proteins: induction of degradation by stress exposures

Many stresses lead to the induction of proteases, which often attack a specific group of proteins, or have dual protein repair and cleavage activities (Table 5.2). Such degradation activities generally target damaged proteins, or specific proteins that are not required at a particular growth stage (e.g. the RpoS protein, which is destroyed by ClpXP, as organisms leave stationary-phase, Li *et al.*, 2000), and would be unlikely to lead to significant lethal effects. Were essential proteins, especially undamaged ones, to be targeted, however, this could lead to lethality.

## Factors which enhance the tolerance of organisms to potentially lethal stresses

### RpoS and inherent stress tolerance

Several factors both inherent and induced influence the levels of stress tolerance. A major study on the influence of inherent factors on stress tolerance was that in Humphrey's laboratory. His group (Humphrey *et al.*, 1995) showed that clinical isolates of *Salmonella enterica* serovar Enteritidis were substantially more resistant than a number of strains from non-clinical sources, and this was shown to result from differences in RpoS. The clinical strains showed greater inherent tolerance to heat, acid and hydrogen peroxide and were better able to survive on surfaces. This work has not been sufficiently emphasized and, accordingly, it is intended here to detail the likely significance of the results. First, the heat tolerance means that

were such tolerant isolates to enter food materials, they would be more likely to survive heating in food production processes and in cooking itself. Such survival might allow them to be ingested. On ingestion, the organisms would be better able to resist acidity in the stomach, than non-clinical isolates. If surviving organisms crossed the intestinal epithelium, even if they were phagocytosed, they might survive because of their inherent acid and $H_2O_2$ tolerance. All in all, the tolerances of these organisms would give them a better chance of reaching the body, surviving in it and causing disease. It is likely that these increased stress tolerances play a major role in another property of these isolates, namely their increased virulence (Humphrey *et al.*, 1996). It is now well established that RpoS plays a major role in determining the level of inherent stress tolerance in isolates; those clinical ones which show increased tolerances have full wild-type *rpoS* genes, whereas inherently less tolerant strains have mutated *rpoS* genes with reduced or abolished activity.

### Growth phase and stress tolerance

It is now clear that stress tolerance is greatly increased as organisms enter stationary phase and the usual explanation is that increased levels of the sigma factor RpoS occur on entry into this phase, and that this sigma factor leads to increased transcription of a large group of genes, with the result that many proteins including tolerance proteins are derepressed. Starvation stress also enhances the translation of the mRNA for RpoS and stabilizes RpoS by stopping its proteolytic cleavage. In log-phase, the *clpP*, *clpX* operon is fully transcribed and the ClpXP protease arises (Li *et al.*, 2000). This protease spe-

**Table 5.2** Dual or multiple roles for chaperones and other heat-shock proteins or for proteins involved in other thermal or low-temperature responses

| Chaperone or other stress-related protein | Cellular location | Activities under non-stressing conditions | Activities under stressing conditions |
|---|---|---|---|
| DnaJ | Cytoplasm | Protein maturation | Protein repair, RpoH destabilization |
| DnaK | Cytoplasm | Protein maturation, secretion and proteolysis | Protein repair, RpoH destabilization, proteolysis |
| GroEL | Cytoplasm | Protein maturation, secretion and proteolysis | Protein repair, RpoH destabilization. Holding proteins in protected state, proteolysis |
| GroEL | Envelope[1] | Cytotoxicity, promotion of cellular invasion | Cytotoxicity, promotion of cellular invasion, likely role in protection from thermal and other stresses |
| GrpE | Cytoplasm | Protein maturation | Protein repair, RpoH destabilization |
| DegP | Periplasm | Protein maturation | Protein repair (low stress), protease (high stress) |
| ClpXP | Cytoplasm | Protein maturation | Protein repair and proteolysis |
| ClpXA | Cytoplasm | Protein maturation | Protein repair and proteolysis |
| Lon | OM and cytoplasm | ? | Protein repair and proteolysis |
| FtsH | Cytoplasm | Protein maturation | Protein repair and maturation |
| CspA (cold shock) | Cytoplasm | ? | RNA chaperone, transcriptional and translational activator |
| ESC/EIC pair[2] | External medium | Heat sensing[3] (ESC) | Thermotolerance induction (EIC) |

[1]Some GroEL appears external or envelope attached in some bacteria, the proportion increasing in heat stress.
[2]Although the thermal response ECs are shown as an ESC/EIC pair, ability of these two components to be readily and reversibly inter-converted, shows they are two forms of the same protein.
[3]Or sensing acidity, alkalinity, UV irradiation or $Cu^{2+}$.

cifically cleaves both RpoS and many of the proteins which have been translated from mRNAs produced only from operons, transcribed by RNA polymerase plus RpoS. In stationary-phase, the ClpXP protease does not arise in appreciable quantities, and RpoS is not cleaved. Note that the ClpXP can also function as a chaperone, to repair damaged proteins (Table 5.2).

The above functioning of RpoS explains a part of the stress tolerance that arises in stationary-phase, but there is another, generally overlooked, component.

## Termination of chromosome replication and stress tolerance

Growth into stationary-phase leads to termination of chromosome replication, i.e. to the production of completed chromosomes. Such chromosomes, by definition, have no replication forks and, accordingly, no single-stranded regions. Many stresses have major effects on DNA, with this being the case, for example, for UV irradiation, for acid and alkali stresses and for stress by heat and hydrogen peroxide (Walker, 1984; Sinha, 1986; Raja et al., 1991; Rowbury, 1994; Ananthaswamy and Eisenstark, 1977), and it is believed that these lethal effects are almost exclusively at single-stranded regions. Accordingly, it would be expected that organisms growing into stationary phase would become stress-tolerant due simply to the fact that single-stranded regions of DNA would disappear in this growth-phase. That replication termination does lead to stress tolerance has been shown by Rowbury (1972). Growth of a DNA synthesis initiation mutant at restrictive temperature led to UV tolerance, mimicking the so-called SIRE effect (starvation-induced resistance enhancement), which leads to UV tolerance on starvation. These findings show that stress-tolerance in stationary-phase results, partially at least, from termination of chromosome replication.

## Attachment of organisms to surfaces and stress tolerance

Organisms attached to surfaces are more tolerant than free organisms to a huge range of chemical, physical and biological agents. Attachment can not only occur in the aquatic environment, but also in foods, with organisms being attached or becoming attached to solid components (Humphrey et al., 1997). If organisms attach to particles in natural waters, this may permit them to survive potentially lethal stress exposures in water or purification or sterilisation procedures. The surviving organisms might then enter food processing or production systems in the sterilized water. Agents or conditions which occur in natural waters or in purification processes and to which attached organisms are resistant, include acid, alkali, heat, chlorine, bacteriophages, colicins, acrylate and metal ions (Hicks and Rowbury, 1986; 1987; Whiting, 1990; Rowbury, 2002b). In foods, among the agents or conditions to which attached organisms would be resistant are heat, weak acid preservatives (such as sorbic and propionic acids), inorganic acid and alkali (Humphrey et al., 1997; Rowbury, 2002b), as shown in Table

5.3. It is likely that attached organisms are resistant to the lethal effects of many other agents, and it should be noted that such resistance extends to a range of antibiotics and other antibacterials (Table 5.3) which have therapeutic uses (Whiting, 1990; Rowbury and Whiting, unpublished observations). This finding suggests that ingested pathogens, which are attached to food particles, may resist lethal effects of antibacterials, and other inhibitory agents such as bile salts, in the intestine. Also, pathogens which become attached to surfaces in the body may show resistance to antibacterials.

## Inducible stress tolerance and sensitization processes

The major factors governing the stress tolerance of organisms relate to the levels of inducible stress responses. Three major classes of such response occur. The first are those involving induction of tolerance to a stress, following low level exposures to the same stress. Thus, micromolar levels of $H_2O_2$ induce tolerance to millimolar concentrations (Demple and Halbrook, 1983), mild heat induces thermotolerance (Mackey and Derrick, 1986; Humphrey et al., 1993) and exposures to extremes of pH induce acid and alkali tolerance (Rowbury et al., 1989). Several other tolerance responses are induced by chemical and physical stresses, such as potentially lethal metal ions, alkylating agents and UV light.

Exposure of bacteria to chemical and physical stresses also leads to cross-tolerance responses. Exposure to acidity, for example, induces tolerances to heat, UV irradiation (Leyer and Johnson, 1993) as well as to acidity and exposure to heat induces tolerance to acid, alkali and UV irradiation, as well as thermotolerance (Rowbury, 2003). Other stresses have similar effects with UV irradiation leading to at least four cross-tolerances (Rowbury, 2003) and starvation leading to at least six (Matin, 1991).

Although the ability of stresses to induce tolerance responses is well known, there is generally little knowledge of cross-sensitisation responses, which may be almost as common. For this reason, some emphasis will be given here to some cross-sensitisation processes. A number of these have been studied at UCL, and three of the major ones are indicated below. There are three acid sensitisation processes; exposure of organisms to any of the inducing conditions that lead to these sensitizations will, for example, make the organisms less able later to resist exposure to potentially lethal proton concentrations in water, in foods or in the animal or human body.

*Response a.* Acid sensitivity induction is a well-studied phenomenon, that occurs on transfer of E. coli to pH 9.0. Almost all strains of E. coli show this response, but *phoE* mutants show more sensitisation than other strains (Rowbury et al., 1993), a property associated with their inherent acid tolerance. Sensitisation involves two processes, induction of only one needing protein synthesis. It is believed that sensitisation involves induction or modification of an OM pore, but this has not been identified.

**Table 5.3** Surface-attached *E. coli* are more resistant to a range of inhibitors than are free organisms

| Inhibitory agent (concentration) | % survival or growth* (± SEM) after challenge of | |
| --- | --- | --- |
| | Free organisms | Surface-attached organisms |
| Kanamycin (20 µg/mL) | 2.6 ± 0.9 | 68.4 ± 6.9 |
| Gentamicin (10 µg/mL) | 2.0 ± 0.19 | 70.5 ± 4.8 |
| Streptomycin (30 µg/mL) | 4.7 ± 0.1 | 40.5 ± 4.0 |
| Rifampicin (30 µg/mL) | 43.5 ± 1.8 | 87.0 ± 11.1 |
| Nalidixic acid (50 µg/mL) | 39.2 ± 4.4 | 68.8 ± 7.3 |
| Chloramphenicol (100 µg/mL) | 26.4 ± 1.8 | 82.3 ± 2.3 |
| Polymyxin B (10 µg/ml$^{-1}$) | 0 ± 0 | 27.8 ± 5.1 |
| Novobiocin (50 µg/mL) | 61.3 ± 2.9 | 90.3 ± 5.3 |
| CTAB (10 µg/mL) | 25.3 ± 0.8 | 101.9 ± 1.1 |
| SDS (100 µg/mL) | 42.5 ± 5.6 | 91.9 ± 3.0 |
| $H_2O_2$ (42 mmol/L) | 52.6 ± 8.4 | 96.7 ± 5.5 |
| Butyric acid (30 mmol/L)* | 59.3 ± 7.1 | 105.7 ± 4.2 |
| Transcinnamic acid (30 mmol/L)* | 0.01 ± 0.01 | 82.5 ± 5.9 |
| Citric acid (30 mmolo/L)* | 34.3 ± 2.6 | 84.3 ± 2.6 |
| Lactic acid (30 mmol/L)* | 6.0 ± 3.0 | 85.6 ± 9.5 |
| Sorbic acid (30 mmol/L)* | 31.0 ± 13.3 | 113.2 ± 4.5 |

*E. coli* strain P678–54 ColV (for tests with weak acids) and strain 1829 ColV (for tests with antibiotics and other agents) were grown in broth at 37°C to midlog-phase and suspensions of free and attached organisms prepared as described by Hicks and Rowbury (1986). Cultures of strain P678-54 ColV were then exposed to pH 3.5 with or without weak acid for 15 min at 37°C, and subsequent growth of washed cells followed at 37°C in broth. Cultures (strain 1829 ColV) were exposed to antibiotics for 2 0 min at 37°C (kanamycin, gentamicin, streptomycin, rifampicin and nalidixic acid) or for 240 min at 37°C (SDS, CTAB, chloramphenicol and novobiocin). Washed organisms were plated after challenge for viable count. All these results are from Whiting (1990) and Whiting and Rowbury (unpublished observations). The ColV used here and in other tables is ColV, I-K94.

*Response b.* Growth with high levels of NaCl leads to acid sensitivity and use of *phoE–lacZ* and *phoE–phoA* mutants (Lazim et al., 1996) showed that this is associated with PhoE induction, i.e. increased levels of this pore in the outer membranes of *E. coli* lead to acid sensitivity. This involvement of PhoE in sensitivity to protons, is in accord with the effects of loss of PhoE (1) in *phoE* mutants, such as *E. coli* AB 1157, which are acid resistant, and (2) by repression, which leads to acid resistance.

*Response c.* Strikingly, organisms that have the third response, L-leucine-induced acid sensitivity, appear to contain an altered OmpA protein, as evidenced by their increased phage K3 resistance (Rowbury et al., 1996). Transfer to L-leucine-containing medium did not induce PhoE synthesis and, in fact, L-leucine reduced PhoE induction by NaCl. Interestingly, mutants of *E. coli* which have lost the OmpA protein do not show acid sensitisation by L-leucine. Some strains with altered forms of OmpA show the L-leucine response, others do not. Studies of *ompA* mutants show that loss of acidic amino acids (e.g. glutamate or aspartate) in the surface loops of OmpA leads to a failure to induce the response (Rowbury, 1999; 2002b).

Other cross-sensitization responses occur. In view of the relevance of alkaline pH to bacterial survival in water (Rowbury et al., 1989), in foods (Humphrey et al., 1991) and in the body (Segal et al., 1981), sensitisation to alkali is of interest. Such sensitisation occurs at pH 5.5–6.0 and is dependent on IHF, H-NS, Lrp and CysB (Rowbury and Hussain, 1996).

## How inducible stress responses are switched on

Inducible responses have been examined for many years, and these researches have included studying the earliest stages in induction, i.e. how such responses are switched on. In contrast, very little work has been undertaken on inducible stress responses and this article will continue to redress this imbalance.

### Initiation of stress responses, as exemplified by acid tolerance induction

Until recently, it has been believed that the induction of stress responses follows essentially the same process as that involved in the switching on of other inducible responses. These processes are believed to involve the response stimulus being detected inside the cell by an intracellular sensor, and with activation of the sensor by its interaction with the stimulus, leading to the formation of an intracellular signaling component able to lead to response induction, often, but not always, by enhancing transcription of the appropriate operon (Neidhardt et al., 1990; Somerville, 1982; Wanner, 1987). It may be that some stress responses, which are induced by intracellular chemical stress stimuli, involve a process like that indicated above (Foster and Moreno, 1999; Kullick et al., 1995), but it is now clear that numerous stress responses, especially those involving extracellular chemicals, but also those switched on by physical stimuli, are quite distinct in their initiation stages. For the switching on of these, extracellular stress sensors function, and the activation of these by stress stimuli leads to formation of extracellular signaling molecules, which subsequently interact with organisms and induce the appropriate response. For most stress responses, therefore, pairs of extracellular components (ECs) function to set in train the earliest stages of the induction process.

*Discovery of role of ECs in acid tolerance induction in Escherichia coli*

Extracellular components which switch on tolerance were specifically looked for (1) by testing the effects of removing ECs from cultures which had been transferred to pH 5.0 to induce the response; removal of ECs by filtration during incubation or by adding proteases during the process, abolished tolerance induction and suggested that the induction components were proteinaceous in nature (Rowbury and Goodson, 1998). Involvement of ECs in the pH 5.0 cultures were also implicated (2) by filtering-off organisms from the cultures and showing that the cell-free filtrates, after neutralization, were able to induce acid tolerance in organisms at pH 7.0. Trivial reasons for the tolerance induction effects by the filtrates were ruled out, and the ECs responsible were shown not to be small metabolites. In fact, the induction components in the filtrates proved to be non-dialyzable proteins, which were relatively heat-stable (Rowbury and Goodson, 1998). The ECs which act as induction components, in the above process, were later termed extracellular induction components (EICs).

*ECs as acidity sensors*

It was argued that if incubation at pH 5.0 leads to acid tolerance due to formation of EICs at this pH, it was highly likely that acidity in the medium was detected by an EC, i.e. that the acidity sensor was extracellular and not intracellular. This proved to be so. Medium filtrates from pH 7.0-grown cultures were unable to induce acid tolerance in pH 7.0 cultures, but if the filtrates were exposed to pH 5.0 for 30 min, in the absence of organisms, the activated filtrates, after neutralization, were able to induce tolerance. Again trivial explanations for the effect were ruled out, and the active agent was shown not to be a small metabolite. The likely explanation, which has been amply confirmed, was that filtrates from pH 7.0-grown cultures contained an EIC-precursor, which was activated to the EIC by acidity, even in the absence of organisms. This EC was, therefore, a component, which detected acidity, and was converted by it to a product (EIC) able to induce tolerance. On this basis, it is an acidity sensor, but highly unusual in being the first extracellular stimulus sensor demonstrated (Rowbury and Goodson, 1999a). Tests show that this sensing component closely resembles the EIC, in being a rather heat-stable non-dialyzable protein, but unlike the EIC cannot induce tolerance without prior activation. This EC sensor arises *de novo* from amino acids in the medium, rather than from peptides, but the synthesis process is unusual, in that some antibiotics which normally target the ribosome do not abolish sensor synthesis

(Rowbury and Goodson, 1999a); small stress-related proteins or peptides seem commonly to involve an unusual synthetic process (Etchegaray and Inouye, 1999). This EC sensor has now been termed an extracellular sensing component (ESC).

*Cross-tolerance responses, which lead to acid tolerance*

It has been known for some time that cultures of enterobacteria grown at 37°C, and shifted to 42–50°C for short periods, become acid tolerant (Humphrey *et al.*, 1993), and we have shown that either exposure to UV irradiation, followed by incubation, or incubation with cupric ions also leads to acid tolerance. The basis for these effects has now been established; the acid tolerance-related ESC can be activated by heat, UV or $Cu^{2+}$ as well as by acidity, the resulting EIC leading to acid tolerance.

*Different forms of the ESC, synthesized at different pH values*

The ESC synthesized at pH 7.0 can be activated to EIC at pH values up to pH 6.0, but not above. In contrast, the ESC formed at pH 9.0 was converted to EIC over the pH range 5.0–7.5, and even at pH 8.0 there was significant activation. Studies of ESCs from cultures grown at intermediate pH values showed that activation occurs appreciably at one pH unit below the ESC synthesis pH. The proposal is that conformationally different forms (or possibly oligomeric forms) of the ESC occur, with those formed at high pH being activated even before the medium becomes acidic; this would protect organisms growing at high pH from a sudden acidification and eventual lethal acidity.

## On the induction of thermotolerance in enterobacteria

The ability to resist heat challenges is of major importance for potentially pathogenic enterobacteria, because if such organisms are present as contaminants in foods or food materials, heat in food preparation or production processes or in cooking, must be resisted if viable organisms are to go on to be ingested and have the potential to cause disease. It is proposed, here, to review how thermal stress is sensed and how inducible thermotolerance processes are switched on. First, it has now been well established that transferring organisms from low temperatures to higher ones induces thermotolerance, but even more strikingly, that cross-tolerance responses can also lead to thermotolerance, when organisms at constant temperature are mildly stressed by other means, e.g. by exposure to acidity, alkalinity, UV or to metal ions (Table 5.4).

*There is no firm direct evidence for intracellular thermal sensors*

For many years, it has been assumed that, as for other inducible responses, the sensor detecting thermal stress and initiating thermotolerance induction, would be an intracellular component, and several membrane and cytoplasmic molecules or structures have been strongly advocated as sensors. In particu-

**Table 5.4** Thermotolerance induction in *E. coli*, by exposure of cultures to mild thermal stress and other stresses

| Culture incubation conditions | % survival ± SEM after exposure of culture to 49°C for 5 min |
|---|---|
| pH 7.0 at 30°C | 0.43 ± 0.07 |
| pH 7.0 at 45°C | 24.9 ± 2.25 |
| pH 5.0 at 30°C | 24.1 ± 1.1 |
| pH 5.5 at 30°C* | 13.1 ± 1.2 |
| pH 5.5 plus $Cu^{2+}$ at 30°C* | 50.5 ± 3.25 |
| pH 9.0 at 30°C | 29.0 ± 1.55 |
| UV 60s then 30°C incubation | 49.0 ± 1.2 |

Cultures of *E. coli* 1829 ColV were grown to log-phase in pH 7.0 broth at the stated temperature, and then incubated further for 60 min under the same conditions or at pH 5.0 or 5.5 or 9.0, or at pH 5.5 with 60 µg/mL $CuSO_4$. One sample was irradiated for 60s with a Philips 6W TUV tube 15 cm from the culture, followed by incubation for 60 min at pH 7.0 and 30°C.
*The culture incubated with $Cu^{2+}$ was dialysed against 50 vols of pH 7.0 broth for 16 h at 4°C (to remove $Cu^{2+}$), after incubation and prior to thermal challenge; the copper -free control was treated in the same way. After incubation the cultures were thermally challenged as stated.

lar, the ribosome has been proposed to act in this way, with the suggestion being that the fall in charged tRNA levels, which may follow temperature up-shifts, would lead to some A sites being empty and heat-shock protein (HSP) synthesis could accordingly be induced (van Bogelen and Neidhardt, 1990). Another favorite candidate for a thermal stress sensor is the DnaK gene product (McCarty and Walker, 1991). This product does show greatly increased activity as the temperature is increased in the range that induces heat-shock, but it is most likely that this has simply evolved as a means of increasing chaperone activity (i.e. so that repair activity for damaged proteins increases with increasing temperature), rather than being related to thermal sensing for response induction. Other cytoplasmic molecules such as DNA have been suggested as thermal sensors, and cytoplasmic and outer membrane components and cellular enzymes, including homoserine-trans-succinylase, have been other favored candidates (Rowbury, 2002b). There has been no appreciable direct evidence for functioning of any of these. In contrast, the view that unfolded proteins (that arise due to partial or complete denaturation by stress damage) may play a role in triggering stress responses (including the heat shock response) has been supported by the finding that mutations and conditions that lead to unfolded protein accumulation, cause HSP synthesis (Lund, 2001), but this would only explain HSP induction at potentially lethal temperatures; the induction that occurs at 40–45°C could not be due to induction by partially denatured proteins, since such denaturation does not occur appreciably at these lower temperatures.

*Possible arguments against extracellular components as thermal sensors*
Tests were made above for functioning of ECs in acid tolerance induction, because potentially lethal acid stresses faced by bacteria are generally extracellular, i.e. the lethal protons are in the medium and, accordingly, responses to them would be more rapid if the sensing components were extracellular. As indicated above, EC sensors and induction components do occur

for acid tolerance induction, as predicted. Assuming that such ECs have evolved to ensure rapid responses to chemical stressing agents or conditions, such as lethal acidity, in the growth medium, the involvement of similar ECs in thermotolerance induction would not be expected, since levels of stress would be equal inside and outside the organism, when it is exposed to thermal challenges.

*Extracellular sensing and induction components switch on thermotolerance*
Although the above argument would appear to suggest that there would be no advantages in ECs functioning in thermotolerance induction, their possible involvement was tested. Cell-free filtrates from 45°C-grown cultures of *E. coli*, were, on addition to organisms growing at 30°C, able to induce the latter to thermotolerance (Table 5.5). All possible trivial explanations were ruled out, and the components in the 45°C filtrates were demonstrated to be protease-sensitive molecules of molecular weight ca 10,000 Da. These components were, therefore, not low molecular weight secreted metabolites, but were apparently proteinaceous thermotolerance-related EICs.

Whereas 45°C filtrates were active in inducing thermotolerance, filtrates from 30°C-grown cultures were not. They did, however, contain thermotolerance related ECs, since on incubation at 45°C, without organisms, 30°C filtrates became able to induce thermotolerance in 30°C cultures. Again trivial explanations were ruled out and tests on the 30°C filtrates, prior to activation at 45°C, showed that the components in the filtrates were relatively low molecular weight proteins. Apparently, interaction of these components with the elevated temperature had converted them to EICs, i.e. a chemical or enzymic reaction had occurred at 45°C. On the basis that the components in the 30°C filtrates behaved as precursors of response induction components, and in view of the fact that they interacted with the stressing condition (heat) to produce the EIC and lead to induction, they can be considered to be EC sensors. By analogy with the acid tolerance induction system, they are thermotolerance-related ESCs. It is believed that they

**Table 5.5** Tolerance to thermal stress induced by filtrates activated by heat and other stresses

| Filtrate from 30°C-grown culture and the stressing conditions used for its activation | % survival (± SEM) after thermal stress for 30°C-grown culture after preincubation with or without filtrate |
|---|---|
| No filtrate | 0.53 ± 0.033 |
| Filtrate, no activation | 0.33 ± 0.09 |
| Filtrate, activated at 45°C | 18.0 ± 0.47 |
| Filtrate, activated at 50°C | 29.8 ± 0.72 |
| Filtrate, activated at 55°C | 44.4 ± 1.25 |
| Filtrate, activated by UV, 60 sec | 20.2 ± 0.83 |
| Filtrate, activated by UV, 90 sec | 34.1 ± 1.5 |
| Filtrate, activated at pH 5.5 | 16.9 ± 1.5 |
| Filtrate, activated at pH 5.0 | 25.0 ± 2.35 |
| Filtrate, activated at pH 8.5 | 21.3 ± 0.58 |
| Filtrate, activated at pH 9.0 | 31.4 ± 0.58 |
| Filtrate, activated with $CuSO_4$ | 44.0 ± 1.22 |

Cultures of *E. coli* 1829 ColV were grown at 30°C and cell-free filtrate prepared. Some samples of filtrate were then activated at the stated elevated temperature. Other samples were activated by incubation at the stated pH value or at pH 5.5 with 60 µg/mL $CuSO_4$, or following UV irradiation. Incubation was in all cases for 30 min. Copper-treated filtrate (and its pH 5.5 control) was then dialysed as for Table 4. Filtrates were then incubated for 60 min at 30°C with 30°C-grown culture of the same strain (1/1 filtrate to culture, the latter containing ca 7–10 × 10$^7$ organisms per mL), prior to thermal challenge (49°C for 5 min).

have evolved so as to be able to diffuse away to non-stressed organisms, and give early warning of thermal stress; on this basis, they are both pheromones and alarmones.

*Cross-tolerance responses leading to thermotolerance induction*

Recent studies have established that exposure of enterobacterial cultures to a number of stresses, other than thermal stress, leads to thermotolerance. Transfer of organisms growing at 30°C and neutral pH to mildly acidic or alkaline conditions leads to such thermotolerance-induction and the same occurs on exposure of such organisms to UV irradiation (followed by incubation) or to $Cu^{2+}$. It is now known (Rowbury, 2002a) that these other stresses act by converting the thermotolerance-related ESCs to EICs, which are able to induce thermotolerance if incubated with 30°C-grown cells (Table 5.5).

### Extracellular stress sensors and induction components and the switching on of UV tolerance

It has been believed, for more than twenty years, that most of the major stages in the switching on and induction of UV tolerance were known, and although the precise ways in which DNA damage switched on the earliest stages of the process were not entirely clear, it seemed rational, and was widely accepted, that DNA was the UV-sensor, the interaction of UV and DNA producing the signal which induced UV tolerance, and all the related processes. The same arguments which suggested that there would not be ECs involved in the switching on of thermotolerance, were also compelling in the case of UV tolerance induction by low doses of UV irradiation, and the widely held belief that DNA was the sensor also discouraged testing for EC involvement. Nonetheless, the ability of filtrates to induce UV tolerance was investigated, at UCL.

First, studies of filtrates from unirradiated cultures showed that they were unable to induce UV tolerance in unirradiated organisms (Table 5.6). They became able to do so, however, if irradiated, in the absence of organisms (Table 5.6). Trivial explanations having been ruled out, it became apparent that exposure of filtrates to UV led to the production of EICs, and that the filtrates, before irradiation, contained EIC precursors, which were converted to these UV tolerance induction components by interaction with UV, i.e. these precursors acted as UV sensors and can be considered UV tolerance-related ESCs. These ESCs prove to be dialyzable and more heat-sensitive than most ESCs. These UV tolerance-related ECs are probably proteins.

*UV tolerance induced by other stresses*

Several stresses other than UV irradiation induce UV tolerance when cultures are exposed to them, and it is now clear that this results from the ability of these stresses to activate the UV tolerance-related ESC. Thus, the UV tolerance EIC arises when filtrates from unirradiated cultures (which contain the UV tolerance-related ESC) are exposed to heat, acid and alkali, since the resulting filtrates will induce UV tolerance when incubated with unstressed cultures. The UV tolerance induction effect of the activated filtrates was greatest for filtrates activated at 55°C, substantial with filtrates activated at pH 5.0, and less but still appreciable by filtrates activated at pH 9.0 (Table 5.6). Strikingly, not only did heat induce UV tolerance, but also the levels of UV tolerance increased with increasing temperature from 40°C to 55°C (Table 5.7).

### Role of ESC/EIC pairs in other stress responses

Responses which lead to changes in alkali tolerance are of interest, because organisms may be exposed to alkaline condi-

**Table 5.6** An extracellular sensor detects UV irradiation, its activation leading to UV tolerance induction

| Filtrate from 37°C-grown culture | Treatment before filtrate activation | Treatment after filtrate activation | % survival (± SEM), after UV, for 37°C culture incubated with filtrate |
|---|---|---|---|
| Not activated | N.A. | N.A. | $0.07 \pm 0.006$ |
| Activated, UV 30 s | None | None | $0.4 \pm 0.06$ |
| Activated, UV 60 s | None | None | $6.5 \pm 0.49$ |
| Activated, UV 90 s | None | None | $6.0 \pm 0.48$ |
| Activated, pH 5, 30 min | None | None | $11.3 \pm 0.49$ |
| Activated, pH 9, 30 min | None | None | $3.5 \pm 0.42$ |
| Activated, UV 90 s | Protease | None | $5.7 \pm 0.35$ |
| Activated, UV 90 s | None | Protease | $0.2 \pm 0.06$ |
| Activated, UV 90 s | 75°C | None | $0.5 \pm 0.06$ |
| Activated, UV 90 s | None | 75°C | $0.6 \pm 0.06$ |
| Activated, UV 90 s | Dialysis | None | $1.3 \pm 0.15$ |
| Activated, UV 90 s | None | Dialysis | $6.1 \pm 0.31$ |

Strain 1829 ColV was grown to midlog-phase in pH 7.0 broth at 37°C and cell-free filtrates prepared. Such filtrates were activated (1) by exposure to UV irradiation (as for Table 5.4), followed by 30 min incubation at 37°C and pH 7.0, or (2) by incubation at pH 5.0, or (3) by incubation at pH 9.0. Where stated, filtrates were treated by the stated methods, before or after activation. Neutralised filtrates were then incubated with mid-log phase cultures of 1829 ColV grown at pH 7.0 and 37°C (1/1 filtrate to culture, with the culture containing $7–10 \times 10^7$ per mL viable organisms), with the incubation being under the same conditions as culture growth. Cultures were then challenged with UV (under the conditions indicated above) for 120 s.

**Table 5.7** An extracellular thermometer detects increasing temperature and induces UV tolerance

| Filtrate from 37°C culture | % survival (± SEM) after UV irradiation, for 37°C-grown culture incubated with filtrate |
|---|---|
| Not activated | $0.0043 \pm 0.00067$ |
| Activated at 40°C | $0.033 \pm 0.012$ |
| Activated at 42°C | $0.26 \pm 0.09$ |
| Activated at 45°C | $1.37 \pm 0.15$ |
| Activated at 50°C | $10.0 \pm 0.86$ |
| Activated at 55°C | $21.8 \pm 1.45$ |
| Activated at 60°C | $15.1 \pm 0.68$ |

Strain 1829 ColV was grown to mid-log phase at 37°C and pH 7.0 and cell-free filtrates prepared. Where appropriate, these filtrates were activated at the stated temperature for 30 min and then incubated for 60 min with the above culture (1/1 filtrate to culture, the latter containing $7–10 \times 10^7$ mL$^{-1}$ viable organisms), before challenge with UV (see Table 5.4 for conditions) for 150 s.

tions in the natural environment, as well as in foods and in the body. Alkali tolerance appears on exposure to non-lethal levels of alkalinity (pH 8.5–9.5) and switching on of this response involves the sensing of alkalinity by ESCs (Rowbury, 2001); the likelihood is that there are two ESCs, one a protein and the other a non-protein component. Exposure of these to alkaline pH, in the absence of organisms, produces a pair of EICs, which are needed for full tolerance induction.

Alkali sensitivity induction occurs at pH 5.5–6.0. The proton sensor is a very heat-stable small extracellular protein or peptide of molecular weight less than 5000 Da (Rowbury, 2001). On activation at pH 5.5, in the absence of organisms, this ESC gives rise to an EIC, which resembles the ESC in size and properties and, on incubation with pH 7.0-grown organisms, induces them to alkali sensitivity.

Alkylhydroperoxide (AHP) tolerance is another response which is induced by pH change, in this case, transfer to pH 9.0. Most ESC/EIC pairs are protein or peptide in nature, but such AHP tolerance induction involves a non-protein pair.

Cultures grown at pH 9.0 contain a small (readily dialyzable) molecule, which will induce AHP tolerance in pH 7.0-grown cultures. This EIC is sensitive to destruction at 75°C but insensitive to protease (Rowbury, 2001). Cultures grown at pH 7.0 do not contain EIC, but they contain a precursor ESC, which is converted to EIC on incubation at pH 9.0, in the absence of organisms.

## The switching on of tolerance by starvation stress; sensing of carbon starvation

Polluting organisms in natural water or in polluted soils may later enter foods, due to the use of polluted water in food production or due to contamination of food components, e.g. vegetables with polluted water or soils. In view of this, starvation responses, to which organisms may be exposed in waters or soils, are relevant to food microbiology. It was established by Matin and his group that carbon starvation induces two major types of response. The first class of response induces enzymes that permit degradation of carbon compounds and, therefore, release organisms from carbon starvation (*cst* genes, regulated by cAMP). Of more relevance here are those responses governed by *pex* genes, which lead to stress tolerance (Jenkins *et al.*, 1988, 1990; Matin, 1991), with carbon starvation inducing salt tolerance, acid tolerance, thermotolerance, osmotic tolerance and oxidative component tolerance.

The proposal is that transcription of *cst* genes is controlled initially by the levels of sugars, a fall in the glucose concentration, for example, leading to a rise in protein III$^{glc}$-phos, followed by increased adenyl cyclase, leading to cAMP synthesis and *cst* gene induction. In this scenario, a CM component, protein III$^{glc}$ is the sensor, i.e. there is an intracellular sensor. I believe that ESC/EIC pairs will prove to be involved, at some stage.

There is so far no direct evidence as to how carbon starvation for *pex* gene induction is sensed, or even the precise nature of the stimulus; certainly, a fall in sugar concentration is unlikely to be the stimulus for starvation-induced stress tolerance, since some tolerance responses are induced by sugars (Rowbury, 1999). It has been assumed that an intracellular sensor occurs. It is proposed, however, to examine whether ESCs and EICs occur here, and this seems likely. Until the nature of the stimulus is determined, however, fruitful studies will be difficult.

## Switching on of cold-shock responses

The critical factor here is how cold is sensed, and several mechanisms involving the internal sensing of cold have been proposed, i.e. these are mechanisms whereby the low temperature directly affects the properties or structures of internal components such as ribosomes or mRNAs (van Bogelen and Neidhardt, 1990; Wolffe, 1995). It has, however, not been previously pointed out that, with such mechanisms, because organisms can only begin to induce the cold-shock response after the stress begins (i.e. after the temperature falls), induction will be very slow. For rapid induction, paradoxically, the response must be switched on <u>before</u> the cold stress begins. The means by which such a paradoxical situation can occur is already known, since the sensing of stresses by extracellular sensing components, ESCs, with interaction of ESC and stress leading to extracellular induction component, EIC, formation (Rowbury, 2001) has already been discussed above. Diffusion of EICs to unstressed regions leads to early warning of impending stress, with response induction beginning, in unstressed cells, *before* stress exposure. Because of the above, it seems highly likely that a cold stress ESC will have evolved and after its conversion by low temperature to a cold stress EIC, cold shock response induction will be able to occur for unstressed organisms *before* exposure to low temperature, and, accordingly, such induction (because it is occurring at higher temperatures) will be much more rapid than for cold-shocked cells. The evolution of an ESC/EIC pair for cold shock responses would, in fact, have been much more likely than for other physical stresses. This is because, responses to other physical stresses (e.g. to thermal stress and UV irradiation stress, for example) will only allow

early warning to be conveyed to unstressed cells. In contrast, in the case of cold-shock, the evolution of a cold-shock ESC/EIC pair will both enable early warning of stress to occur, and allow more rapid response induction in the unstressed cells (which are at a higher temperature, than the stressed ones) following cross-feeding by the EIC, which has diffused from stressed regions. I predict, therefore, that cold will prove to be sensed by a specific ESC, with the EIC that arises by the cold – ESC interaction, inducing cold-shock responses.

## The wider significance of ESCs and EICs

There was no surprise in the finding that ESCs and EICs functioned in the early stages of induction of several stress responses, but they now appear to have greater significance, (1) probably functioning in all stress response induction processes, even those triggered by physical stresses, (2) also acting as early warning alarmones, (3) playing an unexpected role by allowing killed cultures to induce responses, and (4) most strikingly acting (ESCs) as biological thermometers.

## Why are ESCs and EICs involved in switching on responses to physical stresses?

The involvement of ECs in response induction was initially tested for, because it was believed that sensing chemical stresses in the medium might lead to more rapid responses. Physical stresses, such as UV irradiation and heat, however, affect the external milieu and the organisms equally and, therefore, ECs would not lead to a more rapid response. Accordingly, another explanation must be looked for to explain the essential role of ESCs and EICs in the switching on of thermotolerance and other responses to heat and in the switching on of UV tolerance and other responses induced by UV irradiation.

## Pheromonal activity of ECs in stress response induction

It is now clear why ECs have evolved as obligate components in the switching on of responses induced by physical stresses. During early studies of stress response induction, it was shown that EICs were small or medium sized molecules (generally but not always proteins) and that they would, therefore, readily diffuse away from the region of formation. During the early work on acid tolerance induction, for example, it was shown that the acid tolerance-related EIC would actively and rapidly diffuse through membranes, the diffusion leading to the cross-feeding of either unstressed organisms or non-producers (i.e. mutant strains that would not form acid tolerance-related EICs), or strains unable to produce EICs due to inhibition. Such cross-feeding then resulted in acid tolerance induction in the cross-fed cultures (Rowbury, 1999). These three properties of EICs, namely diffusibility, ability to act on unstressed cells and non-producers, and the cross-feeding properties, clearly allow EICs to influence stress responses of other organisms, i.e. to act pheromonally. In the case where the organisms which are cross-fed and induced to tolerance are unstressed ones, such behaviour gives an early warning against stress. For the acid tolerance-related EICs, for example, the unstressed cells are given early warning of impending acidification, and, because the EIC interacts with these unstressed cells, prepared to resist the coming stress. It is now known that the thermotolerance-related EIC and the UV tolerance-related one have very similar properties to the acid tolerance-related one, and so it is clear that ECs have evolved to switch on responses to physical stress, because the pheromonal activity of the EICs both warns the cells of the forthcoming stress and prepares them to resist it. Early warnings, by such intercellular communication, may allow unstressed organisms to survive the impending occurrence of normally lethal physical stress levels.

### Stress response ECs as alarmones

The ability of ESCs and EICs to diffuse away from the site of synthesis to regions where the organisms are unchallenged, means that such ECs can give early warning of impending danger from stress. Diffusing EICs can directly induce tolerance in unstressed cells or in organisms unable for genetical or physiological reasons to produce EICs. Since such EICs readily interact with sensitive cells and induce tolerance, they act as alarmones, in the terminology of Bochner et al., 1984. ESCs can act similarly, if they diffuse to regions where organisms (for genetical or physiological reasons) are not producing ESCs. Both ESCs and EICs, when leading to such intercellular communication, are acting as both *alarmones* and *pheromones* (Rowbury and Goodson, 2001).

### ESCs as biological thermometers

The term biological thermometer refers to cellular components that both sense temperature increases, and are changed by them so as to alter the behaviour of a component or switch on inducible biological responses. McCarty and Walker (1991) referred to the DnaK protein of E. coli as a thermometer, as its activity is increased as temperature is increased. This component has probably evolved simply to ensure that as the extent of protein damage by heat increases, the activity of this major protein repair chaperone increases also, rather than acting to induce a response.

Until very recently, there was little evidence for temperature-sensing components in bacteria, because the growth medium was not being examined. Above, it has been mentioned that the UV tolerance of E. coli cultures is greater at higher temperature than at low ones, and in fact the extent of such tolerance goes up with increasing temperature. It is the basis for this that is of particular interest, namely that as Table 5.7 shows, the ability of ESC-containing filtrates to induce UV tolerance (after conversion to EICs) depends on the activation temperature, the activated filtrates becoming more effective, in UV tolerance induction, as their activation temperature increases, i.e. the UV tolerance-related ESC in the filtrate, acts as a thermometer.

The UV tolerance-related ESC is not unique in its ability to act as a thermometer. Those ESCs involved in sensing stresses that lead to acid tolerance, alkali tolerance and thermotolerance all act in this way, i.e. at least four ESCs behave as biological thermometers in that the ability of heat-activated filtrates to induce responses increases with increasing temperature (Table 5.8).

### Killed cultures can induce tolerance, allowing sensitive organisms to survive potentially lethal stresses

If bacterial cultures are treated so that all living organisms in them are killed or destroyed, it would be expected that any properties that were conferred by the cultures or by individual organisms in them would be abolished, and that they would not be able to influence any other organisms that entered that location subsequently. Indeed, failure for this to happen, would mean that killing contaminating organisms in foods or food materials would not neutralize their baleful effects, which might be transferred to other contaminating organisms that the treated foods came into contact with later. Apart from the ability of the killed culture to possibly alter the pH of a culture that it was subsequently mixed with, or to alter its growth rate positively by supplying extra nutrients or negatively by releasing inhibitory agents or produce other trivial effects, the killed culture would not have been expected to confer other properties, e.g. the ability to resist lethal chemical or physical agents.

In spite of the above, killed cultures have been tested for their ability to confer stress tolerance on unstressed cultures. The original studies were on acid tolerance (Rowbury, 2000) and showed that if acid-tolerant neutralized cultures of E. coli killed by heat, alkali, acid, antibiotic or $Cu^{2+}$, were mixed with unstressed living E. coli, substantial numbers of acid-tolerant organisms arose (i.e. many of the organisms in the incubated mixture survived potentially lethal acidity). The same occurred if acid-sensitive cultures were killed and then activated at pH 5.0; neutralized killed culture on admixture with sensitive living organisms, gave rise to substantial numbers of organisms that were able to survive extreme acidity. All the possible trivial reasons for these results were ruled out (Rowbury, 2000). Thus, killed cultures had virtually no living organisms present, and the possibility that the apparently dead organisms were merely damaged and recovered later was ruled out. To absolutely establish that the acid-tolerant organisms had arisen from the living cultures, rather than from the few living organisms in the killed cultures, genetically marked strains were used; all acid-tolerant organisms in the mixtures had the genotype of the added living originally acid-sensitive organisms. Also, the possibility that the killed culture had reduced the growth rate or lowered the pH of the living culture during incubation (both such changes can increase acid tolerance), was ruled out.

Accordingly, the killed cultures had conferred acid tolerance on the phenotypically acid-sensitive living organisms.

**Table 5.8** Tolerance response ESCs as biological thermometers

| Temperature (°C) for activation of the ESC-containing filtrate | Stress tolerances (mean% survival ± SEM) for unstressed organisms incubated with activated filtrate | | |
|---|---|---|---|
| | Acid tolerance | Alkali tolerance | Thermotolerance |
| 37 | 0.7 ± 0.13 | 0.4 ± 0.055 | 3.1 ± 0.12 |
| 38.5 | NT | NT | 9.3 ± 0.74 |
| 40 | 7.4 ± 1.55 | 0.93 ± 0.033 | 11.2 ± 0.50 |
| 42 | 15.1 ± 0.75 | 2.2 ± 0.11 | 14.4 ± 0.56 |
| 45 | 20.9 ± 1.55 | 5.7 ± 0.24 | 18.5 ± 0.35 |
| 50 | 30.1 ± 0.55 | 11.5 ± 0.48 | 26.4 ± 2.5 |
| 55 | 39.3 ± 1.75 | 6.7 ± 0.74 | 37.1 ± 2.75 |
| 60 | NT | 5.1 ± 0.85 | 32.0 ± 2.3 |

Strains 1829 ColV (for acid tolerance and thermotolerance) or 1829 (for alkali tolerance tests) were grown to midlog-phase at pH 7.0 and 37°C (or 30°C for thermotolerance tests), and cell-free filtrates prepared. Filtrates were activated at the stated temperatures for 30 min at pH 7.0, and then incubated with the culture used to prepare the filtrate, at 1/1 filtrate to culture (the latter containing 7–10 × 10⁷ per mL viable organisms) for 60 min at pH 7.0 and either 30°C (for thermotolerance tests) or 37°C. Cultures were challenged at pH 3.0 for 7 min (acid tolerance), at pH 11.0 for 5 min (alkali tolerance) or at 49°C for 5 min (thermotolerance).

Further studies showed that appropriate neutralized cultures can, after killing, confer alkali tolerance or alkali sensitivity on unstressed cultures (Rowbury, 2000).

Recent work has concentrated on whether thermotolerant cultures can, after killing, confer thermotolerance on thermosensitive organisms. As indicated above, thermotolerance can be induced by stresses other than thermal stress, and these recent experiments (Rowbury and Goodson, 2001) reveal that killed cultures of *E. coli*, activated by heat, by UV irradiation or by exposure to pH 5.0 or pH 9.0, confer thermotolerance on thermosensitive organisms, whether the cultures have been activated before or after killing (Table 5.9).

The original tests on acid tolerance induction strongly suggested that the components conferring acid tolerance in the killed cultures were protease-sensitive, non-dialyzable agents, implicating ESCs and EICs as the likely active agents. The use of a mutant (*cysB*) with a lesion in the functioning of the acid tolerance ECs strongly supported the involvement of these ECs in acid tolerance induction by the killed cultures.

The finding that killed cultures can confer stress responses is of public health significance. Thus, if acid-tolerant organisms in foods or food materials were killed by heat, and the heated food ingested with other organisms, which were acid-sensitive, the acid-tolerance (carried by EICs) could be transferred to the living organisms, and allow the latter to survive subsequent acid challenges in the stomach or in the phagolysosome. Similarly, if thermotolerant organisms in food materials were killed by extreme heat (thermotolerant organisms can still be killed by high enough temperature exposures) and then mixed with other food materials containing living thermosensitive organisms, the living organisms could gain thermotolerance (from the EICs in the heated culture) and subsequently survive exposure to mild heating.

Exposures of polluting organisms in water to stresses could also be of significance in food microbiology or have medical significance. For example, such organisms which had been made thermotolerant by exposure to increasing concentrations of alkalinity in chemically polluted natural waters (Rowbury

**Table 5.9** Killed cultures can induce thermotolerance in 30°C-grown *E. coli*

| Killed culture | Living culture | % survival ± SEM, after heat challenge of living culture preincubated with killed one |
|---|---|---|
| None | 30°C-grown | 0.7 ± 0.06 |
| None | 45°C-grown | 22.5 ± 1.5 |
| 30°C grown, untreated | 30°C-grown | 0.66 ± 0.068 |
| 45°C grown, untreated | 30°C-grown | 25.2 ± 0.83 |
| 45°C grown, protease- treated | 30°C-grown | 12.9 ± 0.58 |
| 30°C grown → 45°C, 30 min | 30°C-grown | 4.1 ± 0.24 |
| 30°C grown → UV, 90 s | 30°C-grown | 20.5 ± 1.26 |
| 30°C grown → pH 9.0, 30 min | 30°C-grown | 16.1 ± 0.85 |

Cultures of strain 1829 ColV were grown at the stated temperature and killed by exposure to 65°C for 15 min. They were then activated either at pH 9.0 for 30 min at 30°C, or at pH 7.0 and 45°C for 30 min or at 30°C/pH 7.0 for 30 min after exposure to UV irradiation (see Table 5.4 for conditions). One sample was exposed to protease (1 mg/mL) for 15 min at 30°C. All killed cultures were incubated with living ones of the same strain grown at pH 7.0 at the stated temperature (1/1 killed culture to living one, the latter containing ca 7–10 × 10⁷ organisms per mL) for 60 min at 30°C, before thermal challenge (49°C for 5 min).

*et al.*, 1989), and subsequently killed as the alkalinity reached lethal levels, would give rise to killed cultures that could confer thermotolerance on contaminating organisms in food materials to which the polluted water had been added. It is likely that killed cultures could confer other stress tolerances, and their ability to make organisms irradiation-resistant is specifically under study.

## Intracellular regulatory processes and stress response induction

Above, the role of ESCs and EICs in thermotolerance has been outlined; these ECs play an obligate role in the switching on of the process, i.e. the thermal sensor and the initial induction component are extracellular. Other processes switched on by thermal stress also show the essential role of an ESC/EIC pair in the earliest stages of induction. It is highly likely that all thermal responses need such EC pairs, but there are other regulatory stages and components in thermotolerance induction and in other thermally induced processes which are intracellular and these will be outlined here. Tolerance responses induced by other stresses also use ESC/EIC pairs for stressor sensing and for switching on the induction processes; all of these have other regulatory stages following interaction of EIC with sensitive cells, and these will also be considered here.

## Responses to thermal stress: intracellular regulatory processes

Numerous responses are induced by heat and, for some, the molecular mechanisms, subsequent to the switching on by ESCs and EICs, have been studied in detail (Marquis *et al.*, 1994). The heat-shock process, which leads to the induction of the heat-shock proteins (HSPs) has been studied in most detail, with respect to the nature of intracellular regulatory components and stages, and this inducible process (Polissi *et al.*, 1995; Yura *et al.*, 2000; Periago *et al.*, 2002) will be considered here.

The levels of HSPs increase very rapidly after a temperature increase (e.g. considering a transfer from 30°C to 42°C), and since most other protein synthesis is reduced or abolished (depending on the protein and the shift temperature), HSPs make-up nearly all of the proteins synthesized after the shift. As with other HSPs, the amount of the RpoH protein increases very rapidly after the temperature shift-up, and for this protein, the increases occur within a few seconds. This suggests that RpoH synthesis at high temperatures could not result from increases in *de novo* synthesis of its mRNA, followed by translation of this increased mRNA amount, because such a process would take too long. In fact the major factors responsible for the above increased RpoH levels relate to stability. First, the mRNA already present at the time of the shift becomes stabilized at the increased temperature, allowing more RpoH synthesis. Second, the pre-existing and newly synthesized RpoH molecules become stabilized, also leading to increased RpoH levels.

These increased numbers of RpoH molecules are then responsible for the increased formation rates for the other HSPs. RpoH is a sigma factor, $\sigma^{32}$, and its binding to the core RNA polymerase leads the latter to detect and attach to heat-shock operon promoters (Grossman *et al.*, 1984); absence of this σ-factor (e.g. in mutants with lesions in *rpoH*), virtually abolishes the heat-shock response.

Whereas high levels of most HSPs remain for some time after a temperature up-shift, this does not apply to RpoH; within a few minutes of the up-shift, RpoH levels fall substantially. There appear to be complex explanations for this disappearance of RpoH, all involving the functioning of chaperones exhibiting dual roles. First, DnaJ, DnaK and GrpE chaperones enhance the activity of proteases that specifically target RpoH, degrading it and second, the same chaperones appear to destabilize the mRNA for RpoH and possibly also affect RpoH activity.

*Structures, properties and locations of molecular chaperones*

There are numerous heat-shock proteins; many are molecular chaperones, which process proteins in both normal and stressing situations. In stressing conditions, they have some protective roles, but their major function is in repair. Several consist of lidded containers, into which, when the lids are open, damaged proteins are permitted to enter. Following entry, typically the protein associates with the inside of the pocket by hydrogen-bonding, and ATP cleavage allows the refolding of damaged hydrophobic regions. Following such repair, the lid opens, and the repaired protein and ADP plus phosphate are released. Many chaperones (although not the periplasmic ones) act, in general, in the above way. In more detail, one of them, DnaK, acts as follows. The 3D structure is folded, so that the main part of the protein forms a hollow vessel. Unlike some other chaperones, however, the lid of the DnaK vessel is not formed from a second protein, as occurs for the GroEL/ES complex, but DnaK itself folds to cover the opening of the vessel. There is another difference between DnaK and the GroEL/ES complex, namely that the hollow channel in the vessel formed by GroEL expands when the GroES lid and ATP attach to it (Ellis, 1996).

DnaK and GroEL are large chaperones (ca 70 kDa and 60 kDa respectively). Several molecules in this class are, however, smaller, and this is the case for the IbpA and IbpB HSPs, which appear to repair stress-damaged proteins (Kitagawa *et al.*, 2000).

The above chaperones are cytoplasmic molecules or complexes. Others occur in the periplasm, although their functioning differs from that of many cytoplasmic chaperones, in being ATP independent. Although many periplasmic proteins are resistant to heat and other stresses (e.g. the PhoA protein), this is not generally true and does not, of course, occur for most OM proteins, which are merely passing through the periplasm, as a stage in their synthesis. Several periplasmic proteins appear to have chaperone functions, and there appear to be two repair pathways (Rizzitello *et al.*, 2001); one pathway can use the DegP protein (Table 5.2) as a chaperone, the other pathway uses SurA; this protein had already been shown to repair damaged proteins, i.e. to have chaperone activity (Behrens *et al.*, 2001).

Many of the chaperones described above, have roles other than in protein maturation and repair (Table 5.2). For example, many can also show protease activity, or present proteins to proteases. There is also evidence for several playing a part in protein secretion, whilst some chaperones can protect proteins during stress, i.e. so that less damage occurs.

## Responses to UV irradiation: intracellular processes regulating response induction

It has been known for many years now that the key to induction of the many UV irradiation responses, lies in the RecA gene product. This component, as with a number of stress-related proteins has dual activities and functions. Like several other such stress response components (e.g. ClpXP, which can have protease or chaperone activity, Li *et al.*, 2000), RecA has potential protease action, as well as its other properties (e.g. in recombination). Its role in the SOS response, involves its activation, leading to protease activity, possibly following its association with SS DNA (resulting from DNA damage produced by, e.g. UV irradiation) and nucleotides. The protease activity of RecA is highly specific, only a few proteins are cleaved by it, the most important being the LexA gene product (Walker, 1984). This protein apparently dimerises onto the so-called SOS box in the operator regions of numerous genes or groups of genes. Such binding prevents transcription taking place, i.e. it has a repressing effect and only as the LexA level falls (following cleavage by activated RecA) do the genes become derepressed. The strength of binding of LexA to the operator regions varies from gene to gene, so that some genes are derepressed early, and others only when LexA has fallen to very low levels. The un-masking of RecA protease activity not only explains derepression of numerous SOS response-related proteins, it also establishes why phages like λ become derepressed after DNA damage; the RecA protease activity destroys the λ repressor.

## Regulation of carbon starvation responses

RpoS is the major component controlling such tolerances, acting as a sigma factor, switching on the induction of several genes and operons. During stationary-phase and starvation, the levels of RpoS rise, probably for two major reasons, namely increased translation of its mRNA and increased stability of RpoS itself. This latter depends, at least mainly, on loss of protease activity by the stress-shock protein complex, ClpXP. This protease specifically targets the RpoS protein and other proteins formed by its activity. This occurs in the log-phase, because the complete ClpXP complex arises in this phase and has protease activity. Accordingly, in the log-phase, RpoS level is low and those stationary phase (or starvation-induced) tolerances, which are induced by it, do not appear. In contrast,

in stationary-phase (or in starvation conditions) a shortened mRNA is formed from the *clpP, clpX* operon (Li *et al.*, 2000), and the ClpXP complex is not formed. The ClpP protein that results does not have protease activity, and so the levels of RpoS (and proteins formed by its activity) increase.

## Regulation of cold shock

Cold shock seems likely to be sensed by a specific ESC. Following activation, a large group of cold shock proteins are induced in bacteria (Wolffe, 1995; Weber and Marahiel, 2003). For example, on transfer of *E. coli* to cold temperatures, e.g. 10°C, at first an initial group of cold-shock proteins is derepressed. Most rapid and marked derepression is for several very low molecular weight (MW) proteins especially CspA (CS7.4), which is derepressed 100-fold (Goldstein *et al.*1990). These are proteins with ca 70 amino acids only. There are several reasons for the increased synthesis of CspA. Thus, the mRNA for this protein shows both increased stability and more efficient translation at low temperatures, the latter resulting from the functioning of a downstream box translational enhancer. Additionally, as soon as appreciable numbers of low MW Csp molecules have accumulated, this will further enhance synthesis from Csp mRNAs, because secondary structure formation in such RNAs, which occurs at low temperatures, will be abolished, because of the ability of the Csp proteins to destabilize such secondary structures (Table 5.2), acting as RNA chaperones (Wolffe, 1995; Jones and Inouye, 1994; Weber and Marahiel, 2003). Similar low MW Csps are induced at low temperatures in, e.g. lactobacilli and in *B. subtilis* and there is evidence that these and the Csp proteins of *E. coli*, play a role in low temperature survival and adaptation (Girgis *et al.*, 2002; Willimsky *et al.*, 1992; Graumann and Marahiel, 1994; Weber and Marahiel, 2003).

Induction of other proteins (e.g. GyrA and H-NS) follows that of the low MW group and, for many, this results from CspA, and the others, acting like Y box proteins, binding to sequences such as ATTGG and enhancing transcription. For example, GyrA and H-NS have such a sequence, and binding of CspA etc enhances synthesis of their mRNAs (La Teana *et al.*, 1991; Wolffe, 1995).

## Biochemical changes that lead to stress tolerance

### Responses to thermal stress: the biochemical and physiological changes which lead to stress tolerance

As indicated above, thermal stress leads to the induction of numerous HSPs, including the DnaJ, DnaK, GrpE, GroEL, GroES, ClpA, ClpB, ClpP, ClpX, GrpE, Lon, HtpG, IbpA and IbpB gene products. Many of these are molecular chaperones, and others have protease activity. This would be expected for HSPs; heat damage to proteins will be repaired by chaperones, whilst if damage is too great, degradation by proteases will occur, allowing re-use of amino acids. Recent work shows that a number of chaperones have a dual function being able at low damage levels to repair proteins, but at high damage levels they gain protease activity (Table 5.2). This applies to the ClpXP protein (Li *et al.*, 2000) and to the periplasmic DegP protein (Spiess *et al.*, 1999).

## Carbon starvation; biochemistry and physiology

As indicated above, RpoS frequently controls carbon starvation responses. Numerous proteins and groups of proteins are induced, and lead to tolerance responses and other related processes. Some of these are as follows (Hengge-Aronis, 1993; Brown *et al.*, 1997; Eichel *et al.*, 1999):

1 Cyclopropane fatty acid (CFA) changes. The levels of CFAs in the CMs of organisms has been shown to increase in stationary-phase and this may be a major factor influencing starvation-induced tolerance responses, since CFA changes are known to influence, e.g. acid tolerance (Brown *et al.*, 1997). These changes in CFA levels, when induced by starvation, are governed by RpoS (Eichel *et al.*, 1999).

2 Increases in HSP levels, especially chaperones like DnaK, GroEL, GroES, GrpE, ClpB, and HtpG.

3 Increases in trehalose synthesis. Trehalose is a sugar with the ability to protect membranes and some proteins from stress damage. This molecule is involved in some stress tolerances, including starvation-induced thermotolerance, and starvation induces synthesis of this sugar, e.g. the Ots A and B gene products are induced.

4 Increases in $H_2O_2$ degradation. Such increases depend on, e.g. induction of RpoS-controlled catalase.

5 Increases in enzymes for DNA damage limitation and DNA repair. Some enzymes that can protect DNA from damage (such as certain DNA-binding proteins) and some which can repair damaged DNA (e.g. certain nucleases) are induced.

6 Protease induction. Certain proteases are induced by starvation, to release amino acids for protein synthesis. Lesions in the genes for such proteases lead to poor resistance to starvation.

Note that on entry into the stationary-phase, RpoS is stabilized by the disappearance of the ClpXP protease (Li *et al.*, 2000). In contrast, on exit from this growth phase, the resynthesis of this protease leads to degradation of both RpoS and numerous proteins which have been induced under its control.

## Biochemical changes in cold-shocked organisms

The biochemical changes that are induced as part of the cold shock response will be those that counteract the inhibitory or even potentially lethal changes that follow a shift to low temperature. Certain changes to physiology, composition and structure occur for all living cells on transfer to low temperatures, and the cold shock response must be able to, at least, partially reverse or counteract these changes. First, on a temperature down-shift, nucleic acids show increased secondary structure and, for DNA, major alterations to superhelicity. Second, membranes exhibit reduced fluidity, which for many organisms can be, even without freezing, lethal. Third, the levels of free radicals and oxidative components markedly increase. Fourth, it is clear that the levels of transcription and translation must be maintained or increased at low temperatures, because strains with lesions in these processes fail to grow at low temperature. Fifth, in many situations, following cold-shock, the temperatures fall below freezing, and accordingly, it would be likely that the cold-shock response would include processes which protect from freezing.

First, it is clear from the studies reported in earlier sections, that cold shock induces components (e.g. CspA) which can reverse the formation of secondary structure, e.g. in RNAs (Wolffe, 1995). In the second stage of the response, amongst the proteins induced are GyrA and H-NS, which would be likely to play a role in reversing changes to superhelicity (La Teana *et al.*, 1991; Jones *et al.*, 1992; Wolffe, 1995).

As stated, low temperatures reduce membrane fluidity. For even reduced growth rate to occur at such temperatures, the membranes need to show, at least, partial fluidity, and only by modulating the membrane fatty acid (FA) compositions do bacteria avoid lethal crystallinity. Changes occurring include increased FA unsaturation, increased proportions of branched-chain FAs, appearance of cyclo-FAs etc. Such changes must be catalyzed by enzymes induced as part of the cold shock response.

It is now well established that free radicals and other oxidative components accumulate at low temperatures. Such an accumulation can occur, for example, in cold-stressed plants and the levels of antioxidant and of enzymes that destroy free radicals increase during cold acclimation (Kaye and Guy, 1995). Strongly supporting the idea that ability to destroy or neutralize oxidative components and radicals is linked to cold tolerance and freezing tolerance are the findings that cold acclimation increases tolerance to oxidative components (Bridger *et al.*, 1994) and that overproduction of, for example, superoxide dismutase correlates with freezing tolerance. It is likely that similar situations will occur in bacteria. One should perhaps ask whether the induction in bacteria, on transfer to low temperatures (Wolffe, 1995), of enzymes containing the powerful anti-oxidant N-acetyl-lipoic acid is significant here.

As indicated above, transcription and translation must remain at high levels on a temperature down-shift, and this is no doubt the reason for the induction at low temperatures of NusA and of the $2\alpha$ and $2\beta$ translation initiation factors.

It is likely that metabolic pathways, especially those related to oxidative metabolism change on cold shock, as indicated by the induction in *E. coli* of pyruvate dehydrogenase and dihydrolipoamide acetyltransferase (Wolffe, 1995).

Cold-shock responses have also often evolved to anticipate the likelihood that freezing will occur, and since freezing leads to desiccation, the accumulation of high levels of cryoprotectants would be anticipated, with the induction of the enzymes that synthesize them being a component part of the cold shock response. This is certainly the case for many living organisms exposed to cold acclimation or cold-shock (Kaye and Guy, 1995; Block, 2003).

## Other physiological properties of cold-shocked bacteria, especially in shellfish

Studies over many years have established that cold-shocked organisms show altered responses to numerous chemical stressors (Strange and Postgate, 1964; Whiting, 1990; Whiting and Rowbury, unpublished observations; Whiting and Rowbury, 1995). As already stated, potential pathogens in shellfish which resist lethal agents in natural waters, could subsequently go on to cause disease if ingested. The bactericidal agent acrylate occurs in shellfish beds (Brown *et al.*, 1977) and tolerance to it would be of Public Health significance (Hicks and Rowbury, 1987). Strikingly, cold-shocked populations of *E. coli* are more tolerant to this agent than control organisms (Table 5.10), and, of particular interest, this applies to plasmid-containing organisms (including strains carrying a virulence plasmid and a resistance plasmid; Table 5.10) as well as plasmid-free ones. Cold-shock also renders organisms more tolerant to $Cu^{2+}$ (Table 5.10), but, in contrast, $ColV^+$ *E. coli* were sensitized to acidic and alkaline pH values and to chlorine by cold-shocking.

## Organisms in shellfish: factors influencing stress tolerance

As indicated above, cold-shocking can influence stress resistance of contaminating organisms in shellfish. In view of the fact that such organisms in shellfish, that have not been cold-shocked, are frequently responsible for serious gastrointestinal disease, even if the shellfish have been disinfected before ingestion, and sometimes even if they have been cooked before ingestion, it seems likely that contaminating organisms in them are either, for some other reason, inherently tolerant to chemical or physical stress, or that the organisms have become inducibly tolerant. One major reason for inherent tolerance is that potential pathogens, especially those containing virulence plasmids, can confer surface attachment (Hicks and Rowbury, 1986; 1987). This leads to inherent stress tolerance, because attached organisms, especially those carrying plasmids, are resistant to a range of stresses (Hicks and Rowbury, 1986; 1987; Humphrey *et al.*, 1997).

In addition, contaminating organisms in shellfish are, when considering induction of stress tolerance, a special case, in that they can be exposed to a range of physical and chemical stresses in natural waters, and this can induce tolerances, including cross-tolerances, and, therefore, influence their ability to survive if the shellfish are disinfected or lightly cooked, and

to resist host defenses and go on to cause disease, if the shellfish are later ingested. For example, shellfish can be subjected, in natural waters, to UV irradiation, thermal stress from hot effluents, acidity from acid rain, mine workings or chemical plant effluents or alkalinity from agricultural or chemical operations. All the above stresses induce UV tolerance, and this could allow organisms to survive the UV irradiation often used for shellfish disinfection (Hicks and Rowbury, 1987). All such exposures can also induce thermotolerance (Tables 5.4 and 5.5), which could allow organisms to survive cooking. Other exposures in natural waters can also induce stress tolerance, e.g. $Cu^{2+}$, which enters natural waters from mine workings and chemical plant effluents, induces very marked thermotolerance in *E. coli* (Rowbury, 2002), as shown in Table 5.4.

Another kind of stress exposure can also affect stress tolerance levels in organisms in shellfish, namely starvation. There will be periods when nutrients are low in aquatic systems from which shellfish are obtained (Elliott and Colwell, 1985), and starvation shock during these can alter stress tolerance. Studies of *E. coli* show that starvation increases the acrylate tolerance of most plasmid-containing strains; tolerance to $Cu^{2+}$ was increased for the $ColV^+$ strain tested, but this strain showed increased sensitivity to acidic and alkaline pH (Table 5.11) after starvation.

## Quorum-sensing in relation to organisms in foods

Quorum-sensing is a process whereby, as a component, secreted by a bacterium, builds-up in a growth medium, the concentration can be sensed by the organisms, and at a specific level, an appropriate response is induced. By definition, the component sensed is in the medium, i.e. is extracellular, although it builds up in the cytoplasm of the organisms as well as in the external milieu. Although such agents are ECs, the concentration changes, which ultimately lead to response induction, are *not* detected in the medium (as for the sensing of stress agents or conditions by an ESC), but there is an *intracellular* sensor.

## Quorum-sensing and the production of growth-inhibitory agents in foods

Naturally occurring lactobacilli in dairy products produce fatty acids and lower the pH in their environment, and this has been believed to be a factor influencing the growth of contaminating organisms in such products. It is now known that such organisms, especially lactobacilli, produce another critical class of inhibitory agents in dairy products, namely bacteriocins, agents that inhibit the growth of other organisms. It is of particular interest here, that this class of bacteriocin is controlled by quorum-sensing, with the positive regulator of bacteriocin synthesis being the bacteriocin itself or a closely related compound (Nes *et al.*, 1996). Generally, bacteriocins are not of great importance, *in vivo*, because they are so specific that they only inhibit a few species, which are generally those closely related to the producer strain. This is not, however, the case for those produced by lactobacilli, which can inhibit quite

**Table 5.10** Cold-shock and tolerance of *E. coli* to lethal agents

| Strain | Lethal agent | Mean % survival (± SEM) for | |
| --- | --- | --- | --- |
| | | Control population | Cold-shocked population |
| P678-54 ColV | Acrylate | 17.9 ± 3.6 | 54.9 ± 3.0 |
| P678-54 | Acrylate | 16.3 ± 5.3 | 62.6 ± 7.0 |
| 1829 ColV | Acrylate | 13.5 ± 3.2 | 60.9 ± 5.3 |
| 1829 | Acrylate | 10.4 ± 5.2 | 49.1 ± 7.6 |
| 1829 R1 | Acrylate | 12.9 ± 2.1 | 74.8 ± 7.6 |
| 1829 F*lac* | Acrylate | 14.2 ± 3.7 | 46.6 ± 3.4 |
| P678-54 ColV | Cu$^{2+}$ | 22.9 ± 5.8 | 78.6 ± 5.3 |
| P678-54 | Cu$^{2+}$ | 8.9 ± 3.9 | 42.3 ± 1.5 |
| 1829 ColV | Cu$^{2+}$ | 13.3 ± 4.6 | 72.9 ± 6.2 |
| 1829 | Cu$^{2+}$ | 34.5 ± 3.3 | 79.5 ± 1.4 |
| 1829 R1 | Cu$^{2+}$ | 11.6 ± 0.5 | 76.7 ± 3.3 |
| 1829 F*lac* | Cu$^{2+}$ | 17.6 ± 5.3 | 64.6 ± 7.6 |

Log-phase cultures grown at 37°C in broth were used throughout. Cold shock was for 60 min at 10°C, with subsequent treatment with lethal agent (4 h with acrylate 1 mg/L[1] or 24 h with Cu$^{2+}$ 3.75 μg/mL) at 20°C.

**Table 5.11** Starvation and stress tolerance in *E. coli*

| Strain | Inhibitory agent | Mean % survival (± SEM) for | |
| --- | --- | --- | --- |
| | | Non-starved population | Starved population |
| P678–54 ColV | Acrylic acid | 6.2 ± 2.3 | 29.5 ± 5.6 |
| 1829 ColV | Acrylic acid | 0.4 ± 0.04 | 35.5 ± 2.6 |
| 1829 R1 | Acrylic acid | 48.3 ± 5.3 | 27.4 ± 1.7 |
| 1829 F*lac* | Acrylic acid | 18.5 ± 4.9 | 46.3 ± 5.5 |
| P678–54 ColV | Cu$^{2+}$ | 16.0 ± 2.2 | 45.8 ± 3.3 |
| P678–54 ColV | pH 3.0 | 53.3 ± 3.0 | 13.6 ± 2.0 |
| P678–54 ColV | pH 11.0 | 0.4 ± 0.04 | 0.06 ± 0.02 |

Starvation (of log-phase 37°C broth-grown cultures) was in river water (which was supplemented with tryptone (20 μg/mL, humic acid, 10 μg/mL, L-threonine (20 μg/mL, L-leucine (20 μg/mL, L-tryptophan (20 μg/mL and D-glucose, 0.2% w/v). Incubation of washed organisms in these media was for 24 h at 20°C, with initial cell populations of 1–2 × 10$^8$/mL. Challenge with inhibitors was in supplemented river water at 20°C with acrylic acid 1mgml$^{-1}$, for 16 h, with CuSO$_4$, 3.75 μg/mL, for 4 h, with acid (pH 3.0) for 15 min, and with alkali (pH 11.0) for 4 min.

a broad range of Gram-positive organisms, and accordingly, such agents may be important in controlling spoilage organisms or potential pathogens in dairy products. Bacteriocins produced by lactobacilli are generally not effective on most Gram-negative bacteria, because of the barrier properties of their OMs (Nikaido and Vaara, 1985; Nikaido, 1994). For example, nisin cannot target most unstressed Gram-negative bacteria, because it fails to penetrate the OM and cannot reach its site of action in the CM. Some foods, however, are acidic in reaction, and acid stress will sensitize contaminating Gram-negative organisms to nisin and other bacteriocins (Boziaris and Adams, 2001).

## Quorum-sensing and the production of bacterial toxins in foods

It is highly likely that many toxins produced in food by pathogens, and responsible for food intoxications on subsequent ingestion, are synthesized in response to the appearance of pheromones, in the food, resulting from a quorum-sensing type control. In particular, we know that the toxins produced by *S. aureus*, including its enterotoxin, are induced, governed by a

quorum-sensing system, following the production of a peptide pheromone (Ji *et al.*, 1995). Since most other tested Gram-positive bacteria show similar quorum-sensed systems, it is likely that other toxins produced in food, e.g. the enterotoxin of *B. cereus*, will be controlled in this way.

## On the relation between stress tolerance and virulence

The question to be asked here is whether enhancement of stress tolerance either as an inherent bacterial property or as an inducible process, leads to increased virulence; the answer is an unequivocal "Yes". First, let us consider inherent stress tolerance. One useful study was that of Humphrey *et al.* (1995), which established that clinical isolates of *Salmonella enterica* Serovar Enteritidis PT4 showed enhanced tolerance to several stresses including heat, acidity, oxidative components and surface exposure compared to strains isolated from non-clinical sources. Obviously, one might expect that the clinical isolates would show increased virulence, because they might be better able to resist, e.g. acidity and oxidative components in the phagolysosome and other locations in the body, and indeed this proved to be the case (Humphrey *et al.*, 1996). In this study, it was established that it was the level of RpoS that determined the isolate virulence.

It is also highly likely that induction of stress tolerance processes aids the infecting organism, in its attempt to establish in the animal or human body, and to go on to cause disease. This can be seen by considering the host factors which reduce the growth or multiplication of potential pathogens or actively attack them. First, in the phagolysosome, any engulfed bacterium, must face a barrage of lethal components. Although there are other lethal agents found in these structures, the major attacking agents are those produced after the oxidative burst, such as high proton levels, and increased concentrations of $H_2O_2$, $O_2^-$, $OH^•$, NO, ClOH etc. Inducible tolerance processes aiding tolerance to acid, hydrogen peroxide and superoxide are well known and such induced tolerance in organisms in the phagolysosome, would be expected to allow increased survival and enhanced ability to cause disease. There is little doubt that induced tolerance to hydroxyl radicals, nitric oxide and chlorine also occurs. Accordingly, it is highly likely that increased levels of stress tolerance, both inherent and induced, lead to increased virulence.

## Quorum-sensing and virulence

Quorum-sensing frequently influences the virulence of bacterial strains. One well-studied example of this phenomenon is in *Staphylococcus aureus*. This species secretes a veritable battery of toxins and other virulence factors, all evolved to ensure that the organism is able, first, to multiply in the host at a specific site and resist the host defenses there, and second, to pass to new sites. These "waves" of virulence factor formation are controlled by quorum-sensing type mechanisms. These ensure that when a strain reaches a high cell density (a characteristic of quorum-sensing controlled processes), the synthesis and secretion of most of the extracellular lethal virulence factors is induced. Synthesis is governed by the Agr group of gene products and the key component in switching on this process is an octapeptide (Ji *et al.*, 1995). As growth occurs, this pheromone builds-up in the organisms and in the medium, and at high cell-density, enough is present to interact with an *intracellular* sensor, to induce RNA III transcript synthesis, and following this, of virulence factors.

Most responses to stress, so far established, are not governed by quorum-sensing, but are switched on following the detection of the stress by an ESC, followed by its conversion to an EIC. Here, one process, regulated by quorum-sensing, will be mentioned, which undoubtedly affects the ability of a bacterium to cause disease. This is the acid tolerance induction process of *Streptococcus mutans*, growing in biofilms, which appears to be controlled by a mechanism involving quorum-sensing (Li *et al.*, 2001).

Quorum sensing means not only that bacteria can regulate their production of toxins and other virulence components, to ensure that they are most effective, but also that such processes, when brought about by commensal organisms, can protect the body from potentially pathogenic bacteria. As indicated above, lactobacilli and related organisms may control the multiplication of undesirable organisms in dairy products. Such organisms may also play a part in limiting the multiplication of potential pathogens in the animal and human intestine, both by leading to acidic pH values and to the production of weak acids, but also by the secretion of bacteriocins; both types of product also play a role in controlling the growth of undesirable organisms in the vagina.

## References

Allwood, M.C., and Russell, A.D. 1968. Thermally induced ribonucleic acid degradation and leakage of substances from the metabolic pool in *Staphylococcus aureus*. J. Bacteriol. 95: 345-349.

Ananthaswamy, H.N., and Eisenstark, A. 1977. Repair of hydrogen peroxide-induced single strand breaks in *Escherichia coli* DNA. J. Bacteriol. 130: 187–191.

Behrens, S., Maier, R., de Cock, H., Schmid, F.X., and Gross, C.A. 2001. The SurA periplasmic PPIase lacking its parvulin domains functions *in vivo* and has chaperone activity. EMBO J. 20: 285–291.

Block, W. 2003. Water or ice? – the challenge for invertebrate survival. In "Temperature Effects on Biological Systems" Sci. Prog. 86: 77–101.

Bochner, B.R., Lee, P.C., Wilson, S.W., Cutler, C.W. and Ames, B.N. 1984. ApppppA and related adenylated nucleotides are synthesised as a consequence of oxidation stress. Cell 37: 227-232.

van Bogelen, R.A. and Neidhardt, F.C. 1990. Ribosomes as sensors of heat and cold shock in *Escherichia coli*. Proc. Nat. Acad. Sci. USA 87: 5589–5593.

Boziaris, I.S. and Adams, M.R. 2001. Temperature shock, injury and transient sensitivity to nisin in Gram-negatives. J. Appl. Microbiol. 91: 715–724.

Bridger, G.M., Yang, W., Falk, D.E., and McKersie, B.D. 1994. Cold acclimation increases tolerance to activated oxygen in winter cereals. J. Plant Physiol. 144: 235–240.

Brown, J.L., Ross, T., McMeekin, T.A. and Nichols, P.D. 1997. Acid habituation of *Escherichia coli* and the potential role of cyclopropane fatty acids in low pH tolerance. Int. J. Food Microbiol. 37: 163–173.

Brown, R.K., McMeekin, T.A., and Balis, C. 1977. Effect of some unicellular algae on *Escherichia coli* populations in seawater and oysters. J. Appl. Bacteriol. 43: 129–136.

Csonka, L.N. 1989. Physiological and genetical responses of bacteria to osmotic stress. Microbiol. Rev. 53: 121–147.

Demple, B., and Halbrook, J. 1983. Inducible repair of oxidative damage in *Escherichia coli*. Nature. 304: 466–468.

Eichel, J., Chang, Y.-Y., Riesenberg, D., and Cronan, J.E. 1999. Effect of ppGpp on *Escherichia coli* cyclopropane fatty acid synthesis is mediated through the RpoS sigma factor ($\sigma^s$). J. Bacteriol. 181: 572–576.

Elliott, E.L., and Colwell, R.R. 1985. Indicator organisms for estuarine and marine waters. FEMS Microbiol. Rev. 32: 61–79.

Ellis, R.J. 1996. The Chaperonins. Academic Press: San Diego, CA.

Etchegaray, J.-P., and Inouye, M. 1999. CspA, CspB and CspG, major cold-shock proteins of *Escherichia coli*, are induced at low temperatures under conditions that completely block protein synthesis. J. Bacteriol. 181: 1827–1830.

Foster, J.W. and Moreno, M. 1999. Inducible acid tolerance mechanisms in enterobacteria. In: Bacterial Responses to pH. Novartis Found. Symp. Vol. 221, p. 55–69.

Girgis, H.S., Smith, J., Luchansky, J.B., and Klaenhammer, T. R. 2002. Stress adaptations of lactic acid bacteria. In: Microbial Stress Adaptation and Food Safety p. 159–211. A.E. Yousef and V.K. Juneja eds. CRC Press LLC; Boca Raton, FL, USA.

Goldstein, J., Pollitt, N.S., and Inouye, M. 1990. Major cold shock protein of *Escherichia coli*. Proc. Nat. Acad. Sci. USA 87: 283–287.

Goodson, M., and Rowbury, R.J. 1990. Habituation to alkali and increased UV-resistance in DNA repair-proficient and -deficient strains of *Escherichia coli* grown at pH 9.0. Letts. Appl. Microbiol. 11: 123–125.

Goodson, M., and Rowbury, R.J. 1991. RecA-independent resistance to irradiation with UV light in acid-habituated *Escherichia coli*. J. Appl. Bacteriol. 70: 177–189.

Grauman, P., and Marahiel, M.A. 1994. The major cold-shock protein of *Bacillus subtilis* CspB binds with high affinity to the ATTGG and CCAAT sequences in single-stranded oligonucleotides. FEBS Lett. 338: 157–160.

Grossman, A.D., Erickson, J.W., and Gross, C.A. 1984. The *htpR* gene of *E. coli* is a sigma factor for heat-shock promoters. Cell. 38: 383–390.

Guilfoyle, D.E., and Hirshfield, I.N. 1996. The survival benefit of short-chain organic acids and the inducible arginine and lysine decarboxylase genes for *Escherichia coli*. Letts. Appl. Microbiol. 22: 393–396.

Hancock, R.E.W. 1984. Alterations in outer membrane permeability. Ann. Revs. Microbiol. 38: 237–264.

Hengge-Aronis, R. 1993. Survival of hunger and stress: the role of *rpoS* in early stationary-phase gene regulation in *Escherichia coli*. Cell. 72: 165–168.

Hicks, S.J., and Rowbury, R.J. 1986. Virulence plasmid-associated adhesion of *Escherichia coli* and its significance for chlorine resistance. J. Appl. Bacteriol. 61: 209–218.

Hicks, S.J. and Rowbury, R.J. 1987. Resistance of attached *Escherichia coli* to acrylic acid and its significance for the survival of plasmid-bearing organisms in water. Ann. Inst. Pasteur. 138: 359–369.

Humphrey, T.J. 1981. The effects of pH and organic matter on the death-rates of salmonellas in chicken scald-tank water. J. Appl. Bacteriol. 51: 27–39.

Humphrey, T.J., Richardson, N.P., Gawler, A.H.L., and Allen, M.A. 1991. Heat resistance in *Salmonella enteritidis* PT4 and the influence of prior exposure to alkaline conditions. Letts. Appl. Microbiol. 12: 258–260.

Humphrey, T.J., Richardson, N.P., Statton, K.M., and Rowbury, R.J. 1993. Effects of temperature shifts on acid and heat tolerance in *Salmonella enteritidis* phage type 4. Appl. Environ. Microbiol. 59: 3120–3122.

Humphrey, T.J., Slater, E., McAlpine, K., Rowbury, R.J., and Gilbert, R.J. 1995. *Salmonella enteritidis* phage type 4 isolates more tolerant of heat, acid or hydrogen peroxide also survive longer on surfaces. Appl. Environ. Microbiol. 61: 3161–3164.

Humphrey, T.J., Williams, A., McAlpine, K., Lever, M.S., Guard-Petter, J., and Cox, J.M. 1996. Isolates of *Salmonella enterica* serovar Enteritidis PT4 with enhanced heat and acid tolerance are more virulent in mice and more invasive in chickens. Epidemiol. Infect. 117: 79–88.

Humphrey, T.J., Wilde, S.J., and Rowbury, R.J. 1997. Heat tolerance of *Salmonella typhimurium* DT104 isolates attached to muscle tissue. Letts. Appl. Microbiol. 25: 265–268.

Jenkins, D.E., Schulz, J.E., and Matin, A. 1988. Starvation-induced cross-protection against heat or $H_2O_2$-challenge in *Escherichia coli*. J. Bacteriol. 170: 3910–3914.

Jenkins, D.E., Chaisson, S.A., and Matin, A. 1989. Starvation-induced cross-protection against osmotic challenge in *Escherichia coli*. J. Bacteriol. 172: 2779–2781.

Ji, G., Beavis, R.C., and Novick, R.P. 1995. Cell density control of staphylococcal virulence by an octapeptide pheromone. Proc. Nat. Acad. Sci. USA 92: 12055–12059.

Jones, P.G., and Inouye, M. 1994. The cold-shock response – a hot topic. Mol. Microbiol. 11: 811–818.

Jones, P.G., Krah, R., Tafuri, S.R., and Wolffe, A.P. 1992. DNA gyrase, CS7.4 and the cold shock response in *Escherichia coli*. J. Bacteriol. 174: 5798–5802.

Kadota, H., Uchida, A., Sako, Y., and Harada, K. 1978. Heat-induced injury in spores and vegetative cells of *Bacillus subtilis*. In *Spores VII*, Eds. G. Chambliss and J.C. Vary. pp 27-30. American Society for Microbiology, Washington.

Katsui, N., Tsuchido, T., Hiramatsu, R., Fujikawa, S., Takano, M., and Shibasaki, I. 1982. Heat-induced blebbing and vesiculation of the outer membrane of *Escherichia coli*. J. Bacteriol. 151: 1523–1531.

Kaye, C., and Guy, C.L. 1995. Perspectives of plant cold tolerance: physiology and molecular responses. Sci. Prog. 78: 271–299.

Kitagawa, M., Matsumara, Y. and Tsuchido, T. 2000. Small heat-shock proteins, IbpA and IbpB, are involved in resistances to heat and $O_2^-$ stress in *Escherichia coli*. FEMS Microbiol. Letts. 184: 165–171.

Kullick, I., Toledano, M.B., Tartaglia, L.A., and Storz, G. 1995. Mutation analysis of the redox-sensitive transcriptional regulator OxyR: regions important for oxidation and transcriptional activation. J. Bacteriol. 177: 1275–1284.

Kwon, Y.M. and Ricke, S.C. 1998. Induction of acid resistance of *Salmonella typhimurium* by exposure to short chain fatty acids. Appl. Environ. Microbiol. 64: 3458–3463.

La Teana, A., Bandi, A., Falconi, M., Sprino, R., Pon, C.L., and Gualerzi, C.O. 1991. Identification of a cold shock transcriptional enhancer of the *Escherichia coli* gene encoding nucleoid protein H-NS. Proc. Nat. Acad. Sci. USA 88: 10907–10911.

Lazim, Z., Humphrey, T.J., and Rowbury, R.J. 1996. Induction of the PhoE porin by NaCl as the basis for salt-induced acid sensitivity in *Escherichia coli*. Letts. Appl. Microbiol. 23: 269-272.

Leyer, G.J. and Johnson, E.A. 1993. Acid adaptation induces cross-protection against environmental stresses in *Salmonella typhimurium*. Appl. Environ. Microbiol. 59: 1842-1847.

Li, C., Tao, Y.P. and Simon, L.D. 2000. Expression of different size transcripts from *clpP-clpX* operon of *Escherichia coli* during carbon deprivation. J. Bacteriol. 182: 6630–6637.

Li, Y.-H., Hanna, M.N., Svensater, G., Ellen, R.P., and Cvitkovitch, D.G. 2001. Cell density modulates acid adaptation in *Streptococcus mutans*: implications for survival in biofilms. J. Bacteriol. 183: 6875–6884.

Lund, P.A. 2001. Microbial molecular chaperones. Adv. Micro. Physiol. 44: 93–140.

Mackey, B.M. 1983. Changes in antibiotic sensitivity and cell surface hydrophobicity in *Escherichia coli* injured by heating, freezing, drying or gamma radiation. FEMS Microbiol. Letts. 20: 395–399.

Mackey, B.M. 1984. Lethal and sub-lethal effects of refrigeration, freezing and freeze-drying on microorganisms. Symp. Soc. Appl. Bacteriol. 12: 45–75.

Mackey, B.M., and Derrick, C.M. 1982. The effect of sub-lethal injury by heating, freezing, drying and gamma radiation on the duration of the lag-phase of *Salmonella typhimurium*. J. Appl. Bacteriol. 53: 243–251.

Mackey, B.M., and Derrick, C.M. 1986. Changes in the heat resistance of *Salmonella typhimurium* during heating at rising temperatures. Letts. Appl. Microbiol. 4: 13–16.

Mackey, B.M., and Seymour, D.A. 1987. The effect of catalase on recovery of heat-injured DNA-repair mutants of *Escherichia coli*. J. Gen. Microbiol. 133: 1601–1610.

Marquis, R.E., Sim, J., and Shin, S.Y. 1994. Molecular mechanisms of resistance to heat and oxidative damage. J. Appl. Bacteriol. 76 Supplement: 40S-48S.

Matin, A. 1991. The molecular basis of carbon starvation-induced general resistance in *Escherichia coli*. Molec. Microbiol. 5: 3–10.

Mattick, K.L., Jorgensen, F., Legan, J.D., Cole, M.B., Porter, J., Lappin-Scott, H.M., and Humphrey, T.J. 2000. The survival and filamentation of *Salmonella enterica* serovar Enteritidis PT4 and *Salmonella enterica* serovar Typhimurium DT104 at low water activity. Appl. Environ. Microbiol. 66: 1274–1279.

Mattick, K.L., Phillips, L.E., Jorgensen, F., Lappin-Scott, H.M., and Humphrey, T.J. 2003a. Filament formation by *Salmonella* spp inoculated into liquid food matrices at refrigeration temperatures, and growth patterns when warmed. J. Food Protect. In the Press.

Mattick, K., Rowbury, R., and Humphrey, T. 2003b. Morphological changes of *Escherichia coli* O157:H7, commensal *E. coli* and *Salmonella* in marginal growth conditions, with special reference to mildly stressing temperatures. In: Temperature Effects on Biological Systems. Sci. Prog. 86: 103–113.

McCarty, J.S., and Walker, G.C. 1991. DnaK as a thermometer: threonine-199 is site of autophosphorylation and is critical for ATPase activity. Proc. Nat. Acad. Sci. USA 88: 9513–9517.

Neidhardt, F.C., Ingraham, J.L., and Schaechter, M. 1990. Physiology of the Bacterial Cell. Sinauer Associates: Sunderland, Mass.

Nes, I.F., Diep, D.B., Harvarstein, L.S., Brurberg, M.B., Eijsink, V., and Holo, H. 1996. Biosynthesis of bacteriocins in lactic acid bacteria. Ant. van Leewenhoek 70: 113–128.

Nikaido, H., and Vaara, M. 1985. Molecular basis of bacterial outer membrane permeability. Microbiol. Revs. 49: 1–32.

Nikaido, H. 1994. Prevention of drug access to bacterial targets: permeability barriers and active efflux. Science. 264: 382–388.

Pauling, C., and Beck, L.A. 1975. Role of DNA ligase in the repair of single strand breaks induced in DNA by mild heating of Escherichia coli. J. Gen. Microbiol. 87: 181–184.

Periago, P.M., van Schaik, W., Abee, T., and Wouters, J.A. 2002. Identification of proteins involved in the heat stress response of Bacillus cereus ATCC 14579. Appl. Environ. Microbiol. 68: 3486–3495.

Polissi, A., Goffin, L., and Georgopoulos, C. 1995. The Escherichia coli heat shock response and bacteriophage lambda development. FEMS Microbiol. Revs. 17: 159–169.

Raja, N., Goodson, M., Smith, D.G., and Rowbury, R.J. 1991. Decreased DNA damage and increased repair of acid-damaged DNA in acid-habituated Escherichia coli. J. Appl. Bacteriol. 70: 507–511.

Rizzitello, A.E., Harper, J.R., and Silhavy, T.J. 2001. Genetic evidence for parallel pathways of chaperone activity in the periplasm of Escherichia coli. J. Bacteriol. 183: 6794–6800.

Ron, E.Z., and Davis, B.D. 1971. Growth rate of Escherichia coli at elevated temperatures: limitation by methionine. J. Bacteriol. 107: 391–396.

Rowbury, R.J. 1972. Observations on starvation-induced resistance enhancement (SIRE) in Salmonella typhimurium. Int. J. Radiat. Biol. 21: 297–302.

Rowbury, R.J. 1994. Inducible enterobacterial responses to environmental pollution by sodium ions and alkalinisation. Sci. Prog. 77: 159–182.

Rowbury, R.J. 1999. Acid tolerance induced by metabolites and secreted proteins, and how tolerance can be counteracted. In "Bacterial responses to pH". Novartis Found. Symp. 221: 93–111.

Rowbury, R.J. 2000. Killed cultures of Escherichia coli can protect living organisms from acid stress. Microbiol. 146: 1759–1760.

Rowbury, R.J. 2001. Cross-talk involving extracellular sensors and extracellular alarmones gives early warning to unstressed Escherichia coli of impending lethal chemical stress and leads to induction of tolerance responses. J. Appl. Bacteriol. 90: 677–695.

Rowbury, R.J. 2002a. Microbial disease: recent studies show that novel extracellular components can enhance microbial resistance to lethal host chemicals and increase virulence. In: Some Insights into Microbial Disease and its Control. Sci. Prog. 85: 1–11.

Rowbury, R.J. 2002b. Physiological and molecular basis of stress adaptation, with particular reference to subversion of adaptation and to the involvement of extracellular components in adaptation. In "Microbial Stress Adaptation and Food Safety" pp 247–302. Ed. A.E.Yousef and V.K. Juneja. CRC Press LLC; Boca Raton, Florida, USA.

Rowbury, R.J. 2003. Extracellular proteins as enterobacterial thermometers. In: Temperature Effects on Biological Systems" Sci. Prog. 86: 139–155.

Rowbury, R.J., Goodson, M., and Whiting, G.C. 1989. Habituation of Escherichia coli to acid and alkaline pH and its relevance for bacterial survival in chemically polluted natural waters. Chem. Ind. 1989: 685–686.

Rowbury, R.J., Goodson, M., and Humphrey, T.J. 1993. Acid sensitivity induction (ASI) at alkaline pH in Escherichia coli involves two major sensitisation components, induction of both being switched on by increased internal pH. Letts. Appl. Microbiol. 17: 272–275.

Rowbury, R.J., and Hussain, N.H. 1996. Exposure of Escherichia coli to acid habituation conditions sensitises it to alkaline stress. Letts. Appl. Microbiol. 22: 57–61.

Rowbury, R.J., Lazim, Z., and Goodson, M. 1996. Involvement of the OmpA protein in L-leucine-induced acid sensitivity. Letts. Appl. Microbiol. 23: 426–430.

Rowbury, R.J., and Goodson, M. 1998. Induction of acid tolerance at neutral pH in log-phase Escherichia coli by medium filtrates from organisms grown at acidic pH. Letts. Appl. Microbiol. 26: 447–451.

Rowbury, R.J., and Goodson, M. 1999. An extracellular acid stress-sensing protein needed for acid tolerance induction in Escherichia coli. FEMS Microbiol. Letts. 174: 49–55.

Rowbury, R.J., and Goodson, M. 2001. Extracellular sensing and signalling pheromones switch on thermotolerance and other stress responses in Escherichia coli. Sci. Prog. 84: 205-233.

Russell, A.D. 1984. Potential sites of damage in micro-organisms exposed to chemical or physical agents. In: The Revival of Injured Microbes, pp 1–18. Eds, Andrew, M.H.E. and Russell, A.D. Academic Press, London.

Salmond, C.V., Kroll, R.G., and Booth, I.R. 1984. The effect of food preservatives on pH homeostasis in Escherichia coli. J. Gen. Microbiol. 130: 2845–2850.

Segal, A.W., Geisow, M., Garcia, R., Harper, A., and Miller, R. 1981. The respiratory burst of phagocytic cells is associated with a rise in vacuolar pH. Nature. 290: 406–409.

Sedgwick, S.G., and Bridges, B.A. 1972. Evidence for indirect production of DNA strand scissions during mild heating of Escherichia coli. J. Gen. Microbiol. 71: 191–193.

Sinha, R.P. 1986. Toxicity of organic acids for repair-deficient strains of Escherichia coli. Appl. Environ. Microbiol. 51: 1364–1366.

Somerville, R.L. 1982. Tryptophan: biosynthesis, regulation and large-scale production. In: Amino Acids and Genetic Regulation. K.M. Hermann and R.L. Somerville, Eds. Adison Wesley, Reading, MA.

Spiess, C., Beil, A. and Ehrmann, M. 1999. A temperature-dependent switch from chaperone to protease in a widely conserved heat-shock protein. Cell. 97: 339–347.

Strange, R.E. and Postgate, J.R. 1964. Penetration of substances into cold-shocked bacteria. J. Gen. Microbiol. 36: 393–403.

Tsuchido, T., Katsui, T., Takeuchi, A., Takano, M. and Shibasaki, I. 1985. Destruction of the outer membrane permeability barrier of Escherichia coli by heat treatment. App. Environ. Microbiol. 50: 298–303.

Tsuchido, T., Aoki, I. and Takano, M. 1989. Interaction of the fluorescent dye 1-N-phenyl-naphthylamine with Escherichia coli cells during heat-stress and recovery from heat stress. J. Gen. Microbiol. 135: 1941–1947.

Walker, G.C. 1984. Mutagenesis and inducible responses to deoxyribonucleic acid damage in Escherichia coli. Microbiol. Revs. 48: 60–93.

Wanner, B.L. 1987. Phosphate regulation of gene expression in Escherichia coli. In Escherichia coli and Salmonella typhimurium: Cellular and Molecular Biology, Vol. 2. Neidhardt, F.C., Ingraham, J.L., Low, K.B., Magasanik, B., Schaechter, M. and Umbarger, H.E. eds. American Society for Microbiology. Washington, D.C. p. 1326–1333.

Weber, M.H.W., and Marahiel, M.A. 2003. Bacterial cold shock responses. In: Temperature Effects on Biological Systems. Sci. Prog. 86: 9–75.

Whiting, G.C. 1990. The effect of plasmid carriage on the survival of Escherichia coli. Ph.D. Thesis: University of London.

Whiting, G.C., and Rowbury, R.J. 1995 Increased resistance of Escherichia coli to acrylic acid and to copper ions after cold-shock. Letts. Appl. Microbiol. 20: 240–242.

Willimsky, G., Bang, H., Fischer, G., and Marahiel, M.A. 1992. Characterization of cspB, a Bacillus subtilis inducible cold shock gene affecting cell viability at low temperatures. J. Bacteriol. 174: 6326–6335.

Wolffe, A.P. 1995. The cold-shock response in bacteria. Sci. Prog. 78: 301–310.

Woodcock, E., and Grigg, G.W. 1972. Repair of thermally induced DNA breakage in Escherichia coli. Nature. 237: 76–79.

Yura, T., Kanemori, M., and Morita, M.T. 2000. The heat shock response: regulation and function. In: Bacterial Stress Responses; Storz, G., and Hengge-Aronis, R. Eds. p. 3–18. ASM Press, Washington, D.C.

# Viable but Nonculturable Bacteria in Food Environments

6

James D. Oliver

## Abstract

Cells in the viable but nonculturable (VBNC) state are alive, but undetectable by routine microbiological methods. The presence of such cells in foods presents a special concern, especially when they are human pathogens. This chapter reviews the biology of the VBNC state and the factors which induce it, the foodborne pathogens that are known to enter this state, how cells resuscitate back to the actively growing state, and the importance of the VBNC state in food microbiology.

## Introduction

The possibility of foods containing bacteria that, due to some environmental stress, cannot be cultured by routine microbiological methods is especially problematic to the food industry. There is growing evidence, however, that such a situation exists. This chapter is intended to introduce the reader to what is currently known about "viable but nonculturable" (VBNC) cells, their ability to exit this "dormancy" state and initiate infection, those studies which have examined the involvement of VBNC cells in foodborne outbreaks, and a description of the VBNC state in a variety of foodborne pathogens.

Because several reviews have appeared in recent years regarding the VBNC state, only a brief summary of this phenomenon will be presented here. The reader is referred to the reviews by Kell *et al.* (1998) and Oliver (2000a; 2000b; 2000c) for detailed coverage of the VBNC state and the bacteria described to enter this condition.

## Definition of the VBNC state

Bacteria in the VBNC state may be defined as those which fail to grow on the routine bacteriological media on which they would normally grow and develop into colonies, but which are in fact alive and capable of renewed metabolic activity (Oliver, 2000b). Cells in the VBNC state appear to be fairly inactive metabolically, but on "resuscitation" are again culturable. Cells in this state differ from "injured" bacteria, in that the latter demonstrate an inability to grow on selective media (Mackey, 2000; McFeters and LeChevalier, 2000). In contrast, VBNC-cells do not grow on any medium, even if non-selective.

## Entry of cells into the viable but nonculturable state

The classic VBNC response is illustrated in Figure 6.1. As demonstrated, exposure to one or more environmental stresses results in a regular decline in colony-forming units ("standard plate counts"). At the same time, "total cell counts" (which do not indicate cell viability but simply the presence of a cell), typically remain quite constant. The critical factor which describes VBNC cells is the "viability" assay. A variety of such assays are currently employed in different laboratories to demonstrate this parameter (see below), but in all instances are designed to provide an estimate of the number of cells that exist in the population which are alive, regardless of their ability to develop into colonies on selective or non-selective media.

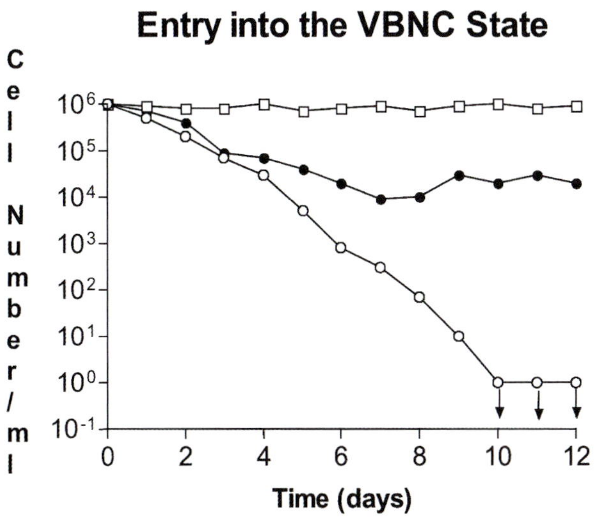

**Figure 6.1** Typical curves demonstrating entrance of bacterial cells into the VBNC state. Shown are total (e.g. acridine orange) cell counts (□), culturable (plate) counts (○), and a direct viability assay (●). Such data indicate that, whereas total cell numbers typically remain high, cells become nonculturable on routine plating media. However, microscopic viability assays indicate a significant portion of the population remains viable.

## Inducers of the VBNC state

Unlike conditions that induce injury (typically agents such as antibiotics, chlorine, and other xenobiotic chemicals; Oliver, 1993), cells enter the VBNC state in response to one or more natural stresses, including nutrient deprivation, temperature up- or down-shifts, elevated osmotic concentrations (e.g. seawater or food brines), oxygen concentration, and exposure to white light (Oliver, 2000c). In all cases, inducers appear to be environmental factors which are potentially injurious or lethal to a given bacterial species. Many of these represent treatments routinely employed in the food industry for reducing bacterial cell numbers and/or for food preservation. The rapidity of the VBNC response to such stresses often depends on the physiological age of the culture (Oliver et al., 1991), with nonculturability times of hours to months being reported. However, significant variation in time to nonculturability occurs between species, and even strains, of a given genus.

## Methods for determining viability

While a total cell count is often obtained through the use of DAPI (4′,6-diamidino-2-phenylindole) or acridine orange staining, viable cell counts are generally determined by the substrate responsive assay of Kogure et al. (1979), hydrolysis of CTC (5-cyano-2,3-ditolyl tetrazolium chloride) or reduction of INT as an indication of metabolic activity, or determination of cytoplasmic membrane integrity (BacLight® or propidium iodide). These methods have been described in detail in several reviews (Oliver, 1993, 2000b; McFeters et al., 1995; Breeuwer and Abee, 2000; Créach et al., 2003). Direct viable count methods such as that of Kogure et al. (1979) are generally useful for all gram-negative bacteria, whereas a study by Regnault et al. (2000) concluded that such methods are not routinely applicable to the various gram-positive bacteria occurring in complex ecosystems (e.g. food samples). Instead, gram-positive bacteria must be examined individually for optimal direct viable count methods. In contrast, the BacLight® method appears to be evolving as a more universal method.

## Bacteria entering the VBNC state

The number of species described to enter the VBNC state constantly increases, with nearly 60 now reported to demonstrate this physiological response. Included are a large number of pathogens of concern to the food industry, including enterohemorrhagic Escherichia coli, Salmonella and Shigella spp., Listeria monocytogenes, and numerous pathogenic vibrios. Species reported to enter this state are shown in Table 6.1.

## Cell biology of the VBNC state

Cells entering the VBNC state typically undergo dwarfing, possibly a result of reductive division, although this aspect has not been investigated to any extent. There are major metabolic changes that occur, including reductions in nutrient transport,

respiration rates, and macromolecular (protein, DNA, RNA) synthesis (Porter et al., 1995; Oliver, 2000a). However, novel starvation and cold shock proteins have been observed to be formed during this time (Morton and Oliver, 1994; McGovern and Oliver, 1995). ATP levels, generally observed to decline rapidly in dead and moribund cells, have generally been found to remain high in VBNC cells (Beumer et al., 1992; Federighi et al., 1998). Further, recent studies have demonstrated the continued expression of selected genes by cells in the VBNC state (Lleò et al., 2000, 2001; Yaron and Matthews, 2002). Extensive modifications in cytoplasmic membrane fatty acid composition has been reported (Day and Oliver, 2004), a response likely necessary for the continued membrane potential which has been reported by some investigators (Porter et al., 1995; Tholozan et al., 1999). Similarly, significant modifications in cell wall peptidoglycan have been reported (Signoretto et al., 2000). Plasmids appear to be retained (Porter et al.; 1995), and their presence or absence has been reported (in several Pseudomonas species) to have dramatic consequences on whether a given factor will induce cells into the VBNC state or not (McDougald et al., 1995). Genomic changes have also been suggested in studies reported by Bej et al. (1997) and Warner and Oliver (1998).

## Evidence for the VBNC state

A few investigators (e.g. Kaprelyants et al., 1993; Barer et al., 1998; Kell et al., 1998; Barer and Harwood, 1999; Begosian and Bourneuf, 2001; Nyström, 2001) have argued against the existence of a VBNC state in bacteria, with much of their concern addressing terminology. However, the overwhelming body of literature from numerous investigators and covering a wide variety of bacterial species argues that this is a novel physiological state induced by a variety of environmental factors. Indeed, several recent papers from investigators that have argued against the VBNC state now offer evidence for it (Mukamolova et al., 1998a, 1998b, 1999) and/or have confirmed what has been previously shown by others (e.g. Begosian et al., 2000). A few of the studies which offer compelling evidence for the VBNC state are noted in this chapter.

Using multi-parameter flow cytometry, Porter et al. (1995) studied E. coli cells as they entered the VBNC state in filter-sterilized lake water. Measurements of membrane potential, membrane integrity, and intracellular enzymatic activity "… provided extensive evidence for the validity of the methods for monitoring cell viability during adoption of the viable-but-non-culturable state in starved E. coli". Beumer et al. (1992) determined that the levels of ATP remained fairly constant within cells of Campylobacter jejuni following incubation in saline at 20°C, conditions that led to complete nonculturability.

Yaron and Matthews (2002) found that a variety of genes, including mobA, rfbE, stx1 and those for 16S rRNA synthesis, were expressed in nonculturable cells of E. coli O157:H7. Similarly, Barrett (1998) and Saux et al. (2002) have reported continued production of message for several genes after cells of

**Table 6.1** Bacteria described to enter the VBNC state

| | | |
|---|---|---|
| *Aeromonas salmonicida* | *Lactobacillus plantarum* | *Serratia marcescens* |
| *Agrobacterium tumefaciens* | *Lactococcus lactis* | *Shigella dysenteriae* |
| *Alcaligenes eutrophus* | *Legionella pneumophila* | *S. flexneri* |
| *Aquaspirillum* sp. | *Listeria monocytogenes* | *S. sonnei* |
| *Burkholderia cepacia* | *Micrococcus flavus* | *Sinorhizobium meliloti* |
| *B. pseudomallei* | *M. luteus* | *Streptococcus faecalis* |
| *Campylobacter coli* | *M. varians* | *Vibrio anguillarum* |
| *C. jejuni* | *Mycobacterium tuberculosis* | *V. campbellii* |
| *C. lari* | *Pasteurella piscida* | *V. cholerae* |
| *Cytophaga allerginae* | *Pseudomonas aeruginosa* | *V. fischeri* |
| *Enterobacter aerogenes* | *P. fluorescens* | *V. harveyi* |
| *E. cloacae* | *P. putida* | *V. mimicus* |
| *Enterococcus faecalis* | *P. syringae* | *V. natriegens* |
| *E. hirae* | *Ralstonia solanacearum* | *V. parahaemolyticus* |
| *E. faecium* | *Rhizobium leguminosarum* | *V. proteolytica* |
| *Escherichia coli* (including EHEC) | *R. meliloti* | *V. shiloi* |
| *Francisella tularensis* | *Rhodococcus rhodochrous* | *V. vulnificus* (types 1 and 2) |
| *Helicobacter pylori* | *Salmonella* Enteritidis | *Xanthomonas campestris* |
| *Klebsiella aerogenes* | *S.* Typhi | |
| *K. pneumoniae* | *S.* Typhimurium | |
| *K. planticola* | | |

*V. vulnificus* were in the VBNC state for as long as 4.5 months. Such findings strongly indicate not only viability, but active involvement in the VBNC process. Similarly, Rahman *et al.* (1994), studying *Shigella dysenteriae* cells that had been in the VBNC state for 4–8 weeks, found that such cells were capable of active uptake of methionine and its incorporation into protein.

Significant, and what might be characteristic biochemical changes in VBNC cells, have also been documented. Signoretto *et al.* (2002), in studying the cell wall peptidoglycan of *E. coli* entering the VBNC state, reported an increase in cross-linking, a 3-fold increase in unusual DAP–DAP (diaminopimelic acid) cross-linking, an increase in muropeptides bearing covalently bound lipoprotein, and a shortening of the average length of glycan strands in comparison to exponentially growing cells. VBNC cells were also found to have an autolytic capability far higher than that measured in exponentially growing cells. Similar findings were reported by Signoretto *et al.* (2000) for *Enterococcus faecalis*.

## Resuscitation from the VBNC state

We have reported cells of *V. vulnificus* to survive in the VBNC state for as long as 5 years (Sides *et al.*, 1999). Ultimately, however, for the VBNC state to be of consequence, the cells must be able to respond to some environmental stimulus and become metabolically active and, as a consequence, culturable. This process is generally referred to as "resuscitation".

The simplest form of resuscitation is that which is observed following removal or reversal of the factor(s) that induced nonculturability. Probably best studied is the case of *V. vulnificus*. This foodborne pathogen is induced into the VBNC state by incubation at temperatures below 10°C (Wolf and Oliver, 1992). This occurs whether or not cells are in a high nutrient environment (Oliver and Wanucha, 1989) or under starvation conditions (Oliver *et al.*, 1991). A simple temperature upshift results in resuscitation of these cells. This has been demonstrated *in vivo* (Oliver and Bockian, 1995), *in situ* (Oliver *et al.*, 1995) as well as *in vitro* (Wolf and Oliver, 1992; Oliver *et al.*, 1991).

One of the persistent questions regarding resuscitation has been the difficulty in showing conclusively that recovery of culturable cells from an otherwise nonculturable population is the result of true "resuscitation" of cells, as opposed to the regrowth of a fully culturable cell which evaded detection when the population was sampled. This problem developed largely from the lack of stringent controls in many early studies conducted on resuscitation (this aspect has been thoroughly reviewed by Kell *et al.*, 1998). However, several studies have now demonstrated conclusively that resuscitation does occur, at least in some bacteria. One of the earliest such studies is that of Whitesides and Oliver (1997), who studied resuscitation of *V. vulnificus* induced into the VBNC state by low temperature incubation. The key to this study was the dilution, up to a thousand-fold, of the nonculturable (< 0.1 CFU/mL) population to an extent where individually inoculated tubes ($n = 10$) each contained ca $10^3$ total (VBNC) cells, but < 0.0001 CFU/mL. Following a 24-hour incubation at room temperature, cells in each of the 10 tubes were observed to resuscitate to the original cell number (Figure 6.2A). An additional study examined the time course of resuscitation, and determined that > 10 generations would have had to occur during a 1-hour period if the

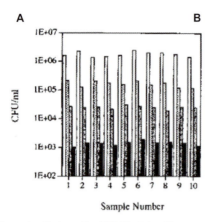 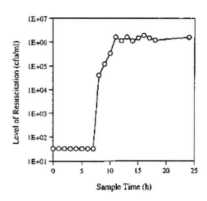

**Figure 6.2** Resuscitation of cells from the VBNC state. Shown are results from a study by Whitesides and Oliver (1997) on resuscitation of *V. vulnificus*. (A) Following dilutions (undiluted, 10-, 100-, and 1000-fold) of the nonculturable (<0.1 CFU/mL) population, cells were exposed to a temperature upshift. Culturability was returned to the population, with CFU/mL values reaching direct viability count values determined prior to the upshift. (B) Plate counts were performed hourly on the nonculturable population during a temperature upshift. Culturability was undetectable (<3.3 × 10$^1$ CFU/mL) for the first 7 hours, but reached 5 × 10$^4$ at 8 hours. Such an increase in culturability, if due to growth of cells, would have required >10 generations, and a generation time of ca. 6 min.

observed increase in culturability had been due to the presence of culturable cells (Figure 6.2B). Such an increase would have required a generation time of 6 minutes, a rate clearly impossible in the absence of any exogenous nutrient, the lack of aeration, and incubation at the suboptimal temperature of 22°C.

While many bacteria have been reported to enter the VBNC state, fewer have been conclusively demonstrated to exit this state and return to the actively metabolizing and culturable state. While bacteria like *V. vulnificus* are easily resuscitated by a simple reversal of the inducing factor, the difficulty in demonstrating resuscitation is exemplified by a study reported by Steinert *et al.* (1997). They examined nonculturability and resuscitation in *Legionella pneumophila*, and while entry into this state was easily induced by nutrient starvation, resuscitation could only be demonstrated following co-incubation of the VBNC cells with the amoeba, *Acanthameoba castellani*. Thus, resuscitation may prove to be the most difficult aspect of studying the VBNC state in some bacteria.

Other "natural" resuscitation conditions have been reported by Pommepuy *et al.* (1996), studying the induction by solar irradiation of the VBNC state in *E. coli*. Using quartz chambers and *in situ* incubation, these authors showed natural cycles of culturability and nonculturability in lake water that correlated with the levels of irradiation. Visible light has also been shown to induce dormancy in *E. coli* in natural river water by Barcina *et al.* (1989).

Reissbrodt *et al.* (2000) reported that the hydroxymate siderophore, ferrioxamine E, allowed resuscitation of *Salmonella enterica* serovar *Typhimurium* from the VBNC state. In a follow-up study, Reissbrodt *et al.* (2002) described the use of "Baxcell", a bacterial growth autoinducer, for the resuscitation of VBNC cells of *Salmonella Typhimurium* and *Escherichia coli* O157:H7. While an autoinducer, Baxcell is not a homoserine lactone nor an AI-2 like quorum-sensing molecule. Baxcell was reported to resuscitate *Escherichia coli* cells, present in water microcosms in the VBNC state for 455 days, which were still able to produce Shiga toxin. Such results exceeded the re-

suscitation ability of Oxyrase®, an oxygen radical inactivator, and of ferrioxamine E.

A novel method of resuscitation was reported for VBNC cells of *Vibrio cholerae* O1 by Wai *et al.* (1996). Cells that had been in the VBNC state for over 70 days were resuscitated by a heat shock of 45°C for 1 minute. This response appeared to require the presence of NH$_4$Cl in the resuscitation medium, but the role this compound played in the resuscitation response is not understood. Similarly, the mechanism by which heat shock allowed the subsequent culture of *V. cholerae* is not understood, although the authors suggested a possible role for protective heat shock proteins.

## Reactive oxygen species (ROS) and the VBNC state

In an important development in the development of methods to allow culture of "nonculturable" cells, Mizunoe *et al.* (1999) determined that EHEC cells which were nonculturable (<10$^0$ CFU/mL) on agar media following incubation in distilled water at 4°C could, in fact, be cultured to a high level (10$^4$–10$^5$ CFU/mL) when plated on the same media containing either catalase, sodium pyruvate or α-ketoglutarate. A similar finding was reported by Sides *et al.* (1999) that same year for both O157:H7 and non-O157:H7 cells of *E. coli*. This finding suggests the metabolic by-product, H$_2$O$_2$, is produced by cells during the VBNC process, and that this toxic agent is neutralized by these anti-reactive oxygen species compounds.

Mizunoe *et al.* (2000) subsequently reported *V. parahaemolyticus* to enter the VBNC state in ca. 12 days following starvation and 4°C incubation. However, when the plating medium was supplemented with catalase or sodium pyruvate, culturability was maintained for a considerably longer period. Similarly, Wai *et al.* (2000) reported cells of *A. hydrophila* to enter the VBNC state in ca. 45 days when incubated under starvation conditions at 4°C. Cells could be cultured, however, when either of these two agents was present in the plating medium. Similar results were reported for *V. vulnificus* by Sides *et al.* (1999) and subsequently by Bogosian *et al.* (2000).

First to suggest that the plating medium itself may be important in the VBNC state is believed to be Whitesides and Oliver (1997). In a study on the resuscitation of *V. vulnificus* from the VBNC state, they stated that "... elevated nutrient might be toxic in some manner to cells in this [the VBNC] state", and "Why resuscitation appears to occur only when little or no nutrient is present is not understood, although this study suggests that HI [heart infusion] broth is not, in itself, toxic to cells in the VBNC state." Shortly thereafter, Bloomfield *et al.* (1998) speculated that, on exposure to high nutrient levels, growth-arrested cells would likely undergo an imbalance in metabolism which would result in a near instantaneous production of super-oxide and other free radicals. In the absence of some pre-adaptation (e.g. starvation-induced stress proteins), such cells would not be able to detoxify these toxic radicals and, as a result, would die. This theory could also account for the inability of VBNC cells to grow on nutrient media.

This hypothesis was quite testable, and several studies (Begosian *et al.*, 2000; Vedeikis and Oliver, 2000; Mizunoe *et al.*, 2000, 2001) were subsequently published indicating that, indeed, the presence of ROS in solid media, or produced by VBNC cells when plated to solid media, might account for entry into the VBNC state. The value of such ROS scavengers is indicated in Figure 6.3, which demonstrates the effect of incorporation of catalase or pyruvate in plated media (HI agar) onto which VBNC cells of *V. vulnificus* are placed. Most recently, we have produced an *oxyR* mutant of *V. vulnificus* which is incapable of producing catalase. These cells are nonculturable even when cultured at room temperature, resembling wild type *V. vulnificus* cells which are nonculturable following incubation

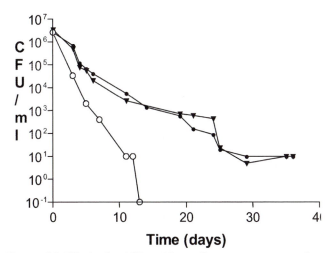

**Figure 6.3** Effect of addition of reactive oxygen scavenging agents to culture media on culturability of "nonculturable" cells. Whereas culturability of 5°C-incubated *V. vulnificus* cells on heart infusion agar was lost by 12 days, addition of catalase or pyruvic acid to the medium resulted in maintenance of significant culturability for over 5 weeks. Open circles represent medium lacking pyruvate or catalase and closed circles and closed triangles represent medium containing catalase or pyruvate, respectively.

at 5°C. Such observations suggest that, at least for some cells, a large basis for the inability to culture cells from the VBNC state resides in their production of peroxide when plated onto solid nutrient media, coupled with their inability to detoxify this lethal metabolite.

Barer (1997) has suggested that, since viability (as classically defined) can only be demonstrated by colony development, the term "viable but nonculturable" is an oxymoron. Further, given the finding that some cells, when present in the VBNC state, may be at least partially culturable under certain conditions (e.g. the presence of ROS-scavengers), Oliver (2000b) also suggested that this phrase is not fully accurate. However, the phrase "viable but nonculturable" has become adopted by the majority of researchers investigating this phenomenon, and the fact remains that certain conditions induce these cells to lose their ability to develop colonies on routine media, and to thus become "nonculturable" by routinely employed methods.

## Virulence of cells in the VBNC state

A key aspect of the VBNC state is, of course, whether or not cells in the VBNC state retain virulence. Cells in the VBNC state generally demonstrate extremely low metabolic rates, and with few exceptions, cells in this near-dormancy state would not be considered to be able to initiate infection. Note, however, that VBNC cells are still capable of gene transcription, and this could result in the production of one or more virulence factors. For example, Saux *et al.* (2002) have recently demonstrated that *V. vulnificus* continues to produce mRNA for its hemolysin/cytolysin even after 4.5 months in the VBNC state. Similarly, Nwoguh *et al.* (1995) demonstrated that nonculturable cells of *E. coli* and *Salmonella* strains retain inducible enzyme activity. Likely more important, however, would be resuscitation under the appropriate conditions that may lead to subsequent infection of a susceptible host.

To date, and depending on the bacterium in question, the evidence on this subject is often contradictory. Certainly, numerous studies have demonstrated the ability of pathogens which, while present in the VBNC state when inoculated into a susceptible host, were able to initiate infection. As examples, Colwell *et al.* (1985) inoculated VBNC cells of *V. cholerae* into ligated rabbit ileal loops, and observed positive responses (i.e. enteropathogenicity) in all samples. Intestinal fluid was also reported to contain culturable *V. cholerae* cells. In a subsequent study, Colwell *et al.* (1996) reported a study wherein two human volunteers ingested VBNC cells of *V. cholerae*. Approximately 48 h after the challenge, one of the volunteers had culturable *V. cholerae* cells in his stool, while at 5 days the second volunteer was stool positive for this pathogen.

As a second example, Oliver and Bockian (1995) demonstrated the ability of *V. vulnificus* cells, present in the VBNC state, to retain virulence against mice. In their study, cells were made VBNC (<0.1 CFU/mL) by incubation at 5°C. These cells were diluted such that, while the preparation contained

$10^5$ cells in the VBNC state, there were <0.04 CFU/mL. Mice inoculated with this preparation died, and culturable *V. vulnificus* cells were recovered from both blood and the peritoneal cavity.

Additional discussions on virulence of cells in the VBNC state are presented below as the various bacterial species entering the VBNC state are described.

## The presence of bacterial pathogens in foods while in the VBNC state

Although concern for the possibility of VBNC cells being present in foods was expressed some time ago (Archer and Young, 1988; Oliver, 1989; Schofield, 1992), relatively few studies have been reported on the presence of such cells in foods. This is likely due to the technical difficulties in determining, of all the bacteria present in the food, which are the target VBNC cells. Methods (e.g. fluorescently labeled monoclonal antibodies) for identifying specific cells do exist, however, but while the possible transmission of pathogens in the VBNC state by food has been suggested (e.g. see van Duynhoven and de Jonge, 2001), this remains a much under-investigated area of food safety.

Despite this, there is strong evidence for the presence of VBNC cells in foodstuffs. Millet and Lonvaud-Funel (2000) studied populations of acetic acid and lactic acid bacteria in wine during its storage. Their results suggested that acetic acid bacteria were induced into the VBNC state by oxygen deprivation, while lactic acid bacteria were induced into this state by sulfites. They noted that, while such cells were nonculturable, they were capable of hydrolyzing esters. Further, at least in the case of the acetic acid bacteria, resuscitation was said to be very rapid when oxygen was re-supplied. As is common for cells entering the VBNC state, these cells underwent dwarfing after several weeks of wine storage, passing 0.45 µm filters, but were again retained upon resuscitation.

Similarly, Gunasekera *et al.* (2002) recently demonstrated that both direct viable counts (using propidium iodide exclusion to demonstrate membrane integrity), and cells demonstrating *de novo* expression of a *gfp* reporter gene, were significantly higher than colony counts for both *E. coli* and *Pseudomonas putida* in commercially available pasteurized milk. They concluded that a "… substantial portion of cells rendered incapable of forming colonies by heat treatment are metabolically active and are able to transcribe and translate genes de novo".

Duffy and Sheridan (1998) examined the correlation between both acridine orange and the *BacLight*® reagent with standard plate counts in samples of processed meats (minced beef, cooked ham, bacon, and beef burgers). They found acridine orange direct counts (i.e. total bacterial cells) not to correlate well with plate counts. However, *BacLight*® counts (i.e. viable cells) correlated well ($r^2 = 0.87–0.93$) with plate counts for all sample types. Such findings again suggest the presence of VBNC cells.

A *Salmonella* outbreak caused by dried, salted squid contaminated with *Salmonella enterica* subsp. *enterica* Oranienburg occurred in Japan in 1999. From this material, Asakura *et al.* (2002) isolated a strain of this species which became nonculturable within 7 days in response to the elevated osmotic concentration induced by 7% NaCl. These cells, however, demonstrated >90% viability by the *Baclight*® assay. That these cells would enter the VBNC state in experimentally infected dried squid was also demonstrated, as well as the ability to recover these cells following a resuscitation step. The authors reported that <20 CFU could be recovered by direct plating from the implicated squid, a level believed to be insufficient to establish the septicemia present in the patients. However, as the authors pointed out, if a large population of VBNC cells were present in the squid, a much higher number of viable cells would have been present than detectable by plate counts, and that this could have caused the outbreak. These authors previously reported (Makino *et al.*, 2000) a food poisoning caused by enterohemorrhagic *E. coli*-contaminated salted salmon roe. Cells isolated from patients also were NaCl sensitive, consistent with the results of their findings with the *Salmolnella*-infected squid.

Makino *et al.* (2000) examined the possibility that a 1998 outbreak of enterohemorrhagic *E. coli* O157 in Japan which resulted from ingestion of salted salmon roe might be due to the presence of these cells in the VBNC state. Because as few as 0.75 CFU of this pathogen was estimated to have caused the infection, such an event seemed feasible. The roe had been soaked in a fermented seasoning with a salt content equivalent to 13% NaCl. Whereas O157 cells obtained from patients lost culturability at this osmotic content, the cells could be resuscitated, and more than 90% were shown to be viable by a fluorescent assay. In contrast, O157 cells obtained directly from the salted roe were able to grow at the elevated salt level, although 20% also appeared to enter the VBNC state. VBNC cells originating from roe, but not the patients, killed mice and could be isolated from their intestines, indicating resuscitation of these cells *in vivo*. The authors thus proposed that O157 cells are able to enter the VBNC state as a result of osmotic shock (e.g. high NaCl concentrations), and then resuscitate when the stress conditions have been eliminated. The authors went on to suggest that "… the VNC state in food is potentially dangerous from a public health viewpoint and may have to be considered at the time of food inspection".

That high salt concentrations might induce *E. coli* into the VBNC state, which later could resuscitate and cause infection, has received strong experimental support by a study reported by Ohtomo and Saito (2001). These authors observed that saline-stressed cells of *E. coli* underwent a significant loss of colony-forming units, which rapidly (within 2 hours) returned to the original culturable levels when the stress was removed. Significantly, this resuscitation occurred in the absence of both DNA synthesis and cell division, indicating that true resuscitation, as opposed to regrowth, had occurred.

What follows is a discussion of foodborne pathogens described to enter the VBNC state, and which thus may be a food safety consideration. The reader is also directed to a recent review by Oliver (2000b) that describes the public health significance of VBNC cells, including those (e.g. *Helicobacter pylori*) not generally transmitted by food.

## *Campylobacter*

*Campylobacter* spp. are responsible for disease in a number of animals, with *C. jejuni* being especially important as a human foodborne pathogen. Indeed, this species is now recognized as one of the most common sources of enteric disease. The major food source of *Campylobacter* infections is poultry, which represents up to 70% of infection, although a wide variety of animals serve as reservoirs for human infection. *Campylobacter* spp. are also frequently isolated from water, which has also been implicated as a vehicle of some outbreaks. Infections are typically of the lower gastrointestinal tract, and vary from asymptomatic to severe diarrhea with blood, lasting more than 1 week (Beumer *et al.*, 1992; Nachamkin, 2001). The role of the VBNC state in survival and epidemiology of this pathogen was recently reviewed by Oliver (2000b), and the reader is directed to that review also.

Probably the first study to demonstrate entry of *C. jejuni* into the VBNC state was that of Rollins and Colwell (1986), who reported that cells held at 4°C in sterile stream water maintained viability for over 4 months, whereas cells incubated at higher temperatures (25 or 37°C) rapidly lost culturability, entering the VBNC state within 28 and 10 days, respectively. A similar finding was reported by Korhonen and Martikainen (1991) who studied the survival of this bacterium in both treated and untreated lake water.

Cells of *C. jejuni* undergo a morphological transition from rod/spiral form to a coccoid form, generally during conditions unfavorable to growth, including during entry into the VBNC state (Rollins and Colwell, 1986). The role of these coccoid forms in the pathogenesis of *C. jejuni* (as well as in the related *Helicobacter pylori*) is not known. While it has been suggested that these are degenerate forms, the membranes remain intact, and such cells demonstrate many of the traits characteristic of cells in the VBNC state. Further, it has been reported that the coccoid forms retain virulence (Oliver, 2000b).

In one of the most comprehensive studies on the VBNC state in *C. jejuni* cells, Tholozan *et al.* (1999) measured cell volume, adenylate energy charge, internal pH, intracellular potassium concentration, and membrane potential values of cells during 30 days of incubation in filter sterilized surface water. They observed reduction in all of these parameters with time in the VBNC state compared to culturable cells, but noted that these results were similar to those reported for other genera undergoing starvation. Further, while the three strains studied became nonculturable in 14–16 days, this apparent inability to maintain internal homeostasis did not result in cell death after the 30 day study.

Beumer *et al.* (1992) determined that, while the number of culturable cells declined to the point that none could be detected after 4 days in saline at 20°C, ATP levels in these cells remained fairly constant during their 3-week period of study. This finding, coupled with DVC (direct viable count) data, support the conclusion that *C. jejuni* enters a VBNC state. Federighi *et al.* (1998) also reported maintenance of ATP levels following entry into the VBNC state in three of the *Campylobacter jejuni* strains they studied.

Lazaro *et al.* (1999) studied the survival of *Campylobacter jejuni* at 4 and 20°C and found that the VBNC cells induced by low temperature incubation had intact DNA after 116 days, and cellular integrity and respiring cells were detected for up to 7 months. Examination of the proteins in these cells by 2-D electrophoresis revealed both up- and down-regulated proteins. Further evidence for low temperature induction into a nonculturable state by *C. jejuni* has also been provided by Ekweozor *et al.* (1998).

The question of whether or not VBNC *C. jejuni* is capable of causing infection is unclear, with studies involving chick, mouse, rat, and human models providing conflicting results. Cappelier *et al.* (1999a) examined the VBNC state in *C. jejuni* and the ability of such cells to infect mice and chicks. Cells from three strains of this pathogen were induced into the VBNC state by incubation at 4°C, then inoculated *per os* into newborn mice and 1-day old chicks. Under these conditions, all three *C. jejuni* strains resuscitated in the murine model, while two revived in the chick model. Their conclusion was that "… the VBNC state should be considered as playing a role in the epidemiology of *Campylobacter* infection." In another study by these authors (1999b), three strains were made VBNC by incubation in sterile surface waters at 15 and 25°C. After 30d, the cells were inoculated into yolk sacs of embryonated chicken eggs. All strains were found to resuscitate under these conditions, and all retained their ability to adhere to HeLa cells. Similarly, Jones *et al.* (1991) reported two of four strains of *C. jejuni* that were made VBNC in water caused death in suckling mice, and Stern *et al.* (1994) found VBNC cells of *C. jejuni* retained the ability to colonize 1 day-old chicks.

Talibart *et al.* (2000), determined that 98% of the 85 strains of *C. jejuni* and *C. coli* they studied became nonculturable within 30 days when incubated at 4°C in sterile water. Of these, 51% could be recovered after injection into 9-day fertilized chicken eggs. The authors concluded that viable but nonculturable forms of *Campylobacter* are a potential public risk to humans and animals, and that embryonic factors may be essential for resuscitation. In contrast, Medema *et al.* (1992) reported that when $1.8 \times 10^5$ cells of a VBNC culture of a *C. jejuni* strain isolated from chicken feces was inoculated into 1-day-old chicks, colonization of the intestinal tract did not occur. Noting that such chicks are a very sensitive and natural animal model for campylobacter colonization, the authors concluded that the viable but nonculturable state appears to have a very limited significance, if any, in the transmission of this genus.

When cocci that developed during the VBNC process were introduced into simulated gastric, ileal, or colon environments by Beumer et al. (1992), no resuscitation was detected. Similarly, when the coccoid forms were administered orally into laboratory animals and human volunteers, no symptoms of campylobacteriosis was observed. Thus, these authors were not able to demonstrate that VBNC cells of this species retain virulence.

Saha et al. (1991) observed that 16 strains of C. jejuni which had been frozen, then inadvertently thawed, were fully nonculturable. However, on inoculation into rat ileal loops, 7 of the 16 were successfully resuscitated and remained toxigenic. That cells could undergo such freeze–thaw and retain viability and virulence should be of special concern to the food industry.

Few studies exist on the VBNC state in C. coli, but Jacob et al. (1993) reported entrance to this dormant state within 48 hours at 37°C, and 14 days at 4°C. In contrast, Rollins and Colwell (1986) reported C. jejuni to become nonculturable in 10 days and 4 months at these two temperatures, respectively. Of the two species, C. coli is less frequently isolated from the environment, and the suggestion by Jacob et al. (1993) that this might be due to a more rapid entry into the VBNC state appears supported by these findings. These authors found no change in whole cell protein or lipooligosaccharide patterns as C. coli became nonculturable, and they concluded that the coccoid forms produced by this species are likely dormant forms (i.e. VBNC), not detectable by conventional microbiological techniques.

## Enterococcus

E. faecalis is the most commonly isolated Enterococcus species from human clinical samples, and E. faecium the second (Lleò et al., 2001). Previously classified as "group D streptococci", the enterococci occur in plants and in the feces of warm- and cold-blooded animals. Most infections are of the urinary tract or blood, and are especially common in patients with catheters and in patients hospitalized for prolonged periods. Infections can be life-threatening, and these bacteria are a leading cause of nosocomial (hospital-acquired) diseases. As a result of poor hygiene or fecal contamination during slaughter, enterococci are also foodborne. Since some of these bacteria are salt, heat, and freeze resistant, they can maintain themselves in foods for long periods (Pierson and Smoot, 2001; Murray et al., 2002).

Entrance by Enterococcus species into the VBNC state, the physiological and biochemical changes that occur during this transition, and the possible public health significance of cells in this state, have largely been described by researchers of the Canepari group at the University of Verona. These researchers (Signoretto et al., 2000) reported that the cell walls of E. faecalis cells in the VBNC state were more resistant to mechanical disruption than cells in the growing state. They found that while the teichoic acid and lipoteichoic acid levels remained unchanged during entry into the VBNC state, the cell wall peptidoglycan underwent an increase in total cross-linking, as

did the penicillin binding proteins (involved in peptidoglycan assembly). The authors suggest that these physiological changes in the cell wall are specific to the VBNC state, and provide indirect confirmation of the viability of such cells.

As further indication of viability in these cells, Lleò et al. (2000) determined the presence of mRNA for one of the penicillin binding proteins, pbp5, in VBNC cells of E. faecalis. The detection of mRNA in cells has been shown to be an indicator of viability, owing to its high instability (Sheridan et al., 1998). In the case of E. faecalis, Lleò et al. (2000) demonstrated a correlation between the presence of pbp5 mRNA and metabolic activity and ability to resuscitate from the VBNC state, indicating the viability of cells which had been in this physiological state for 3 months.

Lleò et al. (2001) subsequently studied entry into, and resuscitation from, the VBNC state in different enterococcal species (E. faecalis, E. hirae, E. faecium) following induction into the VBNC state by incubation in sterile lake water at 4°C. Through examination of pbp5 mRNA levels and by quantifying their resuscitation ability, it was determined that E. faecalis and E. hirae entered the VBNC state within 2 weeks, and remained viable but nonculturable for 3 months. Entry into the VBNC state required 4 weeks in the case of E. faecium, and the period of viability was considerably shorter (2 weeks).

The ability of E. faecalis cells in the VBNC state to adhere to cultured heart and urinary tract epithelial cells has been studied by Pruzzo et al. (2002). When compared to growing cells, they observed a 50–70% decrease in the adherence capability of VBNC cells. However, adherence values were similar to growing cells following resuscitation.

The accumulated data from these studies provide strong evidence for a VBNC state in Enterococcus spp. (especially E. faecalis), and further suggest that this state may be of significance in the epidemiology of infections caused by these bacteria.

## Escherichia coli

E. coli is part of the normal flora in the intestinal tracts of humans and other warm blooded animals, and while most strains do not cause disease, certain strains, notably those of serotype O157:H7, cause hemorrhagic colitis and hemolytic-uremic syndrome, a potentially fatal condition. Such strains are found in cattle, and hence undercooked ground beef is the primary source of human infection (Meng et al., 2001).

Physiological (Porter et al., 1995), molecular (Yaron and Matthews, 2002), and biochemical (Signoretto et al., 2002) studies have provided strong evidence for the presence of a VBNC state in Escherichia coli.

Rigsbee et al. (1997) studied entry of enterohemorrhagic Escherichia coli (EHEC) cells into the VBNC state in river and sea water at both 5 and 25°C. They reported the cells to remain culturable in both water sources at 25°C, but to enter the VBNC state in both waters when incubated at 5°C. Thus, temperature, rather than salt content, appeared to be the primary signal for entering the nonculturable state in this strain.

However, sea water is ca 3.5% salt, and higher NaCl levels have been reported to induce the VBNC state in *E. coli*.

That high NaCl concentrations may induce the VBNC state in *E. coli* had previously been shown by Roth *et al.* (1988). Following incubation in 0.8 mol/L NaCl (4.7%), cells underwent a rapid and large decrease in colony forming ability, but could be resuscitated when the cells were exposed to the osmotic protectant, betaine. Renewed ability to form colonies in the presence of betaine occurred even in the presence of the protein synthesis inhibitor, chloramphenicol, indicating that true resuscitation had occurred. Further, we found (McGovern and Oliver, unpublished) that a mutant for HN-S, a regulatory factor required for dealing with osmotic stress, was not substantially able to recover from such an osmotic stress, even in the presence of betaine.

In Japan, salmon roe is soaked in soy sauce and a variety of seasonings, providing a salt content equal to 13% NaCl. As described in more detail in the introduction to this chapter, an outbreak occurred in Japan in 1998 resulting from salmon roe contaminated with *E. coli* O157:H7. Makino *et al.* (2000) found that, based on plate counts, ca 0.75–1.5 culturable cells would have been responsible for this outbreak, a number considered too low for infection. However, they provide evidence that a larger number of cells likely existed in the VBNC state in this sushi ingredient, and these were the source of the outbreak.

It thus appears that low temperature, and high levels of NaCl, are both inducers of the VBNC state in *E. coli*, and that the presence of such cells in foodstuffs is likely to be of public health concern.

## Listeria

*Listeria* spp. have an unusually wide growth temperature range (0–45°C), with freezing able to preserve the bacterium. It is also capable of growth at unusually high NaCl concentrations (to 12%). Possibly due to these traits, *Listeria monocytogenes* is found in a wide variety of foods, including milk and milk products, cheeses, meat and poultry products, seafood, and a variety of "ready-to-eat" foods. Infections typically occur in certain high-risk groups, including pregnant women, newborns, and immunocompromised individuals. Septicemia, meningitis, and meningoencephalitis, with a mortality rate of 20–25%, are the usual results of infection. Infection during pregnancy may lead to stillbirth or abortion (Swaminathan, 2001). The recognition of *L. monocytogenes* as a foodborne pathogen, its significance, incidence, and survival in foods, has also been reviewed by Schofield (1992) and most recently by Kathariou (2002).

Despite the great significance of *Listeria* as a foodborne pathogen, few studies have been reported on the ability of this organism to enter the VBNC state. Possibly this is due to the fact that so many Gram-positive bacteria fail to respond to the originally described and widely used indicators of viability (e.g. the method of Kogure *et al.*, 1979).

Besnard *et al.* (2002) examined the factors inducing *Listeria monocytogenes* (four strains) into the VBNC state. Factors studied included inoculum size, temperature (4°C and 20°C), NaCl concentration (0 and 7%), pH (5 and 6), and presence of sunlight. Of these factors, higher incubation temperature and salt concentration both decreased the time necessary to enter the VBNC state. The authors concluded that, because of the relevance of these factors to the food industry, the presence of VBNC cells of this pathogen could pose a major public health hazard since the cells can not be detected by routine culture methods. However, Li *et al.* (2003) reported that the combination of cold (4°C) and a carbon dioxide atmosphere, as frequently employed in fresh muscle food and fruit preparation, did not induce the VBNC state in cells of *L. monocytogenes*.

## Salmonella

*Salmonella* are widespread in the natural environment, and occur in a variety of meats, although their presence in poultry and eggs products is the greatest concern. Despite efforts to reduce its presence in foods, *Salmonella* spp. remain a leading cause of foodborne bacterial illness in humans. Outbreaks involving more than 10,000 persons have been reported (D'Aoust *et al.*, 2001).

Roszak *et al.* (1984) appear to be the first to demonstrate entry of a *Salmonella* sp. (*S. enteritidis*) into the VBNC state. Entry appears to have been induced by nutrient deprivation, with cells becoming nonculturable within 48 hours in sterile river water at 25°C. Resuscitation was indicated following addition of nutrient to cells that had been in the VBNC state for up to 21 days. A subsequent study by Roszak and Colwell (1987) employed direct viable counts and microautoradiography (following uptake of radiolabeled glutamate or thymidine) to provide further evidence that these nonculturable cells were, in fact, metabolically active and thus alive. This study also reported data suggesting that these VBNC cells, when inoculated into rabbit ligated ileal loops, could be resuscitated to the culturable state.

In contrast to the studies by Roszak and co-workers (1984, 1987), Chao *et al.* (1987) and Chmielewski and Frank (1995) were unable to demonstrate a VBNC state in *S. enteritidis* when incubated in river water or in PBS. Such conflicting results may be a consequence of the suspending medium, as Cornax *et al.* (1990) reported that, at 18°C, *S. paratyphi* became nonculturable in natural and filtered seawater, but not in artificial seawater or saline solutions.

Smith *et al.* (2002) employed carbon and nitrogen stress in the presence of chloramphenicol to induce *Salmonella enterica* serovar Typhimurium into the VBNC state. These were then inoculated, by both intraperitoneal (IP) and oral routes, into female BALB/c mice. Despite doses of VBNC cells exceeding the oral and IP $LD_{50}$ values by 3.5 and 2 orders of magnitude, respectively, no evidence for infection or colonization was detected. Thus, these authors concluded that VBNC cells of the *Salmonella* strain used should not be considered infective.

A potential problem with such studies, of course, is the possibility that the culture conditions employed do not provide cells which mimic those found in natural, including food, environments. Further, it is possible that the mouse model employed is not appropriate, or that experimental conditions (e.g. the presence of chloramphenicol which might prevent resuscitation of *Salmonella* cells) do not allow resuscitation to culturable and infective cells.

A 1999 *Salmonella* outbreak in Japan, caused by contaminated dried and salted squid (Asakura *et al.*, 2002), has been described in some detail in the introduction. Whether cells in the VBNC state were responsible for this outbreak is impossible to determine with certainty, but the circumstances surrounding the event suggests this is a plausible explanation.

Resuscitation of *Salmonella* spp. was first reported by Roszak *et al.* (1984). They reported cells of *S. Enteritidis*, made VBNC by starvation, to again become culturable following addition of organic nutrient. The time frame within which resuscitation of the cells would occur, however, was brief. Cells in the VBNC state for 4 days, but not 21 days, could be resuscitated. More recently, Reissbrodt *et al.* (2000, 2002) employed an iron-binding siderophore and a bacterial growth autoinducer for resuscitating *S. enterica* (see Introduction, "Resuscitation from the VBNC State" for greater discussion of these studies).

### Shigella

Members of the genus *Shigella* cause bacillary dysentery, or shigellosis. While disease is associated with a wide variety of food types, no specific foods are routinely implicated. It has been estimated that over 400,000 cases of shigellosis occur each year in the United States, making it the third leading cause of foodborne outbreak by bacterial pathogens. Most cases are believed to result from the fecal–oral route, a result of poor personal hygiene. The disease is characterized by the production of bloody diarrhea and by a low infectious dose (as low as 20 cells), traits that differentiate this disease from those caused by most other foodborne pathogens (Lampel and Maurelli, 2001).

Colwell *et al.* (1985) first reported *S. sonnei* and *S. flexneri* to become nonculturable when incubated in estuarine water at 25°C, with *S. flexneri* entering the VBNC state three times faster than *S. sonnei* (which required 2–3 weeks to be fully nonculturable).

Islam *et al.* (1993) found that *S. dysenteriae* Type 1, the leading cause of dysentery worldwide, became nonculturable in Bangladesh pond, river, drain, and lake waters after 2–3 weeks. After 6 weeks, the cells remained in the VBNC state, as shown by the Kogure DVC assay. The VBNC cells could be detected by PCR and fluorescent antibody techniques. These authors concluded that the VBNC state may be important in the epidemiology of shigellosis.

Rahman *et al.* (1994) found *S. dysenteriae* Type 1 to survive more than 6 months in the VBNC state. Such cells retained cytopathogenicity for cultured HeLa cells. In a follow-up study, Rahman *et al.* (1996) reported that VBNC cells of *Shigella dysenteriae* Type 1 not only retained their ability to produce active Shiga toxin, but also the ability to adhere

to intestinal epithelial cells (Henle 407 line). However, such cells appeared to lose their ability to invade Henle cells. The authors concluded that VBNC cells of this species retained virulence factors and remained potentially virulent, posing a public health risk.

Kehoe *et al.* (2002) reported that *Shigella dysenteriae* Type 1 became nonculturable within 50 hours when incubated in water. However, noting that this species lacks catalase activity, they found that the same cells, when plated onto medium supplemented with catalase or pyruvate, remained culturable after 100 hours. They suggested that starvation (as employed for induction of the VBNC state in this study) results in the production of reactive oxygen species (e.g. $H_2O_2$) that lead to cell death. The presence of catalase in the medium, however, allows the culture of at least some of these cells. Thus, they suggested that VBNC cells of *Shigella dysenteriae* Type 1 are likely going undetected during routine analysis of environmental samples.

### Vibrio spp.

A variety of pathogenic vibrios are recognized, most of which are associated with consumption of seafood. These typically cause various degrees of gastroenteritis, although some species (e.g. *V. vulnificus*) produce life-threatening infections. The most significant, worldwide, is certainly *V. cholerae*, the causative agent of cholera, one of the few foodborne illnesses with epidemic and pandemic potential. *V. parahaemolyticus* is the major cause of foodborne disease in some countries (e.g. Japan and Korea), and has caused outbreaks in the United States affecting over a thousand people. *V. vulnificus* is responsible for 95% of all seafoodborne deaths in the United States, a result of consumption of raw or undercooked molluscan shellfish. It carries a fatality rate in excess of 50%, the highest death rate of any foodborne disease in the US. Other seafoodborne vibrios (e.g. *V. fluvialis*, *V. furnissii*, *V. hollisae*, and *V. alginolyticus*) are responsible for fewer infections (Oliver and Kaper, 2001).

### Vibro cholerae

Vibrios are associated with a wide variety of seafoods, and represent the major public health threat following their consumption. The incidence of vibrios in seafood has been recently reviewed by Oliver and Kaper (2001), and the reader is referred to that review for discussions on the isolation, identification, epidemiology, virulence mechanisms and incidence of *V. cholerae* infections following consumption of seafood. The VBNC state of this pathogen has also been recently reviewed by Oliver (2000b).

Chaiyanan *et al.* (2000) studied the VBNC state in both O1 and O139 strains of *V. cholerae* induced into the VBNC state by incubation in 1% Instant Ocean at 4°C for 6 months. The cells, 75–90% of which were found to be viable by the *Baclight*® method, contained a "significant amount" of ATP, retained surface protein determinants, and demonstrated susceptibility to a variety of antimicrobial agents.

The first study to document the entry of *V. cholerae* into the VBNC state was that of Xu *et al.* (1982). The clinical isolate they studied was shown to enter the VBNC state within 6–9 days in natural estuarine water when incubated at ca. 5°C. Research from Colwell's lab subsequently documented that this pathogen was widely present in waters from which it could not be cultured (Xu *et al.*, 1984; Brayton *et al.*, 1987; Huq *et al.*, 1990). Such cells were subsequently shown to be capable of causing human infection (see Introduction to this chapter). Evidence for the continued virulence of *V. cholerae* when in the VBNC state has also been provided by Hasan *et al.* (1992), who were able to show the presence of cholera enterotoxin in diarrheal stool samples from a patient with clinical cholera symptoms, but who was culture negative for *V. cholerae* O1, non-O1, and enterotoxigenic *E. coli*. A number of investigators, including Colwell *et al.* (1985), Ravel *et al.* (1995), Wai *et al.* (1996) and have also suggested that the VBNC state accounts for the seasonal nature of cholera outbreaks observed with this pathogen (Colwell, 1996).

### *Vibrio parahaemolyticus*

The first description of the involvement of *V. parahaemolyticus* in a major foodborne outbreak occurred in 1950. Since then, over 40 outbreaks have been documented in the United States, with several thousands of persons affected. As much as 70% of all foodborne disease is caused by *V. parahaemolyticus* in Japan. Gastroenteritis caused by this *Vibrio* is almost exclusively associated with the consumption of raw, undercooked, or re-contaminated seafood, of which fish, crab, shrimp, lobster, and oysters have all been implicated (Oliver and Kaper, 2001). Only strains of *V. parahaemolyticus* which produce the so-called "Kanagawa hemolysin" are capable of initiating human infections. Interestingly, whereas virtually all clinical isolates are Kanagawa-positive (K+), ca. 99% of all environmental strains are K−. It is now clear that the production of this hemolysin is required for survival in the human host, and the rare K+ strains present in seafoods are selected for in the body.

Although only a single strain of each Kanagawa type was studied, Jiang and Chai (1996) reported that a Kanagawa-negative (K−) strain lost culturability more slowly than a K+ strain at low temperature, suggesting that the hemolysin may play some role in survival. Bates and Oliver (2004) expanded on this study, studying 11 K+ strains and K− strains of *V. parahaemolyticus*. These included mutants in both the *tdh1* and *tdh2* hemolysin-encoding genes, as well as a co-isogenic strain containing a *tdh1/tdh2* encoding plasmid allowing restoration of KP hemolysin activity. Despite employing various nutrient and temperature downshifts and various osmotic concentrations, they found no differences in the ability of these strains to enter the VBNC state. Their study did report the successful resuscitation of the various strains to the fully culturable state following an overnight temperature upshift. However, addition of ROS scavengers was not found to enhance recovery of cells from the VBNC state. This is in contrast to a study reported by Mizunoe *et al.* (2000), which reported that media supplemented with catalase and SOD allowed growth of nonculturable cells of *V. parahaemolyticus*.

### *Vibrio vulnificus*

While worldwide *V. cholerae* is the most important seafood-borne pathogen, *V. vulnificus* carries the highest case-fatality rate of any foodborne pathogen; >50% of all *V. vulnificus* infections are fatal. Most infections follow the ingestion of raw or undercooked molluscan bivalves, particularly raw oysters, and 95% of all seafood-related deaths are due to this pathogen. The Centers for Disease control estimates there are approximately 50 foodborne cases of *V. vulnificus* annually in the United States (Mead *et al.*, 1999). The biology of this pathogen has been the subject of several reviews (Linkous and Oliver, 1999; Strom and Paranjpye, 2000; Oliver and Kaper, 2001).

More is probably known about the VBNC state in *V. vulnificus* than any other bacterium. Because of the number of studies and reviews on this physiological state in this pathogen (Oliver, 2000a; 2000b; 2000c), only a brief summary is provided here.

Whereas *V. vulnificus* is easily isolated from seawater and oysters during warm months, researchers have generally been unable to detect the bacterium when water temperatures are <13°C. This has generally been attributed to entry of these cells into the VBNC state, and both laboratory (Oliver *et al.*, 1991) and *in situ* (Oliver *et al.*, 1995) studies have supported this. Lab studies have revealed that *V. vulnificus* cells dwarf from rods to small (ca. 0.6 μm) cocci. These forms have significantly reduced ribosome and nucleic acid density but maintain an intact cytoplasmic membrane (Linder and Oliver, 1989). Major modification of membrane fatty acids occur as cells enter the VBNC state (Linder and Oliver, 1989; Day and Oliver, 2004), along with drastic and rapid decreases in protein, RNA, and DNA synthesis (Oliver, 2000a). At the same time, novel proteins are synthesized in response to temperature and nutrient downshifts (Morton and Oliver, 1994; McGovern and Oliver, 1995; Paludan-Müller *et al.*, 1996). The time required for *V. vulnificus* to enter the VBNC state is highly dependent on the physiological age of the cells (Oliver *et al.*, 1991), believed to be due to the presence of stress-induced proteins. Genomic changes have also been suggested in a study by Warner and Oliver (1998), and this observation agrees with that of Bej *et al.* (1997).

Resuscitation of *V. vulnificus* from the VBNC state was first shown in 1991 (Nilsson *et al.*, 1991), and subsequently verified in several studies (Oliver *et al.*, 1995; Whiteside and Oliver, 1997). Such cells appear identical to those prior to induction into the VBNC state, and retain virulence (Oliver and Bockian, 1995). The involvement of nonculturable *V. vulnificus* cells in human infection is not known, however, and as for all nonculturable cells, difficult to demonstrate (although see discussions of *Salmonella* roe in the introduction and in the section on *Escherichia coli*). Nevertheless, it is quite feasible that at least part of the reason some persons can consume raw oysters for years with no health consequences, only to develop a fatal *V. vulnificus* infection from another oyster meal, may lie in the presence and numbers of *V. vulnificus* cells present in the oysters in the VBNC state.

## References

Archer, D.L. and Young, F.E. 1988. Contemporary issues: diseases with a food vector. Clin. Microbiol. Rev. 1: 377–398.

Asakura, H., Makino, S.-I., Takagi, T., Kuri, A., Kurazono,T., Watarai, M., and Shirahata, T. 2002. Passage in mice causes a change in the ability of *Salmonella enterica* serovar Oranienburg to survive NaCl osmotic stress: resuscitation from the viable but non-culturable state. FEMS Microbiol. Lett. 212: 87–93.

Barcina, I., González, J.M., Iriberri, J., and Egea L. 1989. Effect of visible light on progressive dormancy of *Escherichia coli* cells during the survival process in natural fresh water. Appl. Environ. Microbiol. 55:246–251.

Barer, M.R. 1997. Viable but non-culturable and dormant bacteria: time to resolve an oxymoron and a misnomer? J. Med. Microbiol. 46: 629–631.

Barer, M.R. and Harwood, C.R. 1999. Bacterial viability and culturability. Adv. Micro. Physiol. 41:93–137.

Barer, M.R., Kaprelyants, A.S., Weichart, D.H., Harwood, C.R., and Kell, D.B. 1998. Microbial stress and culturability: conceptual and operational domains. Microbiology 144: 2009–2010.

Barrett, T. 1998. An investigation into the molecular basis of the viable but non-culturable response in bacteria. Ph.D. thesis, Univ. Aberdeen.

Bates, T.C., and Oliver, J.D. 2004. The viable but nonculturable state of Kanagawa positive and negative strains of *Vibrio parahaemolyticus*. J. Microbiol. 42: 74–79.

Bej, A.K.N, Vickery, F.E., Brasher, C., Jeffreys, A.., Jones, D.D., DePaola, A., and Cook, D.W. 1997. Use of PCR to determine genomic diversity and distribution of siderophore-mediated iron acquisition genes in clinical and environmental isolates of *Vibrio vulnificus*. Abstr. Annu. Meet. Am. Soc. Microbiol. Q177, p. 485.

Besnard, V., Federighi, M., Declerq, E., Jugiau, F., and Cappelier, J.-M. 2002. Environmental and physico-chemical factors induce VBNC state in *Listeria monocytogenes*. Vet. Res. 33: 359–370.

Beumer, R.R., de Vries, J., and Rombouts, F.M. 1992. *Campylobacter jejuni* non-culturable coccoid cells. Int. J. Food Microbiol. 15:153–163.

Bloomfield, S.F., Stewart, G.S.A.B., Dodd, C.E.R., Booth, I.R., and Power, E.G.M. 1998. The viable but nonculturable phenomenon explained? Microbiology 144: 1–3.

Bogosian, G. and Bourneuf, E.V. 2001. A matter of bacterial life and death. EMBO Reports 2: 770–774.

Bogosian, G., Aardema, N.D., Bourneuf, E.V., Morris, P.J.L., and O'Neil, J.P. 2000. Recovery of hydrogen peroxide-sensitive culturable cells of *Vibrio vulnificus* gives the appearance of resuscitation from a viable but non-culturable state. J. Bacteriol. 182: 5070–5075.

Brayton, P., Tamplin, M., Huq, A., and Colwell, R. 1987. Enumeration of *Vibrio cholerae* O1 in Bangladesh waters by fluorescent-antibody direct viable count. Appl. Environ. Microbiol. 53: 2862–2865.

Breeuwer, P. and Abee, T. 2000. Assessment of viability of microorganisms employing fluorescence techniques. Int. J. Food Microbiol. 55: 193–200.

Cappelier, J.M., Magras, C., Jouve, J.L., and Federighi, M. 1999a. Recovery of viable but non-culturable *Campylobacter jejuni* cells in two animal models. Food Microbiol. 16: 375–383.

Cappelier, J.M., Minet, J., Magras, C., Colwell, R.R., and Federeighi, M. 1999b. Recovery in embryonated eggs of viable but nonculturable *Campylobacter jejuni* cells and maintenance of ability to adhere to HeLa cells after resuscitation. Appl. Environ. Microbiol. 65: 5154-5157.

Chao, W.-L., Tai, C.-L., Chen, R.-S., and Ding, R.-J. 1987. Factors affecting the survivals of *Vibrio parahaemolyticus*, *Vibrio cholerae* (non-O1), and *Salmonella enterica* in aquatic environments. N-49, p. 252. Abstr. Annu. Meet. Amer. Soc. Microbiol.

Chaiyanan, S., Chaiyanan, S., Huq, A., and Colwell, R.R. 2000. Viability of nonculturable coccoid form of *Vibrio cholerae* O1 and O139. Abstr. Q-101. Annu. Meet. Am. Soc. Microbiol.

Chmielewski, R.A.N. and Frank, J.F. 1995. Formation of viable but nonculturable *Salmonella* during starvation in chemically defined solutions. Lett. Appl. Microbiol. 20: 380–384.

Colwell, R.R. 1996. Global climate and infectious disease: the cholera paradigm. Science 274: 2025–2031.

Colwell, R.R., Brayton, P.R., Grimes, D.J., Roszak, D.R., Huq, S.A., and Palmer, L.M. 1985. Viable, but non-culturable *Vibrio cholerae* and related pathogens in the environment: implication for release of genetically engineered microorganisms. Bio/Technology 3: 817-820.

Colwell, R.R., Brayton, P.R., Herrington, D., Tall, B.D., Huq, A., and Levine, M.M. 1996. Viable but nonculturable *Vibrio cholerae* O1 revert to a culturable state in human intestine. World J. Microb. Biotechnol. 12: 28–31.

Cornax, R., Moriñigo, M.A., Romero, P., and Borrego, J.J. 1990. Survival of pathogenic microorganisms in seawater. Curr. Microbiol. 20: 293–298.

Créach, V., Baudoux, A.-C., Bertru, G., and Le Rouzic, B. 2003. Direct estimate of active bacteria: CTC use and limitations. J. Microbiol. Meth. 52: 19–28.

D'Aoust, J.-Y., Maurer, J., and Bailey, J.S. 2001. *Salmonella* species. In: Nonculturable Microorganisms in the Environment, R.R. Colwell and D.J. Grimes, eds., Amer. Soc. Microbiol. Press, Washington, D.C., pp. 141–178.

Day, A.P., and Oliver, J.D. 2004. Changes in membrane fatty acid composition during entry of *Vibrio vulnificus* into the viable but nonculturable state. J. Microbiol. 42: 69–73.

Duffy, G. and Sheridan, J.J. 1998. Viability staining in a direct count rapid method for the determination of total viable counts on processed meats. J. Microbiol. Meth. 31: 167–174.

Ekweozor, C.C., Nwoghu, C.E., and Barer, M.R. 1998. Transient increases in colony counts observed in declining populations of *Campylobacter jejuni* held at low temperature. FEMS Microbiol. Lett. 158: 267–272.

Federighi, M., Tholozan, J.L., Cappelier, J.M., Tissier, J.P., and Jouve, J.L. 1998. Evidence of non-coccoid viable but non-culturable *Campylobacter jejuni* cells in microcosm water by direct viable count, CTC-DAPI double staining, and scanning electron microscopy. Food Microbiol. 15:539–550.

Gunasekera, T.S., Sørensen, A., Attfield, P.V., Sørensen, S.J., and Veal, D.A. 2002. Inducible gene expression by nonculturable bacteria in milk after pasteurization. Appl. Environ. Microbiol. 68: 1988–1993.

Hasan, J.A.K., Shahabuddin, M., Huq, A., Loomis, L., and Colwell, R.R. 1992. Polymerase chain reaction for detection of cholera toxin genes in viable but nonculturable *Vibrio cholerae*. D-138, p. 119. Abstr. Annu. Meet. Amer. Soc. Microbiol.

Huq, A., Colwell, R.R., Rahman, R., Ali, A., Chowdhury, M.A.R., Parveen, S., Sack, D.A., and Russek-Cohen, E. 1990. Occurrence of *Vibrio cholerae* in the aquatic environment measured by fluorescent antibody and culture method. Appl. Environ. Microbiol. 56: 2370-2373.

Islam, M.S., Hasasn, M.K., Miah, M.A., Sur, G.C., Felsenstein, A., Venkatesan, M., Sack, R.B., and Albert, M.J. 1993. Use of the polymerase chain reaction and fluorescent-antibody methods for detecting viable but nonculturable *Shigella dysenteriae* Type 1 in laboratory microcosms. Appl. Environ. Microbiol. 59: 536–540.

Jacob, J., Martin, W., and Holler, C. 1993. Characterization of viable but nonculturable state of *Campylobacter coli*, characterized with respect to electron-microscopic findings, whole cell protein and lipooligosaccharide (LOS) patterns. Zentbl. Mikrobiol. 148: 3–10.

Jiang, X. and Chai, T.-J. 1996. Survival of *Vibrio parahaemolyticus* at low temperatures under starvation conditions and subsequent resuscitation of viable, nonculturable cells. Appl. Environ. Microbiol. 62: 1300–1305.

Jones, D.M., Sutcliffe, E.M., and Curry, A. 1991. Recovery of viable but non-culturable *Campylobacter jejuni*. J. Gen. Microbiol. 137: 2477–2482.

Kaprelyants, A.S., Gottschal, J.C., and Kell, D.B. 1993. Dormancy in non-sporulating bacteria. FEMS Microbiol. Rev. 104: 271–286.

Kathariou, S. 2002. *Listeria monocytogenes* virulence and pathogenicity, a food safety perspective. J. Food Prot. 11: 1811–1829.

Kehoe, S.C., Barer, M.R., and McGuigan, K.G. 2002. Hydrogen peroxide sensitivity increases during maintenance in water. Abstr. I-30. Ann. Meet. Am. Soc. Microbiol.

Kell, D.B., Kapreylants, A.S., Weichart, D.H., Harwood, C.L., and Barer, M.R. 1998. Viability and activity in readily culturable bacteria: a review and discussion of the practical issues. Ant. van Leeuwenhoek 73: 169–187.

Kogure, K., Simidu, U., and Taga, N. 1979. A tentative direct microscopic method for counting living marine bacteria. Can. J. Microbiol. 25: 415–420.

Korhonen, L.K. and Martikainen, P.J. 1991. Survival of *Escherichia coli* and *Campylobacter jejuni* in untreated and filtered lake water. J. Appl. Bacteriol. 71: 379–382.

Lampel, K.A. and Maurelli, A.T. 2001. *Shigella* species. In: Nonculturable Microorganisms in the Environment, R.R. Colwell and D.J. Grimes, eds. Amer. Soc. Microbiol. Press, Washington, D.C., pp. 247–261.

Lázaro, B., Cárcamo, J., Audícana, A., Perales, I., and Fernández-Astorga, A. 1999. Viability and DNA maintenance in nonculturable spiral *Campylobacter jejuni* cells after long-term exposure to low temperatures. Appl. Environ. Microbiol. 65: 4677–4681.

Li, J., Kolling, G.L., Matthews, K.R. and Chikindeas, M.L..2003. Cold and carbon dioxide used as multiple-hurdle preservation do not induce appearance of viable but non-culturable *Listeria monocytogenes*. J. Appl. Microbiol. 94: 1–48.

Linder, K. and Oliver, J.D. 1989. Membrane fatty acid and virulence changes in the viable but nonculturable state of *Vibrio vulnificus*. Appl. Environ. Microbiol. 55: 2837–2842.

Linkous, D.A. and Oliver, J.D. 1999. Pathogenesis of *Vibrio vulnificus*. FEMS Microbiol. Lett. 174: 207–214.

Lleò, M.M., Pierobon, S., Tafi, M.C., Signoreto, C., and Canepari, P. 2000. mRNA detection by reverse transcription-PCR for monitoring viability over time in an *Enterococcus faecalis* viable but nonculturable population maintained in a laboratory microcosm. Appl. Environ. Microbiol. 66: 4564–4567.

Lleò, M.M., Bonato, B., Tafi, M.C., Signoretto, C., Boaretti, M., and Canepari, P. 2001. Resuscitation rate in different enterococcal species in the viable but non-culturable state. J. Appl. Microbiol. 91:1095–1102.

Mackey, B.M. 2000. Injured bacteria. In: The Microbiological Safety and Quality of Food. Lund, B.M., A. Baird-Parker, and G.M. Gould, eds. Aspen Publ. Inc., Gaithersburg, MD, p. 315-341.

Makino, S.-I., Kii, T., Asakura, H., Shirahata, T., Ikeda, T., Takeshi, K. and Itoh, K. 2000. Does enterohaemorrhagic *Escherichia coli* O157 enter the VNC state in salmon roe? Appl. Environ. Microbiol. 66: 5536–5539.

McDougald, D., Prosser, J.I., Glover, L.A., and Oliver, J.D. 1995. Effect of temperature and plasmid carriage on nonculturability in organisms targeted for release. FEMS Microbiol. Ecol. 17: 229–238.

McFeters, G.A. and LeChevalier, M.W. 2000. Chemical disinfection and injury of bacteria in water. In: R.R. Colwell and D.J. Grimes, eds., Nonculturable Microorganisms in the Environment. Amer. Soc. Microbiol. Press, Washington, D.C., pp. 255–275.

McFeters, G.A., Yu, F.P., Pyle, B.H., and Stewart, P.S. 1995. Physiological assessment of bacteria using fluorochromes. J. Microbiol. Meth. 21: 1–13.

McGovern, V.P. and Oliver, J.D. 1995. Induction of cold responsive proteins in *Vibrio vulnificus*. J. Bacteriol. 177: 4131–4133.

Mead, P.S., Slutsker, L., Dietz, V., McCraig, L.F., Bresee, J.S., Shapiro, C., Griffin, P.M., and Tauxe, R.B.V. 1999. Food-related illness and death in the United States. Emerg. Infect. Dis. 5: 607–625.

Medema, G.J., Schets, F.M., van de Giessen, A.W., and Havelaar, A.H. 1992. Lack of colonization of 1 day old chicks by viable, non-culturalbe *Campylobacter jejuni*. J. Appl. Bacteriol. 72:512–516.

Meng, J., Doyle, M.P., Zhao, T., and Zhao, S. 2001. Enterohemorrhagic *Escherichia coli*. In: Nonculturable Microorganisms in the Environment, R.R. Colwell and D.J. Grimes, eds., Am. Soc. Microbiol. Press, Washington, D.C., pp. 193–213.

Millet, V. and Lonvaud-Funel, A. 2000. The viable but non-culturable state of wine micro-organisms during storage. Lett. Appl. Microbiol. 30: 136–141.

Mizunoe, Y., Wai, S.N., Ishikawa, T., Takade, A., Yoshida, S.-I. 2000. Resuscitation of viable but nonculturable cells of *Vibrio parahaemolyticus* induced at low temperature under starvation. FEMS Microbiol. Lett. 186: 115–120.

Mizunoe, Y., Wai, S.N., Takade, A., and Yoshida, S.-I. 1999. Restoration of culturability of starvation-stressed and low-temperature-stressed *Escherichia coli* O157 cells by using $H_2O_2$.degrading compounds. Arch. Microbiol. 172:63–67.

Morton, D. and Oliver, J.D. 1994. Induction of carbon starvation proteins in *Vibrio vulnificus*. Appl. Environ. Microbiol. 60: 3653–3659.

Mukamolova, G.V., Yanopolskaya, N.D., Kell, D.B., and Kaprelyants, A.S. 1998a. On resuscitation from the dormant state of *Micrococcus luteus*. Ant. Van Leeuwen. 73: 237-243.

Mukamolova, G.V., Kaprelyants, A.S., Young, D.I., Young, M., and Kell, D.B. 1998b. A bacterial cytokine. Proc. Natl. Acad. Sci. USA 95: 8916–8921.

Mukamolova, G.V., Kormer, S.S., Kell, D.B., and Kaprelyants, A.S. 1999. Stimulation of the multiplication of *Micrococcus luteus* by an autocrine growth factor. Arch Microbiol. 172: 9-14.

Murray, P.R., Rosenthal, K.S., Kobayashi, G.S., and Pfaller, M.A. (eds.). 2002. *Enterococcus* and other gram-positive cocci. In: Medical Microbiology, 4th ed., Mosby, St. Louis, MO, pp. 236–239.

Nachamkin, I. 2001. *Campylobacter jejuni*. In: R.R. Colwell and D.J. Grimes, eds., Nonculturable Microorganisms in the Environment. Amer. Soc. Microbiol. Press, Washington, D.C., pp. 179–192.

Nilsson, L., Oliver, J.D., and Kjelleberg, S. 1991. Resuscitation of *Vibrio vulnificus* from the viable but nonculturable state. J. Bacteriol. 173: 5054–5059.

Nwoguh, C.E., Harwood, C.R., and Barer, M.R. 1995. Detection of induced ⊠-galactosidase activity in individual non-culturable cells in pathogenic bacteria by quantitative cytological assay. Mol. Micrbiol. 17: 545–554.

Nyström, T. 2001. Not quite dead enough: on bacterial life, culturability, senescence, and death. Arch. Microbiol. 176: 159–164.

Ohtomo, R. and Saito, M. 2001. Increase in the culturable cell number of *Escherichia coli* during recovery from saline stress: possible implication for resuscitation from the VBNC state. Microb. Ecol. 42: 208–214.

Oliver, J.D. 1989. *Vibrio vulnificus*. In: Foodborne Bacterial Pathogens. Marcel-Dekker, pp 569-600.

Oliver, J.D. 1993. Formation of viable but nonculturable cells. In: Starvation in Bacteria, S. Kjelleberg, ed., Plenum Press, NY, pp. 239–272.

Oliver, J.D. 2000a. Problems in detecting dormant (VBNC) cells and the role of DNA elements in this response. In: Tracking Genetically Engineered Microorganisms, J.K. Jansson, J.D. van Elsas, and M.J. Bailey, eds. Landes Biosciences, Georgetown, TX, pp. 1–15.

Oliver, J.D. 2000b. The public health significance of viable but nonculturable bacteria. In: R.R. Colwell and D.J. Grimes, eds., Nonculturable Microorganisms in the Environment. Amer. Soc. Microbiol. Press, Washington, D.C., pp. 277–299.

Oliver, J.D. 2000c. The viable but nonculturable state and cellular resuscitation. In: C.R. Bell, M. Brylinsky, and P. Johnson-Green, eds., Microbial Biosystems: New Frontiers. Atlantic Canada Soc. Microb. Ecol., Halifax, Canada, pp. 723–730.

Oliver, J.D. and Wanucha, D. 1989. Survival of *Vibrio vulnificus* at reduced temperatures and elevated nutrient. J. Food Safety 10: 79–86.

Oliver, J.D. and Bockian, R. 1995. *In vivo* resuscitation, and virulence towards mice, of viable but nonculturable cells of *Vibrio vulnificus*. Appl. Environ. Microbiol. 61: 2620–2623.

Oliver, J.D. and Kaper, J.B. 2001. *Vibrio* species. In: Food Microbiology: Fundamentals and Frontiers, M.P. Doyle, *et al.*, ed., 2nd Ed. Amer. Soc. Microbiol. Press, Washington, D.C., pp. 263–300.

Oliver, J.D., Nilsson, L., and Kjelleberg, S. 1991. The formation of nonculturable cells of *Vibrio vulnificus* and its relationship to the starvation state. Appl. Environ. Microbiol. 57: 2640-2644.

Oliver, J.D., Hite, F., McDougald, D., Andon, N.L., and Simpson, L.M. 1995. Entry into, and resuscitation from, the viable but nonculturable state by *Vibrio vulnificus* in an estuarine environment. Appl. Environ. Microbiol. 61: 2624–2630.

Paludan-Müller, C., Weichart, D., McDougald, D. and Kjelleberg, S. 1996. Analysis of starvation conditions that allow for prolonged culturability of *Vibrio vulnificus* at low temperature. Microbiology 142: 1675–1684.

Pierson, M.D. and Smoot, L.M. 2001. Indicator organisms and microbiological criteria. In: Nonculturable Microorganisms in the Environment, R.R. Colwell and D.J. Grimes, eds. Amer. Soc. Microbiol. Press, Washington, D.C., pp. 71–87.

Pommepuy, M., Butin, M., Derrien, A., Gourmelon, M., Colwell, R.R., and Cormier, M. 1996. Retention of enteropathogenicity by viable but nonculturable *Escherichia coli* exposed to seawater and sunlight. Appl. Environ. Microbiol. 62: 4621–4626.

Porter, J., Edwards, C., and Pickup, R.W. 1995. Rapid assessment of physiologicalstatus in *Escherichia coli* using fluorescent probes. J. Appl. Bacteriol. 4: 399–408.

Pruzzo, C., Tarsi, R., Lleò, M. M., Signoretto, C., Zampini, M., Colwell, R.R., and Canepari, P. 2002. Curr. Microbiol. 45: 105–110.

Rahman, I., Shahamat, M., Chowdhury, M.A.R., and Colwell, R.R. 1996. Potential virulence of viable but nonculturable *Shigella dysenteriae* Type 1. Appl. Environ. Microbiol. 62: 115-120.

Rahman, I., Shahamat, M., Kirchman, P.A., Russek-Cohen, E., and Colwell, R.R. 1994. Methionine uptake and cytopathogenicity of viable but nonculturable *Shigella dysenteriae* Type 1. Appl. Environ. Microbiol. 60: 3573–3578.

Ravel, J., Knight, I.T., Monahan, C.E., Hill, R.T., and Colwell, R.R. 1995. Temperature-induced recovery of *Vibrio cholerae* from the viable but nonculturable state: growth or resuscitation? Microbiology 141: 377–383.

Regnault, B., Martin-Delautre, S., and Grimont, P.A.D. 2000. Problems associated with the direct viable count procedure applied to gram-positive bacteria. Intern. J. Food Microbiol. 55: 281–184.

Reissbrodt, R., Heier, H., Tschäpe, H., Kingsley, R.A., and Williams, P.H. 2000. Resuscitation by ferrioxamine E of stressed *Salmonella enterica* serovar *Typhimurium* from soil and water microcosms. Appl. Environ. Microbiol. 66: 4128–4130.

Reissbrodt, R., Romanova, J.M., Freestone, P.P.E., Haigh, R.D., Lyte, M., Tschäpe, H., and Williams, P.H. 2002. Resuscitation of *Salmonella typhimurium* and *Escherichia coli* O15:H7 from viable but non-culturable state by supplementation with the bacterial autoinducer of growth Baxcell. Abstr. Q-117. Ann. Meet. Amer. Soc. Microbiol.

Rigsbee, W., Simpson, L.M., and Oliver, J.D. 1997. Detection of the viable but nonculturable state in *Escherichia coli* O157:H7. J. Food Safety 16: 255–262.

Rollins, D.M. and Colwell, R.R. 1986. Viable but nonculturable stage of *Campylobacter jejuni* and its role in survival in the natural aquatic environment. Appl. Environ. Microbiol. 52: 531–538.

Roth, W.G., Leckie, M.P., and Dietzler, D.N. 1988. Restoration of colony-forming activity in osmotically stressed *Escherichia coli* by betaine. Appl. Environ. Microbiol. 54: 3142–3146.

Roszak, D.B. and Colwell, R.R. 1987. Metabolic activity of bacterial cells enumerated by direct viable count. Appl. Environ. Microbiol. 53: 2889–2893.

Roszak, D.B., Grimes, D.J., and Colwell, R.R. 1984. Viable but non-recoverable stage of *Salmonella enteritidis* in aquatic systems. Can. J. Microbiol. 30: 334–338.

Saha, S.K., Saha, S., and Sanyal, S.C. 1991. Recovery of injured *Campylobacter jejuni* cells after animal passage. Appl. Environ. Microbiol. 57: 3388–3389.

Saux, M.F.-L., Hervio-Heath, D., Loaec, S., Colwell, R.R., and Pommepuy, M. 2002. Detection of cytotoxin-hemolysin mRNA in nonculturable populations of environmental and clinical *Vibrio vulnificus* strains in artificial seawater. Appl. Environ. Microbiol. 68:5641–5646.

Schofield, G.M. 1992. Emerging foodborne pathogens and their significance in chilled foods. J. Appl. Bacteriol. 72:267–273.

Sheridan, G.E.C., Masters, C.I., Shallcross, J.A., and Mackey, B.M. 1998. Detection of mRNA by reverse transcription-PCR as an indicator of viability in *Escherichia coli* cells. Appl. Environ. Microbiol. 64: 1313–1318.

Sides, L., Hite, F. and Oliver, J.D. 1999. Effects of reactive oxygen species (ROS) inhibitors on resuscitation of viable but nonculturable (VBNC) cells of *Vibrio vulnificus*, *Escherichia coli* O157:H7 and non-O157:H7. Q129, p. 558, Annu. Meet. Amer. Soc. Microbiol.

Signoretto, C., Lleò, M.M., and Canepari, P. 2002. Modification of the peptidoglycan of *Escherichia coli* in the viable but nonculturable state. Curr. Microbiol. 44: 125–131.

Signoretto, C., Lleò, M. M., Tafi, M.C., and Canepari, P. 2000. Cell wall chemical composition of *Enterococcus faecalis* in the viable but nonculturable state. Appl. Environ. Microbiol. 66: 1953–1959.

Smith, R.J., Newton, A.T., Harwood, C.R., and Barer M.R. 2002. Active but nonculturable cells of *Salmonella enterica* serovar *Typhimurium* do not infect or colonize mice. Microbiology 148: 2717–2726.

Steinert, M., Emody, L., Amann, R., and Hacker, J. 1997. Resuscitation of viable but nonculturable *Legionella pneumophilia* Philadelphia JR32 by *Acanthamoeba castellani*. Appl. Envrion. Microbiol. 63: 2047–2053.

Stern, N.J., Jones, D.M., Wesley, I.V., and Rollins, D.M. 1994. Colonization of chicks by non-culturable *Campylobacter* spp. Lett. Appl. Microbiol. 18:333–336.

Strom, M.S. and Paranjpye, R.N. 2000. Epidemiology and pathogenesis of *Vibrio vulnificus*. Microbes Infect. 2: 177–188.

Swaminathan, B. 2001. *Listeria monocytogenes*. In: R.R. Colwell and D.J. Grimes, eds., Nonculturable Microorganisms in the Environment. Amer. Soc. Microbiol. Press, Washington, D.C., pp. 383–409.

Talibart, R., Denis, M., Castillo, A., Capperlier, J.M., and Ermel, G. 2000. Survival and recovery of viable but noncultivable forms of *Camplyobacter* in aqueous microcosm. Intern. J. Food Microbiol. 55: 263–267.

Tholozan, J.L., Cappelier, J.M., Tissier, J.P., Delattre, G., and Federighi, M. 1999. Physiological characterization of viable-but-nonculturable *Camplyobacter jejuni* cells. Appl. Environ. Microbiol. 65: 1110–1116.

van Duynhoven, Y.T. and de Jonge, R. 2001. Transmission of *Helicobacter pylori*: a role for food? Bull. WHO 79: 455–460.

Vedeikis, E. and Oliver, J.D. 2000. The role of reactive oxygen species (ROS) in the cultivation of *Vibrio vulnificus* present in the viable but nonculturable state. Ann. Meet. Amer. Soc. Microbiol.

Wai, S.N., Mizunoe, Y., Takade, A., and Yoshida, S. 2000. A comparison of solid and liquid media for resuscitation of starvation- and low-temperature-induced nonculturable cells of *Aeromonas hydrophila*. Arch. Microbiol. 173: 307–310.

Wai, S.N., Moriya, T., Kondo, K., Misumi, H., and Amako, K. 1996. Resuscitation of *Vibrio cholerae* O1 strain TSI-4 from a viable but nonculturable state by heat shock. FEMS Microbiol. Lett. 136: 187–191.

Warner, J.M. and Oliver, J.D. 1998. Randomly amplified polymorphic DNA analysis of starved and viable but nonculturable *Vibrio vulnificus* cells. Appl. Environ. Microbiol. 64: 3025-3028.

Whitesides, M.D. and Oliver, J.D. 1997. Resuscitation of *Vibrio vulnificus* from the viable but nonculturable state. Appl. Environ. Microbiol. 64: 3025–3028.

Wolf, P. and Oliver, J.D. 1992. Temperature effects on the viable but nonculturable state of *Vibrio vulnificus*. FEMS Microbiol. Ecol. 101: 33–39.

Xu, H.-S., Roberts, N., Singleton, F.L., Attwell, R.W., Grimes, D.J., and Colwell, R.R. 1982. Survival and viability of nonculturable *Escherichia coli* and *Vibrio cholerae* in the estuarine and marine environment. Microb. Ecol. 8: 313–323.

Xu, H.-S. Roberts, N.C., Adams, L.B., West, P.A., Siebeling, R.J., Huq, A., Huq, M.I., Rahman, R., and Colwell, R.R. 1984. An indirect fluorescent antibody staining procedure for detection of *Vibrio cholerae* serovar O1 cells in aquatic environmental samples. J. Microbiol. Methods. 2: 221–231.

Yaron, S. and Matthews, K. 2002. A reverse transcriptase-polymerase chain reaction assay for detection of viable *Escherichia coli* O157:H7: investigation of specific target genes. J. Appl. Microbiol. 92: 633–640.

# Modeling Pathogen Behavior in Foods

Mark L. Tamplin

7

## Abstract

Accurately predicting the fate of microbial pathogens in food is a goal of all concerned food safety specialists employed by food companies, government agencies, academic institutions and consulting firms. In this regard, predictive microbiology has emerged as an important scientific discipline for estimating the consequences of diverse food handling and processing operations on the growth, survival and inactivation of microbial pathogens. The development and successful implementation of predictive models involves a series of steps that include high quality experimental designs, sound mathematical algorithms, valid models and effective user interfaces. The net results are tools that are used in HACCP plans to define critical control points and critical limits, as well as safe remedial actions when deviations occur. This chapter provides an overview of current concepts in the field of predictive microbiology and how this discipline can contribute to the production of safer foods, international commerce and increased consumer confidence.

## Introduction

Predictive microbiology has increasingly become an important tool for protecting the safety of the food supply. This scientific field is based on the premise that microbial growth, survival and inactivation can be quantified and expressed with mathematical equations, and that under a specific set of environmental conditions, microbial behavior can be reproduced. An additional but understated assumption is that microorganisms, in general, display "smooth" trends in behavior over a range of environmental variables. Such responses may not always be simple curves at growth/no-growth interfaces, but nevertheless may permit predictions about microbial behavior over a range of conditions, most of which were not experimentally tested.

Models have proven to be valuable resources for companies that strive to reduce uncertainty about the fate of pathogenic bacteria in foods, from the farm to the consumer. Model predictions can be applied to a series of food processing operations, from transport to wholesale and retail outlets, and when foods are handled by consumers. Such knowledge provides food safety managers with knowledge and opportunities to produce more effective pathogen interventions through food safety management systems such as Hazard Analysis and Critical Control Points (HACCP), and through the development of more accurate risk assessments.

This chapter provides an overview of current concepts in the field of predictive microbiology, including a summary of the more common modeling methods. The discussions are primarily limited to models of pathogenic bacteria, rather than that of viruses, fungi, spoilage organisms, and chemical adulterants. In addition, this narrative seeks to describe the discipline in a manner that can be generally understood by scientists in the field of microbiology, without emphasis on mathematical algorithms. For more detailed presentations of modeling techniques, the author recommends readings by Baranyi (2002), Baranyi and Roberts (1994), Baranyi et al. (1993), Ross (1999), Ross and McMeekin (1994), McKellar and Lu (2004), McMeekin et al. (1993), Whiting and Buchanan (2001), and Wilson et al. (2002).

## Phases of bacterial growth and inactivation

Bacterial densities in food are controlled by the type and level of the initial contamination and by a variety of environmental conditions that affect their fate. Bacteria may increase in numbers (grow), decrease in numbers (inactivate or die) or remain at the same level (survive). Predictive models have been developed for each of these types of bacterial behavior.

A survey of the literature reveals that more models have been developed for growth scenarios than for those of inactivation or survival. Also, there are many more models for bacteria in defined microbiological media than in food. As a result, historical advances in the field of predictive microbiology have relied primarily on research of bacterial growth in microbiological broth.

### Growth

Bacteria can exhibit at least four phases of behavior in response to environmental conditions. These include lag, growth, maximum population density (stationary) and death. Within the context of common food processing and storage practices, the first two phases are more relevant for pathogen survival in food.

In general, growth rate and maximum population density are less variable than lag phase because the latter is influenced appreciably by the previous physiological state of the bacterium before it enters a new environment. This situation has much relevance today with the increased production of convenient, ready-to-eat (RTE) foods containing additives to inhibit the growth of pathogens, such as the psychrophile *Listeria monocytogenes*.

*Lag phase*

The physiological status of a bacterial cell at any moment is linked to its response to a new environment. Such transitions require bacteria to make physiological adjustments, a process sometimes referred to as "work-to-be-done." Baranyi and coworkers (1993; 1994) express this physiological adjustment in terms of $h_0$, which is related to the physiological state of the cells ($q_0$), and to the product of the lag phase duration and growth rate. Generally, lag phase increases with decreasing temperature and with exposure to environmental stresses, such as lower water activity and unfavorable pH.

However, not all environmental changes result in a lag phase. For example, bacteria that are transferred from mid-exponential growth in broth to a similar fresh medium do not normally express a lag phase. In these instances, the bacterial physiological state is such that it can immediately grow in the new environment. In contrast, cells in stationary phase typically express a lag phase when introduced into a fresh but similar medium. This likely results from a series of cellular events (i.e., energy expenditures) that are required for the cells to initiate growth.

Exceptions to the introduction of a lag phase as bacteria move among different environments have been reported. For example, Tamplin (2002) showed that a lag phase was not observed when *E. coli* O157:H7 was grown to early stationary phase at 37°C in brain heart infusion (BHI) broth, and then transferred to 6 to 10°C refrigerated sterile raw ground beef. This finding indicates that the physiological state of stationary phase *E. coli* O157:H7 grown in BHI broth results in cells that are prepared for immediate growth in refrigerated ground beef. This same phenomenon may occur for other pathogens moving among food environments.

Other reports show that lag phase may be affected by the initial pathogen density. For example, Métris *et al.* (2003) report that single bacterial cells express a distribution of lag times. At lower cell densities, there is a higher probability of selecting individual cells with longer lag times, versus at higher cell concentrations. These findings indicate that models need to be developed with lower, more realistic, contamination levels similar to what likely occurs when environmental *L. monocytogenes* contaminates RTE products.

Ross (1999) reported that lag time is inversely related to the generation time, i.e., the time for a cell to replicate, sometimes referred to as the doubling time. In addition, he reports that the ratio of lag time to generation time (i.e., relative lag time (*rlt*)) for a specific bacterial species is a unique distribution of responses. For example, when examining the distribution of *rlt* for a given organism under various environmental conditions, a peak in the frequency distribution appeared at *rlt* values ranging from 3 to 6 (Ross, 1999). This observation might be useful for estimating an unknown lag time when the growth rate is known. This relationship between lag time and growth rate is also incorporated into the dynamic Baranyi model (Baranyi *et al.*, 1993, Baranyi and Roberts, 1994).

*Growth phase*

The rate of bacterial growth is affected by the intrinsic and extrinsic properties of the environment, most notably temperature, pH and water activity. For a specific condition, there is a positive relationship between temperature and growth rate. In the field of predictive microbiology, growth rate is commonly expressed as the change in cell number per time interval. Cell numbers can be measured by a variety of methods, including colony-forming units (CFU) and optical density. For growth scenarios, time is usually expressed in hours, and in minutes for acute inactivation. For computational purposes, cell concentration is converted to the natural logarithm and designated as "specific rate" ($h^{-1}$).

*Maximum population density*

The greatest concentration of bacteria that can exist in an environment can be described as the maximum population density. In pure cultures, bacteria typically grow to a density of $\sim 10^9$ to $10^{10}$ $\log_{10}$ CFU per gram or per milliliter of medium. This represents the carrying capacity of the environment. As described later, the maximum population density for a single strain can be reduced in mixed-strain cultures. This effect has been attributed to the production of toxic by-products, pH, bacteriocins, and the depletion of growth-limiting nutrients (Jay, 2000).

## Inactivation

Bacteria are inactivated, or die, when conditions do not support growth. These environmental conditions may cause acute inactivation, as with high temperature, or mild inactivation, as observed with weak organic acid challenges. These conditions may lead to an immediate log-linear reduction in cell numbers or expression of a "shoulder" followed by a linear decrease and possibly a "tail."

*Linear phase*

In the log-linear phase of inactivation, the rate of reduction depends on the number of cell "targets" affected by the effector, such as heat. As the cell concentration declines, the probability of a "hit" on the cell target decreases, resulting in a proportional linear reduction in cell number. The rate of reduction is commonly referred to as the decimal reduction time, or *D*-value. Although *D*-values have been expressed for different levels of reduction, the most common representation is the time for the population to decrease by 90% (10-fold or 1.0 log$_{10}$). A secondary model form of *D* is the *Z*-value. This term describes the change in temperature that results in a 90% (or 10-fold) change in the *D*-value. This term is used for calculating process lethality, especially in thermally processed foods. Process lethality can be expressed as the *F*-value which is an integrated calculation of time-dependent thermal effects on inactivation of cell numbers, and serves to measure the accumulated lethality effects with "come-up" and "come-down" thermal profiles, such as those used in the canning industry.

*"Shoulders and tails"*

The kinetics of both thermal and non-thermal inactivation may display a lag-like period, sometimes referred to as a "shoulder," that preceeds the linear inactivation phase. For thermal inactivation scenarios, this is more commonly observed at lower temperatures and when using higher cell concentrations. It is theorized that this represents a subpopulation of cells that are more thermotolerant, with a greater likelihood of being observed when using higher inocula, such as $\sim 10^{7-9}$ CFU/g-mL. Other scientists speculate that these shoulders may result from inaccurate measurements of the internal temperature of the matrix during temperature "come-up" time, the use of mixed cultures, cell clumping and cell multiple hit mechanisms (Stringer *et al.*, 2000). The Weibull distribution has been increasingly applied to model non-linear inactivation curves (Peleg, 1999; Van Boekel, 2002).

In some instances, the linear phase of inactivation does not intercept the x-axis, but instead transitions to an asymptotic curve referred to as a "tail." Such "tails" are more commonly observed with higher inoculum levels. Investigators theorize that "tails" represent a subpopulation of bacteria that is more resistant to the thermal challenge.

It should be noted that this section on bacterial inactivation has focused more on thermal effects and less on acute inactivation scenarios involving the effects of acidulants, dessicants, and other bacterial antagonists. In this regard, the effects of compounds such as lactic acids and diacetates have received much recent attention due to efforts to control *L. monocytogenes* in RTE foods. These compounds, and other non-thermal treatments, can produce slower rates of inactivation, and may be only bacteriostatic in some applications.

## Models

Models are simplified, imperfect mathematical expressions of the numerous processes that affect bacterial growth, survival and inactivation. In the field of predictive microbiology, models are classified as primary, secondary and tertiary. Primary models reflect changes in cell number as a function of time. Secondary models predict changes in parameters of microbial growth or inactivation as a function of environmental condi-

tion. Tertiary models function as a model interface, whereby values for environmental variables are entered and secondary model predictions are refitted to primary models.

Intuitively, it is desirable to produce primary models with parameters based on cellular mechanisms (i.e., mechanistic models) that affect bacterial behavior. However, to date, the total cellular processes that control growth and inactivation have not been defined; consequently, the majority of models are empirical. Even if all the mechanisms of bacterial growth and inactivation were defined, a model built on these parameters would likely be highly complex and impractical for measurements and routine use.

Models can also be classified as kinetic or stochastic. Kinetic models predict changes in microbial numbers as a function of time whereas stochastic (probabilistic) models predict the likelihood or probability that some event will occur. The latter model form is commonly used when modeling growth/no-growth boundaries, such as predicting the probability-of-growth as a function of temperature and pH.

### Growth models

Two primary models have been extensively used to describe bacterial growth. The Gompertz model was the earliest curve-fitting routine used to model sigmoid-shaped growth curves (Gompertz, 1825). This model was not originally developed for microbial kinetics, but was modified to an empirical form by Gibson *et al.* (1987) to describe bacterial growth (eqn 7.1).

$$\log x(t) = A + \{-C \exp[-B(t-M)]\} \tag{7.1}$$

In eqn 7.1, $x(t)$ is the cell number at time $t$, $A$ is the cell number as $t$ approaches zero, $C$ is the difference between the maximum population density and $A$, and $B$ is the relative growth rate at the time (M) when the growth rate is maximum.

Baranyi and coworkers (Baranyi and Roberts, 1994; Baranyi *et al.*, 1993) describe a more mechanistically based dynamic model for bacterial growth:

$$\frac{dx}{dt} = \frac{q(t)}{q(t)+1} \cdot \mu_{\max}\left(1-\left(\frac{x(t)}{x_{\max}}\right)^m\right)x(t) \tag{7.2}$$

where $x$ is the quantity of cells at time $t$, $q(t)$ is the concentration of a limiting substrate, and $x_{\max}$ is the maximum population density.

The Baranyi model form includes parameters for lag phase based on the initial physiological state $q$ ($q_0$) of the organism. This state is expressed as:

$$h_0 = \ln\left(1+\frac{1}{q_0}\right) = \mu_{\max}\lambda \tag{7.3}$$

where $\mu_{\max}$ is the maximum growth rate and $\lambda$ is the lag time.

By knowing the value of $h_0$, this model can have very practical applications for predicting growth as a function of fluctuating conditions, such as temperature, that are experienced in food processing operations (Bovill et al., 2000; 2001).

## Inactivation models

Historically, the science of predictive microbiology is rooted in the thermal food processing industry, where models were developed to predict the inactivation of bacteria, most notably spores. Over time, regulations have been developed that specify D-values for specific types of processed foods. Simple log-linear models are commonly used to predict first-order inactivation kinetics. This can be expressed as:

$$N = N_0 e^{-kt} \tag{7.4}$$

where $N$ is the concentration of cells, $N_0$ is the initial cell concentration, $k$ is the inactivation rate, and $t$ is the time.

Relatively recently, there has been much discussion and debate about models for "shoulders" and "tails" that are sometimes observed during thermal treatments. Disagreements exist about whether non-linear inactivation curves are a laboratory artifact resulting from high inoculum levels, and if linear models are a more practical and fail-safe approach for the food processing industry. The reader is directed to other publications for more detailed discussions of non-linear inactivation models (McKellar and Lu, 2004; Stringer et al., 2000; Whiting and Buchanan, 2001).

## Growth/no-growth models

Research shows that there is progressively greater variation in bacterial behavior as environmental conditions reach the extremes for supporting growth. Likely, this is related to conformational changes in molecular structures, such as enzymes, that affect vital cellular processes. At certain growth boundaries, such as low temperature, these changes may be reversible whereas at higher temperatures they are destructive.

At the growth/no-growth boundary, primary data can be collected in a binary form by recording "growth" (1) and "no-growth" (0). At this boundary, replicate experiments under a specific set of conditions generate a mixture of growth and no-growth responses, resulting in average values between 0 and 1. These probability data can then be plotted over a range of experimental conditions, where the position of the growth/no-growth boundary can be adjusted to a desired level of probability (Ross and Dalgaard, 2004).

## Steps in the development of predictive models

As described earlier, microbial models permit predictions of pathogen behavior under environmental conditions that have not been tested. Producing a model with satisfactory accuracy involves a series of steps that include the design of the experi- mental conditions, the production of primary, secondary, and tertiary models, and measures of model performance, sometimes referred to as model verification and validation.

## Experimental design

Experimental design is arguably the most important step in the development of a useful model, since the quality of the data gathered will dictate the accuracy of the model. The researcher should ask the question "Ultimately, what environmental conditions should the model be able to predict?" The answer to this question should include the intrinsic and extrinsic environmental conditions that are relevant to the pathogen-food combination of interest. Examples of intrinsic conditions include pH, water activity ($a_w$), and the concentration of NaCl or acidulants, such as potassium lactate and sodium diacetate. The most common extrinsic conditions are temperature and packaging atmosphere. For each of these independent variables, it is necessary to define the potential range of each variable so that the model is robust for the anticipated conditions of prediction.

When only one or two variables are to be modeled, it is relatively simple to define a full factorial experimental design. However, when three or more variables exist, limiting resources usually necessitate application of statistical procedures to define a partial factorial experimental design. For each variable, it is necessary to define the appropriate test values. These values can be set at regular intervals in regions where the responses are expected to occur over a "smooth" surface. However, the test variable intervals may be irregular at growth/no-growth conditions where bacterial behavior is more stochastic (i.e. more variation).

### Strains

It is common to find models that have been developed using a single strain or a mixture (i.e., cocktail) of strains. Single strain models are naturally strain-specific. They provide information about a strain's behavior under defined environmental conditions. There is greater value to single-strain models when it is understood how the strain compares to other strains. For example, does the strain represent a worse-case scenario for growth or thermal inactivation compared to most strains. To have this perspective, however, it is necessary to understand the behavior of the strain in comparison to other strains. If such information is not known, there is higher uncertainty about the model predictions for the species as a whole.

Mixtures of strains can provide liberal, or worse case, estimates of pathogen behavior. For example, in a growth scenario, the strain with the highest growth rate will predominate and be measured. For thermal inactivation experiments, the strain that is most resistant to heat will predominate and be measured. A shortcoming to the use of cocktails is that measurements of growth and inactivation kinetics may show unusual parameters affected by the behavior of multiple strains. Although this could theoretically represent a real-world scenario of mixed strain contamination, there is lower value to the information since individual strains cannot be identified. This effect can be reduced by adjusting all strains in the cocktail to the same initial level.

Another consideration for the experimental design is the source of strains. For example, if you wish to produce a model for the growth of a pathogen in food, then strains isolated directly from foods are preferable. Too commonly, clinical strains linked to foodborne disease are used for experiments in food without knowing whether such strains have changed (i.e., mutated) during passage through humans. Strain characteristics can also mutate with repeated culture, and the odds of perpetuating a mutated strain increase when single colonies are selected with each subculture. Consequently, long-term storage methods should be used to prevent mutations.

*Test matrix*

The majority of published microbial models were developed in microbiological broth to understand how variables, such as temperature, pH and water activity affect microbial behavior. Although defined media such as broth or simulated "foods" offer greater control over matrix variation, they may not reflect microbial behavior in relevant food matrices. A number of studies have shown that bacterial growth rates are high in sterile broth, supporting the belief that these rates are equal to or greater than rates observed in food (Anonymous, 2004). Consequently, broth-based models may serve as worse-case predictions for growth rate, but not for lag phase duration and maximum population density, as described elsewhere in this chapter.

*Microbial flora*

The presence of native microbial flora, including spoilage microbes, should also be considered as a characteristic of the model test matrix. This is especially important if the model is to be applied to food that will contain mixed flora. Numerous reports document the competitive nature of bacteria, and their inhibiting effects on maximum population density. Inhibition of one strain by another strain has been attributed to the production of acidic products, bacteriocins, and/or the carrying capacity of the environment (Jay, 2000). Approaches to experimental design include inoculating the test matrix with one or more spoilage strains or using the undefined spoilage flora found in retail foods. The more recent introduction of irradiated animal foods into the market place indicates that models are needed for pathogens in the presence of lower levels of spoilage organisms.

*Inoculum preparation*

As previously described, the length of the lag phase is affected by cellular "adjustments" that occur before bacteria can resume growth. Therefore, the experimental design should consider what is known about the anticipated previous environment of the pathogen before it contaminates food. For example, a mod-el may need to reflect the physiological state for *E. coli* O157: H7 moving from bovine feces to a beef carcass surface, or *L. monocytogenes* transitioning from the surface of a conveyor belt to a RTE meat. With this said, few models have been developed considering relevant environments of the pathogen.

More commonly, bacteria are grown to stationary phase in bacteriological broth. In a few instances, authors state that this is done assuming that the environmental form of the pathogen is in the stationary phase and that this phase produces an organism more tolerant to subsequent challenges, such as acidic environments. Studies show that bacteria grown to the exponential phase tend not to express a relatively long lag phase, whereas lag phase increases with increasing time in stationary phase. If the goal is to produce a model that reflects greater initial tolerance to an environmental factor, then it is more prudent to previously grow the pathogen in the presence of the challenge factor.

## Primary models

After the experimental protocol is established, time-versus-cell number data are collected for each of the single or multiple independent variable conditions. Next, curve-fitting routines are used to develop a least-squares best-fit algorithm to the data. For growth data, the parameters normally include lag phase duration, growth rate, and maximum population density. For inactivation types of data, parameters may reflect an initial "shoulder", somewhat analogous to the lag phase, a linear reduction in cell count, and possibly a "tail." In cases where probability-of-growth is relevant, such as at growth/no-growth boundaries, data may be scored simply as growth or no-growth.

## Secondary models

Secondary models are produced from primary models to predict the change in primary model parameters as a function of the environment. An example of a secondary model is the prediction of growth rate as a function of temperature, or predictions of growth rate as a function of multiple environmental conditions such as salt, water activity and temperature. The *z*-value is another type of secondary model that describes the change in *D*-value as a function of temperature change. Secondary models can be simple linear regressions or more complex polynomial models that require commercial software.

Various secondary models have been used to model growth and inactivation of bacteria. More commonly, lag time and growth rate have been modeled using square-root, gamma and cardinal approaches. The use of probability models for describing the likelihood of a microbial event in food is increasing in the literature. Applications include modeling growth/no-growth interfaces, the length of the lag phase for pathogens in formulated RTE foods, and the production of microbial toxins. For a thorough discussion of secondary models, the reader is directed to a recent review by Ross and Dalgaard (2004).

Another model form that is increasingly reported is artificial neural networks (ANN). Analogous to human brain neuron communication, ANNs consist of processing nodes that are interconnected to reflect relationships among model parameters. These nodes are assigned specific weights based on "training" and test data. The value of ANNs will likely increase with the application of large databases to complex modeling problems. The reader is directed to publications by Hajmeer and coworkers (1997, 2002), as well as Tu (1996) for more detailed discussions of the production and application of ANNs.

## Tertiary models

The next step of model development involves expressing secondary model predictions through a primary model. This is commonly done in commercial spreadsheets and in stand-alone software, such as the US Department of Agriculture-Agricultural Research Service's *Pathogen Modeling Program* (PMP; www.arserrc.gov/mfs/pathogen.htm) and the UK Institute of Food Research's *Growth Predictor* (http://www.ifr. ac.uk/Safety/GrowthPredictor/default.html). These programs are discussed later in this chapter.

## Measures of model performance

The utility of a model for supporting food safety decisions depends on the validity of the model. Tools for measuring microbial model performance have been described by Baranyi *et al.* (1999), Delignette-Muller *et al.* (1995) and Ross (1996). These methods involve measurements of model bias ($B_f$) and accuracy ($A_f$) factors. Bias is a comparison of model prediction, such as a ratio or least-square measure, to observations. This value can be positive or negative, depending on whether there is over- or underprediction. These discrete values are summed and averaged over the observed conditions, thereby providing a mean measure of model performance. A limitation to this method is that a $B_f$ value could appear perfect or near perfect, yet actually represent a balance between very high over- and underpredictions. Eqn 7.5 has been commonly reported in the literature.

$$B_f = 10^{(\Sigma \log(GT_{predicted}/GT_{observed})/n)} \qquad (7.5)$$

where GT is the generation or doubling time and $n$ is the number of observations.

In contrast, the $A_f$ of a model is the sum of absolute differences between model predictions and observations (eqn 7.6). As such, $A_f$ is a better measure of overall model error. However, as with bias factors, these measurements do not identify specific regions of the model where prediction errors exist. Tamplin (2003) describes "discrete" measurements of model prediction error in the reporting of $B_f$ and $A_f$ for the growth of *E. coli* O157:H7 in sterile raw ground beef. In both tabular or graph forms, these representations of model performance indicate regional assessments of model accuracy.

$$A_f = 10^{(\Sigma |\log(GT_{predicted}/GT_{observed})|/n)} \qquad (7.6)$$

## Enhancing the application of models

In the previous section, the measurement of model performance was discussed. The proven validity of a model for assisting food safety decisions greatly enhances its potential use by food industries and other entities interested in predicting the behavior of pathogens in food. In the US, the US Department of Agriculture's Food Safety Inspection Service issued a notice dated May 4, 2005 giving guidance in the use of microbial models in HACCP plans (Food Safety and Inspection Service, 2005). The directive encourages the use of models as a tool for supporting food safety decisions, but emphasizes that "Before the models could be used in such a manner, the user would have to validate the models for each specific food of interest." Consequently, regulatory agencies and the food industry realize greater value in models that have been validated for foods. This situation will hopefully stimulate more researchers to produce models for commercially relevant foods versus microbiological broths. As described earlier in this chapter, models also gain greater acceptance when (1) the organisms are grown under conditions similar to the suspected route of contamination, (2) food-derived strains are employed, (3) relevant inoculum levels are used, and (4) the effects of native food spoilage organisms are considered.

## Software packages

Advances in computational power have facilitated the execution of complex mathematical calculations that otherwise would be too time-consuming. Computer software programs in the field of predictive microbiology provide an interface between the model's mathematics and the end-user, permitting information to be entered into the model and then predictions to be viewed via graphs and/or tables. Examples of model software packages that have gained use in the food industry and research communities include the *Pathogen Modeling Program*, *Growth Predictor* and the *Seafood Spoilage Predictor*.

### Pathogen modeling program (PMP)

The *PMP* version 7.0 is a package of 39 microbial models that describe growth, survival, inactivation and toxin production of pathogens under various environmental conditions (Buchanan *et al.*, 1993). It is produced by the USDA Agricultural Research Service and can be downloaded at http://www.arserrc. gov/mfs/pathogen.htm. Although the majority of the models are for static environmental conditions, the PMP also includes dynamic temperature models for the growth of *Clostridium perfringens* and *Clostridium botulinum*. These latter models are used by US food industries to meet performance standards for cooked and cooled meat products (Food Safety and Inspection Service, 1999).

The PMP model interface allows users to input conditions for a variety of intrinsic and extrinsic environmental factors, including pH, water activity, atmosphere, temperature, irradiation dose, and concentrations of various acidulants and other bacterial inhibitors. Model predictions are then made for growth/inactivation rate, lag time, and generation time, including upper and lower confidence limits. A helpful feature of the 7.0 version is a link between models and associated full-text refereed publications.

## Growth predictor

The UK Institute of Food Research and the UK Food Standards Agency have produced *Growth Predictor*, a package of models for predicting the growth of bacteria in microbiological broth as a function of temperature, pH, water activity, carbon dioxide and/or acetic acid. The user can also input the inoculum level and a value for the physiological state of bacteria. The latter factor provides the user with control over the prediction of the lag phase. The output includes growth rate/doubling time, and kinetic data in table and graph forms. The software is free and can be downloaded at http://www.ifr.ac.uk/Safety/GrowthPredictor/default.html.

## Other software

The *Seafood Spoilage Predictor* (SSP) is produced by the Danish Institute of Fisheries Research and is a software package that predicts microbial spoilage of fishery products with fixed and changing temperatures. It can be downloaded at http://www.dfu.min.dk/micro/ssp/. Another useful tertiary model is the Food Spoilage Predictor (http://www.arserrc.gov/cemmi/FSPsoftware.pdf).

## Conclusions

Food safety managers strive to produce wholesome and safe foods using the most effective intervention strategies to reduce, eliminate and prevent foodborne hazards. This goal can be best achieved through the use of predictive models that estimate the effects of intervention technologies over a range of critical application limits, as well as process deviations. Presently in the US and other nations, predictive models are used to develop HACCP plans, improve product formulations, and produce risk assessments. As a result, substantial savings in human disease and production costs have been realized, thus increasing consumer confidence in the food supply. To sustain these promising achievements within the context of a global food system, the field of predictive microbiology will increasingly rely on multinational collaborations to produce, share and manage enormous amounts data that are needed to develop microbial models. Ultimately, these efforts will enhance the production of safer foods, promote international trade and increase consumer confidence in the safety and quality of the world's food supply.

## References

Anonymous. 2004. ComBase. www.combase.cc.

Baranyi, J. 2002. Stochastic modeling of bacterial lag phase. Int. J. Food Microbiol. 73: 203–206.

Baranyi, J., Pin, C., and Ross, T. 1999. Validating and comparing predictive models. Int. J. Food Microbiol. 48: 159–166.

Baranyi, J., and Roberts, T.A. 1994. A dynamic approach to predicting bacterial growth in food. Int. J. Food Microbiol. 23: 277–294.

Baranyi, J., Roberts, T.A., and McClure, P. 1993. A non-autonomous differential equation to model bacterial growth. Food Microbiol. 10: 43–59.

Bovill, R., Bew, J., Cook, N., D'Agostino, M., Wilkinson, N., and Baranyi, J. 2000. Predictions of growth for Listeria monocytogenes and Salmonella during fluctuating temperature. Int. J. Food Microbiol. 59: 157–165.

Bovill, R.A., Bew, J., and Baranyi, J. 2001. Measurements and predictions of growth for Listeria monocytogenes and Salmonella during fluctuating temperature. II. Rapidly changing temperatures. Int. J. Food Microbiol. 67: 131–137.

Buchanan R.L., Bagi L.K., Goins R.V., and Phillips J.G. 1993. Response surface models for the growth kinetics of Escherichia coli O157:H7. Food Microbiology 10: 303–315.

Delignette-Muller, M.L., Rosso, L., and Flandrois, J.P. 1995. Accuracy of microbial growth predictions with square root and polynomial models. Int. J. Food Microbiol. 27: 139–146.

Food Safety and Inspection Service (USDA). 1999. Appendix B. Compliance guidelines for cooling heat-treated meat and poultry products (stabilization). http://www.fsis.usda.gov/OA/fr/95033F-b.htm

Food Safety and Inspection Service (USDA). 2005. Use of microbial pathogen computer modeling in HACCP Plans. FSIS Notice 25-05. www.fsis.usda.gov/regulations_&policies/Notice-25-05/index.asp.

Gibson, A.M., Bratchell, N., and Roberts, T.A. 1987. The effect of sodium chloride and temperature on the rate and extent of growth of Clostridium botulinum type A in pasteurized pork slurry. J. Appl. Bacteriol. 62: 479–490.

Gompertz, B. 1825. On the nature of the function expressive of the law of human mortality, and a new mode of determining the value of life contingencies. Philos. Trans. R. Soc. Lond. 115: 513–585.

Hajmeer, M.N., Basheer, I.A., and Najjar, Y.M. 1997. Computational neural networks for predictive microbiology. 2. Applications to microbial growth. Int. J. Food Microbiol. 34: 51–66.

Hajmeer, M.N. and Basheer, I. 2002. A probabilistic neural network approach for modeling and classification of bacterial growth/no-growth data. Int. J. Food Microbiol. 82: 233–243.

Jay, J.M. 2000. Intrinsic and extrinsic parameters of foods that affect microbial growth. In: Modern food microbiology. 6th ed. Aspen Publishers, Inc., Gaithersburg, Maryland p. 35–56.

McKellar, R.C. and Lu, X. 2004. Modeling microbial responses in food. CRC Press, Boca Raton, FL.

McMeekin, T.A., Olley, J.N., Ross, T., and Ratkowsky, D.A. 1993. Predictive microbiology: theory and application. John Wiley and Sons, New York, NY.

Metris, A., George, S.M., and Peck, M.W., and Baranyi, J. 2003. Distribution of turbidity detection times produced by single cell-generated bacterial populations. J. Microbiol. Meth. 55: 821–827.

Peleg, M. 1999. On calculating sterility in thermal and non-thermal preservation methods. Food. Res. Int. 32: 271–278.

Ross, T. 1999. Predictive food microbiology models in the meat industry. North Sydney, Australia, Meat and Livestock Australia.

Ross, T. 1996. Indices for performance evaluation of predictive models in food microbiology. J. Appl. Bacteriol. 81: 501–508.

Ross, T. and Dalgaard, P. 2004. Secondary models. In: Modeling microbial responses in food. CRC Press, Boca Raton, FL. p.63–150.

Ross, T., and McMeekin, T.A. 1994. Predictive microbiology. Int. J. Food Microbiol. 23:241–264.

Stringer, S.C., George, S.M., and Peck, M.W. 2000. Thermal inactivation of Escherichia coli O157:H7. J. Appl. Microbiol. 88:79S-89S.

Tamplin, M. L. 2002. Growth of *Escherichia coli* O157:H7 in raw ground beef stored at 10°C and the influence of competitive bacterial flora, strain variation, and fat level. J. Food Prot. 65: 1535–1540.

Tamplin, M.L., Paoli, G., Marmer, B.S., and Phillips, J.G. 2003. Models of the behavior of *Escherichia coli* O157:H7 in raw sterile ground beef grown at 5 to 46°C. Int. J. Food Microbiol. 100: 335–344.

Tu, J.V. 1996. Advantages and disadvantages of using artificak neural networks versus logistic regression for predicting medical outcomes. J. Clin. Epidemiol. 11: 1225–1231.

Van Boekel, M.A.J.S. 2002. On the use of the Weibull model to describe thermal inactivation of microbial vegetative cells. Int. J. Food Microbiol. 74: 139–159.

Whiting, R.C. and Buchanan, R.L. 2001. Predictive modeling and risk assessment. In: Food microbiology: Fundamentals and frontiers. Doyle, M.P., Beuchat, L.R., and Montville, T.J. eds. ASM Press, Washington, D.C, pp. 813–831.

Wilson, P.D.G., Brocklehurst, T.F., Arino, S., Thuault, D., Jakobsen, M., Lange, M., Farkas, J., Wimpenny, J.W.T., and Van Impe, J.F. 2002. Modelling microbial growth in structured foods: towards a unified approach. Int. J. Food Microbiol. 74: 275–289.

# Foodborne and Waterborne Enteric Viruses

Gary P. Richards

## Abstract

Food- and water-borne viruses contribute to a substantial number of illnesses throughout the world. Among those most commonly known are hepatitis A virus, rotavirus, astrovirus, enteric adenovirus, hepatitis E virus, and the human caliciviruses consisting of the noroviruses and the Sapporo viruses. This diverse group has something in common: they are transmitted by the fecal–oral route, often by ingestion of contaminated food and water. This chapter provides an introduction to the enteric viruses including the diseases they cause, their characteristics, mechanisms of pathogenesis, inactivation procedures, and methods for their detection, analysis, and control.

## Introduction

Enteric virus contamination of food and water is responsible for millions of illnesses each year and an untold number of deaths. Many of the deaths are associated with dehydration, particularly in countries where rehydration therapy is unavailable. The human caliciviruses and rotavirus are especially known to elicit dehydration, whereas, hepatitis A and E viruses are involved in liver disease. All of these pathogens are readily spread by the fecal–oral route and most are highly infectious, even in low numbers. This chapter explores some of the distinguishing features of enteric viruses and the diseases they cause. Mechanisms of pathogenicity, as well as detection, inactivation, and control strategies are also explored.

### Biology, pathogenesis, and structure of food- and water-borne enteric viruses

The hallmark of calicivirus, rotavirus, enteric adenovirus, and astrovirus-induced illness is diarrhea caused by viral infection and resulting damage to the absorptive enterocytes of the villi of the small intestines. Damage results in reduced absorptive and digestive capacity and liquid retention within the intestinal lumen. Fermentation of the undigested material as it passes through the large intestine increases the production of lower molecular mass particles which subsequently increases the osmotic pressures within the large intestines. Once the absorptive capacity of the large intestine is exceeded, diarrhea occurs (Conner and Ramig, 1997). One might speculate that the extent of the diarrhea is proportional to the amount of enterocyte damage; however, this has not been quantitatively determined. Unlike most of the enteric viruses, the principal site of hepatitis A and hepatitis E virus infections is not the intestine, but rather the liver. Enteric viral pathogens are at least partially resistant to stomach acids and bile salts and readily pass from the stomach into the lumen of the small intestines where they initiate infection. It is imperative to study viral pathogenesis if one is to understand the cyclic relationships involving viruses, humans, foods and beverages.

Central to an understanding of enteric virus structure and function is the fact that viruses are not living entities. They do not require food to survive and have no metabolic functions in and of themselves. Enteric viruses consist of either DNA viruses (adenovirus) or RNA viruses (hepatitis A and E viruses, noroviruses, astrovirus, rotavirus) enclosed within a protective protein capsid. Enteric viruses utilize the protein synthetic machinery of host cells to produce capsid proteins which are coded for by viral DNA or RNA, depending on the type of virus. Most of these viruses are highly species specific, infecting humans and other primates. There is some concern over possible zoonotic transmission of hepatitis E virus. The noroviruses contain a broad group of genetically diverse viruses formerly called the Norwalk-like viruses (NLVs). They are highly specific to infecting humans, with only occasional infection of other primates in laboratory challenge studies. The general properties of the enteric viruses are shown in Table 8.1 and are further discussed below.

### Adenovirus

Adenoviruses are 80–110 nm in diameter, nonenveloped, icosahedral-shaped viruses in the *Adenoviridae* family. Adenoviruses cause respiratory, ocular, kidney, neurological, and gastrointestinal illnesses. Human adenoviruses consist of 49 serotypes that are grouped into six subgroups. Enteric adenovirus infection is primarily caused by serotypes 40 and 41 within subgroup F (Ranki *et al.*, 1983; Uhnoo *et al.*, 1984; Brandt *et al.*, 1985; Krajden *et al.*, 1990), although serotypes 1, 3, 5, 7,

**Table 8.1** Properties of enteric viruses

| | Adenovirus | Astrovirus | Hepatitis A | Hepatitis E | Norovirus | Rotavirus |
|---|---|---|---|---|---|---|
| Family | Adenoviridae | Astroviridae | Picornaviridae | Unclassified | Caliciviridae | Reoviridae |
| Capsid | Icosahedral | Circular | Icosahedral | Icosahedral | Icosahedral | Icosahedral |
| Diameter (nm) | 65–80 | 28–30 | 27–32 | 27–34 | 27–40 | 70–75 |
| | 80–110 w/fibers | 41–43 w/spikes | | | | 100 w/spikes |
| Enveloped | No | No | No | No | No | No |
| Genome | ds DNA | ss RNA | ss RNA | ss RNA | ss RNA | ds RNA |
| Size (kb) | 36 | 6.8 | 7.5 | 7.3 | 7.7 | 18.5 |
| Structure | Linear | Linear | Linear | Linear | Linear | Segmented |
| Polarity | NA[1] | Positive | Positive | Positive | Positive | NA |
| Poly (A) | No | Yes | Yes | Yes | Yes | No |
| Cell culture assay | Yes | Yes | Some | No | No | Yes |
| Cytopathic effects | Yes | Yes | Some | No | No | Some |
| Host cell replication in | Nucleus | Cytoplasm | Cytoplasm | Cytoplasm | Cytoplasm | Cytoplasm |
| New cases/year in USA[2] | Unknown[3] | $3.9 \times 10^6$ | $8.0 \times 10^4$ | ND[4] | $2.3 \times 10^7$ | $3.9 \times 10^6$ |
| Incubation period | 2–15 days | 2–4 days | 15–45 days | 2–8 weeks | 10–51 hours | 11 h – 6 days |
| Duration of illness | 3–11 days | < 4 days | ≤ 5 months | ≤ 8 weeks | 1–2 days | 5 days |
| Fecal shedding | | 1 week | ≤ 5 months | ≤ 52 days | 2 weeks | ≤ 2 weeks |

[1]NA, not applicable.
[2]According to estimates by Mead et al. (1999).
[3]Number of enteric adenovirus cases in the USA unknown.
[4]ND, none detected. Hepatitis E is not found in the USA population.

and 31 are occasionally found in stools (Krajden et al., 1990). Acute adenovirus gastroenteritis may occur among children, particularly in developing countries. It is a major cause of infantile diarrhea and is characterized by watery diarrhea, often containing mucus, and may also include fever, vomiting, abdominal pain, dehydration, and it may be accompanied by respiratory illness (Ruuskanen et al., 2002). The incubation period for adenovirus infection ranges from 2–15 days (mean 10 days) (Dudding et al., 1973; Finn et al., 1988; Ruuskanen et al., 1988; Kotloff et al., 1989; Lawler et al., 1994) and the duration of illness is from 3–11 days, although adenovirus 41 persists longer than serotype 40 (Uhnoo et al., 1984). Most cases are mild and transient; however, fatal cases have been described in immunocompromised individuals (Krajden et al., 1990). Most adenoviruses cause respiratory infections, and less is known about enteric adenovirus pathogenesis. Approximately one-third of normal, healthy individuals have latent adenoviruses in their duodenal epithelium (Lawler et al., 1994). It is known that enteric adenoviruses infect mature enterocytes of the villi leading to damage of the apices of the villi of the small intestine.

Serological identification of adenovirus infections is by the detection of a major capsid protein, the hexagon antigen, by enzyme immunoassay (EIA). The IgM is detectable in 20–50% of those infected (Meurman et al., 1983). In one study, 70% of 52 children exhibited an enhanced IgG response, while IgA increased in 37% of the patients (Meurman et al., 1983). In a study of young adult men, IgM, IgG and IgA responses occurred in 39, 89, and 77% of the patients, respectively, within 2 weeks after infection (Julkunen et al., 1986).

The capsid of adenoviruses contain 254 capsomeres: 240 hexons which form the facets of the icosahedron and 12 pentons which are at the vertices of the virus particle. The penton contains two structural proteins: the penton base which affixes the penton to the capsid, and the fiber which forms an elongated spike protruding from the vertices. Figure 8.1 is a cryo-electron micrograph of adenovirus with barely visible spikes. Other structural proteins in the capsid cement the hexons and bind the capsid to the DNA core. The genome consists of a linear, double-stranded chromosome of approximately 36,000 bp and inverted terminal repetitions (ITRs) of 103 bp. Viral replication takes place in the nucleus of host cells causing a lytic infection within 30 hours. Viral entry into the cells is mediated by virus binding to specific cellular receptors and the internalization of clatherin-coated pits into endosomes. The endosomes are acidified releasing the viral DNA which is targeted to the nucleus. Once in the nucleus, viral DNA is replicated by single-stranded displacement (Van der Vliet, 1995) utilizing the ITRs to form partial duplex structures which can initiate second strand synthesis. The viral DNA encodes early transcription units (E1A, E1B, E2A, E2B, E3 and E4) for nonstructural proteins during the early stages of infection and late transcription units (L1-L5) principally for structural proteins during the late stages of infection. These transcription units are coordinately expressed. The first to be activated is E1A which codes for transcriptional regulatory proteins, while E1B codes for proteins which direct the cell away from p53-mediated apoptosis. Regions E2A and E2B code for proteins required for adenovirus replication and E3 codes for proteins to combat host antiviral defense mechanisms. Proteins

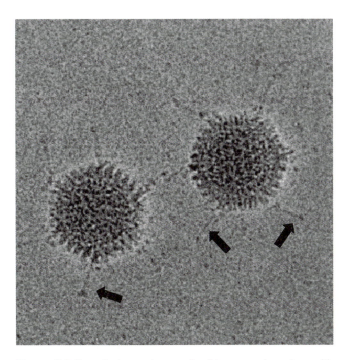

**Figure 8.1** Cryoelectron micrograph of human adenovirus with spikes denoted by arrows. Courtesy of Phoebe L. Stewart, Vanderbilt University Medical Center, Nashville, TN.

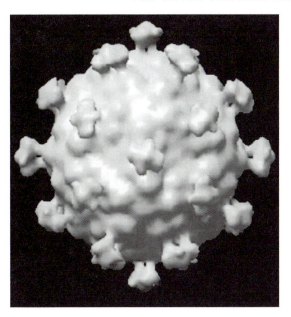

**Figure 8.2** Map of human astrovirus, based on cryoelectron micrograph, viewed along 2-fold axis of symmetry. Courtesy of Mark Yeager, The Scripps Research Institute, La Jolla, CA, and with permission of Fields Virology, 4th ed, Lippincott Williams and Wilkins, Philadelphia, PA, © 2001.

encoded by E4 regulate the transcription of viral and cellular gene expression and are involved in alternative RNA splicing. Late in the infection, after DNA replication, transcription is initiated at the adenovirus major late promoter. Most of the major late mRNAs encode structural proteins of the virion including the penton, hexon and fiber proteins. A protease transcribed from the L3 transcription unit is instrumental in the proteolytic processing of capsid components. In total, about 40 different polypeptides are produced during the early and late translation events.

## Astrovirus

Astroviruses are 28–30 nm diameter, nonenveloped, positive-sense, polyadenylated, single-stranded RNA viruses in the *Astroviridae* family. Electron microscopy reveals a characteristic five- or six-pointed star shape extending over the entire surface in approximately 10% of the virions (Oliver and Phillips, 1988). A cryoelectron micrographic reconstruction reveals information on surface topology as depicted in Figure 8.2. The name, astrovirus, is derived from this distinctive, stellate, surface configuration (Madeley and Cosgrove, 1976). Astroviruses cause a mild and self-limiting illness, usually in children, characterized by mild episodes of vomiting and watery diarrhea which may be accompanied by fever, abdominal pain, and anorexia. More serious symptoms may result in immunocompromised or severely malnourished individuals. Dehydration can become life-threatening, but oral or intravenous fluids can often remedy this condition. The incubation period is generally 24–96 h and the duration of illness is usually less than 4 days. Virus levels in stools are estimated at $10^6$–$10^7$ per gram (Glass *et al.*, 1996), although levels as high as $10^{10}$

per gram have been reported (Kurtz and Lee, 1987). Fecal shedding persists during acute illness and may last for a week or longer in some patients. Immunocompromised individuals can have symptoms of astrovirus illness for months during which time virus particles are shed in the feces (Kurtz and Lee, 1987). Secretion of virus particles by asymptomatic individuals has also been reported. An estimated 3.9 million cases of astrovirus illness occur in the United States annually with only 10 deaths (Mead *et al.*, 1999). The number of cases from food- or water-borne sources is not certain. Antibodies against astroviruses are common among the general population (Konno *et al.*, 1982; LeBaron *et al.*, 1990) and are acquired during early childhood (Kurtz and Lee, 1978). There may be a seasonal predilection to astrovirus infection with the highest incidence in temperate climates during the winter months (Kurtz and Lee, 1987; LeBaron *et al.*, 1990). Astrovirus-induced diarrhea has been observed in a number of animals including humans and in a sheep model (Kurtz, 1994). In sheep, viruses infect mature enterocytes of the villi, whereas in the bovine model, viruses infect the microfold cells overlaying the Peyer's patches (lymph nodes) of the proximal small intestine (Woode *et al.*, 1984). Diarrhea is not associated with the bovine model, perhaps due to the different anatomical structure of the gut.

Human astroviruses may be divided into eight serotypes; however, serotype 1 is the most prevalent (Lee and Kurtz, 1994; Noel *et al.*, 1995; Glass *et al.*, 1996). The RNA is polyadenylated and consists of about 6800 nucleotides (nt) containing three open reading frames, ORF1a, ORF1b, and ORF2. The 5′ portion of the genome contains ORF1a and ORF1b which contain approximately 2700 and 1550 nt, respectively. These regions contain motifs suggestive of functions as nonstructural proteins including a 3C-like serine protease and an RNA-dependent RNA polymerase, respectively. ORF1a also contains four transmembrane alpha-helical regions and a nuclear localization site of unknown function. A 70-nt overlap of ORF1a and ORF1b contains a ribosomal frameshift signal which directs the synthesis of the RNA-dependent RNA polymerase (Lewis and Matsui, 1994, 1996; Marczinke *et al.*, 1994). ORF2 is located in the 3'-third of the genome. It contains approx. 2400 nt and encodes structural proteins involved in capsid formation. Genomic and subgenomic RNAs are produced in Astrovirus-susceptible cells (Monroe *et al.*, 1991; Lewis *et al.*, 1994). The ORF2 of subgenomic RNA encodes a 90-kDa protein precursor which is further processed into variously sized capsid proteins. In infected cells, protein processing results in at least four structural subunits ranging between 20 and 36.5 kDa (Matsui, 2002) and one subunit of 5.2 kDa (Kurtz, 1988), while only 30-kDa subunits were isolated from feces of infected individuals (Midthun *et al.*, 1993).

## Caliciviruses

Human caliciviruses may be divided into two groups, the noroviruses and the Sapporo-like viruses. Both groups encompass a broad range of genetically diverse, nonenveloped, icosahedral viruses, 27–40 nm in diameter. Caliciviruses derive their name from the calyx or cup-shaped surface depressions observed by electron microscopy. The nomenclature for the noroviruses, formerly the Norwalk and Norwalk-like viruses, has evolved steadily over the past 75 years. First recognized as winter vomiting disease in 1929 (Zahorsky, 1929), the proverbial Norwalk virus was identified by immunoelectron microscopy from a 1968 outbreak of gastroenteritis at an elementary school in Norwalk, Ohio (Kapikian *et al.*, 1972) (Figure 8.3). Over the years, noroviruses have been referred to as Norwalk and Norwalk-like viruses, the agent of winter vomiting disease, agents of nonbacterial gastroenteritis, and small round structured viruses, not to be confused with small round viruses of the *Astroviridae*, *Parvoviridae*, and *Picornaviridae* families. To add to the confusion in the nomenclature, agents reported in the literature, like Hawaii agent and Snow Mountain agent, were subsequently identified as noroviruses. Numerous other noroviruses have been identified and named for the locations from which they were obtained (Table 8.2).

Norovirus and Sapporo virus illnesses are characterized by the acute onset of non-bloody diarrhea and/or vomiting, usually accompanied by one or more of the following symptoms: nausea, abdominal cramps, chills, fever, headache, and body ache. Norovirus infection causes a decrease in alkaline phosphatase, sucrase and trehalase activities in brush borders of the small intestines resulting in the maladsorption of carbohydrates and a steatorrhea (Agus *et al.*, 1973). Norovirus gastroenteritis may be the most common enteric viral illness in the United States where an estimated 23 million cases occur annually (Mead *et al.*, 1999). In experimental studies, the incubation period has ranged between 10 and 51 h after virus ingestion (Wyatt *et al.*, 1974; Dolin *et al.*, 1982; Green *et al.*, 2001) and symptoms generally persist for 24–48 h. Norovirus

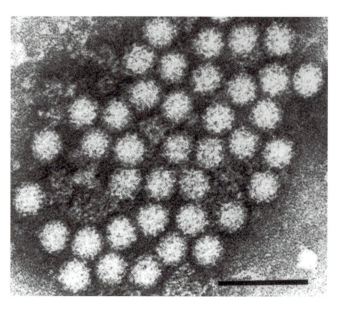

**Figure 8.3** Immune electron micrograph of Norwalk virus from the 1969 outbreak at an elementary school in Norwalk, Ohio (Kapikian *et al.*, 1972). Bar = 100 nm. Courtesy of Albert Z. Kapikian, National Institutes of Health, Silver Springs, MD.

illness is widely under-reported since infected individuals experience rapid remission of symptoms. Virus may be shed in the stools for a week or two after symptoms resolve (Graham *et al.*, 1994; White *et al.*, 1996). Virus levels in stools may reach $10^{10}$ particles per gram or higher (Richards *et al.*, 2004). Virus shedding may also occur in individuals exposed to norovirus but who are asymptomatic (Graham *et al.*, 1994; Richards *et al.*, 2004). The illness spontaneously passes without sequellae in most individuals, although rehydration therapy may be necessary in some cases. Both stool and vomitus contain infectious particles and can serve as a source for secondary spread. An asymptomatic carrier state occurs in some individuals and can be linked to seroconversion and fecal shedding of viruses (Graham *et al.*, 1994; Richards *et al.*, 2004). The Sapporo-like viruses are less prevalent than noroviruses in the United States.

Numerous clinical studies involving norovirus infection have used human volunteers. These viruses have never been propagated in cell or tissue culture, and since they do not infect nonprimate laboratory animals, stools from volunteers have been a valuable source of viruses for research purposes. Volunteer studies have provided viruses and sera/antisera against the viruses. Research on the immune responses elicited by the noroviruses presents a confusing array of results. Volunteer studies have indicated both short- and long-term immunity against norovirus infection (Dolin *et al.*, 1972; Wyatt *et al.*, 1974; Parrino *et al.*, 1977); however, only the short-term immunity follows the traditional pattern where sick individuals rechallenged 6–14 weeks later are resistant to infection. Short-term immunity appears to be genotype specific, since volunteers infected with Hawaii virus (genogroup II, cluster 1) were susceptible to Norwalk virus (genogroup

**Table 8.2** Groupings of human noroviruses and Sapporo-like viruses by genetic relatedness of the major capsid protein[1]

| Genogroup | Genetic cluster | Representative viruses[2] |
|---|---|---|
| **Noroviruses** | | |
| I | 1 | Norwalk/1968/US (M87661); KY-89/1989/JP (L23828); Aichi124/1989/JP (AB031013) |
| I | 2 | Southampton/1991/UK L07418); Whiterose/1996/UK (AJ277610) |
| I | 3 | Desert Shield 395/1990/SA (U04469); Birmingham/1993/UK (AJ277612); Stav/1995/NO (AF145709) |
| I | 4 | Chiba 407/1987/JP (AB022679); Thistlehall/1990/UK (AJ277621); Valetta/1995/Malta (AJ277616) |
| I | 5 | Musgrove/1989/UK (AJ277614) |
| I | 6 | Hesse 3/1997/GE (AF093797); Sindlesham/1995/UK (AJ277615) |
| I | 7 | Winchester/1994/UK (AJ277609) |
| II | 1 | Hawaii/1971/US (U07611); Girlington/1993/UK (AJ277606); Wortley/1990/UK AJ277618); Aichi 76–96 Chitta/1996/JP (AB032758) |
| II | 2 | Snow Mountain/1976/US (U70059, U75682); Melksham/1994/UK (X81879) |
| II | 3 | Toronto 24/1991/CA (U02030); Mexico/1989/MX U22498); OTH-25/1989/JP (L23830); Arg320/1995/AR (AF190817); Rbh/1993/UK (AJ277617); Bham132/1995/UK (AJ277611); Auckland/1994/NZ (U46039) |
| II | 4 | Bristol/1993/UK (X76716); Lordsdale/1993/UK (X86557); Grimsby/1995/UK (AJ004864); Camberwell/1994/AU (AF145896); Parkroyal/1995/UK (AJ277613); Symgreen/1995/UK (AJ277619) |
| II | 5 | Hillingdon/1990/UK (AJ277607) |
| II | 6 | Seacroft/1990/UK (AJ277620) |
| II | 7 | Leeds/1990/UK (AJ277608) |
| **Sapporo-like viruses** | | |
| I | | Sapporo/1982/JP (U65427); Manchester/1993/UK (X86560); Plymouth/1993/UK (X86559); Houston/1986/US (U95643) |
| II | | Houston 27/1990/US (U95644); Parkville/1994/US (U73124); Potsdam/2000/GE (AF294739); London 29845/1992/UK (U95645); Stockholm 318/1997/SE (AF194182) |

[1]Table adapted from Green *et al.* (2001).
[2]Each virus is listed by common name/year of outbreak/country of outbreak (GenBank accession number).

I, cluster 1) (Wyatt *et al.*, 1974). It is not known whether the immunoresponse is also cluster specific. Infection can lead to significant increases in IgA and IgG within 8–11 days, followed shortly thereafter by IgM (Brinker *et al.*, 1998; Gray *et al.*, 1994). Preexisting antibody titer does not appear to protect volunteers from reinfection by norovirus (Blacklow *et al.*, 1972, 1979; Greenberg *et al.*, 1981; Johnson *et al.*, 1990a, 1990b; Graham *et al.*, 1994).

Long-term immunity is more variable. Volunteers challenged with Norwalk virus exhibited different levels of resistance when rechallenged 27–42 months later. In one group of 12 volunteers, six who became ill after an initial challenge, also became ill after the second challenge, whereas the six who did not experience symptoms after the initial challenge also remained symptom-free after the second challenge (Parrino *et al.*, 1977). Interestingly, volunteers who did not become ill after the first challenge had no significant seroresponse after the first or second challenge with the same virus (Parrino *et al.*, 1977). Ill volunteers who elicited an antibody response after the first challenge had a significantly higher response after the second challenge. This phenomenon has been seen in other volunteer studies and in individuals naturally exposed to noroviruses.

In spite of the prevalence of norovirus infection, relatively little has been published about its replication in humans.

Proximal intestinal biopsies were taken from volunteers (Agus *et al.*, 1973; Schreiber *et al.*, 1974; Dolin *et al.*, 1975; Widerlite *et al.*, 1975) and histological studies showed broadening and blunting of intestinal villi, hyperplasia of the crypt cells, the formation of vacuoles within the cytoplasm, and mononuclear cell infiltration into the lamina propria. Since only jejunal tissues were studied, the site of virus replication has not been determined. Viruses were not detected within jejunal cells.

The norovirus genome consists of a single-strand of linear, polyadenylated, positive-sense RNA approximately 7300–7700 nt long, excluding the poly (A) tract containing a 5′-terminal GU cap (Jiang *et al.*, 1993; Hardy and Estes, 1996; Green *et al.*, 2001). Sequence diversity leads to varying length RNAs. Based on nucleotide sequence, they are classified into genogroups I and II which can be further divided into genetic clusters. There are new norovirus strains identified each year and the genetic clusters continue to expand. Traditionally, these viruses have been named for the location from which the first documented outbreak occurred. Representative members of some norovirus clusters are shown in Table 8.2. Among some notable genogroup I viruses are the Norwalk virus (Kapikian *et al.*, 1972), cruise ship virus (Noel *et al.*, 1997a), Desert Shield virus (Lew *et al.*, 1994a), and the Southampton agent (Lambden *et al.*, 1993). Genogroup II noroviruses include the Camberwell virus (Cauchi *et al.*, 1996), Grimsby virus (Maguire *et al.*, 1999), Hawaii agent (Thornhill *et al.*, 1977; Lew *et al.*, 1994b), Lordsdale virus (Dingle *et al.*, 1995), Mexico virus (Jiang *et al.*, 1995), Snow Mountain agent (Dolin *et al.*, 1982), and the Toronto virus (Lew *et al.*, 1994c), to name only a few. Another group of human Caliciviruses that do not fall within the *Norovirus* genus are the Sapporo-like viruses (SLV) which are further divided into SLV genogroups I and II and include the Sapporo virus (Chiba *et al.*, 1979), Houston virus (Jiang *et al.*, 1997) London virus (Jiang *et al.*, 1997), Manchester virus (Liu *et al.*, 1995), Parkville virus (Noel *et al.*, 1997b), Plymouth virus (Liu *et al.*, 1995), and Stockholm virus (Vinje *et al.*, 2000) (Table 8.2).

The classic Norwalk virus genome contains three open reading frames (ORF1 through ORF3) from the 5′ to the 3′ termini of the genome, respectively. The proteins coded for by ORF1 through ORF3 contain 1789, 530 and 212 amino acid residues, respectively (Atmar and Estes, 2001). ORF1 encodes a polyprotein with nonstructural motifs characteristic of a helicase, a protease, and an RNA-dependent RNA polymerase, similar to the *Picornaviridae* family of viruses. The genome organization is similar for the Southampton virus. For the Southampton virus, cleavage of the polyprotein produces a 48-kDa protein, a 41-kDa protein (helicase), and a 113-kDa protein (Liu *et al.*, 1996) that is further cleaved into 22-kDa, 16-kDa, 19-kDa (3C protease), and 57-kDa (RNA-dependent RNA polymerase) proteins (Liu *et al.*, 1999; Clarke and Lambden, 2000). ORF1 and ORF 2 sequences overlap for genogroup I and II noroviruses. The major capsid protein is derived from ORF2 and, when expressed in a baculovirus expression system, ORF2 produces capsid-like proteins that self-assemble into virus-like particles (VLPs). A minor capsid protein is derived from ORF3.

For the Sapporo-like virus (SLV) genome, strains contain three ORFs, with ORF1 being somewhat longer and coding for a polyprotein similar to that of ORF1-coded polyprotein from the noroviruses, except it also codes for the capsid protein that is coded for by ORF2 in the noroviruses (Atmar and Estes, 2001). ORF2 of the SLVs is analogous to the ORF3 of noroviruses, is located on the 3′- terminus of the genome, and appears to encode a minor capsid protein. ORF3 is present in at least some of the SLVs; however, it is currently unclear whether an ORF3-encoded protein is actually produced.

## Hepatitis A virus

Hepatitis A virus is a member of the family *Picornaviridae* which includes the genera hepatovirus, enterovirus, rhinovirus, aphthovirus, cardiovirus, and parechovirus (Anonymous, 2000). Although previously classified as an enterovirus, because of its similarity to the classic enteroviruses (polio-, echo- and coxsackie-viruses), it was reclassified as the sole member of the genus *Hepatovirus*. Hepatitis A virus is nonenveloped with a capsid 27–32 nm in diameter and with apparent icosahedral symmetry.

Hepatitis A virus causes hepatitis A, an acute, self-limiting infection generally restricted to the liver. The disease is often asymptomatic but occasionally becomes fulminant and life-threatening. The incubation period for hepatitis A is generally 15 to 45 days, although the presence of virus particles in the blood and feces occurs a few days after exposure, well before the onset of clinical symptoms. Fecal shedding of virus and viremia may persist for many weeks. During the incubation period, virus levels in the blood and stool increase until symptoms become evident. Symptoms include nausea, vomiting, anorexia, malaise, myalgia, fever, and abdominal pain in the upper right quadrant. Liver damage results in a disruption of normal liver function and increases in the serum titer of liver-derived enzymes, particularly alanine aminotransferase (ALT), alkaline phosphatase (AST), γ-glutamyl transpeptidase (GGTP), and bilirubin. Urine takes on a dark, Coca-Cola-like appearance, stool becomes pale clay-colored, and jaundice becomes apparent. Exposure confers immunity mediated by a rapid IgM response and a delayed IgG response. Although the IgM antibody dissipates within several months after the illness, the IgG persists and confers lifetime immunity.

The severity of illness is correlated with age and demographics. Young children often experience asymptomatic illness, whereas the elderly, who were not previously exposed to hepatitis A virus, can develop fulminant infection and have a greater chance for hepatic failure and other complications. In underdeveloped countries, young children often acquire asymptomatic hepatitis A and have circulating antibodies to hepatitis A which provides life-long immunity to further infection. In more developed countries, where exposure to hepatitis A virus is less frequent, individuals tend to develop hepatitis A at a later stage in life resulting in more severe symptoms. Two complications of hepatitis A include cholestatic hepatitis A and relapsing hepatitis A. In cholestatic hepatitis A, jaundice gives way to pruritus, anorexia and weight loss and may persist for weeks to months. Relapsing hepatitis A may occur 1 to 3 months after the cessation of clinical symptoms leading to an elevation of serum enzymes and potential fecal shedding of viruses.

After entering the stomach via contaminated food, water, or direct person-to-person uptake, hepatitis A virus migrates to the small intestines where it has been observed in the crypts of the jejunum and ileum of experimental animals (Karayiannis *et al.*, 1986; Asher *et al.*, 1995). It is speculated that some virus replication occurs in intestinal epithelial cells. Virus uptake may also occur in the oropharynx and by parenteral injection of infectious virus particles. Regardless of the mode of entry, virus migrates to the liver, which is the primary site of virus replication. The mechanism of this migration has not been elucidated, but initial replication in the intestines might be the springboard for virus transfer to the blood or lymphatic system and subsequently to the liver (Feinstone, 1986). The lymphatic system could play a role in transporting hepatitis A virus to hepatic cells since the lymph ultimately discharges into the circulatory system. As hepatitis A virus replicates in the liver, parenchymal cells become necrotic and Kuffer cells (phagocytic cells that contribute to the formation of the endothelial lining of liver sinusoids) proliferate (Corey, 1990). Viruses propagated in hepatocytes enter the bile and flow through the bile duct and into the stomach where they spread throughout the intestinal tract and into the feces. The number of virions in feces has been reported as high as $10^8$ per gram (Purcell et al., 1984; Lemon et al., 1990) and the shedding of viral RNA has been reported for up to 5 months (Rosenblum et al., 1991; Robertson et al., 2000).

Hepatitis A virus contains a single strand of linear RNA approximately 7.5 kb in length and of positive polarity. The 5′ end of the genome contains an untranslated 738-nt region followed by a single, long open reading frame, and a 63-nt untranslated region adjacent to a 3′ poly(A) tail which varies in length. The open reading frame is 6681 nt long and encodes a 2227-amino acid polyprotein which has been arbitrarily divided into three sections referred to as P1, P2 and P3. P1 is subsequently cleaved into four capsid proteins, VP1-VP4. P2 contains three proteins (2A, 2B, and 2C). Protein 2A has no known function in HAV; however 2B and 2C are believed to be membrane proteins involved in replication (Teterina et al., 1997) and 2C may have NTPase and helicase activity and may contain a nucleoside triphosphate binding domain (Totsuka and Moritsugu, 1999). The P3 portion contains four proteins, 3A–3D. The 3A protein may serve to anchor the 3A and 3B proteins to the cellular membrane. Protein 3B is the Vpg protein which binds to the 5′ portion of the RNA and may be involved in priming of RNA synthesis or in cleavage of the RNA. Protein 3C is the virus-encoded proteinase responsible for most of the cleavages to the polyprotein and protein 3D is RNA polymerase.

## Hepatitis E virus

Hepatitis E virus is an icosahedral, nonenveloped, positive polarity, single-stranded RNA virus with a diameter of approximately 32 nm. It was initially assigned to the *Caliciviridae* family; however, it was recently removed from this family and listed as "unclassified" (Green et al., 2000). Hepatitis E virus is the causative agent of hepatitis E, formerly known as enterically transmitted non-A, non-B hepatitis. The virus causes an acute, generally mild, infection, except in pregnant women where mortality rates range between 15% and 25% (Mast and Krawczynski, 1996).

The incidence of hepatitis E in the United States is nearly negligible; however, hepatitis E is a major cause of morbidity in developing countries. Hepatitis E outbreaks have been reported from water-borne sources (Naik et al., 1992; Tucker et al., 1996; Corwin et al., 1996, 1997; Benjelloun et al., 1997; Rab et al., 1997). Early (prodromal) symptoms of the illness often include malaise, fatigue, vomiting, anorexia, and low-grade fever. Symptoms are indistinguishable from other forms of acute viral hepatitis. Pale clay-colored stools and Coca-Cola-colored urine are present in hepatitis E infection and may persist during the prodromal period.

Like all the other hepatitis viruses, hepatitis E infects the liver causing jaundice in some individuals. Liver involvement

may lead to pain in the upper right quadrant. Spleen enlargement has been reported in 10–15% of patients (Anderson and Shrestha, 2002). The incubation period for hepatitis E appears to depend on the dose of virus received (Tsarev et al., 1994) ranging from 2–8 weeks (Li et al., 1994; Tsarev et al., 1994; Anderson and Shrestha, 2002). The duration of prodromal symptoms is 1–10 days; however, total clinical recovery may take 4–8 weeks. The IgM and IgG responses to infection nearly parallel each other with antibody detection after 1 month, maximum antibody levels detected after three months, and a rapid decline thereafter. The IgM levels disappear by about 6 months; however, IgG levels persist at a low level, thus conferring long-term immunity to further infection. IgA is also produced in response to hepatitis E virus, but is short-lived. Neither cytokine nor cellular immune responses have been studied for hepatitis E infections. Since hepatitis E virus can not be propagated in cell or tissue culture, no quantitative data is available on the levels of virus excreted in feces. Using reverse transcription-PCR (RT-PCR), virus has been detected in feces of patients for 2 weeks (Clayson et al., 1995) and for 52 days (Nanda et al., 1995) after onset. A potential zoonotic transmission of hepatitis E virus between swine and humans has been reported (reviewed by Smith, 2001).

Four genotypes of hepatitis E virus have been proposed (Wang et al., 1999), represented by the following prototypical strains: genotype 1, Burmese strain; genotype 2, Mexican strain; genotype 3, swine strain, and genotype 4, China T1 strain. The swine strain appears to cross species lines (Meng et al., 1998) and is closely related to some human strains (Schlauder et al., 1998). The hepatitis E virus genome contains approximately 7250 nt with a 3′ polyA tail of variable length. The 5′ end has a methylguanosine cap (Kabrane-Lazizi et al., 1999). Both the 5′ and 3′ ends contain highly conserved untranslated regions of 35 and 68–75 nt, respectively, which may be involved in RNA replication and encapsidation. The genome contains three open reading frames with ORF1 at the 5′ end, ORF2 at the 3′ end, and ORF3 towards the center with an overlap of ORF2 and ORF3. ORF1 encodes PORF1, a 186-kDa polyprotein containing methyltransferase, papain-like protease, RNA helicase, and RNA-dependent RNA polymerase motifs. ORF2 encodes PORF2, a 660-amino-acid capsid protein. ORF3 encodes PORF3, a 13-kDa highly basic protein (pI ≈12.5) of unknown function. All mechanisms of hepatitis E virus replication occur in the cytoplasm of host cells. Intact virus particles are released into the bile canaliculium and then flow into the deodenum, through the intestines, and are released in the feces.

## Rotavirus

Rotaviruses are classified within the *Reoviridae* family. They are icosahedral, nonenveloped, 70–75 nm diameter particles containing double-stranded RNA. Worldwide, rotavirus infection is a common illness in infants and young children who make up the majority of the estimated 800,000 deaths annually (Parashar et al., 1998). In the United States alone, the annual incidence of rotavirus illness is estimated at 2.7–3.9 million cases (Mead *et al.* 1999), 49,000–50,000 hospitalizations (Anonymous, 1998), and 30 deaths (Anonymous, 1998).

Clinical signs of human rotavirus infection include diarrhea, anorexia, dehydration, depression, and occasional vomiting (Kapikian and Chanock, 1990; Bishop, 1994; Saif *et al.*, 1994). Childhood mortality is commonly due to dehydration in regions of the world where rehydration therapy is lacking but is uncommon in developed countries. The incubation period is generally short, ranging from 11 hours to 6 days (Kapikian and Chanock, 1990; Bishop, 1994; Saif *et al.*, 1994). Vomiting lasts 1–2 days and diarrhea persists for about 5 days and may be accompanied by several days of fever and dehydration (Uhnoo *et al.*, 1986). Fecal shedding of viruses has been shown for up to 57 days (Richardson *et al.*, 1998), or longer, particularly in children with immunodeficiencies (Saulsbury *et al.*, 1980).

Rotaviruses are ubiquitous pathogens present in feces at levels approximating $10^9$–$10^{10}$ particles per gram (Conner and Ramig, 1997). Consequently, nearly 100% of young children will become infected with rotavirus. This makes the immune response to rotavirus infection difficult to track. In animal models, rotavirus infection elicits rapid humoral immunity with the production of IgM, followed by IgG and IgA in the intestine and serum. Reinfection occurs frequently but with increasingly milder symptoms. Only the primary infection is usually life-threatening, suggesting a partial immunity after the initial exposure, while complete immunity occurs after repeated exposure (Coulson *et al.*, 1992).

Rotaviruses are acid labile in most adult stomachs, which maintain a pH of around 2.0, but they survive longer in infants, in whom the pH of the stomach is often around 3.2; therefore, infants are at particular risk of acquiring rotavirus infections (Weiss and Clark, 1985). The use of antacids by adults may predispose some individuals to rotavirus infection.

Rotavirus targets nondividing enterocytes on the surface of the villi of the small intestine. Proteolytic cleavage of the VP4 spike protein appears to enhance the severity of the infection. Much of the pathophysiology of rotavirus infection has been determined from extensive characterization in animal models including the bovine, ovine, porcine, lapine, and murine models (Conner and Remig, 1997). Rotavirus replication occurs primarily in the mature enterocytes along the tip of the villi of the small intestine. Duodenal biopsies reveal irregular architecture over portions of the mucosal surface. Irregularities include blunting of the villi, increased numbers of mononuclear cells in the lamina propria, increased crypt depth, and a tendency for epithelial cells to appear more cuboidal (Bishop *et al.*, 1973; Davidson *et al.*, 1979; Wyatt *et al.*, 1978). Viral particles are present in the cytoplasm of the duodenal mucosa, but not in other cell types. In severe cases, loss of villi can be observed (Davidson and Barnes, 1979).

Rotaviruses are icosahedral particles containing three layers: the outer viral capsid, the inner viral capsid, and the core. They also have spike proteins extending out from the outer capsid. Although their diameter is generally considered 70–75 nm, the diameter including the spike proteins is about 100 nm. The rotavirus genome is found within the core and consists of 11 segments of double-stranded RNA ranging from about 670–3300 bp (~18,500 bp total). Each segment contains an open reading frame and a short noncoding region at the 5'- and 3'-termini.

The RNA codes for six structural proteins, three in the core (VP1, VP2, VP3), one in the inner shell (VP4), and two in the outer shell (VP6 and VP7), as well as six nonstructural proteins (NSP1–NSP6). VP1 is an RNA-dependent RNA polymerase, VP2 is an inner capsid structural protein, VP3 is a guanylyltransferase involved in the virion transcription complex, VP4 forms the outer capsid spike and is cleaved by trypsin into VP5 and VP8. VP5 permeabilizes the cellular membrane and may enhance virus infectivity (Denisova *et al.*, 1999), while VP8 is involved in virus attachment to cells. VP6 forms the major structural protein for the inner viral capsid, while VP7 is a major structural protein of the outer viral capsid (reviewed in Franco and Greenberg, 2002). The function of NSP1 is uncertain; however, NSP2 may be involved in single-stranded RNA binding. NSP3 binds the 3' nonpolyadenylated end of the rotavirus mRNA and is involved in transcriptional regulation. NSP4 functions as an enterotoxin to induce diarrhea and appears to be involved in virus morphogenesis and virus entry into the endoplasmic reticulum (reviewed in Franco and Greenberg, 2002). NSP5 is the product of the 11th double-stranded RNA segment, is believed to have an autocatalytic kinase activity, and may be involved in viral replication. Finally, NSP6 is the product of the second ORF of gene segment 11 and interacts with NSP5. It has a putative function in viral replication.

Rotaviruses consist of seven serogroups (serogroups A to G) differing in antigen composition, but similar in morphology (Saif and Jiang, 1994). The antigenic regions associated with serogroups A-C are on the VP7 structural protein located on the surface of the outer capsid. Serogroup A is the most prevalent worldwide, although epidemics of serogroup B and C rotaviruses have been reported in China and Japan in both children and adults (Hung *et al.*, 1984; Ushijima *et al.*, 1989). Serogroups D, E, F, and G are nonhuman animal pathogens. Further characterization may be made by virtue of neutralizing antibodies directed against the surface proteins VP7 and VP4. Neutralization with a VP7 antibody indicates the presence of the G serotype (glycosylated protein serotype) and neutralization with an antibody against VP4 indicates the P serotype (protease sensitive protein). The VP4 and VP7 proteins are important not only in determining serotype, but also in inducing neutralizing antibodies and protective immunity.

## Virus extraction and assay from food and water

Current extraction methods for enteric viral pathogens may be classified into two groups: (i) methods for extracting viable viruses for detection in cell culture or animal models, and (ii) methods for extracting viral nucleic acids for use in molecular biological assays. The latter detects infectious as well as non-infectious viruses.

## Molecular methods

### Extraction and assay

Molecular biological methods have become popular techniques for the assay of enteric viruses in food and water. Methods have been developed to extract and assay viral DNA and RNA for subsequent analysis by PCR or RT-PCR, respectively. Procedures are available for the extraction and RT-PCR detection of noroviruses from water (Haflinger *et al.*, 2000; Huang *et al.*, 2000; Beuret, 2003), meats (Schwab *et al.*, 2000, Sair *et al.*, 2002), fruits and vegetables (Dubois *et al.*, 2002; Goswami *et al.*, 2002; Sair *et al.*, 2002) and molluscan shellfish (Atmar *et al.*, 1995; Le Guyanger *et al.*, 1996; Schwab *et al.*, 1998; Shieh *et al.*, 1999; Kingsley and Richards, 2001). Hepatitis A virus extraction and assay methods are available for water (Gilgen *et al.*, 1997; Jothikumar *et al.*, 2000; Carducci *et al.*, 2003), shellfish (Cromeans *et al.*, 1997; Croci *et al.*, 1999; Kingsley and Richards, 2001; Mullendore *et al.*, 2001; Goswami *et al.*, 2002), and produce (Jean *et al.*, 2001; Dubois *et al.*, 2002; Goswami *et al.*, 2002; Sair *et al.*, 2002). Methods were developed for the concentration of astroviruses in water (Abad, 1997; Taylor *et al.*, 2001) and the extraction of astroviruses from shellfish (Traore *et al.*, 1998; Legeay *et al.*, 2000; Le Guyader *et al.*, 2000; Yokoi *et al.*, 2001). Most astrovirus RT-PCR methods were designed for the analysis of clinical specimens (Lewis *et al.*, 1994; Jonassen *et al.*, 1995; Noel *et al.*, 1995) and led to the development of primers useful for food and water analyses. Rotavirus may be detected by PCR of water concentrates (Jothikumar *et al.*, 1995; Gilgen *et al.* 1997) and shellfish extracts (Barardi *et al.*, 1999; Le Guyader *et al.*, 2000). Molecular techniques have also been described for adenovirus detection in water (Cho *et al.*, 2000) and shellfish (Muniain-Mujika *et al.*, 2000) and for hepatitis E virus detection in water (Jothikumar *et al.*, 1993, 1995; Grimm and Fout, 2002).

### PCR and RT-PCR

Procedures are available to: (i) extract some viruses or their genetic materials from food and water, (ii) reduce or eliminate materials inhibitory to PCR or RT-PCR that might be carried over into the extracts, (iii) amplify the genetic materials with high sensitivity and specificity, and (iv) verify that the PCR amplicon is indeed the desired product. Procedures potentially applicable to food and water testing include RT-multiplex PCR (Egger *et al.*, 1995), nested and seminested PCR (Haflinger *et al.*, 1997; Le Guyader *et al.*, 1994; Severini *et al.*, 1993; Castignolles *et al.*, 1998; Gantzer *et al.*, 1997; Gilgen *et al.*, 1997), immunocapture PCR (Beaulieux *et al.*, 1997), magnetic immunoseparation PCR (Jothikumar *et al.*, 1998; Monceyron and Grinde, 1994), antigen capture PCR (Deng *et al.*, 1994; Divizia *et al.*, 1998), indirect antibody capture RT-PCR

(Schwab *et al.*, 1996), nucleic acid capture RT-PCR (Regan and Margolin, 1997) combined or integrated cell culture-RT-PCR (Reynolds *et al.*, 1996; Chapron *et al.*, 2000), and real-time RT-PCR (Kageyama *et al.*, 2003; Nishida *et al.*, 2003; Richards *et al.*, 2004) to name a few. Quantitative real-time PCR can be used if some method of enumeration is available to generate standard curves. None of the methods are to be considered standard procedures since they all require collaborative testing to verify their reliability. Such methods have great potential for identifying virus presence in food and water.

A wide variety of primer sets can be identified for potential use from the current literature. Many primer pairs and reaction conditions have evolved for the molecular detection of the genetically diverse noroviruses (Matsui *et al.*, 1991; Ando *et al.*, 1995, Atmar and Estes, 2001; Le Guyander *et al.*, 1996; Noel *et al.*, 1997b; Kageyama *et al.*, 2003). Some are broadly reactive to detect a wide variety of viruses within a genus, like primers for the enteroviruses (Rotbart *et al.*, 1988) and noroviruses (Jiang *et al.*, 1999; Kageyama *et al.*, 2003; Nishida *et al.*, 2003). However, many of the primers described in the literature are specific for particular viruses and do not amplify closely related species. Examples would be primer sets that only recognize one genogroup of norovirus or only some clusters of noroviruses within a genogroup. Other viruses, like hepatitis A and E viruses, rotaviruses, astroviruses, and enteric adenoviruses are more easily assayed using genus-specific primer sets, since they lack the genetic diversity of the noroviruses.

## Culturable viruses

### Extraction methods

The isolation and detection of infectious viruses in water and food can be a formidable challenge. The pH, conductivity, turbidity, presence of particulates, organic acids, and many other attributes can directly affect the efficiency of virus recovery from water. On the other hand, foods represent an even more complex matrix from which viruses must be extracted. Because molluscan shellfish cause an appreciable number of enteric virus illnesses, most of the extraction efforts for foods have focused on molluscan shellfish. A plethora of methods have been reported for the extraction of enteric viruses from shellfish with considerably fewer methods developed for fruits and vegetables (reviewed by Richards and Cliver, 2001). Many of the methods were developed using vaccine strains of poliovirus as a model, since poliovirus was considered by many as an indicator of fecal pollution. Viruses in water must be concentrated before they can be assayed. Concentration may be accomplished for poliovirus, hepatitis A and rotavirus using various filtration techniques (Sobsey and Glass, 1980; Smith and Gerba, 1982; Raphael *et al.*, 1985; Sobsey *et al.*, 1985; Toranzos and Gerba, 1989). Poliovirus extraction methods have been developed for Eastern oysters, *Crassostrea virginica*; Pacific oysters, *Crassostrea gigas*; hard-shell clams, *Mercenaria mercenaria*; soft-shell clams, *Mya arenaria*, Japanese clams, *Tapes japonica*; and mussels, *Mytilus edulis* (reviewed in Richards and Cliver, 2001). Procedures are also available for the extraction of viruses from the surfaces of fruits, vegetables, and mushrooms (Kostenbader and Cliver, 1973; Konowalchuk and Speirs, 1975, 1977; Badawy et al., 1985; Sullivan et al., 1986). Shellfish and comminuted foods may contain viruses internally and are usually subjected to homogenization and a series of differential centrifugations to purify the viral contaminants. For virus analysis, surface contamination may be removed from fruits and vegetables by vigorous rinsing followed by concentration and assay.

## Infectivity assays

### Cell lines used

Historically, much of the work on the development of cell culture-based assays has focused on enteroviruses (polio-, echo- and coxsackie-viruses). Mammalian tissue and cell cultures have been instrumental for the propagation and assay of many enteric viruses. Primary cell cultures are considered more sensitive to virus infection, but some are very difficult to obtain and they can be contaminated by unknown adventitious agents. Several established cell lines are sensitive to a range of viruses and they have been used with some success. The use of multiple cell lines for the assay of virus-containing samples has proven useful to increase the likelihood of virus detection. Although primary or secondary human embryonic kidney tissues and human embryonic lung fibroblast cells (Flehmig et al., 1981; Graff et al., 1994, 1997; Kok et al., 1998) are highly sensitive to some enteric viruses, there are practical limits on the use of these cells including limits based on the potential presence of adventitious agents, cost, and availability of such tissues. Coupled with the ethical issues concerning the use of fetal tissues, most researchers rely on cell lines that have been demonstrated as practical for routine virus testing.

Among the most commonly used cell lines for the assay of enteric viruses are the Buffalo green monkey kidney (BGM) cells derived from the African green monkey (Barron et al., 1970; Dahling et al., 1974; Schmidt et al., 1976; Dahling and Wright, 1986). These cells are recommended in the US Environmental Protection Agency's (USEPA) ICR for the detection of total culturable viruses in surface and drinking waters (Fout et al., 1996). Detailed methods for the maintenance and use of BGM cells are provided in the USEPA Manual of Methods for Virology (Berg et al., 1987a). The BS-C-1 cell line, also derived from the African green monkey kidney (Day et al., 1992; Zhang et al., 1995); MA-104 cells from the fetal rhesus monkey kidney (Whitaker and Hayward, 1985); rhabdomyosarcoma (RD) cells from a human cancer (McAllister et al., 1969; Schmidt et al., 1975, 1978); and HEp-2 cells derived from a HeLa cell line (Pal et al., 1963; Chen, 1988) can be used to propagate some of the enteric viruses. The human cervical carcinoma cell line (HeLa) (Lund and Hedstrom, 1969; Irving and Smith, 1981) has also been employed for virus assay. Fetal rhesus monkey kidney (FRhK-4, Wallace et al., 1973) cells are used for enterovirus detection and have been particularly useful for the assay of cell culture-adapted strains of hepatitis A virus (Flehmig, 1981; Wheeler et al., 1986; Cromeans et al., 1987; Zou and Chaudhary, 1991; Reiner et al., 1992). FRhK-4 cells can support the growth of some environmentally derived strains of hepatitis A virus, but only with extended incubations. Rotavirus are frequently propagated in MA-104 monkey kidney cells (Ramia and Sattar, 1979; Urasawa et al., 1981; Ward et al., 1984) while enteric adenoviruses can be propagated in HEp-2 cells containing 100 µg/mL guanidine hydrochloride to inhibit the replication of contaminating enteroviruses (Hurst et al., 1988) or in A-549 human lung carcinoma cells (Giard et al., 1973), MRC-5 human lung fibroblast cells (Jacobs et al., 1970), HeLa cells, and in human embryonic kidney cell line transformed with adenovirus type 5 DNA (293 cells) (Graham et al., 1977). Astroviruses were propagated in human embryonic kidney (HEK) cells (Lee and Kurtz, 1981), and in a human colon carcinoma cell line (CaCo-2, Fogh et al., 1977), rhesus kidney cell line (LLCMK2), and a human hepatoma cell line (PLC/PRF/5) (Taylor et al., 1997). In spite of many years of intensive effort, human caliciviruses (noroviruses and Sapporo viruses) have not been propagated in cell or tissue cultures. Hepatitis E virus has also defied propagation efforts. Their detection in food or water must be based on molecular biological techniques until cell culture systems are developed.

### Plaque assays

For lytic viruses capable of growth in cell or tissue culture, plaque assays have been the gold-standard for many years (Berg et al., 1987b). The assay of poliovirus by the plaque technique is simple and semiquantitative. Discrete foci of virus-infected (dead) cells (plaques) are visible on stained cell culture monolayers after several days and may be enumerated. Plaque counts provide a measure of the number of infectious viruses present in the sample inoculum. This measure is variable as the sensitivity of the cells to virus binding, penetration, and replication varies with the cell lines or tissues used and the conditions of the assay. Cell culture-adapted strains of hepatitis A virus, and enteroviruses including poliovirus, may be assayed, with varying degrees of success, by the plaque technique. In addition, plaque assays may be performed on some strains of group A rotavirus, enteric adenovirus (adenovirus types 40 and 41), and astrovirus; however, trypsin treatment of the cell or tissue culture may be required to enhance virus uptake by the cells (Ramia and Sattar, 1979, 1980; Agbalika et al., 1984; Taylor et al., 1997). Trypsin may be incorporated into the agar overlay medium, although treatment of both virus and cells may enhance virus adsorption (Jourdan et al., 1995). The reproducibility of these assays varies greatly from one test to another. It is speculated that varying exposure periods and sensitivities of the cells to the trypsin may be responsible for some of the fluctuations in assay sensitivity and reproducibility.

The plaque assay entails the preparation and use of monolayer cultures in tissue culture flasks or dishes. There are a number of variations in plaque assay techniques that should

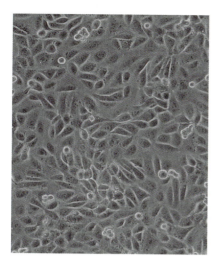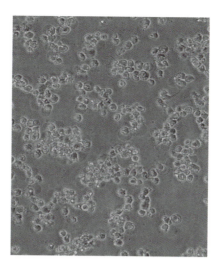

**Figure 8.4** Microscopic comparison of normal Buffalo green monkey kidney cell monolayer with cells exhibiting classical cytopathic effects 3 days post infection with poliovirus.

**Normal cells**          **Virus-infected cells**

be optimized for the particular virus to be tested. Stock viruses obtained from a culture repository may be used to determine the sensitivity of the cells to the particular virus and the optimal conditions of the assay. Some of the conditions which should be optimized include the type of diluent used, the volume of inoculum used per flask, the duration of adsorption, whether adsorption is conducted under rocked or static conditions, the preferred overlay medium, and the duration of incubation. Food extracts or water concentrates may contain compounds cytotoxic to cell culture monolayers and may be diluted to reduce the level of toxicity. Likewise, cytotoxic materials bound to cell monolayers may be removed, in part, by rinsing the monolayers after virus adsorption; however, some virus loss may be experienced during the rinsing process. The composition of the sample, the manner of dilution, and the cell handling procedures may seriously alter results and must be optimized to the extent possible. As a rule of thumb, the volume of inoculum should not exceed 1.0, 3.0, and 6.0 mL per 25, 75, and 150 $cm^2$ tissue culture flask, respectively. The larger the volume, the longer the time required for virus adsorption. In all cases, the inoculum should be of sufficient volume to ensure that the cells do not dry out during the adsorption procedure. Inoculum may also be removed from one flask after the adsorption period and added to a monolayer of different cells and these cells also used for plaque assay to enhance the likelihood of virus detection. Monolayers may be rocked during virus adsorption or allowed to remain static. Rocking increases plaque counts (Richards and Weinheimer, 1985). A variety of enteroviruses may be quantitated using the plaque assay technique. The USEPA has published detailed methods for the plaque assay of human enteroviruses in BGM cells (Berg et al., 1987b).

When testing food extracts or water concentrates for virus presence by the plaque assay technique, it is possible that cytotoxic materials in the extract may kill portions of the monolayer, thus leaving holes or gaps that may be mistaken as plaques.

Plaques should be confirmed by picking to susceptible cell cultures and observing microscopically for cytopathic effects (CPE). Depending on the virus, CPE may include rounding or shortening of cells, disintegration of cells, release of cells from the surface, cell clumping, and/or differences in light refractivity (Figure 8.4). Detailed instructions for picking plaques for confirmation may be found in the *USEPA Manual of Methods for Virology* (Berg et al., 1988a; 1988b).

*Quantal assays*
Quantal assays involve the detection of infectious viruses in an all or nothing approach, where the culture is infected (all) or not infected (nothing). It can use a mathematical model to determine viral density. Quantal assays can provide the most probable number of cytopathic units (MPNCU) or the median tissue culture infectious doses ($TCID_{50}$) (Chang et al., 1958; Sobsey, 1976). The number of viruses in these assays is defined statistically within a 95% confidence interval based on the number of positive and negative replicates within a virus dilution series. An MPN procedure for the semi-quantitation of viruses in drinking water and surface water is provided in the *USEPA ICR Microbial Laboratory Manual* (Fout et al., 1996). A determination of whether or not the cells are infected may be by microscopic examination for cytopathic effects or by other means.

## Other assays

*Enzyme and immunoassays*
For the nonculturable viruses, like noroviruses, immunoelectron microscopy (IEM) (Kapikian et al., 1972; Thornhill et al., 1977), radioimmunoassay (RIA) (Greenberg et al., 1978), and enzyme immunoassays (EIA) (Herrmann et al., 1985, Madore et al., 1986) were initially developed, but these methods relied on antisera that were not commercially available. For the noro-

viruses, antisera availability is still generally limited to research settings. The low sensitivity of IEM, RIA and EIA make them more appropriate for detecting viruses in clinical specimens or after cell culture propagation, where virus levels may be high. Several types of immunoassays can be used to detect virus-antibody: immunofluorescent assays (IF) (Riggs, 1979; Payment, 1997); immunoenzymmatic assays such as immunoperoxidase assays (Payment, 1997; Payment and Trudel, 1985) and enzyme-linked immunosorbent assays (ELISA) (Guttman-Bass et al., 1987; Nasser and Metcalf, 1987); biotin-avidin immunoassays (Gary et al., 1985; Herrmann et al., 1985); and radioimmunoassays (RIA) (Grabow et al., 1983; Dolin et al., 1986).

Viruses that do not produce visible CPE but do infect the cultures may be tested by the radioimmunofocus assay (RIFA) using $^{125}$I-labeled antibody against the virus (Lemon et al., 1983; Liu et al., 1984) or by the luminescent immunofocus assay (LIFA) using enhanced chemiluminescence detection (Richards and Watson, 2001). The RIFA assay is similar to the plaque assay and has been instrumental in the identification and enumeration of hepatitis A virus and rotavirus in infected cell cultures (Lemon et al., 1983; Liu et al., 1984). Foci of virus-infected cells on a cell culture monolayer are comparable to plaques but cannot be seen directly. Viruses within the foci must first be reacted with radiolabeled antibody against the virus and then exposed to X-ray film. Areas of virus infection will produce visible spots on the film and the spots may be counted. Like plaques, each spot represents a region of the monolayer where one virus infects a cell and progeny virus spreads to adjoining cells to produce a focus of virus-infected cells.

An alternative to the RIFA is the recently described LIFA procedure which uses enhanced chemiluminescence instead of radiometric detection to identify and enumerate non-cytopathic hepatitis A virus, rotavirus and other viruses that are capable of infecting cell cultures (Richards and Watson, 2001). Assays may be performed with commercially available monoclonal antibodies against hepatitis A virus and rotavirus. The viruses are cultured in susceptible cells under an agarose overlay. After suitable incubation, the agar is removed, the plate is disinfected under ultraviolet light, and the cells and viruses are then lifted onto nitrocellulose membranes. The membranes are blocked in milk, rinsed, and virus-specific antibody is added followed by rinsing and the addition of a second antibody (anti-rabbit, anti-mouse, etc.) conjugated to horse radish peroxidase. After rinsing, a luminol-based detection reagent is added and the membranes are wrapped in plastic and placed against light sensitive photographic film for up to 10 min. Resulting foci are comparable to the foci observed in the RIFAs. The LIFA has been instrumental in efforts to enumerate non-cytopathic strains of HAV and rotavirus propagated in cell culture monolayers, without the need for radioactive substances and detection can be performed in minutes compared to days as required with radiometric methods.

Immunofluorescent assays have been used to detect viruses in infected cells (Riggs, 1979; Smith and Gerba, 1982; Gutman-Bass et al., 1987; Oragui and Mara, 1989; Patti et al., 1990; Payment and Trudel, 1993; Payment, 1997). This procedure requires virus-specific antibodies conjugated to a fluorogenic compound such as fluorescein isothiocyanate. The labeled antibody reacts with the viruses within the cells and unbound antibody is then washed away. Infected cells must be visualized microscopically under ultraviolet illumination for fluorescent regions within the cytoplasm of some or all of the cells. This procedure can be tedious and microscopic counting of individual foci of virus infection is not always possible.

Immunoenzymatic assays may be conducted by reacting virus-infected cells with a virus-specific antibody followed by a second antibody that has been conjugated to an enzyme such as peroxidase (immunoperoxidase assay). The peroxidase-labeled antibody binds to the first antibody which is attached to virus particles within the cells. Cells are washed thoroughly to remove unbound second antibody and then reacted with a substrate that produces a colored product when cleaved by the enzyme. Only cells containing virus will produce the color which can be observed microscopically. Immunoperoxidase assays have been successfully used for the detection of enteric viruses in water and wastewater (Smith and Gerba, 1982; Payment and Trudel, 1985; Guttman-Bass et al., 1987; Payment et al., 1988; Oragui and Mara, 1989, Patti et al., 1990).

Other immunoassays used for the detection of enteric viruses include radioimmunoassays (RIA) and enzyme-linked immunosorbent assays (ELISAs). Rotavirus may be detected by ELISA or by means of commercially available latex agglutination assay kits (Lipson and Zelinski-Papez, 1989; Dahling et al., 1993). Several latex agglutination kits have been comparatively tested (Lipson and Zelinsky-Papez, 1989; Dahling et al., 1993; Thomas et al., 1994). Hepatitis A virus antigen has been directly detected in shellfish (Zivanovic-Marinkovic et al., 1984; Desenclos et al., 1991) and water (Nasser and Metcalf, 1987) by ELISAs and in water by RIA (Biziagos et al., 1988). The sensitivity of RIAs and ELISAs is insufficient to directly detect low to moderate numbers of viruses in food extracts or water concentrates; however, these methods can be readily used to screen for viruses replicated in cell cultures or in clinical specimens.

## Controls and intervention strategies

### Pre-harvest contamination controls

There are numerous intervention strategies that can be used to curb enteric virus contamination of foods which may become contaminated either pre- or post-harvest (Richards, 2002). Pre-harvest sources of contamination include nonpotable irrigation waters, fertilization with sewage sludge, and fecal pollution of the areas from which the foods are obtained (Sadovski et al., 1978; Oron et al., 1995; Beuchat and Ryu, 1997). The use of wastewater for crop irrigation requires adequate treatment of the water to eliminate microbial contaminants and may be accomplished by primary and secondary treatment followed by adequate chlorination or ultraviolet light disinfection. Irrigation of food crops with wastewater should be discontinued several weeks prior to harvest to permit the natural inactivation of any residual viruses. The practice of fertilizing crops with sewage sludge may be common in some countries and can provide a ready supply of viruses along with undesirable heavy metals. Care should be taken to fertilize with sludge that is free from industrial wastes and low in chemical and biological contaminants. Fertilization with acceptable sludges should take place before planting so that the viruses may become inactivated before crop harvest. After irrigation with wastewater or fertilization with sewage sludge, produce should be washed thoroughly, especially if it is to be consumed raw or only lightly cooked. Peeling may be employed to physically remove surface contamination from some fruits and vegetables. Commercially available rinses are also available to disinfect the surfaces of fruits and vegetables; however, these treatments may not be effective on all viruses.

Pre-harvest contamination of molluscan shellfish has led to numerous outbreaks of hepatitis A and norovirus illness (Richards, 1985). Oysters, clams and mussels are filter feeders, capable of concentrating contaminants from the water column within their edible tissues, making them a highly publicized source of gastrointestinal illness. Adding insult to injury is the fact that molluscan shellfish are often eaten raw or only lightly cooked. Enteric viruses are inactivated by heat, so thorough cooking would provide greater assurances of product safety. Practical testing procedures for viruses in shellfish have not been developed or have not been standardized for routine use. Further research in this area is needed.

## Post-harvest contamination controls

A potential source of post-harvest contamination of produce is via the hands of the harvesters and handlers. Hand contamination was believed responsible for outbreaks of hepatitis A from lettuce (Rosenblum et al., 1990), raspberries (Reid and Robinson, 1987), and strawberries (Niu et al., 1992; Anonymous, 1997; Hutin et al., 1999); norovirus gastroenteritis from raspberries (Gaulin et al., 1999) and shellfish (Kohn et al., 1995; McDonnell et al., 1997); and rotavirus illness from lettuce (Hernandez et al., 1997). Sanitary hand washing facilities are essential and food harvesters and handlers should be encouraged to use them. Of particular concern are field-workers, who may not have access to proper toilet facilities or clean water. Sick food handlers are likely to be the leading cause of food contamination and have been linked to numerous outbreaks of enteric illness (White et al., 1996; Patterson et al., 1997a, 1997b; Daniels et al., 2000). Food handlers should not be allowed to work with foods or in restaurants while actively sick. Since virus shedding may continue for weeks after recovery (see Table 8.1), it is a good precaution to restrict sick employees from handling foods, dishes or utensils for at least 2 weeks after recovery.

Cross-contamination of clean products with contaminated products should be avoided. For instance, workers at seafood counters in grocery stores should not contaminate cooked fish or shellfish with raw seafoods. Even the smallest amount of "juice" dripped from raw product onto cooked, ready-to-eat product may be sufficient to elicit an outbreak. For example, the cross-contamination of salad by seafood caused one outbreak of norovirus illness (Griffin et al., 1982). Many foods are contaminated after they are cooked or processed. Once again, the hands of food preparers are often to blame. Bakery products, held in the hand while frosting or icing are applied, have been implicated in viral illness (Warburton et al., 1991; Weltman et al., 1996). Effective sanitation is essential in minimizing cross-contamination of food products. There is inadequate information on what chemicals inactivate enteric viruses, so in many processing environments, hot disinfection solutions sprayed on equipment offer the advantage of heat and chemical inactivation, as well as dilution, to combat a host of microbial contaminants. Disinfection and sanitation do not only apply to large equipment, but also apply to containers and bowls used to contain foods, work surfaces used for cutting and mixing foods, and utensils used for mixing and serving. Silverware and plates should be cleaned in approved dishwashers with water temperatures adequate for disinfection.

The lack of common sense is also a major factor in eliminating food contamination. An outbreak of Norwalk virus was attributed to the mixing of potato salad in a sink in which a kitchen worker had previously vomited (Patterson et al., 1997a). Likewise, fishermen contaminated their own catch as they discharged stool from sick workers overboard (Kohn et al., 1995; Berg et al., 2000). Businesses that allow or condone sick workers to handle foods can face stiff civil and criminal penalties for negligence when enteric virus-contaminated food causes illness or death.

Self-serve buffets and salad bars offer many opportunities for the spread of enteric viruses. Serving utensils are often too short and do not extend out of the food tray. As a result, patrons must reach into the tray to retrieve the utensils, thus contaminating the food in the process. Excessive handling of foods increases the opportunity of contamination and may be responsible for some of the norovirus outbreaks associated with foods. In recent outbreaks of norovirus illness on board cruise lines, fruits and vegetables are carved into creative forms and this extensive handling may contribute to the outbreaks.

Insects and vermin can also contribute to food contamination; however, there are no data available showing the extent to which insect or animal contamination leads to enteric virus illness. Food processing plants, restaurants, and grocery stores should make an effort to reduce the incidence of undesirable pests within the workplace. Unsanitary restrooms attract unwanted pests which may lead to product contamination.

## Disinfection

A number of studies were performed to evaluate potential chemical disinfectants for inactivating enteric viruses. Viruses may be inactivated by disruption of the capsid or denaturation of capsid proteins required for binding to specific cellular receptors. Inactivation may also be realized by damaging viral RNA or DNA; however, such damage is usually secondary to damage to the capsid proteins. The most effective method to measure virus inactivation is through infectivity studies. Hence, most studies have been performed with easily cultivatable viruses, like the picornaviruses (polio-, echo-, and coxsackieviruses), and cell culture-adapted strains of rotavirus or hepatitis A virus. Unfortunately, the noroviruses cannot be grown in cell culture, making infectivity studies impossible. It is not possible to differentiate infectious from inactivated viruses by PCR or RT-PCR, since chemically treated viruses may be inactivated by denaturation of the capsid proteins without any adverse effects on the DNA or RNA (Richards, 1999). Since binding of capsid proteins to specific cellular receptors is an integral part of the virus replication process, chemical or physical damage to capsid proteins is sufficient to render the virus inactive.

The high number of potential chemicals and the inability to assay some viruses have hampered a thorough investigation for effective disinfectants, thus leaving broad gaps in the inactivation data. In screening 27 chemical disinfectants for the reduction of rotavirus on surfaces, only nine were found effective: chloramine-T, chlorhexidine gluconate, glutaraldehyde, hydrochloric acid, isopropyl alcohol, peracetic acid, povidone iodine, quaterinary ammonium compound, and sodium-o-benzyl-p-chlorophenate (Lloyd-Evans et al., 1986). In another study, rotavirus was sensitive to o-phenylphenol and ethanol and chlorine bleach (Sattar et al., 1994). Rotavirus was readily inactivated by chlorine dioxide at alkaline pH (Chen and Vaughn, 1990), and by chlorine (Vaughn et al., 1986) and ozone (Vaughn et al., 1987), especially at neutral to acidic pH. A number of enteroviruses were inactivated by chlorine, chlorine dioxide, peracetic acid, and ozone (Harakeh and Butler, 1984). Iodine and chlorine dioxide were effective at alkaline pH in inactivating poliovirus (Alvarez and O'Brien, 1982). The pH can play a crucial role in inactivation effectiveness. For instance, a number of picornaviruses were 99% inactivated within 4.5 min by ~0.5 mg/L of free chlorine at pH 6.0 and 7.8, but inactivation at pH 10 required 21–96 min (Engelbrecht et al., 1980).

## Heating and cooking

Perhaps the most effective means to inactivate enteric viruses in foods is by thorough cooking which destabilizes viral capsid proteins causing breakdown of the capsid and release of viral nucleic acids. Times and temperatures of virus inactivation vary widely depending on the food matrix, the cooking method and, to a lesser extent, on the virus type. Overall, relatively little is known about the thermal stability of the enteric viruses in various foods. Studies with a cell culture-adapted strain of hepatitis A virus in skim milk, homogenized milk, and table cream showed a 5-log reduction in titer at 80°C after 0.68 min for the skim and homogenized milk and 1.24 min

for the cream, suggesting that the higher fat content of cream provides the virus with enhanced thermal stability (Bidawid et al., 2000). Early studies on hamburger showed that high fat content and protein levels confer thermal stability to poliovirus (Filippi and Banwart, 1974). The method of cooking can also influence virus inactivation. Studies performed on oysters spiked with $> 1.0 \times 10^4$ poliovirus per gram demonstrated: 87% inactivation of viruses after frying for 8 min to an internal temperature of 100°C, 87% inactivation after baking for 29 min to 90°C, 93% inactivation after steaming for 30 min to 97°C, and 90% inactivation after stewing for 8 min to 75°C (DiGirolamo et al., 1970).

## Irradiation

Three forms of irradiation are practiced in food processing: gamma, ultraviolet (UV), and microwave irradiation. Gamma irradiation is effective for inactivating enteric viruses in foods (Patterson, 1993; Monk et al., 1995; Farkus, 1998), although data are limited on the effects of irradiation on most of the viruses. Studies with hepatitis A, rotavirus, and poliovirus in shellfish demonstrated 90% inactivation with ≤ 3.1 kilograys (kGy) (Mallet et al., 1991). Higher dosages (6.8 kGy) were required to inactivate coxsackievirus from ground beef (Sullivan et al., 1973). A 90% inactivation was observed for poliovirus in water with just 1.92 kGy (Kaupert et al., 1999). Gamma irradiation penetrates the food to inactivate microorganisms throughout the product and can be used for products already packaged. Since foods can become contaminated with high virus levels, reports showing a 1 $\log_{10}$ reduction in infectious virus titer is insufficient to conclude the effectiveness of gamma irradiation as a food processing strategy. Another form of disinfection is UV irradiation which can be used to inactivate viruses in water and on surfaces (Chang et al., 1985; Slade et al., 1986; Ojeh et al., 1995; Gerba et al., 2002; Thurston-Enriquez et al., 2003). The most effective wavelengths are generally between the range of 245–285 nm, with 254 nm considered optimum. UV is also effective for inactivating viruses on the surfaces of food and processing equipment. The dose of UV to inactivate rotavirus and poliovirus is 3–4 times greater than is required for the inactivation of E. coli (Chang et al., 1985). Microwave energy is another form of irradiation effective for inactivating viruses due to the heat produced in the process and the general heat denaturation of proteins. Hepatitis A virus was inactivated in microwaved foods (Mishu et al., 1990), while poliovirus was inactivated in infant formula (Kindle et al., 1986). It is uncertain whether the microwave energy has any direct effect on the viral nucleic acids other than the heating effect common in microwave processing.

## Dessication and dehydration

Limited information is available on the effects of dehydration or freeze drying on enteric virus infectivity. One study from 1969 indicated the inactivation of 3–4 $\log_{10}$ of poliovirus in potato salad, salmon salad, beef pot roast, beef and vegetables, and chicken with gravy (Heidelberg and Giron, 1969). An evaluation of virus inactivation on fomites, hands, and in the air indicated a general tendency of viruses to persist longer in high relative humidity environments and at low temperatures (Ijaz, 1985; Abad *et al.*, 1994; 2001).

## Depuration and relaying

The origin of depuration dates back to late nineteenth-century France, where the term "degorgeoirs" was used (Herdman and Scott, 1896; Herdman and Boyce, 1899) to describe the first depuration process. Depuration, also called "controlled purification," is a commercial process where shellfish are placed in tanks of clean seawater and permitted to purge contaminants for 3–4 days (reviewed by Richards, 1988, 1991). The tank water can be continuously replaced with fresh seawater in a flow-through system, replaced at intervals with clean seawater in what is termed a batch-process, or it can be recirculated. Recirculated systems rely on disinfection of the water using UV irradiation, ozone, and occasionally chlorine. As microbial contaminants are purged from the gut into the water, they are effectively removed by any of these disinfection means. Depuration is widely practiced throughout Europe, but it is relatively uncommon in the United States, where an ample supply of clean shellfish is available. There have been many studies demonstrating the effectiveness of depuration in reducing some bacterial pathogens from shellfish (reviewed by Richards 1988, 1991); however, depuration is less effective for reducing enteric viruses from shellfish. Most of the enteric viruses tested to date, such as poliovirus, hepatitis A virus, and noroviruses, resist depuration in oysters and clams (DiGirolamo *et al.*, 1970; Grohmann *et al.*, 1981; Gill *et al.*, 1983; Ang 1998; Schwab *et al.*, 1998; Formiga-Cruz *et al.*, 2002) and in mussels (Power and Collin, 1989; Franco *et al.*, 1990; Enriquez *et al.*, 1992; De Medici *et al.*, 2001; Chironna *et al.*, 2002). Studies have shown that viruses migrate from the gut into the epithelial cells lining the gut and into other tissues (Hay and Scotti, 1986; Schwab *et al.*, 1998), thus minimizing the effectiveness of depuration. Although the depuration process reduces virus levels as some viruses pass through the digestive tract of shellfish, viruses that become integrated into the tissues are refractive to depuration and pose a potential threat to the shellfish consumer.

Shellfish relaying is the practice of relocating shellfish from contaminated growing waters to areas that meet acceptable water quality standards (reviewed by Richards, 1988). Relayed shellfish are allowed to purge contaminants in the natural environment, generally for 10–14 days. Shellfish may be spread over the ocean floor or, for simplification of harvest, placed in baskets, racks, or other types of containers which are either placed on the ocean bottom or suspended in the water column. Shellfish naturally purge some of the microbial contaminants if the water remains clean and the environmental conditions are suitable during the duration of the relay operation. The practice of relaying has the potential of significantly reducing the levels of viral contaminants if the water quality is maintained throughout the period and shellfish are permitted sufficient time to purge the contaminants. Unfortunately,

enteric viruses may persist in molluscan shellfish for extended periods (Sobsey *et al.*, 1987; Richards, 1988; Enriquez *et al.*, 1992; Kingsley and Richards, 2003), thus rendering relaying less than fully effective in eliminating all virus particles.

## High hydrostatic pressure processing

High hydrostatic pressure processing is effective for inactivating rotaviruses (Pontes *et al.*, 1997; Khadre and Yousef, 2002), hepatitis A virus, and the norovirus surrogate, feline calicivirus (Kingsley *et al.*, 2002). This method has been commercially successful for inactivating vibrios and most vegetative organisms from oysters and other foods by subjecting the food to about 30,000 psi of pressure [which equals 207 megapascals (MPa), 2040 atmospheres, 2110 kg/cm$^2$] for 5 min. Recently, it was shown that 275 MPa for 5 min was sufficient to inactivate 7-log$_{10}$/mL of feline calicivirus, but higher pressures (450 MPa for 5 min) were required to eliminate 7 log$_{10}$/mL of hepatitis A virus (Kingsley *et al.*, 2002). Eight log$_{10}$ TCID$_{50}$/mL of human rotaviruses were inactivated with 300 MPa for 2 min, although some rotaviruses were resistant to pressures as high as 800 MPa for 10 min (Khadre and Yousef, 2002). Large-scale trials are needed to determine the effectiveness of commercial high pressure processing in eliminating enteric viruses from shellfish and other foods.

## Conclusion

Although there are many food- and water-borne enteric viruses, our ability to control them is limited. The incidence of enteric disease may be combated by enhanced vigilance and better education of food handlers. Improved analytical methods are being developed each day, allowing health and regulatory authorities to better monitor the food supply. In time, more practical, cost-effective, and rapid procedures will make monitoring of foods and water more routine. Until then, the tenets of good sanitation and food processing techniques can reduce the incidence of human illness. Restricting ill workers from handling foods would also have a major impact in reducing outbreaks of enteric illness.

## References

Abad, F.X., Pinto, R.M., and Bosch, A. 1994. Survival of enteric viruses on environmental fomites. Appl. Environ. Microbiol. 60: 3704–3710.

Abad, F.X., Pinto, R.M., Villena, C., Gajardo, R., and Bosch, A. 1997. Astrovirus survival in drinking water. Appl. Environ. Microbiol. 63: 3119–3122.

Abad, F.X., Villena, C., Guix, S., Caballero, S., Pinto, R.M., and Bosch, A. 2001. Potential role of fomites in the vehicular transmission of human astroviruses. Appl. Environ. Microbiol. 67: 3904–3907.

Agbalika, F., Hartemann, P., and Foliguet, J.M. 1984. Trypsin-treated Ma-104: a sensitive cell line for isolating enteric viruses from environmental samples. Appl. Environ. Microbiol. 47: 378–380.

Agus, S.G., Dolin, R., Wyatt, R.G., Tousimis, A.J., and Northrup, R.S. 1973. Acute infectious nonbacterial gastroenteritis: intestinal histopathology. Histologic and enzymatic alterations during illness produced by the Norwalk agent in man. Ann. Intern. Med. 79: 18–25.

Alvarez, M.E., and O'Brien, R.T. 1982. Mechanisms of inactivation of poliovirus by chlorine dioxide and iodine. Appl. Environ. Microbiol. 44: 1064–1071.

Anderson, D.A., and Shrestha, I.I. 2002. Hepatitis E virus. In: Clinical Virology, 2nd ed. D.D. Richman, R.J. Whitney, and F.G. Hayden, eds. ASM Press, Washington, D.C. p. 1061–1074.

Ando, T., Monroe, S.S., Gentsch, J.R., Jin, Q., Lewis, D.C., and Glass, R.I. 1995. Detection and differentiation of antigenically distinct small round-structured viruses (Norwalk-like viruses) by reverse transcription-PCR and Southern hybridization. J. Clin. Microbiol. 33: 64–71.

Ang, L. H. 1998. An outbreak of viral gastroenteritis associated with eating raw oysters. Commun. Dis. Public Health 1: 38–40.

Anonymous. 1997. Hepatitis A associated with consumption of frozen strawberries – Michigan, March 1997. Morbid. Mortal. Wkly. Rept. 46: 288, 295.

Anonymous. 1998. Prevention of rotavirus disease: guidelines for use of rotavirus vaccine. American Academy of Pediatrics. Pediatrics. 102: 1483–1491.

Anonymous. 2000. International Committee on Taxonomy of Viruses and International Union of Microbiological Societies. Virology Division. Family: Picornaviridae. In: Virus Taxonomy: Classification and Nomenclature of Viruses, Seventh Report of the International Committee on Taxonomy of Viruses. Academic Press, San Diego, CA. p. 657–678.

Asher, L.V., Binn, L.N., Mensing, T.L., Marchwicki, R.H., Vassell, R.A., and Young, G.D. 1995. Pathogenesis of hepatitis A in orally inoculated owl monkeys (Aotus trivirgatus). J. Med. Virol. 47: 260–268.

Atmar, R.L., and Estes, M.K. 2001. Diagnosis of non-cultivatable gastroenteritis viruses, the human caliciviruses. Clin. Microbiol. Rev. 14: 15–37.

Atmar, R.L., Neill, F.L., Romalde, J.L., Le Guyader, F., Woodley, C.M., Metcalf, T.G., and Estes, M.K. 1995. Detection of Norwalk virus and hepatitis A virus in shellfish tissues with PCR. Appl. Environ. Microbiol. 61: 3014–3018.

Badawy, A.S., Gerba, C.P., and Kelly, L.M. 1985. Development of a method for recovery of rotavirus from the surface of vegetables. J. Food Prot. 48: 261–264.

Barardi. C.R., Yip, H., Emsile, K.R., Vesey, G., Shanker, S.R., and Williams, K.L. 1999. Flow cytometry and RT-PCR for rotavirus detection in artificially seeded oyster meat. Int. J. Food. Microbiol. 49: 9–18.

Baron, A.L., Olshevsky, C., and Cohen, M.M. 1970. Characteristics of the BGM line of cells from African green monkey kidney. Brief report. Arch. Ges. Virusforsch. 32: 389–392.

Beaulieux, F., See, D.M., Leparc-Goffart, I., Aymard, M., and Lina, B. 1997. Use of magnetic beads versus guanidium thiocyanate-phenol-chloroform RNA extraction followed by polymerase chain reaction for the rapid, sensitive detection of enterovirus RNA. Res. Virol. 148: 11–15.

Benjelloun, S., Bahbouhi, B., Bouchrit, N., Cherkaoui, L., Hda, N., Mahjour, J., and Benslimane, A. 1997. Seroepidemiological study of an acute hepatitis E outbreak in Morocco. Res. Virol. 148: 279–287.

Berg, G.E., Kohn, M.A., Farley, T.A., and McFarland, L.M. 2000. Multi-state outbreaks of acute gastroenteritis traced to fecal-contaminated oysters harvested in Louisiana. Infect. Dis., Suppl. 2, 181: S381-S386.

Berg, G., Safferman, R.S., Dahling, D.R., D. Berman, D., and Hurst, C.J. 1987a. USEPA Manual of Methods for Virology. Cell Culture Preparation and Maintenance, EPA/600/4–013 (R9), Environmental Monitoring and Support Laboratory, US Environmental Protection Agency, Cincinnati, OH. Chapter 9, p. 1–8.

Berg, G., Safferman, R.S., Dahling, D.R., D. Berman, D., and Hurst, C.J. 1987b. USEPA Manual of Methods for Virology. Cell Culture Procedures for Assaying Plaque-Forming Viruses, EPA/600/4–013 (R10), Environmental Monitoring and Support Laboratory, US Environmental Protection Agency, Cincinnati, OH. Chapter 10, p. 1–11.

Berg, G., Safferman, R.S., Dahling, D.R., D. Berman, D., and Hurst, C.J. 1988a. USEPA Manual of Methods for Virology. Virus Plaque Confirmation Procedure, EPA/600/4–013 (R11), Environmental Monitoring and Support Laboratory, US Environmental Protection Agency, Cincinnati, OH. Chapter 11, p. 1–3.

Berg, G., Safferman, R.S., Dahling, D.R., Berman, D., and Hurst, C.J. 1988b. USEPA Manual of Methods for Virology. Identification of Enteroviruses, EPA/600/4–013 (R12), Environmental Monitoring and Support Laboratory, US Environmental Protection Agency, Cincinnati, OH. Chapter 12, pp. 1–4.

Beuchat, R.L., and Ryu, J.H.. 1997. Produce handling and processing practices. Emerg. Infect. Dis. 3: 459–465.

Beuret, C. 2003. A simple method for isolation of enteric viruses (noroviruses and enteroviruses) in water. J. Virol. Methods 107: 1–8.

Bidawid, S, Farber, J.M., Satter, S.A., and Hayward, S. 2000. Heat inactivation of hepatitis A virus in dairy foods. J. Food Prot. 63: 522–528.

Bishop, R.F. 1994. Natural history of rotavirus infections. In: Viral Infections of the Gastrointestinal Tract. A.Z. Kapikian, ed., 2nd ed., New York, Marcel Dekker, p. 131–167.

Bishop, R.F., Davidson, G.P., Holmes, I.H., and Ruck, B.J. 1973. Virus particles in epithelial cells of duodenal mucosa from children with acute nonbacterial gastroenteritis. Lancet 2: 1281–1283.

Biziagos, E., Passagot, J., Crance, J.M., and Deloince, R. 1988. Long-term survival of hepatitis A virus and poliovirus type 1 in mineral water. Appl. Environ. Microbiol. 54: 2705–2710.

Blacklow, N.R., Cukor, G., Bedegian, M.K., Echeverria, P., Greenberg, H.B., Schreiber, D.S., and Trier, J.S. 1979. Immune response and prevalence of antibody to Norwalk enteritis virus as determined by radioimmunoassay. J. Clin. Microbiol. 10: 903–909.

Blacklow, N.R., Dolin, R., Fedson, D.S., DuPont, H., Northrup, R.S., Hornick, R.B., and Chanock, R.M. 1972. Acute infectious nonbacterial gastroenteritis: etiology and pathogenesis. Ann. Intern. Med. 76: 993–1008.

Brandt, C.D., Kim, H.W., Rodriguez, W.J., Arrobio, J., Jeffries, B.C., Stalling, E.P., Lewis, C., Miles, A.J., Gardner, M.K., and Parrot, R.H. 1985. Adenoviruses and pediatric gastroenteritis. J. Infect. Dis. 151: 437–443.

Brinker, J.P., Blacklow, N.R., Estes, M.K., Moe, C.L., Schwab, K.J., and Herrmann, J.E. 1998. Detection of Norwalk virus and other genogroup 1 human caliciviruses by a monoclonal antibody, recombinant-antigen-based immunoglobulin M capture enzyme immunoassay. J. Clin. Microbiol. 36: 1064–1069.

Carducci, A., Casini, B., Bani, A., Rovini, E., Verani, M., Mazzoni, F., and Giuntini, A. 2003. Virological control of groundwater quality using bimolecular tests. Water Sci. Technol. 47: 261–266.

Castignolles, N., Petit, F., Mendel, I., Simon, I., Cattolico, L., and Buffet-Janvresse, C. 1998. Detection of adenovirus in waters of the Seine River estuary by nested-PCR. Mol. Cell. Probes 12: 175–180.

Cauchi, M.R., Doultree, J.C., Marshall, J.A., and Wright, P.J. 1996. Molecular characterization of Camberwell virus and sequence variation in ORF3 of small round-structured (Norwalk-like) viruses. J. Med. Virol. 49: 70–76.

Chang, J.C.H., Ossoff, S.F., Lobe, D.C., Dorfman, M.H., Dumais, C.M., Qualls, R.G., and Johnson, J.D. 1985. UV inactivation of pathogenic and indicator microorganisms. Appl. Environ. Microbiol. 49: 1361–1365.

Chang, S.L., Berg, G., Busch, K.A., Stevenson, R.E., Clarke, N.E., and Kabler, P.W. 1958. Application of the most probable number method for estimating concentrations of animal viruses by the tissue culture technique. Virology 6: 27–42.

Chapron, C.D., Ballester, N.A., Fontaine, J.H., Frades, C.N., and Margolin, A.B. 2000. Detection of astroviruses, enteroviruses, and adenovirus types 40 and 41 in surface waters collected and evaluated by the information collection rule and an integrated cell culture-nested PCR procedure. Appl. Environ. Microbiol. 66: 2520–2525.

Chen, T.R. 1988. Re-evaluation of HeLa, HeLa S3, and Hep-2 karyotypes. Cytogenet. Cell Genet. 48: 19–24.

Chen, Y.S., and Vaughn, J.M. 1990. Inactivation of human and simian rotaviruses by chlorine dioxide. Appl. Environ. Microbiol. 56: 1363–1366.

Chiba, S., Sakuma, Y., Kogasaka, R., Akihara, M., Horino, K., Nakao, T., and Fukui, S. 1979. An outbreak of gastroenteritis associated with calicivirus in an infant home. J. Med. Virol. 4: 249–254.

Chironna, M., Germinario, C., De Medici, D., Fiore, A., Di Pasquale, S., Quarto, M., and Barbuti, S. 2002. Detection of hepatitis A virus in mussels from different sources marketed in Puglia region (South Italy). Int. J. Food Microbiol. 75: 11–18.

Cho, H.B., Lee, S.H., Cho, J.C., and Kim, S.J. 2000. Detection of adenoviruses and enteroviruses in tap water and river water by reverse transcription multiplex PCR. Can. J. Microbiol. 46: 417–424.

Clarke, I.N., and Lambden, P.R. 2000. Organization and expression of Calicivirus genes. J. Infect. Dis. 181: S309-S316.

Clayson, E.T., Myint, K.S., Snitbhan, R., Vaughn, D.W., Innis, B.L., Chan, L., Cheung, P., and Shrestha, M.P. 1995. Viremia, fecal shedding and IgM and IgG responses in patients with hepatitis E. J. Infect. Dis. 172: 927–933.

Conner, M.E., and Ramig, R.F. 1997. Viral enteric diseases. In: Viral Pathogenesis. N. Nathanson, ed. Lippincott-Raven Publishers, New York. p. 713–743.

Corey, L. 1990. Hepatitis viruses. In: Medical Microbiology. An Introduction to Infectious Diseases. J.C. Sherris, ed. Elsevier, New York, p. 547–558.

Corwin, A.L., Khiem, H.B., Clayson, E.T., Pham, K.S., Vo, T.T., Vu, T.Y., Cao, T.T., Vaughn, D., Merven, J., Richie, T.L., Putri, M.P., He, J., Graham, R., Wignall, F.S., and Hyams, K.C. 1996. A water-borne outbreak of hepatitis E virus transmission in southwestern Vietnam. Am. J. Trop. Med. Hyg. 54: 559–562.

Corwin, A., Putri, M.P., Winarno, J., Lubis, I., Suparmanto, S., Sumardiati, A., Laras, K., Tan, R., Master, J., Warner, G., Wignall, F.S., Graham, R., and Hyams, K.C. 1997. Epidemic and sporadic hepatitis E virus transmission in West Kalimantan (Borneo), Indonesia. Am. J. Trop. Med. Hyg. 57: 62- 65.

Coulson, B.S., Grimwood, K., Hudson, I.L., Barnes, G.L., Bishop, R.F. 1992. Role of coproantibody in clinical protection of children during reinfection with rotavirus. J. Clin. Microbiol. 30: 1678–1684.

Croci, L., De Medici, D., Morace, G., Fiore, A., Scalfaro, C., Beneduce, F., and Toti, L. 1999. Detection of hepatitis A virus in shellfish by nested reverse transcription-PCR. Int. J. Food Microbiol. 48: 67–71.

Cromeans, T.L., Nainan, O.V., and Margolis, H.S. 1997. Detection of hepatitis A virus RNA in oyster meat. Appl. Environ. Microbiol. 63: 2460–2463.

Cromeans, T., Sobsey, M.D., and Fields, H.A.. 1987. Development of a plaque assay for a cytopathic, rapidly replicating isolate of hepatitis A virus. J. Med. Virol. 22: 45–56.

Dahling, D.R., and Wright, B.A. 1986. Optimization of the BGM cell line culture and viral assay procedures for monitoring viruses in the environment. Appl. Environ. Microbiol. 51: 790-812.

Dahling, D.R., Berg, G., and Berman, D. 1974. BGM, a continuous cell line more sensitive than primary rhesus and African green kidney cells for the recovery of viruses from water. Health Lab. Sci. 11: 275–282.

Dahling, D.R., Wright, B.A., and Williams F.P. Jr. 1993. Detection of viruses in environmental samples: suitability of commercial rotavirus and adenovirus test kits. J. Virol. Methods 45: 137–147.

Daniels, N.A., Bergmire-Sweat, D.A., Schwab, K.J., Hendricks, K.A., Reddy, S., Rowe, S.M., Fankhauser, R.L., Monroe, S.S., Atmar, R.L., Glass, R.I., and Mead, P. 2000. A foodborne outbreak of gastroenteritis associated with Norwalk-like viruses: First molecular trace back to deli sandwiches contaminated during preparation. J. Infect. Dis. 181: 1467–1470.

Davidson, G.P., and Barnes, G.L. 1979. Structural and functional abnormalities of the small intestine in infants and young children with rotavirus gastroenteritis. Acta. Paediatr. Scand. 68: 181–186.

Day, S.P., Murphy, P., Brown, E.A., and Lemon, S.M. 1992. Mutations within the 5′ nontranslated region of hepatitis A virus RNA which enhances replication in BS-C-1 cells. J. Virol. 66: 6533–6540.

De Medici, D., Ciccozzi, M., Fiore, A., Di Pasquale, S., Parlato, A., Ricci-Bitti, P., and Croci, L. 2001. Closed-circuit system for the depuration of mussels experimentally contaminated with hepatitis A virus. J. Food Prot. 64: 877–880.

Deng, M.Y., Day, S.P., and Cliver, D.O. 1994. Detection of hepatitis A virus in environmental samples by antigen-capture PCR. Appl. Environ. Microbiol. 60: 1927–1933.

Denisova, E., Dowling, W., LaMonica, R., Shaw, R., Scarlata, S., Ruggeri, F., and Mackow, E.R. 1999. Rotavirus capsid protein VP5* permeabilizes membranes. J. Virol. 73: 3147–3153.

Desenclos, J.C., Klontz, K.C., Wilder, M.H., Nainan, O.V., Margolis, H.S., and Gunn, R.A. 1991. A multistate outbreak of hepatitis A caused by the consumption of raw oysters. Am. J. Public Health 81: 1268–1272.

DiGirolamo, R., Liston, J., and Marches, J.R. 1970. Survival of virus in chilled, frozen and processed oysters. Appl. Microbiol. 20: 58–63.

Dingle, K.E., Lambden, P.R., Caul, E.O., and Clarke, I.N. 1995. Human enteric Caliciviridae: the complete genome sequence and expression of virus-like particles from a genetic group II small round structured virus. J. Gen. Virol. 76: 2349–2355.

Divizia, M., Ruscio, V., Degener, A.M., and Pana, A. 1998. Hepatitis A virus detection in wastewater by PCR and hybridization. New Microbiol. 21: 161–167.

Dolin, R., Blacklow, N.R., DuPont, H., Buscho, R.F., Wyatt, R.G., Kasel, J.A., Hornick, R., and Chanock, R.M. 1972. Biological properties of Norwalk agent of acute infectious nonbacterial gastroenteritis. Proc. Soc. Exp. Biol. Med. 140: 578–583.

Dolin, R., Levy, Wyatt, R.G., Thornhill, T.S., and Gardner, J.D. 1975. Viral gastroenteritis induced by the Hawaii agent: jejunal histopathology and seroresponse. Am. J. Med. 59: 761-769.

Dolin, R., Reichman, R.C., Roessner, K.D., Tralka, T.S., Schooley, R.T., Gary, W., and Morens, D. 1982. Detection by immune electron microscopy of the Snow Mountain agent of acute viral gastroenteritis. J. Infect. Dis. 146: 184–189.

Dolin, R., Roessner, K.D., Treanor, J.J., Reichman, R.C., Phillips, M., and Madore, H.P. 1986. Radioimmunoassay for detection of the Snow Mountain Agent of viral gastroenteritis. J. Med. Virol. 19: 11–18.

Dubois, E., Agier, C., Traore, O., Hennechart, C., Merle, G., Cruciere, C., and Laveran, H. 2002. Modified concentration method for the detection of enteric viruses on fruits and vegetables by reverse transcriptase-polymerase chain reaction or cell culture. J. Food Prot. 65: 1962–1969.

Dudding, B.A., Top, F.H., Winter, P.E., Buescher, E.L., Lamson, T.H., and Leibovitz, A. 1973. Acute respiratory disease in military trainees. Am. J. Epidemiol. 97: 187–198.

Egger, D., Pasamontes, L., Ostermayer, M., and Bienz, K. 1995. Reverse transcription multiplex PCR for differentiation between polio- and enteroviruses from clinical and environmental samples. J. Clin. Microbiol. 33: 1442–1447.

Engelbrecht, R.S., Weber, M.J., Salter, B.L., and Schmidt, C.A. 1980. Comparative inactivation of viruses by chlorine. Appl. Environ. Microbiol. 40: 249–256.

Enriquez, R., Frosner, G.G., Hochstein-Mintzel, V., Riedemann, S., and Reinhardt, G. 1992. Accumulation and persistence of hepatitis A virus in mussels. J. Med. Virol. 37: 174–179.

Farkas, J. 1998. Irradiation as a method for decontaminating food. a review. Int. J. Food Microbiol. 44: 189–204.

Feinstone, S. 1986. Hepatitis A. Prog. Liver Dis. 8: 299–310.

Filippi, J.A., and Banwart, G.J. 1974. Effect of the fat content of ground beef on the heat inactivation of poliovirus. J. Food Sci. 39: 865–868.

Finn, A., Anday, E., and Talbot, G.H. 1988. An epidemic of adenovirus 7a infection in a neonatal nursery: course, morbidity, and management. Infect. Control Hosp. Epidemiol. 9: 398–404.

Flehmig, B. 1981. Hepatitis A virus in cell culture. II. Growth characteristics of hepatitis A virus in FRhK-4/R cells. Med. Microbiol. Imunol. (Berl) 170: 73–81.

Flehmig, B., Vallbracht, A., and Wurster, G. 1981. Hepatitis A virus in cell culture. III. Propagation of hepatitis A virus in human embryo kidney cells and human embryo fibroblast strains. Med. Microbiol. Immunol. (Berl.) 170: 83–89.

Fogh, J., Fogh, J.M., and Orfeo, T. 1977. One hundred and twenty-seven human tumor cell lines producing tumors in nude mice. J. Natl. Cancer Inst. 59: 221–226.

Formiga-Cruz, M., Tofino-Quesada, G., Bofill-Mas, S., Lees, D.N., Henshilwood, K., Allard, A.K., Conden-Hansson, A.C., Hernroth, B.E., Vantarakis, A., Tsibouxi, A., Papapetropoulou, M., Furones, M.D., and Girones, R. 2002. Distribution of human virus contamination in shellfish from different growing areas in Greece, Spain, Sweden, and the United Kingdom. Appl. Environ. Microbiol. 68: 5990–5998.

Fout, G.S., Schaefer F.W. III, Messer, J.W., Dahling, D.R., and Stetler, R.E. 1996. ICR Microbial Laboratory Manual. EPA/600/R-95/178. National Exposure Research Laboratory, Office of Research and Development, US Environmental Protection Agency, Cincinnati, OH.

Franco, M.A., and Greenberg, H.B. 2002. Rotaviruses. In: Clinical Virology. D.D. Richman, R.J. Whitley, and F.G. Hayden, eds. ASM Press, Washington, D.C. p. 743–762.

Franco, E., Toti, L., Gabrieli, R., Croci, L., De Medici, D., and Pana, A. 1990. Depuration of Mytilus galloprovincialis experimentally contaminated with hepatitis A virus. Int. J. Food Microbiol. 11: 321–327.

Gary, G.W. Jr., Kaplan, J.E., Stine, S.E., and Anderson, L.J. 1985. Detection of Norwalk virus antibodies and antigen with a biotin-avidin immunoassay. J. Clin. Microbiol. 22: 274–278.

Gantzer, C., Senouci, S., Maul, A., Levi, Y., and Schwartzbrod, L. 1997. Enterovirus genomes in wastewater: concentration on glass wool and glass powder and detection by RT-PCR. J. Virol. Methods 65: 265–271.

Gaulin, C.D., Ramsay, D., Cardinal, P., and D'Halevyn, M.A.. 1999. Epidemic of gastroenteritis of viral origin associated with eating imported raspberries. Can. J. Public Health 90: 37–40.

Gerba, C.P., Gramos, D.M., and Nwachuka, N. 2002. Comparative inactivation of enteroviruses and adenovirus 2 by UV light. Appl. Environ. Microbiol. 68: 5167–5169.

Giard, D.J., Aaronson, S.A., Todaro, G.J., Srnstein, P., Kersey, J.H., Dosik, H, and Parks, W.P. 1973. In vitro cultivation of human tumors: establishment of cell lines derived from a series of solid tumors. J. Natl. Cancer Inst. 51: 1417–1423.

Gilgen, M., Germann, D., Luthy, J., and Hubner, P. 1997. Three-step isolation method for sensitive detection of enterovirus, rotavirus, hepatitis A virus, and small round structured viruses in water samples. Int. J. Food Microbiol. 37: 189–199.

Gill, O.N., Cubitt, W.D., McSwiggan, D.A., Watney, B.M., and Bartlett, C.L. 1983. Epidemic of gastroenteritis caused by oysters contaminated with small round structured viruses. Br. Med. J. (Clin. Res. Ed.) 287: 1532–1534.

Glass, R.I., Noel, J., Mitchell, D., Herrmann, J.E., Blacklow, N.R., Pickering, L.K., Dennehy, P., Ruiz-Palacios, G., de Guerrero, M.L., and Monroe, S.S. 1996. The changing epidemiology of astrovirus-associated gastroenteritis: a review. Arch. Virol. Suppl. 12: 287–300.

Goswami, B.B., Kulka, M., Ngo, D., Istafanos, P., and Cebula, T.A. 2002. A polymerase chain reaction-based method for the detection of hepatitis A virus in produce and shellfish. J. Food Prot. 65: 393–402.

Grabow, W.O., Gauss-Muller, V., Prozesky, O.W., and Deinhardt, F. 1983. Inactivation of hepatitis A virus and indicator organisms in water by free chlorine residues. Appl. Environ. Microbiol. 46: 619–624.

Graff, J., Kasang, C., Normann, A., Pfisterer-Hunt, M., Feinstone, S.M., and Flehmig, B. 1994. Mutational events in consecutive passages of hepatitis A virus strain GBM during cell culture adaptation. Virol. 204: 60–68.

Graff, J., Normann, A., and Flehmig, B. 1997. Influence of the 5' noncoding region of hepatitis A virus strain GBM on its growth in different cell lines. J. Gen. Virol. 78: 1841–1849.

Graham, D.Y., Jiang, X., Tanaka, T., Opekun, A.R., Madore, H.P., and Estes, M.K. 1994. Norwalk virus infection of volunteers: new insights based on improved assays. J. Infect. Dis. 170: 34–43.

Graham, F.L., Smiley, J., Russell, W.C., and Nairn, R. 1977. Characteristics of a human cell line transformed by DNA from human adenovirus type 5. J. Gen. Virol. 36: 59–74.

Gray, J.J., Cunliffe, C., Ball, J., Graham, D.Y., Desselberger, U., and Estes, M.K. 1994. Detection of immunoglobulin M (IgM), IgA, and IgG Norwalk virus-specific antibodies by indirect enzyme-linked immunosorbent assay with baculovirus-expressed Norwalk virus capsid antigen in adult volunteers challenged with Norwalk virus. J. Clin. Microbiol. 32: 3059–3063.

Green, K.Y., Ando, T., Balayan, M.S., Berke, T., Clarke, I.N., Estes, M.K., Matson, D.O., Nakata, S., Neill, J.D., Studdert, M.J., and Thiel, H.J. 2000. Taxonomy of the caliciviruses. J. Infect. Dis. 181: S322-S330.

Green, K.Y., Chanock, R.M., and Kapikian, A.Z. 2001. Human caliciviruses. In: Fields Virology. D.M. Knipe and P.M. Howley, eds. Vol. 1. Lippincott Williams & Wilkins, Philadelphia, PA. p. 841–874.

Greenberg, H.B., Wyatt, R.G., Kalica, A.R., Yolken, R.H., Black, R., Kapikian, A.Z., and Chanock, R.M. 1981. New insights in viral gastroenteritis. Prospect. Virol. 11: 163–187.

Greenberg, H.B., Wyatt, R.G., Valdesuso, J., Kalica, A.R., London, W.T., Chanock, R.M., and Kapikian, A.Z. 1978. Solid-phase microtiter radioimmunoassay for detection of the Norwalk strain of acute nonbacterial epidemic gastroenteritis virus and its antibodies. J. Virol. Methods 2: 97–108.

Griffin, M.R., Surowiec, J.J., McCloskey, D.I., Capuano, B., Pierzynski, B., Quinn, M., Wojnarski, R., Parkin, W.E., Greenberg, H., and Gary, G.W. 1982. Foodborne Norwalk virus. Am. J. Epidemiol. 115: 178–184.

Grimm, A.C., and Fout, G.S. 2002. Development of a molecular method to identify hepatitis E virus in water. J. Virol. Methods. 101: 175–188.

Grohmann, G.S., Murphy, A.M., Christopher, P.J., Auty, E., and Greenberg, H.B. 1981. Norwalk virus gastroenteritis in volunteers consuming depurated oysters. Aust. J. Exp. Biol. Med. Sci. 59 (Pt. 2): 219–228.

Guttman-Bass, N., Tchorsk Y., and Marva, E. 1987. Comparison of methods for rotavirus detection in water and results of a survey of Jerusalem wastewater. Appl. Environ. Microbiol. 53: 761–767.

Haflinger, D., Gilgen, M., Luthy, J., and Hubner, P. 1997. Seminested RT-PCR systems for small round structured viruses and detection of enteric viruses in seafoods. Int. J. Food Microbiol. 37: 27–36.

Harakeh, M., and Butler, M. 1984. Inactivation of human rotavirus SA11 and other enteric viruses in effluent by disinfectants. J. Hyg. (Camb.) 93: 157–163.

Hardy, M.E., and Estes, M.K. 1996. Completion of the Norwalk virus genome sequence. Virus Genes 12: 289–292.

Hay, B., and Scotti, P. 1986. Evidence for intracellular adsorption of virus by the Pacific oyster, Crassostrea gigas. N.Z. J. Marine Freshwater Res. 20: 655–659.

Heidelberg, N.D. and Giron, D.J. 1969. Effects of processing on recovery of polio virus from inoculated foods. J. Food Sci. 34: 239–241.

Herdman, W. A, and Boyce, R. 1899. Oysters and disease. An account of certain observations upon the normal and pathological histology and bacteriology of the oyster and other shellfish. Lancashire Sea-Fisheries Memoir No. 1, London, p. 35–40.

Herdman, W. A., and Scott, A. 1896. Report on the investigation carried on in 1895 in connection with the Lancashire Sea-Fisheries Laboratory at the University College, Liverpool. Proc. Trans. Liverpool Biol. Soc. 10: 103–174.

Herrmann, J.E., Nowak, N.A., and Blacklow, N.R. 1985. Detection of Norwalk virus in stools by enzyme immunoassay. J. Med. Virol. 17: 127–133.

Hernandez, F., Monge, R., Jimenez, C., and Taylor, L. 1997. Rotavirus and hepatitis A virus in market lettuce. Int. J. Food Microbiol. 37: 221–223.

Huang, P.W., Laborde, D., Land, V.R., Matson, D.O., Smith, A.W., and Jiang, X. 2000. Concentration and detection of caliciviruses in water samples by reverse transcription-PCR. Appl. Environ. Microbiol. 2000. 66: 4383–4388.

Hung, T., Chen, G., Wang, C., Yao, H.L., Fang, Z.Y., Chao, T.X., Chou, Z.Y., Ye, W., Chang, X.J., and Den, S.S. 1984. Water-borne outbreak of rotavirus diarrhea in adults in China caused by a novel rotavirus. Lancet 2: 1139–1142.

Hurst, C.J., McClellan, K.A., and Benton, W.H. 1988. Comparison of cytopathogenicity, immunofluorescence and in situ DNA hybridization as methods for the detection of adenoviruses. Water Res. 22: 1547–1552.

Hutin, Y.J., Pool, V., Cramer, E.H., Nainan, O.V., Weth, J., Williams, I.T., Goldstein, S.T., Gensheimer, K.F., Bell, B.P., Shapiro, C.N., Alter, M.J., and Margolis, H.S. 1999. A multistate, foodborne outbreak of hepatitis A. National Hepatitis A Virus Investigation Team. N. Engl. J. Med. 340: 595–602.

Ijaz, M.K., Sattar, S.A., Johnson-Lussenburg, C.M., and Springthorpe, V.S. 1985. Comparison of the airborne survival of calf rotavirus and poliovirus type 1 (Sabin) aerosolized as a mixture. Appl. Environ. Microbiol. 49: 289–293.

Irving, L.G., and Smith, F.A. 1981. One-year survey of enteroviruses, adenoviruses and reoviruses isolated from effluent at an activated sludge purification plant. Appl. Environ. Microbiol. 41: 51–59.

Jacobs, J.P., Jones, C.M., and Baille, J.P. 1970. Characteristics of a human diploid cell designated MRC-5. Nature 227: 168–170.

Jean, J., Blais, B., Darveau, A., and Fliss, I. 2001. Detection of hepatitis A virus by the nucleic acid sequence-based amplification technique and comparison with reverse transcription-PCR. Appl. Environ. Microbiol. 67: 5593–5600.

Jiang, X., Cubitt, W.D., Berke, T., Zhong, W., Dai, X., Nakata, S., Pickering, L.K., and Matson, D.O. 1997. Sapporo-like human caliciviruses are genetically and antigenically diverse. Arch. Virol. 142: 1813–1827.

Jiang, X., Huang, P.W., Zhong, W.M., Farkas, T., Cubitt, D.W., and Matson, D.O. 1999. Design and evaluation of a primer pair that detects both Norwalk- and Sapporo-like caliciviruses by RT-PCR. J. Virol. Methods 83: 145–154.

Jiang, X., Matson, D.O., Ruiz-Palacios, G.M., Hu, J., Treanor, J., and Pickering, L.K. 1995. Expression, self-assembly and antigenicity of a Snow Mountain agent-like calicivirus capsid protein. J. Clin. Microbiol. 33: 1452–1455.

Jiang, X., Wang, M., Wang, K., and Estes, M.K. 1993. Sequence and genomic organization of Norwalk virus. Virology 195: 51–61.

Johnson, P.C., Hoy, J., Mathewson, J.J., Ericsson, C.D., and DuPont, H.L. 1990a. Occurrence of Norwalk virus infections among adults in Mexico. J. Infect. Dis. 162: 389–393.

Johnson, P.C., Mathewson, J.J., Dupont, H.L., Greenberg, H.B. 1990b. Multiple challenge study of host susceptibility to Norwalk gastroenteritis in US adults. J. Infect. Dis. 161:18–21.

Jonassen, T.O., Monceyron, C., Lee, T.W., Kurtz, J.B., and Grinde, B. 1995. Detection of all serotypes of human astrovirus by the polymerase chain reaction. J. Virol. Methods 52: 327–334.

Jothikumar, N., Aparna, K., Kamatchiammal, S., Paulmurugan, R., Saravanadevi, S., and Khanna, P. 1993. Detection of hepatitis E virus in raw and treated wastewater with the polymerase chain reaction. Appl. Environ. Microbiol. 59: 2558–2562.

Jothikumar, N., Cliver, D.O. and Mariam, T.W. 1998. Immunomagnetic capture PCR for rapid concentration and detection of hepatitis A virus from environmental samples. Appl. Environ. Microbiol. 64: 504–508.

Jothikumar, N., Khanna, P., Paulmurugan, R., Kamatchiammal, S., and Padmanabhan, P. 1995. A simple device for the concentration and detection of enterovirus, hepatitis E virus and rotavirus from water samples by reverse transcription-polymerase chain reaction. J. Virol. Methods 55: 401–415.

Jothikumar, N., Paulmurugan, R., Padmanabhan, P., Sundari, R.B., Kamatchiammal, S., and Rao, K.S. 2000. Duplex RT-PCR for simultaneous detection of hepatitis A and hepatitis E virus isolated from drinking water samples. J. Environ. Monit. 2: 587–590.

Jourdan, N., Cotte Laffitte, J., Forestier, F., Servin, A.L., and Quero, A.M. 1995. Infection of cultured intestinal cells by monkey RRV and hu-man Wa rotavirus as a function of intestinal epithelial cell differentiation. Res. Virol. 146: 325–331.

Julkunen, I., Lehtomäki, K., and Hovi, T. 1986. Immunoglobulin class-specific serological responses to adenovirus in respiratory infections of young adult men. J. Clin. Microbiol. 24: 212–215.

Kabrane-Lazizi, Y., Meng, X.J., Purcell, R.H., and Emerson, S.U. 1999. Evidence that the genomic RNA of hepatitis E virus is capped. J. Virol. 73: 8848–8850.

Kageyama, T., Kojima, S., Shinohara, M., Uchida, K., Fukushi, S., Hoshino, F.B., Takeda, N., and Katayama, K. 2003. Broadly reactive and highly sensitive assay for Norwalk-like viruses based on real-time quantitative reverse transcription-PCR. J. Clin. Microbiol. 41: 1548-1557.

Kapikian, A.Z., and Chanock, R.M. 1990. Rotaviruses. In: Virology. B.N. Fields, and D.M. Knipe, eds. 2nd ed. Raven Press, New York, NY. p. 1353–1404.

Kapikian, A.Z., Wyatt, R.G., Dolin, R., Thornhill, T.S., Kalica, A.R., and Chanock, R.M. 1972. Visualization by immune electron microscopy of a 27-nm particle associated with acute infectious nonbacterial gastroenteritis. J. Virol. 10: 1075–1081.

Karayiannis, P., Jowett, T., Enticott, M., Moore, D., Pignatelli, M., Brenes, F., Scheuer, P.J., and Thomas, H.C. 1986. Hepatitis A virus replication in tamarins and host immune response in relation to pathogenesis of liver cell damage. J. Med. Virol. 18: 261–276.

Kaupert, N., Burgi, E., and Scolaro, L. 1999. Inactivation of poliovirus by gamma irradiation of wastewater sludges. Rev. Argent. Microbiol. 31: 49–52.

Khadre, M.A., and Yousef, A.E. 2002. Susceptibility of human rotavirus to ozone, high pressure, and pulsed electric field. J. Food Prot. 65: 1441–1446.

Kindle, G., Busse, A., Kampa, D., Meyer-Konig, U., and Daschner, F.D. 1986. Killing activity of microwaves in milk. J. Hosp. Infect. 33: 273–278.

Kingsley, D.H., Hoover, D.G., Papafragkou, E., and Richards, G.P. 2002. Inactivation of hepatitis A virus and a calicivirus by high hydrostatic pressure. J. Food Prot. 65: 1605–1609.

Kingsley, D.H., and Richards, G.P. 2001. Rapid and efficient extraction method for reverse transcription-PCR detection of hepatitis A and Norwalk-like viruses in shellfish. Appl. Environ. Microbiol. 67: 4152–4157.

Kingsley, D.H., and Richards, G.P. 2003. Persistence of hepatitis A virus in oysters. J. Food Prot. 66: 331–334.

Kohn, M.A., Farley, T.A., Ando, T., Curtis, M., Wilson, S.A., Jin, Q., Monroe, S.S., Baron, R.C., McFarland, L.M., and Glass, R.I. 1995. An outbreak of Norwalk virus gastroenteritis associated with eating raw oysters. Implications for maintaining safe oyster beds. J. Am. Med. Assoc. 273: 466–471.

Kok, T.W., Pryor, T., and Payne, L. 1998. Comparison of rhabdomyosarcoma, buffalo green monkey kidney epithelial, A549 (human lung epithelial) cells and human embryonic lung fibroblasts for isolation of enteroviruses from clinical samples. J. Clin. Virol. 11: 61–65.

Konno, T., Suzuki, H., Ishida, N., Chiba, R., Mochizuki, K., and Tsunoda, A. 1982. Astrovirus-associated epidemic gastroenteritis in Japan. J. Med. Virol. 9: 11–17.

Konowalchuk, J., and Speirs, J.I. 1975. Survival of enteric viruses on fresh vegetables. J. Milk Food Technl. 38: 469–472.

Konowalchuk, J., and Speirs, J.L. 1977. Virus detection on grapes. Can. J. Microbiol. 23: 1301-1303.

Kostenbader, K.D. Jr., and Cliver, D.O. 1973. Filtration methods for recovering enteroviruses from foods. Appl. Microbiol. 26: 149–154.

Kotloff, K.L., Losonsky, G.A., Morris, J.G. Jr., Wasserman, S.S., Singh-Naz, N., and Levine, M.M. 1989. Enteric adenovirus infection and childhood diarrhea: an epidemiological study in three clinical settings. Pediatrics 84: 219–225.

Krajden, M., Brown, M., Petrasek, A, and Middleton, P.J. 1990. Clinical features of adenovirus enteritis: a review of 127 cases. Pediatr. Infect. Dis. J. 9: 636–641.

Kurtz, J.B. 1994 Astroviruses. In: Viral Infections of the Gastrointestinal Tract. A.Z. Kapakian, ed. 2nd ed, Marcel Dekker, New York, p. 569–580.

Kurtz, J.B. 1988. Astroviruses. In: Viruses and the Gut. Proceedings of the Ninth British Society of Gastroenterology. M.J.G. Farthing, ed. Smith Kline and French Laboratories, Ltd., Windsor, United Kingdom. p. 84–87.

Kurtz, J. and Lee, T. 1978. Astrovirus gastroenteritis age distribution of antibody. Med. Microbiol. Immunol. 166: 227–230.

Kurtz, J., and Lee, T. 1987. Astroviruses: human and animal. Ciba Found. Symp. 128: 92–107.

Lamden, P.R., Caul, E.O., Ashley, C.R., and Clarke, I.N. 1993. Sequence and genome organization of a human small round-structured (Norwalk-like) virus. Science 259: 516–519.

Lawler, M., Humphries, P., O'Farrelly, C., Hoye, H., Sheils, O., Jeffers, M., O'Brian, D.S., and Kelleher, D. 1994. Adenovirus 12 E1A gene detection by polymerase chain reaction in both the normal and coeliac duodenum. Gut 35: 1226–1232.

LeBaron, C.W., Furutan, N.P., Lew, J.F., Allen, J.R., Gouvea, V., Moe, C., and Monroe, S.S. 1990. Viral agents of gastroenteritis. Morbid. Mortal. Wkly. Rep. 39(RR-5): 1–24.

Lee, T.W., and Kurtz, J.B. 1981. Serial propagation of astrovirus in tissue culture with the aid of trypsin. J. Gen. Virol. 57: 421–424.

Lee, T.W., and Kurtz, J.B. 1994. Prevalence of human astrovirus serotypes in the Oxford region 1976–1992, with evidence for two new serotypes. Epidemiol. Infect. 112: 187–193.

Legeay, O., Caudrelier, Y., Cordevant, C., Rigottier-Gois, L., and Lange, M. 2000. Simplified procedure for detection of enteric pathogenic viruses in shellfish by RT-PCR. J. Virol. Methods 90: 1–14.

Le Guyader, F., Dubois, E., Menard, D., Pommepuy, M. 1994. Detection of hepatitis A virus, rotavirus, and enterovirus in naturally contaminated shellfish and sediment by reverse transcription-seminested PCR. Appl. Environ. Microbiol. 60: 3665–3671.

Le Guyader, F., Estes, M.K., Hardy, M.E., Neill, F.H., Green, J., Brown, D., and Atmar, R.L. 1996. Evaluation of a degenerate primer for the PCR detection of human caliciviruses. Arch. Virol. 141: 2225–2235.

Le Guyader, F., Haugarreau, L., Miossec, L., Dubois, E., and Pommepuy, M. 2000. Three-year study to assess human enteric viruses in shellfish. Appl. Environ. Microbiol. 66: 3241–3248.

Lemon, S.M., Binn, L.N., and Marchwicki, R.H. 1983. Radioimmunofocus assay for quantitation of hepatitis A virus in cell cultures. J. Clin. Microbiol. 17: 834–839.

Lemon, S.M., Binn, L.N., Marchwicki, R., Murphy, P.C., Ping, L.H., Jansen, R.W., Asher, L.V., Stapleton, J.T., Taylor, D.G., LeDuc, J.W. 1990. In vivo replication and reversion to wild-type of a neutralization-resistant variant of hepatitis A virus. J. Infect. Dis. 161: 7–13.

Lew, J.F., Kapikian, A.Z., Jiang, X., Estes, M.K., and Green, K.Y. 1994a. Molecular characterization and expression of the capsid protein of a Norwalk-like virus recovered from a Desert Shield troop with gastroenteritis. Virology. 200: 319–325.

Lew, J.F., Kapikian, A.Z., Valdesuso, J., and Green, K.Y. 1994b. Molecular characterization of Hawaii virus and other Norwalk-like viruses: evidence for genetic polymorphisms among human caliciviruses. J. Infect. Dis. 170: 535–542.

Lew, J.F., Petric, M., Kapikian, A.Z., Jiang, X., Estes, M.K., and Green, K.Y. 1994c. Identification of "mini-reovirus" as a Norwalk-like virus in pediatric patients with gastroenteritis. J. Virol. 68: 3391–3396.

Lewis, T.L., Greenberg, H.B., Herrmann, J.E., Smith, L.S., and Matsui, S.M. 1994. Analysis of astrovirus serotype 1 RNA, identification of the virus RNA-dependent RNA polymerase motif, and expression of a viral structural protein. J. Virol. 68: 77–83.

Lewis, T.L., and Matsui, S.M. 1994. An astrovirus frameshift signal induces ribosomal frameshifting in vitro. Arch. Virol. 140: 1127–1135.

Lewis, T.L., and Matsui, S.M. 1996. Astrovirus ribosomal frameshifting in an infection-transfection transient expression system. J. Virol. 70: 2869–2875.

Li, F., Zhuang, H., Kolivas, S., Locarnini, S., and Anderson, D. 1994. Persistent and transient antibody responses to hepatitis E virus detected by Western immunoblot using open reading frame 2 and 3 and glutathione S-transferase fusion proteins. J. Clin. Microbiol. 32: 2060–2066.

Lipson, S.M., and Zelinsky-Papez, K.A. 1989. Comparison of four latex agglutination (LA) and three enzyme-linked immunosorbent assays (ELISA) for the detection of rotavirus in fecal specimens. Am. J. Clin. Pathol. 92: 637–643.

Liu, S., Birch, C., Coulepis, A., and Gust, I. 1984. Radioimmunofocus assay for detection and quantitation of human rotavirus. J. Clin. Microbiol. 20: 347–350.

Liu, B.L., Clarke, I.N., Caul, E.O., and Lambden, P.R. 1995. Human enteric caliciviruses have a unique genome structure and are distinct from the Norwalk-like viruses. Arch. Virol. 140: 1345–1356.

Liu, B.L., Clarke, I.N., Caul, E.O., and Lambden, P.R. 1996. Polyprotein processing in Southampton virus: identification of 3C-like protease cleavage sites by in vitro mutagenesis. J. Virol. 70: 2605–2610.

Liu, B.L., Viljoen, G.J., Clarke, I.N., and Lambden, P.R. 1999. Identification and further proteolytic cleavage sites in the Southampton calicivirus polyprotein by expression of the viral protease in E. coli. J. Gen. Virol. 80: 291–296

Lloyd-Evans, N., Springthorp, V.S., and Sattar, S.A. 1986. Chemical disinfection of human rotavirus-contaminated inanimate surfaces. J. Hyg. 97: 163–173.

Lund, E., and Hedstrom, C.E. 1969. A study on sampling and isolation methods for the detection of virus in sewage. Water Res. 3: 823–832.

Madeley, C.R., and Cosgrove, B.P. 1976. Caliciviruses in man. Lancet: 1: 199–200.

Madore, H.P., Treanor, J.J., Pray, K.A., and Dolin, R. 1986. Enzyme-linked immunosorbent assays for Snow Mountain and Norwalk agents of acute viral gastroenteritis. J. Clin. Microbiol. 24: 456–459.

Maguire, A.J., Green, J., Brown, D.W.G., Desselberger, U., and Gray, J.J. 1999. Molecular epidemiology of outbreaks of gastroenteritis associated with small round-structured viruses in East Anglia, United Kingdom, during the 1996–1997 season. J. Clin. Microbiol. 37: 81-89.

Mallett, J.C., Beghian, L.E., Metcalf, T.G., and Kaylor, J.D. 1991. Potential of irradiation technology for improving shellfish sanitation. J. Food Safety 11: 231–245.

Marczinke, B., Bloys, A.J., Brown, T.D.K., Willcocks, M.M., Carter, M.J., and Brierley, I. 1994. The human astrovirus RNA-dependent RNA polymerase coding region is expressed by ribosomal frameshifting. J. Virol. 68: 5588–5595.

Mast, E.E., and Krawczynski, K. 1996. Hepatitis E: an overview. Annu. Rev. Med. 47: 257–266.

Matsui, S.M. 2002. Astrovirus. In: Clinical Virology. D.D. Richman, R.J. Whitley, and F.G. Hayden, eds. 2nd ed. ASM Press, Washington, D.C. p. 1075–1086.

Matsui, S.M., Kim, J.P., Greenberg, H.B., Su, W., Sum, Q., Johnson, P.C., DuPont, H.L., Oshiro, L.S., and Reyes, G.R. 1991. The isolation and characterization of a Norwalk virus-specific cDNA. J. Clin. Investig. 87: 1456–1461.

McAllister, R.M., Melnyk, J., Finkelstein, J.Z., Adams, E.C. Jr., and Gardner, M.B. 1969. Cultivation in vitro of cells derived from a human rhabdomyosarcoma. Cancer 24: 520–526.

McDonnell, S., Kirkland, K.B., Hlady, W.G., Aristeguieta, C., Hopkins, R.S., Monroe, S.S., and Glass, R.I. 1997. Failure of cooking to prevent shellfish-associated viral gastroenteritis. Arch. Intern. Med. 157: 111–116.

Mead, P.S., Slutsker, L., Dietz, V., McCraig, L.F., Bresee, J.S., Shapiro, C, Griffin, P.M., and Tauxe, R.V. 1999. Food-related illness and death in the United States. Emerging Infect. Dis. 5: 607–625.

Meng, X.J., Halbur, P.G., Shapito, M.S., Govindarajan, S., Bruna, J.D., Mushahwar, I.K., Purcell, R.H., and Emerson, S.U. 1998. Genetic and experimental evidence for cross-species infection by swine hepatitis E virus. J. Virol. 72: 9714–9721.

Meurman, O., Ruuskanen, O., and Sarkkinen, H. 1983. Immunoassay diagnosis of adenovirus infections in children. J. Clin. Microbiol. 18: 1190–1195.

Midthun, K., Greenberg, H.B., Kurtz, J.B., Gary, G.W., Lin, F.C., and Kapikian, A.Z. 1993. Characterization and seroepidemiology of a type 5 astrovirus associated with an outbreak of gastroenteritis in Marin County, California. J. Clin. Microbiol. 31: 955–962.

Mishu, B., Hadler, S.C., Boaz, V.A., Hutcheson, R.H., Horan, J.M., and Schaffner, W. 1990. Foodborne hepatitis A: evidence that microwaving reduces risk. J. Infect. Dis. 162: 655–658.

Monceyron, C., and Grinde, B. 1994. Detection of hepatitis A virus in clinical and environmental samples by immunomagnetic separation and PCR. J. Virol. Methods 46: 157–166.

Monk, J.D., Beuchat, L.R., and Doyle, M.P. 1995. Irradiation inactivation of foodborne microorganisms. J. Food Prot. 58: 197–208.

Monroe, S.S., Stine, S.E., Gorelkin, L., Herrmann, J.E., Blacklow, N.R., and Glass, R.I. 1991. Temporal synthesis of proteins and RNAs during human astrovirus infection of cultured cells. J. Virol. 65: 641–648.

Mullendore, J.L., Sobsey, M.D., and Shieh, C.Y. 2001. Improved method for the recovery of hepatitis A virus from oysters. J. Virol. Methods 94: 25–35.

Muniain-Mujika, I., Girones, R., and Lucena, F. 2000. Viral contamination of shellfish: evaluation of methods and analyses of bacteriophage and human viruses. J. Virol. Methods 89: 109–118.

Naik, S.R., Aggarwal, R., Salunke, P.N., and Mehrotra, N.N. 1992. A large water-borne viral hepatitis E epidemic in Kanpur, India. Bull. World Health Organ. 70: 597–604.

Nanda, S.K., Ansari, I.H., Acharya, S.K., Jameel, S., and Panda, S.K. 1995. Protracted viremia during acute sporadic hepatitis E virus infection. Gastroenterology 108: 225–230.

Nasser, A.M., and Metcalf, T.G. 1987. An A-ELISA to detect hepatitis A virus in estuarine samples. Appl. Environ. Microbiol. 53: 1192–1195.

Nishida, T., Kimura, H., Saitoh, M., Schnohara, M., Kato, M., Fukuda, S., Munemura, T., Mikami, T., Kawamoto, A., Akiyama, M., Kato, Y., Nishi, K., Kozawa, K., and Nishio, O. 2003. Detection, quantitation, and phylogenetic analysis of noroviruses in Japanese oysters. Appl. Environ. Microbiol. 69: 5782–5786.

Niu, M.T., Polish, L.B., Robertson, B.H., Khanna, B.K., Woodruff, B.A., Shapiro, C.N., Miller, M.A., Smith, J.D., Gedrose, J.K., and Alter, M.J. 1992. Multistate outbreak of hepatitis A associated with frozen strawberries. J. Infect. Dis. 166: 518–524.

Noel, J.S., Ando, T., Leite, J.P., Green, K.Y., Dingle, K.E., Estes, M.K., Seto, Y., Monroe, S.S., and Glass, R.I. 1997a. The correlation of patient immune responses with genetically characterized small round-structured viruses involved in outbreaks of nonbacterial acute gastroenteritis in the United States, 1990 to 1995. J. Med. Virol. 53: 372–383.

Noel, J.S, Lee, T.W., Kurtz, J.B., Glass, R.I., and Monroe, S.S. 1995. Typing of human astroviruses from clinical isolates by enzyme immunoassays and nucleotide sequencing. J. Clin. Microbiol. 33: 797–801.

Noel, J.S., Liu, B.L., Humphrey, C.D., Rodriguez, E.M., Lambden, P.R., Clarke, I.N., Dwyer, D.M., Ando, T., Glass, R.I., and Monroe, S.S. 1997b. Parkville virus: a novel genetic variant of human calicivirus in the Sapporo virus clade, associated with an outbreak of gastroenteritis in adults. J. Med. Virol. 52: 173–178.

Ojeh, C.K., Cusack, T.M., and Yolken, R.H. 1995. Evaluation of the effects of disinfectants on rotavirus RNA and infectivity by the polymerase chain reaction and cell-culture methods. Mol. Cell Probes 9: 341–346.

Oliver, A.R., and Phillips, A.D. 1988. An electron microscopical investigation of faecal small round viruses. J. Med. Virol. 24: 211–218.

Oragui, J.I., and Mara, D.D. 1989. Simple method for the detoxification of wastewater ultrafiltration concentrates for rotavirus assay by indirect immunofluorescence. Appl. Environ. Microbiol. 55: 401–405.

Oron, G., Goemans, M., and Manor, Y. 1995. Poliovirus distribution in the soil-plant system under reuse of secondary wastewater. Water Res. 29: 1069–1078.

Pal, S.R., McQuillin, J., and Gardner, P.S. 1963. A comparative study of susceptibility of primary monkey kidney cells, Hep 2 cells and HeLa cells to a variety of faecal viruses. J. Hyg., Camb. 61: 493–498.

Parashar, U.D., Bresse, J.S., Gentsch, J.R., and Glass, R.I. 1998. Rotavirus. Emerging Infect. Dis. 4: 561–570.

Parrino, T.A., Schreiber, D.S., Trier, J.S., Kapikian, A.Z., and Blacklow, N.R. 1977. Clinical immunity in acute gastroenteritis caused by Norwalk agent. N. Engl. J. Med. 297: 86–89.

Patterson, M.F. 1993. Food irradiation and food safety. Rev. Med. Microbiol. 4: 151–158.

Patterson, W, Haswell, P., Fryers, P.T., and Green, J. 1997a. Outbreak of small round structured virus gastroenteritis after kitchen worker vomited. Commun. Dis. Rep. CDR Rev. 7: R101-103.

Patterson, T., Hutchings, P., and Palmer, S. 1997b. Outbreak of SRSV gastroenteritis at an international conference traced to food handling by a post-symptomatic caterer. Epidemiol. Infect. 111: 157–162.

Patti, A.M., Aulicino, F.A., De Pilippis, P., Gabrieli, R., Volterra, L., and Pana, A. 1990. Identification of enteroviruses isolated from sea-water: indirect immunofluorescence (IIF). Boll. Soc. Ital. Biol. Sper. 66: 595–600.

Payment, P. 1997. Cultivation and assay of viruses. Manual of Environmental Microbiology. American Society for Microbiology, Washington, D.C. p. 72–77.

Payment, P., and Trudel, M. 1985. Immunoperoxidase method with human immune serum globulin for broad-spectrum detection of cultivable human enteric viruses: application to enumeration of cultivable viruses in environmental samples. Appl. Environ. Microbiol. 50: 1308–1310.

Payment, P., and Trudel, M. 1993. Methods and Techniques in Virology. Marcel Dekker Inc., New York, NY.

Payment, P., Affoyon, F., and Trudel, M. 1988. Detection of animal and human enteric viruses in water from the Assomption River and its tributaries. Can. J. Microbiol. 34: 967–973.

Pontes, L., Fornells, I.A., Giongo, V., Araujo, J.R.V., Sepulveda, A., Villas-Boas, M., Bonate, C. F.S., and Silva, J L. 1997. Pressure inactivation of animal viruses: potential biotechnological applications. In: High Pressure Research in the Biosciences and Biotechnology. K. Heremans, ed. Leuven University Press, Leuven, Belgium p. 91–94.

Power, U.F., and Collins, J.K. 1989. Differential depuration of poliovirus, *Escherichia coli*, and a coliphage by the common mussel, *Mytilus edulis*. Appl. Environ. Microbiol. 55: 1386–1390.

Purcell, R.H., Feinstone, S.M., Ticehurst, J.R., Daemer, R.J., and Baroudy, B.M. 1984. Hepatitis A virus. In: Viral Hepatitis and Liver Disease. G.N. Vyas, J.L. Dienstag, and J.H. Hoofnagle, eds. Grune & Stratton, Orlando, FL. p. 9–22.

Rab, M.A., Bile, M.K., Mubarik, M.M., Asghar, H., Sami, Z., Siddiqi, S., Dil, A.S., Barzgar, M.A., Chaudhry, M.A., and Burney, M.I. 1997. Water-borne hepatitis E virus epidemic in Islamabad, Pakistan: a common source outbreak traced to the malfunction of a modern treatment plant. Am. J. Trop. Med. Hyg. 57: 151–157.

Ramia, S., and Sattar, S.A. 1979. Simian rotavirus SA-11 plaque formation in the presence of trypsin. J. Clin. Microbiol. 10: 609–614.

Ramia, S., and Sattar, S.A. 1980. Proteolytic enzymes and rotavirus SA-11 plaque formation. Can. J. Comp. Med. 44: 232–235.

Ranki, M., Virtanen, M., Palva, A., Laaksonen, M., Pettersson, R., Kääriäinen, L., Halonen, P., and Söderlund, H. 1983. Nucleic acid sandwich hybridization in adenovirus diagnosis. Curr. Top. Microbiol. Immunol. 104: 307–318.

Raphael, R.A., Sattar, S.A., and Springthorp, V.S. 1985. Rotavirus concentration from raw water using positively charged filters. J. Virol. Methods 11: 131–140.

Regan, P.M., and Margolin, A.B. 1997. Development of a nucleic acid capture probe with reverse transcriptase-polymerase chain reaction to detect poliovirus in groundwater. J. Virol. Methods 64: 65–72.

Reid, T.M., and Robinson, H.G. 1987. Frozen raspberries and hepatitis A. Epidemiol. Infect. 98: 109–112.

Reiner, P., Reinerova, M., and Veselovska, Z. 1992. Comparison of two defective hepatitis A virus strains adapted to cell culture. Acta. Virol. 36: 245–252.

Reynolds, K.A., Gerba, C.P., and Pepper, I.L. 1996. Detection of infectious enteroviruses by integrated cell culture-PCR procedure. Appl. Environ. Microbiol. 62: 1424–1427.

Richards, G. P. 1985. Outbreaks of shellfish-associated enteric virus illness in the United States: requisite for development of viral guidelines. J. Food Prot. 48: 815–823.

Richards, G.P. 1988. Microbial purification of shellfish: a review of depuration and relaying. J. Food Prot. 51: 218–251.

Richards, G.P. 1999. Limitations of molecular biological techniques for assessing the virological safety of foods. J. Food Prot. 62: 691–697.

Richards, G.P. 1991. Shellfish depuration. In: Microbiology of Marine Food Products. D.S. Ward, and C. Hackney, eds. Van Nostrand Reinhold, New York. p. 395–428.

Richards, G.P. 2002. Enteric virus contamination of foods through industrial practices: a primer on intervention strategies. J. Indust. Microbiol. Biotechnol. 27: 117–125.

Richards, G.P., and Cliver, D.O. 2001. Foodborne viruses. In: Compendium of Methods for the Microbiological Examination of Foods. F.P. Downes and K. Ito, eds. American Public Health Assoc., Washington, DC. p. 447–461.

Richards, G.P., and Watson, M.A. 2001. Immunochemiluminescent focus assays for the quantitation of hepatitis A virus and rotavirus in cell cultures. J. Virol. Methods 94: 69–80.

Richards, G.P., Watson, M.A., and Kingsley, D.H. 2004. A SYBR green, real-time RT-PCR method to detect and quantitate Norwalk virus in stools. J. Virol. Methods 116: 63–70.

Richards, G.P., and Weinheimer, D.A. 1985. Influence of adsorption time, rocking, and soluble proteins on the plaque assay of monodispersed poliovirus. Appl. Environ. Microbiol. 49: 744–748.

Richardson, S., Grimwood, K., Gorrell, R., Palombo, Barnes, G., and Bishop, R. 1998. Extended excretion of rotavirus after severe diarrhea in young children. Lancet 351: 1844–1848.

Riggs, J.L. 1979. Immunofluorescent staining. In: Diagnostic Procedures for Viral, Rickettsial and Chlamydial Infections. E.H. Lennette and N.J. Schmidt. eds. 5th ed. Amer. Pub. Health Assn., Washington, D.C., p. 141–145.

Robertson, B.H., Averhoff, F., Cromeans, T.L., Han, X., Khoprasert, B., Nainan, O.V., Rosenberg, J., Paikoff, L., DeBess, E., Shapiro, C.N., and Margolis, H.S. 2000. Genetic relatedness of hepatitis A virus isolates during a community-wide outbreak. J. Med. Virol. 62: 144–150.

Rosenblum, L.S., Mirkin, I.R., Allen, D.T., Safford, S., and Hadler, S.C. 1990. A multifocal outbreak of hepatitis A traced to commercially distributed lettuce. Am. J. Public Health 80: 1075–1079.

Rosenblum, L.S., Villarino, M.E., Nainan, O.V., Melish, M.E., Hadler, S.C., Pinsky, P.P., Jarvis, W.R., Ott, C.E., and Margolis, H.S. 1991. Hepatitis A outbreak in a neonatal intensive care unit: risk factors for transmission and evidence of prolonged viral excretion among preterm infants. J. Infect. Dis. 164: 476–482.

Rotbart, H.A., Eastman, P.S., Ruth, J.L., Hirata, K.K., and Levin, M.J. 1988. Nonisotopic oligomeric probes for the human enteroviruses. J. Clin. Microbiol. 26: 2669–2671.

Ruuskanen, O., Mertsola, J., and Meurman, O. 1988. Adenovirus infections in families. Arch. Dis. Hild. 63: 1250–1253.

Ruuskanen, O., Meurman, O., and Akusjärvi, G. 2002. Adenoviruses. In: Clinical Virology, D.D. Richman, R.J. Whitley, and F.G. Hayden, eds. 2nd ed. ASM Press, Washington, D.C. p. 515-535.

Sadovski, A.Y., Fattal, B., Goldberg, D., Katzenelson, E., and Shuval, H.I. 1978. High levels of microbial contamination of vegetables irrigated with wastewater by the drip method. Appl. Environ. Microbiol. 36: 824–830.

Saif, L.J., and Jiang, B. 1994. Nongroup A rotaviruses of humans and animals. Curr. Top. Microbiol. Immunol. 185: 339–371.

Saif, L.J., Rosen, B.I., and Parwani, A.V. 1994. Animal rotaviruses. In: Viral Infections of the Gastrointestinal Tract. A.Z. Kapikian, ed. 2nd ed., Marcel Dekker, New York, p. 279–367.

Sair, A.I., D'Souza, Moe, C.L., and Jaykus, L.A. 2002. Improved detection of human enteric viruses in food by RT-PCR. J. Virol. Methods 100: 57–69.

Sattar, S.A., Jacobsen, H., Rahman, H., Cusack, T.M., and Rubino, J.R. 1994. Interruption of rotavirus spread through chemical disinfection. Infect. Control Hosp. Epidemiol. 15: 751-756.

Saulsbury, F.T., Winkelstein, J.A., and Yolken, R.H. 1980. Chronic rotavirus infection in immunodeficiency. J. Pediatr. 97: 61–65.

Schlauder, G.G., Dawson, G.J., Erker, J.C., Kwo, P.Y., Knigge, M.F., Smalley, D.L., Rosenblatt, J.E., Desai, S.M., and Mushahwar, I.K. 1998. The sequence and phylogenetic analysis of a novel hepatitis E virus isolated from a patient with acute hepatitis reported in the United States. J. Gen. Virol. 79: 447–456.

Schmidt, N.J., Ho, H.H., and Lennette, E.H. 1975. Propagation and isolation of group A coxsackievirus in RD cells. J. Clin. Microbiol. 2: 183–185.

Schmidt, N.J., Ho, H.H., and Lennette, E.H. 1976. Comparative sensitivity of the BGM cell line for the isolation of enteric viruses. Health Lab. Sci. 13: 115–117.

Schmidt, N.J., Ho, H.H., Riggs, J.L., and Lennette, E.H. 1978. Comparative sensitivity of various cell culture systems for isolation of viruses from wastewater and fecal samples. Appl. Environ. Microbiol. 36: 480–486.

Schreiber, D.S., Blacklow, N.R., and Trier, J.S. 1974. The small intestinal lesion induced by Hawaii agent acute infectious nonbacterial gastroenteritis. J. Infect. Dis. 129: 705–708.

Schwab, K.J., De Leon, R., and Sobsey, M.D. 1996. Immunoaffinity concentration and purification of water-borne enteric viruses or detection by reverse transcriptase PCR. Appl. Environ. Microbiol. 62: 2086–2094.

Schwab, K.J., Neill, F.H., Estes, M.K., Metcalf, T.G., and Atmar, R.L. 1998. Distribution of Norwalk virus within shellfish following bioaccumulation and subsequent depuration by detection using RT-PCR. J. Food Prot. 61: 1674–1680.

Schwab, K.J., Neill, F.H., Fankhauser, R.L., Daniels, N.A., Monroe, S.S., Bergmire-Sweet, D.A., Estes, M.K., and Atmar, R.L. 2000. Development of methods to detect "Norwalk-like viruses" (NLVs) and hepatitis A virus in delicatessen foods: application to foodborne NLV outbreak. Appl. Environ. Microbiol. 66: 213–218.

Severini, G.M., Mestroni, L., Falaschi, A., Camerini, F., and Giacca, M. 1993. Nested polymerase chain reaction for high-sensitivity detection of enteroviral RNA in biological samples. J. Clin. Microbiol. 31: 1345–1349.

Shieh, Y.C., Calci, K.R., and Baric, R.S. 1999. A method to detect low levels of enteric viruses in contaminated oysters. Appl. Environ. Microbiol. 65: 4709–4714.

Slade, J.S., Harris, N.R., and Chisholm, R.G. 1986. Disinfection of chlorine resistant enteroviruses in ground water by ultraviolet irradiation. Water Sci. Technol. 18: 115–123.

Smith, E.M., and Gerba, C.G. 1982. Development of a method for detection of human rotavirus in water and sewage. Appl. Environ. Microbiol. 43: 1440–1450.

Smith, J.L. 2001. A review of hepatitis E virus. J. Food Prot. 64: 572–586.

Sobsey, M.D. 1976. Field monitoring techniques and data analysis. In: Virus Aspects of Applying Municipal Waste to Land. L.B. Baldwin, J.M. Davidson, and J.F. Gerber, eds. Univ. Florida, Gainsville, FL.

Sobsey, M.D., Davis, A.L., and Rullman, V.A. 1987. Persistence of hepatitis A virus and other viruses in depurated Eastern oysters. Proc. Oceans 5: 1740–1745.

Sobsey, M.D., and Glass, J.S. 1980. Poliovirus concentration from tap water with electropositive absorbent filters. Appl. Environ. Microbiol. 40: 201–210.

Sobsey, M.D., Oglesbee, S.E., and Wait, D.A. 1985. Evaluation of methods for concentrating hepatitis A virus from drinking water. Appl. Environ. Microbiol. 50: 1457–1463.

Sullivan, R., Scarpino, P.V., Fassolitis, A.C., Larkin, E.P., and Peeler, J.T. 1973. Gamma radiation inactivation of coxsackie B-2. Appl. Microbiol. 26: 14–17.

Sullivan, R., Peeler, J.T., and Larkin, E.P. 1986. A method for recovery of poliovirus 1 from a variety of foods. J. Food Prot. 49: 226–228.

Taylor, M.B., Cox, N., Vrey, M.A., and Grabow, W.O. 2001. The occurrence of hepatitis A and astroviruses in selected river and dam waters in South Africa. Water Res. 35: 2653–2660.

Taylor, M.B., Grabow, W.O., and Cubitt, W.D. 1997. Propagation of human astrovirus in the PLC/PRF/5 hepatoma cell line. J. Virol. Methods 67: 13–18.

Teterina, N.L., Bienz, K., Egger, D., Gorbalenya, A.E., and Ehrenfeld, E. 1997. Induction of intracellular membrane rearrangements by HAV proteins 2C and 2BC. Virology 237: 66–77.

Thomas, E.E., Roscoe, D.L., Book, L., Bone, B., Browne, L., and Mah, V. 1994. The utility of latex agglutination assays in the diagnosis of pediatric viral gastroenteritis. Am. J. Clin. Pathiol. 101: 742–746.

Thornhill, T.S., Wyatt, R.G., Kalica, A.R., Dolin, R., Chanock, R.M., and Kapikian, A.Z. 1977. Detection by immune electron microscopy of 26–27 nm virus-like particles associated with two family outbreaks of gastroenteritis. J. Infect. Dis. 135: 20–27.

Thurston-Enriquez, J.A., Haas, C.N., Jacangelo, J., Riley, K., and Gerba, C.P. 2003. Inactivation of feline calicivirus and adenovirus type 40 by UV irradiation. Appl. Environ. Microbiol. 69: 577–582.

Toranzos, G.A., and C.P. Gerba. 1989. An improved method for the concentration of rotaviruses from large volumes of water. J. Virol. Methods 24: 131–140.

Totsuka, A., and Moritsuga, Y. 1999. Hepatitis A virus proteins. Intervirology 42: 63–68.

Traore, O., Arnal, C., Mignotte, B., Maul, A., Laveran, H., Billaudel, S., and Schwartzbrod, L. 1998. Reverse transcriptase PCR detection of astrovirus, hepatitis A virus, and poliovirus in experimentally contaminated mussels: comparison of several extraction and concentration methods. Appl. Environ. Microbiol. 64: 3118–3122.

Tsarev, S.A., Tsarev, T.S., Emerson, S.U., Yarbough, P.O., Legters, L.J., Moskal, T., and Purcell, R.H. 1994. Infectivity titration of a prototype strain of hepatitis E virus in cynomolgus monkeys. J. Med. Virol. 43: 135–142.

Tucker, T.J., Kirsch, R.E., Louw, S.J., Isaacs, S., Kannemeyer, J., and Robson, S.C. 1996. Hepatitis E in South Africa: evidence for sporadic spread and increased seroprevalence in rural areas. J. Med. Virol. 50: 117–119.

Uhnoo, I., Wadell, G., Svensson, L., and Johansson, M.E. 1984. Importance of enteric adenoviruses 40 and 41 in acute gastroenteritis in infants and young children. J. Clin. Microbiol. 20: 365–372.

Uhnoo, I., Olding, S.E., and Kreuger, A. 1986. Clinical features of acute gastroenteritis associated with rotavirus, enteric adenoviruses, and bacteria. Arch. Dis. Child. 61: 732–738.

Urasawa, T., Urasawa, S., and Taniguchi, K. 1981. Sequential passages of human rotavirus in MA-104 cells. Microbiol. Immunol. 25: 1025–1035.

Ushijima, H., Honma, H., Mukoyama, A., Shinozaki, T., Fujita, Y., Kobayashi, M., Ohseto, M., Morikawa, S., and Kitamura, T. 1989. Detection of group C rotaviruses in Tokyo. J. Med. Virol. 27: 299–303.

Van der Vliet, P.C. 1995. Adenovirus DNA replication Curr. Top. Microbiol. Immnol. 199: 1–30.

Vaughn, J.M., Chen, Y.S., Lindburg, K., and Morales, D. 1987. Inactivation of human and simian rotaviruses by ozone. Appl. Environ. Microbiol. 53: 2218–2221.

Vaughn, J.M., Chen, Y.S., and Thomas, M.Z. 1986. Inactivation of human and simian rotaviruses by chlorine. Appl. Environ. Microbiol. 51: 391–394.

Vinje, J., Deijl, H., van der Heide, R., Lewis, D., Hedlund, K.O., Svensson, L., and Koopmans, M.P.G. 2000. Molecular detection and epidemiology of Sapporo-like viruses. J. Clin. Mcrobiol. 38: 530–536.

Wallace, R.E., Vasington, P.J., Petricciani, J.C., Hopps, H.E., Lorenz, D.E., and Kadanka, Z. 1973. Development and characterization of cell lines from subhuman primates. In vitro 8: 333–341.

Wang, Y., Ling, R., Erker, C., Zhang, H., Li, H., Desai, S., Mushahwar, I.K., and Harrison, T.J. 1999. A divergent genotype of hepatitis E virus in Chinese patients with acute hepatitis. J. Gen. Virol. 80: 169–177.

Warburton, A.R., Wreghitt, T.G., Ramplng, A., Buttery, R., Ward, K.N., Perry, K.R., and Parry, J.V. 1991. Hepatitis A outbreak involving bread. Epidemiol. Infect. 106: 199–202.

Ward, R.I., Knowlton, D.R., and Pierce, M.J. 1984. Efficiency of human rotavirus propagation in cell culture. J. Clin. Microbiol. 19: 748–753.

Weiss, C., and Clark, H.F. 1985. Rapid inactivation of rotavirus by exposure to acid buffer or acidic gastric juice. J. Gen. Virol. 66: 2725–2730.

Weltman, A.C., Bennett, N.M., Ackman, D.A., Misage, J.H., Campana, J.J., Fine, L.S., Doniger, A.S., Balzano, G.J., and Birkhead, G.S. 1996. An outbreak of hepatitis A associated with a bakery, New York, 1994: the 1968 "West Branch, Michigan" outbreak repeated. Epidemiol. Infect. 117: 333–341.

Whitaker, A.M., and Hayward, C.J. 1985. The characterization of three monkey kidney cell lines. Dev. Biol. Stand. 60: 125–131.

White, K.E., Osterholm, M.T., Mariotti, J.A., Korlath, J.A., Lawrence, D.H., Ristinen, T.L., and Greenberg, H.L. 1996. A foodborne outbreak of Norwalk virus gastroenteritis: evidence for post-recovery transmission. Am. J. Epidemiol. 124: 120–126.

Wheeler, C.M., Fields, H.A., Schable, C.A., Meinke, W.J. and Maynard, J.E. 1986. Adsorption, purification, and growth characteristics of hepatitis A virus strain HAS-15 propagated in fetal rhesus monkey kidney cells. J. Clin. Microbiol. 23: 434–440.

Widerlite, L., Trier, J.S., Blacklow, N.R., and Schreiber, D.S. 1975. Structure of the gastric mucosa in acute infectious bacterial gastroenteritis. Gastroenterology 68: 425–430.

Woode, G.N., Pohlenz, J.F., Kelso-Gourley, N.E., and Fagerland, J.A. 1984. Astrovirus and bredavirus infection of dome cell epithelium of bovine ileum. J. Clin. Microbiol. 19: 623-630.

Wyatt, R.G., Dolin, R., Blacklow, N.R., DuPont, H.L., Buscho, R.F., Thornhill, T.S., and Kapikian, A.Z. 1974. Comparison of three agents of acute infectious nonbacterial gastroenteritis by cross-challenge in volunteers. J. Infect. Dis. 129: 709–714.

Wyatt, R.G., Kalica, A.R., Mebus, C.A., Kim, H.W., London, W.T., Chanock, R.M., and Kapikian, A.Z. 1978. Reovirus-like agents (rotaviruses) associated with diarrhea illness in animals and man. Perspect. Virol. 10: 121–145.

Yokoi, H., Kitahashi, T., Tanaka, T., and Utagawa, E. 2001. Detection of astrovirus RNA from sewage works, seawater and native oysters samples in Chiba City, Japan using reverse transcription-polymerase chain reaction. Kansenshogaku Zasshi. 75: 263–269.

Zahorsky, J., 1929. Hyperemesis hiemis or the winter vomiting disease. Arch. Pediatr. 46: 391-395.

Zhang, H., Chao, S.F., Ping, L.H., Grace, K., Clarke, B., and Lemon, S.M. 1995. An infectious cDNA clone of a cytopathic hepatitis A virus: genomic regions associated with rapid replication and cytopathic effect. Virol. 212: 686–697.

Zivanovic-Marinkovic, V., Cobeljic, M., Parabucki, S., Stankovic, D., Krstic, L., and Birtasevic, B. 1984. Antigen of hepatitis A virus detected in shellfish using the ELISA test. Vojnosanit Pregl. 41: 263–265.

Zou, S., and Chaudhary, R.K. 1991. Kinetic study of the replication of a cell-culture-adapted hepatitis A virus. Res. Virol. 142: 381–385.

# Foodborne and Waterborne Protozoan Parasites

Ynes Ortega

## Abstract

Protozoan parasites have been associated with food and waterborne outbreaks causing illness in humans. Although parasites are more commonly found in developing countries, developed countries have also experienced several foodborne outbreaks. Contaminants may be inadvertently introduced to the foods by inadequate handling practices, either on the farm or during processing of ready-to-eat foods. In some instances, this contamination has occurred in endemic regions and is carried to non-endemic areas, where an outbreak is initiated. Other protozoan parasites can be found worldwide, either infecting wild animals or in an environment such as water, and eventually finding its way to crops grown for human consumption. Parasites can infect immunocompetent individuals, however the clinical presentation can be much more severe and prolonged in immunocompromised individuals.

## Introduction

Parasites are a very large group of organisms which cause disease in humans and animals. They can be acquired via vectors, person to person transmission, or contact with animals, arthopods, and insects. They can also be acquired by ingestion of contaminated foods and water. Based on their morphological characteristics, parasites have been classified into two large groups, the helminths and protozoa. Helminths are classified as nematodes (round worms), cestodes (flat worms) and trematoda (flukes). Protozoa have been classified based on their morphological and molecular characteristics. Protozoa associated with transmission via food and water will be covered in this chapter: coccidia, flagellates, ciliates, and amoebae. Microsporidia will be discussed as a separate group.

Parasites are unique in their complex life cycles. Most require a vertebrate host to complete their life cycle and produce infectious and environmentally resistant forms, which are found contaminating water, fresh produce, and fruits. Throughout the years, we have learned much from the mechanisms of transmission, pathogenesis and their molecular classification.

Coccidia such *Cryptosporidium*, *Isospora* and *Cyclospora* can cause diarrheal illness in susceptible individuals, and is almost always restricted to the gastrointestinal tract; whereas other parasites such as *Toxoplasma* can infect other tissues and produce birth defects, blindness, encephalitis, and chorioretinitis. *Toxoplasma* also infects other animals and the disease can be acquired by ingestion of raw meats containing viable tissue cysts.

Once parasites are ingested by a susceptible host, the motile forms of the parasites are released from the cysts or oocysts and may initiate colonization of the host's intestinal cells, feeding on the nutrients and cellular debris of the intestinal epithelium. *Giardia lamblia*, a flagellate originally considered a commensal, is now well recognized as an etiological agent for acute and chronic diarrhea but it does not invade the cells of the intestinal epithelium. *Balantidium coli*, a ciliate, also causes diarrhea in humans and if not treated, can cause ulcerative colitis. Various species of amoebae can infect humans; however, *E. histolytica* is the only amoeba pathogenic to humans, causing diarrhea, dysentery, or amebomas if not treated.

Microsporidia are spore forming microorganisms and have been recognized as a cause of illness in humans. They are suspected to be transmitted via food, water, and person to person. Five genera have been implicated in human diseases: *Encephalitozoon*, *Enterocytozoon*, *Septata*, *Pleistophora*, and *Vittaforma*. None of these species is host, tissue, or organ specific with the exception of *Enterocytozoon bieneusii*, which appears to infect only the human intestinal tract. Microscopic images of several of the parasites discussed in this chapter are shown in Figure 9.1.

## *Cryptosporidium*

*Cryptosporidium* has been identified in the gastrointestinal or respiratory tract of most species of animals, including mammals, reptiles, birds, and fish. Cryptosporidiosis was described in immunocompetent populations, particularly animal handlers and travelers (Black, 1986; Rahman *et al.*, 1985). In the early 1980s, *Cryptosporidium* was identified in AIDS patients

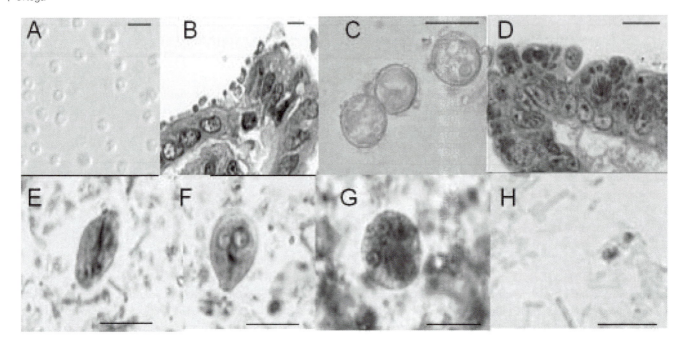

**Figure 9.1** Protozoan parasites observed by Nomarski microscopy (A, C) or by bright field microscopy of stained fecal smears (E, F, G, H). Histological sections were stained with hematoxoline and eosine or silver stain (B and D). (A) *Cryptosporidium parvum* oocysts. (B) Intestinal tissue showing parasitic vacuoles of *Cryptosporidium parvum*. (C) Sporulated and unsporulated oocysts of *Cyclospora cayetanensis*. (D) Intestinal tissue containing parasitic vacuoles of *Toxoplasma gondii*. (E) *Giardia lamblia* cyst. (F) *Giardia lamblia* trophozoites. (G) *Entamoeba histolytica* cyst. (H) *Microsporidia* spores. Bar =10 µm.

presenting with severe enteritis (Anand *et al.*, 1996; Garrido Davila and Ramirez Ronda, 1991). As diagnostic techniques were developed and improved, the number of reported cases continued to increase. *Cryptosporidium* is now recognized as a common cause of enteritis for both immunocompetent and immunocompromised hosts worldwide. The mechanisms of transmission have been mostly fecal–oral (day-care centers), water, and foodborne (Chacin-Bonilla, 1995; Chick *et al.*, 2001).

## Characteristics

*Cryptosporidium* belongs to the phylum Apicomplexa, class Sporozoa, subclass Coccidiasina. *Cryptosporidium* oocysts measure between 4 and 6 µm in diameter.

Infection starts when the susceptible host ingests the oocysts. When in contact with digestive enzymes and bile salts, oocysts excyst and infective sporozoites are released and invade the epithelial cells. Merogony (asexual replication) initiates, resulting in the formation of type I and type II meronts (schizonts) that contain eight and four merozoites respectively. Asexual multiplication can continue or sexual differentiation can occur producing the gametocytes. Microgametes (male) fertilize the macrogametes (female) leading to the formation of oocysts. Oocysts can either reinfect the host (thin walled oocysts) or be excreted to the environment (thick-walled oocysts) (Tzipori and Griffiths, 1998).

*Cryptosporidium* species have been described based on their morphological characteristics and genetic make-up. *Cryptosporidium* infects various hosts: *C. nasorum* in fish, *C.*

*serpentis* in reptiles, *C. saurophilum* in lizards, *C. meleagridis, C. galli,* and *C. baileyi* infect birds. *Cryptosporidium muris, C. hominis, C. meleagridis, C. felis, C. canis, C wrairi, C. andersoni,* and *C. parvum* have been found to infect mammals (Morgan *et al.*, 2000). *Cryptosporidium parvum* was considered to be the only species affecting humans; however, molecular studies led to the differentiation of *C. parvum* genotype I or human genotype and *C. parvum* genotype II or bovine genotype (Widmer *et al.*, 1998; Widmer *et al.*, 2000). Now the human genotype has been renamed as *C. hominis* and is morphologically similar to *C. parvum* (Morgan-Ryan *et al.*, 2002). *Cryptosporidium parvum* infects mammals and can also infect humans. Other *Cryptosporidium* species isolated in humans include *C. muris, C. meleagridis, C. canis,* and *C. felis* (Lindergard *et al.*, 2003; Palmer *et al.*, 2003; Pedraza-Diaz *et al.*, 2001). As researchers examine the genetic make up of *Cryptosporidium* isolates, the names of the species are being reevaluated and new species are now being described. Another factor considered to classify the species is the definite host where the parasite was isolated, susceptibility of infection by a particular host, and the grade of homology of DNA sequences of various protein genes and SS-rRNA (Gasser *et al.*, 2001; Gobet and Toze, 2001; Widmer, 1998; Xiao *et al.*, 2000; Xiao *et al.*, 2001). The correlation of these *Cryptosporidium* species and the clinical presentation in humans still needs to be determined.

## Epidemiology

*Cryptosporidium* can be acquired via human to human transmission or through contaminated foods and water. Studies in human volunteers have demonstrated that 16–100 oocysts of different isolates are sufficient to cause human illness in immunocompetent individuals.

Most of the *Cryptosporidium* outbreaks have been waterborne. In 1993, a large outbreak of cryptosporidiosis involved approximately 400,000 individuals in Milwaukee. The outbreak was caused by contaminated municipal drinking water (Mackenzie *et al.*, 1995b). During the following 2 years of the outbreak, cryptosporidiosis was considered an underlying or contributing cause of death in immunocompromised individuals (Hoxie *et al.*, 1997). Genotyping analysis demonstrated that C. hominis was responsible for this outbreak, and further seroepidemiological studies suggested that the outbreak was larger than reported (Zhou *et al.*, 2003). *Cryptosporidium* has also been implicated in five other major outbreaks in the United States. Filtration of municipal water can reduce oocyst contamination, but protection is not absolute. The municipal water systems that were implicated in outbreaks included filtration treatment, but small numbers of oocysts have been detected in filtered municipal water in 27% to 54% of municipalities tested. Cryptosporidiosis can also be acquired by swimming in fresh surface water (i.e., lakes or rivers). *Cryptosporidium* oocysts have been identified in 65% to 97% of surface water bodies tested. The most recent water-borne outbreaks have been associated with swimming pools contaminated with *Cryptosporidium* oocysts (Anonymous, 1994a; Anonymous, 1994b; Anonymous, 2001; Mackenzie *et al.*, 1995a; Puech *et al.*, 2001; Wilberschied, 1995).

The first documented foodborne cryptosporidial transmission occurred in 1993, in Maine, where contamination was traced to consumption of C. parvum contaminated apple cider. (Millard *et al.*,1994). In Minnesota, in 1995, *Cryptosporidium* was associated with cases of acute gastroenteritis among attendees of a social event. This outbreak was epidemiologically associated with contaminated chicken salad (anonymous, 1996). In 1996, apple cider was again associated with outbreaks of cryptosporidiosis in New York (anonymous, 1997; Millard *et al.*, 1994). In 1998, in Spokane Washington, another foodborne outbreak affected about 50 people, but it could not be traced to a specific type of food (anonymous, 1998).

Transmission has been reported via human to human or fecal–oral transmission in day-care centers, and among household contacts, hospitalized patients, and health care workers. Individuals at risk of acquiring infection include animal handlers, particularly those who work with calves and other mammals. Travelers and tourists are also at an increased risk (Addiss *et al.*, 1991; Alpert *et al.*, 1986; anonymous, 1984; Combee *et al.*, 1986; Cordell and Addiss, 1994; Garcia and Castano, 1991; Garcia-Rodriguez *et al.*, 1990; Heijbel *et al.*, 1987; Jelinek *et al.*, 1997; Mahdi and Ali, 2002; Nwanyanwu *et al.*, 1989; Rahman *et al.*, 1985; Taylor *et al.*, 1985; Walters *et al.*, 1988).

Cryptosporidiosis has been described worldwide. Its true prevalence is currently unknown. In developed countries, the prevalence has been from 0.6% to 20%, and 4% to 32% in de-

veloping countries. In endemic areas, asymptomatic cryptosporidiosis is relatively frequent. Higher infection rates appear to be associated with a younger age and a warm, humid climate. Various serologic surveys have revealed greater than expected rates of seropositivity, suggesting that active or recent infection is common in the general population (Isaac-Renton *et al.*, 1999; Kuhls *et al.*, 1994; McReynolds *et al.*, 1999).

AIDS patients are more susceptible to the infection. In1986, the Centers for Disease Control and Prevention estimated that 3% to 4% of all AIDS patients had cryptosporidiosis. In later studies, *Cryptosporidium* was identified in 15% of patients with AIDS and diarrhea at the National Institutes of Health. In Haiti and Africa, up to 50% of AIDS patients are infected. Asymptomatic cryptosporidiosis has been described and the overall frequency is not known. In developed countries, cryptosporidiosis has been reported less frequently in HIV patients, as this population is educated in preventive measures of acquiring opportunistic infections.

## Pathology and pathogenesis

The *Cryptosporidium* parasitic vacuoles are located in the apical surfaces of epithelial cells of the small intestine. The vacuoles' localization is intracellular but extracytoplasmatic, making the parasite very resistant to drug treatments. The parasite communicates with the host cell selectively via a "feeder organelle", allowing the passage of nutrients as needed (Kosek *et al.*, 2001).

*Cryptosporidium* can also infect the epithelial ligning of the esophagus, stomach, small and large intestine, gallbladder and bile and pancreatic ducts; within colonic submucosal vessels, and in the respiratory tract. Histopathological changes include blunting or complete loss of villi, elongation of crypts, and infiltration of the lamina propria with polymorphonuclear leukocytes, lymphocytes, and plasma cells.

Animals with cryptosporidiosis develop diminished glucose, electrolyte, and water absorption in the small intestinal (Argenzio *et al.*, 1990). Enterotoxic or humoral factors may play a role, but to date have not been identified. The pathogenic mechanisms of *Cryptosporidium* need to be identified.

## Clinical manifestations

Cryptosporidiosis in immunocompetent and immunocompromised hosts presents as voluminous watery diarrhea, abdominal pain, weight loss, flatulence, and malaise. Nausea, vomiting, myalgias, and fever are less common. The incubation period for cryptosporidiosis is typically 2 to 14 days. The severity and duration of the illness varies and is related to immune competency of the host. Usually symptoms may last 10 to 14 days. In immunocompromised persons the disease presents more aggressively. As immune competence lowers, clinical symptoms frequently worsen. Patients may experience voluminous (1 to 25 L daily) watery diarrhea, profound weight loss, electrolyte imbalance, and severe dehydration, requiring hospitalization for months; often until they die if not treated (Flanigan and Soave, 1993; Soave, 1988).

Cryptosporidial cholecystitis or cholangitis is a frequent finding in immunocompromised patients with cryptosporidiosis. There are marked histopathologic changes in the biliary tract and gallbladder, ranging from acute inflammation to gangrenous necrosis. Coinfection of the biliary tract with cytomegalovirus has been reported. Pancreatitis can also occur in association with cryptosporidiosis in both immunocompetent and immunocompromised patients (Farman et al., 1994; Hawkins et al., 1987; Kaup et al., 1994; Norby et al., 1998). In immunocompromised patients, Cryptosporidium has also been isolated from sputum, tracheal aspirates, bronchoalveolar lavage fluid, and lung tissues (Dupont et al., 1996; Goodstein et al., 1989; Kocoshis et al., 1984; Lopez-Velez et al., 1995; Pellicelli et al., 1998).

## Diagnosis

Diagnosis of cryptosporidiosis is done by the detection of oocysts in fecal or intestinal biopsy specimens. The modified acid fast stains are being used to detect the parasite in stool specimens. Concentration techniques have been included to increase the sensitivity.

Diagnostic testing for Cryptosporidium spp. has been developed for clinical settings; however, as more environmental and food samples are involved in outbreaks many of these assays are being modified to facilitate the detection, identification and speciation of the parasite. Immunofluorescence assays (IFA) and enzyme-linked immunosorbent assay (ELISA) are available commercially (Baveja, 1998; Garcia et al., 1988). In water samples, Merifluor® DFA and IFA (Meridian Diagnostics, Inc., Cincinnati, Ohio) were reported to have a 100% specificity and sensitivity. Other kits have been developed for clinical samples such as Crypto-Cell IF-DFA ((Cellabs, Brookvale, Australia) and ProSpect®-EIA (Alexon, Inc., Ramsey, MN). Water and environmental sample kits are also available, including Cryp-a-GloTM Kits and reagents (Water-borne Inc. New Orleans,LA), Dynabeads® anti- Cryptosporidium (Dynal Biotech, Inc. Lake Success, NY), and Hydrofluor™ Combo ((Meridian Diagnostics, Inc., Cincinnati, OH) (Ortega and Arrowood, 2003). Some of these kits are being adapted to be used with food samples. PCR has been described, but its possible role and use in a clinical laboratory has not yet been established; however, it may be valuable when evaluating environmental and food specimens, where classical diagnostic tools may not be sensitive enough.

Although shellfish have not been implicated in Cryptosporidium outbreaks, infectious Cryptosporidium parvum oocysts have been isolated in molluscs. Because molluscs can filter large volumes of water, it has been suggested that mollusks could be used as environmental monitors. In Chesapeake Bay, contamination seems to be more intense during the winter months of November thru January. Oocysts viability in clam shells and in seawater with salinities of 10–30 ppt at 10°C, to 10 ppt at 20°C was demonstrated up to 40 days' storage. In experimental settings oocysts are infectious even after a period of 12 months. Oocysts can also be found in the gills and hemocytes for as long as one month. This suggests that enteric parasites could remain in the environment for larger periods of time and retain their infectivity. Therefore, molluscs should be considered as a potential source for acquiring the disease (Fayer et al., 2003; Freire-Santos et al., 2000; Gomez-Couso et al., 2003; Graczyk et al., 1998).

## Treatment

Cryptosporidiosis in immunocompetent hosts is self-limited and management is generally limited to supportive measures. Cryptosporidiosis in AIDS is chronic and unremitting; support therapy is often necessary. Antiviral drug therapy controls the viral loads, and as the cellular immune response increases cryptosporidial infection resolves.

To date there is no effective therapy for human cryptosporidiosis. Various animal models, including neonates and immunosupressed mice, gnotobiotic pigs, and in vitro cultivation using tissue culture systems (HCT-8, Caco-2, HT29, MDCK) are being used for screening drugs. A wide variety of antimicrobial and immunomodulating agents as well as special diets have been tested, particularly to control infection in AIDS patients with cryptosporidiosis (Allam and Shehab, 2002; Armitage et al., 1992; Blanshard et al., 1997; Castro Hermida et al., 2000; Hewitt et al., 2000; Kadappu et al., 2002).

Few drugs have proven limited success. The macrolide azithromycin demonstrated a decrease in stool oocyst counts in patients. Conflicting results describing the efficacy of paromomycin, a nonabsorbable aminoglycoside, in AIDS related cryptosporidiosis have been reported. In one double blind, placebo controlled trial, partial symptomatic and parasitologic responses were obtained, consistent with other anecdotal observations. However, in an AIDS Clinical Trials Group-sponsored multicenter, double blind, placebo-controlled trial, paromomycin did not show significant benefit relative to placebo (Hammel et al., 1992; Hewitt et al., 2000; Tzipori et al., 1995). In other instances, clinical response may not correlate with reduction in parasite shedding, response may be transient, and eradication may not always be accomplished.

Nitazoxanide, a nitrothiazole benzamide compound, reduces the diarrheal frequency and/or reduction of oocyst shedding in stool samples (Bowers, 1998). Zidovudine therapy may show improvement, probably explained by improved immune function rather than specific anticryptosporidial activity (Garrido Davila and Ramirez Ronda, 1991). Other therapies, including hyperimmune bovine colostrum, bovine colostral immunoglobulins, cow's milk globulin, and cocktails of monoclonal antibodies have resulted in both success and failure in humans and animals when they were administered orally (Fayer et al., 1989; Jenkins et al., 1999; Nord et al., 1990). The use of bovine transfer factor also appeared promising, but more investigation is required, including identification of the active component (Fayer et al., 1987). Clinical investigation of novel agents, including specific egg yolk antibodies to

*Cryptosporidium*, has also been evaluated (Cama and Sterling, 1991).

Cryptosporidiosis in patients with weakened immunity due to exogenous factors such as chemotherapy may resolve if immunoreductive therapy is reduced or interrupted. In those patients with HIV infections, the CD4[+] cell count is the best marker for the ability of the immune system to resolve the infection (Brink *et al.*, 2002; Navin *et al.*, 1999). In developed countries, patients with CD4 counts 180 cells/mm$^3$ or more usually develop a self-limiting cryptosporidiosis, while those with counts below 180 cells/mm$^3$ usually develop chronic and profuse diarrhea, which is exacerbated by the lack of an effective therapy.

## Prevention

Persons at risk should be advised that certain activities will increase their chances of acquiring cryptosporidiosis and should be avoided. These include unprotected physical contact with infected persons or animals, swimming in community pools, and consumption of or swimming in freshwater. Risk of transmission can be eliminated by boiling water or by drinking bottled water from safe sources, or using water filtration units.

The *Cryptosporidium* oocyst is resistant to many disinfectants used in hospitals and laboratories, including 3% hypochlorite solution, iodophor, cresylic acid, benzalkonium chloride, and 5% formaldehyde. Oocysts appear to be temperature sensitive and inactivation is achieved by exposure of oocysts to temperatures above 73°C for 1 minute and reduced by prolonged exposure to undiluted bleach or 5% ammonia, peroxide and iodine. (Barbee *et al.*, 1999; Biswas *et al.*, 2003; Chauret *et al.*, 2001; Drescher *et al.*, 2001; Venczel *et al.*, 1997).

## Cyclospora

*Cyclospora* is the newest coccidian described in humans. In 1979, Ashford described an *Isospora*-like organism affecting humans in Papua New Guinea. Other reports described this organism by various names such as "large *Cryptosporidium*", coccidian like body, cyanobacterium (blue green algae), cyanobacterium like body, or *Cryptosporidium*-like body (Long *et al.*, 1991; Shlim *et al.*, 1991).

In 1993, the conclusive identification of CLB's demonstrated that it was in fact a coccidian parasite of the genus *Cyclospora*. The species name *cayetanensis* was later coined, as initial studies were done at the Cayetano Heredia University in Lima, Peru (Ortega *et al.*, 1993; Ortega *et al.*, 1994).

*Cyclospora* belongs to the family Eimeriidae, subphylum Apicomplexa. *Cyclospora* species have been described in moles, rodents, insectivores, and snakes. To date *C. cayetanensis* seems to be host specific and infects humans exclusively (Ortega *et al.*, 1993). In the past 5 years, three new species of *Cyclospora* have been described in non-human primates. These parasites are morphologically similar to *C. cayetanensis* but phylogenetically distinct (Eberhard *et al.*, 1999).

## Characteristics

Immature and undifferentiated *Cyclospora* oocysts are excreted in the feces of infected individuals. Under optimal environmental conditions the oocysts sporulate and differentiate into two sporocysts, each containing two sporozoites. *Cyclospora* oocysts are typically round and of 8 to 10 μm in diameter, distinct from oocysts of *Cryptosporidium* (also round, but 4 to 6 μm) and *Isospora* (elliptical, 28 × 13 μm). Parasitic vacuoles containing asexual (meronts type I and II) and sexual stages have been observed in the cytoplasm of epithelial cells of the duodenum and jejunum. In spite of various reports suggesting that other animals may be reservoirs of *Cyclospora*, no *in vivo* or *in vitro* system is yet available, making research with this parasite difficult (Ortega *et al.*, 1997a).

## Epidemiology

Even before the true identify of *Cyclospora* was unveiled, epidemiological information on the parasite was being collected worldwide. In Nepal, the epidemiology and clinical description of the disease in adult expatriates was studied and later, in Nepalese children (Shlim *et al.*, 1991; Shlim *et al.*, 1999). In endemic areas of Peru, children under 10 years of age were most susceptible to infection. Asymptomatic cases suggest prior exposure to the parasite and development of protective immunity. Adults living in non endemic regions as well as foreign travelers develop symptomatic cyclosporiasis (Bern *et al.*, 2002; Ortega *et al.*, 1993).

Foodborne outbreaks in the US were recognized in 1995 and later epidemiologically associated with the consumption of contaminated fresh produce such as berries, lettuce or basil (Herwaldt and Ackers, 1997). *Cyclospora* oocysts, as with other parasitic diseases, were not recovered or detected in the implicated produce (Caceres *et al.*, 1998; Connor and Shlim, 1995; Fleming *et al.*, 1998).

Human *Cyclospora* appears to be globally distributed. Cyclosporiasis has been reported in travelers from North America, Central America, South America, the Caribbean Islands, Eastern Europe, India, Nepal, Bangladesh, and Southeast Asia. Incidence is seasonal, with most cases reported from Nepal during the warm and rainy months, whereas in Peru the parasite can be found affecting people during the warmer months of the year, but in regions were there is little or no precipitation. In the United States cases were observed between May and July (Crowley *et al.*, 1996; Lontie *et al.*, 1995; Rijpstra and Laarman, 1993; Verweij *et al.*, 2003) which coincides with the seasonality of the implicated product's country of origin.

Symptomatic infection occurs in all ages and in immunocompetent and immunocompromised hosts alike. In endemic areas, *Cyclospora* was identified in 6% to 18% of children aged 1 to 24 months in shanty towns in Lima, Peru, and in 12% of Nepalese children with diarrhea, aged 18 to 60 months. There were no Nepalese or Haitian children younger than 18 months with diarrhea identified with *Cyclospora* (Bern *et al.*, 2002;

Hoge *et al.*, 1995; Jelinek *et al.*, 1997; Ortega *et al.*, 1994). Among Haitian adults with chronic diarrhea, *Cyclospora* was detected in 11% of human immunodeficiency virus (HIV) seropositive, but none of the HIV seronegative patients tested positive (Lopez *et al.*, 2003; Pape *et al.*, 1994). In the United States and England, *Cyclospora* was identified in 0.1% to 0.5% of stool specimens received in three clinical laboratories. Most of these were from returning travelers.

Routes of transmission for *Cyclospora* are still being studied, although the fecal–oral route, either directly or via water and food, is probably the primary one. In 1995, nearly 1450 sporadic and cluster-related cases were reported in 20 states, Washington, DC, and two Canadian provinces. Epidemiological investigation demonstrated an association with eating fresh raspberries imported from Guatemala, particularly where case clustering was identified. Basil and lettuce has also been associated with outbreaks of *Cyclospora* in the USA. In other instances, *Cyclospora* oocysts have been isolated from lettuce, and a wedding cake filled with cream and raspberries (Caceres *et al.*, 1998; Herwaldt and Ackers, 1997; Katz *et al.*, 1999; Steele *et al.*, 2003).

The first water-borne outbreak occurred in Chicago in 1990. Physicians living in a hospital residence apparently contracted cyclosporiasis from contaminated water. It was epidemiologically associated with tap water from unprotected reservoir tanks that served the building and which had a broken water pump (Huang *et al.*, 1995). Another report described a waterborne outbreak in Nepal where municipal water containing acceptable levels of chlorine contained *Cyclospora* oocysts, suggesting that these oocysts are highly resistant to chemical disinfectants, particularly chlorine (Rabold *et al.*, 1994). Sporadic cases of cyclosporiasis have been connected to acquisition of the infection after exposure to water. In Utah, a man became infected after cleaning his basement that was flooded with runoff from a nearby farm following heavy rains (Sterling and Ortega, 1999).

Direct transmission from animals or person to person has not been documented. *Cyclospora* like oocysts have been reported in animals (chicken, duck and dogs). Whether these were transient or true infections remain to be determined. To date, in spite of several experimental studies attempting to infect animal species, including those reported in the literature, modes of transmission remain inconclusive (Eberhard *et al.*, 2000; Garcia-Lopez *et al.*, 1996; Yai *et al.*, 1997; Zerpa *et al.*, 1995).

Imported raspberries were one of the commodities most frequently associated to *Cyclospora*. Contamination in the farm could have occurred when berries were sprayed with insecticide, possibly diluted with contaminated surface water. Analysis of rivers used for raspberry crop irrigation demonstrated the presence of *Cyclospora* oocysts, thus supporting this idea. In Peru and Nepal, *Cyclospora* and *Cryptosporidium* oocysts have been isolated from vegetables purchased from markets in endemic areas. *Cyclospora* oocysts experimentally

inoculated on vegetables demonstrated that washing with water does not remove all the oocysts (Ortega *et al.*, 1997b). The minimum infectious dose of oocysts is unknown, but is suspected to be low. Oocyst sporulation rate and survival under different environmental conditions needs to be determined.

Analysis of the intervening transcribed spacer-1 (ITS1) region sequence of *Cyclospora* isolates from a foodborne outbreak demonstrated that all were identical, suggesting that a single source of contamination caused the infection. Variability within isolates of endemic areas may suggest mixed infections (Adam *et al.*, 2000).

## Pathology and pathogenesis

Small bowel injury is prominent. Duodenal erythema may be observed in some patients. Histologic features on duodenal and jejunal biopsies include villous atrophy, crypt hyperplasia, and epithelial disarray, with acute and chronic inflammation (Connor *et al.*, 1999; Ortega *et al.*, 1997a). Extensive lymphocytic infiltration into the surface epithelium is present, especially at the tips of the shortened villi (Ortega *et al.*, 1997a).

Parasitic vacuoles can be observed in the cytoplasm at the luminal end of the epithelial cells. Merogony and gametogony occur within the parasitophorous vacuoles in the cytoplasm of intestinal cells. Gametogony and oocyst formation occur in the cells. Unsporulated oocysts are then excreted in the feces of patients with cyclosporiasis.

The inflammatory response does not correlate with the number of parasites present in the tissues. The specific mechanism of intestinal injury remains to be elucidated. Extraintestinal infection (cholangitis) has been reported in patients with AIDS (de Gorgolas *et al.*, 2001; Sifuentes-Osornio *et al.*, 1995; Zar *et al.*, 2001).

## Clinical manifestations

Cyclosporiasis is characterized by mild to severe nausea, anorexia, abdominal cramping, mild fever, and watery diarrhea. Diarrhea alternating with constipation has been commonly reported. Some patients present with flatulent dyspepsia and less frequently, joint pain and night sweats. Onset of illness is usually sudden in patients and symptoms last an average of 7 weeks if untreated. Malabsorption has been demonstrated in a small number of patients. In AIDS patients, untreated *Cyclospora* infection and its symptoms are often chronic or relapsing (Deodhar *et al.*, 2000; Sterling and Ortega, 1999).

## Diagnosis

*Cyclospora* infection is determined by identifying the oocysts in stool samples. *Cyclospora* oocysts are round and 8 to 10 μm in diameter; on unstained wet preparations they are nonrefractile and contain globular inclusions. Modified acid fast staining greatly facilitates detection of *Cyclospora*, although oocysts stain highly variably. A modified Safranin method using microwave heat, improves the homogeneity of the stained oocysts. Without using specific stains, the oocysts autofluoresce blue green under epifluorescence microscopy. This procedure is highly sensitive but not specific (Sterling and Ortega, 1999).

*Cyclospora cayetanensis* oocysts require 2 weeks to sporulate and become infectious at 23°C. Nested PCR targeted to amplify the 18S rDNA can also be used with these preparations. When examining environmental samples, further testing is imperative using restriction fragment length polymorphism (RFLP). The described PCR also amplifies a same size fragment in *Eimeria* species which are infectious to animals, but not to humans, and can be readily found in the environment (Jinneman *et al.*, 1998; Relman *et al.*, 1996).

An oligo-ligation assay has been proposed, as it would simplify identification in the laboratories (Jinneman *et al.*, 1999). No viability assay is available for *Cyclospora*, although sporulation of immature oocysts and electrorotation have been described (Dalton *et al.*, 2001).

## Treatment

Trimethoprim-sulfamethoxazole (TMP-SMX) is a very effective treatment for cyclosporiasis. Alternative treatment with ciprofloxacin, although not as effective, may be prescribed for Sulfa-sensitive patients. AIDS patients tend to relapse and prolonged treatment may be required to eradicate infection. The prevalence of *Cyclospora* in HIV patients is not different from immunocompetent populations, probably due to the frequent use of TMP-SMX for *Pneumocystis carinii* prophylaxis among HIV patients (Madico *et al.*, 1993; Pape *et al.*, 1994; Verdier *et al.*, 2000).

## *Isospora belli*

### Characteristics

The *Isospora belli* oocysts are elliptical and substantially larger than *Cyclospora* and *Cryptosporidium*. When excreted to the environment they are unsporulated. It takes between 12 and 24 h to fully sporulate, differentiate, and form two sporocysts. Inside each sporocyst are four sporozoites. Isosporiasis can be acquired by ingestion of the mature (sporulated) oocyst. The parasite invades host intestinal epithelium and, once in the enterocyte cytoplasm, asexual (merogony) and sexual (gametogony) multiplication occurs. This is followed by production of oocysts which are excreted in the feces and sporulate outside the host to become infectious (Goodgame, 1996b).

### Epidemiology

The prevalence of *I. belli* in humans is not known. It is distributed throughout the world but is more common in tropical and subtropical climates. Endemic areas include Latin America, the Caribbean, Africa, Australia, and Southeast Asia. In the United States, *I. belli* has been implicated in several institutional outbreaks of diarrhea and as a cause of traveler's diarrhea in World War II veterans returning from the Pacific. Isosporiasis has been documented in 0.2% to 1.0% of all AIDS patients in the United States, and in 5% to 19% of AIDS patients in Haiti and Africa (Anand *et al.*, 1998; Atzori *et al.*, 1993; Dieng *et al.*, 1994; Ravera *et al.*, 1996; Sauda *et al.*, 1993; Sorvillo *et al.*, 1995). The relative infrequency of clinical *I. belli* infection among AIDS patients in the United States may also be due to the use of TMP SMX for prophylaxis. The mode of transmission is not well understood and acquisition from infected animals and humans, and through contaminated water, is suspected but not confirmed.

### Pathology and pathogenesis

Patients with isosporiasis present with atrophic mucosa, shortened villi, hypertrophic crypts, and infiltration of the lamina propria with inflammatory cells, particularly eosinophils. Parasites are found within cytoplasmic vacuoles of enterocytes. Extraintestinal isosporiasis is well documented in cats, but rarely described in humans. In one AIDS patient with acalculous cholecystitis, *I. belli* was identified in the lymph nodes and in the gallbladder lumen (Benator *et al.*, 1994; French *et al.*, 1995).

### Clinical manifestations

Isosporiasis is characterized by watery diarrhea without blood or inflammatory cells, cramping abdominal pain, anorexia, and weight loss. Low grade fever may be present. Fat malabsorption is common; peripheral eosinophilia has been documented in some cases. Immunocompetent adults usually have a self-limited diarrheal illness, but there have been case reports of prolonged illness. AIDS patients and immunocompetent infants and children often have chronic or relapsing isosporiasis. *Isospora belli* infects the entire intestine and produces severe intestinal disease. Deaths from overwhelming infections have been reported, especially in immunocompromised patients. The disease may persist for months and even years (DeHovitz *et al.*, 1986).

### Diagnosis

Diagnosis of isosporiasis is done by identifying *Isospora* oocysts in the stools of infected patients. *Isospora belli* oocysts (10–19 × 20–30 μm) can be easily detected in direct wet mounts in heavy infections, or by a modified acid fast stain. *Isospora* organisms may also be identified with a fluorescent auramine stain. Most infections are not heavy, and like most parasites, shedding of oocysts may be intermittent, making it necessary to examine a series of stool samples. Concentration procedures may contribute to detection of low number of oocysts (Soave and Johnson, Jr., 1988).

### Treatment

Trimethropim sulfamethoxazole is an effective treatment for isosporiasis, usually eliminating the clinical symptoms and oocysts shedding within 3 days. Immunocompetent patients may relapse, but will respond to re-treatment with TMP SMX. Pyrimethamine, roxithromycin and diclazuril have also been used in isosporiasis, particularly in sulfa sensitive patients (Pape *et al.*, 1989).

## *Toxoplasma gondii*

*Toxoplasma gondii* is a coccidian parasite that infects a variety of warm blooded hosts. Cats are the definitive hosts and other warm-blooded animals can serve as intermediate hosts. Cats excrete oocysts in their feces. Oocysts are environmentally resistant and can survive several years in moist shaded conditions. Infections are acquired principally by ingestion of contaminated food, water containing oocysts, ingestion of animal tissues containing cystic forms (bradyzoites), or by transplacental transmission.

### Characteristics

The unsporulated oocysts take about 24 h to sporulate outside the host and become infectious. When oocysts are ingested by the intermediate host, the oocyst walls are ruptured and the sporozoites are released. They invade epithelial cells and rapidly multiply asexually (tachyzoites). Tachyzoites multiply by endodyogeny, a process in which the mother tachyzoite is consumed by the formation of two daughter zoites. These tachizoites will encyst in the brain, liver, skeletal muscle, and cardiac muscle. These cysts contain bradyzoites which are slow multiplying forms. Cysts persist for the duration of the life of the host (Sun and Teichberg, 1988). By encysting, the parasite evades the host's immune response and ensures its own viability. Most of the *Toxoplasma* found in humans belong to three clonal lineages. The public health implications of these groups and their pathogenicity remain to be determined (Ajzenberg *et al.*, 2002; Boothroyd and Grigg, 2002).

When tissues of other animal species are ingested by felines, proteolytic enzymes digest the cyst wall and the bradyzoites are released and begin the enteroepithelial cycle and sexual multiplication. Macro and microgametocytes are produced. After fertilization, the zygote differentiates into oocysts which are passed in the feces.

Although most infections occur by ingestion of contaminated meat, foods, and water, they can also be acquired by organ transplantation or by blood transfusion. Disseminated toxoplasmosis may occur in those who received organ transplants and are receiving immunosuppressive therapy, and those individuals whose immune systems are compromised (Wittner *et al.*, 1993).

### Clinical manifestations

Toxoplasmosis can also be acquired vertically by transplacental transmission when a pregnant woman gets infected. After multiplying in the placenta, tachyzoites spread into the fetal tissues. Infection can occur at any stage of the pregnancy, but the fetus is affected the greatest when infection occurs during the first months of pregnancy. Most infected children do not show any signs of the disease until later in life when they may present with chorioretinitis and mental retardation (Gilbert *et al.*, 2001).

The overall prevalence in humans and animals varies according to the eating habits and lifestyle. The prevalence of *T. gondii* in swine is highest if the animals are not contained. Confined housing reduces the exposure to cat feces or infected rodents (Lunden *et al.*, 2002; Nalbantoglu *et al.*, 2002; Wang *et al.*, 2002).

### Treatment

Pyrimethamine in combination with folinic acid or trisulfapyrimidine is the treatment of choice for acute infections. Trimethoprim-sulfamethoxazole is effective and frequently used to prevent recurrence of acute infections in AIDS patients (Haberkorn, 1996; Romand *et al.*, 1995).

### Prevention

Toxoplasmosis can be acquired by ingestion of lamb, poultry, horse, and wild game animals. Cooking, freezing or gamma irradiation will kill the *Toxoplasma* cysts and oocysts. Temperatures of 61°C or higher for 3.6 min will inactivate the parasites and freezing at −13°C will result in non-viable cysts (Gamble, 1997).

## *Giardia lamblia*

*Giardia* is a protozoan flagellate that belongs to the Phylum Zoomastigophora. *Giardia* can cause diarrhea and malabsorption and infects millions of people throughout the world in both epidemic and sporadic forms. Most human infections result from the ingestion of contaminated water or food, or by direct fecal−oral transmission such as would occur in person-to-person contact in childcare centers, and in male homosexual activity (Adam, 2001).

### Characteristics

Three species of *Giardia* have been described based on differences discernible in cysts and trophozoites by light microscopy; *G. agilis* from amphibians; *G. microti* of wolves and muskrats; *G. muris* from rodents; and *G. lamblia* (also called *G. intestinalis* or *G. duodenalis*) from various mammals, including humans. Two additional species which are indistinguishable from *G. lamblia* by light microscopy; *G. ardeae* (heron) and *G. psittaci* (psittacine birds) have been identified based on ultrastructural morphologic differences. *Giardia lamblia* does not appear to be host-restricted and wild animals, such as beavers and muskrats, have been implicated in water-borne outbreaks. *Giardia* has been grouped into genotypes A-1, A-2 and B. Both infect humans, but A-1 is less restrictive to the number of hosts that it can infect. Mayhofer has grouped *Giardia* where assemblages A and B infect humans and C-G infect other animal species. Assemblages C and D are specific to dogs and seems to have little or no potential for zoonotic transmission. More recently, molecular classification using small subunit ribosomal RNA has placed *Giardia* as one of the most primitive eukaryotic organisms (Adam, 2000).

*Giardia* can be observed in two forms: the trophozoite and the cyst. The cyst is the infectious and environmentally resistant form. After cysts are ingested, excystation occurs in the duodenum after exposure to the acidic gastric pH and pancreatic enzymes, chemotrypsin and trypsin. Each cyst releases two vegetative trophozoites. The trophozoites replicate by asexual fission in the crypts of the duodenum and upper jejunum. Some of the trophozoites then encyst in the ileum, possibly as a result of exposure to bile salts or from cholesterol starvation. The trophozoites and cysts are excreted in the feces.

Cysts are round or oval shaped. Each measure 11–14 × 7–10 μm, has four nuclei, and contain axonemes and median bodies. Trophozoites have the shape of a teardrop (viewed dorsally or ventrally) and measure 10–20 μm in length by 5–15 μm in width. The trophozoite has a concave sucking disk with four pairs of flagella, two axonemes, two median bodies and two nuclei. The ventral disks act as suction cups, allowing mechanical attachment to the surface of the intestine. Infections may result from the ingestion of 10 or fewer *Giardia* cysts. Boiling is very effective in inactivating *Giardia* cysts, but cysts can survive after freezing for a few days (Backer, 2000; Gillin *et al.*, 1996).

### Epidemiology

*Giardia lamblia* is prevalent worldwide, and is especially common in areas where poor sanitary conditions and insufficient water treatment facilities prevail. Seasonality has been reported during late summer in the United Kingdom, United States, and Mexico. The majority of cases of *Giardia* are asymptomatic, but they can present as chronic diarrhea. Travelers to endemic areas are at high risk for developing symptomatic giardiasis. In Leningrad (now St. Petersburg), Russia, 95% of travelers developed symptomatic giardiasis. Hikers and campers are also at increased risk because *Giardia* of animal origin can be found in freshwater lakes and streams. Prevalence of *Giardia* can be as high as 35% in children attending childcare centers. Although these children are frequently asymptomatic, they may infect other family members who may develop symptomatic giardiasis (Kettlewell *et al.*, 1998; Levesque *et al.*, 1999; Maltezou *et al.*, 2001; Thompson, 2000).

*Giardia* has been associated with water-borne transmission. This results from inadequate water treatment or sewage contamination of drinking, well, or surface water. Giardiasis has also been associated with exposure to contaminated recreational water, such as swimming pools. *Giardia* cysts are susceptible to inactivation by ozone and halogens; however, the concentration of chlorine used for drinking water may not inactivate *Giardia* cysts. Inactivation by chlorine requires prolonged contact time and filtration is the recommended means for purifying water (Abbaszadegan *et al.*, 1997; Finch and Belosevic, 2001; Fricker and Crabb, 1998).

### Clinical manifestations

Clinical signs of giardiasis include loose, foul-smelling stools, and increased fat and mucus in fecal samples. Flatulence, abdominal cramps, bloating, and nausea are common, as are anorexia, malaise and weight loss. Individuals may present with

fever at the beginning of the infection. Patients present with malabsorption of fats, carbohydrates, and vitamins. Reduced intestinal disaccharidase activity may persist even after *Giardia* is eradicated. Lactase deficiency is the most common residual deficiency and occurs in 20–40% of cases (D'Anchino *et al.*, 2002; Homan and Mank, 2001).

Giardiasis may resolve spontaneously, however the illness frequently lasts for several weeks if left untreated, and sometimes for months. Chronic giardiasis is characterized by profound malaise, and diffuse epigastric and abdominal discomfort. Diarrhea and constipation may intercalate. Persistent infection may be related to the ability of *Giardia* trophozoites to change their surface antigens (VSP) avoiding the antibody mediated immune response (Adam, 2001; Nash, 1997; Yang and Adam, 1995).

Villous blunting and lymphocytic infiltration can be observed in biopsy samples of symptomatic cases. No tissue invasion occurs and high numbers of trophozoites may be present in the crypts without obvious pathology. The presence of a toxin has not been demonstrated, and no other potential mechanisms by which *Giardia* causes diarrhea have been identified (Ebert, 1999).

### Diagnosis

*Giardia* can be diagnosed by finding cysts or less commonly, trophozoites, in fecal specimens. Immunoassays that have been developed against *Giardia* include enzyme immunoassays, indirect and direct immunofluorescent assays using monoclonal antibodies Merifluor® DFA and IFA (Meridian Diagnostics, Inc., Cincinnati, Ohio) or PCR. Water and environmental sample kits are also available, including HydrofluorTM–Combo ((Meridian Diagnostics, Inc., Cincinnati, Ohio) or PCR (Ortega and Arrowood, 2003). All of these procedures are highly sensitive and specific for environmental and stool samples. In some instances, patients with chronic diarrhea and malabsorption present negative stool examinations despite ongoing suspicion of giardiasis (Bisoffi *et al.*, 1995; Ghosh *et al.*, 2000; Jelinek *et al.*, 1996).

### Treatment

Effective treatment for patients with symptomatic giardiasis is mainly a single treatment course with metronidazole. In refractory cases, multiple or combination courses have occasionally been required. Tinidazole is widely used throughout the world for treatment of giardiasis (Abbaszadegan *et al.*, 1997; Cabello *et al.*, 1997; Carroccio *et al.*, 2001; Doumbo *et al.*, 1997; Gardner and Hill, 2001; Goodgame, 1996a; Martinez and Caumes, 2001; Nash, 2001; Nash *et al.*, 2001).

Foodborne outbreaks have also been associated with *Giardia*. To date, molluscs have not been involved in a foodborne outbreak in spite of the fact that shellfish can also concentrate *Giardia* cysts. This makes shellfish a potential source for foodborne outbreaks, especially when eaten raw (Graczyk *et al.*, 1998).

## Balantidium

Balantidium coli is a ciliate parasite that although found worldwide, is not highly prevalent. It is a commensal parasite of pigs. The trophozoites reside in the large intestine and multiply by binary fission. In humans, it can cause ulcerative colitis and diarrhea. The ulcers differ from those caused by Entamoeba histolytica in that the epithelial surface is damaged, but with more superficial lesions compared to those caused by amoebae. The Balantidium cyst is excreted in the feces of the infected individual. The cyst is large and oblong, 45–65 μm in diameter and environmentally resistant. Both cyst and trophozoite contain two nuclei. Trophozoites move via their cilia and rotate on their longitudinal axis. Treatment is preferentially done with tetracycline and, alternatively, with iodoquinol and metronidazole (Garcia, 1999).

## Amoebiasis

Various amoebae can infect humans as commensals; however, Entamoeba. histolytica is the only pathogenic to humans. Entamoeba dispar, which is morphologically similar to E. histolytica, is not pathogenic. Trophozoites and the environmentally resistant E. histolytica cysts are excreted in the feces of infected individuals. Once cysts are ingested, the trophozoite excysts, colonizes the large intestine, and multiplies by binary fission, followed by encystations (Ravdin, 1989).

### Clinical presentation

Most patients are asymptomatic even when shedding cysts in their feces. In other instances, the parasites can invade the mucosa and cause an ulceration from the luminal surface of the intestine, through the lamina propria and to the muscularis mucosa. The parasite then spreads laterally, forming a flask shaped ulcer. The trophozoites feed on cell debris and red blood cells. Infection can progress, producing ulcerative amoebic colitis and causing perforation of the intestinal wall. Patients complain of diarrhea with stools containing blood and mucus, back pain, tenesmus, dehydration, and abdominal tenderness. Fulminant colitis is characterized by severe bloody diarrhea, fever, and abdominal tenderness due to transmural necrosis of the bowel. Another presentation of amoebiasis is ameboma, which resembles a carcinoma and is not necessarily associated with pain. Amoebae can also disseminate to the liver. Complications with amoebiasis are observed when the liver parenchyma is gradually replaced with necrotic debris, inflammatory cells and trophozoites. Patients present with hepatomegaly, weight loss and anemia (Salles et al., 2003).

### Treatment

Asymptomatic amoebiasis can be treated with iodoquinol, paromomycin, or diloxanide. If mild to severe and hepatic abcesses are present, amoebiasis can be treated with metronidazole or tinidazole followed with iodoquinol to treat asymptomatic amoebiasis (Bassily et al., 1987; Blessmann and Tannich, 2002; Chunge et al., 1989; Cooperstock et al., 1992).

## Dientamoeba fragilis

Dientamoeba fragilis seems to be present in various parts of the world, although very little is known of this parasite. Dientamoeba does not have a cyst form, it has an ameboid form that measures 5 to 12 μm in diameter. It is binucleate, with central nuclear granules and cytoplasmic vacuoles. Dientamoeba belongs to the phylum Sarcodina, although it is now considered to be more closely related to the flagellate genera Histomonas and Trichomonas.

### Epidemiology

Dientamoeba has a worldwide distribution, but its true prevalence is not known. In various surveys in the United States, the prevalence of the parasite has ranged from 1.4% to 18.6%.

The mode of transmission of D. fragilis is not known. Dientamoebiasis rarely occurs in conjunction with the other parasitic enteritis that are typically transmitted by contaminated food and water. Some researchers have suggested that D. fragilis may be transmitted by the ova of the human pinworm Enterobius vermicularis.

Dientamoba colonizes the caecum and proximal large intestine. Invasion of the epithelial tissues has not been observed. Symptoms are nonspecific including intermittent diarrhea, cramping, and bloating. Clinical disease has been reported in 50% of infected adults and 90% of infected children (Schwartz and Nelson, 2003). Peripheral eosinophilia has been documented principally in children. Dientamoebiasis is not known to be more severe or more common in immunocompromised hosts. Biliary involvement may occur (Cuffari et al., 1998).

### Diagnosis

Diagnosis of D. fragilis can be done by identifying trophozoites in stool samples. Specimens are optimal if they are submitted fresh for immediate processing, or preserved in polyvinyl alcohol, sodium acetate acetic acid formalin, or Schaudinn fixative. Trophozoites stain well using the hematoxylin Kinyoun or trichrome staining of stool smears, facilitating detection. As with other parasites, multiple specimens obtained on different days may enhance the chances of finding this organism (Sawangjaroen et al., 1993; Silard et al., 1979; Yang and Scholten, 1977).

### Treatment

Iodoquinol, tetracycline, paromomycin, diiodohydroxyquin and metronidazole have demonstrated effectiveness in controlling the infection (Chan et al., 1994; Cuffari et al., 1998; Preiss et al., 1991; Spencer et al., 1979).

## Microsporidia

Microsporidial organisms are obligate intracellular spore-forming parasites long recognized as pathogens in mammals, fish, crustaceans, and insects. Until the AIDS epidemic, association of human disease with microsporidial infection was rare. Human microsporidial infection has a varied clinical presentation, predominantly among HIV-infected persons, but it can also be observed in immunocompetent individuals (Mathis, 2000).

## Characteristics

Microsporidia is the term used to refer to members of the phylum Microspora, order Microsporida. The phylum contains more than 1000 species within approximately 100 genera. Six genera have been described in human infection: *Encephalitozoon* (including *Encephalitozoon intestinalis*, formerly *Septata intestinalis*, *Enterocytozoon*, *Pleistophora*, *Nosema*, *Vittaforma*, and *Trachipleistophora*. Those that have been not been characterized well enough to assign to a genus have been named *Microsporidium* sp. (Weiss, 2001).

The Microsporidia are primitive eukaryotes lacking mitochondria and Golgi apparatus. Spores contain a coiled polar filament and sporoplasm, which consists of cytoplasm and one or two nuclei. In the appropriate environment, spores extrude their filament which penetrates a host cell and sporoplasm is transferred into the host cytoplasm. Merogony and sporogony follow. The resultant new spores are released when the host cell ruptures, remaining within the host or passing into the environment.

Spores are typically ovoid or piriform. Dimensions vary by species and range from 1 to 20 μm in diameter. Microsporidial spores found in humans are relatively small at 1 to 2 μm. Shape, position of nuclei, and number of nuclei and coils of the polar filament also vary. After entering the cell, the parasite multiplies asexually and eventually forms spores lysing the host cell and invading neighboring cells. Classification of microsporidian species into various genera is determined by morphologic features and the mode of replication within the host cell (Marshall *et al.*, 1997).

## Epidemiology

The majority of microsporidiosis cases have been reported in AIDS patients, including nearly all cases of *Enterocytozoon bieneusi*, *Encephalitozoon hellem* and *E. intestinalis*. Recent reports suggest that also immunocompetent individuals and children can be infected with microsporidia.

Human microsporidiosis has a worldwide distribution, with cases in HIV-infected patients reported from North and South America, Europe, Africa, Asia, and Australia. The actual prevalence of microsporidial infection in humans is unknown. Microsporidia have been identified in 7.5% to 50% of AIDS patients with previously unexplained diarrhea. Most, but not all, studies reveal a strong correlation between intestinal microsporidiosis and clinical enteritis. This suggests that microsporidia, particularly *E. bieneusi*, are an important cause of AIDS related diarrhea, but asymptomatic cases may also be observed in individuals with HIV infection. Immunocompetent individuals and children can have asymtomatic microsporidia. *E. bienusi* can also be isolated from pigs and non-human primates (Conteas *et al.*, 1998; Escobedo and Nunez, 1999; Weber *et al.*, 1999).

The modes by which humans become infected with microsporidia are unknown. Spores are environmentally resistant and may be shed in feces, urine, or respiratory secretions of infected humans. Fecal oral and sexual modes of transmission from human to human have been proposed but not documented. Microsporidial spores have been identified in surface water and it has been suggested that it may be present in foods and drinking water as well (Fournier *et al.*, 2000; Fournier *et al.*, 2002; Thurston-Enriquez *et al.*, 2002).

## Pathology and pathogenesis

Tissue distribution, histopathologic features, and propensity to disseminate vary widely among species of microsporidia recognized in human disease. Human infection with *E. bieneusi* can be found in intestinal and biliary tissue, and respiratory tract (delAguila *et al.*, 1997; Marshall *et al.*, 1997).

Human infection with *Encephalitozoon* species is frequently disseminated. Gastroenteritis has been associated with *E. intestinalis*. Small bowel villous atrophy, enterocyte sloughing, and acute inflammatory infiltrates may be present (Magaud *et al.*, 1997). Spores have been identified in the urine or respiratory secretions of several patients with enteritis, suggesting that disseminated infection is under recognized (Chu and West, 1996; Marshall *et al.*, 1997).

The other species of Microsporidia, *E. hellem* has been identified in superficial corneal and conjunctival epithelia. Disseminated infection with *E. hellem* was demonstrated in a patient with advanced AIDS and *E. cuniculi* has been identified in hepatocytes of AIDS patients with granulomatous hepatitis and peritonitis, respectively. Myositis has been reported associated with *Pleistophora* and *Trachipleistophora* infection in which organisms were demonstrated among atrophic muscle fibers.

## Clinical manifestations

Clinical features of microsporidiosis include chronic diarrhea, keratoconjunctivitis, myositis, nephritis, hepatitis, sinusitis, and pneumonia. Enteritis due to *E. bieneusi* or *E. intestinalis* is the most common clinical manifestation of microsporidiosis in patients with AIDS. Diarrhea is typically chronic and intermittent, loose to watery, and non-bloody. Fecal leukocytes are usually absent. Anorexia, weight loss, and dehydration are common, and findings on D-xylose testing are often abnormal, consistent with malabsorption. Abdominal cramping, nausea, and vomiting may occur. The density of *E. bieneusi* infestation and associated histopathologic appearance may not correlate with severity of enteritis, and chronic asymptomatic carriage has been documented. Co-infection has been observed with *Cryptosporidium* or cytomegalovirus.

## Diagnosis

Diagnosis of microsporidiosis is done by demonstration of spores in tissue, stool, or body fluids. With use of chromotrope based stains (trichrome blue stain, Weber chromotrope stain) spores appear pink and refractile, and some have a darker staining band along the shorter axis, making them distinct from other stained elements. Chemofluorescence (Uvitex 2B, calcofluor white 2MR) stain spore walls nonspecifically when observed under fluorescence microscopy. Concentration of body fluid specimens may improve yield but has not been consistently helpful for stool specimens.

Identification of microsporidia in the environment presents an additional challenge: most species of the animal kingdom can be parasitized by some other species of microsporidia which could be confused with those infectious to humans. New tests using specific monoclonal antibodies and molecular tools such as PCR are being developed to aid in the diagnosis of *Microsporidia* in tissue, fecal and environmental samples.

## Treatment

To date, albendazole (at doses of 400 mg orally twice daily for a minimum of 4 weeks) has been the most successful agent used in the treatment of AIDS related microsporidiosis. *E. intestinalis* and *E. cuniculi* are more susceptible and *E. bieneusi* seems much less responsive. Even in responders, microsporidia may not be eradicated, and chronic maintenance therapy may be required. Supportive treatment to control chronic diarrhea, fluid and electrolyte loss, and malabsorption may be required.

## Conclusion

Foodborne disease outbreaks caused by parasites are being recognized with increased frequency. The consumption of fresh fruits and vegetables in developed countries has increased and local production cannot meet this demand. Thus, importation of products from countries where sanitary standards are not controlled makes it even more difficult to control food related diseases.

Enrichment media are not available for parasites; therefore, isolation and identification methodologies need to be extremely sensitive and specific to detect small numbers of contaminants that may be present in foods. Molecular tools such as PCR, RFLP, and variations of these techniques are being developed to improve the sensitivity and specificity of detection and identification processes.

Consumption of raw shellfish is a practice that has increased throughout the years. Although there has not been a reported outbreak associated with consumption of shellfish, it is known that parasites can be concentrated and filtered from contaminated water.

Parasite control and inactivation have been challenging tasks. Good agricultural practices and water quality are key strategies to prevent contamination of fresh produce with parasites. Common disinfectants and sanitizers have demonstrated limited effect on parasites. Chlorination of drinking water at permissible concentrations is relatively ineffective at inactivating spores, cysts, and oocysts, and cannot be recommended as a sole method for water treatment. Boiling of water inactivates cysts, oocysts, and spores. Irradiation treatment of produce has also been investigated. *Toxoplasma* cysts can be effectively inactivated using this method; however, irradiation effects on other parasites have not been evaluated.

## References

Abbaszadegan, M., Hasan, M.N., Gerba, C.P., Roessler, P.F., Wilson, E.R., Kuennen, R., and VanDellen, E. 1997. The disinfection efficacy of a point-of-use water treatment system against bacterial, viral and protozoan water-borne pathogens. Water Res. 31: 574–582.

Adam, R.D. 2000. The *Giardia lamblia* genome. Int. J. Parasitol. 30: 475–484.

Adam, R.D. 2001. Biology of *Giardia lamblia*. Clin. Microbiol. Rev. 14: 447–475.

Adam, R.D., Ortega, Y.R., Gilman, R.H., and Sterling, C.R. 2000. Intervening transcribed spacer region 1 variability in *Cyclospora cayetanensis*. J. Clin. Microbiol. 38: 2339–2343.

Addiss, D.G., Stewart, J.M., Finton, R.J., Wahlquist, S.P., Williams, R.M., Dickerson, J.W., Spencer, H.C., and Juranek, D.D. 1991. *Giardia lamblia* and *Cryptosporidium* infections in child day-care centers in Fulton County, Georgia. Pediatr. Infect. Dis..J. 10: 907–911.

Ajzenberg, D., Cogne, N., Paris, L., Bessieres, M.H., Thulliez, P., Filisetti, D., Pelloux, H., Marty, P., and Darde, M.L. 2002. Genotype of 86 *Toxoplasma gondii* isolates associated with human congenital toxoplasmosis, and correlation with clinical findings. J. Infect. Dis. 186: 684–689.

Allam, A.F. and Shehab, A.Y. 2002. Efficacy of azithromycin, praziquantel and mirazid in treatment of cryptosporidiosis in school children. J. Egypt. Soc. Parasitol. 32: 969–978.

Alpert, G., Bell, L.M., Kirkpatrick, C.E., Budnick, L.D., Campos, J.M., Friedman, H.M., and Plotkin, S.A. 1986. Outbreak of cryptosporidiosis in a day-care center. Pediatrics 77: 152 157.

Anand, L., Brajachand, N.G., and Dhanachand, C.H. 1996. Cryptosporidiosis in HIV infection. J. Commun. Dis. 28: 241–244.

Anand, L., Dhanachand, C., and Brajachand, N. 1998. Prevalence and epidemiologic characteristics of opportunistic and non-opportunistic intestinal parasitic infections in HIV positive patients in Manipur. J. Commun. Dis. 30: 19–22.

Anonymous. 1984. *Cryptosporidiosis* among children attending day-care centers–Georgia, Pennsylvania, Michigan, California, New Mexico. MMWR 33: 599–601.

Anonymous. 1994a. *Cryptosporidium* infections associated with swimming pools–Dane County, Wisconsin, 1993. MMWR. 43: 561–563.

Anonymous. 1994b. *Cryptosporidium* infections associated with swimming pools–Dane County, Wisconsin, 1993. JAMA 272: 914–915.

Anonymous. 1996. Foodborne outbreak of diarrheal illness associated with *Cryptosporidium parvum*–Minnesota, 1995. MMWR. 45: 783–784.

Anonymous. 1997. Outbreaks of *Escherichia coli* O157:H7 infection and cryptosporidiosis associated with drinking unpasteurized apple cider–Connecticut and New York, October 1996. MMWR. 46: 4–8.

Anonymous. 1998. Foodborne outbreak of cryptosporidiosis–Spokane, Washington, 1997. MMWR. 47: 565–567.

Anonymous. 2001. Prevalence of parasites in fecal material from chlorinated swimming pools–United States, 1999. MMWR. 50: 410–412.

Argenzio, R.A., Liacos, J.A., Levy, M.L., Meuten, D.J., Lecce, J.G., and Powell, D.W. 1990. Villous atrophy, crypt hyperplasia, cellular infiltration, and impaired glucose-Na absorption in enteric cryptosporidiosis of pigs. Gastroenterology 98: 1129–1140.

Armitage, K., Flanigan, T., Carey, J., Frank, I., MacGregor, R.R., Ross, P., Goodgame, R., and Turner, J. 1992. Treatment of cryptosporidiosis with paromomycin. A report of five cases. Arch. Intern. Med. 152: 2497–2499.

Atzori, C., Bruno, A., Chichino, G., Cevini, C., Bernuzzi, A.M., Gatti, S., Comolli, G., and Scaglia, M. 1993. HIV-1 and parasitic infections in rural Tanzania. Ann. Trop. Med. Parasitol. 87: 585–593.

Backer, H.D. 2000. *Giardiasis* – An elusive cause of gastrointestinal distress. Phys. Sports Med. 28: 46-+.

Barbee, S.L., Weber, D.J., Sobsey, M.D., and Rutala, W.A. 1999. Inactivation of *Cryptosporidium parvum* oocyst infectivity by disinfection and sterilization processes. Gastroint..Endosc. 49: 605–611.

Bassily, S., Farid, Z., el Masry, N.A., and Mikhail, E.M. 1987. Treatment of intestinal *E. histolytica* and G. lamblia with metronidazole, tinidazole and ornidazole: a comparative study. J. Trop. Med. Hyg. 90: 9–12.

Baveja, UK 1998. Acid fast staining versus ELISA for detection of *Cryptosporidium* in stool. J. Commun. Dis. 30: 241–244.

Benator, D.A., French, A.L., Beaudet, L.M., Levy, C.S., and Orenstein, J.M. 1994. *Isospora belli* infection associated with acalculous cholecystitis in a patient with AIDS. Ann. Intern. Med. 121: 663–664.

Bern, C., Ortega, Y., Checkley, W., Roberts, J.M., Lescano, A.G., Cabrera, L., Verastegui, M., Black, R.E., Sterling, C., and Gilman, R.H. 2002. Epidemiologic differences between cyclosporiasis and cryptosporidiosis in Peruvian children. Emerg. Infect. Dis. 8: 581–585.

Bisoffi, Z., diTommaso, M., Ricciardi, M.L., Majori, S., and Campello, C. 1995. Evaluation of a commercially available ELISA for detection of *Giardia lamblia* antigen in faeces: Preliminary results in unconventional samples. Eur. J. Epidemiol. 11: 703–705.

Biswas, K., Craik, S., Smith, D.W., and Belosevic, M. 2003. Synergistic inactivation of *Cryptosporidium parvum* using ozone followed by free chlorine in natural water. Water Res. 37: 4737–4747.

Black, R.E. 1986. Pathogens that cause travelers' diarrhea in Latin America and Africa. Rev. Infect. Dis. 8: Suppl 2: S131-S135.

Blanshard, C., Shanson, D.C., and Gazzard, B.G. 1997. Pilot studies of azithromycin, letrazuril and paromomycin in the treatment of cryptosporidiosis. Int. J. STD AIDS 8: 124–129.

Blessmann, J. and Tannich, E. 2002. Treatment of asymptomatic intestinal Entamoeba histolytica infection. N. Engl. J. Med. 347: 1384.

Boothroyd, J.C. and Grigg, M.E. 2002. Population biology of *Toxoplasma gondii* and its relevance to human infection: do different strains cause different disease? Curr. Opin. Microbiol. 5: 438–442.

Bowers, M. 1998. Nitazoxanide for cryptosporidial diarrhea. BETA. 30–31.

Brink, A.K., Mahe, C., Watera, C., Lugada, E., Gilks, C., Whitworth, J., and French, N. 2002. Diarrhea, CD4 counts and enteric infections in a community-based cohort of HIV-infected adults in Uganda. J. Infect. 45: 99–106.

Cabello, R.R., Guerrero, L.R., Garcia, M.D.M., and Cruz, A.G. 1997. Nitazoxanide for the treatment of intestinal protozoan and helminthic infections in Mexico. Trans. Roy. Soc. Trop. Med. Hyg. 91: 701–703.

Caceres, V.M., Ball, R.T., Somerfeldt, S.A., Mackey, R.L., Nichols, S.E., Mackenzie, W.R., and Herwaldt, B.L. 1998. A foodborne outbreak of cyclosporiasis caused by imported raspberries. J. Fam. Pract. 47: 231–234.

Cama, V.A. and Sterling, C.R. 1991. Hyperimmune hens as a novel source of anti-*Cryptosporidium* antibodies suitable for passive immune transfer. J. Protozool. 38: 42S-43S.

Carroccio, A., Cavataio, F., Montalto, G., Paparo, F., Troncone, R., and Iacono, G. 2001. Treatment of *Giardiasis* reverses 'active' coeliac disease to 'latent' coeliac disease. Eur. J. Gastroent. Hepatol. 13: 1101–1105.

Castro Hermida, J.A., Freire, S.F., Oteiza Lopez, A.M., Vergara Castiblanco, C.A., and Ares-Mazas, M.E. 2000. *In vitro* and *in vivo* efficacy of lasalocid for treatment of experimental cryptosporidiosis. Vet. Parasitol. 90: 265–270.

Chacin-Bonilla, L. 1995. Cryptosporidiosis in humans. Invest. Clin. 36: 207–250.

Chan, F.T., Guan, M.X., Mackenzie, A.M., and Diaz-Mitoma, F. 1994. Susceptibility testing of Dientamoeba fragilis ATCC 30948 with iodoquinol, paromomycin, tetracycline, and metronidazole. Antimicrob. Agents. Chemother. 38: 1157–1160.

Chauret, C.P., Radziminski, C.Z., Lepuil, M., Creason, R., and Andrews, R.C. 2001. Chlorine dioxide inactivation of *Cryptosporidium parvum* oocysts and bacterial spore indicators. Appl. Environ. Microbiol. 67: 2993–3001.

Chick, S.E., Koopman, J.S., Sooorapanth, S., and Brown, M.E. 2001. Infection transmission system models for microbial risk assessment. Sci. Total Environ. 274: 197–207.

Chu, P.G. and West, A.B. 1996. *Encephalitozoon (Septata) intestinalis* – Cytologic, histologic, and electron microscopic features of a systemic intestinal pathogen. Am. J. Clin. Pathol. 106: 606–614.

Chunge, C.N., Estambale, B.B., Pamba, H.O., Chitayi, P.M., Munanga, P.N., and Kang'ethe, S. 1989. Comparison of four nitroimidazole compounds for treatment of symptomatic amoebiasis in Kenya. East Afr. Med. J. 66: 724–727.

Combee, C.L., Collinge, M.L., and Britt, E.M. 1986. Cryptosporidiosis in a hospital-associated day care center. Pediatr. Infect. Dis. 5: 528–532.

Connor, B.A., Reidy, J., and Soave, R. 1999. Cyclosporiasis: clinical and histopathologic correlates. Clin. Infect. Dis. 28: 1216–1222.

Connor, B.A. and Shlim, D.R. 1995. Foodborne transmission of *Cyclospora*. Lancet 346: 1634.

Conteas, C.N., Berlin, O.G.W., Lariviere, M.J., Pandhumas, S.S., Speck, C.E., Porschen, R., and Nakaya, T. 1998. Examination of the prevalence and seasonal variation of intestinal microsporidiosis in the stools of persons with chronic diarrhea and human immunodeficiency virus infection. Am. Trop. Med. Hyg. 58: 559–561.

Cooperstock, M., DuPont, H.L., Corrado, M.L., Fekety, R., and Murray, D.M. 1992. Evaluation of new anti-infective drugs for the treatment of diarrhea caused by Entamoeba histolytica. Infectious Diseases Society of America and the Food and Drug Administration. Clin. Infect. Dis. 15: Suppl 1: S254-S258.

Cordell, R.L. and Addiss, D.G. 1994. Cryptosporidiosis in child care settings: a review of the literature and recommendations for prevention and control. Pediatr. Infect. Dis. J. 13: 310–317.

Crowley, B., Path, C., Moloney, C., and Keane, C.T. 1996. *Cyclospora* species–a cause of diarrhoea among Irish travellers to Asia. Ir. Med. J. 89: 110–112.

Cuffari, C., Oligny, L., and Seidman, E.G. 1998. Dientamoeba fragilis masquerading as allergic colitis. J. Pediatr. Gastroenterol. Nutr. 26: 16–20.

D'Anchino, M., Orlando, D., and De Feudis, L. 2002. *Giardia lamblia* infections become clinically evident by eliciting symptoms of irritable bowel syndrome. J. Infect. 45: 169–172.

Dalton, C., Goater, A.D., Pethig, R., and Smith, H.V. 2001. Viability of *Giardia* intestinalis cysts and viability and sporulation state of *Cyclospora cayetanensis* oocysts determined by electrorotation. Appl. Environ. Microbiol. 67: 586–590.

de Gorgolas, M., Fortes, J., and Fernandez Guerrero, M.L. 2001. *Cyclospora cayetanensis* Cholecystitis in a patient with AIDS. Ann. Intern. Med. 134: 166.

DeHovitz, J.A., Pape, J.W., Boncy, M., and Johnson, W.D., Jr. 1986. Clinical manifestations and therapy of *Isospora belli* infection in patients with the acquired immunodeficiency syndrome. N. Engl. J. Med. 315: 87–90.

delAguila, C., LopezVelez, R., Fenoy, S., Turrientes, C., Cobo, J., Navajas, R., Visvesvara, G.S., Croppo, G.P., DaSilva, A.J., and Pieniazek, N.J. 1997. Identification of *Enterocytozoon* bieneusi spores in respiratory samples from an AIDS patient with a 2-year history of intestinal microsporidiosis. J. Clin. Microbiol. 35: 1862–1866.

Deodhar, L., Maniar, J.K., and Saple, D.G. 2000. *Cyclospora* infection in acquired immunodeficiency syndrome. J. Assoc. Phys. India 48: 404–406.

Dieng, T., Ndir, O., Diallo, S., Coll-Seck, A.M., and Dieng, Y. 1994. [Prevalence of *Cryptosporidium* sp and *Isospora belli* in patients with acquired immunodeficiency syndrome (AIDS) in Dakar (Senegal]. Dakar Med. 39: 121–124.

Doumbo, O., Rossignol, J.F., Pichard, E., Traore, H.A., Dembele, M., Diakite, M., Traore, F., and Diallo, D.A. 1997. Nitazoxanide in the treatment of cryptosporidial diarrhea and other intestinal parasitic infections associated with acquired immunodeficiency syndrome in tropical Africa. Am. J. Trop. Med. Hyg. 56: 637–639.

Drescher, A.C., Greene, D.M., and Gadgil, A.J. 2001. *Cryptosporidium* inactivation by low-pressure UV in a water disinfection device. J. Environ. Health 64: 31–35.

Dupont, C., Bougnoux, M.E., Turner, L., Rouveix, E., and Dorra, M. 1996. Microbiological findings about pulmonary cryptosporidiosis in two AIDS patients. J. Clin. Microbiol. 34: 227–229.

Eberhard, M.L., Da Silva, A.J., Lilley, B.G., and Pieniazek, N.J. 1999. Morphologic and molecular characterization of new *Cyclospora* species from Ethiopian monkeys: C. cercopitheci sp.n., C. colobi sp.n., and C. papionis sp.n. Emerg. Infect. Dis. 5: 651–658.

Eberhard, M.L., Ortega, Y.R., Hanes, D.E., Nace, E.K., Do, R.Q., Robl, M.G., Won, K.Y., Gavidia, C., Sass, N.L., Mansfield, K., Gozalo, A., Griffiths, J., Gilman, R., Sterling, C.R., and Arrowood, M.J. 2000. Attempts to establish experimental *Cyclospora cayetanensis* infection in laboratory animals. J. Parasitol. 86: 577–582.

Ebert, E.C. 1999. *Giardia* induces proliferation and interferon gamma production by intestinal lymphocytes. Gut 44: 342–346.

Escobedo, A.A. and Nunez, F.A. 1999. Prevalence of intestinal parasites in Cuban acquired immunodeficiency syndrome (AIDS) patients. Acta Tropica 72: 125–130.

Farman, J., Brunetti, J., Baer, J.W., Freiman, H., Comer, G.M., Scholz, F.J., Koehler, R.E., Laffey, K., Green, P., and Clemett, A.R. 1994. AIDS-related cholangiopancreatographic changes. Abdom. Imaging 19: 417–422.

Fayer, R., Andrews, C., Ungar, B.L., and Blagburn, B. 1989. Efficacy of hyperimmune bovine colostrum for prophylaxis of cryptosporidiosis in neonatal calves. J. Parasitol. 75: 393–397.

Fayer, R., Klesius, P.H., and Andrews, C. 1987. Efficacy of bovine transfer factor to protect neonatal calves against experimentally induced clinical cryptosporidiosis. J. Parasitol. 73: 1061–1062.

Fayer, R., Trout, J.M., Lewis, E.J., Santin, M., Zhou, L., Lal, A.A., and Xiao, L. 2003. Contamination of Atlantic coast commercial shellfish with *Cryptosporidium*. Parasitol. Res. 89: 141–145.

Finch, G.R. and Belosevic, M. 2001. Controlling *Giardia* spp. and *Cryptosporidium* spp. in drinking water by microbial reduction processes. Can. J. Civil Eng. 28: 67–80.

Flanigan, T.P. and Soave, R. 1993. Cryptosporidiosis. Prog. Clin. Parasitol. 3: 1–20.

Fleming, C.A., Caron, D., Gunn, J.E., and Barry, M.A. 1998. A foodborne outbreak of *Cyclospora cayetanensis* at a wedding: clinical features and risk factors for illness. Arch. Intern. Med. 158: 1121–1125.

Fournier, S., Dubrou, S., Liguory, O., Gaussin, F., Santillana-Hayat, M., Sarfati, C., Molina, J.M., and Derouin, F. 2002. Detection of microsporidia, cryptosporidia and *Giardia* in swimming pools: a one-year prospective study. FEMS Immunol. Med. Microbiol. 33: 209 213.

Fournier, S., Liguory, O., Santillana-Hayat, M., Guillot, E., Sarfati, C., Dumoutier, N., Molina, J.M., and Derouin, F. 2000. Detection of microsporidia in surface water: a one-year follow-up study. FEMS Immunol. Med. Microbiol. 29: 95–100.

Freire-Santos, F., Oteiza-Lopez, A.M., Vergara-Castiblanco, C.A., Ares-Mazas, E., Alvarez-Suarez, E., and Garcia-Martin, O. 2000. Detection of *Cryptosporidium* oocysts in bivalve molluscs destined for human consumption. J. Parasitol. 86: 853–854.

French, A.L., Beaudet, L.M., Benator, D.A., Levy, C.S., Kass, M., and Orenstein, J.M. 1995. Cholecystectomy in patients with AIDS: clinicopathologic correlations in 107 cases. Clin. Infect. Dis. 21: 852–858.

Fricker, C.R. and Crabb, J.H. 1998. Water-borne cryptosporidiosis: Detection methods and treatment options. Adv. Parasitol. – Opportunistic Protozoa in Humans 40: 241–278.

Gamble, H.R. 1997. Parasites associated with pork and pork products. Revue Scientifique et Technique de l'Office International des Epizooties 16: 496–506.

Garcia, L.S. 1999. Flagellates and ciliates. Clin. Lab. Med. 19: 621–38.

Garcia, L.S., Brewer, T.C., and Bruckner, D.A. 1988. Incidence of *Cryptosporidium* in all patients submitting stool specimens for ova and parasite examination: monoclonal antibody IFA method. Diagn. Microbiol. Infect. Dis. 11: 25–27.

Garcia, P. and Castano, M.A. 1991. Importance of food-handlers in the transmission of cryptosporidiosis. Enferm. Infec. Microbiol. Clin. 9: 583–584.

Garcia-Lopez, H.L., Rodriguez-Tovar, L.E., and Medina-De la Garza CE 1996. Identification of *Cyclospora* in poultry. Emerg. Infect. Dis. 2: 356–357.

Garcia-Rodriguez, J.A., Martin-Sanchez, A.M., Canut, B.A., and Garcia Luis, E.J. 1990. The prevalence of *Cryptosporidium* species in children in day care centres and primary schools in Salamanca (Spain): an epidemiological study. Eur. J. Epidemiol. 6: 432–435.

Gardner, T.B. and Hill, D.R. 2001. Treatment of *Giardia*sis. Clin. Microbiol. Rev. 14: 114–128.

Garrido Davila, J.I. and Ramirez Ronda, C.H. 1991. Updates on AIDS cryptosporidiosis: a review. Bol. Assoc. Med. P. R. 83: 65–68.

Gasser, R.B., Zhu, X., Caccio, S., Chalmers, R., Widmer, G., Morgan, U.M., Thompson, R.C., Pozio, E., and Browning, G.F. 2001. Genotyping *Cryptosporidium parvum* by single-strand conformation polymorphism analysis of ribosomal and heat shock gene regions. Electrophoresis 22: 433–437.

Ghosh, S., Debnath, A., Sil, A., De, S., Chattopadhyay, D.J., and Das, P. 2000. PCR detection of *Giardia lamblia* in stool: targeting intergenic spacer region of multicopy rRNA gene. Mol. Cell. Probes 14: 181–189.

Gilbert, R.E., Gras, L., Wallon, M., Peyron, F., Ades, A.E., and Dunn, D.T. 2001. Effect of prenatal treatment on mother to child transmission of *Toxoplasma gondii*: retrospective cohort study of 554 mother-child pairs in Lyon, France. Int. J. Epidemiol. 30: 1303–1308.

Gillin, F.D., Reiner, D.S., and McCaffery, J.M. 1996. Cell biology of the primitive eukaryote *Giardia lamblia*. Ann. Rev. Microbiol. 50: 679–705.

Gobet, P. and Toze, S. 2001. Sensitive genotyping of *Cryptosporidium parvum* by PCR-RFLP analysis of the 70-kilodalton heat shock protein (HSP70) gene. FEMS Microbiol. Lett. 200: 37–41.

Gomez-Couso, H., Freire-Santos, F., Martinez-Urtaza, J., Garcia-Martin, O., and Ares-Mazas, M.E. 2003. Contamination of bivalve molluscs by *Cryptosporidium* oocysts: the need for new quality control standards. Int. J. Food Microbiol. 87: 97–105.

Goodgame, R.W. 1996a. Diagnosis and treatment of gastrointestinal protozoal infections. Curr. Opin. Infect. Dis. 9: 346–352.

Goodgame, R.W. 1996b. Understanding intestinal spore-forming protozoa: cryptosporidia, microsporidia, *Isospora*, and *Cyclospora*. Ann. Intern. Med. 124: 429–441.

Goodstein, R.S., Colombo, C.S., Illfelder, M.A., and Skaggs, R.E. 1989. Bronchial and gastrointestinal cryptosporidiosis in AIDS. J. Am. Osteopath. Assoc. 89: 195–197.

Graczyk, T.K., Farley, C.A., Fayer, R., Lewis, E.J., and Trout, J.M. 1998. Detection of *Cryptosporidium* oocysts and *Giardia* cysts in the tissues of eastern oysters (Crassostrea virginica) carrying principal oyster infectious diseases. J. Parasitol. 84: 1039–1042.

Haberkorn, A. 1996. Chemotherapy of human and animal coccidioses: State and perspectives. Parasitol. Res. 82: 193–199.

Hammel, P., Rogeaux, O., Diouf, M.L., Gouyon, B., Ruszniewski, P., and Bernades, P. 1992. Efficacy of paromomycin sulfate (Humagel) in an AIDS infected patient with intestinal cryptosporidiosis. Gastroenterol. Clin. Biol. 16: 1010–1011.

Hawkins, S.P., Thomas, R.P., and Teasdale, C. 1987. Acute pancreatitis: a new finding in *Cryptosporidium* enteritis. Br. Med. J. (Clin. Res. Ed) 294: 483–484.

Heijbel, H., Slaine, K., Seigel, B., Wall, P., McNabb, S.J., Gibbons, W., and Istre, G.R. 1987. Outbreak of diarrhea in a day care center with spread to household members: the role of *Cryptosporidium*. Pediatr. Infect. Dis. J. 6: 532–535.

Herwaldt, B.L. and Ackers, M.L. 1997. An outbreak in 1996 of cyclosporiasis associated with imported raspberries. The *Cyclospora* Working Group. N. Engl. J. Med. 336: 1548–1556.

Hewitt, R.G., Yiannoutsos, C.T., Higgs, E.S., Carey, J.T., Geiseler, P.J., Soave, R., Rosenberg, R., Vazquez, G.J., Wheat, L.J., Fass, R.J., Antoninievic, Z., Walawander, A.L., Flanigan, T.P., and Bender, J.F. 2000. Paromomycin: no more effective than placebo for treatment of cryptosporidiosis in patients with advanced human immunodeficiency virus infection. AIDS Clinical Trial Group. Clin. Infect. Dis. 31: 1084–1092.

Hoge, C.W., Shlim, D.R., Ghimire, M., Rabold, J.G., Pandey, P., Walch, A., Rajah, R., Gaudio, P., and Echeverria, P. 1995. Placebo-controlled trial of co-trimoxazole for *Cyclospora* infections among travellers and foreign residents in Nepal. Lancet 345: 691–693.

Homan, W.L. and Mank, T.G. 2001. Human *Giardiasis*: genotype linked differences in clinical symptomatology. Int. J. Parasitol. 31: 822–826.

Hoxie, N.J., Davis, J.P., Vergeront, J.M., Nashold, R.D., and Blair, K.A. 1997. Cryptosporidiosis-associated mortality following a massive water-borne outbreak in Milwaukee, Wisconsin. Am. J. Pub. Health 87: 2032–2035.

Huang, P., Weber, J.T., Sosin, D.M., Griffin, P.M., Long, E.G., Murphy, J.J., Kocka, F., Peters, C., and Kallick, C. 1995. The first reported outbreak of diarrheal illness associated with *Cyclospora* in the United States. Ann. Intern. Med. 123: 409–414.

Isaac-Renton, J., Blatherwick, J., Bowie, W.R., Fyfe, M., Khan, M., Li, A., King, A., McLean, M., Medd, L., Moorehead, W., Ong, C.S., and Robertson, W. 1999. Epidemic and endemic seroprevalence of antibodies to *Cryptosporidium* and *Giardia* in residents of three communities with different drinking water supplies. Am. J. Trop. Med. Hyg. 60: 578–583.

Jelinek, T., Lotze, M., Eichenlaub, S., Loscher, T., and Nothdurft, H.D. 1997. Prevalence of infection with *Cryptosporidium parvum* and *Cyclospora cayetanensis* among international travellers. Gut 41: 801–804.

Jelinek, T., Peyerl, G., Loscher, T., and Nothdurft, H.D. 1996. *Giardiasis* in travellers. Evaluation of an antigen-capture ELISA for the detection of *Giardia lamblia*-antigen in stool. Zeit. Gastroenterol. 34: 237–240.

Jenkins, M.C., O'Brien, C., Trout, J., Guidry, A., and Fayer, R. 1999. Hyperimmune bovine colostrum specific for recombinant *Cryptosporidium parvum* antigen confers partial protection against cryptosporidiosis in immunosuppressed adult mice. Vaccine 17: 2453 2460.

Jinneman, K.C., Wetherington, J.H., Hill, W.E., Adams, A.M., Johnson, J.M., Tenge, B.J., Dang, N.L., Manger, R.L., and Wekell, M.M. 1998. Template preparation for PCR and RFLP of amplification products for the detection and identification of *Cyclospora* sp. and *Eimeria* spp. Oocysts directly from raspberries. J. Food Prot. 61: 1497–1503.

Jinneman, K.C., Wetherington, J.H., Hill, W.E., Omiescinski, C.J., Adams, A.M., Johnson, J.M., Tenge, B.J., Dang, N.L., and Wekell, M.M. 1999. An oligonucleotide-ligation assay for the differentiation between *Cyclospora* and *Eimeria* spp. polymerase chain reaction amplification products. J. Food Prot. 62: 682–685.

Kadappu, K.K., Nagaraja, M.V., Rao, P.V., and Shastry, B.A. 2002. Azithromycin as treatment for cryptosporidiosis in human immunodeficiency virus disease. J. Postgrad. Med. 48: 179 181.

Katz, D., Kumar, S., Malecki, J., Lowdermilk, M., Koumans, E.H., and Hopkins, R. 1999. Cyclosporiasis associated with imported raspberries, Florida, 1996. Public Health Rep. 114: 427–438.

Kaup, F.J., Kuhn, E.M., Makoschey, B., and Hunsmann, G. 1994. Cryptosporidiosis of liver and pancreas in rhesus monkeys with experimental SIV infection. J. Med. Primatol. 23: 304–308.

Kettlewell, J.S., Bettiol, S.S., Davies, N., Milstein, T., and Goldsmid, J.M. 1998. Epidemiology of *Giardia*sis in Tasmania: A potential risk to residents and visitors. J. Travel Med. 5: 127 130.

Kocoshis, S.A., Cibull, M.L., Davis, T.E., Hinton, J.T., Seip, M., and Banwell, J.G. 1984. Intestinal and pulmonary cryptosporidiosis in an infant with severe combined immune deficiency. J. Pediatr. Gastroenterol. Nutr. 3: 149–157.

Kosek, M., Alcantara, C., Lima, A.A., and Guerrant, R.L. 2001. Cryptosporidiosis: an update. Lancet Infect. Dis. 1: 262–269.

Kuhls, T.L., Mosier, D.A., Crawford, D.L., and Griffis, J. 1994. Seroprevalence of cryptosporidial antibodies during infancy, childhood, and adolescence. Clin. Infect. Dis. 18: 731–735.

Levesque, B., Rochette, L., Levallois, P., Barthe, C., Gauvin, D., and Chevalier, P. 1999. Descriptive analysis of *Giardiasis* in Quebec

(Canada) and relation with drinking water source. Revue D Epidemiologie et de Sante Publique 47: 403–410.

Lindergard, G., Nydam, D.V., Wade, S.E., Schaaf, S.L., and Mohammed, H.O. 2003. A novel multiplex polymerase chain reaction approach for detection of four human infective *Cryptosporidium* isolates: *Cryptosporidium parvum*, types H and C, *Cryptosporidium* canis, and *Cryptosporidium* felis in fecal and soil samples. J. Vet. Diagn. Invest 15: 262–267.

Long, E.G., White, E.H., Carmichael, W.W., Quinlisk, P.M., Raja, R., Swisher, B.L., Daugharty, H., and Cohen, M.T. 1991. Morphologic and staining characteristics of a cyanobacterium-like organism associated with diarrhea. J. Infect. Dis. 164: 199–202.

Lontie, M., Degroote, K., Michiels, J., Bellers, J., Mangelschots, E., and Vandepitte, J. 1995. *Cyclospora* sp.: a coccidian that causes diarrhoea in travellers. Acta Clin. Belg. 50: 288–290.

Lopez, A.S., Bendik, J.M., Alliance, J.Y., Roberts, J.M., Da Silva, A.J., Moura, I.N., Arrowood, M.J., Eberhard, M.L., and Herwaldt, B.L. 2003. Epidemiology of *Cyclospora cayetanensis* and other intestinal parasites in a community in Haiti. J. Clin. Microbiol. 41: 2047–2054.

Lopez-Velez, R., Tarazona, R., Garcia, C.A., Gomez-Mampaso, E., Guerrero, A., Moreira, V., and Villanueva, R. 1995. Intestinal and extraintestinal cryptosporidiosis in AIDS patients. Eur. J. Clin. Microbiol. Infect. Dis. 14: 677–681.

Lunden, A., Lind, P., Engvall, E.O., Gustavsson, K., Uggla, A., and Vagsholm, I. 2002. Serological survey of *Toxoplasma gondii* infection in pigs slaughtered in Sweden. Scand. J. Infect. Dis. 34: 362–365.

Mackenzie, W.R., Kazmierczak, J.J., and Davis, J.P. 1995a. An outbreak of cryptosporidiosis associated with a resort swimming pool. Epidemiol. Infect. 115: 545–553.

Mackenzie, W.R., Schell, W.L., Blair, K.A., Addiss, D.G., Peterson, D.E., Hoxie, N.J., Kazmierczak, J.J., and Davis, J.P. 1995b. Massive outbreak of water-borne *Cryptosporidium* infection in Milwaukee, Wisconsin: recurrence of illness and risk of secondary transmission. Clin. Infect. Dis. 21: 57–62.

Madico, G., Gilman, R.H., Miranda, E., Cabrera, L., and Sterling, C.R. 1993. Treatment of *Cyclospora* infections with co-trimoxazole. Lancet 342: 122–123.

Magaud, A., Achbarou, A., and Desportes-Livage, I. 1997. Cell invasion by the Microsporidium *Encephalitozoon* intestinalis. J. Eukar. Microbiol. 44: 81S.

Mahdi, N.K. and Ali, N.H. 2002. Cryptosporidiosis among animal handlers and their livestock in Basrah, Iraq. East Afr. Med. J. 79: 550–553.

Maltezou, H.C., Zafiropoulou, A., Mavrikou, M., Bozavoutoglou, E., Liapi, G., Foustoukou, M., and Kafetzis, D.A. 2001. Acute diarrhoea in children treated in an outpatient setting in Athens, Greece. J. Infect. 43: 122–127.

Marshall, M.M., Naumovitz, D., Ortega, Y., and Sterling, C.R. 1997. Water-borne protozoan pathogens. Clin. Microbiol. Rev. 10: 67–85.

Martinez, V. and Caumes, E. 2001. Metronidazole. Ann. Dermatol. Venereol. 128: 903–909.

Mathis, A. 2000. Microsporidia: emerging advances in understanding the basic biology of these unique organisms. Int. J. Parasitol. 30: 795–804.

McReynolds, C.A., Lappin, M.R., Ungar, B., McReynolds, L.M., Bruns, C., Spilker, M.M., Thrall, M.A., and Reif, J.S. 1999. Regional seroprevalence of *Cryptosporidium parvum*-specific IgG of cats in the United States. Vet. Parasitol. 80: 187–195.

Millard, P.S., Gensheimer, K.F., Addiss, D.G., Sosin, D.M., Beckett, G.A., Houck-Jankoski, A., and Hudson, A. 1994. An outbreak of cryptosporidiosis from fresh-pressed apple cider. JAMA 272: 1592–1596.

Morgan, U., Weber, R., Xiao, L., Sulaiman, I., Thompson, R.C., Ndiritu, W., Lal, A., Moore, A., and Deplazes, P. 2000. Molecular characterization of *Cryptosporidium* isolates obtained from human immunodeficiency virus-infected individuals living in Switzerland, Kenya, and the United States. J. Clin. Microbiol. 38: 1180–1183.

Morgan-Ryan, U.M., Fall, A., Ward, L.A., Hijjawi, N., Sulaiman, I., Fayer, R., Thompson, R.C., Olson, M., Lal, A., and Xiao, L. 2002. *Cryptosporidium* hominis n. sp. (Apicomplexa: Cryptosporidiidae) from Homo sapiens. J. Eukar. Microbiol. 49: 433–440.

Nalbantoglu, S., Vatansever, Z., Deniz, A., Babur, C., Cakmak, A., Karaer, Z., and Korudag, E. 2002. Sero prevalance of *Toxoplasma gondii* by the Sabin feldman and indirect flourescent antibody tests in cattle in the Turkish Republic of Northern Cyprus. Turk. J. Vet. Anim. Sci. 26: 825–828.

Nash, T.E. 1997. Antigenic variation in *Giardia lamblia* and the host's immune response. Phil. Trans. Roy. Soc. London Ser. B-Biol. Sci. 352: 1369–1375.

Nash, T.E. 2001. Treatment of *Giardia lamblia* infections. Pediatr. Infect. Dis. J. 20: 193–195.

Nash, T.E., Ohl, C.A., Thomas, E., Subramanian, G., Keiser, P., and Moore, T.A. 2001. Treatment of patients with refractory *Giardia*sis. Clin. Infect. Dis. 33: 22–28.

Navin, T.R., Weber, R., Vugia, D.J., Rimland, D., Roberts, J.M., Addiss, D.G., Visvesvara, G.S., Wahlquist, S.P., Hogan, S.E., Gallagher, L.E., Juranek, D.D., Schwartz, D.A., Wilcox, C.M., Stewart, J.M., Thompson, S.E., III, and Bryan, R.T. 1999. Declining CD4+ T-lymphocyte counts are associated with increased risk of enteric parasitosis and chronic diarrhea: results of a 3-year longitudinal study. J. Acquir. Immune Defic. Syndr. Hum. Retrovirol. 20: 154 159.

Norby, S.M., Bharucha, A.E., Larson, M.V., and Temesgen, Z. 1998. Acute pancreatitis associated with *Cryptosporidium parvum* enteritis in an immunocompetent man. Clin. Infect. Dis. 27: 223–224.

Nord, J., Ma, P., DiJohn, D., Tzipori, S., and Tacket, C.O. 1990. Treatment with bovine hyperimmune colostrum of cryptosporidial diarrhea in AIDS patients. AIDS 4: 581–584.

Nwanyanwu, O.C., Baird, J.N., and Reeve, G.R. 1989. Cryptosporidiosis in a day-care center. Tex. Med. 85: 40–43.

Ortega, Y. and Arrowood, M.J. 2003. *Cryptosporidium*, *Cyclospora*, and *Isospora*. In: Manual of Clinical Microbiology, P.Murray, E.J. Baron, J.H. Jorgensen, M.A. Pfaller, and R.H.Yolken, eds. ASM Press, Washington, D.C. p. 2008–2019.

Ortega, Y.R., Gilman, R.H., and Sterling, C.R. 1994. A new coccidian parasite (Apicomplexa: Eimeriidae) from humans. J. Parasitol. 80: 625–629.

Ortega, Y.R., Nagle, R., Gilman, R.H., Watanabe, J., Miyagui, J., Quispe, H., Kanagusuku, P., Roxas, C., and Sterling, C.R. 1997a. Pathologic and clinical findings in patients with cyclosporiasis and a description of intracellular parasite life-cycle stages. J. Infect. Dis. 176: 1584–1589.

Ortega, Y.R., Roxas, C.R., Gilman, R.H., Miller, N.J., Cabrera, L., Taquiri, C., and Sterling, C.R. 1997b. Isolation of *Cryptosporidium parvum* and *Cyclospora cayetanensis* from vegetables collected in markets of an endemic region in Peru. Am. J. Trop. Med. Hyg. 57: 683–686.

Ortega, Y.R., Sterling, C.R., Gilman, R.H., Cama, V.A., and Diaz, F. 1993. *Cyclospora* species – A new protozoan pathogen of humans. New Engl. J. Med. 328: 1308–1312.

Palmer, C.J., Xiao, L., Terashima, A., Guerra, H., Gotuzzo, E., Saldias, G., Bonilla, J.A., Zhou, L., Lindquist, A., and Upton, S.J. 2003. *Cryptosporidium muris*, a rodent pathogen, recovered from a human in Peru. Emerg. Infect. Dis. 9: 1174–1176.

Pape, J.W., Verdier, R.I., Boncy, M., Boncy, J., and Johnson, W.D., Jr. 1994. *Cyclospora* infection in adults infected with HIV. Clinical manifestations, treatment, and prophylaxis. Ann. Intern. Med. 121: 654–657.

Pape, J.W., Verdier, R.I., and Johnson, W.D., Jr. 1989. Treatment and prophylaxis of *Isospora belli* infection in patients with the acquired immunodeficiency syndrome. N. Engl. J. Med. 320: 1044–1047.

Pedraza-Diaz, S., Amar, C.F., McLauchlin, J., Nichols, G.L., Cotton, K.M., Godwin, P., Iversen, A.M., Milne, L., Mulla, J.R., Nye, K., Panigrahl, H., Venn, S.R., Wiggins, R., Williams, M., and Youngs, E.R. 2001. *Cryptosporidium meleagridis* from humans: molecular analysis and description of affected patients. J. Infect. 42: 243–250.

Pellicelli, A.M., Palmieri, F., Spinazzola, F., D'Ambrosio, C., Causo, T., De Mori, P., Bordi, E., and D'Amato, C. 1998. Pulmonary crypt-

osporidiosis in patients with acquired immunodeficiency syndrome. Minerva Med. 89: 173–175.

Preiss, U., Ockert, G., Broemme, S., and Otto, A. 1991. On the clinical importance of Dientamoeba fragilis infections in childhood. J. Hyg. Epidemiol. Microbiol. Immunol. 35: 27–34.

Puech, M.C., McAnulty, J.M., Lesjak, M., Shaw, N., Heron, L., and Watson, J.M. 2001. A statewide outbreak of cryptosporidiosis in New South Wales associated with swimming at public pools. Epidemiol. Infect. 126: 389–396.

Rabold, J.G., Hoge, C.W., Shlim, D.R., Kefford, C., Rajah, R., and Echeverria, P. 1994. *Cyclospora* outbreak associated with chlorinated drinking water. Lancet 344: 1360–1361.

Rahman, A.S., Sanyal, S.C., Al Mahmud, K.A., and Sobhan, A. 1985. *Cryptosporidium* diarrhoea in calves and their handlers in Bangladesh. Indian J. Med. Res. 82: 510–516.

Ravdin, J.I. 1989. Immunobiology of human infection by Entamoeba histolytica. Pathol. Immunopathol. Res. 8: 179–205.

Ravera, M., Reggiori, A., and Riccioni, G. 1996. Prevalence of *Isospora belli* in AIDS and immunocompetent patients in Uganda. Presse Med. 25: 1170.

Relman, D.A., Schmidt, T.M., Gajadhar, A., Sogin, M., Cross, J., Yoder, K., Sethabutr, O., and Echeverria, P. 1996. Molecular phylogenetic analysis of *Cyclospora*, the human intestinal pathogen, suggests that it is closely related to Eimeria species. J. Infect. Dis. 173: 440–445.

Rijpstra, A.C. and Laarman, J.J. 1993. Repeated findings of unidentified small *Isospora*-like coccidia in faecal specimens from travellers returning to The Netherlands. Trop. Geogr. Med. 45: 280–282.

Romand, S., Bryskier, A., Moutot, M., and Derouin, F. 1995. In-Vitro and In-Vivo Activities of Roxithromycin in Combination with Pyrimethamine Or Sulfadiazine Against *Toxoplasma gondii*. J. Antimicrob. Chemother. 35: 821–832.

Salles, J.M., Moraes, L.A., and Salles, M.C. 2003. Hepatic amebiasis. Braz. J.I nfect. Dis. 7: 96 110.

Sauda, F.C., Zamarioli, L.A., Ebner, F.W., and Mello, L.B. 1993. Prevalence of *Cryptosporidium* sp. and *Isospora belli* among AIDS patients attending Santos Reference Center for AIDS, Sao Paulo, Brazil. J. Parasitol. 79: 454–456.

Sawangjaroen, N., Luke, R., and Prociv, P. 1993. Diagnosis by faecal culture of Dientamoeba fragilis infections in Australian patients with diarrhoea. Trans. R. Soc. Trop. Med. Hyg. 87: 163–165.

Schwartz, M.D. and Nelson, M.E. 2003. Dientamoeba fragilis infection presenting to the emergency department as acute appendicitis. J. Emerg. Med. 25: 17–21.

Shlim, D.R., Cohen, M.T., Eaton, M., Rajah, R., Long, E.G., and Ungar, B.L. 1991. An alga-like organism associated with an outbreak of prolonged diarrhea among foreigners in Nepal. Am. J. Trop. Med. Hyg. 45: 383–389.

Shlim, D.R., Hoge, C.W., Rajah, R., Scott, R.M., Pandy, P., and Echeverria, P. 1999. Persistent high risk of diarrhea among foreigners in Nepal during the first 2 years of residence. Clin. Infect. Dis. 29: 613–616.

Sifuentes-Osornio, J., Porras-Cortes, G., Bendall, R.P., Morales-Villarreal, F., Reyes-Teran, G., and Ruiz-Palacios, G.M. 1995. *Cyclospora cayetanensis* infection in patients with and without AIDS: biliary disease as another clinical manifestation. Clin. Infect. Dis. 21: 1092 1097.

Silard, R., Colea, A., Panaitescu, D., Florescu, P., and Roman, N. 1979. Studies on Dientamoeba fragilis in Romania. I. Dientamoeba fragilis isolated from clinical cases. Problems of diagnosis, incidence, clinical aspects. Arch. Roum. Pathol. Exp. Microbiol. 38: 359–372.

Soave, R. 1988. Cryptosporidiosis and isosporiasis in patients with AIDS. Infect Dis. Clin. North Am. 2: 485–493.

Soave, R. and Johnson, W.D., Jr. 1988. *Cryptosporidium* and *Isospora belli* infections. J. Infect. Dis. 157: 225–229.

Sorvillo, F.J., Lieb, L.E., Seidel, J., Kerndt, P., Turner, J., and Ash, L.R. 1995. Epidemiology of isosporiasis among persons with acquired immunodeficiency syndrome in Los Angeles County. Am. J. Trop. Med. Hyg. 53: 656–659.

Spencer, M.J., Garcia, L.S., and Chapin, M.R. 1979. Dientamoeba fragilis. An intestinal pathogen in children? Am. J. Dis. Child 133: 390–393.

Steele, M., Unger, S., and Odumeru, J. 2003. Sensitivity of PCR detection of *Cyclospora cayetanensis* in raspberries, basil, and mesclun lettuce. J. Microbiol. Methods 54: 277–280.

Sterling, C.R. and Ortega, Y.R. 1999. *Cyclospora*: an enigma worth unraveling. Emerg. Infect. Dis. 5: 48–53.

Sun, T. and Teichberg, S. 1988. Protozoal infections in the acquired immunodeficiency syndrome. J. Electron Microsc. Tech. 8: 79–103.

Taylor, J.P., Perdue, J.N., Dingley, D., Gustafson, T.L., Patterson, M., and Reed, L.A. 1985. Cryptosporidiosis outbreak in a day-care center. Am. J. Dis. Child 139: 1023–1025.

Thompson, R.C.A. 2000. *Giardiasis* as a re-emerging infectious disease and its zoonotic potential. Int. J. Parasitol. 30: 1259–1267.

Thurston-Enriquez, J.A., Watt, P., Dowd, S.E., Enriquez, J., Pepper, I.L., and Gerba, C.P. 2002. Detection of protozoan parasites and microsporidia in irrigation waters used for crop production. J. Food Prot. 65: 378–382.

Tzipori, S., Griffiths, J., and Theodus, C. 1995. Paromomycin treatment against cryptosporidiosis in patients with AIDS. J. Infect. Dis. 171: 1069–1070.

Tzipori, S. and Griffiths, J.K. 1998. Natural history and biology of *Cryptosporidium parvum*. Adv. Parasitol. 40: 5–36.

Venczel, L.V., Arrowood, M., Hurd, M., and Sobsey, M.D. 1997. Inactivation of *Cryptosporidium parvum* oocysts and Clostridium perfringens spores by a mixed-oxidant disinfectant and by free chlorine. Appl. Environ. Microbiol. 63: 1598–1601.

Verdier, R.I., Fitzgerald, D.W., Johnson, W.D., Jr., and Pape, J.W. 2000. Trimethoprim-sulfamethoxazole compared with ciprofloxacin for treatment and prophylaxis of *Isospora belli* and *Cyclospora cayetanensis* infection in HIV-infected patients. A randomized, controlled trial. Ann. Intern. Med. 132: 885–888.

Verweij, J.J., Laeijendecker, D., Brienen, E.A., van Lieshout, L., and Polderman, A.M. 2003. Detection of *Cyclospora cayetanensis* in travellers returning from the tropics and subtropics using microscopy and real-time PCR. Int. J.Med. Microbiol. 293: 199–202.

Walters, I.N., Miller, N.M., van den, E.J., Dees, G.C., Taylor, L.A., Taynton, L.F., and Bennett, K.J. 1988. Outbreak of cryptosporidiosis among young children attending a day-care centre in Durban. S. Afr. Med. J. 74: 496–499.

Wang, C.H., Diderrich, V., Kliebenstein, J., Patton, S., Zimmerman, J., Hallam, A., Bush, E., Faulkner, C., and McCord, R. 2002. *Toxoplasma gondii* levels in swine operations: differences due to technology choice and impact on costs of production. Food Cont. 13: 103 106.

Weber, R., Ledergerber, B., Zbinden, R., Altwegg, M., Pfyffer, G.E., Spycher, M.A., Briner, J., Kaiser, L., Opravil, M., Meyenberger, C., and Flepp, M. 1999. Enteric infections and diarrhea in human immunodeficiency virus-infected persons – Prospective community-based cohort study. Arch. Int. Med. 159: 1473–1480.

Weiss, L.M. 2001. Microsporidia: emerging pathogenic protists. Acta Tropica 78: 89–102.

Widmer, G. 1998. Genetic heterogeneity and PCR detection of *Cryptosporidium parvum*. Adv. Parasitol. 40: 223–239.

Widmer, G., Akiyoshi, D., Buckholt, M.A., Feng, X., Rich, S.M., Deary, K.M., Bowman, C.A., Xu, P., Wang, Y., Wang, X., Buck, G.A., and Tzipori, S. 2000. Animal propagation and genomic survey of a genotype 1 isolate of *Cryptosporidium parvum*. Mol. Biochem. Parasitol. 108: 187–197.

Widmer, G., Tzipori, S., Fichtenbaum, C.J., and Griffiths, J.K. 1998. Genotypic and phenotypic characterization of *Cryptosporidium parvum* isolates from people with AIDS. J. Infect. Dis. 178: 834–840.

Wilberschied, L. 1995. A swimming-pool-associated outbreak of cryptosporidiosis. Kans. Med. 96: 67–68.

Wittner, M., Tanowitz, H.B., and Weiss, L.M. 1993. Parasitic infections in AIDS patients. Cryptosporidiosis, isosporiasis, microsporidiosis, cyclosporiasis. Infect. Dis. Clin. North Am. 7: 569–586.

Xiao, L., Alderisio, K., Limor, J., Royer, M., and Lal, A.A. 2000. Identification of species and sources of *Cryptosporidium* oocysts in storm waters with a small-subunit rRNA-based diagnostic and genotyping tool. Appl. Environ. Microbiol. 66: 5492–5498.

Xiao, L., Bern, C., Limor, J., Sulaiman, I., Roberts, J., Checkley, W., Cabrera, L., Gilman, R.H., and Lal, A.A. 2001. Identification of 5 types of *Cryptosporidium* parasites in children in Lima, Peru. J. Infect. Dis. 183: 492–497.

Yai, L.E., Bauab, A.R., Hirschfeld, M.P., de Oliveira, M.L., and Damaceno, J.T. 1997. The first two cases of *Cyclospora* in dogs, Sao Paulo, Brazil. Rev. Inst. Med. Trop. Sao Paulo 39: 177–179.

Yang, J. and Scholten, T. 1977. Dientamoeba fragilis: a review with notes on its epidemiology, pathogenicity, mode of transmission, and diagnosis. Am. J. Trop. Med. Hyg. 26: 16–22.

Yang, Y.M. and Adam, R.D. 1995. A group of *Giardia lamblia* variant-specific surface protein (VSP) genes with nearly identical 5′ regions. Mol. Biochem. Parasitol. 75: 69–74.

Zar, F.A., El Bayoumi, E., and Yungbluth, M.M. 2001. Histologic proof of acalculous cholecystitis due to *Cyclospora cayetanensis*. Clin. Infect. Dis. 33: E140-E141.

Zerpa, R., Uchima, N., and Huicho, L. 1995. *Cyclospora cayetanensis* associated with watery diarrhoea in Peruvian patients. J. Trop. Med. Hyg. 98: 325–329.

Zhou, L., Singh, A., Jiang, J., and Xiao, L. 2003. Molecular surveillance of *Cryptosporidium* spp. in raw wastewater in Milwaukee: implications for understanding outbreak occurrence and transmission dynamics. J. Clin. Microbiol. 41: 5254–5257.

# Foodborne Mycotoxins: Chemistry, Biology, Ecology, and Toxicology

10

Maribeth A. Cousin, Ronald T. Riley, and James J. Pestka

## Abstract

Molds produce mycotoxins, which are secondary metabolites that can cause acute or chronic diseases in humans when ingested from contaminated foods. Although the mechanisms and health effects of most mycotoxins are not completely resolved, potential diseases include cancers and tumors in different target organs (heart, liver, kidney, nerves), gastrointestinal disturbances, alteration of the immune system, and reproductive problems. Species of *Aspergillus*, *Fusarium*, *Penicillium*, and *Claviceps* grow in agricultural commodities or foods and elaborate the major mycotoxins, namely aflatoxins, deoxynivalenol, ochratoxin A, fumonisins, ergot alkaloids, T-2 toxin, and zearalenone, and minor mycotoxins such as cyclopiazonic acid and patulin. Mycotoxins occur mainly in cereal grains (barley, maize, rye, wheat), coffee, dairy products, fruits, nuts and spices. Control of mycotoxins in foods has focused on minimizing mycotoxin production in the field, during storage or destruction once produced; however, most control approaches have not been totally successful. Monitoring foods for mycotoxins is important to manage strategies such as regulations and guidelines, which are used by 77 countries, and for developing exposure assessments essential for accurate risk characterization. Chromatographic methods and immunoassays are most commonly used to detect mycotoxins. Research will continue to elucidate different aspects of mycotoxins and their importance to human health.

## Introduction

Mycotoxins are secondary fungal metabolites that are produced from primary metabolites and secreted into the microenvironment around the mold and when consumed or absorbed by animals and humans can cause illness or behavioral changes. This definition distinguishes between "fungal metabolites", "toxic fungal metabolites", and "mycotoxins". There are probably tens of thousands of fungal metabolites; however, the foodborne-fungal metabolites that are suspected or known to cause disease in animals and humans number in the hundreds (Riley nd Norred, 1996; Riley, 1998). The number of known mycotoxins that pose a measurable health risk to animals and humans is quite limited for several reasons. Firstly, a basic tenet of toxicology is that "the dose makes the poison". This means that although animals and humans are exposed every day to mycotoxins, the doses are insufficient to make them acutely poisonous. Secondly, while there are thousands of publications documenting poisonous effects of fungal metabolites in laboratory experiments, the levels and routes of exposure do not model the exposure that occurs when farm animals or humans are exposed to naturally contaminated feeds and foods. Thus, the potential for toxicity must not be confused with the documented and confirmed toxic effects in field situations. Nonetheless, the knowledge derived from *in vitro* studies and with laboratory animals serve as a "red flag" for the possible contribution of mycotoxins in altering immune function (Bondy and Pestka, 2000), contributing to unexplained farm animal and human diseases, and feed-associated performance problems in farm animals (Osweiler, 2000). Finally, for humans, some groups are at much higher risk due to cultural and socioeconomic conditions. For example, in less developed countries, subsistence farmers consuming potentially contaminated commodities may be exposed to much higher levels of mycotoxins just before harvest of the year's crops at a time when the food supply is not sufficiently diverse. In addition, nutritional deficiencies can increase susceptibility to mycotoxin-associated disease.

The mycotoxins that present the greatest risk to humans and farm animals are those that occur in commodities that are consumed in large amounts. Affected commodities include, for example, maize, wheat, sorghum, barley, millet, rye, peanuts, and to a much lesser extent rice. In addition, mycotoxins in airborne dusts, hay and forage grasses can cause diseases in farm animals or humans. The genera of molds of greatest concern for foodborne illnesses are *Aspergillus*, *Fusarium*, *Penicillium*, and *Claviceps* with *Neotyphodium* being of concern for grazing animals and *Stachybotrys* for animals that consume hay and for humans due to skin contact and/or inhalation. The mycotoxins of greatest concern from food consumption are aflatoxins $B_1$ and $M_1$, ochratoxin A, fumonisins $B_1$ and $B_2$, deoxynivalenol, T-2 toxin, zearalenone, ergot alkaloids, ergot-like alkaloids, and macrocyclic trichothecenes. Since 1961 when the first convincing evidence for the effects of aflatoxins in animals was reported, there have been thousands of papers and numerous reviews written on mycotoxins. As an example, since 1990 there have been many reviews published on aspects of mycotoxins produced by *Aspergillus* species (Coulombe, 1991; Wood, 1991; Bilgami and Sinha, 1992; Strømer, 1992; Ellis *et al.*, 1991; 1994a; Scott, 1994; Trail *et al.*, 1995; Galvano *et al.*, 1996a; Smith, 1997; Benford *et al.*, 2001; Henry *et al.*, 2001; Varga *et al.*, 2001; Bayman *et al.*, 2002;), *Claviceps* species (Buchta and Cvak, 1999), *Fusarium* species (Sharma and Kim, 1991; Mirocha *et al.*, 1992; Scott, 1993; Marasas, 1995; D'Mello *et al.*, 1997; Shier and Abbas, 1999; Shier *et al.*, 2000; World Health Organization, 2000b; Bolger *et al.*, 2001; Canady *et al.*, 2001a; 2001b; Marasas, 2001a,b; Miller, 2001), and *Penicillium* species (Scott, 1994; Abramson, 1997). Some general reviews on mycotoxins published since 1990 include those by United Nations Environment Programme, 1990; Sharma and Salunkhe, 1991; Frisvad, 1995; Miller, 1995; Moss, 1998; Pittet, 1998; Steyn, 1998; Sweeney and Dobson, 1998; Bodine, 1999; Fink-Germmels, 1999; Steyn and Stander, 1999; Sweeney and Dobson, 1999; Pitt, 2000; World Health Organization, 2000a; Trucksess, 2001; Deshpande, 2002; Frisvad and Thrane, 2002; and CAST, 2003. This review will be limited to those mycotoxins for which exposure is known to be high and those that are either known to cause disease in animals or humans, are associated with human disease, or are suspected to be modifying factors in disease processes. Biochemical mechanisms of action that could modulate disease processes will also be reviewed.

## Characteristics of the major mycotoxins

Mycotoxins represent a varied group of chemicals that cause toxicities in animals and humans. The chemical structures and properties are often important determinants for the health effects seen in animals and humans. The chemical structure of all of the most important mycotoxins has been published in a recent report by the Council for Agricultural Science and Technology (CAST, 2003). Some of the chemical/physical properties (name, formula, molecular weight, melting point and fluorescence/UV absorbance) of the mycotoxins of the greatest concern in foods and feeds are listed in Table 10.1. Recently a three-volume collection containing chemical/physical/structural/spectral data on approximately 1200 fungal metabolites has been published (Cole and Schweikert, 2003). All mycotoxins have carbon, hydrogen and oxygen and many, including cyclopiazonic acid, ergot alkaloids, fumonisins and ochratoxin A, also contain nitrogen and other elements. For the mycotoxins of greatest concern, molecular weights range from 154 for patulin to 722 for fumonisins with most around 300 to 400. Melting points also vary from a low of 96°C for ochratoxin A to a high of 289°C for aflatoxin $B_2$. All of those listed in Table 10.1, except fumonisins and T-2 toxin, absorb light in the UV range and some have unique UV spectra and, in the case of the aflatoxins and ochratoxin A, emit fluorescence when ex-

cited by UV light. Mycotoxins have many different chemical structures, which are presented in the section on health effects along with the proposed mechanisms of pathogenicity. Many of the major mycotoxins contain five- and six-membered ring structures: aflatoxins (see Figure 10.1), cyclopiazonic acid (see Figure 10.2), ochratoxin A (see Figure 10.3), trichothecenes such as deoxynivalenol and T-2 toxin (see Figure 10.5), zearalenone (see Figure 10.6) and patulin (see Figure 10.7). Fumonisins lack the ring structures that are common in the other major mycotoxins (see Figure 10.4). There are about 70 ergot alkaloids produced by *Claviceps* species, which were discussed by Buchta and Cvak (1999), and chemical structures based on lysergic acids or derivates were given for these ergot alkaloids. Generally, mycotoxins can be classified by their biosynthetic pathways rather than their specific organic chemical structures (Smith and Moss, 1985). The primary pathways for mycotoxin biosynthesis are presented in Table 10.2.

Aflatoxins are heterocyclic compounds that have coumarin rings linked to di/tetrahydrofuran sections (Smith, 1997). Aflatoxins are synthesized by the condensation of units of acetate with malonate minus carbon dioxide and successively reacted with seven malonate units in the polyketide pathway, which then produces cyclic rings to give norsolorinic acid followed by a series of intermediates and ultimately aflatoxin (Hocking, 2001). Aflatoxins $B_1$ and $B_2$ contain a cyclopentenone ring, but aflatoxins $G_1$ and $G_2$ have a δ-lactone (Smith, 1997). Both aflatoxins $B_1$ and $G_1$ have a double bond in the terminal furan ring, which causes greater toxicity (Smith, 1997). Aflatoxin M1 is secreted in milk from mammals after ingestion of aflatoxin $B_1$; therefore, it is a derivative of aflatoxin $B_1$ with a hydroxyl group at the 4-position (Henry *et al.*, 2001). Aflatoxin $M_1$ has a similar formula and molecular weight to aflatoxin $G_1$ (Henry *et al.*, 2001).

Ergot alkaloids are produced by *Claviceps purpurea* and *Claviceps fusiformis*, particularly from the sclerotia (compact hyphae) that have different amino acids, carbohydrates, lipids (triglycerides with high concentration of ricinoleic acid), pigments and the active alkaloids (United Nations Environment Programme, 1990). The basic chemical structures of the ergot alkaloids are 3,4-substituted indole derivatives in tetracyclic ergoline rings (Keller, 1999). ʟ-Tryptophan and mevalonic acid are the precursors for the ergoline ring with methionine contributing to the N-methyl group (United Nations Environment Programme, 1990; Buchta and Cvak, 1999; Keller, 1999). There are about 70 different ergot alkaloids produced by *Claviceps* species; however, derivatives from both lysergic acid and clavines are the two basic types of ergot alkaloids (Buchta and Cvak, 1999; Jegorov, 1999; Keller, 1999). Sclerotia in *C. purpurea* produce color (dark purple) and contain active alkaoids (United Nations Environment Programme, 1990).

Fumonisins belong to the amino-polyalcoholic classification and were named for their producer *Fusarium moniliforme*, which now has been named *Fusarium verticilioides* (Desjardins *et al.*, 1996). There are now 28 different fumonisin analogs that can be separated into A, B, C and P series (Rheeder *et*

**Table 10.1** Chemical properties of the major mycotoxins and some minor mycotoxins[a]

| Mycotoxin | Chemical name | Formula | Molecular weight | Melting point (°C) | Fluorescence/UV absorbance |
|---|---|---|---|---|---|
| **Aflatoxins** | | | | | |
| Aflatoxin B$_1$ | Heterocyclic compounds with di/tertra-hydrodifurano linked to a substituted coumarin | C$_{17}$H$_{12}$O$_6$ | 312 | 268–269 | Bright blue UV 360 nm |
| Aflatoxin B$_2$ | | C$_{17}$H$_{14}$O$_6$ | 314 | 286–289 | Bright blue UV 360 nm |
| Aflatoxin G$_1$ | | C$_{17}$H$_{12}$O$_7$ | 328 | 244–246 | Blue-green UV 360 nm |
| Aflatoxin G$_2$ | | C$_{17}$H$_{14}$O$_7$ | 330 | 230 | Blue-green UV 360 nm |
| Aflatoxin M$_1$ | 4-Hydroxy derivative of aflatoxin B$_1$ | C$_{17}$H$_{12}$O$_7$ | 328 | 299 | Blue-violet UV 360 nm |
| Cyclopiazonic acid | Indole tetramic acid | C$_{20}$H$_{20}$O$_3$N$_2$ | 336 | 245–246 | depends on solvent |
| Deoxynivalenol | 12,13-Epoxy-3α,7α,15-trihydroxy trichothec-9-ene-8-one | C$_{15}$H$_{20}$O$_6$ | 296 | 151–153 | UV spectrum not characteristic |
| Ergot alkaloids | Lysergic acid derivatives | About 70 compounds[b] | About 70 compounds[b] | 162–230 | 235–310 nm – depends on the alkaloid analyzed |
| | Isolysergic acid derivatives | | | 196–243 | |
| | Clavine alkaloids | | | 206–249 | |
| **Fumonisins (28 fumonisins with B$_1$, B$_2$ and B$_3$ most important)** | | | | | |
| Fumonisin B$_1$ | Diester of propane-1,2,3, tricarboxylic acid with backbone of 2-amino-12,16-dimethyl-3,5,10,14,15-pentahydroxyicosane | C$_{34}$H$_{59}$NO$_{15}$ | 722 | Powder | No absorption in UV |
| Ochratoxin A | 3,4-Dihydromethylisocoumarin linked at 7-carboxy to L-β-phenyl alanine by amide bond | C$_{20}$H$_{18}$O$_6$NCl | 403 | 94–96 | Green UV 360 nm |
| Patulin | [4-Hydroxy-4-H-furo-[3,2-c] pyran-2(6H)-one | C$_7$H$_6$O$_4$ | 154 | 110–111 | UV 276 maximum |
| T-2 toxin | 4-β, 15-Diacetoxy-3α-hydroxy-8α-(3-methylbutryloxy)12,13-epoxytrichothec-9-ene | C$_{24}$H$_{34}$O$_9$ | 466 | 151–152 | No absorption in UV |
| Zearalenone | 6-(10-Hydroxy-6-oxo-trans-1-undecenyl)-β-resorcyclic acid μlactone | C$_{18}$H$_{22}$O$_5$ | 318 | 164 | Blue-green UV 254 nm |

[a]Based on information from Betina (1993), Deshpande (2002), Jegorov (1999), Joint FAO/WHO Expert Committee on Food Additives (2001), Kozakiewicz (2001), Marasas (2001b), Mirocha *et al.* (1992), Rheeder *et al.* (2002).

[b]See Butcha and Cvak (1999) for structures.

**Table 10.2** Major biosynthetic pathways used for mycotoxin production[a]

| Pathway | Mycotoxin |
|---|---|
| Polyketide | Aflatoxin, fumonisin, ochratoxin, patulin, zearalenone |
| Terpenoid | Trichothecenes (deoxynivalenol, nivalenol, T-2 toxin) |
| Cyclic peptides | Ergot |
| Sphingoid base | Fumonisins (also polyketide) |

[a]Smith and Moss (1985), Desjardins *et al.* (1996), ApSimon (2001).

*al.*, 2002). Fumonisins of the B series are considered the most important with regards to food and feed safety and fumonisins B$_1$, B$_2$ and B$_3$ are the most abundant on maize. Fumonisin B$_1$ (FB$_1$) is chemically a propane-1,2,3-tricarboxylic diester of 2-amino-12,16-dimethyl-3-,5,10,14,15-pentahydroxyeicosane (Desjardins *et al.*, 1996). The other B series fumonisins are similar; however, fumonisin B$_2$ is not hydroxylated at the carbon 10 (C-10) position, fumonisin B$_3$ has no hydroxyl group at C$_5$, and fumonisin B$_4$ has no hydroxyl groups at C-5 and C-10 (ApSimon, 2001). Biosynthesis occurs as follows (ApSimon, 2001): condensation with an 18-carbon polyketide chain results in alanine's direct incorporation into the C-1 and C-2 positions of the fumonisin backbone; the hydroxyl groups at C-3, C-5, and C-15 and later hydroxylation at C-10 and C-14 are consistent with a carbonyl derivation; methylation at

C-12 and C-16 is methionine-derived and the esterification of the tricarballylic groups at C-14 and C-15 is believed to occur last (ApSimon, 2001).

Ochratoxin A is a 5-choloro-3,4-dihydro-3-methyl-iso-coumarin that is linked at C-8 to a phenylalanine moiety (Steyn, 1998). The isocoumarin moiety is polyketide (pentaketide) in origin; whereas, the biosynthesis of L-phenylalanine is via the shikimate pathway (United Nations Environment Programme, 1990; Steyn, 1998). Ochratoxin A is a colorless crystal, soluble in organic solvents (polar) and slightly soluble in water, which is stable in ethanol, to heat (melting point of 94–96°C) and to irradiation but not to fluorescent light (Scott, 1994; Kozakiewicz, 2001). Ochratoxin A has a molecular weight of 430 based on an emperical formula of $C_{20}H_{18}O_6NCl$ and at UV 360 nm fluoresces green (Kozakiewicz, 2001). Further information on the chemistry of ochratoxin A can be found in Scott (1994). Ochratoxin A was named after the mold from which it was first isolated, *Aspergillus ochraceus* (United Nations Environment Programme, 1990). Other molds that produce ochratoxin A are listed in Table 10.3. Ochratoxin A is not found in ruminants because it is cleaved; however, in monogastric animals, it accumulates in the kidneys (United Nations Environment Programme, 1990).

Patulin is stable to acid but not to alkaline environments and is soluble in polar solvents and water (Scott, 1994). Patulin is stable in acidic conditions as found in apple and grape juices and to heat in the presence of sucrose, but it decreases with addition of vitamin C and when irradiation at 0.35 kGy (Scott, 1994). Patulin reacts with sulfhydryl amino acids, which may cause instability in foods (Deshpande, 2002). Additional information on the chemistry of patulin can be found in Scott (1994). Cyclopiazonic acid is an indole tetramic acid that is produced by *A. flavus* and *Penicillium* species, especially *P. commune* and *P. camemberti* (Hocking, 2001; Kozakiewicz, 2001; Pitt, 2001a; 2002). It co-occurs with aflatoxins, which may affect the toxicity (Deshpande, 2002). Other information on the chemistry of cyclopiazonic acid was reviewed by Scott (1994).

Trichothecenes have a tetracyclic 12,13-epoxytrichothecene backbone and can be grouped into four classes based on the functional groups (United Nations Environment Programme, 1990). Classes are:

1 Type A have a non-ketone at the C-8 position (T-2 toxin, diacetoxyscirpenol)
2 Type B have a carbonyl group at the C-8 position (deoxy-nivalenol and nivalenol)
3 Type C have another epoxide at C-7,8 or C-9,10.
4 Type D have a macrocyclic ring attached by two ester linkages from C-4 to C-15.

Type A trichothecenes are colorless crystals that do not absorb in the UV range and are soluble in polar solvents such as acetone, chloroform and ethyl acetate; however, Type B require polar solvents such as aqueous solutions of acetonitrile or methanol (United Nations Environment Programme, 1990). Zearalenone is a lactone with β-resorcylic acid (D'Mello *et al.*, 1997). While zearalenone biosynthesis follows a polyketide pathway, T-2, nivalenol and diacetoxyscirpenol are terpenoid mycotoxins with mevalonic acid as the starting point in their biosynthesis (Steyn, 1998).

## Taxonomy of major mycotoxin-producing molds

The taxonomy of the major mycotoxin producing mold genera, *Aspergillus*, *Fusarium*, and *Penicillium*, have gone through changes in recent years because of new studies in genetics and physiology (Frisvad and Filtenborg, 1989; Windels, 1991; Pitt and Hocking, 1997; Samson and Pitt, 2000; Leslie *et al.*, 2001). The current taxonomy for these three genera is briefly presented here. *Aspergillus* Fr.:Fr. and *Penicillium* Link:Fr. species are classified in the subdivision Deuteromycotinia or Fungi Imperfecti in the class Hyphomycetes and have teleomorphs that belong to the Ascomycetes with *Emericella* Berk. *Eurotium* Link Fr., *Neosartorya* Malloch & Cain for *Aspergillus* and *Eupenicillium* F. Ludw., *Geosmithia* Pitt, and *Talaromyces* C.R. Benj. for *Penicillium* (Pitt and Hocking, 1997; Samson and Pitt, 2000). *Aspergillus* and *Penicillium* species produce phialides that are borne on stipes that terminate in a vesicle (called an aspergillum) for *Aspergillus* species and are in clusters on the septate stipes or on metulae and/or rami (called a penicillus) for *Penicillium* species (Pitt and Hocking, 1997). Aflatoxins are produced by *Aspergillus flavus* Link, *Aspergillus parasiticus* Speare and *Aspergillus nomius* Kurtzman, horn and Hesseltine (Frisvad, 1995; Pitt and Hocking, 1997). There are two strains of *A. flavus* based on sclerotia and aflatoxin production: the S strain has small sclerotia of < 300 to 400 μm and produces large amounts of either aflatoxin B or both aflatoxins B and G; the L strain has sclerotia > 400 μm and lower amounts of aflatoxin (Egel *et al.*, 1994; Cotty and Cardwell, 1999). Cyclopiazonic acid is produced by *Penicillium griseofulvum* Dierckx, *Penicillium commune* Thom, *A. flavus* and *Aspergillus tamarii* Kita (Frisvad, 1995). All except *P. commune* (meat and cheese) are found on cereals; however, fermentation cultures of *Aspergillus oryzae* (Ahlburg) Cohn (domesticated *A. flavus*) and *Penicillium camemberti* Thom (domesticated *P. commune*) also produce cyclopiazonic acid (Frisvad, 1995). Cyclopiazonic acid occurs naturally in maize, peanuts, sunflower seeds and millet (CAST, 2003). Ochratoxin A is produced by *Penicillium verrucosum* Dierckx (other penicillia reidentified after misidentification) and *Aspergillus ochraceus* Wilhelm and *Aspergillus ostianus* Wehmer (not produced sensu stricto). The mis-identification of *Penicillium* species that produce mycotoxins has caused great confusion over the years and some of this misinformation still appears in the literature (Pitt, 2001a, b). In recent years, some isolates of *Aspergillus carbonius* (Bainier) Thom (Pitt and Hocking, 1997; Joosten *et al.*, 2001; Urbano *et al.*, 2001; Cabañes *et al.*, 2002; Abarca *et al.*, 2003; Taniwaki *et al.*, 2003) and *Aspergillus niger* van Tieghem (Urbano *et al.*, 2001; Samson *et al.*, 2002) have produced ochratoxin

A in some agricultural commodities. Ochratoxin A has been found in processed foods including coffee, beer, wine and pork products and cyclopiazonic acid has been found in cheese and fermented meat products (Benford *et al.*, 2001; CAST, 2003). Although several mycotoxins are produced by *Penicillium* species, only ochratoxin A, cyclopiazonic acid, and patulin are linked to toxicities in animals and, potentially in humans, because the molds that produce these mycotoxins commonly occur in foods worldwide (Table 10.3).

*Fusarium* Link ex Fr. species are classified in the subdivision Deuteromycotinia or Fungi Imperfecti in the class Hyphomycetes and have perithecial states or teleomorphs that belong to *Gibberella* and *Nectria* in the class Hypocreales (Windels, 1991; Burgess *et al.*, 1994). *Fusarium* species are generally defined by the production of crescent-shaped macroconidia with or without a foot-shaped cell and apical beak (Windels, 1991; Pitt and Hocking, 1997). Macroconidia can either be produced in sporodochia (pustules) or pionnotes (slimy masses) and some species also produce one- to two-celled microconidia and chlamydoconidia (Pitt and Hocking, 1997). *Fusarium* species produce many different colors when grown on agar and these are used in identification schemes (Windels, 1991; Burgess *et al.*, 1994; Pitt and Hocking, 1997). Common morphological properties are used to place *Fusarium* species into the different sections (Windels, 1991). Several taxonomic systems have been developed over the years using both microscopic and macroscopic characteristics (Windels 1991; Burgess *et al.*, 1994; Pitt and Hocking, 1997). Trichothecenes (T-2 toxin and deoxynivalenol) are produced by *Fusarium poae* Wollenw, *Fusarium sporotrichioides* Sherb, *Fusarium acuminatum* Ell. And Everh., *Fusarium equiseti* (Corda) Sacc., *Fusarium crookwellense* Burgess, Nelson, and Toussoun, *Fusarium culmorum* (W.G. Smith) Sacc., *Fusarium graminearum* Schwabe, and *Fusarium sambucinum* Fuckel. Zearalenone is produced by *Fusarium sporotrichioides, Fusarium pallidoroseum* (Cooke) Sacc., *F. crookwellense, F. equiseti, F. culmorum, F. graminearum.* Fifteen *Fusarium* species have been reported to produce fumonisins (Rheeder *et al.*, 2002). Most are in the Section *Liseola* or the closely related Section *Dlaminia.* Of the 15 producers, *Fusarium verticillioides* (Sacc.) Nirenberg (formerly *Fusarium moniliforme* Sheldon) and *Fusarium proliferatum* (Matsushima) Nirenberg are the most important because of their widespread distribution, ability to produce high levels of fumonisins in maize, and association with farm animal diseases (Rheeder *et al.*, 2002). Although several *Fusarium* species produce mycotoxins, Marasas (2001b) noted that three species were most important, namely, *F. graminearum, F. sporotrichioides*, and *F. verticillioides* (formerly *F. moniliforme*).

*Claviceps purpurea* (Fries ex Fries), which is in the family Clavicipitaceae and the order Hypocreales, produces ergot alkaloids (Pažoutová and Parbery, 1999; Tenberge, 1999). *Claviceps* species are identified by the color, size and shape of the ascospores, asci, capitula, conidia, perithecia, sclerotia, and stipes; and by whether or not the stroma has loose hyphae (Pažoutová and Parbery, 1999). *C. purpurea*, which occurs in temperate regions of the world, generally has purple-black sclerotia with beige-toned capitula; however, the sclerotia can vary depending on the climate and the host floret (Pažoutová and Parbery, 1999). Although other species of *Claviceps* produce ergot alkalois, *C. purpurea* is dominant ((Pažoutová and Parbery, 1999; Tenberge, 1999).

The microscopic and colonial appearances of these molds and the mycotoxins that they produce are presented in Table 10.3. Some reviews since 1990 that include information on toxigenic molds are: *Aspergillus* species (Smith and Ross, 1991; Pitt and Hocking, 1997; Bhatnagar and García, 2001; Hocking, 2001; Kozakiewicz, 2001), Claviceps species (Mantle, 1991; Křen and Cvak, 1999; Pažoutová and Parbery, 1999; Tenberge, 1999; Kuldau and Bacon, 2001); *Fusarium* species (Marasas, 1991; Leslie, 1996; Pitt and Hocking, 1997; Chelkowski, 1998; Bullerman, 2001a,b; Marasas, 2001b, Rheeder *et al.*, 2002), and *Penicillium* species (Pitt and Leistner, 1991; Pitt and Hocking, 1997; Pitt, 2001a,b; 2002).

## Ecology of mycotoxin production in agricultural commodities

### Ecology of mold growth and mycotoxin production

Mycotoxins can be produced while the plant is growing in the field, during the storage of the plant-based material after harvest, and in some cases, during both storage and processing of foods. There are no specific conditions that can be established in common for the growth of all molds and subsequent production of mycotoxins because there are many species of molds that have different biochemistries, ecologies and morphologies (CAST, 2003). Six genera (*Aspergillus, Claviceps, Fusarium, Penicillium* with *Neotyphodium* associated with plant tissues and *Stachybotrys* associated with building materials) that produce mycotoxins and the conditions under which they are produced were summarized in the recent CAST Task Force Report (CAST, 2003). Bilgrami and Choudhary (1998) reviewed the major ecological factors that determine whether molds infect plants and produce mycotoxins and stressed that the environmental factors, such as where the crops are grown, have a major impact. Factors such as size of inoculum of the mycotoxin-producing molds, susceptibility of the plant to mold invasion, climatic factors (drought and temperature), insect damage to plants allowing molds to enter and agronomic practices that make plants susceptible to mold invasion were reviewed (Bilgrami and Choudhary, 1998). Factors that affect the production of mycotoxins by different species/strains of the molds include environmental factors (pH, temperature and water activity); nutrients; interactions between plants, molds, and microorganisms; and chemicals used on crops and in the storage of cereals (Moss, 1991; Kozakiewicz, 2001). The major mycotoxins and the agricultural commodities in which they appear are listed in Table 10.4. The major mycotoxins that can be produced in agricultural commodities during the growing

**Table 10.3** Taxonomy of major molds that produce mycotoxins[a]

| Genus and species | Section | Microscopic appearance | Colonial appearance[b] | Mycotoxins |
|---|---|---|---|---|
| Aspergillus clavatus Desm. | Clavi | Uniserate clavate-shaped vesicles with closed packed phialides and elliptical smooth conidia | Rapid growth, floccose white mycelium (radially furrowed) with reverse uncolored to dull yellow to brown, dull green to turquoise conidia | Patulin |
| Aspergillus flavus Link | Flavi | Biserate columnar or radiate, globose vesicle, coarsely rough conidiophores with globose conidia that are finely rough; reddish-brown to black spherical sclerotia | Floccose in center and velutinous at margins, white mycelium with reverse uncolored or dull brown or orange, olive to parrot green or yellow-green conidia, dark brown to black sclerotia | Aflatoxin $B_1$, $B_2$, cyclopiazonic acid, patulin |
| Aspergillus parasiticus Speare | Flavi | Mostly uniserate that radiate from rough conidiophores and globose conidia, similar to A. flavus but smaller and elongate sclerotia when produced | Velutinous to floccose white mycelium with reverse uncolored to dull yellow to pinkish-red, dark olive to deep jade green conidia, sometimes brown to black sclerotia | Aflatoxin $B_1$, $B_2$, $G_1$, $G_2$ |
| Aspergillus niger van Tieghem | Nigri | Closely packed biserate, spherical vesicles, smooth heavy-walled conidiophores, black rough spherical conidia | Rapid growth, white mycelium with reverse uncolored to yellow, covered by black conidia | Ochratoxin A |
| Aspergillus carbonarius (Bainier) Thom | Nigri | Closely packed biserate over entire metulae, spherical vesicles, smooth to finely rough walled conidiophores, large black rough with spikes globose conidia | White mycelium with reverse pale to yellow, covered by olive-black to black conidia | Ochratoxin A |
| Aspergillus ochraceus Wilhelm | Cicumdati | Biserate, globose vesicles and small globose conidia, slightly rough with rough-walled conidiophores | Floccose to low sulcate, white mycelium with reverse dull yellow to gray-red or brown, wheat to ochraceus conidia, pink to purple sclerotia if present | Ochratoxin A |
| Claviceps purpurea | NA | Perithecia in stromata, filiform ascospores, long cylindrical asci | Brown to black sclerotia and purple capitulum | Ergot alkaloids |
| Fusarium semitectum sensu Wollenw. | Arthrosporiella | Apical pores to polyphialides, no microconidia, macroconidia (fusiform to curved with beaked apical cell and footcell, 3–5 septate), sparse globose chlamydospores | Floccose/powdery, white or salmon or pale brown, reverse pale yellow, to yellowish-brown | Zearalenone |
| Fusarium culmorum (W.G. Smith) Sacc. | Discolor | Branched conidiophores, thick-walled macroconidia (fusiform to sickle-shaped, pointed apex and footcell, five-septate), orange sporodochia, chlamydospores | Dense felty, floccose, pale red to yellow-brown, reverse pale to deep red | Deoxynivalenol, type A trichothecenes, zearalenone |
| Fusarium graminearum Schwabe | Discolor | Monophialides, micro (ovoid to clavate, truncate base, chains and false heads)/macroconidia (curved or straight, basal cell pedicel, long apex, curved, 5- or 6-septate), orange sporodochia, sparse chlamydospores | Floccose, white to carmine red with yellow, reverse carmine red | Zearalenone, deoxynivalenol, nivalenol |
| Fusarium crookwellense Burgess, Nelson, and Toussoun | Discolor | Unbranched to branched conidiophores with mono- and polyphialides, absent microconidia, macroconidia (sickle-shaped with curved apical cell and footcell, five-septate), chlamydospores, orange to red-brown sporodochia | Floccose, white to yellow or pink and ochraceous to brownish-red with age, reverse red to purple or brown | Trichothecenes, zearalenone |
| Fusarium acuminatum Ellis and Ev.erhart Sensu Gordon | Gibbosum | Conidiophores with lateral to branching phialides, absent to sparse microconidia, macroconidia (curved, sickle-shaped with tapering apical cell and footcell base, 3–5 septate), salmon to bright orange sporodochia, chlamydospores in chains or intercalary | Felty, white to grayish rose to ruby red ith yellow brown arial mycelium, reverse pale to grayish red to reddish brown | Trichothecenes, especially T-2 and HT-2 toxins, zearlenone |

| Species | Section/Genus | Conidiophores/morphology | Colony (colors on agar)[b] | Mycotoxins |
|---|---|---|---|---|
| *Fusarium equiseti* (Corda) Sacc. | Gibbosum | Unbranched to branched conidiophores with mono- and polyphialides, absent to rare microconidia, macroconidia (sickle-shaped with footcell, 3–5 or 7 septate), abundant pale brown chlamydospores | Floccose, white but yellow-brown with age, reverse pale to salmon | Trichothecenes types A – T-2 and HT-2 toxins and B – deoxynivalenol and nivalenol, zeararlenone |
| *Fusarium proliferatum* (*matsushima*) Nirenberg | Liseola | Unbranched to branched conidiophores with mono- and polyphialides, single-celled microconidia in chains or false heads (ellipsoidal to clavate with flat base), rare macroconidia (sickle-shaped to straight with round ends, 3–5 septate), orange sporodochia | Floccose, white or pale pink or orange or grey violet, reverse orange or red or violet | Fumonisins (fusaric acid, moniliformin) |
| *Fusarium verticillioides* (Sacc.) Nirenberg (*F. moniliforme* Sheldon) | Liseola | Monophialides, macroconidia (straight or curved, basal cell pedicel, curved apex, tapers, 3–5 to 7 septate), orange sporodochia | Floccose/powdery, white to purple, reverse purple | Fumonisins (fusaric acid, fusarin C) |
| *Fusarium poae* (Peck) Wollenw. | Sporotrchiella | Branched conidiophores with monophialides, single-celled microconidia (spherical to lemon-shaped), sparce macroconidia (slightly curved with footcell, 2–3 septate) | Cottony white to pale pink and red to violet near agar, reverse yellow to rose to deep red | Trichothecenes – nivalenol, T-2 and HT-2 toxins |
| *Fusarium sporotrichioides* Sherbakoff | Sporotrchiella | Mono-/polyphialides, micro (spindle or globose)/macroconidia (curved, basal cell nondistinct, curved apex, tapers, 3 –5 septate), chlamydospores, orange sporodochia | Floccose, white to carmine red, reverse carmine red | T-2 toxin and diacetoxyscirpenol |
| *Penicillium camemberti* Thom | Penicillium | Conidiophores with rough walls and two-staged branched, terverticillate, ampulliform phialides, smooth subspherical to spherical conidia in disordered chains | Floccose, white mycelium and with age yellow to greenish-gray and reverse pale to yellow-brown, conidia absent to light | Cyclopiazonic acid |
| *Penicillium commune* Thom | Penicillium | Conidiophores with rough walls and two-staged branched, terverticillate, ampulliform phialides, smooth subsherical to spherical conidia in disordered chains | Velutinous to floccose, white mycelium with pale to yellow reverse, gray-green to greenish-turquoise conidia | Cyclopiazonic acid |
| *Penicillium expansum* Link | Penicillium | Conidiophores with smooth walls in fascicles or coremia, terverticillate, ampulliform to cylindrical phialides, smooth ellipsoidal conidia in dense irregular chains | Tufted (coremial) with velutinous to floccose areas, white mycelium, dull green to greenish-turquoise conidia | Patulin, citrinin |
| *Penicilium griseofulvum* Dierckx | Penicillium | Conidiophores with smooth walls in fascicles, ter- to quaterverticillate, short cylindrical to tapering phialides, smooth ellipsoidal conidia in dense disordered chains | Restricted dense with granular surface, white mycelium with reverse pale to yellow to brown, sometimes reddish brown pigment, clear to yellow exudate, gray-green conidia | Cyclopiazonic acid, patulin |
| *Penicillium verrucosum* Dierckx | Penicillium | Conidiophores with rough walls and two-staged branched, terverticillate, ampulliform phialides, smooth globose to subglobose conidia in disordered chains | Velutinous to floccose to fasciculate, white mycelium with reverse yellow brown to deep brown, clear to yellow exudates, gray-green to dull green conidia | Ochratoxin A, citrinin |

[a]Bullerman (2001a; 2001b), Klich and Pitt (1988), Kozakiewicz (2001), Kuldau and Bacon (2001), Kurtzman *et al.* (1987), Nelson *et al.* (1983), Marasas (2001b), Pitt and Hocking (1997), Samson *et al.* (2002).

[b]Depends on agar (colors on Czapek yeast extract agar) and growth conditions.

season include aflatoxins, cyclopiazonic acid, and ochratoxin A by *Aspergillus* species (Wilson *et al.*, 2002); ergot alkaloids by *Claviceps* species (Pažoutová and Parbery, 1999); deoxynivalenol, fumonisins, T-2 toxin, zearalenone by *Fusarium* species (Miller, 2002); and ochratoxin A and patulin by *Penicillium* species (Pitt, 2002).

Widstrom (1992) reviewed the association of *Aspergillus flavus* with maize with an emphasis on the development of ear infection after the silking phase. Plants in the field can be inoculated with *A. flavus* from air, insects or soil that carry conidia (spores), hyphae, sclerotia or combinations of these mold propagules (Wilson and Payne, 1994). *A. flavus* conidia, mycelia or sclerotia originate in soil and plant debris and become airborne or carried by insects (corn earworms and corn borers) to enter the maize at silk initiation and proceed down to the kernels and then, if moisture (> 18% but best at 35 to 40%) and temperature (about 30 to 38°C) are acceptable, the spores can germinate, grow and produce aflatoxin (Bilgrami and Choudhary, 1998; Dowd, 1998; Payne, 1998). Temperatures of 7.5 to 48°C (best at 25 to 42°C), water activity above 0.85, oxygen, sugars and starches in maize were described as best for aflatoxin production by *A. flavus* and *A. parasiticus* (Widstrom, 1992; Payne, 1998). Other factors include stress, insect damage to allow easy entry into the kernel, stress of the plant due to drought, plant genetics, imbalance of nutrients, and competition from other molds (other *Aspergillus* species or species of *Cladosporium*, *Fusarium* (*F. verticillioides* competes with *A. flavus* for colonization of the maize), *Rhizopus*, and *Trichoderma*) for nutrients (Widstrom, 1992; Wilson and Payne, 1994; Wicklow, 1995; Abramson, 1998; Bilgrami and Choudhary, 1998; Payne, 1998).

If a plant is infected with *A. flavus*, the plant and surrounding environmental conditions will determine whether the mold will grow and produce aflatoxin, especially temperature and drought stress in maize. Agronomic practices that have an impact on aflatoxin production include choosing fields where aflatoxin production is not likely, cultivars that are less susceptible to *A. flavus*, less dense planting in fields, irrigation practices, and planting date to take advantage of rain at times when *A. flavus* would not normally infect plants (Wilson and Payne, 1994; Bilgrami and Choudhary, 1998). Peanuts, cottonseed and mustard are the oilseeds where aflatoxins can be produced during growth (Bilgrami and Choudhary, 1998). Although peanut plants are mainly infected by *A. flavus* or *A. parasiticus* in the soil and insect damage by mites and lesser corn stalk borer allow *A. flavus* and *A. parasiticus* to contaminate the peanut shell or nut, contamination of flowers and pegs cannot be totally ignored (Bilgrami and Choudhary, 1998; Dowd, 1998; Payne, 1998). *A. flavus* usually outcompeted *A. parasiticus* in peanuts as it did in maize (Payne, 1998). Soil temperatures between 25 to 32°C with a mean of 30.5°C, drought stress of 20 to 30 days late in the growing season, and concentration of phytoalexins were important for aflatoxin production in peanuts with calcium deficiencies and shell damage playing

some role (Wilson and Payne, 1994; Bilgrami and Choudhary, 1998).

Failure to rotate crops, the type of soil (sandy or loam soils hold less water), and minerals (calcium) may also affect aflatoxin production in peanuts (Bilgrami and Choudhary, 1998). In addition, planting time and other crops planted in the area (competition of fungi) affected aflatoxin production in peanuts (Bilgrami and Choudhary, 1998). In cottonseed, *A. flavus* infects the bolls from windborne soil conidia, but the point of entry is not clear and may be leaf scars, nectar or insect (corn earworms or bollworms) damaged cotton bolls that allow molds to move into the bolls as the plant grows (Bilgrami and Choudhary, 1998; Dowd, 1998; Payne, 1998). For cottonseed, temperatures between 25 to 34°C during the day were important for aflatoxin production as was humidity but more data is needed in this area (Wilson and Payne, 1994; Bilgrami and Choudhary, 1998). Mustard seed is grown in climates with day temperatures ranging from 28 to 31°C and relative humidities from 50 to 94% (Bilgrami and Choudhary, 1998). Insect injury by the navel orange larvae of tree nuts (almonds, pistachios) is important for colonization by *A. flavus* and *A. parasiticus* that come from the soil around the trees; however, *A. niger* is in greater percentages than either of the aflatoxin producers (Payne, 1998). Dowd (1998) discussed the insects involved in allowing fungal entry and included caterpillars (worms and borers) plus beetles, aphids and other insects that damage cereals and nuts; therefore, management of insect contamination by use of insecticides, plant breeding (Bt gene makes them less susceptible to insect damage), and similar measures may decrease the amount of mycotoxins in cereal grains and nuts.

Most *Fusarium* mycotoxins are produced in the field and remain in the stored grain (Frisvad, 1995). *Fusarium* mycotoxins are produced due to complex interactions of biological, chemical, and physical factors and have been found in several African and European countries, India, Japan, and North America (D'Mello *et al.*, 1997). Climatic conditions affect the growth of *Fusarium* species with temperature and moisture (at silking for maize and anthesis for wheat and barley) being the most important (Abramson, 1998). *Fusarium* species cause various rots in barley, maize, and wheat; head blights (scabs) in oats, rye, triticale and wheat; and also produce several mycotoxins in these grains, especially in barley, maize, and wheat (Miller, 1994; D'Mello *et al.*, 1997; Abramson, 1998; Bilgrami and Choudhary, 1998). Head blight is a function of the time moisture is available rather than the amount of water because moisture at anthesis is most important for barley and wheat (Miller, 1994; Abramson, 1998). Gibberella or pink rot is found in northern climates in wet years when *Gibberella zeae* and *F. culmorum* grow; however, *Fusarium* ear rot or kernel rot is associated with warm climates that are dry where insects create an entry for *F. proliferatum*, *F. subglutinans*, and *F. verticillioides* (Miller, 1994; Abramson, 1998).

In addition, heavy rainfall has been associated with head blight and deoxynivalenol production with both time and water activity associated with fumonisin $B_1$ production (D'Mello et al., 1997). Infection usually starts at the ear tip after the maize silk forms and spreads down the ear, especially if damaged by insects or birds (Abramson, 1998; Bilgrami and Choudhary, 1998). F. culmorum (spread by macroconidia) and F. graminearum (spread by macroconidia and ascospores) produce deoxynivalenol at the end of head blight (Miller, 1994). Temperatures above 5°C with optimum temperatures ranging from 25 to 32°C and moisture contents > 35% promoted growth of Fusarium species in maize (Bilgrami and Choudhary, 1998). F. graminearum usually is found in cereals when the daytime temperatures go above 30°C and F. culmorum, F. nivale, F. poae, and F. sporotrichioides normally prefer cooler cereal growing regions in northern Europe (Miller, 1994; Abramson, 1998). In maize, temperature controls the different Fusarium species, for example, F. graminearum, F. subglutinans, and F. verticillioides are common in Canada, parts of Europe and the United States; however, F. sporotrichiodes is common in colder climates (Miller, 1994). Inadequate crop rotation, presence of plant debris and weeds, and type of cultivar can also lead to Fusarium invasion and mycotoxin production (Abramson, 1998; Bilgrami and Choudhary, 1998). Crop rotation is important to prevent the spread of inoculum by using nonhosts such as sorghum or soybeans (Abramson, 1998). In wheat, infection starts around anthesis in humid weather and can be affected by the type of fertilizer (ammonium nitrate > urea), soil conditions (pH and contents of minerals), fungicide application, crop rotation and tillage practices, weed control, and other agronomic practices (D'Mello et al., 1997; Bilgrami and Choudhary, 1998). The chemical composition of the substrate may affect production of mycotoxins because deoxynivalenol is produced when carbohydrates are depleted and T-2 toxin is produced in greater quantities when sorbic acid is present (D'Mello et al., 1997).

Little has been published in recent years about the ecology of mycotoxin production by Penicillium and Claviceps species. Ochratoxin A is produced by Penicillium verrucosum in cereals in temperate climates (optimum 20°C) and by Aspergillus ochraceus in tropical climates (Pitt, 2001b). Penicillium species, especially Penicillium commune in cheese, that produce cyclopiazonic acid have an optimum temperature of 25°C and can grow down to water activities of 0.85 (Pitt, 2001b). Claviceps purpurea invade the floral ovary before pollination where the ascospores and conidia germinate and the mycelium forms destroying the ovarian cells and producing hardened sclerotia that fall onto the ground when mature and remain dormant until the next planting season (Kuldau and Bacon, 2001). Infection is encouraged in climates that are cool at flowering because pollination takes longer and Claviceps sclerotia can invade the plant because pollinated plants are resistant to Claviceps species (Kuldau and Bacon, 2001).

Growth of molds and mycotoxin production during storage of cereal grains and oilseeds results from a combination of factors: damage to commodity (insect, mechanical, etc.), temperature, environmental gases (oxygen and carbon dioxide) and humidity, moisture (drying conditions and regain of moisture), mold spore inoculum, storage time, and presence of other microorganisms (Ominski et al., 1994; Abramson, 1998). Species of Alternaria, Cladosporium and Fusarium are molds that occur in the fields but generally are not competitive during storage because they require high water activities (Table 10.4). Spores of A. flavus can enter stored maize via grain dust or air (Wicklow, 1995). Both the water inside the grains, the water vapor around the grain and the presence of carbon dioxide, nitrogen and oxygen within the grain are important in mold growth (FAO, 2001), In addition, insects and arthropods can damage grain, increase internal temperature and water activity and allows molds to enter and grow (FAO, 2001; Miller, 2002). Aflatoxin production during storage of grains, oilseeds, and nuts can result if there is insufficient drying (water activity > 0.85), warehousing problems that increase temperature to between 25 and 32°C and allow aeration, and damage due to insects or rodents that create points of entry for A. flavus (Wilson and Payne, 1994). Although aflatoxins are produced better in tropical and subtropical countries, they can be produced in cereals in temperature countries if the humidity is high and if treated with formic or propionic acids (Frisvad, 1995). Aspergillus and Penicillium species normally grow during storage because they can tolerate lower water activities (Table 10.4). In stored grains, aflatoxins are most important in warmer climates and ochratoxin A in cooler climates (Frisvad, 1995). Other molds such as xerophilic Wallemia species or molds that require higher water activities (Mucor and Rhizopus species) can grow under certain conditions (Ominski et al., 1994), but these molds do not produce mycotoxins.

## Factors involved in production of mycotoxins

There are many factors that will promote mold growth in grains, nuts and plants. These include pest damage and favorable temperature (−5 to 55°C) and water activity ($a_w$). Under field conditions, mold growth and mycotoxin production are favored by $a_w$ (> 0.88); whereas, storage molds can grow well at $a_w$ levels below this (Pitt and Hocking, 1997; FAO, 2001). The general physiological conditions involved in the production of mycotoxins by species of Aspergillus, Fusarium, and Penicillium with emphasis on nutrition (carbon, nitrogen, and minerals), pH, temperature, water activity ($a_w$), energy charge and redox potential have been reviewed (Payne and Brown, 1998; Sweeney and Dobson, 1998). Relative humidity (RH) affected aflatoxin production on maize incubated at 30°C because none was produced below 80% RH, very little was produced between 80 to 84.5% RH, and an increase was found above 91% RH (Guo et al., 1996). Some key factors that affect mold growth and mycotoxin production are presented in Table 10.5. When maize kernels were kept at high humidity over water for 3 days before inoculation with A. flavus, maize germination was ≥ 93% compared to ≤ 17.5% for no humidity equilibration. In humidity-equilibrated kernels in Louisiana (USA), aflatoxin production was reduced by 68% to 96% in all

**Table 10.4** Major mycotoxins and their occurrence in agricultural commodities[a]

| Mycotoxin | Agricultural commodities |
|---|---|
| Aflatoxins | Major: cottonseed, ground and tree nuts, maize; sometimes: barley, cassava, millet, oats, rice, rye, sorghum, wheat, spices[b] |
| Cyclopiazonic acid | Maize, millet, peanuts |
| Deoxynivalenol | Barley, maize, wheat |
| Ergot alkaloids | Millet, rye, wheat, fescue |
| Fumonisins | Maize, (sometimes rice) |
| Ochratoxin A | Barley, maize, oats, rice, rye, sorghum, wheat, coffee beans, grapes |
| Patulin | Apples |
| T-2 toxin | Barley, maize, oats, safflower seeds, sorghum, wheat |
| Zearalenone | Barley, maize, oats, rye, sorghum, wheat |

[a]Sharma and Salunhke (1991), Shotwell (1991), Strange (1991), Yoshizawa (1991), WHO (2001).
[b]Especially in tropical countries

**Table 10.5** Factors that affect growth of some mycotoxigenic molds and the mycotoxin that they produce[a]

| Mold | Mycotoxin | Growth of mold | | | Mycotoxin production | | |
|---|---|---|---|---|---|---|---|
| | | °C (optimum) | $a_w$ (optimum) | pH (optimum) | °C (optimum) | $a_w$ (optimum) | pH (optimum) |
| *Aspergillus flavus* | Aflatoxins $B_1$, $B_2$ | 10–43 (33) | > 0.78 (0.99) | 2.1–11.2 (3.4–10.0) | 13–37 (16–31) | > 0.82 (0.95-.0.99) | 3.5–8.0 (6.0) |
| *Aspergillus parasiticus* | Aflatoxins $B_1$, $B_2$, $G_1$, $G_2$ | 12–42 (32) | > 0.80 (0.99) | 2.4–10.5 (3.5–8.0) | 12–40 (15–25) | > 0.86 (0.99) | 3.0–8.0 (6.0) |
| *Aspergillus ochraceus* | Ochratoxin A | 8–37 (24–31) | > 0.77 (0.95–0.99) | 2.2–10.0 | 12–37 | > 0.80 (0.95–0.99) | ND[b] |
| *Penicillium verrucosum* | Ochratoxin A | 0–31 (20) | >0.80 | 2.1–10.0 (6.0–7.0) | 0–31 (20) | > 0.86 (0.92) | (5.6) |
| *Fusarium graminearum* | Deoxynivalenol Nivalenol Zearalenone | (24–26) | >0.90 | 2.4–9.5[c] (6.7–7.2) | Similar to growth | Similar to growth | Similar to growth |
| *Fusarium proliferatum,* | Fumonisin $B_1$ | 4–35 (25–30) | > 0.88 (0.96–0.995) | (5.5) | 25–30 | > 0.92 | ND |
| *Fusarium verticillioides* | | 4–35 (30–37) | > 0.88 (0.96–0.98) | (7.0) | 13–28 (20) | > 0.92 | ND |

[a]LeBars *et al.* (1994), Cahagnier *et al.* (1995), Marín *et al.* (1995; 1996a; 1996b), Pitt and Hocking (1997), Sweeney and Dobson (1998).
[b]Not determined.
[c]Temperature dependent.

but one hybrid possibly due to production of antifungal proteins or other chemical inhibitors (Guo *et al.*, 1996). In controlled laboratory media, *Aspergillus flavus*, *A. parasiticus*, and *A. oryzae* had minimum $a_w$ of 0.80, 0.81, and 0.83 at 37, 30, and 25°C, respectively; however, *A. nomius* had a minimum $a_w$ of 0.81 at 37°C and 0.83 at both 30 and 25°C (Pitt and Miscamble, 1995). Rainfall and humidity did not significantly affect aflatoxin production in maize grown in Costa Rica in 1992–1993; however, temperatures of 20.6–25.5°C, minerals (zinc, magnesium, and calcium), and glucose/mannose all significantly affected aflatoxin production (Viquez *et al.*, 1994).

Environmental conditions can affect whether *A. flavus* or *A. parasiticus* will produce aflatoxin in grains. Scott and Zummo (1994) reported that for plots in Mississippi (USA), kernel infection with *A. flavus* increased over harvest dates from 46 to 62 days after midsilk for sensitive but not resistant maize hybrids and that the harvest dates may be important in controlling aflatoxin accumulation in maize. Another

concern with maize was the amount of oil in the hybrid because Severns *et al.* (2003) reported that more aflatoxin was produced in high oil (241 ng/g) than normal oil (178 ng/g) maize grown in Illinois (USA). In August and September of 1998, the low rainfall and warmer temperatures favored aflatoxin production (average 238 ng/g); however, in 1999, opposite conditions of high rainfall and low temperatures decreased aflatoxin production (average 182 ng/g) in the maize grown in Mississippi in the USA (Severns *et al.*, 2003). Plant stress promotes mycotoxin production in the field.

Maize is frequently infected with *Fusarium* species, especially *F. verticillioides* and *F. proliferatum*, which produce fumonisins. Some maize hybrids appear to be more resistant to mold growth and fumonisin production within specific geographical areas to which they are adapted. For example, Pascale *et al.* (1997) studied 14 maize hybrids for 1–4 seasons in plots in Warsaw, Poland for production of fumonisins from ears inoculated 1 week after silk appeared. They reported that inoculated kernels developed ear rots and had fumonisin isolated from them with differences between hybrids and also between seasons that could not be correlated to environmental conditions. One of their hybrids showed low accumulation of fumonisins regardless of the season and could be used by farmers to decrease this mycotoxin in maize. Miedaner *et al.* (2001) also could not correlate temperature and moisture to the degree of deoxynivalenol production in wheat, rye and triticale in Germany. Also, the amount of ergosterol at harvest could not be used to predict the amount of deoxynivalenol in wheat and rye. The genotype-location-year interactions were highly significant; therefore, choice of resistance to *Fusarium* infection could reduce deoxynivalenol production in these three grains (Miedaner *et al.*, 2001).

When *F. moniliforme* and *F. proliferatum* were grown on maize in the presence of *Aspergillus* and *Penicillium* species, there was mutual inhibition at all conditions studied except for $a_w$ of 0.98 at 15°C where the two *Fusarium* species out competed all but *Aspergillus niger*, which inhibited fumonisin production at all other conditions but enhanced it at $a_w$ of 0.98 and 15°C (Marín *et al.*, 1998). With *Aspergillus ochraceus*, fumonisin production by *F. moniliforme* was stimulated at $a_w$ of 0.98 and 15 or 25°C and at $a_w$ of 0.95 and 25°C but for *F. proliferatum* only $a_w$ of 0.98 and 15°C showed more fumonisin (Marín *et al.*, 1998). More fumonisins were produced by *F. proliferatum* at all temperatures and $a_w$ except 15°C and 0.93, respectively, when *A. flavus* was present; however, for *F. moniliforme* only 15°C at $a_w$ of 0.93 or 0.95 and 25°C at $a_w$ of 0.98 had more fumonisin production in the presence of *A. flavus* (Marín *et al.*, 1998). *F. proliferatum* produced more fumonisin at $a_w$ of 0.95 and 0.98 at both 15 and 25°C when *Penicillium implicatum* was present; however, *F. moniliforme* only produced more at $a_w$ of 0.98 and 25°C (Marín *et al.*, 1998). At $a_w$ of 0.93 and 0.95, there was less infection by *F. moniliforme* and *F. proliferatum* due to competition from the *Aspergillus* and *Penicillium* species (Marín *et al.*, 1998). By using scanning electron microscopy to study temperature and water activity of mold growth on maize, Torres *et al.* (2003) found that *A. ochraceus* did not germinate at $a_w$ of 0.98 and 30°C when *F. verticillioides* and *Alternaria alternata* also were inoculated onto maize kernels; however, *F. verticillioides* grew faster probably to be competitive with the rapidly growing *A. alternata* and grew 48.9% slower when *A. ochraceus* was present.

McMullen *et al.* (1997) speculated that the re-emergence of scab or *Fusarium* head blight by *F. graminearum* in barley and wheat in the United States and Canada from 1991 to 1996 was due to weather conditions where high rainfall occurred during flowering and grain development. Other factors that probably contributed to the scab problem were use of reduced soil tillage, too short of rotation periods for susceptible crops, and large amount of acres planted (McMullen *et al.*, 1997). The stage of development significantly affected the production of fumonisin $B_1$ by *F. verticillioides* in maize hybrids grown in Davis, CA (USA), with the most occurring at the dent stage and the least at the blister stage; however, if

moisture were controlled at 45%, then more fumonisin $B_1$ was produced at the dough stage than dent stage but the least was still at the blister stage (Warfield and Gilchrist, 1999). More fumonisin $B_1$ was produced by *F. verticillioides* and *F. proliferatum* on rice under laboratory conditions when temperatures were cycled between 10 and 25°C and 5 and 20°C every 12 h for 6 weeks, respectively (Ryu *et al.*, 1999b); however, *F. graminearum* produce less deoxynivalenol and zearalenone on rice when temperatures were cycled every 12 h between 15 and 30°C but produced the greatest amount of these two mycotoxins at 25°C for 2 weeks followed by incubation at 15°C for an additional 4 weeks (Ryu and Bullerman, 1999). Temperature seems to be an important factor in the production of fumonisin $B_1$, deoxynivalenol and zearalenone in cereal grains.

The production of *Fusarium* mycotoxins in the laboratory may not agree with what happens in the field. Desjardins *et al.* (1998) noted that fumonisin production in the laboratory may not agree with production on maize growing in fields (Kansas, USA). *F. verticillioides* produced more fumonisin $B_1$ at 25°C on maize than at either 20 or 30°C (Alberts *et al.*, 1990). *F. verticillioides* transgenic isolates that expressed green fluorescent protein were used to determine the interaction of the mold with maize, which was grown in greenhouses in Israel (Oren *et al.*, 2003). They found that soil infection of *F. verticillioides* produced more infection than did using infected maize seeds. The infection started in the roots and mesocotyl and then proceeded to cell walls of stems and leaves with greater infection when grown in low rather than high light conditions (Oren *et al.*, 2003). Miller (2001) concluded that generally *F. graminearum* grew best between 26 to 28°C, low oxygen tension and needed rain at silking for ear rot to develop and produce deoxynivalenol, but high oxygen, cool temperatures and even dead tissues allowed zearalenone to be produced. In contrast, *F. verticillioides* grew above 28°C at low pH when organic acids were produced from starch metabolism and under drought and insect damage produced fumonisins (Miller, 2001). It may not be possible to develop laboratory experiments to simulate field conditions; therefore, observations of environmental conditions when mycotoxins are produced will be critical to further understanding of mycotoxin production.

Practices used in the field during growing of crops can affect mycotoxin production. The use of lodging with a drum roller 3 weeks after ear emergence to flatten the crop caused more deoxynivalenol to be produced in oats and barley in Norway; however, natural lodging due to heavy rains and wind did not always increase the level of deoxynivalenol (Langseth and Sabbetorp, 1996). Temperature and rainfall in late summer contributed to more ear rot by *F. graminearum* and, subsequently, more deoxynivalenol in Canadian maize, especially in hybrids that were sensitive to disease (Reid *et al.*, 1996). Weather again was considered important for deoxynivalenol production in wheat in Ontario, Canada because 73% of variation in this mycotoxin concentration could be explained by temperature and rainfall (Hooker *et al.*, 2002).

HT-2 and T-2 toxin were detected in oats in the years of 1996–1998 in Norway (Torp and Langseth, 1999). A "powdery *Fusarium poae*" was isolated from Norwegian oats, barley, and wheat contaminated with T-2 and HT-2 toxins (Torp and Langseth, 1999). This "powdery *Fusarium poae*" was different from *F. poae* because it lacked the fruity odor and aerial mycelium on czapek-dox iprodione dichloran agar (CZID) and from *F. sporotrichioides* because it lacked chlamydospores, macroconidia and polyphialides (Torp and Langseth, 1999). Mateo *et al.* (2002) studied the production of type A trichothecenes (T-2 toxin, HT-2 toxin, diacetoxyscirpenol and neosolaniol) by *F. sporotrichioides* in maize, rice, and wheat in culture flasks at water activities ($a_w$) of 0.990, 0.995 and 0.999 and temperatures of 20, 26, and 33°C over 3 weeks. The type of cereal, temperature and $a_w$ combination affected the production of type A trichothecenes. For example, more T-2 toxin was produced as the $a_w$ increased in maize and the temperature decreased; however, in wheat, T-2 toxin decreased as $a_w$ increased but increased as temperature decreased. In rice, the greatest amount of T-2 toxin was produced at 0.990 and 26°C (Mateo *et al.*, 2002).

The molecular biology and genetics of mycotoxin biosynthesis is beyond the scope of this chapter, but they have been reviewed recently by Payne and Brown (1998) and Sweeney and Dobson (1999). Reviews and recent studies of the genes responsible for the production and regulation of aflatoxin, fumonisin, and trichothecenes have been published (Trail *et al.*, 1995; Bhatnagar *et al.*, 2003; Peplow *et al.*, 2003; Proctor *et al.*, 2003).

## Occurrence in foods

Mycotoxins occur worldwide in many different agricultural commodities and foods because molds are common contaminants in soil, air, water and vegetation (Table 10.4). Many molds grow on various substrates under many different environmental conditions; however, the presence of molds does not mean that they will produce mycotoxins. Also, if no molds are present, it does not mean that mycotoxins are not present because the molds could have died after producing mycotoxins.

Sharma and Salunkhe (1991) and Yoshizawa (1991) reviewed the information before 1990 on the occurrence of mycotoxins, especially aflatoxins, ochratoxins, trichothecenes and zearlaenone in cereals with some information on ergot alkaloids, cyclopiazonic acid and other minor mycotoxins of concern for human health. Shotwell (1991) reviewed the mycotoxins found in maize before 1990 and Strange (1991) reviewed the mycotoxins in groundnuts and cottonseed before 1990. More information is available in these early reviews for aflatoxins than for any other mycotoxin with information for trichothecenes coming in second. Some general reviews on the occurrence of mycotoxins in agricultural commodities and foods have been published in recent years (Rustom, 1997; Pittet, 1998; Deshpande, 2002; CAST, 2003). An excellent summary of the occurrence of DON, fumonisin, ochratoxin A and T-2 toxin in commodities can be found in the recent safety evaluation by the World Health Organization/Food and Agricultural Organization (WHO, 2001).

### Cereal grains

Mycotoxins can be produced in cereals during production of the commodities in the field, during storage, in processing of some foods such as beer, cheese and fermented meats, and in foods during distribution and consumer storage. Although many chemical metabolites produced by molds have been identified as mycotoxins, only a few are important in human and animal health. Miller (1995) reviewed the mycotoxins in grains and reported that the five most important in cereal crops are aflatoxin, deoxynivalenol (nivalenol), fumonisin, ochratoxin A, and zearalenone. In the recent review published by the Council for Agricultural Science and Technology (CAST, 2003), these five mycotoxins plus ergot alkaloids and the trichothecene, T-2 toxin were also deemed to be of concern. Minor mycotoxins that can occur in cereal grains or co-occur with the major mycotoxins include citrinin, cyclopiazonic acid, fusaric acid, moniliformin, patulin, penitrems, and sterigmatocystin (CAST, 2003).

Maize and cottonseed are very susceptible to contamination by *A. flavus* because insects may carry *A. flavus* to the silk and ears of the maize and both soil and air with the help of insects may be involved in *A. flavus* contamination of cottonseed (Wilson and Payne, 1994). Maize in the southeastern USA is more susceptible to *A. flavus* than maize grown in the Midwest because of the differences in climatic factors (Wilson and Payne, 1994).

Wheat and wheat-based products have been surveyed for the presence of *Fusarium* mycotoxins, particularly, the trichothecenes. Deoxynivalenol at the advisory level of $\geq 1$ µg/g was detected in 5%, 10%, 12%, 16% of wheat byproducts (gluten, germ, flakes), white flour, wheat bran, and whole wheat flour, respectively (Trucksess *et al.*, 1996). In Portugal, 78% of the wheat-based breakfast cereals had deoxynivalenol in concentrations from 103 to 6040 µg/kg (Martins and Martins, 2001). Wheat flour from southwestern Germany had 98%, 38%, and 12% contamination by deoxynivalenol, zearalenone, and nivalenol, respectively, with all other trichothecenes having < 7% contamination (Schollenberger *et al.*, 2002). Although no mycotoxins were included in the analyses, 32 mycotoxin-producing mold species belonging to *Alternaria*, *Aspergillus*, and *Penicillium* were identified in wheat flours in Germany (Weidenbörner *et al.*, 2000).

There have been numerous surveys of maize-based foods worldwide to determine the occurrence of fumonisins and in some case, the molds that produce them, especially *Fusarium verticillioides* and *Fusarium proliferatum* (Bullerman and Tsai, 1994; Shepherd *et al.*, 1996; WHO, 2000; 2001). Sydenham *et al.* (1991) analyzed commercial maize-based foods from five countries and found fumonisins in foods from all countries with the lowest level in Canada and the highest in Egypt; however, only two maize-meal samples were collected from these two areas. In the United States and South Africa where more samples were collected that included maize-meals, flakes, grits, mixes and breakfast cereals, fumonisin $B_1$ and $B_2$ ranged from 0 to 2545 and 0 to 475 ng/g, respectively (Sydenham *et al.*, 1991). Fumonisin $B_1$ in levels from 55 to 260 ng/g was detected in 38.8% of maize-based foods (grits, flakes, meals, sweet corn) analyzed by high-performance liquid chromatography (HPLC) in a Swiss study (Pittet *et al.*, 1992), and similar levels from 50 to 200 ng/g were detected in 15% of the maize grits, flakes, flour and snacks analyzed in a Spanish study (Sanchis *et al.*, 1994). Fumonisins $B_1$, $B_2$, and $B_3$ were detected in maize-meals in levels ranging from 33 to 840 µg/g with the highest levels being fumonisin $B_1$, but only fumonisin $B_1$ and hydrolyzed fumonisin $B_1$ (17–320 ng/g) were detected in canned maize, masa and tortilla chips by HPLC (Hopmans and Murphy, 1993). Maize-based products that were collected in Arizona, Maryland, and Nebraska had from < 75 to 1565; < 75 to 5916; and < 75 to 1927 ng fumonisin $B_1$/g, respectively, by HPLC (Castelo *et al.*, 1998b). Food-grade maize had from 0 to 1642 µg/kg of fumonisin $B_1$ and from 0 to 774 µg/kg of moniliformin and similar levels were detected in maize-based foods (grits, flours, masa, meals, and muffin/bread mixes) with 0 to 2,676 and 0 to 858 µg/kg of fumonisin $B_1$ and moniliformin, respectively (Gutema *et al.*, 2000).

Ergot alkaloids are produced by *Claviceps purpurea* and although most cereals are susceptible to *C. purpurea*, rye is most commonly infected (Scott, 1992; Kuldau and Bacon, 2001). Over 92% of rye flour in a 6-year Canadian study had from 70 to 414 ng/g; however, rye-containing breads and crackers had lower levels of 4.8–100 ng/g (Scott, 1992). In the same study, over 92% of triticale flour had from 46–283 ng/g ergot alkaloids, but 74% of wheat flour and 83% of bran/bran cereals had from 12 to 88 ng/g ergot alkaloids (Scott, 1992). These studies can be used to determine the human intake of ergot alkaloids.

Co-occurrence of mycotoxins complicates the evaluation of toxicity of individual mycotoxins. Aflatoxins and cyclopiazonic acid were found together in 51% of maize and 90% of peanut samples from a mycotoxin-monitoring survey in the USA (Urano *et al.*, 1992a). Positive correlations between aflatoxin $B_1$ and cyclopiazonic acid production were noted for *A. flavus* isolated from soils of peanut growing regions from Virginia to Texas in the southern USA (Horn and Dorner, 1999). L strains (sclerotia > 400 µm in diameter) of *A. flavus* showed variable production of aflatoxin $B_1$ and cyclopiazonic acid; whereas, most S strains (numerous sclerotia < 400 µm in diameter) of *A. flavus* produced both mycotoxins (Horn and Dorner, 1999). In Venezuelan maize destined for human consumption, there was no correlation between the development of aflatoxin $B_1$ and fumonisin $B_1$ because the former was detected in only 17% of the samples compared to the latter, in about 84% of the samples (Medina-Martínez and Martínez, 2000). Similar results were found in Chinese maize from regions where esophageal cancer is high because levels of fumonisin $B_1$ were high but aflatoxins were low in these samples;

however, high levels of type A trichothecenes (T-2 toxin, HT-2 toxin, isoneosolaniol and hydrolyzed products) and type B trichothecenes (deoxynivalenol-derivatives), which are produced by *Fusarium* species, also were detected (Chu and Li, 1994). In maize from Ghana, all samples had fumonisins and over half of them also had aflatoxin (Kpodo *et al.*, 2000). Ochratoxin A and citrinin were isolated from almost 90% of stored wheat and barley in the United Kingdom and occasionally viomellein, vioxanthin and xanthomegnin were isolated (Scudamore *et al.*, 1993). Patulin and cytochalasin E were produced during the malting of barley and wheat in the laboratory and about 20% of these two mycotoxins survived 80°C for 24 h (Lopez-Diaz and Flannigan, 1997a, b). Abouzied *et al.* (1991) surveyed several grain-based foods (breakfast cereals, crackers, cookies, flours, mixes and snack foods) for mycotoxins in a year after drought and found only 1 buckwheat flour with aflatoxin $B_1$ at 12 ng/g, 50% of samples with 1.2–19 µg/g deoxynivalenol, and 26% of samples with 5–100 ng/g zearalenone. Deoxynivalenol was present in between 52 to 93% of cereal-based baby foods, breads, breakfast cereals, noodles, rice, and similar foods; however, only 38, 28, 21, 13 and 7% of these foods contained HT-2, 15-acetyl-deoxynivalenol, 3- acetyl-deoxynivalenol, and nivalenol, respectively (Schollenberger *et al.*, 1999). Since several foods contain more than one mycotoxin, the additive effects of these mycotoxins on human health need to be further assessed.

*Fusarium* species are found in many soils and are common plant pathogens that cause blights, rots and wilts; therefore, trichothecenes are common metabolic contaminants (United Nations Environment Programme, 1990). D'Mello *et al.* (1997) reviewed the occurrence of *Fusarium* mycotoxins in several countries and noted that deoxynivalenol occurred in wheat in Canada, Germany, Japan, Poland, The Netherlands and the United States; in oats in Finland and The Netherlands; in maize in Italy; and in barley and rye in The Netherlands. In addition, the occurrence or co-occurrence of other *Fusarium* mycotoxins was reported, such as diacetoxyscirpenol, nivalenol, T-2 toxin, zearalenone, and their derivatives. Also, maize was contaminated with fumonisins (D'Mello *et al.*, 1997). Chelkowski (1998) mentioned that although about 20 species of *Fusarium* can be found on small grain cereals worldwide, only six species are significant because of the mycotoxins they produce: *F. culmorum* and *F. graminearum* produce deoxynivalenol or vomitoxin, nivalenol, and zearalenone; *F. poae* produces nivalenol; and *F. avenaceum* produces moniliformin and *F. verticillioides* and *F. proliferatum* produce fumonisins in maize. Fumonisins were present in high levels in maize in areas of South Africa (Sydenham *et al.*, 1990) and of China (Yoshizawa *et al.*, 1994) where esophageal cancer is prevalent. Maize from Illinois, Iowa and Wisconsin in the midwestern USA had fumonisins $B_1$, $B_2$, and $B_3$ in concentrations from 0.2 to 3.3 µg/g over a 4-year period (Murphy *et al.*, 1993). About 85% of maize and its meal in Botswana, Africa contained from 20 to 1270 µg/kg of fumonisin $B_1$; however, 15% of sorghum only had 20 to 60 µg/kg (Siame *et al.*, 1998). Mexican masa and tortillas had 0.21 to 1.8 µg/g of fumonisin $B_1$ compared to 0.4 to 1.29 µg/g in the United States (Dombrink-Kurtzman and Dvorak, 1999). Over 90% of freshly harvested or stored maize in Brazil had fumonisins $B_1$ and $B_2$ in concentrations from 0.9 to 49.3 µg/g (Orsi *et al.*, 2000). Barley can also be contaminated with molds that produce fumonisins and beer can be contaminated due the growth of these molds during malting (Scott and Lawrence, 1995). Beattie *et al.* (1998) showed that storage for 7 months at temperatures from −20 to 24°C with or without forced air did not affect the deoxynvalenol concentration; however, *Fusarium* species decreased over time at 24°C in forced air conditions. Similar results for deoxynvalenol in wheat were shown by Homdork *et al.* (2000b); however, both deoxynivalenol and zearalenone increased in warm and humid conditions (25°C and 90% relative humidity) of storage. Trichothecenes (deoxynivalenol, HT-2 toxin, T-2 toxin, nivalenol and traces of other derivatives) were found in barley, wheat and oats in Norway (Langseth and Rundberget, 1999). In this study, oats for human consumption generally showed the highest incidence and concentrations of the four mycotoxins with more contamination for HT-2 and T-2 toxins with mean levels of 115 µg/kg and 60 µg/kg, respectively. Fungal infection varied from year to year in cereal grains grown in The Netherlands due to weather conditions in each region and although several toxigenic molds were isolated from the grains, only 3% of wheat had deoxynivalenol and 1% of barley had zearalenone (de Nijs *et al.*, 1996).

Ochratoxin A occurred in contaminated Canadian grains (Scott, 1994) and grains from temperate climates (Varga *et al.*, 2001). In a survey in the USA, Truckress *et al.* (1999) found that 56 of 383 wheat samples had from 0.03 to 31.4 ng ochratoxin A/g with only four samples being > 5 ng/g and 11 of 103 barley samples had from 0.1 to 17 ng/g with only one sample being > 5 ng/g. A comprehensive compilation of survey data for ochratoxin A in food commodities, including cereals, was produced by the World Health Organization (WHO, 2001). In addition to ochratoxin A, cyclopiazonic acid, which is produced by *Penicillium aurantiogriseum*, was isolated from stored grain in Canada (Scott, 1994).

### Coffee

Ochratoxin A has been a concern in coffee from the cherries through the roasting and brewing processes ever since its presence was first reported about 30 years ago. Ochratoxin-producing molds have been isolated from coffee and coffee processing environments in Brazil, India, Indonesia Kenya, Uganda, Vietnam, and Zaire (Mantle and Chow, 2000; Urbano *et al.*, 2001; Varga *et al.*, 2001). *Aspergillus niger* and *Aspergillus ochraceus* made up 22.9 and 10.3% of *Aspergillus* species isolated from coffee beans, respectively; however, of the strains tested, only 11.5% and 88.1%, respectively, could produce ochratoxin A (Urbano *et al.*, 2001). In Thailand, *A. ochraceus* and *Aspergillus carbonarius* were isolated from one and seven coffee cherry samples, respectively, with all *A. carbonarius* producing ochratoxin A (Joosten *et al.*, 2001). When Robusta coffee was stud-

ied during three seasons in Thailand, overripe coffee cherries were more susceptible to ochratoxin A with most being found in the husks; therefore, dehulling procedures are important to prevent ochratoxin A from entering the final roasted coffee (Bucheli *et al.*, 2000). In Brazil, *A. ochraceus* was the dominant ochratoxin producer isolated with 75% of isolates producing ochratoxin A; however, *A. niger* was isolated twice as often but only 3% could produce ochratoxin A, and a few *A. carbonarius* were isolated from one region with 77% being able to produce ochratoxin A (Taniwaki *et al.*, 2003). Struder-Rohr *et al.* (1995) found ochratoxin A in about 50% of the commercial green coffee beans and brewed coffee analyzed. Roasting green coffee beans at 250°C for 150 sec caused some decrease in ochratoxin A; however, the lack of homogeneity in contamination level in the beans gave high standard deviations that made estimation of an average level difficult (Struder-Rohr *et al.*, 1995). When 162 samples of green coffee from Africa, Asia and Central or South America were analyzed for ochratoxin A, 106 samples had up to 48 µg ochratoxin A/kg with coffees from Africa having the highest levels of ochratoxin A (Romani *et al.*, 2000). In all 76/84, 11/18, and 19/60 samples from Africa, Asia and Central or South America had detectable levels of ochratoxin A (Romani *et al.*, 2000). In a Brazilian study of coffee cherries and beans, only 39 of 135 samples had > 0.2 µg ochratoxin A/kg with 10 of these samples from the drying yard or storage areas having levels >5 µg/kg (Taniwaki *et al.*, 2003). In a survey of green and roasted coffee from the market in the USA, lower amounts of ochratoxin A were detected with 9 of 19 samples and 9 of 13 samples having from 0.1–4.6 ng/g and 0.1–1.2 ng/g of ochratoxin A, respectively (Trucksess *et al.*, 1999). Bucheli *et al.* (1998) found no growth of ochratoxin-producing molds or ochratoxin A in green robusta coffee stored in Thailand for up to 8 months, suggesting that ochratoxin A production was a pre-storage concern. From all these studies, it appears that conditions during harvesting, drying, and storage of coffee beans in the different tropical growing regions may be important in preventing mold growth and ochratoxin production.

### Dairy products

Mycotoxins can be present in milk and dairy products from either indirect carryover from animal feeds or by direct growth of molds in or on the dairy product. Similar to the effect of consuming aflatoxin (Galvano *et al.*, 1996a), ewes that consumed cyclopiazonic acid, showed an increase of this mycotoxin in milk as the ingestion increased and the level decreased after consumption ended (Dorner *et al.*, 1994). When cows were fed diets containing up to 12 mg/kg deoxynivalenol, none was detected in the milk (Charmley *et al.*, 1993). No detectable residues of fumonisin were found in the milk of dairy cows dosed orally or intravenously with pure fumonisin B1 (Scott *et al.*, 1994). Galvano *et al.* (1996a) reviewed the presence of aflatoxin $M_1$ in milk and dairy products and concluded that contamination is seen worldwide but at levels that are not health concerns except for breast milk in tropical and subtropical countries. Aflatoxin $M_1$ was detected in 45.7% of Spanish cheeses made from cow, ewe, goat or combinations at levels from 20 to 200 ng/g (Barrios et al., 1996). In Argentina, aflatoxin $M_1$ was low in cheese, with levels from 0.20 to 0.33 µg/L, probably because 40% of the aflatoxin $M_1$ from milk partitioned with the casein and 60% with the whey (López et al., 2001). No mycotoxins were detected in 36 samples of cheese sold in Brazil, although mycotoxin-producing molds were isolated (Taniwaki and van Dender, 1992). A comprehensive survey of aflatoxin $M_1$ in foods can be found in the WHO/FAO safety evaluation of mycotoxins (WHO, 2001). Cyclopiazonic acid was produced by *Penicillium camemberti*, which was used to produce mold-ripened cheese (Scott, 1994). Although several isolates of *Penicillium aurantiogriseum*, *Penicillium commune*, and *Penicillium expansum*, which could produce cyclopiazonic acid were isolated from Italian Taleggio cheese, less than 0.25 mg/kg cyclopiazonic acid was detected in cheese (Finoli et al., 1999).

*Fruits and wine*
Some fruits, especially pome and vine varieties, have been susceptible to mycotoxigenic mold growth and subsequent production of mycotoxins. Pome fruits such as apples and pears can be contaminated with *Penicillium expansum*, which produces patulin. Apple and apple products can be contaminated with patulin as noted in papers on measures to reduce patulin (Taniwaki et al., 1992; Sydenham et al., 1995b; Gökmen and Acar, 1998; Huebner et al., 2000; Bissessur et al., 2001; Kryger, 2001; Kadakal and Nas, 2002). Patulin was detected in 57.7% of apple, pear and mixed fruit juices with 28.3% exceeding 50 µg/L and in 38.9% of apple and mixed fruit products with none exceeding 50 µg/L (Burda, 1992). Vine fruits are a concern for ochratoxin A because they can be used for table fruits, dried fruits or wine production. Ochratoxin A was detected in 16.2% of commercial red wines, especially those from Argentina and Spain, and only in 3.9% of commercial white wines; however, all but two samples had levels below 0.1 µg/L (Soleas et al., 2001). Since 96.7% of all *Aspergillus carbonarius* and only 0.6% of *Aspergillus niger* var. *niger* isolated from dried vine fruits (currants, raisins and sultanas) were able to produce ochratoxin A, it was concluded that these were the main source of this mycotoxin in dried fruits sold in Spain (Abarca et al., 2003). Only *A. carbonarius* produced ochratoxin A in moldy grapes in a vineyard in Spain (Cabañes et al., 2002) and in grapes and musts where levels of ochratoxin A ranged from <10 to 461 ng/mL from French vineyards (Sage et al., 2002). For Greek wines, ochratoxin A ranged from <0.02 to 3.2 ng/mL with more in red than white wines and generally higher levels in sweet versus dry wines (Soufleros et al., 2003). Also, wines from the eastern region of Greece had the highest levels of ochratoxin A suggesting that both variety and environmental factors could be important in mold growth (Soufleros et al., 2003). No aflatoxin or ochratoxin A was detected in figs collected from orchards in Turkey and only one sample from the drying process had aflatoxin $B_1$ of 30 µg/kg (Özay et al.,

1995). For figs from California orchards infected with *A. parasiticus*, 83% had aflatoxin > 100 ng/g; however, only 32% of figs infected with *A. flavus* did (Doster et al., 1996). Ochratoxin A was detected in a few figs at levels from 19 to 9,600 ng/g; however, 40% of the samples had trace amounts of <10 ng/g (Doster et al., 1996).

*Meat and eggs*
There is little new information on the presence of mycotoxins in meat and eggs because these usually are not a good substrate for mold growth and mycotoxin production. Prelusky (1994) reviewed the literature on the accumulation of ochratoxin, trichothecenes (DON, T-2 toxin), and zearalenone in tissues of poultry, ruminants and swine and by-products (eggs and milk). It was concluded that when animals consumed feeds containing high levels of mycotoxins, detectable levels could be found in tissues; however, once the mycotoxins were removed from the diets of the animals, the levels decreased. Cyclopiazonic acid preferentially accumulated in egg albumin with less being detected in the yolks when chickens were fed this mycotoxin but this mycotoxin decreased when intake decreased (Dorner et al., 1994). Fumonisins are poorly absorbed and rapidly eliminated and only very low levels are retained in tissues, milk or eggs (Smith and Thacker, 1996; WHO, 2001). Low levels of ochratoxin A are sometimes found in products made from pig kidney or blood but the levels are relatively low compared to wheat or other cereals (WHO, 2001). Mycotoxigenic molds have been isolated from dry-cured (Núñez et al., 1996) and fermented (López-Díaz et al., 2001) meats; however, mycotoxins were not analyzed in any of these products because the purpose was to find nontoxic strains to use as starter cultures in the processing of these meats. Overall, meat and eggs are less likely than other foods to be sources of mycotoxicoses.

*Nuts*
Both tree and groundnuts can be contaminated with molds that produce aflatoxins. Peanuts are most susceptible to contamination by *A. flavus* or *A. parasiticus* because these molds are commonly found in the soil and on plant debris (Wilson and Payne, 1994). Peanuts in many countries around the world are commonly contaminated with *A. flavus* and *A. parasiticus* that produce aflatoxin $B_1$ in concentrations ranging from 1 to 6450 µg/kg; however, regulations for total aflatoxins in peanuts are 1–20 µg/kg depending on the country (Schatzki and Pan, 1996; Rustom, 1997). All four aflatoxins ($B_1$, $B_2$, $G_1$, $G_2$) were detected in both 52% of peanuts (3.2–48 µg/kg) and 71% of peanut butter (1.6–64 µg/kg) in Botswana, Africa (Siame et al., 1998). Also, pistachio nuts have been contaminated with *A. flavus* as noted by reports from California, USA (Doster and Michailides, 1994; Schatzki, 1995a, b; Schatzki and Pan, 1996, 1997) and Turkey (Heperkan et al., 1994). Greater that 99% of the aflatoxin detected in pistachio nuts was in the early split samples that would be removed during processing because the physical characteristics distinguish them from good quality nuts (Doster and Michailides, 1994). Schatzki (1995a,

b) found that aflatoxin levels in pistachio nuts harvested between 1981 and 1991dropped from 10 to 1.5 ng/g based on a probability distribution model (Schatzki, 1995b). Further research to study the distribution in the pistachio process stream showed that 90% of the aflatoxin was in low-quality product (<5% of total) and that removal of this product would reduce aflatoxin from 1.2 to 0.12 ng/g (Schatzki and Pan, 1996). In addition, small nuts had increased aflatoxin contents with an average of 8 ng/g; therefore, sorting nuts by size could reduce aflatoxin levels (Schatzki and Pan, 1997). In contrast, there were no aflatoxins detected in a Turkish study by thin layer chromatography, although 35% of the *A. flavus* isolates from pistachios could produce aflatoxins on a synthetic medium (Heperkan *et al.*, 1994). Similarly, almonds and Brazil nuts are sometimes contaminated with aflatoxin $B_1$ (Trucksess *et al.*, 1994; Schatzki, 1996; Pittet, 1998). In 1993, for about 78% of the almond crop in California, only 0.7% had greater than 20 ng of aflatoxin/g (from one processor) and the average was 0.67 ng/g (Schatzki, 1996). Control of the amount of aflatoxin in nuts is done through regulations and banning of imports if nuts have unacceptably high levels.

*Miscellaneous foods*
Bosch and Mirocha (1992) found that zearalenone was present in sugar beets in concentrations ranging from 12 to 391 ng/g and in beet fibers from 13 to 4,650 ng/g possibly because *Fusarium equiseti* was the dominant mold isolated from both freshly harvested and stored sugar beets. Because many toxigenic fungi are either plant pathogens or saprophytes, mycotoxin contamination is possible in many agricultural products. The list of agricultural products contaminated with mycotoxins is extensive (WHO, 2001; CAST, 2003). For example, in addition to the agricultural products mentioned above, aflatoxins have been reported in apricots, candy, cookies, lotus seeds, marzipan, melon seeds, spices, pumpkin seeds and sunflower seeds (CAST, 2003). Ochratoxin A has been reported, for example, in beer, many breads, chocolate, currants, figs, olives, pasta, polenta, pulses, rice, and spices (WHO, 2001). Similar lists could be made for other mycotoxins.

## Human diseases
Aflatoxins, ergot alkaloids, fumonisins, ochratoxin A, trichothecenes (deoxynivalenol and T-2 toxin), and zearalenone are the most common mycotoxins that naturally contaminate agricultural commodities, feeds, and foods. Each of these groups of mycotoxins has been shown to cause natural outbreaks of disease in farm animals and are known or suspected causes of disease in humans. Table 10.6 lists the major molds, the mycotoxins produced that are known or suspected to affect human health, and the common sources for these mycotoxins. The linking of consumption of mycotoxins to animal and human disease is complex and several recent reviews have been written about different aspects of toxicity. Bodine (1999); Steyn and Stander (1999); and Hussein and Brasel (2001) have written general reviews on toxicity of mycotoxins. The

World Health Organization and the Food and Agriculture Organization recently have completed a comprehensive review of the toxicity of aflatoxin M1, fumonisins, ochratoxin A, and some trichothecenes (DON, T-2 toxin, HT-2 toxin) (WHO, 2001). Eaton and Ramsdell (1992) reviewed the early literature on the differences between species and how they react to aflatoxins in relation to their diet. Galtier (1999) reviewed the biotransformations that mycotoxins undergo in humans and animals. Reviews of mycotoxins that produce cancers (Castegnaro and McGregor, 1998), estrogenic effects (Shier, 1998); genotoxic effects (Dirheimer, 1998); haemotoxicity (Parent-Massin and Parchment, 1998), and immunotoxicity (Oswald and Coméra, 1998) have been published within recent years. Many *Fusarium* species occur simultaneously in cereal grains. For example, deoxynivalenol, fumonisins, zearalenone, fusaric acid, deoxynivalenol, and other trichothecenes may co-occur on maize (D'Mello *et al.*, 1997). Fusaric acid by itself is not very toxic; however, it has been postulated that it may increase the toxicity of other *Fusarium* mycotoxins with which it co-occurs (D'Mello *et al.*, 1997). However, in a subchronic feeding study in rats, fusaric acid at 400 ppm with fumonisin at (3.4, 18.4 and 437 ppm) in the diet, exerted no synergistic, additive or antagonistic effects (Voss *et al.*, 1999). Also, co-occurrence of aflatoxins and fumonisins in maize could contribute to the high incidence of hepatocellular cancer in parts of China (Li *et al.*, 2001; Ueno *et al.*, 1997). Hence, the link of individual or combined mycotoxins to human disease will continue to be researched for many years to come. This review on human disease will focus on the major mycotoxins and the known or suspected effects to human health based on circumstantial or epidemiological evidence. A summary of the mycotoxins, mold producers and suspected human health effects are presented in Table 10.6.

## *Aspergillus* mycotoxins
Economically important mycotoxins produced by *Aspergillus* species include aflatoxins, cyclopiazonic acid, ochratoxins, and patulin. Aflatoxin $B_1$ ($AFB_1$) is the most troubling, and also the most frequently encountered, *Aspergillus* mycotoxin in food. Aflatoxins were first identified in the early 1960s as the cause of a mysterious outbreak in England, dubbed 'Turkey X Disease', which killed over 100,000 turkeys that consumed imported feed heavily contaminated with *A. flavus* (Blount, 1961). Other mycotoxins may have been involved in this outbreak, most notably cyclopiazonic acid. $AFB_1$ was identified as a potent liver carcinogen in 1964 (Butler and Barnes, 1964) and shortly thereafter liver cancer epizootics in farm raised rainbow trout, which had been recurring since 1936, were determined to be due to the feeding of diets prepared with $AFB_1$ contaminated cotton seed (Jackson, *et al.*, 1968). Since their discovery, aflatoxins have caused outbreaks of farm animal disease worldwide and have been implicated in human disease outbreaks in areas of the world where subsistence farmers consume contaminated crops. A recent example is the report from Kenya's Meru North district that attributed 12 deaths

**Table 10.6** Major mycotoxins involved in human health and common molds that produce them[a]

| Mycotoxins | Known or suspected human health effects of mycotoxins |
|---|---|
| Aflatoxin $B_1$, $B_2$ | Clearly plays role in liver cancer etiology, especially in carriers of hepatitis B surface antigen; acute hepatitis; suppression of immune system, increased susceptibility to infections, growth retardation in children |
| Aflatoxin $G_1$, $G_2$ | Same as above |
| Cyclopiazonic acid | Suspected but not confirmed – fat degeneration and necrosis in kidney and liver |
| Deoxynivalenol (nivalenol) = type B trichothecenes | Drunken bread syndrome, akakabi-byo (red mold disease, scabby grain intoxication); suspected but not confirmed – suppression of immune system and glomerulonephritis |
| Ergot alkaloids | Ergotism or St. Anthony's fire (burning sensation and gangrene of extremities), nausea, vomiting, sleepiness |
| Fumonisin $B_1$, $B_2$ | Circumstantially linked to esophageal cancer, disruption of sphingolipid metabolism may cause toxicities in liver and kidneys (both not resolved) |
| Ochratoxin A | Nephrotoxicities such as Balkan endemic nephropathy and urinary tract tumors and potential effect on liver are not conclusive |
| Patulin | No reports on toxicity in humans |
| T-2 toxin = type A trichothecenes | Suspected to cause alimentary toxic aleukia |
| Zearalenone | Estrogenic effects in animals including monkeys but no information for humans |

[a]D'Mello *et al.* (1997), Frisvad and Thrane (2002), Henry and Bosch (2001), FAO (2001), Pitt (2000), Samson *et al.* (2002), United Nations Environment Programme (1990), IARC (2002)

to consumption of aflatoxin-contaminated maize (The Nation (Nairobi), 3 October, 2001).

Exposure to aflatoxins is determined by food intake surveys, analysis of food, and assay of biological specimens (Hall and Wild, 1994). Autrup and Autrup (1992) reviewed the early literature on the monitoring of human body fluids for the presence of aflatoxin in an effort to determine the relationship to liver cancer and other potential diseases. Since analysis of foods for aflatoxins is not always an accurate measure of exposure, methods that check the blood, urine and other tissues are viewed as more reliable (Hall and Wild, 1994). The link between aflatoxin ingestion from foods and the possibility of liver cancer was discussed for areas of Africa and Asia (Coulombe, 1991; Bilgrami and Sinha, 1992; Hall and Wild, 1994). Aflatoxins are ingested from contaminated food, absorbed from the gastrointestinal tract and go to the liver where they are metabolized by oxidases and enzymes with the toxicity from the most to least toxic being $B_1 > G_1 > B_2 > G_2$ (Smith, 1997). Formation of a reactive epoxide on the terminal furan at the 8,9 position of aflatoxin $B_1$ is probably responsible for the carcinogenicity and mutagenicity because the epoxide can bind to macromolecules (Smith, 1997). Lee *et al.* (2001) showed that some dietary components of foods may inhibit the biotransforamation of aflatoxins to their active epoxide forms because some flavonoids, coumarins, and anthraquinones prevented formation of $AFB_1$–8,9 epoxide in rat livers.

The toxicity of $AFB_1$ is attributed to its ability to undergo mixed function oxidation by cytochrome P450 to form a number of metabolites, some of which react readily with nucleic acids and proteins causing cell death or tumor initiation (Figure 10.1). Coulombe (1991) discussed the biotransformation of aflatoxin $B_1$ by cytochromes P450 and the formation of active compounds including aflatoxin $M_1$. Although some species are

more sensitive to the liver toxicity and tumorigenicity of $AFB_1$, risk factors contributing to an individuals sensitivity to aflatoxin-induced liver tumors include level of exposure to aflatoxin, expression of aflatoxin activation/detoxification pathways, nutritional status, and most importantly, chronic infection with hepatitis B or C virus (Henry *et al.*, 1999). $AFB_1$ induction of liver tumors in laboratory animals is closely correlated with mutations to specific genes that are known to control tumorigenicity. These same genetic alterations were shown in humans consuming large amounts of $AFB_1$-contaminated food, particularly if exposure to hepatitis B or C virus has occurred (Henry *et al.*, 1999). Infection with hepatitis B or C virus acts synergistically with $AFB_1$ to induce mutations known to cause tumors. The nature of the mutations in rats caused by aflatoxin exposure, and the incidence of similar mutations in humans with hepatocellular carcinoma, implicate this mycotoxin as a human carcinogen in Group 1 (IARC, 2003). The best strategy for reducing aflatoxin liver cancer risk is through implementation of a universal hepatitis B virus vaccination program in high-risk areas (Henry *et al.*, 1999). In addition, several other risk factors such as consumption of alcohol and contaminated water, malnutrition, oral contraceptive use, presence of liver flukes, and smoking, have been linked to liver cancer (Henry and Bosch, 2001). Dietary factors are also extremely important. For example, prophylactic intervention with chlorophyllin (a semisynthetic water soluble chlorophyll derivative) reduced urinary levels of the aflatoxin–N-7-guanine adduct in populations at high risk for hepatocellular carcinoma in China (Egner *et al.*, 2001). Furthermore, it was hypothesized that diets rich in chlorophyll would prevent the development of hepatocellular carcinoma from other environmental carcinogens. However, increased liver cancer is not the only effect of dietary aflatoxin exposure in humans because Gong *et al.* (2002) have

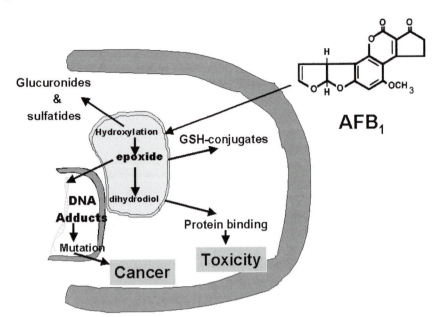

**AFB₁**

**Figure 10.1** Mechanism of action of aflatoxin. The pathways of aflatoxin B₁ (AFB₁) metabolism lead to cancer and cytotoxicity (after Eaton and Gallager, 1994). Briefly, aflatoxin B₁ enters the cell and is either metabolized via mono-oxygenases in the endoplasmic reticulum to hydroxylated metabolites, which are further metabolized to glucuronide and sulfate conjugates or is oxidized to the reactive epoxide, which undergoes spontaneous hydrolysis to the AFB₁–8,9-dihydrodiol which can bind to proteins resulting in cytotoxicity. The epoxide can react with DNA or protein, or be detoxified by an inducible gluta-thione S-transferase to the glutathione (GSH)-conjugate.

clearly shown an association between exposure to aflatoxin and impaired growth in children in West Africa.

Sabbioni and Sepai (1998) reviewed the human exposure to aflatoxins and reported that there are possible links between aflatoxin ingestion and liver cancer; however, the use of bio-markers to determine exposure is critical to determine actual human risks. Biologically active markers, such as aflatoxin B₁-guanine and aflatoxin B₁–albumin adducts, have potential use in studying the exposure and risks to aflatoxin in the diet; however, some of these markers only track a few days to weeks exposure (Sabbioni and Sepai, 1998). When the dietary intake of aflatoxins, ochratoxin A, patulin, and trichothecenes were evaluated for Swedish populations, only some Brazil nuts and pistachios had high levels of aflatoxin above the recommended daily intake (Thuvander *et al.*, 2001a). To be carcinogenic, af-latoxin B₁ must be converted to 8,9-dihydro-8,9-expoxy-afla-toxin B₁, which is termed aflatoxin B₁-8,9-epoxide (Massey *et al.*, 1995; McLean and Dutton, 1995; Sabbioni and Sepai, 1998). Cytochrome P450 monooxygenase enzymes and pos-sibly lipid hydroperoxide-dependent microsomal prostaglan-din H synthase and cytosolic lipoxygenase are involved in the activation of aflatoxin B₁ and have been reviewed by Massey *et al.* (1995). Lewis *et al.* (1999) used human cell lines that expressed cytochrome P450 and showed that aflatoxins B₁ and G₁ were very toxic. In addition to the aflatoxin B₁-8,9-epoxide, Massey *et al.* (1995) reviewed research on aflatoxi-col and aflatoxin B₁-8,9-dihydrodiol, which may be involved in toxicity. Besides being cytotoxic, AFB₁ was able to suppress the immune system as demonstrated by low phagocytic activ-ity and inability to kill *Candida albicans* in human monocytes (Cusumano *et al.*, 1996). A brief review of the immunologi-cal, hormonal, mutagenic and carcinogenic effects of aflatoxins

was done by McLean and Dutton (1995) with the pre-1990 literature reviewed by Sharma (1991).

Aflatoxins have also been implicated in Kwashiorkor dis-ease and Reye's syndrome (Wild and Hall, 1996); however, more evidence is needed before it can be accepted that afla-toxins cause these diseases. While the evidence for aflatoxin causing these diseases is not strong, the possible involvement of aflatoxin cannot be ruled out since higher levels of aflatoxin in tissues of affected individuals and hepatic involvement is apparent (CAST, 2003; Wild and Hall, 1996). The early lit-erature on the ingestion of aflatoxin from contaminated foods that have been linked to Reye's syndrome, childhood cirrhosis, and mental retardation was reviewed by Bilgrami and Sinha (1992); however, Henry and Bosch (2001) have pointed out the lack of good epidemiological studies to document the limited observations. Coulombe (1991) suggested that AFB₁ was probably not the cause of Reye's syndrome in the USA. Aflatoxin M₁, a metabolite of AFB₁, is found in milk of dairy cows and other animals that consume aflatoxin-contaminated feeds (JECFA 56th, 2001). Wood (1991) stated that the rela-tionship between aflatoxin M₁ and cancer in humans was not known; however, the potential for exposure of children to afla-toxin M₁ is carefully regulated in many countries.

Cyclopiazonic acid (CPA) is produced by several molds; however, in maize and peanuts, the primary producers are *Aspergillus* species, including *A. flavus*, which produces afla-toxins (Abramson, 1997). In certain mold ripened cheeses and meats, CPA is produced by several *Penicillium* species, particularly by *Penicillium camemberti* in Camembert cheese (Abramson, 1997). Although CPA is not as toxic as aflatoxin, and apparently not carcinogenic, concern over its presence in foods stems primarily from its co-occurrence with AFB₁ in peanuts and maize and its mechanism of action, which is similar to that of the known tumor promoter thapsigargin (Riley et al., 1995). CPA has been implicated as a causative agent in several cases of field intoxications in pigs, cattle, quail, and humans (Bryden, 1991). In India, millet contaminated with CPA was suggested as the cause of "kodua poisoning" in humans that had symptoms of fatigue, nausea, slurred speech and tremors (Abramson, 1997). Acute doses of CPA administered to rats caused toxic lesions in the liver, spleen, gastrointestinal tract and skeletal muscle (Riley et al., 1995). CPA is a potent inhibitor of calcium ($Ca^{2+}$) uptake and $Ca^{2+}$-dependent ATPase activity in both sarcoplasmic and endoplasmic reticulum (Riley et al., 1995) (Figure 10.2). While CPA is a useful biochemical tool for the study of sarcoplasmic and endoplasmic reticulum function, there is little evidence to implicate CPA in human disease (Wild and Hall, 1996).

Ochratoxin A, like CPA, is produced by several molds: by *Aspergillus ochraceus* in tropical to subtropical climates and by *Penicillium verrucosum* in temperate climates (Abramson, 1997). In northern Europe and elsewhere, ochratoxins are the cause of farm animal disease, most notably porcine nephropathy (Fink-Gremmels, 1999). Ochratoxin A is a renal carcinogen in mice and can induce proximal tubular dysfunction in many animals (CAST, 2003). The mechanism of action of ochratoxin is unclear; however, it has structural similarity to phenylalanine (Zanic-Grubisic et al., 2000) and inhibits many enzymes and processes that are phenylalanine dependent, which strongly suggest that ochratoxin A acts by disrupting phenylalanine metabolism (Figure 10.3). Whether or not ochratoxin A is the cause of any human disease, including Balkan Endemic Nephropathy (a kidney disease commonly attributed to ochratoxin A exposure), is much debated (JECFA 56th, 2001). Balkan endemic nephropathy and the possible relationship to tumors of the urinary tract for people in parts of Bulgaria, Romania and Yugoslavia that lie in the Danube Basin were reviewed by the United Nations Environment Programme (1990) and Størmer (1992). Ochratoxin A may not be the sole factor in Balkan endemic nephropathy and urinary tract tumors; therefore, more research is needed to confirm the associations to date and study other risk factors (Beardall and Miller, 1994; Henry and Bosch, 2001).

Ochratoxin A was highly toxic to both immature and mature rat brain cells and inhibited protein synthesis but had no effect on reactive oxygen species at nanomolar concentrations (Monnet-Tschudi et al., 1997). Conversely, ochratoxin B and the 3S-epimer form of ochratoxin A produced under thermal conditions had 19- and 10-fold lower toxicities in embryonic chick meninges than ochratoxin A, respectively (Bruinink et al., 1997). Since ochratoxin A affects the proximal tubules in the kidney, Babu et al. (2002) found that a human anionic transporter was most active on the apical side of the proximal tubule in mouse kidney cells. Some cytochrome P450 expressing cell lines showed increased toxicity to ochratoxin A (Lewis et al., 1999). Because ochratoxin A is a known cause of renal disease in farm animals, causes renal tumors in rodents, and can be found in human blood in both Europe and Africa (Peraica et al., 1999; Jonsyn-Ellis, 2001), ochratoxin A is a mycotoxin worthy of concern. A provisional tolerable weekly intake (PTWI) of 100 ng ochratoxin A/kg of body weight has been recommended (JECFA 56th, 2001).

There have been many surveys worldwide to determine the level of ochratoxin A in both human blood and in foods. Ochratoxin A was found in human blood, especially in people from Eastern Europe, Bulgaria and the former Yugoslavia, where it is linked to Balkan endemic nephropathy (Abramson, 1997) as well as people living in Canada, the Czech Republic, Germany, Italy, Japan, and Poland (Abramson, 1997; Ueno et al, 1998). In an area of Bulgaria where Balkan endemic nephropathy occurs, both ochratoxin A and citrinin were isolated

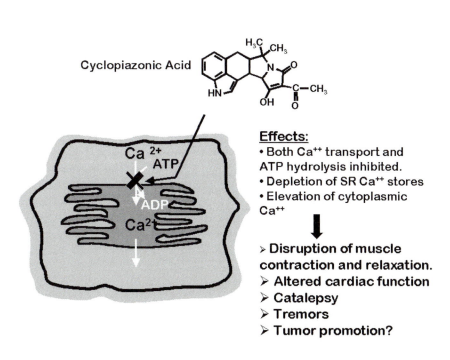

**Cyclopiazonic Acid**

**Effects:**
- Both $Ca^{++}$ transport and ATP hydrolysis inhibited.
- Depletion of SR $Ca^{++}$ stores
- Elevation of cytoplasmic $Ca^{++}$

➢ **Disruption of muscle contraction and relaxation.**
➢ **Altered cardiac function**
➢ **Catalepsy**
➢ **Tremors**
➢ **Tumor promotion?**

**Figure 10.2** Biochemical mechanism of action of cyclopiazonic acid. The chemical structures of cyclopiazonic acid and a schematic showing inhibition of sarcoplasmic or endoplasmic reticulum calcium-dependent ATPases (modified from Norred and Riley, 2001).

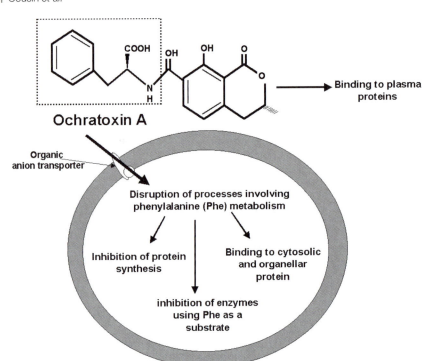

**Figure 10.3** Hypothesized mechanism of action of ochratoxin A, which is transported into cells via a multispecific organic anion transporter (Kobayashi *et al.*, 2002). Ochratoxin's chemical structure contains an isocumarin moity linked to phenylalanine (inside box). Ochratoxin A can alter processes that require phenylalanine and many of the biological effects of ochratoxin can be at least partially prevented by supplementation with phenylalanine or phenylalanine analogs (see Schwerdt *et al.*, 1999; Zanic-Grubisic *et al.*, 2000; Baudrimont *et al.*, 2001).

from wheat and wheat bran, with neither isolated from maize, and only ochratoxin A being isolated from oats (Vrabcheva *et al.*, 2000). Ochratoxin A was found in 14% of bovine milk, 58% of human milk, and 100% of human blood in Sweden (Breitholtz-Emanuelsson *et al.*, 1993) and in 215 of the human milks collected in Norway, where woman who had higher than normal intakes of liver pastes, fruit juices, and cakes and cookies had high levels of ochratoxin A (Skaug *et al.*, 2001). High plasma levels of ochratoxin A in Scandinavian countries were somewhat related to consumption of beer, cereal products, pork, and wine (Thuvander *et al.*, 2001b). All samples of plasma in a Canadian study contained ochratoxin A (Scott *et al.*, 1998). In Croatia, ochratoxin A was found in most plasma collected and in almost all cereals and beans analyzed over 10 years (Radić *et al.*, 1997); however, in an earlier study in SR Croatia, the isolation of ochratoxin A-producing *Aspergillus ochraceus* was not statistically different between nephropathic and nonnephropathic areas (Cvetnić and Pepeljnjak, 1990). In Japan, 85% of the plasma analyzed over 4 years contained ochratoxin A; however, the levels (average of 61 to 98 pg/mL) were lower than those reported from Europe and Canada (Ueno *et al*, 1998). More research is needed to determine if ochratoxin A is toxic to humans; however, many humans worldwide are exposed to this mycotoxin as determined by these surveys of levels in human blood and in common foods.

Sterigmatocystin, "tremorgens", and patulin are some other mycotoxins produced by *Aspergillus* species and can be found on cereal grains or other commodities likely to be consumed by humans or farm animals. Sterigmatocystin is a precursor of aflatoxin and is a liver carcinogen. Fungal metabolites that elicit tremors in animals are defined as tremorgens (Cole and Cox, 1981) and are produced by many molds. Those pro-

duced by *Aspergillus fumigatus* have been implicated in farm animal diseases associated with improperly handled maize silage (CAST, 2003). Patulin is produced by *Aspergillus* and *Penicillium* species and other molds; therefore, the health risks will be described in the section on *Penicillium* toxins.

## *Fusarium* mycotoxins

Economically important mycotoxins produced by *Fusarium* include fumonisins, deoxynivalenol, T-2 toxin, and zearalenone. Aflatoxin $B_1$ and fumonisin $B_1$ are now considered to be the most important foodborne mycotoxins worldwide with 2001 being an especially important year from an international regulatory perspective. For example, in addition to the publication of the Food and Drug Administration's Final Guidance for Industry, the Codex Committee on Food Additives requested that the 56th Joint Expert Committee on Food Additives (JECFA) (WHO, 2001) evaluate the health risks from fumonisin $B_1$, $B_2$ and $B_3$ ($FB_1$, $FB_2$ and $FB_3$) in foods. As a result of the evaluation, the 56th JECFA (2001) proposed a maximum tolerable daily intake (PMTDI) of $FB_1$, $FB_2$ and $FB_3$ of 2 μg/kg of body weight (bw) per day. In most developed countries, consumption of fumonisins fall well below the proposed maximum tolerable daily intake; however, in countries where maize consumption is high (Central America, Africa, China), large numbers of people could easily exceed the proposed PMTDI.

## *Fumonisins*

Several reviews on the health effects of fumonisins have been published (Riley *et al.*, 1994b; Marasas, 1995; Riley *et al.*, 1998; Shier and Abbas, 1999; Shier, 2000). Equine leucoencephalomalacia (ELEM) and porcine pulmonary edema

(PPE) are two farm animal diseases caused by consumption of fumonisin-contaminated feeds (WHO, 2000). ELEM is a fatal disease for horses and other equines with horses being most sensitive to fumonisin ingestion and the only species that develop brain lesions in response to this mycotoxin. PPE is a fatal disease that develops in pigs and is characterized by pulmonary edema, hydrothorax, cardiovascular dysfunction, and liver toxicity. In addition to the species-specific ELEM and PPE, fumonisins are hepatotoxic in all species tested and nephrotoxic in many species. Fumonisins have been isolated from contaminated maize eaten by humans in high esophageal cancer areas of Southern Africa and China and may be one factor involved in human carcinogenesis (Sydenham et al., 1990; Rheeder et al., 1992; Beardall and Miller, 1994; Groves et al., 1999; Chelule et al., 2001). $FB_1$ is a non-genotoxic renal and liver carcinogen in male rats (Gelderblom et al., 1991; NTP Technical Report, 2002), a hepatocarcinogen in female mice (NTP Technical Report, 2002), and a promoter of $AFB_1$ liver tumors in rainbow trout and rats (Carlson et al., 2001; Gelderblom et al., 2002a). Recently, $FB_1$ was evaluated by the International Agency for Research on Cancer (IARC) and the conclusion was that there was sufficient evidence in experimental animals for the carcinogenicity of $FB_1$ and the possibility that $FB_1$ was a Group 2B carcinogen for humans (IARC, 2003).

Fumonisins have been associated with cancer, reproductive toxicity (neural tube defects), and acute disease outbreaks in humans where low quality maize is consumed as a staple (WHO, 2000; WHO, 2001). In all cases, there is only limited evidence for fumonisins being the cause of these diseases in humans (WHO, 2001; IARC, 2003). The association of fumonisins to esophageal cancer has not been demonstrated epidemiologically; however, fumonisins were reported to be involved in an outbreak of foodborne illness in humans consuming moldy maize and sorghum in India in 1995 (Bhat et al., 1997). Ueno et al. (1997) found high levels of $FB_1$ and $AFB_1$ in maize over a 3-year period in Haimen, China where primary liver cancer is a concern and suggested that $FB_1$ and aflatoxin $B_1$ may be risk factors for the high incidence of liver cancer. Fumonisins target different organs in each species; thus, the esophageal cancer possibly could be a species specific outcome in humans; however, other factors such as vitamin and mineral deficiencies, drinking and smoking also may be involved (Marasas, 2001b). Nonetheless, exposure in some developing countries at levels that can exceed the recommended provisional maximum tolerable daily intake (WHO, 2001), its co-occurrence with $AFB_1$ (and possibly ochratoxin A), and its unusual mechanism of action make it a mycotoxin of considerable concern in countries where low-quality maize can constitute a significant portion of the total caloric intake (WHO, 2000).

There is considerable evidence that the underlying mechanism by which fumonisins cause toxicity to animals is disruption of lipid metabolism (Riley et al., 2001; Merrill et al., 2001; Gelderblom et al., 2002b). Most notably, fumonisins are spe-

cific inhibitors of ceramide synthase (sphinganine and sphingosine N-acyltransferase) (Merrill et al., 2001), a key enzyme in the pathway leading to formation of ceramide and more complex sphingolipids (Figure 10.4). The structural similarity between sphinganine and $FB_1$ led to the hypothesis that the mechanism of action of this mycotoxin was via disruption of sphingolipid metabolism or a function of sphingolipids (Wang et al., 1991). In every cell line, animal, plant or fungus in which it has been tested, $FB_1$ inhibits the CoA-dependent acylation of sphinganine (Sa) and sphingosine (So) via the enzyme ceramide synthase. The enzyme ceramide synthase recognizes both the amino group (sphingoid binding domain) and the tricarboxylic acid side-chains (fatty acyl-CoA domain) of $FB_1$ (Merrill et al., 2001). While removal of the tricarboxylic acid side-chains reduces the ability of $FB_1$ to inhibit ceramide synthase, N-acylation of $FB_1$ completely abolishes the inhibitory activity (Norred et al., 2001). When ceramide synthase is completely inhibited, either in vitro or in vivo, the intracellular sphinganine and sometimes sphingosine concentration increases rapidly (Wang et al., 1991; Yoo et al., 1992; Wang et al., 1992; Riley et al., 1993; 1994b; 1999b). In vivo, there is a close relationship between sphinganine accumulation and the expression of toxicity in liver and kidney (Delongchamp and Young, 2001; Riley et al., 2001). Accumulated, free sphingoid bases can persist in tissues (especially kidney) much longer than $FB_1$ (Shephard and Snijman, 1999; Enongene et al., 2000; Garren et al., 2001). In urine from rats fed $FB_1$, > 95% of the free sphinganine was recovered in dead cells (Riley et al., 1994a). A subthreshold dose in rats or mice can prolong the elevation of free sphinganine in urine or kidney caused by a higher dose (Wang et al., 1999; Enongene et al., 2001). $FB_1$-induced elevation of free sphingoid bases and toxicity are both reversible, although the elimination of free sphinganine from the liver is more rapid than from the kidney (Enongene et al., 2000; Enongene et al., 2001; Garren et al., 2001).

Inhibition of ceramide synthase results in the redirection of substrates and metabolites into other pathways (Merrill et al., 2001; Riley et al., 2001). For example, when sphinganine accumulates, it is metabolized to sphinganine 1-phosphate, which breakdowns into a fatty aldehyde and ethanolamine phosphate. Both products are redirected into other biosynthetic pathways, in particular into an increased biosynthesis of phosphatidylethanolamine (Badiani et al., 1996). Disrupted sphingolipid metabolism leads to imbalances in phosphoglycerolipid, fatty acid metabolism, and cholesterol metabolism via free sphingoid-base and sphingoid base 1-phosphate induced alterations in phosphatidic acid phosphatase and monoacylglycerol acyltransferase (Merrill et al., 2001). Thus, $FB_1$ inhibition of ceramide synthase can cause a wide spectrum of changes in lipid metabolism and associated lipid-dependent signaling pathways leading to the altered cell growth, differentiation, and cell injury observed both in vitro and in vivo (WHO, 2000b; Merrill et al., 2001; Riley et al., 2001; WHO, 2001; IARC, 2003)

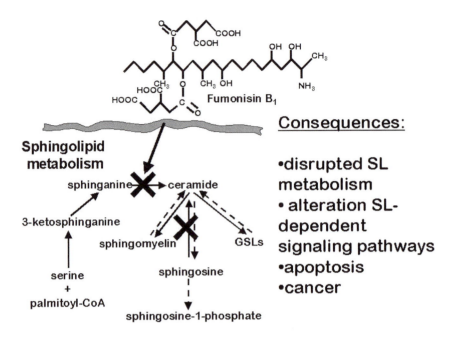

**Figure 10.4** Biochemical mechanism of action of fumonisin. The structure of fumonisin B₁ and inhibition of sphinganine (sphingosine) N-acyltransferase results in a blockage of complex sphingolipid biosynthesis, accumulation of sphinganine (and sometimes sphingosine), and diversion of sphinganine to sphinganine 1-phosphate, which is further broken down to a fatty aldehyde and ethanolamine phosphate (modified from Norred and Riley, 2001). Fumonisins also block the reacylation of dietary sphingosine and the sphingosine that is released by the turnover of more complex sphingolipids. As a consequence of disruption of sphingolipid metabolism, many bioactive lipid intermediates are created and some are lost. Some examples of the biological activity of these intermediates are increases in intracellular free sphingoid bases which can inhibit protein kinase C and other kinases, increase in sphingosine 1-phosphate that triggers endoplasmic reticulum calcium release, increase in free sphinganine and sphinganine 1-phosphate, which can affect cell proliferation, cell cycle progression, and initiate apoptosis or necrotic cell death, and depletion of complex sphingolipids (glycosphingolipids=GSLs and sphingomyelin), which are integral components of pathways that regulate cell growth and cell cycle progression and vitamin transport (for example folic acid) (see Merrill *et al.*, 2001).

In short-term studies with rats, rabbits and mice, disruption of sphingolipid metabolism usually occurred at or below the fumonisin dosages that caused liver or kidney lesions (WHO, 2001). For example, in a long-term feeding study with Fischer-344/N Nctr BR rats (NTP TR 496), FB₁ induced an increase in the Sa/So ratio in kidney tissue and urine, which correlated with increased non-neoplastic and neoplastic kidney lesions (WHO, 2001). However, elevation of free sphinganine and the Sa/So ratio were only increased after 3 and 9 weeks at 50 and 80 mg FB₁/kg diet, respectively, and these were doses that also induced liver adenoma and carcinoma of female B6C3F₁/Nctr BR mice (NTP TR 496) (WHO, 2001). In Sprague Dawley rats, hepatoxicity was associated with free sphinganine levels > 12 nmol/g fresh weight based on *in vivo* studies (Voss *et al.*, 1996b) and in mouse liver, increased apoptosis was associated with free sphinganine levels greater than 12 nmole/g fresh weight (Delongchamp and Young, 2001). In the rat, nephrotoxicity was associated with higher sphinganine levels; however, renal free sphinganine increased more dramatically and at lower FB₁ doses than did hepatic free sphinganine concentration, which were 15 ppm FB₁ and 50 ppm FB₁ for kidney and liver, respectively (Voss *et al.*, 1996b). This is similar to the free sphinganine levels associated with significantly increased nephropathy and hepatopathy in male

BALBc mice (Tsunoda *et al.*, 1998). FB₁ was not a complete carcinogen in the rainbow trout model; however, there was a close correlation between the elevation of free sphinganine in liver and FB₁ promotion of aflatoxin B₁ hepatocarcinogenicity (Carlson *et al.*, 2001). When pure FB₁ in AIN-76 diets was fed to Sprague-Dawley rats, the no-observed-effect level (NOEL) for elevation of urinary free sphinganine was > 1 mg/kg diet to < 5 mg/kg diet (equivalent to >0.1 mg/kg bw/day to < 0.5 mg/kg/bw day) (Wang *et al.*, 1999).

Evidence of disruption of sphingolipid metabolism is an early indicator of fumonisin exposure in horses fed diets containing fumonisins (Wang *et al.*, 1992; Riley *et al.*, 1997; Constable, 2000a,b). For example, a pony showed large increases in serum free sphinganine when fed a diet containing maize screenings naturally contaminated with fumonisins at 22 ppm. The elevation in serum sphinganine was reversible and the increase in free sphinganine and the Sa/So ratio occurred before there were increases in serum transaminase activity and clinical signs of ELEM. In pigs fed diets prepared from naturally contaminated maize screenings, there was a dose–response relationship between liver pathology, the ratio of free sphinganine to free sphingosine in serum or liver, and the dose of fumonisin (Riley *et al.*, 1993). After 14 days, statistically significant increases in the serum ratio of free sphinganine to free sphingosine were observed at feed concentrations as low as 5 mg/kg total fumonisins (equivalent to 0.2 mg/kg bw/day). Free sphingoid bases were also significantly elevated in the kidney and the lung at dosages that showed no signs of toxicity in these organs. In pigs fed fumonisin in maize cultural material, significant effects on cardiovascular function were associated with significant increases in free sphingoid bases in heart tissues (Haschek et al., 2001). Damage to pig alveolar endothelial cells, in vivo, was preceded by accumulation of free sphingoid bases in lung tissue; therefore, the cardiovascular alterations leading to acute left-sided heart failure may be a consequence of sphingoid-base-induced inhibition of L-type calcium channels (Smith et al., 1999; Constable et al., 2000b; Smith et al., 2001). The minimum oral dose needed to induce PPE has not been clearly established; however, Smith et al. (1999) have predicted that when the free sphinganine and free sphingosine in plasma are greater than or equal to 2.2 µmol/L or 1 µmol/L, respectively, the onset of hemodynamic changes will occur.

There have been numerous studies done with agriculturally important species using pure fumonisins, contaminated maize screenings or maize cultural material of F. verticillioides (WHO, 2000; 2001). These studies have been done with catfish, cattle, goats, lambs, mink, poultry, and rabbits. In all cases where toxicity was evident, it involved liver and/or kidney, or heart or homologous organs and where measured, elevation in free sphinganine in tissues, serum, or urine. The use of elevated free sphinganine in human urine or blood has been examined with mixed success. For example, van der Westhuizen et al. (1999) found no relationship between urine or serum sphingoid base levels and low dietary FB$_1$ intake. In China, Qui and Liu (2001) found that the free sphinganine to free sphingosine ratio was significantly greater in male urine from households where the estimated daily FB$_1$ intake was greater than 110 µg/ kg body weight/day.

In both in vivo and in vitro studies, a good correlation between disruption of sphingolipid metabolism and FB$_1$ exposure has been found; however, this correlation is not proof that disruption of sphingolipid metabolism is the cause of the toxicity and carcinogenicity by FB$_1$. Nonetheless, using cultured cells, sphingolipid-dependent mechanisms for inducing apoptosis have been demonstrated (WHO, 2000; WHO, 2001; Riley et al., 2001). For example, accumulation of excess ceramide, glucosylceramide (Korkotian et al., 1999) or sphingoid bases, or depletion of ceramide, or more complex sphingolipids induced apoptotic or oncotic cell death (WHO, 2000; Merrill et al., 2001; Riley et al., 2001; WHO 2001). Conversely, the balance between sphingosine 1-phosphate and ceramide is critical for signaling proliferation or cell survival (Spiegel, 1998). There also will be a diversity of alterations in cellular regulation resulting from FB disruption of sphingolipid metabolism, which identify cell processes that are ceramide mediated (WHO, 2000; 2001). Many of these processes are relevant to understanding the toxicity and carcinogenicity of FB$_1$, especially, the ability of FB$_1$ to protect oxidant damaged cells from apoptosis

and to alter the proliferative response (WHO, 2000; 2001). Perhaps the best evidence for a cause and effect relationship between disruption of sphingolipid metabolism and the toxic effects of FB$_1$ are the studies conducted using inhibitors of sphinganine accumulation to reverse the increased apoptosis and altered cell growth induced by FB1 treatment (Schroeder et al., 1994; Yoo et al., 1996; Schmelz et al., 1998; Riley et al., 1999a; Tolleson et al., 1999; Kim et al., 2001; Yu et al., 2001; He et al. 2002).

The loss of complex sphingolipids also plays a role in the abnormal behavior, altered morphology, and altered proliferation of fumonisin-treated cells (WHO, 2000b; WHO, 2001). The ability of FB$_1$ to alter the function of specific glycosphingolipids and lipid rafts (membrane associations of sphingolipids, ceramide-anchored proteins, and other lipids) is an example as are the functions such as vitamin and toxin transporters and cell–cell and cell–substratum contact (Sandvig et al., 1996; Stevens and Tang, 1997; Pelagalli et al., 1999). The evidence for fumonisin-induced disruption of sphingolipid metabolism in target tissues has been demonstrated repeatedly in many independent studies (Riley et al., 2001). Nonetheless, the precise mechanism by which disrupted sphingolipid metabolism contributes to the increased organ toxicity in rodents is unclear. The current understanding of the sphingolipid signaling pathways (Merrill et al., 2001; Riley et al., 2001) indicates that the balance between the intracellular concentration of sphingolipid effectors that protect cells from apoptosis (decreased ceramide, increased sphingosine 1-phosphate) and the effectors that induce apoptosis (increased ceramide, increased free sphingoid bases, increased fatty acids) will determine the observed cellular response. Cells sensitive to the proliferative effect of decreased ceramide and increased sphingosine 1-phosphate will be selected to survive and proliferate. Conversely, when the rate of increase in free sphingoid bases exceeds a cell's ability to convert sphinganine/sphingosine to dihydroceramide/ceramide or their sphingoid base 1-phosphate, then free sphingoid bases will accumulate to toxic levels. In this case, cells that are sensitive to sphingoid base-induced growth arrest will cease growing and insensitive cells will survive. Thus, the kinetics of fumonisin elimination (rapid), affinity of FB$_1$ for ceramide synthase (competitive and reversible), and kinetics of sphinganine elimination (persistent but reversible) could influence the time course, amplitude, and frequency of spikes in intracellular ceramide, sphingoid base-1 phosphates, and free sphinganine concentration in tissues of animals consuming fumonisins. This is important, because the balance between the rates of apoptosis and proliferation are critical determinants in the process of hepato- and nephrotoxicity and tumorigenesis in animal models (Dragan et al., 2001; Howard et al., 2001; Voss et al., 2001b). At the cellular level, apoptotic necrosis should be considered similar to oncotic necrosis and both will lead to a regenerative process involving sustained cell proliferation and increased cancer risk (Hard et al., 2001; Dragan et al., 2001).

*Trichothecenes*

The trichothecenes are a group of over 180 sesquiterpenoid mycotoxins produced by *Fusarium* and other fungal species (*Myrothecium*, *Trichoderma*, *Stachybotrys* and others) that include some of the most potent translational inhibitors known (Ueno, 1983; Grove, 1988; 1993; 2000). Sharma and Kim (1991) reviewed the pre-1990 research on effects of trichothecenes on human health and noted that T-2 toxin was associated with alimentary toxic aleukia, acute reactions (skin, diarrhea, vomiting and other effects), and potential effects on the immune system. Deoxynivalenol (DON or "vomitoxin"), nivalenol (NIV), T-2 toxin and diacetoxyscirpenol (DAS) are among the trichothecenes detected in cereal grains and, thus, enter human and animal foods (Ueno, 1987; Rotter *et al.*, 1996). These compounds have been associated with human and animal toxicoses that are sometimes fatal (Joffe, 1978, 1983; Coté *et al.*, 1984; Ueno, 1987; Bhat *et al.*, 1989). DON, the most commonly encountered worldwide, has a broad spectrum of effects on gastrointestinal homeostasis, growth, neuroendocrine function and immunity. There are strong species differences with pigs being the most sensitive followed by rodents > dogs > cats > poultry > ruminants. This spectrum of sensitivities relates to differences in toxicokinetics and feeding habits of different species. DON does not accumulate in tissues or appear at high concentrations as residues in animal foods. The potential for DON to cause chronic effects such as anorexia, reduced weight gain, or immunotoxicity in humans is of obvious concern. Epidemiological studies have not yet targeted these possibilities. The PMTDI for DON is 1 µ/kg bw and for T-2/HT, it is 60 ng/kg bw (JECFA 56th, 2001). Many of the reports of trichothecenes causing human disease have been complicated by isolation of more than one mycotoxin and *Fusarium* species; however, based on epidemiological studies, Henry and Bosch (2001) have concluded that DON may be involved in human toxicities.

Scabby grain intoxication (SGI) or "drunken bread" or "red mold disease" (akakabi-byo in Japan) is due to consumption of grains (barley, oats, rice, rye, wheat) that are infected by *F. graminearum* and *F. culmorum*, which cause scab and head blight (Beardall and Miller, 1994; Marasas, 2001b). This mild, nonfatal illness causes GI symptoms such as abdominal pain, chills, diarrhea with bloody stools, headache, nausea, throat irritation, vomiting, weight loss, and weakness within 2 h after consumption of contaminated food; however, the feeling of fullness prevents overeating the bread (Marasas, 2001b). The trichothecenes, DON and NIV, have been implicated in SGI (Beardall and Miller, 1994; Marasas, 2001b). Mycotoxins such as DON and its derivatives, fusarenon-X, NIV, and zearalenone, are possible causes, but more research is needed (Beardall and Miller, 1994). Also, gastrointestinal symptoms have been reported in China after consumption of moldy maize and wheat from which *F. graminearum* and DON were isolated and this led to a suggested limit of 1.0 ppm DON in wheat (Beardall and Miller, 1994). In India, DON in wheat produced diarrheal symptoms and throat irritation

in people; hence, a tolerance for 34 ppb of DON in wheat was suggested based on the level in the food of 0.34 µg/g for the 30 to 150 g of wheat consumed (Beardall and Miller, 1994). DON at low concentrations suppressed cytokine production in human lymphocytes (Meky *et al.*, 2001); therefore, a urinary biomarker using β-glucuronidase treatment was developed to estimate the daily intake of deoxynivalenol in people (Meky *et al.*, 2003).

Alimentary toxic aleukia (ATA) is caused by T-2 toxin and maybe DAS that are produced mainly by *Fusarium sporotrichioides* and, possibly, *F. poae*, which can grow and produce toxins at 6–12°C and survive under the snow in overwintered grains (Beardall and Miller, 1994; Marasas, 2001b). Symptoms of ATA include gastrointestinal (GI) symptoms (diarrhea, nausea, vomiting) plus dermatitis, necrosis from the mouth through the entire GI tract, and hemorrhages in different organs (brain, heart, gastrointestinal tract, lungs, muscles), and even death (Beardall and Miller, 1994; Marasas, 2001b). Since T-2 toxin affects the central nervous system, Wang *et al.* (1998a) found that this toxin affected some monoamines such as increased 5-hydroxy 3-indoleacetic acid and serotonin in the brain of rats. T-2 toxin may cause neurochemical imbalances in protein synthesis and monoamine oxidase in the rat brain (Wang *et al.*, 1998b). From these studies, it can be postulated that T-2 toxin may affect the blood brain barrier in addition to changing the transport of amino acids into the brain.

Experimentally, low dose or chronic exposure of trichothecenes mycotoxin can cause anorexia and vomiting (Bondy and Pestka, 2000). Acute oral exposure to trichothecene doses causes severe damage to actively dividing cells in bone marrow, lymph nodes, spleen thymus and intestinal mucosa. Depending on dose and exposure regimen, trichothecenes can be both immunosuppressive and immunostimulatory (Bondy and Pestka, 2000). Both innate and acquired immune function can be affected. One of the most dramatic effects of the chronic DON ingestion in mice is a pronounced elevation in serum IgA and concurrent depression in IgM and IgG (Forsell *et al.*, 1986; Pestka *et al.*, 1989; 1990a,b,c; Dong *et al.*, 1991). The immunopathology associated with DON-induced dysregulation of IgA production is very similar to human IgA nephropathy (Berger's disease), the most common glomerulonephritis worldwide (D'Amico, 1987). These effects can persist long after withdrawal of DON from the mouse diet (Dong and Pestka, 1993).

The seemingly paradoxical effects of trichothecenes can be explained by their cellular and molecular modes of action (Figure 10.5). Exposure to low levels of trichothecenes appears to promote expression of a diverse array of cytokines and proinflammatory genes *in vitro* and *in vivo* (Dong *et al.*, 1994; Warner *et al.*, 1994; Azcona-Olivera 1995a,b; Zhou *et al.*, 1997; Wong *et al.*, 1998; Zhou *et al.*, 1998, 1999). These genes have the potential to modulate many immune and inflammation-associated sequelae. For example, DON-induced aberrant IgA production may be mediated through superinduction of IL-6 gene expression in T helper (TH) cells and

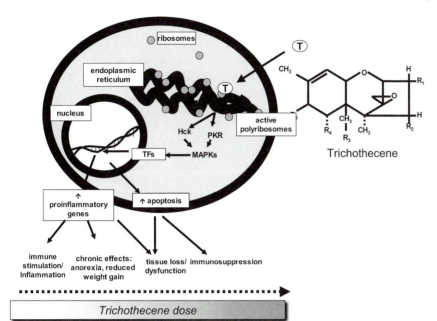

**Figure 10.5** Molecular mode of action for deoxynivalenol and other trichothecenes. Deoxynivalenol enters cell via diffusion and binds to active ribosomes, which transduces signal to RNA-activated protein kinase (PKR) and Src kinase (HCK). Subsequent phosphorylation of mitogen–activated protein kinases (MAPKs) drives transcription factor activation and the resultant chronic and immuntoxic effects.

macrophages (Yan *et al.*, 1997, 1998). Up-regulation of proinflammatory genes appears to relate to the capacity to increase transcription and mRNA stability (Ouyang *et al.*, 1996; Li *et al.*, 1997c). In contrast to these stimulatory effects, high doses of trichothecenes promote rapid onset of leukocyte apoptosis and this will undoubtedly be manifested as immunosuppression (Pestka *et al.* 1994b; Ueno *et al.*, 1995; Islam *et al.*, 1998a; 1998b). Trichothecene-induced apoptosis also occurred in tissues of the gastrointestinal tract including the gastric mucosa, gastric glandular epithelium, and intestinal crypt cell epithelium (Li and Shimizu, 1997; Li *et al.*, 1997a, 1997b). Apoptotic loss in these tissues might result in the breakdown of non-specific mucosal defense mechanisms such as the epithelial barrier and mucus secretion and, thus, results in increased translocation of gut bacteria and endotoxin. The latter effects of trichothecenes appear to involve the activation of p53, a well-known apoptotic gene (Zhou and Pestka, 2003).

Trichothecenes bind to eukaryotic ribosomes (Bamburg, 1983) and, thus, inhibition of translation is an obvious outcome of this interaction. However, in addition, cell signalling appeared to be rapidly affected via a "ribotoxic stress response" that had been demonstrated for trichothecenes and other translation inhibitors such as anisomycin, Shiga toxin, ricin, and α-sarcin (Iordanov *et al.*, 1997). In this model, alteration of 28s rRNA by these inhibitors was postulated to be an initiation signal for activation of stress-activated protein kinases/cJun $NH_2$-terminal kinases (SAPK/JNK), which is a mitogen-activated protein kinase (MAPK) (Figure 10.5). Relatedly, Yang *et al.* (2000) observed a close correlation in protein synthesis inhibition capacity, MAPK activation and apoptosis in macrophages among a number of trichothecenes examined. Increased MAPK activity might also drive activation of transcription factors that promote expression of proinflammatory and stress-related genes as well as induce apopto-

sis. Zhou *et al.* (2003a) have demonstrated the rapid sequential activation of MAPKs transcription factors cytokine genes in mice given a single oral exposure to DON. Recently, it has been shown that trichothecenes induce the phosphorylation of double-stranded RNA-activated protein kinase (PKR) and Src kinase (HCK), which are two kinases that are upstream of the MAPKs (Zhou and Pestka, 2001; Zhou *et al.*, 2003b). Both PKR and HCK are likely to transduce the signal induced upon ribosomal binding to downstream MAPK-driven pathways (Figure 10.5). The transition between these two effects occurs with increasing concentration/dose of trichothecene with the net effects being immune stimulation or suppression, respectively. Future research on trichothecene risk assessment should, therefore, focus on the role of cell signalling deregulation in evoking effects on leukocytes *in vitro* and *in vivo* as well as development of quantitative structure activity relationships among these toxins.

*Zearalenone and other Fusarium mycotoxins*
Zearalenone is an estrogenic mycotoxin that often co-occurs with DON on scabby wheat and on maize that shows signs of *Fusarium* (*Gibberella*) ear rot. Both zearalenone and DON are produced by *F. graminearium*. There is no doubt that zearalenone is a cause of farm animal performance problems. Zearalenone causes vulvovaginitis and estrogenic responses in adult pigs and estrogenism in suckling piglets exposed during lactation via the sow's milk (CAST, 2003). Zearalenone acts by binding to the estradiol receptor; however, it binds with less affinity and is much less potent than estradiol (Figure 10.6) (Osweiler, 2000). Nonetheless, zearalenone can cause reproductive problems in swine. Zearalenone and zearalanol, a zearalenone metabolite used as a growth promoter in swine, were suggested as being involved in precocious development in children in Puerto Rico and other countries; however, the evidence

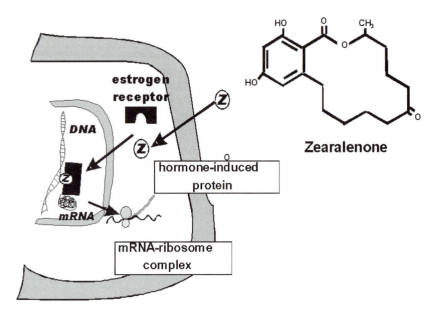

**Zearalenone**

**Figure 10.6** Mechanism of action of zearalenone. Interaction of zearalenone (Z) with the cytosolic estrogenic receptor illustrating hormonal mimicry (modified from Riley and Norred, 1996). Zearalenone (Z)-like phytoestrogens (PE) and environmental estrogens, passively cross the cell membrane and bind to the cytosolic estrogen receptor. The receptor-Z complex (or PE complex) is rapidly transferred into the nucleus where it binds to specific nuclear receptors and generates estrogenic responses via gene activation resulting in the production of mRNAs that code for proteins that are normally expressed by receptor-estrogen complex binding.

is inconclusive (CAST, 2003). Climatic conditions favoring fungal growth and production of *Fusarium* toxins, including zearalenone, may be risk factors for hormone-dependent cancer and adverse reproductive outcomes in farmers' families (Kristensen *et al.*, 2000). Although there are no epidemiological studies that specifically link zearalenone to human diseases because in many cases it has occurred with other mycotoxins and other molds (Henry and Bosch, 2001), a PMTDI of 0.5 µg/kg bw was set for zearalenone by the JECFA 53rd (2000).

Other mycotoxins produced by *Fusarium* species and found on cereal grains (primarily maize) or other crops likely to be consumed by humans or farm animals include nivalenol, fusarenon X, fusarochromanone, moniliformin, fusaric acid, and fusarin C (CAST, 2003). These mycotoxins probably are of little or no risk to consumers of cereal grains. Fusarochromanone has been suggested to play a role in a disease of poultry called tibial dyschondroplasia (CAST, 2003). Fusarin C, which can co-occur with fumonisins, is genotoxic and has been suggested to be involved in the rodent carcinogenicity of *F. verticillioides* cultural material (CAST, 2003). Wortmannin is a *Fusarium* toxin that has been suggested as a possible agent in onyalai disease in south-central Africa; a human disease of unknown origin characterized by hemorrhaging in the mouth, tongue, palate, skin and gastrointestinal tract (Wild and Hall, 1996).

### *Penicillium* mycotoxins

Economically important mycotoxins produced by *Penicillium* include ochratoxins, cyclopiazonic acid (described in the *Aspergillus* toxins section) and patulin (apple and pear juices). Of lesser importance are rubratoxin (maize), citreoviridin (rice), citrinin (moldy feeds) and penitrim A (moldy foods). Compared to the toxins produced by *Aspergillus* and *Fusarium*, the farm animal and human health risks from *Penicillium* toxins in foods (with the exception of ochratoxin A) are much more

difficult to document. Patulin is of some concern since it is often found in apple juice where it has been reported to reach concentrations as high has 45 mg/L (Friedman, 1990); however, concentrations between 1 µg/L and 1 mg/L are the norm. There is no evidence that apple products contaminated with patulin are a cause of farm animal or human disease (Friedman, 1990). Patulin's mechanism of action appears to involve its potent interaction with sulfhydral groups and in particular non-protein sulfhydrals such as glutathione (Figure 10.7). There is no human disease documented for patulin; however, there were effects noted in several animal species (McKinley and Carlton, 1991; Henry and Bosch, 2001). A PMTDI of 0.4 µg/kg bw has been set by JECFA (1995), probably because infants and children consume apple juice and related products.

Citreoviriden is now believed to have caused acute cardiac beri-beri in Japan (Yellow Rice Toxic Syndrome); however, this disease has not been seen since the passage of laws to prevent moldy rice from entering the marketplace in Japan (CAST, 2003). Rubratoxin and citrinin are suspected of having caused or contributed to farm animal diseases but this has not been proven. Penitrim A may cause sickness in domestic animals and humans (CAST, 2003). Abramson (1997) included patulin, penicillic acid, rubratoxin B, secalonic acid D, viomellein and xanthomegnin as mycotoxins produced by *Penicillium* species that cause oral toxicities in humans and animals. More research needs to be done to determine how much the mycotoxins produced by *Penicillium* species are involved in adverse human health effects.

### *Claviceps* mycotoxins

Ergotism is the name given to the acute diseases associated with consumption of cereals (primarily rye, barley, and millet) that are contaminated with sclerotia (ergot) from several

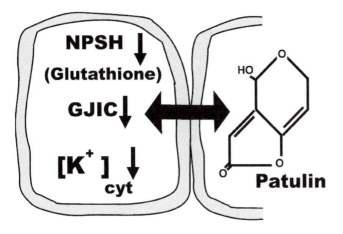

**Figure 10.7** Hypothesized mechanism of patulin toxicity showing the cellular consequences of exposure to patulin in the absence of antioxidants. Patulin-induced depletion of non-protein sulfhydrals (NPSH), primarily reduced glutathione, results in simultaneous suppression of gap junction-mediated intercellular communication (GJIC) in clone 9 rat liver cells and simultaneous potassium efflux in LLC-PK$_1$ pig kidney epithelial cells (modified from Riley, 1998).

*Claviceps* species. The ergot (sclerotia) alkaloids, which consist of 3,4-substituted indole derivatives, can be either clavine or peptide (acid amide derivatives of lysergic acid) ergot alkaloids with the latter causing the most severe disease (Kuldau and Bacon, 2001). If the ergot is not removed from the cereal grain, then humans consuming the contaminated cereal will exhibit symptoms of the acute diseases known as ergotism with the most recent outbreak occurring in the mid to late 1970s (Wild and Hall, 1996). There are two major types of ergotism that affect humans: convulsive (mild convulsions of the diaphragm and esophagus yielding vomiting and mild convulsions of the face and vocal cords) and gangrenous (burning sensation with swelling of extremities that become black, shrunken and dried leading to amputations and death) (Beardall and Miller, 1994; Kuldau and Bacon, 2001). Ergot probably occurs when there is at least 1% ergot in the grain with 7% leading to death (Kuldau and Bacon, 2001). It was estimated that the Swiss daily intake of ergot alkaloids was 5.1 µg/person after analyzing wheat and rye flours and the products made from them (Henry and Bosch, 2001) and in The Netherlands, 7.8 µg/kg body weight was estimated as the maximal daily intake; however, there was no effect in rats feed 10 mg/kg ergometrinemaleate during a 4 week study (Peters-Volleberg *et al.*, 1996). In the 1975 ergot outbreak from millet, there was no toxic effect below about 28 µg/kg bw (Beardall and Miller, 1994). Ergot poisoning from *Claviceps purpurea* infected feeds and forage grasses is also seen in pigs, cattle, and other farm animals (Osweiler, 2000). Tall fescue and perennial ryegrass also can be endophytically colonized with *Neotyphodium* species that produce "ergot-like" alkaloids (CAST, 2003). The acute toxic effects of ergot- and ergot-like poisoning in farm animals are similar and often involve central nervous system (CNS) and "gangrenous" effects that result from the potent vasoconstrictive activity of some of the alkaloids.

## *Stachybotrys* mycotoxins

Several species of *Stachybotrys* cause the disease known as "stachybotryotoxicosis" in farm animals and humans exposed to *Stachybotrys* contaminated feeds (CAST, 2003). The causative agents are believed to be macrocyclic trichothecenes which are probably the most acutely toxic of all the mycotoxins. Like T-2 toxin, they are potent immunomodulators and inhibitors of protein synthesis (Bondy and Pestka, 2000). *Stachybotrys* toxins are currently the focus of considerable interest because they have been associated with what is called "sick building" syndrome (Etzel *et al.*, 1998). Typically, the "syndrome" involves respiratory illness in people inhabiting buildings that have been water damaged and in which there is extensive fungal contamination (Trout *et al.*, 2001). There is considerable debate over whether or not mycotoxins are involved in these illnesses (Etzel *et al.*, 1998). These toxins have caused dermal or inhalation health problems in humans but have not been implicated in foodborne disease (CAST, 2003).

## Miscellaneous mycotoxins

Other mycotoxins that are known or suspected of causing farm animal or human diseases include sporidesmin, slaframine and toxins produced by *Alternaria* species. Sporidesmin is produced by *Pithomyces chartarum* and is found in dead and decaying plants in pastures (Le Bars and Le Bars, 1990). Consumption of sporidesmin causes liver toxicity in many grazing animals and intoxication is characterized by sunlight-induced "eczema" in light-skinned farm animals (typically in the facial region). Slaframine is a mycotoxin found in red clover and produced by *Rhizoctonia leguminicola*, which causes "slobbers" or excess salivation in cattle and other farm animals (CAST, 2003). Toxins produced by *Alternaria* on cereal grains (barley) have been implicated in contributing to Kashin-Beck disease in Tibet, a multifactorial disease characterized by joint deformities (Suetens *et al.*, 2001). Kashin-Beck disease has also been attributed to the *Fusarium* toxin, fusarochromanone (Wild and Hall, 1996). Beardall and Miller (1994) discussed the possible links of mycotoxins to three crippling bone diseases: Kashin-Beck disease (KBD), which is an osteoarthritis that results in shortening of the extremities and skeletal malformations; Mseleni joint disease (MJD), which is an osteoarthropathic disorder of hips, knees and ankles in blacks in southern Africa; and endemic familial arthritis of Malnad (EFAM), which affects the hips of people in southern India. More research is needed to determine whether mycotoxins are involved in these human diseases.

There has not been enough research on whether the interactions of mycotoxins with other mycotoxins and with organisms or the continual ingestion of low levels of mycotoxins can affect human health. Koshinsky *et al.* (2001) reviewed the interactions of mycotoxins with other mycotoxins and organisms, plants, metals and other chemicals and noted that viruses, flukes, and yeasts may be found in association with mycotoxins such as aflatoxin B$_1$ and ochratoxin A and that their interaction may increase the toxicity of the mycotoxin. In addition, combinations of mycotoxins have occurred in the same foods, such as aflatoxin $B_1$ and fumonisins; aflatoxin $B_1$ and zearalenone; ochratoxin A and citrinin; deoxynivalenol and nivalenol; and trichothecenes, fumonisins and moniliformin. Another area that needs more research is the effect of several mycotoxins on the immune system. Pestka and Bondy (1994) reviewed the effects of individual mycotoxins (aflatoxins, fumonisin, ochratoxin A, patulin, tichothecenes (DON, T-2 toxin), and zearalenone) on the immune system of animals, such as suppression that results in infection or stimulation that can produce an autoimmune response. Although they reported that there was some comparison of the DON immune suppression on the kidneys in animals to the human IgA nephropathy and that ochratoxin A inhibited interferon, which may relate to renal cancer, there is not sufficient evidence to extrapolate that mycotoxins can affect the immune system in humans. However, the studies in animals for aflatoxins, ochratoxin A and trichothecenes suggest concerns for toxin-mediated immune changes in humans as well. More research on the combination of mycotoxins and their effects on the human immune system should help to answer some of these concerns.

## Detection, identification, and analysis of mycotoxins

Mycotoxins are a diverse group of chemical secondary metabolites produced by molds; therefore, there is no one method that can be successfully used to qualitatively or quantitatively isolate, detect and identify all mycotoxins. The analytical methods used for mycotoxins must consider their chemical structures, molecular weights and functional groups (Wilson *et al.*, 1998). Samplings and subsampling plans are critical to insure that the results from the analytical technique are reliable because mycotoxins are unevenly distributed in grains, kernels, nuts, seeds, and foods (Chu, 1991; Scott, 1995; Smith, 1997; Gilbert, 2000). Wilson *et al.* (1998) reviewed the methods used to extract (Table 10.7) and analyze (Table 10.8) the following mycotoxins: aflatoxin, citrinin, deoxynivalenol, fumonisins, ochratoxin, patulin, and zearalenone. There are five essential steps in the analysis of mycotoxins: sampling of the food or agricultural commodity, preparation of the sample, extraction of the mycotoxin, analytically measuring the amount of mycotoxins, and confirmation of the chemical detected. In addition to these steps, filtration and cleanup of extracts followed by concentration may be needed depending on the method of analysis (Chu, 1991; Steyn *et al.*, 1991). Some of the information generated since 1990 on sampling, extraction, cleanup, and analytical methods is presented in the following sections.

### Sampling for mycotoxins

A major problem with analyzing agricultural commodities and foods for mycotoxins has been inadequate sampling. Mycotoxins are not uniformly distributed in commodities and, thus, sample size is critical for obtaining reproducible and accurate results (Whitaker and Park, 1994). Coker (1998) described

the sampling sequence (purpose of sample, place where collected, sampling method and collection, preparation of sample for analysis) as being essential to the collection of reliable samples for mycotoxin analysis. The sample preparation is important because there may be only a few kernels or nuts in a lot that are contaminated; therefore, many 100 g up to 200 g or more samples may need to be analyzed (Scott, 1995; Coker, 1998). The type of mill used to grind samples can affect the final mycotoxin level as shown by Dorner and Cole (1993) who reported that four common mills used to comminute peanuts for aflatoxin analysis produced different results. Also, sampling plans can be based on probability or statistics with special concerns taken for problem commodities (Coker, 1998). Whitaker *et al.* (1994) reported that sampling variance was greater than the amount of aflatoxin in the lot and increased as the amount of aflatoxin in the samples increased for farmer's stock peanuts. In this study, sampling accounted for 92.7% of the variability with sample preparation (7.2%) and the analyses (0.1%) being less important in variability. When wheat was analyzed by the Romer FluoroQuant™ analytical method for deoxynivalenol using a 0.454 kg sample that was ground in a Romer mill and subdivided into 25 g samples, the variation was 6.3% for sampling, 10.0% for sampling preparation and 6.3% for analysis (Whitaker *et al.*, 2000). To overcome the distribution problem of only a few kernels contaminated with variable levels of aflatoxin that is randomly distributed in the lots, Giesbrecht and Whitaker (1998) proposed models to determine the distribution of aflatoxin in peanuts; however, the use of more than 120 data sets and 50 determinations would be needed to determine whether the compound gamma or negative binomial would be the best model. Although these authors did not indicate a contamination level based on the models, they speculated that 1 peanut out of 1,000 would be contaminated based on previous experience. Shelled maize from 18 lots that were tested for aflatoxin by liquid chromatography (LC) had from 6 to 667 ppb aflatoxin and showed that the greatest variance was for sampling at 77.8% followed by 20.5% for sample preparation and only 1.7% for method of analysis (Johansson *et al.*, 2000c). To further understand the problem with sampling variability of shelled maize, the compound gamma model that was used to determine the distribution of aflatoxin in the maize estimated that for 20 ppb, there would be six contaminated kernels in a batch of 10,000 (Johansson *et al.*, 2000a). For this model the 18 observed values were compared to 15 theoretical distributions and the compound gamma model was chosen over the negative binomial, three-parameter log normal and truncated normal models (Johansson *et al.*, 2000a). Further research was done with the gamma compound model by evaluating 6 sampling plans for determination of aflatoxin in shelled maize with three evaluating sample size (5, 10, or 20 kg) and three evaluating acceptance at 10, 20 or 30 ppb (Johansson *et al.*, 2000b). More research needs to be done to choose the best model for maize and other commodities. Freese *et al.* (2000) studied the variability of using commercial enzyme linked immunosorbent assay (ELISA) test kits to determine deoxyni-

**Table 10.7** Extraction methods used to separate mycotoxins from foods and feeds[a]

| Mycotoxin | Functional groups | Solvents | Additional methods/cleanup | Problems |
|---|---|---|---|---|
| Aflatoxin | | Polar (acetone, acetonitrile, methanol) Supercritical carbon dioxide | Salt for animal tissues, silica gel for cleanup, dilution in water for immunoassay, immunoaffinity columns | Emulsions, incomplete extractions, co-extraction |
| Citrinin | Carboxylic acid | Acid or base | | Unstable |
| Deoxynivalenol | Epoxytrichothecenone | Aqueous acetonitrile, ethyl acetate, methanol, water Supercritical carbon dioxide | Activated carbon, alumina, n-hexane, solid-phase extraction, immunoaffinity columns | Need for clean-up |
| Fumonisins | Polar diester of propane Tricarboxylic acid and amino-dimethylpolyhydroxy-eicosanes | Water:methanol (1:1) Water:acetonitrile (1:3) | Solid-phase extraction, immunoaffinity columns | Food matrix affects recovery, "hidden" fumonisins[b] |
| Ochratoxin | Carboxylic acid | Acid or base | Solid-phase extraction, immunoaffinity columns | Variable recovery |
| Patulin | Hydroxy-furo-pyranone lactone | Ethyl acetate, diphasic dialysis | Solid-phase extraction (silica gel), sodium carbonate partitioning | |
| Zearalenone | Resorcyclic acid lactone | Acetonitrile/methanol/water Supercritical carbon dioxide | Defat with hexane, dilute in water, immunoaffinity columns | Variable recovery |

[a]Trucksess and Pohland (2001), Wilson et al. (1998).
[b]Kim et al. (2003).

**Table 10.8** Methods used to separate and analyze mycotoxins from foods and feeds[a]

| Primary methods | Addition to primary method | Purpose of methods | Mycotoxins |
|---|---|---|---|
| Thin-layer chromatography (TLC) | | Identify, quantify | Aflatoxins, deoxynivalenol, ergot alkaloids, fumonisins, ochratoxin A, patulin, zearalenone |
| | HPLC, MS | Prepare, quantify | Fumonisins |
| Column chromatography | TLC, HPLC, MS | Isolate, purify, quantify | Fumonisins |
| High-performance liquid chromatography (HPLC) | Ultraviolet (UV) or fluorescence | Identify, quantify | Aflatoxins, deoxynivalenol, ergot alkaloids, fumonisins, ochratoxin, patulin, zearalenone |
| | TLC | Identify, quantify | Fumonisins, zearalenone |
| Gas chromatography (GC) | MS, electron capture (EC) | Detect, quantify | Deoxynivalenol, fumonisins, patulin, zearalenone |
| Mass spectrometry (MS) – electron impact (EI), chemical ionization (CI), electrospray ionization (ESI), tandem (MS/MS) | Fast-atom bombardment, electrospray, GC | Identify, quantify | Aflatoxins, ergot alkaloids, fumonisins, ochratooxins, patulin, zearalenone Deoxynivalenol |
| Immunoassays | | Screen (semi-quantify) | Aflatoxins, cyclopiazonic acid, fumonisins, ochratooxina A, trichothecenes, zearalenones |
| | HPLC, TLC, capillary electrophoresis | Identify, quantify | Fumonisins |
| | Biosensors (beads, fiberoptic, surface plasmon resonance) | Identify, quantify | Aflatoxins, fumonisins, T-2 toxin |

[a]Chu (2001); Porter (2001), Wilson et al. (1998).

valenol in barley and noted that increasing the sample size did not change the variability; however, there was variation among samples from the same lot, from ground subsamples and from extracts measured more than once. These authors speculated that deoxynivalenol was stratified within lots because of mixing and that no solutions to variability could be given without further research. Also, analysis of mycotoxins does not always agree between laboratories; therefore, if the levels in foods are too close to the limits for detection by the method, there can be problems (Smith, 1997). The sampling technique and choice of sampling methods are essential to collect good data for mycotoxin analysis.

## Extraction of mycotoxins

The presence of mycotoxins in diverse foods and feeds presents challenges for extraction. Extraction methods to remove mycotoxins from other constituents are important because naturally contaminated samples can differ from laboratory prepared samples, the matrix may interfere with the extraction, the chemical structure of the mycotoxin determines the extraction liquids, and special cleanup may be needed for mycotoxins after extraction (Scott, 1995; Gilbert, 2000). Extraction usually involved various solvents because mycotoxins were not soluble in water; therefore, filtration removes solid materials (Chu, 1991; Steyn *et al.*, 1991). Porter (2001) has reviewed the liquid-liquid and solid phase extraction procedures, solvent cleanup methods and chromatography that are supplemented with standards for mycotoxin analysis. Examples of extraction solvents and potential problems are presented in Table 10.7.

Fumonisins were extracted from maize with a 50/50 mixture of acetonitrile and water that was shaken for 1 h (Bennett and Richard, 1994) and with 3 parts methanol to 1 part water that was blended for 5 min by Sydenham *et al.* (1992) and Thiel *et al.* (1993). Extraction with sodium EDTA added to the methanol/acetonitrile solvents improved the recovery of fumonisins from white rice flour, cornstarch and cornmeal; however, there still seemed to be the possibility of the fumonisins binding to the food (Kim *et al.*, 2002). Meredith *et al.* (1998) analyzed test diets for fumonisin $B_1$ and used an ether/acetone extraction step to remove lipids before the high-pressure liquid chromatography (HPLC) was done. The following are some examples of solvent extraction strategies used for various mycotoxins. Like the fumonisins, trichothecenes were also extracted with acetonitrile and water but in an 84/16 mixture that included Celite and shaking for 30 min (Mossoba *et al.*, 1996). Ochratoxin A was extracted from barley and maize by a 1/10 solution of phosphoric acid and chloroform followed by filtration through diatomaceous earth (Nesheim *et al.*, 1992). For extraction of cyclopiazonic acid from maize and peanuts, Urano *et al.* (1992b) used 7 parts methanol and 3 parts sodium bicarbonate in a blender for 3 min.

Supercritical fluid extraction (SFE) also has been evaluated to eliminate the need for hazardous solvents; however, supercritical carbon dioxide (SC-$CO_2$) is best used for extraction of nonpolar lipophilic compounds and other chemicals are needed for best recovery of polar mycotoxins. Holcomb *et al.* (1996) extracted aflatoxins from maize with SC-$CO_2$ and found that $B_1$ was best at >34.2 MPa pressure and > 55°C; however, recoveries were poor for the aflatoxins $B_2$, $G_1$ and $G_2$. The extraction of $B_1$, $B_2$, $G_1$ and $G_2$ from maize at 51.7 MPa SC-$CO_2$ plus methanol and 65°C gave recoveries of 77.3, 82.9, 75.4 and 80.3%, respectively, when the levels of aflatoxin ranged from 3 to 11 ppb (Holcomb *et al.*, 1996). Taylor *et al.* (1993)

used acetonitrile-methanol with SC-$CO_2$ to extract aflatoxin $B_1$ from maize and further refined the method for extraction of aflatoxin $M_1$ from beef liver (Taylor *et al.*, 1997). A combination of SC-$CO_2$ and methanol was used by Wu *et al.* (1995) to extract aflatoxins $B_1$, $B_2$, $G_1$ and $G_2$ from peanuts. Forty times more fumonisin $B_1$ was extracted from maize and maize dust with SC-$CO_2$ at 15 mL liquid $CO_2$, 1200 psi, 60°C, 20 min and addition of 750 µLof 5% acetic acid/g sample than was achieved by solvent extraction with ethyl acetate and methanol-water (Selim *et al.*, 1996). SFE and polar solvents (acetonitrile-water, hexane, methanol, trichloromethane) were used to extract deoxynivalenol from rolled oats; however, the co-extraction of other compounds from the oats interfered with the assay and made additional purification necessary (Järvenpää *et al.*, 1997). Research is continuing on the development of good extraction methods for removing mycotoxins from foods and feeds.

## Cleanup in mycotoxin analysis

Cleanup is done in some mycotoxin methods to separate other substances that are co-extracted and to further concentrate the mycotoxin; however, it can be omitted if the screening method is sensitive and specific for the mycotoxin (Gilbert, 2000). Cleanup removes interfering substances by methods such as precipitation with anionic liquids, separation with chromatographic or immunoaffinity chromatographic columns, dialysis, proprietary packed columns, and partitioning by different solvents (Steyn *et al.*, 1991; Scott, 1995). Rice and Ross (1994) used a $C_{18}$-solid phase extraction (SPE) column rinsed with solutions of potassium chloride and acetonitrile to cleanup maize-based foods before using HPLC to detect fumonisins. Trucksess and Tang (1999) used a commercial hydrophilic (divinylbenzene) and lipophilic (N-vinylpyrrolidone) macroporus copolymer column as the SPE to clean up ethyl ether-acetonitrile extracts of patulin from apple juice before being separated with LC. Bennett and Richard (1994) found that a reverse-phase $C_{18}$ column recovered 83–88% of 10 µg fumonisins $B_1$ and $B_2$, and a strong anion exchange column recovered 92–95% of the fumonisins during cleanup; however, the latter column operated at 1 mL/min and pH 6–7 and could not recover fumonisin hydrolysis products. Stack (1998) also used a $C_{18}$ column as a clenup for fumonisin $B_1$ analysis of tortillas. Sydenham *et al.* (1992) and Thiel *et al.* (1993) used strong anion exchange columns with 3 parts methanol to 1 part water for cleanup of maize and feed before fumonisin analysis.

Scott and Truckress (1997) reviewed the use of immunoaffinity columns for cleanup in the analysis of aflatoxins, deoxynivalenol, fumonisins, ochratoxin A, and zearalenone. The advantages of using immunoaffinity columns were that they were specific, used less solvents, were rapid, and lent themselves to both automation and reuse (Scott and Truckress, 1997). However, they had a limited shelf life and needed to be refrigerated. The disadvantages of these columns were that only a few were commercially available and not for all mycotoxins, they were expensive, and some needed precolumn cleanup methods

(Scott and Truckress, 1997). Immunoaffinity column cleanup using commercial products has been used for aflatoxin M1 from cheese (Dragacci and Fremy, 1996), ochratoxin A from coffee and alcoholic beverages (Pittet et al., 1996; Visconti et al., 2001), for zearalenone from cereals (Kruger et al., 1999; Fazekas and Tar, 2001), and for fumonisins from maize, maize products and other food matrices (Scott and Trucksess, 1997). These methods used the commercial immunoaffinity column to cleanup the sample before using liquid chromatography to detect the mycotoxins. There were from 11.9 to 26.1% differences between laboratories for the results using an OchraTest™ immunoaffinity column for determination of ochratoxin A from beer and wine in a collaborative study between 18 laboratories (Visconti et al., 2001). The commercial Zearala-Test™ for immunoaffinity cleanup gave a recovery of 93% zearalenone and from 77% to 93% recovery of resorcyclic acid lactone derivatives of zearalenone (Kruger et al., 1999). An immunoaffinity column that was made by linking a cyclopiazonic acid monoclonal antibody to CNBr-activated Sepharose 4B was used for cleanup of cyclopiazonic acid from maize and peanuts (Yu et al., 1998). This column was able to remove impurities before analysis by ELISA that resulted in detection of 10 to 20 ng/g, which was 20 times more sensitive than doing the ELISA without cleanup.

## Methods for mycotoxin detection

Gilbert (2000) briefly reviewed the following three major types of analytical approaches that are used for mycotoxin analysis: (1) research methods involving the latest analytical tools because they are not limited by instrumentation or sample costs; (2) screening methods that are rapid and qualitative or quantitative and methods that have acceptable levels of false positive reactions; and (3) validated official methods that have undergone interlaboratory validation for reproducibility, repeatability, detection limits and peer review and are accepted as reference methods by the Association of Official Analytical Chemists (AOAC International), European Committee for Standardization (CEN), or similar official bodies. The Official Methods of the AOAC International should be used for foods or feeds that are sold in the market place (Wilson et al., 1998). Various reviews on mycotoxin detection in foods and feeds have been published (Steyn et al., 1991; Richard et al., 1993; Scott, 1995; Gilbert, 2000; Porter, 2001). Several of the major methods used in aflatoxin analysis are presented in Table 10.8.

### Chromatographic methods for mycotoxin detection and quantification

Trucksess and Wood (1994) in their review of methods used to detect aflatoxins included solid-phase extraction with silica gel or similar particles followed by detection and quantification with thin-layer chromatography (TLC), LC, and immunochemical methods. TLC methods have been developed for most mycotoxins (Betina, 1993); however, the results are more variable than either GC or HPLC. GC is best for de-

oxynivalenol; whereas, HPLC can be used for the other major mycotoxins (Wilson et al., 1998). Most research in recent years has focused on improvements in LC, HPLC, GC, and combinations with other methods for separation or detection such as mass spectrometry, capillary zone electrophoresis, and electron capture. Bennett and Richard (1994) found that a 25 cm, 5 μm reversed-phase $C_{18}$ column separated fumonisins $B_1$ and $B_2$ plus their hydrolysis products in 30 min when a gradient of acetonitrile (99 v/v):acetate (1 v/v) and water (99 v/v): acetate (1 v/v) was used. LC methods have been used in collaborative studies to detect ochratoxin A in grains (Larsson and Möller, 1996; Nesheim et al., 1992) and fumonisins $B_1$ and $B_2$ in maize (Thiel et al., 1993). Sydenham et al. (1992) optimized the sample purification and LC method used to detect fumonisins $B_1$ and $B_2$ in maize-based foods and feeds to also detect fumonisin $B_3$.

Combinations of methods have improved the detection of some mycotoxins. Urano et al. (1992b) showed that cyclopiazonic acid in maize and peanuts could be separated by LC and confirmed by tandem mass spectrometry (MS/MS). Mass spectrometry (MS) is used to confirm the mycotoxin based on its structure as determined by its unique fragmentation pattern or its molecular weight and charge (Wilson et al., 1998). Trichothecenes were detected in grains by a combination of gas chromatography, matrix isolation, Fourier transform infrared spectroscopy (GC/MI/FTIR) and GC/MS (Mossoba et al., 1996). The GC/MI/FTIR spectroscopy was used to characterize the mycotoxins (deoxynivalenol, nivalenol, T-2 toxin and derivatives) because the fingerprints were unique at cryogenic temperatures and the GC/MS of the negative ion chemical ionization spectrum was used to quantify deoxynivalenol in sweet corn (Mossoba et al., 1996). Walker and Meier (1998) used LC with a diode array detector to simultaneously determine the presence of nivalenol, deoxynivalenol and two dexoynivalenol derivatives in wheat flour. They used a GC electron capture detection method when mycotoxin levels were below 1 ppm. Yu and Chu (1998a) used LC and a reverse phase $C_{18}$ column for samples extracted from maize that omitted cleanup and derivatization and the fractions eluted from the column were analyzed by an ELISA to detect and quantify fumonisins. This method resulted in an average of 85% recovery of fumonisin $B_1$ added to maize at 50 to 1000 ng/g quantities.

Most mycotoxins can be separated by high-performance liquid chromatography (HPLC) and some (patulin, trichothecenes, zearalenone) can be separated by gas–liquid chromatography (GLC). Electrospray ionization (ES) mass spectrometry (MS) has been used for large mycotoxins that are not very volatile and lack UV chromophores such as fumonisins and AAL (Alternaria alternata f. sp. lycospersici) toxins (Trucksess and Pohland, 2001). Since fumonisins are composed of alkylamine compounds, a HPLC method using detection by fluorescence was developed (Shephard et al., 1990). HPLC has been used to detect and quantify mycotoxins that have fluorescent properties; however, fumonisins do not have these properties and must undergo derivatization with fluorescent compounds before HPLC can be used. This has led to the linking of other methods to the HPLC system to adequately detect fumonisins. HPLC was used to separate derivatives of fumonisins B$_1$ and B$_2$ in grains and this was combined with fast atom bombardment mass spectrometry (FABMS) to detect low levels of the fumonisins (Holcomb *et al.*, 1993). Lukacs *et al.* (1996) used HPLC to separate fumonisins that were then detected by electrospray-ionization (protonated ions), tandem mass spectrometry. Miyahara *et al.* (1996) used tandem solid phase extraction with C$_{18}$ and strong anion exchange columns to clean up fumonisins extracted from maize before derivatizing with *o*-phthaldialdehyde/N-acetyl-L-cysteine and analyzing with ion pairing chromatographic separation to eliminate interfering substances and to be able to detect fluorescence with HPLC. These authors speculated that this method could be used to screen foods for fumonisins. HPLC has been compared to other methods for detecting and quantifying mycotoxins. Maragos (1995) compared HPLC to capillary zone electrophoresis (CZE) with laser-induced fluorescence (LIF) detection because fumonisins contain charged tricarballylic acid groups that can be separated by electrophoresis. This CZE-LIF detection was more sensitive than HPLC for quantifying low fumonisin levels (pg versus ng) and produced 100 times less solvent waste. Radová *et al.* (2001) found that a HPLC (reversed phase C$_{18}$ column) that was equipped with fluorescence detection gave better repeatability for detection of zearalenone in wheat than did an ELISA test kit; however, the ELISA was more sensitive than the HPLC. Dawlatana *et al.* (1998) reported that a high performance thin layer chromatographic method showed linearity of detection of zearalenone in maize over a 10 to 80 µg/kg range but not at higher levels. They proposed this method as an alternative to HPLC for countries where the electrical supply is not constant.

GC-MS has been used to detect several mycotoxins. Patulin from apple juice was detected by GC-MS in laboratories in Spain (Llovera *et al.*, 1999), Taiwan (Sheu and Shyu, 1999), and the USA (Rupp and Turnipseed, 2000). Since HPLC showed the same retention time for more than one peak, Llovera *et al.* (1999) used GC-MS to confirm but not quantify the presence of patulin in underivatized apple juice. Rupp and Turnipseed (2000) used GC-MS to confirm that patulin derivatized with bis (trimethylsilyl) trifluoroacetamide was present in apple juice. In order to quantify patulin in apple juice, Sheu and Shyu (1999) used both dialysis and acylation to extract and derivatize patulin, which was detected and quantified by GC-MS. This method was linear from 10 to 250 µg/L with a recovery of 77 to 109% for 10 apple juice samples. Also, deoxynivalenol and its derivatives were detected by GC-MS for single kernels to 100 g samples of wheat and barley (Mirocha *et al.*, 1998). This method compared well with a GC-electron capture (EC) method that was done on the same samples by another laboratory (Mirocha *et al.*, 1998). Croteau *et al.* (1994) used a GC-EC detection to screen maize for 13 trichothecenes and noted that the method was more sensitive

if there were more hydroxyl groups for derivatization; however, the method was able to quantify between 50 and 200 µg toxin/kg maize depending on the mycotoxin. In an attempt to develop a rapid method to quantify ochratoxin A and deoxynivalenol in barley, Olsson *et al.* (2002) evaluated volatile compounds with an electronic nose and GC-MS. Volatiles such as isooctylacetate, methylpyrazine, pentane, 3-pentanone, and 3-octene-2-ol correlated with deoxynivalenol and could predict its presence by both GC-MS and the electronic nose, but ketones were better predictors of ochratoxin A above the 5 µg/kg maximal allowable level in barley (Olsson *et al.*, 2002). The GC-MS method had fewer false positives (3 of 37) compared to the electronic nose (7 of 37) for prediction of ochratoxin A levels (Olsson *et al.*, 2002).

Capillary electrophoresis (CE) has been used to separate many different charged compounds in an aqueous system. Pragsongsidh *et al.* (1997b) compared CE to reversed phase liquid chromatography (RPLC) for detection of cyclopiazonic acid in milk and found that there was no difference between high levels (200 to 500 ppb) of cyclopiazonic acid; however, at levels below 50 ppb, CE was able to detect cyclopiazonic acid but RPLC could not. Micellar electokinetic capillary electrophoresis was used to detect patulin in apple juice because it could detect neutral or mixed charged and neutral compounds (Tsao and Zhou, 2000).

*Immunoassays for mycotoxin detection*
In an effort to rapidly detect mycotoxins in foods, immunoassays were developed. Some of their potential advantages are that they can be developed for many different types of compounds, are sensitive to low levels of contaminants, and usually can be very specific. Chu (1991; 1992; 1994) reviewed the early research on development and use of immunoassays to detect aflatoxins, cyclopiazonic acid, ochratoxin A, patulin trichothecenes (T-2 toxin, deoxynivalenol, and related toxins), zearalenone, and several other mycotoxins in foods and feeds. In a more recent review, Chu (2001) discussed the immunological detection of mycotoxins focusing on the development of antibodies, different types of assays, combining assays with other methods, and new immunochemical methods.

Most mycotoxins are too small (molecular weight < 800) to be imunogenic. A molecular weight of 3000–4000 is minimum; therefore, most mycotoxins must be conjugated to molecules that have protein or peptides that act as carriers (Pestka *et al.*, 1995; Chu, 2001). The general protocol for antibody production that uses these carrier proteins was reviewed by Pestka *et al.* (1995). Radioimmunoassays (RIA) were the earliest immunoassay developed, which use a radioactive marker; therefore, they had limited use (Pestka *et al.*, 1995; Chu, 2001). The enzyme immunoassay (EIA) uses enzymes as the marker and an amplifier to make the system more sensitive than RIA (Chu, 2001) and eliminates the need for radioactivity (Pestka *et al.*, 1995). The enzyme-linked immunosorbent assay (ELISA) has either the antibody or antigen bound to a solid material and can be used as direct competitive (dc-

ELISA) where the antibody is bound to the solid support or indirect competitive (idc-ELISA) where a mycotoxin protein conjugate is bound to the solid support for mycotoxin analysis (Chu, 2001). These assays usually can be done within 1 to 2 h, respectively, and detect as little as 0.5 pg of mycotoxin (Chu, 2001). Immunoscreening using commercial test kits can be done in < 30 min if one is only trying to detect levels above or below a certain limit (Pestka et al., 1995; Newton, 1999; Chu, 2001).

Most of the work over the last few years on new antibody production has focused on fumonisins because they are the newest mycotoxins of concern to human health. Usleber et al. (1994) used keyhole limpet hemocyanin (KLH) conjugated to fumonisin $B_1$ with glutaraldehyde to produce the immunogen for rabbit immunization. In this research a dc-ELISA generally gave higher readings than the corresponding HPLC analysis of the same samples. Yu and Chu (1996) produced a polyclonal antibody to fumonisin $B_1$ by using carbodiimide to conjugate KLH or bovine serum albumin (BSA) to fumonisin. In these experiments, KLH was the best carrier and the affinity of the antibody was better than previous studies with other conjugates in a dc-ELISA. When this dc-ELISA was compared to HPLC, this ELISA was similar to that of Usleber et al. (1994) because it showed more recovery of fumonisins than did the HPLC method. In an effort to produce antibodies from less toxic material, affinity purified monoclonal antibodies produced in reaction to aflatoxin (Hsu and Chu, 1994) or to fumonisin $B_1$ (Chu et al., 1995) were used to produce anti-idiotype antibodies in rabbits or anti-anti-idiotype antibodies in mice. These purified antibodies were mirror images of the mycotoxins and were able to produce antibodies that were used in indirect ELISA to determine aflatoxin (Hsu and Chu, 1994) and fumonisin $B_1$ (Chu et al., 1995) in maize. When 10–100 ppb of aflatoxin $B_1$ were added to white and yellow maize, this ELISA recovered 84.1 and 85.4%, respectively (Hsu and Chu, 1994). The ELISA for fumonisin $B_1$ in naturally contaminated maize only showed a correlation coefficient of 0.31 with the HPLC analysis of the same samples (Chu et al., 1995). Whitaker et al. (1996) studied the variability for collaborative studies for three analytical methods (ELISA, LC, TLC) and noted that ELISA had greater variability than LC for analysis of aflatoxin in 11 commodities and that among laboratory variabilities were about two times those of with-in laboratories. A dc-ELISA was developed to detect cyclopiazonic acid in maize, feed, and peanuts with 93 to 98% recovery rate (Yu and Chu, 1998b).

Other formats for immunoassay include use of membranes (on cards, dipsticks or in cups); immunoblot with ELISA combined with high-performance thin layer chromatography in a system called ELISAGRAM; and immunoaffinity columns (Pestka et al., 1995). A line immunoblot system was used to simultaneously detect aflatoxin $B_1$, fumonisin $B_1$, and zearalenone in maize within 30 min and the results were recorded on a computer that used image analysis (Abouzied and Pestka,

1994). Sibanda et al. (2000) developed a membrane-based flow-through enzyme immunoaasy for T-2 detection in cereal grain. This method had a detection limit of 50 ng T-2 toxin/g grain and was suggested as a screening tool when laboratory analysis is not possible. A collaborative study was done with these membrane-based flow-through enzyme immunoassays for ochratoxin A and T-2 toxin in wheat and false positives rates were < 5% suggesting that the method could be used to screen these mycotoxins (De Saeger et al., 2002). A dipstick immunoassay was developed to detect a minimum of 12 ng of T-2 toxin/g of wheat within 45 min (De Saeger and van Peteghiem, 1996). Park and Chu (1996) combined LC with ELISA to detect trichothecenes in maize and noted that the combined methods were more effective at detecting the mycotoxins than the use of immunoassays alone because they could simultaneously detect multiple toxins. The combination of fluorescence polarization with immunoassays, where competition between the mycotoxin and the mycotoxin-fluorescein tracer for a monoclonal antibody site formed the basis of the test, was developed to detect aflatoxins and deoxynivalenol in cereal grains (Maragos and Plattner, 2002; Nasir and Jolley, 2002). New immunoassays likely will be developed because they are rapid, sensitive and specific.

An important area for method development to detect mycotoxins is the comparison of methods for accuracy and reproducibility by interlaboratory collaborative studies and for repeatability in intralaboratory studies (Scott, 1995). There have been a few studies where ELISA was compared to other methods for mycotoxin analysis. Beaver et al. (1991) compared an ELISA for the detection of aflatoxins in maize with a LC method and found that the two methods were comparable when levels were greater than 5 ng aflatoxin/g, but false negatives occurred below 5 ng/g. When naturally contaminated maize and feed were analyzed by ELISA, TLC and LC for aflatoxin $B_1$, Hongyo et al. (1992) found that there was good correlation between the three methods because the monoclonal antibody was specific and reproducible. Pestka et al. (1994a) found that the monoclonal ELISA method for fumonisin $B_1$ detection in grain-based foods gave higher levels of the mycotoxin than did HPLC and GC-MS; therefore, this ELISA could only be used as a preliminary screen. When a polyclonal-based ELISA was compared to HPLC for the determination of fumonisins in maize, Sydenham et al. (1996) also reported that the ELISA gave higher values than the HPLC and could be used as a screening tool. Maragos and Richard (1994) found that a commercial ELISA was reproducible but had lower sensitivity than did the LC method for determining fumonisins $B_1$ and $B_2$ in milk. Therefore, it has been proposed that immunoassays be used for initial screening or to determine whether mycotoxins fall within specific concentrations (Wilson et al., 1998). From all these studies, it seems that ELISA methods can be used for initial screening for mycotoxins, but more work is needed before they can be used as the sole quantitation method.

*Biological methods for mycotoxin detection*

Chu (1991) reviewed the biological methods that have been used for mycotoxin analysis and these included bioassays in laboratory animals, antimicrobial methods, cytotoxic assays in eukaryotes or cell lines, and phytotoxicity tests in plants; however, most of these methods were neither highly specific nor sensitive for routine use. The problems with some of the current methods for mycotoxin detection have led to a search for new methods that include bioassays. Mitterbauer *et al.* (2003) developed a yeast bioassay to monitor grains, foods and feeds for zearalenone and its derivatives because this *Saccharomyces cerevisiae* was genetically modified to make it dependent on zearalenone for growth. The minimum detectable level of zearalenone needed for growth of this yeast was 1 μg/liter of liquid and took up to 48 h to detect. A bioassay for fumonisin-like activity in maize, maize products, and fungal culture materials was developed based on the ability of fumonisins to specifically inhibit the enzyme ceramide synthase (Norred *et al.*, 1999; Riley *et al.*, 1999a; Palencia *et al.*, 2003). New methods for mycotoxin detection will continue to be developed because of the need for accuracy, specificity, sensitivity, and speed.

## Quality assurance for mycotoxin analysis

The increase in regulatory limits for mycotoxins and the sampling protocol used plus the need for method validation are all leading to more concern about quality assurance in mycotoxin analysis. The European Union has developed compliance guidelines for laboratories that do analytical work, which can be found in the ISO Guide 25 or EN45000 Series Norms (Brera and Miraglia, 1996). These guidelines include information on staffing the laboratory, sampling the commodity or food, using standard operation procedures, use of validated methods, inclusion of reference materials in analyses, cooperating in interlaboratory trials, and reporting and storing results. Romer and Maune (1993) stressed effective sampling and sample preparation (grinding, mixing and subsampling) and the use of reference samples with known levels of mycotoxin in each analysis for good quality control in mycotoxin analysis. To this end, the use of certified reference mycotoxins and pure reference mycotoxins have improved the detection and confirmation of the mycotoxins (Smith, 1997). Gilbert (1999) stressed the importance of sampling, use of validated methods by official organizations (European Committee for Standardization and Association of Official Analytical Chemists International), and use of reference materials. It was stressed that proficiency testing and accreditation of laboratory personnel be done on a regular basis by sending samples with known mycotoxin levels for good quality assurance for mycotoxin analysis.

## Detection of mycotoxin-producing molds

Mycotoxins can be produced during storage of agricultural commodities and ingredients or during processing of foods; therefore, the detection of these molds may be important in preventing mycotoxins from being formed. In an effort to develop rapid methods to detect the presence of mycotoxin-producing molds two major methods have been used, enzyme-linked immunosorbent assays (ELISA) and methods based on detecting genetic components of the mold. Several ELISAs have been developed to detect mycotoxin-producing molds in grain, nuts and laboratory media. Tsai and Yu (1997) reported on the development of an ELISA using mycelial extracts from *A. parasiticus* to detect *A. flavus* and *A. parasiticus* and *A. flavofrucatis* and *A. sojae* if *A. parasiticus* were not present to outcompete them. Yong and Cousin (2001) used the extracellular antigen to *A. parasiticus* to produce an ELISA that detected *A. parasiticus* in maize and peanuts before aflatoxin could be detected by TLC. This assay detected *A. parasiticus* and *A. flavus* plus their domesticated phenotypes *A. sojae* and *A. oryzae*, respectively, which do not produce aflatoxins and are generally only found with fermented foods (Yong and Cousin, 2001).

Both soluble mycelial and extracellular antigens from *Fusarium graminearum*, *F. sporotrichioides*, and *F. poae* were used to develop ELISAs to *Fusarium* species (Gan *et al.*, 1997). The *F. poae* antibody raised against the extracellular antigen was species specific; however, the other two cross-reacted to some degree with all *Fusarium* species analyzed. The antibodies raised against the extracellular antigen were genus specific; however, those raised against the mycelial extract cross-reacted with several genera of molds (Gan *et al.*, 1997). In an effort to develop more specific antibodies to *Fusarium* species, Iyer and Cousin (2003) used proteins extracted from *F. graminearum* and *F. verticillioides* to immunize rabbits. The antibodies produced showed reaction with *Fusarium* species plus two molds that are not common in grains or foods, a *Monascus* species and *Phoma exigua*; therefore, they were used to produce an ELISA that detected *F. graminearum* and *F. verticillioides* in cornmeal at a sensitivity of $10^2$ to $10^3$ CFU/g (Iyer and Cousin, 2003).

There have been many advances in the understanding of the genes involved in mycotoxin biosynthesis that can be used to detect mycotoxin-producing molds. Shapira *et al.* (1996) developed three primers to two enzymatic genes (*ver-1* encodes versicolorin A dehydrogenase and *omt-1* encodes sterigmatocystin-o-methyltransferase) and a regulatory gene (*apa-2* regulates aflatoxin biosynthesis) to detect *A. flavus* and *A. parasiticus* in inoculated maize. The *ver-1* probe was most sensitive in this study because it detected $10^2$ spores of *A. parasiticus* after a 24 h enrichment. Based on this research, Shapira *et al.* (1997) produced two chimeric proteins for genes *ver-1* and *apa-2* in *Escherichia coli* and used these to produce polyclonal antibodies for the detection of aflatoxin-producing molds. These polyclonal antibodies cross-reacted with all *Aspergillus* species and to a lesser extent with all *Penicillium* species but not with *Fusarium* species or the maize substrate. Polymerase chain reaction (PCR) assays have been developed to detect either the production of mycotoxins or the molds that produce mycotoxins for species of *Aspergillus* and *Fusarium*. A PCR reaction based on three genes for enzymes involved in aflatoxin biosynthesis, namely, *nor-1* (encodes norsolorinic acid reductase), *omt-1*, and *ver-1* was used by Färber *et al.* (1997) to detect *A. flavus* in figs where the DNA was detected but the sensitivity was lower than for pure mold DNA. These three genes were used to develop a multiplex PCR that detected the presence of these genes in *A. flavus, A. parasiticus,* and *A. versicolor* (produces strerigmatocystin but not aflatoxin), but only *omt-A* and *ver-1* were detected in the nonaflatoxigenic fermentation strains of *A. oryzae* and *A. sojae* (Geisen, 1996). Mayer *et al.* (2003) developed real-time PCR against genes involved in aflatoxin biosynthesis and it detected *A. flavus, A. parasiticus,* and *A. oryzae,* which contained the *nor-1* gene but not other molds that did not contain the genes. This research contradicts that published earlier (Geisen, 1996) for detection of *A. oryzae* with the *nor-1* gene (Mayer *et al.,* 2003). When this assay was used to detect *A. flavus* inoculated into pepper, paprika and maize, the number of gene copies of *nor-1* correlated with the number of colony-forming units; however, the copy numbers were higher than the mold count (Mayer *et al.,* 2003).

Several PCR-based assays have been developed recently to detect *Fusarium* species that produce trichothecenes and fumonisins. Niessen and Vogel (1998) developed a PCR assay to the *tri-5* gene that encodes trichodiene synthase to detect *Fusarium* species that produce trichothecenes. When wheat was analyzed for DNA, deoxynivalenol and numbers of *Fusarium graminearum,* the intensity of the DNA signal increased as the amount of mycotoxin and mold increased (Niessen and Vogel, 1998). This research group further developed the method into real-time detection by using a LightCycler™-PCR combined with SYBR®Green I fluorescent dye (Schnerr *et al.,* 2001). This method detected *Fusarium* species in 29 of 30 wheat samples with 2% to 78% infected kernels based on microbiological analysis and did not detect *Fusarium* species in noninfected samples (Schnerr *et al.,* 2001). In an analysis of 300 wheat samples, there was a good correlation ($r = 0.96$) between detection of *Fusarium* DNA and deoxynivalenol (Schnerr *et al.,* 2002). The internal transcribed spacer (ITS) region of several *Fusarium* species were examined to identify regions that were specific to species that produced fumonisins (Grimm and Geisen, 1998). For this research, two primers were identified and used to make a PCR assay, which was combined with a commercial ELISA kit to detect species of *Fusarium* that produced fumonisins. This method only detected pure cultures of molds that produced fumonisin; however, it was not used with grain or food samples (Grimm and Geisen, 1998). A Multiplex PCR was developed to detect species of *Fusarium* and those that produce trichothecenes and fumonisins by making primers to conserved ITS1 and ITS2 regions of *Fusarium* rDNA and to genes involved in mycotoxin production *TRI6* (trichothecenes), and *FUM5* (fumonisins), respectively (Bluhm *et al.,* 2002). All of these primers were highly specific for their intended detections and were able to detect $8 \times 10^4$ *F. graminearum* and $4 \times 10^4$ *F. verticillioides* CFU/g of cornmeal after 24 h (Bluhm *et al.,* 2002). Research is continuing on the rapid detection of mycotoxin-producing molds in raw agricultural commodities and in foods by PCR, multiplex PCR and related methods.

Research was done to compare exoantigens from *F. graminearum* and *F. sporotrichioides* by analyzing for ergosterol and deoxynivalenol to determine which method would be best to detect *Fusarium* head blight in preharvest wheat (Abramson *et al.,* 1998). They reported that there were 0.80 and 0.76 correlations between the exoantigen from *F. sporotrichioides* and deoxynivalenol for hard red spring and soft white winter wheats, respectively; whereas, the correlations between the exoantigen from *F. sporotrichioides* and ergosterol were 0.66 and 0.81 correlations for hard red spring and soft white winter wheats, respectively (Abramson *et al.,* 1998). Hence, they suggested that measuring the level of *Fusarium* exoantigens may be an early way to predict deoxynivalenol in grains (Abramson *et al.,* 1998).

## Control and inactivation of mycotoxins

Control of mycotoxins can occur at three different stages: preharvest, post-harvest or storage, and food processing. The major approaches to preharvest control of mycotoxins before 1990 involved (1) breeding of the agricultural commodity for resistance to insects and environmental stress; (2) determination of the genetics of mycotoxin production; and (3) changing agricultural practices (irrigation, crop rotation, and application of fungicides and pesticides); and (4) use of pesticides (Lisker and Lillehoj, 1991). For post-harvest or storage control before 1990, various chemical preservatives and treatments were studied (ammonia, mold inhibitors approved for food use, volatile fatty acids, and other chemicals); however over the 20 years between 1970–1990 these chemicals had limited use (Kiessling and Pettersson, 1991). Physical methods of removing mycotoxins from the agricultural commodities researched before 1990 included sorting to remove contaminated fractions (bran, fiber, germ, hull), kernels or seeds; use of heat (bake, cook or roast) or radiation (ionizing or ultraviolet); use of visible light adsorption or extraction; and use of chemicals such as acids or bases, ascorbic acid, bisulfite, and oxidizing agents; and washing (Beaver, 1991; Pemberton and Simpson, 1991; Samarajeewa, 1991; West and Bullerman, 1991). Methods that showed promise were sorting of nuts for aflatoxin, use of alkali to remove aflatoxins from oils extracted from cereals, and ammonia treatment of feeds (Pemberton and Simpson, 1991; West and Bullerman, 1991). Bhatnagar *et al.* (1991) reported that various microorganisms could detoxify some of the mycotoxins: *Flavobacterium aurantiacum* for aflatoxin $B_1$ and $M_1$ and *Saccharomyces cerevisiae* for patulin. Some of recent findings on control of mycotoxins after 1990 are presented below.

### Preharvest control of mycotoxins

Management of mycotoxins preharvest has been reviewed for different mycotoxins and commodities (Munkvold and Desjardins, 1997; Lopez-Garcia *et al.,* 1999; Riley and Norred, 1999). Management pre-harvest can focus on pest management, crop rotation, soil conditions, irrigation, and use of resistant seeds

(McMullen *et al.*, 1997; Lopez-Garcia *et al.*, 1999). Strategies to prevent mycotoxin production in maize pre-harvest have included controlling plant stresses (insects, lack of nutrients and water), minimizing crop residues that harbor molds, using cultivars resistance to molds, and applying chemicals to prevent mold infection (Riley and Norred, 1999). Agricultural practices to control aflatoxin production in the field have not been effective for maize, cottonseed and peanuts (Smith, 1997). Brenneman *et al.* (1993) reported that the fungicide, diniconazole, was not effective in either reducing the numbers of *A. flavus* nor aflatoxin production in peanuts. To control fumonisins in maize, Munkvold and Desjardins (1997) suggested using maize cultivars resistant to *Fusarium verticillioides* and improving grain handling to prevent growth of *F. verticillioides*, and removing infected kernels. New strategies being researched involve developing genetically modified seeds to resist infection by insects and fungi or to interfere with mycotoxin production if infected, genetically modifying plants to allow them to turn off mycotoxin synthesis after a select signal is produced, modifying plants to produce enzymes to detoxify mycotoxins *in situ* to prevent accumulation, and using biocompetition by other microorganisms or non-toxigenic molds (Munkvold and Desjardins, 1997; Riley and Norred, 1999; Duvick, 2001).

Biological control of plant infection by mycotoxin-producing strains has been suggested because some nontoxigenic fungi and bacteria can colonize plant tissue and successfully out compete the mycotoxin-producer. In 2001, nearly 20,000 acres of Arizona cotton fields were treated with the atoxigenic biocontrol strain AF36. The level of aflatoxin in cottonseed was reduced by 80% compared to untreated fields (Robens and Riley, 2002). This approach also is proving successful for cotton in South Texas and is now being explored for maize and peanuts and has promise for pistachio nuts and figs (Robens and Riley, 2002). The use of other biocontrol agents such as saprophytic yeast also is effective against colonization by toxigenic fungi in almond and pistachio (Robens and Riley, 2002). *Trichoderma viride* was isolated from maize root tissue and decreased growth of *F. verticillioides* by 46% at day 6 and 91% at day 14 with > 80% reduction in fumonisin $B_1$ production over 12 weeks (Yates *et al.*, 1999). Bacon *et al.* (2001) suggested that a *Trichoderma* species tentatively identified as *T. koningii* be used as a biological control agent to reduce fumonisins in maize kernels, which will be used in animal feeds, during storage. The endophytic stages of colonization of maize tissues by a patented *Bacillus subtilis* strain and *F. verticillioides* were studied and this bacterium completely prevented the mold growth on laboratory agar and prevented colonization of the maize plant and reduced fumonisin production if kernels were infected with *F. verticillioides* (Bacon *et al.*, 2001). Field trials are still in progress on the use of *B. subtilis* (Bacon *et al.*, 2001). Competitive exclusion of toxigenic strains of *A. flavus* and *A. parasiticus* with nontoxigenic strains was successful in test plots for peanuts and cottonseed; however, the survival of these strains in nature is not known (Dorner *et al.*, 1992; Smith, 1997). As long as the biological control microorganisms do not negatively affect the plant or the resulting maize, then they may be a form of control to further investigate.

Mesterházy *et al.* (1999) studied the resistance of wheat to *Fusarium* head blight and deoxynivalenol over 6 years and noted that there was much variation between the genotypes; however, the kernel resistance to *Fusarium* infection did not always correlate with reduced dexoynivalenol concentrations. Chelkowski *et al.* (1999) suggested that the PCR assay could be used to estimate resistance of grain cultivars to *Fusarium* species because the species-specific primers could distinguish between *F. avenaceum*, *F. crookwellense*, *F. culmorum*, and *F. graminearum*. Murillo *et al.* (1998) suggested that the PCR assay developed to *F. verticillioides* could be used to analyze soil for contamination and maize hybrids for resistance to *Fusarium* species. Nicholson *et al.* (1998) developed species-specific PCR to quantify *F. culmorum* and *F. graminearum* during colonization of wheat and suggested that this method could be used to study the *Fusarium*-plant interactions, which need to be understood before control measures are applied. A PCR to quantify trichothecene-producing *Fusarium* species was produced to the *Tri5* gene that encodes trichodiene synthase and was used to determine if fungicides such as azoxystrobin, metconazole, and tebuconazole were effective against *Fusarium* head blight in winter wheat (Edwards *et al.*, 2001). Only metconazole and tebuconazole were effective against *F. culmorum* and *F. graminearum* as determined by the decrease in both PCR and deoxynivalenol detected in these samples of wheat (Edwards *et al.*, 2001). When 200 μg of glufosinate-ammonium herbicide/mL were used as the immersion liquid for maize kernels, there was a 76.8% reduction in the production of aflatoxin $B_1$ by *A. flavus*. The fungicide tebuconazole reduced the production of deoxynivalenol by *F. culmorum* in wheat by 53.5–68.8% if treated 5 days after or 3 days before infection, respectively; however, plots that were treated with this fungicide both before and after infection by *F. culmorum* showed a 79.8% decrease in deoxynivalenol (Homdork *et al.*, 2000a). Boyacioglu *et al.* (1992) found that the fungicides triadimefon and propiconazole, which inhibit ergosterol biosynthesis, decreased deoxynivalenol production by 65.0–78% and 34.4–77.9%, respectively; however, triabendazole had no effect on infection by *F. graminearum* but deoxynivalenol was decreased by 83.4% when it was sprayed 2 days before inoculation with the mold spores.

Head blight of wheat is caused by *Fusarium graminearum* Schwabe along with its perfect state (*Gibberella zeae*) and *Fusarium culmorum* (Wm.G.Sm.) Sacc. (Snijders, 1994). Various *Fusarium* species (*F. graminearum*, *F. verticillioides*, *F. culmorum*, *F. subglutinans* (Wollenweb. and Reinking) Nels, Toussoun and Marasas, *F. equiseti* (Corda) Sacc. Sensu Gordon, and *F. proliferatum* (Matsushima) Nirenberg) cause seedling blight and rots of the root, stalk and ear in maize (Snijders, 1994). Brown *et al.* (1998) mentioned that preharvest control of mycotoxins produced by *Fusarium* species are just beginning because not enough is understood about their interactions with the plants, biosynthesis and genetic control.

Classical resistance to *Fusarium* infection was done by selecting varieties that showed the least damage by *Fusarium* species and planting those varieties the next year (Brown *et al.*, 1998). Mesterházy *et al.* (1999) theorized that resistance of wheat to *Fusarium* species involved resistance to (1) kernel infection or invasion, (2) spreading, (3) toxin accumulation, and (4) mold tolerance. Although the study of resistance to *Fusarium* infection or spread in plants is complex, some resistance to *F. culmorum* and *F. graminearum* was shown in wheat and to *F. graminearum* and *F. verticillioides* in maize and several genes involved in the resistance have been identified (Snijders, 1994). Šrobárová and Pavolová (2001) reported that the accumulation of free proline in wheat seeds that were germinating and had deoxynivalenol, diacetoxyscirpenol or culture filtrate of *F. culmorum* added after 4 d to determine the seedling resistance supported these four theories of resistance to the mold and mycotoxin. Also, there is variation to resistance in spring and winter wheat with some genotypes of spring wheat being resistant to both *Fusarium culmorum* and *Fusarium graminearum*, but not all genotypes were resistant to both species (Snijders, 1994).

For maize, cultivars reacted differently to infection by species of *Fusarium* because both stalk and ear infections were important in maize and there was no direct relationship between responses to *Fusarium* species (Snijders, 1994; Duvick, 2001). Correlations between ear rot and aflatoxin B$_1$ were $r = 0.54$ but between insect damage and ear rot were 0.73 (Tubajika and Damann, 2001). For maize where ear rot is the problem, insect or bird damage plus viral infection allow *Fusarium verticillioides* and *Fusarium subglutinans* to invade the kernels (Snijders, 1994; Duvick, 2001). Resistance to insects helped to improve resistance to *Fusarium* species as shown by Bakan *et al.* (2002) because Bt maize had 0.05–0.3 ppm fumonisin B$_1$ compared to 0.4–9.0 ppm on isogenic maize. In some cases, Bt maize hybrids showed lower levels of fumonisins than non-Bt maize hybrids; however, fumonisin levels were usually not significantly different between the two types of maize (Dowd, 2001; Robens and Riley, 2002).

Some genes involved in resistance to *Fusarium* species have been identified for ear or stalk rot; however, not all genes involved have been identified (Snijders, 1994). Resistance to *Fusarium* infection is complex and many more genes need to be identified to develop resistant seeds because there is not a strong correlation between the infection by *Fusarium* species and the amount of mycotoxin produced or accumulated in the plant material (Snijders, 1994). The major ways presented to achieve resistance to *Fusarium* infection of maize and wheat were to focus on genes involved in resistance to initial penetration of the wheat and to spreading once infection was initiated; however, complete resistance to *Fusarium* species has not been shown (Snijders, 1994). Miedaner (1997) reviewed the methods used to breed rye and wheat for resistance to *Fusarium* species and concluded that more research is needed on host resistance and aggressiveness of pathogenicity. The resistance to mycotoxin production may be another area to investigate.

Naturally occurring and pesticidal chemicals have been investigated to determine if they can prevent mycotoxin production in the field. Two anthocyanins, pelargonidin and delphinidin (Norton, 1999), and the carotenoid α-carotene (Norton, 1997) inhibited aflatoxin B$_1$ production early in the biosynthetic pathway before norsolorinic acid; hence, compounds naturally present in plant tissues could serve as toxin inhibitors if the plant were modified to have sufficient quantities. Molyneux *et al.* (2000) tried to identify phytochemicals in tree nuts that would prevent aflatoxin production by *A. flavus*; however, they were not totally effective for control. Hua *et al.* (1999) reported that three phenolic compounds (acetosyringone > syringaldehyde > sinapinic acid) also inhibited aflatoxin B$_1$ biosynthesis before norsolorinic acid. Glufosinate-ammonium, an herbicide, inhibited both growth of *A. flavus* and aflatoxin production when maize kernels were immersed in 2 to 200 µg/mL (Tubajika and Damann, 2002). In cottonseed, the triglyceride storage lipids allowed *A. flavus* to produce aflatoxin B$_1$; however, if they were removed, then aflatoxin B$_1$ decreased by 800-fold (Mellon *et al.*, 2000). In cottonseed, *A. flavus* preferentially used the carbohydrates, especially raffinose, to produce biomass and begin aflatoxin production at which time fatty acids began to appear (Mellon *et al.*, 2000). Gluconeogenesis may be related to aflatoxin biosynthesis; therefore, breeding to control raffinose may reduce the amount of aflatoxins in cottonseed (Mellon *et al.*, 2000).

Molecular pre-harvest reduction of aflatoxins could involve blocking the plant-mold interaction to prevent aflatoxin production or preventing the mold from producing the aflatoxins (Smith, 1997). The identification of the genes involved in aflatoxin biosynthesis means that traditional plant breeding or genetic manipulation could be used to control aflatoxin biosynthesis (Smith, 1997). There are about 16 genes in major gene clusters that have been identified in aflatoxin biosynthesis that could be used to prevent the production of aflatoxin (Smith, 1997). Various genes have been studied to determine if mycotoxins can be controlled during plant growth. Flaherty *et al.* (1995) fused the *ver1* gene to the *uidA* gene (codes for β-glucuronidase) of *E. coli* to make a reporter that could be used as a probe to determine if aflatoxin is induced in maize. This technique could be used to control aflatoxin biosynthesis by developing transgenic plants with resistance genes. Mayer *et al.* (2003) developed a real-time reverse transcription (RT)-polymerase chain reaction (PCR) to measure the *nor-1* gene for aflatoxin production by *A. flavus* in wheat. The *nor-1* mRNA was detected on day 4 and increased rapidly (Mayer *et al.*, 2003). This was used to study the progression of aflatoxin production. During this time, aflatoxin B$_1$ could be detected by day 6, increased until day 9, and then was constant (Mayer *et al.*, 2003). If these genes could be turned off, then aflatoxin production could be controlled. Development of resistance through plant breeding; genetic alterations of the plant to control fungal invasion, growth or mycotoxin biosynthesis; competition by other organisms; and combinations of these approaches were reviewed for the control of aflatoxin production (Brown *et al.*, 1998).

Infection of cereal grains by *Claviceps* species can be controlled by farm management practices, such as weed control (many grasses are susceptible to *C. purpurea*), crop rotation, plowing 1 foot or greater in infected fields, and using cultivars that are resistant to *Claviceps* infection (Kuldau and Bacon, 2001). Use of pesticides may or may not help because they must contact the ovary at the susceptible time for infection by *Claviceps* species (Kuldau and Bacon, 2001). Some of these measures may not completely control the spread of the ascospores and conidia because they are blown by wind over large areas (Kuldau and Bacon, 2001). It is hard to predict if grains will be contaminated with ergot alkaloids because of changes in agricultural practices, global climate and worldwide distribution of food and disease agents (Kuldau and Bacon, 2001).

Analytical methods for mycotoxins are now able to detect very low levels of contamination. As a result, it has become apparent that some control methods will not completely destroy or prevent mycotoxin production in some agricultural commodities. Nonetheless, reducing mycotoxin levels in the field through the use of effective control strategies at the pre-harvest stage will alleviate the need for post-harvest destruction of mycotoxins.

## Post-harvest control of mycotoxins

Several approaches have been tried to prevent the production of or to destroy mycotoxins in agricultural commodities after harvest, namely (1) drying and proper storage of agricultural commodities to prevent mold growth; (2) inactivation of mycotoxins by heat, irradiation or chemical degradation; (3) biological inactivation of mycotoxins; (4) chemical absorption or extraction of mycotoxins; and (5) modification of animal or human diets to protect consumers from mycotoxins (Phillips *et al.*, 1994; Smith, 1997). Some of the recent reviews on reduction of mycotoxins post-harvest are briefly summarized here. Charmley and Prelusky (1994) reviewed the methods used to decontaminate cereal grains of mycotoxins produced by *Fusarium* species. These included (1) physical methods of cleaning and washing, separation of contaminated kernels, dehulling and milling, dilution by blending, and heat – success depended on amount of contamination and where it was located in the grain; (2) chemical methods such as calcium hydroxide, ammonia, sodium bisulfite, chlorine gas, and sulfur dioxide – success depended on mycotoxin and its concentration, application time, moisture and amount of mycotoxin – no single treatment was effective for all mycotoxins; and (3) biological methods such as adding mycotoxin-binders and adding to feeds microorganisms that detoxify mycotoxins. Saunders *et al.* (2001) reviewed the effects of alkaline processing for masa, canning and cooking, dry- and wet-milling, and extrusion on fumonisins. Sinha (1998) reviewed the methods used to remove mycotoxins from grains and nuts that included: (1) physical separation by sorting, sieving, floatation and similar techniques; (2) filtration of oils and absorption to clays, sili-

cates, activated charcoal and other materials; (3) milling where the mycotoxin is concentrated in one or more fraction; and (4) solvent extraction. Sinha (1998) reviewed the methods used to detoxify mycotoxins including: (1) physical methods involving thermal and irradiation treatments; (2) chemical methods that used ammonia, chlorine gases, hydrogen peroxide, sulfites, and hydroxides; and (3) biological methods using bacteria and molds with only *Flavobacterium aurantiacum* showing promise for some foods (maize, milk, oil, peanuts and peanut butter). Although no economical method exists to detoxify feeds or foods, the removal of contaminated grains and nuts from the food chain should be continued (Smith, 1997; Sinha, 1998).

A major concern with the presence of mycotoxins in foods is that most types of processing have not been effective in destroying all mycotoxins in foods as noted in early reviews (Doyle *et al.*, 1982; Scott, 1984; Samarajeewa *et al.*, 1990; Park and Liang, 1993). More recent reviews have focused on the destruction of *Fusarium* mycotoxins (Bennett and Richard, 1996; Saunders *et al.*, 2001). López-García and Park (1998) reviewed the methods used to prevent mycotoxins from entering food by stressing a four-point program: (1) establishment of regulatory limits; (2) developing effective monitoring programs; (3) implementing process controls; and (4) decontaminating ingredients used in food processing. For this latter control, cleaning and segregation of clean from contaminated; partitioning of mycotoxins during processing such as wet and dry milling; thermal and irradiation inactivation; adsorption by chemicals; biological control; and chemical inactivation were discussed. Although total inactivation of mycotoxins is not possible; some of these methods may reduce the levels of mycotoxins in foods (López-García and Park, 1998).

### *Adsorption of mycotoxins*

Various materials have been studied to determine if they bind mycotoxins and remove them from feed or food before further processing. Activated or natural carbons, charcoal, and clays have been used to remove mycotoxins from contaminated grains. Phillips *et al.* (2002) reviewed the clay-based adsorbents used to remove aflatoxins. Galvano *et al.* (1996b) found that when they added activated carbon and aluminosilicates to feed used for dairy cows that they bound aflatoxin $B_1$ and prevented its absorption in the intestines. Activated carbon reduced aflatoxin $B_1$ in feed 40.6% to 73.6% and aluminosilicate reduced aflatoxin $B_1$ in the feed by 59.2% (Galvano *et al.*, 1996b). This produced 27% to 50% reduction of carryover of aflatoxin $B_1$ into milk. Activated carbons reduced the amounts of deoxynivalenol (1.83% to 98.9% depending on the carbon used) and ochratoxin A (0.8% to 99.9% depending on the compound used) in solutions but no work was done in feeds (Galvano *et al.*, 1998). Kadakal and Nas (2002) reported that patulin could be reduced from 62.3 ppb to 26.7 ppb after a 30 min exposure to 3 g activated charcoal/liter of apple juice. Similarly, Bissessur *et al.* (2001) reported that adsorption by bentonite resulted in 77% loss of patulin from apple juice compared to 89% for pressing followed by centrifugation at 6551

× g for 10 min. In an attempt to develop a continuous process for patulin removal, ultrafine activated carbon was bond to granular quartz to produce a fixed-bed adsorbent system that effectively removed patulin from apple juice; however, the color of the apple juice was deceased by the process (Huebner et al., 2000). To transfer the use of absorption to other mycotoxins, activated carbon and potassium caseinate showed the highest removal of ochratoxin A from red wine when several commercial fining agents were evaluated (Castellari et al., 2001).

Another way to remove mycotoxins is to have them bound in the intestine by a nonnutritive adsorbent material. Kerkadi et al. (1999) used cholestyramine to bind ochratoxin A in the diets of rats fed ochratoxin A. Cholestyramine binds bile salts by electrostatic charges; however, more ochratoxin A was bound to cholestyramine because of the hydrophobic adsorption and ion-exchange reactions (Kerkadi et al., 1999). Other agents used to bind ochratoxin A were reviewed by Varga et al. (2001). Kensler et al. (1994) suggested using chemoprotective measures to reduce the incidence of liver cancer from aflatoxins or other causes. Consumption of a semisynthetic water-soluble chlorophyll derivative reduced levels of aflatoxin adducts in populations at high risk for hepatocellular carcinoma in China (Egner et al., 2001). Another chemical agent that has been successfully used to protect farm animals from aflatoxin-induced hepatotoxicity, growth retardation and reproductive effects is hydrated sodium calcium aluminosilicates (Smith, 1997; Phillips et al., 2002). Adsorbing agents may, therefore, be useful in both the food and the intestines to prevent mycotoxins from causing undesirable health effects in animals and humans.

*Physical methods for mycotoxin removal*
Physical methods such as washing, peeling, milling, sieving, soaking, and related methods have been used to decrease mycotoxins in grains, nuts and fruits before further processing into foods. Reduction of patulin in apples and apple juice has been important in recent years because of the concern for consumption of these products by children. Moodley et al. (2002) found that low density polyethylene packaging was superior to polypropylene in preventing growth of *Penicillium expansum* (68% reduction) and patulin production (99.5% reduction in normal air and 99.97% reduction in 88% $CO_2$:12% $N_2$) during storage of apples. Patulin was reduced in apple juice by washing and removal of rotted apples before processing (Sydenham et al., 1995b). Taniwaki et al. (1992) found that most patulin in apples migrated about 1 cm around the lesion; therefore, just removing the rotted tissue would not prevent all patulin from getting into the juice after pressing. Washing of grains with water or water with added chemicals has been evaluated to reduce the level of mycotoxins. Trenholm et al. (1992) found that washing with water reduced the levels of deoxynivalenol and zearalenone on barley, maize and wheat kernels; however, the addition of 0.1 mol/L sodium carbonate to the water and allowing the grains to soak for 72 h removed the most deoxynivalenol and zearalenone. Clavero et al. (1993)

reported that the use of 0.08% hydrogen peroxide in the water to wash peanuts resulted in a 90% reduction in total aflatoxins after 1.0 min.

Milling of grains can reduce the level of mycotoxins in some of the end-products. Bennett and Richard (1996) reviewed research on wet and dry milling of grains and their effects on levels of *Fusarium* mycotoxins. Trenholm et al. (1991) reported that dehulling grains (barley, maize, rye, wheat) followed by sieving to remove particles smaller than + 16 mesh resulted in removing from 67 to 83% of deoxynivalenol and zearalenone depending on the grain; however, these processes also removed protein and other grain components. Dexter et al. (1996) found that flour contained 50% of the original deoxynivalenol from Canadian wheat with *Fusarium* head blight damage. After milling wheat, deoxynivalenol and zearalenone were found in bran, shorts and flour in decreasing order of concentration and fractions from some samples had amounts greater than the 1 ppm advisory level (Trigo-Stockli et al., 1996). Fumonisins were detected in all dry-milled fractions of maize with the highest concentrations in bran, fines and germ, which are used for animal feeds, and lower concentrations in the flaking grits that are used for human foods (Katta et al., 1997). Grinding to make cornmeal did not reduce the level of fumonisins; however, roasting the cornmeal reduced the level by 25% to 33% and frying reduced the levels even more (Shier et al., 2000). Ergot alkaloids were retained in the late reduction flours and shorts after wheat milling; however, lower levels were found in patent and break flours plus the bran (Fajardo et al., 1995). Although physical methods can reduce levels of mycotoxins, the size of the reduction depends on the mycotoxin, the raw ingredients, and the method used.

*Chemical destruction or alteration of mycotoxin*
Various chemicals have been used in an attempt to remove both the molds and the mycotoxins that they produce from raw agricultural commodities before they are produced into feeds or foods. Montville and Shih (1991) found that 1–2% ammonium bicarbonate reduced the natural microflora in maize by 5–6 log CFU/g; however, 1–2% sodium bicarbonate only reduced the natural microflora by 1–2 log CFU/g. In this same research, 1% sodium bicarbonate caused a 100-fold reduction in ochratoxin A and no toxin was detected for the 2% solution or for either concentration of ammonium bicarbonate. Although ammonia detoxifies maize and cottonseed in relation to aflatoxin, fumonisin $B_1$ was only reduced by 30–35% after treatment with 1–5% ammonium hydroxide (Norred et al., 1991). These chemicals would not be acceptable for food use; however, alkaline treatments are used in some maize processing. Alkaline hydrolysis with lime and heat (nixtamalization) is used in the manufacture of tortillas; hence, several studies have been done to determine if this alkaline treatment reduced fumonisin $B_1$. Hendrich et al. (1993) reported that nixtamalization hydrolyzed fumonisin $B_1$ and a more toxic product was formed because the feeding of this hydrolyzed corn with 8 to 11 mg/kg of hydrolyzed fumoni-

sin $B_1$ produced toxic symptoms in rats that were similar to feeding 45 to 48 mg/kg fumonisin $B_1$. When ground maize was treated with 0.1 mol/L calcium hydroxide, 70–80% of fumonisin $B_1$ was in the aqueous fraction and 10–25% remained in the maize fraction; however, only 5.1% remained with the maize kernel when the pericarp was removed before treatment (Sydenham et al., 1995a). Higher amounts of fumonisin $B_1$ were found in tortillas from Santa Maria de Jesus, Sacatepequez than from Patzicia, Chimaltenango when two regions of Guatemala were studied (Meredith et al., 1999). A concern is that fumonisin $B_1$ can be converted to other toxic forms or be bound to maize as seen in nixtamalization or lime treatment for tortilla manufacture where radiolabeled fumonisin $B_1$ bound to protein or starch; therefore, there was concern about whether it could be released by digestive enzymes and become active in the gastrointestinal tract (Shier et al., 2000). Alkaline hydrolysis reduced the level of fumonisins and their toxicity (Voss et al., 1996a; Palencia et al., 2003) but did not eliminate them completely.

Different chemicals and food additives have been studied for their ability to reduce the level of mycotoxins during processing of foods. Oxidizing agents (L-ascorbic acid and potassium bromate) had no effect on reducing the level of deoxynivalenol during the baking of bread; however, reducing agents (L-cystine and sodium bisulfite) and ammonium phosphate reduced the level by 38 to 46% (Boyacioğlu et al., 1993). Similarly, Tabata et al. (1994) reported that potassium bromate had no effect on degradation of aflatoxins; however, sodium chlorite and ammonium peroxodisulfate completely degraded aflatoxins added to maize grits. Hydrogen peroxide at 0.05% for 30 min detoxified stock solutions of citrinin but not ochratoxin A (Fouler et al., 1994). Treatment with 0.4% bisulfite or 0.1% hydrogen peroxide + lactoperoxidase decreased aflatoxin $M_1$ by 45% to 50% and 5% bentonite reduced aflatoxin $M_1$ by 89%; however, these methods are not acceptable for dairy product manufacture (van Egmond, 1994; Henry et al., 2001). Some chemicals have been studied for their inhibition of mycotoxin production by molds. Monolaurin, eugenol, and isoeugenol at 100 ppm in yeast extract dextrose broth inhibited aflatoxin $B_1$ production by A. parasitcus by 25%, 58% and 57%, respectively (Mansour et al., 1996). When sclerotia of C. purpurea were treated with 1% chlorine at 150–200°C for 4 h, a 90% reduction of ergot alkaloids occurred, but autoclaving at 121°C for 30 min only caused about a 25% decrease (United Nations Environment Programme, 1990). Yazici and Velioglu (2002) reported that use of 625 to 2500 mg/kg of B-vitamins (panthothenate, pyridoxine and thiamine) reduced the level of patulin in apple juice by 55.5% to 67.7% after 6 months at 4°C without destruction of the juice quality. The use of chemicals to detoxify mycotoxins in ingredients or foods will depend on the safety of the chemical, the food in which it is used, the mycotoxin, the processing conditions, and other factors.

### Thermal stability of mycotoxins

The heat generated during baking, frying, roasting, and other processing has been evaluated to determine the effects on different mycotoxins. Most of the recent work has been directed toward destruction of fumonisins. The production of masa from maize in commercial tortilla manufacture reduced the fumonisin level by 80% or greater due to the nixtamalization step (cooking and steeping in aqueous alkali such as lime or calcium hydroxide); however, the subsequent processing steps of preparing the dough, cutting, baking and frying did not affect the fumonisin levels (Dombrink-Kurtzman et al., 2000; Voss et al., 2001a). Canning of maize-based products at 121°C for 22–87 min resulted in 0–15% decrease in fumonisins and baking maize muffins/bread at 204°C (internal about 100°C) for 20 min resulted in no loss of fumonisins; however, roasting dry cornmeal at 218°C (internal from 182–192°C) for 15 min almost totally decreased fumonisins (Castelo et al., 1998c). Baking maize muffins at 200°C (150°C surface and 115°C internal) and 175°C (135°C surface and 107°C internal) resulted in only 28% and 16% loss of fumonisin $B_1$, respectively (Jackson et al., 1997). Similar results were reported by Dupuy et al. (1993) who found that 25% and 13% fumonisin $B_1$ remained in dry maize after 80 min at 125°C and 40 min at 150°C, respectively. In an aqueous buffer solution, fumonisin $B_1$ and $B_2$ decrease was dependent on temperature, pH and time because less fumonisins were recovered at alkaline (pH 10) or acid pH (pH 4) compared to neutral (pH 7) and at high temperatures from 100–235°C for longer times, especially 60 min or more (Jackson et al., 1996a; b). Frying maize-based products and the loss of fumonisins depended on temperature and time because at < 180°C, there was no loss, but as temperature approached 190°C, losses were seen at 8 min or longer (Jackson et al., 1997). There was a decrease of 67% fumonisin $B_1$ at 190°C for 15 min; however, the maize chips were too dark in color for good consumer acceptance (Jackson et al., 1997). The losses in the different thermal processing methods may be due to nonenzymatic browning involving the amines of fumonisins and reducing sugars because fumonisins react with the reducing sugars, glucose and fructose, during heating in aqueous solutions to form a stable hydrolyzed Schiff's base termed N-carboxymethyl fumonisin $B_1$ that may or may not alter toxicity depending on the animal model used (Howard et al., 1998; Lu et al., 2002; Poling et al., 2002). Although several researchers have reported that N-carboxymethyl fumonisin $B_1$ is formed in the presence of reducing sugars, Castelo et al. (2001) found that only 13–16% of the original fumonisin $B_1$ in maize muffins were converted to this hydrolyzed product. Based on these results, canning, baking, frying, and boiling will not sufficiently reduce fumonisins in foods; however, alkaline treatments with heat significantly reduced fumonisins in maize-based foods (Dombrink-Kurtzman et al., 2000; Voss et al, 2001a; Palencia et al., 2003).

Although most research over the past few years has been done on fumonisin thermostability, some other mycotoxins have been studied as well. When deoxynivalenol was heated in a buffer at pH 4.0, 7.0, or 10.0 and temperatures of 100, 120, and 170°C for 15, 30, or 60 min, it was reported that pH 10.0 was most detrimental with partial loss at 100°C for 60 min and total loss at 120°C for 30 min and 170°C for 15 min (Wolf and Bullerman, 1998). Deoxynivalenol was stable to pH 4.0 and only showed partial loss at 170°C for 60 min; whereas, at pH 7.0, there was more loss (Wolf and Bullerman, 1998). Autoclaving cream-style maize at 121°C for 72 min resulted in only 13% reduction of deoxynivalenol but no reduction was noted for dog food autoclaved at 121°C for 87 min (Wolf-Hall et al., 1999). In breadmaking, deoxynivalenol was reduced by 21.6% during the fermentation process and by 28.9% during the baking process (Neira et al., 1997). In contrast to these results, there was no loss of ergot alkaloids during bread fermentation or in the crumb; however, the crust showed a 25–55% reduction (Fajardo et al., 1995), which contradicts the 50 to 100% decrease of ergolines in wheat and rye bread depending on the flour and the process (United Nations Environment Programme, 1990). There was little change in the ergot alkaloids after processing flour into noodles and pasta and cooking the noodles and pasta (Fajardo et al., 1995). When cyclopiazonic acid in milk was analyzed, it was stable to most heat treatments given to milk and even canning at 120°C for 30 min only reduced the level by 33 to 36% (Prasongsidh et al., 1998b). If milk were held at 100°C for 2 h, 50% cyclopiazonic acid still remained; however, storage of milk at 4°C resulted in from 19–62% loss depending on the previous heat treatment (Prasongsidh et al., 1998b). Evaporating and spray-drying of milk resulted in little to no loss of cyclopiazonic acid but concentrating sweetened condensed milk with steam injection caused almost 40% loss (Prasongsidh et al., 1997a). Storage of raw milk at 4°C for 4 days and of pasteurized milk for 21 days caused < 2% and 5.8% loss of cyclopiazonic acid, respectively; however, in frozen or freeze-dried milk stored at −18°C, the loss was < 12% after 140 days (Prasongsidh et al., 1997a).

There have been questions about the stability of ochratoxin A in coffee for about 30 years. Ochratoxin A survived the drying of coffee cherries with more in damaged or overripe cherries (Bucheli et al., 2000) and roasting of Robusta green coffee beans from Cote d'Ivoire resulted in a 69% reduction of ochratoxin A (van der Stegen et al., 2001) and in Thai robusta green coffee beans, a reduction of 84% occurred (Blanc et al., 1998). Most of the ochratoxin A is in the husks and good agricultural and manufacturing practices can control it getting into the final coffee beans (Viani, 2002). Soluble coffee extraction, concentration and spray drying resulted in the final coffee powder having 13% of the initial ochratoxin A (Blanc et al., 1998). Ochratoxin A was reduced by 20% in ground dry wheat at 100°C for 80 min and 150°C for 32 min, by 88.3% at 200°C for 48 min and by 93.5% at 250°C for 16 min; however, when the wheat was wet, > 50% ochratoxin A was destroyed at 100 and 150°C for 120 min but at 200°C, only 61% was destroyed after 32 min (Boudra et al., 1995). The use of hot hexane to extract peanut oil from peanut cake left a protein residue with 52% aflatoxin $B_1$ because it is insoluble in hexane and probably binds to the hydrophobic amino acids of the peanut protein (Sashidhar, 1993). Patulin in apple aroma or distillate was 250 times less than that of the starting apple juice after 2.5–3 h of constant boiling; therefore, commercial apple distillate should be free of patulin although the juice is not (Kryger, 2001). These results show that unless there are very high temperatures for long times or alkaline pH that mycotoxins will remain in the processed foods.

### Extrusion and mycotoxins

Extrusion is a process where temperature and screw speed are used to make dry breakfast cereals and snack foods from maize, wheat, rice and other grains. Most of the research with extrusion processing has been done on the mycotoxins produced by *Fusarium* species. Castelo et al. (1998a) reported that more fumonisin $B_1$ was lost from maize grits with the mixing screw (29–69% by ELISA and 31–68% by HPLC) than nonmixing screw (13–54% by ELISA and 20–47% by HPLC) extruders because of greater sheer of mixing and greater residence time in the machine. These conditions may help to explain why there was greater loss of fumonisin $B_1$ at 120°C rather than 140°C but not why only 18% and not 22 or 26% moisture affected the reduction (Castelo et al., 1998a). These authors speculated that the extraction of fumonisin $B_1$ with 70:30 methanol:water may have affected the recovery (Castelo et al., 1998a). The barrel temperature (160–200°C) and screw speed (120–160 rpm) affected the amount of fumonisin $B_1$ remaining in maize, which showed reductions of 46–76% (Katta et al., 1999). The temperature plus high pressure and shear may account for the reduction of fumonisin $B_1$ in extruded maize (Katta et al., 1999). In addition to the high temperatures and slow screw speeds, glucose at 10% caused greater losses of fumonisin $B_1$ in extruded maize (Castelo et al., 2001). Extrusion of maize and dog food resulted in 53 and 21% reduction of deoxynivalenol, respectively (Wolf-Hall et al., 1999). Cazzaniga et al. (2001) reported a 95% reduction of deoxynivalenol in extruded maize but only a 10–25% reduction of aflatoxin $B_1$. In contrast to these two reports, common extrusion conditions did not statistically lower the level of deoxynivalenol in wheat; however, the process reduced the moisture by 8% and also the odor (Accerbi et al., 1999). More zearalenone was destroyed by extrusion processing at 120–160°C than by other thermal processing methods with 66 to 83% reduction using mixing screws and 65% to 77% reduction with nonmixing screws (Ryu et al., 1999a). From these results, extrusion processing will decrease the mycotoxin depending on the temperature, moisture, screw speed, food, and mycotoxin; however, complete reduction may not be possible.

### Fate of mycotoxins during fermentation

Food fermentation is a process where microorganisms alter a substrate or food to produce a new food; therefore, microorganisms used for fermentations would come into contact with mycotoxins. Some research has been done to determine how mycotoxins are affected during dairy and alcoholic fermentations. Hassanin (1994) found that 83–89% aflatoxin $M_1$ was recovered from yogurt, yogurt cheese, and acidified milk; however, the level decreased to 41%, 71%, and 34%, respectively after storage at 4°C for 13 days. When yogurt and Cheddar cheese were made from milk containing cyclopiazonic acid, there was a 70% reduction of cyclopiazonic acid in yogurt on day 1, but 21% was still in yoghurt after 21 days at 4°C (Prasongsidh *et al.*, 1998a) and cyclopiazonic acid was concentrated about 2-fold in cheese where the amounts increased and decreased over the 10-month aging but still was about 93% of the concentrated value (Prasongsidh *et al.*, 1999a). The fate of *Fusarium* mycotoxins during alcoholic fermentation has been studied (Bothast *et al.*, 1992; Bennett and Richard, 1996; Boeira *et al.*, 1999a; b; 2000). Deoxynivalenol, fumonisins, and zearalenone have been recovered from beer (Bennett and Richard, 1996; Wolf-Hall and Schwarz, 2002). No fumonisin $B_1$ occurred in distilled ethanol because it partitioned into distillers grains and solubles and in thin stillage when contaminated maize was fermented (Bothast *et al.*, 1992). Fusarium mycotoxins affected the growth of brewers yeasts if present in high concentrations: 50 µg zearalenone/mL slowed growth of both ale and lager *Saccharomyces cerevisiae* and 50 µg of fumonisin $B_1$ slowed ale strains but 10 µg/mL slowed lager strains (Boeira *et al.*, 1999a); 50 µg deoxynivalenol or nivalenol/mL slowed the growth of ale strains and 50 µg nivalenol/mL slowed lager strains but 100 µg deoxynivalenol/mL were needed to slow lager strains (Boeira *et al.*, 1999b). When deoxynivalenol, fumonisin $B_1$ and zearalenone were combined in low concentrations, there was no effect on yeast growth; however, when concentrations higher than would be used normally in beer manufacture were evaluated, there was a synergistic effect on yeast growth (Boeira *et al.*, 2000). Fermentation can reduce the level of mycotoxins in some fractions but may increase them in others. Thus, the level of concern will depend on which fraction is consumed.

*Degradation of mycotoxins by microorganisms*
Cells of bacteria and molds plus the enzymes that they produce have been evaluated for their ability to degrade mycotoxins either in foods or in the intestinal tract. One bacterium that could degrade aflatoxin $B_1$ and has been researched in two laboratories is *Flavobacterium aurantiacum*, an orange pigmented Gram-negative rod (Line *et al.*, 1994; Line and Bracket, 1995 a, b; D'Souza and Bracket, 1998, 2000; Smiley and Draughon, 2000; D'Souza and Bracket, 2001). *F. aurantiacum* degraded aflatoxin $B_1$ to water soluble components (Line *et al.*, 1994) with live cells that were 72 h old being more effective than 24 h cells at populations of $10^9$ CFU/mL or greater (Line and Brackett, 1995a). *F. aurantiacum* did not grow if aflatoxin $B_1$ were the only carbon source (Line and Brackett, 1995b) and aflatoxin degradation was decreased if trace metals (cobalt, copper, manganese and zinc) were present because they could be affecting a reducing enzyme (D'Souza and Bracket, 1998);

however, aflatoxin degradation increased in the presence of divalent cations (calcium and magnesium) because they are important in enzymatic reactions in the glycolytic pathway (D'Souza and Bracket, 2000). Smiley and Draughon (2000) further supported the enzymatic nature of the degradation of aflatoxin $B_1$ by *F. aurantiacum* when they isolated crude protein extracts that degraded aflatoxin $B_1$ but did not if heat treated. Research to determine if reducing conditions affected aflatoxin $B_1$ degradation by *F. aurantiacum* was not conclusive (D'Souza and Bracket, 2001). More research on the enzymes involved in aflatoxin $B_1$ degradation by *F. aurantiacum* are needed before they can be used in foods or feeds; also the orange pigmentation of the bacterium could limit its direct use in foods or feeds.

Lactic acid bacteria and other probiotic bacteria such as *Bifidobacterium* and *Propionibacterium* species have been evaluated to determine if they could inhibit growth of *A. flavus* and aflatoxin production or if they could bind to or detoxify aflatoxins. *Lactobacillus acidophilus* and *Lactobacillus plantarum* caused *A. flavus* to disappear after 4 days; however, *Lactobacillus bulgaricus* only reduced the mold counts when grown in laboratory media but when grown on maize or rice, these lactic acid bacteria did not inhibit growth or aflatoxin production (Karunaratne *et al.*, 1990). A commercial silage inoculum containing a mixture of *Lactobacillus* species inhibited mold growth and aflatoxin production in laboratory media and the cell-free supernatant from these *Lactobacillus* species inhibited aflatoxin production but not mold growth (Gourama and Bullerman, 1995). *Lactobacillus rhamnosus* strains GG and LC-705 were able to bind more aflatoxin $B_1$ after treatment with 1 mol/L HCl and heat at 100 to 121°C but not after treatment with 70% ethanol or propanol, sodium hydroxide, sonication, and UV irradiation (El-Nezami *et al.*, 1998). *L. rhamnosus* strains GG and LC-705 removed more aflatoxin M1 from buffer than did the other dairy lactic acid bacteria; however, for strain GG, more aflatoxin $M_1$ was removed by the heat killed than by viable cells (Pierides *et al.*, 2000). The reaction was just the opposite for strain LC-705, which was more effective as viable rather than heat killed cells in both skim and full fat milks (Pierides *et al.*, 2000). When strain GG was used to bind aflatoxin $M_1$ in the Caco-2 cell cultural model, there was from 5% to 30% reduced ability to adhere to these cells suggesting that these bacteria can increase the excretion of aflatoxins and prevent their accumulation in the intestinal tract (Kankaanpää *et al.*, 2000).

El-Nezami *et al.* (2000) noted that when *L. rhamnosus* GG, *L. rhamnosus* LC-705, and *Propionibacterium freudenreichii* spp. *shermanii* JS were present, uptake of aflatoxin $B_1$ by the chicken intestinal tissues was reduced by 74%, 37%, and 63%, respectively. Species of *Bifidobacterium*, *Lactococcus* and other species of *Lactobacillus* bound 18 to 48.7%, 5.6 to 41.1%, and 5.8 to 59.7% aflatoxin $B_1$, respectively (Peltonen *et al.*, 2000, 2001). *Lactobacillus* species weakly bound aflatoxin $B_1$ because 17.4 to 94.4% of the cells were released when the cell-aflatoxin complexes were washed with phosphate buffered saline

(Peltonen *et al.*, 2001). Washing heat-killed *Bifidobacterium* species bound to aflatoxin $B_1$ also showed that the binding was reversible (Oatley *et al.*, 2000). In addition to binding aflatoxins, *L. rhamnosus* GG and *L. rhamnosus* LC-705 also bound zearalenone and α-zearalenol if at least $10^9$ cells/mL were present because no degradation products were detected after growth of these bacteria (El-Nezami *et al.*, 2002). More research is needed to determine if common bacteria used in dairy and other fermentations can be used on either immobilization columns to remove mycotoxins from liquids or in foods to prevent mycotoxins from adhering to intestinal cells.

Limited research has been done on the use of enzymes extracted from microorganisms to degrade mycotoxins. Stander *et al.* (2000) isolated a lipase from *Aspergillus niger* that degraded ochratoxin A to the nontoxic ochratoxin α and phenylalanine. Similarly, a proposed esterase from *Acinetobacter calcoaceticus* degraded ochratoxin A (Hwang and Draughon, 1994). *Clonostachys rosea* produced a laconohydrolase that degraded zearalenone into a compound that was less estrogenic (Takahashi-Ando *et al.*, 2002). The gene for this enzyme was cloned in the hopes of potentially using it to modify wheat to detoxify mycotoxins in the plant (Takahashi-Ando *et al.*, 2002). Shima *et al.* (1997) suggested that an extracellular enzyme produced by a soil bacterium from the *Agrobacterium-Rhizobium* group may be responsible for the degradation of deoxynivalenol to the less immunosuppressive (> 90%) 3-keto-deoxynivalenol by oxidizing the 3-OH group. In the future, other enzymes may be discovered that degrade mycotoxins and more research may be done to determine both their mechanisms of action and how they can be used to make safer foods and feeds.

*Degradation of mycotoxins by miscellaneous methods*
Miscellaneous methods have been researched for their ability to reduce the level of mycotoxins in different foods. Destruction of aflatoxin $M_1$ and cyclopiazonic acid in milk and dairy products has been a concern in recent years. Pasteurization, drying, refrigeration/freezing, and fermentation did not affect the aflatoxin $M_1$ content of milk (van Egmond, 1994; Henry *et al.*, 2001). Aflatoxin $M_1$ is semipolar and concentrated in the aqueous phase; therefore, it does not appear in cream but does concentrate in casein and cheese (van Egmond, 1994). In the manufacture of butter, 92.1% of cyclopiazonic acid partitioned into the buttermilk with only 4.8% recoverable from the butter (Prasongsidh *et al.*, 1999b), but >27% of the cyclopiazonic acid from milk was present in ice-cream stored at −20°C for 9 weeks (Prasongsidh *et al.*, 1999c). Ultrafiltration and diafiltration removed 43.8% and 25% of aflatoxin $M_1$ from milk, respectively; however, the use of both methods together removed 68.5% of the aflatoxin $M_1$ originally present in milk (Higuera-Ciapara *et al.*, 1995).

Park *et al.* (1999) and FAO (2001) suggested that Hazard Analysis and Critical Control Points (HACCP) could be used to reduce mycotoxins in animal feeds, cereal grains, fruits, nuts, and oilseeds from pre-harvest through post-harvest handling of these agricultural commodities, foods and feeds by controlling mold infestation, growth and mycotoxin production. The controls presented included agricultural practices used to produce these agricultural commodities (insect and mold control, irrigation and tillage activities, crop rotation, etc.); appropriate harvest time, temperature and moisture; storage conditions and environmental controls; and product testing for mycotoxigenic molds and mycotoxins (Park *et al.*, 1999). A HACCP system has been identified for mycotoxins and some examples are given for aflatoxin and patulin (FAO, 2001). Since mycotoxins are produced in both pre- and postharvest of plant materials, storage and use of these materials will determine what control and preventive measures are implemented (FAO, 2001).

*Packaging foods to control mold growth and mycotoxin production*
There have been limited studies on the use of packaging and modified atmospheres to control mold growth and mycotoxin production. Ellis *et al.* (1994a) found that the use of films that had high barriers to oxygen ($O_2$) and carbon dioxide ($CO_2$) transmission, with or without a modified atmosphere of 65% $CO_2$ to 35% $O_2$, limited mold growth and aflatoxin production by *A. flavus* in peanuts. When oxygen absorbers were used with high gas barrier films, then the growth and aflatoxin production by *A. parasiticus* was controlled in peanuts (Ellis *et al.*, 1994b, c). When mycotoxin-producing molds (*A. flavus* and *P. commune*) were grown on Cheddar cheese at 25°C in modified atmospheres of 20% and 40% $CO_2$ with 1 to 5% oxygen, 0.03 to 4.0 μg aflatoxin $B_1$/10 g and 0.02 to 0.07 μg cyclopiazonic acid/10 g were produced compared to 5850 μg aflatoxin $B_1$/10 g and 34 μg cyclopiazonic acid/10 g in air alone (Taniwaki *et al.*, 2001). Also, a 7-to 20-fold decrease in production of roquefortine C by *P. roqueforti* in Cheddar cheese at 25°C was noted in the modified atmospheres (Taniwaki *et al.*, 2001). As new packaging materials and techniques are developed in the future, more research will undoubtedly be done on mycotoxin control in foods.

*Control of mycotoxins in foods through regulations*
In an effort to control the amount of mycotoxins in foods and feeds, regulatory action and guidance levels for different mycotoxins have been established in many countries. Regulations before 1990 have been reviewed by Gilbert (1991) and van Egmond (1991a) for the European Union (EU) and the United States of America (USA) and by van Egmond (1991a, b) for Africa and Asia. Aflatoxin was the major mycotoxin of concern with regulatory levels ranging from 0 to 50 μg/kg depending on the food and the country; however, there were some countries that regulated deoxynivalenol, ochratoxin A, patulin, T-2 toxin, and zearalenone (Gilbert, 1991; van Egmond, 1991a; b). In 1995, 77 countries around the world had some regulations for mycotoxins (FAO, 1997); however, 13 countries had no regulations and no data were available from 50 other countries (CAST, 2003). Boutrif and Canet (1998) and CAST (2003) list the countries with the mycotoxins they regulate. Currently

regulated mycotoxins include aflatoxins, deoxynivalenol, fumonisins, HT-2 toxin, ochratoxin, patulin, and zearalenone in products such as baby foods, barley, beans, dried fruits, fruit juices, grains, maize, milk and dairy products, nuts, rice, and wheat (FAO, 1997; Boutrif and Canet, 1998; CAST, 2003). Six factors that affect the establishment of regulations for the level of mycotoxins in foods are: production of mycotoxins in food commodities; distribution of these mycotoxins; methods to analyze foods for mycotoxins; data on toxicology; legislation in countries where food trade exists; and availability of food to feed people (Wood and Trucksess, 1998; van Egmond, 2002). Park (1995) noted that an effective food safety program for mycotoxin control includes not only regulations but also monitoring agricultural commodities pre-, during and post-harvest and testing final products for mycotoxins.

Although the control of aflatoxins $B_1$, $B_2$, $G_1$, and $G_2$ in peanuts was regulated in the United States of America (USA) by establishing acceptable limits with an action level of 30 µg/kg; this level was lowered in 1969 to 20 µg/kg for foods and feeds (Park and Troxell, 2002). However, levels for some feeds now range from 100 to 300 µg/kg and 0.5 µg/kg was set for milk (Nesheim and Wood, 1995). The Food and Durg Administration (FDA) sets action levels for aflatoxin as guidelines to enhance food safety and not as laws subject to court actions (Wood and Trucksess, 1998). The maximum levels for aflatoxin $B_1$ in animal feeds in the EU are much lower than those in the USA, ranging from 5 to 50 µg/kg depending on the animal species (Rustom, 1997). For human food, the level set by the Joint Food and Agriculture Organization/World Health Organization (FAO/WHO) Expert Committee was 0.05 µg/kg for $M_1$ in milk, 10 µg/kg for combined $B_1$, $B_2$, $G_1$, and $G_2$ in processed peanut products and 15 µg/kg for combined $B_1$, $B_2$, $G_1$, and $G_2$ in raw peanuts (Rustom, 1997). Blanc (2001) discussed the problems with regulations in the EU for aflatoxins in peanuts and pistachio nuts because there is variation in methods and lack of repeatability that caused a high rejection rate for these nuts among member states.

The FDA established guidelines for fumonisins (total of $B_1$, $B_2$ $B_3$) in maize products intended for human consumption that range from 2 to 4 mg/kg based on the health concerns noted in animals and for patulin in apple juice products at 50 µg/kg based on risk assessment and consumption by children under age 2 (Trucksess, 2001; USA Food and Drug Administration (US FDA), 2001a,b,c). Levels ranging from 5 to 100 mg/kg for fumonisins were set for animal feeds depending on the sensitivity of the animal species to this mycotoxin (Trucksess, 2001; USA FDA, 2001a). Although the FDA noted that there are no conclusive health effects in humans from fumonisins, the effects in animals were sufficient to set these guidelines (USA FDA, 2001a, b). Other countries have regulatory limits for deoxynivalenol, ergot alkaloids, ochratoxin A, patulin, and zearalenone (FAO, 2001). Based on this information, the FDA or other associations has made recommendations for deoxynivalenol, fumonisins, and patulin (Table 10.9). The co-occurrence of the *Fusarium* mycotoxins

(deoxynivalenol, fumonisins, fusaric acid, fusarin, moniliformin, zearalenone and other toxins) in grains complicates the evaluation of tolerances and potential regulatory limits (D'Mello *et al.*, 1997). Before further regulatory action can be taken, more information is needed on the levels of mycotoxins in foods, the human exposure to mycotoxins, and the toxicological effects in animals (Park, 1995; Wood and Trucksess, 1998). In addition, good analytical methods must be developed to survey foods.

### Control of mycotoxins through risk assessment and management

The hazards of mycotoxins for human health are now being determined using risk assessment (hazard identification, hazard characterization, assessment of exposure, and characterization of risk). Within the last 10 to 15 years, the WHO and FAO have led these activities related to risk assessment for mycotoxins in foods (Kuiper-Goodman, 1999). The FAO/WHO Joint Expert Committee on Food Additives (JECFA) has scientifically evaluated aflatoxins, patulin, fumonisins, trichothecenes, and ochratoxin A and is studying zearalenone (Boutrif and Canet, 1998; Herrman and Walker, 1999; JECFA 56th, 2001). Generally, hazard characterization results in a provisional maximal tolerable daily intake (PMTDI) when the mechanism of the toxicity shows a dose and effect at a threshold level (Kuiper-Goodman, 1999). The PMTDI value is extrapolated from animals to humans as the "no observed adverse effect level" (NOAEL), which is divided by 100 for added safety (Kuiper-Goodman, 1999). For genotoxic carcinogens (patulin), the levels "as low as is reasonable" (ALAR) are used to determine risk (Kuiper-Goodman, 1999). One problem with developing risks is the determination of exposure to mycotoxins because the epidemiology and use of reliable and accurate measurement for mycotoxins are lacking (Kuiper-Goodman, 1999). De Nijs *et al.* (1997) identified trichothecenes (deoxynivalenol, nivalenol, T-2 toxin and acetyl deoxynivalenol), zearalenone, and fumonisin $B_1$ plus 12 secondary metabolites produced by *Fusarium* species in raw materials that are used to make foods and proposed that these mycotoxins be used in all risk assessments. The agreement on what compounds are hazardous may create problems in doing risk assessment. Also, models for risk will need to be re-evaluated as new information is generated.

The hazards of the mycotoxins patulin and ochratoxin A have been determined by the FAO/WHO Joint Expert Committee on Food Additives (JECFA) and Provisional Tolerable Weekly Intake (PTWI) that include safety factors have been identified (van Egmond, 2002). JECFA has suggested As Low As Reasonably Achievable (ALARA) levels for aflatoxins because they are linked to cancer in humans (van Egmond, 2002). In addition, fumonisins, deoxynivalenol (T-2 and HT-2), and aflatoxin $M_1$ have been evaluated (JECFA 56th, 2001). Estimates of the daily intake of mycotoxins in different EU countries have been done for ochratoxin A with levels to be assessed for beer, cereals, coffee, and wine, and plans

**Table 10.9** Regulatory guidelines for some major mycotoxins set by the Food and Drug Administration (FDA) or other associations[a]

| Mycotoxin | Human foods | Animal feeds |
|---|---|---|
| Aflatoxin $B_1$ | 20 µg/kg | Corn/peanut for breeding animals 100 µg/kg |
| | | Corn/peanut for finishing swine 200 µg/kg |
| | | Corn/peanut for finishing beef 300 µg/kg |
| | | Cottonseed meal 300 µg/kg |
| | | Corn for young and all other feed 20 µg/kg |
| Aflatoxin $M_1$ | 0.5 µg/kg | |
| Deoxynivalenol | 1 µg/g – wheat products | Swine 5 µg/g |
| | | Beef 10 µg/g |
| | | Poultry 10 µg/g |
| | | Other animals 10 µg/g |
| Fumonisin[b,c] (total $B_1$ + $B_2$ + $B_3$) | 2 µg/g –degermed dry milled corn products | Horses ≤ 5 ppm |
| | 4 µg/g – whole or partially degermed dry milled corn products | Swine ≤10 ppm |
| | 4 µg/g – dry milled corn bran | Poultry or beef ≤ 50 ppm |
| | 4 µg/g – clean corn for masa production | Equids and Rabbits – 5 µg/g |
| | 3 µg/g –clean corn for popcorn | Swine and catfish – 20 µg/ |
| | | Breeding ruminants, poultry, and mink – 30 µg/g |
| | | Ruminants ≥ 3 months old raised for slaughter and mink being raised for pelt production – 60 µg/g |
| | | Poultry being raised for slaughter – 100 µg/g |
| | | All other livestock and pets – 10 µg/g |
| Ochratoxin A | None set by FDA | |
| Patulin | 50 µg/kg | |
| | 30–50 µg/kg (worldwide) | |

[a]Wood and Trucksess (1998).
[b]Suggested by American Association of Veterinary Laboratory Diagnosticians.
[c]Anonymous (2000) Guidance for Industry: Fumonisin levels in human food and animal feed. Final Guidance USFD, CFSAN, CVM. URL: http://www.cfsan.fda.gov/~dms/fumongu2.html.

are under way for deoxynivalenol, fumonisins, and zearalenone (van Egmond, 2002). Tolerable level of risk (TLR) can differ among countries. The TLR for aflatoxin in peanuts may be 0.04 cancers/$10^6$ population in Europe, but in Indonesia may be 100 cancers/$10^6$ population because there is higher exposure to aflatoxin due to greater consumption of contaminated peanuts (ICMSF, 2002)

Risk assessments have been done in some countries using either actual consumption data or estimates of consumption patterns. Gauchi and Leblanc (2002) used the Monte Carlo simulation, which included the nonparametric-nonparametric method, the parametric method of simulation and bootstrap confidence intervals, to determine the exposure of the French population to ochratoxin A from 8 groups of foods (cereals, coffee, fruit juices, pork, poultry, raisins, other dried fruits, wines) over one week. Food consumption and contamination data in this study were variable because not everyone consumed all foods and the contamination levels were taken from a previous project and not analyzed in this research. The calculated risk for nephrotoxicity was 1.8 ng/kg body weight/day; however, a previous risk analysis for ochratoxin A done in France was 1.3 ng/kg body weight/day (Gauchi and Leblanc, 2002). It was estimated by de Nijs et al. (1998) that half of the people of the Netherlands were possibly exposed to 1000 ng fumonisin $B_1$/day if they ate at least 3 g maize and people with celiac or Dürhring's disease were more at risk because they ate

more maize daily. Humphreys et al. (2001) did quantitative risk assessment for fumonisins $B_1$ and $B_2$ in maize consumed in the United States by using FDA surveillance data for the mycotoxins, USDA food intake surveys, dose-response based on rats and nephrotoxicity, extrapolation to humans, and model simulations. They concluded that risk management of the amount of maize eaten/day would be better than trying to decrease the level of fumonisins in the maize.

The intake of mycotoxins can be determined by direct analysis of levels in body fluids or tissues or by analyzing for a marker that accompanies the mycotoxin. To assess dietary intake of fumonisins, the ratio of sphinganine (Sa) to sphingosine (So) in blood or urine has been proposed because fumonisins inhibit sphingolipid biosynthesis (Crews et al., 2001). Since sphingolipids play a role in cell membrane function, the Sa/So biomarkers can monitor the biological effect in the body. Biomarkers (aflatoxin $M_1$ and aflatoxin $B_1$-$N^7$-guanine adducts) have been used to study toxicities of aflatoxin $B_1$ for quantitative risk assessment and determination of dietary intake by analyzing urine and milk for these markers (Crews et al., 2001). Recently, a urinary biomarker has been used to assess DON exposure (Meky et al., 2003). Turner et al. (1999) discussed the use of biomarkers such as free fumonisins in urine and the Sa/So ratio in urine and blood to determine the health risks of fumonisins in humans. Also, the use of human urine combined with analysis of duplicate diets gave better estimates of the human exposure to ochratoxin A than did blood levels and dietary intake (Gilbert *et al.*, 2001). Sewram *et al.* (2001) suggested using hair to determine the exposure to fumonisins. Biomarkers must be easily measured, sensitive to low levels of the compound, specific to a given mycotoxin, and not affected by environmental factors such as diet, infections, or chronic disease states (Turner *et al.*, 1999).

WHO and FAO have been concerned about food safety and have done some risk analysis on mycotoxins through the Codex Alimentarius Commission, which was established to protect human health and foster fair trade (Moy, 1998). The Joint FAO/WHO Expert Committee on Food Additives (JECFA) and the International Agency for Research on Cancer (IARC) have studied the risks and hazards of mycotoxins with only aflatoxin being designated as a carcinogen for humans (Moy, 1998). The exposure to mycotoxins is being surveyed by 70 countries that belong to the Global Environmental Monitoring System/Food Contamination Monitoring and Assessment Program (GEMS/Food) with most efforts focused on aflatoxin (Moy, 1998). Work has been done on risk characterization, especially for ochratoxin A and nephropathy in humans and pigs, patulin in apple juice, and other mycotoxins (Moy, 1998). The Codex Committee on Food Additives and Contaminants (CCFAC) tried to determine maximum limits for aflatoxins; however, there was not agreement among nations (Moy, 1998). Also, for aflatoxins, the presence of either hepatitis B or C virus may be necessary for liver cancer because worldwide it has been estimated that 50 to 100% of liver cancer is associated with infection from hepatitis B virus in developing countries and from hepatitis C in developed countries (ICMSF, 2002). The relationship between aflatoxin, hepatitis B and primary liver cancer were studied by Bowers *et al.* (1993) who concluded from the multistage model that the lifetime cancer potency was 230 $(mg/kg/day)^{-1}$ for populations with hepatitis B and 9 $(mg/kg/day)^{-1}$ for populations without hepatitis B. Gorelick *et al.* (1994) reviewed risk assessment for aflatoxins based on animal data and estimated a virtually safe dose as 0.16 ng/kg body weight/day or a life risk of $1.0 \times 10^{-5}$ based on studies with rats. For liver cancer in the southeast USA, it was estimated that exposure was 110 ng aflatoxin/kg body weight/day and that translated into a lifetime risk of $6.88 \times 10^{-2}$ and 98 liver cancers/100,000/year compared to the current rates of 3.4 for males and 1.3 for females (Gorelick *et al.*, 1994). These differences could be due to the differences between species in metabolism and activation of aflatoxin. If the risk is based on epidemiological data, then the value becomes $2.17 \times 10^{-6}$ times the nanograms of aflatoxin/kg body weight/day (Gorelick *et al.*, 1994). More information is needed before risk assessment can be used for other mycotoxins.

Kuiper-Goodman (1994) has proposed that determining the risk to humans of a given mycotoxin for the Canadian population and then managing that mycotoxin in food could prevent health effects in humans. Nuts and products made from them were of most concern for aflatoxin $B_1$ in Canada;

therefore, sorting, roasting, and ingredient blending for peanut butter all reduced the level of aflatoxin $B_1$ (Kuiper-Goodman, 1994). The estimated risk for intake of aflatoxin $B_1$ in a 1- to 11-year-old child was 1–2 ng/kg body weight/day; however, this was higher than the total dietary intake (TDI) of 0.11 to 0.19 ng/kg body weight/day with a carcinogenic risk of 1 in 100,000 calculated for Asia and Africa (Kuiper-Goodman, 1994). The link of aflatoxin and hepatitis B to liver cancer may mean that a higher intake of aflatoxin in Canada may not be a health risk because hepatitis B is not a concern (Kuiper-Goodman, 1994). For zearalenone, there are no guidelines or regulation in Canada; however, the dietary exposure was assessed at 50 –100 ng/kg body weight/day, which was within the estimated TDI of 100 ng/kg body weight/day (Kuiper-Goodman, 1994). For ochratoxin A, the exposure for children from pork and cereal foods was 3.5 ng/kg body weight/day and for the overall population, ranged from 1.5 to 5.7 ng/kg body weight/day with a proposal by the Health Protection Branch to try to limit intake to 4 ng/kg body weight/day (Kuiper-Goodman, 1994). Concern for DON in wheat lead the Canadian government to establish guidelines for uncleaned wheat at 2.0 μg and for infant foods at 1 μg to achieve a TDI of 1.5 μg/kg body weight/day for children and 3.0 μg/kg body weight/day for adults (Kuiper-Goodman, 1994). Studies have been done for the exposure of the population to fumonisins in Canada (Kuiper-Goodman *et al.*, 1996).

JECFA (2002) calculated that there would be 29 cancers/1000 people/year in populations with 1% hepatitis B viral infection when the maximum acceptable level of aflatoxin $M_1$ is 0.5 μg/kg instead of 0.05 μg/kg. The Committee recommended that the vaccination of populations susceptible to hepatitis B virus could reduce incidences of liver cancer. The Committee recommended more research on fumonisins before further risk assessment can be done, especially on the relationship between intake of nixtamalized maize products and human diseases such as neural tube defects, hepatic and renal diseases, and nasopharyngeal and esophageal cancers. A PMTDI of 2 μg/kg of body weight of total fumonisins was suggested by this committee based on renal toxicity seen in rats. The Committee did not change the PWTI recommendation of 100 ng/kg of body weight for ochratoxin A because of its link to renal carcinogenicity and nephrotoxicity; however, they recommended more research on the occurrence of ochratoxin A worldwide and on the ecology of the fungi that produce it and on the mechanisms that cause the renal diseases (JECFA, 2002). Although a 2-year study with mice fed deoxynivalenol did not show that it was carcinogenic; a PMTDI of 2 μg/kg of body weight was suggested because deoxynivalenol can cause acute illness in humans (JECFA, 2002). A TDI of 1.1 μg/kg of body weight of deoxynivalenol was proposed by Pieters *et al.* (2002) who studied dietary intake in The Netherlands. Although this TDI would limit the concentration of deoxynivalenol in wheat to 129 μg/kg, levels of 446 μg/kg was the average amount of deoxynivalenol in wheat in The Netherlands from 1998–2000 (Pieters *et al.*, 2002). Since humans are exposed to deoxynivalenol in foods, it was recommended that studies be done on the presence in foods, effects of processing, dietary intake, combination of mycotoxins and toxicity, and study of carcinogenicity in other laboratory species (JECFA, 2002). T-2 toxin was immunotoxic and haematotoxic based on several short-term dietary intakes of T-2 toxin in several animal species; therefore, a PMTDI of 60 ng/kg of body weight was suggested for the T-2 toxin and its metabolite HT-2 toxin, either alone or combined (JECFA, 2002). Long-term studies in other animals, especially pigs, were recommended for the T-2 and HT-2 toxins plus combinations with other trichothecenes (JECFA, 2002). In addition, more information on intake of T-2 and HT-2 toxins is needed from non-European countries to determine if the PMTDI is reasonable (JECFA, 2002). Thuvander et al. (1999) reported that there were both additive toxicities and inhibition in human lymphocytes depending on the combinations of trichothecene mycotoxins studied; therefore, total intake of different mycotoxins may be needed when determining risks. D'Mello et al. (1997) also mentioned the co-occurrence of the *Fusarium* mycotoxins in grains and the need to evaluate combinations of mycotoxins in any risk assessment because co-occurrence complicates the determination of health effects. Continued research is needed on mycotoxins and human health to improve risk assessment.

There is concern for ochratoxin A intake because it is slowly eliminated from the human body with a half-life of about 35 days, possibly because this mycotoxin is either reabsorbed or bound to protein (Petzinger and Weidenbach, 2002). Kuiper-Goodman (1996) updated the risk assessment for ochratoxin A and noted that hazard assessment based on animal studies is a concern because humans usually consume lower doses of the mycotoxin but over longer times than do animals. There is worldwide exposure to ochratoxin A as determined by the high incidence of this mycotoxin in human sera due to its long half-life of 840 h (Kuiper-Goodman, 1996). The provisional tolerable dietary intake (PTDI) of 3.7 ng/kg body weight/day presents an incremental risk of 1 renal cancer/100,000 population when large safety factors are used to account for toxic properties of ochratoxin A (Kuiper-Goodman, 1996). Gilbert et al. (2001) stated that ochratoxin A occurred in food in less than nanogram quantities per gram of several foods (grains, beans, dried fruits, pig kidneys, milk, etc.), which makes exposure determinations based on levels in foods difficult. The daily exposure levels to ochratoxin A were calculated as 0.26 to 3.54 ng/kg body weight/day using the food intake and body weights of the test subjects (Gilbert et al., 2001). Zimmerli and Dick (1996) found a median of 10 pg of ochratoxin A/mL of wine and calculated an intake in men of 0.7 ng ochratoxin A/kg body weight; hence, normal wine consumption should not increase the risk for renal cancer. Ochratoxin A was detected in red grape juice at an estimated 240 pg/mL; therefore, daily consumption by infants or children could result in an intake of 3–3.6 ng/kg body weight (Zimmerli and Dick, 1996). Based on these worst case estimates, it was suggested that the

amount of ochratoxin A in red grape juice may need to be decreased to ≤100 pg/mL (Zimmerli and Dick, 1996).

Expert Committees were formed by the Codex Committee on Food Additives and Contaminants to do quantitative risk assessment for several mycotoxins. For aflatoxin $M_1$, the committee was charged with determining the risks if the maximum level were set at 0.05 μg/kg for Western Europe and 0.5 μg/kg for the United States (Henry et al., 2001). After reviewing the biological and toxicological data, analytical methods, effects of processing, food contamination, dietary intake estimates, prevention and control, dose-response and potential cancer, the committee concluded that at the 0.5 μg/kg level and contamination of all milk that only 29 additional cancers/1000 million persons/year would occur in a population with a 1% hepatitis B rate of infection (Henry et al., 2001). Rats fed pure fumonisins $B_1$ developed renal toxicity; therefore, the committee recommended a maximum tolerable intake of total fumonisins ($B_1$, $B_2$, $B_3$) of 2 μg/kg body weight/day based upon a no observed effect level (NOEL) of 0.2 mg/kg body weight/day (Bolger et al., 2001). The committee also recommended more studies to determine if fumonisins $B_1$ could be linked to nasopharyngeal and esophageal cancers in adults and neural tube defects in infants in areas of the world where maize consumption was high (Bolger et al., 2001). After evaluating all new information and data on ochratoxin A, the committee concluded that it needed more information about the mechanisms for renal carcinogenicity and nephrotoxicity before it could suggest a maximum for cereal products (Benford et al., 2001). When the committee evaluated the safety of deoxynivalenol for human health, it established a PMTDI of 1 μg/kg body weight (Canady et al., 2001a). Additional recommendations were made to study toxicities in different species with deoxynivalenol and combinations of trichothecenes, to evaluate areas of the world where moldy maize or scabby wheat are endemic, to develop better analytical methods, to determine dietary intake by generating more information on effects of processing, and to develop ways to prevent *Fusarium* infection in grains (Canady et al., 2001a). T-2 was evaluated for the first time and a PMTDI of 60 ng/kg body weight/day was proposed based on a dietary study in pigs because short periods (3 weeks) of intake resulted in immunotoxic and haematotoxic effects (Canady et al., 2001b). Further recommendations for T-2 toxin were to study the carcinogenic potential in several species including pigs, to use combinations of trichothecene mycotoxins, and to obtain more information on dietary intake (Canady et al., 2001b).

Risk analysis can be complicated by other compounds or defects in the agricultural commodity. Hicks et al. (2000) discussed the implication of calculating risks for mycotoxins that included those for pesticide residues. They calculated risks for deoxynivalenol and benomyl by using a comparative risk calculation from the ratio of the two chemicals and noted that the presence of the fungicide reduced the risk for the mycotoxin. Johnson et al. (2001) examined the risk in merchandising grain when it had defects, such as vomitoxin in wheat. Many factors need to be considered when doing risk analysis on mycotoxins.

Trucksess and Pohland (2001) have edited a book on protocols for mycotoxin analysis in an effort to provide the scientific basis for determining the exposure data that can be used in developing risk assessments for mycotoxins. This collection includes information on sampling and isolation of mycotoxins, preparation of standards, and methods for the major mycotoxins produced by *Aspergillus, Fusarium, Penicillium* and other genera. The methods included in this book are based on capillary electrophoresis, chromatography (ESMS, GC, GC-ECD, GC-MS, GLC, HPLC, LC, TLC), fluorometry, and immunochemistry.

## Conclusions

In conclusion, many mycotoxins have been or will be associated with different human or animal diseases. However, when dealing with foodborne chronic diseases, it is very difficult to establish causation because unlike acute diseases, the onset of symptoms seldom occurs concurrent with exposure. It also is certain that most mycotoxin-associated diseases are probably multifactorial and that in many cases, mycotoxins are probably not the prime causes but could be factors that increase susceptibility to the disease. Finally, since mycotoxins are often suspected agents in farm animal disease outbreaks and production problems in developed countries, this should serve as a warning for the potential of mycotoxins to be agents of disease of unknown etiology in humans consuming low quality commodities in large amounts, as is sometimes the case in many less developed countries. All these uncertainties about mycotoxins and their effects on animal and human health mean that research on the control of mold growth and mycotoxin production in agricultural commodities must continue. Also, monitoring foods for mycotoxins will continue to be necessary to develop scientifically sound regulations and appropriate risk assessments. These activities, in turn, require that new, more sensitive and reliable methods be developed for mycotoxin detection.

## References

Abarca, M.L., Accensi, F., Bragulat, M.R., Castellá, G., and Cabañes, F.J. 2003. *Aspergillus carbonarius* as the main source of ochratoxin A contamination in dried vine fruits from the Spanish market. J. Food Prot. 66: 504–506.

Abouzied, M.M. and Pestka, J.J. 1994. Simultaneous screening of fumonisin B$_1$, aflatoxin B$_1$, and zearalenone by line immunoblot: A computer-assisted multianalyte assay system. J. AOAC Int. 77: 495–500.

Abouzied, M.M., Azcona, J.I., Braselton, W.E., and Pestka, J.J. 1991. Immunochemical assessment of mycotoxins in 1989 grain foods: Evidence for deoxynivalenol (vomitoxin) contamination. Appl. Environ. Microbiol. 57: 672–677.

Abramson, D. 1997. Toxicants of the genus *Penicillium*. In: Handbook of Plant and Fungal Toxicants. J.P.F. D'Mello, ed. CRC Press, Boca Raton, FL, pp. 303–317.

Abramson, D. 1998. Mycotoxin formation and environmental factors. In: Mycotoxins in Agriculture and Food Safety. K. K Sinha and D. Bhatnagar, ed. Marcel Dekker, Inc., New York, pp. 225–277.

Abramson, D., Gan, Z., Clear, R.M., Gilbert, J., and Marquardt, R. R. 1998. Relationships among deoxynivalenol, ergosterol and *Fusarium* exoantigens in Canadian hard and soft wheat. Int. J. Food Microbial.45: 217–224.

Accerbi, M., Rinaldi, V.E.A., and Ng, P.K.W. 1999. Utilization of highly deoxynivalenol-contaminated wheat via extrusion processing. J. Food Prot. 62: 1485–1487.

Alberts, J.F., Gelderbloom, W.C.A., Thiel, P.G., Marasas, W.F.O., van Schalkwyk, D.J., andBehrend, Y. 1990. Effects of temperature and incubation period on production of fumonisin B$_1$ by *Fusarium moniliforme*. Appl. Environ. Microbiol. 56: 1729–1733.

ApSimon, J. W. 2001. Structure, synthesis, and biosynthesis of fumonisin B$_1$ and related compounds. Environ. Health Perspect. 109: 245–249.

Autrup, H. and Autrup, J.L. 1992. Human exposure to aflatoxins – biological monitoring. In: Handbook of Applied Mycology. Volume 5: Mycotoxins in Ecological Systems. D. Bhatnagar, E.B. Lillehoj, and D.K. Arora, ed. Marcel Dekker, Inc., New York, pp. 213–230.

Azcona-Olivera, J.I., Ouyang, Y., Murtha, J., Chu, F.S. and Pestka, J.J. 1995a. Induction of cytokine mRNAs in mice after oral exposure to the trichothecene vomitoxin deoxynivalenol: relationship to toxin distribution and protein synthesis inhibition. Toxicol. Appl. Pharmacol. 133: 109–120.

Azcona-Olivera, J.I., Ouyang, Y.L., Warner, R.L., Linz, J.E. and Pestka, J.J. 1995b. Effects of vomitoxin deoxynivalenol and cycloheximide on IL-2,4, 5, and 6 secretion and mRNA levels in murine CD4$^+$ cells. Fd. Chem. Toxicol. 35: 433–441.

Babu, E., Takeda, M., Narikawa, S., Kobayashi, Y., Enomoto, A., Tojo, A., Cha, S. H., Sekine, T., Sakthisekaran, D., and Endou, H. 2002. Role of human organic ion transporter 4 in the transport of ochratoxin A. Biochim. Biophys. Acta 1590: 64–75.

Bacon, C.W., Yates, I.E., Hinton, D.M., and Meredith, F. 2001. Biological control of *Fusarium moniliforme* in maize. Environ. Health Perspect. 109: 325–332.

Badiani, K., Byers, D.M., Cook, H.W., and Ridgway, N.D. 1996. Effect of fumonisin B$_1$ on phosphatidylethanolamine biosynthesis in Chinese hamster ovary cells. Biochim. Biophys. Acta 1304: 190–196.

Bakan, B., Melcion, D., Richard-Molard, D., and Cahagnier, B. 2002. Fungal growth and *Fusarium* mycotoxin content in isogenic traditional maize and genetically modified maize grown in France and Spain. J. Agric. Food Chem. 50: 728–731.

Bamburg, J.R. 1983. Biological and biochemical actions of trichothecene mycotoxins. Prog. Mol. Subcell. Biol. 8: 41–110.

Barrios, M.J., Gualda, M. J., Cabanas, J.M., Medina, L.M., and Jordano, R. 1996. Occurrence of aflatoxin M$_1$ in cheeses from the south of Spain. J. Food Prot. 59: 898–900.

Baudrimont, I., Sostaric, B., Yenot, C., Betbeder, A.M., Dano-Djedje, S., Sanni, A., Steyn, P.S., and Creppy, E.E. 2001. Aspartame prevents the karyomegaly induced by ochratoxin A in rat kidney. Arch. Toxicol. 75: 176–183.

Bayman, P., Baker, J.L., Doster, M.A., Michailides, T.J., and Mahoney, N.E. 2002. Ochratoxinproduction by the *Aspergillus ochraceus* group and *Aspergillus alliaceus*. Appl. Environ.Microbiol. 68: 2326–2329.

Beardall, J.M. and Miller, J.D. 1994. Diseases in humans with mycotoxins as possible causes. In: Mycotoxins in Grain. Compounds Other Than Aflatoxin. J.D. Miller and H.L. Trenholm. ed. Eagan Press, St. Paul, MN, pp. 487–539.

Beattie, S., Schwarz, P.B., Horsley, R., Barr, J., and Casper, H.H. 1998. The effect of grain storage conditions on the viability of *Fusarium* and deoxynivalenol production in infested malting barley. J. Food Prot. 61: 103–106.

Beaver, R. W. 1991. Decontamination of mycotoxin-containing foods and feedstuffs. Trends Food Sci. Technol. 2: 170–173.

Beaver, R.W., James, M.A., and Lin, T.Y. 1991. Comparison of an ELISA-based screening test with liquid chromatography for the determination of aflatoxins in corn. J. Assoc. Off. Anal. Chem 74: 827–829.

Benford, D., Boyle, C., Dekant, W., Fuchs, R., Gaylor, D.W., Hard, G., McGregor, D.B., Pitt, J.I., Plestina, R., Shephard, G., Solfrizzo, M., Verger, P.J.P., and Walker, R. 2001. Ochratoxin A. In: Safety

Evaluation of Certain Mycotoxins in Food. WHO Food Additives Series: 47and FAO Food and Nutrition Paper 74. Joint FAO/WHO Expert Committee on Food Additives, ed. International Programme on Chemical Safety, World Health Organization, Geneva, Switzerland, pp. 281–415

Bennett, G.A. and Richard, J.L. 1994. Liquid chromatographic method for analysis of the naphthalene dicarboxaldehyde derivative of fumonisins. J. AOAC Int. 77: 501–506.

Bennett, G.A. and Richard, J.L. 1996. Influence of processing on *Fusarium* mycotoxins in contaminated grains. Food Technol.50(5): 235–238.

Betina, V. 1993. Thin-layer chromatography of mycotoxins. In: Chromatography of Mycotoxins. Techniques and Applications. V. Betina, ed. Elsevier, New York, pp. 141–251

Bhat, R.V., Beedu, S.R., Ramakrishna, Y., and Munshi, K.L. 1989. Outbreak of trichothecene mycotoxicosis associated with consumption of mould-damaged wheat production in Kashmir Valley, India. Lancet 1: 35–37.

Bhat, R.V., Prathaphumar, H.S., Rao, P.A., and Rao, V.S. 1997. A foodborne disease outbreak due to the consumption of moldy sorghum and maize containing fumonisin mycotoxins. Clin. Tox. 35: 249–255.

Bhatnagar, D. and García, S. 2001. *Aspergillus*. In Guide to Foodborne Pathogens. R.G. Labbé and S. García, ed. Wiley-Interscience, New York, pp. 35–50.

Bhatnagar, D., Ehrlich, K.C. and Cleveland, T.E. 2003. Molecular genetic analysis and regulation of aflatoxin biosynthesis. Appl. Microbiol. Biotech. 61: 83–93

Bhatnagar, D., Lillehoj, E.B., and Bennett, J.W. 1991. Biological detoxification of mycotoxins. In: Mycotoxins and Animal Foods. J.E. Smith and R.S. Henderson, ed. CRC Press, Boca Raton, FL, pp. 815–826.

Bilgrami, K.S. and Choudhary, A.K. 1998. Mycotoxins in preharvest contamination of agricultural crops. In: Mycotoxins in Agriculture and Food Safety. K.K. Sinha and D.Bhatnagar, ed. Marcel Dekker, Inc., New York, pp. 1–43.

Bilgrami, K.S. and Sinha, K.K. 1992. Aflatoxins: Their biological effects and ecological significance. In: Handbook of Applied Mycology. Volume 5: Mycotoxins in Ecological Systems. D. Bhatnagar, E.B. Lillehoj, and D.K. Arora, ed. Marcel Dekker, Inc., New York. p. 59–86.

Bissessur, J., Permaul, K., and Odhav, B. 2001. Reduction of patulin during apple juice clarification. J. Food Prot. 64: 1216–1219.

Blanc, M. 2001. Législation communautaire sur les aflatoxines: incidences sur le commerce de l'arachide de bouche et de la pistache. Food Nutr. Agric. 28: 16–25

Blanc, M., Pittet, A., Muñoz-Box, R., and Viani, R. 1998. Behavior of ochratoxin A during green coffee roasting and soluble coffee manufacture. J. Agric. Food. Chem. 46: 673–675.

Blount, W.P. 1961. Turkey "X" Disease. Turkeys 9: 52–77.

Bluhm, B.H., Flaherty, J.E., Cousin, M.A., and Woloshuk, C.P. 2002. Multiplex polymerase chain reaction assay for the differential detection of trichothecene- and fumonisin-producing species of *Fusarium* in cornmeal. J. Food Prot. 65: 1955–1961.

Bodine, A.B. 1999. The menace of mycotoxins: From nature's bounty …nature's bane. Assoc. Food Drug Off. 63: 42–59.

Boeira, L.S., Bryce, J.H., Stewart, G.G., and Flannigan, B. 1999a. Inhibitory effect of *Fusarium* mycotoxins on growth of brewing yeasts. 1. Zearalenone and fumonisin B$_1$. J. Inst. Brew.105: 366–374.

Boeira, L.S., Bryce, J.H., Stewart, G.G., and Flannigan, B. 1999b. Inhibitory effect of *Fusarium* mycotoxins on growth of brewing yeasts. 2. Deoxynivalenol and nivalenol. J. Inst. Brew. 105: 376–381.

Boeira, L.S., Bryce, J.H., Stewart, G.G., and Flannigan, B. 2000. The effect of combinations of *Fusarium* mycotoxins (deoxynivalenol, zearalenone and fumonisin B$_1$) on growth of brewing yeasts. J. Appl. Microbiol. 88: 388–403.

Bolger, M., Coker, R.D., DiNovi, M., Gaylor, D., Gelderblom, W., Olsen, M., Paster, N., Riley, R.T., Shephard, G., and Speijers, G.J.A. 2001. Fumonisins. In: Safety Evaluation of Certain Mycotoxins in Food. WHO Food Additives Series: 47and FAO Food and Nutrition Paper 74. Joint FAO/WHO Expert Committee on Food Additives, ed. International Programme on Chemical Safety, World Health Organization, Geneva, Switzerland, pp. 103–279.

Bondy, G.S. and Pestka, J.J. 2000. Immunomodulation by fungal toxins. J. Toxicol. Environ. Health, part B 3: 109–143

Bosch, U. and Mirocha, C.J. 1992. Toxin production by *Fusarium* species from sugar beets and natural occurrence of zearalenone in beets and beet fibers. Appl. Environ. Microbiol. 58: 3233–3239.

Bothast, R.J., Bennett, G.A., Vancauwenberge, J.E., and Richard, J.L. 1992. Fate of fumonisin B$_1$ in naturally contaminated corn during ethanol fermentation. Appl. Environ. Microbiol. 58: 233–236.

Boudra, H., Le Bars, P., and Le Bars, J. 1995. Thermostability of ochratoxin A in wheat under two moisture conditions. Appl. Environ. Microbiol. 61: 1156–1158.

Boutrif, E., and Canet, C. 1998. Mycotoxin prevention and control: FAO programmes. Revue Méd. Vét. 149: 681–694.

Bowers, J., Brown, B., Springer, J., Tollefson, L., Lorentzen, R., and Henry, S. 1993. Risk assessment for aflatoxin: An evaluation based on the multistage model. Risk Anal. 13: 637-642.

Boyacioğlu, D., Hettiarachchy, N.S., and D'Appolonia, B.L. 1993. Additives affect deoxynivalenol (vomitoxin) flour during breadbaking. J. Food Sci. 58: 416–418.

Boyacioglu, D., Hettiarachchy, N.S., and Stack, R.W. 1992. Effect of three systemic fungicides on deoxynivalenol (vomitoxin) production by *Fusarium graminearum* in wheat. Can. J. Plant Sci. 72: 93–101.

Breitholtz-Emanuelsson, A., Olsen, M., Oskarsson, A., Palminger, I., and Hult, K. 1993.Ochratoxin A in cow's milk and in human milk with corresponding human blood samples. J. AOAC Int. 76: 842–846.

Brenneman, T.B., Wilson, D.M., and Beaver, R.W. 1993. Effects of diniconazole on *Aspergillus* populations and aflatoxin formation in peanut under irrigated and nonirrigated conditions.Plant Dis. 77: 608–612.

Brera, C. and Miraglia, M. 1996. Quality assurance in mycotoxin analysis. Microchem. J. 54: 465–471.

Brown, R.L., Bhatnagar, D., Cleveland, T.E., and Cary, J.W. 1998. Recent advances in preharvest prevention of mycotoxin contamination. In: Mycotoxins in Agriculture and Food Safety. K.K. Sinha and D. Bhatnagar, ed. Marcel Dekker, Inc., New York, pp. 351–379.

Bruininki, A., Rasonyi, T., and Sidler, C. 1997. Reduction of ochratoxin A toxicity by heat-induced epimerization. *In vitro* effects of ochratoxins on embryonic chick meningeal and other cell cultures. Toxicology. 118: 205–210.

Bryden, W.L. 1991. Occurrence and biological effects of cyclopiazonic acid. In: Emerging Food Safety Problem Resulting from Microbial Contamination. K. Mise and J.L. Richard, ed. Ministry of Health and Welfare, Tokyo, pp. 127–147.

Bucheli, P., Kanchanomai, C., Meyer, I., and Pittet, A. 2000. Development of ochratoxin A during Robusta (*Coffea canephora*) coffee cherry drying. J. Agric. Food Chem. 48: 1358-1362.

Bucheli, P., Meyer, I., Pittet, A., Vuataz, G., and Viani, R. 1998. Industrial storage of green Robusta coffee under tropical conditions and its impact on raw material quality and ochratoxin A content. J. Agric. Food Chem. 46: 4507–4511.

Buchta, M. and Cvak, L. 1999. Ergot alkaloids and other metabolites of the genus *Claviceps*. In: Ergot. The Genus *Claviceps*. V. Křen and L. Cvak, ed. Hardwood Academic Publishers, Amsterdam, The Netherlands, pp. 173–200.

Bullerman, L. B. 2001a. Fusaria and Toxigenic Molds Other than Aspergilli and Penicillia. In:Food Microbiology. Fundamentals and Frontiers. 2$^{nd}$ edition. M. P. Doyle, L. R. Beuchat, and T. J. Montville, ed. ASM Press, Washington, D. C. p. 481–497.

Bullerman, L.B. 2001b. *Fusarium*. In: Guide to Foodborne Pathogens. R.G. Labbé and S, García, ed. Wiley-Interscience, New York, pp. 87–98.

Bullerman, L.B. and Tsai, W.-Y. J. 1994. Incidence and levels of *Fusarium moniliforme*, *Fusarium proliferatum* and fumonisins in corn and corn-based foods and feeds. J. Food Prot. 57: 541–546.

Burda, K. 1992. Incidence of patulin in apple, pear, and mixed fruit products marketed in New South Wales. J. Food Prot. 55: 796–798.

Burgess, L.W., Summerell, B.A., Bullock, S., Gott, K.P., and Backhouse, D. 1994. Laboratory Manual for *Fusarium* Research, 3rd edition. University of Sydney, Sydney, Australia.

Butler, W.H. and Barnes, J.M. 1964. Toxic effects of groundnut meal containing aflatoxin to rats and guinea-pigs. Brit. J. Cancer 17: 699–703.

Cabañes, F.J., Accensi, F., Bragulat, M.R., Abarca, M.L., Castellá, G., Minguez, S., and Pons, A. 2002. What is the source for ochratoxin A in wine? Int. J. Food Microbiol. 79: 213–215.

Cahagnier, B., Melcion, D., and Richard-Molard, D. 1995. Growth of *Fusarium moniliforme* and its biosynthesis of fumonisin B1 on maize grain as a function of different water activities. Lett. Appl. Microbiol. 20:247–251.

Canady, R.A., Coker, R.D., Egan, S.K., Krska, R., Kuiper-Goodman, T., Olsen, M., Pestka, J., Resnik, S., and Schlatter, J. 2001a. Deoxynivalenol. In: Safety Evaluation of Certain Mycotoxins in Food. WHO Food Additives Series: 47and FAO Food and Nutrition Paper 74. Joint FAO/WHO Expert Committee on Food Additives, ed. International Programme on Chemical Safety, World Health Organization, Geneva, Switzerland, pp. 419–555.

Canady, R.A., Coker, R.D., Egan, S.K., Krska, R., Olsen, M., Resnik, S., and Schlatter, J. 2001b. T-2 and HT-2 toxins. In: Safety Evaluation of Certain Mycotoxins in Food. WHO Food Additives Series: 47and FAO Food and Nutrition Paper 74. Joint FAO/WHO Expert Committee on Food Additives, ed. International Programme on Chemical Safety, World Health Organization, Geneva, Switzerland, pp. 557–680.

Carlson, D.B., Williams, D.E., Spitsbergen, J.M., Ross, P.F., Bacon, C.W., Meredith, F.I., and Riley, R.T. 2001. Fumonisin B1 promotes aflatoxin B1 and N-methyl-N'-nitro-nitrosoguanidine initiated liver tumors in rainbow trout. Toxicol. Appl. Pharmacol. 172: 29–36.

CAST (Council for Agricultural Science and Technology). 2003. Mycotoxins: Risks in Plant and Animal Systems, Task Force Report 139. Council for Agricultural Science and Technology, Ames, Iowa.

Castegnaro, M. and McGregor, D. 1998. Carcinogenic risk assessment of mycotoxins. Revue Méd. Vét. 149: 671–678.

Castellari, M., Versari, A., Fabiani, A., Parpinello, G.P., and Galassi, S. 2001. Removal of ochratoxin A in red wines by means of adsorption treatments with commercial fining agents. J. Agric. Food Chem. 49: 3917–3921.

Castelo, M.M., Jackson, L.S., Hanna, M.A., Reynolds, B.H., and Bullerman, L.B. 2001. Loss of fuminosin B1 in extruded and baked corn-based foods with sugars. J. Food Sci. 66: 416–421.

Castelo, M.M., Katta, S.K., Sumner, S.S., Hanna, M.A., and Bullerman, L.B. 1998a. Extrusion cooking reduces recoverability of fumonisin B1 from extruded corn grits. J. Food Sci. 63: 696–698.

Castelo, M.M., Sumner, S.S., and Bullerman, L. B. 1998b. Occurrence of fumonisins in corn-based foods. J. Food Prot. 61: 704–707.

Castelo, M.M., Sumner, S.S., and Bullerman, L. B. 1998c. Stability of fumonisins in thermally processed corn products. J. Food Prot. 61: 1030–1033.

Cazzaniga, D., Basílico, J.C., González, R.J., Torres, R.L., and de Greef, D.M. 2001. Mycotoxins inactivation by extrusion cooking of corn flour. Lett. Appl. Microbiol. 33: 144–147.

Charmley, L.L. and Prelusky, D.B. 1994. Decontamination of *Fusarium* mycotoxins. In: Mycotoxins in Grain. Compounds Other Than Aflatoxin. J.D. Miller and H.L. Trenholm, ed. Eagan Press, St. Paul, MN, pp. 421–435.

Charmley, E., Trenholm, H.L., Thompson, B.K., Vudathala, D., Nicholson, J.W.G., Prelusky, D.B., and Charmley, L.L. 1993. Influence of level of deoxynivalenol in the diet of dairy cows on feed intake, milk production, and its composition. J. Dairy Sci. 76: 3580–3587.

Chelkowski, J. 1998. Distribution of *Fusarium* species and their mycotoxins in cereal grains. In: Mycotoxins in Agriculture and Food Safety. K.K. Sinha and D. Bhatnagar, ed. Marcel Dekker, Inc., New York, pp. 45–64.

Chelkowski, J., Bateman, G.L., and Mirocha, C.J. 1999. Identification of toxigenic *Fusarium* species using PCR assays. J. Phytopathol. 147: 307–311.

Chelule, P. K., Gqaleni, N., Dutton, M.F., and Chuturgoon, A.A. 2001. Exposure of rural and urban populations in KwaZulu Natal, South Africa, to fumonisin B1 in maize. Environ. Health Perspect. 109: 253–256.

Chu, F.S. 1991. Detection and determination of mycotoxins. In: Mycotoxins and Phytoalexins. R.P. Sharma and D.K. Salunkhe, ed. CRC Press, Boca Raton, FL, pp. 33–79.

Chu, F.S. 1992. Development and use of immunoassays in the detection of ecologically important mycotoxins. In: Handbook of Applied Mycology. Volume 5: Mycotoxins in Ecological Systems. D. Bhatnagar, E.B. Lillehoj, and D.K. Arora, ed. Marcel Dekker, Inc., New York, pp. 87–136.

Chu, F.S. 1994. Development of antibodies against aflatoxins. In: The Toxicology of Aflatoxins.Human Health, Veterinary, and Agricultural Significance. D.L. Eaton and J.D. Groopman, ed. Academic Press, New York, pp. 451–490.

Chu, F.S. 2001. Mycotoxin analysis: Immunological techniques. In: Foodborne Disease Handbook. Volume 3: Plant Toxicants. 2nd edition. Y.H. Hui, R.A. Smith, and D.G. Spoerke, Jr., ed. Marcel Dekker, Inc., New York, pp. 683–713.

Chu, F.S. and Li, G.Y. 1994. Simultaneous occurrence of fumonisin B1 and other mycotoxins in moldy corn collected from the People's Republic of China in regions with high incidences of esophageal cancer. Appl. Environ. Microbiol. 60: 847–852.

Chu, F.S., Huang, X., and Maragos, C.M. 1995. Production and characterization of anti-idiotype and anti-anti-idiotype antibodies against fumonisin B1. J. Agric. Food Chem. 43: 261–267.

Clavero, M.R.S., Hung, Y-C., Beuchat, L.R., and Nakayama, T. 1993. Separation of aflatoxin-contaminated kernels from sound kernels by hydrogen peroxide treatment. J. Food Prot. 56: 130–133.

Coker, R.D. 1998. Design of sampling plans for determination of mycotoxins in foods and feeds. In: Mycotoxins in Agriculture and Food Safety. K.K. Sinha and D. Bhatnagar, ed. Marcel Dekker, Inc., New York, pp. 109–133.

Cole, R.J. and Cox, R.H. 1981. Handbook of Toxic Fungal Metabolites. Academic Press, New York.

Cole, R.J. and Schweikert, M.A. 2003. Handbook of Secondary Fungal Metabolites. Volumes I and II. Academic Press, Boston, MA.

Constable, P.D., Foreman, J.H., Waggoner, A.L., Smith, G.W., Eppley, R.M., Tumbleson, M.E., and Haschek, W.M. 2000a. The mechanism of fumonisin mycotoxicosis in horses. Draft Report on USDA-CSREES Grant #928–39453, pp. 1–28.

Constable, P.D., Smith, G.W., Rottinghaus, G.E., and Haschek, W.M. 2000b. Ingestion of fumonisin B1-containing culture material decreases cardiac contractility and mechanica efficiency in swine. Toxicol Appl. Pharmacol. 162: 151–160.

Coté, L.M., Reynolds, J.D., Vesonder, R.F., Buck, W.B., Swanson, S.P., Coffey, R.T., and Brown, D.C. 1984. Survey of vomitoxin-contaminated feed grains in midwestern United States, and associated health problems in swine. J. Am. Vet. Med. Assoc. 184: 189–192.

Cotty, P.J. and Cardwell, K.F. 1999. Divergence of West African and North American communities of *Aspergillus* section *flavi*. Appl. Environ. Microbiol. 65: 2264–2266.

Coulombe, R.A., Jr. 1991. Aflatoxins. In: Mycotoxins and Phytoalexins. R.P. Sharma and D.K. Salunkhe, ed. CRC Press, Boca Raton, FL, pp. 103–143.

Crews, H., Alink, G., Andersen, R., Braesco, V., Holst, B., Maiani, G., Ovesen, L., Scotter, M., Solfrizzo, M., van den Berg, R., Verhagen, H., and Williamson, G. 2001. A critical assessment of some biomarker approaches linked with dietary intake. Brit. J. Nutr. 86 (Suppl. 1): S5-S35.

Croteau, S.M., Prelusky, D.B., and Trenholm, H.L. 1994. Analysis of trichothecene mycotoxins by gas chromatography with electron capture detection. J. Agric. Food Chem. 42: 928–933.

Cusumano, V. Rossano, F., Merendino, R.A., Arena. A., Costa, G.B., Mancuso, G., Baroni, A., and Losi, E. 1996. Immunological activities of mould products: functional impairment of human monocytes exposed to aflatoxin B1. Res. Microbiol. 147: 385–391.

Cvetnić, Z. and Pepeljnjak, S. 1990. Ochratoxinogenicity of *Aspergillus ochraceus* strains from nephropathic and non- nephropathic areas of Yugoslavia. Mycopathologia 110: 93–99.

D'Amico, G.D. 1987. The commonest glomerulonephritis in the world: IgA nephropathy. Q. J. Med. 247: 709–727.

Dawlatana, M., Coker, R.D., Nagler, M.J., Blunden, G., and Oliver, G.W.O. 1998. An HPTLC method for the quantitative determination of zearalenone in maize. Chromatographia 47: 215–218.

Delongchamp, R.R. and Young, J.F. 2001. Tissue sphinganine as a biomarker of fumonisin-induced apoptosis. Food Additiv. Contam. 18: 255–261.

de Nijs, M., Soentoro, P., Delfgou-van Asch, E., Kamphuis, H., Rombouts, F. M., and Notermans, S.H.W. 1996. Fungal infection and presence of deoxynivalenol and zearalenone in cereals grown in The Netherlands. J. Food Prot. 59: 772–777.

de Nijs, M., van Egmond, H.P., Nauta, M., Rombouts, F. M., and Notermans, S.H.W. 1998. Assessment of human exposure to fumonisin B$_1$. J. Food Prot. 61: 879–884.

de Nijs, M., van Egmond, H.P., Rombouts, F.M., and Notermans, S.H.W. 1997. Identification of hazardous *Fusarium* secondary metabolites occurring in food raw materials. J. Food Safety.17: 161–191.

Desjardins, A.E., Plattner, R.D. and Proctor, R.H. 1996. Genetic and biochemical aspects of fumonisin production. In: Fumonisins in Food. L.S Jackson, J.W. DeVries, and L.B. Bullerman. Plenum Press, New York, pp. 165–173.

Desjardins, A.E., Plattner, R.D., Lu, M., and Claflin, L.E. 1998. Distribution of fumonisins in maize ears infected with strains of *Fusarium moniliforme* that differ in fumonisin production.Plant Dis. 82: 953–958.

De Saeger, S. and van Peteghem, C. 1996. Dipstick enzyme immunoassay to detect *Fusarium* T-2 toxin in wheat. Appl. Environ. Microbiol. 62: 1880–1884.

De Saeger, S., Sibanda, L., Desmet, A., and van Peteghem, C. 2002. A collaborative study tovalidate novel field immunoassay kits for rapid mycotoxin detection. Int. J. Food Microbiol. 75: 135–142.

Deshpande, S.S. 2002. Handbook of Food Toxicology. Marcel Dekker, Inc., New York.

Dexter, J.E., Clear, R.M., and Preston, K.R. 1996. Fusarium head blight: Effect on the milling and baking of some Canadian wheats. Cereal Chem. 73: 695–701.

Dirheimer, G. 1998. Recent advances in the genotoxicity of mycotoxins. Revue Méd. Vét. 149: 605–616.

D'Mello, J.P.F., Porter, J.K., MacDonald, A.M.C., and Placinta, C.M. 1997. *Fusarium* mycotoxins. In: Handbook of Plant and Fungal Toxicants. J.P.F. D'Mello, ed. CRC Press, Boca Raton, FL, pp. 287–301.

Dombrink-Kurtzman, M.A. and Dvorak, T.J. 1999. Fumonisin content in masa and tortillas from Mexico. J. Agric. Food Chem. 47: 622–627.

Dombrink-Kurtzman, M.A., Dvorak, T.J., Barron, M.E., and Rooney, L.W. 2000. Effect of nixtamalization (alkaline cooking) on fumonisin-contaminated corn for production of masa and tortillas. J. Agric. Food Chem. 48: 5781–5786.

Dong, W., and Pestka, J.J. 1993. Persistent dysregulation of IgA production and IgA nephropathy in the B6C3F1 mouse following withdrawal of dietary vomitoxin (deoxynivalenol). Fundam. Appl. Toxicol. 20: 38–47.

Dong, W., Sell, J.E., and Pestka, J.J. 1991. Quantitative assessment of mesangial immunoglobulin A (IgA) accumulation, elevated circulating IgA immune complexes, and hematuria during vomitoxin-induced IgA nephropathy. Fundam. Appl. Toxicol. 17: 197-207.

Dong, W., Azcona Olivera, J.I., Brooks, K.H., Linz, J.E., and Pestka, J.J. 1994. Elevated gene expression and production of interleukins 2, 4, 5, and 6 during exposure to vomitoxin (deoxynivalenol) and cycloheximide in the EL-4 thymoma. Toxicol. Appl. Pharmacol. 127: 282–290.

Dorner, J.W. and Cole, R.J. 1993. Variability among peanut subsamples prepared for aflatoxin analysis with four mills. J. AOAC Int. 76: 983–987.

Dorner, J.W., Cole, R.J., and Blankenship, P.D. 1992. Use of a biocompetitive agent to control preharvest aflatoxin in drought stressed peanuts. J. Food Prot. 55: 888–892.

Dorner, J. W., Cole, R.J., Erlington, D.J., Sukksupath, S., McDowell, G.H., and Bryden, W.L. 1994. Cyclopiazonic acid residues in milk and eggs. J. Agric. Food Chem. 42: 1516–1518.

Doster, M.A. and Michailides, T.J. 1994. *Aspergillus* molds and aflatoxins in pistachio nuts in California. Phytopathology 84: 583–590.

Doster, M.A., Michailides, T.J., and Morgan, D.P. 1996. *Aspergillus* species and mycotoxins in figs from California orchards. Plant Dis. 80: 484–489.

Dowd, P.F. 1998. Involvement of arthropods in the establishment of mycotoxigenic fungi under field conditions. In: Mycotoxins in Agriculture and Food Safety. K.K. Sinha and D. Bhatnagar, ed. Marcel Dekker, Inc., New York, pp. 307–350.

Dowd, P.F. 2001. Biotic and abiotic factors limiting efficacy of Bt corn in indirectly reducing mycotoxin levels in commercial fields. J. Econ. Entomol. 94: 1067–1074.

Doyle, M.P., Applebaum, R.S., Bracket, R.E., and Marth, E.H. 1982. Physical, chemical, and biological degradation of mycotoxins in foods and agricultural commodities. J. Food Prot. 45:964–971.

Dragacci, S. and Fremy, J.M. 1996. Application of immunoaffinity column cleanup to aflatoxin M$_1$ determination and survey in cheese. J. Food Prot. 59: 1011–1013.

Dragan, Y.P., Bidlack, W.R., Cohen, S.M., Goldsworthy, T.L., Hard, G.C., Howard, P.C., Riley, R.T., and Voss, K.A..2001. Implications of apoptosis for toxicity, carcinogenicity and risk assessment: fumonisin B1 as an example. Toxicol. Sci. 61: 6–17.

D'Souza, D.H. and Brackett, R.E. 1998. The role of trace metal ions in aflatoxin B$_1$ degradation by *Flavobacterium aurantiacum*. J. Food Prot. 61: 1666–1669.

D'Souza, D.H. and Brackett, R.E. 2000. The influence of divalent cations and chelators on aflatoxin B$_1$ degradation by *Flavobacterium aurantiacum*. J. Food Prot. 63: 102–105.

D'Souza, D.H. and Brackett, R.E. 2001. Aflatoxin B$_1$ degradation by *Flavobacterium aurantiacum* in the presence of reducing conditions and seryl and sulfhydryl group inhibitors. J. Food Prot. 64: 268–271.

Dupuy, J., Le Bars, P., Boudra, H., and Le Bars, J. 1993. Thermostability of fumonisin B$_1$, a mycotoxin from *Fusarium moniliforme*, in corn. Appl. Environ. Microbiol. 59: 2864–2867.

Duvick, J. 2001. Prospects for reducing fumonisin contamination of maize through genetic modification. Environ. Health Perspect. 109: 337–342.

Eaton, D.L. and Gallagher, E.P. 1994. Mechanisms of aflatoxin carcinogenesis. Ann. Rev. Pharmacol. Toxicol. 34: 135–172

Eaton, D.L. and Ramsdell, H.S. 1992. Species- and diet-related differences in aflatoxin biotransformation. In: Handbook of Applied Mycology. Volume 5: Mycotoxins in Ecological Systems. D. Bhatnagar, E.B. Lillehoj, and D.K. Arora, ed. Marcel Dekker, Inc., New York, pp. 157–182.

Edwards, S.G., Pirgozliev, S.R., Hare, M.C., and Jenkinson, P. 2001. Quantification oftrichothecene-producing *Fusarium* species in harvested grain by competitive PCR todetermine efficacies of fungicides against *Fusarium* head blight of winter wheat. Appl. Environ. Microbiol. 67: 1575–1580.

Egel, D.S., Cotty, P.J. and Elias, K.S. 1994. Relationships among isolates of *Aspergillus* sect. *flavi* that vary in aflatoxin production. Phytopathology. 84: 906–912.

Egner, P.A., Wang, J.B., Zhu, Y.R., Zhang, B.C., Wu, Y., Zhang, Q.N., Qian, G.S., Kuang, S.Y., Gange, S.J., Jacobson, L.P., Helzlsouer, K.J., Bailey, G.S Groopman, J.D., and Kensler, T.W. 2001. Chlorophyllin intervention reduces aflatoxin-DNA adducts in individuals at high risk for liver cancer. Proc. Natl. Acad. Sci. USA 98: 14601–14606.

Ellis, W.O., Smith, J.P., Simpson, B.K., and Oldham, J.H. 1991. Aflatoxins in food: Occurrence, biosynthesis, effects on organisms, detection, and methods of control. Crit. Rev. Food Sci. Nutr. 30: 403–439.

Ellis, W.O., Smith, J.P., Simpson, B.K., Ramaswamy, H., and Doyon, G. 1994a. Growth of and aflatoxin production by *Aspergillus flavus* in peanuts stored under modified atmosphere packaging (MAP) conditions. Int. J Food Microbiol. 22: 173–187.

Ellis, W.O., Smith, J.P., Simpson, B.K., Ramaswamy, H., and Doyon, G. 1994b. Effect of gas barrier characteristics of films on aflatoxin production by *Aspergillus flavus* in peanuts packaged under modified atmosphere packaging (MAP) conditions. Food Res. Int. 27: 505-512.

Ellis, W.O., Smith, J.P., Simpson, B.K., Ramaswamy, H., and Doyon, G. 1994c. Novel techniques for controlling the growth of and aflatoxin production by *Aspergillus parasiticus* in packaged peanuts. Food Microbiol. 11: 357–368.

El-Nezami, H., Kankaanpää, P., Salminen, S., and Ahokas, J. 1998. Physicochemical alterations enhance the ability of dairy strains of lactic acid bacteria to remove aflatoxin from contaminated media. J. Food Prot. 61: 466–468.

El-Nezami, H., Mykkänen, H., Kankaanpää, P., Salminen, S., and Ahokas, J. 2000. Ability of *Lactobacillus* and *Propionibacterium* strains to remove aflatoxin B$_1$ from the chicken duodenum. J. Food Prot. 63: 549–552.

El-Nezami, H., Polychronaki, N., Salminen, S., and Mykkänen H. 2002. Binding rather than metabolism may explain the interaction of two food-grade *Lactobacillus* strains with zearalenone and its derivative α-zearalenol. Appl. Environ. Microbiol. 68: 3545–3549.

Enongene, E.N., Sharma, R.P., Neetesh, B., Meredith, F.I., Voss, K.A., and Riley, R.T. 2001.Time and dose-related changes in mouse sphingolipid metabolism following gavage fumonisin B1 administration. Toxicol Sci. 60: 408 (abstract).

Enongene, E.N., Sharma, R.P., Voss, K.A., and Riley, R.T. 2000. Subcutaneous fumonisin administration disrupts sphingolipid metabolism in mouse digestive epithelia, liver, and kidney. Food Chem. Toxic. 38: 793–799.

Etzel, R.A., Montana, E., Sorenson, W.G., Kullman, G.J., Allan, T.M., Dearborn, D.G., Olson, D.R., Jarvis, B.B., and Miller, J.D. 1998. Acute pulmonary haemorrhage in infants associated with exposure to *Stachybotrys atra* and other fungi. Arch. Pediatr. Adolesc. Med. 152: 757–762

Fajardo, J.E., Dexter, J.E., Roscoe, M.M., and Nowicki, T.W. 1995. Retention of ergot alkaloids in wheat during processing. Cereal Chem. 72: 291–298.

FAO (Food and Agriculture Organization of the United Nations). 1997. Worldwide Regulations for Mycotoxins 1995. A Compendium. FAO, Rome, Italy.

FAO (Food and Agriculture Organization). 2001. Manual on the Application of the HACCP System in Mycotoxin Prevention and Control. FAO Food and Nutrition Paper 73. International Atomic Energy Agency. Food and Agricultural Organization of the United Nations. Food and Agricultural Organization, Rome, Italy.

Färber, P., Geisen, R., and Holzapfel, W. H. 1997. Detection of aflatoxigenic fungi in figs by a PCR reaction. Int. J. Food Microbiol. 36: 215–220.

Fazekas, B. and Tar, A. 2001. Determination of zearalenone content in cereals and feedstuffs by immunoaffinity column coupled with liquid chromatography. J. AOAC Int. 84: 1453–1459.

Fink-Gremmels, J. 1999. Mycotoxins: their implications for human and animal disease. Vet. Quart. 21: 115–120

Finoli, C., Vecchio, A., Galli, A., and Franzetti, L. 1999. Production of cyclopiazonic acid by molds isolated from Taleggio cheese. J. Food Prot. 62: 1198–1202.

Flaherty, J.E., Weaver, M.A., Payne, G.A., and Woloshuk, C.P. 1995. A β-glucuronidase reporter gene construct for monitoring aflatoxin biosynthesis in *Aspergillus flavus*. Appl. Environ. Microbiol. 61: 2482–2486.

Forsell, J.H., Witt, M.F., Tai, J.H., Jensen, R., and Pestka, J.J. 1986. Effects of 8-week exposure of the B6C3F1 mouse to dietary deoxynivalenol (vomitoxin) and zearalenone. Food Chem. Toxicol. 24:213–219.

Fouler, S.G., Trivedi, A.B., and Kitabatake, N. 1994. Detoxification of citrinin and ochratoxin A by hydrogen peroxide. J. AOAC Int. 77: 631–637.

Freese, L., Friedrich, R., Kendall, D., and Tanner, S. 2000. Variability of deoxynivalenol measurements in barley. J. AOAC Int. 83: 1259–1263.

Friedman, L. 1990. Patulin: Mycotoxin or fungal metabolite? (Current state of knowledge). In: Biodeterioration Research 3. G.C. Llewellyn and C.E. O'Rear, ed. Plenum Press, New York, pp. 21–54.

Frisvad, J. C. 1995. Mycotoxins and mycotoxigenic fungi in storage. In: Stored-Grain Ecosystems. D.S. Jayas, N.D.G. White, and W.E. Muir, ed. Marcel Dekker, Inc., New York, pp. 251–288.

Frisvad, J.C. and Filtenborg, O. 1989. Terverticillate penicillia: Chemotaxonomy and mycotoxin production. Mycologia 81: 837–861.

Frisvad, J. C. and U. Thrane. 2002. Mycotoxin production by common filamentous fungi. In: Introduction to Food- and Airborne Fungi. 6th edition. R.A. Samson, E.S. Hoekstra, J.C. Frisvad, and O. Filtenborg, ed. Centraalbureau voor Schimmelcultures, Utrecht, The Netherlands, pp. 321–331.

Galtier, P. 1999. Biotransformation and fate of mycotoxins. J. Toxicol. Toxin Rev. 18: 295–312.

Galvano, F., Galofaro, V., and Galvano, G. 1996a. Occurrence and stability of aflatoxin M$_1$ in milk and milk products: A worldwide review. J. Food Prot. 59: 1079–1090.

Galvano, F., Pietri, A., Bertuzzi, T., Fusconi, G., Galvano, M., Piva, A. and Piva, G. 1996b. Reduction of carryover of aflatoxin from cow feed to milk by addition of activated carbons. J. Food Prot. 59: 551–554.

Galvano, F., Pietri, A., Bertuzzi, T., Piva, A., Chies, L., and Galvano, M. 1998. Activated carbons: *In vitro* affinity for ochratoxin A and deoxynivalenol and relation of adsorption ability to physicochemical parameters. J. Food Prot. 61: 469–475.

Gan, Z., Marquardt, R. R., Abramson, D., and Clear, R.M. 1997. The characterization of chicken antibodies raised against *Fusarium* spp. by enzyme-linked immunosorbent assay and immunoblotting. Int. J. Food Microbial 38: 191–200.

Garren, L., Galendo, D., Wild, C.P., and Castegnaro, M. 2001. The induction and persistence of altered sphingolipid biosynthesis in rats treated with fumonisin B1. Food Addit. Contam. 18: 850–856.

Gauchi, J.-P. and Leblanc, J.-C. 2002. Quantitative assessment of exposure to the mycotoxinochratoxin A in food. Risk Anal. 22: 219–234.

Gelderblom W.C.A., Kriek N.P.J., Marasas W.F.O, and Thiel P.G. 1991. Toxicity and carcinogenicity of the *Fusarium moniliforme* metabolite, fumonisin B$_1$, in rats. Carcinogenesis 12: 1247-1251.

Gelderblom, W.C. A., Marasas, W.F.O., Lebepe-Mazur, S., Swanevelder, S., Vessy, C.J., and de la M Hall, P. 2002a. Interaction of fumonisin B1 and aflatoxin B1 in a short-term carcinogenesis model in rat liver. Toxicology. 171: 161–173.

Gelderblom, W.C. A., Moritz, W., Swanevelder, S., Smuts, C.M., and Abel, S. 2002b. Lipids and delta 6-desaturase activity alterations in rat liver microsomal membranes induced by fumonisin B$_1$. Lipids 37: 869–877.

Geisen, R. 1996. Multiplex polymerase chain reaction for the detection of potential aflatoxin and sterigmatocystin producing fungi. System. Appl. Microbiol. 19: 388–392.

Giesbrecht, F.G. and Whitaker, T.B. 1998. Investigations of the problems of assessing aflatoxin levels in peanuts. Biometrics 54: 739–753.

Gilbert, J. 1991. Regulatory aspects of mycotoxins in the European Community and USA. In: Fungi and Mycotoxins in Stored Products. B.R. Champ, E. Highley, A.D. Hocking, and J.I. Pitt, ed. Australian Centre for International Agricultural Research (AICAR), Canberra, Australia, pp. 194–197.

Gilbert, J. 1999. Quality assurance in mycotoxin analysis. Food Nutr. Agric. 23: 33–37.

Gilbert, J. 2000. Overview of mycotoxin methods, present status and future needs. Nat. Toxins 7: 347–352.

Gilbert, J., Brereton, P., and MacDonald, S. 2001. Assessment of dietary exposure to ochratoxin A in the UK using duplicate diet approach and analysis of urine and plasma samples. Food Addit. Contam. 18: 1088–1093.

Gökmen, V. and Acar, J. 1998. An investigation on the relationship between patulin and fumaric acid in apple juice concentrates. Lebensm. Wiss. u. Technol. 31: 480–483.

Gong, Y.Y., Cardwell, K., Hounsa, A., Egal, S., Turner, P.C., Hall, A.J., and Wild, C.P. 2002. Dietary aflatoxin exposure and impaired growth in young children from Benin and Togo: cross sectional study. Brit. Med. J. 325: 20–21.

Gorelick, N.J., Bruce, R.D. and Hoseyni, M.S. 1994. Human risk assessment based on animal data: Inconsistencies and alternatives. In: The Toxicology of Aflatoxins. Human Health, Veterinary, and Agricultural Significance. D.L. Eaton and J.D. Groopman, ed. Academic Press, New York, pp. 493–511.

Gourama, H. and Bullerman, L. B. 1995. Inhibition of growth and aflatoxin production of Aspergillus flavus by Lactobacillus species. J. Food Prot. 58: 1249–1256.

Grimm, C. and Geisen, R. 1998. A PCR-ELISA for the detection of potential fumonisin producing Fusarium species. Lett. Appl. Microbiol. 26: 456–462.

Grove, J.F. 1988. Non-macrocyclic trichothecenes. Nat. Prod. Rep. 5, 187–209.

Grove, J.F. 1993. Macrocyclic trichothecenes. Nat. Prod. Rep. 10: 429–448.

Grove, J.F. 2000. Non-macrocyclic trichothecenes. Part 2. Prog. Chem. Org. Nat. Prod. 69: 1-70.

Groves, F.D., Zhang, L., Chang, Y.-S., Ross, P.F., Casper, H., Norred, W.P., You, W.-C., and Fraumeni, J.F., Jr. 1999. Fusarium mycotoxins in corn and corn products in a high-risk area for gastric cancer in Shandong Province, China. J. AOAC Int. 82: 657–662.

Guo, B. Z., Russin, J.S., Brown, R.L., Cleveland, T.E., and Widstrom, N.W. 1996. Resistance to aflatoxin contamination in corn as influenced by relative humidity and kernel germination. J. Food Prot. 59: 276–281.

Gutema, T., Munimbazi, C., and Bullerman, L. B. 2000. Occurrence of fumonisins and moniliformin in corn and corn-based food products of USA origin. J. Food Prot. 63: 1732–1737.

Hall, A.J. and Wild, C.P. 1994. Epidemiology of aflatoxin-related disease. In: The Toxicology of Aflatoxins. Human Health, Veterinary, and Agricultural Significance. D.L. Eaton and J.D.Groopman, ed. Academic Press, New York, pp. 233–258.

Hard, G.C., Howard, P.C., Kovatch, R.M., and Bucci, T.J. 2001. Rat kidney pathology induced by chronic exposure to fumonisin B1 includes rare variants of renal tubule tumor. Toxicol. Pathol. 29: 379–386.

Haschek, W.M., Gumprecht, L.A., Smith, G., Tumbleson, M.E., and Constable, P.D. 2001. Fumonisin toxicosis in swine: An overview of porcine pulmonary edema and current perspectives. Environ. Health Perspect. 109: 251–257.

Hassanin, N.I. 1994. Stability of aflatoxin $M_1$ during manufacture and storage of yoghurt, yoghurt-cheese and acidified milk. J. Sci. Food Agric. 65: 31–34.

He, Q., Riley, R.T., and Sharma, R.P. 2002. Pharmacological antagonism of fumonisin $B_1$ cytotoxicity in porcine renal epithelial cells (LLC-$PK_1$): A model for reducing fumonisin-induced nephrotoxicity In Vivo. Pharmacol. Toxicol. 90: 268–277

Hendrich, S., Miller, K.A., Wilson, T.M., and Murphy, P.A. 1993. Toxicity of Fusarium proliferatum-fermented nixtamalized corn-based diets fed to rats: Effect of nutritional status. J. Agric. Food Chem. 41: 1649–1654.

Henry, S.H., and F.X. Bosch. 2001. Foodborne disease and mycotoxin epidemiology. In Foodborne Disease Handbook. Volume 3: Plant Toxicants. 2nd edition. Y.H. Hui, R.A. Smith, and D.G. Spoerke, Jr., ed. Marcel Dekker, Inc., New York, pp. 593–626.

Henry, S., Bosch, F.X., Troxell, T.C., and Bolger, P.M. 1999. Reducing liver cancer – global incidence of aflatoxin. Science. 286: 2453–2454.

Henry, S.H., Whitaker, T., Rabbani, I., Bowers, J., Park, D., Price, W., Bosch, F.X., Pennington, J., Verger, P., Yoshizawa, van Egmond, H., Jonker, M.A., and Coker, R. 2001. Aflatoxin $M_1$. In: Safety Evaluation of Certain Mycotoxins in Food. WHO Food Additive Series: 47 and FAO Food and Nutrition Paper 74. Joint FAO/WHO Expert Committee on Food Additives, ed. International Programme on Chemical Safety, World Health Organization, Geneva, Switzerland, pp. 1–102.

Heperkan, D., Aran, N., and Ayfer, M. 1994. Mycoflora and aflatoxin contamination in shelledpistachio nuts. J. Sci. Food Agric. 66: 273–278.

Herrman, J.L. and Walker, R. 1999. Risk analysis of mycotoxins by the joint FAO/WHO expert committee on food additives (JECFA). Food Nutr. Agric. 23: 17–24.

Hicks, L.R., Brown, D.R., Storch, R.H., and Bushway, R.J. 2000. Need to determine the relative developmental risks of Fusarium mycotoxin deoxynivalenol (DON) and benomyl (BEN) in wheat. Hum. Ecol. Risk Assess. 6: 341–354.

Higuera-Ciapara, I. Esqueda-Valle, M., and Nieblas, J. 1995. Reduction of aflatoxin $M_1$ from artificially contaminated milk using ultrafiltration and diafiltration. J. Food Sci. 60: 645–647.

Hocking, A.D. 2001. Toxigenic Aspergillus species. In: Food Microbiology. Fundamentals and Frontiers. 2nd edition. M. P. Doyle, L. R. Beuchat, and T. J. Montville, ed. ASM Press, Washington, D. C. p. 451–465.

Holcomb, M., Sutherland, J.B., Chiarelli, M.P., Korfmacher, W.A., Thompson, H.C. Jr., Lay, J.O. Jr., Hankins, L.J., and Cerniglia, C.E. 1993. HPLC and FAB mass spectrometry analysis of fumonisins $B_1$ and $B_2$ produced by Fusarium moniliforme on food substrates. J. Agric. Food Chem. 41: 357–360.

Holcomb, M., Thompson, H.C.Jr., Cooper, W.M., and Hopper, M.L. 1996. SFE extraction of aflatoxins ($B_1$, $B_2$, $G_1$, and $G_2$) from corn and analysis by HPLC. J. Super. Fluids 9: 118–121.

Homdork, S., Fehrmann, H., and Beck, R. 2000a. Effects of field application of tebuconazole on yield, yield components and the mycotoxin content of Fusarium-infected wheat grain. J. Phytopathol. 148: 1–6.

Homdork, S., Fehrmann, H., and Beck, R. 2000b. Influence of different storage conditions on the mycotoxin production and quality of Fusarium-infected wheat grain. J. Phytopathol. 148: 7–15.

Hongyo, K.-I., Ithoh, Y., Hifumi, E., Takeyasu, A., and Uda, T. 1992. Comparison of monoclonal antibody-based enzyme-linked immunosorbent assay with thin-layer chromatography and liquid chromatography for aflatoxin $B_1$ determination in naturally contaminated corn and mixed feed. J. AOAC Int. 75: 307–312.

Hooker, D.C., Schaafsma, A.W., and Tamburic-Ilincic, L. 2002. Using weather variables pre- and post-heading to predict deoxynivalenol content in winter wheat. Plant Dis. 86: 611–619.

Hopmans, E. C. and Murphy, P. A. 1993. Detection of fumonisins $B_1$, $B_2$, and $B_3$ and hydrolyzed fumonisin $B_1$ in corn-containing foods. J. Agric. Food Chem. 41: 1655–1658.

Horn, B.W. and Dorner, J.W. 1999. Regional differences in production of aflatoxin $B_1$ and cyclopiazonic acid by soil isolates of Aspergillus flavus along a transect within the United States. Appl. Environ. Microbiol. 65: 1444–1449.

Howard, P.C., Churchwell, M.I., Couch, L.H., Marques, M.M., and Doerge, D.R. 1998. Formation of N-(carboxymethyl)fumonisin B1, following the reaction of fumonisin B1 with reducing sugars. J. Agric. Food Chem. 46: 3546–3557.

Howard, P.C., Warbritton, A., Voss, K.A., Lorentzen, R.J., Thurman, J.D., Kovach, R.M., and Bucci, T.J. 2001. Compensatory regeneration as a mechanism for renal tube carcinogenesis of fumonisin B1 in the F344/N/Nctr BR rat. Environ. Health Perspect. 109: 309–314.

Hsu, K.-H. and Chu, F.S. 1994. Production and characterization of anti-idiotype and anti-anti-idiotype antibodies from a monoclonal antibody against aflatoxin J. Agric. Food Chem. 42: 2353–2359.

Hua, S.-S.T., Grosjean, O.-K., and Baker, J.L. 1999. Inhibition of aflatoxin biosynthesis by phenolic compounds. Lett. Appl. Microbiol. 29: 289–291.

Huebner, H.J., Mayura, K., Pallaroni, L., Ake, C.L., Lemke, S.L., Herrera, P., and Phillips, T. D. 2000. Development and characterization of a carbon-based composite material for reducing patulin levels in apple juice. J. Food Prot. 63: 106–110.

Humphreys, S. H., Carrington, C., and Bolger, M. 2001. A quantitative risk assessment for fumonisins $B_1$ and $B_2$ in US corn. Food Addit. Contam. 18: 211–220.

Hussein, H.S. and Brasel, J.M. 2001. Toxicity, metabolism, and impact of mycotoxins on humans and animals. Toxicology 167: 101–134.

Hwang, C.-A., and Draughon, F. A. 1994. Degradation of ochratoxin A by *Acinetobacter calcoaceticus*. J. Food Prot. 57: 410–414.

IARC (International Agency for Research on Cancer). 2003. IARC Monographs on the Evaluation of Carcinogenic Risks of Chemicals to Humans, Vol. 82, Some Traditional Herbal Medicines, Some Mycotoxins, Naphthalene and Styrene. International Agency for Research on Cancer, Lyon, France.

ICMSF (International Commission on the Microbiological Specifications for Food). 2002. Microorganisms in Foods 7. Microbiological Testing in Food Safety Management. Kluwer Academic/Plenum Publishers, New York.

Iordanov, M.S., Pribnow, D., Magun, J.L., Dinh, T.H., Pearson, J.A., Chen, S.L., and Magun, B.E. 1997. Ribotoxic stress response: activation of the stress-activated protein kinase JNK1 by inhibitors of the peptidyl transferase reaction and by sequence-specific RNA damage to the alpha-sarcin/ricin loop in the 28S rRNA. Mol. Cell Biol. 17: 3373–3381.

Islam, Z., Nagase, M., Yoshizawa, T., Yamauchi, K., and Sakato, N. 1998a. T-2 toxin induces thymic apoptosis *in vivo* in mice. Toxicol. Appl. Pharmacol. 148: 205–214.

Islam, Z., Nagase, M., Ota, A., Ueda, S., Yoshizawa, T., and Sakato, N. 1998b. Structure-function relationship of T-2 toxin and its metabolites in inducing thymic apoptosis *in vivo* in mice. Biosci. Biotechnol. Biochem. 62: 1492–1497.

Iyer, M.S. and Cousin, M.A. 2003. Immunological detection of *Fusarium* species in cornmeal. J. Food Prot. 66: 451–456.

Jackson, E.W., Wolf, H., and Sinnhuber, R.O. 1968. The relationship of hepatoma in rainbow trout to aflatoxin contamination and cottonseed meal. Cancer Res. 28: 987–997.

Jackson, L.S., Hlywka, J.J., Senthil, K.R., and Bullerman, L.B. 1996a. Effects of thermal processing on the stability of fumonisin $B_2$ in an aqueous system. J. Agric. Food Chem. 44: 984–1987.

Jackson, L.S., Hlywka, J.J., Senthil, K.R., Bullerman, L.B., and Musser, S.M. 1996b. Effects of time, temperature, and pH on the stability of fumonisin B1 in an aqueous model system. J. Agric. Food Chem. 44: 906–912.

Jackson, L.S., Katta, S.K., Fingerhut, D.D., DeVries, J.W., and Bullerman, L.B. 1997. Effects of baking and frying on the fumonisin $B_1$ content of corn-based foods. J. Agric. Food Chem. 45: 4800–4805.

Järvenpää, E.P., Taylor, S.L., King, J.W., and Huopalahti, R. 1997. The use of supercritical fluid extraction for the determination of 4-deoxynivalenol in grains: The effect of the sample clean-up and analytical methods on quantitative results. Chromatographia 46: 33- 39.

JECFA (Joint FAO/WHO Expert Committee on Food Additives). 1995. Evaluation of Certain Food Additives and Contaminants. World Health Organization Technical Report Series, No. 859. World Health Organization, Geneva, Switzerland.

JECFA 53th (Joint FAO/WHO Expert Committee on Food Additives 53th). 2000. Safety Evaluation of Certain Food Additives. World Health Organization Food Additives Series 44. World Health Organization, Geneva, Switzerland.

JECFA 56th (Joint FAO/WHO Expert Committee on Food Additives 56th). 2001. Safety Evaluation of Certain Mycotoxins in Food. Food and Agriculture Organization of the UnitedNations, paper 74. World Health Organization Food Additives Series 47. World Health Organization, Geneva, Switzerland.

JECFA (Joint FAO/WHO Expert Committee on Food Additives). 2002. Evaluation of Certain Mycotoxins in Food. WHO Technical Report Series 906. World Health Organization, Geneva, Switzerland.

Jegorov, A. 1999. Analytical chemistry of ergot alkaloids. In: Ergot. The Genus *Claviceps*. V. Křen and L. Cvak, ed. Hardwood Academic Publishers, Amsterdam, The Netherlands. p. 267–301.

Joffe, A. 1978. *Fusarium poae* and *F. sporotrichoides* as principal causal agents of alimentary toxic aleukia. In: Mycotoxic Fungi, Mycotoxins, Mycotoxicoses. T. Wyllie and L. Morehouse, ed. Marcel Dekker, New York. p. 21–86.

Joffe, A.Z. 1983. Foodborne disease: Alimentary toxic aleukia. In: Handbook of Foodborne Diseases of Biological Origin. M. Rochigle, ed. CRC Press, Boca Raton, FL, pp. 353–495.

Johansson, A.S., Whitaker, T.B., Giesbrecht, F. G., Hagler, W. M. Jr., and Young, J.H. 2000a. Testing shelled corn for aflatoxin, Part II: Modeling the observed distribution of aflatoxin test results. J. AOAC Int. 83: 1270–1278.

Johansson, A.S., Whitaker, T.B., Giesbrecht, F. G., Hagler, W. M.Jr., and Young, J.H. 2000b. Testing shelled corn for aflatoxin, Part III: Evaluating the performance of aflatoxin sampling plans. J. AOAC Int. 83: 1279–1284.

Johansson, A.S., Whitaker, T.B., Hagler, W. M. Jr., Giesbrecht, F. G., Young, J.H., and Bowman, D.T. 2000c. Testing shelled corn for aflatoxin, Part I: Estimation of variance components. J. AOAC Int. 83: 1264–1269.

Johnson, D.D., Wilson, W.W., and Diersen, M.A. 2001. Quality uncertainty, procurement strategies, and grain merchandising risk: Vomitoxin in spring wheat. Rev. Agr. Econ. 23: 102–119.

Jonsyn-Ellis, F.E. 2001. Seasonal variation in exposure frequency and concentration levels of aflatoxin and ochratoxins in urine samples of boys and girls. Mycopathologia. 152: 35–40.

Joosten, H.M.L.J., Goetz, J., Pittet, A., Schellenberg, M., and Bucheli, P. 2001. Production of ochratoxin A by *Aspergillus carbonarius* on coffee cherries. Int. J. Food Microbiol. 65: 39-44.

Kadakal, C. and Nas, S. 2002. Effect of activated charcoal on patulin, fumaric acid and some other properties of apple juice. Nahrung 46: 31–33.

Kankaanpää, P., Tuomola, E., El-Nezami, H., Ahokas, J., and Salminen, S.J. 2000. Binding of aflatoxin $B_1$ alters the adhesion properties of *Lactobacillus rhamnosus* strain GG in caco-2 model. J. Food Prot. 63: 412–414.

Karunaratne, A., Wezenberg, E., and Bullerman, L.B. 1990. Inhibition of mold growth and aflatoxin production by *Lactobacillus* spp. J. Food Prot. 53: 230–236.

Katta, S.K., Cagampang, A.E., Jackson, L.S., and Bullerman, L.B. 1997. Distribution of *Fusarium* molds and fumonisins in dry-milled corn fractions. Cereal Chem. 74: 858–863.

Katta, S.K., Jackson, L.S., Sumner, S.S., Hanna, M.A., and Bullerman, L.B. 1999. Effect of temperature and screw speed on stability of fumonisin $B_1$ in extrusion-cooked corn grits. Cereal Chem. 76: 16–20.

Keller, U. 1999. Biosynthesis of ergot alkaloids. In: Ergot. The Genus *Claviceps*. V. Křen and L. Cvak, ed. Hardwood Academic Publishers, Amsterdam, The Netherlands, pp. 95–163.

Kensler, T.W., Davis, E.F., and Bolton, M.G. 1994. Strategies for chemoprotection against aflatoxin-induced liver cancer. In: The Toxicology of Aflatoxins. Human Health, Veterinary, and Agricultural Significance. D.L. Eaton and J.D. Groopman, ed. Academic Press, New York, pp. 281–306.

Kerkadi, A., Barriault, C., Marquardt, R.R., Frohlich, A.A., Yousef, I.M., Zhu, X.X., and Tuchweber, B. 1999. Cholestyramine protection against ochratoxin A toxicity: Role of ochratoxin A sorption by the resin and bile acid enterohepatic circulation. J. Food Prot. 62: 1461–1465.

Kiessling, K.-H. and Pettersson, H. 1991. Chemical preservatives. In: Mycotoxins and Animal Foods. J.E. Smith and R.S. Henderson, ed. CRC Press, Boca Raton, FL, pp. 765–775.

Kim, E-K, Scott, P.M., and Lau, B.P-Y. 2003. Hidden fumonisins in corn flakes. Food Addit. Cont. 20:161–169.

Kim, E-K, Scott, P.M., Lau, B.P-Y, and Lewis, D.A. 2002. Extraction of fumonisins $B_1$ and $B_2$ from the white rice flour and their stability in white rice flour, cornstarch, cornmeal, and glucose. J. Agric. Food Chem. 50: 3614–3620.

Kim, M.S., Lee, D-Y., Wang, T., and Schroeder, J.J. 2001. Fumonisin $B_1$ induces apoptosis in LLC-PK1 renal epithelial cells via a sphinganine- and calmodulin-dependent pathway. Toxicol. Appl. Pharmacol. 176: 118–126

Klich, M.A. and Pitt, J.I. 1988. A Laboratory Guide to Common *Aspergillus* Species and Their Teleomorphs. Commonwealth Scientific and Industrial Research Organization, Division of Food Processing, North Ryde, N.S.W., Australia.

Kobayashi, Y., Ohshiro, N., Shibusawa, A., Sasaki, T., Tokuyama, S., Sekine, T., Endou, H., and Yamamoto, T. 2002. Isolation, characterization and differential gene expression of multispecific anion transporter in mice. Mol. Pharmacol. 62: 7–14.

Korkotian, E., Schwartz, A., Pelled, D., Schwarzmann, G., Segal, M., and Futerman, A.H. 1999. Elevation of intercellular glucosylceramide levels results in an increase in endoplasmic reticulum density and in functional calcium stores in cultured neurons. J. Biol. Chem. 274: 21673–21678.

Koshinsky, H.A., Woytowich, and Khachatourians, G.G. 2001. Mycotoxicoses: The effects of interactions with mycotoxins. In: Foodborne Disease Handbook. Volume 3: Plant Toxicants. 2nd edition. Y.H. Hui, R.A. Smith and D.G. Spoerke, Jr., ed. Marcel Dekker, Inc., New York, p. 627.

Kozakiewicz, Z. 2001. Aspergillus. In: Foodborne Disease Handbook. Volume 3: Plant Toxicants. 2nd edition. Y.H. Hui, R.A. Smith and D.G. Spoerke, Jr., ed. Marcel Dekker, Inc., New York, pp. 471–501.

Kpodo, K., Thrane, U. and Hald, B. 2000. Fusaria and fumonisins in maize from Ghana and their co-occurrence with aflatoxins. Int. J. Food Microbiol. 61: 147–157.

Křen, V. and Cvak, L. (ed.) 1999. Ergot. The Genus Claviceps. Harwood Academic Publishers, Amsterdam, The Netherlands.

Kristensen, P., Andersen, A., and Irgens, L.M. 2000. Hormone-dependent cancer and adverse reproductive outcomes in farmers' families–effects of climatic conditions favoring fungal growth in grain. Scand. J. Work Environ. Health 26: 331–337.

Kruger, S.C., Kohn, B., Ramsey, C. S., and Prioli, R. 1999. Rapid immunoaffinity-based method for determination of zearalenone in corn by fluorometry and liquid chromatography. J. AOAC Int. 82: 1364–1368.

Kryger, R.A. 2001. Volatility of patulin in apple juice. J. Agric. Food Chem. 49: 4141–4143.

Kuiper-Goodman, T. 1994. Prevention of human mycotoxicosis through risk assessment and risk management. In: Mycotoxins in Grain. Compounds Other Than Aflatoxin. J.D. Miller and H.L. Trenholm, ed. Eagan Press, St. Paul, MN, pp. 439–469.

Kuiper-Goodman, T. 1996. Risk assessment of ochratoxin A: an update. Food Addit. Contam. 13(Suppl): 53–57.

Kuiper-Goodman, T. 1999. Approaches to the risk analysis of mycotoxins in the food supply. Food Nutr. Agric. 23: 10–16.

Kuiper-Goodman T., Scott P.M., McEwen N.P., Lombaert G.A., and Ng, W. 1996. Approaches to the risk assessment of fumonisins in corn-based foods in Canada. Adv. Exp. Med. Biol. 392:369–393.

Kuldau, G.A. and Bacon, C.W. 2001. Claviceps and related fungi. In: Foodborne Disease Handbook. Volume 3: Plant Toxicants. 2nd edition. Y.H. Hui, R.A. Smith and D.G. Spoerke, Jr., ed. Marcel Dekker, Inc., New York, pp. 503–534.

Kurtzman, C.P., Horn, B.W., and Hesseltine, C.W. 1987. Aspergillus nomius, a new aflatoxin-producing species related to Aspergillus flavus and Aspergillus tamarii. Antonie van Leeuwenhoek 53: 147–158.

Langseth, W. and Rundberget, T. 1999. The occurrence of HT-2 toxin and other trichothecenes in Norwegian cereals. Mycopathologia 147: 157–165.

Langseth, W. and Stabbetorp, H. 1996. The effect of lodging and time of harvest ondeoxynivalenol contamination in barley and oats. Phytopathology. 144: 241–245.

Larsson, K. and Möller, T. 1996. Liquid chromatographic determination of ochratoxin A in barley, wheat bran, and rye by the AOAC/IUPAC/NMKL method: NMKL collaborative study. J. AOAC Int. 79: 1102–1105.

Le Bars, J. and Le Bars, P. 1990. Etiologie des troubles divers consecutifs au developpement de champignons dans les fourrages. Association Francaise des Techniciens de L'Alimentation Animale 90: 40–44.

Le Bars, J., Le Bars, P., Dupuy, J., Boudra, H., and Cassini, R. 1994. Biotic and abiotic factors in fumonsins $B_1$ production and stability. J. AOAC Int. 77: 517–521.

Lee, S.-E., Campbell, B. C., Molyneux, R.J., Hasegawa, S., and Lee, H.-S. 2001. Inhibitory effects of naturally occurring compounds on aflatoxin $B_1$ biotransformation. J. Agric. Food Chem. 49: 5171–5177.

Leslie, J.F. 1996. Introductory biology of Fusarium moniliforme. In: Fumonisins in Food. L.J. Jackson, J.W. DeVries, and L.B. Bullerman, ed. Plenum Press, New York, pp. 153–164.

Leslie, J.F., Zeller, K.A., and Summerell, B.A. 2001. Icebergs and species in populations of Fusarium. Physiol. Mol. Pathol. 59: 107–117.

Lewis, C.W., Smith, J.E., Anderson, J.G., and Freshney, R.I. 1999. Increased cytotoxicity of foodborne mycotoxins toward human cell lines in vitro via enhanced cytochrome p450 expression using the MTT bioassay. Mycopathologia 148: 97–102.

Li, F.-Q., Yoshizawa, T., Kawamura, O., Luo, X.-Y, and Li, Y.-W. 2001. Aflatoxins and fumonisins in corn from the high-incidence area for human hepatocellular carcinoma in Guangxi, China. J. Agric. Food Chem. 49: 4122–4126.

Li, G., Shinozuka, J., Uetsuka, K., Nakayama, H., and Doi, K. 1997a. T-2 toxin-induced apoptosis in intestinal crypt epithelial cells of mice. Exp. Toxicol. Pathol. 49: 447–450.

Li, J. and Shimizu, T. 1997. Course of apoptotic changes in the rat gastric mucosa caused by oral administration of fusarenon-X. J. Vet. Med. Sci. 59: 191–199.

Li, J., Shimizu, T., Miyoshi, N., and Yasuda, N. 1997b. Rapid apoptotic changes in the gastric glandular epithelium of rats administered intraperitoneally with fusarenon-X. J. Vet. Med. Sci. 59: 17–22.

Li, S. G., Ouyang, Y.L., Dong, W., and Pestka, J.J. 1997c. Superinduction of IL2 geneexpression by vomitoxin deoxynivalenol involves increased mRNA stability. Toxicol. Appl. Pharmacol. 147: 331–342.

Line, J.E., and Brackett, R.E. 1995a. Factors affecting aflatoxin $B_1$ removal by Flavobacterium aurantiacum. J. Food Prot. 58: 91–94.

Line, J.E., and Brackett, R.E. 1995b. Role of toxin concentration and second carbon source in microbial transformation of aflatoxin $B_1$ by Flavobacterium aurantiacum. J. Food Prot. 58: 1042–1044.

Line, J.E., Brackett, R.E., and Wilkinson, R. E. 1994. Evidence for degradation of aflatoxin $B_1$ by Flavobacterium aurantiacum. J. Food Prot. 57: 788–791.

Lisker, N., and Lillehoj, E.B.1991. Prevention of mycotoxin contamination (principally aflatoxins and Fusarium toxins) at the preharvest stage. In: Mycotoxins and Animal Foods. J.E. Smith and R.S. Henderson, ed. CRC Press, Boca Raton, FL, pp. 689–719.

Llovera, M., Viladrich, R., Torres, M., and Canela, R. 1999. Analysis of underivatized patulin by a GC-MS technique. J. Food Prot. 62: 202–205.

López, C., Ramos, L., Ramadán, S., Bulacio, L., and Perez, J. 2001. Distribution of aflatoxin $M_1$ in cheese obtained from milk artificially contaminated. Int. J. Food Microbiol. 64: 211–215.

Lopez-Diaz, T.-M. and Flannigan, B. 1997a. Mycotoxins of Aspergillus clavatus: Toxicity of cytochalasin E, patulin, and extracts of contaminated barley malt. J. Food Prot. 60: 1381–1385.

Lopez-Diaz, T.M., and Flannigan, B. 1997b. Production of patulin and cytochalasin E by Aspergillus clavatus during malting of barley and wheat. Int. J. Food Microbiol. 35: 129–136.

López-Díaz, T.M., Santos, J-A, Garca-López, M-L, and Otero, A. 2001. Surface mycoflora of a Spanish fermented meat sausage and toxigenicity of Penicillium isolates. Int. J. Food Microbiol. 68: 69–74.

López-García, R. and Park, D. L. 1998. Effectiveness of postharvest procedures in management of mycotoxin hazards. In: Mycotoxins in Agriculture and Food Safety. K.K. Sinha and D. Bhatnagar, ed. Marcel Dekker, Inc., New York, pp. 407–433.

Lopez-Garcia, R., Park, D.L., and Phillips, T.D. 1999. Integrated mycotoxin management systems. Food Nutr. Agric. 23: 38–48.

Lu, Y., Clifford, L., Hauck, C.C., Hendrich, S., Osweiler, G., and Murphy, P.A. 2002. Characterization of fumonisin $B_1$-glucose reaction kinetics and products. J. Agric. Food Chem. 50: 4726–4733.

Lukacs, Z., Schaper, S., Herderich, M., Schreier, P., and Humpf, H.-U. 1996. Identification and determination of fumonisin $FB_1$ and $FB_2$ in corn and corn-products by high-performance liquid chromatography – electrospray-ionization tandem mass spectrometry (HPLC-ESI-MS-MS). Chromatographia. 43: 124–128.

Mansour, N., Yousef, A.E., and Kim, J-G. 1996. Inhibition of surface growth of toxigenic and nontoxigenic aspergilli and penicillia by eugenol, isoeugenol, and monolaurin. J. Food Safety. 16: 219–229.

Mantle, P.G. 1991. Miscellaneous toxigenic fungi. In: Mycotoxins and Animal Foods. J.E. Smith and R.S. Henderson, ed. CRC Press, Boca Raton, FL. p. 141–152.

Mantle, P.G., and Chow, A.M. 2000. Ochratoxin formation in *Aspergillus ochraceus* with particular reference to spoilage of coffee. Int. J. Food Microbiol. 56: 105–109.

Maragos, C.M. 1995. Capillary zone electrophoresis and HPLC for the analysis of fluorescein isothiocyanate-labeled fumonisin B₁. J. Agric. Food Chem. 43: 390–394.

Maragos, C.M., and Plattner, R.D. 2002. Rapid fluorescence polarization immunoassay for the mycotoxin deoxynivalenol in wheat. J. Agric. Food Chem. 50: 1827–1832.

Maragos, C.M., and Richard J.L. 1994. Quantitation and stability of fumonisins B₁ and B₂ in milk. J. AOAC Int. 77: 1162–1167.

Marasas, W.F.O. 1991. Toxigenic fusaria. In: Mycotoxins and Animal Foods. J.E. Smith and R.S. Henderson, ed. CRC Press, Boca Raton, FL, pp. 119–139.

Marasas, W.F.O. 1995. Fumonisins: Their implications for human and animal health. Nat. Tox. 3: 193–198.

Marasas, W.F.O. 2001a. Discovery and occurrence of fumonisins: A historical perspective. Environ. Health Perspect. 109: 239–243.

Marasas, W.F.O. 2001b. *Fusarium*. In: Foodborne Disease Handbook. Volume 3: Plant Toxicants. 2nd edition. Y.H. Hui, R.A. Smith, and D.G. Spoerke, Jr., ed. Marcel Dekker, Inc., New York, pp. 535–580.

Marín, S., Sanchis, V., and Magan, N. 1995. Water activity, temperature and pH effects on growth of *Fusarium moniliforme* and *Fusarium proliferatum* isolates from maize. Can. J. Microbiol. 41: 1063–1070.

Marín, S., Sanchis, V., Rull, F., Ramos, A.J., and Magan, N. 1998. Colonization of maize grain by *Fusarium moniliforme* and *Fusarium proliferatum* in the presence of competing fungi and their impact on fumonisin production. J. Food Prot. 61: 1489–1496.

Marín, S., Sanchis, V., Teixido, A., Saenz, R., Ramos, A.J., Vinas, I., and Magan, N. 1996a. Water and temperature relations and microconidial germination of *Fusarium moniliforme* and *Fusarium proliferatum* from maize. Can. J. Microbiol. 42: 1045–1050.

Marín, S., Sanchis, Vinas, I., Canela, R., and Magan, N. 1996b. Effect of water activity and temperature on growth and fumonisin B₁ and B₂ production by *Fusarium moniliforme* and *Fusarium proliferatum* on maize grain. Lett. Appl. Microbiol. 21: 298–301.

Martins, M.L. and Martins, H.M. 2001. Determination of deoxynivalenol in wheat-based breakfast cereals marketed in Portugal. J. Food Prot. 64: 1848–1850.

Massey, T.E., Stewart, R.K., Daniels, J.M., and Liu, L. 1995. Biochemical and molecular aspects of mammalian susceptibility to aflatoxin B1 carcinogenicity. Proc. Soc. Exp. Biol. Med. 208: 213–227.

Mateo, J.J., Mateo, R., and Jiménez, M. 2002. Accumulation of type A trichothecenes in maize, wheat and rice by *Fusarium sporotrichioides* isolates under diverse culture conditions. Int. J. Food Microbiol. 72: 115–123.

Mayer, Z., Färber, P., and Geisen, R. 2003. Monitoring the production of aflatoxin B₁ in wheat by measuring the concentration of *nor-1* mRNA. Appl. Environ. Microbiol. 69: 1154–1158.

McKinley, E.R., and Carlton, W.W. 1991. Patulin. In: Mycotoxins and Phytoalexins. R.P. Sharma and D.K. Salunkhe, ed. CRC Press, Boca Raton, FL. p. 191–236.

McLean, M., and Dutton, M.F. 1995. Cellular interactions and metabolism of aflatoxin: An update. Pharmac. Ther. 65: 163–192.

McMullen, M., Jones, R., and Gallenberg, D. 1997. Scab of wheat and barley: A re-emerging disease of devastating impact. Plant Dis. 81: 1340–1348.

Medina-Martínez, M.S., and Martínez, A.J. 2000. Mold occurrence and aflatoxin B₁ and fumonisin B₁ determination in corn samples in Venezuela. J. Agric. Food Chem. 48: 2833-2836.

Meky, F.A., Hardie, L.J., Evans, S.W., and Wild, C.P. 2001. Deoxynivalenol-induced immunomodulation of human lymphocyte proliferation and cytokine production. Food Chem. Toxicol. 39: 827–836.

Meky, F.A., Turner, P.C., Ashcroft, A.E., Miller, J.D., Qiao, Y.-L., Roth, M.J., and Wild, C.P. 2003. Development of a urinary biomarker of human exposure to deoxynivalenol. Food Chem. Toxicol. 41: 265–273.

Mellon, J.E., Cotty, P.J., and Dowd, M.K. 2000. Influence of lipids with and without other cottonseed reserve materials on aflatoxin B₁ production by *Aspergillus flavus*. J. Agric. Food Chem. 48: 3611–3615.

Meredith, F.I., Riley, R.T., Bacon, C.W., Williams, D.E., and Carlson, D.B. 1998. Extraction, quantification, and biological availability of fumonisin B₁ incorporated into the Oregon test diet and fed to rainbow trout. J. Food Prot. 61: 1034–1038.

Meredith, F.I., Torres, O.R., Saenz de Tejada, S., Riley, R.T., and Merrill, A.H.Jr. 1999. Fumonisin B₁ and hydrolyzed fumonisin B₁ (AP₁) in tortillas and nixtamalized corn (*Zea mays* L.) from two different geographic locations in Guatemala. J. Food Prot. 62: 1218–1222.

Merrill, A.H., Jr. Sullards, M.C., Wang, E., Voss, K.A., and Riley, R.T. 2001. Sphingolipid metabolism: roles in signal transduction and disruption by fumonisins. Environ. Health Perspect. 109: 283–289.

Mesterházy, A., Bartók, T., Mirocha, G.G., and Komoróczy, R. 1999. Nature of wheat resistance to *Fusarium* head blight and the role of deoxynivalenol for breeding. Plant Breeding 118: 97–110.

Miedaner, T. 1997. Breeding wheat and rye for resistance to *Fusarium* diseases. Palnt Breeding 116: 201–220.

Miedaner, T., Reinbrecht, C., Lauber, U., Schollenberger, M., and Geiger, H.H. 2001. Effects of genotype and genotype−environment interaction on deoxynivalenol accumulation and resistance to *Fusarium* head blight in rye, triticale, and wheat. Plant Breeding 120: 97–105.

Miller, J.D. 1994. Epidemiology of *Fusarium* ear diseases of cereals. In: Mycotoxins in Grain. Compounds Other Than Aflatoxin. J.D.Miller and H.L. Trenholm. ed. Eagan Press, St. Paul, MN, pp. 19–36

Miller, J.D. 1995. Fungi and mycotoxins in grain: Implications for stored product research. J. Stored Prod. Res. 31: 1–16.

Miller, J.D. 2001. Factors that affect the occurrence of fumonisin. Environ. Health Perspect. 109: 321–324.

Miller, J.D. 2002. Aspects of the ecology of *Fusarium* toxins in cereals. In: Mycotoxins and Food Safety. J.W. DeVries, M.W. Trucksess, and L.S. Jackson, ed. Kluver Academic/Plenum Publishers, New York, pp. 19–27.

Mirocha, C.J., Gilchrist, D.G., Shier, W.T., Abbas, H.K., Wen, Y., and Vesonder, R.F. 1992. AAL toxins, fumonisins (biology and chemistry) and host-specificity concepts. Mycopathologia 117: 47–56.

Mirocha, C.J., Kolaczkowski, E., Xie, W., Yu, H., and Jelen, H. 1998. Analysis of deoxynivalenol and its derivatives (batch and single kernel) using gas chromatography/mass spectrometry. J. Agric. Food Chem. 46: 1414–1418.

Mitterbauer, R., Weindorfer, H., Safaie, N., Krska, R., Lemmens, M., Ruckenbauer, P., Kuchler, K., and Adam, G. 2003. A sensitive and inexpensive yeast bioassay for the mycotoxin zearalenone and other compounds with estrogenic activity. Appl. Environ. Microbiol. 69: 805–811.

Miyahara, M., Akiyama, H., Toyoda, M., and Saito, Y. 1996. New procedure for fumonisins B₁ and B₂ in corn and corn products by ion pair chromatography with *o*-phthaldialdehyde postcolumn derivatization and fluorometric detection. J. Agric. Food Chem. 44: 842–847.

Molyneux, R.J., Mahoney, N., and Campbell, B.C. 2000. Anti-aflatoxigenic constituents of *Pistacia* and *Juglans* species. In: Natural and Selected Synthetic Toxins. Biological Implications. A.T. Tu and W. Gaffield, ed. American Chemical Society, Washington, D.C, pp. 43–53.

Monnet-Tschudi, F., Sorg, O., Honegger, P., Zurich, M.-G., Huggett, A.C., and Schilter, B. 1997. Effects of the naturally occurring food mycotoxin ochratoxin A on brain cells in culture. NeuroToxicology 18: 831–840.

Montville, T.J., and Shih, P-L. 1991. Inhibition of mycotoxigenic fungi in corn by ammonium and sodium bicarbonate. J. Food Prot. 54: 295–297.

Moodley, R.S., Govinden, R., and Odhav, B. 2002. The effect of modified atmospheres and packaging on patulin production in apples. J. Food Prot. 65: 867–871.

Moss, M.O. 1991. The environmental factors controlling mycotoxin formation. In Mycotoxins and Animal Foods. J. E. Smith and R.S. Henderson, ed. CRC Press, Boca Raton, FL, pp. 37-56.

Moss, M.O. 1998. Recent studies of mycotoxins. J. Appl. Microbiol. Symp. Suppl. 84: 62S-76S.

Mossoba, M.M., Adams, S., Roach, J.A.G., and Trucksess, M. W. 1996. Analysis of trichothecene mycotoxins in contaminated grains by gas chromatography/matrix isolation/fourier transform infrared spectroscopy and gas chromatography/mass spectrometry. J. AOAC Int. 79: 1116–1123.

Moy, G.G. 1998. Roles of national governments and international agencies in the risk analysis of mycotoxins. In: Mycotoxins in Agriculture and Food Safety. K.K. and D. Bhatnagar, ed. Marcel Dekker, Inc., New York, pp. 483–496.

Munkvold, G.P., and Desjardins, A. E. 1997. Fumonisins in maize. Can we reduce their occurrence? Plant Dis. 81: 556–565.

Murillo, I., Cavallarin, L., and San Segundo, B. 1998. The development of a rapid PCR assay for detection of *Fusarium moniliforme*. Eur. J. Plant Pathol. 104: 301–311.

Murphy, P.A., Rice, L.G., and Ross, P.F. 1993. Fumonisin B$_1$, B$_2$, and B$_3$ content of Iowa, Wisconsin, and Illinois corn and corn screenings. J. Agric Food Chem. 41: 263–266.

Nasir, M.S. and Jolley, M.E. 2002. Development of a fluorescence polarization assay for the determination of aflatoxins in grains. J. Agric. Food Chem. 50: 3116–3121.

Neira, M.S., Pacin, A.M., Martínez, E.J., Moltó, G., and Resnik, S.L. 1997. The effects of bakery processing on natural deoxynivalenol contamination. Int. J. Food Microbiol. 37: 21–25.

Nelson, P.E., Toussoun, T.A., and Marasas, W.F.O. 1983. *Fusarium* Species: An Illustrated Manual for Identification. Pennsylvania State University Press, University Park, PA.

Nesheim, S., and Wood, G.E. 1995. Regulatory aspects of mycotoxins in soybean and soybean products. J. Am. Oil Chem. Soc. 72: 1421–1423.

Nesheim, S., Stack, M.E., Trucksess, M.W., Eppley, R.M., and Krogh, P. 1992. Rapid solvent-efficient method for liquid chromatographic determination of ochratoxin A in corn, barley, and kidney: Collaborative study. J. AOAC Int. 75: 481–487.

Newton, B. 1999. Managing vomitoxin in the cereal processing industry. Cereal Foods World. 44: 338–341.

Nicholson, P., Simpson, D.R., Weston, G., Rezanoor, H.N., Lees, A.K., Parry, D.W., and Joyce, D. 1998. Detection and quantification of *Fusarium culmorum* and *Fusarium graminearum* in cereals using PCR assays. Physiol. Mol. Plant Pathol. 53: 17–37.

Niessen, M. L., and Vogel, R.F. 1998. Group specific PCR-detection of potential trichothecene-producing *Fusarium*-species in pure cultures and cereal samples. System. Appl. Microbiol. 21: 618–631.

Norred W.P., and Riley, R.T. 2001. Toxicology – mode of action of mycotoxins. In: Mycotoxins and Phycotoxins in Perspective at the Turn of the New Millennium. W.J. de Koe, R.A. Samson, H.P. van Egmond, J. Gilbert, and M. Sabino, eds. W.J. de Koe, Hazekamp2, Wageningen, The Netherlands, pp. 211–222.

Norred, W.P., Bacon, C.W., Riley, R.T., Voss, K.A., and Meredith, F.I. 1999. Screening of fungal species for fumonisin production and fumonisin-like disruption of sphingolipid biosynthesis. Mycopathologia 146: 91–98.

Norred, W.P., Riley, R.T., Meredith, F.I., Poling, S.M., and Plattner, R.D. 2001. Instability of N-acetylated fumonisin B1 (FA1) and the impact on inhibition of ceramide synthase in rat liver slices. Food Chem. Toxicol. 39: 1071–1078.

Norred, W.P., Voss, K.A., Bacon, C.W., and Riley, R.T. 1991. Effectiveness of ammonia treatment in detoxification of fumonisin-contaminated corn. Food Chem. Toxicol. 29: 815–818.

Norton, R.A. 1997. Effect of carotenoids on aflatoxin B$_1$ synthesis by *Aspergillus flavus*. Phytopathology 87: 814–821.

Norton, R.A. 1999. Inhibition of aflatoxin B$_1$ biosynthesis in *Aspergillus flavus* by anthocyanidins and related flavonoids. J. Agric. Food Chem. 47: 1230–1235.

NTP Technical Report 2002. Toxicology and Carcinogenesis Studies of Fumonisin B$_1$ in F344/N rats and B6C3F1 mice. NIH Publication No. 99–3955.

Núñez, F., Rodríguez, M.M., Bermúdez, M.E., Córdoba, J.J., and Asensio, M.A. 1996. Composition and toxigenic potential of the mould population and dry-cured Iberian ham. Int. J. Food Microbiol. 32: 185–197.

Oatley, J.T., Rarick, M.D., Ji, G.E., and Linz, J.E. 2000. Binding of aflatoxin B$_1$ to bifidobacteria in vitro. J. Food Prot. 63: 1133–1136.

Olsson, J., Börjesson, T., Lundstedt, T., and Schnürer, J. 2002. Detection and quantification ofochratoxin A and deoxynivalenol in barley grains by GC-MS and electronic nose. Int. J. Food Microbiol. 72: 203–214.

Ominski, K.H., Marquardt, R.R., Sinha, R.N., and Abramson, D. 1994. Ecological aspects of growth and mycotoxin production by storage fungi. In: Mycotoxins in Grain. Compounds Other Than Aflatoxin. J.D. Miller and H.L. Trenholm, ed. Eagan Press, St. Paul, MN, pp. 287-312.

Oren, L., Ezrati, S., Cohen, D., and Sharon, A. 2003. Early events in the *Fusarium verticillioides*-maize interaction characterized by using a green fluorescent protein-expressing transgenic isolate. Appl. Environ. Microbiol. 69: 1695–1701.

Orsi, R.B., Corrêa, B., Possi, C.R., Schammass, E.A., Nogueira, J.R., Dias, S.M.C., and Malozzi, M.A.B. 2000. Mycoflora and occurrence of fumonisins in freshly harvested and stored hybrid maize. J. Stored Prod. Res. 36: 75–87.

Oswald, I.P., and Coméra, C. 1998. Immunotoxicity of mycotoxins. Revue Méd. Vét. 149: 585–590.

Osweiler, G.D. 2000. Mycotoxins: Contemporary issues of food animal health and productivity. Vet. Clin. North Am. Food Anim. Pract. 16: 511–530.

Ouyang, Y.L., Li, S., and Pestka, J.J. 1996. Effects of vomitoxin (deoxynivalenol) on transcription factor NF-kappa B/Rel binding activity in murine EL-4 thymoma and primary CD4+ T cells. Toxicol. Appl. Pharmacol. 140: 328–336.

Özay, G., Aran, N., and Pala, M. 1995. Influence of harvesting and drying techniques on microflora and mycotoxin contamination of figs. Nahrung 39: 156–165.

Palencia, E., Torres, O., Hagler, W., Meredith, F.I., Williams, L.D., and Riley, R.T. 2003. Total fumonisins are reduced in tortillas using the traditional nixtamalization method of Mayan communities. J. Nutr. 133: 3200–3203.

Parent-Massin, D., and Parchment, R.E. 1998. Haematotoxicity of mycotoxins. Revue Méd. Vét. 149: 591–598.

Park, D.J., and Liang, B. 1993. Perspectives on aflatoxin control for human food and animal feed. Trends Food Sci. Technol. 4: 334–342.

Park, D.L. 1995. Surveillance programmes for managing risks from naturally occurring toxicants. Food Addit. Contam. 12: 361–371.

Park, D.L., and Troxell, T.C. 2002. USA perspective on mycotoxin regulatory issues. In: Mycotoxins and Food Safety. J.W. DeVries, M.W. Trucksess, and L.S. Jackson, ed. Kluwer Academic/Plenum Publishers, New York, pp. 277–285.

Park, D.L., Njapau, H., and Boutrif, E. 1999. Minimizing risks posed by mycotoxins utilizing the HACCP concept. Food Nutr. Agric. 23: 49–56.

Park, J.J. and Chu, F.S. 1996. Assessment of immunochemical methods for the analysis of trichothecene mycotoxins in naturally occurring moldy corn. J. AOAC Int. 79: 465–471.

Pascale, M., Visconti, A., Prończuk, M., Wiśniewska, H., and Chelkowski, J. 1997. Accumulation of fumonisins in maize hybrids inoculated under field conditions with *Fusarium moniliforme* Sheldon. J. Sci. Food Agric. 74: 1–6.

Payne, G.A. 1998. Process of contamination of aflatoxin-producing fungi and their impact on crops. In: Mycotoxins in Agriculture and Food Safety. K.K. Sinha and D. Bhatnagar, ed. Marcel Dekker, Inc., New York. pp. 279–306.

Payne, G.A., and Brown, M.P. 1998. Genetics and physiology of aflatoxin biosynthesis. Ann. Rev. Phytopathol. 36: 329–362.

Pažoutová, S., and Parbery, D. P. 1999. In: Ergot. The Genus *Claviceps*. V. Křen and L. Cvak, ed. Harwood Academic Publishers, Amsterdam, The Netherlands, pp. 57–93.

Pelagalli, A., Belisario, M.A., Squillacioti, C., Morte, R.D., d'Angelo, D., Tafuri, S., Lucisano, A., and Staiano, N. 1999. The mycotoxin fumonisin B1 inhibits integrin-mediated cell-matrix adhesion. Biochimie 81: 1003–1008.

Peltonen, K., El-Nezami, H., Haskard, C., Ahokas, J., and Salminen, S. 2001. Aflatoxin B$_1$ binding by dairy strains of lactic acid bacteria and bifidobacteria. J. Dairy Sci. 84: 2152–2156.

Peltonen, K.D., El-Nezami, H.S., Salminen, S.J., and Ahokas, J.T. 2000. Binding of aflatoxin B$_1$ by probiotic bacteria. J. Sci. Food Agric. 80: 1942–1945.

Pemberton, A.D., and Simpson, T.J. 1991. The chemical degradation of mycotoxins. In: Mycotoxins and Animal Foods. J.E. Smith and R.S. Henderson, ed. CRC Press, Boca Raton, FL, pp. 797–813.

Peplow, A.W., Tag, A.G., Garifullina, G.F., and Beremand, M.N. 2003. Identification of new genes positively regulated by Tri10 and a regulatory network for trichothecene mycotoxin production. Appl. Environ. Microbiol. 69: 2731–2736

Peraica, M., Radic, B., Lucic, A., and Pavlovic, M. 1999. Toxic effects of mycotoxins in humans. Bull. World Health Org. 77: 754–766

Pestka, J.J., and Bondy, G.S. 1994. Immunotoxic effects of mycotoxins. In: Mycotoxins in Grain. Compounds Other Than Aflatoxin. J.D. Miller and H.L. Trenholm, ed. Eagan Press, St. Paul, MN, pp. 339–358.

Pestka, J.J., Abouzeid, M.N., and Sutikno. 1995. Immunological assays for mycotoxin detection. Food Technol. 49(2): 120–128.

Pestka, J.J., Azcona-Olivera, J.I., Plattner, R.D., Minervini, F., Doko, M.B., and Visconti, A. 1994a. Comparative assessment of fumonisin in grain-based foods by ELISA, GC-MS, and HPLC. J. Food Prot. 57: 169–172.

Pestka, J.J., Dong, W., Warner, R.L., Rasooly, L., Bondy, G.S., and Brooks, K.H. 1990a. Elevated membrane IgA$^+$ and CD4$^+$-T helper populations in murine Peyer's patch and splenic lymphocytes during dietary administration of the trichothecene vomitoxin deoxynivalenol. Food Chem. Toxicol. 28: 409–420.

Pestka, J.J., Moorman, M.A., and Warner, R. 1989. Dysregulation of IgA production and IgA nephropathy induced by the trichothecene vomitoxin. Food Chem. Toxicol. 27: 361–368.

Pestka, J.J., Moorman, M.A., and Warner, R. 1990b. Altered serum immunoglobulin response to model intestinal antigens during dietary exposure to vomitoxin deoxynivalenol. Toxicol. Lett. 50: 75–84.

Pestka, J.J., Warner, R.L., Dong, W., Rasooly, L., and Bondy, G.S. 1990c. Effects of dietary administration of the trichothecene vomitoxin deoxynivalenol on IgA and IgG secretion by Peyer's patch and splenic lymphocytes. Food Chem. Toxicol. 28: 693–699.

Pestka, J. J., Yan, D., and King, L.E. 1994b. Flow cytometric analysis of the effects of *in vitro* exposure to vomitoxin (deoxynivalenol) on apoptosis in murine T, B and IgA$^+$ cells. Food Chem. Toxic. 32: 1125–1136.

Peters-Volleberg, G.W.M., Beems, R.B., and Speijers, G.J.A. 1996. Subacute toxicity of ergometrine maleate in rats. Food Chem. Toxicol. 34: 951–958.

Petzinger, E., and Weidenbach, A. 2002. Mycotoxins in the food chain: the role of ochratoxins. Livest. Prod. Sci. 76: 245–250.

Phillips, T.D., Clement, B.A., and Park, D.L. 1994. Approaches to reduction of aflatoxins in foods and feeds. In: The Toxicology of Aflatoxins. Human Health, Veterinary, and Agricultural Significance. D.L. Eaton and J.D. Groopman, ed. Academic Press, New York, pp. 383–406.

Phillips T.D., Lemke S.L., and Grant P.G. 2002. Characterization of clay-based enterosorbents for the prevention of aflatoxicosis. Adv. Exp. Med. Biol. 504: 157–171.

Pierides, M., El-Nezami, H., Peltonen, K., Salminen, S., and Ahokas, J. 2000. Ability of dairy strains of lactic acid bacteria to bind aflatoxin M$_1$ in a food model. J. Food Prot. 63: 645–650.

Pieters, M.N., Freijer, J., Baars, B.-J., Fiolet, D.C.M., van Klaveren, J., and Slob, W. 2002. Risk assessment of deoxynivalenol in food: concentra-

tion limits, exposure and effects. In: Mycotoxins and Food Safety. J.W. DeVries, M.W. Trucksess, and L.S. Jackson, ed. KluverAcademic/Plenum Publishers, New York, pp. 235–248.

Pitt, J.I. 2000. Toxigenic fungi and mycotoxins. Br. Med. Bull. 56: 184–192.

Pitt, J.I. 2001a. *Penicillium*. In: Foodborne Disease Handbook. Volume 3: Plant Toxicants. 2nd edition. Y.H. Hui, R.A. Smith, and D.G. Spoerke, Jr., ed. Marcel Dekker, Inc., New York, pp. 581–591.

Pitt, J.I. 2001b. Toxigenic *Penicillium* species. In: Food Microbiology. Fundamentals and Frontiers. 2ND edition. M.P. Doyle, L.R. Beuchat, and T.J. Montville, ed. ASM Press, Washington, D.C. p. 467–480.

Pitt, J.I. 2002. Biology and ecology of toxigenic *Penicillium* species. In: Mycotoxins and Food Safety. J.W. DeVries, M.W. Trucksess, and L.S. Jackson, ed. Kluver Academic/Plenum Publishers, New York, pp. 29–41.

Pitt, J.I., and Hocking, A.D. 1997. Fungi and Food Spoilage. 2nd edition. Blackie Academic & Professional (Chapman and Hall), New York.

Pitt, J.I., and Leistner, L. 1991. Toxigenic *Penicillium* species. In: Mycotoxins and Animal Foods. J.E. Smith and R.S. Henderson, ed. CRC Press, Boca Raton, FL, pp. 81–99.

Pitt, J.I., and Miscamble, B.F. 1995. Water relations of *Aspergillus flavus* and closely related species. J. Food Prot. 58: 86–90.

Pittet, A.1998. Natural occurrence of mycotoxins in foods and feeds: an updated review. Revue Méd. Vét. 149: 479–492.

Pittet, A., Parisod, V., and Schellenberg, M. 1992. Occurrence of fumonisins B$_1$ and B$_2$ in corn-based products from the Swiss market. J. Agric. Food Chem. 40: 1352–1354.

Pittet, A., Tornare, D., Huggett, A., and Viani, R. 1996. Liquid chromatographic determination of ochratoxin A in pure and adulterated soluble coffee using an immunoaffinity column cleanup procedure. J. Agric. Food Chem. 44: 3564–3569.

Poling, S.M., Plattner, R.D., and Weisleder, D. 2002. N-(1-deoxy-D-fructos-1-yl) fumonisin B$_1$, the initial reaction product of fumonisin B$_1$ and D-glucose. J. Agric. Food Chem. 50: 1318–1324.

Porter, J.K. 2001. Analytical methodology for mycotoxins. In: Foodborne Disease Handbook. Volume 3: Plant Toxicants. 2nd edition. Y.H. Hui, R.A. Smith, and D.G. Spoerke, Jr., ed. Marcel Dekker, Inc., New York, pp. 653–682.

Prasongsidh, B.C., Kailasapathy, K., Skurray, G.R., and Bryden, W.L. 1997a. Stability of cyclopiazonic acid during storage and processing of milk. Food Res. Int. 30: 793–798.

Prasongsidh, B.C., Kailasapathy, K., Skurray, G.R., and Bryden, W.L. 1997b. Analysis of cyclopiazonic acid in milk by capillary electrophoresis. Food Chem. 61: 515–519.

Prasongsidh, B.C., Kailasapathy, K., Skurray, G.R., and Bryden, W.L. 1998a. Behaviour of cyclopiazonic acid during the manufacture and storage of yogurt. Aust. J. Dairy Technol. 53: 152–155.

Prasongsidh, B.C., Kailasapathy, K., Skurray, G.R., and Bryden, W.L. 1998b. Kinetic study of cyclopiazonic acid during the heat-processing of milk. Food Chem. 62: 467–472.

Prasongsidh, B.C., Sturgess, R., Skurray, G.R., and Bryden, W.L. 1999a. Fate of cyclopiazonic acid in Cheddar cheese. Milchwissenschaft 54: 200–203.

Prasongsidh, B.C., Sturgess, R., Skurray, G.R., and Bryden, W.L. 1999b. Partition behaviour of cyclopiazonic acid in cream and butter. Milchwissenschaft 54: 272–274.

Prasongsidh, B.C., Kailasapathy, K., Sturgess, R., Skurray, G.R., and Bryden, W.L. 1999c. Influence of manufacturing and storing of ice-cream on cyclopiazonic acid. Milchwissenschaft 54: 141–143.

Prelusky, D.B. 1994. Residues in food products of animal origin. In: Mycotoxins in Grain. Compounds Other Than Aflatoxin. J.D. Miller and H.L. Trenholm, ed. Eagan Press, St. Paul, MN, pp. 405–419.

Proctor, R.H., Brown, D.W., Plattner, R.D., and Desjardin, A.E. 2003. Co-expression of 15 contiguous genes delineates a fumonisin biosynthetic cluster in *Gibberella moniliformis*. Fungal Genet. Biol. 38: 237–249.

Qiu, M. and Liu, X. 2001. Determination of sphinganine, sphingosine and Sa/So ratio in urine of humans exposed to dietary fumonisin B$_1$. Food Addit. Contam. 18: 263–269.

Radíc, B., Fuchs, R., Peraica, M., and Lucíc, A. 1997. Ochratoxin A in human sera in the area with endemic nephropathy in Croatia. Toxicol. Lett. 91: 105–109.

Radová, Z., Hajšlová, J., and Králová, J. 2001. Analysis of zearalenone in wheat using high-performance liquid chromatography with fluorescence detection and/or enzyme-linked immunosorbent assay. Cereal Res. Comm. 29: 435–442.

Reid, L.M., Stewart, D.W., and Hamilton, R.I. 1996. A 4-year study of the association between Gibberella ear rot severity and deoxynivalenol concentration. J. Phytopathol. 144: 431–436.

Rheeder, J.P. Marasas, W.F.O., and Vismer, H.F. 2002. Production of fumonisin analogs by Fusarium species. Appl. Environ. Microbiol. 68: 2101–2105.

Rheeder, J.P., Marasas, W.F.O., Thiel, P.G., Sydenham, E.W., Shephard, G.S., and van Schalkwyk, D. J. 1992. Fusarium moniliforme and fumonisins in corn in relation to human esophageal cancer in Transkei. Phytopathology 82: 353–357.

Rice, L.G. and Ross, P.F. 1994. Methods for detection and quantitation of fumonisins in corn, cereal products and animal excreta. J. Food Prot. 57: 536–540.

Richard, J.L., Bennett, G.A., Ross, P.F., and Nelson, P.E. 1993. Analysis of naturally occurring mycotoxins in feedstuffs and food. J. Anim. Sci. 71: 2563–2574.

Riley, R.T. 1998. Mechanistic interaction of mycotoxins: Theoretical considerations. In: Mycotoxins in Agriculture and Food Safety. K.K. Sinha and D. Bhatnagar, ed. Marcel Dekker, New York, NY, pp. 227–223.

Riley, R.T., and Norred, W.P. 1996. Mechanistic toxicology of mycotoxins. In: The Mycota VI. Human and Animal Relationships. Howard, D.H., and Miller, J.D., eds. Springer-Verlag, Berlin, pp. 193–211.

Riley, R.T., and Norred, W.P. 1999. Mycotoxin prevention and decontamination – a case study on maize. Food Nutr. Agric. 23: 25–32.

Riley, R.T., An, N-H., Showker, J.L, Yoo, H-S., Norred, W.P., Chamberlain, W.J., Wang, E., Merrill, A.H. Jr., Motelin, G., Beasley, V.R., and Haschek, W.M. 1993. Alteration of tissue and serum sphinganine to sphingosine ratio: An early biomarker of exposure to fumonisin-containing feeds in pigs. Toxicol. Appl. Pharmacol. 118: 105–112.

Riley, R.T., Enongene, E., Voss, K.A., Norred, W.P., Meredith, F.I., Sharma, R.P., Williams, D.E., Carlson, D.B., Spitsbergen, J., and Merrill, A.H., Jr. 2001. Sphingolipid perturbations as mechanisms for fumonisin carcinogenesis. Environ. Health Perspect. 109: 301–308.

Riley, R.T., Goeger, D.E., and Norred, W.P. 1995. Disruption of calcium homeostasis: The cellular mechanism of cyclopiazonic acid toxicity in laboratory animals. In: Molecular Approaches to Food Safety. M. Eklund, J.L. Richard, and K. Mise, ed. Alaken, Inc., Fort Collins, CO, pp. 461–480.

Riley, R.T., Hinton, D.M., Chamberlain, W.J., Bacon, C.W., Wang, E., Merrill, A.H., Jr., and Voss, K.A. 1994a. Dietary fumonisin $B_1$ induces disruption of sphingolipid metabolism in Sprague–Dawley rats: A new mechanism of nephrotoxicity. J. Nutr. 124: 594–603.

Riley, R.T., Norred, W.P., Wang, E., and Merrill, A.H., Jr. 1999a. Alteration in sphingolipid metabolism: bioassay for fumonisin- and ISP-I-like activity in tissues, cells, and other matrices. Natural Toxins 7: 407–414.

Riley, R.T., Showker, J.L., Owens, D.L., and Ross, P.F. 1997. Disruption of sphingolipid metabolism and induction of equine leudoencephalomacia by Fusarium proliferatum culture material containing $FB_2$ or $FB_3$. Environ. Toxicol. Pharmacol. 3: 221–228.

Riley, R.T., Voss, K.A., Norred, W.P., Sharma, R.P., Wang, E., and Merrill, A.H., Jr. 1998. Fumonisins: mechanism of mycotoxicity. Revue Méd. Vét. 149: 617–626.

Riley, R.T., Voss, K.A., Norred, W.P., Bacon C.W., Meredith F.I., and Sharma, R.P. 1999b. Serine palmitoyltransferase inhibition reverses antiproliferative effects of ceramide synthase inhibition in cultured renal cells and suppresses free sphingoid base accumulation in kidney of BALBc mice. Environ Toxicol Pharmacol. 7: 109–118.

Riley, R.T., Voss, K.A., Yoo, H.-S., Gelderblom, W.C.A., and Merrill, A.H., Jr. 1994b. Mechanism of fumonisin toxicity and carcinogenesis. J. Food Prot. 57: 638–645.

Robens, J. and Riley, R.T. (ed.). 2002. Proceedings of the 1st Fungal Genomics, 2nd Fumonisin Elimination and 14th Aflatoxin Elimination Workshops, Mycopathologia 155: 1–164.

Romani, S., Sacchetti, G., López, C.C., Pinnavaia, G.G., and Rosa, M.D. 2000. Screening on the occurrence of ochratoxin A in green coffee beans of different origins and types. J. Agric. Food Chem. 48: 3616–3619.

Romer, T., and Maune, C. 1993. A practical approach to mycotoxin quality control. Cereal Foods World 38: 349–352.

Rotter, B.A., Prelusky, D.B., and Pestka, J.J. 1996. Toxicology of deoxynivalenol (vomitoxin). J. Toxicol. Environ. Health 48: 1–34.

Rupp, H.S., and Turnipseed, S.B. 2000. Confirmation of patulin and 5-hydroxymethylfurfural in apple juice by gas chromatography/mass spectrometry. J. AOAC Int. 83: 612–620.

Rustom, I.Y.S. 1997. Aflatoxin in food and feed: occurrence, legislation and inactivation by physical methods. Food Chem. 59: 57–67.

Ryu, D., and Bullerman, L.B. 1999. Effect of cycling temperatures on the production of deoxynivalenol and zearalenone by Fusarium graminearum NRRL 5883. J. Food Prot. 62: 1451–1455.

Ryu, D., Hanna, M.A., and Bullerman, L.B. 1999a. Stability of zearalenone during extrusion of corn grits. J. Food Prot. 62: 1482–1484.

Ryu, D., Munimbazi, C., and Bullerman, L.B. 1999b. Fumonisin $B_1$ production by Fusarium moniliforme and Fusarium graminearum as affected by cycling temperatures. J. Food Prot. 62: 1456–1460.

Sabbioni, G., and Sepai, O. 1998. Determination of human exposure to aflatoxins. In: Mycotoxins in Agriculture and Food Safety. K.K. Sinha and D. Bhatnagar, ed. Marcel Dekker, Inc., New York, pp. 183–226.

Sage, L., Krivobok, S., Delbos, E., Seigle-Murandi, F., and Creppy, E.E. 2002. Fungal flora and ochratoxin A production in grapes and musts from France. J. Agric. Food Chem. 50: 1306–1311.

Samarajeewa, U. 1991. In situ degradation of mycotoxins by physical methods. In: Mycotoxins and Animal Foods. J.E. Smith, and R.S. Henderson, ed. CRC Press, Boca Raton, FL, pp. 785–796.

Samarajeewa, U., Sen, A.C., Cohen, M.D., and Wei, C.I. 1990. Detoxification of aflatoxins in foods and feeds by physical and chemical methods. J. Food Prot. 53: 489–501.

Samson, R.A. and Pitt, J.I. (ed.) 2000. Integration of Modern Taxonomic Methods for Penicillium and Aspergillus Classification. Harwood Academic Publishers, Amsterdam, The Netherlands.

Samson, R.A., Hoekstra, E.S., Frisvad, J.C., and Filtenborg, O. (ed.) 2002. Introduction to Food- and Airborne Fungi. 6th edition. Centraalbureau voor Schimmelcultures, Utrecht, The Netherlands.

Sanchis, V., Abadias, M., Oncins, L., Sala, N., Viñas, I., and Canela, R. 1994. Occurrence of fumonisins $B_1$ and $B_2$ in corn-based products from the Spanish market. Appl. Environ. Microbiol. 60: 2147–2148.

Sandvig, K., Garred, O., Van Helvoort, A., Van Meer, G., and Van Deurs, B. 1996. Importance of glycolipid synthesis for butyric acid-induced sensitization to Shiga toxin and intracellular sorting of toxin in A431 cells. Mol Biol Cell 7: 1391–1404.

Sashidhar, R.B. 1993. Fate of aflatoxin $B_1$ during the industrial production of edible defattedpeanut protein flour from raw peanuts. Food Chem. 48: 349–352.

Saunders, D.S., Meredith, F.I., and Voss, K.A. 2001. Control of fumonisin: Effects of processing. Environ. Health Perspect. 109: 333–336.

Schartzki, T.F. 1995a. Distribution of aflatoxin in pistachios. 1. Lot distributions. J. Agric. Food Chem. 43: 1561–1565.

Schartzki, T.F. 1995b. Distribution of aflatoxin in pistachios. 2. Distribution in freshly harvested pistachios. J. Agric. Food Chem. 43: 1566–1569.

Schartzki, T.F. 1996. Distribution of aflatoxin in almonds. J. Agric. Food Chem. 44: 3595–3597.

Schartzki, T.F., and Pan, J.L. 1996. Distribution of aflatoxin in pistachios. 3. Distribution in pistachio process streams. J. Agric. Food Chem. 44: 1076–1084.

Schartzki, T.F., and Pan, J.L. 1997. Distribution of aflatoxin in pistachios. 4. Distribution in small pistachios. J. Agric. Food Chem. 45: 205–207.

Schmelz, E-M., Dombrink-Kurtzman, M.A., Roberts, P.C., Kozutsumi, Y., Kawasaki, T., and Merrill, A.H., Jr. 1998. Induction of apoptosis by fumonisin B$_1$ in HT29 cells is mediated by the accumulation of endogenous free sphingoid bases. Toxicol. Appl. Pharmacol. 148: 252–260.

Schnerr, H., Niessen, L., and Vogel, R.F. 2001. Real time detection of the *tri5* gene in *Fusarium* species by LightCycler$^{TM}$ -PCR using SYBR$^{®}$ Green I for continuous fluorescence monitoring. Int. J. Food Microbiol. 71: 53–61.

Schnerr, H., Vogel, R.F., and Niessen, L. 2002. Correlation between DNA of trichothecene-producing *Fusarium* species and deoxynivalenol concentrations in wheat-samples. Lett. Appl. Microbiol. 35: 121–125.

Schollenberger, M., Jara, H.T., Suchy, S., Drochner, W., and Müller, H.-M. 2002. *Fusarium* toxins in wheat flour collected in an area in southwest Germany. Int. J. Food Microbiol. 72: 85–89.

Schollenberger, M., Suchy, S., Jara, H.T., Drochner, W., and Müller, H-M. 1999. A survey of *Fusarium* toxins in cereal-based foods marketed in an area of southwest Germany. Mycopathologia 147: 49–57.

Schroeder, J.J., Crane H.M., Xia J., Liotta, D.C., and Merrill, A.H., Jr. 1994. Disruption of sphingolipid metabolism and stimulation of DNA synthesis by fumonisin B$_1$. A molecular mechanism for carcinogenesis associated with *Fusarium moniliforme*. J. Biol. Chem. 269: 3475–3481.

Schwerdt, G., Freudinger, R., Silbernagl, S., and Gekle, M. 1999. Ochratoxin A-binding proteins in rat organs and plasma and in different cell lines of the kidney. Toxicology 135: 1–10.

Scott, G.E., and Zummo, N. 1994. Kernel infection and aflatoxin production in maize by *Aspergillus flavus* relative to inoculation and harvest dates. Plant Dis. 78: 123–125.

Scott, P.M. 1984. Effects of food processing on mycotoxins. J. Food Prot. 47: 489–499.

Scott, P.M. 1992. Ergot alkaloids in grain foods sold in Canada. J. AOAC Int. 75: 773–779.

Scott, P.M. 1993. Fumonisins. Int. J. Food Microbiol. 18: 257–270.

Scott, P.M. 1994. *Penicillium* and *Aspergillus* toxins. In: Mycotoxins in Grain. Compounds Other Than Aflatoxin. J.D. Miller and H.L. Trenholm, ed. Eagan Press, St. Paul, MN, pp. 261–285.

Scott, P.M. 1995. Mycotoxin methodology. Food Addit. Contam. 12: 395–403.

Scott, P.M., and Lawrence, G.A. 1995. Analysis of beer for fumonisins. J. Food Prot. 58: 1379–1382.

Scott, P.M., and Trucksess, M.W. 1997. Application of immunoaffinity columns to mycotoxin analysis. J. AOAC Int. 80: 941–949.

Scott, P.M., Delgado T., Prelusky, D.B., Trenholm, H.L., and Miller, J. D. 1994. Determination of fumonisins in milk. J. Environ. Sci. Health. B29: 989–998.

Scott, P. M., Kanhere, S.R., Lau, B.P.-Y., Lewis, D.A., Hayward, S., Ryan, J.J., and Kuiper-Goodman, T. 1998. Survey of Canadian human blood plasma for ochratoxin A. Food Addit. Contam. 15: 555–562.

Scudamore, K.A., Clarke, J.H., and Hetmanski, M.T. 1993. Isolation of *Penicillium* strains producing ochratoxin A, citinin, xanthomegnin, viomellein, and vioxanthin from stored cereal grains. Lett. Appl. Microbiol. 17: 82–87.

Selim, M.I., El-Sharkawy, S.H., and Popendorf, W.J. 1996. Supercritical fluid extraction of fumonisin B$_1$ from grain dust. J. Agric. Food Chem. 44: 3224–3229.

Severns, D.E., Clements, M.J., Lambert, R.J., and White, D.G. 2003. Comparison of *Aspergillus* ear rot and aflatoxin contamination in grain of high-oil and normal-oil corn hybrids. J. Food Prot. 66: 637–643.

Sewram, V., Nair, J.J., Nieuwoudt, T.W., Gelderblom, W.C., Marasas, W.F., and Shephard, GS. 2001. Assessing chronic exposure to fumonisin mycotoxins: the use of hair as a suitable noninvasive matrix. J. Anal. Toxicol. 25: 450–455

Shapira, R., Paster, N., Eyal, O., Menasherov, M., Mett, A., and Salomon, R. 1996. Detection of aflatoxigenic molds in grains by PCR. Appl. Environ. Microbiol. 62: 3270–3273.

Shapira, R., Paster, N., Menasherov, M., Eyal, O., Mett, A., Meiron, T., Kuttin, E., and Salomon, R. 1997. Development of polyclonal antibodies for detection of aflatoxigenic molds involving culture filtrate and chimeric proteins expressed in *Escherichia coli*. Appl. Environ. Microbiol. 63: 990–995.

Sharma, R.P. 1991. Immunotoxic effects of mycotoxins. In: Mycotoxins and Phytoalexins. R.P. Sharma and D.K. Salunkhe, ed. CRC Press, Boca Raton, FL, pp. 81–99.

Sharma, R.P., and Kim, Y.-W. 1991. Trichothecenes. In: Mycotoxins and Phytoalexins. R.P. Sharma and D.K. Salunkhe, ed. CRC Press, Boca Raton, FL, pp. 339–359.

Sharma, R.P., and Salunkhe, D.K. 1991. Occurrence of mycotoxins in foods and feeds. In: Mycotoxins and Phytoalexins. R.P. Sharma and D.K. Salunkhe, ed. CRC Press, Boca Raton, FL, pp. 13–32.

Shephard, G.S., and Snijman, P.W. 1999. Elimination and excretion of a single dose of the mycotoxin fumonisin B$_2$ in a non-human primate. Food Chem. Toxic. 37: 111–116.

Shephard, G.S., Sydenham, E.W., Thiel, P.G., and Gelderblom, W.C.A. 1990. Quantitative determination of fumonisins B$_1$ and B$_2$ by high-performance liquid chromatography with fluorescence detection. J. Liq. Chromatogr. 13: 2077–2087.

Shephard, G.S., Thiel, P.G., Stockenström, S., and Sydenham, E.W. 1996. Worldwide survey of fumonisin contamination of corn and corn-based products. J. AOAC Int. 79: 671–687.

Sheu, F., and Shyu, Y.T. 1999. Analysis of patulin in apple juice by diphasic dialysis extraction with *in situ* acylation and mass spectrometric determination. J. Agric. Food Chem. 47: 2711–2714.

Shier, W.T. 1998. Estrogenic mycotoxins. Revue Méd. Vét. 149: 599–604.

Shier, W.T. 2000. The fumonisin paradox: A review of research on oral bioavailability of fumonisin B$_1$, a mycotoxin produced by *Fusarium moniliforme*. J. Toxicol. Toxin Rev. 19: 161–187.

Shier, W.T., and Abbas, H.K. 1999. Current issues in research on fumonisins, mycotoxins which may cause nephropathy. J. Toxicol. Toxin Rev. 18: 323–335.

Shier, W.T., Tiefel, P.A., and Abbas, H.K. 2000. Current research on mycotoxins: Fumonisins. In: Natural and Selected Synthetic Toxins. Biological Implications. A.T. Tu and W. Gaffield, ed. American Chemical Society, Washington, D,C, pp. 54–66.

Shima. J., Takase, S., Takahashi, Y., Iwai, Y., Fujimoto, H., Yamazaki, M., and Ochi, K. 1997. Novel detoxification of trichothecene mycotoxin deoxynivalenol by a soil bacterium isolated by enrichment culture. Appl. Environ. Microbiol. 63: 3825–3830.

Shotwell, O.L. 1991. Natural occurrence of mycotoxins in corn. In: Mycotoxins and Animal Foods. J.E. Smith and R.S. Henderson, ed. CRC Press, Boca Raton, FL, pp. 325–340.

Siame, B.A., Mpuchane, S.F., Gashe, B.A., Allotey, J., and Teffera, G. 1998. Occurrence of aflatoxins, fumonisin B$_1$ and zearalenone in foods and feeds in Botswana. J. Food Prot. 61: 1670–1673.

Sibanda, L., De Saeger, S., van Peteghem, C., Grabarkiewicz-Szczesna, J., and Tomczak, M. 2000. Detection of T-2 toxin in different cereals by flow-through enzyme immunoassay with a simultaneous internal reference. J. Agric. Food Chem. 48: 5864–5867.

Sinha, K.K. 1998. Detoxification of mycotoxins and food safety. In: Mycotoxins in Agriculture and Food Safety. K.K. Sinha and D. Bhatnagar, ed. Marcel Dekker, Inc., New York, pp. 381-405.

Skaug, M.A., Helland, I., Solvoll, K., and Saugstad, O.D. 2001. Presence of ochratoxin A in human milk in relation to dietary intake. Food Addit. Contam. 18: 321–327.

Smiley, R.D., and Draughon, F.A. 2000. Preliminary evidence that degradation of aflatoxin B$_1$ by *Flavobacterium aurantiacum* is enzymatic. J. Food Prot. 63: 415–418.

Smith, G.W., Constable, P.D., Foreman, J.H., Eppley, R.M., Waggoner, A.L., Tumbleson, M.E., and Haschek, W.M. 2001. Intravenous fumonisin B1 decreases cardiovascular function in horses. Toxicol. Sci. 60: 14 (abstract).

Smith, G.W., Constable, P.D., Tumbleson, M.E., Rottinghaus, G.E., and Haschek, W.M. 1999. Sequence of cardiovascular changes leading to pulmonary edema in swine fed fumonisin-containing culture material. Am. J. Vet. Res. 60: 1292–1300.

Smith, J.E. 1997. Aflatoxins. In: Handbook of Plant and Fungal Toxicants. J.P.F. D'Mello, ed. RC Press, Boca Raton, FL, pp. 269–285.

Smith, J.E. and Moss, M.O. 1985. Mycotoxins. Formation, Analysis and Significance. John Wiley and Sons, New York.

Smith, J.E., and Ross, K. 1991. The toxigenic aspergilli. In: Mycotoxins and Animal Foods. J.E. Smith and R.S. Henderson, ed. CRC Press, Boca Raton, FL, pp. 101–118.

Smith, J. S. and Thakur, R.A. 1996. Occurrence and fate of fumonisins in beef. In: Fumonisins in Food. L.J. Jackson, J.W. DeVries, and L.B. Bullerman, ed. Plenum Press, New York, pp. 39–55.

Snijders, C.H.A. 1994. Breeding for resistance to *Fusarium* in wheat and maize. In: Mycotoxins in Grain. Compounds Other Than Aflatoxin. J.D. Miller and H.L. Trenholm, ed. Eagan Press, St. Paul, MN, pp. 37–58.

Soleas, G.J., Yan, J., and Goldberg, D.M. 2001. Assay of ochratoxin A in wine and beer by high-pressure liquid chromatography photodiode array and gas chromatography mass selective detection. J. Agric. Food Chem. 49: 2733–2740.

Soufleros, E.H., Tricard, C., and Bouloumpasi, E.L. 2003. Occurrence of ochratoxin A in Greek wine. J. Sci. Food Agric. 83: 173–179.

Spiegel, S. 1998. Sphingosine 1-phosphate: a prototype of a new class of second messenger. J. Leuk. Biol. 65: 341–344.

Šrobárová, A., and Pavlová, A. 2001. Toxicity of secondary metabolites of the fungus *F. culmorum* in relation to resistance of winter wheat cultivars. Cereal Res. Commun. 29: 101–108.

Stack, M.E. 1998. Analysis of fumonisin $B_1$ and its hydrolysis product in tortillas. J. AOAC Int. 81: 737–740.

Stander, M.A., Bornscheuer, U.T., Henke, E., and Steyn, P.S. 2000. Screening of commercial hydrolases for the degradation of ochratoxin A. J. Agric. Food Chem. 48: 5736–5739.

Stevens, V.L., and Tang, J. 1997. Fumonisin $B_1$-induced sphingolipid depletion inhibits vitamin uptake via the glycosylphosphatidylinositol-ancored folate uptake. J Biol. Chem. 272: 18020–18025.

Steyn, P. S. 1998. The biosynthesis of mycotoxins. Revue Méd. Vét. 149: 469–478.

Steyn, P.S. and Stander, M.A. 1999. Mycotoxins as causal factors of diseases in humans. J. Toxicol. Toxin Rev. 18: 229–243.

Steyn, P.S., Thiel, P.G., and Trinder, D.W. 1991. Detection and quantification of mycotoxins by chemical analysis. In: Mycotoxins and Animal Foods. J.E. Smith and R.S. Henderson, ed. CRC Press, Boca Raton, FL, pp. 165–221.

Størmer, F.C. 1992. Ochratoxin A – a mycotoxin of concern. In: Handbook of Applied Mycology. Volume 5: Mycotoxins in Ecological Systems. D. Bhatnagar, E.B. Lillehoj, and D.K. Arora, ed. Marcel Dekker, Inc., New York, pp. 403–432.

Strange, R.N. 1991. Natural occurrence of mycotoxins in groundnuts, cottonseed, soya, and cassava. In: Mycotoxins and Animal Foods. J.E. Smith and R.S. Henderson, ed. CRC Press, Boca Raton, FL, pp. 341–362.

Struder-Rohr, I., Dietrich, D. R., Schlatter, J., and Schlatter, C. 1995. The occurrence of ochratoxin A in coffee. Food Chem. Toxic. 33: 341–355.

Suetens, C., Moreno-Reyes, R., Chasseur, C., Mathieu, F., Begaux, F., Haubruge, E., Durand, M.C., Nève, J., and Vanderpas, J. 2001. Epidemiological support for a multifactorial aetiology of Kashin-Beck disease in Tibet. Int. Orthop. 25: 180–187.

Sweeney, M.J., and Dobson, A.D.W. 1998. Mycotoxin production by *Aspergillus*, *Fusarium* and *Penicillium* species. Int. J. Food Microbiol. 43: 141–158.

Sweeney, M.J., and Dobson, A.D.W. 1999. Molecular biology of mycotoxin biosynthesis. FEMS Microbiol. Lett. 175: 149–163.

Sydenham, E.W., Shephard, G.S., and Thiel, P.G. 1992. Liquid chromatographic determination of fumonisins $B_1$, $B_2$, and $B_3$ in foods and feeds. J. AOAC Int. 75: 313–318.

Sydenham, E.W., Shephard, G.S., Thiel, P.G., Marasas, W.F.O., and Stockenström, S. 1991. Fumonisin contamination of commercial corn-based human foodstuffs. J. Agric. Food Chem. 39: 2014–2018.

Sydenham, E.W., Stockenström, S., Thiel, P.G., Rheeder, J.P., Doko, M.B., Bird, C., and Miller, B.M. 1996. Polyclonal antibody-based ELISA and HPLC methods for the determination of fumonisins in corn: A comparative study. J. Food Prot. 59: 893–897.

Sydenham, E.W., Stockenström, S., Thiel, P.G., Shephard, G.S., Koch, K.R., and Marasas, W.F.O. 1995a. Potential of alkaline hydrolysis for the removal of fumonisins fromcontaminated corn. J. Agric. Food Chem. 43: 1198–1201.

Sydenham, E.W., Thiel, P.G., Marasas, W.F.O., Shephard, G.S., van Schalkwyk, D.J., and Koch, K.R. 1990. Natural occurrence of some *Fusarium* mycotoxins in corn from low and high esophageal cancer prevalence areas of the Transkei, southern Africa. J. Agric. Food Chem. 38: 1900–1903.

Sydenham, E.W., Vismer, H.F., Marasas, W.F.O., Brown, N., Schlechter, M., van der Westhuizen, L., and Rheeder, J.P. 1995b. Reduction of patulin in apple juice samples – influence of initial processing. Food Control 6: 195–200.

Tabata, S., Kamimura, H., Ibe, A., Hashimoto, H., and Tamura, Y. 1994. Degradation of aflatoxins by food additives. J. Food Prot. 57: 42–47.

Takahashi-ando, N., Kimura, M., Kakeya, H., Osada, H., and Yamaguchi, I. 2002. A novel lactonohydrolase responsible for the detoxification of zearalenone: enzyme purification and gene cloning. Biochem. J. 365: 1–6.

Taniwaki, M.H., and van Dender, A.G.F. 1992. Occurrence of toxigenic molds in Brazilian cheese. J. Food Prot. 55: 187–191.

Taniwaki, M.H., Hocking, A.D., Pitt, J.I., and Fleet, G.H. 2001. Growth of fungi and mycotoxinproduction on cheese under modified atmospheres. Int. J. Food Microbiol. 68: 125–133.

Taniwaki, M.H., Hoenderboom, C.J.M., Vitali, A.D.A.., and Eiroa, M.N.U. 1992. Migration of patulin in apples. J. Food Prot. 55: 902–904.

Taniwaki, M.H., Pitt, J.I., Teixeira, A.A., and Iamanaka, B.T. 2003. The source of ochratoxin A in Brazilian coffee and its formation in relation to processing methods. Int. J. Food Microbiol. 82: 173–179.

Taylor, S.L., King, J.W., Greer, J.I., and Richard, J.L. 1997. Supercritical fluid extraction of aflatoxin $M_1$ from beef liver. J. Food Prot. 60: 698–700.

Taylor, S.L., King, J.W., Richard, J.L., and Greer, J.I. 1993. Analytical-scale supercritical fluid extraction of aflatoxin $B_1$ from field-inoculated corn. J. Agric. Food Chem. 41: 910–913.

Tenberge, K.B. 1999. In: Ergot. The Genus *Claviceps*. V. Křen and L. Cvak, ed. Harwood Academic Publishers, Amsterdam, The Netherlands, pp. 25–56.

Thiel, P.G., Sydenham, E.W., Shephard, G.S., and van Schalkwyk, D.J. 1993. Study of reproducibility characteristics of a liquid chromatographic method for the determination of fumonisins $B_1$ and $B_2$ in corn: IUPAC collaborative study. J. AOAC Int. 76: 361–366.

Thuvander, A., Möller, T., Barbieri, H.E., Jansson, A., Salomonsson, A.-C., and Olsen, M. 2001a. Dietary intake of some important mycotoxins by the Swedish population. Food Addit. Contam. 18: 696–706.

Thuvander, A., Paulsen, J.E., Axberg, K., Johansson, N., Vidnes, A., Enghardt-Barbieri, H., Trygg, K., Lund-Larsen, K., Jahrl, S., Widenfalk, A., Bosnes, V., Alexander, J., Hult, K., and Olsen, M. 2001b. Levels of ochratoxin A in blood from Norwegian and Swedish blood donors and their possible correlation with food consumption. Food Chem. Toxicol. 39: 1145–1151.

Thuvander, A., Wikman, C., and Gadhasson, I. 1999. *In vitro* exposure of human lymphocytes to trichothecenes: Individual variation in sensitivity and effects of combined exposure on lymphocyte function. Food Chem. Toxicol. 37: 639–648.

Tolleson, W.H., Couch, L.H., Melchior, W.B., Jr, Jenkins, G.R., Muskhelishvili, M., Muskhelishvili, L., McGarrity, L.J., Domon, O.E., Morris, S.M., and Howard, P.C. 1999. Fumonisin $B_1$ induces apoptosis in cultured human keratinocytes through sphinganine accumulation and ceramide depletion. Int. J. Oncol. 14: 833–843.

Torp, M., and Langseth, W. 1999. Production of T-2 toxin by a *Fusarium* resembling *Fusariumpoae*. Mycopathologia 147: 89–96.

Torres, M.R., Ramos, A.J., Soler, J., Sanchis, V., and Marín, S. 2003. SEM study of water activity and temperature effects on the initial growth of *Aspergillus ochraceus*, *Alternaria alternata*, and *Fusarium verticillioides* on maize grain. Int. J. Food Microbiol. 81: 185–193.

Trail, F., Mahanti, N., and Linz, J. 1995. Molecular biology of aflatoxin biosynthesis. Microbiology 141: 755–765.

Trenholm, H.L., Charmley, L.L., Prelusky, D.B., and Warner, R.M. 1991. Two physical methods for the decontamination of four cereals contaminated with deoxynivalenol and zearalenone. J. Agric. Food Chem. 39: 356–360.

Trenholm, H.L., Charmley, L.L., Prelusky, D.B., and Warner, R.M. 1992. Washing procedures using water or sodium carbonate solutions for the decontamination of three cereals contaminated with deoxynivalenol and zearalenone. J. Agric. Food Chem. 40: 2147–2151.

Trigo-Stockli, D.M., Deyoe, C.W., Satumbaga, R.F., and Pedersen, J.R. 1996. Distribution of deoxynivalenol and zearalenone in milled fractions of wheat. Cereal Chem. 73: 388–391.

Trout, D., Bernstein, J., Martinez, K., Biagini, R., and Wallingford, K. 2001. Bioaerosol lung damage in a worker with repeated exposure to fungi in a water-damaged building. Environ.Health Perspect. 109: 641–644.

Trucksess, M.W. 2001. Mycotoxins. J. AOAC Int. 84: 202–211.

Trucksess, M.W., and Pohland, A.E. (ed.). 2001. Mycotoxin Protocols. Humana Press, Totowa, NJ.

Trucksess, M.W., and Tang, Y. 1999. Solid-phase extraction method for patulin in apple juice and unfiltered apple juice. J. AOAC Int. 82: 1109–1113.

Trucksess, M.W., and Wood, G.E. 1994. Recent methods of analysis in foods and feeds. In: The Toxicology of Aflatoxins. Human Health, Veterinary, and Agricultural Significance. D.L. Eaton and J.D. Groopman, ed. Academic Press, New York, pp. 409–431.

Trucksess, M.W., Giler, J., Young, K., White, K. D., and Page, S.W. 1999. Determination andsurvey of ochratoxin A in wheat, barely, and coffee – 1997. J. AOAC Int. 82: 85–89.

Trucksess, M.W., Ready, D.E., Pender, M.K., Ligmond, C.A., Wood, G.E., and Page, S.W.1996. Determination and survey of deoxynivalenol in white flour, whole wheat flour, and bran. J. AOAC Int. 79: 883–887.

Trucksess, M.W., Stack, M.E., Nesheim, S., Albert, R.H., and Romer, T.R. 1994. Multifunctional column coupled with liquid chromatography for determination of aflatoxins B$_1$, B$_2$, G$_1$, and G$_2$ in corn, almonds, Brazil nuts, peanuts, and pistachio nuts: Collaborative study. J.AOAC Int. 77: 1512–1521.

Tsai, G.-J., and Yu, S.-C. 1997. An enzyme-linked immunosorbent assay for the detection of *Aspergillus parasiticus* and *Aspergillus flavus*. J. Food Prot. 60: 978–984.

Tsao, R., and Zhou, T. 2000. Micellar electrokinetic capillary electrophoresis for rapid analysis of patulin in apple cider. J. Agric. Food Chem. 48: 5231–5235.

Tsunoda, M., Sharma, R.P., and Riley, R.T. 1998. Early fumonisin B, toxicity in relation to disrupted sphingolipid metabolism in male BALB/c mice. J. Biochem. Molec. Toxicol. 12: 281–289.

Tubajika, K.M., and Damann, K.E. 2001. Sources of resistance to aflatoxin production in maize. J. Agric. Food Chem. 49: 2652–2656.

Tubajika, K.M., and Damann, K.E., Jr. 2002. Glufosinate-ammonium reduces growth and aflatoxin B$_1$ production by *Aspergillus flavus*. J. Food Prot. 65: 1483–1487.

Turner, P.C., Nikiema, P., and Wild, C.P. 1999. Fumonisin contamination of food: progress in development of biomarkers to better assess human health risks. Mut. Res. 443: 81–93.

Ueno, Y. 1983. General toxicology In: Trichothecenes: Chemical, biological and toxicological aspects. Y. Ueno, ed. Elsevier, New York, pp. 135–146.

Ueno, Y. 1987. Trichothecenes in food. In: Mycotoxins in Food. P. Krogh, ed. Academic Press, New York, pp. 123–147.

Ueno, Y., Iijima, K., Wang, S.-D., Sugiura, Y., Sekijima, M., Tanaka, T., Chen, C., and Yu, S.-Z. 1997. Fumonisins as a possible contributory risk factor for primary liver cancer: A 3-year study of corn harvested in Haimen, China, by HPLC and ELISA. Food Chem. Toxicol. 35: 1143–1150.

Ueno, Y., Maki, S., Lin, J., Furuya, M., Sugiura, Y., and Kawamura, O. 1998. A 4-year study of plasma ochratoxin A in a selected population in Tokyo by immunoassay and immunoaffinity column-linked HPLC. Food Chem. Toxicol. 36: 445–449.

Ueno, Y., Umemori, K., Niimi, E., Tanuma, S., Nagata, S., Sugamata, M., Ihara, T., Sekijima, M., Kawai, K., Ueno I., and Tashiro, F. 1995. Induction of apoptosis by T-2 toxin and other natural toxins in HL-60 human promyelotic leukemia cells. Nat. Toxins 3: 129–137.

United Nations Environment Programme, International Labor Organisation, and World Health Organization. 1990. Environmental Health Criteria105. Selected Mycotoxins: Ochratoxins, Trichothecenes, Ergot. World Health Organization, Geneva, Switzerland.

U. S. Food and Drug Administration (FDA). 2001a. Guidance for industry. Fumonisin levels in human foods and animal feeds. http://www.cfsan.fda.gov/~dms/fumongu2.html (accessed 1/3/03).

U. S. Food and Drug Administration (FDA). 2001b. Background paper in support of fumonisin levels in corn and corn products intended for human consumption. http://www.cfsan.fda.gov/~dms/fumongu2.html (accessed 1/3/03).

U. S. Food and Drug Administration (FDA). 2001c. Patulin in apple juice, apple juice concentrates and apple juice products. http://www.cfsan.fda.gov/~dms/patubck2.html (accessed 1/3/03).

Urano, T., Trucksess, M.W., Beaver, R.W., Wilson, D.M., Dorner, J.W., and Dowell, F.E. 1992a. Co-occurrence of cyclopiazonic acid and aflatoxins in corn and peanuts. J. AOAC Int. 75: 838–841.

Urano, T., Trucksess, M.W., Matusik, J., and Dorner, J.W. 1992b. Liquid chromatographic determination of cyclopiazonic acid in corn and peanuts. J. AOAC Int. 75: 319–322.

Urbano, G.R., Taniwaki, M.H., de F. Leitão, and Vicentini, M.C. 2001. Occurrence of Ochratoxin A-producing fungi in raw Brazilian coffee. J. Food Prot. 64: 1226–1230.

Usleber, E., Straka, M., and Terplan, G. 1994. Enzyme immunoassay for fumonisin B$_1$ applied to corn-based food. J. Agric Food Chem. 42: 1392–1396.

van der Stegen, G.H.D., Essens, P.J.M., and van der Lijn, J. 2001. Effect of roasting conditions on reduction of ochratoxin A in coffee. J. Agric. Food Chem. 49: 4713–4715.

van der Westhuizen, L., Brown, N.L., Marasas, W.F.O., Swanevelder, S., and Shephard, G.S. 1999. Sphinganine/sphingosine ratio in plasma and urine as a possible biomarker for fumonisin exposure in humans in rural areas of Africa. Food Chem. Toxicol. 37: 1153–1158.

van Egmond, H. P. 1991a. Limits and regulations for mycotoxins in raw materials and animal feeds. In: Mycotoxins and Animal Foods. J.E. Smith, and R.S. Henderson, ed. CRC Press, Boca Raton, FL, pp. 423–436.

van Egmond, H.P. 1991b. Regulatory aspects of mycotoxins in Asia and Africa. In: Fungi and Mycotoxins in Stored Products. B.R. Champ, E. Highley, A.D. Hocking, and J.I. Pitt, ed. Australian Centre for International Agricultural Research (AICAR), Canberra, Austarlia. p. 198–204.

van Egmond, H.P. 1994. Aflatoxins in milk. In: The Toxicology of Aflatoxins. Human Health, Veterinary, and Agricultural Significance. D.L. Eaton and J.D. Groopman, ed. Academic Press, New York, pp. 365–381.

van Egmond, H.P. 2002. Worldwide regulations for mycotoxins. In: Mycotoxins and Food Safety. J.W. DeVries, M.W. Trucksess, and L.S. Jackson, ed. Kluwer Academic/Plenum Publishers, New York, pp. 257–269.

Varga, J., Rigo, K., Teren, J., and Mesterhazy, A. 2001. Recent advances in ochratoxin research I. Production, detection and occurrence of ochratoxins. Cereal Res. Commun. 29: 85–99.

Viani, R. 2002. Effect of processing on ochratoxin A (OTA) content of coffee. In: Mycotoxins and Food Safety. J.W. DeVries, M.W. Trucksess, and L.S. Jackson, ed. Kluver Academic/Plenum Publishers, New York, pp. 189–193.

Viquez, O.M, Castell-Perez, M.E., Shelby, R.A., and Brown, G. 1994. Aflatoxin contamination in corn samples due to environmental conditions, aflatoxin-producing strains, and nutrients in grain grown in Costa Rica. J. Agric. Food Chem. 42: 2551–2555.

Visconti, A., Pascale, M., and Centonze, G. 2001. Determination of ochratoxin A in wine and beer by immunoaffinity column cleanup and liquid chromatographic analysis with fluorometric detection: Collaborative study. J. AOAC Int. 84: 1818–27.

Voss, K.A., Bacon, C.W., Meredith, F.I., and Norred, W.P. 1996a. Comparative subchronictoxicity studies of nixtamalized and water-extracted *Fusarium moniliforme* culture material. Food Chem. Toxicol, 34: 623–632.

Voss, K.A., Poling, S.M., Meredith, F.I., Bacon, C.W., and Saunders, D.S. 2001a. Fate of fumonisins during the production of fried tortilla chips. J. Agric. Food Chem. 49: 3120-3126.

Voss, K.A., Porter, J.K., Bacon, C.W., Meredith, F.I., and Norred, W.P. 1999. Fusaric acid and modification of subchronic toxicity to rats of fumonisins in *F. moniliforme* culture material. Food Chem. Toxicol. 37: 853–861.

Voss, K.A., Riley, R.T., Bacon, C.W., Chamberlain, W.J., and Norred, W.P. 1996b. Subchronic toxic effects of *Fusarium moniliforme* and fumonisin B$_1$ in rats and mice. Nat. Toxins 4: 1623.

Voss, K.A., Riley, R.T., Norred, W.P., Bacon, C.W., Meredith, F.I., Howard, P.C., Plattner, R.D., Collins, T., Hansen, D.K., and Porter, J.K. 2001b. An overview of rodent toxicities: liver, and kidney effects of *Fusarium moniliforme* and fumonisins. Environ. Health Perspect. 109: 259–266.

Vrabcheva, T., Usleber, E., Dietrich, R., and Märtlbauer, E. 2000. Co-occurrence of ochratoxin A and citrinin in cereals from Bulgarian villages with a history of Balkan endemic nephropathy. J. Agric. Food Chem. 48: 2483–2488.

Walker, F. and Meier, B. 1998. Determination of the *Fusarium* mycotoxins nivalenol, deoxynivalenol, 3-acetyl deoxynivalenol, and 15-O-acetyl-4- deoxynivalenol in contaminated whole wheat flour by liquid chromatography with diode array detection and gas chromatography with electron capture detection. J. AOAC Int. 81: 741–748.

Wang, E., Norred, W.P., Bacon, C.W., Riley, R.T., and Merrill, A.H., Jr. 1991. Inhibition of sphingolipid biosynthesis by fumonisins: implications for diseases associated with *Fusarium moniliforme*. J. Biol. Chem. 266: 14486–14490.

Wang, E., Riley, R.T., Meredith, F.I., and Merrill, A.H., Jr. 1999. Fumonisin B$_1$ consumption by rats causes reversible, dose-dependent increases in urinary sphinganine and sphingosine. J. Nutr. 129: 214–220.

Wang, E., Ross, P.F., Wilson, T.M., Riley, R.T., and Merrill, A.H., Jr. 1992. Increases in serum sphingosine and sphinganine and decreases in complex sphingolipids in ponies given feed containing fumonisins, mycotoxins produced by *Fusarium moniliforme*. J. Nutr. 122: 1706–1716.

Wang, J., Fitzpatrick, D. W., and Wilson, J.R. 1998a. Effects of the trichothecene mycotoxin T-2 toxin on neurotransmitters and metabolites in discrete areas of the rat brain. Food Chem. Toxicol. 36: 947–953.

Wang, J., Fitzpatrick, D. W., and Wilson, J.R. 1998b. Effect of T-2 toxin on blood-brain barrier permeability monoamine oxidase activity and protein synthesis in rats. Food Chem. Toxicol. 36: 955–961.

Warfield, C.Y., and Gilchrist, D.G. 1999. Influence of kernel age on fumonisin B$_1$ production in maize by *Fusarium moniliforme*. Appl. Environ. Microbiol. 65: 2853–2856.

Warner, R.L., Brooks, K., and Pestka, J.J. 1994. *In vitro* effects of vomitoxin deoxynivalenol on lymphocyte function: enhanced interleukin production and help for IgA secretion. Food Chem. Toxicol. 32: 617–625.

Weidenbörner, M., Wieczorek, C., Appel, S., and Kunz, B. 2000. Whole wheat and white wheat flour – the mycobiota and potential mycotoxins. Food Microbiol. 17: 103–107.

West, D.I., and Bullerman, L. B. 1991. Physical and chemical separation of mycotoxins from agricultural products. In: Mycotoxins and Animal Foods. J.E. Smith and R.S. Henderson, ed. CRC Press, Boca Raton, FL, pp. 777–783.

Whitaker, T., Horwitz, W., Albert, R., and Nesheim, S. 1996. Variability associated with analytical methods used to measure aflatoxin in agricultural commodities. J. AOAC Int. 79: 476–485.

Whitaker, T.B., and Park, D.L. 1994. Problems associated with accurately measuring aflatoxin in food and feeds: Errors associated with sampling, sample preparation, and analysis. In: The Toxicology of Aflatoxins. Human Health, Veterinary, and Agricultural Significance. D.L. Eaton and J.D. Groopman, ed. Academic Press, New York. p. 433–450.

Whitaker, T.B., Dowell, F.E., Hagler, W.M., Jr., Giesbrecht, F.G., and Wu, J. 1994. Variability associated with sampling, sample preparation, and chemical testing for aflatoxin in farmers' stock peanuts. J. AOAC Int. 77: 107–116.

Whitaker, T.B., Hagler, W.M.Jr., Giesbrecht, F.G., and Johansson, A.S. 2000. Sampling, sample preparation, and analytical variability associated with testing wheat for deoxynivalenol. J. AOAC Int. 83: 1285–1292.

WHO (World Health Organization). 2000. Environmental Health Criteria 219: Fumonisin B$_1$, W.H.O. Marasas, J.D. Miller, R.T. Riley, and A. Visconti, ed. International Programme on Chemical Safety, United Nations Environmental Programme, The International Labor Organization, and the World Health Organization, Geneva, Switzerland.

WHO (World Health Organization). 2001. Safety Evaluation of Certain Mycotoxins in Food. Fifty-sixth report of the Joint FAO/WHO Expert Committee on Food Additives. WHO Food Additives Series 47, FAO Food and Nutrition Paper 74, pp 701 WHO Geneva.

Wicklow, D.T. 1995. The mycology of stored grain: An ecological perspective. In: Stored-Grain Ecosystems. D.S. Jayas, N.D.G. White, and W.E. Muir, ed. Marcel Dekker, Inc., New York, pp. 197–249.

Widstrom, N.W. 1992. Aflatoxin in developing maize: Interactions among involved biota and pertinent econiche factors. In: Handbook of Applied Mycology. Volume 5: Mycotoxins in Ecological Systems. D. Bhatnagar, E.B. Lillehoj, and D.K. Arora, ed. Marcel Dekker, Inc., New York, pp. 23–58.

Wild, C.P., and Hall, A.J. 1996. Epidemiology of mycotoxin-related disease. In: The Mycota VI. Human and Animal Relationships. D.H. Howard, and J.D. Miller, ed. Springer-Verlag, Berlin, pp. 213–227.

Wilson, D.M., and Payne, G.A. 1994. Factors affecting *Aspergillus flavus* group infection and aflatoxin contamination of crops. In: The Toxicology of Aflatoxins. Human Health, Veterinary, and Agricultural Significance. D.L. Eaton and J.D. Groopman, ed. Academic Press, New York, pp. 309–325.

Wilson, D.M., Mubatanhema, W., and Jurjevic, Z. 2002. Biology and ecology of mycotoxigenic *Aspergillus* species as related to economic and health concerns. In: Mycotoxins and Food Safety. J.W. DeVries, M.W. Trucksess, and L.S. Jackson, ed. Kluver Academic/Plenum Publishers, New York, pp. 3–17.

Wilson, D.M., Sydenham, E.W., Lombart, G.A., Trucksess, M. W., Abramson, D., and Bennett, G.A. 1998. Mycotoxin analytical techniques. In: Mycotoxins in Agriculture and Food Safety. K.K. Sinha and D. Bhatnagar, ed. Marcel Dekker, Inc., New York, pp. 135–182.

Windels, C.E. 1991. Current status of *Fusarium* taxonomy. Phytopathology 81: 1048–1051.

Wolf, C.E., and Bullerman, L.B. 1998. Heat and pH alter the concentration of deoxynivalenol inan aqueous environment. J. Food Prot. 61: 365–367.

Wolf-Hall, C.E., and Schwarz, P.B. 2002. Mycotoxins and fermentation – beer production. In: Mycotoxins and Food Safety. J.W. DeVries, M.W. Trucksess, and L.S. Jackson, ed. Kluver Academic/Plenum Publishers, New York, pp. 217–226.

Wolf-Hall, C.E., Hanna, M.A., and Bullerman, L.B. 1999. Stability of deoxynivalenol in heat-treated foods. J. Food Prot. 62: 962–964.

Wong, S.S., Zhou, H.R., Marin-Martinez, M.L., Brooks, K., and Pestka, J.J. 1998. Modulation of IL-1beta, IL-6 and TNF-alpha secretion and mRNA expression by the trichothecene vomitoxin in the RAW

264.7 murine macrophage cell line. Food Chem. Toxicol. 36: 409-419.

Wood, G.E. 1991. Aflatoxin M₁. In: Mycotoxins and Phytoalexins. R.P. Sharma and D.K. Salunkhe, ed. CRC Press, Boca Raton, FL, pp. 145–164.

Wood, G.E., and Trucksess, M.W. 1998. Regulatory control programs for mycotoxin-contaminated food. In: Mycotoxins in Agriculture and Food Safety. K.K. Sinha and D. Bhatnagar, ed. Marcel Dekker, Inc., New York, pp. 459–481.

Wu, P.-Y., Yen, Y.-H., and Chiou, R.Y.-Y. 1995. Extraction of aflatoxin from peanut meal and kernels by $CO_2$-methanol using a highly pressurized fluid extraction process. J. Food Prot. 58: 800–803.

Yan, D., Zhou, H.R., Brooks, K.H., and Pestka, J.J. 1997. Potential role for IL-5 and IL-6 in enhanced IgA secretion by Peyer's patch cells isolated from mice acutely exposed to vomitoxin. Toxicology 122: 145–158.

Yan, D., W.K. Rumbeiha, and Pestka, J.J. 1998. Experimental murine IgA nephropathy following passive administration of vomitoxin-induced IgA monoclonal antibodies. Food Chem. Toxicol. 36: 1095–1106.

Yang, G.H., Jarvis, B.B., Chung, Y.J., and Pestka, J.J. 2000. Apoptosis induction by the satratoxins and other trichothecene mycotoxins: relationship to ERK, p38 MAPK and SAPK/JNK activation. Toxicol. Appl. Pharmacol. 164: 149–160.

Yates, I.E., Meredith, F., Smart, W., Bacon, C.W., and Jaworski, A.J. 1999. *Trichoderma viride* suppresses fumonisin B₁ production by *Fusarium moniliforme*. J. Food Prot. 62: 1326–1332.

Yazici, S., and Velioglu, Y.S. 2002. Effect of thiamine hydrochloride, pyridine hydrochloride and calcium d-panthothenate on the patulin content of apple juice concentrate. Nahrung 46: 256–257.

Yong, R.K., and Cousin, M.A. 2001. Detection of moulds producing aflatoxins in maize and peanuts by an immunoassay. Int. J. Food Microbiol. 65: 27–38.

Yoo, H.-S., Norred, W.P., Showker, J.L., and Riley, R.T. 1996. Elevated sphingoid bases and complex sphingolipid depletion as contributing factors in fumonisin-induced cytotoxicity. Toxicol. Appl. Pharmacol. 138: 211–218.

Yoo, H.-.S., Norred, W.P., Wang, E., Merrill, A.H., Jr., and Riley, R.T. 1992. Fumonisin inhibition of *de novo* sphingolipid biosynthesis and cytotoxicity are correlated in LLC-PK₁ cells. Toxicol. Appl. Pharmacol. 114: 9–15.

Yoshizawa, T. 1991. Natural occurrence of mycotoxins in small grain cereals (wheat, barley, rye, oats, sorghum, millet, rice). In: Mycotoxins and Animal Foods. J.E. Smith and R.S. Henderson, ed. CRC Press, Boca Raton, FL, pp. 301–324.

Yoshizawa, T., Yamashita, A., and Luo, Y. 1994. Fumonisin occurrence in corn from high- and low-risk areas for human esophageal cancer in China. Appl. Environ. Microbiol. 60: 1626–1629.

Yu, C.-H., Lee, Y.-M., Yum, Y.-P., and Yoo, H.-S. 2001. Differential effects of fumonisin B₁ on cell death in cultured cells : the significance of elevated sphinganine. Arch. Pharmacol. Res. 24: 136–143.

Yu, F.-Y., and F. S. Chu. 1996. Production and characterization of antibodies against fumonisin B₁. J. Food Prot. 59: 992–997.

Yu, F.-Y., and F. S. Chu. 1998a. Analysis of fumonisins and *Alternaria alternata* toxin by liquid chromatography – enzyme-linked immunosorbent assay. J. AOAC Int. 81: 749–756.

Yu, W. and F. S. Chu. 1998b. Improved direct competitive enzyme-linked immunosorbent assay for cyclopiazonic acid in corn, peanuts, and mixed feed. J. Agric. Food Chem. 46: 1012–1017.

Yu, W., Dorner, J.W., and F. S. Chu. 1998. Immunoaffinity column as cleanup tool for a direct competitive enzyme-linked immunosorbent assay of cyclopiazonic acid in corn, peanuts, and mixed feed. J. AOAC Int. 81: 1169–1175.

Zanic-Grubisic, T., Zrinski, R., Cepelak, I., Petrik, J., Radic, B., and Pepeljnjak, S. 2000. Studies of ochratoxin A-induced inhibition of phenylalanine hydroxylase and its reversal by phenylalanine. Toxicol. Appl. Pharmacol. 167: 132–139.

Zhou, H.R., and Pestka, J.J. 2001. Essential role of non-receptor tyrosine kinase Hck in vomitoxin-induced phosphorylation of JNK, ERK and p38 mitogen-activated protein kinase. FASEB J. 15: 5.

Zhou, H.R. and Pestka, J.J. 2003. Deoxynivalenol-induced apoptosis mediated by p38 MAPK-dependent p53 gene induction in raw 264.7 macrophages. Toxicologist 72: 330.

Zhou, H.R., Harkema, J.R., Yan, D., and Pestka, J.J. 1999. Amplified proinflammatory cytokineexpression and toxicity in mice coexposed to lipopolysaccharide and the trichothecene vomitoxin (deoxynivalenol). J. Toxicol. Environ. Health 57: 115–136.

Zhou, H.R., Islam, Z., and Pestka, J.J. 2003a. Rapid, sequential activation of mitogen-activated protein kinases and transcription factors precedes proinflammatory cytokine mRNA expression in spleens of mice exposed to the trichothecene vomitoxin. Toxicol. Sci. 72: 130–142.

Zhou, H.R., Lau, A.S., and Pestka, J.J. 2003b. Role of double-stranded RNA-activated protein kinase R (PKR) in deoxynivalenol-induced ribotoxic stress response. Toxicol. Sci. 74: 335-344.

Zhou, H.R., Yan, D., and Pestka, J.J. 1997. Differential cytokine mRNA expression in mice after oral exposure to the trichothecene vomitoxin deoxynivalenol: dose–response and time course. Toxicol. Appl. Pharmacol. 144: 294–305.

Zhou, H.R., Yan, D., and Pestka, J.J. 1998. Induction of cytokine gene expression in mice after repeated and subchronic oral exposure to vomitoxin deoxynivalenol: differential toxin-induced hyporesponsiveness and recovery. Toxicol. Appl. Pharmacol. 151: 347–358.

Zimmerli, B., and Dick, R. 1996. Ochratoxin A in table wine and grapejuice: occurrence and risk assessment. Food Addit. Contam. 13: 655–668.

# Yersinia enterocolitica

Truls Nesbakken

## 11

## Abstract

*Yersinia enterocolitica* includes well-established pathogens and environmental strains that are ubiquitous in terrestrial and fresh water ecosystems. Evidence from large outbreaks of yersiniosis and from epidemiological studies of sporadic cases has shown that *Y. enterocolitica* is a foodborne pathogen, and that in many cases pork is implicated as the source of infection. The pig is the only animal consumed by man that regularly harbours pathogenic *Y. enterocolitica*. An important property of the bacterium is its ability to multiply at temperatures near to 0°C, and therefore in many chilled foods. The pathogenic serovars (mainly O:3, O:5,27, O:8 and O:9) show different geographical distribution. However, the appearance of strains of serovars O:3 and O:9 in Europe, Japan in the 1970s, and in North America by the end of the 1980s, is an example of a global pandemic. The possible risk of reactive arthritis following infection with *Y. enterocolitica* has led to further attention being paid to this microbe.

## Introduction

*Yersinia enterocolitica* is an important cause of gastroenteritis in humans, especially in temperate countries (Mollaret *et al.*, 1979; World Health Organization, 1983; World Health Organization, 1987). The consequences of yersiniosis are severe, and include prolonged acute infections, pseudoappendicitis, and long-term sequelae such as reactive arthritis and erythema nodosum, particularly in northern Europe where the prevalence of the HLA-B27 histocompatibility type is high (Ahvonen, 1972b; Winblad, 1975; Black and Slome, 1988; Cover and Aber, 1989; Sæbø and Lassen, 1991; Kapperud and Slome, 1998). These consequences make *Y. enterocolitica* infection a public health and economic problem of greater magnitude than the actual number of recorded cases would suggest (Ostroff *et al.*, 1992; Ostroff, 1995).

In many instances, attempts to isolate *Y. enterocolitica* from foods implicated in cases of disease in humans have been unsuccessful. However, evidence from large outbreaks of yersiniosis in the USA, Canada and Japan (Tauxe *et al.*, 1987; Cover and Aber, 1989) and from epidemiological studies of sporadic cases (Ostroff *et al.*, 1994;) has shown that *Y. enterocolitica* is a foodborne pathogen, and that in many cases pork is implicated as the source of infection (Morris and Feeley, 1963; Hurvell, 1981; Tauxe *et al.*, 1987; Lee *et al.*, 1991; Ostroff *et al.*, 1994). An important property of the bacterium is its ability to multiply at temperatures near to 0°C, and therefore in many chilled foods.

The emergence of yersiniosis is probably also related to changes that have occurred in livestock farming, food technology and the food industry. Of greatest importance are changes in the meat industry, where meat production has shifted from small-scale slaughterhouses with limited distribution patterns, to large facilities that process thousands of pigs each day and distribute their products nationally and internationally. Farm sizes have increased and animal husbandry methods have also become more intensive. While many modern slaughter techniques reduce the risk of meat contamination, opportunities for animal-to-animal transmission of the organism, and for cross-contamination of carcasses and meat products, exist on a scale that was unthinkable decades ago. In addition, advances in packaging and refrigeration now allow industry and consumers to store foods for much longer periods, a significant factor with regard to a cold-adapted pathogen such as *Y. enterocolitica*.

Due to the relative lack of information on *Y. pseudotuberculosis* as a foodborne pathogen (Schiemann, 1989), this bacterium will be considered less extensively than *Y. enterocolitica*. *Y. pseodotuberculosis* mainly causes epizootic disease, especially in rodents, with necrotizing granulomatous disease of liver, spleen and lymph nodes (Aleksic and Bockemühl, 1990; Aleksic *et al.*, 1995). In humans it may cause acute abdominal disease, septicaemia, arthritis and erythema nodosum (Knapp, 1958; Ahvonen, 1972a).

## Characteristics and taxonomy of *Yersinia enterocolitica*

The genus *Yersinia* of the family *Enterobacteriaceae* includes three well-established pathogens (*Yersinia pestis*, *Yersinia pseudotuberculosis* and *Yersinia enterocolitica*) and several non-pathogens (Mollaret *et al.*, 1979; Kapperud and Bergan, 1984). *Y. pestis* was isolated by Alexandre Yersin in 1894 (Yersin, 1894). The most important *Yersinia* infection, plague, caused by *Y. pestis*, is one of the most ancient recognized human diseases. Although infections caused by *Y. pseudotuberculosis* and *Y. enterocolitica* have only been reported more recently, it is nevertheless likely that these infections have also occurred for many years. Disease due to *Y. pseudotuberculosis* (first described in 1884) has been recognized since the beginning of the 20th century, and *Y. enterocolitica* was first shown to be associated with human disease in 1939 (Mollaret, 1995).

The current interest in *Y. enterocolitica* started in 1958 following a number of epizootics among chinchillas and hares (Mollaret *et al.*, 1979; Hurvell, 1981), and after the establishment of a causal relationship with abscedizing lymphadenitis in man. The similarity between the human and animal isolates was established in 1963 (Knapp and Thal, 1963), and in 1964 the species name *Y. enterocolitica* was formally proposed by Frederiksen (1964). During the past thirty years, the bacterium has been found with increasing frequency as a cause of human disease, and from animals and inanimate sources.

The genus *Yersinia* was proposed in 1944 by Van Loghem (1944) for bacteria that were related to the genus *Pasteurella*, and that were pathogenic. Thal (1954) drew attention to evidence relating *Yersinia* to the *Enterobacteriaceae*. A general numerical taxonomic study from 1958 placed *Yersinia* between *Klebsiella* and *Escherichia coli* (Sneath and Cowan, 1958). The allocation of *Yersinia* to the family *Enterobacteriaceae* was further supported by Frederiksen (1964).

## Differentiation of *Y. enterocolitica* from other *Yersinia* spp.

A range of strains of *Yersinia* variants has been isolated from animals, water and food (Mollaret *et al.*, 1979; Hurvell, 1981; Lee *et al.*, 1981; Kapperud, 1991). Many of these bacteria have characteristics that deviate considerably from the typical pattern shown by *Y. enterocolitica*, but can be classified as belonging to the genus *Yersinia* (Mollaret *et al.*, 1979; Kapperud and Bergan, 1984). Such *Y. enterocolitica*-like bacteria have now been divided on a genetic basis into seven new species (Bercovier *et al.*, 1980a; Bercovier *et al.*, 1980b, Brenner *et al.*, 1980a; Brenner *et al.*, 1980b; Ursing *et al.*, 1980; Brenner, 1981; Bercovier *et al.*, 1984; Aleksic *et al.*, 1987; Wauters *et al.*, 1988b): *Yersinia frederiksenii, Yersinia kristensenii, Yersinia intermedia, Yersinia aldovae, Yersinia rohdei, Yersinia mollaretii* and *Yersinia bercovieri*.

The four last-mentioned are not included as separate species in Bergey's Manual of Systematic Bacteriology (Bercovier and Mollaret, 1984). For practical reasons, the common term "*Y. enterocolitica*-like bacteria" is applied to these seven species. The species name *Y. enterocolitica* is reserved for the most typical strains, and it is only within this species that pathogenic strains have been found. It is therefore important to be able to differentiate the seven new species from *Y. enterocolitica*. This can be achieved by applying nine biochemical tests, and the fermentation of rhamnose, melibiose and sucrose being key characteristics (Bercovier *et al.*, 1980a; Bercovier *et al.*, 1980b; Brenner, 1981; Brenner *et al.*, 1980a; Brenner *et al.*, 1980b; Ursing *et al.*, 1980; Bercovier *et al.*, 1984; Aleksic *et al.*, 1987; Wauters *et al.*, 1988b). A description of the characteristics of *Y. enterocolitica* is given in Table 11.1.

*Y. enterocolitica* is a Gram-negative, oxidase-negative, catalase-positive, nitrate reductase-positive, facultative anaerobic rod (occasionally coccoid), 0.5–0.8 × 1–3 μm in size (Bercovier and Mollaret, 1984). It does not form a capsule or spores. It is non-motile at 35–37°C, but motile at 22–25°C with relatively few, peritrichous flagellae. Some human pathogenic strains of serovar O:3 are, however, non-motile at both temperatures. In addition, the bacterium is urease-positive, $H_2S$-negative, ferments mannitol, and produces acid, but not gas, from glucose.

The genus *Yersinia* also includes two further species that will not be considered further here, namely (i) *Y. pestis*, and (ii) *Y. ruckeri*, which is a fish pathogen.

Both phenotypic typing and genotyping methods have been used to differentiate *Y. enterocolitica* strains.

## Phenotypic characterization

### Biotyping

The bacteria that are currently classified as *Y. enterocolitica* do not constitute a homogeneous group. Within the species there is a wide spectrum of biochemical variants (Mollaret *et al.*, 1979; Bercovier *et al.*, 1980a; Swaminathan *et al.*, 1982; Kapperud and Bergan, 1984). Such variations form the basis for dividing *Y. enterocolitica* into biovars. A number of different biotyping schemes have been described (Wauters, 1970; Brenner, 1981; Swaminathan *et al.*, 1982; Kapperud and Bergan, 1984). However, Wauters *et al.* (1987) described a revised biotyping scheme (Table 11.2) that differentiates between pathogenic (biovars 1B, 2, 3, 4, 5) and non-pathogenic (only biovar 1A) variants. In this chapter, the term biovar has been used in accordance with the latest edition of Bergey's Manual (Bercovier and Mollaret, 1984). The proposed biovar 6 (Wauters *et al.*, 1987) is re-classified into two new species: *Y. bercovieri* and *Y. mollaretii* (Wauters *et al.*, 1988b).

### Serotyping using O- and H-antigens

*Y. enterocolitica* can be divided into serovars using O-antigens. So far, 76 different O-factors have been described in both *Y. enterocolitica* and *Y. enterocolitica*-like bacteria (Wauters, 1981; Wauters *et al.*, 1991). A few strains, however, cannot be typed by this system, and the number of described antigen factors is, therefore, likely to increase in the future. Fifty-four H-factors have been recognized (Wauters *et al.* 1991; S. Aleksic, personal communication, 1995), but H-antigen determination is rarely carried out and most studies are limited to the O-anti-

**Table 11.1** Characteristics of *Y. enterocolitica* (Brenner, 1981; Kapperud and Bergan, 1984)

| Parameter | | Reaction |
|---|---|---|
| Acid production from | Adonitol | – |
| | D-Cellobiose | + |
| | Dulcitol | – |
| | D-Glucose | + |
| | Malonate | – |
| | D-Mannitol | + |
| | D-Mannose | + |
| | D-Melibiose | – |
| | Mucate | – |
| | Raffinose | – |
| | L-Rhamnose | – |
| | Sucrose | + |
| | D-Sorbitol | + |
| | D-Trehalose | + |
| Other tests | Arginine dihydrolase | – |
| | Motility, 22–25°C | v* |
| | Motility, 35–37°C | – |
| | Metabolism of citrate (Simmons') | – |
| | Phenylalanine deaminase | – |
| | Gas from glucose | – |
| | Breakdown of gelatin, 22°C | – |
| | Production of $H_2S$ | – |
| | Production of indole | v |
| | KCN | – |
| | Lysine decarboxylase | – |
| | Methyl red | v |
| | Nitrate reduction | + |
| | Oxidase | – |
| | Ornithine decarboxylase | + |
| | Pigment | – |
| | Urease | + |
| | Voges–Proskauer reaction, 22–25°C | + |
| | Voges–Proskauer reaction, 35–37°C | – |

The tests are performed at 28°C, unless stated otherwise.
*Some of the human pathogenic variants of serovar O:3/biovar 4 are non-motile or only slightly motile at 22–25°C.
+, more than 90% of strains showing a positive reaction; –, less than 10% of strains showing a positive reaction. v, between 10% and 90% of strains showing a positive reaction.
Source: Nesbakken (1992).

gens. Most isolates of *Y. enterocolitica* possess H-antigens consisting of several subfactors. The H-antigens are monophasic and remain stable after repeated subculture and after storage as stab cultures (Aleksic *et al.*, 1986). The great variability in H-antigens allows differentiation of strains that cannot be distinguished by other phenotyping methods employed (Kapperud *et al.*, 1990b). Even though the preparation and standardization of specific antisera is both time-consuming and costly, H-antigen typing can be a valuable supplement to O-antigen typing and biochemical characterization of *Y. enterocolitica* in epidemiological investigations.

## Correlation between biovars, serovars, ecology, and pathogenicity

Although some antigenic factors are linked to pathogenic strains, for example O:3, O:9, O:8 etc., serotyping alone cannot be used to indicate pathogenicity because these antigenic factors also occur in non-pathogenic species or biovars. For instance, factor O:3 is common in *Y. frederiksenii*, and may also

be encountered in other species and in biovar 1A of *Y. enterocolitica*. Factor O:8 occurs in biovar 1A and in *Y. bercovieri*, and factor O:9 in *Y. frederiksenii*. Hence, antigenic typing of the strains should always be done after appropriate biochemical characterization.

The relationships between the biovars, the O-antigens and the ecology of *Y. enterocolitica* and related species are presented in the paragraphs below (Wauters, 1991):

Biovar 1A includes a large number of serovars which are found in the environment, in food and occasionally in the digestive tract of animals and humans. They do not possess virulence properties and are of little or no clinical significance for man.

Biovars 1B, 2, 3, 4 and 5 are potential pathogens for man or animals. They exhibit pathogenic properties and, when freshly isolated, usually harbour the virulence plasmid. Strains of biovar 1B belong to a small number of pathogenic serovars, the most frequent being O:8, O:21, O:13a and O:13b, whereas O:4, O:18 and O:20 occur less frequently. They have been

**Table 11.2** Revised biotyping scheme for *Y. enterocolitica* (Wauters *et al.* (1987). A proposed biovar 6 was reclassified into two new species (Wauters *et al.*, 1988b)

| Parameter | Biovar | | | | | |
|---|---|---|---|---|---|---|
| | 1 A* | 1 B | 2 | 3 | 4 | 5 |
| Lipase (Tween-esterase) | + | + | − | − | − | − |
| Esculin/salicin 24 h | +/− | − | − | − | − | − |
| Formation of indole | + | + | (+) | − | − | − |
| Acid from xylose | + | + | + | + | − | v |
| Acid from trehalose/NO$_3$ reduction | + | + | + | + | + | − |
| Pyrazinamidase | + | − | − | − | − | − |
| β-D-glucosidase | + | − | − | − | − | − |
| Voges–Proskauer reaction | + | + | + | + | + | (+) |
| Proline peptidase | v | − | − | − | − | − |

Incubation at 28°C.
*Nonpathogenic variants.
v, variable; +/−, delayed positive; (+), weakly positive.

isolated mainly in the United States and were therefore called "American strains". Since the beginning of the nineties, a few strains have been isolated outside North-America; in Europe, Japan, India and Chile (Wauters, 1991).

Biovar 2 only includes two serovars, O:9 and O:5,27, which are pathogenic for man. O:9 is important in Belgium, Holland and France, while O:5,27 is quite common in the United States (Mollaret *et al.*, 1979; World Health Organization, 1983; Ministere des Affaires Sociales, de la Sante Publique et de l'Environnement Institut d'Hygiene et d'Epidemiologie, 1995).

Biovar 3 includes serovar O:1,2,3, which has been isolated mainly from chinchilla and rodents and rarely from man (Mollaret *et al.*, 1979; Hurvell, 1981). There are also a few serovars O:5,27 in this biovar (Wauters, 1991). As the beginning of the 1990s, a Voges-Proskauer negative variant of biovar 3, belonging to serovar O:3 was isolated in Japan and East Asia and has become a prominent pathogenic for man in these countries (Wauters 1991, Wauters *et al.*, 1991).

Biovar 4 contains only one serovar, O:3, which is the main pathogenic serovar for man, distributed world-wide. It is also isolated regularly from healthy pigs (Wauters, 1970; Wauters, 1979; Christensen, 1980; Doyle *et al.*, 1981; Hurvell, 1981; Schiemann and Fleming, 1981; Nesbakken and Kapperud, 1985).

Biovar 5 is of no importance for humans. It includes serovar O:2,3, which is found in hares and some other animals (Krogstad, 1974; Mollaret *et al.*, 1979; Hurvell, 1981; Lanada, 1990; Slee and Button, 1990; Wauters, 1991).

## Other phenotypic typing methods

### Antibiogram typing

Antimicrobial susceptibility testing is a highly standardized method employed in a large number of laboratories (Cornelis, 1981; Baker and Farmer, 1982). However, susceptibility patterns may be influenced by local antibiotic usage, which may make differentiation of outbreak strains from epidemiologically unrelated strains difficult. Of the six typing methods that were compared in one study, antimicrobial susceptibility testing showed the greatest variability in all serovars investigated (Kapperud *et al.*, 1990b). On the basis of the investigation of strains from three outbreaks, Baker and Farmer (1982) concluded that even though antimicrobial susceptibility testing contributed to the differentiation of the isolates, results were less reliable than those obtained by serotyping and phage typing.

### Phage typing

Phage typing requires a battery of phages and indicator strains. Two European phage typing systems are described, but not used in many laboratories (Mollaret and Nicolle, 1965; Niléhn, 1969; Nicolle, 1973). The French system (Mollaret and Nicolle, 1965; Nicolle, 1973) is more often used than the Swedish one (Niléhn, 1969). A bacteriophage typing system that allows greater differentiation of American O:8 strains has also been described (Baker and Farmer, 1982).

## Genotyping

Methods based on the characterization of the genotype include plasmid profile analysis, restriction enzyme analysis of plasmid and chromosomal DNA (DNA fingerprinting), pulsed-field gel electrophoresis (Buchrieser *et al.*, 1994; Najdenski *et al.*, 1994; Saken *et al.*, 1994) and the use of DNA or RNA probes (Wachsmuth, 1985; Mayer, 1988; Tenover, 1988; Andersen and Saunders, 1990).

### Restriction endonuclease analysis of the plasmid (REAP)

Pathogenic strains of *Y. enterocolitica* harbour a family of virulence plasmids showing a DNA sequence homology of 55–100% (Portnoy and Martinez, 1985). REAP yields specific patterns for each sero-/biovar combination. Within some sero-/biovar combinations like serovar O:8/biovar 1B there are a number of different patterns, while within serovar O:3/biovar 4, the diversity of patterns are limited (Nesbakken *et al.*, 1987). Generally, the discriminatory power increases if more than one restriction enzyme is used (Heesemann *et al.*, 1983; Nesbakken *et al.*, 1987).

*Restriction endonuclease analysis of the chromosome (REAC)*

REAC patterns varied clearly between sero-/biovar combinations. Just as REAP, this method shows the greatest discriminatory power with regard to serovar O:8. Compared with O:8, serovars O:3 and O:9 were relatively homogeneous with regard to REAC patterns (Kapperud *et al.*, 1990b). It may be some problems to interpret complex restriction patterns consisting of hundreds of bands.

*Ribotyping*

The problems with complex REAC patterns may be avoided by the use of DNA probes that hybridize to specific DNA sequences. In conclusion, a close relationship exists between the ribotypes and the sero-/biovar combinations. As for REAP and REAC, the diversity among serovar O:3/biovar 4 is limited (Andersen and Saunders, 1990).

*Pulsed-field gel electrophoresis (PFGE) typing*

This technique has been applied for comparing *Y. enterocolitica* strains belonging to the same sero-/biovar combination (Buchrieser *et al.*, 1994; Najdenski *et al.* 1994; Saken *et al.* 1994). Some studies demonstrate clearly the high discriminative power of PFGE (Fenwick, 1998; Harnett *et al.*, 1998; Frederiksson-Ahomaa *et al.* 1999). Though several pulsotypes are found among O:3/biovar 4 strains, most of the strains belong to one or two dominating pulsotypes (Buchrieser *et al.* 1994; Najdenski *et al.* 1994; Saken *et al.* 1994; Asplund *et al.*, 1998; Frederiksson-Ahomaa *et al.* 1999).

## Comparison of typing methods

In a comparison of six methods for typing *Y. enterocolitica* (Kapperud *et al.*, 1990b), antimicrobial susceptibility had the greatest overall discriminatory power, followed by REAC, H-antigen typing, restriction enzyme analysis of plasmid DNA (REAP), phage typing, and biotyping (in decreasing order of discrimination) (Kapperud *et al.*, 1990b). The use of multilocus enzyme electrophoresis to investigate allelic variations in the chromosomal genomes of the same strains indicates that this method is a less suitable epidemiological tool for *Y. enterocolitica* (Caugant *et al.*, 1989). The discriminatory power of the genotypic methods was greatest within serovar O:8. Since this serovar is homogeneous with regard to biovar and phagevar, there has long been a need for methods that permit a finer degree of differentiation in epidemiological investigations (Baker and Farmer, 1982). The results also show that restriction enzyme analysis is an effective method for the analysis of serovar O:8. The high discriminative power of PFGE is demonstrated in studies during the last 10 years (Fenwick, 1998; Harnett *et al.*, 1998; Fredriksson-Ahomaa *et al.*, 1999). However, a comparison between PFGE and other epidemiological typing methods is still missing.

## Characteristics of diseases caused by *Yersinia enterocolitica*

### Clinical symptoms of *Yersinia enterocolitica* infection

Gastroenteritis is by far the most common symptom of *Y. enterocolitica* infection (yersiniosis) in humans (Bottone, 1977; Mollaret *et al.*, 1979; Wormser and Keusch, 1981; Cover and Aber, 1989). The clinical picture is usually one of self-limiting diarrhea associated with mild fever and abdominal pain (Wormser and Keusch, 1981). Nausea and vomiting occur, but less frequently. The portion of the gastrointestinal tract usually involved is the ileocaecal region (Sandler *et al.*, 1982). The colon may also be affected and the infection may simulate Crohn's disease, which has a different prognosis (Vantrappen *et al.*, 1977; Gleason and Patterson, 1982). Occasionally the infection is limited to the right fossa iliaca in the form of terminal ileitis or mesenteriel lymphadenitis, with symptoms that can be confused with those of acute appendicitis. In several studies of patients with the appendicitis-like syndrome, *Y. enterocolitica* has been found in up to 9% of patients (Niléhn and Sjöström, 1967; Ahvonen, 1972a; Jebsen *et al.*, 1976; Pai *et al.*, 1982; Samadi *et al.*, 1982; Attwood *et al.*, 1987; Megraud, 1987). Infections with serovars O:3 or O:9 are, in some patients, followed by reactive arthritis (Aho *et al.*, 1981), which is most common in patients possessing the tissue type HLA-B27. Often, although not always, the patient has shown prior gastrointestinal symptoms. Other complications seen with *Y. enterocolitica* infection are reactive skin complaints, erythema nodosum being the most common. Many such patients have no history of prior gastrointestinal involvement. Septicaemia due to *Y. enterocolitica* is seen almost exclusively in individuals with underlying disease (Bottone, 1977), while those with cirrhosis and disorders associated with excess iron are particularly predisposed to infection and increased mortality.

Clinical symptoms are influenced by the age of the patient (Bottone, 1977; Wormser and Keusch, 1981). Gastroenteritis dominates in children and young people, while various forms of reactive arthritis are most common in young adults, and most patients with skin manifestations are adult females (Wormser and Keusch, 1981). In Scandinavia, there is a relatively high incidence of both reactive arthritis (10–30%) (Winblad, 1975) and erythema nodosum (30%) (Ahvonen, 1972b), caused by serovars O:3 and O:9. These forms of the disease have been almost totally absent in the USA, where O:8, historically, has been the most common cause of yersiniosis. This situation may change, however, as serovar O:3 is on the increase in the USA (Bissett *et al.*, 1990).

Transient carriage and excretion of both pathogenic and non-pathogenic *Y. enterocolitica* may occur following exposure to the bacteria. High rates of asymptotic carriage have been reported in connection with outbreaks (Tacket *et al.*, 1985), although in surveys unrelated to outbreaks, *Y. enterocolitica* has been detected in the stools of fewer than 1% of individuals (Niléhn and Sjöström, 1967). In patients with *Y. enterocolitica* enteritis, the organism may be excreted in the stools for lengthy periods after symptoms have resolved. In a study of Norwegian patients, convalescent carriage of *Y. enterocolitica* O:3 was detected in the stools of 47% of 57 patients for a median period of 40 days (range 17–116 days) (Ostroff *et al.*, 1992).

## Mechanisms of pathogenesis

### Pathogenesis and immunity

Human infection due to *Y. enterocolitica* is most often acquired by the oral route. The minimal infectious dose required to cause disease is unknown. In one volunteer, ingestion of 3.5 × 10$^9$ organisms was sufficient to produce illness (Szita *et al.*, 1973). The incubation period is uncertain, but has been estimated as being between 2 and 11 days (Szita *et al.*, 1973; Ratnam *et al.*, 1982).

Enteric infection leads to proliferation of *Y. enterocolitica* in the lumen of the bowel and in the lymphoid tissue of the intestine. Adherence to and penetration into the epithelial cells of the intestinal mucosa are essential factors in the pathogenesis of *Y. enterocolitica* infection (Portnoy and Martinez, 1985; Cornelis *et al.*, 1987; Miller *et al.*, 1988; Bliska and Falkow, 1994). When the bacteria reach the lymphoid tissues in the terminal ileum, a massive inflammatory response takes place in the Peyer's patches. Reactive arthritis and erythema nodosum appear to be delayed immunologic sequelae of the original intestinal infection.

In humans, infection with pathogenic strains of *Y. enterocolitica* stimulates development of specific antibodies. It is not known whether specific serum antibody protects against reinfection with *Y. enterocolitica* organisms of the same or different serovars. The immunological response can be measured by a variety of techniques, including tube agglutination, indirect haemagglutination, enzyme-linked immunosorbent assay (ELISA), and solid-phase radioimmunoassay (Atwood *et al.*, 1987). An indirect immunofluorescent-antibody assay has also been used (Cafferkey and Buckley, 1987). However, ELISA is probably the most suitable and extensively applied method in the world today. Agglutinating antibodies appear soon after the onset of illness and persist for from 2 to 6 months. Some serovars may be associated with illness without eliciting a detectable serological response (Ahvonen, 1972a; Toma, 1973).

The serological diagnosis of *Y. enterocolitica* infection may be complicated by the existence of cross-reactions between *Y. enterocolitica*, most notably serovar O:9, and such organisms as *Y. pseudotuberculosis*, *Brucella*, *Vibrio*, *Salmonella*, *Proteus* and *Escherichia coli* (Wauters, 1981; Nielsen *et al.*, 1996; Weynants *et al.*, 1996b). The interpretation may also be confounded by high prevalence of seropositive individuals in the healthy population. Of 813 Norwegian military recruits selected at random, 67 (8.2%) had antibodies against *Y. enterocolitica* O:3 (Nesbakken *et al.*, 1991b). Patients with thyroiditis of an immunological aetiology have an unexplained increased frequency of cross-reacting antibodies to *Y. enterocolitica* (Shenkman

and Bottone, 1976). Detection of antibodies to plasmid-encoded proteins, *Yersinia* outer membrane proteins (Yops), by immunoblots, has been suggested as a highly specific means of demonstrating previous *Y. enterocolitica* infection (Ståhlberg *et al.*, 1989; Hoogkamp-Korstanje *et al.*, 1992). Demonstration of specific circulating IgA to the Yops is indicative of recent or persistent infection and is strongly correlated with the presence of virulent *Y. enterocolitica* in the intestinal lymphatic tissue of patients with reactive arthritis.

### Virulence factors

#### The virulence plasmid

Human pathogenic strains of *Y. enterocolitica* possess a special plasmid 40–50 megadaltons in size (Brubaker, 1979; Portnoy and Martinez, 1985). The presence of this plasmid is an essential, though not sufficient, prerequisite for the bacterium to be able to induce disease. Properties associated with the chromosome are also necessary for virulence. Corresponding plasmids are found in *Y. pseudotuberculosis* and *Y. pestis*, and there is thus a family of related virulence plasmids within the genus *Yersinia* (Portnoy and Martinez, 1985). The way in which the plasmid contributes to virulence has not been fully elucidated. The presence of this virulence plasmid has been associated with several properties, most of which are phenotypically expressed only at elevated growth temperatures of 35–37°C (Portnoy and Martinez, 1985). The list of such plasmid-mediated and temperature-regulated properties includes: Ca$^{2+}$-dependent growth (Gemski *et al.*, 1980; Perry and Brubaker, 1983; Portnoy *et al.*, 1984), production of V and W antigens (Perry and Brubaker, 1983), spontaneous autoagglutination (Laird and Cavanaugh, 1980; Skurnik *et al.*, 1984), mannose-resistant haemagglutination (Kapperud *et al.*, 1987), serum resistance (Pai and DeStephano, 1982), binding of Congo red dye (Prpic *et al.*, 1985), hydrophobicity (Lachica *et al.*, 1984), mouse virulence (Ricciardi *et al.*, 1978; Pai and DeStephano, 1982; Schiemann and Devenish, 1982; Nesbakken *et al.*, 1987), and production of a number of proteins (Portnoy *et al.*, 1984; Bölin *et al.*, 1985; Portnoy and Martinez, 1985), of which one is a true outer membrane protein (YadA, previously termed Yop1) (Michiels *et al.*, 1990). This true outer membrane protein forms a fibrillar matrix on the bacterial surface and mediates cellular attachment and entry (Bliska and Falkow, 1994). It also confers resistance to the bactericidal effect of normal human serum and inhibition of the anti-invasive effect of interferon.

The mechanism of virulence is known to vary even between different serovars within *Yersinia enterocolitica*. Plasmid-containing strains of O:8, O:13a, O:13b, and O:21 are lethal to orally infected mice, whereas O:3, O:9 and O:5,27 are not (Ricciardi *et al.*, 1978; Pai and DeStephano, 1982; Schiemann and Devenish, 1982; Nesbakken *et al.*, 1987). These latter serovars, however, are capable of maintaining intestinal colonization for at least 1 week after orogastric challenge (Nesbakken *et al.*, 1987).

## The chromosome

Elements encoded by the chromosome are also necessary for maximum virulence. The pathogenic yersiniae share at least two chromosomal loci, *inv* and *ail*, that play a role in their entry into eukaryotic cells (Miller *et al.*, 1988). The *inv* and *ail* gene products can be classified as adhesins since they mediate adherence to the eukaryotic surface. Unlike other previously characterized bacterial adhesins, they also mediate entry into a variety of mammalian cells.

### Virulence assays

Two of the above-mentioned characteristics, Ca$^{2+}$-dependent growth and spontaneous autoagglutination, have often been employed as indicators of the pathogenic properties of the bacterium. These tests depend on the presence of the plasmid. However, the plasmid may be lost in the course of a prolonged isolation procedure. Other situations where loss of the plasmid may occur are: (i) cultivation on magnesium oxalate agar at 37°C (Perry and Brubaker, 1983), and (ii) extended storage at refrigeration temperatures (Bhaduri and Turner-Jones, 1983). A test for pyrazinamidase activity has been included in the new biotyping scheme of Wauters *et al.* (1987) (Table 11.2). The test is simple, reproducible, and shows a striking correlation between potential pathogenicity and lack of pyrazinamidase activity, independent of the presence of the virulence plasmid. The reliability of this test has been documented following studies of more than 1000 *Y. enterocolitica* isolates representing a broad range of serovars and different geographical origins (Kandolo and Wauters, 1985). Pyrazinamidase activity is commonly found in all environmental strains, which never harbour the virulence plasmid. There are a few exceptions in biovar 1A strains that lack pyrazinamidase activity. Furthermore, in *Y. kristensenii*, the reaction is often weak. Pathogenic strains, whether or not the plasmid is still present, show a negative reaction. Also the invasiveness in HeLa cells is chromosomally encoded (Schiemann and Devenish, 1982). Only pyrazinamidase negative, pathogenic strains are invasive for HeLa cells, although they do not multiply within the cells (Devenish and Schiemann, 1981).

### Enterotoxin production

Many strains of *Y. enterocolitica* and related species produce a heat-stable enterotoxin (YEST) when the bacteria are cultured at 20–30°C (Pai *et al.*, 1978; Kapperud, 1982; Nesbakken, 1985a). Certain strains, especially within the species *Y. kristensenii*, are also able to produce YEST at 4 and 37°C (Pai *et al.*, 1978; Kapperud, 1982; Nesbakken, 1985a). This property is regulated by chromosomal genes and is independent of the virulence plasmid.

The toxin of *Y. enterocolitica* resembles the heat-stable enterotoxin (ST) of *E. coli* in pH stability, activity in suckling mice and the rabbit ileal loop (Boyce *et al.*, 1979), and the stimulation of guanylate cyclase activity (Inoue *et al.*, 1983). However, unlike the ST of *E. coli*, the YEST is found only in the methanol-soluble fraction (Velin and Emody, 1982). The toxin is not inactivated by exposure to 121°C for 30 min or by storage at 4°C for at least 5 months (Boyce *et al.*, 1979). It is also highly resistant to enzymatic degradation (Okamoto *et al.*, 1983), but sensitive to treatment with 2-mercaptoethanol (Velin and Emody, 1982). Purification of YEST reveals a protein with a molecular weight of about 3000, and a high degree of amino acid homology with *E. coli* ST (Takao *et al.*, 1984).

Although there is no evidence to support the involvement of YEST in the pathogenesis of *Y. enterocolitica* enteritis, the possibility still remains that enterotoxigenic strains may produce foodborne intoxication by means of preformed enterotoxins (Kapperud and Langeland, 1981; Kapperud, 1982). This assumption is based on the fact that YEST is able to resist gastric acidity as well as temperatures used in food processing and storage, without losing activity (Boyce *et al.*, 1979). The ability to produce YEST at room and refrigeration temperature may give the "environmental" yersiniae a new significance in food hygiene as potential agents of foodborne intoxication.

## Reservoirs

### Food animals

#### Pigs

Healthy pigs are often carriers of strains of *Y. enterocolitica* that are pathogenic to humans, in particular strains of serovar O:3/biovar 4 and serovar O:9/biovar 2) (Hurvell, 1981; Schiemann, 1989; Kapperud, 1991). The organisms are present in the oral cavity, especially the tongue and tonsils, and in the intestine and faeces. Shiozawa *et al.* (1991) reported that O:3 strains were isolated from 85% of oral swabs from 40 freshly slaughtered, healthy pigs and presented evidence that the organism colonized the pigs' tonsils. In this study 24.3% of 140 pigs were carriers of the organism in the caecum, with counts ranging from fewer than 300 to 110,000 *Y. enterocolitica*/g of caecal contents. Strains of O:3 have been found frequently on the surface of freshly slaughtered pig carcasses (in frequencies up to 63.3%) (Andersen, 1988; Nesbakken, 1988). This is probably the result of spread of the organism via faeces and intestinal contents during slaughter and dressing operations. The association between yersiniosis in humans and the consumption of raw pork in Belgium (Tauxe *et al.*, 1987) and the apparently rare incidence of the infection in Moslem countries (Samadi *et al.*, 1982), where consumption of pork is restricted, point to pork as a source of infection with *Y. enterocolitica*.

Other pathogenic strains do not appear to be as closely associated with pigs, and may have a different ecology. In western Canada, O:8 and O:5,27 strains have been found most commonly in humans, but only O:5,27 strains were found in the throats of slaughter-age pigs (Schiemann, 1989). In the USA, O:5,27 strains were isolated from the caecal contents and faeces of two out of 50 pigs at slaughter (Kotula and Sharar, 1993). Serovar O:8/biovar 1B, until recently considered to be the most common human pathogenic strain of *Y. enterocolitica* in the USA (Ostroff, 1995) and in western Canada (Toma and Lafleur, 1981), has seldom been reported in pigs.

Healthy pigs have been found to be infected with *Y. enterocolitica* O:3 in frequencies up to 85% (Wauters, 1970; Wauters, 1979; Christensen, 1980; Doyle *et al.*, 1981; Hurvell, 1981; Schiemann and Fleming, 1981; Nesbakken, 1985b; Nesbakken and Kapperud, 1985; Nesbakken, 1988; Shiozawa *et al.*, 1991).

### Cattle

Positive tests in serological control programs for brucellosis in cattle, have in some cases proved to be cross-reactions against *Y. enterocolitica* serovar O:9 (Wauters, 1981; Nielsen *et al.*, 1996; Weynants *et al.*, 1996b; Danish Zoonosis Centre, Copenhagen). However, cattle are generally not considered to be carriers of human pathogenic *Y. enterocolitica*.

### Sheep and goats

In Norway, Krogstad (1974) demonstrated outbreaks of *Y. enterocolitica* infection in goat herds in which serovar O:2/biovar 5 was implicated. He also described a case in which an animal attendant was infected by the same serovar. Biovar 5 has also been isolated from goats in New Zealand (Lanada, 1990). Enteritis in sheep and goats due to infection of *Y. enterocolitica* O:2,3, biovar 5 is also seen in Australia (Slee and Button, 1990). Serovar O:3 was isolated from the rectal contents in two (3.0%) of 66 lambs in New Zealand (Bullians, 1987).

### Poultry

Stengel (1985) isolated *Y. enterocolitica* serovars O:3 (*n* =3), O:9 (*n* = 3), and non-virulent *Y. enterocolitica* (*n* = 13) from 130 samples of poultry. This is probably the first time these virulent serovars have been isolated from poultry, and there was no obvious opportunity for cross-contamination from pigs or pork. Nevertheless, according to other investigations, poultry has not been proven to be carrier of pathogenic *Y. enterocolitica* (Leistner *et al.*, 1975; Guthertz *et al.*, 1976; Inoue and Nagao, 1976; Capriolli *et al.*, 1978; Norberg, 1981; Cox *et al.*, 1990; Nesbakken *et al.*, 1991b).

### Deer

Surveys in New Zealand have found deer to carry both O:5,27/biovar 2 and O:9/biovar 2 (S. Fenwick, personal communication, 1996).

## Other animals

Dogs, cats and rodents, such as rats, may also occasionally be faecal carriers of O:3 and O:9 (Hurvell; 1981; Fukushima *et al.*, 1984b; Fenwick *et al.*, 1994; Hayashidani *et al.*, 1995). The relative intimate contact between man and pets suggests a potential reciprocal transmission, although such an epidemiological link has not been clearly confirmed (Nesbakken *et al.*, 1991b; Fenwick *et al.*, 1994; Ostroff *et al.*, 1994). O:3 has been isolated from one (0.9%) of 117 crows in Japan, and O:9 from one (0.7%) of 156 Japanese serows (Kato *et al.*, 1985).

It has been suggested that rodents may be reservoirs of O:8 and O:21 strains in North America (Schiemann, 1989), and evidence has been published that this is indeed the case in Japan (Hayashidani *et al.*, 1995). Serovars O:8 and O:21 are closely related in many ways (biochemical profile, H antigens, and animal virulence). Serovar 21 ("O:Tacoma") has been isolated from wild rodent fleas in the western United States (Quan *et al.*, 1974).

As is the case with human pathogenic variants, the animal pathogenic strains also belong to particular combinations of serovars and biovars (Mollaret *et al.*, 1979; Hurvell, 1981). Serovar O:1/biovar 3 was responsible for widespread outbreaks in chinchilla both in Europe and in the USA during the period 1958–1964 (Mollaret *et al.*, 1979; Hurvell, 1981). During the same period, epizootics were observed among hares along the French-Belgium frontier caused by serovar O:2/biovar 5 (Mollaret *et al.*, 1979; Hurvell, 1981).

## Environmental reservoirs and water

Although a range of different serovars and biovars has been isolated from the environment (e.g. soil and water) (Mollaret *et al.*, 1979; Kapperud, 1981; Kapperud *et al.*, 1981; Kapperud, 1991), most belong to biovar 1A, according to the revised biotyping scheme described by Wauters *et al.* (1987) (Table 11.2). As regards the so-called *Y. enterocolitica*-like bacteria, which are also common in the same environments, no clinical significance is currently attributed to most of the strains within biovar 1 A (Bottone, 1977; Mollaret *et al.*, 1979). However, such environmental strains have occasionally caused disease in immunocompromised patients (Bottone, 1977; Mollaret *et al.*, 1979).

Shallow wells in particular, and also rivers and lakes, are susceptible to contamination with by surface runoff from rain or snow melt. Such runoff may become fecally contaminated by wild or domestic animals, or by leakage from septic tanks or open latrines in the surrounding areas. Water is a significant reservoir of *Y. enterocolitica* (Lassen, 1972; Harvey *et al.* 1976; Kapperud and Jonsson, 1978; Saari and Jansen, 1979; Langeland, 1983; Brennhovd, 1991). However, most isolates of *Y. enterocolitica* and *Y. enterocolitica*-like bacteria obtained from water are variants with no known pathogenic significance to man.

Evidence that water supplies can be a source of pathogenic *Y. enterocolitica* has been found in certain outbreaks associated with contamination from water sources, and in a few sporadic cases, as well as case–control studies.

## Incidence of *Yersinia enterocolitica* in foods

### Pork

In contrast to the frequent occurrence of the bacterium in pigs and on freshly slaughtered carcasses, pathogenic *Y. enterocolitica* have only exceptionally been found from pork products at the retail sale stage, with the exception of fresh tongues (Delmas and Vidon, 1985; Nesbakken *et al.*, 1985; DeBoer *et al.*, 1986; Ternström and Molin, 1987; World Health Organization, 1987; Wauters *et al.*, 1988a; Schiemann, 1989; Nesbakken *et al.*, 1991a).

This phenomenon might be explained by the lack of proper selective methodology for the isolation of pathogenic strains. However, Wauters *et al.* (1988a) described a method which allowed *Y. enterocolitica* O:3 to be recovered from as many as 12 (24%) of 50 ground pork samples. Genetic probes can also be used in DNA colony hybridization to demonstrate virulent *Y. enterocolitica* strains. The results of one such investigation support the supposition that traditional culture methods lead to underestimation of the presence of virulent *Y. enterocolitica* in pork products. A significantly higher detection rate (60%) was achieved when two isolation procedures (Nordic Committee on Food Analysis, 1987; Wauters *et al.*, 1988a) were combined with colony hybridization, than when the isolation procedures were employed alone (18%) (Nesbakken *et al.*, 1991a). In this study, counts of virulent yersiniae in pork sausage meat varied from 50 to 2500/g, and in pork cuts from 50 to 300/g.

Ostroff *et al.* (1994) showed in a case–control study that persons with *Y. enterocolitica* infection reported having eaten significantly more pork or sausages than their matched controls. This phenomenon might be explained by recontamination of sausages after pasteurization, bacterial multiplication during storage, and insufficient heat treatment in the kitchen.

### Beef and mutton

With a few exceptions (Bullians, 1987; Weynants *et al.*, 1996b; Danish Zoonosis Centre, Copenhagen), cattle and sheep are generally not considered to be carriers of human pathogenic *Y. enterocolitica*. Nevertheless, the possibility exists of cross-contamination to beef or lamb from pig carcasses and pork in slaughterhouses, cutting plants, meat processing establishments, and butchers' shops. Beef is sometimes consumed after little or no heat treatment, and some concern has therefore been expressed about *Y. enterocolitica* contamination in beef. According to the case–control study of Ostroff *et al.* (1994), persons with yersiniosis were also more likely than controls to report a preference for eating meat prepared raw or rare.

In an extensive survey conducted in Germany by Leistner *et al.* (1975), four strains of *Y. enterocolitica* and six *Y. enterocolitica*-like bacteria were isolated from 37 samples of beef. Two of the beef isolates belonged to biovars that can cause typical human illness, but their virulence was not tested (Leistner *et al.*, 1975). According to other investigations, beef and lamb has not been proven to be carrier of pathogenic *Y. enterocolitica* (Inoue and Kurose; 1975; Hanna *et al.*, 1976; Ternström and Molin, 1987; Ibrahim and MacRae, 1991).

### Milk and dairy products

World-wide studies indicate that *Y. enterocolitica* is fairly common in raw milk (Lee *et al.*, 1981), and several outbreaks of yersiniosis due to serovars O:3, O:6,30, O:8, O:10K and O:13,18 have been associated with the consumption of milk or milk products (Zen-Yoji, 1973; Black *et al.*, 1978; Morse *et al.*, 1984; Tacket *et al.*, 1984; Greenwood and Hooper, 1990; Alsterlund *et al.*, 1995; Centers for Disease Control and Prevention, 1995). *Y. enterocolitica* was isolated from ice-cream (Mollaret *et al.*, 1972) and pasteurized milk (Sarrouy, 1972; Zen-Yoji, 1973) as early as 1970.

Reported levels of raw milk contamination with *Y. enterocolitica* vary: Japan, 4% (Fukushima *et al.*, 1984a); Denmark, 10% (Christensen, 1982); Australia, 16% (Ibrahim and MacRae, 1991); Northern Ireland, 22% (Walker and Gilmour, 1986); Morocco, 30% (Hamama *et al.*, 1992); Italy, 37% (Franzin *et al.*, 1984), Ireland, 39% (Rea *et al.*, 1992) and France, 81% (Vidon and Delmas, 1981). Almost all reported isolates from raw milk samples have been categorized as nonvirulent environmental strains.

According to Greenwood *et al.* (1990), strains of *Yersinia* spp. could be isolated from a sample of milk taken immediately after pasteurization, but before bottling, even though the time/temperature conditions applied during pasteurization appeared to be adequate. The presence of *Y. enterocolitica* has also been demonstrated in pasteurized milk in other studies (Delmas, 1983; Moustafa *et al.*, 1983; Stengel, 1984; DeBoer *et al.*, 1986). As *Y. enterocolitica* has the capacity for growth in milk at refrigeration temperatures (Stern *et al.*, 1980b), consumption of pasteurized milk contaminated with *Yersinia* spp. constitutes a potential health risk. However, pasteurized milk, cheese made from pasteurized milk, and milk products such as ice-cream prepared from pasteurized mixes should normally not pose a health problem unless recontaminated with *Y. enterocolitica*, because the pasteurization process is usually adequate to destroy *Y. enterocolitica*. In the study of Greenwood *et al.* (1990) it also proved that the bottles into which the pasteurized milk was filled, were probably also a source of contamination. None of the strains found were virulent.

DeBoer *et al.* (1986) found four positive samples (4.5%) out of 89 Brie and Camembert cheeses, and one positive sample (2.0%) out of 50 blue-veined cheeses. None of the isolated strains proved to be virulent. In another study (Hamama *et al.*, 1992) seven (7.4%) of 94 cheeses were contaminated with nonvirulent *Yersinia* spp., all the positive cheeses being made of raw milk.

In conclusion, it is almost solely in connection with outbreaks caused by contaminated pasteurized milk (Tacket *et al.*, 1984; Greenwood and Hooper, 1990; Alsterlund *et al.*, 1995), reconstituted powdered milk (Morse *et al.*, 1984), contaminated chocolate milk (Black *et al.*, 1978) that one has been able to find the pathogenic strains. If not found in the product itself, epidemiological investigations have incriminated the product. In contrast, the screening of "normal" samples has most often only revealed nonvirulent strains. Nevertheless, the consumption of raw milk or dairy products made from raw milk may pose a potential public health problem.

*Seafood*

Some seafood may be consumed raw, and thus the presence of virulent strains of *Y. enterocolitica* due to faecal contamination of water or contamination during processing, could cause foodborne illness. The presence of environmental strains of *Y. enterocolitica* has been reported in fish (Rakovsky *et al.*, 1973; Alonso *et al.*, 1975; Kapperud and Jonsson, 1976; DeBoer *et al.*, 1986), mussels (Spadaro and Infortuna, 1968) and oysters in Canada (Toma, 1973) and in the USA (Lee *et al.*, 1977). *Y. enterocolitica*-like bacteria have been isolated from shrimps and crab from the Gulf of Mexico (Peixotto *et al.*, 1980).

*Vegetables*

Raw vegetables might be contaminated with faeces from wild animals or from the use of animal or human wastes as fertilizer, and may constitute a health risk (DeBoer *et al.*, 1986). Contamination might also occur in a food processing plant when vegetables are washed and prepared. However, it should be mentioned that vegetables are usually contaminated with large numbers of other bacteria and yeasts, and as *Yersinia* is a poor competitor (Stern *et al.*, 1980b), extensive outgrowth of the organism is unlikely.

In France, *Y. enterocolitica* has been recovered from carrots, tomatoes, and green salads from hospital foods (Louiseau-Marolleau and Alonso, 1976). Some isolation have been made from vegetables in Czechoslovakia (Aldova *et al.*, 1975). Thirty (46.1%) of 65 samples in a French study (Delmas and Vidon, 1985), and 37 (43%) of 87 samples in one study from The Netherlands (DeBoer *et al.*, 1986) were reported to be *Yersinia* spp. positive. None of the isolated strains proved to be virulent.

## Epidemiology

Our understanding of the epidemiology of yersiniosis is still incomplete. World-wide surveillance data show that great changes have occurred over the last two decades. The importance of *Y. enterocolitica* as the cause of a number of clinical syndromes is unclear in many areas of the world. Standardized disease surveillance is needed within and across national boundaries so that data from each location are comparable. Improved screening of stools and other specimens for *Y. enterocolitica* is necessary to further elucidate the epidemiology of the disease.

### Sporadic cases

*Y. enterocolitica* has been isolated from humans in many countries of the world, but it seems to be found most frequently in cooler climates (North America, the western coast of South America, Europe, northern, central and eastern Asia, Australia, New Zealand and South Africa) (Mollaret *et al.*, 1979; World Health Organization, 1983; Kapperud and Bergan, 1984; World Health Organization, 1987; Aleksic and Bockemühl, 1990). The widespread nature of *Y. enterocolitica* has been well documented; by the mid-1970s Mollaret *et al.* (1979)

had compiled reports of isolates from 35 countries on six continents. *Y. enterocolitica* infections are an important cause of gastroenteritis in the developed world, occurring particularly as sporadic cases in northern Europe (Black and Slome, 1988; Cover and Aber, 1989), where a clustering of cases during autumn and winter has been reported (World Health Organization, 1983).

There are appreciable geographic differences in the distribution of the different phenotypes of *Y. enterocolitica* isolated from man (Mollaret *et al.*, 1979; Wauters, 1991). There is also a strong correlation between the serovars isolated from humans and pigs in the same geographical area (Esseveld and Godzwaard, 1973; Pedersen, 1979; Wauters, 1979; Bercovier *et al.*, 1980a; Schiemann and Fleming, 1981). Serovar O:3 is widespread in Europe, Japan, Canada, Africa and Latin America. Sometimes, but not always, phage typing makes it possible to distinguish between European, Canadian and Japanese strains (Mollaret *et al.*, 1979; Kapperud *et al.*, 1990b). Serovar O:3 seem to be responsible for more than 90% of the cases in Denmark, Norway, Sweden and New Zealand, and as many as 78.8% of the cases in Belgium (Table 11.3). In general, the data in Table 11.3 originating from different surveillance programs, national statistics and even estimates are not directly comparable. Serovar O:9/biovar 2 is the second most common in Europe, but its distribution is uneven; while it still accounts for a relatively high percentage of the strains isolated in France, Belgium and the Netherlands, only a few strains have been isolated in Scandinavia (World Health Organization, 1983). However, recent data shows that serovar O:9 is on decrease in Belgium. In 1979–1981 this serovar was implicated in 23.3% of the cases, while in 1994 it was responsible for only 5.9% of the cases (Ministere des Affaires Sociales, de la Sante Publique et de l'Environnement. Institut d'Hygiene et d'Epidemiologie, 1995). A similar reduction has also been seen in France and The Netherlands (L. de Zutter, personal communication, 1996). Until recently, the most frequently reported serovars in the United States were O:8 followed by O:5,27 (Mollaret *et al.*, 1979; Bisset *et al.*, 1990; Ostroff, 1995; World Health Organization, 1995). In recent years, serovar O:3 has been on the increase in the United States; O:3 now accounts for the majority of sporadic *Y. enterocolitica* isolates in California (Bisset *et al.*, 1990). In 1989, the estimated cost of yersiniosis in the United States was 138 millions of dollars (World Health Organization, 1995). Principal foodborne infections, as estimated for 1997, are ranked by estimated number of cases caused by foodborne transmission each year in the United States. *Y. enterocolitica* is number ten in the list (among the bacteria in the list, *Y. enterocolitica* is number seven) (Mead *et al.*, 1999) (Table 11.3). The appearance of strains of serovars O:3 and O:9 in Europe, Japan in the 1970s (Anon., 1976), and in North America by the end of the 1980s (Lee *et al.*, 1990; Lee *et al.*, 1991), is an example of a global pandemic (Tauxe, 2002).

The first Japanese case of *Y. enterocolitica* O:8 infection was linked to consumption of imported raw pork (Ichinohe

**Table 11.3** Verified and estimated cases of yersiniosis in some countries

| Country | Total number of cases (year) | Cases per 100,000 inhabitants | References |
|---|---|---|---|
| Belgium | 829[1] (1994) | 8.5 | Ministere des Affaires Sociales, de la Sante Publique et de l'Environnement. Institut d'Hygiene et d'Epidemiologie (1995) |
| Denmark | 286[2] (2001) | 5.3 | Danish Zoonosis Centre, Copenhagen |
| Finland | 728 (2001) | 14.1 | National Public Health Institute, Helsinki |
| Germany | 7113 (2001) | 8.7 | RKI (2002) |
| New Zealand | 3000[2,3] (1994) | 84 | Wright *et al.* (1995) |
| Norway | 123[2] (2001) | 2.7 | National Institute of Public Health, Oslo |
| Sweden | 579[2] (2001) | 6.5 | Swedish institute for Infectious Disease Control |
| Switzerland | 95 (1993) | 1.4 | Swiss National Reference Laboratory for Foodborne Diseases, Berne |
| The European Union | 7385[4] (2000) | | European Commission, Health and Consumer Protection Directorate-General |
| United States | 87,000[3] (1997) | 33.4 | Mead *et al.* (1999) |

[1]Serovar O:3: 78.8%; serovar O:9: 5.9%.
[2]Serovar O:3: >90%.
[3]Estimates.
[4]Figures from nine countries in the European Union.

*et al.*, 1991), although O:8 infection from raw water have also occurred in Japan (Hayashidani *et al.*, 1995).

The incidence of *Y. enterocolitica* infection in patients with acute endemic enterocolitis ranges from 0 to 4%, depending on the geographic location, study method, and population (Kapperud and Slome, 1998). In Australia, Canada, Denmark, Germany, New Zealand, Norway and Sweden (Lassen and Kapperud, 1984; Aleksic and Bockemühl, 1990; Nesbakken, 1992; Swedish Institute for Infectious Disease Control, 1995; Wright *et al.*, 1995, Danish Zoonosis Centre, Copenhagen; National Institute of Public Health, Oslo), *Y. enterocolitica* has surpassed *Shigella*, and now rivals *Salmonella* and *Campylobacter* as a cause of acute bacterial gastroenteritis. World-wide surveillance data show great changes over the last two decades. There appears to have been a real and generalized increase in incidence (World Health Organization, 1983; Ostroff, 1995; World Health Organization, 1995), even though there is still much under-reporting. Nevertheless, improvements in detection and reporting systems may have contributed a great deal to the increase in reported incidence.

Only a few epidemiological studies have been performed to investigate the sources of sporadic human infections. A 1985 study of *Y. enterocolitica* in Belgium identified consumption of raw pork as a risk factor for disease (Tauxe *et al.*, 1987). The following variables were found to be independently related to an increased risk of yersiniosis in a case–control study conducted in Norway: Drinking untreated water, general preference for meat to be prepared raw or rare, and frequency of consumption of pork and sausages (Ostroff *et al.*, 1994). The infrequent occurrence of infections with *Y. enterocolitica* in some areas of the world may be due in part to avoidance of certain risk factors, such as lack of consumption of pork in Muslim countries (Samadi *et al.*, 1982). Two seroepidemiolog-

ical studies have indicated that occupational exposure to pigs may be a risk factor (Merilahti-Palo *et al.*, 1991; Nesbakken *et al.*, 1991b), although in these two studies, confounders could not be excluded.

### Outbreaks

In the United States, chocolate milk (Black *et al.*, 1978), pasteurized milk ((Tacket *et al.*, 1984), soybean curd (tofu) (Tacket *et al.*, 1985), and bean sprouts (Aber *et al.*, 1982) have been implicated as sources in outbreaks of *Y. enterocolitica* infection. These outbreaks, all of which occurred before 1983, were caused by *Y. enterocolitica* serovars which have been infrequently associated with human disease (serovars O:13, O:18) or which no longer predominate in the United States (serovar O:8). More recently, the preparation of raw pork intestines (chitterlings) was associated with an outbreak of *Y. enterocolitica* O:3 infections among black US infants in Georgia (Lee *et al.*, 1990); the organism was isolated from samples of the pork intestines. Also, in outbreaks in Buffalo, New York, 1994–1996 (Kondracki *et al.*, 1996) chitterlings were the vehicle.

The milk-borne outbreak in Sweden in 1988 (Alsterlund *et al.*, 1995) was probably caused by recontamination of pasteurized milk because of lack of chlorination of the water supply. In the multistate outbreak in 1982 (Tacket *et al.*, 1984), milk cartons were contaminated with mud from a pig farm (Aulisio *et al.*, 1982). In the case of the outbreak described by Greenwood and Hooper (1990), post-pasteurization contamination may have occurred from bottles. Previous studies have shown that milk-associated *Y. enterocolitica* outbreaks have been linked to the addition of ingredients after pasteurization (Black *et al.*, 1978; Morse *et al.*, 1984).

In 1981, an outbreak of infection due to *Y. enterocolitica* O:8 in Washington State occurred in association with the consumption of tofu packed in untreated spring water (Tacket *et al.*, 1985). The outbreak serovar was isolated from the spring water samples. Another outbreak caused by serovar O:8 was traced to ingestion of contaminated water used in manufacturing or preparation of food (Schiemann, 1989). Two other *Yersinia* outbreaks have been associated with well water. One occurred among members of a Pennsylvania girl scout troop after they ate bean sprouts grown in contaminated well water (Aber *et al.*, 1982); the other was a familial outbreak of yersiniosis in Canada (Thompson and Gravel, 1986).

The epidemiology of yersiniosis in the United States seems to have evolved into a pattern similar to the picture in Europe (Bottone *et al.*, 1987; Bisset *et al.*, 1990; Ostroff, 1995), where foodborne *Yersinia* outbreaks are rare, and where serovar 3 predominates (Mollaret *et al.*, 1979; Prentice *et al.*, 1991; Verhaegen *et al.*, 1991). Although yersiniosis appears to be more common in Europe than in the United States, only five foodborne outbreaks have been reported in Europe (Toivanen *et al.*, 1973; Olsovsky, *et al.*, 1975; Greenwood and Hooper, 1990; Alsterlund *et al.*, 1995; Swedish Institute for Infectious Disease Control, 1995).

In Japan, several outbreaks connected to schools (Zen-Yoji *et al.*, 1973; Maruyama, 1987) and communities (Asakawa *et al.*, 1973) are reported. In all cases the vehicle is unknown. Often few hundred patients were ill out of several hundreds were at risk. Serovar O:3 has been the agent involved in all cases. In one outbreak in China, caused by serovar O:3 from pickled vegetables, 351 persons were ill (Anon., 1987).

## Susceptibility to physical and chemical agents and factors influencing survival and growth

*Y. enterocolitica* is a facultative organism able to multiply in both aerobic and anaerobic conditions.

### Temperature

The ability of *Y. enterocolitica* to multiply at low temperatures is of considerable concern to food producers. The reported growth range is −2 to 42°C (Bercovier and Mollaret, 1984; Gill and Reichel, 1989). Optimum temperature is 28–29°C (Bercovier and Mollaret, 1984). *Y. enterocolitica* can multiply in foods such as meat and milk at temperatures approaching and even below 0°C (Hanna *et al.*, 1977a; Stern *et al.*, 1980a; Stern *et al.*, 1980b; Lee *et al.*, 1981). It is important to recognize the rate at which *Y. enterocolitica* can multiply, which is considerably greater than that for *L. monocytogenes* (Bhaduri *et al.*, 1994). Results show that, in a food with a neutral pH stored at 5°C, *Y. enterocolitica* counts may increase from e.g. 10/mL to $2.8 \times 10^7$/mL in 5 days.

### pH

The minimum pH for growth has been reported at being between 4.2 and 4.4 (Kendall and Gilbert, 1980), while in a medium in which the pH had been adjusted with HCl, growth occurred at pH 4.18 and 22°C (Brocklehurst and Lund, 1990). The presence of organic acids will reduce the ability of *Y. enterocolitica* to multiply at low pH, acetic acid being more inhibitory per gram mol. at a given pH than lactic and acetic acids (Brocklehurst and Lund, 1990).

## Growth and survival in foods

The ability to propagate at refrigeration temperature in vacuum-packed foods with a prolonged shelf-life (Hanna *et al.*, 1976; Hanna *et al.*, 1979) is of considerable significance in food hygiene. *Y. enterocolitica* may survive in frozen foods for long periods (Schiemann, 1989).

*Y. enterocolitica* is not able to grow at pH < 4.2 or > 9.0 (Kendall and Gilbert, 1980; Stern *et al.*, 1980a) or at salt concentrations greater than 7% ($a_w$ < 0.945) (Stern *et al.*, 1980a). The organism does not survive pasteurization or normal cooking, boiling, baking, and frying temperatures. Heat-treatment of milk and meat products at 60°C for 1–3 minutes effectively inactivates *Y. enterocolitica* (Lee *et al.*, 1981). D-values determined in scalding water were 96, 27 and 11 seconds at 58, 60 and 62°C, respectively (Sörqvist and Danielsson-Tham, 1990).

The literature is contradictory regarding the multiplication of *Y. enterocolitica* in meat during conventional cold storage (Hanna *et al.*, 1977a; Stern *et al.*, 1980a; Stern *et al.*, 1980b; Lee *et al.*, 1981; Fukushima and Gomyoda, 1986; Schiemann, 1989; Kleinlein and Untermann, 1990; Lindberg and Borch, 1994; Borch and Arvidsson, 1996; Bredholt *et al.*, 1999). A comparison of published (Hanna *et al.*, 1977b) and predicted generation times (GT) (Sutherland and Bayliss, 1994) for *Y. enterocolitica* in raw pork at 7°C, 0.5% NaCl (w/v) and pH 5.5–6.5 shows GTs of 8.4–12.4 hours (published) and 8.15–5.05 hours (predicted). However, according to many reports, the ability of *Y. enterocolitica* to compete with other psychotrophic organisms normally present in food may be poor (Stern *et al.*, 1980b; Fukushima and Gomyoda, 1986; Schiemann, 1989; Kleinlein and Untermann, 1990). In contrast, a number of studies have shown that *Y. enterocolitica* is able to multiply in foods kept under chill storage and might even compete successfully (Hanna *et al.*, 1977a; Stern *et al.*, 1980a; Grau, 1981; Lee *et al.*, 1981; Gill and Reichel, 1989; Lindberg and Borch, 1994; Borch and Arvidsson, 1996; Bredholt *et al.* 1999). The study of Bredholt *et al.*(1999) indicate that $10^4$ CFU/g of *Y. enterocolitica* is able to grow well at 8°C in vacuum-packaged cooked ham and servelat sausage in the presence of $10^{4-5}$ CFU/g of lactic acid bacteria (LAB). These LAB cultures inhibited growth of *Listeria monocytogenes* and *Escherichia coli* O157:H7 in the same experiment. The effect of lactic acid (concentration range of 0.1 to 1.1% v/v within a pH range of 3.9 to 5.8 at 4°C) on growth of *Y. enterocolitica* O:9 is greater under anaerobic than aerobic conditions, although the bacterium has proved to be more tolerant of low-pH conditions under anaerobic atmosphere than under an aerobic atmosphere in the absence of lactic acid (El-Ziney *et al.*, 1995).

Pig carcasses are often held in chilling rooms for 2–4 days after slaughter prior to cutting. Pre-packaged raw meat products may remain in retail chill cabinets for more than a week, depending on the product, packaging, package atmosphere, and rate of turnover. Pathogenic variants of *Y. enterocolitica* might propagate considerably during the course of this relatively long storage period.

As a facultative organism, the growth of *Y. enterocolitica* is drastically affected by a gaseous atmosphere. Under anaerobic conditions, *Y. enterocolitica* is unable to grow in beef at pH 5.4–5.8, whereas growth occurs at pH 6.0 (Grau, 1981). One hundred percent $CO_2$ is reported to inhibit the growth of *Y. enterocolitica* (Molin, 1986; Gill and Reichel, 1989). In the study of Gill and Reichel (1989), *Y. enterocolitica* was inoculated into high pH (> 6.0) beef DFD (dark firm dry)-meat. Samples were packaged under vacuum or in an oxygen-free $CO_2$ atmosphere maintained at atmospheric pressure after the meat had been saturated with the gas and stored at –2, 0, 2, 5 or 10°C. In vacuum packs, *Y. enterocolitica* grew at all storage temperatures at rates similar or faster than those of the spoilage flora. In $CO_2$ packs, the bacterium grew at both 5 and 10°C, but not at lower temperatures. Growth of *Y. enterocolitica* was nearly totally inhibited both at 4 and 10°C in a 60% $CO_2$/0.4% CO mixture, while the bacterial numbers in samples packed in high $O_2$ mixture (70% $O_2$/30% $CO_2$) increased from about $5 \times 10^2$ bacteria/g at day 0 to about $10^4$ at day 5 at 4°C and to $10^5$ at 10°C. Growth in chub packs (stuffed in plastic casings) was even higher (Nissen *et al.*, 2000).

Mohammad and Draughon (1987) investigated the growth characteristics of *Y. enterocolitica* strains in pasteurized milk at 4°C. Pasteurized milk was inoculated with 10 or 1000 cells/mL of *Y. enterocolitica*. *Y. enterocolitica* competed well with the background microflora and reached levels of log 5.0 to 7.0/mL after 7 days. However, a study of Stern *et al.* (1980b) indicated that while *Y. enterocolitica* has the capacity for growth in milk at refrigeration temperatures, it is a poor competitor with common spoilage organisms.

## Survival and inactivation in water

### Survival in untreated water

The 3-log reduction (99.9% inactivation) time for *Y. enterocolitica* in lake water has been reported to be 17–18 days at 4°C and 14–15 days at 10°C, the bacterium possibly surviving for a longer time in cold, clean surface waters than *E. coli* (Lund, 1991). A Polish study described an inactivation time ($T_{100}$) of 38 days at spring temperatures and seven days under summer conditions in unfiltered water, in contrast to 197 and 184 days, respectively, in filtered water (Dominowska and Malottke, 1971). Some strains of *Y. enterocolitica* are even able to grow in water at low temperatures (4°C) (Highsmith *et al.*, 1977).

### Inactivation in chlorinated water

In the study of Lund (1991), autoclaved tap water (pH 6.5) was chlorinated according to conventional water-treatment practices, resulting in a free residual level of approximately 0.05 mg/L after a contact time of 30 min. A 3-log reduction for *Y. enterocolitica* and *E. coli* exposed to 0.2 mg/L $Cl_2$ was obtained in 20–180 and 20–25 s, respectively, depending on bacterial strain, plasmid content (*Y. enterocolitica* O:3 harbouring a 40–50 MDa virulence plasmid exhibits enhanced resistance to chlorine) and temperature (Lund, 1991).

### Inactivation by ultraviolet (UV) radiation

The UV radiation dose at 254 nm required for a 3-log reduction (99.9% inactivation) of *Y. enterocolitica* O:3 and *E. coli*, is 2.7 and 5.0 mWs/cm$^2$, respectively. Using *E. coli* as the basis for comparison, it appears that *Y. enterocolitica* O:3 is more sensitive to UV than many of the pathogens associated with water-borne disease outbreaks and can be readily inactivated in most commercially available UV reactors (Butler *et al.*, 1987). In contrast, Carlson *et al.* (1985) demonstrated that avirulent strains were more sensitive to UV radiation (3-log dose of 7–10 mWs/cm$^2$) than a virulent *Y. enterocolitica* O:8 strain, which contained plasmids of 41 and 73 MDa (3-log dose of 22 mWs/cm$^2$).

## Analysis and detection

### General principles

In order to isolate *Y. enterocolitica* from foodstuffs, methods other than those used for isolation from patients with acute intestinal infection are in part required (Mehlman *et al.*, 1978; Schiemann, 1982; Schiemann, 1989; Kapperud, 1991). This is mainly due to four factors:

1 The background flora in foodstuffs varies considerably, both with regard to number and species distribution.

2 The *Yersinia* count in foods is usually far less than in clinical samples. The inclusion of an enrichment step in the isolation procedure is therefore required.

3 The bacteria may be sublethally damaged after exposure to cold, heat or chemicals. This may necessitate a resuscitation step.

4 A broad range of different *Yersinia* is usually found in foods, most of which are of no particular clinical significance (Kapperud, 1991; Lee *et al.*, 1981). An isolation method that as far as possible specifically selects pathogenic variants is therefore desirable.

The individual biovars and serovars of *Y. enterocolitica* differ in their tolerance to selective agents and to the conditions encountered during the isolation process. The choice of procedure will thus have a considerable influence on the range of bacteria isolated. Differences in tolerance occur between the different pathogenic serovars (O:3, O:8 and O:9) (Schiemann, 1989). A method for the isolation of pathogenic *Y. enterocolitica* strains that works well for one serovar, may not necessarily be effective for another ones.

The analytical methods available today for the determination of pathogenic *Y. enterocolitica* suffer from limitations such as insufficient selectivity, and, in particular inadequate differentiation between pathogenic and non-pathogenic strains. These analytical shortcomings will hopefully be solved by the new generation of analytical techniques now being developed.

## Specific procedures for isolation

A three-step method based on a combination of cold enrichment in a non-selective medium with subsequent inoculation onto a highly selective medium has been developed for the Nordic Committee on Food Analysis (1987). The method, a reference method that can be employed for the qualitative demonstration of *Y. enterocolitica* in all types of foods, comprises three steps: (i) 3 h resuscitation at 20–25°C in phosphate buffered saline (pH 7.6) with 2% sorbitol and 0.15% bile salts (PSB), (ii) 8 days' pre-enrichment in PSB at 4°C followed by selective enrichment in a modified Rappaport broth (MRB) (four days at 20–25°C), and (iii) 3 weeks' cold enrichment at 4°C in PSB. After each step, culture is carried out on cefsulodin-irgasan-novobiocin (CIN) agar, a highly selective and differential medium that is more effective than conventional enteric media for the recovery of *Y. enterocolitica* (Head *et al.*, 1982). After incubation at 28°C for 18 to 20 h *Y. enterocolitica* appears as 0.5- to 1.0-mm-diameter colonies with a dark red "bull's eye" and transparent border. CIN agar permits the growth of all pathogenic serovars.

Wauters *et al.* (1988a) developed a method for isolation of serovars O:3 and O:9 from meat and meat products. The procedure is based on a 2- to 3-day selective enrichment period in irgasan–ticarcillin–potassium chlorate (ITC) enrichment broth at room temperature, and is therefore very time-saving compared with the method described above. Culture is carried out on modified *Salmonella–Shigella* agar with 1% sodium deoxycholate and 0.1% CaCl$_2$ (SSDC). Under practical conditions, the method has proved effective for the detection of serovars O:3 and O:9, but not particularly satisfactory with regard to *Y. enterocolitica* serovar O:8 and environmental strains (Wauters *et al.*, 1988a). However, De Zutter *et al.*(1994) have shown that ITC is not optimal for recovery of serovar O:9.

Both *Y. enterocolitica* and *Y. pseudotuberculosis* seem to be more tolerant of alkaline conditions than most other *Enterobacteriaceae*, and treatment of food enrichments with potassium hydroxide (KOH) may be used to selectively reduce the level of background flora (Aulisio *et al.*, 1980). Tolerance of high pH is influenced by the suspending medium, temperature, and phase of growth (Schiemann, 1983). However, despite some improvements in the conditions provided during exposure to high pH, a method based on treatment with KOH proved inferior to selective enrichment in bile-oxalate-sorbose broth for recovering pathogenic strains of *Y. enteroco-*

*litica* from inoculated foods (Schiemann, 1983). Elements of the methods from the Nordic Committee on Food Analysis (1987), Schiemann (1982), Wauters *et al.* (1988a), and KOH treatment (Schiemann, 1983) are incorporated into the International Organization for Standardization (ISO) method (ISO 10273) (Figure 11.1) (International Organization for Standardization, 1994).

Methods for the isolation of *Y. enterocolitica* from water usually involve concentration by filtration followed by procedures similar to those used for isolation from food (Nordic Committee on Food Analysis, 1987; Wauters *et al.*, 1988a; International Organization for Standardization, 1994).

## Detection by DNA colony hybridization

Genetic probes can also be used in DNA colony hybridization to demonstrate virulent *Y. enterocolitica* strains (Wachsmuth, 1985; Tenover, 1988; Kapperud *et al.*, 1990a). Probes based on cloned fragments that encode for virulence factors, and synthetic oligonucleotide probes based on sequence analysis of the cloned fragments, are available (Hill *et al.*, 1983; Jagow and Hill, 1986; Fukushima *et al.*, 1990; Kapperud *et al.*, 1990a; Nesbakken *et al.*, 1991a; Goverde *et al.*, 1993). A study (Nesbakken *et al.*, 1991a) in which gene probes and colony hybridization were used following isolation either by the method of Wauters *et al.* (1988a) or the Nordic Committee on Food Analysis (1987) method, showed that a substantial increase in the rate of detection of pathogenic *Y. enterocolitica* could be achieved. Isolation plus hybridization increased the detection rate from 16% to 38% for the method according to Wauters *et al.* (1988b) and from 10% to 48% for the Nordic method. The results of this investigation (Nesbakken *et al.*, 1991a) support the supposition that conventional culture methods lead to underestimation of virulent *Y. enterocolitica* in pork products.

## Detection by polymerase chain reaction (PCR)

Since Wren and Tabaqchali (1990) developed a simple polymerase chain reaction (PCR) assay using oligonucleotide primers based on nucleotide sequences from two independent virulence genes that readily distinguish between virulent and non-virulent *Y. enterocolitica*, a number of PCR assays have been developed for detection of pathogenic *Y. enterocolitica* in natural samples. These methods use often primers targeting the *virF* (Thisted-Lambertz *et al.*, 1996; Weynants *et al.*, 1996a) or *yadA* (Kapperud *et al.*, 1993) gene, but also the *IcrE* gene (Viitanen *et al.*, 1991) and the *yopT* gene (Arnold *et al.*, 2001) from the virulence plasmid have been used.

*Y. enterocolitica* may loose the virulence plasmid during culture, subculture or storage (Blais and Philippe, 1995). Accordingly, PCR methods based on chromosomal virulence genes, often the *ail* gene, have been developed. Often a combination of genes from the virulence plasmid and the chromosome are used. A common gene combination in such a multiplex PCR assay is the *virF* and *ail* genes (Kaneko *et al.*, 1995; Nilsson *et al.*, 1998).

## Test portion (x g)

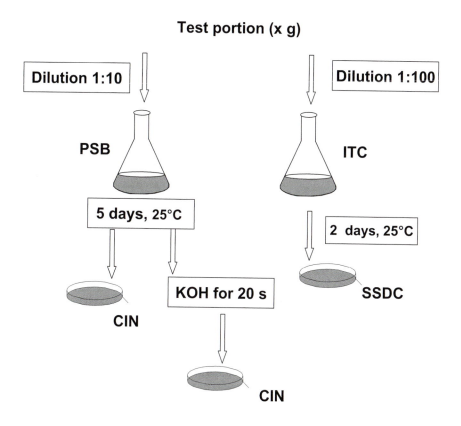

**Figure 11.1** Method for recovery of *Y. enterocolitica* from foods according to the International Organization for Standardization (1994). The "right side" of the method was developed by Wauters (1988a) and is recommended for serovar O:3 in particular. The "left side" of the method has elements from the methods of Nordic Committee on Food Analysis (1987), Schiemann (1982), and Schiemann (1983). PSB, phosphate-buffered saline and sorbitol, 2%; bile salts, 0.15%. ITC, irgasan–ticarcillin–potassium chlorate enrichment broth; KOH, KOH, 0.25% (w/v); NaCl, 0.85% (w/v) in water. CIN, cefsulodin–irgasan–novobiocin agar; SSDC = *Salmonella–Shigella* + sodium deoxycholate, 1%; CaCl$_2$, 0.1%.

Rasmussen *et al.* (1995) detected *Y. enterocolitica* O:3 in faecal samples and tonsil swabs from pigs using IMS and PCR based on the *inv* gene. O:3 cells were detected after pre-enrichment, but direct detection needed further optimization of the sample preparation procedures. By combining *inv*, *virF* and *ail* genes in a multiplex-PCR assay, Weynants *et al.* (1996a) could differentiate between *Y. pseudotuberculosis*, virulent *Y. enterocolitica*, and *Y. enterocolitica* O:3.

## Control

### At the farm level

New-born piglets are easily colonized and become long-term healthy carriers of *Y. enterocolitica* in the oral cavity and intestines (Schiemann, 1989). In a recent study (Skjerve *et al.*, 1998), an enzyme-linked immunosorbent assay (ELISA) (Nielsen *et al.*, 1996) was used to detect IgG antibodies against *Y. enterocolitica* O:3 in sera from 1605 slaughter pigs from 321 different herds. Positive titers were found in 869 (54.1%) of the samples. In the final epidemiological study 182 (63.4%) of 287 herds were defined as positive. Among the positive herds, there were significantly fewer combined herds of piglets and fatteners than fattening herds. Among the risk factors were using an own farm vehicle for transport of slaughter pigs to abattoirs, daily observations of a cat with kittens at the farm, and using straw bedding for slaughter pigs. In conclusion, the epidemiological data suggest that it is possible to reduce the herd prevalence of *Y. enterocolitica* O:3 by minimizing contact between infected and non-infected herds. Further, attempts to reduce the prevalence at the top levels of the breeding pyramids may be beneficial for the industry as a whole. The meat industry may use serological tests as a tool to lower the prevalence in the pig population by limiting the contact between seropositive and seronegative herds.

### During slaughter, meat inspection, and deboning

Because of the high prevalence of *Y. enterocolitica* in pig herds, a strict slaughter hygiene will remain an important means to reduce carcass contamination with *Y. enterocolitica* as well as other pathogenic micro-organisms (Skjerve *et al.*, 1998). However, it is not possible to sort out pigs contaminated with *Y. enterocolitica* at post-mortem meat inspection. Pig slaughter is an open process with many opportunities for the contamination of the pork carcass with *Y. enterocolitica*, and it does not contain any point where hazards are completely eliminated (Borch *et al.*, 1996). Contamination of the carcass with *Y. enterocolitica* during pig slaughter is most likely to arise from faecal and pharyngeal sources (Borch *et al.*, 1996; Nesbakken, 1988). HACCP (Hazard Analysis Critical Control Point) and GMP (Good Manufacturing Practice) in pig slaughter must be focused on limiting this spread. As a guide, attention should be given to the possible establishment of critical points (CPs) or critical control points (CCPs) at the following specific steps during slaughter and dressing: (i) lairage, (ii) killing, (iii) scalding, (iv) dehairing, (v) singeing/flaming, (vi) scraping, (vii) circumanal incision and removal of the intestines, (viii) excision of the tongue, pharynx, and in particular the tonsils, (ix) splitting, (x) post mortem meat inspection procedures, and (xi) deboning of the head. Some of the above-mentioned critical control points will be discussed in more detail below.

During lairage, pathogenic *Y. enterocolitica* may spread from infected to non-infected pigs (Fukushima *et al.*, 1990). If possible, herds should be handled separately, and cleaning and disinfection of the lairage facilities should be performed between herds, since some herds are free from this pathogen (Skjerve *et al.*, 1998).

A working procedure that is employed in many slaughterhouses, and that can be introduced immediately in all slaughterhouses, is the two-knife method. This can be used to interrupt the path of infection from oral cavity and intestine to other parts of the carcass. The two-knife method involves the installation in the slaughter hall of knife decontaminators, with running water held at a temperature of approximately 82°C. When an unclean working operation has been performed, for example, in the region around the rectum or oral cavity, the knife is rinsed before being placed in the decontaminator. The operator should then wash his hands before the other knife is used for clean working operations. The two-knife method should be used both by operators and meat inspection personnel.

Meat inspection procedures concerning the head also seem to represent a cross-contamination risk: Incision of the submaxillary lymph nodes is a compulsory procedure according to the EU regulations (European Commission, 1995). In a screening of 97 animals, 5.2% of samples from the submaxillary lymph nodes were positive and by the sampling of 24 these lymph nodes in a follow-up study, 12.5% of the samples were positive (Nesbakken *et al.*, 2003). This may, however, result in the bacterium being transported from the medial neck region to other parts of the carcass by the knives and hands of the meat inspection personnel (Nesbakken, 1988; Nesbakken *et al.*, 2003). In view of the fact that the incidence of tuberculosis in pigs and humans has been reduced to a very low level in many parts of the world, it may be possible to re-consider regulations that require incision of the submaxillary lymph nodes by meat inspectors.

Cutting and removal of head-meat in pigs should be carried out on a separate work table, preferably in a separate room (European Commission, 1995). This room should be considered to be an unclean area. Knives and equipment must not be used for cutting and deboning other parts of the carcass without prior cleaning and disinfection. Current EU regulations that require removal of pig head-meat to be carried out in a separate department are unfortunately not complied with in all abattoirs.

The possibility of decapitation early on in the carcass dressing procedure has been considered, with the head, including tongue and tonsils, then being removed on a separate line for heat-treatment and cutting (Christensen, 1987). The results presented by Andersen (1988) indicate, however, that it is even more important to modify procedures for removal of

the guts, in order to avoid contamination of the carcass from the rectum. Technological solutions have already been found which allow removal of the rectum without soiling of the carcass. This can be done, inter al., by insertion of a pre-frozen plug into the anus prior to rectum-loosening and gut removal (Danish Meat Research Institute, 1989). The sealing off of the rectum with a plastic bag immediately after it has been freed, can significantly reduce the spread of *Y. enterocolitica* to pig carcasses (Andersen (1988; Nesbakken *et al.*, 1994). According to data from the Norwegian National Institute of Public Health, the occurrence of human yersiniosis has dropped by about 30% after the introduction of the plastic bag technique in about 90% of the pig slaughterhouses in Norway. A similar drop in yersiniosis has also been seen in Sweden after the introduction in this technique (Swedish Institute for Infectious Disease Control, 1995).

## During the processing of meat products

In addition to the slaughter hall, and the cutting and deboning departments, the sausage-making department must also be considered to be a contaminated area. It must be assumed that raw materials such as pig head-meat and pork cuts, and consequently also sausage meat, are likely to be contaminated with pathogenic *Y. enterocolitica*. Strict cleaning and disinfecting requirements must therefore also apply here. It is important to maintain an effective separation between sausage preparation and packing, so as to avoid recontamination after heat-treatment.

## During food processing, sale, and food preparation

### Refrigeration

*Y. enterocolitica* is able to propagate at temperatures approaching 0°C. While a refrigeration of food does not prevent the multiplication of *Y. enterocolitica*, the rate at which this takes place will be reduced.

### Separation of raw meats from other foods

Raw meats (in particular pork) should be separated from other foods. Cross-contamination from raw meat to heat-treated end products must be avoided in meat-processing establishments, butchers' shops, meat departments in retail food stores, and in kitchens in institutions, restaurants and homes.

### Cleaning and disinfection

Knives, equipment, and machines used to cut or process raw meat products must be cleaned and disinfected before being used for handling other foods. All surfaces that have been in contact with raw meat must be cleaned and disinfected with appropriate and effective agents.

Adequate cooking of meat. A case–control study carried out in Norway revealed that inadequate heat-treatment of meat is a risk factor for human yersiniosis Ostroff *et al.*, 1994). Consumption of undercooked pork should be discouraged.

### Pasteurization of milk

Well-controlled pasteurization of milk is necessary to safeguard against *Y. enterocolitica* as well as many other pathogens.

### Kitchen hygiene

Information to the public on the consequences of eating raw and lightly cooked foods, and of the need for proper kitchen hygiene, will thus be a most important prophylactic measure. Good hygiene is particularly necessary because *Y. enterocolitica* is able to propagate at refrigeration temperature.

### Personal hygiene

Precautions aimed at preventing fecal–oral spread of the pathogens should be taken. Hand-washing and proper stool disposal must be employed in household, day-care, and hospital settings.

### Animal contact

Avoidance of contact with excreta from pigs or domestic pets may reduce transmission. Domestic animals should be kept away from food preparation areas.

### Water supplies

Evidence discussed earlier indicates that wild and domestic animals can often carry strains of *Y. enterocolitica* that are pathogenic to humans. It is therefore important (i) not to drink from raw water supplies that are liable to contamination by animals, (ii) not to allow such water supplies to be used in food production, and (iii) to ensure that drinking water supplies are treated effectively so that *Y. enterocolitica*, and the multitude of other pathogens, are inactivated or eliminated.

## References

Aber, R.C., Mc Carthy, M.A., Berman, R., DeMelfi, T., and Witte, E. 1982. An outbreak of *Yersinia enterocolitica* illness among members of a Brownie troop in Centre County, Pennsylvania. Program and Abstracts of the 22nd Interscience Conference on Antimicrobial Agents and Chemotherapy. Miami Beach, American Society for Microbiology, Washington, D.C.

Aho, K., Ahvonen, P., Laitinen, O., and Leirisalo, M. 1981. Arthritis associated with *Yersinia enterocolitica* infection. In: *Yersinia enterocolitica*. E.J. Bottone, ed. CRC Press, Inc., Boca Raton, FL, pp. 113–124.

Ahvonen, P. 1972a. Human yersiniosis in Finland. I. Bacteriology and serology. Ann. Clin. Res. 4: 30–38.

Ahvonen, P. 1972b. Human yersiniosis in Finland. II. Clinical features. Ann. Clin. Res. 4: 39-48.

Aldova, E., Cerna, J., Janeckova, M., and Pegrimkova, J. 1975. *Yersinia enterocolitica* and its demonstration in foods. Czech. Hyg. 20: 395.

Aleksic, S., Bockemühl, J., and Lange, F. 1986. Studies on the serology of flagellar antigens of *Yersinia enterocolitica* and related *Yersinia* species. Zbl. Bakteriol. Mikrobiol. Hyg. Ser. A 261: 299–310.

Aleksic, S., and Bockemühl, J. 1990. Mikrobiologie und Epidemiologie der Yersiniosen. Immun. Infekt. 18: 178–185.

Aleksic, S., Bockemühl, J., and Wuthe, H.H. 1995. Epidemiology of *Y. pseudotuberculosis* in Germany, 1983–1993. Contrib. Microbiol. Immunol. 13: 55–58.

Aleksic, S., Steigerwalt, A.G., Bockemühl, J., Huntley-Carter, G.P., and Brenner, D.J. 1987. *Yersinia rohdei* sp. nov. isolated from human and dog feces and surface water. Int. J. Syst. Bact. 37: 327–332.

Alonso, J.M., Bejot, J., Bercovier, H., and Mollaret, H.H. 1975. Sur un groupe de souches de *Yersinia enterocolitica* fermentant le rhamnose. Intérêt diagnostique et particularitès écologiques. Med. Mal. Infect. 5: 490.

Alsterlund, R., Danielsson-Tham, M.-L., Edén, T., de Jong, B., Lyxell, G., Nilsson, P.O., and Ransjö, U. 1995. *Yersinia enterocolitica*-utbrott på Bjärehalvön. Risker med kylda matvaror. Svensk Vet. Tidsskr. 47: 257–260.

Andersen, J.K. 1988. Contamination of freshly slaughtered pig carcasses with human pathogenic *Yersinia enterocolitica*. Int. J. Food Microbiol. 7: 193–202.

Andersen, J.K., and Saunders, N.A. 1990. Epidemiological typing of *Yersinia enterocolitica* by analysis of restriction fragment length polymorphisms with a cloned ribosomal RNA gene. J. Med. Microbiol. 32: 179–187.

Anon. 1976. Worldwide spread of infections with *Yersinia enterocolitica*. WHO Chronicle 30: 494–496.

Anon. 1987. First report of an outbreak caused by *Yersinia enterocolitica* serotype O:3 in Shengyang. Chin. J. Epid. 8: 264–267.

Arnold, T., Hensel, A., Hagen, R., Aleksic, S., Neubauer, H., and Scholz, H.C. 2001. A highly specific one-step PCR-assay for the rapid discrimination of enteropathogenic *Yersinia enterocolitica* from pathogenic *Yersinia pseudotuberculosis* and *Yersinia pestis*. Syst. Appl. Microbiol. 24: 285–289.

Asakawa, Y., Akahane, S., Kagata, N., Noguchi, M., Sakazaki, R., and Tamura, K. 1973. Two community outbreaks of human infection with *Yersinia enterocolitica*. J. Hyg. 71: 715–723.

Asplund, K., Johansson, T., and Siitonen, A. 1998. Evaluation of pulsed field gel electrophoresis of genomic restriction fragments in the discrimination of *yersinia enterocolitica* O:3. Epidemiol. Infect. 121: 579–586.

Attwood, S.E.A., Mealy, K., Cafferkey, M.T., Buckley, T.F., West, A.B., Boyle, N., Healy, E., and Keane, F.B.V. 1987. *Yersinia* infection and acute abdominal pain. Lancet I: 529–533.

Aulisio, C.C.G., Lanier, J.M., and Chappel, M.A. 1982. *Yersinia enterocolitica* and O:13 associated with outbreaks in three southern states. J. Food. Protect. 45: 1263.

Aulisio, C.C.G., Mehlman, I.J., and Sanders, A.C. 1980. Alkali method for rapid recovery of *Yersinia enterocolitica* and *Yersinia enterocolitica* from foods. Appl. Environ. Microbiol. 39: 135–140.

Baker, P.M., and Farmer, J.J. III. 1982. New bacteriophage typing system for *Yersinia enterocolitica*, *Yersinia kristensenii*, *Yersinia frederiksenii*, and *Yersinia intermedia*: correlation with serotyping, biotyping, and antibiotic susceptibility. Clin. Microbiol. 15: 491–502.

Bercovier, H., Brenner, D.J., Ursing, J., Steigerwalt, A.G., Fanning, G.R., Alonsio, J.M., Carter, G.P., and Mollaret, H.H. 1980a. Characterization of *Yersinia enterocolitica* sensu stricto. Curr. Microbiol. 4: 201–206.

Bercovier, H., and Mollaret, H.H. 1984. Genus XIV. *Yersinia*. In: Bergey's manual of systematic bacteriology, vol. 1. N.R. Krieg, ed. Williams & Wilkins, Baltimore, pp. 498-506.

Bercovier, H., Steigerwalt, A.G., Guiyoule, A., Carter, G.P., and Brenner, D. 1984. *Yersinia aldovae* (formerly *Yersinia enterocolitica*-like group X2): a new species of *Enterobacteriaceae* isolated from aquatic ecosystems. Int. J. Syst. Bacteriol. 34: 166–172.

Bercovier, H., Ursing, J., Brenner, D.J., Steigerwalt, A.G., Fanning, G.R., Alonso, J.M., Carter, G.P., and Mollaret, H.H. 1980b. *Yersinia kristensenii*: a new species of *Enterobacteriaceae* composed of sucrose negative strains (formerly called atypical *Yersinia enterocolitica* or *Yersinia enterocolitica*-like). Curr. Microbiol. 4: 219–224.

Bhaduri, S., and Turner-Jones, C. 1983. The effect of anaerobic atmospheres on the stability of the virulence-related characteristics in *Yersinia enterocolitica*. Food Microbiol. 10: 239–242.

Bhaduri, S., Turner-Jones, C.O., Buchanan, R.L., and Phillips, J.G. 1994. Response surface model of the effect of pH, sodium chloride and sodium nitrite on growth of *Yersinia enterocolitica* at low temperatures. Int. J. Food Microbiol. 23: 333–343.

Bissett, M.L., Powers, C., Abbott, S.L., and Janda, J.M. 1990. Epidemiologic investigations of *Yersinia enterocolitica* and related species: sources, frequency, and serogroup distribution. J. Clin. Microbiol. 28: 910–912.

Black, R.E., Jackson, R.J., Tsai, T., Medvesky, M., Shayegani, M., Feeley, J.C., MacLeod, K.I.E., and Wakelee, A.M.1978. Epidemic *Yersinia enterocolitica* infection due to contaminated chocolate milk. N. Engl. J. Med. 298: 76–79.

Black, R.E., and Slome, S. 1988. *Yersinia enterocolitica*. Infect. Dis. Clin. N. Am. 2: 625-641.

Bliska, J.B., and Falkow, S. 1994. Interplay between determinants of cellular entry and cellular disruption in the enteropathogenic *Yersinia*. Curr. Opin. Infect. Dis. 7: 323–328.

Blais, B.W., and Philippe, L.M. 1995. Comparative analysis of *yadA* and *ail* polymerase chain reaction methods for virulent *Yersinia enterocolitica*. Food Control. 6: 211–214.

Borch, E., and Arvidsson, B. 1996. Growth of *Yersinia enterocolitica* O:3 in pork. In: Proceedings: Food Associated Pathogens, the International Union of Food Science and Technology. Uppsala, Sweden, p. 202–203.

Borch, E., Nesbakken, T., and Christensen, H. 1996. Hazard identification in swine slaughter with respect to foodborne bacteria. Int. J. Food. Microbiol. 30: 9–25.

Bottone, E.J. 1977. *Yersinia enterocolitica*: a panoramic view of a charismatic microorganism. Crit. Rev. Microbiol. 5: 211–241.

Bottone, E.J., Gullans, C.R., and Sierra, M.F. 1987. Disease spectrum of *Yersinia enterocolitica* serogroup O:3, the predominant cause of human infection in New York City. Contrib. Microbiol. Immunol. 9: 56–60.

Boyce, J.M., Evans, D.J., Evans, D.G., and DuPont, H.L. 1979. Production of heat-stable, methanol-soluble enterotoxin by *Yersinia enterocolitica*. Infect. Immun. 25: 532–537.

Bredholt, S., Nesbakken, T., and Holck, A. 1999. Protective cultures inhibit growth of *Listeria monocytogenes* and *Escherichia coli* O157: H7 in cooked, sliced vacuum- and gas-packaged meat. Int. J. Food Microbiol. 53: 43–52.

Brenner, D.J. 1981 Classification of *Yersinia enterocolitica*. In: *Yersinia enterocolitica*. E.J. Bottone, ed. CRC Press, Inc., Boca Raton, FL, pp. 1–8.

Brenner, D.J., Bercovier, H., Ursing, J., Alonso, J.M., Steigerwalt, A.G., Fanning, G.R., Carter, G.P., and Mollaret, H.H.1980a. *Yersinia intermedia*: a new species of *Enterobacteriaceae* composed of rhamnose-positive, melibiose-positive, raffinose-positive strains (formerly called *Yersinia enterocolitica* or *Yersinia enterocolitica*-like). Curr. Microbiol. 4: 207–212.

Brenner, D.J., Ursing, J., Bercovier, H., Steigerwalt, A.G., Fanning, G.R., Alonso, J.M., and Mollaret, H.H. 1980b. Deoxyribonucleic acid relatedness in *Yersinia enterocolitica* and *Yersinia enterocolitica*-like organisms. Curr. Microbiol. 4: 195–200.

Brennhovd, O. 1991. Termotolerante *Campylobacter* spp. og *Yersinia* spp. i noen norske vannforekomster. Thesis. Norwegian College of Veterinary Medicine, Oslo.

Brocklehurst, T.F., and Lund, B.M. 1990. The influence of pH, temperature and organic acids on the initiation of growth of *Yersinia enterocolitica*. J. Appl. Bacteriol. 69: 390–397.

Brubaker, R.R. 1979. Expression of virulence in yersiniae. In: Microbiology-1979. D. Schlessinger, ed. American Society for Microbiology, Washington, D.C., p. 168–171.

Buchrieser, C., Weagant, S.D., and Kaspar, C.W. 1994. Molecular characterization of *Yersinia enterocolitica* by pulsed-field gel electrophoresis and hybridization of DNA fragments to *ail* and pYV probes. Appl. Environ. Microbiol. 60: 4371–4379.

Bullians, J.A. 1987. *Yersinia* species infection of lambs and cull cows at an abattoir. New Zealand Vet. J. 35: 65–67.

Butler, R.C., Lund, V., and Carlson, D.A. 1987. Susceptibility of *Campylobacter jejuni* and *Yersinia enterocolitica* to UV radiation. Appl. Environ. Microbiol. 53: 375–378.

Bölin, I., Portnoy, D.A., and Wolf-Watz, H. 1985. Expression of the temperature-inducible outer membrane proteins of yersiniae. Infect. Immun. 48: 234–240.

Cafferkey, M.T., and Buckley, T.F. 1987. Comparison of saline agglutination, antibody to human gammaglobulin, and immunofluorescence tests in the routine serological diagnosis of yersiniosis. J. Infect. Dis. 156: 845–848.

Capriolli, T., Drapeau, A.J., and Kasatiya, S. 1978. *Yersinia enterocolitica*: serotypes and biotypes isolated from humans and the environment in Quebec, Canada. Canada. J. Clin. Microbiol. 8: 7.

Carlson, D.A., Seabloom, R.W., DeWalle, F.B., Wetzler, T.F., Engeset, J., Butler, R., Wangsuphachart, S., and Wang, S. 1985. Ultraviolet disinfection of water for small water supplies. Publication no. EPA/600/52–85/092. USA Environmental Protection Agency, Water Engineering Research Laboratory, Cincinnati, Ohio.

Caugant, D.A., Aleksic, S., Mollaret, H.H., Selander, R.K., and Kapperud, G. 1989. Clonal diversity and relationships among strains of *Yersinia enterocolitica*. J. Clin. Microbiol. 27: 2678–2683.

Centers for Disease Control and Prevention. 1995. Outbreak of *Yersinia enterocolitica* infections in Upper Valley of Vermont and New Hampshire. Department of Health and Human Services, Memorandum, Epi-aid 96–5 trip report, December 5, 1995, Atlanta, GA, 9 pp.

Christensen, S.G. 1980. *Yersinia enterocolitica* in Danish pigs. J. Appl. Bacteriol. 48: 377–382.

Christensen, S.G. 1982. The prevalence of *Yersinia enterocolitica* in slaughter animals, water and raw milk in Denmark. In: Psychotrophic microorganisms in spoilage and pathogenicity. T.A. Roberts, B. Hobbs, J.H.B. Christian, and Skovgaard, N., eds. Academic Press, London.

Christensen, S.G. 1987. The *Yersinia enterocolitica* situation in Denmark. Contrib. Microbiol. Immunol. 9: 93–97.

Cornelis, G. 1981. Antibiotic resistance in *Yersinia enterocolitica*. In: *Yersinia enterocolitica*. E.J. Bottone, ed. CRC Press, Inc., Boca Raton, FL, pp. 55–71.

Cornelis, G., Laroche, Y., Balligand, G., Sory, M.-P., and Wauters, G. 1987. *Yersinia enterocolitica*, a primary model for bacterial invasiveness. Rev. Infect. Dis. 9: 64–87.

Cover, T.L., and Aber, R.C. 1989. *Yersinia enterocolitica*. New Engl. J. Med. 321: 16–24.

Cox, N.A., del Corral, F., Bailey, J.S., Shotts, E.B., and Papa, C.M. 1990. The presence of *Yersinia enterocolitica* and other *Yersinia* species on the carcasses of market broilers. Poultry Sci. 69: 482–485.

Danish Meat Research Institute. 1989. Annual Report. Roskilde, 36 pp.

DeBoer, E., Seldam, W.M., and Oosterom, J. 1986. Characterization of *Yersinia enterocolitica* and related species isolated from foods and porcine tonsils in the Netherlands. Int. J. Food Microbiol. 3: 217–224.

Delmas, C.L. 1983. La contamination du lait par *Yersinia enterocolitica*. Med. Nut. T. 19: 208–210.

Delmas, C.L., and Vidon, D.J.-M. 1985. Isolation of *Yersinia enterocolitica* and related species from foods in France. Appl. Environ. Microbiol. 50: 767–771.

Devenish, J.A., and Schiemann, D. 1981. HeLa cell infection by *Yersinia enterocolitica*: evidence for lack of intracellular multiplication and development of a new procedure to quantitative expression of infectivity. Infect. Immun. 32: 48–55.

De Zutter, L., Le Mort, L., Janssens, M., and Wauters, G. 1994. Shortcoming of irgasan ticarcillin chlorate broth for the enrichment of *Y. enterocolitica* biotype 2, serotype O:9 from meat. Int. J. Food Microbiol. 23: 231–237.

Dominowska, C., and Malottke, R. 1971. Survival of *Yersinia* in water samples originated from various sources. Bull. Inst. Med. Gdansk 22: 173.

Doyle, M.P., Hugdahl, M.B., and Taylor, S.L. 1981. Isolation of *Yersinia enterocolitica* from porcine tongues. Appl. Environ. Microbiol. 42: 661–666.

El-Ziney, M.G., DeMeyer, H., and Debevere, J.M. 1995. Kinetics of interactions of lactic acid, pH and atmosphere on the growth and survival of *Yersinia enterocolitica* IP 383 O:9 at 4°C. Int. J. Food Microbiol. 27: 229–244.

Esseveld, H., and Goudzwaard, C. 1973. On the epidemiology of *Y. enterocolitica* infections: pigs as the source of infections in man. Contrib. Microbiol. Immunol. 2: 99–101.

European Commission. 1995. Council directive 64/433/EEC on health condition for the production and marketing of fresh meat. Brussels, 34 pp.

Fenwick, S.G. 1998. *Yersinia enterocolitica* infections in animals and people in New Zealand. Nederlands Tijdsch. Med. Microbiologie 6: Supplement II, 12.

Fenwick, S.G., Madie, P., and Wicks, C.R. 1994. Duration of carriage and transmission of *Yersinia enterocolitica* biotype 4, serotype O:3 in dogs. Epidemiol. Infect. 113: 471–477.

Franzin, L., Fantino, P., and Vidotto, V. 1984. Isolation of *Yersinia enterocolitica* and *Yersinia enterocolitica*-like organisms from raw milk in Italy. Curr. Microbiol. 10: 357-362.

Frederiksen, W. 1964. A study of some *Yersinia pseudotuberculosis*-like bacteria ("*Bacterium enterocoliticum*" and "*Pasteurella X*"). In: Proc. XIV Scand. Congr. of Path. Microbiol. Universitetsforlagets Trykningssentral, Oslo, pp. 103–104.

Fredriksson-Ahomaa, M., Autio, T., and Korkeala, H. 1999. Efficient subtyping of *Yersinia enterocolitica* bioserotype 4/O:3 with pulsed-field gel electrophoresis. Lett. Appl. Microbiol. 29: 308–312.

Fukushima, H., and Gomyoda, M. 1986. Inhibition of *Yersinia enterocolitica* serotype O3 by natural microflora of pork. Appl. Environ. Microbiol. 51: 990–994.

Fukushima, H., Maruyama, K., Omori, I., Ito, K., and Iorihara, M. 1990. Contamination of pigs with *Yersinia* at the slaughterhouse. Fleischwirtsch. 70: 1300–1302.

Fukushima, H., Saito, K., Tsubokura, M., Otsuki, K., and Kawaoka, Y. 1984a. Significance of milk as a possible source of infection for human yersiniosis. I. Incidence of *Yersinia* organisms in raw milk in Shimane perfecture. Jpn. Vet. Microbiol. 9: 139–146.

Fukushima, H., Tsubokura, M., Otsuki, K., and Kawaoka, Y. 1984b. Biochemical heterogenity of serotype O3 strains of 700 *Yersinia* strains isolated from humans, other mammals, flies, animal feed, and river water. Curr. Microbiol. 11: 149–154.

Gemski, P., Lazere, J.R., and Casey, T. 1980. Plasmid associated with pathogenicity and calcium dependency of *Yersinia enterocolitica*. Infect. Immun. 27: 682–685.

Gill, C., and Reichel, M. 1989. Growth of cold-tolerant pathogens *Yersinia enterocolitica*, *Aeromonas hydrophila* and *Listeria monocytogenes* on high-pH beef packaged under vacuum or carbon dioxide. Food Microbiol. 6: 223–230.

Gleason, T.H., and Patterson, S.D. 1982. The pathology of *Yersinia enterocolitica* ileocolitis. Am. J. Surg. Pathol. 6: 347–355.

Goverde, R.L.J., Jansen, W.H., Brunings, H.A., Huis in 't Veld, J.H.J., and Mooi, F.R. 1993. Digoxigenin-labeled *inv*- and *ail*-probes for the detection and identification of pathogenic *Yersinia enterocolitica* in clinical specimens and naturally contaminated pig samples. J. Appl. Bacteriol. 74: 301–313.

Grau, F.H. 1981. Role of pH, lactate, and anaerobiosis in controlling the growth of some fermentative gram-negative bacteria on beef. Appl. Environ. Microbiol. 42: 1043–1050.

Greenwood, M.H., and Hooper, W.L. 1990. Excretion of *Yersinia* spp. associated with consumption of pasteurized milk. Epidemiol. Infect. 104: 345–350.

Greenwood, M.H., Hooper, W.L., and Rodhouse, J.C. 1990. The source of *Yersinia* spp. in pasteurized milk: an investigation at a dairy. Epidemiol. Infect. 104: 351–360.

Guthertz, L.S., Fruin, J.T., Spicer, D., and Fowler, J.L. 1976. Microbiology of fresh comminuted turkey meat. J. Milk Food Technol. 39: 823.

Hamama, A., El Marrakchi, A., and El Othmani, F. 1992. Occurrence of *Yersinia enterocolitica* in milk and dairy products in Morocco. Int. J. Food Microbiol. 16: 69–77.

Hanna, M.O., Stewart, J.C., Zink, D.L., Carpenter, Z.L., and Vanderzant, C. 1977a. Development of *Yersinia enterocolitica* on raw and cooked beef and pork at different temperatures. J. Food Sci. 42: 1180–1184.

Hanna, M.O., Stewart, J.C., Carpenter, Z.L., Zink, D.L., and Vanderzant, C. 1979. Isolation and characteristics of *Yersinia enterocolitica*-like bacteria from meats. Contr. Microbiol. Immunol. 5: 234–242.

Hanna, M.O., Zink, D.L., Carpenter, Z.L., and Vanderzant, C. 1976. *Yersinia enterocolitica*-like organisms from vacuum packaged beef and lamb. J. Food Sci. 41: 1254–1256.

Hanna, M.O., Zink, D.L., Carpenter, Z.L., and Vanderzant, C. 1977b. Effect of heating, freezing and pH on *Yersinia enterocolitica*-like organisms from meat. J. Food Protect. 40: 689–692.

Harnett, N., Hu, G., Wan, J., Brunins, V., Borcyk, A., and Jamieson, F. 1998. Epidemiology of *Yersinia enterocolitica* infections in Canada. Nederlands Tijdsch. Med. Microbiologie 6: Supplement II, 12.

Harvey, S., Greenwood, M.J., Pickett, M.J., and Robert, A.M. 1976. Recovery of *Yersinia enterocolitica* from streams and lakes of California. Appl. Environ. Microbiol. 32: 352-354.

Hayashidani, H., Ohtomo, Y., Toyokawa, Y., Saito, M., Kaneko, K.-I., Kosuge, J., Kato, M., Ogawa, M., and Kapperud, G.1995. Potential sources of sporadic human infection with *Yersinia enterocolitica* serovar O:8 in Aomori prefecture, Japan. J. Clin. Microbiol. 33: 1253–1257.

Head, C.B., Whitty, D.A., and Ratnam, S. 1982. Comparative study of selective media for recovery of *Yersinia enterocolitica*. J. Clin. Microbiol. 16: 615–621.

Heesemann, J., Keller, C., Morawa, R., Schmidt, N., Siemens, H.J., and Laufs, R. 1983. Plasmids of human strains of *Yersinia enterocolitica*: Molecular relatedness and possible importance of pathogenesis. J. Infect. Dis. 147: 107–115.

Highsmith, A.K., Feeley, J.C., Skaley, P.S., Wells, J.G., and Wood, B.T. 1977. Isolation of *Yersinia enterocolitica* from well water and growth in distilled water. Appl. Environ. Microbiol. 34: 745–750.

Hill, W.E., Payne, W.L., and Aulisio, C.C.G. 1983. Detection and enumeration of virulent *Yersinia enterocolitica* in food by DNA colony hybridization. Appl. Environ. Microbiol. 46: 636–641.

Hoogkamp-Korstanje, J.A.A., de Koning, J., Heesemann, J., Festen, J.J., Houtman, P.M., and vanOyen, P.L.1992. Influence of antibiotics on IgA and IgG response and persistence of *Yersinia enterocolitica* in patients with *Yersinia*-associated spondylarthropathy. Infection 20: 53–57.

Hurvell, B. 1981. Zoonotic *Yersinia enterocolitica* infection: host range, clinical manifestations, and transmission between animals and man. In: *Yersinia enterocolitica*. E.J. Bottone, ed. CRC Press, Inc., Boca Raton, Fla., p. 145–59.

Ibrahim, A., and MacRae, I.C. 1991. Isolation of *Yersinia enterocolitica* and related species from red meat and milk. J. Food Sci. 56: 1524–1526.

Ichinohe, H., Yoshioka, M., Fukushima, H., Kaneko, S., and Maruyama, T. 1991. First isolation of *Yersinia enterocolitica* serotype O:8 in Japan. J. Clin. Microbiol. 29: 846–847.

Inoue, M., and Kurose, M. 1975. Isolation of *Yersinia enterocolitica* from cow's intestinal contents and beef meat. Jpn. J. Vet. Sci. 37: 91.

Inoue, M., and Nagao, H. 1976. Isolation of *Yersinia enterocolitica* from bovine cecal contents, commercial meat, and meat shops. J. Jpn. Vet. Med. Assoc. 29: 612.

Inoue, M., Okamoto, K., Moriyama, T., Takahashi, T., Shimizu, K., and Miyama, 1983. Effect of *Yersinia enterocolitica* ST on cyclic guanosine 3',5'- monophosphate levels in mouse intestines and cultured cells. Microbiol. Immunol. 27: 159–166.

International Organization for Standardization. 1994. Microbiology – General Guidance for the Detection of Presumptive Pathogenic Yersinia enterocolitica (ISO 10273). International Organization for Standardization, Genève, Switzerland, 16 pp.

Jagow, J., and Hill, W.E. 1986. Enumeration by DNA colony hybridization of virulent *Yersinia enterocolitica* colonies in artificially contaminated food. Appl. Environ. Microbiol. 52: 441–443.

Jebsen, O.B., Korner, B., Lauritsen, K.B., Hancke, A.B., Andersen, L., Henrichsen, S., Brenoe, E., Christiansen, P.M., and Johansen, A.1976. *Yersinia enterocolitica* infection in patients with acute surgical abdominal disease. A prospective study. Scand. J. Infect. Dis. 8: 189–194.

Kandolo, K., and Wauters, G. 1985. Pyrazinamidase activity in *Yersinia enterocolitica* and related organisms. J. Clin. Microbiol. 21: 980–982.

Kaneko, S., Ishizaki, N., and Kokubo, Y. 1995. Detection of pathogenic *Yersinia enterocolitica* and *Yersinia pseudotuberculosis* from pork using polymerase chain reaction. Contr. Microbiol. Immunol. 13: 153–155.

Kapperud, G. 1981. Survey on the reservoirs of *Yersinia enterocolitica* and *Yersinia enterocolitica*-like bacteria in Scandinavia. Acta Pathol. Microbiol. Immunol. Scand. Sect. B 89: 29–35.

Kapperud, G. 1982. Enterotoxin production at 4°, 22°, and 37°C among *Yersinia enterocolitica* and *Yersinia enterocolitica*-like bacteria. Acta Pathol. Microbiol. Immunol. Scand. Sect. B 90: 185–189.

Kapperud, G. 1991. *Yersinia enterocolitica* in food hygiene. Int. J. Food Microbiol. 12: 53-66.

Kapperud, G., Bergan, T., and Lassen, J. 1981. Numerical taxonomy of *Yersinia enterocolitica* and *Yersinia enterocolitica*-like bacteria. Int. J. Syst. Bacteriol. 31: 401–419.

Kapperud, G., and Bergan, T. 1984. Biochemical and serological characterization of *Yersinia enterocolitica*. In: Methods in Microbiology, vol. 15. T. Bergan and J.R. Norris, eds. Academic Press, London, p. 295–344.

Kapperud, G., Dommarsnes, K., Skurnik, M., and Hornes, E. 1990a. A synthetic oligonucleotide probe and a cloned polynucleotide probe based on the *yopA* gene for detection and enumeration of virulent *Yersinia enterocolitica*. Appl. Environ. Microbiol. 56: 17–23.

Kapperud, G., and Jonsson, B. 1976. *Yersinia enterocolitica* in brown trout (*Salmo trutta* L.) from Norway. Acta Microbiol. Scand. 84B: 66.

Kapperud, G., and Jonsson, B. 1978. *Yersinia enterocolitica* et bactéries apparentées isolées à partir d'écosystèmes d'eau douce en Norvège. Med. Mal. Infect. 8: 500–506.

Kapperud, G. and Langeland, G. 1981. Enterotoxin production at refrigeration temperature by *Yersinia enterocolitica* and *Yersinia enterocolitica*-like bacteria. Curr. Microbiol. 5: 119-122.

Kapperud, G., Namork, E., Skurnik, M., and Nesbakken, T. 1987. Plasmid-mediated surface fibrillae of *Yersinia pseudotuberculosis* and *Yersinia enterocolitica*: relationship to the outer membrane protein YOP1 and possible importance for pathogenesis. Infect. Immun. 55: 2247–2254.

Kapperud, G., Nesbakken, T., Aleksic, S., and Mollaret, H.H. 1990b. Comparison of restriction endonuclease analysis and phenotypic typing methods for differentiation of *Yersinia enterocolitica* isolates. J. Clin. Microbiol. 28: 1125–1131.

Kapperud, G., and Slome, S.B. 1998. *Yersinia enterocolitica* infections. In: Bacterial Infections of Humans, 3rd. edition. A. Evans, and P.F. Brachman, eds. Plenum Medical Book Company, New York, pp. 859–873.

Kapperud, G., Vardund, T., Skjerve, E., Hornes, E., and Michaelsen, T.E. 1993. Detection of pathogenic *Yersinia enterocolitica* in foods and water by immunomagnetic separation, nested polymerase chain reactions, and colorimetric detection of amplified DNA. Appl. Environ. Microbiol. 59: 2938–2944.

Kato, Y., Ito, K., Kubokura, Y., Maruyama, K., Kaneko, I., and Ogawa, M. 1985. Occurrence of *Yersinia enterocolitica* in wild-living birds and Japanese serows. Appl. Environ. Microbiol. 49: 198–200. 107.

Kendall, M., and Gilbert, R.J. 1980. Survival and growth of *Yersinia enterocolitica* in media and in food. In: Microbial Growth and Survival in Extremes of Environment, Society for Applied Bacteriology Technical Series, No 15. G.W. Gould and J.E.L. Corry, eds. Academic Press, London, pp. 215–226.

Kleinlein, N., and Untermann, F. 1990. Growth of pathogenic *Yersinia enterocolitica* strains in minced meat with and without protective gas with consideration of the competitive background flora. Int. J. Food Microbiol. 10: 65–72.

Knapp, W. 1958. Mesenteric adenitis due to *Pasteurella pseudotuberculosis* in young people. N. Engl. J. Med. 259: 776–778.

Knapp, W., and Thal. E. 1963. Untersuchungen über die kulturellbiochemischen, serologischen, tierexperimentellen und immunologischen Eigenschaften einer vorläufig "Pasteurella X" benannten Bakterienart. Zbl. Bakteriol. Orig. A 190: 472–484.

Kondracki, S., Balzano, G., Schwartz, J., Kiehlbauch, J., Ackman, D., and Morse, D. 1996. Recurring outbreaks of yersiniosis associated with pork chitterlings. Abstract. 36th Interscience Conference on Antimicrobial agents and Chemotherapy, New Orleans. p. 259.

Kotula, A.W., and Sharar, A.K. 1993. Presence of *Yersinia enterocolitica* serotype O:5,27 in slaughter pigs. J. Food Protect. 56: 215–218.

Krogstad, O. 1974. *Yersinia enterocolitica* infection in goat. A serological and bacteriological investigation. Acta Vet. Scand. 15: 597–608.

Lachica, R.V., Zink, D.L., and Ferris, W.R. 1984. Association of fibril structure formation with cell surface properties of *Yersinia enterocolitica*. Infect. Immun. 46: 272–275.

Laird, W.J., and Cavanaugh, D.C. 1980. Correlation of autoagglutination and virulence of yersiniae. J. Clin. Microbiol. 11: 430–432.

Lanada, E.B. 1990. The Epidemiology of Yersinia Infections in Goat Flocks. Thesis. Massey University, Palmerston North, New Zealand.

Langeland, G. 1983. *Yersinia enterocolitica* and *Yersinia enterocolitica*-like bacteria in drinking water and sewage sludge. Acta Pathol. Microbiol. Immunol. Scand. Sect. B 91: 179–185.

Lassen, J. 1972. *Yersinia enterocolitica* in drinking-water. Scand. J. Infect. Dis. 4: 125–127.

Lassen, J., and Kapperud, G. 1984. Epidemiological aspects of enteritis due to *Campylobacter* spp. in Norway. J. Clin. Microbiol. 19: 153–156.

Lee, L.A., Gerber, A.R., Lonsway, D.R., Smith, J.D., Carter, G.P., Pohr, N.D., Parrish, C.M., Sikes, R.K., Finton, R.J., and Tauxe, R.V. 1990. *Yersinia enterocolitica* O:3 infections in infants and children, associated with the household preparation of chitterlings. N. Engl. J. Med. 322: 984–987.

Lee, L.A., Taylor, J., Carter, G.P., Quinn, B., Farmer, J.J.III, and Tauxe, R.V. 1991. *Yersinia enterocolitica* O:3: an emerging cause of pediatric gastroenteritis in the United States. J. Infect. Dis. 163: 660–663.

Lee, W.H., McGrath, P.P., Carter, P.H., and Eide, E.L. 1977. The ability of some *Yersinia enterocolitica* strains to invade HeLa cells. Can. J. Microbiol. 23: 1714.

Lee, W.H., Vanderzant, C., and Stern, N. 1981. The occurrence of *Yersinia enterocolitica* in foods. in *Yersinia enterocolitica*. E.J. Bottone, ed. CRC Press, Inc., Boca Raton, FL, p. 161–171.

Leistner, L., Hechelmann, H., Kashiwazpaki, M., and Albertz, R. 1975. Nachweis von *Yersinia enterocolitica* in Faeces und Fleisch von Schweinen, Rindern und Geflügel. Fleischwirtsch. 55: 1599.

Lindberg, C.W., and Borch, E. 1994. Predicting the aerobic growth of *Yersinia enterocolitica* O:3 at different pH values, temperatures and *L*-lactate concentrations using conductance measurements. Int. J. Food Microbiol. 22: 141–153.

Louiseau-Marolleau, M.L., and Alonso, J.M. 1976. Isolement de *Yersinia enterocolitica* lors d'une étude systematique des aliments en milieu hospitalier. Med. Mal. Infect. 6: 373.

Lund, V. 1991. Drinking Water Disinfection Processes. Effect on Microorganisms and Organic Substances in Water. Thesis. Norwegian College of Veterinary Medicine, Oslo, 121 pp.

Maruyama, T. 1987. *Yersinia enterocolitica* infection in humans and isolation of the microorganism from pigs in Japan. Contrib. Microbiol. Immunol. 9: 48–55.

Mayer, L.W. 1988. Use of plasmid profiles in epidemiologic surveillance of disease outbreaks and in tracing the transmission of antibiotic resistance. Clin. Microbiol. Rev. 1: 228–243.

Mead, P.S., Slutsker, L., Dietz, V., McCaig, L.F., Bresee, J.S., Shapiro, C., Griffin, P.M., and Tauxe, R.V. 1999. Food-related illness and death in the United States. Emerging Infectious Diseases 5: 607–625.

Megraud, F. 1987. *Yersinia* infection and acute abdominal pain. Lancet I: 1147.

Mehlman, I.J., Aulisio, C.C.G., and Sanders, A.C. 1978. Problems in the recovery and identification of *Yersinia* from food. J. Assoc. Off. Anal. Chem. 61: 761–771.

Merilahti-Palo, R., Lahesmaa, R., Granfors, K., Gripenberg-Lerche, C., and Toivanen, P. 1991. Risk of *Yersinia* infection among butchers. Scand. J. Infect. Dis. 23: 55–61.

Michiels, T., Wattiau, P., Brasseur, R., Ruysschaert, J.M., and Cornelis, G. 1990. Secretion of *Yop* proteins by yersiniae. Infect. Immun. 58: 2840–2849.

Miller, V.L., Finlay, B.B., and Falkow, S. 1988. Factors essential for the penetration of mammalian cells by *Yersinia*. Curr. Top. Microbiol. Immunol. 138: 15–39.

Ministere des Affaires Sociales, de la Sante Publique et de l'Environnement. Institut d'Hygiene et d'Epidemiologie. 1995. Surveillance van Infectieuze Aandoeningen door een Netwerk van Laboratoria voor Microbiologie 1994 + Retrospectieve 1983 – 1993, Brussels, p.58.

Mohammad, K.A., and Draughon, F.A. 1987. Growth characteristics of *Yersinia enterocolitica* in pasteurized skim milk. J. Food Protect. 50: 849–852.

Molin, G. 1986. The resistance to carbon dioxide of some food related bacteria. Eur. J. Appl. Microbiol. Biotechnol. 18: 214–217.

Mollaret, H.H. 1995. Fifteen centuries of yersiniosis. Contrib. Microbiol. Immunol. 13: 1–4.

Mollaret, H.H., Bercovier, H., and Alonso, J.M. 1979. Summary of the data received at the WHO Reference Center for *Yersinia enterocolitica*. Contrib. Microbiol. 5: 174–184.

Mollaret, H.H., and Nicolle, P. 1965. Sur la fréquence de la lysogénie dans l'espèce nouvelle *Yersinia enterocolitica*. C. R. Acad. Sci. 260: 1027–1029.

Mollaret, H.H., Nicolle, P., Brault, J., and Nicolas, R. 1972. Importance actuelle des infections a *Yersinia enterocolitica*. Bull. Acad. Nat. Med. Paris 156: 704.

Morris, G.K., and Feeley, J.C. 1976. *Yersinia enterocolitica*: a review of its role in food hygiene. Bull. World Health Organ. 54: 79–85.

Morse, D.L., Shayegani, M., and Gallo, R.J. 1984. Epidemiologic investigation of a *Yersinia* camp outbreak linked to a food handler. Am. J. Public Health 74: 589–592.

Moustafa, M.K, Ahmed, A.A.H., and Marth, E.H. 1983. Occurrence of *Yersinia enterocolitica* in raw and pasteurized milk. J. Food Protect. 46: 276–278.

Najdenski, H., Iteman, I., and Carniel, E. 1994. Efficient subtyping of pathogenic *Yersinia enterocolitica* strains by pulsed-field gel electrophoresis. J. Clin. Microbiol. 32: 2913–2920.

Nesbakken, T. 1985a. Enterotoxin production at 4, 22, and 37°C by *Yersinia enterocolitica* and *Yersinia enterocolitica*-like bacteria isolated from porcine tonsils and pork products. *Acta Vet. Scand.* 26: 13–20.

Nesbakken, T. 1985b. Comparison of sampling and isolation procedures for recovery of *Yersinia enterocolitica* serotype O:3 from the oral cavity of slaughter pigs. Acta Vet. Scand. 26: 127–135.

Nesbakken, T. 1988. Enumeration of *Yersinia enterocolitica* O:3 from the porcine oral cavity, and its occurrence on cut surfaces of pig carcasses and the environment in a slaughterhouse. Int. J. Food. Microbiol. 8: 287–293.

Nesbakken, T. 1992. Epidemiological and food hygienic aspects of *Yersinia enterocolitica* with special reference to the pig as a suspected source of infection. Thesis. Norwegian College of Veterinary Medicine, Oslo, 114 pp.

Nesbakken, T., Eckner, K., Høidal, H.K., and Røtterud, O.-J. 2003. Occurrence of *Yersinia enterocolitica* and *Campylobacter* spp. in slaughter pigs and consequences for meat inspection, slaughtering, and dressing procedures. Int. J. Food. Microbiol. 80: 231–240.

Nesbakken, T., Gondrosen, B., and Kapperud, G. 1985. Investigation of *Yersinia enterocolitica*, *Yersinia enterocolitica*-like bacteria, and thermotolerant campylobacters in Norwegian pork products. Int. J. Food Microbiol. 1: 311–320.

Nesbakken, T., and Kapperud, G. 1985. *Yersinia enterocolitica* and *Yersinia enterocolitica*-like bacteria in Norwegian slaughter pigs. Int. J. Food Microbiol. 1: 301–309.

Nesbakken, T., Kapperud, G., Sørum, H., and Dommarsnes, K. 1987. Structural variability of 40–50 MDa virulence plasmids from *Yersinia enterocolitica*. Geographical and ecological distribution of plasmid variants. Acta Path. Microbiol. Immunol. Scand. Sect. B. 95: 167–173.

Nesbakken, T., Kapperud, G., Dommarsnes, K., Skurnik, M., and Hornes, E. 1991a. Comparative study of a DNA hybridization method and two isolation procedures for detection of *Yersinia enterocolitica* O:3 in naturally contaminated pork products. Appl. Environ. Microbiol. 57: 389–394.

Nesbakken, T., Kapperud, G., Lassen, J., and Skjerve, E. 1991b. *Yersinia enterocolitica* antibodies in slaughterhouse employees, veterinarians, and military recruits. Occupational exposure to pigs as a risk factor for yersiniosis. Contrib. Microbiol. Immunol. 12: 32–39.

Nesbakken, T., Nerbrink, E., Røtterud, O.-J., and Borch, E. 1994. Reduction of *Yersinia enterocolitica* and *Listeria* spp. on pig carcasses by enclosure of the rectum during slaughter. Int. J. Food Microbiol. 23: 197–208.

Nicolle, P. 1973. *Yersinia enterocolitica*. In: Lysotypie und andere spezielle epidemiologische Laboratoriums-methoden. H. Rische, ed. VEB Gustav Fischer Verlag, Jena, p. 377–387.

Nielsen, B., Heisel, C., and Wingstrand, A. 1996. Time course of the serological response to *Yersinia enterocolitica* O:3 in experimentally infected pigs. Vet. Microbiol. 48: 293–303.

Niléhn, B. 1969. Studies on *Yersinia enterocolitica* with special reference to bacterial diagnosis and occurrence in human acute enteric disease. Acta. Pathol. Microbiol. Scand. Contrib. Microbiol. Immunol. Suppl. 206: 1.

Niléhn, B., and Sjöström, B. 1967. Studies on *Yersinia enterocolitica*: occurrence in various groups of acute abdominal disease. Acta Pathol. Microbiol. Scand. 71: 612–628.

Nilsson, A., Lambertz, S.T., Stålhandske, P., Norberg, P., and Danielsson-Tham, M.L. 1998. Detection of *Yersinia enterocolitica* in food by PCR amplification. Lett. Appl. Microbiol. 26: 140–141.

Nissen, H., Alvseike, O., Bredholt, S., and Nesbakken, T. 2000. Comparison between growth of *Yersinia enterocolitica*, *Listeria monocytogenes*, *Escherichia coli* O157:H7 and *Salmonella* spp. in ground beef packed by three commercially used packaging techniques. Int. J. Food Microbiol. 59: 211–220.

Norberg, P. 1981. Enteropathogenic bacteria in frozen chicken. Appl. Environ. Microbiol. 42: 32–34.

Nordic Committee on Food Analysis. 1987. *Yersinia enterocolitica. Detection in Food.* Method no. 117. 2nd ed. Nordic Committee on Food Analysis, Esbo, 12 pp.

Okamoto, K., Inoue, T., Ichikawa, H., Hara, S., and Miyama, A. 1983. Partial purification and characterization of heat-stable enterotoxin produced by *Yersinia enterocolitica*. Infect. Immun. 31: 544–559.

Olsovsky, Z., Olsakova, V., Chobot, S., and Sviridov, V. 1975. Mass occurrence of *Yersinia enterocolitica* in two establishments of collective care of children. J. Hyg. Epidemiol. Immunol. 19: 22–29.

Ostroff, S.M. 1995. *Yersinia* as an emerging infection: epidemiologic aspects of yersiniosis. Contrib. Microbiol. Immunol. 13: 5–10.

Ostroff, S.M., Kapperud, G., Lassen, J., Aasen, S., and Tauxe, R.V. 1992. Clinical features of sporadic *Yersinia enterocolitica* infections in Norway. J. Infect. Dis. 166: 812–817.

Ostroff, S.M., Kapperud, G., Hutwagner, L.C., Nesbakken, T., Bean, N.H., Lassen, J., and Tauxe, R.V. 1994. Sources of sporadic *Yersinia enterocolitica* infections in Norway: a prospective case–control study. Epidemiol. Infect. 112: 133–141.

Pai, C.H., and DeStephano, L. 1982. Serum resistance associated with virulence in *Yersinia enterocolitica*. Infect. Immun. 35: 605–611.

Pai, C.H., Gillis, F., and Marks, M.I. 1982. Infection due to *Yersinia enterocolitica* in children with abdominal pain. J. Infect. Dis. 146: 705.

Pai, C.H., Mors, V., and Toma, S. 1978. Prevalence of enterotoxigenicity in human and nonhuman isolates of *Yersinia enterocolitica*. Infect. Immun. 22: 334–338.

Pedersen, K.B. 1979. Occurrence of *Yersinia enterocolitica* in the throat of swine. Contrib. Microbiol. Immunol. 5: 253–256.

Peixotto, S.S., Finne, G., Hanna, M.O., and Vanderzant, C. 1980. Presence, growth, and survival of *Yersinia enterocolitica* in oysters, shrimp, and crab. J. Food. Prot. 42: 972.

Perry, R.D., and Brubaker, R.R. 1983. Vwa[+] phenotype of *Yersinia enterocolitica*. Infect. Immun. 40: 166–171.

Portnoy, D.A., and Martinez, R.J. 1985. Role of a plasmid in the pathogenicity of *Yersinia* species. Curr. Topics Microbiol. Immunol. 118: 29–51.

Portnoy, D.A., Wolf-Watz, H., Bölin, I., Beeder, A.B., and Falkow, S. 1984. Characterization of common virulence plasmids in *Yersinia* species and their role in the expression of outer membrane proteins. Infect. Immun. 43, 108–114.

Prentice, M.B., Cope, D., and Swann, R.A. 1991. The epidemiology of *Yersinia enterocolitica* infection in the British isles 1983–1988. Contrib. Microbiol. Immunol. 12: 17–25.

Prpic, J.K., Robins-Browne, R.M., and Davey, R.B. 1985. *In vitro* assessment of virulence in *Yersinia enterocolitica* and related species. J. Clin. Microbiol. 22: 105–110.

Quan, T.J., Meek, J.L., Tsuchiya, K.R., Hudson, B.W., and Barnes, A.M. 1974. Experimental pathogenicity of recent North American isolates of *Yersinia enterocolitica*. J. Infect. Dis. 129: 341–344.

Rakovsky, J., Pauckova, V., and Aldova, E. 1973. Human *Yersinia enterocolitica* infections in Czechoslovakia. Contrib. Microbiol. Immunol. 2: 93.

Rasmussen, H.N., Rasmussen, O.F., Christensen, H., and Olsen, J.E. 1995. Detection of *Yersinia enterocolitica* O:3 in faecal samples and tonsil swabs from pigs using IMS and PCR. J. Appl. Bacteriol. 78: 563–568.

Ratnam, S., Mercer, E., Pico, B., Parsons, S., and Butler, R. 1982. A nosocomial outbreak of diarrheal disease due to *Yersinia enterocolitica* serotype O:5 biotype 1. J. Infect. Dis. 145: 242–247.

Rea, M.C., Cogan, T.M., and Tobin, S. 1992. Incidence of pathogenic bacteria in raw milk in Ireland. J. Appl. Bacteriol. 73: 331–336.

Ricciardi, I.D., Pearson, A.D., Suckling, W.G., and Klein, C. 1978. Long-term fecal excretion and resistance induced in mice infected with *Yersinia enterocolitica*. Infect. Immun. 21: 342–344.

RKI. 2002. Epidemiologisches Bulletin No. 29. Robert Koch Institute, Berlin, Germany.

Saari, T.N., and Jansen, G.P. 1979. Water-borne *Yersinia enterocolitica* in the midwest United States. Contr. Microbiol. Immunol. 5: 185–196.

Saken, E., Roggenkamp, A., Aleksic, S., and Heesemann, J. 1994. Characterisation of pathogenic *Yersinia enterocolitica* serogroups by pulsed-field gel electrophoresis of genomic *Not*I restriction fragments. J. Med. Microbiol. 41: 329–338.

Samadi, A.R., Wachsmuth, K., Huq, M.I., Mahbub, M., and Agbonlahor, D.E. 1982. An attempt to detect *Yersinia enterocolitica* infection in Dacca, Bangladesh. Trop. Geogr. Med. 34: 151–154.

Sandler, M., Girdwood, A.H., Kottler, R.E., and Marks, I.N. 1982. Terminal ileitis due to *Yersinia enterocolitica*. S. Afr. Med. J. 62: 573–576.

Sarrouy, J. 1972. Isolement d'une *Yersinia enterocolitica* a partir du lait. Med. Mal. Infect. 2: 67.

Schiemann, D.A. 1982. Development of a two-step enrichment procedure for recovery of *Yersinia enterocolitica* from food. Appl. Environ. Microbiol. 43: 14–27.

Schiemann, D.A. 1983. Alkalotolerance of *Yersinia enterocolitica* as a basis for selective isolation from food enrichments. Appl. Environ. Microbiol. 46: 22–27.

Schiemann, D.A. 1989. *Yersinia enterocolitica* and *Yersinia pseudotuberculosis*. In: Foodborne bacterial pathogens. M.P. Doyle, ed. Marcel Dekker, Inc., New York, pp. 601–672.

Schiemann, D.A., and Devenish, J.A. 1982. Relationship of HeLa cell infectivity to biochemical, serological and virulence characteristics of *Yersinia enterocolitica*. Infect. Immun. 35: 497–506.

Schiemann, D.A., and Fleming, C.A. 1981. *Yersinia enterocolitica* from throats of swine in eastern and western Canada. Can. J. Microbiol. 27: 1326–1333.

Shenkman, L., and Bottone, E.J. 1976. Antibodies to *Yersinia enterocolitica* in thyroid disease. Ann. Intern. Med. 85: 735–739.

Shiozawa, K., Nishina, T., Miwa, Y., Mori, T., Akahane, S., and Ito, K. 1991. Colonization in the tonsils of swine by *Yersinia enterocolitica*. Contrib. Microbiol. Immunol. 12: 63–67.

Skjerve, E., Lium, B., Nielsen, B., and Nesbakken, T. 1998. Control of *Yersinia enterocolitica* in pigs at herd level. Int. J. Food Microbiol. 45: 195–203.

Skurnik, M., Bölin, I., Heikkinen, H., Phia, S., and Wolf-Watz, H. 1984. Virulence plasmid-associated autoagglutination in *Yersinia* spp. J. Bacteriol. 158: 1033–1036.

Slee, K.J., and Button, C. 1990. Enteritis in sheep and goats due to *Yersinia enterocolitica* infection. Austr. Vet. J. 67: 396–398.

Sneath, P.H.A., and Cowan, S.T. 1958. An electro-taxonomic survey of bacteria. J. Gen. Microbiol. 19: 551.

Spadaro, M., and Infortuna, M. 1968. Isolamenta di *Yersinia enterocolitica* in *Mitilus galloprovincialis* lamk. Boll. Soc. Ital. Biol. Sper. 44: 1896.

Stengel, G. 1984. Vorkommen von *Yersinia enterocolitica* in Milch und Milcherzeugnissen. Arch. Lebensm. Hyg. 35: 91–95.

Stengel, G. 1985. *Yersinia enterocolitica*. Vorkommen und Bedeutung in Lebensmitteln. Fleischwirtsch. 65: 1490–1495.

Stern, N.J., Pierson, M.D., and Kotula, A.W. 1980a. Effects of pH and sodium chloride on *Yersinia enterocolitica* growth at room and refrigeration temperature. J. Food Sci. 45: 64-67.

Stern, N.J., Pierson, M.D., and Kotula, A.W. 1980b. Growth and competitive nature of *Yersinia enterocolitica* in whole milk. J. Food Sci. 45: 972–974.

Ståhlberg, T.H., Heesemann, J., Granfors, K., and Toivonen, P. 1989. Immunoblot analysis of IgM, IgG, and IgA responses to plasmid encoded released proteins of *Yersinia enterocolitica* in patients with or without yersinia triggered reactive arthritis. Ann. Rheum. Dis. 48: 577–581.

Sutherland, J.P., and Bayliss, A.J. 1994. Predictive modelling of growth of *Yersinia enterocolitica*: the effects of temperature, pH and sodium chloride. Int. J. Food Microbiol. 21: 197–215.

Swaminathan, B., Harmon, M.C., and Mehlman, I.J. 1982. A review – *Yersinia enterocolitica*. J. Appl. Bact. 52: 151–183.

Swedish Institute for Infectious Disease Control. 1995. Smittskyddsinstitutets Epidemiologiska Årsrapport 1994, Stockholm, 27 pp.

Szita, M.I., Káli, M., and Rédey, B. 1973. Incidence of *Yersinia enterocolitica* infection in Hungary. Contrib. Microbiol. Immunol. 2: 106–110.

Sæbø, A. and Lassen, J. 1991. A survey of acute and chronic disease associated with *Yersinia enterocolitica* infection. A Norwegian 10-year follow-up study on 458 hospitalized patients. Scand. J. Infect. Dis. 23: 517–527.

Sörqvist, S., and Danielsson-Tham, M.L. 1990. Survival of *Campylobacter, Salmonella* and *Yersinia* spp. in scalding water used at pig slaughter. Fleischwirtsch. 70: 1460–1466.

Tacket, C.O., Ballard, J., Harris, N., Allard, J., Nolan, C., Quan, T., and Cohen, M.L. 1985. An outbreak of *Yersinia enterocolitica* infections caused by contaminated tofu (soybean curd). Am. J. Epidemiol. 121: 705–711.

Tacket, C.O., Narain, J.P., Sattin, R., Löfgren, J.P., Königsberg, C.Jr., Rendtorff, R.C., Rausa, A., Davis, B.R., and Cohen, M.L. 1984. A multistate outbreak of infections caused by *Yersinia enterocolitica* transmitted by pasteurized milk. J. Am. Med. Assoc. 251: 483–486.

Takao, T., Tomiunaga, N., Shimonishi, Y., Hara, S., Inoue, T., and Miyama, A. 1984. Primary structure of heat-stable enterotoxin produced by *Yersinia enterocolitica*. Biochem. Biophys. Res. Commun. 125: 845–851.

Tauxe, R.V. 2002. Emerging foodborne pathogens. Int. J. Food Microbiol. 78: 31–42.

Tauxe, R.V., Wauters, G., Goossens, V., vanNoyen, R., Vandepitte, J., Martin, S.M., de Mol, P., and Thiers, G. 1987. *Yersinia enterocolitica* infections and pork: the missing link. Lancet I, 1129–1132.

Tenover, F.C. 1988. Diagnostic deoxyribonucleic acid probes for infectious diseases. Clin. Microbiol. Rev. 1: 82–101.

Ternström, A., and Molin, G. 1987. Incidence of potential pathogens on raw pork, beef and chicken in Sweden, with special reference to *Erysipelothrix rhusiopathiae*. J. Food Protect. 50: 141–146.

Thal, E. 1954. Untersuchungen über *Pasteurella pseudotuberculosis*. Thesis, University of Lund, Lund.

Thisted Lambertz, S., Ballagi-Pordány, A., Nilsson, A., Nordberg, P., and Danielsson-Tham, M.L. 1996. a comparison between a PCR method and a conventional culture method for detection of pathogenic *Yersinia enterocolitica* in foods. J. Appl. bacteriol. 81: 303–308.

Thompson, J.S., and Gravel, M.J. 1986. Family outbreak of gastroenteritis due to *Yersinia enterocolitica* serotype O:3 from well water. Can. J. Microbiol. 32, 700–701.

Toivanen, P., Toivanen, A., Olkkonen, L., and Aantaa, S. 1973. Hospital outbreak of *Yersinia enterocolitica* infection. Lancet I: 1801–1803.

Toma, S. 1973. Survey on the incidence of *Yersinia enterocolitica* in the province of Ontario. Can. J. Publ. Health 64: 477.

Toma, S. and Lafleur, L. 1981. *Yersinia enterocolitica* infections in Canada 1966 to August 1978. In: *Yersinia enterocolitica*. E.J. Bottone, ed. CRC Press, Inc., Boca Raton, FL, pp. 183–191.

Ursing, J., Brenner, D.J., Bercovier, H., Fanning, G.R., Steigerwalt, A.G., Brault, J., and Mollaret, H.H. 1980. *Yersinia frederiksenii*: a new species of *Enterobacteriaceae* composed of rhamnose-positive strains (formerly called atypical *Yersinia enterocolitica* or *Yersinia enterocolitica*-like). Curr. Microbiol. 4: 213–217.

Van Loghem, J.J. 1944. The classification of plague-*Bacillus*. Antonie van Leeuwenhoek 10: 15.

Vantrappen, G., Agg, H.O., Ponette, E., Geboes, K., and Bertrand, P. (1977) *Yersinia* enteritis and enterocolitis: Gastroenterological aspects. Gastroenterology 72: 220–227.

Velin, D., and Emody, L. 1982. The stability of enterotoxin production in *Yersinia enterocolitica* and the methanol solubility of heat-stable enterotoxin. Acta Microbiol. Acad. Sci. Hung. 29: 227–233.

Verhaegen, J., Dancsa, L., Lemmens, P., Janssens, M., Verbist, L., Vandepitte, J., and Wauters, G. 1991. *Yersinia enterocolitica* surveillance in Belgium (1979–1989). Contrib. Microbiol. Immunol. 12:11–16.

Vidon, D.J.M., and Delmas, C.L. 1981. Incidence of *Yersinia enterocolitica* in raw milk in Eastern France. Appl. Environ. Microbiol. 41: 355–359.

Viitanen, A.M, Arstila, P., Lahesmaa, R., Granfors, K., Skurnik, M., and Toivonen, P. 1991. Application of the polymerase chain reaction and immunofluorescence techniques to the detection of bacteria in *Yersinia*-triggered reactive arthritis. Arthritis Rheum. 34: 89–96.

Wachsmuth, K. 1985. Genotypic approaches to the diagnosis of bacterial infections: Plasmid analyses and gene probes. Infect. Control. 6: 100–109.

Walker, S.J., and Gilmour, A. 1986. The incidence of *Yersinia enterocolitica* and *Yersinia enterocolitica*-like organisms in raw and pasteurized milk in Northern Ireland. J. Appl. Bacteriol. 60: 133–138.

Wauters, G. 1970. Contribution à l'étude de *Yersinia enterocolitica*. Thèse d'agrégé. Université Catholique de Louvain. Vander, Louvain, 165 pp.

Wauters, G. 1979. Carriage of *Yersinia enterocolitica* serotype 3 by pigs as a source of human infection. Contrib. Microbiol. Immunol. 5: 249–252.

Wauters, G. 1981. Antigens of *Yersinia enterocolitica*. In: *Yersinia enterocolitica*. E.J. Bottone, ed. CRC Press, Inc., Boca Raton, Fla., pp. 41–53.

Wauters, G. 1991. Taxonomy, identification and epidemiology of *Yersinia enterocolitica* and related species. In: I Problemi della Moderna Biologia: Ecologia Microbica, Analitica di Laboratorio, Biotecnologia, Vol. 1. H.Grimme, E. Landi, and S. Dumontet, eds. Atti del IV Convegno Internazionale, Sorrento, pp. 93–101.

Wauters, G., Aleksic, S., Charlier, J., and Schulze, G. 1991. Somatic and flagellar antigens of *Yersinia enterocolitica* and related species. Contrib. Microbiol. Immunol. 12: 239–243.

Wauters, G., Goossens, V., Janssens, M., and Vandepitte, J. 1988a. New enrichment method for isolation of pathogenic *Yersinia enterocolitica* O:3 from pork. Appl. Environ. Microbiol. 54: 851–854.

Wauters, G., Janssens, M., Steigerwalt, A.G., and Brenner, D.J. 1988b. *Yersinia mollaretii* sp. nov., formerly called *Yersinia enterocolitica* biogroups 3A and 3B. Int. J. Syst. Bacteriol. 38: 424–429.

Wauters, G., Kandolo, K., and Janssens, M. 1987. Revised biogrouping scheme of *Yersinia enterocolitica*. Contrib. Microbiol. Immunol. 9: 14–21.

Weynants, V., Jadot, V., Denoel, P.A., Tibor, A., and Letesson, J.-J. 1996a. Detection of *Yersinia enterocolitica* serogroup O:3 by a PCR method. J. Clin. Microbiol. 34: 1224–1227.

Weynants, V., Tibor, A., Denoel, P.A., Saegerman, C., Godfroid, J., Thiange, P., and Letesson, J.-J. 1996b. Infection of cattle with *Yersinia enterocolitica* O:9 a cause of the false positive serological reactions in bovine brucellosis diagnostic tests. Vet. Microbiol. 48: 101–112.

Winblad, S. 1975. Arthritis associated with *Yersinia enterocolitica* infections. Scand. J. Infect. Dis. 7: 191–195.

World Health Organization. 1983. Yersiniosis: report on a WHO meeting, Paris 1981. WHO Regional Office for Europe, Euro reports and studies 60, Copenhagen, 31 pp.

World Health Organization. 1987. Report of the round table conference on veterinary public health aspects of *Yersinia enterocolitica*, Orvieto 1985. WHO Collaborative Centre for Research and Training in Veterinary Public Health, Instituto Superiore di Sanità, Rome, 30 pp.

World Health Organization. 1995. Report of the WHO Consultation on Emerging Foodborne Diseases, Berlin, 25 pp.

Wormser, G.P., and Keusch, G.T.1981. *Yersinia enterocolitica*: clinical observations. In: *Yersinia enterocolitica*. E.J. Bottone, ed. CRC Press, Inc., Boca Raton, FL, pp. 83–93.

Wren, B.W., and Tabaqchali, S. 1990. Detection of pathogenic *Yersinia enterocolitica* by the polymerase chain reaction. Lancet 336: 693.

Wright, J., Fenwick, S., and McCarthy, M. 1995. Yersiniosis: an emerging problem in New Zealand. The New Zealand Public Health Report 2: 65–66.

Yersin, A. 1894. La peste bubonique a Hong Kong. Ann. Inst. Pasteur, Paris 8: 662–667.

Zen-Yoji, H., Maruyama, T., Sakai, S., Kimura, S., Mizuno, T., and Momose, T. 1973. An outbreak of enteritis due to *Yersinia enterocolitica* occurring at a junior high school. Jpn. J. Microbiol. 17: 220–222.

# *Vibrio* species

Mitsuaki Nishibuchi and Angelo DePaola

12

## Abstract

*Vibrio* species are prevalent in estuarine and marine environments and seven species can cause seafoodborne infections. *Vibrio cholerae* O1 and O139 serotypes produce cholera toxin and are agents of cholera. However, fecal–oral route infections in the terrestrial environment are responsible for epidemic cholera. *V. cholerae* non-O1/O139 strains may cause gastroenteritis through production of known toxins or unknown mechanism. *Vibrio parahaemolyticus* strains capable of producing thermostable direct hemolysin (TDH) and/or TDH-related hemolysin are most important cause of gastroenteritis associated with seafood consumption. *Vibrio vulnificus* is responsible for seafoodborne primary septicemia and its infectivity depends primarily on the risk factors of the host. *V. vulnificus* infection has the highest case fatality rate (50%) of any foodborne pathogen. Four other species (*Vibrio mimicus*, *Vibrio hollisae*, *Vibrio fluvialis*, and *Vibrio furnissii*) that have potential to cause gastroenteritis have been reported. Some strains of these species produce known toxins but the pathogenic mechanism is largely not understood. The ecology of and detection and control methods for all seafoodborne *Vibrio* pathogens are essentially similar.

## Overview of foodborne *Vibrio* pathogens

The bacteria belonging to the genus *Vibrio* are facultatively anaerobic, gram-negative rods motile by one or more polar flagella. They are usually oxidase positive, sensitive to the vibriostatic agent O/129 and generally require sodium ions for growth. Vibrios are one of the most dominant bacterial populations in estuarine and marine environments. They are sensitive to low pH but grow well at high pH. In general, vibrios can be readily isolated from coastal waters when water temperature is around 15°C or above.

*Vibrio* species can establish intestinal or extraintestinal infection in humans. As summarized in previous reviews (Tison and Kelly, 1984; Morris and Black 1985; Janda *et al.*, 1988; Oliver and Kaper, 2001), *Vibrio cholerae*, *Vibrio parahaemolyticus*, *Vibrio mimicus*, *Vibrio hollisae*, *Vibrio fluvialis*, and *Vibrio furnissii* were solely or mainly isolated from gastroenteritis cases and *Vibrio vulnificus*, *Vibrio alginolyticus*, *Vibrio damsela*, *Vibrio metschnikovii*, and *Vibrio cincinnatiensis* were primarily reported from extraintestinal infections (septicemia, wounds, etc.). Of the latter group primary septicemia due to *V. vulnificus* infection was often associated with consumption of seafood. Therefore, *V. cholerae*, *V. parahaemolyticus*, *V. mimicus*, *V. hollisae*, *V. fluvialis*, *V. furnissii*, and *V. vulnificus* are considered to have potential to establish infection in humans through seafood consumption. Of these, *V. cholerae* and *V. parahaemolyticus* have long been established as important pathogens in various parts of the world. Unlike other *Vibrio* spp., which occur naturally in seafood, *V. cholerae* primarily contaminates various types of food through fecal and water-borne routes in endemic and epidemic areas since this organism, unlike most other *Vibrio* species, can survive in the freshwater environment. The pathogenic mechanisms of epidemic *V. cholerae* and *V. parahaemolyticus* have been studied well. It is now possible to differentiate environmental strains of these species into virulent and avirulent ones based on their ability or inability to produce their major virulence factors. Infection by *V. vulnificus* is characterized by high fatality rates and thus, is considered an important pathogen. About 50 foodborne *V. vulnificus* septicemia cases are estimated in the US annually (Mead *et al.*, 1999). The pathogenic mechanism of *V. vulnificus* has been studied fairly extensively but it has not been clearly elucidated. Establishment of *V. vulnificus* infection is considered to depend primarily on the risk factors of the host. Ecology including the "viable but nonculturable (VNBC)" state has been studied mainly for the three important *Vibrio* species. Infections by other *Vibrio* species are rare and they have not been well studied. Information on the pathogenic mechanism, epidemiology, ecology, detection, and control of the three important vibrios are summarized and available information for the other seafoodborne *Vibrio* pathogens is also included in this chapter.

## Taxonomy and differentiation of pathogenic *Vibrio* species

Taxonomy of the bacteria belonging to the genus *Vibrio* has been in a state of flux. Numerical taxonomy was followed by

DNA–DNA hybridization-based taxonomy for classification of vibrios and the results obtained by the two methods agreed fairly well (Colwell, 1970). This established the basis for the current classification of vibrios. Their phylogentic relationship has been further clarified by the analysis of the 16S rRNA gene sequence (Dorsch *et al.*, 1992; Kita-Tsukamoto *et al.*, 1993; Ruimy *et al.*, 1994). In *Bergey's Manual of Determinative Bacteriology*, ninth edition, 34 species belonging to the genus *Vibrio* are described, and a table listing 12 species reported from clinical specimens and summarizing the biochemical characteristics to differentiate these species is given (Holt *et al.*, 1994). However, the strains isolated from environmental samples (seawater, sediments, seafood, etc.) may exhibit atypical biochemical characteristics and identification based on biochemical tests is slow, resource intensive, and not always definitive. For this reason, identification methods based on genetic techniques such as DNA probes and polymerase chain reaction (PCR) have been developed for clinically important species (described below).

## *Vibrio cholerae*

### Characteristics of the organism and the diseases

Kaper *et al.* (1995) and Wachsmuth *et al.* (1994) wrote extensive reviews on *V. cholerae* and diseases caused by this organism. Readers are referred to these reviews for details. Only essential characteristics are summarized in this section.

*V. cholerae* and a closely related *Vibrio*, *V. mimicus*, are atypical *Vibrio* species in that they can survive and grow without added salts. This feature allows *V. cholerae* causing, in addition to seafoodborne infection, water-borne outbreaks and epidemics in terrestrial environments. The type of O antigen is an important marker for epidemic potential of *V. cholerae*. Epidemic cholera used to be associated exclusively with the O1 serotype. Serotype O139 was established as an epidemic strain in association with the cholera epidemic that emerged in the Bengal area in 1992 and O139 strains have been regarded as another causative agent of cholera since then. Strains belonging to O1 and O139 serotypes generally possess the *ctx* gene encoding cholera toxin (CT) and CT is responsible for the cholera symptoms (profuse, watery diarrhea often accompanied by vomiting leading to severe dehydration and electrolyte deficiencies). Damage to intestinal tissue was not observed in cholera cases. O1 strains can be classified into the classical or El Tor biotype based on biochemical characteristics. The O1 strains can also be classified into Ogawa, Inaba, and Hikojima subserotypes for epidemiological investigation. Unlike O1 strains, O139 strains produce a capsule. Some strains belonging to the O serotypes other than O1 and O139 (referred as non-O1/O139) cause foodborne diarrhea that is milder than cholera. In the US non-O1/O139 *V. cholerae* is the second leading cause of seafood associated bacterial gastroenteritis after *V. parahaemolyticus*. *V. cholerae* non-O1 was associated

with gastroenteritis twice as frequently as extraintestinal infections in Florida (Hlady and Klontz, 1996).

The infectious dose of *V. cholerae* O1 El Tor varies with the host conditions. When given in buffered saline, the inoculum that consistently induced diarrhea in healthy North American volunteers was $10^{11}$ colony-forming units, whereas $10^6$ organisms caused diarrhea in volunteers when gastric acid was neutralized with sodium bicarbonate or when the organisms were ingested with fish and rice. The incubation period became longer and stool volume decreased with smaller inocula (Kaper *et al.*, 1995). The inoculum in natural infection in Bangladesh was estimated to be $10^2$ to $10^3$ (Kaper *et al.*, 1995).

### Virulence mechanisms

Various virulence-associated factors and regulators of *V. cholerae* have been studied at the molecular genetic level to date (Table 12.1). Viable cells of *V. cholerae* O1 and O139 may reach the small intestine after ingestion if they can survive in the acidic environment of the stomach. The cells then colonize the epithelial surface and produce CT. A bundle-forming pilus termed toxin-coregulated pilus (TCP) is considered to play an important role in colonization and CT is the determinant of the cholera symptoms. CT is composed of one A and five B subunits. The B subunits mediate binding of CT (holotoxin) to the $GM_1$ ganglioside receptor on the intestinal cell. The A subunit is cleaved into $A_1$ and $A_2$ peptides after CT enters the cell. The $A_1$ peptide catalyzes ADP-ribosylation of $Gs\alpha$ protein and results in activation of adenylate cyclase. Increase in cAMP level leads to activation of protein kinase A and subsequently alteration of ion transport that is responsible for profuse, watery diarrhea. The CT and TCP are thus considered two major virulence factors of O1 and O139. The *ctx* operon and *tcp* gene clusters encoding, respectively, CT and TCP are usually unique to the O1 and O139 strains. The *ctx* operon composed of the *ctxA* and *ctxB* genes exists in a transposon-like genetic unit called CTX element that also contains two other toxin genes (*zot*, *ace*). The CTX genetic element is a part of the lysogenic phage called CTXΦ (Waldor and Mekalanos 1996). The *tcp* gene cluster is contained in a 40-kb DNA segment termed a pathogenicity island (VPI), and VPI is reported to be encoded on a lysogenic filamentous phage VPIΦ, and the TCP serves as a receptor for CTXΦ (Karaolis *et al.*, 1999). This suggests that an ancestor of O1 and O139 serotypes acquired the ability to produce CT and TCP by phage conversion during evolution and that O139 serotype derived from an O1 El Tor strain (Mekalanos *et al.*, 1997). The finding that the difference in the O antigenicity between O1 and O139 serotypes is due to replacement of the *rfb* region in the O1 strains possibly by recombination supports the above hypothesis. The O139 *rfb* region appears to have resulted from replacement of the 22-kb O1 *rfb* region by the 35-kb nucleotide sequence containing *otnAB* genes associated with capsule synthesis and the genes related to the O antigen synthesis (Bik *et al.*, 1996; Comstock *et al.*, 1996).

**Table 12.1** Virulence factors (gene) and regulators (gene) of foodborne *Vibrio* pathogens studied at molecular levels

| Organism | Virulence factor (gene) or regulator (gene) | Reference |
|---|---|---|
| V. cholerae | *Major virulence factors* | |
| | CTX genetic element | Faruque *et al*. (1998) |
| | Cholera toxin (*ctx*) | Kaper and Levine (1981), Pearson and Mekalanos 1982) |
| | Zonula occludens toxin (*zot*) | Baudry *et al*. (1992) |
| | Accessory cholera enterotoxin (*ace*) | Trucksis *et al*. (1993) |
| | Pathogenicity island (VPI) | Karaolis *et al*. (1998) |
| | Toxin-coregulated pilus (*tcpA*) | Taylor *et al*. (1987) |
| | Accessory colonization factor (*acf*) | Peterson and Mekalanos (1988) |
| | *Other virulence factors* | |
| | Hemolysin (*hly*) | Brown and Manning 1985 |
| | RTX (repeats in toxin) toxin | Lin *et al*. (1999) |
| | Mannose-fucose-resistant hemagglutinin (*mfr*) | Franzon *et al*. (1993) |
| | Iron acquisition systems (*irgA*, *viuA*, *vib*, *hut*) | Goldberg *et al*. (1990), Butterton, *et al*. (1992), Wyckoff *et al*. (1997) |
| | | Butterton *et al*. (2000), Henderson and Payne 1994), Occhino *et al*. (1998) |
| | Hemagglutinin/protease (*hap*) | Hase and Finkelstein (1991) |
| | Neuraminidase (*nanH*) | Galen *et al*. (1992) |
| | Motility (*motAB*, *motY*, *flaA*) | Lee *et al*. (2001) |
| | Chemotaxis (*cheA*, *cheY*) | Lee *et al*. (2001) |
| | Lipopolysaccharide (*galU*) | Nespar *et al*. (2001) |
| | Thermostable direct hemolysin (*tdh*) | Baba *et al*. (1991) |
| | Heat-stable enterotoxin (*stn, sto*) | Ogawa *et al*. (1990), Ogawa and Takeda (1993) |
| | *Virulence regulators* | |
| | ToxR (*toxR*) | Miller and Mekalanos (1984) |
| | ToxT (*toxT*) | DiRita (1992) |
| | cAMP receptor protein (*crp*) | Skorupski and Taylor (1997a), Skorupski and Taylor (1997b) |
| | TcpP–TcpH complex (*tcpP*, *tcpH*) | Cotter and DiRita (2000) |
| | LysR-family transcriptional regulator (*aphB*) | Kovacikova and Skorupski (1999) |
| | Heat shock protein (*dnaK*) | Chakrabarti *et al*. (1999) |
| | Histone-like nucleoid structuring protein (*hns*) | Nye *et al*. (2000) |
| | Multifunctional protein (*pepA*) | Behari *et al*. (2001) |
| | Virulence-associated regulator (*varA*) | Wong *et al*. (1998) |
| | Quorum-sensing regulator (*luxO*, *hapR*) | Zhu *et al*. (2002) |
| | Phosphate response regulator (*phoR*, *phoB*) | von Krüger *et al*. (1999) |
| | Histidine kinase and response regulators (*vieSAB*) | Lee *et al*. (1998a) |
| | Hemagglutinin/protease gene regulator (*hapR*) | Jobling and Holmes (1997) |
| | Hemolysin gene regulator (*hlyU*) | Williams *et al*. (1993) |
| | Iron acquisition regulators (*fur*, *irgB*) | Litwin *et al*. (1992), Butterton *et al*. (1992), Goldberg *et al*. (1991) |
| V. parahaemolyticus | *Major virulence factors* | |
| | Thermostable direct hemolysin (*tdh*) | Nishibuchi *et al*. (1992) |
| | TDH-related hemolysin (*trh*) | Xu *et al*. (1994) |
| | *Other virulence factors* | |
| | Chemotaxis and motility | Kim and McCarter (2000), McCarter and Wright (1993) |
| | Iron acquisition systems (*pvuA*, *psuA*) | Funahashi *et al*. (2002) |
| | Pilus | Nakasone and Iwanaga (1990) |
| | Cell-associated hemagglutinin | Nagayama *et al*. (1995) |
| | Thermostable hemolysin | Taniguchi *et al*. (1990 ) |
| | Thermolabile hemolysin (*tlh*) | Taniguchi *et al*. (1986), Shinoda *et al*. (1991) |
| | Metalloproteases (*prtVp*, *vppC*) | Lee *et al*. (1995c), Kim *et al*. (2002) |

**Table 12.1** Continued

| Organism | Virulence factor (gene) or regulator (gene) | Reference |
|---|---|---|
| | *Virulence regulators* | |
| | ToxR (*toxR*) | Lin *et al*. (1993) |
| | Regulator of iron acquisition system (*fur*) | Yamamoto *et al*. (1997) |
| *V. vulnificus* | *Virulence factors* | |
| | Iron acquisition systems (*hupA*, *vvuA*) | Litwin and Byrne (1998), Webster and Litwin (2000) |
| | Capsule (epinerase gene and *wza*) | Zuppardo and Siebeling (1998), Wright *et al*. (2001) |
| | Type IV pilus and type II secretion (*vvpC*, vvpD) | Paranjpye *et al*. (1998) |
| | Hemolysins (*vvh*, *vllY*) | Wright *et al*. (1985), Yamamoto *et al*. (1990), Chang *et al*. (1997) |
| | Zinc metalloprotease (*vvp*, *empV*, *vvpE*) | Cheng *et al*. (1996), Chuang *et al*. 1997), Jeong *et al*. (2000) |
| | *Virulence regulator* | |
| | Iron acquisition regulators (*fur, hupR*) | Litwin and Calderwood (1993), Litwin and Quackenbush (2001) |
| | ToxR (*toxR*) | Lee *at al*. (2000) |
| | LuxR homologue (*smcR*) | Shao and Hor (2001) |
| *V. mimicus* | *Possible major virulence factors* | |
| | CTX genetic element | Boyd *et al*. (2000) |
| | Cholera toxin (*ctx*) | Bi *et al*. (2001), Shi *et al*. (1998) |
| | Zonula occludens toxin (*zot*) | Chowdhury *et al*. (1994), Shi *et al*. (1998) |
| | Accessory cholera enterotoxin (*ace*) | Shi *et al*. (1998) |
| | Pathogenicity island (VPI) | |
| | Toxin-coregulated pilus (*tcpA, tcpP, toxT*) | Boyd *et al*. (2000), Bi *et al*. (2001) |
| | Heat-stable enterotoxin (*st*) | Shi *et al*. (1998) |
| | Thermostable direct hemolysin (*tdh*) | Terai *et al*. (1990) |
| | *Other virulence factors* | |
| | Hemolysin (*vmhA*) | Kim *et al*. (1997) |
| | Phospholipase (*phl*) | Kang *et al*. (1998) |
| | Metalloprotease (*vmc*) | Lee *et al*. (1998c) |
| | Siderophore | Okujo and Yamamoto (1994) |
| | Hemagglutinins | Alam *et al*. (1996) |
| | *Virulence regulators* | |
| | ToxR (*toxR*) | Bi *et al*. (2001) |
| | ToxT (*toxT*) | Bi *et al*. (2001) |
| *V. hollisae* | *Possible major virulence factor* | |
| | Thermostable direct hemolysin (*tdh*) | Yamasaki *et al*. (1991), Nishibuchi *et al*. (1996) |
| | *Other virulence factors* | |
| | Siderophore | Okujo and Yamamoto (1994) |
| | Chinese hamster ovary cell elongation factor | Kothary *et al*. (1995) |
| *V. fluvilais* | *Virulence factors* | |
| | Chinese hamster ovary cell elongation factor | Lockwood *et al*. (1982) |
| | Chinese hamster ovary cell-killing factor | Wall *et al*. (1984) |
| | Siderophores | Yamamoto *et al*. (1993) |

Virulence factors other than CT and TCP probably play auxiliary roles in pathogenesis. Some of the studies supporting this idea are listed in Table 12.1. Motility and chemotaxis help *V. cholerae* cells penetrate through the mucus gel layer and reach epithelial cells before the bacterial cells colonize the intestine. The *acf*-encoded accessory colonization factor and the mannose-fucose-resistant hemagglutinin may play some roles in intestinal colonization. Lipopolysaccharide can render *V. cholerae* the ability to colonize the intestine and to resist serum killing. Neuraminidase destroys the neuraminic acid of the intestinal cell surface and can help CT bind to the intestinal epithelial cells. The zonula occludens toxin may contribute to diarrheal symptoms to some degree by affecting the structure of the intercellular tight junctions (zonula occludens) of intestinal cells. The *ace*-encoded accessory cholera enterotoxin may also cause fluid secretion in the intestine. RTX toxin (repeats in toxin: a toxin containing repeats of glycine and aspartic acid residues at the C-terminus) is toxic to cultured animal cells. The gene cluster encoding RTX is physically linked to the cholera toxin (CTX) element but its exact role is not known. The hemolysin encoded by the *hly* gene is cytolytic to various erythrocytes and the *hly* gene is distributed in almost all strains of *V. cholerae* regardless of the O serotype. Therefore, the pathogenic role of this hemolysin, if any, is not restricted to a particular serotype(s). Various systems to acquire iron that is needed for bacterial growth *in vivo* were reported (Table 12.1). Expression of one of the iron acquisition components, the outer membrane protein encoded by the *irgA* gene, was shown to be associated with colonization ability and lethality in an animal model.

Of non-O1/O139 strains isolated from the patients with diarrhea, some strains were shown to carry the *ctx* gene (Yamamoto *et al.*, 1995), the *tdh* gene similar to the *V. parahaemolyticus tdh* gene encoding the thermostable direct hemolysin (TDH), or the *stn* gene related to the *Eschericia coli* gene encoding the heat-stable enterotoxin. A similar heat-stable enterotoxin gene (*sto*) was also detected in a *V. cholerae* O1 strain. The genetic structures surrounding these toxin genes are associated with gene mobilization and suggest that these toxin genes were acquired possibly by horizontal transfer mechanisms (Nishibuchi and Kaper, 1995; Ogawa and Takeda 1993). While these toxins are likely to play important pathogenic roles in these strains, most clinical strains lack these genes and identification of good virulence markers for *V. cholerae* non-O1/O139 is an important research area.

Expression of major virulence genes in O1 and O139 strains are regulated by a complex system. The *toxR* and *toxS* genes form an operon and ToxR, a transmembrane protein, is stabilized in the membrane by ToxS, another transmembrane protein. The ToxR–ToxS complex positively regulates expression of the *toxT* gene present in the *tcp* gene cluster in response to various environmental signals (temperature, pH, salt concentration, and nutrient). ToxT, as the second regulator, then regulates virulence-associated genes (*ctx*, *tcp*, *acf*), and some other genes. This regulatory system is called ToxR regulon (DiRita 1992). Recently, the regulators encoded by the *crp*, *tcpP*, *tcpH*, *aphB*, *dnaK*, *hns*, *pepA*, *varA*, *luxO*, and *hapR* genes were shown to influence expression of the genes in the ToxR regulon (Table 12.1) and thus the ToxR regulon seems to form a very complex cascade. The *toxR* gene is distributed in various pathogenic and non-pathogenic species of *Vibrio*. It appears that the ancestral role of ToxR is regulation of the genes needed for adaptation to environmental change and that ToxR has been appropriated for the regulation of virulence genes in only several pathogenic *Vibrio* species (Okuda *et al.*, 2001). The regulators encoded by the *phoR* and *phoB* genes are involved in expression of a colonization-related gene(s) but its exact mechanism is not understood. The *vieSAB* genes are expressed during infection in an infant mouse model. The *hapR*, *hlyU*, *fur*, and *irgB* gene products are the regulators of some of the auxiliary virulence factors (Table 12.1).

## Epidemiology

Seven pandemics of cholera have been described since 1823; the first six originated from the Indian subcontinent and the seventh from Indonesia (Pollizer, 1959; Blake, 1994a). Strains from the seventh pandemic, which began in 1961 and still persists, are different than available strains from previous pandemics in that they are hemolytic and are referred to as the El Tor biotype whereas the nonhemolytic strains are called the classical biotype. In general, illness by the classical biotype is more severe than that by the El Tor biotype but the exact reason is not known. The persistence of the current pandemic may be related to the ability of the El Tor strains to survive in the natural environment. *V. cholerae* O139, which emerged in India and Bangladesh in 1992, appears to be closely related to the seventh pandemic strains as described above. *V. cholerae* O139 was prevalent in this region for several years but has not become pandemic.

## Ecology

### Distribution

While *V. cholerae* is indigenous to fresh and brackish water environments in tropical, subtropical and temperate areas worldwide, the threat of epidemic cholera is confined primarily to developing countries with warm climates (Wachsmuth *et al.* 1994). *V. cholerae* non-O1/O139 strains are most prevalent at warmer water temperatures and low to intermediate salinities but, unlike *V. cholerae* O1 and O139, are not correlated with fecal coliform density (Kaper *et al.*, 1979; DePaola *et al.*, 1984; DePaola *et al.*, 1983; Motes *et al.*, 1994).

### Reservoirs

Cholera is exclusively a human disease and human feces from infected individuals is the primary source of infections in cholera epidemics. Long-term carriers have been documented but are rare. Drinking water and food may be directly contaminated with feces. Alternatively, contamination of food production environments or wash water can indirectly introduce *V.*

*cholerae* into foods. This fecal oral route of infection is prevalent with the 7th pandemic strain in developing countries in Asia, Africa and South America. The recovery of free-living toxigenic *V. cholerae* in remote rivers in northeast Australia and in coastal areas along the USA Gulf Coast suggests the existence of natural reservoirs (Blake, 1994b). *V. cholerae* produce chitinase and have been shown to attach and multiply on zooplankton such as copepods (Huq *et al.*, 1983).

Epidemic cholera can be introduced from abroad by infected travelers, imported foods (Mintz *et al.*, 1994) and through the ballast water of cargo ships (McCarthy and Khambaty, 1994). However, legally imported foods have seldom been implicated in cholera outbreaks and extensive sampling of shrimp and other seafoods produced in countries with endemic cholera and exported has detected toxigenic *V. cholerae* but the frequency was very low (Minami *et al.*, 1991; DePaola *et al.* 1992). Toxigenic *V. cholerae* O1 was detected in migratory aquatic birds in Colorado indicating their potential to introduce cholera from abroad (Ogg *et al.*, 1989). *V. cholerae* has been shown to enter a VBNC state when subjected to stress such as extended exposure to cold seawater and this has been hypothesized as a mechanism for overwintering (Colwell and Huq, 1994).

## Detection

Methods for culturing *V. cholerae* from foods and clinical samples in most countries are quite similar and have changed little in decades (Centers for Disease Control and Prevention, 1998a; DePaola *et al.*, 1988b; Elliot *et al.*, 1998; Kaysner and Depaola, 2001). Foods are homogenized and placed into an enrichment broth, typically 1:10 in alkaline peptone water (APW), and incubated at either 35 or 42°C. Incubation of APW at 42°C has been shown to significantly improve *V. cholerae* recovery from oysters by reducing growth of background flora (DePaola *et al.*, 1987; DePaola *et al.*, 1988b). After incubation for 6 to 8h and/or overnight, a loopful of the surface pellicle (area of higher *V. cholerae* cell density) is streaked for isolation onto thiosulfate-citrate-bile salts-sucrose (TCBS) agar and incubated overnight at 35°C for the selective and differential isolation of *V. cholerae*. Flat yellow (sucrose positive) colonies with translucent peripheries are typical of *V. cholerae* and several colonies should be purified for further biochemical identification techniques. Key characteristics for identification of a suspect isolate as *V. cholerae* include positive tests for production of catalase, oxidase, string (cell lysis in sodium desoxychollate), growth in tryptone broth without added NaCl, production gelatinase, decarboxylation of lysine and ornithine and fermentation of glucose, sucrose and mannitol (Elliot *et al.*, 1998). Isolates identified as *V. cholerae* should be tested for agglutination in O1 polyvalent and O139 antisera and tested for enterotoxigenicity to determine their epidemic potential. Production of CT can be determined phenotypically using tissue culture such as the Y-1 mouse adrenal cell assay, or immunologically (i.e. reverse passive latex agglutination) (Elliot *et al.*, 1998).

Various molecular detection methods alternative to the conventional methods have been devised since late 1980. The PCR method targeting the 16S–23S rRNA intergenic spacer region (Chun *et al.*, 1999), the *toxR* gene (Ghosh *et al.*, 1997; Rivera *et al.*, 2001), or the *ompW* gene (Nandi *et al.*, 2000) were reported to allow specific identification of *V. cholerae*. O1 and O139 strains can be identified by PCR methods targeting the respective *rfb* gene (Keasler and Hall., 1993; Guhathakurta *et al.*, 1999). Classical and El Tor biotypes may be differentiated by the PCR methods targeting the biotype-specific *tcp* operons (Keasler and Hall, 1993; Rivera *et al.*, 2001). The *ctx* gene can be detected by DNA probe methods (Kaper *et al.* 1989; Wright *et al.* 1992) or by PCR methods (Shirai *et al.*, 1991; Fields *et al.*, 1992; Koch *et al.*, 1993; Elliot *et al.* 1998). A multiplex PCR method for simultaneous examination of the O1 and O139 antigens and the *ctx* gene allows detection of toxigenic O1 and O139 strains using a single PCR reaction (Hoshino *et al.*, 1998). DNA probe methods (Takeda *et al.*, 1991; Hoge *et al.*, 1990) or PCR methods (Guglielmetti *et al.*, 1994; Rivera *et al.*, 2001) are useful alternatives to cumbersome suckling mouse assays to detect the ability of *V. cholerae* non-O1/O139 to produce the heat-stable enterotoxins. The results obtained by the molecular detection methods are usually definitive, but the chance that oligonucleotide probe and PCR methods may miss a base change(s) in the target nucleotide sequence cannot be ruled out. Therefore, polynucleotide DNA probes may be employed when appropriate. For example, a DNA probe study carried out in Japan revealed that only 26.6% of the *V. cholerae* O1 strains isolated from imported seafood and none of the *V. cholerae* O1 strains isolated from the natural water carried the *ctx* gene (Minami *et al.*, 1991). On the other hand, a small percentage of environmental *V. cholerae* non-O1 strains were shown to carry the *ctx* gene in a study using a DNA probe (Nair *et al.*, 1988).

Another approach for detection of epidemic cholera that eliminates the need to isolate colonies is to test the APW enrichment directly for O1 and O139 *V. cholerae* or the *ctx* gene using ELISA or PCR, respectively (Oliver and Kaper, 2001; Koch *et al.*, 1993; DePaola, and Hwang, 1995; Karnasagar *et al.*, 1995). These methods are more efficient, rapid and sensitive than culture methods but are generally recommended for screening. Positive results in food samples are infrequent and the above culture technique can be employed to obtain an isolate for confirmation. The prevalence and range of *V. cholerae* and other potentially pathogenic vibrios in the environment and the food supply may be much higher than previously reported because of the availability of more efficient and sensitive techniques. For example a study in the early 1990s using a 42°C enrichment method found a similar incidence of *V. cholerae* non-O1/O139 in 0.1g portions of Gulf Coast oysters as a study a decade earlier, which examined 100 g oyster portions from another Gulf Coast area using a 35°C enrichment method. The real-time PCR method to detect the *hlyA* gene (Lyon, 2001) appears to be useful for rapid and quantitative detection of *V. cholerae*, regardless of serotype and possession of the *ctx* gene, in raw oyster and seawater. With the employment of powerful new methods such as real time PCR we may obtain a much more complete understanding of the distribution and ecology of pathogenic vibrios in the environment and the food supply. This knowledge should greatly aid current efforts to assess and control the risk that these organisms present to public health.

## Control

Protecting the food supply from contamination of *V. cholerae* employs many of the same hygienic approaches used with other foodborne pathogens but there are noteworthy considerations due to differences in the ecology of strains with epidemic potential. Prevention of foodborne infections of epidemic cholera relies primarily on breaking the fecal/oral route. Family contacts and health care workers of cholera patients should be especially diligent with hand washing and other hygienic measure to prevent person to person spread and contamination of water or food. Food processors in addition to insuring good hygienic practices among food handlers should also screen workers for signs of potential cholera infections. Food products such as vegetables or seafood that are likely to have been exposed to human feces (i.e. irrigation, fertilization with night soil, or harvest from polluted waters) should be avoided or properly cooked. Chlorination or other disinfection should be used with all potable water used for ice or food processing. Avoiding contact between cooked food products and raw seafood is important for preventing contamination of both epidemic cholera and *V. cholerae* non-O1/O139. Since nearly all *V. cholerae* non-O1/O139 infections are associated with seafood consumption, primarily raw molluscan shellfish, control measures should focus on this commodity. *V. cholerae* non-O1/O139 occurs naturally in warm and temperate esturarine environments independently of fecal pollution, thus traditional shellfish sanitation measures are ineffective in mitigating the risk of infection from consumption of raw shellfish. The risk of illness from consumption of raw shellfish from cooler waters is substantially reduced compared to those harvested from waters > 20°C. Proper cooking of shellfish and other seafood products readily inactivates *V. cholerae* even in highly contaminated products. *V. cholerae* does not survive well at low pH and is seldom associated with high-acid foods (Mintz *et al.*, 1994). Cerviche is a popular dish in Latin America that is prepared from raw seafood products marinated in lime juice. While survival of *V. cholerae* in cerviche depends on the level and location of contamination, the pH and the time marinated, cerviche was not considered an important vehicle in the spread of epidemic cholera in Latin America during the 1991 epidemic. Post-harvest treatment of raw shellfish by mild heat treatments (Shultz *et al.*, 1984), freezing combined with frozen storage (Reily and Hackney, 1985), and high hydrostatic pressure (Berlin *et al.* 1999) can reduce *V. cholerae* densities by many magnitudes and practically eliminate the risk of infection.

## *Vibrio parahaemolyticus*

## Characteristics of the organism and the diseases

Strains of *V. parahaemolyticus* and *V. vulnificus* usually form green colonies when grown on TCBS agar. Biochemical characteristics of these two species are similar and environmental strains may show somewhat atypical biochemical patterns. Differentiation of the strains belonging to these species, particularly those isolated from environmental sources, may require additional tests. An O:K serotyping scheme has been established for examination of clinical strains of *V. parahaemolyticus* and it can be useful for confirmation of identification as well as epidemiological investigation. However, the O and K antigens of the environmental strains of *V. parahaemolyticus* are more diversified and they have not been included in the typing scheme. PCR methods for identification at the species level are now available (described below).

*V. parahaemolyticus* is characterized by its fast growth. Its generation time can be as short as 9 minutes in culture medium and 12 minutes in seafood (Miwatani and Takeda, 1976; Oliver and Kaper, 2001). Therefore, this species is an important cause of gastroenteritis associated with consumption of raw or undercooked seafood in tropical and temperate countries. In addition, pandemic spread of the infections by a new clone started around 1996 (Matsumoto *et al.*, 2000).

Most clinical strains induce beta-type hemolysis when grown on a special blood agar, Wagatsuma agar. This phenomenon is called Kanagawa phenomenon (KP) and KP is caused by TDH. Only 1 to 2% of the environmental strains exhibit KP (Sakazaki *et al.*, 1968; Miyamoto *et al.*, 1969) and thus KP has long been considered a marker for virulent strains. However, some KP-negative strains capable of producing a TDH-related hemolysin (TRH) but not TDH were isolated from patients with diarrhea in late 1980s (Honda *et al.*, 1988). Subsequent molecular genetic studies established that the strains capable of producing TDH, TRH, or both are virulent strains (explained in the virulence mechanism section below).

Consumption of seafood and other foods contaminated with virulent strains causes gastroenteritis. The main symptom of the patients with gastroenteritis is diarrhea. Diarrhea is usually watery and bloody diarrhea is observed less frequently. Various other symptoms were also reported from outbreaks of gastroenteritis. These included abdominal pain, vomiting, fever, headache, nausea, and chill. Average incubation period ranged from 15 to 24 hours and illness generally lasted for 4 to 7 days (Miwatani and Takeda, 1976; Oliver and Kaper, 2001). Patients with gastroenteritis died in very rare cases. Extraintestinal infection is rare.

Human volunteer studies showed that $2 \times 10^5$ to $10^9$ organisms of virulent strains when protected against gastric acid by antacid or food could induce abdominal discomfort or diarrhea and that $2 \times 10^{10}$ cells of a KP-negative strain did not cause any symptom (Aiso and Fujiwara, 1963; Sanyal and Sen, 1974).

## Virulence mechanisms

TDH and/or TRH are considered major virulence factors of *V. parahaemolyticus*. TDH is composed of two identical subunits. It is hemolytic to erythrocytes and cytotoxic to various animal cells, increases vascular permeability in rabbit skin, and stimulates fluid accumulation in the rabbit ileal loop (Takeda, 1983). Heating at 100°C does not inactivate the hemolytic activity of TDH and the hemolytic activity is not enhanced by addition of lectin (direct action) and thus this hemolysin was named thermostable direct hemolysin. TDH binds to intestinal cells through ganglioside receptors, $GT_1$ and forms a pore in the membrane (Honda *et al.*, 1992; Huntley *et al.*, 1993). TDH stimulates ion flux in myocardial tissue, rat erythrocytes, and human erythrocytes (Takashi *et al.*, 1982; Seyama *et al.*, 1977; Huntley *et al.*, 1993). TDH was shown to use calcium ion as an intracellular second messenger and induce intestinal chloride ion secretion in a rabbit model (Raimondi *et al.*, 1995). The *tdh* gene encodes a TDH subunit. KP-positive strains carry two *tdh* genes (*tdh1* and *tdh2* sharing 97% identity) and the *tdh2* gene is expressed at a much high level than the *tdh1* gene and is responsible for KP (Nishibuchi and Kaper, 1995; Okuda and Nishibuchi, 1998). The strong promoter plays a major role and ToxR regulator has a minor role in the high-level *tdh2* expression (Lin *et al.*, 1993; Okuda and Nishibuchi, 1998). Comparison in rabbit ileal loop and the Ussing chamber models of a KP-positive strain and its isogenic TDH-deficient mutant strain where both *tdh1* and *tdh2* genes were knocked out established that TDH is the major enterotoxin of KP-positive strain (Nishibuchi *et al.*, 1992). KP-negative strains do not carry the *tdh* gene except in rare cases when only one *tdh* gene is present and production of TDH is much smaller than with KP-positive strains (Nishibuchi and Kaper, 1995). Production of TDH from the strains carrying both the *tdh* and *trh* genes is generally at low levels (Nakaguchi *et al.*, 2004). The *tdh* gene is associated with insertion sequence-like genetic structure. *V. parahaemolyticus* probably acquired the *tdh* gene from another organism by horizontal transfer and subsequent genetic rearrangement may have led to variation of the *tdh* gene (Nishibuchi and Kaper, 1995).

The *trh* gene that can be classified into two subtypes, *trh1* and *trh2* sharing 84% sequence identity, encodes TRH. Most clinical strains carry the *tdh* gene, the *trh* gene (*trh1* or *trh2*), or both genes whereas only a small proportion of environmental strains have these genes (Nishibuchi and Kaper, 1995). TRH is 62% to 63% identical to TDH at the amino acid sequence level and shares structural similarity and biological activities with TDH except that TRH is thermolabile (Honda *et al.*, 1988; Nishibuchi *et al.*, 1989; Kishishita *et al.*, 1992). Experiment to evaluate the pathogenic role of TRH by using a mutagenesis was attempted but clear results have not been obtained (Xu *et al.*, 1994).

Other factors may play some auxiliary roles in pathogenesis of *V. parahaemolyticus* although they are not associated only with clinical strains (Table 12.1). Chemotaxis, motility by means of polar and lateral flagella, and the systems needed for iron acquisition have been studied at the genetic level. *V. para-* haemolyticus was shown to adhere to various animal cells and tissues. A pilus and a cell-associated hemagglutinin that mediate adherence to rabbit intestinal tissue and a human colonic cell line, respectively, have been characterized. The gene encoding a thermolabile hemolysin, which is distinct from TRH and found to be an atypical phospholipase and the gene encoding the thermostable hemolysin, termed delta-VPH and different from TDH, have been reported. The genes encoding two different metalloproteases have also been studied. Other factors including a lethal factor, Shiga-like toxin, and a Chinese hamster ovary (CHO) cell elongation factor were reported in the literature but they have not been characterized.

## Epidemiology

Nearly all *V. parahaemolyticus* illnesses are associated with seafoods that are consumed raw, undercooked or contaminated after cooking. In countries where consumption of raw seafood is popular, *V. parahaemolyticus* is a major cause of foodborne gastroenteritis and has been reported to be the leading cause of foodborne illness in some Asian countries such as Japan and Taiwan (Kudo *et al.*, 1974; Chiou *et al.* 2000). It was discovered and first associated with seafood consumption (sardines) in the Shirasu outbreak in Japan during 1950 (Fujino *et al.*, 1953). During the 1960s and early 1970s numerous outbreaks were reported throughout Asia (Joseph *et al.*, 1983). The first outbreak in the USA was documented in 1971 and was followed by 12 more by the end of 1972; nearly all were associated with cooked products such as crabs, lobster and shrimp (Barker, 1974). Since 1973, *V. parahaemolyticus* outbreaks have been less frequent and there has been an increasing association of raw molluscan shellfish as the vehicle of infection (Daniels *et al.*, 2000b). *V. parahaemolyticus* is the leading cause of bacterial gastroenteritis reported in the USA associated with seafood consumption (Hlady and Klontz, 1996; Daniels *et al.*, 2000a; Altekruse *et al.*, 2000).

In early 1996 a sharp increase in the incidence of *V. parahaemolyticus* in Calcutta, India, was associated with a new clone of the O3:K6 serotype and this clone became the most prevalent strain in many Asian countries (Okuda *et al.*, 1997). The O3:K6 clone elevated to pandemic status in 1998 when it caused oyster associated outbreaks in Texas and New York (Matsumoto *et al.*, 2000). This clone appears to be endemic in much of Asia but has not been associated with additional outbreaks in North America or other countries outside of Asia until recently. Outbreaks in Chile and infection in Spain due to this O3:K6 clone have occurred recently (Gonzalez-Escalona *et al.*, 2005; Martinez-Urtaza, unpublished data). However, few countries have active surveillance programs or capabilities for serotyping. Prevalence of infection by the pandemic clone including serotype variants was demonstrated in southern Thailand (Laohaprertthisan *et al.*, 2003). The strains belonging to this clone were isolated from bivalves in the same area and their epidemiological relationship with clinical strains were demonstrated by using DNA fingerprinting techniques (Vuddhakul *et al.*, 2000a; Vuddhakul *et al.*, unpublished data).

# Ecology

## Distribution

Densities of *V. parahaemolyticus* in the environment and seafoods vary greatly by season, location, fecal pollution, sample type, and analytical methodology (Earle and Crisley, 1975; DePaola *et al.*, 1990; DePaola *et al.*, 1988a; DePaola *et al.*, 2000; Cook *et al.*, 2002a; Kaneko and Colwell, 1976; Watkins and Cabelli, 1985; Kaysner *et al.*, 1990). Seawater temperature is the most important factor controlling environmental levels of *V. parahaemolyticus* with densities increasing from 10 to 30°C (Anonymous, 2001; DePaola *et al.*, 1990; Kiiyukia *et al.*, 1989; Watkins and Cabelli, 1985; Kaneko and Colwell, 1978). *V. parahaemolyticus* can be detected over a broad range of salinities (5 to 35 ppt) with an optimal salinity of 22 ppt (Anonymous, 2001; DePaola *et al.*, 1990). Most studies have found that fecal coliforms do not reliably predict *V. parahaemolyticus* levels (DePaola *et al.*, 1990; Kaysner *et al.*, 1990; Kaneko and Colwell, 1975b). An indirect link between fecal pollution and *V. parahaemolyticus* levels was reported in a Rhode Island estuary that may have resulted from biostimulation of the microfauna associated with *V. parahaemolyticus* attachment (Watkins and Cabelli, 1985) and this may occur in other estuaries (Venkateswaran *et al.*, 1990). Association of *V. parahaemolyticus* with zooplankton, especially copepods, is well documented and has been attributed to their chitinase activity and affinity to chitin (Kaneko and Colwell, 1978; Venkateswaran *et al.*, 1990; Watkins and Cabelli, 1985; Kaneko and Colwell, 1975a). Typically, *V. parahaemolyticus* levels in molluscan shellfish and sediments are 1 to 3 logs higher than in water (DePaola *et al.*, 1990; Kaneko and Colwell, 1978; Ogawa *et al.*, 1989). *V. parahaemolyticus* doubles approx. every hour in live oysters held at typical ambient temperatures (26°C) along the USA Gulf Coast (Gooch *et al.*, 2002) and are 1 to 2 logs higher in oysters at retail than at harvest (Cook *et al.*, 2002a). *V. parahaemolyticus* levels can fluctuate over 3 logs among individual oysters harvested at the same time and location (Kaufman *et al.*, 2003). Most environmental and seafood surveys have not employed quantitative methods and lack of common methodology complicates data comparisons between the few quantitative studies available. For example, mean quantitative recoveries of *V. parahaemolyticus* in USA seawater and oysters varied 4 and 9-fold respectively, using 4 previously reported methods (DePaola *et al.*, 1988a).

Much less is known about the distribution of pathogenic *V. parahaemolyticus* in the environment compared to total *V. parahaemolyticus*. Strains possessing *tdh*, *trh* or both are generally considered to be pathogenic even though these genes may not be expressed in some strains, and occasionally, strains without either of these genes are isolated from the stools of patients with diarrhea. Most studies have found that fewer than 1% of the *V. parahaemolyticus* isolates from seafoods or environmental sources possess *tdh* or are KP positive and few studies have examined isolates for the presence of *trh* (Cook *et al.*, 2002a; DePaola *et al.*, 1990; DePaola *et al.*, 2000; Thompson and Vanderzant, 1976; Kiiyukia *et al.*, 1989; Kaysner *et al.*, 1996; Kaysner *et al.*, 1990; Cook *et al.*, 2002b). Recent advances in DNA colony hybridization and PCR (McCarthy *et al.*, 2000; Blackstone *et al.*, 2003) (see section on genetic methods) have facilitated detection of pathogenic *V. parahaemolyticus*. While the prevalence of *tdh*+ isolates in total isolates of *V. parahaemolyticus* was similar to that reported in earlier studies (< 1%), approximately 50% of oyster samples (10–12 oysters pooled) or individual oysters were *tdh*+ by either PCR or DNA colony hybridization (Kaufman *et al.*, 2003; Nordstrom and DePaola, 2003; Blackstone *et al.*, 2003). In the recent surveys carried out in the USA, it was found that nearly all the food and environmental *V. parahaemolyticus* strains possessing *tdh* also possess *trh* (DePaola *et al.*, 2000; DePaola *et al.*, 2003). Along the USA Gulf Coast the proportion of *V. parahaemolyticus* strains possessing *tdh* were approximately 10 times more (4.8%) when water temperature was <15°C than at warmer temperatures (DePaola *et al.* 2003). A higher proportion of *tdh*+ isolates (3.2%) has also been reported in the USA Pacific Northwest which also has cool water temperatures (Anonymous, 2001). These observations may explain the higher reported *V. parahaemolyticus* illness rates associated with raw oyster consumption in the Pacific Northwest (Centers for Disease Control and Prevention, 1998b) and in the winter along the Gulf Coast (Klontz *et al.*, 1994) than expected (Anonymous, 2001). Densities of pathogenic *V. parahaemolyticus* are generally <10/g (Cook *et al.*, 2002b; DePaola *et al.*, 2000) but levels of up to 140/g have been reported in oysters at harvest (Anonymous, 2001) and 500/g after 24 h storage at 26°C (Kaufman *et al.*, 2003). Incidence of *tdh*+ and *trh*+ strains in the *V. parahaemolyticus* strains isolated from the seafood imported into Japan mostly from other Asian countries was 0.67% and 2%, respectively (Nishibuchi, unpublished data).

## Reservoirs

The primary reservoir for *V. parahaemolyticus* is the estuarine environment including seawater, sediment and fauna. While *V. parahaemolyticus* typically is undetectable in seawater after prolonged periods below 10°C, it could be cultured from Chesapeake Bay sediments throughout the year at temperatures as low as 1°C (Kaneko and Colwell, 1978). Long-term human carriers have not been reported, but *V. parahaemolyticus* has been isolated from the intestinal contents of wild raccoons in Florida. Passage in warm blooded vertebrates could select for more virulent strains and act as a reservoir although the duration is probably brief (Dutta *et al.*, 1963). The new clone of the O3:K6 serotype may have additional reservoirs as it has remained endemic in many Asian countries since 1996 and has demonstrated pandemic spread (Okuda *et al.*, 1997; Matsumoto *et al.*, 2000). It has been suggested that ballast water from cargo ships may be responsible for pandemic spread as Galveston Bay, Texas is a busy international port (Daniels *et al.*, 2000b).

## Detection

A variety of methods have been described for culturing *V. parahaemolyticus* from seafood and environmental samples but the most popular methods are quite similar to those described for *V. cholerae* (Kaysner and DePaola, 2001; Elliot *et al.*, 1998). Seafood samples are homogenized and generally enriched 1:10 in APW overnight at 35°C. In general, TCBS agar is employed as the selective isolation medium with large green colonies being typical of *V. parahaemolyticus*. Salt polymyxin broth is a second popular enrichment broth and a newly developed selective isolation medium, CHROMagar, appears to be very effective (Hara-Kudo *et al.*, 2001). Several colonies or more are selected for identification using standard biochemical assays such as API 20E. It is a general practice to follow the recommendation that the NaCl concentration in test media is adjusted to 1% or 2–3% (Janda, *et al.*, 1988; Kaysner and DePaola, 2001). PCR methods that can be used in place of biochemical tests for identification were reported. These PCR targets the *tlh* gene (Bej *et al.*, 1999), *gyrB* gene (Venkateswaran *et al.*, 1998), *toxR* gene (Kim *et al.*, 1999), or 0.76-kb nucleotide sequence of unknown function (Lee *et al.*, 1995a; 1995b). Tada *et al.* (1992) reported PCR protocols for detection of the major virulence genes (*tdh* and *trh*) in the isolated strains. Bej *et al.* (1999) reported a multiplex PCR method to detect the *tlh*, *tdh* and *trh* genes at once so that shellfish samples can be examined for total as well as virulent strains of *V. parahaemolyticus*. A real time PCR assay was recently described that detected *tdh* in approximately 50% of oyster enrichments in an environmental survey of Alabama oysters (Blackstone *et al.*, 2003). PCR methods for specific detection of the strains belonging to the pandemic clone were developed. These PCR methods target the nucleotide bases in the *toxR* gene (Matsumoto *et al.*, 2000) or an open reading flame in a lysogenic filamentous phage (Nasu *et al.*, 2000) that appear to be unique to the pandemic strains.

For quantification, serial 10-fold sample dilutions are enriched in an MPN format prior to isolation on TCBS agar. Use of biochemical utilization profiles for identification of *V. parahaemolyticus* in an MPN format is extremely laborious and DNA colony hybridization targeting the species specific *tlh* gene has been substituted for biochemical testing in several studies (Ellison *et al.*, 2001; Cook *et al.*, 2002a; Gooch *et al.*, 2001). Another approach for *V. parahaemolyticus* enumeration using hydrophobic grid membrane filters gave higher *V. parahaemolyticus* recoveries from seawater and oysters samples from the USA than 3 other test methods (two MPN methods and a membrane filter method) but has not been widely accepted (DePaola *et al.* 1988a). A PCR method to detect the 0.76-kb nucleotide sequence of Lee *et al.* (1995a) was combined with an MPN technique to enumerate the total number of *V. parahaemolyticus* cells in mussels (Croci *et al.*, 2002). Alam *et al.* (2002) utilized the *toxR*-targeted PCR method of Kim *et al.* (1999) and the multiplex PCR method of Bej *et al.* (1999) in MPN analyses and showed that the PCR-based MPN methods were more sensitive than the conventional culture-based MPN methods in examination of seawater and marine organic material.

## Control

The control measures described previously for *V. cholerae* non-O1/O139 are also effective in preventing foodborne *V. parahaemolyticus* illnesses. Particularly, *V. parahaemolyticus* grows much faster than other foodborne pathogens under favorable conditions. Therefore, a cold treatment of raw seafood during transportation and marketing is a very effective control measure. Prompt refrigeration of molluscan shellfish is strongly recommended to minimize post harvest multiplication of *V. parahaemolyticus* (Gooch *et al.*, 2002; Anonymous, 2001).

---

## *Vibrio vulnificus*

### Characteristics of the organism and the diseases

Three biogroups (biogroups 1, 2, and 3) of *V. vulnificus* have been reported and they can usually be differentiated by biochemical characteristics. The biogroup 1 is pathogenic to humans and its details are described below in this chapter. The biogroup 2 is an eel pathogen (Tison *et al.*, 1982) and an exceptional isolation of this biogroup from a human wound, possibly transmitted by direct contact of open wounds with infected eels, was reported (Veenstra *et al.*, 1992). Recently, the biogroup 3 was reported from an outbreak in Israel where 62 cases of wound infection and bacteremia were found among fish-handling people (Bisharat *et al.*, 1999).

Biogroup 1 infects humans through seafood consumption or skin lesions. It can cause wound infections, primary septicemia, or gastroenteritis with the latter occurring less frequently. The patients had underlying chronic illnesses such as liver and renal diseases and diabetes or those who use antacids or immunosuppressive agents. The symptoms of gastroenteritis include abdominal pain, vomiting, and diarrhea and those of primary septicemia are fever, chills, nausea, and hypotension. Secondary lesion may occur on extremities and develop into necrotizing fasciitis or vasculitis. The infection by biogroup 1 is characterized by high fatality rates (Tison and Kelly 1984). Cases of biogroup 1 infection were reported often in the United States (Todd, 1989), Japan (Yamauchi *et al.*, 1994), Korea (Park *et al.*, 1991), and Taiwan (Chuang *et al.*, 1992), and sporadically from Sweden (Melhus *et al.*, 1995), Denmark (Bock *et al.*, 1994), Spain (Garcia Cuevas *et al.*, 1998), Italy (Stabellini *et al.*, 1998), Turkey (Horre *et al.*, 1998), and Australia (Maxwell *et al.*, 1991).

The infectious dose in humans is not known. Lethality in experimentally infected mice has often been used to measure the virulence of *V. vulnificus* strains.

### Virulence mechanisms

Virulence mechanisms of *V. vulnificus* infection have been studied mostly for biogroup 1 and they are summarized below and in Table 12.1. Unlike for *V. cholerae* and *V. parahaemolyticus*, a virulence determinant has not been established for *V. vulnificus* and therefore it is not clear whether only a particular group of the strains are virulent. The host factor (underlying chronic diseases) appears to be the primary determinant for *V. vulnificus* infection although the following virulence factors of *V. vulnificus* have been studied.

The ability to acquire iron is considered essential for virulence expression of *V. vulnificus*. Individuals with elevated serum iron levels are at higher risk for developing a *V. vulnificus* infection (Hlady and Klontz, 1996). Wright *et al.* (1981) demonstrated that addition of ferric ammonium citrate to the inoculum reduced the $LD_{50}$ in mice of a clinical strain from $6 \times 10^6$ to $1 \times 10^0$ organisms. The mechanisms of iron acquisition including a siderophore and its receptor, a heme receptor, and their regulators have been studied at the molecular genetic level (Table 12.1).

An encapsulated phase variant that exhibits opaque colony morphology appears in the culture. Several workers reported that the capsule is associated with virulence in mice, serum resistance, hydrophobicity, and ability to utilize transferrin-bound iron (Yoshida *et al.*, 1985; Simpson *et al.*, 1987; Wright *et al.*, 1990).

Proteins presumably involved in type IV pilus biogenesis and type II secretion are associated with the expression of cell cytotoxicity, adherence, and mouse virulence of *V. vulnificus* (Paranjpye *et al.*, 1998).

The hemolysin encoded by the *vvh* gene and often called cytotoxin–hemolysin because of its cytotoxic and cytolytic activities has been analyzed as a possible virulence factor. A zinc metalloprotease has also been characterized as another possible virulence factor. However, neither the hemolysin nor the metalloprotease was shown to contribute to the virulence of *V. vulnificus* (Wright and Morris 1991; Shao and Hor 2000; Jeong *et al.*, 2000). The vllY-encoded hemolysin resembling the hemolysin (legiolysin) of *Legionella pneumophila*, was reported recently but its exact role in pathogenesis is not clear.

ToxR regulator enhances the expression of the cytotoxin–hemolysin gene (*vvhA*) fivefold in an *E. coli* background. The *smcR* gene, a homologue of the *luxR* gene of *V. harveyi*, is involved in expression control of the metalloprotease gene (*vvp*) and the cytotoxin–hemolysin gene (*vvhA*) but this regulator has no influence on the virulence in mice. Accordingly, these regulators do not seem to significantly contribute to control of virulence in *V. vulnificus*.

A recent study indicates that sequence polymorphism of the 16S rRNA gene may be a virulence marker for *V. vulnificus* as a statistically significant association between clinical strains and a particular 16S rRNA genotype was demonstrated (Nilsson *et al.*, 2003).

## Epidemiology

Primary septicemia, in which the intestine is the portal of bacterial entry into the bloodstream, is clearly the most important foodborne disease syndrome caused by *V. vulnificus* (Shapiro *et al.*, 1998; Strom and Paranjpye, 2000). It is responsible for nearly all seafood associated deaths in the US and the 61% fatality rate among primary septicemia patients in the US (Shapiro *et al.*, 1998) is the highest of any foodborne pathogen (Mead *et al.*, 1999). Similar fatality rates have also been reported in Asia (Yamauchi *et al.*, 1994; Chuang *et al.* 1992; Park *et al.* 1991). *V. vulnificus* can cause relatively mild gastroenteritis but these infections account for < 5% of reported *V. vulnificus* infections in the US (Shapiro *et al.*, 1998). Foodborne illness from *V. vulnificus* is characterized by sporadic cases and an outbreak has never been reported (Blake *et al.*, 1979; Shapiro *et al.*, 1998; Morris and Black, 1985). Nearly all primary septicemia cases are associated with preexisting chronic illnesses, especially liver disease or alcoholism (Shapiro *et al.*, 1998; Strom and Paranjpye, 2000). Infections are most frequently reported in men over 40 years old. In the USA there have been an average of 32 primary septicemia cases per year reported from 1995 through 2001 and nearly all traceable cases have been associated with consumption of raw oysters from the Gulf Coast (Shapiro *et al.*, 1998; Glatzer, 2001). Approximately 90% of the cases occur from April through November when water temperatures exceed 20°C (Shapiro *et al.*, 1998). A similar seasonal pattern has been reported in Japan (Yamauchi *et al.*, 1994), Korea (Park *et al.* 1991), and Taiwan (Chuang *et al.* 1992); patients typically reported consumption of raw or undercooked seafood within one week of the onset of symptoms but the species of seafood was generally not specified.

## Ecology

The seasonal distribution of *V. vulnificus* in seafood and the environment closely mirrors that of reported illnesses (Motes *et al.*, 1998; Cook *et al.*, 2002a). While *V. vulnificus* can be detected in US Gulf Coast oysters at harvest and at market during each month of the year, the abundance is low (< 10/g) until late March (Motes *et al.*, 1998). By late April densities at harvest usually exceed 1000/g and remain high until water temperatures cool to < 20°C in October or November. *V. vulnificus* multiplies rapidly in oysters if not refrigerated (Cook, 1994; Cook, 1997) and levels at market are often >1 log greater than at harvest (Cook *et al.*, 2002a). *V. vulnificus* is present in all oyster tissues and fluids examined including hemolymph (Tamplin and Capers, 1992) and coexist with high densities of phages lytic to *V. vulnificus* (DePaola *et al.*, 1997a). High levels of *V. vulnificus* are also found in estuarine sediments and fauna with the highest levels in the intestines of near shore bottom feeding fish (DePaola *et al.*, 1994). *V. vulnificus* has been detected outside the Gulf of Mexico from widespread locations under a variety of environmental conditions but these studies were not intended to examine association with foodborne illnesses (Oliver *et al.*, 1982; Oliver *et al.*, 1983; Kaysner *et al.*, 1987; Wright *et al.*, 1996; Höi *et al.*, 1998a; Veenstra *et al.*, 1994; Radu *et al.*, 2000).

## Detection

Early surveys and environmental studies of *V. vulnificus* utilized the TCBS agar designed to detect and differentiate *V. cholerae* and *V. parahaemolyticus* (Oliver *et al.*, 1983; Oliver *et al.*, 1982; Cook and Ruple, 1989; Kaysner *et al.*, 1987). However, TCBS agar inhibits many *V. vulnificus* strains and does not differentiate *V. vulnificus* from *V. parahaemolyticus*, complicating detection and especially enumeration (Höi *et al.*, 1998b). Cellobiose-polymyxin B-colistin agar was proposed as selective plating medium for *V. vulnificus* (Massad and Oliver, 1987) and it was subsequently modified (Elliot *et al.*, 1998; Höi *et al.*, 1998b). These have been extensively employed in ecological studies and seafood surveys (Cook and Ruple, 1992; Oliver *et al.*, 1992; Cook *et al.*, 2002a; Tamplin and Capers, 1992; Motes *et al.*, 1998).

Polymyxin B and colistin are also inhibitory to *V. vulnificus* and agar media containing these antibiotics are not generally recommended as a direct plating medium but rather for isolation of *V. vulnificus* after enrichment (Höi *et al.*, 1998b; Elliot *et al.*, 1998). Thus, an MPN approach is required for enumeration after enrichment, which is extremely resource intensive if biochemical utilization profiles are used for screening and identification. *V. vulnificus* identification was facilitated by the development of a species specific enzyme immunoassay using a monoclonal antibody (Tamplin *et al.*, 1991) Direct plating procedures are amenable for *V. vulnificus* enumeration because of the high densities found in many seafood and environmental samples. Miceli *et al.* (1993) reported a direct plating procedure using VVE, a selective and differential medium, which incorporates the chromogenic lactose analog (X-gal), followed by isolation and biochemical testing of selected representative colonies.

Various workers have reported molecular genetic methods for identification or detection of *V. vulnificus*. These included the methods for detection of the species-specific *vvhA* gene by using a 3.2-kb isotopically labeled DNA probe (Wright *et al.*, 1985), an alkaline phosphatase-labeled oligonucleotide probe (Wright *et al.*, 1993), and PCR protocols (Hill *et al.*, 1991; Aono *et al.*, 1997; Lee *et al.*, 1998b; Coleman *et al.*, 1996), those for detection of the 16S rRNA gene by using oligonucleotide probes (Aznar *et al.* 1994; Cerda-Cuellar *et al.*, 2000), and that for detection of the 23S rRNA gene by a nested PCR method (Arias *et al.*, 1995).

These methods do not differentiate virulent and avirulent strains and can be used in place of biochemical identification or incorporated into MPN protocols for enumeration. The alkaline phosphatase-labeled oligonucleotide probe was superior to a commercial biochemical identification kit in identifying *V. vulnificus* when environmental strains were examined (Dalsgaard *et al.*, 1996). Direct plating on a nonselective medium followed by colony hybridization with the alkaline phosphatase-labeled oligonucleotide probe eliminated the need to identify colonies individually (Wright *et al.*, 1993; Wright *et al.*, 1996). This direct plating approach provided similar recoveries but was more rapid, efficient and precise than the standard MPN method for enumerating *V. vulnificus* in oysters at harvest (DePaola *et al.*, 1997b). However, in market oysters (1–14 days after harvest) direct plating may be less effective due to elevated background flora and possible cold injury to *V. vulnificus* (Cook *et al.*, 2002a). Seafoods other than molluscan shellfish generally require a 1:10 mixture in a diluent to facilitate homogenization and pipetting and this reduces sensitivity to 100/g when a 0.1-mL portion is plated.

## Control

Foodborne *V. vulnificus* illnesses in general can be controlled in most seafoods as previously described for *V. cholerae* and *V. parahaemolyticus*. Since nearly all *V. vulnificus* infections in the US are associated with consumption of raw oysters, a variety of harvest and postharvest treatments to reduce *V. vulnificus* levels have been investigated (Cook and Ruple, 1992; Sun and Oliver, 1994; Ama *et al.*, 1994; Berlin *et al.*, 1999; Parker *et al.*, 1994; Motes and DePaola, 1996). Prompt refrigeration of oysters and other molluscan shellfish prevents postharvest *V. vulnificus* multiplication (Cook, 1997). Mild pasteurization at 50°C for 10 min reduced *V. vulnificus* 6 logs to nondetectable levels (< 3/g) (Cook and Ruple, 1992). Subsequently, this approach was patented and oysters marketed with the label "processed to reduce *V. vulnificus* to nondetectable levels" as described by the National Shellfish Sanitation Program (USA Department of Health and Human Services.Public Health Services.FDA, 1999).

Freezing oysters followed by frozen storage was also shown to reduce *V. vulnificus* by 4 to 5 logs but the natural populations could still be cultured at detectable levels even after 70 days of frozen storage (Cook and Ruple, 1992; Parker *et al.*, 1994). Depuration of oysters is problematic as *V. vulnificus* readily grows in oyster tissues (Tamplin and Capers, 1992) but relaying oysters offshore to high salinity water was shown to usually reduce natural *V. vulnificus* populations (levels at harvest) to nondetectable levels in approximately 2 weeks (Motes and DePaola, 1996). The addition of chemical agents to depuration tanks was investigated and diacetyl was of limited value reducing *V. vulnificus* densities 1 to 2 logs after a 24-h depuration period (Sun and Oliver, 1994). *V. vulnificus* numbers were readily reduced 6 logs in both fish and oyster homogenates with less than 1.0 kGy dose of irradiation and raising the temperature to 40°C increased the lethality of irradiation (Ama *et al.*, 1994).

High hydrostatic pressure (200 MPa for 10 min at 25°C) was reported to reduce both *V. parahaemolyticus* and *V. vulnificus* in inoculated oyster homogenates by approximately 6 logs (Berlin *et al.*, 1999). Some of these postharvest treatments such as freezing and irradiation also may be applicable to other raw seafood products. Special care should be practiced during evisceration of finfish intended for raw consumption to prevent contamination of the edible flesh with the intestinal contents as extremely high *V. vulnificus* levels are typical of near shore bottom feeding fish (DePaola *et al.*, 1994).

## Other vibrios of potential importance

### Vibrio mimicus

*V. minicus* shares many physiologic, phenotypic, and pathogenic characteristics in common with *V. cholerae* non-O1/O139. Both are distributed in freshwater and estuarine environments and they share most biochemical characteristics and some of O antigens except that *V. minicus* does not ferment sucrose. *V. mimicus* was isolated from the patients with diarrhea and much less frequently from extraintestinal infections (Hlady and Klontz, 1996). *V. mimicus* infection appears to be less common than the infection by *V. cholerae* non-O1/O139 (Morris and Black, 1985). Outbreaks of seafoodborne diarrhea were reported but diarrhea cases are usually sporadic (Oliver and Kaper, 2001; Campos *et al.*, 1996).

As it has been reported for *V. cholerae* non-O1/O139, some clinical strains of *V. mimicus* were shown to have the ability to produce virulence factors that are major virulence factors of other enteric pathogens (Table 12.1). The strains carrying two virulence-associated lysogenic phages of *V. cholerae* were found: CTXΦphage containing the CTX genetic element and VPIΦencoding TCP. However, atypical strains lacking the *tcpA* gene were reported among clinical strains although they had *tcpP*, *toxT*, *ctx*, and *toxRS* genes. The *zot* and *ace* genes were detected in *ctx*-positive and -negative strains. Some strains carried the heat-stable enterotoxin gene similar to that of enterotoxigenic *E. coli* or the gene resembling the *tdh* gene of *V. parahaemolyticus*. Most environmental strains lack the ability to produce these toxins and, as hypothesized for *V. cholerae* non-O1/O139, the exceptional strains of *V. mimicus* is likely to have acquired the above toxin genes by horizontal transfer mechanisms. A hemolysin, a phospholipase, and a metalloprotease that may play auxiliary roles in pathogenesis have been studied at the genetic level (Table 12.1).

### Vibrio hollisae

Thirty cases of *V. hollisae* infection were summarized in a review published in 1994 (Abbott and Janda, 1994). Most illness (87%) were gastroenteritis among the otherwise healthy individuals who consumed raw oysters, clams, or shrimp and the rest were septicemia or wound infection. Only two publications reported isolation of *V. hollisae* from environmental samples (fish intestine and cultivated oyster beds) (Nishibuchi *et al.*, 1988; Kelley and Stroh, 1988).

The paucity of *V. hollisae* isolation from both clinical and environmental samples is likely due to inability of this organism to grow or poor growth on selective isolation media for enteric pathogens including TCBS agar and MacConkey agar. In addition, identification by standard biochemical tests is difficult because this organism is biochemically inert and addition of NaCl to test media at a final concentration of 1% is recommended (Janda *et al.*, 1988). PCR methods targeting the *toxR* and *gyrB* genes can be employed as alternative identification tools (Vuddhakul *et al.*, 2000b).

All tested strains of *V. hollisae* including an environmental strain carried the *tdh* gene encoding a hemolysin similar to TDH of *V. parahaemolyticus* (Nishibuchi *et al.* 1996) and thus this seems to be an important virulence factor. Other possible virulence factors that were characterized at molecular levels include a siderophore and a CHO cell elongation factor (Table 12.1).

### Vibrio fluvialis

*V. fluvialis* is similar to *Aeromonas hydrophila* in that arginine dihydrolase is positive and that lysine and ornithine decarboxylases are negative but other important biochemical characteristics of *V. fluvialis* resemble those of the genus *Vibrio*. *V. fluvialis* was called "group F" or "group EF-6" before its taxonomic position was established in 1981. This organism was isolated from more than 500 patients with diarrhea in Bangladesh during a 9-month period (Huq *et al.*, 1980) and from sporadic cases of gastroenteritis in US and other countries (Tacket *et al.*, 1982; Magalhaes *et al.*, 1990; Desenclos *et al.*, 1991; Levine *et al.*, 1993). *V. fluvialis* is widely distributed in aquatic environments (West *et al.*, 1986). Surveillance studies suggested that consumption of raw oyster and other seafood is an important cause of *V. fluvialis*-associated gastroenteritis (Desenclos *et al.*, 1991; Levine *et al.*, 1993). Therefore, *V. fluvilais* is considered a potential pathogen in seafood. *V. fluvialis* was also implicated in a bacteremia case and a wound infection (Albert *et al.*, 1991; Varghese *et al.*, 1996).

Several workers demonstrated that *V. fluvialis* produces various intracellular enzymes, siderophores, a hemolysin, and factors that induce fluid accumulation in rabbit ileal loops and in the intestine of suckling mice or affect CHO cells. Of these possible virulence factors only a CHO cell elongation factor, a CHO cell-killing factor, and siderophores were studied at molecular levels (Table 12.1).

### Vibrio furnissii

Like *V. fluvialis*, *V. furnissii* is positive for the arginine dihydrolase test and negative in lysine and ornithine decarboxylase tests. *V. furnissii* is unique among the genus *Vibrio* in that it produces gas from glucose. It was formally classified as an aerogenic biogroup of *V. fluvialis* and was established as a new species in 1983. This organism was isolated from the patients with gastroenteritis in different parts of the world on rare occasions and from environmental samples (Brenner *et al.*, 1983; Hickman-Brenner *et al.*, 1984; West *et al.*, 1986; Magalhaes *et al.*, 1990; Dalsgaard *et al.*, 1997). The bacterial cultures but not the culture filtrates of *V. furnissii* (aerogenic *V. fluvialis*) strains experimentally caused diarrhea and mortality in suckling mice (Nishibuchi *et al.*, 1983; Chikahira and Hamada, 1988). Bacterial cultures of 50% of tested strains induced fluid accumulation in rabbit ileal loops and a rare strain produced a CHO cell elongation factor (Chikahira and Hamada, 1988). However, strong evidence to support that *V. furnissii* plays a pathogenic role in gastroenteritis is currently unavailable.

## References

Abbott, S.L. and Janda, J.M. 1994. Severe gastroenteritis associated with *Vibrio hollisae* infection: Report of two cases and review. Clin. Infect. Dis. 18: 310–312.

Aiso, K. and Fujiwara, K. 1963. Feeding tests of "pathogenic halophilic bacteria". Ann. Rep. Inst. Food Microbiol. Chiba Univ. 15: 34–38.

Alam, M., Miyoshi, S.I., Tomochika, K.I., and Shinoda, S. 1996. Purification and characterization of novel hemagglutinins from *Vibrio mimicus*: a 39-kilodalton major outer membrane protein and lipopolysaccharide. Infect. Immun. 64: 4035–4041.

Alam, M.J., Tomochika, K., Miyoshi, S. and Shinoda, S. 2002. Environmental investigation of potentially pathogenic *Vibrio parahaemolyticus* in the Seto-Inland Sea, Japan. FEMS Microbiol. Lett. 208: 83–87.

Albert, M.J., Hossain, M.A., Alam, K., Kabir, I., Neogi, P.K., and Tzipori, S. 1991. A fatal case associated with shigellosis and *Vibrio fluvialis* bacteremia. Diagn. Microbiol. Infect. Dis. 14: 509–510.

Altekruse, S.F., Bishop, R.D., Baldy, L.M., Thompson, S.G., Wilson, S.A., Ray, B.J., and Griffin, P.M. 2000. *Vibrio* gastroenteritis in the US Gulf of Mexico region: the role of raw oysters. Epidemiol. Infect. 124: 489–495.

Ama, A.A., Hamdy, M.K., and Toledo, R.T. 1994. Effects of heating, pH and thermoradiation on inactivation of *Vibrio vulnificus*. Food Microbiol. 11: 215–227.

Anonymous. 2001. Draft Risk Assessment on the Public Health Impacts of *Vibrio parahaemolyticus* in Raw Molluscan Shellfish. USA Food and Drug Administration Washington, DC.

Aono, E., Sugita, H., Kawasaki, J., Sakakibara, H., Takahashi, T., Endo, K., and Deguchi, Y. 1997. Evaluation of the polymerase chain reaction method for identification of *Vibrio vulnificus* isolated from marine environments. J. Food Prot. 60: 81–83.

Arias, C.R., Garay, E., and Aznar, R. 1995. Nested PCR method for rapid and sensitive detection of *Vibrio vulnificus* in fish, sediments, and water. Appl. Environ. Microbiol. 61: 3476–3478.

Aznar, R., Ludwig, W., Amann, R.I., and Schleifer, K.H. 1994. Sequence determination of rRNA genes of pathogenic *Vibrio* species and whole-cell identification of *Vibrio vulnificus* with rRNA-targeted oligonucleotide probes. Int. J. Syst. Bacteriol. 44: 330–337.

Baba, K., Shirai, H., Terai, A., Kumagai, K., Takeda, Y., and Nishibuchi, M. 1991. Similarity of the *tdh* gene-bearing plasmids of *Vibrio cholerae* non-01 and *Vibrio parahaemolyticus*. Microb. Pathog. 10: 61–70.

Barker, W.H., Jr. 1974. *Vibrio parahaemolyticus* outbreaks in the United States. In: International Symposium on *Vibrio parahaemolyticus*. T. Fujino, G. Sakaguchi, R. Sakazaki, and Y. Takeda, ed. Saikon Publishing Co., Tokyo, pp. 47–52.

Baudry, B., Fasano, A., Ketley, J., and J.B. Kaper. 1992. Cloning of a gene (*zot*) encoding a new toxin produced by *Vibrio cholerae*. Infect. Immun. 60: 428–434.

Behari, J., Stagon, L., and Calderwood, S.B. 2001. *pepA*, a gene mediating pH regulation of virulence genes in *Vibrio cholerae*. J. Bacteriol. 183: 178–188.

Bej, A.K., Patterson, D.P., Brasher, C.W., Vickery, M.C., Jones, D.D., and Kaysner, C.A. 1999. Detection of total and hemolysin-producing *Vibrio parahaemolyticus* in shellfish using multiplex PCR amplification of *tlh*, *tdh*, and *trh*. J. Microbiol. Methods 36: 215–225.

Berlin, D.L., Herson, D.S., Hicks, D.T., and Hoover, D.G. 1999. Response of pathogenic *Vibrio* species to high hydrostatic pressure. Appl. Environ. Microbiol. 65: 2776–2780.

Bi, K., Miyoshi, S.I., Tomochika, K.I., and Shinoda, S. 2001. Detection of virulence associated genes in clinical strains of *Vibrio mimicus*. Microbiol. Immunol. 45: 613–616.

Bik, E.M., Bunschoten, A.E., Willems, R.J.L., Chang, A.C.Y., and Mooi, F.R. 1996. Genetic organization and functional analysis of the *otn* DNA essential for cell-wall polysaccharide synthesis in *Vibrio cholerae* O139. Mol. Microbiol. 20: 799–811.

Bisharat, N., Agmon, V., Finkelstein, R., Raz, R., Ben-Dror, G., Lerner, L., Soboh, S., Colodner, R., Cameron, D.N., Wykstra, D.L., Swerdlow, D.L., and Farmer., J. J., III. 1999. Clinical, epidemiological, and microbiological features of *Vibrio vulnificus* biogroup 3 causing outbreaks of wound infection and bacteraemia in Israel. Lancet 354: 1421–1424.

Blackstone, G.M., Nordstrom, J.L., Vickery, M.C.L., Bowen, M.D., Meyer, R.F., and DePaola, A. 2003. Detection of pathogenic *Vibrio parahaemolyticus* in oyster enrichments by real time PCR. J. Microbiol. Meth. 53:149–155.

Blake, P.A. 1994a. Historical perspectives on pandemic cholera. In: *Vibrio cholerae* and Cholera: Molecular to Global Perspectives. D.C. Wachsmuth, I.K., Blake, P.A. Olsvik, Ö., ed. ASM Press, Washington, D.C. p. 293–295.

Blake, P.A. 1994b. Endemic cholera in Australia and the United States. In: *Vibrio cholerae* and Cholera: Molecular to Global Perspectives. D.C. Wachsmuth, I.K., Blake, P.A. Olsvik, Ö., ed. ASM Press, Washington, D.C. p. 309–319.

Blake, P.A., Merson, M.H., Weaver, R.E., Hollis, D.G., and Heublein, P.C. 1979. Disease caused by a marine *Vibrio*: clinical characteristics and epidemiology. New Engl. J. Medicin. 300: 1–5.

Bock, T., Christensen, N., Eriksen, H.N., Winter, S., Rygaard, H., and Jorgensen, F. 1994. The first fatal case of *Vibrio vulnificus* infection in Denmark. APMIS 102: 874–876.

Boyd, E.F., Moyer, K.E., Shi, L., and Waldor, M.K. 2000. Infectious CTXΦ and the *Vibrio* pathogenicity island prophage in *Vibrio mimicus*: evidence for recent horizontal transfer between *V. mimicus* and *V. cholerae*. Infect. Immun. 68: 1507–1513.

Brenner, D.J., Hickman-Brenner, F.W., Lee, J.V., Steigerwalt, A.G., Fanning, G.R., Hollis, D.G., Farmer, J.J., III, Weaver, R.E., Joseph, S.W., and Seidler, R.J. 1983. *Vibrio furnissii* (formerly aerogenic biogroup of *Vibrio fluvialis*), a new species isolated from human feces and the environment. J. Clin. Microbiol. 18: 816–824.

Brown, M.H., and Manning, P.A. 1985. Haemolysin genes of *Vibrio cholerae*: presence of homologous DNA in non-haemolytic O1 and haemolytic non-O1 strains. FEMS Microbiol. Lett. 30: 197–201.

Butterton, J.R., Choi, M.H., Watnick, P.I., Carroll, P.A., and Calderwood, S.B. 2000. *Vibrio cholerae* VibF is required for vibriobactin synthesis and is a member of the family of nonribosomal peptide synthetases. J. Bacteriol. 182: 1731–1738.

Butterton, J.R., Stoebner, J.A., Payne, S.M., and Calderwood, S.B. 1992. Cloning, sequencing, and transcriptional regulation of viuA, the gene encoding the ferric vibriobactin receptor of *Vibrio cholerae*. J. Bacteriol. 174: 3729–3738.

Campos, E., Bolanos, H., Acuna, M.T., Diaz, G., Matamoros, M.C., Raventos, H., Sanchez, L.M., Sanchez, O., and Barquero, C. 1996. *Vibrio mimicus* diarrhea following ingestion of raw turtle eggs. Appl. Environ. Microbiol. 62: 1141–1144.

Centers for Disease Control and Prevention. 1998a. Laboratory Methods for the Diagnosis of *Vibrio cholerae*. Centers for Disease Control, Atlanta, GA.

Centers for Disease Control and Prevention. 1998b. Outbreak of *Vibrio parahaemolyticus* infections associated with eating raw oysters – Pacific Northwest, 1997. MMWR 47: 457–462.

Cerda-Cuellar, M., Jofre, J., and Blanch, A.R. 2000. A selective medium and a specific probe for detection of *Vibrio vulnificus*. Appl. Envrion. Microbiol. 66: 855–859.

Chakrabarti, S., Sengupta, N., and Chowdhury, R. 1999. Role of DnaK in *in vitro* and *in vivo* expression of virulence factors of *Vibrio cholerae*. Infect. Immun. 67: 1025–1033.

Chang, T.M., Chuang, Y.C., Su, J.H., and Chang, M.C. 1997. Cloning and sequence analysis of a novel hemolysin gene (*vllY*) from *Vibrio vulnificus*. Appl. Environ. Microbiol. 63: 3851–3853.

Cheng, J.-C., Shao, C.-P., and Hor, L.-I. 1996. Cloning and nucleotide sequencing of the protease gene of *Vibrio vulnificus*. Gene. 183: 255–257.

Chikahira, M., and Hamada, K. 1988. Enterotoxigenic substance and other toxins produced by *Vibrio fluvialis* and *Vibrio furnissii*. Nippon Juigaku Zasshi (Jpn. J. Vet. Sci.) 50: 865–873.

Chiou, C.S., Hsu, S.Y., Chiu, S.I., Wang, T.K., and Chao, C.S. 2000. *Vibrio parahaemolyticus* serovar O3:K6 as cause of unusually high incidence of foodborne disease outbreaks in Taiwan from 1996 to 1999. J.Clin.Microbiol. 38: 4621–4625.

Chowdhury, M.A., Hill, R.T., and Colwell, R.R. 1994. A gene for the enterotoxin zonula occludens toxin is present in *Vibrio mimicus* and *Vibrio cholerae* O139. FEMS Microbiol. Lett. 119: 377–380.

Chuang, Y.C., Chang, T.M., and Chang, M.C. 1997. Cloning and characterization of the gene (*empV*) encoding extracellular metalloprotease from *Vibrio vulnificus*. Gene 189: 163–168.

Chuang, Y.C., Yuan, C.Y., Liu, C.Y., Lan, C.K., and Huan, A.H. 1992. *Vibrio vulnificus* infection in Taiwan: report of 28 cases and review of clinical manifestations and treatment. Clin. Infect. Dis. 15: 271–276.

Chun J., Huq, A., and Colwell, R. R. 1999. Analysis of 16S–23S rRNA intergenic spacer regions of *Vibrio cholerae* and *Vibrio mimicus*. Appl. Environ. Microbiol. 65: 2202–2208.

Coleman, S.S., Melanson, D.M., Biosca, E.G., and Oliver, J.D. 1996. Detection of *Vibrio vulnificus* biotypes 1 and 2 in eels and oysters by PCR amplification. Appl. Environ. Microbiol. 62: 1378–1382.

Colwell, R.R. 1970. Polyphasic taxonomy of the genus *Vibrio*: numerical taxonomy of *Vibrio cholerae*, *Vibrio parahaemolyticus*, and related *Vibrio* species. J. Bacteriol. 104: 410–433.

Colwell, R.R., and Hug, A. 1994. Vibrios in the environment: viable but nonculturable *Vibrio cholerae*. In: *Vibrio cholerae* and Cholera: Molecular to Global Perspectives. I.K. Wachsmuth, P.A.Blake, Ö. Olsvik, ed. ASM Press, Washington, D,C, pp. 117–125.

Comstock, L.E., Johnson, J.A., Michalski, J.M., Morris, J.G., Jr., and Kaper, J.B. 1996. Cloning and sequence of a region encoding a surface polysaccharide of *Vibrio cholerae* O139 and characterization of the insertion site in the chromosome of *Vibrio cholerae* O1. Mol. Microbiol. 19: 815–826.

Cook, D.W. 1994. Effect of time and temperature on multiplication of *Vibrio vulnificus* in postharvest Gulf coast shellstock oysters. Appl. Environ. Microbiol. 60: 3483–3484.

Cook, D.W. 1997. Refrigeration of oyster shellstock: conditions which minimize the outgrowth of *Vibrio vulnificus*. J. Food Prot. 60: 349–352.

Cook, D.W., Bowers, J.C., and DePaola, A. 2002b. Density of total and pathogenic (*tdh*+) *Vibrio parahaemolyticus* in Atlantic and Gulf Coast molluscan shellfish at harvest. J.Food Prot. 65: 1873–1880.

Cook, D.W., O'Leary, P., Hunsucker, J.C., Sloan, E.M., Bowers, J.C., Blodgett, R.J., and DePaola, A. 2002a. *Vibrio vulnificus* and *Vibrio parahaemolyticus* in USA retail shell oysters: A national survey June 1998 to July 1999. J. Food Prot. 65: 79–87.

Cook, D.W., and Ruple, A.D. 1989. Indicator bacteria and *Vibrionaceae* multiplication in post-harvest shellstock oysters. J. Food Prot. 52: 343–349.

Cook, D.W., and Ruple, A.D. 1992. Cold storage and mild heat treatment as processing aids to reduce the numbers of *Vibrio vulnificus* in raw oysters. J. Food. Prot. 55: 985–989.

Cotter, P.A., and DiRita, V.J. 2000. Bacterial virulence gene regulation: an evolutionary perspective. Ann. Rev. Microbiol. 54: 519–565.

Croci, L., Suffredini, E., Cozzi L., and Toti, L. 2002. Effects of depuration of molluscs experimentally contaminated with *Escherichia coli*, *Vibrio cholerae* O1, and *Vibrio parahaemolyticus*. J. Appl. Microbiol. 92: 460–465.

Dalsgaard, A., Dalsgaard, I., Hoi, L., and Larsen J.L. 1996. Comparison of a commecial biochemical kit and an oligonucleotide probe for identification of environmental isolates of *Vibrio vulnificus*. Lett. Appl. Microbiol. 22: 184–188.

Dalsgaard, A., Glerup, P., Hoybye, L.L., Paarup, A.M., Meza, R., Bernal, M., Shimada, T., and Taylor, D.N. 1997. *Vibrio furnissii* isolated from humans in Peru: a possible human pathogen? Epidemiol. Infect. 119: 143–149.

Daniels, N., MacKinnon, L., Bishop, R., Altekruse, S., Ray, B., Hammond, R., Thompson, S., Wilson, S., Bean, N., Griffin, P., and Slutsker, L. 2000a. *Vibrio parahaemolyticus* infections in the United States, 1973–1998. J. Infect. Dis. 181: 1661–1666.

Daniels, N.A., Ray, B., Easton, A., Marano, N., Kahn, E., McShan, A.L., Del Rosario, L., Baldwin, T., Kingsley, M.A., Puhr, N.D., Wells, J.G.,

and Angulo, F.J. 2000b. Emergence of a new *Vibrio parahaemolyticus* serotype in raw oysters. JAMA 284: 1541–1545.

DePaola, A., Capers, G.M., and Alexander, D. 1994. Densities of *Vibrio vulnificus* in the intestines of fish from the USA Gulf Coast. Appl. Environ. Microbiol. 60: 984–988.

DePaola, A., Hopkins, L.H., and McPhearson, R.M. 1988a. Evaluation of four methods for enumeration of *Vibrio parahaemolyticus*. Appl. Environ. Microbiol. 54: 617–618.

DePaola, A., Hopkins, L.H., Peeler, J.T., Wentz, B., and McPhearson, R.M. 1990. Incidence of *Vibrio parahaemolyticus* in USA coastal waters and oysters. Appl. Environ. Microbiol. 56: 2299–2302.

DePaola, A., and Hwang. G.C. 1995. Effect of dilution, incubation time, and temperature of enrichment on cultural and PCR detection of *Vibrio cholerae* obtained from the oyster *Crassostrea virginica*. Mol. Cell. Probes 9: 75–81.

DePaola, A., Kaysner, C.A., Bowers, J.C., and Cook, D.W. 2000. Environmental investigations of *Vibrio parahaemolyticus* in oysters following outbreaks in Washington, Texas, and New York (1997, 1998). Appl. Environ. Microbiol. 66: 4649–4654.

DePaola, A., Kaysner, C.A., and McPhearson, R.M. 1987. Elevated temperature method for recovery of *Vibrio cholerae* from oysters (*Crassostrea gigas*). Appl. Environ. Microbiol. 53: 1181–1182.

DePaola, A., Mcleroy, S., and Mcmanus, G. 1997a. Distribution of *Vibrio vulnificus* phage in oyster tissue and other eustarine habitats. Appl. Environ. Microbiol. 63: 2464–2467.

DePaola, A., Motes, M.L., Cook, D.W., Veazey, J., Garthright, W.E., and Blodgett, R. 1997b. Evaluation of alkaline phosphatase-labeled DNA probe for enumeration of *Vibrio vulnificus* in Gulf Coast oysters. J. Microbiol. Meth. 29: 115–120.

DePaola, A., Motes, M.L., and McPhearson, R.M. 1988b. Comparison of APHA and elevated temperature enrichment methods for recovery of *Vibrio cholerae* from oysters. J. Assoc. Off. Anal. Chem. 71: 584–589.

DePaola, A., Nordstrom, J.L., Bowers, J.C., Wells, J.G., and Cook, D.W. 2003. Seasonal abundance of total and pathogenic *Vibrio parahaemolyticus* in Alabama oysters. Appl. Enviro. Microbiol. 69: 1521–1526.

DePaola, A., Presnell, M.W., Becker, R.E., Motes, M.L.Jr., Zywno, S.R., Musselman, J.F., Taylor, J., and Williams, L. 1984. Distribution of *Vibrio cholerae* in the Apalachicola (Florida) bay estuary. J. Food. Prot. 47: 549–553.

DePaola, A., Presnell, M.W., Motes, M.L. Jr., McPhearson, M.R., Twedt, R.M., Becker, R.E., and Zywno, S. 1983. Non-O1 *Vibrio cholerae* in shellfish, sediment and waters of the USA Gulf coast. J. Food Prot. 46: 802–806.

DePaola, A., Rivadeneyra, C., Gelli, D.S., Zuazua, H., and Grahn, M. 1992. Peruvian cholera epidemic: Role of seafood. In: Proceedings of the 16th Annual Tropical and Subtropical Fisheries Technological Conference of the Americas. S.W. Otwell, ed. University of Florida. Gainesville, FL, pp. 28–32.

Desenclos, J.C., and Klontz, K.C., Wolfe, L.E., and Hoecherl, S. 1991. The risk of *Vibrio* illness in the Florida raw oyster eating population, 1981–1988. Am. J. Epidemiol. 134: 290–297.

DiRita, V.J. 1992. Co-ordinate expression of virulence genes by ToxR in *Vibrio cholerae*. Mol. Microbiol. 6: 451–458.

Dorsch, M., Lane, D., and Stackebrandt, E. 1992. Towards a phylogeny of the genus *Vibrio* based on 16S rRNA sequences. Int. J. Syst. Bacteriol. 42: 58–63.

Dutta, N.K., Panse, M.V., and Jhala, H.I. 1963. Choleragenic property of certain strains of El Tor, nonagglutinable, and water vibrios confirmed experimentally. Br. Med. J. 1: 1200–1203.

Earle, P.M. and, Crisley, F.D. 1975. Isolation and characterization of *Vibrio parahaemolyticus* from Cape Cod soft-shell clams (*Mya arenaria*). Appl. Microbiol. 29: 635–640.

Elliot, E.L., Kaysner, C.A., Jackson, L., and Tamplin, M.L. 1998. *Vibrio cholerae*, *V. parahaemolyticus*, *V. vulnificus*, and other *Vibrio* spp. In: USA Food and Drug Administration Bacteriological Analytical Manual, 8th edn. Anonymous. A.O.A.C International, Gaithersburg, MD, pp. 9.01–9.27.

Ellison, R.K., Malnati, E., DePaola, A., Bowers, J.C., and Rodrick, G.E. 2001. Populations of *Vibrio parahaemolyticus* in retail oysters from Florida using two methods. J. Food Prot. 64: 682–686.

Faruque, S.M., Albert, M.J., and Mekalanos, J.J. 1998. Epidemiology, genetics, and ecology of toxigenic *Vibrio cholerae*. Microbiol. Mol. Biol. Rev. 62: 1301–1314.

Fields, P.I., Popovic, T., Wachsmuth, K., and Olsik, O. 1992. Use of the polymerase chain reaction for detection of toxigenic *Vibrio cholerae* O1 strains from the Latin American cholera epidemic. J. Clin. Microbiol. 30: 2118–2121.

Food and Drug Administration. 1984. Bacteriological Analytical Manual, 6th ed. Association of Official Analytical Chemists, Arlington, Va.

Franzon, V.L., Barker, A., and Manning, P.A. 1993. Nucleotide sequence encoding the mannose-fucose-resistant hemagglutinin of *Vibrio cholerae* O1 and construction of a mutant. Infect. Immun. 61: 3032–3037.

Fujino, T., Okuno, Y., Nakada, D., Aoyama, A., Fukai, K., Mukai, T., and Ueho, T. 1953. On the bacteriological examination of shirasu-food poisoning. Med. J. Osaka Univ. 4: 299–304.

Funahashi, T., Moriya, K., Uemura, S., Miyoshi, S., Shinoda, S., Narimatsu, S., and Yamamoto, S. 2002. Identification and characterization of pvuA, a gene encoding the ferric vibrioferrin receptor protein in *Vibrio parahaemolyticus*. J. Bacteriol. 184: 936–946.

Galen, J.E., Ketley, J.M., Fasano, A.S., Richardson, H., Wasserman, S.S., and Kaper, J. B. 1992. Role of *Vibrio cholerae* neuraminidase in the function of cholera toxin. Infect. Immun. 60: 406–415.

Garcia Cuevas, M, Collazos Gonzalez, J., Martinez Gutierrez, E., and Mayo Suarez, J. 1998. *Vibrio vulnificus* septicemia in Spain. An. Med. Interna. 15: 485–486. (In Spanish.)

Ghosh, C., Nandy, R. K., Dasgupta, S. K., Nair, G. B., Hall, R. H., and Ghose, A. C. 1997. A search for cholera toxin (CT), toxin coregulated pilus (TCP), the regulatory element ToxR and other virulence factors in non-O1/non-O139 *Vibrio cholerae*. Microb. Pathog. 22: 199–208.

Glatzer, M. 2001. *Vibrio vulnificus* shellfish cases file from 1989 to 2000. USA Food and Drug Administration.

Goldberg, M.B., Boyko, S.A., and Calderwood, S.B. 1991. Positive transcriptional regulation of an iron-regulated virulence gene in *Vibrio cholerae*. Proc. Natl. Acad. Sci. USA 88: 1125–1129.

Goldberg, M.B., and DiRita, V.J., and Calderwood, S.B. 1990. Identification of an iron-regulated virulence determinant in *Vibrio cholerae*, using TnphoA mutagenesis. Infect. Immun. 58: 55–60.

Gonzalez-Escalona, N., Cachicas, V., Acevedo, C., Rioseco, M.L., Vergara, J.A., Cabello, F., Romero, J., and Espejo, R.T. 2005. *Vibrio parahaemolyticus* diarrhea, Chile, 1998 and 2004. Emerg. Infect. Dis. 11: 129–131.

Gooch, J.A., DePaola, A., Bowers, J.C., and Marshall, D.L. 2002. Growth and survival of *Vibrio parahaemolyticus* in postharvest American oysters. J. Food Prot. 65: 970–974.

Gooch, J.A., DePaola, A., Kaysner, C.A., and Marshall, D.L. 2001. Evaluation of two direct plating methods using nonradioactive probes for enumeration of *Vibrio parahaemolyticus* in oysters. Appl. Environ. Microbiol. 67: 721–724.

Guglielmetti, P., Bravo, L., Zanchi, A., Monte, R., Lombardi, G., and Rossolini, G.M. 1994. Detection of the *Vibrio cholerae* heat-stable enterotoxin gene by polymerase chain reaction. Mol. Cell. Probes. 8: 39–44.

Guhathakurta, B., Sasmal, D., Pal, S., Chakraborty, S., Nair, G.B., and Datta, A. 1999. Comparative analysis of cytotoxin, hemolysin, hemagglutinin and exocellular enzymes among clinical and environmental isolates of *Vibrio cholerae* O139 and non-O1, non-O139. FEMS Microbiol. Lett. 179:401–407.

Hara-Kudo, Y., Nishina, T., Nakagawa, H. Konuma, H., Hasegawa, J., and Kumagai, S. 2001. Improved method for detection of *Vibrio parahaemolyticus* in seafood. Appl. Environ. Microbiol. 67: 5819–5823.

Hase, C.C., and Finkelstein, R.A. 1991. Cloning and nucleotide sequence of the *Vibrio cholerae* hemagglutinin/protease (HA/protease) gene and construction of an HA/protease-negative strain. J. Bacteriol. 173: 3311–3317.

Henderson, D.P., and Payne, S.M. 1994. Characterization of the *Vibrio cholerae* outer membrane heme transport protein HutA: sequence of the gene, regulation of expression, and homology to the family of TonB-dependent proteins. J. Bacteriol. 176: 3269–3277.

Hickman-Brenner, F.W., Brenner, D.J., Steigerwalt, A.G., Schreiber, M., Holmberg, S.D., Baldy, L.M., Lewis, C.S., Pickens, N.M., and Farmer, J.J., III. 1984. *Vibrio fluvialis* and *Vibrio furnissii* isolated from a stool sample of one patient. J. Clin. Microbiol. 20: 125–127.

Hill, W.E., Keasler, S.P., Trucksess, M.W., Feng, P., Kaysner, C. A., and Lampel, K. A. 1991. Polymerase chain reaction identification of *Vibrio vulnificus* in artificially contaminated oysters. Appl. Environ. Microbiol. 57: 707–711.

Hlady W.G., and Klontz, K.C. 1996. The epidemiology of *Vibrio* infection in Florida, 1981–1993. J. Infect. Dis. 173: 1176–1183.

Hoge, C.W., Sethabutr, O., Bodhidatta, L., Echeverria, P., Robertson, D.C., Morris, J.G., Jr. 1990. Use of a synthetic oligonucleotide probe to detect strains of non-serovar O1 *Vibrio cholerae* carrying the gene for heat-stable enterotoxin (NAG-ST). J. Clin. Microbiol. 28: 1473–1476.

Höi, L., Dalsgaard, I., and Dalsgaard, A. 1998b. Improved isolation of *Vibrio vulnificus* from seawater and sediment with cellobiose-colistin agar. Appl. Environ. Microbiol. 64: 1721–1724.

Höi, L., Larsen, J.L., Dalsgaard, I., and Dalsgaard, A. 1998a. Occurrence of *Vibrio vulnificus* biotypes in Danish marine environments. Appl. Environ. Microbiol. 64: 7–13.

Holt, J.G., Krieg, N.R., Sneath, P.H.A., Staley, J.T., and Williams, S.T. 1994. Bergey's Manual of Determinative Bacteriology 9th ed., Williams & Wilkins, Baltimore, MD.

Honda, T., Ni, Y., and Miwatani, T. 1988. Purification and characterization of a hemolysin produced by a clinical isolate of Kanagawa phenomenon-negative *Vibrio parahaemolyticus* and related to the thermostable direct hemolysin. Infect. Immun. 56: 961–965.

Honda, T., Ni, Y., Miwatani, T., and Kim, J. 1992. The thermostable direct hemolysin of *Vibrio parahaemolyticus* is a pore-forming toxin. Can. J. Microbiol. 38: 1175–1180.

Horre, R., Becker, S., Marklein, G., Shimada, T., Stephan, R., Steuer, K., Bierhoff, E., and Schaal, K.P. 1998. Necrotizing fasciitis caused by *Vibrio vulnificus*: first published infection acquired in Turkey is the second time a strain is isolated in Germany. Infection 26: 399–401.

Hoshino, K., Yamasaki, S., Mukhopadhyay, A.K., Chakraborty, S., Basu, A., Bhattacharya, S.K., Nair, G.B., Shimada, T., and Takeda, Y. 1998. Development and evaluation of a multiplex PCR assay for rapid detection of toxigenic *Vibrio cholerae* O1 and O139. FEMS Immunol. Med. Microbiol. 20: 201–207.

Huntley, J.S., Hall, A.C., Sathyamoorthy, V., and Hall, R.H. 1993. Cation flux studies of the lesion induced in human erythrocyte membranes by the thermostable direct hemolysin of *Vibrio parahaemolyticus*. Infect. Immun. 61: 4326–4332.

Huq, M.I., Alam, A.K., Brenner, D.J., and Morris, G.K. 1980. Isolation of *Vibrio*-like group, EF-6, from patients with diarrhea. J. Clin. Microbiol. 11: 621–624.

Huq, A., Small, E.B., West, P.A., Huq, M.I., Rahman, R., and Colwell, R.R. 1983. Ecological relationships between *Vibrio cholerae* and planktonic crustacean copepods. Appl. Environ. Microbiol. 45: 275–283.

Janda, J.M., Powers, C., Bryant, R., and Abbott, S.L. 1988. Current perspectives on the epidemiology and pathogenesis of clinically significant *Vibrio* spp. Clin. Microbiol. Rev. 1: 245–267.

Jeong, K.C., Jeong, H.S., Rhee, J.H., Lee, S.E., Chung, S.S., Starks, A.M., Escudero, G.M., Gulig, P.A., and Choi, S.H. 2000. Construction and phenotypic evaluation of a *Vibrio vulnificus* vvpE mutant for elastolytic protease. Infect. Immun. 68: 5096–5106.

Jobling, M.G., and Holmes, R.K. 1997. Characterization of hapR, a positive regulator of the *Vibrio cholerae* HA/protease gene hap, and its identification as a functional homologue of the *Vibrio harveyi* luxR gene. Mol. Microbiol. 26: 1023–1034.

Joseph, S.W., Colwell, R.R., and Kaper, J.B. 1983. *Vibrio parahaemolyticus* and related halophilic vibrios. Crit. Rev Microbiol. 10: 77–123.

Kaneko, T., and Colwell, R.R. 1975a. Adsorption of *Vibrio parahaemolyticus* onto chitin and copepods. Appl. Microbiol. 29: 269–274.

Kaneko, T., and Colwell, R.R. 1975b. Ecology of *Vibrio parahaemolyticus* in Chesapeake Bay. Appl. Microbiol. 30: 251–257.

Kaneko, T., and Colwell, R.R. 1976. Incidence of *Vibrio parahaemolyticus* in Chesapeake Bay. Appl. Microbiol. 30: 251–257.

Kaneko, T., and Colwell, R.R. 1978. The annual cycle of *Vibrio parahaemolyticus* in Chesapeake Bay. Microb. Ecol. 4: 135–155.

Kang, J.H., Lee, J.H., Park, J.H., Huh, S.H., and Kong, I.S. 1998. Cloning and identification of a phospholipase gene from *Vibrio mimicus*. Biochim. Biophys. Acta 1394: 85–89.

Kaper, J.B., and Levine, M.M. 1981. Cloned cholera enterotoxin genes in study and prevention of cholera. Lancet. 2: 1162–1163.

Kaper, J., Lockman, H., Colwell, R.R. and Joseph, S.W. 1979. Ecology, serology, and enterotoxin production of *Vibrio cholerae* in Chesapeake Bay. Appl. Environ. Microbiol. 37: 91–103.

Kaper, J.B., Morris, J.G., Jr., and Levine, M.M. 1995. Cholera. Clin. Microbiol. Rev. 8: 48–86.

Kaper, J.B., Morris, J.G., Jr., and Nishibuchi, M. 1989. DNA probes for pathogenic *Vibrio* species. In: DNA Probes for Infectious Diseases. F.C. Tenover, ed. CRC Press, Boca Raton, Florida. p. 65–77.

Karaolis, D.K., Johnson, R.J.A., Bailey, C.C., Boedeker, E.C., Kaper, J.B., and Reeves, R.R. 1998. A *Vibrio cholerae* pathogenicity island associated with epidemic and pandemic strains. Proc. Natl. Acad. Sci. USA 95: 3134–3139.

Karaolis, D.K., Somara, S., Maneval, D.R., Jr., Johnson, J.A., and Kaper, J.B. 1999. A bacteriophage encoding a pathogenicity island, a type-IV pilus and a phage receptor in cholera bacteria. Nature 399: 375–379.

Karunasagar, I., Sugumar, G., and Karunasagar, I. 1995. Rapid detection of *Vibrio cholerae* contamination of seafood by polymerase chain reaction. Mol. Mar. Biol. Biotechnol. 4: 365–368.

Kaufman, G.E., Bej, A.K., Bowers, J.C., and DePaola, A. 2003. Oyster-to-oyster variability in levels of *Vibrio parahaemolyticus*. J. Food Prot. 66: 125–129.

Kaysner, C.A., Abeyta, C., Jr., Stott, R.F., Krane, M.H., and Wekell, M.M. 1996. Enumeration of *Vibrio* species, including *V. cholerae* from samples of an oyster growing area, Grays Harbor, Washington. J. Food Prot. 53: 300–302.

Kaysner, C.A., Abeyta, C., Jr., Stott, R.F., Lilja, J.L., and Wekell, M.M. 1990. Incidence of urea-hydrolyzing *Vibrio parahaemolyticus* in Willapa Bay, Washington. Appl. Environ. Microbiol. 56: 904–907.

Kaysner, C.A., Abeyta, C., Jr., Wekell, M.M., DePaola, A., Stott, R.F., and Leitch, J.M. 1987. Virulent strains of *Vibrio vulnificus* isolated from estuaries of the United States West coast. Appl. Environ. Microbiol. 53: 1349–1351.

Kaysner, C.A., and DePaola, A., Jr. 2001. *Vibrio*. In: Compendium of Methods for the Microbiological Examination of Foods. 4th ed. F.P. Downes, and K. Ito, eds. American Public Health Association, Washington, D.C. p. 405–420.

Keasler, S.P., and Hall, R.H. 1993. Detection and biotyping *Vibrio cholerae* O1 with multiplex polymerase chain reaction. Lancet. 341: 1661.

Kelly, M.T., and Stroh, E.M.D. 1988. Occurrence of *Vibrionaceae* in natural and cultivated oyster populations in the Pacific Northwest. Diagn. Microbiol. Infect. Dis. 9: 1–5.

Kiiyukia, C., Venkateswaran, K., Navarro, I.M., Nakano, H., Kawakami, H., and Hashimoto, H. 1989. Seasonal distribution of *Vibrio parahaemolyticus* serotypes along the oyster beds in Hiroshima coast. J. Fac. Appl. Biol. Sci. 28: 49–61.

Kim, G.T., Lee, J.Y., Huh, S.H., Yu, J.H., and Kong, I.S. 1997. Nucleotide sequence of the *vmhA* gene encoding hemolysin from *Vibrio mimicus*. Biochim. Biophys. Acta 1360: 102–104.

Kim, Y.-K., and McCarter, L.L. 2000. Analysis of the polar flagellar gene system of *Vibrio parahaemolyticus*. J. Bacteriol. 182: 3693–3704.

Kim, Y. B., Okuda, J., Matsumoto, C., Takahashi, N., Hashimoto, S., and Nishibuchi, M. 1999. Identification of *Vibrio parahaemolyticus* at the species level by PCR targeted to the *toxR* gene. J. Clin. Microbiol. 37: 1173–1177.

Kim, S.K., Yang, J.Y., and Cha, J. 2002. Cloning and sequence analysis of a novel metalloprotease gene from *Vibrio parahaemolyticus*. Gene. 283: 277–286.

Kishishita, M., Matsuoka, N., Kumagai, K., Yamasaki, S., Takeda, Y., and Nishibuchi, M. 1992. Sequence variation in the thermostable direct hemolysin-related hemolysin (*trh*) gene of *Vibrio parahaemolyticus*. Appl. Environ. Microbiol. 58: 2449–2457.

Kita-Tsukamoto, K., Oyaizu, H., Nanba, K., and Shimidu, U. 1993. Phylogenetic relationships of marine bacteria, mainly members of the family *Vibrionaceae*, determined on the basis of 16S rRNA sequences. Int. J. Syst. Bacteriol. 43: 8–19.

Klontz, K.C., Williams, L., Baldy, L.M., and Campos, M. 1994. Raw oyster-associated *Vibrio* infections: linking epidemiologic data with laboratory testing of oysters obtained from a retail outlet. J. Food Prot. 56: 977–979.

Koch, W.H., Payne, W.L., Wentz, B.A., and Cebula, T.A. 1993. Rapid polymerase chain reaction method for detection of *Vibrio cholerae* in foods. Appl. Environ. Microbiol. 59: 556–560.

Kothary M.H., Claverie, E.F., Miliotis, M.D., Madden, J.M., and Richardson, S.H. 1995. Purification and characterization of a Chinese hamster ovary cell elongation factor of *Vibrio hollisae*. Infect. Immun. 63: 2418–2423.

Kovacikova, G., and Skorupski, K. 1999. A *Vibrio cholerae* LysR homologue, AphB, cooperates with AphA at the tcpPH promoter to activate expression of the ToxR virulence cascade. J. Bacteriol. 181: 4250–4256.

Kudo, Y., Sakai, S., Zen-Yoji, H., and LeClair, R.A. 1974. Epidemiology of food poisoning due to *Vibrio parahaemolyticus* occurring in Tokyo during the last decade. In: International Symposium on *Vibrio parahaemolyticus*. T. Fjino, G. Sakaguchi, R. Sakazaki, and Y. Takeda, eds. Saikon Publishing Co., Tokyo. p. 9–13.

Laohaprertthisan, V., Chowdhury, A., Kongmuang, U., Kalnauwakul, S., Ishibashi, M., Matsumoto, C., and Nishibuchi, M. 2003. Prevalence and serodiversity of the pandemic clone among the clinical strains of *Vibrio parahaemolyticus* isolated in southern Thailand. Epidemiol. Infect. 130: 1–12.

Lee, S.H., Angelichio, M.J., Mekalanos, J.J., and Camilli, A. 1998a. Nucleotide sequence and spatiotemporal expression of the *Vibrio cholerae vieSAB* genes during infection. J. Bacteriol. 180: 2298–2305.

Lee, S.H., Butler, S.M., and Camilli, A. 2001. Selection for *in vivo* regulators of bacterial virulence. Proc. Natl. Acad. Sci. USA 98: 6889–6894.

Lee, C.Y., Chen, C.H., and Chou, Y.W. 1995a. Characterization of a cloned pR72H probe for *Vibrio parahaemolyticus* detection and development of a nonisotopic colony hybriziation assay. Microbiol. Immunol. 39: 177–183.

Lee, S.E., Kim, S.Y., Kim, S.J., Kim, H.S., Shin, J.H., Choi, S.H., Chung, S.S., and Rhee, J.H. 1998b. Direct identification of *Vibrio vulnificus* in clinical specimens by nested PCR. J. Clin. Microbiol. 36: 2887–2892.

Lee, J.H., Kim, G.T., Lee, J.Y., Jun, H.K., Yu, J.H., and Kong, I.S. 1998c. Isolation and sequence analysis of metalloprotease gene from *Vibrio mimicus*. Biochim. Biophys. Acta 1384: 1–6.

Lee, C.Y., Pan, S.F., and Chen, C.H. 1995b. Sequence of a cloned pR72H fragment and its use for detection of *Vibrio parahaemolyticus* in shellfish with the PCR. Appl. Environ. Microbiol. 61: 1311–1317.

Lee, S.E., Shin, S.H., Kim, S.Y., Kim, Y.R., Shin, D.H., Chung, S. S., Lee, Z.H., Lee, J.Y., Leong, K.C., Choi, S.H., and Rhee, J.H. 2000. *Vibrio vulnificus* has the transmembrane transcription activator *toxRS* stimulating the expression of the hemolysin gene *vvhA*. J. Bacteriol. 182: 3405–3415.

Lee, C.Y., Su, S.C., and Liaw, R.B. 1995c. Molecular analysis of an extracellular protease gene from *Vibrio parahaemolyticus*. Microbiology 141: 2569–2576.

Levine, W.C., and Griffin, P.M. 1993. *Vibrio* infections on the Gulf Coast: results of first year of regional surveillance. Gulf Coast *Vibrio* Working Group. J. Infect. Dis. 167: 479–483.

Lin, W., Fullner, K.J., Clayton, R., Sexton, J.A., Rogers, M.B., Calia, K.E., Calderwood, S.B., Fraser, C., and Mekalano, J.J. 1999. Identification of a *Vibrio cholerae* RTX toxin gene cluster that is tightly linked to the cholera toxin prophage. Proc. Natl. Acad. Sci. USA 96: 1071–1076.

Lin, Z., Kumagai, K., Baba, K., Mekalanos, J.J., and Nishibuchi, M. 1993. *Vibrio parahaemolyticus* has a homolog of the *Vibrio cholerae toxRS* operon that mediates environmentally induced regulation of the thermostable direct hemolysin gene. J. Bacteriol. 175: 3844–3855.

Litwin, C.M., Boyko, S.A., and Calderwood, S.B. 1992. Cloning, sequencing, and transcriptional regulation of the *Vibrio cholerae* fur gene. J. Bacteriol. 174: 1897–1903.

Litwin, C.M., and Calderwood, S.B. 1993. Cloning and genetic analysis of the *Vibrio vulnificus* fur gene and construction of a fur mutant by *in vivo* marker exchange. J. Bacteriol. 175: 706–715.

Litwin, C.M., and Byrne, B.L. 1998. Cloning and characterization of an outer membrane protein of *Vibrio vulnificus* required for heme utilization: regulation of expression and determination of the gene sequence. Infect. Immun. 66: 3134–3141.

Litwin, C.M., and Quackenbush, J. 2001. Characterization of a *Vibrio vulnificus* LysR homologue, HupR, which regulates expression of the haem uptake outer membrane protein, HupA. Microb. Pathog. 31: 295–307.

Lockwood, D.E., Kreger, A.S., and Richardson, S.H. 1982. Detection of toxins produced by *Vibrio fluvialis*. Infect. Immun. 35: 702–708.

Lyon, W.J. 2001. TaqMan PCR for detection of *Vibrio cholerae* O1, O139, non-O1, and non-O139 in pure cultures, raw oysters, and synthetic seawater. Appl. Environ. Microbiol. 67: 4685–4693.

Magalhaes, M., de Silva, G.P., Magalhaes, V., Antas, M.G., Andrade, M.A., Tateno, S. 1990. *Vibrio fluvialis* and *Vibrio furnissii* associated with infantile diarrhea. Rev. Microbiol., sao Paulo 21: 295–298.

Massad, G., and Oliver, J.D. 1987. New selective and differential medium for *Vibrio cholerae* and *Vibrio vulnificus*. Appl. Environ. Microbiol. 53:2262–2264.

Matsumoto, C., Okuda, J., Ishibashi M., Iwanaga M., Garg, P., Rammamurthy, T., Wong, H.-C., Depaola, A., Kim, Y.B., Albert, M.J., and Nishibuchi, M. 2000. Pandemic spread of an O3:K6 clone of *Vibrio parahaemolyticus* and emergence of related strains evidenced by arbitrarily primed PCR and *toxRS* sequence analyses. J. Clin. Microbiol. 38: 578–585.

Maxwell, E.L., Mayall, B.C., Pearson, S.R., and Stanley, P.A. 1991. A case of *Vibrio vulnificus* septicaemia acquired in Victoria. Med. J. Aust. 154: 214–215.

McCarthy, S.A., DePaola, A., Kaysner, C.A., Hill, W.E., and Cook, D.W. 2000. Evaluation of nonisotopic DNA hybridization methods for detection of the *tdh* gene of *Vibrio parahaemolyticus*. J. Food Prot. 63: 1660–1664.

McCarter, L.L., and Wright, M.E. 1993. Identification of genes encoding components of the swarmer cell flagellar motor and propeller and a sigma factor controlling differentiation of *Vibrio parahaemolyticus*. J. Bacteriol. 175: 3361–3371.

McCarthy, S.A., and Khambaty, F.M. 1994. International dissemination of epidemic *Vibrio cholerae* by cargo ship ballast and other nonportable waters. Appl. Environ. Microbiol. 60: 2597–2601.

Mead, P.S., Slutsker, L., Dietz.V., McGaig, L.F., Bresee, J.S., Shapiro, C., Griffin, P.M. and Tauxe, R.V. 1999. Food-related illness and death in the United States. Emerg. Infect. Dis. 5: 607–625.

Mekalanos, J.J., Rubin, E.J., and Waldor, M.K. 1997. Cholera: molecular basis for emergence and pathogenesis. FEMS Immunol. Med. Microbiol. 18: 241–248.

Melhus, A., Holmdahl, T., and Tjernberg, I. 1995. First documented case of bacteremia with *Vibrio vulnificus* in Sweden. Scand J. Infect. Dis. 27: 81–82.

Miceli, G.A., Watkins, W.D., and Rippey, S.R. 1993. Direct plating procedure for enumerating *Vibrio vulnificus* in oysters (Crassostrea virginica). Appl. Environ. Microbiol. 59: 3519–3524.

Miller, V.L., and Mekalanos, J.J. 1984. Synthesis of cholera toxin is positively regulated at the transcriptional level by *toxR*. Proc. Natl. Acad. Sci. USA 81: 3471–3475.

Minami, A., Hashimoto, S., Abe, H., Arita, M., Taniguchi, T., Honda, T., Miwatani, T., and Nishibuchi, M. 1991. Cholera enterotoxin production in *Vibrio cholerae* O1 strains isolated from the environment and from humans in Japan. Appl. Environm. Microbiol. 57: 2152–2157.

Mintz, E.D., Popovic, T., and Blake, P.A. 1994. Transmission of *Vibrio cholerae* O1. In: *Vibrio cholerae* and Cholera: Molecular to Global Perspectives. I.K. Wachsmuth, P.A. Blake, and Ö. Olsvik, eds. ASM Press, Washington, D.C. p. 345–356.

Miwatani, T., and Takeda, Y. 1976 *Vibrio parahaemolyticus*: A Causative Bacterium of Food Poisoning. Saikon Publishing, Tokyo.

Miyamoto, Y., Kato, T., Obara, Y., Akiyama, S., Takizawa, K., Yamai, S. 1969. *In vitro* hemolytic characteristic of *Vibrio parahaemolyticus*: its close correlation with human pathogenicity. J. Bacteriol. 100: 1147–1149.

Morris, J.G., Jr., and Black, R.E. 1985. Cholera and other vibrioses in the United States. N. Engl. J. Med. 312: 343–350.

Motes, M.L., and DePaola, A. 1996. Offshore suspension relaying to reduce levels of *Vibrio* vulnifiucs in oysters (*Crassostrea virginica*). Appl. Environ.Microbiol. 62: 3875–3877.

Motes, M.L., DePaola, A., Cook, D.W., Veazey, J.E., Hunsucker, J.C., Garthright, W.E., Blodgett, R.J., and Chirtel, S.J. 1998. Influence of water temperature and salinity on *Vibrio vulnificus* in northern Gulf and Atlantic Coast oysters (Crassostrea virginica). Appl. Environ. Microbiol. 64: 1459–1465.

Motes, M.L., DePaola, A., Zywno-Van Ginkel, S., and McPhearson, M. 1994. Occurrence of toxigenic *Vibrio cholerae* O1 in oysters in Mobile Bay, Alabama: an ecological investigation. J. Food Prot. 57: 975–980.

Nagayama, K., Oguchi, T., Arita, M., and Honda, T. 1995. Purification and characterization of a cell-associated hemagglutinin of *Vibrio parahaemolyticus*. Infect. Immun. 63: 1987–1992.

Nair, G.B., Oku, Y., Takeda, Y., Ghosh, R.K., Ghosh, A., Chattopadhyyay, S., Pal, S.C., Kaper, J.B., and Takeda, T. 1988. Toxin profiles of *Vibrio cholerae* non-O1 from environmental sources in Calcutta, India. Appl. Microbiol. 54: 3180–3182.

Nakaguchi, Y., Ishizuka, T., Ohnaka, S., Hayashi, T., Yasukawa, K., Ishiguro, T., and Nishibuchi, M. 2004. Rapid and specific detection of *tdh*, *trh1*, and *trh2* mRNA of *Vibrio parahaemolyticus* by transcription–reverse transcription concerted reaction with an automated system. J. Clin. Microbiol. 429: 4284–4292.

Nakasone, N., and Iwanaga, M. 1990. Pili of a *Vibrio parahaemolyticus* strain as a possible colonization factor. Infect. Immun. 58: 61–69,

Nandi, B., Nandy, R.K., Mukhopadhyay, S., Nair, G.B., Shimada, T., and Ghose, A.C. 2000. Rapid method for species-specific identification of *Vibrio cholerae* using primers targeted to the gene of outer membrane protein OmpW. J. Clin. Microbiol. 38: 4145–4151.

Nasu, H., Iida, T., Sugahara, T., Yamaguchi, Y., Park, K.-S., Yokoyama, K., Makino, K., Shinagawa, H., and Honda, T. 2000. A filamentous phage associated with recent pandemic *Vibrio parahaemolyticus* O3: K6 strains. J. Clin. Microbiol. 8: 2156–2161.

Nesper, J., Lauriano, C.M., Klose, K.E., Kapfhammer, D., Kraib, A., and Reidl, J. 2001. Characterization of *Vibrio cholerae* O1 El Tor *galU* and *galE* mutants: influence on lipopolysaccharide structure, colonization, and biofilm formation. Infect. Immun. 69: 435–445.

Nilsson, W.B., Paranjpye, R.N., DePaola, A., and Strom, M.S. 2003. Sequence polymorphism of the 16S rRNA gene of *Vibrio vulnificus* is a possible indicator of strain virulence. J. Clin. Microbiol. 41: 442–446.

Nishibuchi, M., Doke, S., Toizumi, S., Umeda, T., Yoh, M., and Miwatani, T. 1988. Isolation from a coastal fish of *Vibrio hollisae* capable of producing a hemolysin similar to the thermostable direct hemolysin of *Vibrio parahaemolyticus*. Appl. Environ. Microbiol. 54: 2144–2146.

Nishibuchi, M., Fasano, A., Russell, R., and Kaper, J.B. 1992. Enterotoxigenicity of *Vibrio parahaemolyticus* with and without genes encoding thermostable direct hemolysin. Infect. Immun. 60: 3539–3545.

Nishibuchi, M., Janda, J.M., and Ezaki, T. 1996. The thermostable direct hemolysin gene (*tdh*) of *Vibrio parahaemolyticus* is dissimilar in prevalence to and phylogenetically distant from the *tdh* genes of other vibrios: implications in the horizontal transfer of the *tdh* gene. Microbiol. Immunol. 40: 59–65.

Nishibuchi, M., and Kaper, J.B. 1995. Thermostable direct hemolysin gene of *Vibrio parahaemolyticus*: a virulence gene acquired by a marine bacterium. Infect. Immun. 63: 2093–2099.

Nishibuchi, M., Seidler, R.J., Rollins, D.M., and Joseph, S.W. 1983. *Vibrio* factors cause rapid fluid accumulation in suckling mice. Infect. Immun. 40: 1083–1091.

Nishibuchi, M. Taniguchi, T., Misawa, T., Khaeomanee-iam, V., Honda, T., and Miwatani. T., 1989. Cloning and nucleotide sequence of the gene (*trh*) encoding the hemolysin related to the thermostable direct hemolysin of *Vibrio parahaemolyticus*. Infect. Immun. 57: 2691–2697.

Nordstrom, J.L. and DePaola, A. 2003. Improved recovery of pathogenic *Vibrio parahemolyticus* from oysters using colony hybridization following enrichment. J. Microbiol. Methods 52: 273–277.

Nye, M.B., Pfau, J.D., Skorupski, K., and Taylor, R.K. 2000. *Vibrio cholerae* H-NS silences virulence gene expression at multiple steps in the ToxR regulatory cascade. J. Bacteriol. 182: 4295–4303.

Occhino, D.A., Wyckoff, E.E., Henderson, D.P., Wrona, T.J., and Payne, S.M. 1998. *Vibrio cholerae* iron transport: haem-transport genes are linked to one of two sets of *tonB*, *exbB*, *exbD* genes. Mol. Microbiol. 29: 1493–1507.

Ogawa, A., Kato, J.- I., Watanabe, H., Nair, G.B., and Takeda, T. 1990. Cloning and nucleotide sequence of a heat-stable enterotoxin gene from *Vibrio cholerae* non-O1 isolated from a patient with traveler's diarrhea. Infect. Immun. 58: 3325–3329.

Ogawa, A. and Takeda, T. 1993. The gene encoding the heat-stable enterotoxin of *Vibrio cholerae* is flanked by 123-base pair direct repeats. Microbiol. Immunol. 37: 607–616.

Ogawa, H., Tokunou, H., Kishimoto, T., Fukuda, S., Umemura, K., and Takata, M. 1989. Ecology of *Vibrio parahaemolyticus* in Hiroshima Bay. Hiroshima J. Vet. Med. 4: 47–57.

Ogg, J.E., Ryder, R.A. and Smith, and H.L.Jr. 1989. Isolation of *Vibrio cholerae* from aquatic birds in Colorado and Utah. Appl. Environ. Microbiol. 55: 95–99.

Okuda, J., Ishibashi, M., Hayashi, E., Nishino, T., Takeda, Y., Mukhopadhyary, A.K., Garg, S., Bhattacharya, S.K., Nair, B.G. and Nishibuchi, M. 1997. Emergence of a unique O3:K6 clone of *Vibrio parahaemolyticus* in Calcutta, India, and isolation of strains from the same clonal group from southeast Asian travelers arriving in Japan. J. Clin. Microbiol. 35: 3150–3155.

Okuda, J., Nakai, T., Chang, P.S., Oh, T., Nishino, T., Koitabashi, T., and Nishibuchi, M. 2001. The *toxR* gene of *Vibrio* (*Listonella*) *anguillarum* controls expression of the major outer membrane proteins but not virulence in a natural host model. Infect. Immun. 69: 6091–6101.

Okuda, J., and Nishibuchi, M. 1998. Manifestation of the Kanagawa phenomenon, the virulence-associated phenotype, of *Vibrio parahaemolyticus* depends on a particular single base change in the promoter of the thermostable direct haemolysin gene. Mol. Microbiol. 30: 499–511.

Okujo, N., and Yamamoto, S. 1994. Identification of the siderophores from *Vibrio hollisae* and *Vibrio mimicus* as aerobactin. FEMS Microbiol. Lett. 118: 187–192.

Oliver, J.D., Guthrie, K., Preyer, J., Wright, A.C., Simpson, L.M., Siebeling, R.J., and Morris, J.G. 1992. Use of colistin-polymyxin B-cellobiose agar for isolation of *Vibrio vulnificus* from the environment. Appl. Environ. Microbiol. 58: 737–739.

Oliver, J.D., and Kaper, J.B. 2001. *Vibrio* species. In: Food Microbiology: Fundamentals and Frontiers, 2nd ed. M.P. Doyle, L.R. Beuchat, and T.J. Montville, eds. ASM Press, Washington D.C, pp. 228–264.

Oliver, J.D., Warner, R.A., and Cleland, D.R. 1982. Distribution and ecology of *Vibrio vulnificus* and other lactose-fermenting marine vibrios in coastal waters of the Southeastern United States. Appl. Environ. Microbiol. 44: 1404–1414.

Oliver, J.D., Warner, R.A., and Cleland, D.R. 1983. Distribution of *Vibrio vulnificus* and other lactose fermenting vibrios in the marine environment. Appl. Environ. Microbiol. 45: 985–998.

Paranjpye, R.N., Lara, J.C., Pepe, J.C., Pepe, C.M., and Strom, M.S. 1998. The type IV leader peptidase/N-methyltansferase of *Vibrio vulnificus* controls factors required for adherence to Hep-2 cells and virulence in iron-overloaded mice. Infect. Immun. 66: 5659–5668.

Park, S.D., Shon, and H.S. Joh, N J. 1991. *Vibrio vulnificus* septicemia in Korea: clinical and epidemiologic findings in 70 patients. J. Am. Acad. Dermatol. 24: 397–403.

Parker, R.W., Maurer, E.M., Childers, A.B., and Lewis, D.H. 1994. Effect of frozen storage and vaccum-packaging on survival of *Vibrio vulnificus* in Gulf coast oysters (*Crassostrea virginica*). J. Food Protect. 57: 604–606.

Pearson, G.D., and Mekalanos, J.J. 1982. Molecular cloning of *Vibrio cholerae* enterotoxin genes in *Escherichia coli* K-12. Proc. Natl. Acad. Sci. USA 79: 2976–2980.

Peterson, K.M., and Mekalanos, J.J. 1988. Characterization of the *Vibrio cholerae* ToxR regulon: identification of novel genes involved in intestinal colonization. Infect. Immun. 57: 2822–2829.

Pollizer, R. 1959. Cholera. Monograph no. 43. Geneva: World Health Organization.

Radu, S., Yuherman, Rusul, G., Yeang, L.K., and Nishibuchi, M. 2000. Detection and molecular characterization of *Vibrio vulnificus* from coastal waters of Malaysia. Southeast Asian J. Trop. Med. Pub. Health 31: 668–673.

Raimondi, D. Kao, J.P.Y., Kaper, J.B., Guandalini, S., and Fasano, A. 1995. Calcium-dependent intestinal chloride secretion by *Vibrio parahaemolyticus* thermostable direct hemolysin in a rabbit model. Gastroenterol. 109: 381–386.

Ray, B., Hawkins, S.M., and Hackney, C.R. 1978. Method for the detection of injured *Vibrio parahaemolyticus* in seafoods. Appl. Enviro. Microbiol. 35: 1121–1127.

Reily, L.A. and Hackney, C.R. 1985. Survival of *Vibrio cholerae* during cold storage in artificially contaminated seafoods. J. Food Sci. 50: 838–839.

Rivera, T.G., Chun, J., Huq, A., Sack, R.B., and Colwell, R.R. 2001. Genotypes associated with virulence in environmental isolates of *Vibrio cholerae*. Appl. Environ. Microbiol. 67: 2421–2429.

Ruimy, R., Breittmayer, V., Elbaze, P., Lafay, B., Boussemart, O., Gauthier, M., and Christine, R. 1994. Phylogenetic analysis and assessment of the genera *Vibrio*, *Photobacterium*, *Aeromonas*, and *Plesiomonas* deduced from small-subunit rRNA sequences. Int. J. Syst. Bacteriol. 44: 416–426.

Sakazaki, R., Tamura, K., Kato, T., Obara, Y., Yamai, S., and Hobo, K. 1968. Studies of the enteropathogenic, facultatively halophilic bacteria, *Vibrio parahaemolyticus*. III. Enteropathogenicity. Jpn. J. Med. Sci. Biol. 21: 325–331.

Sanyal, S.C., and Sen, P.C. 1974. Human volunteer study on the pathogenicity of *Vibrio parahaemolyticus*. In: International Symposium on *Vibrio parahaemolyticus*. T Fujino, G Sakaguchi, R Sakazaki, and Y Takeda, eds. Saikon Publishing Co., Tokyo, pp. 227–235.

Seyama, I., Irisawa, H., Honda, T., Takeda, Y., and Miwatani, T. 1977. Effect of hemolysin produced by *Vibrio parahaemolyticus* on membrane conductance and mechanical tension of rabbit myocardium. Jpn. J. Physiol. 27: 43–56.

Shao, C.P., and Hor, L.I. 2000. Metalloprotease is not essential for *Vibrio vulnificus* virulence in mice. Infect. Immun. 68: 3569–3573.

Shao, C.-P., and Hor, L.I. 2001. Regulation of metalloprotease gene expression in *Vibrio vulnificus* by a *Vibrio harveyi* LuxR homologue. J. Bacteriol. 183: 1369–1375.

Shapiro, R.L., Altekruse, S., Hutwagner, S., Bishop, R., Hammond, R., Wilson, S., Ray, B., Thompson, S., Tauxe, R.V., Griffin, P.M., and *Vibrio* working group. 1998. The role of Gulf Coast oysters harvested in warmer months in *Vibrio vulnificus* infections in the United States, 1988–1996. J. Infect. Dis. 178: 752–759.

Shi, L., Miyoshi, S., Hiura, M., Tomochika, K., Shimada, T., and Shinoda, S. 1998. Detection of genes encoding cholera toxin (CT), zonula occludens toxin (ZOT), accessory cholera enterotoxin (ACE) and heat-stable enterotoxin (ST) in *Vibrio mimicus* clinical strains. Microbiol. Immunol. 42: 823–828.

Shinoda, S., Matsuoka, H., Tsuchie, T., Miyoshi, S., Yamamoto, S., Taniguchi, H., and Mizuguchi, Y. 1991. Purification and characterization of a lecithin-dependent haemolysin from *Escherichia coli* transformed by a *Vibrio parahaemolyticus* gene. J. Gen. Microbiol. 137: 2705–2711.

Shirai, H., Nishibuchi, M., Ramamurthy, T., Bhattacharya, S.K., Pal, S.C., and Takeda, Y. 1991. Polymerase chain reaction for detection of the cholera enterotoxin operon of *Vibrio cholerae*. J. Clin. Microbiol. 29: 2517–2521.

Shultz, L.M., Rutledge, J.E., Grodner, R.M. and Biede, S.L. 1984. Determination of the thermal death time of *Vibrio cholerae* in blue crabs (*Callinectes sapidus*). J. Food Prot. 47: 4–10.

Simpson, L.M., White, V.K., Zane, S.F., and Oliver, J.D. 1987. Correlation between virulence and colony morphology in *Vibrio vulnificus*. Infect. Immun. 55: 269–272.

Skorupski, K., and Taylor, R.K. 1997a. Sequence and functional analysis of the gene encoding *Vibrio cholerae* cAMP receptor protein. Gene. 198: 297–303.

Skorupski, K., and Taylor, R.K. 1997b. Cyclic AMP and its receptor protein negatively regulate the coordinate expression of cholera toxin and toxin-coregulated pilus in *Vibrio cholerae*. Proc. Natl. Acad. Sci. USA 94: 265–270.

Stabellini, N., Camerani, A., Lambertini, D., Rossi, M.R., Bettoli, V., Virgili, A., and Gilli, P. 1998. Fatal sepsis from *Vibrio vulnificus* in a hemodialyzed patient. Nephron. 78: 221–224.

Sun, Y., and Oliver, J.D. 1994. Effects of GRAS compounds on natural *Vibrio vulnificus* populations in oysters. J. Food Prot. 57: 921–923.

Strom, M.S., and Paranjpye, R.N. 2000. Epidemiology and pathogenesis of *Vibrio vulnificus*. Microbes Infect. 2: 177–188.

Tacket, C.O., Hickman, F., Pierce, G.V., and Mendoza, L.F. 1982. Diarrhea associated with *Vibrio fluvialis* in the United States. J. Clin. Microbiol. 16: 991–992.

Tada, J., Ohashi, T., Nishimura, N., Shirasaki, Y., Ozaki, H., Fukushima, S., Takano, J., Nishibuchi, M., and Takeda, Y. 1992. Detection of thermostable direct hemolysin gene (*tdh*) and the thermostable direct hemolysin-related hemolysin gene (*trh*) of *Vibrio parahaemolyticus* by polymerase chain reaction. Mol. Cell. Probes. 6: 477–487.

Takashi, K., Nishiyama, M., and Kuga, T. 1982. Hemolysis and hyperpotasemia in rat induced by thermostable direct hemolysin of *Vibrio parahaemolyticus*. Jpn. J. Pharmacol. 32: 377–380.

Takeda, Y. 1983. Thermostable direct hemolysin of *Vibrio parahaemolyticus*. Pharmacol. Ther. 19: 123–146.

Takeda, T., Peina, Y., Ogawa, A., Dohi S., Abe, H., Nair G.B., and Pal S.C. 1991. Detection of heat-stable enterotoxin in a cholera toxin gene-positive strain of *Vibrio cholerae* O1. FEMS Microbiol. Lett. 64: 23–27

Tamplin, M.L., and Capers, G.M. 1992. Persistence of *Vibrio vulnificus* in tissues of Gulf Coast oysters, Crassostrea virginica, exposed to seawater disinfected with UV light. Appl. Environ. Microbiol. 58: 1506–1510.

Tamplin, M.L., Martin, A.L., Ruple, A.D., Cook, D.W., and Kaspar, C.W. 1991. Enzyme immunoassay for identification of *Vibrio vulnificus* in seawater, sediment, and oysters. Appl. Environ. Microbiol. 57: 1235–1240.

Taniguchi, H., Hirano, H., Kubomura, S., Higashi, K., and Mizuguchi, Y. 1986. Comparison of the nucleotide sequences of the genes for the thermostable direct hemolysin from *Vibrio parahaemolyticus*. Microb. Pathog. 1: 425–432.

Taniguchi, H., Kubomura, S., Hirano, H., Mizue, K., Ogawa, M., and Mizuguchi, Y. 1990. Cloning and characterization of a gene encoding a new thermostable hemolysin from *Vibrio parahaemolyticus*. FEMS Microbiol. Lett. 67: 339–346.

Taylor, R.K., Miller, V.L., Furlong, D.B., and Mekalanos, J.J. 1987. The use of *phoA* gene fusions to identify a pilus colonization factor coordinately regulated with cholera toxin. Proc. Natl. Acad. Sci. USA 84: 2833–2837.

Terai, A., Shirai, H., Yoshida O., Takeda, Y., and Nishibuchi, M. 1990. Nucleotide sequence of the thermostable direct hemolysin gene (*tdh* gene) of *Vibrio mimicus* and its evolutionary relationship with the *tdh* genes of *Vibrio parahaemolyticus*. FEMS Microbiol. Lett. 71: 319–324.

Tison, D.L., and Kelly, T. 1984. *Vibrio* species of medical importance. Diagn. Micrbiol. Infect. 2: 263–276.

Tison, D.L, Nishibuchi, M., Greenwood, J.G, and Seidler, R.J. 1982. *Vibrio vulnificus* Biogroup 2: A new biogroup pathogenic for eels. Appl. Environ. Microbiol. 44: 640–646.

Todd, E.C.D. 1989. Preliminary estimates of costs of foodborne diseases in the United States. J. Food. Prot. 52: 595–601.

Thompson, C.A., and Vanderzant, C. 1976. Serological and hemolytic characteristics of *Vibrio parahaemolyticus* from marine sources. J. Food Sci. 41: 204–205.

Trucksis, M., Galen, J.E., Michalski, J., Fasano, A., and Kaper, J.B. 1993. Accessory cholera enterotoxin (Ace), the third toxin of a *Vibrio cholerae* virulence cassette. Proc. Natl. Acad. Sci. USA 90: 5267–5271.

USADepartment of Health and Human Services.Public Health Services. FDA. 1999. National Shellfish Sanitation Program Guide for the Control of Molluscan Shellfish. USA Department of Health and Human Services, Washington, DC.

Varghese, M.R., Farr, R.W., Wax, M.K., Chafin, B.J., and Owens, R.M. 1996. *Vibrio fluvialis* wound infection associated with medicinal leech therapy. Clin. Infect. Dis. 22: 709–710.

Veenstra, J., Rietra, P.J., Coster, J.M., Slaats, E., and Dirks-Go, S. 1994. Seasonal variations in the occurrence of *Vibrio vulnificus* along the Dutch coast. Epidemiol. Infect. 112: 285–290.

Veenstra, J., Rietra, P.J., Stoutenbeek, C.P., Coster, J.M., de Gier, H.H., and Dirks-Go, S. 1992. Infection by an indole-negative variant of *Vibrio vulnificus* transmitted by eels. J. Infect. Dis. 166: 209–210.

Venkateswaran, K., Dohmoto, N., and Harayama, S. 1998. Cloning and nucleotide sequence of the *gyrB* gene of *Vibrio parahaemolyticus* and its application in detection of this pathogen in shrimp. Appl. Environmen. Microbiol. 64: 681–687.

Venkateswaran, K., Kiiyukia, C., Nakanishi, K., Nakano, H., Matsuda, O., and Hashimoto, H. 1990. The role of sinking particles in the overwintering process of *Vibrio parahaemolyticus* in a marine environment. FEMS Microbiol.Lett. 73: 159–166.

von Krüger, W.M.A., Humphreys, S., and Ketley, J.M. 1999. A role for the PhoBR regulatory system homologue in the *Vibrio cholerae* phosphate-limitation response and intestinal colonization. Microbiology. 145: 2463–2475.

Vuddhakul, V., Chowdhury, A., Laohaprertthisan, V., Pungrasamee, P., Patararungrong, N., Thianmontri, P., Ishibashi, M., Matsumoto, C., and Nishibuchi, M. 2000a. Isolation of *Vibrio parahaemolyticus* strains belonging to a pandemic O3:K6 clone from environmental and clinical sources in Thailand. Appl. Environ. Microbiol. 66: 2685–2689.

Vuddhakul, V., Nakai, T., Matsumoto, C., Oh, T., Nishino, T., Chen, C.-H., Nishibuchi, M., and Okuda, J. 2000b. Analysis of the *gyrB* and *toxR* gene sequences of *Vibrio hollisae* and establishment of the *gyrB*- and *toxR*-targeted PCR methods for isolation and identification of *V. hollisae* in the environment. Appl. Environ. Microbiol. 66: 3506–3514.

Wachsmuth, I.K., Blake, P.A., and Olsvik, O. 1994. *Vibrio cholerae* and Cholera: Molecular to Global Perspectives. ASM Press, Washington, D.C.

Waldor, M.K., and Mekalanos, J.J. 1996. Lysogenic conversion by a filamentous phage encoding cholera toxin. Science. 272: 1910–1914.

Wall, V.W., Kreger, A.S., and Richardson, S.H. 1984. Production and partial characterization of a *Vibrio fluvialis* cytotoxin. Infect. Immun. 46: 773–777.

Watkins, W.D., and Cabelli, V.J. 1985. Effect of fecal pollution on *Vibrio parahaemolyticus* densities in an estuarine environment. Appl. Environ. Microbiol. 49: 1307–1313.

Webster, A.C., and Litwin, C.M. 2000. Cloning and characterization of vuuA, a gene encoding the *Vibrio vulnificus* ferric vulnibactin receptor. Infect. Immun. 68: 26–534.

West, P.A., Brayton, P.R., Bryant, T.N., and Colwell, R.R. 1986. Numerical taxonomy of vibrios isolated from aquatic environments. Int. J. Syst. Bacteriol. 36: 531–543.

Williams, S.G., Attridge, S.R., and Manning, P.A. 1993. The transcriptional activator HlyU of *Vibrio cholerae*: nucleotide sequence and role in virulence gene expression. Mol. Microbiol. 9: 751–760.

Wong, S.M., Carroll, P.A., Rahme, L.G., Ausubel, F.M. and Calderwood, S. 1998. Modulation of expression of the ToxR regulon in *Vibrio cholerae* by a member of the two-component family of response regulators. Infect. Immun. 66: 5854–5861.

Wright, A.C., Guo, Y., Johnson, J.A., Nataro, J.P., and Morris, J.G., Jr. 1992. Development and testing of a non-radioactive DNA oligonucleotide probe that is specific for *Vibrio cholerae* cholera toxin. J. Clin. Microbiol. 30: 2302–2306.

Wright, A.C., Hill, R.T., Johnson, J.A., Roghman, M.-C., Colwell, R.R., and Morris, J.G. 1996. Distribution of *Vibrio vulnificus* in Chesapeake Bay. Appl. Environ. Microbiol. 62: 717–724.

Wright, A.C., Miceli, G.A., Landry, W.L., Christy, J.B., Watkins, W.D., and Morris, J.G. 1993. Rapid identification of *Vibrio vulnificus* on nonselective media with an alkaline phosphatase-labeled oligonucleotide probe. Appl. Environ. Microbiol. 59: 541–546.

Wright A.C., and Morris, J.G., Jr. 1991. The extracellular cytolysin of *Vibrio vulnificus*: inactivation and relationship to virulence in mice. Infect. Immun. 59: 192–197.

Wright A., Morris, J.G., Jr., Maneval, D.R., Jr., Richardson, K., and Kaper, J.B. 1985. Cloning of the cytotoxin-hemolysin gene of *Vibrio vulnificus*. Infect. Immun. 50: 922–924.

Wright, A.C., Powell, J.L., Kaper, J.B., and Morris, J.G., Jr. 2001. Identification of a group 1-like capsular polysaccharide operon for *Vibrio vulnificus*. Infect. Immun. 69: 6893–68901.

Wright, A.C., Simpson, L.M., and Oliver J.D. 1981. Role of iron in pathogenesis of *Vibrio vulnificus*. Infect. Immun. 34: 503–507.

Wright, A.C., Simpson, L.M., Oliver, J.D., and Morris, J.G., Jr. 1990. Phenotypic evaluation of acapsular transposon mutants of *Vibrio vulnificus*. Infect. Immun. 58: 1769–1773.

Wyckoff, E.E., Stoebner, J.A., Reed, K.E., and Payne, S.M. 1997. Cloning of a *Vibrio cholerae* vibriobactin gene cluster: identification of genes required for early steps in siderophore biosynthesis. J. Bacteriol. 179: 7055–7062.

Xu, M., Yamamoto, K., and Honda, T. 1994. Construction and characterization of an isogenic mutant of *Vibrio parahaemolyticus* having a deletion in the thermostable direct hemolysin-related hemolysin gene. J. Bacteriol. 176: 4757–4760.

Yamamoto, K., Do Valle, G.R. Xu, M., Miwatani, T., and Honda, T. 1995. Amino acids of the cholera toxin from *Vibrio cholerae* O37 strain S7 which differ from those of strain O1. Gene. 163:155–156.

Yamamoto, S., Funahashi, T., Ikai, H., and Shinoda S. 1997. Cloning and sequencing of the *Vibrio parahaemolyticus* fur gene. Microbiol. Immunol. 41: 737–740.

Yamamoto, S., Okujo, N., Fujita, Y., Saito, M., Yoshida, T., and Shinoda, S. 1993. Structures of two polyamine-containing catecholate siderophores from *Vibrio fluvialis*. J. Biochem. (Tokyo) 113: 538–544.

Yamamoto K., Wright, A.C., Kaper, J.B., and Morris, J.G., Jr. 1990. The cytolysin gene of *Vibrio vulnificus*: sequence and relationship to the *Vibrio cholerae* El Tor hemolysin gene. Infect. Immun. 58: 2706–2709.

Yamasaki, S., Shirai, H., Takeda, Y., and Nishibuchi, M. 1991. Analysis of the gene of *Vibrio hollisae* encoding the hemolysin similar to the thermostable direct hemolysin of *Vibrio parahaemolyticus*. FEMS Microbiol. Lett. 80: 259–264.

Yamauchi, Y., Ito, S., Yamane, Y., Matsuo, Y., Kumura, S., and Arai, T. 1994. Clinical characteristics and epidemiology of *Vibrio vulnificus* infection in Japan: A collective review of 58 cases. Ehime Igaku 13: 176–183. (In Japanese.)

Yoshida, S.-I., Ogawa, M., and Mizuguchi, Y. 1985. Relation of capsular materials and colony opacity to virulence of *Vibrio vulnificus*. Infect. Immun. 47: 446–451.

Zhu, J., Miller, M.B., Vance, R.E., Dziejman, M.B., Bassler, L., and Mekalanos, J.J. 2002. Quorum-sensing regulators control virulence gene expression in *Vibrio cholerae*. Proc. Natl. Acad. Sci. USA 99: 3129–3134.

Zuppardo, A.B., and Siebeling, R.J. 1998. An epimerase gene essential for capsule synthesis in *Vibrio vulnificus*. Infect. Immun. 66: 2601–2606.

# Staphylococcus aureus

## 13

George C. Stewart

## Abstract

*Staphylococcus aureus* is a common cause of confirmed bacterial foodborne disease worldwide. Food poisoning episodes are characterized by symptoms of vomiting and diarrhea that occur shortly after ingestion of *S. aureus*-contaminated food. The symptoms arise from ingestion of preformed enterotoxin, which accounts for the short incubation time. Staphylococcal enterotoxins are a family of sequence similar, but serologically distinct proteins. These proteins have the additional property of being superantigens and, as such, have adverse effects on the immune system. The enterotoxin genes are accessory genetic elements in *S. aureus*, meaning that not all strains of this organism are enterotoxin-producing. The enterotoxin genes are found on prophage, plasmids, and pathogenicity islands in different strains of *S. aureus*. Expression of the enterotoxin genes is often under the control of global virulence gene regulatory systems. Although much progress has been made recently in defining enterotoxin structure and superantigenicity properties, much remains to be learned regarding the binding of enterotoxins to receptors in the gastrointestinal tract and how toxin production leads to the symptoms associated with staphylococcal food poisoning.

## Introduction

*Staphylococcus aureus* is a common cause of confirmed bacterial foodborne disease worldwide. For example, from 1993–1997, outbreaks of food poisoning attributed to *S. aureus* accounted for 1.5% of the reported outbreaks and 1.8% of the total cases in the United States (Olsen *et al.*, 2000). It is estimated that there are 185,060 total cases in the USA each year (Mead *et al.*, 1999). A previous estimate was higher with 1,513,000 cases of illness and 1210 deaths annually (Bennett *et al.*, 1987). The difficulty with determining an accurate case number lies in the short duration of illness associated with staphylococcal food poisoning which results in most cases going unreported. Annual human illness cost estimates from *S. aureus* foodborne cases are estimated at $1.2 billion (Buzby *et al.*, 1996).

After an incubation period of 30 minutes to 8 hours, consumption of staphylococcal contaminated foods results in symptoms of vomiting, diarrhea, and abdominal cramping (Holmberg and Blake, 1984). Complete recovery usually occurs within 1–2 days. The condition results from an intoxication, not an infection, which accounts for the short incubation time prior to the onset of symptoms. The protein toxin responsible, the enterotoxin, is produced by *S. aureus* during growth on the contaminated food. Foods often incriminated in staphylococcal food poisoning include meat (especially ham), poultry and egg products, casseroles, bakery products, and milk and dairy products. Implicated foods are often those that require considerable handling during preparation and are kept at slightly elevated temperatures after preparation. The bacteria replicate in the food and elaborate one or more enterotoxins. Food poisoning is usually caused by food which has not been kept hot enough ($\geq 60°C$) or cold enough ($\leq 7.2°C$). The halotolerance of the bacteria, and the relative heat resistance of the enterotoxin, contribute to prevalence of the disease.

Enterocolitis resulting from replication of *S. aureus* in the gastrointestinal tract is a less common condition. Predisposing factors for this condition include age (infants at a higher risk), immune status, and antibiotic therapy. Methicillin-resistant *S. aureus*, because of their resistance to multiple antibiotics, are often implicated in enterocolitis in patients undergoing antibiotic therapy.

## Classification and characteristics of staphylococci

Bacteria from the genus *Staphylococcus* can be found on the skin and mucous membranes of virtually all warm and cold blooded animals. The association between the bacterium and the host has evolved to the point where the parasite of each host is specific to that host, and hence the organism is often characterized as a distinct species of *Staphylococcus*. At other times, this host specificity was considered to be insufficient for unique species status and the number of species was reduced. Therefore, the taxonomy of this bacterium was dependent of the criterion *du jour* and the list of species periodically expanded and contracted. Minimal standards have now been applied to the genus *Staphylococcus* and for species designations (Freney *et al.*, 1999). Assignment of a strain to the genus *Staphylococcus* requires that it be a Gram-positive nonmotile coccus whose cells tend to remain associated after division to form grape-like clusters, produces catalase, has a cell wall structure with peptidoglycan with the appropriate oligopeptide cross-bridging and the presence of teichoic acid, and has a G+C content of DNA in the range of 30–40 mol%. The criteria for species designation include colony morphology and the results of certain conventional tests including pigment production, growth requirements, fermentative and oxidative activity on carbohydrates, novobiocin susceptibility, enzymatic activities (nitrate reductase, alkaline phosphatase, arginine dihydrolase, ornithine decarboxylase, urease, cytochrome oxidase, staphylocoagulase in rabbit plasma, heat-stable nuclease, amidases, oxidases, clumping factor, and hemolytic activity on sheep or bovine blood agar). DNA–DNA hybridization may distinguish between species where there is less than 70% sequence homology. Currently, there are 38 species recognized for the genus *Staphylococcus* (Table 13.1). The three coagulase-positive species (*aureus*, *intermedius*, and *hyicus* [coagulase is a variable trait in this species]) are the primary pathogens whereas the coagulase-negative species may be opportunistic pathogens. *S. aureus* has an extended host range, being found on humans and a wide range of animal species. Although it resembles *S. aureus* in many characteristics, including the array of toxins produced, *S. intermedius* differs in DNA se-

quence and the presence of glycerol teichoic acid rather than ribitol teichoic acid in the cell envelope. *S. intermedius* is the predominant coagulase-positive staphylococcus isolated from dogs and is also found in other carnivores, horses and birds. Certain strains of *S. intermedius* (11.3% in one survey) have been shown to express enterotoxins and there are reports of involvement of this species in food poisoning episodes (Almazan *et al.*, 1987; Hirooka *et al.*, 1988; Khambaty *et al.*, 1994; Edwards *et al.*, 1997; Burkett and Frank, 1998; Becker *et al.*, 2001). *S. hyicus* is an inhabitant of carnivores, poultry, cattle, and especially pigs. Culture supernatants from certain strains of *S. hyicus*, representing both of its subspecies, were shown to induce an emetic response in monkeys, the principle characteristic of staphylococcal enterotoxins. However, no immunoreactive species wee detected in these supernatants with antisera to the major enterotoxins of *S. aureus* (Adesiyun *et al.*, 1984).

## Staphylococcal enterotoxins

Studies of staphylococcal food poisoning (SFP) date back to 1894 and the work of J. Denys. In 1914, Barber was able to produce the symptoms of SFP by consuming milk from a cow with *S. aureus* mastitis. In 1930, it was shown that the symptoms of SFP could be induced with a filtrate of a culture of *S. aureus*, demonstrating the involvement of a toxin (Dack *et al.*, 1930). Staphylococcal enterotoxins are secreted proteins with molecular weights of approximately 26,000 to 29,000.

**Table 13.1** Species of the genus *Staphylococcus*[a]

| | |
|---|---|
| *arlettae* | *hyicus* subsp. *hyicus* |
| *aureus* subsp. *anaerobius* | *intermedius* |
| *aureus* subsp. *aureus* | *kloosii* |
| *auricularis* | *lentus* |
| *capitis* subsp. *capitis* | *lugdunesis* |
| *capitis* subsp. *urealyticus* | *lutrae* |
| *caprae* | *muscae* |
| *carnosus* subsp. *carnosus* | *pasteuri* |
| *carnosus* subsp. *utilis* | *piscifermentans* |
| *caseolyticus* | *pulvereri* |
| *chromogenes* | *saccharolyticus* |
| *cohnii* subsp. *cohnii* | *saprophyticus* subsp. *bovis* |
| *cohnii* subsp. *urealyticus* | *saprophyticus* subsp. *saprophyticus* |
| *condimenti* | *schleiferi* subsp. *coagulans* |
| *delphini* | *schleiferi* subsp. *schleiferi* |
| *epidermidis* | *sciuri* subsp. *carnaticus* |
| *equorum* | *sciuri* subsp. *lentus* |
| *felis* | *sciuri* subsp. *rodentium* |
| *fleurettii* | *sciuri* subsp. *sciuri* |
| *gallinarum* | *simulans* |
| *haemolyticus* | *succinus* |
| *hominis* subsp. *hominis* | *vitulinus* |
| *hominis* subsp. *novobiosepticus* | *warneri* |
| *hyicus* subsp. *chromogenes* | *xylosus* |

[a]Taken from http://www.bacterio.cict.fr/index.html.

The enterotoxins have the following properties: (1) they are emetic; (2) they are polyclonal activators of T-cells; (3) they are relatively heat resistant; and (4) they are relatively protease resistant. The defining property of the enterotoxins is that of inducing emesis and gastroenteritis upon oral administration to primates. Recent studies have also shown that emesis can be induced in ferrets (Wright *et al.*, 2000) and shrew mice (Hu *et al.*, 2003). Thus nonprimate animal models of enterotoxigenicity may be available to researchers. To date, a number of enterotoxins have been identified which are distinguished serologically. These are designated staphylococcal enterotoxin A (SEA) through R (SER), in accordance with an alphabetical nomenclature for the classification of enterotoxins (Betley, *et al.*, 1990). The known enterotoxin genes share 26–78% amino acid sequence identity (Table 13.2). Minor sequence variants of SEC have been reported and have been designated SEC1, SEC2, SEC3, SEC$_{bovine}$, SEC$_{ovine}$, and SEC$_{canine}$ (Marr *et al.*, 1993). A toxin which is a sequence variant of SEG, Seg$_v$, has been identified (Abe, *et al.*, 2000). Toxic shock syndrome toxin-1 (TSST-1) was once referred to as enterotoxin F (Bergdoll *et al.*, 1981). Characterization of TSST-1 revealed that this toxin shared many biological activities with enterotoxins, but did not cause emesis. The SEF designation has been retired. Additional enterotoxin-like sequences have been identified in *S. aureus*. An enterotoxin-like pseudogene, *sezA* has also been reported (Soltis *et al.*, 1990). This determinant is transcribed but is not translated because it lacks an initiation codon. The *seg, sei, sek, sel,* and *sem* determinants are organized as an operon which additionally includes two pseudogenes Ψ*ent1* and Ψ*ent2* on a 3.2-kb DNA element (Jarraud *et al.*, 2001; Monday and Bohach, 2001). The operon has been given the designation "enterotoxin gene cluster (*egc*)" and is present in many clinical isolates of *S. aureus*. Phylogenetic analysis of enterotoxin genes indicated that they all potentially could have arisen from this cluster, identifying *egc* as a putative nursery for the development of new enterotoxin genes.

The incidence pattern of particular enterotoxins in SFP has changed over time. However, it must be kept in mind that most of the enterotoxins were unrecognized during the earlier studies of the role of the enterotoxins in SFP. Early studies indicated that SEA was the most frequently identified enterotoxin in SFP cases, followed by SED and SEC (Holmberg and Blake, 1984). SEB, only rarely implicated in early studies of SFP, is more frequently implicated in more recent outbreaks. Many recent studies have implicated SEC as the major enterotoxin involved in cases of SFP. Enterotoxins have also been implicated in cases of toxic shock syndrome (Jarraud *et al.*, 1999).

## The enterotoxin proteins

The staphylococcal enterotoxins are defined by their ability to induce emesis in primates. They are part of a family of pyrogenic exotoxins produced by staphylococci and *Streptococcus pyogenes* (Bohach *et al.*, 1990). The enterotoxins can induce profound immune system responses by stimulating T cells with particular Vβ elements. The fraction of responding T cells can be orders of magnitude greater than that evoked by conventional antigens. Accordingly, they are termed "superantigens" and the family of proteins the pyrogenic toxin superantigens (PTSAg; Marrack and Kappler, 1990; Dinges *et al.*, 2000). The enterotoxins are functionally bivalent T cell mitogens that bind simultaneously to an MHC II molecule on antigen presenting cells and the variable region of the β chain (Vβ) of the T cell receptor (TCR; Figure 13.1). Unlike conventional protein antigens that are processed by antigen presenting cells and then peptides placed in a peptide binding grove on the MHC II, superantigens are not processed before binding to the MHC II molecule. They bind on the outer surface of the MHC II. These superantigens bind to the Vβ chain of the TCR regardless of the composition of the variable regions. A given type of superantigen can bind to all of the T cells bearing a particular Vβ chain. Superantigens activate a much higher percentage of T cells than can be activated by conventional protein antigens. As a consequence of the superantigen bridging the MHC II on the antigen presenting cell and the TCR, a large amount of proinflammatory cytokines, including tumor necrosis factor alpha (TNFα), interferon gamma (INF-γ), and interleukins-and 2 (IL-1 and -2) are produced and these cytokines produce clinical symptoms that include fever, hypotension, and shock. SEA has also been shown to bind to the MHC class I molecule and produce a cytokine response, although with a lower efficiency than the response seen with cells bearing MHCII (Wright and Chapes, 1999).

Crystal structures have been determined for a number of the enterotoxins, including SEA, SEB, SEC1, SEC2, and SED. Although there is substantial variation in amino acid sequences, their structures are remarkably similar and resemble that of TSST-1.

The emetic properties of the enterotoxins are distinct from the superantigenicity functions. From protease digestion studies, it was concluded that loss of the amino-terminal part of the protein had no effect on induction of emesis in monkeys. For example, the SEC1 enterotoxin which had its 59 N-terminal amino acid residues removed retained the ability to provoke emesis in monkeys (Spero *et al.*, 1976). Most staphylococcal enterotoxins contain two cysteine residues that can potentially form an intramolecular disulfide bond which results in the generation of a constrained loop. Pyrogenic exotoxins which lack emesis inducing activity do not possess the paired cysteine residues, and thus it was thought that the disulfide bond and loop formation are important in inducing emesis. They are not essential, however, in that SEI contains only a single cysteine residue and is emetic, albeit at higher doses (Table 13.3; Munson *et al.*, 1998). The disulfide constrained loops of the various enterotoxins differ in length and amino acid sequence composition and proteolytic nicking of the loop has no effect on emesis-inducing activity. Therefore, the loop sequences appear not to be the critical component. Cysteine residues from SEC1 have been replaced by means of site-directed mutagenesis and the effects on emesis-inducing activity differed depending on

**Table 13.2** Percent amino acid sequence identity between precursor forms of staphylococcal enterotoxins

| | SEA | SEB | SEC1 | SEC2 | SEC3 | SED | SEE | SEG | SEGv | SEH | SEI | SEJ | SEK | SEL | SEM | SEN | SEO | SEP | SEQ | SER |
|---|---|---|---|---|---|---|---|---|---|---|---|---|---|---|---|---|---|---|---|---|
| SEA | 100 | 33 | 30 | 31 | 31 | 50 | 83 | 27 | 27 | 37 | 39 | 64 | 31 | 34 | 35 | 39 | 38 | 77 | 35 | 31 |
| SEB | | 100 | 68 | 67 | 69 | 35 | 32 | 44 | 44 | 33 | 31 | 33 | 30 | 28 | 29 | 32 | 36 | 34 | 31 | 46 |
| SEC1 | | | 100 | 97 | 94 | 31 | 29 | 41 | 41 | 27 | 26 | 30 | 26 | 28 | 26 | 29 | 34 | 30 | 35 | 40 |
| SEC2 | | | | 100 | 96 | 31 | 30 | 42 | 42 | 27 | 28 | 31 | 28 | 28 | 27 | 28 | 34 | 31 | 32 | 41 |
| SEC3 | | | | | 100 | 32 | 29 | 42 | 42 | 26 | 27 | 30 | 26 | 26 | 27 | 29 | 33 | 31 | 34 | 43 |
| SED | | | | | | 100 | 52 | 26 | 28 | 35 | 33 | 51 | 33 | 39 | 41 | 38 | 40 | 48 | 34 | 28 |
| SEE | | | | | | | 100 | 27 | 26 | 35 | 35 | 63 | 31 | 35 | 37 | 39 | 38 | 78 | 38 | 29 |
| SEG | | | | | | | | 100 | 97 | 34 | 28 | 29 | 27 | 28 | 28 | 31 | 31 | 31 | 29 | 60 |
| SEGv | | | | | | | | | 100 | 34 | 33 | 35 | 28 | 28 | 28 | 31 | 33 | 31 | 29 | 58 |
| SEH | | | | | | | | | | 100 | 33 | 34 | 29 | 28 | 32 | 37 | 35 | 33 | 28 | 33 |
| SEI | | | | | | | | | | | 100 | 34 | 67 | 55 | 57 | 31 | 31 | 33 | 59 | 30 |
| SEJ | | | | | | | | | | | | 100 | 32 | 31 | 34 | 37 | 43 | 60 | 36 | 31 |
| SEK | | | | | | | | | | | | | 100 | 56 | 57 | 28 | 30 | 33 | 55 | 29 |
| SEL | | | | | | | | | | | | | | 100 | 54 | 30 | 37 | 34 | 55 | 28 |
| SEM | | | | | | | | | | | | | | | 100 | 28 | 31 | 34 | 60 | 30 |
| SEN | | | | | | | | | | | | | | | | 100 | 42 | 40 | 27 | 33 |
| SEO | | | | | | | | | | | | | | | | | 100 | 38 | 35 | 39 |
| SEP | | | | | | | | | | | | | | | | | | 100 | 35 | 31 |
| SEQ | | | | | | | | | | | | | | | | | | | 100 | 32 |
| SER | | | | | | | | | | | | | | | | | | | | 100 |

Sequence comparison was performed with program BLASTP. Sequences were obtained from GenBank.

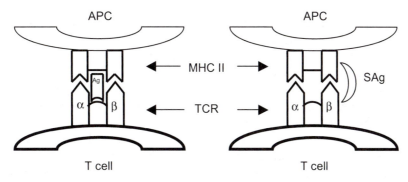

**Figure 13.1** Interactions between antigen-presenting cells (APC) and T cells involving conventional peptide antigens (Ag) as shown on the left and superantigens (SAg) shown on the right. APC process proteins and present peptides to specific subsets of T cell receptors (TCR) in the peptide-binding groove of the MHC class II molecules. Superantigens are not processed but interact with MHC class II molecules directly by binding on the outside of the MHC class II molecule and the variable region of the β chain of the TCR.

the specific amino acid substitution for the cysteine residue. Mutants with a serine substitution were emetic while alanine substitution mutants were inactive (Hovde *et al.*, 1994). Thus residues near the disulfide bond appear to be critical to emesis-induction. Mutagenesis experiments with SEA indicated that residues close to the N terminus can also influence emesis (Harris and Betley, 1995). Glycine substitutions at positions 25, 47, and 48 resulted in reductions in emetic potency of the enterotoxin.

The enterotoxins are resistant to proteolytic digestion with trypsin, chymotrypsin, pepsin, rennin, and papain. Resistance to proteases is an important feature of an ingested enterotoxin. In addition, protease cleavage of an enterotoxin, without separation of the resulting fragments, is apparently insufficient to cause the inactivation of the enterotoxin. The toxins tend to be resistant to denaturation, and if denatured, can spontaneously renature and regain biological activity (Spero *et al.*, 1976; Warren, 1977).

Another characteristic feature of the staphylococcal enterotoxins is that they are relatively heat resistant. In crude preparations of toxin, boiling of the sample for 30 minutes is inadequate to completely inactivate the enterotoxin (Bergdoll, 1983). Lower temperatures, such as those used for pasteurization of milk or temperatures generated by heat lamps used for keeping food warm, have little or no effect on enterotoxin activity. The biological activity of SEB was retained after heating

for 16 hours at 60°C (Schantz *et al.*, 1965). The heat stability of the proteins may even be further enhanced when the toxin is present in food or other complex environments. Because the temperatures required for inactivation of enterotoxins are higher than those needed to kill the enterotoxin-producing bacteria, food involved in cases of SFP may not have culturable organisms at the time the food is served.

## Mode of action

The typical symptoms associated with staphylococcal food poisoning are vomiting with or without diarrhea and abdominal cramping. In severe cases, patients may experience fever and shock. Enterotoxin effects are likely to be limited to the gastrointestinal tract under normal conditions. In studies with rats, orally administered enterotoxins did enter the circulation, but were rapidly removed by the kidneys thus limiting any possible systemic effects (Beery *et al.*, 1984). Enteritis resulting from food poisoning in humans was initially studied by Palmer (1951). Thereafter, many animals including cats, dogs, monkeys, rabbits, chinchillas, rats, and mice have been used to study the pathologic feature of staphylococcal food poisoning. Some early studies showed that injection of enterotoxins induced symptoms in cats similar to those seen in food poisoning cases and repeated daily injection of enterotoxins caused moderate to severe jejunitis in cats and kittens (Tan *et al.*, 1959). Severe gastroenteritis was produced in dogs by intra-

**Table 13.3** Emetic doses for staphylococcal enterotoxins

| Toxin | Estimated ED$_{50}$ | Species | Reference |
|-------|---------------------|---------|-----------|
| SEA | < 25 µg/kg | Rhesus monkey | Harris *et al.* (1993) |
| | 10 µg/kg | Rhesus monkey | Stelma and Bergdoll (1982) |
| | 144 ± 50 ng | Humans | Everson *et al.* (1988) |
| SEB | 0.4 µg/kg | Humans | Raj and Bergdoll (1969) |
| | 0.9 µg/kg | Rhesus monkey | Schantz *et al.* (1965) |
| SEC1 | 0.1–1.0 µg/kg | Pigtail monkey | Schlievert *et al.* (2000) |
| SEG | ~ 80 µg/kg | Rhesus monkey | Munson *et al.* (1998) |
| SEI | >150 µg/kg | Rhesus monkey | Munson *et al.* (1998) |

jejunal instillation of crude staphylococcal filtrates (Warren *et al.*, 1964). Vomiting, diarrhea, shock, and acute gastroenteritis marked by regional edema, hypermenia, mucosal exudation, muscular irritation, and destruction of intestinal villi were observed 2 to3 hours after receiving one or two large doses of enterotoxin. Lymphoid hyperplasia was observed after repeated instillation of enterotoxin filtrates. Monkeys have been proven to be the most susceptible laboratory animals to enterotoxins. Vomiting and diarrhea occurred 2 to 8 hours after oral administration of enterotoxins, though some animals did not develop symptoms until 16 hours after feeding of toxins (Kent, 1966). Acute gastroenteritis was well developed by 2 hours, reached a maximum at 4 to 8 hours, and rapidly regressed after 12 to 72 hours. The lesions found in the small intestine were characterized by epithelial damage, villus distension, and crypt lengthening. Electron microscopy revealed mitochondrial alterations in villus and crypt epithelial cells, as well as in diverse cells of the jejunal mucosa (Merrill and Sprinz, 1968). Though emetic response and the severity of tissue destruction correlated well, vomiting and intestinal lesions did not always occur at the same time. For instance, after repeated daily feeding of enterotoxins, monkeys became refractory to emesis, but intestinal lesions were still observable (Kent, 1966). It was suggested that specific receptors for enterotoxins were located within the gastrointestinal tract, and the sensory stimulus reached the medullary vomiting center *via* vagus and sympathetic nerves (Sugiyama and Hayama, 1964; 1965). However, such receptors have not been identified. Studies with human kidney proximal tubular cells have identified a neutral glycosphingolipid as a putative receptor for enterotoxin (Chatterjee and Jett, 1992). This study may provide clues as to the nature of the receptor on intestinal epithelial cells.

Enterotoxin exposure results in intestinal lesions, suggesting a direct damaging effect on the intestinal epithelium. However, *in vitro* studies have produced conflicting results. Enterotoxins SEA and SEB did not damage the integrity of cultured Henle 407 human intestinal cells as measured by leakage of cytoplasmic constituents, amino acid transport, and macromolecular synthesis (Buxser and Bonventre, 1981). However, SEB was shown to have a direct effect on the barrier function of endothelial cells (Campbell *et al.*, 1997). The damage induced by SEB was prevented using inhibitors of protein tyrosine kinases. The pathogenic effect of enterotoxins on the gastrointestinal tract could also result from production of cytokines by local immune cells, such as macrophages, mast cells, and other MHC II bearing cells lining the gut, and/or T cells located in this area. In this regard, one study has shown that SEB, through its effect of expanding T cell populations, gives rise to T cells which induce apoptosis in syngeneic intestinal epithelial cells (Ito *et al.*, 2000). The autoreactive activity of the T cells was suppressed by the addition of IL-10. The authors suggest that the bacterial toxin has the potential to abrogate self tolerance by stimulating autoreactive T cells which are cytolytic to the target cells.

## Genetics of enterotoxin production

Many strains of *S. aureus* do not carry genes encoding the enterotoxins. Others can produce one or more of these toxins. Therefore, all enterotoxin genes are "accessory elements", genes which get added to the basic genetic capacity of this organism. Enterotoxin genes in *S. aureus* have been found on plasmids, on bacteriophage, and as part of pathogenicity islands. The enterotoxin A determinant (*sea*) is introduced into *S. aureus* strains as part of a bacteriophage (Betley and Mekalanos, 1985). The *sea*-containing phages comprise a family of related, but not identical, phage as evidenced by differences in physical maps. The actual prophage integration site is within the *hlb* locus, encoding the hemolytic sphingomyelinase known as β-toxin. The *sea*-containing phage are referred to as double converting, or triple converting phage (Coleman *et al.*, 1989). Strains acquiring this prophage lose the capacity to produce β-toxin due to insertional inactivation of *hlb* by the integration of the prophage. Doubling converting phage thus lose β-toxin synthesis simultaneously with acquisition of production of SEA. Triply converting phage carry *sak*, the gene for staphylokinase, along with *sea* and, therefore, lysogens produce both SEA and staphylokinase while losing β-toxin production.

The *sezA* pseudogene is carried on a temperate phage which can be induced by exposure of the lysogens to ultraviolet light (Soltis *et al.*, 1990). The gene encoding SEE was also identified on UV-inducible phage which share sequence homology with *sea* containing phage (Couch *et al.*, 1990). Because the UV-induced *see* containing phage were not able to form plaques on various *S. aureus* strains tested, it was suggested that the *see* containing phage may be defective. Based on chromosomal sequence analysis, it is likely that *sep* is also associated with bacteriophage DNA. The *sed*, *sej*, and *ser* determinants are carried on a 27.6 kb penicillinase-type plasmid in *S. aureus* (Bayles and Iandolo, 1989; Zhang *et al.*, 1998; Omoe *et al.*, 2003). The determinants are oriented in opposite directions and the open reading frames are separated by an 895 bp intergenic sequence. Physical mapping of 21 independent *sed* encoding plasmids indicated that *sej* is always co-resident. Thus SED producing staphylococci also produce SEJ. The *seh* determinant, as well as many of the *sec* family of enterotoxins, is found in the chromosome of the respective enterotoxin-producing strains, but the nature of the genetic element on which they reside, bacteriophage, transposon, integrated plasmid, etc., has not been elucidated.

Certain enterotoxin genes have been found to be within pathogenicity islands. Five staphylococcal pathogenicity islands (SaPIs) have recently been described (SaPI1–4, SaPIbov). They are present in the genomes of many staphylococci but absent from closely related strains, are relatively large genomic fragments (> 15 kb) that differ in GC content from the rest of the chromosome, contain virulence determinants, are flanked by short direct repeats likely generated upon insertion of the elements into the genome, and possess genes encoding functions associated with genetic mobility such as integrases.

The first staphylococcal pathogenicity island, SaPI1, was identified and characterized by Lindsay *et al.* (1998) as the genetic element encoding TSST-1. SaPI1 also encodes SEK and part of SEQ. Mobility of SaPI1 has been demonstrated only in the presence of a helper phage, such as 80α. SEB, SEC, SEK, SEL, and SEQ are known to be encoded within pathogenicity islands. This suggests that transfer of enterotoxin determinants can occur between staphylococcal strains. These bacteriophage-mobilized islands are likely to be involved in an ongoing evolution of enterotoxigenic *S. aureus*. There may be some evidence in the literature for movement of the pathogenicity islands. A 56.2-kb plasmid, referred to as pZA10, encoding β-lactamase and heavy metal resistance was identified from an enterotoxigenic clinical isolate (Altboum *et al.*, 1985). The physical map of pZA10 and subsequent co-transformation analysis indicated that *seb* and *sec1* were linked and associated with *bla* and metallic ion resistance genes on an 18.1-kb *Sal*I fragment. Curing of the plasmid resulted in a loss of SEB production. This may have represented the phage-mediated movement of a pathogenicity island into a penicillinase-type plasmid, rather than to a more commonly seen chromosomal location. The plasmid was physically and segregationally unstable, which may be a problem with plasmid carriage of the island and would explain why plasmid copies are not a more common occurrence.

## Enterotoxin expression

The expression of enterotoxins varies among serological types. SEA and SEJ are produced in an apparently unregulated fashion during the exponential phase of growth. However, as much as an eight-fold difference in the amount of SEA produced can be attributed to the specific *sea*-harboring temperate phage present (Betley *et al.*, 1992). In contrast, SEB, SEC, and SED are maximally produced during the transition from exponential to stationary phase of growth (Gaskill and Khan, 1988; Bayles and Iandolo, 1989; Otero *et al.*, 1990). In addition to differences in the kinetics of expression, the amount of toxin synthesized also varies widely. SEA, SED, SEE, and SEJ are often produced at concentrations less than 5 µg/mL of culture supernatant, whereas SEB and SEC are produced in larger quantities, usually on the order of 100 µg/mL of culture supernatant (Bergdoll, 1979). Production of a given enterotoxin may also vary with the host background. For instance, strains S6, DU4916, and COL produce 375, 50, and 12 µg/mL of SEB, respectively (Compagnone-Post *et al.*, 1991). Strains isolated from clinical cases often produce higher amounts of SEC (≥25 µg/mL) than food strains and animal strains (Marr *et al.*, 1993). Most of the bovine and ovine isolates produce less than 5 µg/mL of SEC. Similar information concerning expression levels for the other enterotoxins are not yet available.

The postexponential growth phase stimulation of *seb*, *sec*, and *sed* expression is a characteristic of many staphylococcal extracellular virulence factors which are under the control of the accessory gene regulator (Agr) system. The Agr system is a quorum sensing system in the staphylococci that regulates the expression of a number of virulence-associated genes of *S. aureus* (for a review, see Arvidson and Tegmark, 2001). The *agr* locus consists of two divergent transcripts (Figure 13.2). The *agrACDB* genes encode the two component regulatory system. AgrD is a polypeptide which appears to be processed into a short peptide (i.e. an octapeptide) and exported from the cell by the AgrB protein. The peptide binds to the AgrC protein which is the sensory protein of the two component regulatory system. When a threshold concentration of peptide is bound, the histidine kinase activity of AgrC membrane protein is induced and AgrC becomes phosphorylated. The phosphate moiety is then transferred to the AgrA protein, the response regulator of the two component system. The phosphorylated AgrA is a transcriptional factor which binds to the *agr* promoters and results in increased transcription of RNAII (encoding *agrACDB*) and RNAIII. The latter is a 514 nt transcript which encodes the delta hemolysin (*hld*). However, it has been shown that it is the RNA molecule, not the Hld polypeptide, which is the effector molecule of the Agr system. When the concentration of RNAIII in the cell increases, transcription of a variety of exoproteins (including alpha hemolysin, toxic shock syndrome toxin, and proteases) is increased while the transcription of a variety of cell surface proteins (including protein A and fibrinogen binding protein) is reduced. The mechanism(s) by which the RNAIII mediates this transcription control have not been elucidated. Regulation of virulence factors is even more complicated because there are other transcriptional factors (the Sar family including SarA, SarR, SarS, SarT, and Rot) which can affect the Agr system (SarA for examples up-regulates the expression of the *agr* transcripts). SarA also regulates virulence genes in an Agr-independent fashion (Cheung *et al.*, 1992). Furthermore, additional two component regulatory systems (*sae* and *arl*) have been identified and may contribute to regulation of virulence-associated genes (Giraudo *et al.*, 1997; Fournier and Hooper, 2000).

The *agr*-associated regulation of enterotoxin gene expression has been examined by Iandolo, Betley, Khan and their co-workers (Gaskill and Khan, 1988; Bayles and Iandolo, 1989; Mahmood and Khan, 1990; Regassa *et al.*, 1991). Their experiments demonstrated that loss of *agr* signal transduction system resulted in substantial reductions of enterotoxin protein and mRNA production. The reductions in mRNA levels were 4-fold for *seb*, 5.5-fold for *sed*, and 2–3 fold for *sec*. The reduction in enterotoxin protein production was more dramatic. For example, the amount of SEC was reduced 16 to 32-fold as revealed by western blot analysis (Regassa *et al.*, 1991). The levels of staphylococcal enterotoxin B (SEB) produced by various naturally occurring toxogenic strains of *S. aureus* are highly variable. The *seb* determinants from a high-producer strain, S6, and from DU4916 and COL (medium-and low-level toxin-producer strains, respectively) showed that their open reading frame upstream sequences are identical (Compagnone-Post *et al.*, 1991). The *seb* determinants from these three strains, when cloned into the laboratory 8325 genetic background, each were expressed to the same extent. RNAIII, the regulatory species

# Agr Network

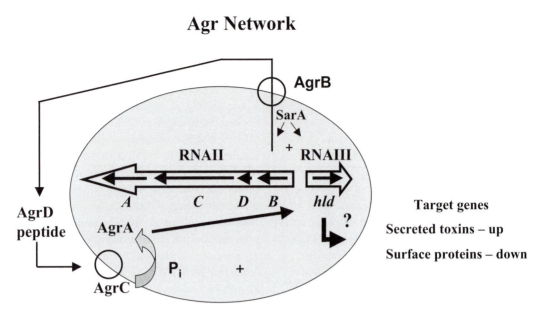

**Figure 13.2** Agr system regulation of virulence genes of *S. aureus*. The polypeptide encoded by *agrD* is processed to a peptide and exported by the AgrB protein. As growth of the cells occurs, the AgrD peptide accumulates in the environment. When a threshold concentration is achieved, the peptide binds to the AgrC membrane protein (the sensory protein of the two-component regulatory system). Binding of the peptide activates the histidine kinase activity of AgrC resulting in an autophosphorylation of AgrC. The phosphate moiety is then transferred to AgrA, the response regulator of the two component regulatory system. When phosphorylated, AgrA is a transcriptional factor which binds to the *agr* promoter region resulting in an increased production of RNAII and RNAIII. Elevation of RNAIII levels in the cells results in an increased transcription and expression of genes encoding a variety of exotoxins, including many of the enterotoxins, and simultaneously effects a reduction in expression of genes encoding certain wall-associated surface proteins. The SarA protein is a transcriptional regulator which is a positive regulator of *agr* transcription and additionally affects expression of certain virulence genes in an Agr-independent fashion. Proper functioning of the Agr quorum sensing system has been shown to be essential for virulence in *S. aureus*.

of the *agr* system, was shown by Northern blot analysis to vary in an identical fashion as *seb* RNA levels in strains S6, DU4916, and COL. These results suggest that differences in the *agr* system are responsible for the differential enterotoxin production by many naturally occurring strains. The plasmid-borne *sed* determinant is regulated two-fold by the Agr system and genetic studies on its promoter have shown that the Rot protein (for repressor of toxin; McNamara *et al.*, 2000) represses *sed* transcription and that the Agr system inactivates this Rot-mediated repression at the post-exponential phase of growth (Zhang and Stewart, 2000; Tseng *et al.*, 2004).

In addition to *agr* and *sar* systems, the expression of enterotoxin genes is also affected by glucose concentration in culture medium. Several early studies suggested that addition of glucose to culture medium resulted in decreased production of SEA, SEB, SEC and other exoproteins. Because these early experiments were carried out without a strictly controlled glucose concentration and pH, it was not known whether the glucose effect was due to a catabolite repression mechanism or is a result of the decreased pH resulting from metabolism of the glucose. The addition of glucose to the culture medium reduced RNAIII expression, an effect which was most pronounced when the broth pH was allowed to become acidic (Regassa *et al.*, 1991). However, repression of *sec* expression by glucose occurred even when the culture pH was maintained at 6.5.

## Food poisoning episodes

Staphylococci are ubiquitous in animals and as such they become likely contaminants of meat and meat products. Staphylococci can also be isolated from raw milk and are found in higher numbers if mastitis is present in the herd. Staphylococci are poor competitors against other bacteria which are normal flora contaminants. Thus, they are not normally present and are usually eliminated by cooking or pasteurization. Involvement of staphylococci as agents in food poisoning requires production of enterotoxin. A toxin dose of less than one microgram in contaminated food will elicit symptoms of food poisoning. This toxin level is reached when the bacterial population exceeds $10^5$ per gram (Food and Drug Administration, 2003). Several species of *Staphylococcus* have been shown to produce enterotoxin and thus have the potential to cause gastroenteritis. However, nearly all cases of staphylococcal food poisoning are attributed to *S. aureus*. Food handlers, and not the food animals, are the usual source of the bacterial contamination. The epidemiology of staphylococcal food poisoning is thus very different from that caused by *Salmonella*, *Campylobacter*, or *Escherichia coli* where the animal or a product from the animal is the usual source of the contamination. In humans, 30% or more of healthy individuals harbor *S. aureus*, especially in the nasal vestibule (Holmberg and Blake, 1984; Lina *et al.*, 2003). Human strains of *S. aureus* are more likely produce enterotoxin than are animal associated strains, and this may also contribute to the usual source of infection being food preparers. However, recent reports have indicated that a significant percentage of bovine and ovine mastitis isolates (28–88%) produce one or more enterotoxins (Kenny *et al.*, 1993; Orden *et al.*, 1994; Adesiyun, 1995; Lee *et al.*, 1998). Substantial geographical variation has been found as isolates from certain geographical locations have not yet been found to produce enterotoxin (Lee *et al.*, 1998). Although human isolates are most commonly associated with SFP episodes, animal isolates have been the sources of some cases of SFP. For example, staphylococcal mastitis can lead to food poisoning cases involving improperly pasteurized dairy products.

The predominant enterotoxin serotypes implicated in food poisoning episodes depend on the geographical locale, the nature of the food or source population examined, and the date of the SFP episode, because reported frequencies have changed somewhat over time. The superantigenic properties of the enterotoxins likely provide the selective pressure to allow new toxin types to emerge and determine which of the enterotoxins will predominate in staphylococci from various animal species. The superantigenicity is important to the immunosuppressive virulence properties of the organism. Different toxins activate different V$_\beta$ subtype populations, and different animal species have characteristic V$_\beta$ subtype patterns.

Enterotoxin A was, and often remains, the predominant toxin identified in cases of SFP, with SED and SEC being the next most common serotypes. With the evolution of more sensitive detection methodologies, the range of enterotoxin serotypes implicated in SFP episodes is likely to increase.

The emetic dose of specific enterotoxins has been shown to vary by serotype (Table 13.3). Because of differences in the emetic dose of a particular serotype of enterotoxin, intrinsic variation in expression levels of enterotoxins, host-specific expression differences, and environmental factors including temperature and the type of food harboring the organism, it is not possible to state with any degree of certainty the numbers of *S. aureus* in food required to cause SFP. However, as the estimated number of cases in the United States of 185,060 will attest, the cell numbers necessary to achieve a toxin load sufficient for SFP episodes is easily achieved. SFP episodes usually result from one or more of the following conditions: improper hygiene (hand washing, sanitation of instruments), inadequate refrigeration, food preparation long time periods in advance of consumption, inadequate cooking or heating of food, or maintaining food warm for long periods of time prior to consumption. Specific examples of outbreaks can be found in the *Staphylococcus aureus* section in the *Bad Bug Book* (Food and Drug Administration, 2003).

Staphylococci are able to grow in a variety of food items. They do not replicate well in raw meat, but do multiply in fermented sausage. They are well suited for growth in different environments. They can utilize a wide range of carbon sources, they have a broad range of growth temperatures (7°C to 47–48°C), pH tolerance (4.5 to 9.3), and can grow under conditions of lower water activity ($a_w$) than other non-halophilic bacteria. Growth has been observed with $a_w$ values as low as 0.83. The organism is thus relatively salt tolerant and able to replicate in certain salt cured meat products.

## Identification of staphylococcal food poisoning episodes

Because of the transient nature of SFP, many occurrences go unreported. In addition, the symptoms are similar to that of food poisoning cause by *Bacillus cereus*. Thus the true incidence of SFP is unknown. In the diagnosis of SFP, proper interviews with the victims and collection and analysis of the food consumed are essential. Incriminated foods should be collected and examined for staphylococci. However, a failure to culture staphylococci does not rule out SFP in that the enterotoxins will survive temperatures which cause loss of viability of the bacteria. Baird–Parker agar is a widely used selective plating medium for isolation of *S. aureus*. In this medium, lithium chloride and potassium tellurite are included as selective agents and egg yolk and pyruvate improve the recovery of stressed or damaged cells. Reduction of the tellurite by *S. aureus* produces characteristic shiny black colonies that are surrounded by a zone of clearing resulting from lecithinase activity on the egg yolk supplement. *S. hyicus* produces colonies resembling those of *S. aureus* on this medium. *S. intermedius*, however, is unable to reduce tellurite and produces white colonies on Baird–Parker agar.

Detection of the enterotoxin in the food is the most conclusive test linking illness with consumption of a particular item of food. A number of serological methods for the detection of staphylococcal enterotoxins in food have been developed. The limitation of this approach is that only antibodies to certain serotypes of the toxins are included in the tests and thus those toxins with more divergent amino acid sequences may not be detected. Reverse passive latex agglutination tests with a sensitivity of 0.5 ng per mL are commercially available. Several enzyme-linked immunosorbent assay (ELISA) methods have also been developed. ELISA tests will detect 0.1–1.0 ng of enterotoxin per gram of food. Specific monoclonal or polyclonal antibodies are adsorbed onto wells of a microtiter plate. Putative enterotoxin-containing samples are added to the wells so that the enterotoxin, if present, can be captured by the antibodies. The unbound material is washed away and the presence of any enterotoxin is detected by the addition of a secondary antibody–enzyme conjugate, followed by addition of a chromogenic substrate for the enzyme. The amount of enterotoxin present in the sample is then determined spectrophotometrically.

If the staphylococci can be isolated from food, it is possible to screen for the presence of enterotoxin genes by PCR. This is important for a number of reasons. First of all, the amount of enterotoxin produced and the amount necessary to induce emesis varies markedly among the different enterotoxin serotypes. Furthermore, the serological detection techniques utilize antibodies to the classic enterotoxin serotypes, which may lack sufficient reactivity with the newer serotypes and thus fail to detect their presence. Recent studies have indicated that the frequency of foodborne *S. aureus* isolates bearing enterotoxins G, H, I, and J was high, with more than half of the strains testing positive by PCR (Rosec and Gigaud, 2002). Multiplex PCR, designed to detect all known enterotoxin determinants, is capable of rapidly assessing if a staphylococcal isolate harbors any of the known enterotoxins and the test is easily expanded as new enterotoxin determinants are identified (Monday and Bohach, 1999).

## Prevention of staphylococcal food poisoning

Proper hygiene by food handlers and holding prepared food at temperatures which limit the growth and enterotoxin production by the staphylococci are the most effective measures that can be taken to prevent SFP episodes. However, nothing is fool-proof because fools are so ingenious. Food will continue to become contaminated by handlers and improper holding conditions will occur. Therefore, it would be desirable to limit either the growth of the staphylococci when introduced into food or to limit the amount of enterotoxins produced. The fact that many of the enterotoxin genes are regulated by the Agr system, provides a possible control point to limit expression of these enterotoxins. Accumulation of the AgrD peptide during growth of the organism triggers an increased synthesis of the Agr-regulated enterotoxin when the concentration of organisms reaches a critical level. However, the stimulation of Agr-regulated toxins only occurs when the homologous AgrD peptide binds to its receptor. Staphylococci have been found to exist in several Agr groups, each differing in the sequence of the AgrD peptide and the AgrC membrane receptor (Ji *et al.*, 1997; Dufour *et al.*, 2002). When the AgrD peptide binds to cells of a different Agr group, there is an inhibition of toxin production, rather than the stimulation seen with binding of the homologous AgrD peptide (Lyon *et al.*, 2002). Because the majority of SFP episodes result from human strains of *S. aureus*, it should be possible to devise an inhibitory peptide from a different species of *Staphylococcus*, which would be unlikely to be involved in food poisoning. This inhibitor could then be incorporated into food which might be at risk for SFP, including food prepared for large groups and kept warm for extended periods of time. The peptide would not prevent the organism from replicating in the food, but might limit the amount of enterotoxin produced by inhibiting activation of the Agr system. This type of approach has been utilized with staphylococci, including inhibition of toxin production by *S. aureus* and prevention of biofilm production by *S. epidermidis* (Otto *et al.*, 1999; Balaban *et al.*, 2003). The small size of these peptides and because they contain an unusual thiol ester-linked cyclic structure (Mayville *et al.*, 1999) should make these compounds relatively heat resistant, stable, and nonimmunogenic.

The staphylococcal enterotoxins, because of the symptoms they induce including shock resulting from their superantigenic properties, their relative stability, and especially because of their activity at low doses, are included on the National Institutes of Health/Centers for Disease Control list of select agents. There is concern that these toxins might be used as biological warfare agents. Because of this, there is ongoing research to develop antagonists that are effective as therapeutic measures following exposure to enterotoxin. Short peptide antagonists have been developed which block the lethal effect of these superantigens by inhibiting the triggering of expression of the T cell cytokines that mediate toxic shock (Kaempfer *et al.*, 2002). The antagonists may prove useful not only as countermeasures in cases of a biological attack with the enterotoxins, but perhaps as therapeutics in the treatment of staphylococcal toxic shock.

## References

Abe, J., Ito, Y., Onimaru, M., Kohsaka, T., and Takeda, T. 2000. Characterization and distribution of a new enterotoxin-related superantigen produced by *Staphylococcus aureus*. Microbiol. Immunol. 44: 79–88.

Adesiyun, A.A. 1995. Characteristics of *Staphylococcus aureus* strains isolated from bovine mastitic milk: bacteriophage and antimicrobial agent susceptibility, and enerotoxigenicity. Zentralbl Veterinarmed [B] 42: 129–139.

Adesiyun, A.A., Tatini, S.R., and Hoover, D.G. 1984. Production of enterotoxin(s) by *Staphylococcus hyicus*. Vet. Microbiol. 9: 487–495.

Almazan, J., de la Fuente, R., Gomez Lucia, E., and Suarez, G. 1987. Enterotoxin production by strains of *Staphylococcus intermedius* and *Staphylococcus aureus* from dog infections. Zentbl. Bakteriol. Mikrobiol. Hyg. A 264: 174–176.

Altboum, Z., Hertman, I., and Sarid, S. 1985. Penicillinase plasmid-linked genetic determinants for enterotoxin B and $C_1$ production in *Staphylococcus aureus*. Infect. Immun. 47: 514–521.

Arvidson, S. and Tegmark, K. 2001. Regulation of virulence determinants in *Staphylococcus aureus*. Int. J. Med. Microbiol. 291: 159–170.

Balaban, N., Giacometti, A., Cirioni, O., Gov, Y., Ghiselli, R., Mocchegiani, F., Vitricchi, C., Del Prete, M.S., Saba, V., Scalise, G., and Dell'Acqua, G. 2003. Use of the quorum-sensing inhibitor RNAIII-inhibiting peptide to prevent biofilm formation *in vivo* by drug-resistant *Staphylococcus epidermidis*. J. Infect. Dis. 187: 625–630.

Bayles, K.W. and Iandolo, J.J. 1989. Genetic and molecular analyses of the gene encoding staphylococcal enterotoxin D. J. Bacteriol. 171: 4799–4806.

Becker, K., Keller, B., von Eiff, C., Bruck, M., Lubritz, G., Etienne, J., and Peters, G. 2001. Enterotoxigenic potential of *Staphylococcus intermedius*. Appl. Environ. Microbiol. 67: 5551–5557.

Beery, J.T., Taylor, S.L., Schlunz, L.R., Freed, R.C., and Bergdoll, M.S. 1984. Effects of staphylococcal enterotoxin A on the rat gastrointestinal tract. Infect. Immun. 44: 234–240.

Bennett, J.V., Holmberg, S.D., Rogers, M.F., and Solomon, S.L. 1987. Infectious and parasitic diseases. In: Closing the Gap: The Burden of Unnecessary Illness. R.W. Amler and H.B. Dull, eds. Oxford University Press, New York, p. 102–114.

Bergdoll, M.S. 1979. Staphylococcal intoxications In: Foodborne Infections and Intoxications, H. Riemann and F.L. Bryan, eds. Academic Press, New York, pp. 443–494.

Bergdoll, M.S. 1983. Enterotoxins. In: Staphylococci and Staphylococcal Infections, vol. 2. C.S.F. Easmon and C. Adlum, eds. Academic Press, London, p. 559–598.

Bergdoll, M.S., Crass, B.A., Reiser, R.F., Robbins, R.N., and Davis, J.P. 1981. A new staphylococcal enterotoxin, enterotoxin F, associated with toxic-shock-syndrome *Staphylococcus aureus* isolates. Lancet 1: 1017–1021.

Betley, M.J., Borst, D.W., and Regassa, L.B. 1992. Staphylococcal enterotoxins, toxic shock syndrome toxin and streptococcal pyrogenic exotoxins: a comparative study of their molecular biology. Chem. Immunol. 55: 1–35.

Betley, M.J. and Mekalanos, J.J. 1985. Staphylococcal enterotoxin A is encoded by phage. Science 229: 185–187.

Betley, M.J., Schlievert, P.M., Bergdoll, M.S., Bohach, G.A., Iandolo, J.J., Khan, S.A., Pattee, P.A., and Reiser, R.R. 1990. Staphylococcal gene nomenclature. ASM News 56: 182.

Bohach, G.A., Fast, D.J., Nelson, R.D., and Schlievert, P.M. 1990. Staphylococcal and streptococcal pyrogenic toxins involved in toxic shock syndrome and related illnesses. Crit. Rev. Microbiol. 17: 251–272.

Burkett, G., and Frank, L.A. 1998. Comparison of production of *Staphylococcus intermedius* exotoxins among clinically normal dogs, atopic dogs with recurrent pyoderma, and dogs with a single episode of pyoderma. J. Am. Vet. Med. Assoc. 213: 232–234.

Buxser, S. and Bonventre, P.F. 1981. Staphylococcal enterotoxins fail to disrupt membrane integrity or synthetic function of Henle 407 intestinal cells. Infect. Immun. 31: 929–934.

Buzby, J.C., Roberts, T., Lin, C.T.J., and MacDonald, J.M. 1996. Bacterial foodborne disease: medical costs and productivity losses. Economic Research Service Report no. 741. USDA Economic Research Service, Washington, D.C.

Campbell, W.N., Fitzpatrick, M., Ding, X., Jett, M., Gemski, P., and Goldblum, S.E. 1997. SEB is cytotoxic and alters EC barrier function through protein tyrosine phosphorylation *in vitro*. Am. J. Physiol. 273: L31–39.

Chatterjee, S. and Jett, M. 1992. Glycosphingolipids: the putative receptor for *Staphylococcus aureus* enterotoxin-B in human kidney proximal tubular cells. Mol. Cell. Biochem. 113: 25-31.

Cheung, A.L., Koomey, J.M., Butler, C.A., Projan, S.J., and Fischetti, V.A. 1992. Regulation of exoprotein expression in *Staphylococcus aureus* by a locus (*sar*) distinct from *agr*. Proc. Ntl. Acad. Sci. USA 89: 6462–6466.

Coleman, D.C., Sullivan, D.J., Russell, R.J., Arbuthnott, J.P., Carey, B.F., and Pomeroy, H.M. 1989. *Staphylococcus aureus* bacteriophages mediating the simultaneous lysogenic conversion of β-lysin, staphylokinase and enterotoxin A: molecular mechanism of triple conversion. J. Gen. Microbiol. 135: 1679–1697.

Compagnone-Post, P., Malyankar, U., and Khan, S.A. 1991. Role of host factors in the regulation of the enterotoxin B gene. J. Bacteriol. 173: 1827–1830.

Couch, J.L., Soltis, M.T., and Betley, M.J. 1988. Cloning and nucleotide sequence of the type E staphylococcal enterotoxin gene. J. Bacteriol. 170: 2954–2960.

Dack, G.M., Cary, W.E., Woolpert, O., and Wiggers, H. 1930. An outbreak of food poisoning proved to be due to a yellow hemolytic staphylococcus. J. Prev. Med. 4: 167–175.

Dinges, M.M., Orwin, P.M., and Schlievert, P.M. 2000. Exotoxins of *Staphylococcus aureus*. Clin. Microbiol. Rev. 13: 16–34.

Dufour, P., Jarraud, S., Vandenesch, F., Greenland, T., Novick, R.P., Bes, M., Etienne, J., and Lina, G. 2002. High genetic variability of the *agr* locus in *Staphylococcus* species. J. Bacteriol. 184: 1180–1186.

Edwards, V.M., Deringer, J.R., Callantine, S.D., Deobald, C.F., Berger, P.H., Kapur, V., Stauffacher, C.V., and Bohach, G.A. 1997. Characterization of the canine type C enterotoxin produced by *Staphylococcus intermedius* pyoderma isolates. Infect. Immun. 65: 2346–2352.

Everson, M.L., Hinds, M.W., Bernstein, R.S., and Bergdoll, M.S. 1988. Estimation of human dose of staphylococcal enterotoxin A from a large outbreak of staphylococcal food poisoning involving chocolate milk. Int. J. Food Microbiol. 7: 311–316.

Food and Drug Administration. 2003. *Staphylococcus aureus*. In: Foodborne Pathogenic Microorganisms and Natural Toxins Handbook (Bad Bug Book), URL: http://vm.cfsan.fda.gov/~mow/intro.html.

Fournier, B. and Hooper, D.C. 2000. A new two-component regulatory system involved in adhesion, autolysis, and extracellular proteolytic activity of *Staphylococcus aureus*. J. Bacteriol. 182: 3955–3964.

Freney, J., Kloos, W.E., Hajek, V., Webster, J.A., *et al.* 1999. Recommended minimal standards for description of new staphylococcal species. Int. J. System. Bacteriol. 49: 489–502.

Gaskill, M.E., and Khan, S.A. 1988. Regulation of the enterotoxin B gene in *Staphylococcus aureus*. J. Biol. Chem. 263: 6276–6280.

Giraudo, A.T., Cheung, A.L., and Nagel, R. 1997. The *sae* locus of *Staphylococcus aureus* controls exoprotein synthesis at the transcriptional level. Arch. Microbiol. 168: 53–58.

Harris, T.O. and Betley, M.J. 1995. Biological activities of staphylococcal enterotoxin type A mutants with N-terminal substitutions. Infect. Immun. 63: 2133–2140.

Harris, T.O., Grossman, D, Kappler, J.W., Marrack, P., Rich, R.R., and Betley, M.J. 1993. Lack of complete correlation between emetic and T-cell-stimulatory activities of staphylococcal enterotoxins. Infect. Immun. 61: 3175–3183.

Hirooka, E.Y., Muller, E.E., Freitas, J.C., Vicente, E., Yoshimoto, Y., and Bergdoll, M.S. 1988. Enterotoxigenicity of *Staphylococcus intermedius* of canine origin. Int. J. Food Microbiol. 7: 185–191.

Holmberg, S.D. and Blake, P.A.1984. Staphylococcal food poisoning in the United States. New facts and old misconceptions. JAMA 251: 487- 489.

Hovde, C.J., Marr, J.C., Hofmann, M.L., Hackett, S.P., Chi, Y.I., Crum, K.K., Stevens, D.L., Stauffacher, C.V., and Bohach, G.A. 1994. Investigation of the role of the disulphide bond in the activity and structure of staphylococcal enterotoxin C1. Mol. Microbiol. 13: 897-909.

Hu, D.L., Omoe, K., Shimoda, Y., Nakane, A., and Shinagawa, K. 2003. Induction of emetic response to staphylococcal enterotoxins in the house musk shrew (*Suncus murinus*). Infect. Immun. 71: 567–570.

Ito, K., Takaishi, H., Yin, Y., Song, F., Denning, T.L., and Ernst, P.B. 2000. Staphylococcal enterotoxin B stimulates expansion of autoreactive T cells that induce apoptosis in intestinal epithelial cells: regulation of autoreactive responses by IL-10. J. Immunol. 164: 2994–3001.

Jarraud, S., Cozon, G., Vandenesch, F., Bes, M., Etienne, J., and Lina, G. 1999. Involvement of enterotoxins G and I in staphylococcal toxic shock syndrome and staphylococcal scarlet fever. J. Clin. Microbiol. 37: 2446–2449.

Jarraud, S., Peyrat, M.A., Lim, A., Tristan, a., Bes, M., Mougel, C., Etienne, J., Vandenesch, F., Bonneville, M., and Lina, G. 2001. *egc*, a highly prevalent operon of enterotoxin gene, forms a putative nursery of superantigens in *Staphylococcus aureus*. J. Immunol. 166 :669-677.

Ji, G. Beavis, R.C., and Novick, R.P. 1997. Bacterial interference caused by autoinducing peptide variants. Science 276: 2027–2030.

Kaempfer, R., Arad, G., Levy, R., and Hillman, D. 2002. Defense against biologic warfare with superantigen toxins. Isr. Med. Assoc. J. 4: 520–523.

Kenny, K., Reiser, R.F., Bastida-Corcuera, F.D., and Norcross, N.L. 1993. Production of enterotoxins and toxic shock syndrome by bovine mammary isolates of *Staphylococcus aureus*. J. Clin. Microbiol. 31: 706–707.

Kent, T.H. 1966. Staphylococcal enterotoxin gastroenteritis in rhesus monkeys. Am. J. Pathol. 48: 387–399.

Khambaty, F.M., Bennett, R.W., and Shah, D.B. 1994. Application of pulsed-field gel electrophoresis to the epidemiological characterization of *Staphylococcus intermedius* implicated in a food-related outbreak. Epidemiol. Infect. 113: 75–81.

Lee, S.U., Quesnell, M., Fox, L.K., Yoon, J.W., Park, Y.H., Davis, W.C., Falk, D., Deobald, C.F., and Bohach, G.A. 1998. Characterization of staphylococcal bovine mastitis isolates using the polymerase chain reaction. J. Food Prot. 61: 1384–1386.

Lina, G., Boutite, F., Tristan, A., Bes, M., Etienne, J., and F. Vandenesch. 2003. Bacterial competition for human nasal cavity colonization: role of staphylococcal *agr* alleles. Appl. Environ. Microbiol. 69: 18–23.

Lindsay, J.A., Ruzin, A., Ross, H.F., Kurepina, N., and Novick, R.P. 1998. The gene for toxic shock toxin is carried by a family of mobile pathogenicity islands in *Staphylococcus aureus*. Mol. Microbiol. 29: 527–543.

Lyon, G.J., Wight, J.S., Christopoulos, A, Novick, R.P., and Muir, T.W. 2002. Reversible and specific extracellular antagonism of receptor-histidine kinase signaling. J. Biol. Chem. 277: 6247–6253.

Mahmood, R. and Khan, S.A. 1990. Role of upstream sequences in the expression of the staphylococcal enterotoxin B gene. J. Biol. Chem. 15: 4652–4656.

Marr, J.C., Lyon, J.D., Robberson, J.R., Lupher, M., Davis, W.C., and Bohach, G.A. 1993. Characterization of novel type C staphylococcal enterotoxins: biological and evolutionary implications. Infect. Immun. 61: 4254–4262.

Marrack, P. and Kappler, J. 1990. The staphylococcal enterotoxins and their relatives. Science 248: 705- 711.

Mayville, P., Ji, G., Beavis, R., Yang, H., Goger, M., Novick, R.P., and Muir, T.W. 1999. Structure-activity analysis of synthetic autoinducing thiolactone peptides from *Staphylococcus aureus* responsible for virulence. Proc. Natl. Acad. Sci. USA 96:1218–1223.

McNamara, P.J., Milligan-Monroe, K.C., Khalili, S., and Proctor R.A. 2000. Identification, cloning, and initial characterization of rot, a locus encoding a regulator of virulence factor expression in Staphylococcus aureus. J. Bacteriol. 182: 3197–3203.

Mead, P.S., Slutsker, L., Dietz, V., McCaig, L.F., Bresee, J.S., Shapiro, C., Griffin, P.M., and Tauxe, R.V. 1999. Food-related illness and death in the United States. Emerg. Infect. Dis. 5: 607–625.

Merrill, T.G., and Sprinz, H. 1968. The effect of staphylococcal enterotoxin on the fine structure of the monkey jejunum. Lab. Invest. 18: 114–123.

Monday S.R. and Bohach, G.A. 1999. Use of multiplex PCR to detect classical and newly described pyrogenic toxin genes in staphylococcal isolates. J. Clin. Microbiol. 37: 3411–3414.

Monday, S.R. and Bohach, G.A. 2001. Genes encoding staphylococcal enterotoxins G and I are linked and separated by DNA related to other staphylococcal enterotoxins. J. Nat. Toxins 10: 1–8.

Munson, S.H., Tremaine, M.T., Betley, M.J., and Welch, R.A. 1998. Identification and characterization of staphylococcal enterotoxin types G and I from *Staphylococcus aureus*. Infect. Immun. 66: 3337–3348.

Olsen, S.J., MacKinon, L.C., Goulding, J.S., Bean, N.H., and Slutsker, L. 2000. Surveillance for foodborne-disease outbreaks – United States, 1993–1997, Morb. Mortal. W. Rep. 49(SS-1): 1–51.

Omoe, K., Hu, D.L., Takashashi-Omoe, H., Nakane, A., and Shinagawa, K. 2003. Identification and characterization of a new staphylococcal enterotoxin-related putative toxin encoded by two kinds of plasmids. Infect. Immun. 71: 6088–6094.

Orden, J.A., Cid, D., Blanco, M.E., Ruiz Santa Quiteria, J.A., Gomez-Lucia, E., and de la Fuente, R. 1992. Enterotoxin and toxic shock syndrome toxin-one production by staphylococci isolated from mastitis in sheep. APMIS 100: 132–134.

Otero, A., Garcia, M.L., Garcia, M.C., Moreno, B., and Bergdoll, M.S. 1990. Production of staphylococcal enterotoxins $C_1$ and $C_2$ and thermonuclease throughout the growth cycle, Appl. Environ. Microbiol. 56: 555–559.

Otto, M., Sussmuth, R., Vuong, C., Jung, G., and Gotz, F. 1999. Inhibition of virulence factor expression in *Staphylococcus aureus* by the *Staphylococcus epidermidis* agr pheromone and derivatives. FEBS Lett. 450: 257–262.

Palmer, E.D. 1951. The morphologic consequences of acute exogenous (staphylococcic) gastroenteritis of the gastric mucosa. Gastroenterology 19: 462–475.

Raj, H.D. and Bergdoll. 1969. Effect of enterotoxin B on human volunteers. J. Bacteriol. 98: 833–834.

Regassa, L.B., Couch, J.L., and Betley, M.J. 1991. Steady-state staphylococcal enterotoxin type C mRNA is affected by a product of the accessory gene regulator (agr) and by glucose, Infect. Immun. 59: 955–962.

Rosec, J.P. and Gigaud, O. 2002. Staphylococcal enterotoxin genes of classical and new types detected by PCR in France. Int. J. Food Microbiol. 77: 61–70.

Schantz, E.J., Roessler, W.G., Wagman, J., Spero, L., Dunnery, D.A., and Bergdoll, M.S. 1965. Purification of staphylococcal enterotoxin B. J. Biochem. 4: 1011–1016.

Schlievert, P.M., Jablonski, L.M., Roggiani, M., Sadler, I., Callantine, S., Mitchell, D.T., Ohlendorf, D.H., and Bohach, G.A. 2000. Pyrogenic toxin superantigen site specificity in toxic shock syndrome and food poisoning in animals. Infect. Immun. 68: 3630–3634.

Soltis, M.T., Mekalanos, J.J., and Betley, M.J. 1990. Identification of a bacteriophage containing a silent staphylococcal variant enterotoxin gene (sezA). Infect. Immun. 58: 1614–1619.

Spero, L., Griffin, B.Y., Middlebrook, J.L., and Metzger, J.F. 1976. Effect of single and double peptide bond scission by trypsin on the structure and activity of staphylococcal enterotoxin C. J. Biol. Chem. 251: 5580–5588.

Stelma, G.N. Jr. and Bergdoll, M.S. 1982. Inactivation of staphylococcal enterotoxin A by chemical modification. Biochem. Biophys. Res. Commun. 105: 121–126.

Sugiyama, H., and Hayama, T. 1964. Comparative resistance of vagotomized monkeys to intravenous vs. intragastric staphylococcal enterotoxin challenges. Proc. Soc. Exp. Biol. Med. 115: 243–246.

Sugiyama, H. and Hayama, T. 1965. Abdominal viscera as site of emetic action for staphylococcal enterotoxin in monkeys. J. Infect. Dis. 115: 330–336.

Tan, T.L., Drake, C.T., Jacobson, M.J., and Prohaska, J.V. 1959. The experimental development of pseudomembranous enterocolitis. Surg. Gynec. Obst. 108: 415–420.

Tseng C.W., Zhang S., and Stewart G.C. 2004. Accessory gene regulator control of staphyloccoccal enterotoxin D gene expression, J. Bacteriol. 186: 1793–1801.

Warren, J.R. 1977. Comparative kinetic stabilities of staphylococcal enterotoxin types A, B, and C1. J. Biol. Chem. 252: 6831–6834.

Warren, S.E., Jacobson, M., Mirany, J., and Prodaska, J.V. 1964. Acute and chronic enterotoxin enteritis. J. Exp. Med. 120: 561–568.

Wright, A., Andrews, P.L., and Titball, R.W. 2000. Induction of emetic, pyrexic, and behavioral effects of *Staphylococcus aureus* enterotoxin B in the ferret, Infect. Immun. 68: 2386–2389.

Wright, A.D. and Chapes, S.K. 1999. Cross-linking staphylococcal enterotoxin A bound to major histocompatibility complex class I is required for TNF-alpha secretion. Cell. Immunol. 197: 129–135.

Zhang, S., Iandolo, J.J., and Stewart, G.C. 1998. The enterotoxin D plasmid of *Staphylococcus aureus* encodes a second enterotoxin determinant (sej). FEMS Microbiol. Lett. 168: 227–233.

Zhang, S. and Stewart, G.C. 2000. Characterization of the promoter elements for the staphylococcal enterotoxin D gene. J. Bacteriol. 182: 2321–2325.

# *Campylobacter* Infections

14

Irving Nachamkin and Patricia Guerry

## Abstract

*Campylobacter* spp., primarily *C. jejuni* subsp. *jejuni*, is one of the major causes of bacterial gastroenteritis in the USA and worldwide. *Campylobacter* infection is primarily a foodborne illness, usually without complications; however, serious sequelae such as Guillain–Barré syndrome occur in a small subset of infected patients. Detection of *C. jejuni* in clinical samples is readily accomplished by culture and non-culture methods, although improvements in diagnostic approaches are needed. A significant body of knowledge exists on the epidemiology of *Campylobacter* infections; however, much less is known about the mechanism of disease, despite over two decades of research.

## Characteristics and taxonomy

The family Campylobacteraceae contains three genera, *Campylobacter*, *Arcobacter*, and *Sulfurospirillum* (On, 2001; Vandamme, 2000). Within *Campylobacter*, there are 16 described species, four species in the genus *Arcobacter* and five species in the genus *Sulfurospirillum*. *C. hominis*, a species related to *C. gracilis* and *C. sputorum*, was isolated from healthy human feces and was recently proposed (Lawson *et al.*, 2001). Another recently proposed species, *Campylobacter lanienae*, isolated from healthy human feces, is closely related to *C. hyointestinalis* (Logan *et al.*, 2000). Some isolates described as *C. gracilis* have been renamed as *Sutterella wadsworthensis* (Wexler *et al.*, 1996). A number of animals serve as reservoirs of human infection for *Campylobacter* spp. (Table 14.1). Morphologically, *Campylobacter* spp. are gram-negative, non-sporeforming curved, s-shaped, or spiral rods, but occasionally form straight rods (i.e. *C. hominis*). Most species are motile and produce a single polar flagellum. *Campylobacter* spp. are usually microaerophillic and some species require increased hydrogen for growth (Vandamme and De Ley, 1991).

## Clinical significance

Campylobacters are recognized as the most common cause of bacterial gastrointestinal infection in the United States and other developed nations (Finch and Riley, 1984; Borczyk *et al.*, 1991; Tauxe, 1992; Friedman *et al.*, 2000). Based on active surveillance data collected through CDC's Foodborne Diseases Active Surveillance Network (FoodNet), it is estimated that 1% of the USA population develops campylobacter infection each year and is similar to data from the United Kingdom and other developed countries (Friedman *et al.*, 2000). However, the incidence of *Campylobacter* infections in the USA appears to be decreasing. The incidence of *Campylobacter* infections in the USA in 2003 was reported to be 12.6 per 100,000 persons overall (range 7.0 to 26.9 among nine reporting sites) compared with an overall incidence of 21.7 during the 1996–1998 reporting period (Centers for Disease Control, 2004). Ingestion of contaminated food from improper handling or cooking is primarily responsible for sporadic infections and occur more often in the summer months (Stern, 1992; Friedman *et al.*, 2000). Several sources account for sporadic infections including ingestion of raw milk, contaminated surface water, exposures during overseas travel, and contact with domestic pets (Table 14.2).

Of the many *Campylobacter* species, *C. jejuni* and *C. coli* are the most recognized causes of gastrointestinal. Infections have the highest incidence in infants/young children and young adults, 20–40 years of age. In contrast to the USA, incidence rates in countries such as Mexico and Thailand are several orders of magnitude higher (Oberhelman and Taylor, 2000). Symptomatic infections occur primarily in infancy and early childhood in developing countries due to increasing immunity in older populations. Travel to developing countries is a major risk factor for acquiring *Campylobacter* infection (Oberhelman and Taylor, 2000; Friedman *et al.*, 2004).

Patients with *Campylobacter* infection may have no symptoms or may be severely ill. In ill patients, signs and symptoms include fever, abdominal cramping and diarrhea lasting days to more than one week. *Campylobacter* infections are generally self-limited, but relapses may occur in a small proportion of untreated patients (Blaser and Allos, 2005). For patients who require antimicrobial therapy, increasing resistance to fluoroquinolones has become a problem in many parts of the world, including the United States, whereas macrolide resistance re-

**Table 14.1** Reservoirs for *Campylobacter* species

| Species | Reservoir |
|---|---|
| *C. jejuni* subsp. *jejuni*, *C. sputorum* biovar *sputorum* | Humans |
| *C. concisus*, *C. curvus*, *C. rectus*, *C. showae*, *C. gracilis* | |
| *C. jejuni* subsp. *jejuni*, *C. fetus* subsp. *fetus/venerealis* | Cattle |
| *C. hyointestinalis*, *C. sputorum* biovars *paraureolyticus*, *faecalis* | |
| *C. fetus* subsp. *fetus*, *C. sputorum* biovar *faecalis* | Sheep |
| *C. coli*, *C. hyointestinalis*, *C. mucosalis* | Pigs |
| *C. jejuni* subsp. *jejuni*, *C. coli*, *C. lari* | Birds |
| *C. jejuni* subsp. *jejuni*, *C. lari*, *C. upsaliensis* | Domestic pets |
| *C. hyointestinalis*, *C. helveticus* | |

References: Skirrow (1994), Vandamme *et al*. (1995), Nachamkn (2001).

**Table 14.2** Sources of sporadic *Campylobacter* infections

| Source | Estimated proportion of infections (%) |
|---|---|
| Milk, raw | 5 |
| Drinking water | 8 |
| Travel overseas | 9 |
| Contact with pets | 6–30 |
| Poultry | 10–70 |

Reprinted with permission from Friedman *et al*. (2000).

mains low (Engberg *et al*., 2001). Poultry have also been implicated as a source for acquisition of fluoroquinolone-resistant *Campylobacter jejuni* infections (Kassenborg *et al*., 2004).

Extraintestinal *Campylobacter* infections occasionally occur including bacteremia, hemolytic-uremic syndrome, peritonitis, and focal infections (Skirrow *et al*., 1993; Skirrow and Blaser, 2000). In the immunocompromised host (e.g. HIV), persistent diarrheal illness and bacteremia may occur and can be difficult to treat (Perlman *et al*., 1988). *Campylobacter* infections rarely result in death but the mortality associated with infection may be underestimated (Friedman *et al*., 2000; Havelaar *et al*., 2000; Helms *et al*., 2003).

Several sequelae from campylobacter infection are recognized; Guillain–Barré syndrome (GBS) and reactive arthritis. Both are immune-mediated responses to gastrointestinal infection. GBS is a reversible, acute flaccid paralysis, affecting the peripheral nervous system. *C. jejuni* exhibit ganglioside-like mimicry in the lipooligosaccharide outer core and the immune response to these structures is thought result in damage to peripheral nerve tissue, rich in gangliosides (Ho *et al*., 1998). Certain serotypes (Penner) appear to be over-represented in some GBS cases including HS:19 and HS:41 (Nachamkin *et al*., 2000).

Reactive arthritis and Reiter's syndrome may occur in some patients following campylobacter infection. Joint swelling and pain usually lasts a few weeks and may be prolonged (Skirrow and Blaser, 2000). An immunogenetic basis for joint involvement, such as HLA B27 positivity, is strongly associated with the condition (Skirrow and Blaser, 2000).

## Pathogenesis and virulence factors

Despite its importance as a human pathogen, surprisingly, little is understood about how *C. jejuni* causes disease. This has been due to lack of both appropriate systems of experimental genetics and small animal models that mimic human disease. Additionally, *C. jejuni* strains undergo phase variation of surface antigens, many of them virulence factors, at high frequency due to a mechanism called slip strand mismatch repair (Parkhill *et al*., 2000). Effectively, the surface of the organism is constantly changing, an occurrence that has likely complicated *in vitro* studies of virulence. Even the availability of the complete genome sequence of one strain of *C. jejuni* failed to provide major new insight into pathogenesis (Parkhill *et al*., 2000). *Campylobacter* has evolved several systems of protein glycosylation and the role of these systems in contributing to virulence is an important area for investigation (Szymanski *et al*., 2003).

The bloody, inflammatory nature of *Campylobacter* damage to the intestine is consistent with a virulence mechanism that includes invasion of intestinal epithelial cells, damage by cytotoxins or a combination of the two. While complete molecular details are far from complete, pathogenesis of *C. jejuni* appears to be multifactorial, and there are several specific virulence factors or processes that have been recognized.

### Motility

The rapid, darting motility imparted by the polar flagella of *C. jejuni* is critical to pathogenesis. *C. jejuni* must be motile to colonize the intestinal tract of animals and humans (Black *et al*., 1988; Nachamkin *et al*., 1993; Wassenaar *et al*., 1993), perhaps allowing the bacterium to penetrate the thick mucus lining of the intestinal tract prior to adherence or invasion, or simply to persist in the mucus layer (Lee *et al*., 1986). However, motility is also necessary for campylobacters to invade intestinal epithelial cells in culture. Thus, mutants lacking flagella (Wassenaar *et al*., 1991; Grant *et al*., 1993) or mutants that synthesize a "paralyzed" flagella filament that does not confer motility (Yao *et al*., 1997) are non-invasive *in vitro*. Although the reasons for this are not totally understood, it may be that flagellin functions as an adhesin, mediating attachment to eukaryotic cells prior to invasion. Additionally, flagella are known to mediate autoagglutination, a phenomenon that may be required for microcolony formation both *in vitro* and *in vivo* (Misawa and Blaser, 2000). Recent data have also suggested that the flagella filament may function as a secretory organelle through which virulence factors pass during infection, as demonstrated for other enteric pathogens (Young *et al.*, 1999) and discussed below.

## Toxins

There are numerous early reports that *C. jejuni* synthesized a cholera-like enterotoxin, but these reports have not been substantiated, nor was any gene encoding an enterotoxin observed in the genome sequence. Although novel toxins may exist that remain to be identified, the only known toxin in *C. jejuni* is the cytolethal distending toxin or CDT. CDT is a unique cytotoxin that inhibits eukaryotic cells at the G2 phase of cell division (Whitehouse *et al.*, 1998). It is a membrane bound toxin (Hickey *et al.*, 2000) composed of three subunits, CdtA, CdtB and CdtC; the active component is CdtB (Lara-Tejero and Galan, 2001). CdtB is taken up by the eukaryotic cell and transported to the nucleus where it nicks the chromosomal DNA (Lara-Tejero and Galan, 2000; Lara-Tejero and Galan, 2001). CDT is highly conserved among *C. jejuni* strains, and is also found in *Shigella dysenteriae*, some diarrheagenic *E. coli*, *Helicobacter hepaticus*, *Hemophilus ducreyii* and *Actinobacillus actinomycetemcomitans* (Lara-Tejero and Galan, 2002).

## Adherence to intestinal epithelial cells (IECs)

*C. jejuni* adheres to both polarized and non-polarized IECs *in vitro* and this is likely an important component of the virulence process and a prerequisite to invasion. Multiple adhesins have been reported, including flagellin (McSweegan and Walker, 1986), an antigen termed PEB1 (de Melo and Pechère, 1990; Pei *et al.*, 1998), and CadF, an outer membrane protein that binds fibronectin (Konkel *et al.*, 1997). Adherence likely precedes invasion and allows for secretion of proteins required for uptake and/or induction of a proinflammatory response (see below).

## Invasion and translocation of intestinal epithelial cells

Invasion of IECs is an important virulence factor for numerous enteropathogens, particularly *Salmonella* and *Shigella*. As mentioned above, *C. jejuni* are generally considered to be invasive for IECs also, although the relative invasion levels are, in general, much lower than that seen with other enteric pathogens and appear to vary considerably among different strains of *C. jejuni* (Kopecko *et al.*, 2001). There also appear to be different pathways for uptake into IECs since some stains appear to utilize a microtubule-dependent pathway and others utilize a microfilament-dependent pathway (Oelschlaeger *et al.*, 1993; Kopecko *et al.*, 2001). In addition to these host cytoskeletal rearrangements, *C. jejuni* invasion requires de novo protein synthesis and stimulates signal transduction pathways. Thus,

invasion results in host protein phosphorylation changes and induces release of host intracellular calcium (Wooldridge *et al.*, 1996; Hu and Kopecko, 1999; Kopecko *et al.*, 2001).

*C. jejuni* appear to be able to translocate through intestinal epithelial cells in culture, which may be an important event in the pathogenesis of disease and the transient bacteremia observed in some infections (Konkel *et al.*, 1992). The translocation may occur by transcytosis through the cytoplasm following invasion or by a paracellular route between intestinal cells (Everest *et al.*, 1992). Importantly, this form of translocation does not result in major effects on tight junction integrity. Passage through M cells has also been suggested to play a role in translocation (Walker *et al.*, 1988).

Importantly, unlike many other invasive pathogens, no "invasion" has been identified that is clearly responsible for mediating uptake of *C. jejuni* into IECs. The only obvious requirement is that the bacteria be flagellated, which may reflect either the need for autoagglutination or the fact that flagella may function as an adhesin. The only other surface structure that has been firmly associated with invasion of *C. jejuni* strain 81–176 is the capsular polysaccharide (Bacon *et al.*, 2001b). This surface carbohydrate, whose existence was only realized subsequent to the genome sequence (Parkhill *et al.*, 2000), is the serodeterminant of the Penner serotyping scheme (Karlyshev *et al.*, 2000; Bacon *et al.*, 2001b). A capsule mutant of *C. jejuni* strain 81–176 was unable to invade intestinal epithelial cells (Bacon *et al.*, 2001b). Expression of capsule in 81–176 also undergoes phase variation at high frequency (Bacon *et al.*, 2001b), indicating that not all cells in the population are equally invasive. Moreover, since capsule structure varies among strains, it may be that particular carbohydrate structures function more efficiently as adhesions or invasions.

## Interactions with monocytes/macrophages

Once *C. jejuni* have translocated the intestinal epithelial cell barrier, they must resist killing by macrophages. Indeed, *C. jejuni* has been reported to be able to survive within macrophages for several days (Kiehlbauch *et al.*, 1985). Macrophage survival might offer the highly serum-sensitive *C. jejuni* a mechanism by which it can disseminate to regional lymph nodes (Blaser *et al.*, 1985).

## Intestinal Inflammatory responses

One of the key clinical features of *C. jejuni* enteritis is an intense inflammatory response in which polymorphonuclear leukocytes infiltrate the intestinal epithelium. Early work has described elevation of cAMP, prostaglandin $E_2$ and leukotrience $B_4$ levels in a rabbit intestinal loop model of infection (Everest *et al.*, 1993). More recently, *C. jejuni* strains have been shown to induce release of proinflammatory cytokines from human IECs (Hickey *et al.*, 1999). Release of IL-8 is due in part to damage resulting from CDT, but there also appears to be an alternate mechanism that requires either adherence and/or invasion of IECs that is independent of CDT (Hickey *et al.*, 2000).

## Specialized protein secretion systems

The cross talk that is generated between *C. jejuni* and eukaryotic cells likely involves protein secretion by the infecting bacterium. Some strains of *C. jejuni* have been shown to secrete a protein called CiaB that is internalized into the eukaryotic target cell (Konkel *et al.*, 1999). This secretion is thought to be through the flagellar organelle (Konkel *et al.*, 1999), since there does not appear to be a specialized type III secretion system similar to those that are involved in secretion of virulence factors in the *Enterobacteriaceae* (Parkhill *et al.*, 2000). The secretion triggers calcium release and signal transduction changes in the target cell (Kopecko *et al.*, 2001).

Some strains of *C. jejuni* contain a plasmid called pVir that encodes a putative type IV secretion system, similar to those found in other pathogens (Christie and Vogel, 2000). In strain 81–176, this plasmid encoded system appears to contribute to the ability of the strain to invade IECs at high levels (Bacon *et al.*, 2000; Bacon *et al.*, 2001a) likely because it is required for secretion of effector proteins. Importantly, the presence of this plasmid in only a subset of *C. jejuni* strains would suggest that there may be major differences in virulence among strains.

## Lipooligosaccharide

*C. jejuni* strains lack the lipopolysaccharide (LPS) found in most other enteric pathogens and instead have variable lipooligosaccharide (LOS) lacking side chains. Importantly, the outer cores of LOS of many strains of *C. jejuni* contain sialic acid in structures that mimic human gangliosides including expression of GM1, GM2, GD1a, GD3, and GT1a like structures (Moran *et al.*, 2000). *C. jejuni* is one of only a few bacteria that are capable of endogenous synthesis of sialic acid. The molecular mimicry of LOS to human gangliosides is thought to be involved in development of the autoimmune disease Guillain–Barré syndrome, in susceptible hosts (Magira *et al.*, 2003). The role of LOS in diarrheal disease remains to be fully examined, however. It has been shown that sialylation of the core increases resistance to normal human serum (Guerry *et al.*, 2000). The structures of LOS cores undergo high frequency phase variations such that the ganglioside mimicry can change. Thus, strain NCTC 11168 has been shown to change from GM2 to GM1 mimicry (Linton *et al.*, 2000) and strain 81-176 can change from GM2 to GM3 mimicry (Guerry *et al.*, 2000; Guerry *et al.*, 2002).

## Stress responses

As a foodborne pathogen, *C. jejuni* must be capable of surviving numerous physiological stresses encountered in food and water for effective transmission. Thus, it is a seeming paradox that this strict microaerophile, which is fastidious in its growth *in vitro*, can survive on chicken skin in the supermarket and in the harsh, strictly anaerobic environment of the intestine. However, there is little information about stress responses in general. There is one report that *C. jejuni* synthesizes >40 new proteins in response to temperature stress (Konkel *et al.*, 1998), although the molecular basis for this response is not clear. Interestingly, the genome sequence revealed that *C. jejuni* lacks the typical sigma factor involved in stress responses in other bacteria (Parkhill *et al.*, 2000).

One of the major stress responses occurs following internalization of *C. jejuni* into either IECs or macrophages. Superoxide dismutase (SOD) has been shown to be one of the bacterium's major defenses against oxidative damage by catalysis of superoxide radicals (Pesci *et al.*, 1994). Similarly, catalase provides resistance to hydrogen peroxide *in vitro* (Day *et al.*, 2000).

The extremely low levels of free iron within mammalian hosts is another stress condition that *C. jejuni* must handle. Unlike other invasive enteric bacteria, *C. jejuni* does not produce siderophores, outer membrane proteins that sequester free iron, but it can bind exogenous siderophores (Field *et al.*, 1986) and possesses an enterochelin uptake system (Richardson and Park, 1995). Moreover, *C. jejuni* synthesizes bacterial ferritin which stores intracellular iron and also functions to protect the bacterium from oxidative stresses (Wai *et al.*, 1996).

## Epidemiology of outbreaks and sporadic infections

Infections caused by *C. jejuni* are mainly foodborne and there are numerous animals reservoirs described; the major animal source of sporadic infections are poultry (Doyle and Jones, 1992; Stern, 1992; Altekruse *et al.*, 1994; Jacobs-Reitsma, 2000) (Table 14.3). A recent case–control study conducted by Friedman and colleagues confirmed the importance of consuming poultry (chicken and turkey), particularly prepared in restaurants, as risk factors for sporadic human *Campylobacter* infection in the USA (Friedman *et al.*, 2004). In addition to animal sources, contaminated vegetables and shellfish may also be sources of infection (Doyle and Jones, 1992; Altekruse *et al.*, 1994; Jacobs-Reitsma, 2000). Contaminated water supplies have been implicated in point-source outbreaks of *Campylobacter* infection. A dormant form referred to as "viable but nonculturable" has been described but the role of these forms as a source for human infection is unclear (Rollins and Colwell, 1986; Medema *et al.*, 1992).

Numerous outbreaks implicating *Campylobacter* have been reported to the CDC and from 1978 to 1996, there were 111 *Campylobacter* outbreaks affecting 9913 individuals (Friedman *et al.*, 2000). Until the later 1980s, most outbreaks involved contaminated water and unpasteurized milk. After 1988 through 1996, water and milk accounted for less than 20% of reported outbreaks (Friedman *et al.*, 2000). Unpasteurized milk has been an important source of outbreaks with high attack rates, particularly in the 1980s. Outbreaks due to raw milk occurred yearly from 1981 to 1990, and children going on field trips to dairy farms were the primary population affected (Wood *et al.*, 1992). The most recent milkborne outbreak occurred mong children in Wisconsin from a cow-lease program (Harrington *et al.*, 2002). Outbreaks due to milkborne and waterborne

**Table 14.3** Rates of isolation of *Campylobacter* from various food sources

| Product | Percent positive samples |
|---------|--------------------------|
| Beef | 0 to 23.6 |
| Chicken | 14 to 98 |
| Cow milk | 0 to 12.3 |
| Duck | 48 |
| Ewe's milk | 0 |
| Goat milk | 0 |
| Goose | 38 |
| Lamb | 0 to 15.5 |
| Mussels | 47 to 69 |
| Offal | 47 |
| Oysters | 6 to 27 |
| Pork | 1 to 23.5 |
| Sheep | 3 |
| Turkey | 3 to 25 |
| Vegetables | < 5 |

Reprinted with permission from Jacobs-Reitsma (2000).

sources usually occur in the spring and fall, whereas sporadic infections are most frequent in the summertime (Friedman *et al.*, 2000) (Figure 14.1).

## Susceptibility to environmental conditions

*C. jejuni* do not survive for extended periods of time outside the host because the organism is not very tolerant of a variety of environmental conditions. *C. jejuni* is thermophillic and doesn't grow at low temperatures, below 30°C. The species is also killed by drying, high-oxygen conditions, and low pH (Doyle and Jones, 1992). *C. jejuni* is also adequately killed in properly cooked foods and at various temperatures (Table 14.4). Freezing reduces the concentration of *Campylobacter* in contaminated poultry, but even after freezing to –20°C, low levels of *Campylobacter* can be recovered (Jacobs-Reitsma, 2000). Gamma irradiation is effective in killing *C. jejuni* but

the rate of killing may vary with the type of food and temperature of the material. The growth phase of organism may also influence susceptibility to irradiation (Radomyski *et al.*, 1994). Irradiation treatment effective against other foodborne pathogens such as *Salmonella* spp. and *Listeria monocytogenes* should also kill *Campylobacter* spp. (Patterson, 1995).

Common disinfectants have good activity against *Campylobacter* spp. Compounds that are effective include sodium hypochlorite (Clorox), *o*-phenylphenol (Amphyl), iodine polyvinylpyrrolidine (Betadine), alkylbenzyl dimethylammonium chloride (Zephiran), glutaraldehyde (Cidex), formaldehyde, and ethanol (Wang *et al.*, 1980). *Campylobacter* spp. are killed at low pH (Blaser *et al.*, 1980). Ascorbic acid will inhibit the organism at 0.05% and has a killing effect at 0.09% (Jacobs-Reitsma, 2000).

## Detection and analysis

Detection of *Campylobacter* by microscopic methods may be difficult because *Campylobacter* spp. are not easily visualized with safranin counterstain commonly used in the Gram stain procedure. A stain containing carbol-fuchsin or 0.1% aqueous basic fuchsin should be used as the counter stain for smears of stools or pure cultures (Sazie and Titus, 1982; Park *et al.*, 1983). Phase contrast and dark-field microscopy have also been used to directly detect motile campylobacters in fresh stool samples (Karmali and Fleming, 1979; Paisley *et al.*, 1982).

For culture isolation, *Campylobacter* species require a microaerobic atmosphere containing approximately 5% $O_2$, 10% $CO_2$ and 85% $N_2$ and this gas mixture can be produced using commercially available gas generating systems. There are some *Campylobacter* species such as *C. sputorum*, *C. concisus*, *C. mucosalis*, *C. curvus*, *C. rectus* and *C. hyointestinalis* that require increased hydrogen levels for primary isolation and growth. These species may not be recovered using the conventional microaerobic conditions since the amount of hydrogen generated in properly used commercial gas-packs is < 2% (Nachamkin, 1995). A gas mixture of 10% $CO_2$, 6% $H_2$ and balanced $N_2$

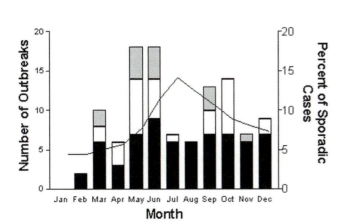

■ Other food
□ Milk
▨ Water
— Sporadic Cases

**Figure 14.1** Foodborne and waterborne outbreaks of *Campylobacter* infections, 1978 to 1996, and distribution of sporadic cases by month, 1982 to 1995, in the United States. From reference (Friedman *et al.*, 2000), reprinted with permission.

**Table 14.4** D-values reported for different food sources

| Source | Temperature (°C) | D-value range (min) |
|---|---|---|
| Skim milk | 48 | 7.2–12.8 |
| | 55 | 0.74–1.0 |
| Red meat | 50 | 5.9–6.3 |
| | 60 | < 1 |
| Ground chicken | 49 | 20 |
| | 57 | < 1 |

Reprinted with permission from Jacobs-Reitsma (2000).

used in an evacuation-replacement jar is sufficient for isolating hydrogen-requiring species (Nachamkin, 1995).

A number of selective media are recommended for isolating *C. jejuni* and *C. coli* from clinical samples and include blood-containing and blood-free (charcoal based) media (Nachamkin, 2003). Several studies suggest that using a combination of media may increase the yield of *Campylobacter* from stool samples by 10–15% over the use of a single medium (Gun Monro *et al.*, 1987; Endtz *et al.*, 1991). A number of enrichment broths have been formulated to enhance the recovery of *Campylobacter* from stool samples (Bolton and Robertson, 1982; Martin *et al.*, 1983; Chan and Mackenzie, 1984; Agulla *et al.*, 1987; Sjogren *et al.*, 1987; Korhonen and Martikainen, 1990;). Enrichment cultures may be useful in situations where low numbers of organisms may be expected, such as with delayed sample transport, during convalescing infections, or in the investigation of GBS following acute *Campylobacter* infection (Taylor *et al.*, 1988; Bowen-Jones, 1989; Nachamkin, 1997).

In addition to biochemical identification of *Campylobacter* spp., commercial systems have been developed as an aid to identifying *Campylobacter* spp. to the genus level. These methods include immunologic and nucleic acid hybridization techniques (Nachamkin, 2003).

*Campylobacter* antigen may also be detected directly in stool samples and is available commercially (ProSpecT *Campylobacter*, Alexon-Trend, Inc., Ramsey, MN). The assay has been evaluated by several investigators and exhibits reasonable performance characteristics (Endtz *et al.*, 2000; Hindiyeh *et al.*, 2000; Tolcin *et al.*, 2000). Polymerase chain reaction (PCR) has also been widely studied to directly detect *Campylobacter* in stool samples (Oyofo *et al.*, 1992; Oyofo *et al.*, 1996; Waegel and Nachamkin, 1996; Lawson *et al.*, 1997; Linton *et al.*, 1997; Takeshi *et al.*, 1997).

## Control

Measures to control and reduce the incidence of *Campylobacter* infection are multifaceted and complex to implement. In order to have an impact on campylobacter infections in humans, a detailed understanding of the epidemiology and sources of human infection are needed. Since poultry is considered to be one of the most important sources for human infection, a significant effort has been placed on studying pre- and post-harvest procedures in poultry processing as a means for devising control and prevention strategies. This topic has been extensively reviewed (Jacobs-Reitsma, 2000; Newell and Wagenaar, 2000; Ransom *et al.*, 2000).

## References

Agulla, A., Merino, F.J., Villasante, P.A., Saz, J.V., Diaz, A., and Velasco, A.C. 1987. Evaluation of four enrichment media for isolation of *Campylobacter jejuni*. J. Clin. Microbiol. 25: 174–175.

Altekruse, S.F., Hunt, J.M., Tollefson, L.K., and Madden, J.M. 1994. Food and animal sources of human *Campylobacter jejuni* infection. J. Am. Vet. Med. Assoc. 204: 57–61.

Bacon, D.J., Alm, R.A., Burr, D.H., Hu, L., Kopecko, D.J., Ewing, C.P. et al. 2000. Involvement of a plasmid in virulence of *Campylobacter jejuni* 81–176. Infect. Immun. 68: 4384–4390.

Bacon, D.J., Alm, R.A., Hu, L., Hickey, T.E., Ewing, C.P., Trust, T.J., and Guerry, P. 2001a. DNA sequence and mutational analyses of the pVir plasmid of *Campylobacter jejuni* 81 176. Infect. Immun. 70: 6242–6250.

Bacon, D.J., Szymanski, C.M., Burr, D.H., Silver, R.P., Alm, R.A., and Guerry, P. 2001b. A phase variable capsule is involved in virulence of *Campylobacter jejuni* 81–176. Mol. Microbiol. 40: 769–777.

Black, R.E., Levine, M.M., Clements, M.L., Hughs, T.P., and Blaser, M.J. 1988. Experimental *Campylobacter jejuni* infections in humans. J Infect. Dis. 157: 472–480.

Blaser, M.J., and Allos, B.M. 2005. *Campylobacter jejuni* and Related Species. In: Principles and Practice of Infectious Diseases. G.L. Mandell, J.E. Bennett, and R. Dolin, eds. Elsevoer, Philadelphia, pp. 2548–2557.

Blaser, M.J., Hardesty, H.L., Powers, B., and Wang,W.L. 1980. Survival of *Campylobacter* fetus subsp jejuni in biological milieus. J. Clin. Microbiol. 11: 309–313.

Blaser, M.J., Smith, P.F., and Kohler, P.A. 1985. Susceptibility of *Campylobacter* isolates to the bactericidal activity in human serum. J. Infect. Dis. 151: 227–235.

Bolton, F.J., and Robertson, L. 1982. A selective medium for isolating *Campylobacter jejuni/coli*. J. Clin. Pathol. 35: 462–467.

Borczyk, A., Rosa, S.D., and Lior, H. 1991. Enhanced recognition of *Campylobacter cryaerophila* in clinical and environmental specimens. Annual Meeting of the American Society for Microbiology. Abstract # C-267, 386.

Bowen-Jones, J. 1989. Infection and cross-infection in a paediatric gastro-enteritis unit. Curationis 12: 30–33.

Centers for Disease Control. 2004. Preliminary FoodNet data on the incidence of infection with pathogens transmitted commonly through food-selected sites, United States, 2003. MMWR 53: 338–343.

Chan, F.T.H., and Mackenzie, A.M.R. 1984. Advantage of using enrichment-culture techniques to isolate *Campylobacter jejuni* from stools. J. Infect. Dis. 149: 481–482.

Christie, P., and Vogel, J.P. 2000. Bacterial type IV secretion: conjugation system adapted to deliver effector molecules to host cells. Trends Microbiol. 8: 354–360.

Day, W.A., Sajecki, J.L., Pitts, T.M., and Joens, L.A. 2000. Role of catalase in *Campylobacter jejuni* intracellular survival. Infect. Immun. 68: 6337–6345.

de Melo,M.A., and Pechëre,J.C. 1990. Identification of *Campylobacter jejuni* surface proteins that bind to Eucaryotic cells *in vitro*. Infect. Immun. 58: 1749–1756.

Doyle, M.P., and Jones, D.M. 1992. Foodborne transmission and antibiotic resistance of *Campylobacter jejuni*. In: *Campylobacter jejuni*: Current Status and Future Trends. I. Nachamkin, M.J. Blaser, and L.S. Tompkins, eds. American Society for Microbiology, Washington, D.C., pp. 45–48.

Endtz, H.P., Ang, C.W., Van Den Braak, N., Luijendijk, A., Jacobs, B.C., de Man, P. *et al.* 2000. Evaluation of a new commercial immunoassay for rapid detection of *Campylobacter jejuni* in stool samples. Eur. J. Clin. Microbiol. Infect. Dis. 19: 794–797.

Endtz, H.P., Ruijs, G.J., Zwinderman, A.H., van der Reijden, T., Biever, M., and Mouton, R.P. 1991. Comparison of six media, including a semisolid agar, for the isolation of various *Campylobacter* species from stool specimens. J. Clin. Microbiol. 29: 1007–1010.

Engberg, J., Aarestrup, F.M., Taylor, D.E., Gerner-Smidt, P., and Nachamkin, I. 2001. Quinolone and macrolide resistance in *Campylobacter jejuni* and *C. coli*: resistance mechanisms and trends in human isolates. Emerg. Infect. Dis. 7: 24–34.

Everest, P.H., Cole, A.T., Hawkey, C.J., Knutton, S., Goossens, H., Butzler, J.P. *et al.* 1993. Role of leukotriene B4, prostaglandin E2, and cyclic AMP in *Campylobacter jejuni* induced intestinal fluid secretion. Infect. Immun. 61: 4885–4887.

Everest, P.H., Goossens, H., Butzler, J.P., Lloyd, D., Knutton, S., Ketley, J.M., and Williams, P.H. 1992. Differentiated Caco-2 cells as a model for enteric invasion by *Campylobacter jejuni* and *C. coli*. J. Med. Microbiol. 37: 319–325.

Field, L.H., Headley, V.L., Payne, S.M., and Berry, L.J. 1986. Influence of iron on growth, morphology, outer membrane protein composition, and synthesis of siderophores in *Campylobacter jejuni*. Infect. Immun. 54: 126–132.

Finch, M.J., and Riley, L.W. 1984. *Campylobacter* infections in the United States: Results of an 11-state surveillance. Arch. Intern. Med. 144: 1610–1612.

Friedman, C.R., Hoekstra, R.M., Samuel, M., Marcus, R., Bender, J., Shiferaw, B. *et al.* 2004. Risk factors for sporadic *Campylobacter* infection in the United States. A case–control study in FoodNet sites. Clin. Infect. Dis. 38 (Suppl. 3): 285–96.

Friedman, C.R., Neimann, J., Wegener, H.C., and Tauxe, R.V. 2000. Epidemiology of *Campylobacter jejuni* Infections in the United States and other Industrialized Nations. In: *Campylobacter*. I. Nachamkin, and M.J. Blaser, eds. ASM Press, Washington, D.C. p. 121 138.

Grant, C.C.R., Konkel, M.E., Cieplak, W., and Tompkins, L.S. 1993. Role of flagella in adherence, internalization, and translocation of *Campylobacter jejuni* in nonpolarized and polarized epithelial cells. Infect. Immun. 61: 1764–1771.

Guerry, P., Ewing, C.P., Hickey, T.E., Prendergast, M.M., and Moran, A.P. 2000. Sialylation of lipooligosaccharide cores affects immunogenecity and serum resistance of *Campylobacter jejuni*. Infect. Immun. 68: 6656–6662.

Guerry, P., Szymanski, C., Prendergast, M.M., Hickey, T.E., Ewing, C.P., Pattarini, D.L., and Moran, A.P. 2002. Phase variation of *Campylobacter jejuni* 81–176 lipooligosaccharide affects ganglioside mimicry and invasiveness *in vitro*. Infect. Immun. 70: 787–793.

Gun-Monro, J., Rennie, R.P., Thornley, J.H., Richardson, H.L., Hodge, D., and Lynch, J. 1987. Laboratory and clinical evaluation of isolation media for *Campylobacter jejuni*. J. Clin. Microbiol. 25: 2274–2277.

Harrington, P., Archer, J., Davis, J.P., Croft, D.R., and Varma, J.K. 2002. Outbreak of *Campylobacter jejuni* infections associated with drinking unpasteurized milk procured through a cow-leasing program, Wisconsin, 2001. MMWR 51: 548–549.

Havelaar, A.H., de Wit, M.A.S., van Koningsveld, R., and van Kempen, E. 2000. Health burden in the Netherlands due to infection with thermophilic *Campylobacter* spp. Epidemiol. Infect. 125: 505–522.

Helms, M., Vastrup, P., Gerner-Smidt, P., and Molbak, K. 2003. Short and long term mortality associated with foodborne bacterial gastrointestinal infections: registry based study. Br. Med. J. 326: 1–5.

Hickey, T.E., Baqar, S., Bourgeois, A.L., Ewing, C.P., and Guerry, P. 1999. *Campylobacter jejuni*-stimulated secretion of interleukin 8 by INT407 cells. Infect. Immun. 67: 88–93.

Hickey, T.E., McVeigh, A.L., Scott, D.A., Michielutti, R.E., Bixby, A., Carroll, S.A. *et al.* 2000. *Campylobacter jejuni* cytolethal distending toxin mediates release of interleukin 8 from intestinal epithelial cells. Infect. Immun. 68: 6535–6541.

Hindiyeh, M., Jense, S., Hohmann, S., Benett, H., Edwards, C., Aldeen, W. *et al.* 2000. Rapid detection of *Campylobacter jejuni* in stool specimens by an enzyme immunoassay and surveillance for *Campylobacter upsaliensis* in the greater Salt Lake City area. J. Clin. Microbiol. 38: 3076–3079.

Ho, T.W., McKhann, G.M., and Griffin, J.W. 1998. Human autoimmune neuropathies. Ann. Rev. Neurosci. 21: 187–226.

Hu, L., and Kopecko, D.J. 1999. *Campylobacter jejuni* 81–176 associates with microtubules and dynein during invasion of human epithelial cells. Infect. Immun. 67: 4171–4182.

Jacobs-Reitsma, W. 2000. *Campylobacter* in the Food Supply. In: *Campylobacter*, 2nd ed. I. Nachamkin, and M.J. Blaser, eds. ASM Press, Washington, D.C., pp. 467–481.

Karlyshev, A.V., Linton, D., Gregson, N.A., Lastovica, A.J., and Wren, B.W. 2000. Genetic and biochemical evidence of a *Campylobacter jejuni* capsular polysaccharide that accounts for Penner serotype specificity. Mol. Microbiol. 35: 529–541.

Karmali, M.A., and Fleming, P.C. 1979. *Campylobacter* enteritis in children. J Pediatr 94: 527 533.

Kassenborg, H.D., Smith, K.E., Vugia, D.J., Rabatsky-Ehr, T., Bates, M.R., Carter, M.A. *et al.* 2004. Fluoroquinolone-resistant *Campylobacter* infections: eating poultry outside of the home and foreign travel are risk factors. Clin. Infect. Dis. 38 (Suppl 3): 279–284.

Kiehlbauch, J.A., Albach, R.A., Baum, L.L., and Chang, K.P. 1985. Phagocytosis of *Campylobacter jejuni* and its intracellular survival in mononuclear phagocytes. Infect. Immun. 48: 446–451.

Konkel, M.E., Garvis, S.G., Tipton, S.L., Anderson, D.E., and Cieplak, W. 1997. Identification and molecular cloning of a gene encoding a fibronectin-binding protein. Mol. Microbiol. 24: 953–963.

Konkel, M.E., Kim, B.J., Klena, J.D., Young, C.R., and Ziprin, R. 1998. Characterization of the thermal stress response of *Campylobacter jejuni*. Infect. Immun. 66: 3666–3672.

Konkel, M.E., Kim, B.J., Rivera-Amill, V., and Garvis, S.G. 1999. Bacterial secreted proteins are required for the internalization of *Campylobacter jejuni* into cultured mammalian cells. Mol. Microbiol. 32: 691–701.

Konkel, M.E., Mead, D.J., Hayes, S.F., and Cieplak, W., Jr. 1992. Translocation of *Campylobacter jejuni* across human polarized epithelial cell monolayer cultures. J. Infect. Dis. 166: 308–315.

Kopecko, D.J., Hu, L., and Zaal, K.J.M. 2001. *Campylobacter jejuni*- microtubule dependent invasion. Trends Microbiol. 9: 389–396.

Korhonen, L.K., and Martikainen, P.J. 1990. Comparison of some enrichment broths and growth media for the isolation of thermophilic campylobacters from surface water samples. J. Appl. Bacteriol. 68: 593–599.

Lara-Tejero, M., and Galan, J.E. 2000. A bacterial toxin that controls cell cycle progression is a deoxyribonuclease I-like protein. Science 290: 354–357.

Lara-Tejero, M., and Galan, J.E. 2001. CdtA, CdtB, and CdtC form a tripartite complex that is required for cytolethal distending toxin activity. Infect. Immun. 69: 4358–4365.

Lara-Tejero, M., and Galan, J.E. 2002. Cytolethal distending toxin: limited damage as a strategy to modulate cellular functions. Trends Microbiol. 10: 147–152.

Lawson, A.J., Linton, D., Stanley, J., and Owen, R.J. 1997. Polymerase chain reaction and speciation of *Campylobacter* upsaliensis and C. helveticus in human faeces and comparison with culture techniques. J. Appl. Microbiol. 83: 375–380.

Lawson, A.J., On, S.L.W., Logan, J.M.J., and Stanley, J. 2001. *Campylobacter hominis* sp. nov., from the human gastrointestinal tract. Int. J. Sys. Evol. Microbiol. 51: 651–660.

Lee, A., O'Rourke, L.L., Barrington, P.J., and Trust, T.J. 1986. Mucus colonization as a determinant of pathogenicity in intestinal infection by *Campylobacter jejuni*: a mouse cecal model. Infect. Immun. 51: 536–546.

Linton, D., Gilbert, M., Hitchen, P.G., Dell, A., Morris, H.R., Wakarchuk, W.W. *et al.* 2000. Phase variation of a B-1,3 galactosyltransferase involved in generation of the ganglioside GM1-like lipo-oligosaccharide of *Campylobacter jejuni*. Mol. Microbiol. 37: 501–514.

Linton, D., Lawson, A.J., Owen, R.J., and Stanley, J. 1997. PCR detection, identification to species level, and fingerprinting of *Campylobacter jejuni* and *Campylobacter coli* direct from diarrheic samples. J. Clin. Microbiol. 35: 2568–2572.

Logan, J.M.J., Burnens, A., Linton, D., Lawson, A.J., and Stanley, J. 2000. *Campylobacter lanienae* sp. nov., a new species isolated from workers in an abattoir. Int. J. Sys. Evol. Microbiol. 50: 865–872.

Magira, E.E., Papaioakim, M., Nachamkin, I., Asbury, A.K., Li, C.Y., Ho, T.W. *et al.* 2003. Differential distribution of HLA-DQb/DRb epitopes in the two forms of Guillain-Barre syndrome, acute motor axonal neuropathy (AMAN) and acute inflammatory demyelinating polyneuropathy (AIDP): identification of DQb epitopes associated with susceptibilty to and protection from AIDP. J. Immunol. 170: 3074–3080.

Martin, W.T., Patton, C.M., Morris, G.K., Potter, M.E., and Puhr, N.D. 1983. Selective enrichment broth for isolation of *Campylobacter jejuni*. J. Clin. Microbiol. 17: 853–855.

McSweegan, E., and Walker, R.I. 1986. Identification and characterization of two *Campylobacter jejuni* adhesins for cellular and mucous substrates. Infect. Immun. 53: 141 148.

Medema, G.J., Schets, F.M., van de Giessen, A.W., and Gavelaar, A.H. 1992. Lack of colonization of 1 day old chicks by viable, non-culturable *Campylobacter jejuni*. J. Appl. Bacteriol. 72: 512–516.

Misawa, N., and Blaser, M.J. 2000. Detection and characterization of autoagglutination activity by *Campylobacter jejuni*. Infect. Immun. 68: 6168–6175.

Moran, A.P., Penner, J.L., and Aspinall, G.O. (2000) *Campylobacter* lipopolysaccharides. In: *Campylobacter*, 2nd ed. I. Nachamkin, and M.J. Blaser, eds. ASM Press, Washington, D.C., pp. 241–257.

Nachamkin, I. 1995. *Campylobacter* and *Arcobacter*. In: Manual of Clinical Microbiology. P.R. Murray, E.J. Baron, M.A. Pfaller, F.C. Tenover, and R.H. Yolken, eds. ASM Press, Washington, D.C., pp. 483–491.

Nachamkin, I. 1997. Microbiologic approaches for studying *Campylobacter* in patients with Guillain–Barré syndrome. J. Infect. Dis. 176 (Suppl. 2): S106–S114.

Nachamkin, I. 2001. *Campylobacter jejuni*. In: Food Microbiology: Fundementals and Frontiers. M.P. Doyle, L.R. Beuchat, and T.J. Montville, eds. ASM Press, Washington, D.C., pp. 179–192.

Nachamkin, I. 2003. *Campylobacter* and *Arcobacter*. In: Manual of Clinical Microbiology, 8th edn. Murray, P.R., Baron, E.J., Jorgensen, J.H., Pfaller, M.A., and Yolken, R.H. eds. ASM Press, Washington, D.C., pp. 902–914.

Nachamkin, I., B.M. Allos, and T.W. Ho. 2000. *Campylobacter jejuni* infection and the association with Guillain–Barre syndrome. In: *Campylobacter* 2nd Edition. I. Nachamkin, and M.J. Blaser eds. ASM Press, Washington, D.C. p. 155–175.

Nachamkin, I., Yang, X.H., and Stern, N.J. 1993. Role of *Campylobacter jejuni* flagella as colonization factors for three-day-old chicks: analysis with flagellar mutants. Appl. Environ. Microbiol. 59: 1269–1273.

Newell, D.G., and Wagenaar, J.A. 2000. Poultry infections and their control at the farm level. In: *Campylobacter*, 2nd Edition. I. Nachamkin, and M.J. Blaser, eds. ASM Press, Washington, D.C., pp. 497–509.

Oberhelman, R.A., and Taylor, D.N. 2000. *Campylobacter* infections in developing countries. In: *Campylobacter*, 2nd Edition. I. Nachamkin, and M.J. Blaser, eds. ASM Press, Washington, D.C., pp. 139–153.

Oelschlaeger, T.A., Guerry, P., and Kopecko, D.J. 1993. Unusual microtubule-dependent endocytosis mechanisms triggered by *Campylobacter jejuni* and *Citrobacter freundii*. Proc. Natl. Acad. Sci. 90: 6884–6888.

On, S.L.W. 2001. Taxonomy of *Campylobacter, Arcobacter, Helicobacter* and related bacteria: current status, future prospects and immediate concerns. J. Appl. Microbiol. 90: 1S–15S.

Oyofo, B.A., Mohran, Z.S., El-Etr, S., Wasfy, M.O., and Peruski, L.F. 1996. Detection of enterotoxigenic *Escherichia coli, Shigella*, and *Campylobacter* spp. by multiplex PCR assay. J. Diarrh. Dis. Res. 14: 207–210.

Oyofo, B.H., Thornton, S.A., Burr, D.H., Trust, T.J., Pavlovskis, O.R., and Guerry, P. 1992. Specific detection of *Campylobacter jejuni* and *Campylobacter coli* by using polymerase chain reaction. J. Clin. Microbiol. 30: 2613–2619.

Paisley, J.W., Mirrett, S., Lauer, B.A., Roe, M., and Reller, L.B. 1982. Dark-field microscopy of human fece for presumptive diagnosis of *Campylobacter fetus* subsp. *jejuni* enteritis. J. Clin. Microbiol. 15: 61–63.

Park, C.H., Hixon, D.L., Polhemus, A.S., Ferguson, C.B., Hall, S.L., Risheim, C.C., and Cook, C.B. 1983. A rapid diagnosis of campylobacter enteritis by direct smear examination. Am. J. Clin. Pathol. 80: 388–390.

Parkhill, J., Wren, B.W., Mungall, K., Ketley, J.M., Churcher, C., Basham, D. *et al.* 2000. The genome sequence of the foodborne pathogen *Campylobacter jejuni* reveals hypervariable sequences. Nature 403: 665–668.

Patterson, M.F. 1995. Sensitivity of *Campylobacter* spp. to irradiation in poultry meat. Lett. Appl. Microbiol. 20: 338–340.

Pei, Z., Burucoa, C., Grignon, B., Baqar, S., Huang, X.-Z., Kopecko, D.J. *et al.* 1998. Mutation in peb1A of *Campylobacter jejuni* reduces interactions with epithelial cells and intestinal colonization of mice. Infect. Immun. 66: 938–943.

Perlman, D.M., Ampel, N.M., Schifman, R.B., Cohn, D.L., Patton, C.M., Aguirre, M.L. *et al.* 1988. Persistent *Campylobacter jejuni* infections in patients infected with human immunodeficiency virus (HIV). Ann. Intern. Med. 108: 540–546.

Pesci, E.C., Cottle, D.L., and Pickett, C.L. 1994. Genetic, enzymatic and pathogenic studies of the iron superoxide dismutase of *Campylobacter jejuni*. Infect. Immun. 62: 2687–2694.

Radomyski, T., Murano, E.A., Olson, D.G., and Murano, P.S. 1994. Eliminaton of pathogens of significance in food by low-dose irradiation: a review. J. Food Prot. 57: 73–86.

Ransom, G.M., Kaplan, B., McNamara, A.M., and Wachsmuth, I.K. 2000. *Campylobacter* prevention and control: the USDA-Food Safety and Inspection Service role and new food safety approaches. In: *Campylobacter*, 2nd Edition. I. Nachamkin, and M.J. Blaser, eds. ASM Press, Washington, D.C., pp. 511–528.

Richardson, P.T., and Park, S.F. 1995. Enterochelin acquisition in *Campylobacter coli*: characterization of components of a binding-protein dependent transport system. Microbiology 141: 3181–3191.

Rollins, D.M., and Colwell, R.R. 1986. Viable but nonculturable stage of *Campylobacter jejuni* and its role in the survival in the natural aquatic environment. Appl. Environ. Microbiol. 52: 531–538.

Sazie, E.S.M., and Titus, A.E. 1982. Rapid diagnosis of *Campylobacter* enteritis. Ann. Intern. Med. 96: 62–63.

Sjogren, E., Lindblom, G.B., and Kaijser, B. 1987. Comparison of different procedures, transport media, and enrichment media for isolation of *Campylobacter* species from healthy laying hens and humans with diarrhea. J. Clin. Microbiol. 25: 1966–1968.

Skirrow, M.B. 1994. Diseases due to *Campylobacter, Helicobacter* and related bacteria. J. Comp. Path 111: 113–149.

Skirrow, M.B., and Blaser, M.J. 2000. Clinical aspects of *Campylobacter* infection. In: *Campylobacter*. 2nd Edition. I. Nachamkin, and M.J. Blaser, eds. ASM Press, Washington, D.C., pp. 69–88.

Skirrow, M.B., Jones, D.M., Sutcliffe, E., and Benjamin, J. 1993. *Campylobacter* bacteremia in England and Wales, 1981–1991. Epidemiol. Infect. 110: 567–573.

Stern, N.J. 1992. Reservoirs for *Campylobacter jejuni* and approaches for intervention in poultry. In: *Campylobacter jejuni*: Current Status and Future Trends. I. Nachamkin, Blaser, M.J., and Tompkins, L.S., eds. American Society for Microbiology, Washington, D.C., pp. 49–60.

Szymanski, C.M., Logan, S.M., Linton, D., and Wren, B.W. 2003. *Campylobacter* – a tale of two protein glycosylation systems. Trends Microbiol. 11: 233–238.

Takeshi, K., Ikeda, T., Kubo, A., Fujinaga, Y., Makino, S., Oguma, K. *et al.* 1997. Direct detection by PCR of *Escherichia coli* O157 and enteropathogens in patients with bloody diarrhea. Microbiol. Immunol. 41: 819–822.

Tauxe, R.V. 1992. Epidemiology of *Campylobacter jejuni* infections in the United States and other industrialized nations. In: *Campylobacter jejuni*: Current Status and Future Trends. I. Nachamkin, M.J. Blaser, and L.S. Tompkins, eds. American Society for Microbiology, Washington, D.C., pp. 9–19.

Taylor, D.N., Echeverria, P., Pitarangsi, C., Seriwatana, J., Bodhidatta, L., and Blaser, M.J. 1988. Influence of strain characteristics and immunity on the epidemiology of *Campylobacter* infections in Thailand. J. Clin. Microbiol. 26: 863–868.

Tolcin, R., LaSalvia, M.M., Kirkley, B.A., Vetter, E.A., Cockerill, F., and Procop, G.W. 2000. Evaluation of the alexon-trend ProSpecT *Campylobacter* microplate assay. J. Clin. Microbiol. 38: 3853–3855.

Vandamme, P. 2000. Taxonomy of the family *Campylobacteraceae*. In: *Campylobacter*, 2nd edn. I. Nachamkin, and M.J. Blaser, eds. ASM Press, Washington, D.C., pp. 3–26.

Vandamme, P., Daneshvar, M.I., Dewhirst, F.E., Paster, B.J., Kersters, K., Goossens, H., and Moss, C.W. 1995. Chemotaxonomic analyses *of Bacteroides gracilis* and *Bacteroides ureolyticus* and reclassification of *B. gracilis* as *Campylobacter gracilis* comb. nov. Int. J. Sys. Bacteriol. 45: 145–152.

Vandamme, P., and De Ley, J. 1991. Proposal for a new family, *Campylobacteraceae*. Int. J. Syst. Bacteriol. 41: 451–455.

Waegel, A., and Nachamkin, I. 1996. Detection and molecular typing of *Campylobacter jejuni* in fecal samples by polymerase chain reactioin. Mol. Cell Probes 10: 75–80.

Wai, S.N., Nakayama, K., Umene, K., Moriya, T., and Amako, K. 1996. Construction of a ferritin-deficient mutant of *Campylobacter jejuni*: contribution of ferritin to iron storage and protection against oxidative stress. Mol. Microbiol. 20: 1127–1134.

Walker, R.I., Schmauder-Chock, E.A., Parker, J.L., and Burr, D. 1988. Selective association and transport of *Campylobacter jejuni* through M cells of rabbit Peyer's patches. Can. J. Microbiol. 34: 1142–1147.

Wang, W.L., Powers, B.W., Luechtefeld, N.W., and Blaser, M.J. 1980. Effects of disinfectants on *Campylobacter jejuni*. Appl. Environ. Microbiol. 45: 1202–1205.

Wassenaar, T.M., Bleumink-Pluym, N.M., and van der Zeijst, B.A. 1991. Inactivation of *Campylobacter jejuni* flagellin genes by homologous recombination demonstrates that flaA but not flaB is required for invasion. EMBO J. 10: 2055–2061.

Wassenaar, T.M., Van Der Zeijst, B.A.M., Ayling, R., and Newell, D.G. 1993. Colonization of chicks by motility mutants of *Campylobacter jejuni* demonstrates the importance of flagellin A expression. J. Gen. Microbiol. 139: 1171–1175.

Wexler, H.M., Reeves, D., Summanen, P.H., Molitoris, E., McTeague, M., Duncan, J. *et al.* 1996. *Sutterella wadsworthensis* gen. nov., sp. nov., bile-resistant microaerophilic *Campylobacter gracilis*-like clinical isolates. Int. J. Syst. Bacteriol. 46: 252–258.

Whitehouse, C.A., Balbo, P.B., Pesci, E.C., Cottle, D.L., Mirabito, P.M., and Pickett, C.L. 1998. *Campylobacter jejuni* cytolethal distending toxin causes a G2-phase cell cycle block. Infect Immun. 66: 1934–1940.

Wood, R.C., MacDonald, K.L., and Osterholm, M.T. 1992. *Campylobacter* enteritis outbreaks associated with drinking raw milk during youth activities. A 10-year review of outbreaks in the United States. JAMA 268: 3228–3230.

Wooldridge, K.G., Williams, P.H., and Ketley, J.M. 1996. Host signal transduction and endocytosis of *Campylobacter jejuni*. Microb. Pathog. 21: 299–305.

Yao, R., Burr, D.H., and Guerry, P. 1997. CheY mediated modulation of *Campylobacter jejuni* virulence. Mol. Microbiol. 23: 1021–1031.

Young, G.M., Schmiel, D.H., and Miller, V.L. 1999. A new pathway for the secretion of virulence factors by bacteria: the flagellar export apparatus functions as a protein secretion system. Proc. Natl. Acad. Sci. USA 96: 6456–6461.

# Listeria monocytogenes

15

George C. Paoli, Arun K. Bhunia, and Darrell O. Bayles

## Abstract

*Listeria monocytogenes* is a Gram-positive foodborne bacterial pathogen and the causative agent of human listeriosis. The organism has served as a model for the study of intracellular pathogenesis for several decades and many aspects of the pathogenic process are well understood. *Listeriae* are acquired primarily through the consumption of contaminated foods including soft cheese, raw milk, deli salads, and ready-to-eat foods such as luncheon meats and frankfurters. Although *L. monocytogenes* infection is usually limited to individuals that are immunocompromised, the high mortality rate associated with human listeriosis makes *L. monocytogenes* the leading cause of death amongst foodborne bacterial pathogens. As a result, tremendous effort has been made at developing methods for the isolation, detection and control of *L. monocytogenes* in foods. Additional research in the area of genomics and proteomics has begun to be applied toward developing a better understanding of how *L. monocytogenes* responds to its environment. These efforts will allow a more complete understanding of the pathogenic process and will aid the design and development of targeted strategies for detection and intervention, leading to improved control of *L. monocytogenes* in foods.

## Introduction

*Listeria monocytogenes* was originally isolated by E.G.D. Murray from laboratory rabbits in 1926 (Murray *et al.*, 1926). Although the organism has been used as a model for the study of intracellular parasitism for decades, *L. monocytogenes* was not considered a significant animal pathogen until the late 1970s and early 1980s when it was recognized as a major foodborne human pathogen. It is now estimated that *L. monocytogenes* is the leading cause of death from foodborne bacteria in the United States (Mead *et al.*, 1999). In recent years, the detailed biology and disease-producing capability of this pathogen has been explored. *Listerial* pathogenesis is a complex process and multiple virulence factors aid this pathogen in evading the host immune system, ultimately resulting in human listeriosis. Over the past few decades, tremendous strides have been made in improving methods for isolation and detection of *L. monocytogenes* from food and environmental samples. The re-

cent determination of the complete genome sequence of four strains of *L. monocytogenes* (Glaser *et al.*, 2001; Nelson *et al.*, 2004) and one strain of *L. innocua* (Glaser *et al.*, 2001) has allowed the comparative analysis of these genomes and should greatly expand our understanding of the biology of this important foodborne pathogen.

## *Listeria* characteristics and taxonomy

The genus *Listeria* is grouped with other Gram-positive non-spore forming bacilli. Members of the genus *Listeria* are generally aerobic or facultatively anaerobic, catalase positive, and oxidase negative. *Listeria* are motile via a few peritrichous flagella when grown at temperatures below 30°C. *L. monocytogenes* is one of six species in the genus *Listeria*. The other species are *L. ivanovii*, *L. innocua*, *L. seeligeri*, *L. welshimeri*, and *L. grayi*. Although *L. ivanovii* is known to be pathogenic to animals, *L. monocytogenes* is the only human pathogen within the genus *Listeria*. Differentiation of the *Listeria* species can be made based on hemolytic activity (sheep or horse blood) and sugar fermentation (Table 15.1), although exceptions have been reported (Wiedmann *et al.*, 1997; Johnson *et al.*, 2004). The hemolytic activity can be difficult to interpret on standard blood agar and can be enhanced by employing the Christie, Atkins, Munch-Petersen test (CAMP test) (Christie *et al.*, 1944). A series of 10 biochemical tests specifically designed to differentiate the members of the genus *Listeria*, the API-*Listeria*® test (bioMerieux, Marcy l'Etoile, France), includes a patented DIM test (Differentiation of Innocua and Monocytogenes) that obviates the need for hemolysis testing (Bille *et al.*, 1992). A 15-test kit (Micro-ID® *Listeria*; Remel, Inc., Lenexa, Kansas, USA), a 12-test kit (Microgen® *Listeria*-ID; Microgen Bioproducts, Camberley, UK), and a 12-test kit that includes a hemolysis assay (Microbact® *Listeria* 12L System; Oxoid, Basingstoke, UK) are also commercially available.

## Listeriosis: disease and epidemiology

*Listeria monocytogenes*, an emerging pathogen since the late 1970s, causes an estimated 2500 cases of serious illness and 500 deaths per year in the United States (Mead *et al.*, 1999).

**Table 15.1** Biochemical differentiation of *Listeria* species (Linnan *et al.*, 1995)

| *Listeria* species | Hemolysis[b] | Acid production from[a] | | |
|---|---|---|---|---|
| | | D-Xylose | L-Rhamnose | D-Mannitol |
| *L. monocytogenes* | + | − | + | − |
| *L. seeligeri* | + | + | − | − |
| *L. ivanovii* | ++ | + | − | − |
| *L. innocua* | − | − | V | − |
| *L. welshimeri* | − | + | V | − |
| *L. grayi* | − | − | V | + |

[a]+, positive reaction; −−, negative reaction; V, strain variable.
[b]*L. ivanovii* is strongly hemolytic while *L. monocytogenes* and *L. seeligeri* are weakly hemolytic.

Although Murray recognized the oral route of infection in his original isolation of *L. monocytogenes* (Murray *et al.*, 1926), the key in recognizing the organism as a foodborne pathogen came nearly 60 years later when an outbreak of listeriosis was epidemiologically linked to the consumption of contaminated coleslaw (Schlech *et al.*, 1983). Since then, listeriosis has been associated with the consumption of a variety of contaminated foods including soft cheese, raw milk, hot dogs, delicatessen meats and salads, seafood, and fresh vegetables. It is now recognized that most cases of listeriosis (85–95%), both sporadic cases (Pinner *et al.*, 1992; Schuchat *et al.*, 1992) and common-source outbreak cases, are caused by the consumption of *L. monocytogenes* contaminated food. A summary of epidemiological information from some foodborne listeriosis outbreaks is presented in Table 15.2.

Listeriosis is a serious health concern because of the severity of the disease, its high mortality rate, and the opportunistic nature of the pathogen in causing disease. *L. monocytogenes* primarily affects individuals with suppressed cellular immunity, including the young, the elderly, and individuals experiencing immunosuppression due to chemotherapeutic treatment for cancer or organ transplant surgery, diabetes, alcoholism, or cardiovascular diseases. Pregnancy is also associated with suppressed cell-mediated immunity as a mechanism to protect the fetus from rejection; therefore, pregnant women are extremely susceptible to listeriosis (Smith, 1999; Abram *et al.*, 2003). In pregnant women, this organism can cause abortion, mainly in the third trimester, resulting in stillbirth (Vazquez-Boland *et al.*, 2001). In some cases infants are born with *Listeria* infection (Vazquez-Boland *et al.*, 2001). In adults, the infection begins with flu-like symptoms that gradually progress to severe headache. Systemic infection may cause meningitis, encephalitis, septicemia, bacteremia, liver abscess, etc. The symptoms of neuropathic listeriosis include fever, malaise, ataxia, seizures, meningitis, and encephalitis (Roberts and Wiedmann, 2003). The mortality rate is about 20–30% resulting in approximately 500 deaths annually, the highest among all the foodborne pathogens (Mead *et al.*, 1999). A number of outbreaks in which most of the infected individuals developed mild symptoms, such as diarrhea, fever, headache and myalgia have been reported (Dalton *et al.*, 1997; Aureli *et al.*,

2000; Frye *et al.*, 2002). These outbreaks of self-limiting listerial gastroenteritis usually involved the ingestion of high doses of *L. monocytogenes* by healthy individuals. *L. monocytogenes* may also be associated with gastrointestinal Crohn's disease (Hugot *et al.*, 2003). Although the levels of *L. monocytogenes* in outbreak associated foods is generally greater than $10^3$ CFU/g, as little as 1 CFU of *L. monocytogenes* per gram of food has been associated with the disease in susceptible individuals (Hubbard and Billy, 2001). Nevertheless, there is a great deal of uncertainty concerning these estimates because the actual level of the pathogen in the serving of food consumed by an individual could vary considerably from that observed in the portion of the food sampled during the subsequent investigation. Furthermore, an infective dose has not been determined in human infection studies and estimates vary from $10^2$–$10^9$ cells depending on the immunological status of the host. The incubation period for the disease also varies from 11–70 days (median of 3 weeks) in humans.

### *Listeria monocytogenes* in food

In addition to being a deadly pathogen, *L. monocytogenes* is well suited for growth and survival in food. The bacterium can survive and even grow at refrigerator temperatures (McClure *et al.*, 1989; Walker *et al.*, 1990; McClure *et al.*, 1991; Phan-Thanh and Gormon, 1995; Bayles *et al.*, 1996; Bayles and Wilkinson, 2000) and is tolerant of both low pH (Davis, 1996; Phan-Thanh *et al.*, 2000) and high salt concentrations (McClure *et al.*, 1989; McClure *et al.*, 1991; Bayles and Wilkinson, 2000; Duché *et al.*, 2002a). *Listeria* bacteria are common in the environment and can be incidentally introduced into food through contact with contaminated surfaces. Food processing facilities can easily become contaminated by soil on worker's shoes, transportation and handling equipment, animal hides, raw plant material, etc. Once introduced into a facility, *Listeria* can become resident in floor drains, in standing water, and can form biofilms on food processing surfaces.

Although *Listeria* are inactivated by heating to 50°C and are easily killed by cooking, they are a significant concern for foods that are minimally processed and purchased ready-to-eat. Surveys of a variety of foods cited a prevalence of *L.*

**Table 15.2** Epidemiological information for some published foodborne listeriosis outbreaks

| Location | Year(s) | Cases | Deaths | Serotype | Food | Reference |
|---|---|---|---|---|---|---|
| Halifax, Canada | 1981 | 41 | 18 | 4b | Coleslaw | Schlech *et al.* (1983) |
| Massachusetts, USA | 1983 | 49 | 14 | 4b | Milk | Fleming *et al.* (1985) |
| Vaud, Switzerland | 1983–87 | 122 | 34 | 4b | Cheese | Bula *et al.* (1995) |
| California, USA | 1985 | 142 | 48 | 4b | Cheese | Linnan *et al.* (1995) |
| United Kingdom | 1989–90 | 300 | 0 | 4b | Pâté | McLauchlin *et al.* (1991) |
| France | 1992 | 279 | 88 | 4b | Pork tongue | Jacquet *et al.* (1995) |
| France | 1993 | 39 | 0 | 4b | Pork pâté | Goulet *et al.* (1995) |
| France | 1995 | 36 | 0 | 4b | Soft cheese | Goulet *et al.* (1995) |
| Multistate, USA | 1998–99 | 40 | 4 | 4b | Deli meats | CDC (1998) |
| Finland | 1988–99 | 25 | 6 | 3a | Butter | Lyytikainen *et al.* (2000) |
| France | 1999 | 29 | 7 | NR[a] | Pork tongue | WHO (2000) |
| Multistate, USA | 2000 | 29 | 4 | 4b | Turkey deli meats | CDC (2002) |
| North Carolina, USA | 2000–01 | 12 | 0 | 4b | Cheese | CDC (2002a) |
| Multistate, USA | 2002 | 46 | 7 | N.R.[a] | Chicken and turkey | CDC (2002b) |
| Quebec, Canada | 2002 | 17 | 0 | NR[a] | Cheese | Gaulin *et al.* (2003) |

[a] Serotype not reported.

*monocytogenes* of between 2% and 10% (Farber and Peterkin, 1991; Hubbard and Billy, 2001) due to incidental contamination. This unique combination of factors has resulted in the US Department of Agriculture and US Food and Drug Administration adopting a "zero tolerance" policy for *L. monocytogenes* in ready-to-eat foods. These ready-to-eat foods provide an ideal environment for the multiplication of *L. monocytogenes* during processing and storage. All establishments that produce ready-to-eat products are required to develop and implement Hazard Analysis and Critical Control Points (HACCP) strategies aimed at reducing or eliminating *L. monocytogenes* from food processing facilities (Hubbard and Billy, 2001; FSIS, 2003)

## Mechanism of pathogenesis

For a foodborne pathogen, bacterial translocation through the intestinal tract is prerequisite for infection. *L. monocytogenes* cells are able to pass through the stomach, in part due to protection imparted by the associated food matrix as well physiological acid resistance (Merrell and Camilli, 2002). Upon arrival in the small intestine, the cells encounter bile salts, and *L. monocytogenes* resistance to bile is mediated by a bile salt hydrolase encoded by the *bsh* gene (Dussurget *et al.*, 2002). During the intestinal phase of infection, *L. monocytogenes* can translocate via two pathways; Peyer's patch (PP)-dependent and PP independent pathways. In the PP-dependent pathway, M cells, naturally phagocytic cells in the PP, facilitate *L. monocytogenes* entry through the lymphatic tissues (Jensen *et al.*, 1998; Havell *et al.*, 1999). *L. monocytogenes* can cause abscesses in Peyer's patches and infiltration in polymorphonuclear leukocytes and macrophages (Marco *et al.*, 1997). Dendritic cells in Peyer's patches carrying *L. monocytogenes* are also important for initial dissemination to peripheral organs (Pron *et al.*, 2001). In the PP-independent pathway, bacteria bind to the epithelial cell lining of the microvilli, actively penetrate, and reach the lamina propria where phagocytic cells (macrophages or dendritic cells) engulf and transport the bacteria to blood. Within 24 hours of infection, about 90% of *Listeria* cells migrate to the liver and 10% to the spleen. Recently, *L. monocytogenes* was also reported to colonize in the gall bladder, which may act as a potential reservoir for this organism in diseased as well as in asymptomatic animals (Hardy *et al.*, 2004). In the progressive infection, *L. monocytogenes* can cross the blood brain or fetoplacental barrier to initiate neurological or pregnancy-related complications.

The cellular mechanism of pathogenesis involves five major events (Figure 15.1).

1   *Adhesion and invasion: Listeria* cells initially adhere to intestinal enterocytes and penetrate the intestinal wall, a process mediated by internalin and/or other adhesion factors. In defense, the host cell traps the bacteria in phagosomes.

2   *Lysis of primary vacuole:* Listeriolysin O (LLO) and phosphatidylinositol-specific phospholipase C (PI-PLC) lyse the phagosome releasing *L. monocytogenes* into host cell's cytosol.

3   *Intracellular growth:* Bacterial propagation in cytosol is mediated by a bacterial hexose phosphate transporter (Hpt) and lipoate protein ligase (LpLA1) that allow *L. monocytogenes* to scavenge mammalian carbon sources for energy.

4   *Cell-to-cell spread:* Following multiplication in the host cytoplasm, *L. monocytogenes* travels across the host cell using an actin polymerization protein (ActA) to polymerize mammalian actin filaments. This actin polymerization results in directed "movement" of the *L. monocytogenes* cells towards the host cell membrane and causes pseudopod-like structures that extend into neighboring cells. This ul-

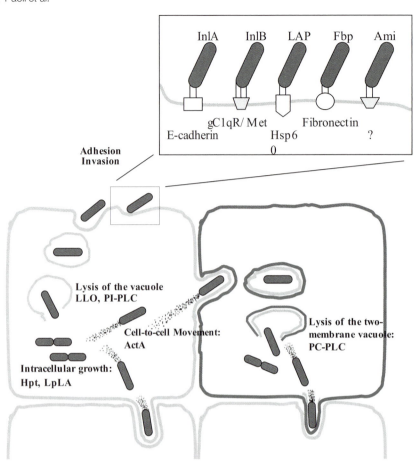

**Figure 15.1** Cellular events in *Listeria monocytogenes* pathogenesis. The intracellular lifestyle of *Listeria monocytogenes* is a complex process and several *L. monocytogenes* virulence factors are essential for pathogenesis. The infection cycle is initiated by the adhesion of *L. monocytogenes* to the host cell which is mediated by a several listerial surface proteins as illustrated in the boxed blow-out. This is followed by invasion of the host cell, lysis of the vacuole, intracellular growth, cell-to-cell movement, and lysis of the two-membrane vacuole. The *L. monocytogenes* virulence factors involved in each of these steps are shown. InlA, internalin A; InlB, internalin B, LAP, *Listeria* adhesion protein; Fbp, fibronectin binding protein; Ami, autolytic amidase; LLO, listeriolysin O, PI-PLC, phosphatidylinositol-specific phospholipase C; PC-PLC, phosphatidylcholine-specific phospholipase C; ActA, actin polymerization protein; Hpt, hexose phosphate transporter, LplA lipoate protein ligase.

timately causes the *L. monocytogenes* cell to move into the neighboring cell within a two-membrane vacuole.

5    *Lysis of the two-membrane vacuole:* A listerial phosphatidylcholine-specific phospholipase C (PC-PLC) along with LLO lyse the two-membrane vacuole and release the *L. monocytogenes* infecting the neighboring cell. Other factors such as the role of innate and cellular immunity during infection are critical but will not be discussed here.

## Adhesion and invasion

The first step in the intracellular infection by *L. monocytogenes* involves the adhesion to and invasion of the host cell by the bacterium and is mediated by the interaction of *L. monocytogenes* surface proteins with macromolecules on the surfaces of host cells. Several *L. monocytogenes* surface proteins interact with a variety of target cells receptors, collectively resulting in the invasion of a wide range of host cell types. Thus, *L. monocytogenes* has the unique ability to cross three major barriers in humans: the intestinal, placental, and the blood–brain barriers leading to the various manifestations of listerial disease.

Internalin (Inl) proteins are a large class of cell surface proteins in *L. monocytogenes* and are encoded by the internalin multigene family. InlA and InlB were the first listerial proteins shown to mediate bacterial adhesion and invasion into host cells (Gaillard *et al.*, 1991; Dramsi *et al.*, 1995). The genes encoding InlA and InlB are cotranscribed and, along with the gene encoding InlC, which is in a separate operon, are regulat-

ed by the positive regulatory factor PrfA. Recently, more than 20 potential internalin genes were identified in the *L. monocytogenes* genome (Glaser *et al.*, 2001), which greatly expanded the internalin protein family. The internalin protein family and several other listerial surface proteins that act as adhesins are discussed below.

## Internalin A

Internalin A (InlA) is involved in the invasion of *L. monocytogenes* into many, but not all, non-professional phagocytic cells. Internalin mutants also bind and cause infection in a murine model, albeit in lower frequency than wild type strains (Dramsi *et al.*, 1995), suggesting that other factors are involved in the adhesion and invasion process. InlA is encoded by the first gene of the *inlAB* operon. It is an 88-kDa protein, rich in threonine and serine, contains 16 leucine-rich repeats (LRRs) and a membrane anchoring LPXTG motif. Sortase A (SrtA), a transamidase, anchors internalin, and other LPXTG proteins, to the bacterial cell wall (Bierne and Cossart, 2002) and a mutation in *srtA* resulted in reduced invasion in epithelial cells. Part of the internalin amino acid sequence is similar to certain cell wall proteins of other bacteria including, protein A and fibronectin-binding protein from *Staphylococcus aureus*; M protein, Fc, and IgA binding proteins from *Streptococcus pyogenes*; and protein G from group G streptococci (Reid *et al.*, 2003).

InlA is thermoregulated at the transcriptional level by both PrfA-dependent and PrfA-independent mechanisms (Lingnau *et al.*, 1995). Internalin production is very high at 37°C and low at 25°C with maximal expression reported during log phase, and minimal expression during stationary phase. Increased internalin production directly correlated with *Listeria* invasion into mammalian cells.

E-cadherin, present on human epithelial cells, acts as a receptor for InlA (Mengaud *et al.*, 1996). E-cadherin is a 135-kDa protein that undergoes post-translation processing to a mature form (120 kDa), and is composed of a highly conserved carboxy terminal cytodomain, a single transmembrane region, and an extracellular domain. The extracellular domain consists of five tandemly repeated cadherin-motif subdomains, each carrying two conserved regions for calcium binding. The matured E-cadherin is directed to the basolateral side of epithelial cells (Beavon, 2000) where it undergoes calcium-dependent dimer formation (*cis*-dimerization), followed by the interaction between dimerized E-cadherins on adjacent cells (*trans*-dimerization). The cytoplasmic domain of E-cadherin is bound to catenins ($\alpha$, $\beta$, and $\gamma$-catenins), which interact with the cytoskeleton. When InlA binds to E-cadherin, the terminal 35 amino acids of E-cadherin react with $\beta$-catenin, which recruits $\alpha$-catenin, which in turn reacts with the actins. $\beta$-catenin is involved in signal transduction events (Peifer, 1997), but the mechanism of actin polymerization during the InlA-dependent entry is not fully understood. However, the involvement of Rac, Arp2/3, myosin VIIa, and vezatin during this process is speculated (Cossart and Sansonetti, 2004). Overall, internalin mediated invasion is described as zipper-like, where bacteria sit on the surface and sink into the cytoplasm after initiating a signal transduction event that results in cytoskeletal rearrangements.

Mouse E-cadherin compared to human E-cadherin has a single amino acid difference at residue 16 that prevents internalin–mediated binding and invasion. Substitution of the critical glutamine with proline in mouse E-cadherin, restored the human E-cadherin sequence and allowed InlA interaction with murine epithelial cells (Lecuit *et al.*, 1999). Later studies using a transgenic mouse model expressing human E-cadherin demonstrated InlA dependent translocation into various organs (Lecuit *et al.*, 2001). Interaction between InlA and E-cadherin is critical for *L. monocytogenes* translocation through the human feto-placental barrier (Lecuit *et al.*, 2004); however, a study by Bakardjiev *et al.*, 2004, contradicts that finding in a guinea-pig model. More study is needed to resolve this discrepancy.

## Internalin B

Internalin B (65 kDa), encoded by *inlB*, is a surface protein responsible for the hepatic phase of infection. It is responsible for the invasion of a variety of cultured cell lines including Vero, HEp-2, HeLa, some hepatocytes, and endothelial cells (Braun *et al.*, 1997; Greiffenberg *et al.*, 1998; Braun *et al.*, 2000; Greiffenberg *et al.*, 2000). InlB can be present as an anchored

molecule or as an extracellular-secreted product. InlB is composed of an amino-terminal signal sequence, eight leucine-rich repeats (LRR) and three carboxy-terminal GW repeats (an 80 amino acid repeated sequence that starts with the amino acids GW). The LRR region is responsible for binding to the tyrosine kinase receptor, c-Met (Shen *et al.*, 2000). The GW domains are required for non-covalent anchoring of InlB to cell wall lipoteichoic acids as well as for potentiation of c-Met activation (Jonquieres *et al.*, 1999; Banerjee *et al.*, 2004). When this interaction takes place, Met is dimerized and phosphorylated to recruit and phosphorylate various signaling and adaptor molecules. Phosphorylation of the adaptor molecules Cb1, Gab1, and Shc induces the recruitment of phosphoinositide 3-kinase (PI 3-K) and the synthesis of phosphatidylinositol 3,4,5-phosphate (PIP3) at the plasma membrane (Cossart and Sansonetti, 2004). The role of PIP3 is not clear. On the other hand, the phosphorylation of Met *also* modulates actin polymerization through Rac, WAVE, and the Arp2/3 complex (Cossart and Sansonetti, 2004).

Gregory *et al.* (1997) indicated that InlB-mediated invasion is dependent on the multiplicity of infection (*Listeria*:target cells) and the overall concentration of bacterial cells, affecting *Listeria* replication in hepatocytes both in *in vivo* and *in vitro* (Gregory *et al.*, 1997). InlB has three known cellular receptors, the globular part of the complement component C1q receptor (gC1qR/p32), proteoglycan (glycosaminoglycans), and Met, a receptor tyrosine kinase (Jonquieres *et al.*, 1999; Braun *et al.*, 2000; Shen *et al.*, 2000). gC1q-R/p32 (33 kDa) is important for entry into non-phagocytic cells (Braun *et al.*, 2000). The most studied InlB receptor is Met (also known as hepatocyte growth factor receptor, HGF-R) that mediates the binding of InlB to hepatocytes in an interaction in which InlB apparently mimics the binding of hepatocyte growth factor (HGF). The interaction between the LRR region of InlB and Met induces *L. monocytogenes* invasion through a signal transduction event mediated by Met. Met activation is enhanced by the calcium-dependent (Marino *et al.*, 1999) interaction between InlB and glycosaminoglycans (Lyon *et al.*, 2002). When this interaction takes place, Met is dimerized and phosphorylated resulting in the recruitment of various signaling and adaptor molecules (Schlessinger, 2000). Cb1, Gab1, and Shc are phosphorylated and phosphoinositide (PI) –3 kinase is activated to generate phosphatidylinositol 3,4,5-triphosphate (PIP3) at the plasma membrane (Cossart and Sansonetti, 2004). The role of PIP3 is not clear. In addition, the phosphorylation of Met modulates actin polymerization through Rac, WAVE, and the Arp2/3 complex (Cossart and Sansonetti, 2004) initiating the formation of the membrane extension that envelopes the bacterium.

## Internalin-independent invasion

Several experimental evidences suggest that InlA-independent translocation occurs during *L. monocytogenes* infection. First, in mouse studies, *L. monocytogenes* was found mainly in Peyer's patches after oral ingestion (MacDonald and Carter, 1980) and translocated via M cells to the organs (Jensen *et al.*, 1998;

Havell *et al.*, 1999). Second, in a rat ligated ileal loop assay, the *inl*AB mutant- and the wild type-strains translocated equally (Pron *et al.*, 1998) suggesting internalin-independent translocation, which possibly occurs through M (microfolds) cells of GALT (gut associated lymphoid tissue) or by other mechanisms. M cells can transport antigens, including bacteria, by non-specific and/or antigen-specific mechanisms (Sansonetti and Phalipon, 1999). In *Salmonella* Typhimurium, one of the fimbriae proteins, Lpf, is responsible for bacterial adhesion to M cells (Baumler *et al.*, 1996). In *Yersinia*, invasion proteins interact with M cells for translocation (Clark *et al.*, 1998), and in *Shigella* the M cells are the major, if not exclusive, entry route, causing membrane ruffling (Sansonetti and Phalipon, 1999). Additional evidence for internalin-independent entry is that InlA does not interact with mouse E-cadherin (Lecuit *et al.*, 2001); thus, the possibility of InlA mediated entry and translocation is excluded in the mouse model.

### p60 protein (cwhA)
A major extracellular protein of 60 kDa, produced by *L. monocytogenes*, was originally reported to be an invasion associated protein (iap) and termed p60. Expression of p60 is necessary for invasion of fibroblasts, hepatocytes, and Caco-2 cells (Kuhn and Goebel, 1989; Ruhland *et al.*, 1993; Wuenscher *et al.*, 1993; Hess *et al.*, 1995). A recombinant *E. coli* expressing p60 showed specific binding to Caco-2 cells leading to speculation that p60 may modulate host cell physiology during invasion (Park *et al.*, 2000). p60 acts as a bacteriolytic protein and was recently renamed a cell wall hydrolase (*cwhA*) (Pilgrim *et al.*, 2003).

p60 protein is uniformly distributed on the surface of all *Listeria* species, and larger amounts of p60 are produced in cells grown at 37°C. About 25% of p60 is cell-associated and the remaining is extracellular. Although present in all *Listeria* spp., the apparent molecular weight of p60 was shown to be heterogeneous among species (Bubert *et al.*, 1992). Amino acid sequence data revealed that 120 residues at the N- and C-terminal ends were conserved in p60 proteins from all *Listeria* spp., whereas the middle part, comprising 240 amino acids, showed considerable variation. *L. monocytogenes* carries two unique sequence regions, which are absent in other *Listeria* spp.

### *Listeria* adhesion protein (LAP)
*Listeria* adhesion protein (LAP) is a 104-kDa (Pandiripally *et al.*, 1999) thermo-regulated surface protein with maximum production observed at 37°C and 42°C in stationary phase (Santiago *et al.*, 1999). Nutrient-rich media and increased glucose concentrations suppress expression of the LAP (Jaradat and Bhunia, 2002). LAP has been recognized as an important adhesion factor during the intestinal phase of infection (Jaradat *et al.*, 2003). LAP-mediated adhesion appeared to be high in human enterocyte-like Caco-2 (colon), HT-29 (colon), and HCT-8 (ileocecal) cells (Jaradat *et al.*, 2003), and low in human cell lines from the upper small intestine (HuTu-80, Int-407),

kidney (A-498), larynx (HEp-2), breast (MCF-7), bladder (UM-UC-3), or ovary (SK-OV-3) (Jaradat *et al.*, 2003). LAP is principally present in the cell cytoplasm with about 10–15% of the protein present on the cell surface (Jaradat and Bhunia, 2002). A LAP-deficient mutant showed a reduced translocation to liver and spleen in a mouse model (Jaradat *et al.*, 2003). Recently, LAP has been characterized as an alcohol acetaldehyde dehydrogenase (Aad). A recombinant *E. coli* expressing Aad also showed adhesion properties similar to a wild-type *L. monocytogenes* strain (A.K. Bhunia, Unpublished).

The receptor for LAP has been identified as eukaryotic Hsp60, and blocking of Hsp60 with a specific antibody caused about 74% reduction in binding of *L. monocytogenes* to Caco-2 cells (Wampler *et al.*, 2004). Hsp60 is present in most cells, acts as a molecular chaperone, regulates folding of mitochondrial proteins, and functions in degradation of misfolded or damaged proteins. Surface expression is essential for Hsp60 to act as a receptor. Heat shock proteins are thought to be present only in the mitochondria of cells; however, they have been found on the surface of cultured cells (Jones *et al.*, 1994; Soltys and Gupta, 1997), and stress induces increased surface expression, possibly acting as a warning signal to the host immune system (Ranford and Henderson, 2002; Wallin *et al.*, 2002). It is speculated that binding of *Listeria* to Hsp60 may initiate a signal transduction event during *L. monocytogenes* infection. Overall, interaction of *L. monocytogenes* with the eukaryotic cell is intricate and multifaceted; however, identification of Hsp60 as a receptor for LAP provides another link in the network of *L. monocytogenes*–host interactions.

*Miscellaneous adhesion proteins*
Many other *L. monocytogenes* surface molecules and proteins function directly or indirectly during adhesion and invasion affecting the overall infection process; thus, they warrant a brief description.

*Autolysin amidase (Ami)*. The autolytic amidase (Ami) is a bacteriolysin able to lyse *L. monocytogenes* (Braun *et al.*, 1997; McLaughlan and Foster, 1998) and *Staphylococcus aureus* (Foster, 1995) cells, and was later identified as an *L. monocytogenes* adhesion factor (Milohanic *et al.*, 2000). Ami attaches to the bacterial cell surface using a GW-repeat similar to those found in InlB (Braun *et al.*, 1997) and this cell-wall anchoring domain mediates *L. monocytogenes* adhesion to cultured human HepG2 liver cells (Milohanic *et al.*, 2001).

*ClpC ATPase*. The common cellular chaperone ClpC ATPase (Schirmer *et al.*, 1996) is used by *L. monocytogenes* to aid in escape from the phagosome during the early stages of infection (Rouquette *et al.*, 1998). A *clpC* mutant strain displayed reduced adhesion, invasion, and survival in cultured TIB73, Caco-2, and murine hepatocytes (Nair *et al.*, 2000). The effect of ClpC ATPase on the infection process may be indirect in that ClpC is needed to maintain maximal cellular levels of other adhesins such as InlA, InlB, and ActA, and may be needed for folding and transport of these virulence proteins (Nair *et al.*, 2000).

*Fibronectin-binding protein.* Fibronectin, a 450-kDa glycoprotein, is present in plasma, in extracellular fluids and on a cell surfaces. Five fibronectin-binding proteins with molecular weights of 55.3, 48.6, 46.7, 42.4, and 26.8 kDa were identified in *L. monocytogenes* (Gilot *et al.*, 1999). Two genes, one encoding a low molecular weight (Gilot *et al.*, 2000) and one encoding a high molecular weight (Dramsi *et al.*, 2004) fibronectin protein, have been cloned and sequenced. The gene products may correspond to the 26.8 and 55.3 kDa proteins identified by Gilot *et al.* (1999) and both proteins are present on the surface of *L. monocytogenes* cells (Gilot *et al.*, 2000; Dramsi *et al.*, 2004). The high molecular weight fibronectin-binding protein, FbpA, is required for efficient colonization of mouse liver, binds to human fibronectin, and contributes to adherence to human epithelial cells (Dramsi *et al.*, 2004). In addition, FbpA, forms a protein complex with InlB and LLO, and post-translationally modulates the intracellular levels of these two proteins (Dramsi *et al.*, 2004); thus, in addition to acting as an adhesion, FbpA may act as a chaperonin.

*Actin polymerization protein (ActA).* Although more commonly known for promoting the cell-to-cell migration of *L. monocytogenes* (see detailed description below), the actin polymerization protein ActA also functions in adhesion of *L. monocytogenes* to eukaryotic cells. ActA binds to heparin sulfate proteoglycans (HSPG) present on the surface of most eukaryotic cells, as demonstrated by the ability of soluble heparin (an HSPG analog) to inhibit *L. monocytogenes* adhesion to and invasion of human enterocyte-like cultured cells (Henry-Stanley *et al.*, 2003). Removal of HSPG from CHO cells or the use of mutant cell lines lacking surface HSPG, significantly decreased adhesion and invasion by *L. monocytogenes* (Alvarez-Dominguez *et al.*, 1997). In addition, the capacity of an *L. monocytogenes actA*-mutant to invade CHO cell-lines was decreased, demonstrating the specificity of the ActA-HSPG interaction (Alvarez-Dominguez *et al.*, 1997).

*Lipoteichoic acid (LTA).* LTA is a structural component of the Gram-positive cell wall and may play a role in *L. monocytogenes* adhesion and virulence. A mutant deficient in D-alanine residues normally found in the LTA (Abachin *et al.*, 2002) displayed reduced adhesion to murine bone marrow macrophages (BMM), Caco-2, and TIB73 cultured cells; however, once inside Caco-2 and TIB73 cells, the growth rate was similar to wild-type *L. monocytogenes*. It was suggested that the altered adhesion in the D-alanine deficient strain may be due to an increase in the electronegativity of the cell surface due to the changes in the cell wall LTA (Abachin *et al.*, 2002). In addition, internalization of *L. monocytogenes* by dendritic cells is dependent on LTA and independent of both InlA and InlB (Kolb-Maurer *et al.*, 2000).

*Lipoprotein promoting entry (LpeA).* One of the most recent additions to the *L. monocytogenes* invasin family is LpeA, which shares significant sequence similarity (~60%) to pneumococcal surface adhesin A (PsaA), a known adhesin from *Streptococcus pneumoniae* (Reglier-Poupet *et al.*, 2003). LpeA is a listerial cell surface-exposed lipoprotein. A LpeA deletion-mutant of *L. monocytogenes* could not invade TIB73 murine hepatocytes or Caco-2 cells *in vitro* but could still infect bone marrow macrophages. Thus, LpeA is required for invasion of nonprofessional phagocytic cells but not macrophages (Reglier-Poupet *et al.*, 2003).

## Lysis of the primary vacuole or phagosome

### Listeriolysin O

Listeriolysin O (LLO) is a 58 to 60 kDa sulfhydryl (SH)-activated, pore-forming hemolysin and is the most-studied *L. monocytogenes* antigen (Geoffroy *et al.*, 1987). It is encoded by the *hly* gene (Mengaud *et al.*, 1989), which is under direct control of PrfA. LLO is a heat-labile extracellular toxin that produces a zone of β-hemolysis on blood agar. LLO binds to cholesterol in the target membrane, and about 30 to 40 molecules of LLO are required to lyse a single erythrocyte. Once a single LLO molecule is bound to the membrane, additional toxin molecules are oligomerized to form a pore, and cell lysis results (Alouf, 2000). LLO produces maximum cytolytic activity at pH 5.5. This pH optimum also helps *L. monocytogenes* to grow and replicate in the acidic environment of the phagosome.

LLO is an essential virulence factor for *L. monocytogenes* (Kathariou *et al.*, 1987). All virulent strains are hemolytic; however, avirulent strains may also produce LLO. Furthermore, heterogeneous forms of LLO may be produced by various serotypes of *L. monocytogenes*. LLO assists in bacterial survival inside the phagosome and produces transmembrane pores, which allow escape of *L. monocytogenes* into the cytoplasm prior to phagolysosomal fusion. LLO mutants can invade host cells but are unable to survive in the phagolysosome. Once *L. monocytogenes* escapes the phagosome, LLO is not required for cytoplasmic bacterial multiplication or actin polymerization (Goetz *et al.*, 2001). LLO is also reported to cause apoptosis or cell death in hepatocytes, dendritic cells, and hybrid B-lymphocytes resulting from membrane damage, intracellular enzyme release, and DNA degradation (see chapter on "animal and cell culture models for foodborne bacterial pathogens").

LLO is a member of the cholesterol-dependent, pore-forming toxin (CDTX) family (Tweten *et al.*, 2001). The LLO is antigenically and genetically related to streptolysin O (SLO), ivanolysin O (ILO), pneumolysin (PLY), alveolysin (ALV), perfringolysin O (PFO), and seeligerolysin with each toxin sharing a conserved C-terminal domain (Mengaud *et al.*, 1987; Geoffroy *et al.*, 1989; Haas *et al.*, 1992). It was suggested, based on high sequence similarity (up to 70%), that all members of this family may be structurally similar to PFO whose crystal structure was delineated (Rossjohn *et al.*, 1997; Dramsi and Cossart, 2002). PFO is composed of four domains, three of which are responsible for toxin oligomerization and membrane disruption with the fourth domain functioning in membrane binding. Cytolytic activity of LLO was believed to be due to a single cysteine residue found in the region of identity shared between hemolysins; however, subsequent study indicated that tryptophan present in this conserved region may actually be responsible for the hemolytic activity (Michel *et al.*, 1990). A PEST (Pro, Glu, Ser, Thr)-like sequence in the N-terminus of LLO appears to facilitate lysis of phagosomes (Decatur and Portnoy, 2000; Lety *et al.*, 2001). The most distinctive feature of LLO is that it loses its cytolytic activity upon phagosomal lysis due to rapid degradation of the toxin in the host cytoplasm by proteases that target the PEST-like sequence (Decatur and Portnoy, 2000). Thus, LLO function is limited to invasion and lysis of the phagosome.

### Phospholipase C (PI-PLC)

*L. monocytogenes* produces two types of phospholipase C (PLC), phosphatidylinositol-specific PLC (PI-PLC) and phosphatidylcholine-specific PLC (PC-PLC). PI-PLC, encoded by *plcA*, works synergistically with LLO to cause lysis of the vacuole (phagosome). PI-PLC is a 33–36 kDa enzyme and the *plcA* gene is regulated by PrfA (Camilli *et al.*, 1991; Mengaud *et al.*, 1991a; Camilli *et al.*, 1993). *plcA*-deficient mutants are unable to grow in murine organs and are avirulent (Camilli *et al.*, 1991). Recombinant *L. innocua* expressing *plcA* multiplied within the primary vacuole (phagosome) but were unable to escape, suggesting that PI-PLC alone is not sufficient for lysis of the host phagosome without the help of LLO (Camilli *et al.*, 1993; Smith *et al.*, 1995). In an LLO-minus *L. monocytogenes*, early and increased expression of PC-PLC, whose primary task is lysis of the two-membrane vacuole during cell-to-cell spread (discussed below), compensates for the lack of LLO, working synergistically with PI-PLC to allow *L. monocytogenes* to escape from the primary vacuole (Grundling *et al.*, 2003).

### Intracellular growth

### Hexose phosphate translocase (Hpt) and lipoate protein ligase (LpLA1)

Proliferation in the mammalian host cell cytosol is an important step in *L. monocytogenes* pathogenesis. Glucose is a preferred carbon source for many bacteria, including *L. monocytogenes*. The PrfA virulence gene regulator; however, induces expression of a homolog of the mammalian hexose phosphate (glucose-6-phosphate) translocase (Hpt) in *L. monocytogenes* upon entry into the host cytosol (Chico-Calero *et al.*, 2002). Expression of Hpt allows pathogenic *L. monocytogenes* to scavenge glucose-1-phosphate, glucose-6-phosphate, fructose-6-phosphate, and mannose-6-phosphate in the host cell cytoplasm (Chico-Calero *et al.*, 2002). In addition, Hpt-mutants were unable to synthesize glucose-6-phosphate following attachment to target cells, thereby affecting virulence by limiting initial growth of adhered bacteria.

For replication in the cytosol, *L. monocytogenes* also utilizes host-derived lipoic acid, a disulfide containing cofactor required for the function of pyruvate dehydrogenase (O'Riordan *et al.*, 2003). *L. monocytogenes* possess lipoate protein ligase (LpLA1) that ligates exogenous lipoic acid to the E2 subunit of the pyruvate dehydrogenase to form E2-lipoamide, which is important during aerobic metabolism. A mutant lacking *lplA1* exhibited defective growth in the cytosol and was less virulent in a mouse bioassay (O'Riordan *et al.*, 2003).

### Cell-to-cell movement

### ActA and actin polymerization

The cell surface protein ActA (90 kDa), encoded by *actA*, is responsible for actin nucleation and tail formation (Tilney and Portnoy, 1989; Pistor *et al.*, 1994; Geese *et al.*, 2002), and is transcriptionally controlled by PrfA. ActA is composed of three functional domains; the N-terminal domain that interacts with the host Actin-related protein (Arp/2) complex to promote actin nucleation, a central proline-rich region that interacts with members of the Enabled (Ena)/vasodilator-stimulated phosphoprotein (VASP) family that aid in speed and directionality of polymerization, and the C-terminal domain that anchors ActA to the bacterial wall (Chakraborty *et al.*, 1995; May *et al.*, 1999; Pistor *et al.*, 2000; Skoble *et al.*, 2000; Skoble *et al.*, 2001; Auerbuch *et al.*, 2003). After escaping the phagosomal compartment, *L. monocytogenes* multiply, move inside the cytoplasm, and spread intracellularly from cell to cell. This movement requires polymerization of actin filaments in the host cytoplasm. *L. monocytogenes* uses the actin tail as a 'diving board' for propulsive movement inside the host cell cytoplasm.

Inside the cytoplasm, the bacteria remain enveloped with a thick layer of F-actin. ActA proteins are evenly distributed on the bacterial cell surface but not present in the actual actin tail (Kocks *et al.*, 1993; Niebuhr *et al.*, 1993). Microfilament cross-linking proteins, such as α-actinin, fimbrin, and villin are found around the bacteria as soon as actin filaments are detected on the bacterial surface. Other proteins associated with the actin tail include tropomyosin, plastin (fimbrin), vinculin, talin, and profilin. Profilin, a G-actin binding protein that is responsible for actin filament elongation (Grenklo *et al.*, 2003), and Ena/VASP co-localize with ActA during polarized actin assembly. ActA protein, especially its proline rich region, nucleates and recruits actin filaments into the comet tail localized at one pole of the bacterial cell wall during directional assembly. Arp/2 initiates actin nucleation and requires the hydrolysis of ATP for bacterial propulsion (Welch *et al.*, 1998; Dayel *et al.*, 2001). Actin depolymerizing factor (ADF) and cofilin are also necessary for motility by directing polarized growth of actin filaments (Loisel *et al.*, 1999).

Disruption of the *actA* gene resulted in strains of *L. monocytogenes* that are unable to accumulate actin, and thus fail to infect adjacent cells. The *actA* mutant strain showed attenuated infection in human (Angelakopoulos *et al.*, 2002) and murine (Darji *et al.*, 2003) models. Microheterogeneity exists in the molecular mass of the ActA polypeptide among different strains of *L. monocytogenes* (Niebuhr *et al.*, 1993; Wiedmann *et al.*, 1997; Jaradat *et al.*, 2002). ActA expression is 200-fold higher when *L. monocytogenes* is present in cytosol of mammalian cells (Shetron-Rama *et al.*, 2003) than when grown in LB broth (Moors *et al.*, 1999).

## Lysis of the double membrane vacuole

*Phosphotidylcholine-specific Phospholipase C (PC-PLC)*

As mentioned above, phosphatidylcholine-specific phospholipase C (PC-PLC) mediates release of *L. monocytogenes* to neighboring cells through lysis of the two-membrane vacuole after cell-to-cell transmission. PC-PLC is a 29-kDa enzyme that expresses both lecithinase and sphingomyelinase activity, requires zinc as a cofactor, and is active at a pH range of 6 to 7 (Geoffroy *et al.*, 1991; Leimeister-Wachter *et al.*, 1991; Goldfine *et al.*, 1993). PC-PLC is encoded by *plcB*, which is under PrfA control (Vazquez-Boland *et al.*, 1992). The main substrate for this enzyme is phosphatidylcholine (PC), hence the name PC-PLC. PC-PLC has weak hemolytic activity but does not lyse sheep red blood cells because these cells are devoid of phosphatidylcholine (Geoffroy *et al.*, 1991). PC-PLC is involved in the lysis of the double membrane vacuoles (outer membrane generated from a newly infected host cell and the inner membrane from a previous host cell) that surround bacteria after cell-to-cell spread (Vazquez-Boland *et al.*, 1992). In the absence of LLO, PC-PLC is required for lysis of the phagosome following primary uptake of the bacteria (Grundling *et al.*, 2003).

It was also shown that PC-PLC is critical for the spread of *L. monocytogenes* from macrophages to endothelical cells, and for cerebral spread (Greiffenberg *et al.*, 1998; Schlüter *et al.*, 1998). Fusion of PC-PLC (*plcB*) with green fluorescence protein showed that PC-PLC expression is initiated upon the entry of *L. monocytogenes* into the cytosol and is dependent on the presence of LLO (Freitag and Jacobs, 1999). PC-PLC is a secreted protein and a signal peptidase (SipZ) was reported to be responsible for secretion. A *sipZ* mutant showed greatly increased $LD_{50}$ value (Bonnemain *et al.*, 2004). Recently, PC-PLC was shown to be able to induce Fas ligands (FasL) expression on the murine splenocytes synergistically when LLO is present (FasL is involved in cell apoptosis and downregulation of immune response), suggesting that the secretory PC-PLC may be important in modulation of immune response upon *L. monocytogenes* infection (Zenewicz *et al.*, 2004).

The *plcB* gene encodes the proenzyme precursor of PC-PLC, which is a 33-kDa protein. Bacteria engaged in intracellular growth in the host maintain a reservoir of PC-PLC precursor, which is activated at low pH (Raveneau *et al.*, 1992; Marquis and Hager, 2000). Production of mature PC-PLC

(29 kDa) is controlled by a zinc metalloprotease, which is encoded by *mpl*, a proximal gene in the lecithinase operon and it is regulated by PrfA (Raveneau *et al.*, 1992; Marquis *et al.*, 1997). However, translocation across the bacterial cell wall has been defined as the rate-limiting step in PC-PLC secretion (Snyder and Marquis, 2003). Mpl is 510 amino acids long with molecular mass of about 57 kDa, and contains an N-terminal signal sequence and a proteolytic cleavage site. Mpl becomes a 35 kDa active protein following proteolytic cleavage. It is stable at various temperatures and shows relatively narrow substrate specificity (Coffey *et al.*, 2000). Mutation in *mpl* caused reduced virulence of *L. monocytogenes* in mice (Raveneau *et al.*, 1992; Poyart *et al.*, 1993).

## Regulation of virulence genes

A majority of the known virulence genes in *L. monocytogenes* are clustered in a 9-kb region (the *Listeria* pathogenicity island) containing *prfA–plcA–hly–mpl–actA–plcB* (Vazquez-Boland *et al.*, 2001), and the genes in this pathogenicity island are regulated by the product of the first gene in the cluster, a positive regulatory factor encoded by *prfA* (Leimeister-Wachter *et al.*, 1990; Chakraborty *et al.*, 1992; Vazquez-Boland *et al.*, 2001) (Figure 15.2). The presence of these genes and their expression is important for listerial hemolytic activity and pathogenicity. This pathogenicity island is not present in the strains of *L. innocua* or *L. welshimeri*, but the genes are present in strains of *L. ivanovii* and *L. seeligeri* (Gouin *et al.*, 1994; Chakraborty *et al.*, 2000). The organization of these genes in the animal pathogen *L. ivanovii* is similar to that in *L. monocytogenes*, but the homologous genes demonstrate significant sequence divergence. The organization of the genes within the nonpathogenic *L. seeligeri* is considerably different from that observed in the pathogenic species due to the presence of several additional open reading frames. Furthermore, one of the additional open reading frames present in the *L. seeligeri* pathogenicity island disrupts the proper expression of *prfA*, preventing the expression of the PrfA regulated genes and the virulence phenotype (Karunasagar *et al.*, 1997). Numerous other studies have also demonstrated the importance of PrfA, and the genes of the PrfA regulon, for virulence.

PrfA is the principal regulator of virulence genes in *L. monocytogenes*. In addition to each of the genes in the *Listeria* pathogenicity island, PrfA positively regulates genes encoding members of the internalin protein family (*inlA*, *inlB*, and *inlC*) (Lingnau *et al.*, 1995; Dramsi *et al.*, 1996; Engelbrecht *et al.*, 1996), a bile salt hydrolase (*bsh*) (Dussurget *et al.*, 2002), and a hexose phosphate transporter (*hpt*) (Chico-Calero *et al.*, 2002), all of which serve important virulence functions (Figure 15.2). Recently two additional genes, *lmo2219* and *lmo0788*, not known to encode virulence factors, were found to be positively regulated by PrfA (Milohanic *et al.*, 2003). In this same study, a group of eight genes (two operons) whose gene products are involved in sugar metabolism were found to be negatively regulated by PrfA.

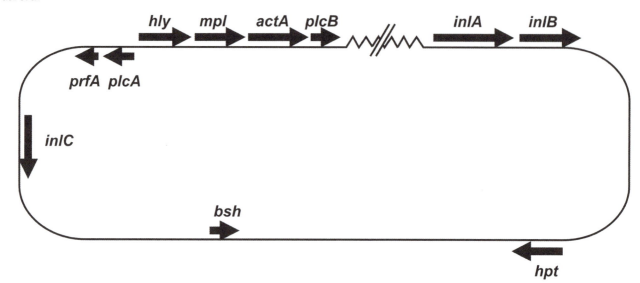

**Figure 15.2** PrfA regulated virulence genes. PrfA is a positive regulatory factor that regulates the expression of several virulence genes in *L. monocytogenes*. A PrfA binding region (PrfA-box) is present upstream of each of the genes shown. Genes depicted on the outside of the circle are transcribed from one strand of the chromosome and genes on the inside of the circle are transcribed from the opposite strand. Together *prfA*, *plcA*, *hly*, *mpl*, *actA*, and *plcB* make up the *Listeria* pathogenicity island. The rest of the genes are scattered around the *L. monocytogenes* chromosome. The jagged line represents the large region of the chromosome between the *Listeria* pathogenicity island and the *inlA* and *inlB* genes. The genes and gene products are as follows: *prfA*, positive regulatory factor for virulence gene expression; *plcA*, phosphatidylinositol-specific phospholipase C *hly*, listeriolysin O; *mpl*, zinc metalloprotease; *actA*, actin polymerization protein; *plcB*, phosphatidylcholine-specific phospholipase C; *inlA*, internalinA; *inlB*, internalinB; *hpt* (*uhpT*), hexose phosphate transporter, *bsh*, bile salt hydrolase; and *inlC*, internalinC.

PrfA is structurally and functionally related to the Crp/Fnr family of global transcription regulators (Lampidis *et al.*, 1994). Crp/Fnr-like regulators are site-specific DNA-binding proteins possessing a C-terminal helix–turn–helix DNA binding motif and a β-roll structure in their N-terminal domain. The protein encoded by *prfA* is 237 amino acids long with a deduced molecular weight of 27 kDa. PrfA binds to a palindromic *prfA* recognition sequence (PrfA box; –TTAACANNTGTTAA–) near position –40 from the transcription start site of each of the PrfA regulated genes/operons (Mengaud *et al.*, 1989; Sheehan *et al.*, 1996). All PrfA regulated virulence genes are precede by a PrfA-box sequence (Vazquez-Boland *et al.*, 2001; Chico-Calero *et al.*, 2002). However, recent microarray analyses have identified several genes that are not preceded by a PrfA-box, but whose transcription is positively influenced by PrfA (Milohanic *et al.*, 2003). Interestingly many of the genes were also regulated by the stress response sigma factor B.

The *prfA* gene is located immediately downstream of *plcA* and two different size *prfA* coding mRNAs were found (Mengaud *et al.*, 1991b), a *prfA*-only transcript and a transcript containing both *plcA* and *prfA*. Three distinct promoters contribute to the *prfA* expression. Two promoters, *prfA*-P1 and *prfA*-P2, are located immediately upstream of *prfA* resulting in the *prfA* only transcripts. The third promoter, *plcA*-P is located upstream of the *plcA* gene and transcription from this promoter accounts for the *plcA*-*prfA* transcripts (Camilli *et al.*, 1993; Freitag and Portnoy, 1994). PrfA positively regulates its own expression through the activation of *plcA* transcription. Increased PrfA synthesis results in the generation of the *prfA*-

*plcA* transcript, which is essential for virulence and spread of the bacteria from cell-to-cell within the host (Camilli *et al.*, 1993; Greene and Freitag, 2003). The bicistronic *plcA*-*prfA* message is regulated by the formation of mRNA secondary structure. In this case, loop formation within the bicistronic transcript increases transcription of *prfA* (Freitag *et al.*, 1993). In addition, the *prfA*-P2 promoter is known to be controlled by alternative sigma factor B (Nadon *et al.*, 2002).

The expression of virulence genes is dependent on various environmental and host factors, and environmental factors that affect *prfA* expression also affect PrfA-regulated genes. The regulation of PrfA synthesis is very complex and is controlled at the level of transcription and translation as well as post-translationally. The cellular levels of PrfA are dependent on growth phase and temperature (Mengaud *et al.*, 1991b). At growth temperatures similar to those encountered during host infection (37°C) PrfA, and as a consequence the virulence gene products of the PrfA regulon, is highly expressed. At low growth temperatures (below 30°C) the levels of PrfA are much lower, although the *prfA* transcript is present even at the lower temperature (Renzoni *et al.*, 1997). This translational-regulation was found to be mediated by a "thermosensor" in the mRNA leader sequence (Johansson *et al.*, 2002). At the lower temperature, the untranslated leader of the *prfA* transcript forms a secondary structure that conceals the ribosome binding site and prevents translation. At higher temperatures, the secondary structure within the *prfA* mRNA leader is relaxed and translation proceeds. In the presence of readily metabolizable sugars like glucose, cellobiose, fructose, mannose, trehalose, β-glucosides, and maltose, the PrfA-dependent genes are negatively regulated. Surprisingly, the expression of PrfA is induced in the presence of cellobiose, suggesting that PrfA activity might be regulated post-translationally (Renzoni *et al.*, 1997; Milohanic *et al.*, 2003). The downregulation of *prfA* and PrfA-dependent virulence genes by sugars is thought to be due to the presence of a common catabolite repressor (CR), spontaneous mutation or specifically to β-glucosides (Milenbachs *et al.*, 1997; Brehm *et al.*, 1999). The addition of charcoal to complex media results in high levels of virulence gene expression, which may be due to the removal of repressive sugars from the growth media (Geoffroy *et al.*, 1989; Ripio *et al.*, 1996). A more complex mechanism for charcoal-mediated derepression of virulence gene expression has recently been demonstrated (Ermolaeva *et al.*, 2004). Ermolaeva *et al.* demonstrated that charcoal sequesters a diffusible autoregulatory compound produced by *L. monocytogenes*. In the absence of charcoal, this autoregulatory substance represses *prfA*-dependent virulence gene expression by inhibiting PrfA activity. Several other environmental factors such as pH (Behari and Youngman, 1998) and iron availability (Conte *et al.*, 1996) have also been found to affect virulence gene expression as well.

## Isolation and identification

### Microbiological methods

Since the original isolation and characterization of *Listeria monocytogenes* from an infected laboratory animal by Murray in 1926, the organism has been difficult to isolate (Donnelly, 1988). Following decades of laborious isolation by direct plating methods, a cold enrichment procedure was introduced in 1948 (Gray *et al.*, 1948). While this procedure took advantage of *L. monocytogenes'* ability to outgrow other organisms at low temperatures, the enrichments often took several weeks or even months. In addition, the cold enrichments were typically done using clinical samples and the media used proved to be insufficiently selective when applied to foods. With the recognition of *L. monocytogenes* as a foodborne pathogen, more rapid and selective methods of enrichment became necessary. More selective media needed to be developed in order to isolate the relatively low levels of *L. monocytogenes* generally found in food samples ($< 10^2$) compared to the relatively high numbers of other bacterial flora (often $> 10^5$). Over the next several decades, a number of basal media were developed utilizing a variety of compounds to suppress the growth of the competitive flora found in food samples (Donnelly, 1988; van Netter *et al.*, 1989; Curtis and Lee, 1995). These media most often utilize acriflavine to inhibit the growth of other Gram-positive bacteria, naladixic acid to inhibit Gram-negative bacteria, and cycloheximide to inhibit fungi. Other often used antimicrobials include the broad spectrum agents ceflazidime and moxalactam as well as carefully controlled concentrations of lithium chloride to inhibit the growth of Gram-negative and certain Gram-positive bacteria. Another important improvement was the recognition that all *Listeria* spp. hydrolyze esculin and that

the inclusion of esculin and ferric iron in enrichment or plating media resulted in the formation of an intense black color (Fraser and Sperber, 1988). This is due to the complexation of the ferric iron with 6,7-dihdroxycoumarin, the product of esculin cleavage by β-D-glucosidase, resulting in a black precipitate. The combination of antimicrobial compounds with the esculin/ferric iron indicator yielded a number of selective chromogenic plating and broth media for the isolation of *Listeria* (Tables 15.3 and 15.4). One disadvantage of the esculin/ferric iron indicator is that the precipitate diffuses throughout the media and non-*Listeria* colonies growing in the precipitation zone can be mistaken for *Listeria* spp. A new medium, Harlequin™ *Listeria*, utilizing a β-D-glucosidase substrate for which the cleavage product is not diffusible but still forms a black precipitate in the presence of ferric iron, has been developed to overcome this problem (Smith *et al.*, 2001). On Harlequin™ *Listeria*, *Listeria* colonies appear black with no precipitation zone. The composition of the most often used broth and plating media is given in Table 15.3.

### Chromogenic media

More recently, chromogenic media have been developed that take advantage of a phosphatidylinositol-specific phospholipase C (PI-PLC) activity that is present in *L. monocytogenes* and *L. ivanovii*, but not the other *Listeria* spp. (Table 15.4). This phospholipase, encoded by the gene *plcA*, contributes to the hemolytic activity of *L. monocytogenes* and *L. ivanovii* and is important for virulence. The first use of PI-PLC for the differentiation of hemolytic and non-hemolytic *Listeria* employed an overlay technique using L-α-phosphatidylinositol (PI) (Notermans *et al.*, 1991). Cleavage of PI by PI-PLC resulted in the production of water insoluble fatty acids and the formation of an opaque white halo-like zone of precipitation around the colonies of the hemolytic species. Later this activity was combined with a chromogenic substrate (5-bromo-4-chloro-3-indolyl-β-D-glucopyranoside, X-gluc) for β-D-glucosidase activity (esculinase) present in all *Listeria* spp. (Ottaviani *et al.*, 1997). In this medium, referred to as ALOA, selective agents are added to reduce the background flora, *Listeria* colonies appear blue due to the cleavage of X-gluc by β-D-glucosidase and the hemolytic species of *Listeria* are surrounded by a white halo due to the cleavage of PI by PI-PLC. Additional variations of these media have been developed (Table 15.4).

In ALOA and other media that include PI as a substrate for PI-PLC, non-hemolytic *Listeria* can grow within the zone of precipitation surrounding the hemolytic strains, making them difficult to distinguish. In addition, these media do not differentiate between the two hemolytic species of *Listeria*, *L. monocytogenes* and *L. ivanovii*. Recently a chromogenic substrate for PI-PLC, 5-bromo-4-chloro-3-indolyl-myo-inositol-1-phosphate (X-IP), was reported (Restaino *et al.*, 1999). The cleavage of X-IP by PI-PLC results in a non-diffusible product that leaves the bacterial colony a blue/turquoise color. X-IP has been employed in chromogenic media for the isolation of *Listeria* and detection of *L. monocytogenes* (Table

**Table 15.3** Media used for conventional enrichment and isolation of *Listeria*

| Medium composition/liter | Reference |
|---|---|
| **Lithium Chloride–Phenylethanol–Moxalactam (LPM)** | Lee and McClain (1986) |
| Phenylethanol agar (35.5 g), glycine anhydride (10 g), LiCl (5 g), moxalactam (2 mL of a 1% stock solution in phosphate buffer, pH 6.0) | |
| **LPM plus Esculin and Ferric Iron** | |
| LPM plus, esculin (1 g) and ferric ammonium citrate (0.5 g) | Hitchins (2003) |
| **University of Vermont (UVM) Broth** | Donnelly and Baigent (1986) |
| Proteose peptone (5 g), tryptone (5 g), Lab-Lemco powder (5 g), yeast extract (5 g), NaCl (20 g), $KH_2PO_4$ (12 g), $Na_2HPO_4$ (1.35 g), esculin (1 g), nalidixic acid (40 mg), acriflavine (12 mg) | |
| **Fraser Broth** | Fraser and Sperber (1988) |
| Proteose peptone (5 g), tryptone (5 g), beef extract (5 g), yeast extract (5 g), NaCl (20g), $KH_2PO_4$ (12 g), $Na_2HPO_4$ (1.35 g), esculin (1 g), nalidixic acid (20 mg), LiCl (3 g), acriflavine (25 mg), ferric ammonium citrate (0.5 g) | |
| **Oxford Agar** | Curtis *et al.* (1989) |
| Columbia agar base (39 g), Esculin (1 g), Ferric ammonium citrate (0.5 g), LiCl (15 g), Cycloheximide (0.4 g), Colistin (20 mg), Acriflavine (5 mg), Cefotetan (2 mg), Fosfomycin (10 mg) | |
| **Modified Oxford (MOX)** | FSIS (2002) |
| Columbia agar base (39 g), Esculin (1 g), ferric ammonium citrate (0.5 g), LiCl (15 g), cycloheximide (0.4 g), colistin (10 mg), sodium moxalactam (20 mg) | |
| **Trypticase Soy Broth (TSBYe)** | Hitchins (2003) |
| Trypticase Soy broth (30 g), yeast extract (6 g) | |
| **Buffered *Listeria* Enrichment Broth (BLEB): TSBYe with Supplements** | Hitchins (2003) |
| TSBYe plus, $KH_2PO_4$ (1.35 g), $Na_2HPO_4$ (9.6 g), Pyruvic acid (11.1 mL of a 10% w/v solution), Acriflavine HCl (10mg), nalidixic acid (40 mg), ceftazidime (50 mg) | |
| **PALCAM Agar** | van Netter *et al.* (1989) |
| Columbia agar (39 g), D-glucose (0.5 g), D-mannitol (10 g), esculin (0.8 g), ferric ammonium citrate (0.5 g), phenol red (80 mg), polymyxin B sulfate (10 mg), acriflavine HCl (5mg), LiCl (15 g), ceftazidime (20 mg) | |

15.4). In X-IP containing media, background flora is reduced by the addition of selective agents, and colonies of non-hemolytic *Listeria* appear white while colonies of *L. monocytogenes* and *L. ivanovii* appear blue. Rapid L'Mono medium (BioRad, Hercules, California, USA) contains xylose and a pH indicator that differentiates *L. monocytogenes* from *L. ivanovii*. Acid production due to xylose fermentation by *L. ivanovii* (Table 15.1) causes a drop in pH and a yellow zone around *L. ivanovii* colonies. This yellowing of the media causes colonies of *L. ivanovii* to appear green/blue on Rapid L'Mono media while colonies of *L. monocytogenes* always appear blue with no yellow halo. The *Listeria* Chromogenic Detection System (R&F Labs, Chicago, IL, USA) is a two component detection system (Restaino *et al.*, 1999) in which X-IP is included in the *Listeria* chromogenic plating medium (LCPM). Colonies that appear blue on the LCPM are picked onto a second confirmatory medium that allow users to distinguish *L. monocytogenes* from *L. ivanovii* based on fluorescence due to the acid production from rhamnose fermentation or the product of an α-mannosidase reaction. Fluorescence in either reaction confirms the identification of *L. monocytogenes*. Regulatory agencies have not yet incorporated these chromogenic media into the standardized

methods for the detection of *L. monocytogenes*, but are currently endorsing their use in conjunction with the existing standardized methods. These chromogenic media may someday replace the current plating media used in the standardized methods described below.

## Standardized isolation methods

Over the years a variety of combinations of enrichment and plating media have been evaluated for the isolation of *L. monocytogenes* from foods. While no single method is ideal for all types of food and the optimal procedure should be determined for each food product, regulatory agencies provide guidance through the publication of standardized methods for the isolation of *L. monocytogenes* from food. Brief descriptions of the conventional methods currently recommended by the United State Department of Agriculture (USDA), the Food and Drug Administration (FDA), the Association Français de Normalisation (AFNOR), and the International Organization for Standardization (ISO) for the isolation and identification of *L. monocytogenes* from foods are given below. The composition of many of the media used in these procedures is given in Table 15.3.

**Table 15.4** Selective chromogenic plating media for *Listeria*

| Plating medium[a] | Diagnostic enzyme and substrate[b] | | Colony appearance | | |
| --- | --- | --- | --- | --- | --- |
| | β-D-Glucosidase | PI-PLC | Nonpathogenic *Listeria* spp. | *L. ivanovii* | *L. monocytogenes* |
| PALCAM[c]/Oxford and MOX | Esculin | NA | Green-gray with black precipitate halo/black with black precipitate halo | Green-gray with black precipitate halo/black with black precipitate halo | Green-gray with black precipitate halo/ Black with black precipitate halo |
| Harlequin™ *Listeria* | CHE-gluc | NA | Black, no halo | Black, no halo | Black, no halo |
| ALOA | X-gluc | PI | Blue, no halo | Blue, surrounded by white precipitate halo | Blue, surrounded by white precipitate halo |
| CHROMAgar® *Listeria* and BBL ® CHROMagar *Listeria* | X-gluc | PI | Blue, no halo | Blue, surrounded by white precipitate halo | Blue, surrounded by white precipitate halo |
| Rapid L'Mono[d] | NA | X-IP | White, with or without a yellow halo | Blue-green, with yellow halo | Blue |
| R&F® Labs LCPM[e] | NA | X-IP | White | Blue, with our without blue halo | Blue, with our without blue halo |

[a]PALCAM, polymyxin B–acriflavin–lithium chloride–ceftazidime–aesculin–mannitol medium; MOX, modified Oxford medium; Harlequin™ *Listeria* (LAB M Limited, Bury, Lancashire, UK); ALOA, Agar *Listeria* Ottavani & Agosti; CHROMAgar® *Listeria* (CHROMagar Microbiology, Paris, France & Becton Dickinson and Company, Franklin Lakes, NJ, USA) Rapid L'Mono (BioRad, Hercules, California, USA); R&F Labs® (West Chicago, Illinois, USA) LCPM, *Listeria monocytogenes* Chromogenic Plating Medium.

[b]PI-PLC, phosphatidylinositol-specific phospholipase C; CHE-gluc, 3,4-cyclohexenoesculetin-D-glucoside; X-gluc, 5-bromo-4-chloro-3-indolyl-β-D-glucopyranoside; PI, L-α-phosphatidylinositol; X-IP, 5-bromo-4-chloro-3-indolyl-myo-inositol-1-phosphate.

[c]PALCAM contains a second indicator system, mannitol and phenol red.

[d]Rapid L'Mono contains xylose and a pH indicator to distinguish *L. ivanovii* (yellow halo) from *L. monocytogenes*.

[e]In addition to LCPM, the R&F Labs *Listeria monocytogenes* Chromogenic Detection System includes a second confirmatory medium to distinguish *L. monocytogenes* from *L. ivanovii* based on fluorescence due to the acid production from rhamnose fermentation or the product of an α-mannosidase reaction. Fluorescence in either reaction confirms the identification of *L. monocytogenes*.

## FDA method

The FDA method (Hitchins, 2003) employs a 48-hour enrichment at 30°C in Buffered *Listeria* Enrichment Broth (BLEB). The sample can be pre-enriched for 4 hours at 30°C prior to the addition of the selective supplements. At 24 and 48 hours the culture is streaked onto selective Oxford, PALCAM, lithium chloride-phenylethanol-moxalactam (LPM) containing esculin and ferric iron, or modified Oxford (MOX) agar. The plates are then incubated at 30/35°C for an additional 24–48 h. *Listeria* colonies appear black with a black halo on the esculin/iron containing selective plating media. Five or more presumptive *Listeria* colonies (black colonies) are streaked onto trypticase soy agar and the plates are incubated at 30°C for 24–48 h. The presumptive *Listeria* are examined using Henry illumination (a light beam is shown upon the bottom of the plate at a 45° angle) causing the *Listeria* colonies to sparkle blue or white and the species determined by further biochemical testing. Serotyping and genetic subtyping by pulse-field gel electrophoresis is required for all *L. monocytogenes* isolates.

## USDA method

The USDA method (FSIS, 2002) utilizes a two-stage enrichment procedure involving a 24–48 hour primary enrichment in University of Vermont (UVM) medium followed by a 24–48 hour secondary enrichment in Fraser broth. The primary enrichment is plated on MOX plates at both 24 and 48 hours. Fraser broth contains the esculin/ferric ion indicator and darkening of the medium due to the formation of the black precipitate is an indicator of a putative *Listeria*-positive culture. If darkening of the Fraser broth is observed at 24 or 48 hours, this secondary enrichment is also plated on MOX. The presence of black colonies on any of the MOX plates is considered a presumptive *Listeria*. Individual black colonies from the MOX plates are further purified, tested for hemolytic activity, and the species identified by biochemical tests.

## ISO/AFNOR method

The ISO/AFNOR methods (Scotter *et al.*, 2001a; Scotter *et al.*, 2001b) are similar to the USDA method, but uses 1/2 Fraser broth (using half the amount of selective agents) in the first enrichment step followed by incubation in full-strength Fraser broth during the secondary enrichment. Both steps are followed by plating on selective media, PALCAM, and/or Oxford agars. *Listeria* colonies appear black with a black halo on Oxford agar. The Palcam agar contains a dual indicator system, esculin/ferric iron, as in the Oxford agar, and D-mannitol with phenol red. On PALCAM, *Listeria* colonies appear gray-green in color with a black halo. Once again putative *Listeria* colonies are identified to the species level by biochemical testing.

## Rapid detection methods

Advances in biological research have lead to the development of more rapid *L. monocytogenes* detection assays that utilize immunological and nucleic acid-based techniques. While the conventional microbiological methods described above are the gold standard for isolating *L. monocytogenes* from food samples and are particularly important for regulatory agencies that desire the axenic bacterial culture as an end result from a positive sample, these methods have several limitations for routine pathogen testing applications of food safety laboratories and the food industry. The conventional methods are laborious and are not generally amenable to automation or high throughput analysis. Although a negative result can be confirmed in 3–4 days, the time to a positive result is usually 5–7 days from sample collection. In the interest of cost and product shelf life, it is often not possible to hold food products for 7 days prior to distribution. Thus, the food industry desires more rapid, less labor intensive, and automated methods for the detection of *L. monocytogenes* and other foodborne pathogens. The immunological and nucleic acid-based rapid methods can reduce the time to result by 24 hours or more, but the time consuming culture enrichments are still necessary to overcome limitations of these rapid methods. The rapid detection methods generally require less than 8 hours to complete, but the low level of *Listeria* cells present in contaminated food coupled with the relatively high limits of detection for immunological methods ($< 10^4$ cells per milliliter) necessitates culture enrichments. Nucleic acid amplification methods are much more sensitive (theoretically as little as a single cell is detectable); but the method is often inhibited by components present in food matrices and cannot distinguish living from dead cells. Thus, the increase in number of cells after culture enrichment allows the dilution of inhibitory substances and ensures that the detection of nucleic acids is from growing cells. Many papers have been published concerning the development and evaluation of rapid methods for the detection of *L. monocytogenes* and other foodborne pathogens, and many commercial kits for rapid testing are available. This information will be discussed briefly below and has been more thoroughly reviewed elsewhere (Batt, 1999; Ivnitski *et al.*, 1999; Kathariou, 2002; Mello and Kubota, 2002; Sellars *et al.*, 2002). In addition, a list of commercially available test kits, performance tested methods, and validated methods is available at the website for the Association of Official Analytical Chemists (AOAC) at www.aoac.org/testkits/testkits.html

### Immunological methods

All immunological detection methods rely on the availability of polyclonal antisera or monoclonal antibody with sufficient binding affinity and specificity to allow the detection of the target of interest without binding to non-target antigens. Most immunoassays work by first capturing the target antigen of interest from a complex mixture followed by detection of the captured antigen. For use in immunoassays, antibody molecules are modified to allow them to function as capture and/or detection reagents. To function as a capture reagent, antibodies are often fixed to a solid surface. As the sample is passed over the antibody coated surface, target antigens are captured from the sample, non-target antigens pass over the surface without being bound, and any non-target antigen that binds non-specifically to the surface is removed by washing. The binding of the target antigen to the antibody is measured directly in some immunoassay methods, but in most cases a secondary detection method is employed. This secondary detection can involve the use of an antibody that has been modified by conjugation to a reporter molecule. The reporter may be a chromophore, fluorophore, or some other molecule that can be detected by eye or through the use of instrumentation. Alternatively, the secondary antibody can be linked to an enzyme that catalyzes the formation of a detectable product. The advantage of the enzyme-linked antibody as a detection reagent is the signal amplification afforded by the multiple rounds enzymatic catalysis. A large number of immunoassays for the detection of bacteria of the genus *Listeria* have been developed and many are commercially available. Until recently, development of immunoassays for the detection of *L. monocytogenes* at the species level has been hindered by the lack of polyclonal antisera or monoclonal antibodies with sufficient specificity.

The simplest system for immunological detection is the agglutination assay. In this assay, antibody solutions or antisera are mixed with the sample. Positive samples are indicated by a formation of visible clumps in the solution caused by the formation of large antibody–antigen aggregates. These agglutination reactions are much easier to read if the antibody is immobilized on the surface of latex beads. At least two latex agglutination tests for the detection of *Listeria* spp. are commercially available: Microscreen *Listeria* (Microgen Bioproducts, Camberley, UK) and Oxoid *Listeria* Test Kit (Oxoid, Basingstoke, UK). Another simple-to-use method is the lateral flow immunoassay. This is the type of disposable detection device that was popularized in the form of the home pregnancy test. Although different formats have been developed, the device always works via the vectorial flow of the sample solution through the various zones. In the simplest version of the assay, a liquid sample is applied through a sample port at one end of the device and flows over the first zone containing chromogenically labeled antibodies that can bind to *Listeria* bacteria or listerial antigen (e.g., heat-killed cells). If *Listeria* is present, an antibody-antigen complex is formed between the *Listeria* in the sample and the labeled antibody. As the solution continues to flow through the device, the *Listeria* and labeled antibody complexes are captured by anti-listerial antibodies in the second zone. The captured antibody–*Listeria* complex forms a colored line in the second zone due to the presence of the chromogenically labeled primary antibody. If *Listeria* is not present in the sample, the chromogenically labeled antibody flows through the second zone without being captured and no colored line is formed. Most of these devices also contain a third zone containing an antibody that captures the chromogenically labeled antibody to form a positive control line. This control line confirms the integrity of the device and ensures flow of the sample across the entire device. A number of single used lateral flow devices for the detection of *Listeria* spp. are available: Oxoid *Listeria* Rapid Test (Oxoid Inc.), Reveal® for *Listeria* (Neogen Corp., Lansing, Michigan, USA), Singlepath®

*Listeria* (Merck KGaA, Darmstadt, Germany), and VIP® for *Listeria* (BioControl Systems Inc., Bellevue, Washington, USA) (Feldsine *et al.*, 1997a; Feldsine *et al.*, 2000; Feldsine *et al.*, 2002a). The agglutination tests and lateral flow devices are technically simple and can be completed in only minutes, but they are not very sensitive. These methods typically require $> 10^5 – 10^6$ *Listeria* cells per milliliter

The most common format for immunological detection of pathogens is the enzyme-linked immunosorbant assay (ELISA) and variations of this technique. The simplest ELISA format involves the use of antibody-coated microtiter plates to capture the pathogen of interest. Following incubation of the sample in the antibody-coated wells of the microplate, unbound and non-specifically bound antigens are removed by a washing. This is followed by the addition of a secondary enzyme-linked antibody and washing steps to remove excess antibody. If the pathogen is present it will be captured by the antibody coating the well and the secondary enzyme-linked antibody will form a sandwich. The addition of substrate for the antibody-conjugated enzyme results in the formation of a detectable product. In variations of this method, the secondary antibody is not linked to an enzyme but to a reporter molecule that is detectable without the conversion of a substrate. A thorough review of the large volume of literature on development of enzyme-linked immunoassays for the detection of *Listeria* spp. is not possible here.

TECRA International Ltd. (Frenchs Forest, Australia) markets two different colorimetric immunoassays. The *Listeria*-VIA® (Visual ImmunoAssay) uses a standard microplate format (Knight *et al.*, 1996; Hughes *et al.*, 2003), and the *Listeria*-Unique® employs a dipstick assay in which the capture antibody is immobilized on a plastic stick that is moved between the sample, wash, and detection solutions. In this assay, a colored line on the dipstick indicates a positive reaction. Diffchamb AB (Västra Frölunda, Sweden) markets two microplate format colorimetric ELISA kits. One kit is for the detection of *Listeria* spp. (Transia Plate *Listeria*) and the other allows the detection of *L. monocytogenes* (Transia Plate *L. monocytogenes*) (Bubert *et al.*, 1994). Another colorimetric ELISA kit, the Assurance® EIA, is available from BioControl Systems Inc., Bellevue, Washington, USA (Feldsine *et al.*, 1997b; Feldsine *et al.*, 2002b). An automated fluorescent ELISA assay for the detection of *Listeria* spp., the VIDAS® LIS, or detection of *L. monocytogenes*, the VIDAS® LMO, are marketed by bioMérieux, Marcy l'Etoile, France (Gangar *et al.*, 2000; Vaz-Velho *et al.*, 2000; Sewell *et al.*, 2003).

One other method that has received considerable attention is the use of antibody-coated immunomagnetic beads (IMBs) for the capture of *Listeria* from food matrices or enrichment cultures (Jackson *et al.*, 1993; Mitchell *et al.*, 1994; Avoyne *et al.*, 1997; Hsih and Hsen, 2001; Kaclikova *et al.*, 2001; Jung *et al.*, 2003). IMBs coated with anti-*Listeria* antibodies (Dynabeads® anti-*Listeria*) are available from Dynal Biotech, Oslo, Norway. The Dynal procedure recommends using the anti-*Listeria* Dynabeads® to capture the *Listeria* cells from enrichment cultures followed by standard methods for plating and identification of *L. monocytogenes*. Three other commercially available kits employ IMB for the capture of *Listeria* spp. The EiaFoss® *Listeria* ELISA Kit (Foss Electric A/S, Hillerød, Denmark) couples the IMB capture to a immunofluorescent detection of *Listeria* spp. (Petersen and Madsen, 2000), the ListerTest® Kit (Vicam, Watertown, MA, USA) involves the direct plating of capture bacteria followed by a colony lift and immunoblot for the colorimetric detection of *L. monocytogenes* (Ben Embarek *et al.*, 1997), and the BioVeris Corporation BV® *Listeria* Test Kit (Gaithersburg, MD, USA, formerly IGEN International, PathIGEN® *Listeria*) detects the IMB captured *Listeria* spp. using a proprietary ruthenium-based electrochemiluminescent system.

*Nucleic acid-based methods*

Nucleic acid-based rapid methods for the detection of *L. monocytogenes* or other foodborne pathogens typically involve nucleic acid hybridization, nucleic acid amplification, or a combination of these techniques.

AccuProbe® *Listeria monocytogenes* culture identification test by Gen-Probe (San Diego, CA, USA) utilizes the chemiluminescent detection of an acridinium ester labeled DNA probe that specifically hybridizes to the 16S rRNA of *L. monocytogenes*. In the absence of target rRNA, the single-stranded labeled DNA is degraded by the addition of hydrogen peroxide, but in the presence of *L. monocytogenes* a luminescent rRNA:DNA hybrid molecule is formed. The requirement that the acridinium labeled DNA hybridize with rRNA ensures the detection of viable *L. monocytogenes*. The Verimicon Identification Technology (VIT®, Vermicon, Munchen, Germany) is a nucleic acid hybridization that allows the simultaneous detection of members of the genus *Listeria* and discrimination of *L. monocytogenes*. In this technique, differently labeled fluorescent probes are hybridized with a sample that has been treated to permeabilize the cells (Stephan *et al.*, 2003). Epifluorescence microscopy is used to distinguish bacteria from the genus *Listeria*, which appear green, and *L. monocytogenes*, which appear red and green. Neogen Corporation's (Lansing, Michigan, USA) GeneTrak and GeneQuench® systems allow solution-based hybridization of an enzyme-linked nucleic acid probe in a test-tube or microtiter plate platform, respectively. Both assay formats are available for the colorimetric detection of *Listeria* spp. or the specific detection *L. monocytogenes*. Neogen Corp. has also developed a GeneTrak®DLP kit that utilizes a directly label probe that further simplifies the detection method by eliminating the need for the addition of substrate.

Nucleic acid amplification methodologies utilize the polymerase chain reaction (PCR) to detect target DNA sequences. The PCR employs a thermostable DNA polymerase to exponentially amplify a target DNA sequence in the presence of two opposing oligonucleotide primers by several successive rounds of replication. To be used as a detection method, the oligonucleotide primers must be carefully selected to allow amplification of the target DNA from only the organism of interest. In many cases a single set of primers is used to amplify a single target DNA molecule, but multiple primer sets can be used to amplify multiple targets. This is referred to as multiplex PCR and can greatly increase the specificity and discriminatory power of PCR-based detection (Lawrence and Gilmour, 1994; Bubert *et al.*, 1999a; Rodriguez-Lazaro *et al.*, 2004). For many years PCR products were detected by separating them on agarose gels followed by staining with the fluorescent intercalating dye ethidium bromide. Today the presence of double-stranded PCR products can be determined in real-time between successive rounds of amplification. In real-time PCR (RT-PCR) the double stranded product can be quantitated by measuring the fluorescence due to the binding of an intercalating dye or due to the binding of a fluorescent hybridization probe using a thermocycler equipped with the necessary optical system. In the latter technique, the hybridization probe, or molecular beacon, is labeled at one end with a fluorescent molecule and at the other end with a molecule capable of quenching that fluorescence via fluorescence resonance energy transfer (FRET). The probe is designed in such a way that in the absence of amplified PCR product, the molecular beacon does not fluoresce due to the close proximity of the two ends of the probe resulting in FRET quenching. If a PCR product is made, the molecular beacon hybridizes to it, the two ends of the probe are sufficiently separated to prevent the FRET, and a fluorescent signal is detected. Three commercial systems are available for the RT-PCR detection of bacteria of the genus *Listeria* or the specific detection of *L. monocytogenes*. The BAX® *L. monocytogenes* Detection System (DuPont-Qualicon, Inc., Wilmington Delaware, USA) (Stewart and Gendel, 1998; Hochberg *et al.*, 2001; Silbernagel *et al.*, 2004) allows the real-time detection of PCR product using the intercalating dye SYBR® Green (Molecular Probes, Eugene, OR, USA) and has recently been adopted by the USA Dept. of Agriculture – Food Safety Inspection Service as a screening method. Two other commercially available RT-PCR-based systems, FoodProof BIOTECON Diagnostics (Hamilton Square, NJ, USA) and GeneVision® Rapid Pathogen Detection System (Warnex Diagnostics, Laval, Quebec, Canada), utilize molecular beacon technology for detection of PCR product.

With the recent determination of the complete nucleotide sequences of the genomes of 4 strains of *L. monocytogenes* (Glaser *et al.*, 2001; Nelson *et al.*, 2004), a more rational approach to the development of detection methods might be possible, in particular the identification of unique genes for PCR detection.

*L. monocytogenes subtyping methods*

A variety of subtyping methods have been devised to differentiate *L. monocytogenes* isolates. These methods play an important role in the surveillance of listeriosis, outbreak detection, and pathogen tracking in food processing environments. Aside from ease and cost, the feasibility of a subtyping methods depends upon its reproducibility, discriminatory power (i.e., how many subgroups can be discerned), and the ability of the method to fit all strains into established subgroups.

Several conventional methods have been developed including phage typing, multilocus enzyme electrophoresis (MLEE), and serotyping. In the phage typing method, *L. monocytogenes* isolates are differentiated by their susceptibility to a standard set of lytic bacteriophage. A number of lytic phage have been isolated for *L. monocytogenes* and the discriminatory power of the phage typing system can be increased by increasing the number of phages used to screen *L. monocytogenes* isolates (Capita *et al.*, 2002). While phage typing has been used to identify at least one outbreak of listeriosis (Jacquet *et al.*, 1995), the value of the method is limited by problems with variability and the fact that many foodborne isolates are untypable by this method (Capita *et al.*, 2002). The MLEE method for typing differentiates strains by examining the variation in the electrophoretic mobility of constitutively expressed enzymes (Bibb *et al.*, 1989). While most *L. monocytogenes* isolates can be typed by MLEE, the method is less discriminatory than some of the other typing methods and is difficult to standardize. *L. monocytogenes* strains can be differentiated into 13 different serotypes based upon the repertoire of somatic (O) and flagellar (H) antigens as determined by the reaction of individual strains with a standard set of O and H antisera (Seeliger and Hohne, 1979). Some of the 13 serotypes are shared by other species of *Listeria* (*L. seeligeri* and *L. innocua*); thus, serotyping alone is not sufficient for speciation of *Listeria*. Ninety-five percent of all foodborne listeriosis can be attributed to *L. monocytogenes* strains of serotypes 1/2a, 1/2b, and 4b, with nearly 90% of the cases being caused by serotype 4b strains (Kathariou, 2002). Nevertheless, serotyping cannot be used as an indicator of potential virulence of a strain. The fact that only three serotypes account for most of the cases of foodborne listeriosis reduces the discriminatory power of this typing method. Recently an ELISA for serotyping *L. monocytogenes* was reported (Palumbo *et al.*, 2003). The ELISA platform significantly reduces reagent cost and increases throughput for serotyping a large collection of strains. To date, serotyping is the most widely used and reported method for typing *L. monocytogenes*; but newer molecular methods of subtyping are rapidly becoming more popular.

Molecular methods for subtyping *L. monocytogenes* include, but are not to, pulsed field gel electrophoresis (PFGE), ribotyping and multilocus sequence typing (MLST). Although technically more difficult, these DNA-based methods provide superior strain discrimination and improved standardization and reproducibility than conventional typing methods. PFGE groups bacteria into "pulsotypes" based upon the pattern of bands separated by agarose gel electrophoresis after digestion of the genomic DNA by an infrequently cutting restriction endonuclease. The use of restriction enzymes results in very large DNA fragments (20–600 kilobases) that are separated on agarose gels subjected to pulsed electric fields. More than one restriction endonuclease can be used in separate reactions to increase the discriminatory power of PFGE. The Centers for Disease Control along with a number of certified state public health departments and food regulatory laboratories have established a national network to catalog and exchange PFGE subtypes for several foodborne pathogens including *L. monocytogenes* (Graves and Swaminathan, 2001) (available on the world-wide web at: http://www.cdc.gov/pulsenet/). Another DNA-based typing method, ribotyping, is based upon differentiating strains due to variability in their ribosomal DNA genes. In this method, genomic DNA is cut with a frequently cutting restriction endonuclease yielding DNA fragments ranging in size from approximately 2–40 kilobases. The DNA fragments are separated by agarose gel electrophoresis, transferred to a membrane (Southern blot), and hybridized with a labeled ribosomal DNA probe. This procedure results in a specific banding pattern that depends on sequence variability within the bacterial strain's ribosomal DNA genes. A fully automated ribotyping instrument, Riboprinter® Microbial Characterization System (Qualicon Inc., Wilmington, Delaware, USA) is commercially available. The high level of standardization and reproducibility of the Riboprinter® has made ribotyping the method of choice for large-scale studies that require the molecular subtyping of many isolates. MLST involves the determination of nucleotide sequences for a number of genes or gene fragments for comparison to other strains. The genes most often used for MLST are those encoding housekeeping genes. These genes are present in all strains and provide a high degree of discriminatory power because they are highly conserved. In *L. monocytogenes*, the PrfA regulated genes represent a group of highly conserved genes unique to *L. monocytogenes* that might be ideal targets for MLST subtyping. Studies on MLST for *L. monocytogenes* have recently been published (Salcedo *et al.*, 2003; Meinersmann *et al.*, 2004; Revazishvili *et al.*, 2004; Zhang *et al.*, 2004) but more widespread use of this technique will depend upon the selection of a standard set of genes, an established database of gene sequences for comparison, and more automated methods. Two other molecular techniques, nucleic acid microarrays and PCR, are beginning to be used as methods for typing strains of *L. monocytogenes*. Nucleic acid microarrays are most often used to examine genome-wide physiological responses in gene expression for cells grown under different culture conditions by measuring changes in the levels of mRNA, but the analysis of genomic DNA by means of microarrays has recently received attention as a rapid method for detection (Volokhov *et al.*, 2002; Hong *et al.*, 2004) and differentiation of *L. monocytogenes* strains (Call *et al.*, 2003; Zhang *et al.*, 2003). PCR methods for the "molecular serotyping" of *L. monocytogenes* strains have been developed (Borucki and Call, 2003; Doumith *et al.*, 2004). Although the PCR methods do not yet match the discriminatory power of the standard serotyping methods, they are an attractive alternative to the conventional serotyping method due to the relative ease of conducting and standardizing PCR protocols.

## Proteomics of *Listeria monocytogenes*

Determining the complete genetic and protein makeup of *L. monocytogenes* and unraveling the interconnected functions of the cellular architecture of *L. monocytogenes* is central to increasing our understanding of the survival and growth of this foodborne pathogen in the natural environment, in foods, and during active cases of listeriosis.

Studies involving the total genome or proteome (cellular protein complement) can be categorized several ways. For genomic studies, the scientific literature can be divided into studies that probe the entire genome and those that probe a selected subset of genes. Genomic studies can be further differentiated based on whether or not the studies used whole-genome data released as part of the major *L. monocytogenes* sequencing projects (Glaser *et al.*, 2001; Nelson *et al.*, 2004). The genome-wide analyses and comparative genomics studies released with the whole genome sequencing projects, and some subsequent studies, have largely been *in silico* analyses that have provided new insights into the genetic makeup, evolution, and pathogenicity of *L. monocytogenes* and have served as a framework for contextualizing previous work (Glaser *et al.*, 2001; Cabanes *et al.*, 2002; Buchrieser *et al.*, 2003; Nelson *et al.*, 2004). Readers interested in reviews of the available *Listeria* whole-genome literature will find these broad works an excellent starting point. Since protein separation technologies (O'Farrell, 1975) predate the ability to sequence entire genomes, total cellular protein analyses have a longer history than total genome analyses. Unfortunately, the ability to positively identify the proteins of interest in cellular protein analyses has only recently begun to be approached on a large-scale basis. The inability to identify proteins of interest was in part due to the insensitivity of protein sequencing methods, which did not match the level of detection sensitivity routinely achieved with separation techniques like two-dimensional gel electrophoresis (2DE), and in part due to the lack of the complete genome sequences needed to populate protein sequence databases. Advances in the mass spectrometry of peptides have been at the forefront of recent technical improvements which are just now beginning to allow scientists to identify large numbers of proteins in the cellular proteome. Similarly, advances in the availability and searchability of large databases are also improving our ability to identify proteins of interest.

The goal here will be to summarize the proteome studies performed on *L. monocytogenes*. Proteomic studies of *L. monocytogenes* have focused on both general stress responses of the organism and on more specific stresses that are important for the survival and growth of *L. monocytogenes* under conditions related to foods, biofilms, and the infection process. Food preservation techniques often involve the implementation of multiple sublethal stresses. A greater understanding of the physiological responses and cross-protections brought about by individual or multiple stresses should allow the implementation and combination of more appropriate food preservation methods. In summarizing the available literature, we wish to provide a framework for contextualizing this diverse body of information and to provide some insights into how *L. monocytogenes* responds to various conditions of importance for food safety. The data presented are summarized in Table 15.5.

## Cold or heat stress

The ability of *L. monocytogenes* to survive and grow at refrigeration temperatures makes it important to determine the mechanisms by which this foodborne pathogen accomplishes that activity. Two early *L. monocytogenes* proteomic studies focused on the response of the organism to heat shock (Phan-thanh and Gormon, 1995) and cold shock (Phan-thanh and Gormon, 1995; Bayles *et al.*, 1996). Phan-thanh and Gormon found 32 proteins were induced 2-fold or greater following a heat shock from 25°C to 49°C. They also found a cold shock from 25°C to 4°C induced 38 proteins 2-fold or greater. Two proteins were induced 3-fold or greater by both heat shock and cold shock, indicating they might be general stress response proteins. Bayles, *et al.* (1996) conservatively identified 12 proteins induced upon cold shock from 37°C to 5°C. Additionally, four of these proteins remained highly expressed during balanced growth at 5°C, indicating that these proteins also functioned as cold acclimation proteins. While neither of these studies reported the identity of the cold-expressed proteins, two of the cold shock proteins found in one study (Bayles *et al.*, 1996) were later identified as ribosomal protein S6 and elongation factor Tu (EF-Tu) (Bayles, unpublished). The increased expression of EF-Tu indicates that *L. monocytogenes* increased the production of translation factors in an effort to overcome the effects of cold shock. The low-temperature induction of EF-Tu in *L. monocytogenes* may reflect a chaperone type role for EF-Tu similar to what has been observed in *Escherichia coli* (Caldas *et al.*, 1998). Increased expression of ribosomal protein S6 may be indicative of a cellular response to translation inefficiency, or possibly ribosomal protein S6 may assist ribosome function at low temperature. While ribosomal protein S6 has been implicated in the response of bacteria to other environmental stresses (see salt stress), it is interesting to note that three ribosomal proteins, one of which was S6, were induced by cold treatment in soybeans (Kim *et al.*, 2004), indicating that this cold stress response is beneficial for both prokaryotic and eukaryotic organisms. Given the food safety importance of *L. monocytogenes* growth at low temperature, it may be time for a thorough reanalysis of the *Listeria* cold stress response.

## Acid stress

A number of groups have used proteomics to explore the response of *L. monocytogenes* to conditions of acid stress. One early study focused on the development of the exponential phase acid tolerance response (ATR) (Davis *et al.*, 1996). This study showed that treating *L. monocytogenes* to a sublethal pH of between 4.0 and 6.0 (pH adjusted using HCl), for up to one hour, induced an ATR and provided measurable protection against a killing at pH 3.0. The addition of the protein synthesis inhibiting antibiotics chloramphenicol or tetracycline substantially reduced the development of the ATR indicating that

**Table 15.5** Summary of proteomic studies involving *L. monocytogenes*

| Treatment condition | Number of protein spots newly expressed or upregulated 2-fold or more | Number of up-regulated proteins identified | Reference |
|---|---|---|---|
| Heat shock 25°C to 49°C | 32 | None | Phan-thanh and Gormon (1995) |
| Cold shock 25°C to 4°C | 38 | None | Phan-thanh and Gormon (1995) |
| Cold shock 37°C to 5°C | 12 | 1 | Bayles *et al.* (1996) |
| pH 5.8 (HCl) | 11 | None | Davis *et al.* (1996) |
| pH 5.5 (lactic acid) | 17 | None | O'Driscoll *et al.* (1997) |
| pH 5.0 (HCl) | 30 | None | O'Driscoll *et al.* (1997) |
| pH 3.5 (HCl) | 49 | 10 | Phan-thanh and Mahouin (1999) |
| pH 5.5 (HCl) | 39 | 10 | Phan-thanh and Mahouin (1999) |
| pH 4.0 (HCl) | 21 | None | Phan-thanh and Gormon (1997) |
| pH 10 (NaOH) | 27 | None | Phan-thanh and Gormon (1997) |
| 0.015% SDS | 17 | None | Phan-thanh and Gormon (1997) |
| 0.03% sodium deoxycholate | 31 | None | Phan-thanh and Gormon (1997) |
| 4% ethanol | 37 | None | Phan-thanh and Gormon (1997) |
| 3.5% NaCl | 6 | None | Esvan *et al.* (2000) |
| 6.5% NaCl | 1 | None | Esvan *et al.* (2000) |
| 3.5% NaCl for 30 min | 6 | 5 | Duché *et al.* (2002a) |
| 3.5% NaCl for 1 h | 9 | 7 | Duché *et al.* (2002a) |
| 3.5% NaCl in minimal media or 3.5% or 6% NaCl in BHI | 56 | 4 | Duché *et al.* (2002b) |
| Minimal medium compared to BHI | 34 increased in minimal media, 8 increased in BHI | 3 | Duché *et al.* (2002b) |
| Biofilm | 22 | 8 | Trémoulet *et al.* (2002) |
| Biofilm with glucose | 26 | 9 | Helloin *et al.* (2003) |
| Biofilm without glucose | 14 | 6 | Helloin *et al.* (2003) |
| Glucose starvation in minimal medium | 35 | 17 | Helloin *et al.* (2003) |
| Entry into stationary phase in defined medium | 206 | 10[a] | Weeks *et al.* (2004) |
| Stationary phase in BHI | 106 | 17 | Folio *et al.* (2004) |
| *rpoN* mutation | Not given | 7 | Arous *et al.* (2004) |
| Divercin V41 sensitive and resistant | 9 and 8[b] | 2[b] | Duffes *et al.* (2000) |
| Multiple strains spontaneously resistant to bacteriocins | Not given | 2 | Gravesen *et al.* (2002) |
| Strain EGDe and multiple food isolates | 13 to 28%[c] | 33 | Ramnath *et al.* (2003) |

[a]Includes the 10 most up- and downregulated proteins.
[b]Only protein spots not seen in the comparison gels were reported.
[c]The percentage represents the approximate percentage of protein spots, major and minor, that were present in food isolates and were not matched in *L. monocytogenes* EGDe.

newly synthesized proteins were involved in the development of the ATR. Using 2DE, the protein profiles of non-adapted cells (pH 6.5 to 7.0) and acid adapted cells (pH 5.8 exposure for 1 h) were compared. This revealed that approximately 11 proteins were induced upon exposure to pH 5.8 conditions. These findings were subsequently confirmed and extended in a study that compared the ATR response of *L. monocytogenes* to pH 5.5 lactic acid or pH 5.0 HCl (O'Driscoll *et al.*, 1997). Adaptation with pH 5.5 lactic acid for 1 h induced expression of 17 proteins while adaptation with pH 5.0 HCl for 1 h induced expression of 30 proteins, six of which were not previously detected in the unstressed controls. Twelve proteins were induced in response to both lactic acid and HCl, indicating a high degree of commonality in the acid stress responses. Similarly, another group found that *L. monocytogenes* grown in minimal medium and exposed to pH 4.0 (pH adjusted with HCl) resulted in the expression of 21 proteins that were either not detected in the controls or were expressed at least 2-fold higher at the lower pH (Phan-thanh and Gormon, 1997). A subsequent study by this same group provided the first successful attempts to identify several of the proteins that were up-regulated or newly synthesized in response to either pH 3.5 or pH 5.5 acidic conditions. Seven of the 13 identified proteins were common to the acid stress response at both pH 3.5 and pH 5.5. Protein identifications based on cross-species peptide mass footprints revealed known stress proteins like GroEL and chaperonin as well as potential transcriptional regulators. Further studies need to be conducted using *L. monocytogenes* specific databases and MS/MS sequence information to more completely define the acid stress proteins.

## Salt stress

*Listeria monocytogenes* has the ability to tolerate conditions of high osmolarity. The organism is capable of growing in the presence of 10% NaCl (McClure *et al.*, 1989) and has been isolated from foods that have a high salt content. Studies profiling the expressed proteins of salt-stressed *L. monocytogenes* have the potential of pointing out key cellular proteins essential for survival of the organism when salt is present. A 30 min exposure to 3.5% NaCl resulted in the production of the general stress proteins DnaK and Ctc and the enzymes alanine dehydrogenase, glyceraldehyde-3-phosphate dehydrogenase (Gap), and CysK, which is involved in biosynthesis of cysteine (Duché *et al.*, 2002a). This indicates that a NaCl-shock initially induces a general stress response, biosynthesis of amino acids, and portions of the metabolic pathway leading to the production of pyruvate. After being exposed to 3.5% NaCl for 60–90 min, proteins involved in longer term acclimation to high osmotic conditions were expressed. These proteins, sometimes referred to as salt acclimation proteins, include GbuA, EF-Tu, GuaB, CcpA, a mannose-specific PTS, PdhA, and PdhD. The last two are subunits of the pyruvate dehydrogenase complex. Glycine betaine has been shown to be one of the most effective compounds for providing *L. monocytogenes* protection against NaCl stress (Bayles and Wilkinson, 2000), so it is not surprising that one of the highly induced proteins was GbuA, which is part of the glycine betaine transport system. The translation factor EF-Tu was induced by both NaCl stress and cold stress, as mentioned above. These results are consistent with observations that the EF-Tu of *E. coli* has chaperone properties (Caldas *et al.*, 1998) and indicate that similar chaperone activities may be provided by the EF-Tu of *L. monocytogenes* in response to various stresses. The remaining identified salt-stress acclimation proteins are involved in purine synthesis (GuaB) and various aspects of carbon metabolism (CcpA, mannose-specific PTS, PdhA, and PdhD) (Duché *et al.*, 2002a). A second study by this same group (Duché *et al.*, 2002b) compared protein expression in minimal medium with 3.5% NaCl to protein expression in brain heart infusion medium (BHI) with 3.5% or 6% NaCl. The glycine betaine transport protein, GbuA, and the general stress protein Ctc were highly expressed in response to salt stress in BHI. An acetate kinase (AckA) and the ribosomal protein S6 were induced by salt stress in complex medium, but were not induced by salt stress in minimal medium. The expression of ribosomal proteins S6 and Ctc may be a cellular response to translation deficiencies following stress. Ribosomal protein S6 was also induced by cold stress (noted above). A subunit of pyruvate dehydrogenase, PdhD, which is induced in response to salt stress in minimal medium, was repressed in BHI with 6% NaCl. The commonality between the salt- and cold-stress responses is reinforced by the expression of EF-Tu and S6 under both stress conditions. This begins to confirm at the proteome level the observation by many researchers that the *L. monocytogenes* salt- and cold-stress responses are highly linked (Ko *et al.*, 1994; Becker *et al.*, 1998; Gerhardt *et al.*, 2000; Becker *et al.*, 2000; Bayles and Wilkinson, 2000).

## Biofilms

The development and persistence of biofilms containing *L. monocytogenes* is an important area of research since biofilms are prevalent in food processing plants and since biofilms can reduce the effectiveness of sanitation treatments designed to eliminate *Listeria* and other microorganisms. There are currently two published studies examining the protein response of *L. monocytogenes* in biofilms (Trémoulet *et al.*, 2002; Helloin *et al.*, 2003). Of the eight up-regulated proteins identified in a *L. monocytogenes* biofilm (Trémoulet *et al.*, 2002), PdhD and CysK were previously identified as salt-stress proteins in defined medium (Duché *et al.*, 2002a), and PdhD and YvyD were proteins that had greater expression in minimal medium than in BHI with or without 6% added NaCl (Duché *et al.*, 2002b). The finding that PdhD, YvyD, and CysK are highly expressed in both a minimal medium and in a biofilm grown on a complex medium agar indicates that the cells in the biofilm are responding to nutrient limitation similar to cells in a minimal medium. Other proteins whose level increased in a biofilm were 6-phosphofructokinase, ribosomal protein S2, superoxide dismutase (SOD), RecO, and DivIVA (Trémoulet *et al.*, 2002). The expression of 6-phosphofructokinase, ribosomal protein S2, and SOD respectively parallels PdhD, YvyD, and CysK in regard to central carbon metabolism, ribosomal proteins, and proteins involved in oxidative stress. The role of the DNA repair protein RecO and the role of the cell division protein DivIVA remain unclear for *L. monocytogenes* in a biofilm. One down-regulated protein that was identified from biofilm cultures was the flagellin protein FlaA, which indicates the need to suppress flagellin production while in a biofilm.

In a complementary *L. monocytogenes* study, proteins from planktonic cells grown in defined medium with glucose were compared to proteins from cells growing in biofilms both with glucose and without glucose (Helloin *et al.*, 2003). Twenty-six proteins were induced when *L. monocytogenes* was grown in a biofilm with glucose. Of these 26 proteins, nine were identified by MS sequencing and seven could be assigned a function via similarity to known proteins. The seven proteins identified were: ClpP, an ATP-dependent serine protease; phosphoglycerate mutase; DivIVA, a cell division protein; a transcription elongation factor similar to GreA; EF-Tu, a translation elongation factor; CodY, a GTP-sensing transcriptional regulator; and peptidyl-polyl cis-trans isomerase, a general stress protein involved in protein repair. The ClpP protein and the phosphoglycerate mutase were also induced when planktonic cultures were glucose-starved, reinforcing the point that cells in a biofilm respond in a fashion similar to planktonic cells experiencing nutrient limitation. The induction of EF-Tu again points out the key role that this translation factor plays in response to a wide variety of stressful conditions. The DivIVA protein was induced in biofilms in both studies (Trémoulet *et al.*, 2002; Helloin *et al.*, 2003), indicating the needed for modifications to the cell division process when cells are in a biofilm. Also similar in the two biofilm studies was the finding that flagellin production was reduced. The reduction in flagellin during nutrient limitation may be mediated through CodY, which was found to be more expressed under nutrient limited conditions (Helloin *et al.*, 2003). The *codY* gene has previously been shown to be essential for repression of flagellin production in *Bacillus subtilis* when the bacterium is experiencing nutrient limitation (Mirel *et al.*, 2000).

*L. monocytogenes* protein profiles from planktonic cells grown with glucose were also compared to glucose-starved cells in a biofilm. This comparison revealed that the cells in a glucose-starved biofilm increased the expression of 14 proteins, six of which could be identified, and five of which were assigned a function based on amino acid sequence similarity. The five proteins assigned a function were: desoxyribose phosphate aldolase; an aminotransferase of ramified amino acids; FtsE, a hydrophilic nucleotide-binding protein possibly involved in translocating potassium transport proteins and in cell division; SOD, superoxide dismutase; and Upp, a nucleotide metabolism protein. The induced proteins identified from glucose-starved biofilm cells generally indicate the induction of metabolic and catabolic proteins required for survival in a nutrient deprived state. It is interesting that none of the proteins found to be induced in biofilms (Helloin *et al.*, 2003), 29

in biofilms with glucose and 14 in biofilms without glucose, were common to both conditions. This indicates that while the responses of *L. monocytogenes* to nutrient limitation and to a biofilm have some commonality, the availability of glucose profoundly alters the protein species that dominate the protein profile.

## Batch culture conditions

The entry into stationary phase demarks the point at which cells leave balanced growth and begin the transition that prepares the cells for long term survival under growth limiting conditions. The entry of *L. monocytogenes* into stationary phase is mediated by a variety of regulatory mechanisms including the alternative sigma factor $\sigma^B$, which promotes transcription of specific genes including some transcriptional regulators (Christiansen *et al.*, 2004). Since the ability of *L. monocytogenes* to survive stressful conditions is greatly enhanced when the cells are in stationary phase, an understanding of the mechanism by which this enhanced stress tolerance is developed is important. There is a large change in the types and amounts of proteins produced when a batch culture of *L. monocytogenes* makes the transition to stationary phase (Weeks *et al.*, 2004). In *L. monocytogenes*, the total number of resolved proteins decreased from approximately 975 proteins in exponential phase to 701 proteins in stationary phase. Of the 701 proteins observed in stationary phase, 66 were up-regulated proteins that were detected in exponential phase and 140 were proteins not detected in exponential phase. The ten proteins with the greatest up- or down-regulation during the period of transition from exponential to stationary phase were identified by MALDI–TOF peptide mass footprints (Weeks *et al.*, 2004). Three proteins, GrpE, SOD, and a LacI-family transcriptional regulator, began to be induced near the start of the transition to stationary phase and continued to be increasingly expressed throughout the transition period. Two other proteins, ATP synthase ε-chain and chromosome partition protein smc, were expressed during the transition period, but were not detected in stationary phase cell cultures. An AraC-family transcriptional regulator was only expressed during stationary phase. There were four proteins that were down-regulated during the transition, an electron transfer flavoprotein (β-subunit), elongation factor Ts, DNA polymerase III, and a phospho-carrier protein HPr. Additionally, the electron transfer flavoprotein subunit and elongation factor Ts were not present in detectable amounts in stationary phase. The up-regulated proteins include expected stress proteins like GrpE and SOD, and regulatory proteins like the LacI-family transcriptional regulator and the AraC-family transcriptional regulator, which appear to be involved in adaptation to stressful environments. The down-regulated proteins ATP synthase, the electron transfer flavoprotein subunit, elongation factor Ts, and DNA polymerase, illustrate that overall the population of cells is decreasing metabolism and growth, consistent with the stationary phase.

A recent study designed to provide a 2DE proteome database for *L. monocytogenes* EGDe has also provided some information regarding protein expression changes occurring in complex medium (BHI) in response to stationary phase (Folio *et al.*, 2004). These authors used 2DE coupled with MALDI-TOF-MS to identify 17 proteins having increased expression in stationary phase and eight proteins having decreased expression in stationary phase. A number of the increasingly expressed proteins, such as GlpD, PdhD, Lmo1372, Lmo2696, and Lmo2743, are involved in cellular metabolism. The increased expression of pyruvate dehydrogenase (PdhD) has already been noted in conjunction with salt stress and biofilms of *L. monocytogenes* further confirming the role of this protein in adaptations to multiple types of stress. The heat shock protein GroES was up-regulated; however, the authors comment that this may be a consequence of the medium pH shifting from 7.5 to 5.5 under the culturing conditions used. If that is the case, it is not clear to what extent other proteins may have been induced by the pH change. The levels of the ribosomal proteins S6 (RpsF) and L10 (RplJ) were reduced in stationary phase. The reduction in cell growth occurring in stationary phase likely suppresses the production of ribosomal proteins; however, when exponentially growing *L. monocytogenes* cultures were subjected to either cold stress or salt stress, ribosomal protein S6 was up-regulated. Five genes corresponding to four up-regulated proteins, *lmo2696*, *lmo1372*, *rmpE*, and *fri*, and one of the down-regulated proteins, *rplJ*, had $\sigma^B$ consensus sequences in their promoter regions. This indicated that $\sigma^B$ may be responsible for altering the expression of the corresponding proteins during stationary phase. One interesting finding was that stationary phase increased the level of listeriolysin O (LLO). This was attributed to the alteration of the PrfA level that occurs when cells enter stationary phase (Mengaud *et al.*, 1991). No other proteins whose genes were under the control of PrfA were identified as having increased expression in stationary phase. The authors speculate that a basic protein p*I* or possibly the cellular localization of virulence proteins may explain the absence of other PrfA-controlled virulence proteins from among the more highly expressed proteins. At least in the case of InlA, a basic pI is likely not the reason since the predicted pI for InlA is 4.6. We propose an alternative interpretation based on studies that show differential regulation of virulence genes under the control of PrfA. At times, PrfA can induce expression of the virulence gene *hly* and *actA*, while not inducing the expression of *inlA* (Bohne *et al.*, 1996). Furthermore, there is good evidence that differential regulation of at least *hly* and *inlA* is strongly dependent on the level of iron available to the cells and that some additional factor influences the *in vivo* activity of PrfA with its target sequence (Böckmann *et al.*, 1996; Dickneite *et al.*, 1998).

Studies performed using mammalian host cells have added weight to the role of iron (Conte *et al.*, 2000) and have shown that in cell culture PrfA differentially regulates virulence genes (Bubert *et al.*, 1999b). When iron became limiting, a PrfA activating factor (Paf) formed a complex favoring the binding of PrfA to the *hly* promoter region while at the same time inhibiting the binding of PrfA to the *inlA* promoter region (Dickneite

*et al.*, 1998). This begs the question, what is Paf? We propose that Paf may be the *L. monocytogenes* ferritin. Ferritin was one of the proteins identified as being up-regulated in stationary phase (Folio *et al.*, 2004). If ferritin is synonymous with Paf, then conditions that induce the expression of ferritin could be expected to induce expression of *hly*, leading to production of LLO while at the same time suppressing the expression of *inlA*. Recent studies have shown that the *Listeria* ferritin is induced by iron limitation and by stationary phase (Polidoro *et al.*, 2002). Even though the *L. monocytogenes* ferritin appears to bind iron, the observation that ferritin production is negatively regulated by iron concentration provides evidence that iron sequestration is not the sole purpose of this protein (Polidoro *et al.*, 2002). The ferritin of *Listeria* appears to be somewhat unique among all other characterized ferritins since the *Listeria* ferritin is the only known example of a non-heme iron binding ferritin that is related to the DNA-binding protein (Dps) previously noted in starved *E. coli* cells (Almirón *et al.*, 1992; Bozzi *et al.*, 1997). This lends support for a role of the *L. monocytogenes* ferritin in functioning as a protein factor able to bind DNA and influence gene regulation. Taken together, this provides a nice model for how *L. monocytogenes* might sense iron limitation and use ferritin to differentially alter the expression of PrfA-mediated virulence genes throughout the infection process. When iron is available, PrfA would more strongly induce *inlA*, leading to expression of internalin. As *L. monocytogenes* became internalized and iron became limiting, ferritin production would increase and complexes of ferritin with PrfA at the promoter sites for virulence genes like *hly*, *actA*, and possibly *mpl* would increase, while expression of *inlA* would decrease.

It is interesting to note that the *L. monocytogenes* ferritin was both up-regulated in three proteins spots and down-regulated in one spot (Folio *et al.*, 2004). These spots appear to be isoforms with different pI and could be the result of changes in the phosphorylation level of the protein. This might reflect one way that the DNA-binding activities of this Dps-like ferritin could be modulated in response to environmental stimuli like growth phase and iron availability. Also, the *L. monocytogenes* ferritin has been shown to be involved in the cold shock response (Hébraud and Guzzo, 2000).

### Other stresses

The role of alternative sigma 54 ($\sigma^{54}$) factor in global gene regulation of *L. monocytogenes* been studied using a *rpoN* mutant ($\sigma^{54}-$) and a proteomic approach (Arous *et al.*, 2004). There appeared to be no difference in the protein content of a *rpoN* mutant compared to the wild-type *L. monocytogenes* EGDe strain for proteins with a pI between 6 and 10; therefore, the authors focused on proteins resolved in the first dimension pH 3–6 range where seven up-regulated protein spots and two down-regulated protein spots were identified. Based on a companion mRNA expression analysis, only the repressed protein MptA, similar to a mannose PTS subunit, was found to have its corresponding gene directly repressed due to the *rpoN* mutation. The seven up-regulated proteins, three of which were subunits of the pyruvate dehydrogenase complex, were thought to be indirectly induced by absence of $\sigma^{54}$. Three of the induced proteins, Pdh, Lmo1579, and Ldh, are all involved in pyruvate metabolism. It is possible that a change in the pyruvate to phosphoenolpyruvate ratio may alter gene expression for these proteins or that in the case of Pdh there may be post-translational modulation of activity via phosphorylation (Arous *et al.*, 2004). Both Pdh and the alanine dehydrogenase proteins have been identified as up-regulated proteins in salt-stressed *L. monocytogenes* (Duché *et al.*, 2002a).

Bacteriocins from lactic acid bacteria have been applied to foods to inhibit the growth of *L. monocytogenes* and subsequently improve food safety. There has been one proteomic study designed to examine the protein differences observed in *L. monocytogenes* strains that are either sensitive or resistance to the bacteriocin divercin V41, which is produced by *Carnobacterium divergens* (Duffes *et al.*, 2000). The study did not report alterations in the expression ratio of proteins produced by both sensitive and resistant strains; rather, only the proteins observed to be unique to each strain were reported. There were nine protein spots unique to the divercin V41 sensitive strain and eight protein spots unique to the resistant strain. Two protein spots were positively identified, one unique to the sensitive strain and one unique to the resistant strain. The unique protein identified in the sensitive strain was ferritin (*fri* gene product), while the unique protein identified in the resistant strain was flagellin (*flaA* gene product). It was not possible to directly link increased flagellin expression to the diversin V41 resistance phenotype; therefore, the authors speculated that the change in flagellin synthesis may be an indirect effect of a more global shift in gene expression. Given the proposal that ferritin may interact with PrfA to mediate differential gene expression of PrfA-regulated genes; it may also be possible that ferritin can modulate PrfA regulation of *flaA*. This could explain why the diversin V41 sensitive strain, which showed up-regulated expression of ferritin, did not have observable amounts of flagellin.

A separate study analyzed eight strains of *L. monocytogenes* that were spontaneously resistant to high levels of class IIa bacteriocins produced by either *Pediococcus acidilactici* PA-2 (PA-1), *Leuconostoc gelidum* UAL 187–22 (leucocin A), *Leuconostoc carnosum* 4010, and *Carnobacterium piscicola* A9b (Gravesen *et al.*, 2002). Using 2DE, the eight bacteriocin resistant mutants were compared to the five wild type *Listeria* strains from which the resistant mutants were derived. All eight mutants were found to be missing a protein that was present in all the wild type strains. The protein was identified as the MptA subunit of the mannose-specific PTS, EII$_t$^Man. This provides strong evidence that repression of EII$_t$^Man is a general requirement for development of class IIa bacteriocin resistance. Since the expression of *mpt* in *L. monocytogenes* can be blocked by mutations in *rpoN*, *manR*, or *mpt*, *L. monocytogenes* EGDe mutants insertionally inactivated for *manR*, *mptA*, or *mptD* were analyzed by 2DE. These *mpt* deficient strains were also missing the MptA subunit of EII$_t$^Man and were resistant to class IIa bacteriocins. The results confirmed that class IIa bacteriocin resistance is mediated by a common mechanism in *L. monocytogenes* and possibly in Gram-positive bacteria in general, which had been previously proposed (Ramnath *et al.*, 2000; Hechard and Sahl, 2002; Gravesen *et al.*, 2002).

It is sometimes advantageous to have a reference map when making multiple proteome comparisons, and a reference map has recently been generated for *L. monocytogenes* EGDe (Ramnath *et al.*, 2003). A total of 33 proteins comprising the four major functional classes of proteins were identified for the reference strain. The reference map was subsequently compared to the proteome of three *L. monocytogenes* serotype 1/2a food isolates and one serotype 1/2b food isolate. This comparison revealed that the food-isolate strains had an average of 13% unmatched major protein spots and 28% unmatched minor protein spots when comparisons were made to the reference map. Of the 33 proteins identified in the reference map, only two corresponded to proteins that were missing in one or more of the food-isolate proteomes. The GAPDH protein spot was not found in one of the food-isolate strains, while the phosphomethylpyrimidine kinase spot was not found in three of the food-isolates. The PCR carried out using primers internal to the genes encoding GAPDH or phosphomethylpyrimidine kinase confirmed that the food-isolate strains carried the genes corresponding to these proteins. This may indicate that there are differences in the expression of these genes in some food isolates, that the genes may be non-functional, or that there may be posttranslational modifications that cause the proteins to not match the reference map.

In the course of examining protein expression in biofilms of *L. monocytogenes*, a comparison was made that showed 35 proteins were induced by glucose starvation of free-living cells (Helloin *et al.*, 2003). Seventeen of these induced proteins were positively identified from their MS/MS spectra. Eleven of the seventeen identified proteins were assigned a probable function based on similarities to proteins indicated in the *L. monocytogenes* EGDe genome annotation (Glaser *et al.*, 2001). The suite of proteins induced during glucose starvation included several proteins that could be expected in cells experiencing carbon starvation. For example, enzymes for the degradation of proteins (ClpP), catabolism of nucleosides (deoxyribosephosphate aldolase), and amelioration of oxidative stress (SOD) were induced in glucose-starved cells. Glucose starvation also caused the cells to increase the amounts of the glycolytic enzyme phosphoglycerate mutase and a pyruvate dehydrogenase subunit, PdhB. Pyruvate dehydrogenase has also been found to be induced by a variety of conditions including minimal media (Duché *et al.*, 2002b), salt stress (Duché *et al.*, 2002a), stationary phase (Folio *et al.*, 2004), and in a *rpoN* mutant strain (Arous *et al.*, 2004). Another protein that was expressed in response to glucose starvation was a protein that is part of the mannose-specific PTS. Similar to Pdh, proteins that are part of the mannose-specific PTS, EII$_t$^Man, are up-regulated in response to salt stress (Duché *et al.*, 2002a) and *rpoN* mutation (Arous *et al.*, 2004). Expression of the EII$_t^{Man}$ complex is also responsible for sensitivity to the class IIa bacteriocins, and at some level appears to be tied to the regulation of other carbohydrate PTS systems in *Listeria* (Gravesen *et al.*, 2002).

The available *L. monocytogenes* proteome studies make it apparent that the stress responses of *L. monocytogenes* can at times be diverse, yet do demonstrate significant overlap. Additional proteomic studies need to be conducted to define the response of *L. monocytogenes* to the stresses that impact the growth and survival of the bacterium. There is also a need to periodically reevaluate and update older studies that were limited in scope due to technical limitations. As the capability to identify the proteins in whole proteomes improves, defining the areas of difference and overlap will generate opportunities for creating both specific and general interventions that increase our control of *L. monocytogenes*.

## Conclusions and future scopes

*Listeria monocytogenes* is now recognized as a foodborne pathogen causing serious illnesses in immunosuppressed hosts and as one of the leading causes of death due to bacterial food poisoning. *Listerial* pathogenesis is a complex process and as noted above, several bacterial and host factors are pivotal in the infection process. Understanding how environmental factors, such as those encountered in food or the potential host, influence virulence gene expression is critical to further unraveling detailed pathogenic mechanisms. A more complete and integrated understanding of *L. monocytogenes* response to stresses, particularly those encountered in foods, will be important for developing rational intervention strategies. Despite tremendous advances in our understanding of listeriosis and *L. monocytogenes* physiology, the organism remains an elusive pathogen and additional efforts are needed to more fully understand and control this pathogen. Recent advances in listerial genomics, proteomics, and transcriptome analysis will certainly lead to a rapid increase in our understanding of listerial pathogenesis, virulence, physiology, and responses to the environmental. These advances in understanding will be used to take a more rational approach to the development of improved intervention strategies and detection methods to safeguard the food supply.

## Acknowledgments

Sincere thanks to Jennifer Wampler, Kwang-Pyo Kim and B.P. Padmapriya for assistance in preparation of this document.

## References

Abachin, E., Poyart, C., Pellegrini, E., Milohanic, E., Fiedler, F., Berche, P., and Trieu-Cuot, P. 2002. Formation of D-alanyl-lipoteichoic acid is required for adhesion and virulence of *Listeria monocytogenes*. Mol. Microbiol. 43: 1–14.

Abram, M., Schluter, D., Vuckovic, D., Wraber, B., Doric, M., and Deckert, M. 2003. Murine model of pregnancy-associated *Listeria* monocytogenes infection. FEMS Immunol. Med. Microbiol. 35: 177–182.

Almiron, M., Link, A.J., Furlong, D., and Kolter, R. 1992. A novel DNA-binding protein with regulatory and protective roles in starved *Escherichia coli*. Genes Dev. 6: 2646–2654.

Alouf, J.E. 2000. Cholesterol-binding cytolytic protein toxins. Int. J. Med. Microbiol. 290: 351–356.

Alvarez-Dominguez, C., Vazquez-Boland, J.A., Carrasco-Marin, E., Lopez-Mato, P., and Leyva-Cobian, F. 1997. Host cell heparan sulfate proteoglycans mediate attachment and entry of *Listeria monocytogenes*, and the listerial surface protein ActA is involved in heparan sulfate receptor recognition. Infect. Immun. 65: 78–88.

Angelakopoulos, H., Loock, K., Sisul, D.M., Jensen, E.R., Miller, J.F., and Hohmann, E.L. 2002. Safety and shedding of an attenuated strain of *Listeria monocytogenes* with a deletion of *actA/plcB* in adult volunteers: a dose escalation study of oral inoculation. Infect. Immun. 70: 3592–3601.

Arous, S., Buchrieser, C., Folio, P., Glaser, P., Namane, A., Hebraud, M., and Hechard, Y. 2004. Global analysis of gene expression in an *rpoN* mutant of *Listeria monocytogenes*. Microbiology 150: 1581–1590.

Auerbuch, V., Loureiro, J.J., Gertler, F.B., Theriot, J.A., and Portnoy, D.A. 2003. Ena/VASP proteins contribute to *Listeria monocytogenes* pathogenesis by controlling temporal and spatial persistence of bacterial actin-based motility. Mol. Microbiol. 49: 1361–1375.

Aureli, P., Fiorucci, G.C., and Caroli, D. 2000. An outbreak of febrile gastroenteritis associated with corn contaminated by *Listeria monocytogenes*. N. Engl. J. Med. 342: 1236–1241.

Avoyne, C., Butin, M., Delaval, J., and Bind, J.-L. 1997. Detection of *Listeria* spp. in food samples by immunomagnetic capture: ListerScreen method. J. Food Protect. 60: 377–384.

Bakardjiev, A.I., Stacy, B.A., Fisher, S.J., and Portnoy, D.A. 2004. Listeriosis in the pregnant guinea-pig: a model of vertical transmission. Infect. Immun. 72: 489–497.

Banerjee, M., Copp, J., Vuga, D., Marino, M., Chapman, T., van der Geer, P., and Ghosh, P. 2004. GW domains of the *Listeria monocytogenes* invasion protein InlB are required for potentiation of Met activation. Mol. Microbiol. 52: 257–271.

Batt, C.A. 1999. Rapid methods for detection of *Listeria*. In: *Listeria, Listeriosis, and Food Safety*. E. T. Ryser and E. H. Marth, eds. 2nd ed. Marcel Dekker, Inc., New York, pp. 261–278.

Baumler, A.J., Tsolis, R.M., and Heffron, F. 1996. The lpf fimbrial operon mediates adhesion of *Salmonella typhimurium* to murine Peyer's patches. Proc. Natl. Acad. Sci. USA 93: 279–283.

Bayles, D., Annous, B., and Wilkinson, B. 1996. Cold stress proteins induced in *Listeria monocytogenes* in response to temperature downshock and growth at low temperatures. Appl. Environ. Microbiol. 62: 1116–1119.

Bayles, D.O., and Wilkinson, B.J. 2000. Osmoprotectants and cryoprotectants for *Listeria monocytogenes*. Lett. Appl. Microbiol. 30: 23–27.

Beavon, I.R.G. 2000. The E-cadherin-catenin complex in tumour metastasis: structure, function and regulation. Eur. J. Cancer. 36: 1607–1620.

Becker, L.A., Cetin, M.S., Hutkins, R.W., and Benson, A.K. 1998. Identification of the gene encoding the alternative sigma factor σB from *Listeria monocytogenes* and its role in osmotolerance. J. Bacteriol. 180: 4547–4554.

Becker, L.A., Evans, S.N., Hutkins, R.W., and Benson, A.K. 2000. Role of σB in adaptation of *Listeria monocytogenes* to growth at low temperature. J. Bacteriol. 182: 7083–7087.

Behari, J., and Youngman, P. 1998. A homolog of CcpA mediates catabolite control in *Listeria monocytogenes* but not carbon source regulation of virulence genes. J. Bacteriol. 180: 6316–6324.

Ben Embarek, P.K., Hansen, L.T., Enger, O., and Huss, H.H. 1997. Occurrence of *Listeria* spp. in farmed salmon and during subsequent slaughter: comparison of Listertest Lift and the USDA method. Food Microbiol. 14: 39–46.

Bibb, W.F., Schwartz, B., Gellin, B.G., Plikaytis, B.D., and Weaver, R.E. 1989. Analysis of *Listeria monocytogenes* by multilocus enzyme electrophoresis and application of the method to epidemiologic investigations. Int. J. Food Microbiol. 8: 233–239.

Bierne, H., and Cossart, P. 2002. InIB, a surface protein of *Listeria monocytogenes* that behaves as an invasin and a growth factor. J. Cell Sci. 115: 3357–3367.

Bille, J., Catimel, B., Bannerman, E., Jacquet, C., Yersin, M., Caniaux, I., Monget, D., and Rocourt, J. 1992. API *Listeria*, a new and promising one-day system to identify *Listeria* isolates. Appl. Environ. Microbiol. 58: 1857–1860.

Böckmann, R., Dickneite, C., Middendorf, B., Goebel, W., and Sokolovic, Z. 1996. Specific binding of the *Listeria monocytogenes* transcriptional regulator PrfA to target sequences requires additional factor(s) and is influenced by iron. Mol. Microbiol. 22: 643–653.

Bohne, J., Kestler, H., Uebele, C., Sokolovic, Z., and Goebel, W. 1996. Differential regulation of the virulence genes of *Listeria monocytogenes* by the transcriptional activator PrfA. Mol. Microbiol. 20: 1189–1198.

Bonnemain, C., Raynaud, C., Reglier-Poupet, H., Dubail, I., Frehel, C., Lety, M.-A., Berche, P., and Charbit, A. 2004. Differential roles of multiple signal peptidases in the virulence of *Listeria monocytogenes*. Mol. Microbiol. 51: 1251–1266.

Borucki, M.K., and Call, D.R. 2003. *Listeria monocytogenes* serotype identification by PCR. J. Clin. Microbiol. 41: 5537–5540.

Bozzi, M., Mignogna, G., Stefanini, S., Barra, D., Longhi, C., Valenti, P., and Chiancone, E. 1997. A novel non-heme iron-binding ferritin related to the DNA-binding proteins of the Dps family in *Listeria innocua*. J. Biol. Chem. 272: 3259–3265.

Braun, L., Dramsi, S., Dehoux, P., Bierne, H., Lindahl, G., and Cossart, P. 1997. InIB: An invasion protein of *Listeria monocytogenes* with a novel type of surface association. Mol. Microbiol. 25: 285–294.

Braun, L., Ghebrehiwet, B., and Cossart, P. 2000. gC1q-R/p32, a C1q-binding protein, is a receptor for the InIB invasion protein of *Listeria monocytogenes*. EMBO J. 19: 1458–1466.

Brehm, K., Ripio, M.-T., Kreft, J., and Vazquez-Boland, J.A. 1999. The bvr locus of *Listeria monocytogenes* mediates virulence gene repression by beta-glucosides. J. Bacteriol. 181: 5024–5032.

Bubert, A., Kuhn, M., Goebel, W., and Kohler, S. 1992. Structural and functional properties of the p60 proteins from different *Listeria* species. J. Bacteriol. 174: 8166–8171.

Bubert, A., Schubert, P., Kohler, S., Frank, R., and Goebel, W. 1994. Synthetic peptides derived from the *Listeria monocytogenes* p60 protein as antigens for the generation of polyclonal antibodies specific for secreted cell-free *L. monocytogenes* p60 proteins. Appl. Environ. Microbiol. 60: 3120–3127.

Bubert, A., Hein, I., Rauch, M., Lehner, A., Yoon, B., Goebel, W., and Wagner, M. 1999a. Detection and differentiation of *Listeria* spp. by a single reaction based on multiplex PCR. Appl. Environ. Microbiol. 65: 4688–4692.

Bubert, A., Sokolovic, Z., Chun, S.K., Papatheodorou, L., Goebel, W., Simm, A., and Bubert, A. 1999b. Differential expression of *Listeria monocytogenes* virulence genes in mammalian host cells. Mol. Gen. Genet. 261: 323–336.

Buchrieser, C., Rusniok, C., Kunst, F., Cossart, P., and Glaser, P. 2003. Comparison of the genome sequences of *Listeria monocytogenes* and *Listeria innocua*: clues for evolution and pathogenicity. FEMS Immunol. Med. Microbiol. 35: 207–213.

Bula, C.J., Bille, J., and Glauser, M.P. 1995. An epidemic of foodborne listeriosis in western Switzerland: description of 57 cases involving adults. Clin. Infect. Dis. 20: 66–72.

Cabanes, D., Dehoux, P., Dussurget, O., Frangeul, L., and Cossart, P. 2002. Surface proteins and the pathogenic potential of *Listeria monocytogenes*. Trends Microbiol. 10: 238–245.

Caldas, T.D., El Yaagoubi, A., and Richarme, G. 1998. Chaperone properties of bacterial elongation factor EF-Tu. J. Biol. Chem. 273: 11478–11482.

Call, D.R., Borucki, M.K., and Besser, T.E. 2003. Mixed-genome microarrays reveal multiple serotype and lineage-specific differences among strains of *Listeria monocytogenes*. J. Clin. Microbiol. 41: 632–639.

Camilli, A., Goldfine, H., and Portnoy, D.A. 1991. *Listeria monocytogenes* mutants lacking phosphatidylinositol-specific phospholipase C are avirulent. J. Exp. Med. 173: 751–754.

Camilli, A., Tilney, L.G., and Portnoy, D.A. 1993. Dual roles of *plcA* in *Listeria monocytogenes* pathogenesis. Mol. Microbiol. 8: 143–157.

Capita, R., Alonso-Calleja, C., Mereghetti, L., Moreno, B., and Garcia-Fernandez, M.d.C. 2002. Evaluation of the international phage typing set and some experimental phages for typing of *Listeria monocytogenes* from poultry in Spain. J. Appl. Microbiol. 92: 90–96.

CDC. 1998. Multistate outbreak of listeriosis – United States, 1998. Morb. Mortal. Weekly Rep. 47: 1085–1086.

CDC. 2000. Multistate outbreak of listeriosis – United States, 2000. Morb. Mortal. Weekly Rep. 49: 1129–1130.

CDC. 2002a. Outbreak of listeriosis associated with homemade Mexican-style cheese – North Carolina, October 2000-January 2001. Morb. Mortal. Weekly Rep. 50: 560–562.

CDC. 2002b. Public Health Dispatch: Outbreak of Listeriosis – Northeastern United States, 2002. Morb. Mortal. Weekly Rep. 15: 950–951.

Chakraborty, T., Leimeister-Wachter, M., Domann, E., Hartl, M., Goebel, W., Nichterlein, T., and Notermans, S. 1992. Coordinate regulation of virulence genes in *Listeria monocytogenes* requires the product of the *prfA* gene. J. Bacteriol. 174: 568–574.

Chakraborty, T., Ebel, F., Domann, E., Niebuhr, K., Gerstel, B., Pistor, S., Temmgrove, C.J., Jockusch, B.M., Reinhard, M., Walter, U., and Wehland, J. 1995. A focal adhesion factor directly linking intracellularly motile *Listeria monocytogenes* and *Listeria ivanovii* to the actin-based cytoskeleton of mammalian cells. EMBO J. 14: 1314–1321.

Chakraborty, T., Hain, T., and Domann, E. 2000. Genome organization and the evolution of the virulence gene locus in *Listeria* species. Int. J. Med. Microbiol. 290: 167–74.

Chico-Calero, I., Suarez, M., Gonzalez-Zorn, B., Scortti, M., Slaghuis, J., Goebel, W., and Vazquez-Boland, J.A. 2002. Hpt, a bacterial homolog of the microsomal glucose-6-phosphate translocase, mediates rapid intracellular proliferation in *Listeria*. Proc. Natl. Acad. Sci. USA 99: 431–436.

Christiansen, J.K., Larsen, M.H., Ingmer, H., Sogaard-Andersen, L., and Kallipolitis, B.H. 2004. The RNA-binding protein Hfq of *Listeria monocytogenes*: role in stress tolerance and virulence. J. Bacteriol. 186: 3355–3362.

Christie, N.E., Atkins, E., and Munch-Petersen, E. 1944. A note on a lytic phenomenon shown by group B streptococci. Aust. J. Exp. Biol. Med. Sci. 22: 197–200.

Clark, M.A., Hirst, B.H., and Jepson, M.A. 1998. M-cell surface beta 1 integrin expression and invasin-mediated targeting of *Yersinia pseudotuberculosis* to mouse Peyer's patch M cells. Infect. Immun. 66: 1237–1243.

Coffey, A., van den Burg, B., Veltman, R., and Abee, T. 2000. Characteristics of the biologically active 35-kDa metalloprotease virulence factor from *Listeria monocytogenes*. J. Appl. Microbiol. 88: 132–141.

Conte, M., Longhi, C., Polidoro, M., Petrone, G., Buonfiglio, V., Di Santo, S., Papi, E., Seganti, L., Visca, P., and Valenti, P. 1996. Iron availability affects entry of *Listeria monocytogenes* into the enterocyte-like cell line Caco-2. Infect. Immun. 64: 3925–3929.

Conte, M.P., Longhi, C., Petrone, G., Polidoro, M., Valenti, P., and Seganti, L. 2000. Modulation of *actA* gene expression in *Listeria monocytogenes* by iron. J. Med. Microbiol. 49: 681–683.

Cossart, P., and Sansonetti, P.J. 2004. Bacterial invasion: the paradigms of enteroinvasive pathogens. Science 304: 242–248.

Curtis, G.D.W., Mitchell, R.G., King, A.F., and Griffen, E.J. 1989. A selective differential medium for the isolation of Listiera monocytogenes. Lett. Appl. Microbiol. 8: 95–98.

Curtis, G.D.W., and Lee, W.H. 1995. Culture media and methods for the isolation of *Listeria monocytogenes*. Int. J. Food Microbiol. 26: 1–13.

Dalton, C.B., Austin, C.C., Sobel, J., Hayes, P.S., Bibb, W.F., Graves, L.M., Swaminathan, B., Proctor, M.E., and Griffin, P.M. 1997. An outbreak of gastroenteritis and fever due to *Listeria monocytogenes* in milk. N. Engl. J. Med. 336: 100–106.

Darji, A., Mohamed, W., Domann, E., and Chakraborty, T. 2003. Induction of immune responses by attenuated isogenic mutant strains of *Listeria monocytogenes*. Vaccine. 21: S102–S109.

Davis, M.J., Coote, P.J., and O'Byrne, C.P. 1996. Acid tolerance in *Listeria monocytogenes*: the adaptive acid tolerance response (ATR) and growth-phase-dependent acid resistance. Microbiology 142: 2975–2982.

Dayel, M.J., Holleran, E.A., and Mullins, R.D. 2001. Arp2/3 complex requires hydrolyzable ATP for nucleation of new actin filaments. Proc. Natl. Acad. Sci. USA 98: 14871–14876.

Decatur, A.L., and Portnoy, D.A. 2000. A PEST-like sequence in listeriolysin O essential for *Listeria monocytogenes* pathogenicity. Science. 290: 992–995.

Dickneite, C., Bockmann, R., Spory, A., Goebel, W., and Sokolovic, Z. 1998. Differential interaction of the transcription factor PrfA and the PrfA-activating factor (Paf) of *Listeria monocytogenes* with target sequences. Mol. Microbiol. 27: 915–928.

Donnelly, C.W., and Baigent, G.J. 1986. Method for flow cytometric detection of *Listeria monocytogenes* in milk. Appl. Environ. Microbiol. 52: 689–695.

Donnelly, C.W. 1988. Historical perspectives on methodology to detect *Listeria monocytogenes*. J. Assoc. Off. Anal. Chem. Int. 71: 644–6.

Doumith, M., Buchrieser, C., Glaser, P., Jacquet, C., and Martin, P. 2004. Differentiation of the major *Listeria monocytogenes* serovars by multiplex PCR. J. Clin. Microbiol. 42: 3819–3822.

Dramsi, S., Biswas, I., Maguin, E., Braun, L., Mastroeni, P., and Cossart, P. 1995. Entry of *Listeria monocytogenes* into hepatocytes requires expression of InlB, a surface protein of the internalin multigene family. Mol. Microbiol. 16: 251–261.

Dramsi, S., Lebrun, M., and Cossart, P. 1996. Molecular and genetic determinants involved in invasion of mammalian cells by *Listeria monocytogenes*. In: Bacterial Invasiveness. V. L. Miller, ed. Springer-Verlag, New York, NY, pp. 61–77.

Dramsi, S., and Cossart, P. 2002. Listeriolysin O: a genuine cytolysin optimized for an intracellular parasite. J. Cell Biol. 156: 943–946.

Dramsi, S., Bourdichon, F., Cabanes, D., Lecuit, M., Fsihi, H., and Cossart, P. 2004. FbpA, a novel multifunctional *Listeria monocytogenes* virulence factor. Mol. Microbiol. 53: 639–649.

Duché, O., Trémoulet, F., Glaser, P., and Labadie, J. 2002a. Salt stress proteins induced in *Listeria monocytogenes*. Appl. Environ. Microbiol. 68: 1491–1498.

Duché, O., Trémoulet, F., Namane, A., and Labadie, J. 2002b. A proteomic analysis of the salt stress response of *Listeria monocytogenes*. FEMS Microbiol. Lett. 215: 183–188.

Duffes, F., Jenoe, P., and Boyaval, P. 2000. Use of two-dimensional electrophoresis to study differential protein expression in divercin V41-resistant and wild-type strains of *Listeria monocytogenes*. Appl. Environ. Microbiol. 66: 4318–4324.

Dussurget, O., Cabanes, D., Dehoux, P., Lecuit, M., Buchrieser, C., Glaser, P., and Cossart, P. 2002. *Listeria monocytogenes* bile salt hydrolase is a PrfA-regulated virulence factor involved in the intestinal and hepatic phases of listeriosis. Mol. Microbiol. 45: 1095–1106.

Engelbrecht, F., Chun, S.K., Ochs, C., Hess, J., Lottspeich, F., Goebel, W., and Sokolovic, Z. 1996. A new PrfA-regulated gene of *Listeria monocytogenes* encoding a small, secreted protein which belongs to the family of internalins. Mol. Microbiol. 21: 823–837.

Ermolaeva, S., Novella, S., Vega, Y., Ripio, M.T., Scortti, M., and Vazquez-Boland, J.A. 2004. Negative control of *Listeria monocytogenes* virulence genes by a diffusible autorepressor. Mol. Microbiol. 52: 601–611.

Esvan, H., Minet, J., Laclie, C., and Cormier, M. 2000. Proteins variations in *Listeria monocytogenes* exposed to high salinities. Int. J. Food Microbiol. 55: 151–155.

Farber, J.M., and Peterkin, P.I. 1991. *Listeria monocytogenes*, a foodborne pathogen. Microbiol. Rev. 55: 476–511.

Feldsine, P.T., Lienau, A.H., Forgey, R.L., and Calhoon, R.D. 1997a. Visual immunoprecipate assay (VIP) for *Listeria monocytogenes* and related *Listeria* species detection in selected foods: collaborative study. J. Assoc. Off. Anal. Chem. Int. 80: 791–805.

Feldsine, P.T., Lienau, A.H., Forgey, R.L., and Calhoon, R.D. 1997b. Assurance polyclonal enzyme immunoassay for detection of *Listeria*

*monocytogenes* and related *Listeria* species in selected foods: collaborative study. J. Assoc. Off. Anal. Chem. Int. 80: 775–90.

Feldsine, P.T., Mui, L.A., Forgey, R.L., and Kerr, D.E. 2000. Equivalence of Visual Immunoprecipate Assay (VIP) for *Salmonella* for the detection of motile and nonmotile *Salmonella* in all foods to AOAC culture method: collaborative study. J. Assoc. Off. Anal. Chem. Int. 83: 888–902.

Feldsine, P.T., Lienau, A.H., Leung, S.C., and Mui, L.A. 2002a. Method extension study to validate applicability of AOAC Official Method 997.03 visual immunoprecipate assay (VIP) for *Listeria monocytogenes* and related *Listeria* spp. from environmental surfaces: collaborative study. J. Assoc. Off. Anal. Chem. Int. 85: 470–478.

Feldsine, P.T., Lienau, A.H., Leung, S.C., and Mui, L.A. 2002b. Method extension study to validate applicability of AOAC Official Method 996.14 Assurance polyclonal enzyme immunoassay for detection of *Listeria monocytogenes* and related *Listeria* spp. from environmental surfaces: collaborative study. J. Assoc. Off. Anal. Chem. Int. 85: 460–9.

Fleming, D.W., Cochi, S.L., MacDonald, K.L., Brondum, J., Hayes, P.S., Plikaytis, B.D., Holmes, M.B., Audurier, A., Broome, C.V., and Reingold, A.L. 1985. Pasteurized milk as a vehicle of infection in an outbreak of listeriosis. N. Engl. J. Med. 312: 404–407.

Folio, P., Chavant, P., Chafsey, I., Belkorchia, A., Chambon, C., and Hebraud, M. 2004. Two-dimensional electrophoresis database of *Listeria monocytogenes* EGDe proteome and proteomic analysis of mid-log and stationary growth phase cells. Proteomics 4:3187–3201.

Foster, S.J. 1995. Molecular characterization and functional analysis of the major autolysin of *Staphylococcus aureus* 8325/4. J. Bacteriol. 177: 5723–5725.

Fraser, J.A., and Sperber, W.H. 1988. Rapid detection of *Listeria* spp. in food and environmental sampes by esculin hydrolysis. J. Food Prot. 51: 762–765.

Freitag, N.E., Rong, L., and Portnoy, D.A. 1993. Regulation of the *prfA* transcriptional activator of *Listeria monocytogenes*: multiple promoter elements contribute to intracellular growth and cell-to-cell spread. Infect. Immun. 61: 2537–2544.

Freitag, N.E., and Portnoy, D.A. 1994. Dual promoters of the *Listeria monocytogenes prfA* transcriptional activator appear essential *in vitro* but are redundant *in vivo*. Mol. Microbiol. 5: 845–853.

Freitag, N.E., and Jacobs, K.E. 1999. Examination of *Listeria monocytogenes* intracellular gene expression by using the green fluorescent protein of *Aequorea victoria*. Infect. Immun. 67: 1844–1852.

Frye, D.M., Zweig, R., Sturgeon, J., Tormey, M., LeCavalier, M., Lee, I., Lawani, L., and Mascola, L. 2002. An outbreak of febrile gastroenteritis associated with delicatessen meat contaminated with *Listeria monocytogenes*. Clin. Infect. Dis. 35: 943–949.

FSIS. 2002. Microbiology Laboratory Guidebook, Chapter 8: Isolation and identification of *Listeria monocytogenes* from red meat, poultry, egg, and environmental samples. [Online]. Available from United Stated Department of Agriculture – Food Safety Inspection Service, http://www.fsis.usda.gov/ophs/microlab/mlg8.03.pdf (verified December, 2004).

FSIS. 2003. Control of *Listeria monocytogenes* in ready-to-eat meat and poultry products. Federal Register. 68: 34208–34254.

Gaillard, J.-L., Berche, P., Frehel, C., Gouln, E., and Cossart, P. 1991. Entry of *L. monocytogenes* into cells is mediated by internalin, a repeat protein reminiscent of surface antigens from Gram-positive cocci. Cell 65: 1127–1141.

Gangar, V., Curiale, M.S., D'Onorio, A., Schultz, A., Johnson, R.L., and Atrache, V. 2000. VIDAS enzyme-linked immunoflourescent assay for detection of *Listeria* in foods: collaborative study. J. Assoc. Off. Anal. Chem. Int. 83: 903–18.

Gaulin, C., Ramsay, D., Ringuette, L., and Ismail, J. 2003. First documented outbreak of *Listeria monocytogenes* in Quebec, 2002. Can. Commun. Dis. Rep. 29: 181–6.

Geese, M., Loureiro, J.J., Bear, J.E., Wehland, J., Gertler, F.B., and Sechi, A.S. 2002. Contribution of Ena/VASP proteins to intracellular motility of *Listeria* requires phosphorylation and proline-rich core but not F-actin binding or multimerization. Mol. Biol. Cell. 13: 2383–2396.

Geoffroy, C., Gaillard, J.L., Alouf, J.E., and Berche, P. 1987. Purification, characterization, and toxicity of the sulfhydryl-activated hemolysin listeriolysin O from *Listeria monocytogenes*. Infect. Immun. 55: 1641–1646.

Geoffroy, C., Gaillard, J.L., Alouf, J.E., and Berche, P. 1989. Production of thiol-dependent hemolysins by *Listeria monocytogenes* and related species. J. Gen. Microbiol. 135: 481–487.

Geoffroy, C., Raveneau, J., Beretti, J.L., Lecroisey, A., Vazquez-Boland, J.A., Alouf, J.E., and Berche, P. 1991. Purification and characterization of an extracellular 29-kiloDalton phospholipase C from *Listeria monocytogenes*. Infect. Immun. 59: 2382–2388.

Gerhardt, P.N.M., Tombras Smith, L., and Smith, G.M. 2000. Osmotic and chill activation of glycine betaine porter II in *Listeria monocytogenes* membrane vesicles. J. Bacteriol. 182: 2544–2550.

Gilot, P., Andre, P., and Content, J. 1999. *Listeria monocytogenes* possesses adhesins for fibronectin. Infect. Immun. 67: 6698–6701.

Gilot, P., Jossin, Y., and Content, J. 2000. Cloning, sequencing and characterisation of a *Listeria monocytogenes* gene encoding a fibronectin-binding protein. J. Med. Microbiol. 49: 887–896.

Glaser, P., Frangeul, L., Buchrieser, C., Rusniok, C., Amend, A., *et al.* 2001. Comparative genomics of *Listeria* species. Science 294: 849–852.

Goetz, M., Bubert, A., Wang, G., Chico-Calero, I., Vazquez-Boland, J.A., Beck, M., Slaghuis, J., Szalay, A.A., and Goebel, W. 2001. Microinjection and growth of bacteria in the cytosol of mammalian host cells. Proc. Natl. Acad. Sci. USA 98: 12221–12226.

Goldfine, H., Johnston, N.C., and Knob, C. 1993. Nonspecific phospholipase-C of *Listeria monocytogenes*: activity on phospholipids in Triton X-100-mixed micelles and in biological membranes. J. Bacteriol. 175: 4298–4306.

Gouin, E., Mengaud, J., and Cossart, P. 1994. The virulence gene cluster of *Listeria monocytogenes* is also present in *Listeria ivanovii*, an animal pathogen, and *Listeria seeligeri*, a nonpathogenic species. Infect. Immun. 62: 3550–3553.

Goulet, V., Jacquet, C., Vaillant, V., Rebiere, I., Mouret, E., Lorente, C., Maillot, E., Stainer, F., and Rocourt, J. 1995. Listeriosis from consumption of raw-milk cheese. Lancet 345: 1581–1582.

Graves, L.M., and Swaminathan, B. 2001. PulseNet standardized protocol for subtyping *Listeria monocytogenes* by macrorestriction and pulsed-field gel electrophoresis. Int. J. Food Microbiol. 65: 55–62.

Gravesen, A., Ramnath, M., Rechinger, K.B., Andersen, N., Jänsch, L., Héchard, Y., Hastings, J.W., and Knøchel, S. 2002. High-level resistance to class IIa bacteriocins is associated with one general mechanism in *Listeria monocytogenes*. Microbiology 148: 2361–2369.

Gray, M.L., Stafseth, H.J., Jr., T.F., Sholl, L.B., and Riley, W.F.J. 1948. A new technique for isolating Listerellae from the bovine brain. J. Bacteriol. 55.

Greene, S.L., and Freitag, N.E. 2003. Negative regulation of PrfA, the key activator of *Listeria monocytogenes* virulence gene expression, is dispensable for bacterial pathogenesis. Microbiology 149: 111–120.

Gregory, S.H., Sagnimeni, A.J., and Wing, E.J. 1997. Internalin B promotes the replication of *Listeria monocytogenes* in mouse hepatocytes. Infect. Immun. 65: 5137–5141.

Greiffenberg, L., Goebel, W., Kim, K.S., Weiglein, I., Bubert, A., Engelbrecht, F., Stins, M., and Kuhn, M. 1998. Interaction of *Listeria monocytogenes* with human brain microvascular endothelial cells: InlB-dependent invasion, long-term intracellular growth, and spread from macrophages to endothelial cells. Infect. Immun. 66: 5260–5267.

Greiffenberg, L., Goebel, W., Kim, K.S., Daniels, J., and Kuhn, M. 2000. Interaction of *Listeria monocytogenes* with human brain microvascular endothelial cells: an electron microscopic study. Infect. Immun. 68: 3275–3279.

Grenklo, S., Geese, M., Lindberg, U., Wehland, J., Karlsson, R., and Sechi, A.S. 2003. A crucial role for profilin-actin in the intracellular motility of *Listeria monocytogenes*. EMBO Rep. 4: 523–529.

Grundling, A., Gonzalez, M.D., and Higgins, D.E. 2003. Requirement of the *Listeria monocytogenes* broad-range phospholipase PC-PLC during infection of human epithelial cells. J. Bacteriol. 185: 6295–6307.

Haas, A., Dumbsky, M., and Kreft, J. 1992. Listeriolysin genes: complete sequence of ILO from *Listeria ivanovii* and of LSO from *Listeria seeligeri*. Biochim. Biophys. Acta 1130: 81–84.

Hardy, J., Francis, K.P., DeBoer, M., Chu, P., Gibbs, K., and Contag, C.H. 2004. Extracellular replication of *Listeria monocytogenes* in the murine gall bladder. Science 303: 851–853.

Havell, E.A., Beretich, G.R., and Carter, P.B. 1999. The mucosal phase of *Listeria* infection. Immunobiology 201: 164–177.

Hébraud, M. and Guzzo, J. 2000. The main cold shock protein of *Listeria monocytogenes* belongs to the family of ferritin-like proteins. FEMS Microbiol. Lett. 190: 29–34.

Hechard, Y. and Sahl, H.G. 2002. Mode of action of modified and unmodified bacteriocins from Gram-positive bacteria. Biochimie 84: 545–557.

Helloin, E., Jansch, L., and Phan-Thanh, L. 2003. Carbon starvation survival of *Listeria monocytogenes* in planktonic state and in biofilm: a proteomic study. Proteomics 3: 2052–2064.

Henry-Stanley, M.J., Hess, D.J., Erickson, E.A., Garni, R.M., and Wells, C.L. 2003. Role of heparan sulfate in interactions of *Listeria monocytogenes* with enterocytes. Med. Microbiol. Immunol. 192: 107–115.

Hess, J., Gentschev, I., Szalay, G., Ladel, C., Bubert, A., Goebel, W., and Kaufmann, S.H.E. 1995. *Listeria monocytogenes* p60 supports host cell invasion by and *in vivo* survival of attenuated *Salmonella typhimurium*. Infect. Immun. 63: 2047–2053.

Hitchins, A.D. 2003. Bacteriological Analytical Manual Online, Chapter 10: Detection and enumeration of *Listeria monocytogenes* in foods [Online]. Available from United States Food and Drug Administration, http://www.cfsan.fda.gov/~ebam/bam-10.html#authors (verified December, 2004).

Hochberg, A.M., Roering, A., Gangar, V., Curiale, M., Barbour, W.M., and Mrozinski, P.M. 2001. Sensitivity and specificity of the BAX for screening *Listeria monocytogenes* assay: internal validation and independent laboratory study. J. Assoc. Off. Anal. Chem. Int. 84: 1087–97.

Hong, B.-X., Jiang, L.-F., Hu, Y.-S., Fang, D.-Y., and Guo, H.-Y. 2004. Application of oligonucleotide array technology for the rapid detection of pathogenic bacteria of foodborne infections. J. Microbiol. Methods 58: 403–411.

Hsih, H.-Y., and Hsen, H.-Y. 2001. Combination of immunomagnetic separation and polymerase chain reaction for the simultaneous detection of *Listeria monocytogenes* and *Salmonella* spp. in food samples. J. Food Prot. 64: 1744–1750.

Hubbard, W.K., and Billy, T.J. 2001. Draft assessment of the relative risk to public health from foodborne *Listeria monocytogenes* among selected categories of ready-to-eat foods. Federal Register. 66: 5515–5517.

Hughes, D., Dailianis, A., Duncan, L., Briggs, J., McKintyre, D.A., and Silbernagel, K. 2003. Modification of enrichment protocols for TECRA *Listeria* Visual Immunoassay method 995.22: collaborative study. J. Assoc. Off. Anal. Chem. Int. 86: 340–54.

Hugot, J.P., Alberti, C., Berrebi, D., Bingen, E., and Cezard, J.P. 2003. Crohn's disease: the cold chain hypothesis. Lancet. 362: 2012–2015.

Ivnitski, D., Abdel-Hamid, I., Atanasov, P., and Wilkins, E. 1999. Biosensors for detection of pathogenic bacteria. Biosensors Bioelectron. 14: 599–624.

Jackson, B.J., Brookins, A.M., Tetreault, D., and Costello, K. 1993. Detection of *Listeria* in food and environmental samples by immunomagnetic bead capture and by culture methods. J. Rapid Methods Autom. Microbiol. 2: 39–54.

Jacquet, C., Catimel, B., Brosch, R., Buchrieser, C., Dehaumont, P., Goulet, V., Lepoutre, A., Veit, P., and Rocourt, J. 1995. Investigations related to the epidemic strain involved in the French listeriosis outbreak in 1992. Appl. Environ. Microbiol. 61: 2242–2246.

Jaradat, Z.W., and Bhunia, A.K. 2002. Glucose and nutrient concentrations affect the expression of a 104-kilodalton *Listeria* adhesion protein in *Listeria monocytogenes*. Appl. Environ. Microbiol. 68: 4876–4883.

Jaradat, Z.W., Schutze, G.E., and Bhunia, A.K. 2002. Genetic homogeneity among *Listeria monocytogenes* strains from infected patients and meat products from two geographic locations determined by phenotyping, ribotyping and PCR analysis of virulence genes. Int. J. Food Microbiol. 76: 1–10.

Jaradat, Z.W., Wampler, J.L., and Bhunia, A.K. 2003. A *Listeria* adhesion protein-deficient *Listeria monocytogenes* strain shows reduced adhesion primarily to intestinal cell lines. Med. Microbiol. Immunol. 192: 85–91.

Jensen, V.B., Harty, J.T., and Jones, B.D. 1998. Interactions of the invasive pathogens *Salmonella typhimurium*, *Listeria monocytogenes*, and *Shigella flexneri* with M cells and murine Peyer's patches. Infect. Immun. 66: 3758–3766.

Johansson, J., Mandin, P., Renzoni, A., Chiaruttini, C., Springer, M., and Cossart, P. 2002. An RNA thermosensor controls expression of virulence genes in *Listeria monocytogenes*. Cell 110: 551–561.

Johnson, J., Jinneman, K., Stelma, G., Smith, B.G., Lye, D., Messer, J., Ulaszek, J., Evsen, L., Gendel, S., Bennett, R.W., Swaminathan, B., Pruckler, J., Steigerwalt, A., Kathariou, S., Yildirim, S., Volokhov, D., Rasooly, A., Chizhikov, V., Wiedmann, M., Fortes, E., Duvall, R.E., and Hitchins, A.D. 2004. Natural atypical *Listeria innocua* strains with *Listeria monocytogenes* pathogenicity island 1 genes. Appl. Environ. Microbiol. 70: 4256–4266.

Jones, M., Gupta, R.S., and Englesberg, E. 1994. Enhancement in amount of P1 (Hsp60) in mutants of Chinese hamster ovary (CHO-K1) cells exhibiting increases in the A-system of amino acid transport. Proc. Natl. Acad. Sci. USA 91: 858–862.

Jonquieres, R., Bierne, H., Fiedler, F., Gounon, P., and Cossart, P. 1999. Interaction between the protein InIB of *Listeria monocytogenes* and lipoteichoic acid: a novel mechanism of protein association at the surface of Gram-positive bacteria. Mol. Microbiol. 34: 902–914.

Jung, Y.S., Frank, J.F., and Brackett, R.E. 2003. Evaluation of Antibodies for Immunomagnetic Separation Combined with Flow Cytometry Detection of *Listeria monocytogenes*. J. Food Prot. 66: 1283–1287.

Kaclikova, E., Kuchta, T., Kay, H., and Gray, D. 2001. Separation of *Listeria* from cheese and enrichment media using antibody-coated microbeads and centrifugation. J. Microbiol. Methods. 46: 63–67.

Karunasagar, I., Lampidis, R., Goebel, W., and Kreft, J. 1997. Complementation of *Listeria seeligeri* with the *plcA-prfA* genes from *L. monocytogenes* activates transcription of seeligerolysin and leads to bacterial escape from the phagosome of infected mammalian cells. FEMS Microbiol. Lett. 146: 303–310.

Kathariou, S., Metz, P., Hof, H., and Goebel, W. 1987. Tn916-induced mutations in the hemolysin determinant affecting virulence of *Listeria monocytogenes*. J. Bacteriol. 169: 1291–1297.

Kathariou, S. 2002. *Listeria monocytogenes* virulence and pathogenicity, a food safety perspective. J. Food Prot. 65: 1811–1829.

Kim, K.Y., Park, S.W., Chung, Y.S., Chung, C.H., Kim, J.I., and Lee, J.H. 2004. Molecular cloning of low-temperature-inducible ribosomal proteins from soybean. J. Exp. Bot. 55: 1153–1155.

Knight, M.T., Newman, M.C., Benzinger, M.J., Jr., Agin, J.R., Ash, M., Sims, P., and Hughes, D. 1996. TECRA *Listeria* Visual Immunoassay (TLVIA) for detection of *Listeria* in foods: collaborative study. J. Assoc. Off. Anal. Chem. Int. 79: 1083–94.

Ko, R., Smith, L.T., and Smith, G.M. 1994. Glycine betaine confers enhanced osmotolerance and cryotolerance on *Listeria monocytogenes*. J. Bacteriol. 176: 426–431.

Kocks, C., Hellio, R., Gounon, P., Ohayon, H., and Cossart, P. 1993. Polarized distribution of *Listeria monocytogenes* surface protein ActA at the site of directional actin assembly. J. Cell Sci. 105: 699–710.

Kolb-Maurer, A., Gentschev, I., Fries, H.W., Fiedler, F., Brocker, E.B., Kampgen, E., and Goebel, W. 2000. *Listeria monocytogenes*-infected human dendritic cells: Uptake and host cell response. Infect. Immun. 68: 3680–3688.

Kuhn, M., and Goebel, W. 1989. Identification of an extracellular protein of *Listeria monocytogenes* possibly involved in intracellular uptake by mammalian cells. Infect. Immun. 57: 55–61.

Lampidis, R., Gross, R., Sokolovic, Z., Goebel, W., and Kreft, J. 1994. The virulence regulator protein of *Listeria ivanovii* Is highly homologous to PrfA from *Listeria monocytogenes* and both belong to the Crp-Fnr family of transcription regulators. Mol. Microbiol. 13: 141–151.

Lawrence, L., and Gilmour, A. 1994. Incidence of *Listeria* spp. and *Listeria monocytogenes* in a poultry processing environment and in poultry products and their rapid confirmation by multiplex PCR. Appl. Environ. Microbiol. 60: 4600–4604.

Lecuit, M., Dramsi, S., Gottardi, C., Fedor-Chaiken, M., Gumbiner, B., and Cossart, P. 1999. A single amino acid in E-cadherin responsible for host specificity towards the human pathogen *Listeria monocytogenes*. EMBO J. 18: 3956–3963.

Lecuit, M., Vandormael-Pournin, S., Lefort, J., Huerre, M., Gounon, P., Dupuy, C., Babinet, C., and Cossart, P. 2001. A transgenic model for listeriosis: role of internalin in crossing the intestinal barrier. Science 292: 1722–1725.

Lecuit, M., Nelson, D.M., Smith, S.D., Khun, H., Huerre, M., Vacher-Lavenu, M.-C., Gordon, J.I., and Cossart, P. 2004. Targeting and crossing of the human maternofetal barrier by *Listeria monocytogenes*: role of internalin interaction with trophoblast E-cadherin. Proc. Natl. Acad. Sci. USA 101: 6152–6157.

Lee, W.H., and McClain, D. 1986. Improved *Listeria monocytogenes* selective agar. Appl. Environ. Microbiol. 52: 1215–7.

Leimeister-Wachter, M., Haffner, C., Domann, E., Goebel, W., and Chakraborty, T. 1990. Identification of a gene that positively regulates expression of listeriolysin, the major virulence factor of *Listeria monocytogenes*. Proc. Natl. Acad. Sci. USA 87: 8336–8340.

Leimeister-Wachter, M., Domann, E., and T., C. 1991. Detection of a gene encoding a phosphatidylinositol-specific phospholipase C that is co-ordinately expressed with listeriolysin in *Listeria monocytogenes*. Mol. Microbiol. 5: 361–366.

Lety, M.-A., Frehel, C., Dubail, I., Beretti, J.-L., Kayal, S., Berche, P., and Charbit, A. 2001. Identification of a PEST-like motif in listeriolysin O required for phagosomal escape and for virulence in *Listeria monocytogenes*. Mol. Microbiol. 39: 1124–1139.

Lingnau, A., Domann, E., Hudel, M., Bock, M., Nichterlein, T., Wehland, J., and Chakraborty, T. 1995. Expression of the *Listeria monocytogenes* EGD *inlA* and *inlB* genes, whose products mediate bacterial entry into tissue culture cell lines, by PrfA-dependent and PrfA-independent mechanisms. Infect. Immun. 63: 3896–3903.

Linnan, M.J., Mascola, L., Lou, X.D., Goulet, V., May, S., Salminen, C., Hird, D.W., Yonekura, M.L., Hayes, P., Weaver, R., and others. 1988. Epidemic listeriosis associated with Mexican-style cheese. N. Engl. J. Med. 319: 823–828.

Loisel, T.P., Boujemaa, R., Pantaloni, D., and Carlier, M.F. 1999. Reconstitution of actin-based motility of *Listeria* and *Shigella* using pure proteins. Nature 401: 613–616.

Lyon, M., Deakin, J.A., and Gallagher, J.T. 2002. The mode of action of heparan and dermatan sulfates in the regulation of hepatocyte growth factor/scatter factor. J. Biol. Chem. 277: 1040–1046.

Lyytikainen, O., Autio, T., Maijala, R., Ruutu, P., Honkanen-Buzalski, T., Miettinen, M., Hatakka, M., Mikkola, J., Anttila, V.J., Johansson, T., Rantala, L., Aalto, T., Korkeala, H., and Siitonen, A. 2000. An outbreak of *Listeria monocytogenes* serotype 3a infections from butter in Finland. J Infect Dis. 181: 1838–1841.

MacDonald, T.T., and Carter, P.B. 1980. Cell-mediated immunity to intestinal infection. Infect. Immun. 28: 516–523.

Marco, A.J., Altimira, J., Prats, N., Lopez, S., Dominguez, L., Domingo, M., and Briones, V. 1997. Penetration of *Listeria monocytogenes* in mice infected by the oral route. Microb. Pathog. 23: 255–263.

Marino, M., Braun, L., Cossart, P., and Ghosh, P. 1999. Structure of the InIB leucine-rich repeats, a domain that triggers host cell invasion by the bacterial pathogen *L. monocytogenes*. Mol. Cell. 4: 1063–1072.

Marquis, H., Goldfine, H., and Portnoy, D.A. 1997. Proteolytic pathways of activation and degradation of a bacterial phospholipase C during intracellular infection by *Listeria monocytogenes*. J. Cell Biol. 137: 1381–1392.

Marquis, H., and Hager, E.J. 2000. pH-regulated activation and release of a bacteria-associated phospholipase C during intracellular infection by *Listeria monocytogenes*. Mol. Microbiol. 35: 289–298.

May, R.C., Hall, M.E., Higgs, H.N., Pollard, T.D., Chakraborty, T., Wehland, J., Machesky, L.M., and Sechi, A.S. 1999. The Arp2/3 complex is essential for the actin-based motility of *Listeria monocytogenes*. Curr. Biol. 9: 759–762.

McClure, P.J., Roberts, T.A., and Oguru, P.O. 1989. Comparison of the effect of sodium chloride, pH, and temperature on the growth of *Listeria monocytogenes* on gradient plates and in liquid-medium. J. Appl. Microbiol. 9: 95–99.

McClure, P.J., Kelly, T.M., and Roberts, T.A. 1991. The effects of temperature, pH, sodium chloride and sodium nitrite on the growth of *Listeria monocytogenes*. Int. J. Food Microbiol. 14: 77–91.

McLauchlin, J., Hall, S.M., Velani, S.K., and Gilbert, R.J. 1991. Human listeriosis and pate: a possible association. Brit. Med. J. 303: 773–5.

McLaughlan, A.M., and Foster, S.J. 1998. Molecular characterization of an autolytic amidase of *Listeria monocytogenes* EGD. Microbiology 144: 1359–1367.

Mead, P.S., Slutsker, L., Dietz, V., McCaig, L.F., Bresee, J.S., Shapiro, C., Griffin, P.M., and Tauxe, R.V. 1999. Food-related illness and death in the United States. Emerg. Infect. Dis. 5: 607–625.

Meinersmann, R.J., Phillips, R.W., Wiedmann, M., and Berrang, M.E. 2004. Multilocus sequence typing of *Listeria monocytogenes* by use of hypervariable genes reveals clonal and recombination histories of three lineages. Appl. Environ. Microbiol. 70: 2193–2203.

Mello, L.D., and Kubota, L.T. 2002. Review of the use of biosensors as analytical tools in the food and drink industries. Food Chem. 77: 237–256.

Mengaud, J., Chenevert, J., Geoffroy, C., Gaillard, J.L., and Cossart, P. 1987. Identification of the structural gene encoding the SH-activated hemolysin of *Listeria monocytogenes* listeriolysin O is homologous to streptolysin O and pneumolysin. Infect. Immun. 55: 3225–3227.

Mengaud, J., Vicente, M.F., and Cossart, P. 1989. Transcriptional mapping and nucleotide sequence of the *Listeria monocytogenes* hlyA region reveal structural features that may be involved in regulation. Infect. Immun. 57: 3695–3701.

Mengaud, J., Braunbreton, C., and Cossart, P. 1991a. Identification of phosphatidylinositol-specific phospholipase-C activity in *Listeria monocytogenes* – a novel type of virulence factor. Mol. Microbiol. 5: 367–372.

Mengaud, J., Dramsi, S., Gouin, E., Vazquez-Boland, J.A., Milon, G., and Cossart, P. 1991b. Pleiotropic control of *Listeria monocytogenes* virulence factors by a gene that is autoregulated. Mol. Microbiol. 5: 2273–2283.

Mengaud, J., Lecuit, M., Lebrun, M., Nato, F., Mazie, J.C., and Cossart, P. 1996. Antibodies to the leucine-rich repeat region of internalin block entry of *Listeria monocytogenes* into cells expressing E-cadherin. Infect. Immun. 64: 5430–5433.

Merrell, D.S., and Camilli, A. 2002. Acid tolerance of gastrointestinal pathogens. Curr. Opin. Microbiol. 5: 51–55.

Michel, E., Reich, K.A., Favier, R., Berche, P., and Cossart, P. 1990. Attenuated mutants of the intracellular bacterium *Listeria monocytogenes* obtained by single amino acid substitutions in listeriolysin O. Mol. Microbiol. 4: 2167–2178.

Milenbachs, A.A., Brown, D.P., Moors, M., and Youngman, P. 1997. Carbon-source regulation of virulence gene expression in *Listeria monocytogenes*. Mol. Microbiol. 23: 1075–1085.

Milohanic, E., Pron, B., Berche, P., and Gaillard, J.L. 2000. Identification of new loci involved in adhesion of *Listeria monocytogenes* to eukaryotic cells. Microbiology 146: 731–739.

Milohanic, E., Jonquieres, R., Cossart, P., Berche, P., and Gaillard, J.L. 2001. The autolysin Ami contributes to the adhesion of *Listeria monocytogenes* to eukaryotic cells via its cell wall anchor. Mol. Microbiol. 39: 1212–1224.

Milohanic, E., Glaser, P., Coppee, J.-Y., Frangeul, L., Vega, Y., Vazquez-Boland, J.A., Kunst, F., Cossart, P., and Buchrieser, C. 2003. Transcriptome analysis of *Listeria monocytogenes* identifies three

groups of genes differently regulated by PrfA. Mol. Microbiol. 47: 1613–1625.

Mirel, D.B., Estacio, W.F., Mathieu, M., Olmsted, E., Ramirez, J., and Marquez-Magana, L.M. 2000. Environmental regulation of *Bacillus subtilis* sigma D-dependent gene expression. J. Bacteriol. 182: 3055–3062.

Mitchell, B.A., Milbury, J.A., Brookins, A.B., and Jackson, B.J. 1994. Use of immunomagnetic capture on beads to recover *Listeria* from environmental samples. J. Food Prot. 57: 743–745.

Moors, M.A., Levitt, B., Youngman, P., and Portnoy, D.A. 1999. Expression of listeriolysin O and ActA by intracellular and extracellular *Listeria monocytogenes*. Infect. Immun. 67: 131–139.

Murray, E.G.D., Webb, R.A., and Swann, H.B.R. 1926. A disease of rabbits characterized by a laarge mononuclear leucocytosis caused by a hitherto undescribed *Bacillus* Bacterium monocytognes (n. sp.). J. Path. Bacteriol. 29: 407–439.

Nadon, C.A., Bowen, B.M., Wiedmann, M., and Boor, K.J. 2002. Sigma B contributes to PrfA-mediated virulence in *Listeria monocytogenes*. Infect. Immun. 70: 3948–3952.

Nair, S., Milohanic, E., and Berche, P. 2000. ClpC ATPase is required for cell adhesion and invasion of *Listeria monocytogenes*. Infect. Immun. 68: 7061–7068.

Nelson, K.E., Fouts, D.E., Mongodin, E.F., Ravel, J., DeBoy, R.T., *et al.* 2004. Whole genome comparisons of serotype 4b and 1/2a strains of the foodborne pathogen *Listeria monocytogenes* reveal new insights into the core genome components of this species. Nuclic Acids. Res. 32: 2386–2395.

Niebuhr, K., Chakraborty, T., Rohde, M., Gazlig, T., Jansen, B., Kollner, P., and Wehland, J. 1993. Localization of the ActA polypeptide of *Listeria monocytogenes* in infected tissue culture cell lines: ActA is not associated with actin comets. Infect. Immun. 61: 2793–2802.

Notermans, S.H., Dufrenne, J., Leimeister-Wachter, M., Domann, E., and Chakraborty, T. 1991. Phosphatidylinositol-specific phospholipase C activity as a marker to distinguish between pathogenic and nonpathogenic *Listeria* species. Appl. Environ. Microbiol. 57: 2666–70.

O'Driscoll, B., Gahan, C.G.M., and Hill, C. 1997. Two-dimensional polyacrylamide gel electrophoresis analysis of the acid tolerance response in *Listeria monocytogenes* LO28. Appl. Environ. Microbiol. 63: 2679–2685.

O'Farrell, P.H. 1975. High resolution two-dimensional electrophoresis of proteins. J Biol. Chem. 250: 4007–4021.

O'Riordan, M., Moors, M.A., and Portnoy, D.A. 2003. *Listeria* intracellular growth andvirulence require host-derived lipoic acid. Science 302: 462–464.

Ottaviani, F., Ottaviani, M., and Agosti, M. 1997. Esperienze su un agar selettivo e differenziale per *Listeria monocytogenes* (Differential agar medium for *Listeria monocytogenes*). Industrie Alimentari 36: 888–895.

Palumbo, J.D., Borucki, M.K., Mandrell, R.E., and Gorski, L. 2003. Serotyping of *Listeria monocytogenes* by enzyme-linked immunosorbent assay and identification of mixed-serotype cultures by colony immunoblotting. J. Clin. Microbiol. 41: 564–571.

Pandiripally, V.K., Westbrook, D.G., Sunki, G.R., and Bhunia, A.K. 1999. Surface protein p104 is involved in adhesion of *Listeria monocytogenes* to human intestinal cell line, Caco-2. J. Med. Microbiol. 48: 117–124.

Park, J.H., Lee, Y.S., Lim, Y.K., Kwon, S.H., Lee, C.U., and Yoon, B.S. 2000. Specific binding of recombinant *Listeria monocytogenes* p60 protein to Caco-2 cells. FEMS Microbiol. Lett. 186: 35–40.

Peifer, M. 1997. Beta-catenin as oncogene – the smoking gun. Science. 275: 1752–0.

Petersen, L., and Madsen, M. 2000. *Listeria* spp. in broiler flocks: recovery rates and species distribution investigated by conventional culture and the EiaFoss method. Int. J. Food Microbiol. 58: 113–116.

Phan-Thanh, L., and Gormon, T. 1995. Analysis of heat and cold shock proteins in *Listeria* by two-dimensional electrophoresis. Electrophoresis 16: 444–450.

Phan-thanh, L. and Gormon, T. 1997. Stress proteins in *Listeria monocytogenes*. Electrophoresis 18: 1464–1471.

Phan-thanh, L. and Mahouin, F. 1999. A proteomic approach to study the acid response in *Listeria monocytogenes*. Electrophoresis 20: 2214–2224.

Phan-Thanh, L., Mahouin, F., and Aligé, S. 2000. Acid responses of *Listeria monocytogenes*. Int. J. Food Microbiol. 55: 121–126.

Pilgrim, S., Kolb-Maurer, A., Gentschev, I., Goebel, W., and Kuhn, M. 2003. Deletion of the gene encoding p60 in *Listeria monocytogenes* leads to abnormal cell division and loss of actin-based motility. Infect. Immun. 71: 3473–3484.

Pinner, R.W., Schuchat, A., Swaminathan, B., Hayes, P.S., Deaver, K.A., Weaver, R.E., Plikaytis, B.D., Reeves, M., Broome, C.V., and Wenger, J.D. 1992. Role of foods in sporadic listeriosis. II. Microbiologic and epidemiologic investigation. The *Listeria* Study Group. J. Am. Med. Assoc. 267: 2046–50.

Pistor, S., Chakraborty, T., Niebuhr, K., Domann, E., and Wehland, J. 1994. The ActA protein of *Listeria monocytogenes* acts as a nucleator inducing reorganization of the actin cytoskeleton. EMBO J. 13: 758–763.

Pistor, S., Grobe, L., Sechi, A.S., Domann, E., Gerstel, B., Machesky, L.M., Chakraborty, T., and Wehland, J. 2000. Mutations of arginine residues within the 146-KKRRK-150 motif of the ActA protein of *Listeria monocytogenes* abolish intracellular motility by interfering with the recruitment of the Arp2/3 complex. J. Cell Sci. 113: 3277–3287.

Polidoro, M., De Biase, D., Montagnini, B., Guarrera, L., Cavallo, S., Valenti, P., Stefanini, S., and Chiancone, E. 2002. The expression of the dodecameric ferritin in *Listeria* spp. is induced by iron limitation and stationary growth phase. Gene. 296: 121–128.

Poyart, C., Abachin, E., Razafimanantsoa, I., and Berche, P. 1993. The zinc metalloprotease of *Listeria monocytogenes* is required for maturation of phosphatidylcholine phospholipase C: direct evidence obtained by gene complementation. Infect. Immun. 61: 1576–1580.

Pron, B., Boumaila, C., Jaubert, F., Sarnacki, S., Monnet, J.P., Berche, P., and Gaillard, J.L. 1998. Comprehensive study of the intestinal stage of listeriosis in a rat ligated ileal loop system. Infect. Immun. 66: 747–755.

Pron, B., Boumaila, C., Jaubert, F., Berche, P., Milon, G., Geissmann, F., and Gaillard, J.L. 2001. Dendritic cells are early cellular targets of *Listeria monocytogenes* after intestinal delivery and are involved in bacterial spread in the host. Cell. Microbiol. 3: 331–340.

Ramnath, M., Beukes, M., Tamura, K., and Hastings, J.W. 2000. Absence of a putative mannose-specific phosphotransferase system enzyme IIAB component in a leucocin A-eesistant strain of *Listeria monocytogenes*, as shown by two-dimensional sodium dodecyl sulfate-polyacrylamide gel electrophoresis. Appl. Environ. Microbiol. 66: 3098–3101.

Ramnath, M., Rechinger, K.B., Jansch, L., Hastings, J.W., Knochel, S., and Gravesen, A. 2003. Development of a *Listeria monocytogenes* EGDe partial proteome reference map and comparison with the protein profiles of food isolates. Appl. Environ. Microbiol. 69: 3368–3376.

Ranford, J.C., and Henderson, B. 2002. Chaperonins in disease: mechanisms, models, and treatments. J. Clin. Pathol.-Mol. Pathol. 55: 209–213.

Raveneau, J., Geoffroy, C., Beretti, J., Gaillard, J., Alouf, J., and Berche, P. 1992. Reduced virulence of a *Listeria monocytogenes* phospholipase-deficient mutant obtained by transposon insertion into the zinc metalloprotease gene. Infect. Immun. 60: 916–921.

Reglier-Poupet, H., Pellegrini, E., Charbit, A., and Berche, P. 2003. Identification of LpeA, a PsaA-like membrane protein that promotes cell entry by *Listeria monocytogenes*. Infect. Immun. 71: 474–482.

Reid, S.D., Montgomery, A.G., Voyich, J.M., DeLeo, F.R., Lei, B.F., Ireland, R.M., Green, N.M., Liu, M.Y., Lukomski, S., and Musser, J.M. 2003. Characterization of an extracellular virulence factor made by group A Streptococcus with homology to the *Listeria monocytogenes* internalin family of proteins. Infect. Immun. 71: 7043–7052.

Renzoni, A., Klarsfeld, A., Dramsi, S., and Cossart, P. 1997. Evidence that PrfA, the pleiotropic activator of virulence genes in *Listeria monocytogenes*, can be present but inactive. Infect. Immun. 65: 1515–1518.

Restaino, L., Frampton, E.W., Irbe, R.M., Spitz, H., and Schabert, G. 1999. Isolation and detection of *Listeria monocytogenes* using fluorogenic and chromogenic substrates for phosphatidylinositol-specific phospholipase C. J. Food Prot. 62: 244–251.

Revazishvili, T., Kotetishvili, M., Stine, O.C., Kreger, A.S., Morris, J.G., Jr., and Sulakvelidze, A. 2004. Comparative analysis of multilocus sequence typing and pulsed-field gel electrophoresis for characterizing *Listeria monocytogenes* strains isolated from environmental and clinical sources. J. Clin. Microbiol. 42: 276–285.

Ripio, M.-T., Dominguez-Bernal, G., Suarez, M., Brehm, K., Berche, P., and Vazquez-Boland, J.A. 1996. Transcriptional activation of virulence genes in wild-type strains of *Listeria monocytogenes* in response to a change in the extracellular medium composition. Res. Microbiol. 147: 371–384.

Roberts, A.J., and Wiedmann, M. 2003. Pathogen, host and environmental factors contributing to the pathogenesis of listeriosis. Cell. Mol. Life Sci. 60: 904–918.

Rodriguez-Lazaro, D., Hernandez, M., and Pla, M. 2004. Simultaneous quantitative detection of *Listeria* spp. and *Listeria monocytogenes* using a duplex real-time PCR-based assay. FEMS Microbiol. Lett. 233: 257–267.

Rossjohn, J., Feil, S.C., McKinstry, W.J., Tweten, R.K., and Parker, M.W. 1997. Structure of a cholesterol-binding, thiol-activated cytolysin and a model of its membrane form. Cell 89: 685–692.

Rouquette, C., de Chastellier, C., Nair, S., and Berche, P. 1998. The ClpC ATPase of *Listeria monocytogenes* is a general stress protein required for virulence and promoting early bacterial escape from the phagosome of macrophages. Mol. Microbiol. 27: 1235–1245.

Ruhland, G.J., Hellwig, M., Wanner, G., and Fiedler, F. 1993. Cell-surface location of *Listeria*-specific protein p60 detection of *Listeria* cells by indirect immunofluorescence. J. Gen. Microbiol. 139: 609–616.

Salcedo, C., Arreaza, L., Alcala, B., de la Fuente, L., and Vazquez, J.A. 2003. Development of a multilocus sequence typing method for analysis of *Listeria monocytogenes* clones. J. Clin. Microbiol. 41: 757–762.

Sansonetti, P.J., and Phalipon, A. 1999. M cells as ports of entry for enteroinvasive pathogens: mechanisms of interaction, consequences for the disease process. Semin. Immunol. 11: 193–203.

Santiago, N.I., Zipf, A., and Bhunia, A.K. 1999. Influence of temperature and growth phase on expression of a 104-kilodalton *Listeria* adhesion protein in *Listeria monocytogenes*. Appl. Environ. Microbiol. 65: 2765–2769.

Schirmer, E.C., Glover, J.R., Singer, M.A., and Lindquist, S. 1996. HSP100/Clp proteins: A common mechanism explains diverse functions. Trends Biochem. Sci. 21: 289–296.

Schlech, W.F., 3rd, Lavigne, P.M., Bortolussi, R.A., Allen, A.C., Haldane, E.V., Wort, A.J., Hightower, A.W., Johnson, S.E., King, S.H., Nicholls, E.S., and Broome, C.V. 1983. Epidemic listeriosis: evidence for transmission by food. N. Engl. J. Med. 308: 203–6.

Schlessinger, J. 2000. Cell signaling by receptor tyrosine kinases. Cell. 103: 211–25.

Schlüter, D., Domann, E., Buck, C., Hain, T., Hof, H., Chakraborty, T., and Deckert-Schlüter, M. 1998. Phosphatidylcholine-specific phospholipase C from *Listeria monocytogenes* is an important virulence factor in murine cerebral listeriosis. Infect. Immun. 66: 5930–5938.

Schuchat, A., Deaver, K.A., Wenger, J.D., Plikaytis, B.D., Mascola, L., Pinner, R.W., Reingold, A.L., and Broome, C.V. 1992. Role of foods in sporadic listeriosis. I. Case-control study of dietary risk factors. The *Listeria* Study Group. J. Am. Med. Assoc. 267: 2041–5.

Scotter, S.L., Langton, S., Lombard, B., Lahellec, C., Schulten, S., Nagelkerke, N., in't Veld, P.H., and Rollier, P. 2001a. Validation of ISO method 11290: Part 2. Enumeration of *Listeria monocytogenes* in foods. Int. J. Food Microbiol. 70: 121–129.

Scotter, S.L., Langton, S., Lombard, B., Schulten, S., Nagelkerke, N., In't Veld, P.H., Rollier, P., and Lahellec, C. 2001b. Validation of ISO method 11290 Part 1 – Detection of *Listeria monocytogenes* in foods. Int. J. Food Microbiol. 64: 295–306.

Seeliger, H.P.R., and Hohne, K. 1979. Serotyping of *Listeria monocytogenes* and related species. Meth. Microbiol. 13: 31–49.

Sellars, M.J., Hall, S.J., and Kelly, D.J. 2002. Growth of *Campylobacter jejuni* supported by respiration of fumarate, nitrate, nitrite, trimethylamine-N-oxide, or dimethyl sulfoxide requires oxygen. J. Bacteriol. 184: 4187–4196.

Sewell, A.M., Warburton, D.W., Boville, A., Daley, E.F., and Mullen, K. 2003. The development of an efficient and rapid enzyme linked fluorescent assay method for the detection of *Listeria* spp. from foods. Int. J. Food Microbiol. 81: 123–129.

Sheehan, B., Klarsfeld, A., Ebright, R., and Cossart, P. 1996. A single substitution in the putative helix-turn-helix motif of the pleiotropic activator PrfA attenuates *Listeria monocytogenes* virulence. Mol. Microbiol. 20: 785–797.

Shen, Y., Naujokas, K., Park, M., and Ireton, K. 2000. InlB-dependent internalization of *Listeria* is mediated by the Met receptor tyrosine kinase. Cell 103: 501–510.

Shetron-Rama, L.M., Mueller, K., Bravo, J.M., Bouwer, H.G.A., Way, S.S., and Freitag, N.E. 2003. Isolation of *Listeria monocytogenes* mutants with high-level *in vitro* expression of host cytosol-induced gene products. Mol. Microbiol. 48: 1537–1551.

Silbernagel, K., Jechorek, R., Barbour, W.M., Mrozinski, P., Alejo, W., *et al.* 2004. Evaluation of the BAX system for detection of *Listeria monocytogenes* in foods: collaborative study. J. Assoc. Off. Anal. Chem. Int. 87: 395–410.

Skoble, J., Portnoy, D.A., and Welch, M.D. 2000. Three regions within ActA promote Arp2/3 complex-mediated actin nucleation and *Listeria monocytogenes* motility. J. Cell Biol. 150: 527–537.

Skoble, J., Auerbuch, V., Goley, E.D., Welch, M.D., and Portnoy, D.A. 2001. Pivotal role of VASP in Arp2/3 complex-mediated actin nucleation, actin branch-formation, and *Listeria monocytogenes* motility. J. Cell Biol. 155: 89–100.

Smith, G.A., Marquis, H., Jones, S., Johnston, N.C., Portnoy, D.A., and Goldfine, H. 1995. The two distinct phospholipases C of *Listeria monocytogenes* have overlapping roles in escape from a vacuole and cell-to-cell spread. Infect. Immun. 63: 4231–4237.

Smith, J.L. 1999. Foodborne infections during pregnancy. J. Food Prot. 62: 818–829.

Smith, P.A., Mellors, D., Holroyd, A., and Gray, C. 2001. A new chromogenic medium for the isolation of *Listeria* spp. Lett. Appl. Microbiol. 32: 78–82.

Snyder, A., and Marquis, H. 2003. Restricted translocation across the cell wall regulates secretion of the broad-range phospholipase C of *Listeria monocytogenes*. J. Bacteriol. 185: 5953–5958.

Soltys, B.J., and Gupta, R.S. 1997. Cell surface localization of the 60 kDa heat shock chaperonin protein (hsp60) in mammalian cells. Cell Biol. Int. 21: 315–320.

Stephan, R., Schumacher, S., and Zychowska, M.A. 2003. The VIT(R) technology for rapid detection of *Listeria monocytogenes* and other *Listeria* spp. Int. J. Food Microbiol. 89: 287–290.

Stewart, D., and Gendel, S.M. 1998. Specificity of the BAX polymerase chain reaction system for detection of the foodborne pathogen *Listeria monocytogenes*. J. Assoc. Off. Anal. Chem. Int. 81: 817–22.

Tilney, L., and Portnoy, D. 1989. Actin filaments and the growth, movement, and spread of the intracellular bacterial parasite, *Listeria monocytogenes*. J. Cell Biol. 109: 1597–1608.

Trémoulet, F., Duché, O., Namane, A., Martinie, B., The European *Listeria* Genome Consortium, and Labadie, J.C. 2002. Comparison of protein patterns of *Listeria monocytogenes* grown in biofilm or in planktonic mode by proteomic analysis. FEMS Microbiol. Lett. 210: 25–31.

Tweten, R.K., Parker, M.W., and Johnson, A.E. 2001. The cholesterol-dependent cytolysins. In: Pore-Forming Toxins. F. G. van der Goot, ed. Springer-Verlag, New York, NY, pp. 15–33.

van Netter, P., Perales, I., van de Moosdijk, A., Curtis, G.D.W., and Mossel, D.A.A. 1989. Liquid and solid selective differential media for the detection and enumeration of *L. monocytogenes* and other Listera spp. Int. J. Food Microbiol. 8: 299–316.

Vazquez-Boland, J.A., Kocks, C., Dramsi, S., Ohayon, H., Geoffroy, C., Mengaud, J., and Cossart, P. 1992. Nucleotide sequence of the lecithinase operon of *Listeria monocytogenes* and possible role of lecithinase in cell-to-cell spread. Infect. Immun. 60: 219–230.

Vazquez-Boland, J.A., Kuhn, M., Berche, P., Chakraborty, T., Dominguez-Bernal, G., Goebel, W., Gonzalez-Zorn, B., Wehland, J., and Kreft, J. 2001. *Listeria* pathogenesis and molecular virulence determinants. Clin. Microbiol. Rev. 14: 584–640.

Vaz-Velho, M., Duarte, G., and Gibbs, P. 2000. Evaluation of mini-VIDAS rapid test for detection of *Listeria monocytogenes* from production lines of fresh to cold-smoked fish. J. Microbiol. Methods 40: 147–151.

Volokhov, D., Rasooly, A., Chumakov, K., and Chizhikov, V. 2002. Identification of *Listeria* Species by Microarray-Based Assay. J. Clin. Microbiol. 40: 4720–4728.

Walker, S.J., Archer, P., and Banks, J.G. 1990. Growth of *Listeria monocytogenes* at refrigeration temperatures. J. Appl. Bacteriol. 68: 157–62.

Wallin, R.P.A., Lundqvist, A., More, S.H., von Bonin, A., Kiessling, R., and Ljunggren, H.G. 2002. Heat shock proteins as activators of the innate immune system. Trends Immunol. 23: 130–135.

Wampler, J.L., Kim, K.P., Jaradat, Z., and Bhunia, A.K. 2004. Heat shock protein 60 acts as a receptor for the *Listeria* adhesion protein in Caco-2 cells. Infect. Immun. 72: 931–936.

Weeks, M.E., James, D.C., Robinson, G.K., and Smales, C.M. 2004. Global changes in gene expression observed at the transition from growth to stationary phase in *Listeria monocytogenes* ScottA batch culture. Proteomics 4: 123–135.

Welch, M.D., Rosenblatt, J., Skoble, J., Portnoy, D.A., and Mitchison, T.J. 1998. Interaction of human Arp2/3 complex and the *Listeria monocytogenes* ActA protein in actin filament nucleation. Science 281: 105–108.

WHO. 2000. Outbreak of listeriosis linked to the consumption of pork tongue in jelly in France. WHO surveillance progamme for control of foodborne infections and intoxicants in Europe. 64: 3.

Wiedmann, M., Bruce, J., Keating, C., Johnson, A., McDonough, P., and Batt, C. 1997. Ribotypes and virulence gene polymorphisms suggest three distinct *Listeria monocytogenes* lineages with differences in pathogenic potential. Infect. Immun. 65: 2707–2716.

Wuenscher, M.D., Kohler, S., Bubert, A., Gerike, U., and Goebel, W. 1993. The iap gene of *Listeria monocytogenes* is essential for cell viability, and its gene product, p60, has bacteriolytic activity. J. Bacteriol. 175: 3491–3501.

Zenewicz, L.A., Skinner, J.A., Goldfine, H., and Shen, H. 2004. *Listeria monocytogenes* virulence proteins induce surface expression of Fas ligand on T lymphocytes. Mol. Microbiol. 51: 1483–1492.

Zhang, C., Zhang, M., Ju, J., Nietfeldt, J., Wise, J., Terry, P.M., Olson, M., Kachman, S.D., Wiedmann, M., Samadpour, M., and Benson, A.K. 2003. Genome diversification in phylogenetic lineages I and II of *Listeria monocytogenes*: identification of segments unique to lineage II populations. J. Bacteriol. 185: 5573–5584.

Zhang, W., Jayarao, B.M., and Knabel, S.J. 2004. Multi-virulence-locus sequence typing of *Listeria monocytogenes*. Appl. Environ. Microbiol. 70: 913–920.

# Salmonella Species

Helene L. Andrews and Andreas J. Bäumler

16

## Abstract

*Salmonella* serotypes continue to be a prominent threat to food safety in the United States. Infections are commonly acquired by animal to human transmission though consumption of undercooked food products derived from livestock or domestic fowl. The second half of the twentieth century saw the emergence of *Salmonella* serotypes that became associated with new food sources (i.e. chicken eggs) and the emergence of *Salmonella* serotypes with resistance against multiple antibiotics. This review provides an overview over epidemiology, pathogenesis and control of this important foodborne disease in the United States.

## Characteristics and taxonomy

### Characteristics of the genus

The genus *Salmonella* is named after the American veterinarian Daniel E. Salmon, who first described Bacterium *choleraesuis* (now called *Salmonella enterica* serotype *Choleraesuis*) as an isolate from pigs in 1885 (Salmon and Smith, 1885). The members of this genus are gram negative facultative-anaerobic and peritrichously flagellated rods. *Salmonella* serotypes are distinguished from members of other genera of the *Enterobacteriaceae* by a series of biochemical reactions including the production of hydrogen sulfide, the reduction of nitrates, the decarboxylation of lysine and ornithine and the utilization of citrate.

### Serotyping

Members of the genus *Salmonella* are traditionally classified into serotypes based on expression of two surface antigens, lipopolysaccharide (LPS) and flagella. Typing sera recognizing H (flagellar) antigens are directed against a protein, termed flagellin, the major component of the flagellar filament. Many members of the genus *Salmonella* contain two flagellin genes encoding two distinct flagellins, termed H1 and H2 antigen. Flagellin expression in *S. enterica* serotype Typhimurium oscillates between H1 and H2 expression states (Andrews, 1922). This observation represents the first description of the phenomenon of phase variation (Andrews, 1922). Flagellar phase variation is mediated by a heritable molecular switching mechanism composed of a promoter located on an invertible DNA controlling element. If the invertible DNA element is in the phase H2 orientation the promoter drives the expression of two open reading frames, one encoding H2 phase flagellin and the second encoding a repressor of phase H1 flagellin expression (H2 phase). Inversion of the controlling DNA element silences both open reading frames, resulting in expression of H1 flagellin (H1 phase) (Silverman and Simon, 1980).

Typing sera specific for O (somatic) antigen recognize epitopes in the O-antigen repeat unit of LPS. For example, *S.* Typhimurium expresses the O-antigens 1, 4, 5 and 12, the H1-antigen i and the H2-antigens 1 and 2, and is thus defined by its antigenic formula (1,4,5,12:i:1,2). A repeat unit of the *S.* Typhimurium O-antigen consists of a trisaccharide backbone (representing the O12-antigen) carrying two branching sugars, the O1 and O4 antigens (Figure 16.1). The glucose branch represents the O1 antigen and the abequose branch represents the O4 antigen. The O5 antigen is an O-acetylation of the abequose residue. Based on variations in the O-antigen of *S.* Typhimurium several serological variants can be differentiated within this serotype. The O1 antigen is not expressed by all *S.* Typhimurium isolates and can be transferred by lysogenic conversion with bacteriophage P22 (Zinder, 1957; Stocker, 1958; Staub and Girard, 1965). *S.* Typhimurium isolates may also lack the O5-antigen, a serological variation known as var. Copenhagen (Kauffmann, 1934). Thus, the antigenic formulas (1,4,5,12:i:1,2), (1,4,12:i:1,2), (4,5,12:i:1,2) and (4,12:i:1,2) are serological variants of a single serotype: *S.* Typhimurium.

### Serogrouping

When antiserum is raised against LPS from one serotype, one epitope ordinarily elicits the production of high-affinity antibodies and is termed the immunodominant antigen (Lüderitz et al., 1966). The O4 antigen (i.e. the abequose branch) is the immunodominant antigen of the *S.* Typhimurium O-antigen repeat unit (Figure 16.1). This immunodominant antigen is used to classify *Salmonella* serotypes into serogroups. All serotypes expressing the O4 antigen, such as *S.* Typhimurium

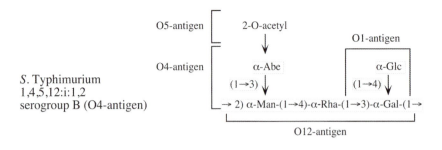

**Figure 16.1** Chemical composition of the O-polysaccharide repeating unit of LPS from *S.* Typhimurium (top) and *S.* Enteritidis (bottom). Abe, abequose; Gal, galactose; Glc, glucose; Man, manose; Rha, rhamnose; Tyv, tyvelose.

or *S. enterica* serotype Heidelberg, belong to serogroup B. The O-antigen of *S. enterica* serotypes Enteritidis, Typhi and Gallinarum, on the other hand, contain the O9 antigen, which is the immunodominant antigen of serogroup D1 (Figure 16.1). Serogrouping is of practical importance since the majority (95.7%) of isolates from humans or food animals are members of only five serogroups, including B (O4-antigen), C1 (O7-antigen), C2 (O8-antigen), D1 (O9-antigen) and E1 (O10-antigen) (Kelterborn, 1967). Therefore, serogrouping performed with only 5 antisera (i.e. antisera specific for serogroups B, C1, C2, D1 and E1) can be used for confirming the biochemical identification of a member of the genus *Salmonella* with a reasonable degree of confidence.

## Phage typing

To further differentiate between isolates of the same serotype for epidemiologic analysis, phage typing schemes have been developed for the most common serotypes. Phage typing distinguishes different isolates of the same serotype based on their susceptibility to a set of bacteriophages. *S.* Typhimurium isolates are commonly characterized by a phage typing system developed by Anderson, that distinguishes more than 200 definitive phage types (DT) (Anderson *et al.*, 1977). The phage type is a stable property, that provides an opportunity to trace the spread of an epidemic clone through the food chain and within its animal reservoirs (Anderson *et al.*, 1978).

## Current nomenclature

In the original typing scheme developed by Kauffmann, each different serotype was considered to be equivalent with a new species defined as "a group of related sero-fermentative phage types" and was designated by a latin binomial (e.g. *Salmonella typhimurium*) (Kauffmann, 1960). However, due to the close genetic relatedness of *Salmonella* serotypes (Crosa *et al.*, 1973), the one serotype–one species concept was later abandoned in favor of a classification system that recognizes only two species in the genus *Salmonella* (Reeves *et al.*, 1989). This nomenclature is still evolving (Brenner *et al.*, 2000). The system currently in use by the Centers for Disease Control and Prevention (CDC) groups the 2,463 *Salmonella* serotypes in two species, *Salmonella bongori* (containing 20 different serotypes) and *Salmonella enterica* (containing 2,443 different serotypes). *S. enterica* is further subdivided into six subspecies that are referred to by a roman numeral or a name (subsp. I or subsp. *enterica*, subsp. II or subsp. *salamae*, subsp. IIIa or subsp. *arizonae*, subsp. IIIb or subsp. *diarizonae*, subsp. IV or subsp. *houtenae*, subsp. VI or subsp. *indica*) (Figure 16.2) (Brenner *et al.*, 2000). According to this nomenclature, the Kauffmann species *Salmonella typhimurium* is now referred to as *Salmonella enterica* subsp. I serotype Typhimurium. However, the subspecies designation is commonly omitted when referring to a serotype (e.g. *S. enterica* serotype Typhimurium instead of *S. enterica* subsp. I serotype Typhimurium) and a short form (i.e. *S.* Typhimurium) is ordinarily used as an abbreviation.

## Characteristics of the disease, mechanisms of pathogenesis and virulence factors

### Human disease syndromes

Members of *S. enterica* subsp. I account for more than 99% of human cases of disease (Figure 16.2) (Aleksic *et al.*, 1996) and are associated with three distinct clinical syndromes, typhoid fever, bacteremia and enterocolitis.

Typhoid fever is a systemic infection caused by strictly human-adapted serotypes, including *S. enterica* serotypes Typhi, Paratyphi A, Paratyphi B, and Paratyphi C. Patients develop fever within 5–9 days post infection and about one-third may subsequently develop diarrhea (Hornick *et al.*, 1970). Bacteria

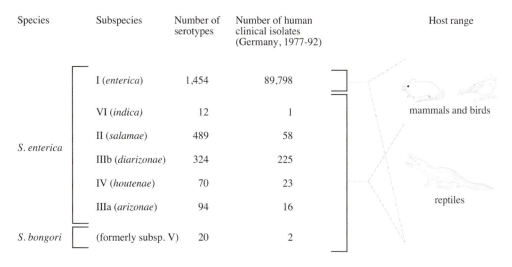

| Species | Subspecies | Number of serotypes | Number of human clinical isolates (Germany, 1977-92) | Host range |
|---|---|---|---|---|
| *S. enterica* | I (*enterica*) | 1,454 | 89,798 | mammals and birds |
| | VI (*indica*) | 12 | 1 | |
| | II (*salamae*) | 489 | 58 | |
| | IIIb (*diarizonae*) | 324 | 225 | |
| | IV (*houtenae*) | 70 | 23 | |
| | IIIa (*arizonae*) | 94 | 16 | reptiles |
| *S. bongori* | (formerly subsp. V) | 20 | 2 | |

**Figure 16.2** Current taxonomy of the genus *Salmonella* used by the CDC. The designation of species and subspecies is shown on the left. The number of *Salmonella* serotypes distinguished within each species and subspecies has been reported previously (Brenner *et al.*, 2000) and is shown in the adjacent column. The frequency of isolating serotypes belonging to each species and subspecies from cases of human disease in Germany has been reported previously (Aleksic *et al.*, 1996) and is shown in the adjacent column. The host range of *Salmonella* serotypes has been reported previously (Popoff and Le Minor, 1992) and is shown on the right.

spread via the circulation and can be cultured from internal organs that become enlarged and granulomatous (Levy and Gaehtgens, 1908). A diffuse infiltrate dominated by monocytes develops within 3 days post infection in the ileum (Sprinz *et al.*, 1966; Kraus *et al.*, 1999). In the absence of antibiotic treatment Peyer's patches of the terminal ileum may become necrotic within 2 weeks after infection, leading to intestinal perforation that is usually observed in the third week (Bitar and Tarpley, 1985).

Bacteremia is a systemic infection caused by the porcine-adapted *S.* Choleraesuis and the bovine-adapted *S. enterica* serotype Dublin (Saphra and Wassermann, 1954; Werner *et al.*, 1979; Fang and Fierer, 1991). Both organisms are highly invasive causing a high spiking fever and septicemia often without local manifestations (Saphra and Wassermann, 1954; Cherubin *et al.*, 1969; Fang and Fierer, 1991). Only about one-third of bacteremia patients develop diarrhea and, in contrast to typhoid fever patients, enteric pathology is rarely encountered (Saphra and Wassermann, 1954).

Enterocolitis is a localized infection caused by a large number of *Salmonella* serotypes, however, *S. Typhimurium* and *S. Enteritidis* are encountered most frequently. Enterocolitis patients characteristically develop nausea, vomiting and diarrhea within 12 to 72 hours after infection. An acute inflammation with predominantly polymorphonuclear leukocytes (PMN) in the terminal ileum and colon leads to necrosis of the upper mucosa (Day *et al.*, 1978; McGovern and Slavutin, 1979). Septicemia is uncommon and transient in this syndrome(Mandal and Brennand, 1988), but may develop as a complication that is more likely to occur in infants, elderly or immuno-compromised patients (Altekruse *et al.*, 1994; Ruiz-Contreras *et al.*, 1995; Tumbarello *et al.*, 1995; Kovacs *et al.*, 1997).

## Models to study human disease syndromes

Mechanisms of *Salmonella* pathogenesis can be elucidated using a variety of animal and cell culture models (Santos *et al.*, 2001b). The perhaps most widely used animal model to study virulence of *Salmonella* serotypes is *S.* Typhimurium infection of mice. The distribution of bacteria in organs and the pathology in the intestine observed during *S.* Typhimurium infection of mice closely resemble typhoid fever (Santos *et al.*, 2001b). An obvious shortcoming of this typhoid fever model is that *S.* Typhimurium is associated with a different disease syndrome in humans, namely enterocolitis. Furthermore, it has been argued that *S.* Typhimurium infection of mice may represent an animal model to study bacteremia rather than typhoid fever (Fierer and Guiney, 2001). Despite these shortcomings, the mouse continues to be an important model because the only alternative to study typhoid fever in an animal model is *S.* Typhi infection of higher primates (e.g. chimpanzees) (Geoffrey *et al.*, 1960). Because mice infected with *S.* Typhimurium do not develop diarrhea, this model is not suited to study the pathogenesis of enterocolitis. Instead, the pathogenesis of enterocolitis can be studied using one of several alternative animal models, including experimental infection of rhesus monkeys, calves or rabbit ligated ileal loops (Zhang *et al.*, 2003). Tissue culture models have been developed to study isolated aspects of *Salmonella* pathogenesis, such as invasion of the intestinal epithelium (Finlay, 1988; Galán and Curtiss, 1989), induction of PMN influx (McCormick *et al.*, 1995) or survival in macrophages (Fields *et al.*, 1986).

## Virulence factors

The availability of good disease models in combination with the development of elegant methods to genetically manipulate *Salmonella* serotypes has led to the identification of a large number of virulence determinants (Fields *et al.*, 1986; Gulig and Curtiss, 1987; Galán and Curtiss, 1989; Mahan, 1993; Hensel *et al.*, 1995; Valdivia and Falkow, 1997). Broadly defined, any factor may be considered to be a virulence determinant that is required by a *Salmonella* serotype to be a successful pathogen. That is, any determinant that enables a *Salmonella* serotype to enter a host, to find a unique niche to multiply, to avoid or subvert the host defenses, to cause disease and to be transmitted to the next susceptible host may be considered a virulence determinant. With the exception of the ability to cause disease, these properties are also characteristic of bacteria present in the normal flora. Therefore, it is not surprising that many genes absolutely required for virulence of *Salmonella* serotypes in an animal model are also present in non-pathogenic relatives, such as commensal *Escherichia coli*. This group of virulence determinants is of considerable importance as it includes a number of genes whose inactivation provides optimal attenuation for converting a *Salmonella* serotype into a live attenuated vaccine strain, including *aroA, phoP, htrA, dam, ompR* and *cya/crp* (Hoiseth and Stocker, 1981; Curtiss *et al.*, 1988; Dorman *et al.*, 1989; Strahan *et al.*, 1992; Miller *et al.*, 1993; Heithoff *et al.*, 1999). But to understand what makes *Salmonella* serotypes different from closely related pathogens or commensal organisms, it is useful to apply a narrower definition of virulence determinants. According to the molecular Koch's postulates, the definition of virulence genes can be narrowed down to those that are "associated with pathogenic members of a genus or pathogenic strains of a species" (Falkow, 1988). By this definition, these *Salmonella*-specific virulence determinants should be absent from closely related commensal organisms, such as *E. coli* K12. Genetic elements harboring such *Salmonella*-specific virulence determinants include plasmids, bacteriophages, pathogenicity islands, and pathogenicity islets (Kingsley and Baumler, 2002).

## Pathogenesis of enterocolitis

Mechanisms by which *S.* Typhimurium causes enterocolitis have recently been elucidated using tissue culture models and the neonatal calf model. These studies suggest that the ability of *S.* Typhimurium to elicit severe PMN infiltration in the intestinal mucosa is important for causing disease and contributes to fluid loss during diarrhea (Zhang *et al.*, 2003). The *Salmonella*-specific virulence determinant most important for causing diarrhea and PMN influx in the intestinal mucosa is the invasion-associated type III secretion system (TTSS-1) (Galán and Curtiss, 1989; Hobbie *et al.*, 1997; Watson *et al.*, 1998; Wood *et al.*, 1998; Ahmer *et al.*, 1999; Tsolis *et al.*, 1999a; Tsolis *et al.*, 1999b). The TTSS-1 was originally identified as a factor required for entry of *S. Typhimurium* into intestinal epithelial cell lines (Galán and Curtiss, 1989). Elegant work with cell culture models shows that the main function of this secretion system is the delivery of proteins, termed effectors, into host cells (Galán, 2001). While genes encoding the TTSS-1 are clustered on *Salmonella* pathogenicity island 1 (SPI-1) (Mills *et al.*, 1995), genes encoding secreted effectors

lie scattered around the genome on bacteriophages (Hardt *et al.*, 1998b; Figueroa-Bossi *et al.*, 2001), pathogenicity islets (Tsolis *et al.*, 1999b) or pathogenicity islands (Hueck *et al.*, 1995; Kaniga *et al.*, 1995; Hardt and Galán, 1997; Wood *et al.*, 1998). Six TTSS-1 secreted effectors, including SipA (SspA), SopA, SopB (SigD), SopD, SopE and SopE2, are required for eliciting fluid accumulation and PMN influx in bovine ligated ileal loops (Galyov *et al.*, 1997; Jones *et al.*, 1998; Wood *et al.*, 2000; Santos *et al.*, 2001a; Zhang *et al.*, 2002a; Zhang *et al.*, 2002b). Several of these effectors, including SopB, SipA, SopE and SipE2, elicit nuclear responses in epithelial cell lines resulting in the production of proinflammatory signals that may be involved in attracting PMN (Hardt *et al.*, 1998a; Lee *et al.*, 2000; Steele-Mortimer *et al.*, 2000; Stender *et al.*, 2000; Friebel *et al.*, 2001). Experiments using the calf model suggest that the increase in vascular permeability accompanying PMN extravasation in conjunction with PMN-mediated injury to the intestinal epithelium results in the flow of liquid and solutes from the blood to the intestinal lumen, thereby contributing to the severe fluid loss observed during *S.* Typhimurium-induced diarrhea (Santos *et al.*, 2001a; Santos *et al.*, 2002). SopB hydrolyzes phosphatidylinositol 3,4,5-trisphosphate, an inhibitor of chloride secretion, suggesting that a secretory mechanism may also contribute to fluid loss (Norris *et al.*, 1998).

## Pathogenesis of systemic disease

Mechanisms involved in causing systemic infections (i.e. typhoid fever or bacteremia) have been elucidated using tissue culture models and one animal model, the mouse. The ability of *S.* Typhimurium to survive and multiply in macrophages is of central importance for causing lethal systemic infection in mice (Fields *et al.*, 1986). *Salmonella*-specific virulence determinants contributing to this ability include genes that confer the ability to resist macrophage killing mechanisms (*sodC1*), to obtain nutrients in the intracellular environment (*mgtBC*), to interfere with actin polymerization (*spvB*) and to alter vesicular trafficking (SPI-2) (Roudier *et al.*, 1990; Ochman *et al.*, 1996; Blanc-Potard and Groisman, 1997; DeGroote *et al.*, 1997; Gulig *et al.*, 1998). The sodCI gene is located on a bacteriophage and encodes a periplasmic superoxide dismutase that is required for full mouse virulence, macrophage survival and protection of *S.* Typhimurium from phagocyte-derived reactive oxygen and reactive nitrogen species (DeGroote *et al.*, 1997; Fang *et al.*, 1999). The genes mgtB and mgtC are located on SPI-3 and encode a high-affinity transport system required for sequestering magnesium during growth in macrophages and for full virulence in mice (Blanc-Potard and Groisman, 1997; Blanc-Potard *et al.*, 1999). The *spvRABCD* operon is located on large virulence plasmids present in some *S. enterica* subsp. I serotypes, including *S.* Typhimurium, *S.* Enteritidis, *S.* Dublin, *S.* Choleraesuis, *S.* Paratyphi C, *S.* Abortusovis and *S.* Gallinarum (Popoff *et al.*, 1984; Poppe *et al.*, 1989; Woodward, 1989; Roudier *et al.*, 1990). SpvB encodes a mono(ADP ribosyl)transferase which ADP-ribosylates actin, thereby interfering with actin polymerization and contributing to full mouse virulence (Otto *et al.*, 2000; Lesnick *et al.*, 2001; Tez-can-Merdol *et al.*, 2001). Finally, SPI-2 encodes a second type III secretion system (TTSS-2) present in *S. enterica* serotypes (Hensel *et al.*, 1995; Ochman and Groisman, 1996; Ochman *et al.*, 1996; Hensel *et al.*, 1997). Inactivation of the TTSS-2 causes a complete loss of *S.* Typhimurium virulence for mice (Hensel *et al.*, 1995) and a reduction in its ability to survive in murine macrophages *in vitro* (Ochman *et al.*, 1996). Several effectors that are translocated into host cells by the TTSS-2 have been identified, including SifA, SifB, SlrP, SseI, SseJ, SspH2, SopD2, and PipB (Stein *et al.*, 1996; Miao *et al.*, 1999; Tsolis *et al.*, 1999b; Uchiya *et al.*, 1999; Miao and Miller, 2000; Worley *et al.*, 2000; Brumell *et al.*, 2001; Knodler *et al.*, 2002; Brumell *et al.*, 2003). Although the precise mechanism by which TTSS-2 allows *S.* Typhimurium to thrive within mac-rophages remains to be worked out, current evidence suggests that interference with intracellular trafficking plays an impor-tant role in this process. That is, the TTSS-2 allows *S. Ty-phimurium* to avoid macrophage-derived reactive oxygen spe-cies by interfering with the trafficking of NADPH oxidase-containing vesicles to the *Salmonella* containing phagosome (SCV) (Vazquez-Torres *et al.*, 2000). The TTSS-2 is required for inhibiting fusion of the SCV with lysosomes and endo-somes in murine macrophage cell lines (Uchiya *et al.*, 1999; Yu *et al.*, 2002). Furthermore, several TTSS-2 effectors, including SifA, SopD2 and PipB, localize to the SCV in infected host cells *in vitro* (Brumell *et al.*, 2002; Knodler *et al.*, 2002; Bru-mell *et al.*, 2003). One of these effectors, SifA, associates with microtubules at sites of interaction between lysosomal glyco-protein-containing vesicles and the SCV, thereby suggesting a role in vesicular trafficking (Brumell *et al.*, 2002).

## Epidemiology, vehicles of infection, and outbreaks

### Importance of human disease syndromes
*Salmonella* serotypes are the leading cause of foodborne disease outbreaks with known etiology in the United States (Olsen *et al.*, 2000). Non-typhoidal *Salmonella* serotypes are also the single most common cause of death from foodborne illnesses associated with viruses, parasites or bacteria in the United States, causing an estimated 550 fatal cases each year (Mead *et al.*, 1999). *Salmonella*-induced enterocolitis is furthermore the second most common cause of bacterial foodborne disease of known etiology in the United States. According to a recent CDC estimate, *Salmonella* serotypes cause approximately 1.4 million cases of illness annually (Mead *et al.*, 1999), thereby producing between $0.5 billion to $2.3 billion in annual costs for medical care and lost productivity (Frenzen *et al.*, 1999). In contrast, typhoid fever and bacteremia are currently rare in the United States and will thus be discussed in less detail. How-ever, typhoid fever is still a significant disease in Asia, Africa and South America, where an estimated 16 million illnesses

and approximately 600,000 deaths occur annually (Merican, 1997).

### Outbreak investigations
Typhoid fever is transmitted from human to human through food or water. In contrast, non typhoidal salmonellosis is com-monly transmitted from animals to humans via food. Knowl-edge about the sources of infection is derived to a large part from outbreak investigations, since it is often difficult to iden-tify vehicles of infection from isolated cases. Outbreaks often provide the opportunity to trace their origin through the food chain directly to an animal reservoir. The first time a source of an foodborne outbreak was identified in this fashion is the analysis by Gärtner of 57 cases of meat poisoning occurring in Frankenhausen, Germany (Gärtner, 1888). Gärtner reported that a cow developed acute diarrhea on May 7th, 1888, and was slaughtered on the evening of May 9th. A veterinarian in-spected the carcass and noted redness (hemorrhage) in some areas of the small intestine. However, no spleno- or hepato-megaly was observed and the smell and appearance of the meat were normal. After the veterinarian declared the meat fit for consumption, it was sold on May 11th. By May 18th, 93 peo-ple had consumed meat from the infected cow; 57 developed symptoms, including nausea, vomiting and diarrhea, and one died. While only 45 of 81 people who had consumed cooked meat or soup became ill, all 12 persons who had consumed raw meat developed disease. The deceased, a 21-year-old healthy worker, had consumed the largest amount of raw meat (ap-proximately 800 g), heavily seasoned with salt and pepper. He fell ill with diarrhea and vomiting 2 hours later in the evening of May 11th and died in the morning of May 13th. His moth-er was the only person in this outbreak that had not consumed meat from the cow but developed the same symptoms. Gärt-ner states that the bed linens of the deceased were heavily con-taminated with liquid feces that could have been a source of infection for his mother. The autopsy of the deceased revealed pathological changes in the terminal ileum and changes in the consistency of his blood. Gärtner isolated a previously uniden-tified organism from meat of the cow and the spleen of the de-ceased, that he termed *Bacillus enteritidis* (now *S.* Enteritidis). Gärtner ended his investigation by showing that the isolated organism caused an enteric disease in mice, guinea-pigs, rab-bits and goats (Gärtner, 1888).

Since the seminal report by Gärtner, numerous outbreak investigations have traced infections with non-typhoidal *Salmonella* serotypes back to their food vehicles. The food ve-hicles most commonly implicated in outbreaks of enterocoli-tis in the United States today are chicken eggs, chicken, beef and turkey (Olsen *et al.*, 2000). Only five *S. enterica* subsp. I serotypes accounted for 61% of human cases reported to CDC between 1987 and 1997 (Olsen *et al.*, 2001). These in-clude *S.* Typhimurium (23% of human cases), *S.* Enteritidis (21% of human cases), *S.* Heidelberg (8% of human cases), *S.* Newport (5% of human cases), and *S.* Hadar (4% of hu-man cases). Each of these serotypes is associated with one or several of the food vehicles most commonly implicated in outbreaks. S. Typhimurium is the serotype most frequently isolated from beef carcasses and the third most common isolate from chicken carcasses and raw ground beef in the United States (Schlosser et al., 2000). The majority of S. Enteritidis infections are caused by consumption of raw or undercooked chicken eggs (Angulo and Swerdlow, 1998; St. Louis et al., 1988). S. Heidelberg is the Salmonella serotype most frequently isolated from chicken carcasses and raw ground chicken and the second most common isolate from turkey carcasses in the United States (Schlosser et al., 2000). Outbreaks caused by S. Newport are frequently traced back to beef originating from dairy cattle (Fontaine et al., 1978; Spika et al., 1987; Zansky et al., 2002). Finally, S. Hadar is the serotype most frequently cultured from turkey carcasses and ground raw turkey in the United States (Schlosser et al., 2000).

## Recent changes in the epidemiology of non-typhoidal salmonellosis

The two most important changes in the epidemiology of non-typhoidal salmonellosis in recent decades are the emergence S. Enteritidis as an egg associated pathogen and the emergence of multiple antibiotic-resistant strains (Rabsch et al., 2001). The first report of an outbreak caused by a multidrug-resistant Salmonella serotype is the description by Anderson of an S. Typhimurium epidemic among cattle and humans that occurred in England and Wales between 1964 and 1966 (Anderson, 1968). The 1970s and 1980s saw a slow increase in the incidence of multiple antibiotic-resistant strains (Threlfall et al., 1993). However, the fraction of multidrug-resistant isolates rose considerably in the 1990s with the appearance of S. Typhimurium phage type DT104 (Threlfall et al., 1997; Glynn et al., 1998; Akkina et al., 1999; Threlfall et al., 2000). Unlike previous multi drug-resistant isolates, DT104 carries antibiotic resistance determinants encoded on its chromosome (Threlfall et al., 1994). Integration into the chromosome increases the probability that these genetic determinants are maintained in the absence of antibiotic selection which may have contributed to the rapid increase in the fraction of multiple antibiotic resistant isolates observed in the 1990s. More recently, the incidence of infections caused by multi drug-resistant S. Newport, has increased considerably in the United States (Zansky et al., 2002). The emergence of multiple antibiotic-resistant strains is likely a consequence of the use of antimicrobial agents in agriculture (Anderson, 1968; Cohen and Tauxe, 1986; Angulo and Griffin, 2000), that was initiated in the 1940s and 1950s (Smith and Crabb, 1956; 1957).

S. Enteritidis was not isolated from foods containing raw or undercooked eggs in the United States in the first half of the 20th century (Edwards and Bruner, 1943). Since the 1960s, S. Enteritidis steadily increased in frequency in the United States from being the sixth most common serotype to becoming the serotype most frequently isolated from humans in 1990 (Aserkoff et al., 1970; Mishu et al., 1994). The incidence of

human S. Enteritidis cases also increased in other countries in Europe and the Americas during this time period (Rodrigue et al., 1990). In the United States the epidemic peaked in 1995 and since then the incidence of human S. Enteritidis infections has decreased slowly (Olsen et al., 2001). During this epidemic, the majority of S. Enteritidis outbreaks in Europe and the Americas were associated with a new food source, namely foods containing raw or undercooked chicken eggs (Coyle et al., 1988; St. Louis et al., 1988; Cowden et al., 1989a; Cowden et al., 1989b; Henzler et al., 1994). Analysis of outbreaks in Germany reveals an association of S. Enteritidis with foods containing chicken eggs as early as 1974 (Rabsch et al., 2001). Retrospective analysis of epidemiological data suggests that the emergence of S. Enteritidis as an egg-associated pathogen was triggered by the elimination in the 1970s of a competitor, S. Gallinarum, from poultry populations in the United States and Europe (Bäumler et al., 2000; Rabsch et al., 2000).

## Reservoirs

Salmonella serotypes can be isolated from a variety of vertebrates, including reptiles, mammals and birds. While members of S. bongori and S. enterica subsp. II–VI are almost exclusively isolated from reptiles (Iveson et al., 1969; Habermalz and Pietzsch, 1973; Roggendorf and Muller, 1976; Chiodini and Sundberg, 1981; Woodward et al., 1997), S. enterica subsp. I serotypes are also frequently isolated from a variety of mammalian and avian species (Figure 16.2). The presence of S. enterica subsp. I serotypes in livestock and domestic fowl results in their subsequent introduction into the derived food products, that are also the most common vehicles implicated in outbreaks among humans (Olsen et al., 2000). An exception to this are the typhoidal S. enterica subsp. I serotypes, including S. Typhi, S. Paratyphi A, and S. Paratyphi B, which are associated exclusively with a human reservoir. The scarcity of S. bongori and S. enterica subsp. II–VI serotypes within populations of mammals and birds, on the other hand, is likely responsible for the fact that these pathogens account for less than 1% of human clinical Salmonella isolations (Figure 16.2) (Aleksic et al., 1996).

Numerous studies suggest that the most important mechanism for persistence of S. enterica subspecies I serotypes within populations of livestock and domestic fowl is animal to animal transmission through fecal contamination of the environment. Housing large numbers of animals in confined areas, extensive animal trade and insufficient decontamination of holding areas can compound this problem (Stevens et al., 1967; Anderson, 1968; Threlfall et al., 1978, 1979; Davies, 1979; Wray et al., 1987; McLaren and Wray, 1991; Wray et al., 1991; Dahl et al., 1997; Davies et al., 1997; Hurd et al., 2001). In addition, vertical transmission through milk (Giles et al., 1989), contamination with rodent feces (Sato et al., 1970; Krabisch and Dorn, 1980; Henzler and Opitz, 1992; Kinde et al., 1996; Guard-Petter et al., 1997) or fecal contamination of animal feed (Crump et al., 2002) may occasionally contribute to the spread of infection. The importance of fecal contamination for the spread of *Salmonella* serotypes among poultry and livestock illustrates that persistent intestinal carriage of the organism is highly relevant for food safety. However, the molecular mechanisms allowing *S. enterica* subsp. I serotypes to persist in the intestines of healthy food animals are poorly understood. Recent work suggests that adherence factors, such as the non-fimbrial adhesin ShdA, may be important for this trait (Kingsley *et al.*, 2000; Kingsley *et al.*, 2002; Kingsley *et al.*, 2003).

*Salmonella* serotypes are associated with a variety of febrile diseases in food animals characterized by enterocolits, bacteremia or abortion. Some of these diseases are not longer endemic in the United States. For instance, the avian-adapted serotype *S. Gallinarum* causes fowl typhoid and pullorum disease in chickens and was associated with high morbidity and mortality among poultry flocks during the first half of the 20th century (Bullis, 1977). The implementation of a control program (Poultry Stock Improvement Plan) by the United States Department of Agriculture (USDA) in 1935 triggered a steady decline in the incidence of *S. Gallinarum* in poultry and led to eradication of this pathogen in the 1970s (Bullis, 1977). Other diseases, such as *S. Typhimurium* or *S. Dublin* infection of cattle, continue to pose a threat to animal health and in the United States (Rothenbacher, 1965; Wells *et al.*, 1998). However, *Salmonella* serotypes are currently a food safety problem not because they cause disease in animals, but because a considerable proportion of infected animals are asymptomatic carriers. Apparently healthy carriers may lay contaminated eggs or introduce the organism into abattoirs or poultry processing plants. Intestinal carriage or an infected organ may result in contamination of equipment surfaces or workers hands, thereby leading to contamination of carcasses and processed foods (Gronstol *et al.*, 1974; Samuel *et al.*, 1979; Moo *et al.*, 1980; Samuel *et al.*, 1981). Thus, carriage of *Salmonella* serotypes by apparently healthy food animals is directly responsible for their subsequent introduction into the derived food products.

## Susceptibility to physical and chemical agents and factors influencing survival

Growth of *Salmonella* serotypes in food vehicles is important for food-safety as it may allow these pathogens to reach numbers sufficient to cause disease. *Salmonella* serotypes can grow at temperatures between 5°C and 47°C at a pH range of 4.5 to 9.0, with optimal growth occurring between 35°C and 37°C at neutral pH (6.5–7.5) in moist environments with low osmolarity. The organism is killed by ordinary household cooking but a risk for *Salmonella* food poisoning exists when raw or undercooked foods are consumed. Growth in food can be prevented by storage below 5°C or by acidification below pH4.5. The organism is killed in drinking water by chlorination.

## Detection and analysis

Animals infected with some of the more invasive serotypes, such as the avian-adapted *S. Gallinarum* can be readily identified using simple serological techniques. Agglutination tests for the detection of birds infected with *S. Gallinarum* were developed at the beginning of the 20th century (Jones, 1913; Schaffer and MacDonald, 1931) and were used successfully to eradicate this pathogen in the United States by the test-and-slaughter method of disease control (Bullis, 1977). Agglutination tests detect the O-antigen of *S. Gallinarum*, which is identical to that of *S. Enteritidis* (Figure 16.1). However, the simple agglutination tests used for the detection of birds exposed to *S. Gallinarum* do not detect birds infected with *S. Enteritidis*, since the latter elicits very weak antibody responses in chickens (Barrow *et al.*, 1992). Similarly, weak antibody responses in apparently healthy food animals carrying other serotypes important for food safety make it difficult to develop simple and inexpensive methods for detection.

Enzyme-linked immunosorbent assays (ELISAs) are more sensitive than agglutination tests and provide an effective means to identify animals infected with *Salmonella* serotypes (Barrow *et al.*, 1989; Chart *et al.*, 1990; Nicholas and Cullen, 1991; van Zijderveld *et al.*, 1992). However, it is difficult to use these techniques for large scale screening on the farm. Animals infected with a *Salmonella* serotype shed the organism intermittently with their feces, and it can be cultivated by enrichment (e.g. tetrathionate broth and rappaport broth) followed by selective differential plating (e.g. bismuth sulfite agar, brilliant green agar or xylose lysine deoxycholate agar). The identity of an isolate suspected to be a *Salmonella* serotype can then be confirmed by agglutination with antisera specific for serogroups B, C1, C2, D1 and E1. However, since shedding is not continuous and these techniques are relatively complicated detection by these methods is not well suited for identifying individual carriers in large operations.

## Control

Typhoid fever has been eradicated in the United States by proper sanitation of drinking water, immunization of persons at risk and the elimination of chronic carriers (typhoid Marys). Today, the small number of typhoid fever cases occurring in the United States annually are commonly associated with foreign travel to endemic areas (Mead *et al.*, 1999). A second success story is the eradication of *S. Gallinarum* from poultry flocks in the United States by culling seropositive birds (Bullis, 1977). Eradication of these diseases was facilitated by the fact that both *S. Typhi* and *S. Gallinarum* are strictly host-adapted serotypes, thereby preventing their continuous reintroduction from a wild animal reservoir subsequent to their elimination from populations of humans and chickens, respectively. In contrast, non-typhoidal serotypes currently posing a threat to food safety in the United States have a broader host range and this makes eradication an unrealistic goal.

Control measures to reduce the prevalence of non-ty-phoidal serotypes at the pre-harvest level include vaccination of animals and blocking routes of transmission by improving sanitation. There is currently no federal initiative to implement vaccination to reduce the prevalence of *Salmonella* serotypes in food animals. However, vaccination has been used successfully in England and The Netherlands to reduce the prevalence of *S.* Enteritidis among poultry (Feberwee *et al.*, 2000; Ward *et al.*, 2000; Feberwee *et al.*, 2001). Vaccination commonly confers protection against *Salmonella* serotypes belonging to the same serogroup, while little or no cross-protection between members of different serogroups is observed (Kingsley and Bäumler, 2000).

Between 1998 and 2000 new federal control programs were adopted to reduce the prevalence of *Salmonella* serotypes at the post-harvest level. In 1998, USDA implemented pathogen reduction/hazard analysis critical control point (PR/HACCP) systems regulations in the meat and poultry slaughter and processing plants. PR/HACCP implementation between 1998 and 2000 was accompanied by a decline in the prevalence of *Salmonella* serotypes in raw meat and poultry products (Rose *et al.*, 2002) and a decline in human cases of non-typhoidal salmonellosis in the United States (CDC, 2002).

The Food and Drug Administration (FDA) has implemented an Egg Safety Action Plan in 1999 with the goal of eliminating *S.* Enteritidis illnesses associated with the consumption of eggs by 2010, and a 50% reduction in egg-associated illness by 2005. However, the number of *S.* Enteritidis cases reported to CDC did not decline between 1999 and 2001, and outbreaks associated with shell eggs continue to occur (CDC, 2003). Thus, it is questionable whether measures to improve egg production, shell egg processing, egg products processing, egg distribution, and egg handling and preparation implemented under the Egg Safety Action Plan will accomplish the goal of eradicating *S.* Enteritidis from this food-source. A possible reason for the limited success in reducing the contamination of shell eggs by minimizing the spread of *S.* Enteritidis in the farm-to-table continuum is the fact that compared to meat products, the contamination rate in eggs is already relatively low and likely originates from infection of a small fraction of eggs on the farm before they are laid (Okamura *et al.*, 2001). The USDA baseline study performed prior to PR/HACCP implementation revealed that between 1% and 50% of raw meat and raw ground meat samples tested positive for *Salmonella* serotypes (Schlosser *et al.*, 2000). In contrast, USDA estimates that *S.* Enteritidis can be isolated from only 0.005% of shell eggs sold in the United States (Ebel and Schlosser, 2000). It may indeed be difficult to reduce the already low prevalence of *S.* Enteritidis in this food source by methods that were successful in reducing the considerably higher contamination rates in meat products. The fact that *S.* Enteritidis became only recently associated with chicken eggs suggest that the number of human cases could be reduced using an alternative strategy. While current control measures mainly target the symptoms (i.e. contamina-

tion of eggs) produced by the current *S.* Enteritidis epidemic, an alternative approach would be to devise control measures designed to eliminate its cause (i.e. factors that triggered the epidemic). One of the factors that may have triggered the *S.* Enteritidis epidemic is the eradication of *S.* Gallinarum from poultry flocks in the United States (Bäumler *et al.*, 2000). Eradication of *S.* Gallinarum resulted in the loss of population wide immunity against other *Salmonella* serotypes belonging to serogroup D1 (Figure 16.1), thereby opening an ecological niche that was subsequently filled by *S.* Enteritidis (Rabsch *et al.*, 2000). This analysis suggest that vaccination would be an effective strategy to exclude *S.* Enteritidis from chicken flocks because this strategy would eliminate one of the risk factors (loss of flock immunity against *Salmonella* serotypes belonging to serogroup D1), which likely contributed to the emergence of *S.* Enteritidis as an egg-associated pathogen.

## Acknowledgments
Work in A.J.B.'s laboratory is supported by USDA/NRICGP grant #2002-35204-12247 and Public Health Service grants AI40124 and AI44170. H.A. is currently supported by Public Health Service grant AI052250.

## References

Ahmer, B.M., van Reeuwijk, J., Watson, P.R., Wallis, T.S., and Heffron, F. 1999. *Salmonella* SirA is a global regulator of genes mediating enteropathogenesis. Mol. Microbiol. 31: 971–982.

Akkina, J.E., Hogue, A.T., Angulo, F.J., Johnson, R., Petersen, K.E., Saini, P.K., Fedorka-Cray, P.J., and Schlosser, W.D. 1999. Epidemiologic aspects, control, and importance of multiple drug resistant *Salmonella* Typhimurium DT104 in the United States. J. Am. Vet. Med. Assoc .214: 790–798.

Aleksic, S., Heinzerling, F., and Bockemühl, J. 1996. Human infection caused by *Salmonellae* of subspecies II to VI in Germany, 1977–1992. Zbl. Bakt. 283: 391–398.

Altekruse, S., Hyman, F., Klontz, K., Timbo, B., and Tollefson, L. 1994. Foodborne bacterial infections in individuals with the human immunodeficiency virus. South. Med. J. 87: 169–173.

Anderson, E.S. 1968. Drug resistance in *Salmonella typhimurium* and its implications. Br. Med. J. 3: 333–339.

Anderson, E.S., Ward, L.R., Saxe, M.J., and de Sa, J.D. 1977. Bacteriophage-typing designations of *Salmonela typhimurium*. J. Hyg .(Lond.) 78: 297–300.

Anderson, E.S., Ward, L.R., De Saxe, M.J., Old, D.C., Barker, R., and Duguid, J.P. 1978. Correlation of phage type, biotype and source in strains of *Salmonella typhimurium*. J. Hyg. (Lond.) 81: 203–217.

Andrews, F.W. 1922. Studies in group-agglutination. I. The *Salmonella* group and its antigenic structure. J. Pathol. Bacteriol. 25: 515–521..

Angulo, F.J., and Swerdlow, D.L. 1998. *Salmonella enteritidis* infections in the United States. J Am Vet Med Assoc 213: 1729–1731.

Angulo, F.J., and Griffin, P.M. 2000. Changes in antimicrobial resistance in *Salmonella* enterica serovar Typhimurium. Emerg. Infect. Dis .6: 436–438.

Aserkoff, B., Schroeder, S.A., and Brachman, P.S. 1970. Salmonellosis in the United States–a five-year review. Am. J. Epidemiol. 92: 13–24.

Barrow, P., Hassan, J.O., Mockett, A.P., and McLeod, S. 1989. Detection of *Salmonella* infection by ELISA. Vet. Rec. 125: 586.

Barrow, P.A., Berchieri, A., Jr., and al-Haddad, O. 1992. Serological response of chickens to infection with *Salmonella gallinarum-S. pullorum* detected by enzyme-linked immunosorbent assay. Avian Dis. 36: 227–236.

Bäumler, A.J., Hargis, B.M., and Tsolis, R.M. 2000. Tracing the origins of *Salmonella* outbreaks. Science 287: 50–52.

Bitar, R., and Tarpley, J. 1985. Intestinal perforation and typhoid fever: a historical and state-of-the-art review. Rev. Infect. Dis. 7: 257.

Blanc-Potard, A.-B., and Groisman, E.A. 1997. The *Salmonella selC* locus contains a pathogenicity island mediating intramacrophage survival. EMBO J. 16: 5376–5385.

Blanc-Potard, A.B., Solomon, F., Kayser, J., and Groisman, E.A. 1999. The SPI-3 pathogenicity island of *Salmonella* enterica. J. Bacteriol. 181: 998–1004.

Brenner, F.W., Villar, R.G., Angulo, F.J., Tauxe, R., and Swaminathan, B. 2000. *Salmonella* nomenclature. J .Clin. Microbiol. 38: 2465–2467.

Brumell, J.H., Rosenberger, C.M., Gotto, G.T., Marcus, S.L., and Finlay, B.B. 2001. SifA permits survival and replication of *Salmonella typhimurium* in murine macrophages. Cell. Microbiol. 3: 75–84.

Brumell, J.H., Goosney, D.L., and Finlay, B.B. 2002. SifA, a type III secreted effector of *Salmonella typhimurium*, directs *Salmonella*-induced filament (Sif) formation along microtubules. Traffic 3: 407–415.

Brumell, J.H., Kujat-Choy, S., Brown, N.F., Vallance, B.A., Knodler, L.A., and Finlay, B.B. 2003. SopD2 is a novel type III secreted effector of *Salmonella typhimurium* that targets late endocytic compartments upon delivery into host cells. Traffic 4: 36–48.

Bullis, K.L. 1977. The history of avian medicine in the USA II. Pullorum disease and fowl typhoid. Avian Dis. 21: 422–429.

CDC. 2002. Preliminary FoodNet data on the incidence of foodborne illnesses–selected sites, United States, 2001. MMWR 51: 325–329.

CDC. 2003. Outbreaks of *Salmonella* serotype *enteritidis* infection associated with eating shell eggs–United States, 1999–2001. MMWR 51: 1149–1152.

Chart, H., Rowe, B., Baskerville, A., and Humphrey, T.J. 1990. Serological analysis of chicken flocks for antibodies to *Salmonella enteritidis*. Vet. Rec. 127: 501–502.

Cherubin, C.E., Fodor, T., Denmark, L.I., Master, C.S., Fuerst, H.T., and Winter, J.W. 1969. Symptoms, septicemia and death in salmonellosis. Am. J. Epidemiol. 90: 285–291.

Chiodini, R.J., and Sundberg, J.P. 1981. Salmonellosis in reptiles: a review. Am. J. Epidemiol. 113: 494–499.

Cohen, M.L., and Tauxe, R.V. 1986. Drug-resistant *Salmonella* in the United States: an epidemiologic perspective. Science 234: 964–969.

Cowden, J.M., Chisholm, D., O'Mahony, M., Lynch, D., Mawer, S.L., Spain, G.E., Ward, L., and Rowe, B. 1989a. Two outbreaks of *Salmonella enteritidis* phage type 4 infection associated with the consumption of fresh shell-egg products. Epidemiol. Infect. 103: 47–52.

Cowden, J.M., Lynch, D., Joseph, C.A., O'Mahony, M., Mawer, S.L., Rowe, B., and Bartlett, C.L. 1989b. Case-control study of infections with *Salmonella enteritidis* phage type 4 in England. Br. Med. J. 299: 771–773.

Coyle, E.F., Palmer, S.R., Ribeiro, C.D., Jones, H.I., Howard, A.J., Ward, L., and Rowe, B. 1988. *Salmonella enteritidis* phage type 4 infection: association with hen's eggs. Lancet 2: 1295–1297.

Crosa, J.H., Brenner, D.J., Ewing, W.H., and Falkow, S. 1973. Molecular relationship among the *Salmonellae*. J. Bacteriol. 115: 307–315.

Crump, J.A., Griffin, P.M., and Angulo, F.J. 2002. Bacterial contamination of animal feed and its relationship to human foodborne illness. Clin. Infect. Dis. 35: 859–865.

Curtiss, R., Goldschmidt, R.M., Fletchall, N.B., and Kelly, S.M. 1988. Avirulent *Salmonella typhimurium* delta cya delta crp oral vaccine strains expressing a streptococcal colonization and virulence antigen. Vaccine 6: 155–160.

Dahl, J., Wingstrand, A., Nielsen, B., and Baggesen, D.L. 1997. Elimination of *Salmonella typhimurium* infection by the strategic movement of pigs. Vet. Rec. 140: 679–681.

Davies, G. 1979. Chloramphenicol-resistant S *typhimurium*. Vet. Rec. 104: 128.

Davies, P.R., Morrow, W.E., Jones, F.T., Deen, J., Fedorka-Cray, P.J., and Gray, J.T. 1997. Risk of shedding *Salmonella* organisms by market-age hogs in a barn with open-flush gutters. J. Am. Vet. Med. Assoc. 210: 386–389.

Day, D.W., Mandal, B.K., and Morson, B.C. 1978. The rectal biopsy appearances in *Salmonella* colitis. Histopathology 2: 117–131.

DeGroote, M.A., Ochsner, U.A., Shiloh, M.U., Nathan, C., McCord, J.M., Dinauer, M.C., Libby, S.J., Vazquez, T.A., Xu, Y., and Fang, F.C. 1997. Periplasmic superoxide dismutase protects *Salmonella* from products of phagocyte NADPH-oxidase and nitric oxide synthase. Proc. Natl. Acad. Sci. USA 94: 13997–14001.

Dorman, C.J., Chatfield, S., Higgins, C.F., Hayward, C., and Dougan, G. 1989. Characterization of porin and *ompR* mutants of a virulent strain of *Salmonella typhimurium*: *ompR* mutants are attenuated *in vivo*. Infect. Immun. 57: 2136–2140.

Ebel, E., and Schlosser, W. 2000. Estimating the annual fraction of eggs contaminated with *Salmonella enteritidis* in the United States. Int. J. Food Microbiol. 61: 51–62.

Edwards, P.R., and Bruner, D.W. 1943. The occurrence and distribution of *Salmonella* types in the United States. J. Infect. Dis. 72: 58–67.

Falkow, S. 1988. Molecular Koch's postulates applied to microbial pathogenicity. Rev. Infect. Dis. 10 Suppl. 2: S274–276.

Fang, F.C., and Fierer, J. 1991. Human infection with *Salmonella dublin*. Medicine (Baltimore) 70: 198–207.

Fang, F.C., DeGroote, M.A., Foster, J.W., Bäumler, A.J., Ochsner, U., Testerman, T., Bearson, S., Giard, J.C., Xu, Y., Campbell, G., and Laessig, T. 1999. Virulent *Salmonella typhimurium* has two periplasmic Cu, Zn-superoxide dismutases. Proc. Natl. Acad. Sci. USA 96: 7502–7507.

Feberwee, A., de Vries, T.S., Elbers, A.R., and de Jong, W.A. 2000. Results of a *Salmonella enteritidis* vaccination field trial in broiler-breeder flocks in The Netherlands. Avian Dis. 44: 249–255.

Feberwee, A., de Vries, T.S., Hartman, E.G., de Wit, J.J., Elbers, A.R., and de Jong, W.A. 2001. Vaccination against *Salmonella enteritidis* in Dutch commercial layer flocks with a vaccine based on a live *Salmonella* gallinarum 9R strain: evaluation of efficacy, safety, and performance of serologic *Salmonella* tests. Avian Dis. 45: 83–91.

Fields, P.I., Swanson, R.V., Haidaris, C.G., and Heffron, F. 1986. Mutants of *Salmonella typhimurium* that cannot survive within the macrophage are avirulent. Proc. Natl. Acad. Sci. USA 83: 5189–5193.

Fierer, J., and Guiney, D.G. 2001. Diverse virulence traits underlying different clinical outcomes of *Salmonella* infection. J. Clin. Invest. 107: 775–780.

Figueroa-Bossi, N., Uzzau, S., Maloriol, D., and Bossi, L. 2001. Variable assortment of prophages provides a transferable repertoire of pathogenic determinants in *Salmonella*. Mol. Microbiol. 39: 260–272.

Finlay, B.B., Starnbach, M.N., Francis, C.L., Stocker, B.A.D., Chatfield, S., Dougan, G., and Falkow, S 1988. Identification and characterization of TnphoA mutants of *Salmonella* that are unable to pass through a polarized MDCK epithelial cell monolayer. Mol. Microbiol. 2: 757–766.

Fontaine, R.E., Arnon, S., Martin, W.T., Vernon, T.M., Jr., Gangarosa, E.J., Farmer, J.J., 3rd, Moran, A.B., Silliker, J.H., and Decker, D.L. 1978. Raw hamburger: an interstate common source of human salmonellosis. Am. J. Epidemiol. 107: 36–45.

Frenzen, P., Riggs, T., Buzby, J., Breuer, T., Roberts, T., Voetsch, D., Reddy, S., and Group, t.F.W. 1999. *Salmonella* cost estimate update using FoodNet data. Food Review 22: 10–15.

Friebel, A., Ilchmann, H., Aepfelbacher, M., Ehrbar, K., Machleidt, W., and Hardt, W.D. 2001. SopE and SopE2 from *Salmonella typhimurium* activate different sets of RhoGTPases of the host cell. J. Biol. Chem. 276: 34035–34040.

Galán, J.E. 2001 *Salmonella* interactions with host cells: type III secretion at work. Annu. Rev. Cell. Dev. Biol .17: 53–86.

Galán, J.E., and Curtiss III, R. 1989. Cloning and molecular characterization of genes whose products allow *Salmonella typhimurium* to penetrate tissue culture cells. Proc. Natl. Acad. Sci. USA 86: 6383–6387.

Galyov, E.E., Wood, M.W., Rosqvist, R., Mullan, P.B., Watson, P.R., Hedges, S., and Wallis, T.S. 1997. A secreted effector protein of *Salmonella dublin* is translocated into eukaryotic cells and mediates inflammation and fluid secretion in infected ileal mucosa. Mol. Microbiol. 25: 903–912.

Gärtner, A. 1888. Ueber die Fleischvergiftung in Frankenhausen a. Kyffh. und den Erreger derselben. Correspondenz-Blätter des allgemeinen ärztlichen Vereins von Thüringen 17: 573–600.

Geoffrey, E., Gaines, S., Landy, M., Tigertt, W.D., Sprintz, H., Trapani, R.-J., Mandel, A.D., and Benenson, A.S. 1960. Studies on infection and immunity in experimental typhoid fever: Typhoid fever in chimpanzees orally infected with Salmonella typhosa. J. Exp. Med. 112: 143–166.

Giles, N., Hopper, S.A., and Wray, C. 1989. Persistence of S. Typhimurium in a large dairy herd. Epidemiol. Infect .103: 235–241.

Glynn, M.K., Bopp, C., Dewitt, W., Dabney, P., Mokhtar, M., and Angulo, F.J. 1998. Emergence of multidrug-resistant Salmonella enterica serotype Typhimurium DT104 infections in the United States. N. Engl. J. Med. 338: 1333–1338.

Gronstol, H., Osborne, A.D., and Pethiyagoda, S. 1974. Experimental Salmonella infection in calves. 1. The effect of stress factors on the carrier state. J. Hyg. (Lond.) 72: 155–162.

Guard-Petter, J., Henzler, D.J., Rahman, M.M., and Carlson, R.W. 1997. On-farm monitoring of mouse-invasive Salmonella enterica serovar Enteritidis and a model for its association with the production of contaminated eggs. Appl. Environ. Microbiol. 63: 1588–1593.

Gulig, P.A., and Curtiss, R. 1987. Plasmid-associated virulence of Salmonella typhimurium. Infect. immun. 1987: 2891–2901.

Gulig, P.A., Doyle, T.J., Hughes, J.A., and Matsui, H. 1998. Analysis of host cells associated with the Spv-mediated increased intracellular growth rate of Salmonella typhimurium in mice. Infect. Immun. 66: 2471–2485.

Habermalz, D., and Pietzsch, O. 1973. [Identification of Arizona bacteria. A contribution to the problem of Salmonella infections among reptiles and amphibians in zoological gardens Zentralbl. Bakteriol. [Orig A] 225: 323–342.

Hardt, W.D., and Galán, J.E. 1997. A secreted Salmonella protein with homology to an avirulence determinant of plant pathogenic bacteria. Proc. Natl. Acad. Sci. USA 94: 9887 9892.

Hardt, W.D., Chen, L.M., Schuebel, K.E., Bustelo, X.R., and Galán, J.E. 1998a. S. Typhimurium encodes an activator of Rho GTPases that induces membrane ruffling and nuclear responses in host cells. Cell 93: 815–826.

Hardt, W.D., Urlaub, H., and Galan, J.E. 1998b. A substrate of the centisome 63 type III protein secretion system of Salmonella typhimurium is encoded by a cryptic bacteriophage. Proc. Natl. Acad. Sci. USA 95: 2574–2579.

Heithoff, D.M., Sinsheimer, R.L., Low, D.A., and Mahan, M.J. 1999. An essential role for DNA adenine methylation in bacterial virulence. Science 284: 967–970.

Hensel, M., Shea, J.E., Gleeson, C., Jones, M.D., Dalton, E., and Holden, D.W. 1995. Simultaneous identification of bacterial virulence genes by negative selection. Science 269: 400–403.

Hensel, M., Shea, J.E., Bäumler, A.J., Gleeson, C., Blattner, F., and Holden, D.W. 1997. Analysis of the boundaries of Salmonella pathogenicity island 2 and the corresponding chromosomal region of Escherichia coli K-12. J. Bacteriol. 179: 1105–1111.

Henzler, D.J., and Opitz, H.M. 1992. The role of mice in the epizootiology of Salmonella enteritidis infection on chicken layer farms. Avian Dis 36: 625–631.

Henzler, D.J., Ebel, E., Sanders, J., Kradel, D., and Mason, J. 1994. Salmonella enteritidis in eggs from commercial chicken layer flocks implicated in human outbreaks. Avian Dis. 38: 37–43.

Hobbie, S., Chen, L.M., Davis, R.J., and Galan, J.E. 1997. Involvement of mitogen-activated protein kinase pathways in the nuclear responses and cytokine production induced by Salmonella typhimurium in cultured intestinal epithelial cells. J. Immunol .159: 5550–5559.

Hoiseth, S.K., and Stocker, B.A.D. 1981. Aromatic-dependent Salmonella typhimurium are non virulent and effective as live oral vaccines. Nature 291: 238–239.

Hornick, R.B., Greisman, S.E., Woodward, T.E., DuPont, H.L., Dawkins, A.T., and Snyder, M.J. 1970. Typhoid fever: pathogenesis and immunologic control. N. Engl. J. Med. 283: 686–691.

Hueck, C.J., Hantman, M.J., Bajaj, V., Johnston, C., Lee, C.A., and Miller, S.I. 1995. Salmonella typhimurium secreted invasion determinants are homologous to Shigella Ipa proteins. Mol. Microbiol. 18: 479–490.

Hurd, H.S., Gailey, J.K., McKean, J.D., and Rostagno, M.H. 2001. Rapid infection in market weight swine following exposure to a Salmonella typhimurium-contaminated environment. Am. J. Vet. Res. 62: 1194–1197.

Iveson, J.B., Mackay-Scollay, E.M., and Bamford, V. 1969. Salmonella and Arizona in reptiles and man in Western Australia. J Hyg (Lond) 67: 135–145.

Jones, F.S. 1913. The value of the macroscopic agglutination test in detecting fowls that are harboring Bact. pullorum. J. Med. Res. 27: 481–495.

Jones, M.A., Wood, M.W., Mullan, P.B., Watson, P.R., Wallis, T.S., and Galyov, E.E. 1998. Secreted effector proteins of Salmonella dublin act in concert to induce enteritis. Infect. Immun. 66: 5799–5804.

Kaniga, K., Tucker, S., Trollinger, D., and Galán, J.E. 1995 Homologs of the Shigella IpaB and IpaC invasins are required for Salmonella typhimurium entry into cultured epithelial cells. J. Bacteriol. 177: 3965–3971.

Kauffmann, F. 1934. Über die serologische und kulturelle Varianten der Paratyphus D- und Mäusetyphus-Bacillen. Z. Hyg. 116: 368–384.

Kauffmann, F. 1960. Two biochemical subdivisions of the genus Salmonella. Acta Path. Microbiol. Scand. 49: 393.

Kelterborn, E. 1967. Salmonella-species. First isolations, names and occurrence. Karl Marx Stadt: S. Hirzel Verlag Leipzig.

Kinde, H., Read, D.H., Ardans, A., Breitmeyer, R.E., Willoughby, D., Little, H.E., Kerr, D., Gireesh, R., and Nagaraja, K.V. 1996. Sewage effluent: likely source of Salmonella enteritidis, phage type 4 infection in a commercial chicken layer flock in southern California. Avian Dis .40: 672–676.

Kingsley, R.A., and Bäumler, A.J. 2000. Host adaptation and the emergence of infectious disease: the Salmonella paradigm. Mol. Microbiol. 36: 1006–1014.

Kingsley, R.A., van Amsterdam, K., Kramer, N., and Baumler, A.J. 2000. The shdA gene is restricted to serotypes of Salmonella enterica subspecies I and contributes to efficient and prolonged fecal shedding. Infect. Immun. 68: 2720–2727.

Kingsley, R.A., and Baumler, A.J. 2002. Pathogenicity islands and host adaptation of Salmonella serovars. Curr. Top. Microbiol. Immunol .264: 67–87.

Kingsley, R.A., Santos, R.L., Keestra, A.M., Adams, L.G., and Bäumler, A.J. 2002. Salmonella enterica serotype Typhimurium ShdA is an outer membrane fibronectin-binding protein that is expressed in the intestine. Mol. Microbiol. 43: 895–905.

Kingsley, R.A., Humphries, A.D., Weening, E., de Zoete, M., Winter, S., Papaconstantinopoulou, A., Dougan, G., and Bäumler, A.J. 2003. Molecular and phenotypic analysis of the CS54 island of Salmonella enterica serotype Typhimurium. Infect. Immun. 71: 629–640.

Knodler, L.A., Celli, J., Hardt, W.D., Vallance, B.A., Yip, C., and Finlay, B.B. 2002. Salmonella effectors within a single pathogenicity island are differentially expressed and translocated by separate type III secretion systems. Mol. Microbiol. 43: 1089–1103.

Kovacs, A., Leaf, H.L., and Simberkoff, M.S. 1997. Bacterial infections. Med Clin North Am 81: 319–343.

Krabisch, P., and Dorn, P. 1980. Epidemiologic significance of live vectors in the transmission of Salmonella infections in broiler flocks. Berl. Munch. Tierarztl. Wochenschr. 93: 232–235.

Kraus, M.D., Amatya, B., and Kimula, Y. 1999. Histopathology of typhoid enteritis: morphologic and immunophenotypic findings. Mod .Pathol. 12: 949–955.

Lee, C.A., Silva, M., Siber, A.M., Kelly, A.J., Galyov, E., and McCormick, B.A. 2000. A secreted Salmonella protein induces a proinflammatory response in epithelial cells, which promotes neutrophil migration. Proc .Natl. Acad. Sci. USA 97: 12283–12288.

Lesnick, M.L., Reiner, N.E., Fierer, J., and Guiney, D.G. 2001 The Salmonella spvB virulence gene encodes an enzyme that ADP-ribosylates actin and destabilizes the cytoskeleton of eukaryotic cells. Mol. Microbiol. 39: 1464–1470.

Levy, E., and Gaehtgens, W. 1908. Über die Verbreitung der Typhusbazillen in den Lymphdrüsen bei Typhusleichen. Arb. Kaiserl. Gesundh. 28: 168–171.

Lüderitz, O., Staub, A.M., and Westphal, O. 1966. Immunochemistry of O and R antigens of *Salmonella* and related *Enterobacteriaceae*. Bacteriol. Rev. 30: 192–255.

Mahan, M.J., Slauch, J.M., and Mekalanos, J.J. 1993. Selection of bacterial virulence genes that are specifically induced in host tissues. Science 259: 686–688.

Mandal, B.K., and Brennand, J. 1988. Bacteraemia in salmonellosis: a 15 year retrospective study from a regional infectious diseases unit. Br. Med. J. 297: 1242–1243.

McCormick, B.A., Miller, S.I., Carnes, D., and Madara, J.L. 1995. Transepithelial signaling to neutrophils by *Salmonellae*: a novel virulence mechanism for gastroenteritis. Infect. Immun. 63: 2302–2309.

McGovern, V.J., and Slavutin, L.J. 1979. Pathology of *Salmonella* colitis. Am. J. Surg. Pathol. 3: 483–490.

McLaren, I.M., and Wray, C. 1991. Epidemiology of *Salmonella typhimurium* infection in calves: persistence of *Salmonellae* on calf units. Vet. Rec. 129: 461–462.

Mead, P.S., Slutsker, L., Dietz, V., McCaig, L.F., Bresee, J.S., Shapiro, C., Griffin, P.M., and Tauxe, R.V. 1999. Food-related illness and death in the United States. Emerg. Infect. Dis. 5: 607–625.

Merican, I. 1997. Typhoid fever: present and future. Med. J. Malaysia 52: 299–308; quiz 309.

Miao, E.A., Scherer, C.A., Tsolis, R.M., Kingsley, R.A., Adams, L.G., Baumler, A.J., and Miller, S.I. 1999. *Salmonella typhimurium* leucine-rich repeat proteins are targeted to the SPI1 and SPI2 type III secretion systems. Mol. Microbiol. 34: 850–864.

Miao, E.A., and Miller, S.I. 2000. A conserved amino acid sequence directing intracellular type III secretion by *Salmonella typhimurium*. Proc. Natl. Acad .Sci .USA 97: 7539–7544.

Miller, S.I., Loomis, W.P., Alpuche-Aranda, C., Behlau, I., and Hohmann, E. 1993. The PhoP virulence regulon and live oral *Salmonella* vaccines. Vaccine 11: 122–125.

Mills, D.M., Bajaj, V., and Lee, C.A. 1995. A 40kb chromosomal fragment encoding *Salmonella typhimurium* invasion genes is absent from the corresponding region of the *Escherichia coli* K-12 chromosome. Mol. Microbiol. 15: 749–759.

Mishu, B., Koehler, J., Lee, L.A., Rodrigue, D., Brenner, F.H., Blake, P., and Tauxe, R.V. 1994. Outbreaks of *Salmonella enteritidis* infections in the United States, 1985–1991. J. Infect. Dis. 169: 547–552.

Moo, D., O'Boyle, D., Mathers, W., and Frost, A.J. 1980. The isolation of *Salmonella* from jejunal and caecal lymph nodes of slaughtered animals. Australian Vet. J. 56: 181–183.

Nicholas, R.A., and Cullen, G.A. 1991. Development and application of an ELISA for detecting antibodies to *Salmonella enteritidis* in chicken flocks. Vet. Rec. 128: 74–76.

Norris, F.A., Wilson, M.P., Wallis, T.S., Galyov, E.E., and Majerus, P.W. 1998. SopB, a protein required for virulence of *Salmonella dublin*, is an inositol phosphate phosphatase. Proc. Natl. Acad. Sci. USA 95: 14057–14059.

Ochman, H., and Groisman, E.A. 1996 Distribution of pathogenicity islands in *Salmonella* spp. Infect. Immun. 64: 5410–5412.

Ochman, H., Soncini, F.C., Solomon, F., and Groisman, E.A. 1996. Identification of a pathogenicity island for *Salmonella* survival in host cells. Proc. Natl. Acad. Sci. USA 93: 7800–7804.

Okamura, M., Kamijima, Y., Miyamoto, T., Tani, H., Sasai, K., and Baba, E. 2001. Differences among six *Salmonella* serovars in abilities to colonize reproductive organs and to contaminate eggs in laying hens. Avian Dis. 45: 61–69.

Olsen, S.J., MacKinnon, L.C., Goulding, J.S., Bean, N.H., and Slutsker, L. 2000. Surveillance for foodborne-disease outbreaks–United States, 1993–1997. MMWR CDC Surveill. Summ 49: 1–62.

Olsen, S.J., Bishop, R., Brenner, F.W., Roels, T.H., Bean, N., Tauxe, R.V., and Slutsker, L. 2001. The changing epidemiology of *Salmonella*: trends in serotypes isolated from humans in the United States, 1987–1997. J. Infect. Dis. 183: 753–761.

Otto, H., Tezcan-Merdol, D., Girisch, R., Haag, F., Rhen, M., and Koch-Nolte, F. 2000. The *spvB* gene-product of the *Salmonella* enterica virulence plasmid is a mono(ADP-ribosyl)transferase. Mol. Microbiol. 37: 1106–1115.

Popoff, M.Y., Miras, I., Coynault, C., Lasselin, C., and Pardon, P. 1984. Molecular relationships between virulence plasmids of *Salmonella* serotypes *typhimurium* and *dublin* and large plasmids of other *Salmonella* serotypes. Ann. Microbiol. 135A: 389–398.

Popoff, M.Y., and Le Minor, L. 1992. Antigenic formulas of the *Salmonella* serovars. Paris: WHO Collaborating Center for Reference and Research on *Salmonella*, Institute Pasteur.

Poppe, C., Curtiss III, R., Gulig, P.A., and Gyles, C.L. 1989. Hybridization with a DNA probe derived from the virulence region of the 60 MDa plasmid of *Salmonella typhimurium*. Can. J. Vet. Res. 53: 378.

Rabsch, W., Hargis, B.M., Tsolis, R.M., Kingsley, R.A., Hinz, K.H., Tschäpe, H., and Bäumler, A.J. 2000. Competitive exclusion of *Salmonella enteritidis* by *Salmonella* gallinarum from poultry. Emerg. Infect. Dis. 6: 443–448.

Rabsch, W., Tschape, H., and Baumler, A.J. 2001. Non-typhoidal salmonellosis: emerging problems. Microbes Infect. 3: 237–247.

Reeves, M.W., Evins, G.M., Heiba, A.A., Plikaytis, B.D., and Farmer III, J.J. 1989. Clonal nature of *Salmonella* typhi and its genetic realtedness to other *Salmonellae* as shown by multilocus enzyme electrophoresis, and proposal of *Salmonella bongori* comb. nov. J. Clin. Microbiol. 27: 313–320.

Rodrigue, D.C., Tauxe, R.V., and Rowe, B. 1990. International increase in *Salmonella enteritidis*: a new pandemic? Epidemiol. Infect. 105: 21–27.

Roggendorf, M., and Muller, H.E. 1976. Enterobacteria from reptiles. *Zentralbl. Bakteriol. Parasitenk. Hyg. Abt. I Orig. A* 236: 22–35.

Rose, B.E., Hill, W.E., Umholtz, R., Ransom, G.M., and James, W.O. 2002. Testing for *Salmonella* in raw meat and poultry products collected at federally inspected establishments in the United States, 1998 through 2000. J. Food Prot. 65: 937–947.

Rothenbacher, H. 1965. Mortality and morbidity in calves with salmonellosis. J. Am. Vet. Med. Assoc. 147: 1211–1214.

Roudier, C., Krause, M., Fierer, J., and Guiney, D.G. 1990. Correlation between the presence of sequences homologous to the vir region of *Salmonella dublin* plasmid pSDL2 and the virulence of twenty-two *Salmonella* serotypes in mice. Infect. Immun. 58: 1180–1185.

Ruiz-Contreras, J., Ramos, J.T., Hernandez-Sampelayo, T., Gurbindo, M.D., Garcia de Jose, M., De Miguel, M.J., Cilleruelo, M.J., and Mellado, M.J. 1995. Sepsis in children with human immunodeficiency virus infection. The Madrid HIV Pediatric Infection Collaborative Study Group. Pediatr. Infect. Dis. J. 14: 522–526.

Salmon, D.E., and Smith, T. 1885. Report on swine plague. Washington, DC: United States Department of Agriculture, pp. 184–246.

Samuel, J.L., O'Boyle, D.A., Mathers, W.J., and Frost, A.J. 1979. Isolation of *Salmonella* from mesenteric lymph nodes of healthy cattle at slaughter. Res. Vet. Sci. 28: 238–241.

Samuel, J.L., Eccles, J.A., and Francis, J. 1981. *Salmonella* in the intestinal tract and associated lymph nodes of sheep and cattle. J. Hyg. 87: 225–232.

Santos, R.L., Tsolis, R.M., Zhang, S., Ficht, T.A., Bäumler, A.J., and Adams, L.G. 2001a. *Salmonella*-induced cell death is not required for enteritis in calves. Infect. Immun. 69: 4610 4617.

Santos, R.L., Zhang, S., Tsolis, R.M., Kingsley, R.A., Adams, L.G., and Bäumler, A.J. 2001b. Animal Models of *Salmonella* Infections: Enteritis vs. Typhoid Fever. Mircrob. Infect. 3: 237–247.

Santos, R.L., Tsolis, R.M., Bäumler, A.J., and Adams, L.G. 2002. Dynamics of Hematologic and Blood Chemical Changes in *Salmonella typhimurium* Infected Calves. Am. J. Vet. Res. 63: 1145–1150.

Saphra, I., and Wassermann, M. 1954. *Salmonella Cholerae suis*. A clinical and epidemiological evaluation of 329 infections identified bewteen 1940 and 1954 in the New York *Salmonella* Center. Am. J. Med. Sci. 228: 525–533.

Sato, G., Miyamae, T., and Miura, S. 1970. A long term epizootiological study of chicken salmonellosis on a farm with reference to elimination of paratyphoid infection by cloacal swab culture. Jap. J. Vet. Res. 18: 47–62.

Schaffer, J.M., and MacDonald, A.D. 1931. A stained antigen for the rapid whole blood test for pullorum disease. J. Am. Vet. Med. Assoc. 32: 236–240.

Schlosser, W., Hogue, A., Ebel, E., Rose, B., Umholtz, R., Ferris, K., and James, W. 2000. Analysis of *Salmonella* serotypes from selected carcasses and raw ground products sampled prior to implementation of the Pathogen Reduction: Hazard Analysis and Critical Control Point Final Rule in the US. Int. J. Food Microbiol. 58: 107–111.

Silverman, M., and Simon, M. 1980. Phase variation: genetic analysis of switching mutants. Cell 19: 845–854.

Smith, H.W., and Crabb, W.E. 1956. The sensitivity to chemotherapeutic agents of a further series of strains of Bacterium coli from cases of white scours: the relationship between sensitivity and response to treatment. Vet. Rec. 68: 274–277.

Smith, H.W., and Crabb, W.E. 1957. The effect of the continuous administration of diets containing low levels of tetracyclines on the incidence of drug-resistant Bacterium coli in the feaces of pigs and chickens: the sensitivity of the Bact. coli to other chemotherapeutic agents. Vet. Rec. 69: 24–30.

Spika, J.S., Waterman, S.H., Hoo, G.W., St Louis, M.E., Pacer, R.E., James, S.M., Bissett, M.L., Mayer, L.W., Chiu, J.Y., Hall, B., and *et al*. 1987. Chloramphenicol-resistant *Salmonella newport* traced through hamburger to dairy farms. A major persisting source of human salmonellosis in California. N. Engl. J. Med. 316: 565–570.

Sprinz, H., Gangarosa, E.J., Williams, M., Hornick, R.B., and Woodward, T.E. 1966. Histopathology of the upper small intestines in typhoid fever. Biopsy study of experimental disease in man. Am. J. Dig. Dis. 11: 615–624.

St. Louis, M.E., Morse, D.L., Potter, M.E., DeMelfi, T.M., Guzewich, J.J., Tauxe, R.V., and Blake, P.A. 1988. The emergence of grade A eggs as a major source of *Salmonella enteritidis* infections. New implications for the control of salmonellosis. JAMA 259: 2103 2107.

Staub, A.M., and Girard, R. 1965. Immunochemical study on *Salmonella*. Analysis of factors 1 in groups B, E4 and G: their relationship with factors 1–12, 19 and 37. Bull. Soc. Chim. Biol. (Paris) 47: 1245–1268.

Steele-Mortimer, O., Knodler, L.A., Marcus, S.L., Scheid, M.P., Goh, B., Pfeifer, C.G., Duronio, V., and Finlay, B.B. 2000. Activation of Akt/ Protein kinase B in epithelial cells by the *Salmonella typhimurium* effector sigD. J. Biol. Chem. 275: 37718–37724.

Stein, M.A., Leung, K.Y., Zwick, M., Garcia-del Portillo, F., and Finlay, B.B. 1996. Identification of a *Salmonella* virulence gene required for formation of filamentous structures containing lysosomal membrane glycoproteins within epithelial cells. Mol. Microbiol. 20: 151–164.

Stender, S., Friebel, A., Linder, S., Rohde, M., Mirold, S., and Hardt, W.D. 2000. Identification of SopE2 from *Salmonella typhimurium*, a conserved guanine nucleotide exchange factor for Cdc42 of the host cell. Mol. Microbiol. 36: 1206–1221.

Stevens, A.J., Gibson, E.A., and Hughes, L.E. 1967. Salmonellosis: the present position in man and animals. 3. Recent observations on field aspects. Vet. Rec. 80: 154–161.

Stocker, B.A.D. 1958. Lysogenic conversion by the A phages of *Salmonella typhimurium*. J. Gen. Microbiol. 18: 9.

Strahan, K., Chatfield, S.N., Tite, J., Dougan, G., and Hormaeche, C.E. 1992. Impaired resistance to infection does not increase the virulence of *Salmonella htrA* live vaccines for mice. Microb. Pathog. 12: 311–317.

Tezcan-Merdol, D., Nyman, T., Lindberg, U., Haag, F., Koch-Nolte, F., and Rhen, M. 2001. Actin is ADP-ribosylated by the *Salmonella enterica* virulence-associated protein SpvB. Mol. Microbiol. 39: 606–619.

Threlfall, E.J., Ward, L.R., and Rowe, B. 1978. Epidermic spread of a chloramphenicol-resistant strain of *Salmonella typhimurium* phage type 204 in bovine animals in Britain. Vet. Rec. 103: 438–440.

Threlfall, E.J., Ward, L.R., and Rowe, B. 1979. Chloramphenicol-resistant S typhimurium. Vet. Rec. 104: 60–61.

Threlfall, E.J., Rowe, B., and Ward, L.R. 1993. A comparison of multiple drug resistance in *Salmonellas* from humans and food animals in England and Wales, 1981 and 1990. Epidemiol. Infect. 111: 189–197.

Threlfall, E.J., Frost, J.A., Ward, L.R., and Rowe, B. 1994. Epidemic in cattle and humans of *Salmonella typhimurium* DT104 with chromosomally integrated multiple drug resistance. Vet. Rec. 134: 577.

Threlfall, E.J., Ward, L.R., Skinner, J.A., and Rowe, B. 1997. Increase in multiple antibiotic resistance in nontyphoidal *Salmonellas* from humans in England and Wales: a comparison of data for 1994 and 1996. Microb. Drug Resist. 3: 263–266.

Threlfall, E.J., Ward, L.R., Skinner, J.A., and Graham, A. 2000. Antimicrobial drug resistance in non-typhoidal *Salmonellas* from humans in England and Wales in 1999: decrease in multiple resistance in *Salmonella* enterica serotypes Typhimurium, Virchow, and Hadar. Microb. Drug Resist. 6: 319–325.

Tsolis, R.M., Adams, L.G., Ficht, T.A., and Baumler, A.J. 1999a. Contribution of *Salmonella typhimurium* virulence factors to diarrheal disease in calves. Infect. Immun. 67: 4879–4885.

Tsolis, R.M., Townsend, S.M., Miao, E.A., Miller, S.I., Ficht, T.A., Adams, L.G., and Bäumler, A.J. 1999b. Identification of a putative *Salmonella* enterica serotype Typhimurium host range factor with homology to IpaH and YopM by signature-tagged mutagenesis. Infect. Immun. 67: 6385–6393.

Tumbarello, M., Tacconelli, E., Caponera, S., Cauda, R., and Ortona, L. 1995. The impact of bacteraemia on HIV infection. Nine years experience in a large Italian university hospital. J. Infect. 31: 123–131.

Uchiya, K., Barbieri, M.A., Funato, K., Shah, A.H., Stahl, P.D., and Groisman, E.A. 1999. A *Salmonella* virulence protein that inhibits cellular trafficking. EMBO J. 18: 3924–3933.

Valdivia, R.H., and Falkow, S. 1997. Fluorescence-based isolation of bacterial genes expressed within host cells. Science 277: 2007–2011.

van Zijderveld, F.G., van Zijderveld-van Bemmel, A.M., and Anakotta, J. 1992. Comparison of four different enzyme-linked immunosorbent assays for serological diagnosis of *Salmonella enteritidis* infections in experimentally infected chickens. J. Clin. Microbiol. 30: 2560–2566.

Vazquez-Torres, A., Xu, Y., Jones-Carson, J., Holden, D.W., Lucia, S.M., Dinauer, M.C., Mastroeni, P., and Fang, F.C. 2000 *Salmonella* pathogenicity island 2-dependent evasion of the phagocyte NADPH oxidase. Science 287: 1655–1658.

Ward, L.R., Threlfall, J., Smith, H.R., and O'Brien, S.J. 2000. *Salmonella* enteritidis epidemic. Science 287: 1754.

Watson, P.R., Galyov, E.E., Paulin, S.M., Jones, P.W., and Wallis, T.S. 1998. Mutation of invH, but not stn, reduces *Salmonella*-induced enteritis in cattle. Infect. Immun. 66: 1432–1438.

Wells, S.J., Ott, S.L., and Seitzinger, A.H. 1998. Key health issues for dairy cattle–new and old. J Dairy Sci. 81: 3029–3035.

Werner, S.B., Humphrey, G.L., and Kamei, I. 1979. Association between raw milk and human *Salmonella dublin* infection. Br. Med. J. 2: 238–241.

Wood, M.W., Jones, M.A., Watson, P.R., Hedges, S., Wallis, T.S., and Galyov, E.E. 1998. Identification of a pathogenicity island required for *Salmonella* enteropathogenicity. Mol. Microbiol. 29: 883–891.

Wood, M.W., Jones, M.A., Watson, P.R., Siber, A.M., McCormick, B.A., Hedges, S., Rosqvist, R., Wallis, T.S., and Galyov, E.E. 2000. The secreted effector protein of *Salmonella dublin*, SopA, is translocated into eukaryotic cells and influences the induction of enteritis. Cell Microbiol. 2: 293–303.

Woodward, D.L., Khakhria, R., and Johnson, W.M. 1997. Human salmonellosis associated with exotic pets. J. Clin. Microbiol. 35: 2786–2790.

Woodward, M.J., McLaren, I., and Wray, C. 1989. Distribution of virulence plasmids within *Salmonellae*. J. Gen. Microbiol. 135: 503–511.

Worley, M.J., Ching, K.H., and Heffron, F. 2000. *Salmonella* SsrB activates a global regulon of horizontally acquired genes. Mol. Microbiol .36: 749–761.

Wray, C., Todd, J.N., and Hinton, M. 1987. Epidemiology of *Salmonella typhimurium* infection in calves: excretion of S. Typhimurium in the faeces of calves in different management systems. Vet. Rec. 121: 293–296.

Wray, C., Todd, N., McLaren, I.M., and Beedell, Y.E. 1991. The epidemiology of *Salmonella* in calves: the role of markets and vehicles. Epidemiol. Infect. 107: 521–525.

Yu, X.J., Ruiz-Albert, J., Unsworth, K.E., Garvis, S., Liu, M., and Holden, D.W. 2002. SpiC is required for secretion of *Salmonella* Pathogenicity Island 2 type III secretion system proteins. Cell. Microbiol. 4: 531–540.

Zansky, S., Wallace, B., Schoonmaker-Bopp, D., Smith, P., Ramsey, F., Painter, J., Gupta, A., Kalluri, P., and Noviello, S. 2002. From the Centers for Disease Control and Prevention. Outbreak of multidrug resistant *Salmonella* Newport–United States, January–April 2002. JAMA 288: 951–953.

Zhang, S., Santos, R.L., Tsolis, R.M., Mirold, S., Hardt, W.-D., Adams, L.G., and Bäumler, A.J. 2002a. Phage mediated horizontal transfer of the sopE1 gene increases enteropathogenicity of *Salmonella* enterica serotype *Typhimurium* for calves. FEMS Microbiol. Lett. 217: 243 247.

Zhang, S., Santos, R.L., Tsolis, R.M., Stender, S., Hardt, W.-D., Bäumler, A.J., and Adams, L.G. 2002b. SipA, SopA, SopB, SopD and SopE2 act in concert to induce diarrhea in calves infected with *Salmonella* enterica serotype *Typhimurium*. Infect. Immun. 70: 3843–3855.

Zhang, S., Kingsley, R.A., Santos, R.L., Andrews-Polymenis, H., Raffatellu, M., Figueiredo, J., Nunes, J., Tsolis, R.M., Adams, L.G., and Bäumler, A.J. 2003. Molecular pathogenesis of *Salmonella* enterica serotype *typhimurium*-induced diarrhea. Infect. Immun. 71: 1–12.

Zinder, N. (1957) Lysogenic conversion in *Salmonella typhimurium*. Science 126: 1237.

# *Shigella* Species

Keith A. Lampel

**17**

## Abstract

*Shigella* species are members of the family *Enterobacteriacae* and are Gram-negative, non-motile rods. Four subgroups exist based on O-antigen structure and biochemical properties; *S. dysenteriae* (subgroup A), *S. flexneri* (subgroup B), *S. boydii* (subgroup C) and *S. sonnei* (subgroup D). Clinical manifestations include mild to severe diarrhea with or without blood, fever, tenesmus, and abdominal pain. Further complications of the disease may be seizures, toxic megacolon, reactive arthritis and hemolytic uremic syndrome. Transmission of the pathogen is by the fecal–oral route, commonly through food and water. The infectious dose ranges from 10 to 100 organisms. *Shigella* spp. have a sophisticated pathogenic mechanism to invade colonic epithelial cells of the host, man and higher primates, and the ability to multiply intracellularly and spread from cell to adjacent cell via actin polymerization. Shigellae are one of the leading causes of bacterial foodborne illnesses and can spread quickly within a population.

## Characteristics of the organism

### Historical

In 1898, Shiga described an organism that caused dysentery in a patient that was distinct from amoebic dysentery (Shiga, 1906; see Bensted (1956) for a historical review). This new species was designated as *Bacillus dysenteriae*. Flexner (Flexner, 1900), 2 years later, as did Strong and Musgrave (Strong and Musgrave, 1900), identified a serologically related organism from patients in the Philippine Islands with dysentery. In that same year, Kruse isolated a nearly identical etiological agent from many cases of dysentery in Germany (Kruse, 1900); a slight controversy erupted since the latter described the *Bacillus* as being non-motile and the other reports suggested that the organism was slightly motile. In a spirit of compromise, the bacterium was described as non-motile and designated as the Shiga-Kruse *Bacillus*. Another serotype was isolated and identified in India in 1929 by Boyd (Boyd, 1931). The Shigella Commission at the Congress of the International Association of Microbiologists in 1950 (Enterobacteriaceae report 1954),

with input based on the work of Ewing (Ewing, 1949) adopted a new genus, *Shigella*, and the prototypical *Bacillus* was renamed *Shigella dysenteriae*.

## Taxonomy

The genus *Shigella* are members of the family *Enterobacteriacea*, tribe *Escherichiaeae*. There are four subgroups, *S. dysenteriae* (Group A), *S. flexneri* (Group B), *S. boydii* (Group C) and *S. sonnei* (Group D) and are subdivided into a total of 43 serotypes and subtypes; *S. sonnei* has only 1 serotype (Table 17.1). Genetically, these species are closely related evolutionarily with the genus *Escherichia coli* and are considered to be clones of this group. Historically, the shigellae were distinguished from *E. coli* based on their medical differences. However, proponents of reclassifying *Shigella* spp. within *E. coli* posit that since DNA-DNA reassociation studies and transconjugants examined from conjugation experiments between *Shigella* and *E. coli* demonstrated that these bacteria are closely related (Luria and Burrous, 1957; Falkow *et al.*, 1963; Brenner, 1969). In order to examine more closely the relatedness of these bacteria, recent analysis using multilocus enzyme electrophoresis, ribotyping, and DNA sequencing of select genetic loci, e.g. the malate dehydrogenase (*mdh*) gene, provided additional support that *Shigella* are, in essence, strains of *E. coli* (Lan and Reeves, 2002.). Accordingly, *Shigella* species have been grouped as *E. coli* strains and have been placed within ECOR groups A or B1 (Johnson, 1999). Further evidence of the relatedness of the *E. coli* and *Shigella* chromosome was confirmed by data generated from the sequence of the chromosome from *S. flexneri* 2a (Jin *et al.*, 2002; accession no. AE005674).

*Shigella* are short (1–3 μm), non-motile, non-pigmented, non-encapuslated, non-spore forming, faculatively anaerobic rods. Important biochemical traits are lysine decarboxylase negative (*S. boydii* serotype 13 is positive), lactose nonfermenting (with the exception of *S. sonnei* after prolonged incubation), do not produce gas from glucose utilization (*S. flexneri* and *S. boydii* serotypes 13 and 14 produce gas), do not produce hydrogen sulphide ($H_2S$), are ornithine decarboxylase negative (*S. sonnei* is positive) and salicin, adonitol, and inositol are not fermented (Edwards and Ewing, 1972). Biochemical reac-

**Table 17.1** Characteristics of *Shigella* spp.

| Species | Serogroup | Serotypes and subtypes | Geographic distribution | Distinguishing charcteristics |
|---|---|---|---|---|
| *S. dysenteriae* | A | 15 serotypes | Indian subcontinent, Africa, Asia, Central America | Type I produces Shiga toxin, causes most severe dysentery, high mortality rate if untreated |
| *S. flexneri* | B | Eight serotypes, nine subtypes | Most common isolate in developing countries | Elicits less severe dysentery |
| *S. boydii* | C | 19 serotypes | Indian subcontinent, rarely isolated in developed countries | Biochemically identical to *S. flexneri*, distinguished by serology |
| *S. sonnei* | D | one[a] serotype | Most common isolate in developed countries | Produces mildest form of shigellosis |

[a]Forms I and II are serotypically distinguishable.

tions are used to separate *Shigella* from *E. coli* and to differentiate the four species from each other (Tables 17.2 and 17.3). Utilization of mannitol and tests for ornithine decarboxylase and *o*-nitrophenyl-β-D-galactopyranosidase activity are useful for the latter.

Commensal *E. coli* strains are easily differentiated from shigellae with the notable exception of enteroinvasive *E. coli* (EIEC). Serotypes of this strain produce the same clinical symptoms, carry the same genetic information for pathogenesis as *Shigella* and in some instances, serologically cross-react with certain serotypes of shigellae (Cheasty and Rowe, 1983).

## Clinical manifestations

Bacillary dysentery is a highly communicable disease. The rapid spread of the etiological agent in some cases is due to poor sanitary conditions within a crowded situation. Another contributing factor is the low number of organisms that are required to cause disease. Studies with volunteers have shown that ingestion of fewer than 200 virulent cells is necessary to elicit clinical symptoms (Dupont *et al.*, 1989).

Clinical presentations range from mild diarrhea to severe dysenteric syndrome, the latter consisting of abdominal pain,

**Table 17.2** Tests to differentiate *Shigella* spp. from *E. coli*

| Test | Reaction | |
|---|---|---|
| | *Shigella* | *E. coli* |
| Motility | − | +[a] |
| Gas from glucose | −[b] | +[c] |
| Lysine decarboxylase | − | +[d] |
| Christensen's citrate | − | + |
| Acetate | − | + |
| Mucate | − | + |

[a]Most positive, some negative (enteroinvasive *E. coli* are nonmotile).
[b]Some strains of *S. flexneri* 6 produce small amounts of gas from glucose.
[c]Some exceptions.
[d]Enteroinvasive *E. coli* are also negative.

tenesmus, and bloody, mucoid stools. Fever is common and a copious amount of watery diarrhea is accompanied by large numbers of leukocytes. This may be a reflection of the anatomical events of bacterial destruction of the colonic mucosa and the inability of the host to reabsorb fluids manifesting in diarrhea (Rout *et al.*, 1975; Kinsey *et al.*, 1976). The presence of high numbers of leukocytes and shigellae in stool specimens occurs in the early stages of the disease. Later in infection, the bacteria spread from cell to cell, incurring more destruction and sloughing of the colonic mucosal cells as evident by the presence of blood, pus and mucous in stools. Shigellosis is a self-limiting disease, with the notably exception of *S. dysenteriae*, but can be fatal in immunocompromised individuals and malnourished people, particularly children under the age of 4 and the elderly.

Onset of illness usually occurs within 3 days after a 1–7 day incubation period and persists for 1–2 weeks. *S. dysenteriae* 1 carries the genetic information for a potent enterotoxin, Shiga toxin, and causes the most severe form of the disease and accounts for deadly epidemics of bacillary dysentery. This organism is typically isolated from patients in third world countries where poverty, crowded and unsanitary conditions exist. The mildest form of the disease is associated with *S. sonnei*, and is endemic in developed countries; infections caused by *S. flexneri* and *S. boydii* could be mild or severe. Further complications of shigellosis can be toxic megacolon, dehydration, intestinal perforation, seizures, reactive arthritis, and hemolytic uremic syndrome (HUS; Bennish, 1991). As with other gram-negative bacteria, reactive arthritis is associated with individuals of the HLA-B27 histocompatibility group (Brewerton *et al.*, 1973; Simon *et al.*, 1981; Bunning *et al.*, 1988). HUS, more commonly associated with *E. coli* 0157:H7, is a rare post infection sequela of *Shigella* and the sole etiological species is *S. dysenteriae* 1 (Raghupathy *et al.*, 1978). Shiga toxin, as with Shiga-like toxin of *E. coli* 0157:H7, may be involved with HUS by damaging the vascular endothelial cells of the kidney (Karmali *et al.*, 1985; Lopez *et al.*, 1989). Septicemia is not a usual complication from shigellosis although cases have been reported from around the world.

**Table 17.3** Biochemical differentiation of *Shigella* species

| Test | S. dysenteriae | S. flexneri | S. boydii | S. sonnei |
|---|---|---|---|---|
| β-Galactosidase | −[a] | − | − | + |
| Ornithine decarboxylase | − | − | − | + |
| Indole production | +/−[b,c] | +/−[c,d] | +/−[c] | − |
| | | | | |
| Acid from | | | | |
| Dulcitol | −[e] | −[f] | − | − |
| Lactose | − | − | −[g] | +[h] |
| D-Mannitol | − | + | + | + |
| Raffinose | − | +/−[c] | − | +[h] |
| Sucrose | − | − | − | +[h] |
| D-Xylose | − | − | +/−[c] | − |
| Melibiose | − | +/−[c] | +/−[c] | − |
| D-Sorbitol | +/−[c] | +/−[c] | +/−[c] | − |

[a]*S. dysenteriae* 1 are positive and *S. flexneri* 2a and *S. boydii* 9 have also been reported a as positive.
[b]*S. dysenteriae* 1 is negative; *S. dysenteriae* 2 is positive.
[c]Reaction is variable.
[d]*S. flexneri* 6 is negative.
[e]*S. dysenteriae* 1 may be positive.
[f]*S. flexneri* 6 may be positive.
[g]*S. boydii* may be positive.
[h]Positive reactions may take 24 hours or longer.
+/−, variable reaction.

## Pathogenesis

### Invasion of epithelial cells

In a highly orchestrated movement, *Shigella* species invade the host cell using their own genetic machinery and exploit the host inflammatory response to facilitate the invasion process. After ingestion, shigellae transit the acidic environment of the stomach and ultimately attach to target cells, currently thought to be the follicle-associated epithelium found on the mucosa-associated lymph nodes (LaBrec *et al.*, 1964). Entry is initially via engulfment by M cells through a macropinocytic event entailing cytoskeletal rearrangements on the apical surface. Attachment of the bacteria to the host cell induces the formation of filopodia and lamellipodia with the recruitment and polymerization of actin by the interaction of the *Shigella* protein IpaA and host vinculin (Tran van Nhieu *et al.*, 1997). These microfilament rearrangements eventually lead to engulfment of the bacterial cell by pseudopods allowing the entry into the host cell (see reviews P.J. Sansonetti, 1999; Gao and Kwaik, 2000).

Translocation across the epithelial barrier involves the lysis of endocytic vacuole. After translocation via M cells, *Shigella* cells are engulfed by macrophages. Subsequent to release from macrophages, bacterial cells invade the basolateral layer of the epithelial cells (Mounier *et al.*, 1992). *Shigella* spp. survive phagocytosis by inducing apoptotic killing of macrophages (Zychlinsky *et al.*, 1992). Apoptosis induced by *Shigella* generated proteins, notably IpaB, also solicits the immune system to initiate an inflammatory response to the bacterial invasion and is proposed to be an integral segment in assisting in apoptosis

(Zychlinsky *et al.*, 1994). Capase I appears to have a dual functions as it is involved with apoptosis and host inflammation. It binds with IpaB and this complex induces programmed cell death of macrophages and leads to the cleavage of pro-IL-1β (Hilbi *et al*, 1997, 1998). The release of chemokines, such as IL-1β and IL-8, are partially responsible for attracting polymorphic mononucleocytes (PMNs) to the site of inflammation, a characteristic of shigellosis. One possible role of PMNs is that once stimulated by the release of pro-IL-1β, these cells migrate to the basolateral membrane of epithelial cells, disturb the integrity of the lining, and facilitate the invasion of *Shigella* into the epithelium (Perdomo *et al.*, 1994).

### Intracellular multiplication/intercellular spread

Post invasion of *Shigella* into epithelial cells, the bacteria are able to multiply. Although *Shigella* are non-motile and do not posses flagella, they are capable of movement inside epithelial cells by a process of actin polymerization. IcsA (VirG) is an outermembrane protein that localizes at one pole of the bacterium and catalyzes the polymerization of host actin monomers (Bernardini *et al.*, 1989; Goldberg *et al.*, 1993). This unipolar localization of IcsA activity propels the bacterium through the host cell cytoplasm and enables the cell to penetrate an adjacent cell. The polymerized actin can be seen as a "comet" extending from the pole of the bacterium. As the bacterial cell enters the adjacent cell in a cytoplasmic protrusion, escape from these membrane structures is accomplished by the lytic action of an unknown bacterial protein.

## Genetics

*Plasmid-associated virulence genes*

In many pathogenic bacteria, genetic factors for virulence are encoded on a plasmid. In the case of all *Shigella*, a 180–220 kilobase pair plasmid carries most of the genes necessary for invasion, intracellular replication, and intracellular and intercellular spread as well as some genetic factors to regulate expression of the former genes (see Table 17.4). Even though plasmid size may vary between *Shigella* species and EIEC, the genes for virulence and their spatial arrangement remain conserved. The entire sequence of the virulence plasmid of *S. flexneri* 5a has been reported (Buchrieser *et al.*, 2000; Venkatesan *et al.*, 2001; accession numbers AL391753 and AF348706, respectively). The involvement of plasmid-borne genes was first demonstrated in 1981 (Sansonetti *et al.*, 1981; 1982). Subsequently, studies have shown that mutants of *S. flexneri* and *S. sonnei* that lack the plasmid do not display any invasive phenotype. Genetic introduction of the virulence plasmid into plasmid-cured strains restores the ability to invade mammalian tissue culture cells (Sansonetti *et al.*, 1983). Figure 17.1 depicts the invasion of *S. flexneri* into HeLa cells.

Most of the genetic factors required for virulence reside within a 37-kb region of the virulence plasmid (Maurelli *et al.*, 1985). A striking feature is that these genes are divided into two transcriptional clusters in opposing orientation. The *ipaBCDA* (invasion plasmid antigen) genes, which encode the immunodominant proteins recognized by sera of convalescent patients (Buysee *et al.*, 1987; Baudry *et al.*, 1987.), are located

in one cluster as are the *ipgCBA* (invasion plasmid gene) genes. The Ipa proteins are involved with invasion, and furthermore, Ipa B, C, and A may form a complex on the outer surface of the bacteria (Ménard *et al.*, 1994). As such, these proteins are considered necessary for signaling the uptake of the bacteria into the host cell. There is no apparent signal sequence on the Ipa proteins for secretion to the outer milieu of the bacterial cell. The Ipa proteins, as are IpgD and VirA, are translocated from within the bacterial cell to the outer surface by means of the Type III secretory pathway (Page and Parsot, 2002). The *mxi/spa* (membrane expression of invasion plasmid antigens/surface presentation of Ipa antigens) genes are contained in the other cluster and encode for the Type III secretion pathway. In addition to the translocators (*Shigella* proteins that are inserted into the host cell membrane), effectors (proteins secreted from the bacterium into the host cell cytoplasm) and secretion apparatus, this secretion system also consists of chaperones of the translocators and effectors and transcriptional regulators (*virB* and *mxiE*).

This system exists in other animal bacterial pathogens, e.g. *Salmonella enterica* serovar Typhimurium, enterohemorrhagic and enteropathogenic *E. coli*, *Yersinia* and in some plant pathogens such as *Pseudomonas solanacearum*, *Erwinia carotovaora*, and *Xanthomonas campestris* (Galán and Sansonetti, 1996; Hueck *et al.*, 1998). Furthermore, strong homology exists at the nucleotide level of the *mxi* genes amongst these pathogens. The *spa* genes have high homology with flagella genes of *Salmonella*, *Bacillus subtilis* and *Caulobacter crescentus* (Galán and Sansonetti, 1996). The *mxi/spa* genes are essential for virulence; mutations in one of these genes negates the ability of the pathogen to invade mammalian tissue culture cells. Proteins encoded by the *mxi/spa* loci have been shown by electron microscopy to form a "needle complex" with a basal portion (Tamano *et al.*, 2000). The entire complex spans the inner and outer membrane of the bacterial cell and acts as a conduit for the effector proteins, e.g. Ipa proteins, to the cell surface.

The *ipg* genes, located in both clusters, encode chaperones of the Ipa proteins. Mutants that lack IpgC degrade IpaB and IpaC rapidly and IpgE appears to be a chaperone for IpgD, a protein that is also secreted by the Type III secretory apparatus (Niebuhr *et al.*, 2000).

Other genes, not within either of these two clusters, have significant roles in pathogenesis. The gene product of the *icsA* (*virG*) gene, IcsA, as described above, is critical for cell movement within the epithelial cell. This protein catalyzes the polymerization of host actin from one pole of the bacterial cell enabling the organism to be propelled from one epithelial cell to an adjacent cell. A new class of exported proteins was identified from the sequence of the virulence plasmid. Designated *osp* (outer *Shigella* proteins), the function of these genes has not been determined (Buchrieser *et al.*, 2000).

VirB controls the expression of the *ipa*, *mxi* and *spa* genes (Adler *et al.*, 1989). Temperature is a critical factor affecting the activity of the *virB* gene. At 30°C, shigellae are avirulent; upon a shift to 37°C, the pathogen is able to express its viru-

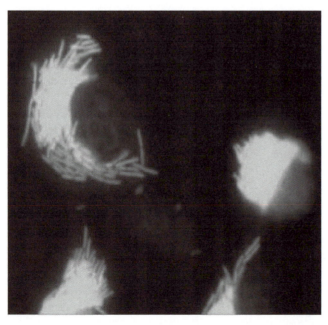

**Figure 17.1** Invasion of *S. flexneri* 2457T, modified by carrying a plasmid expressing GFP (green fluorescent protein), into HeLa cells stained with Hoechst dye. Anita Verma, Uniformed Services University of the Health Sciences, Bethesda, MD, kindly provided this photo.

lence phenotype as measured by assays in tissue culture cells. Although regulation of virulence gene expression is temperature dependent, the exact mechanism of this control has not been elucidated. The transcription of the *virB* gene is regulated by another plasmid-encoded gene, *virF*, a transcriptional activator in the AraC family (Tobe *et al.*, 1993). The VirF is a DNA-binding protein and, as such, binds to a region upstream of the *virB* gene where it regulates the expression of *icsA*.

## Chromosomal-encoded genes

Chromosomal-encoded virulence genes are listed in Table 17.4. Biosynthesis of the O-antigen, an important component of the lipopolysaccharide (LPS) structure, is dependent on the genes located in the *rfb* cluster. The LPS molecule is composed of the Lipid A moiety, which attaches the LPS structure to the bacterial outer membrane, a succession of saccharide units that lies between the Lipid A and the O-antigen, which extends from the bacterial surface. The O-antigen composition is responsible for the antigenic differences exhibited by *Shigella* as with other Gram-negative organisms. Mutations that affect the synthesis of the oligosaccharides that comprise the O-antigen or its assembly in *Shigella* spp. lead to a "rough" phenotype indicative of an avirulent strain (Sandlin *et al.*, 1995, 1996). Proper LPS structure is required for the correct unipolar orientation of IcsA on the cell surface and the genes *rfa* and *rfb*,

have been found by mutation analysis to be essential for this activity (Hong and Payne, 1997). Although most genes responsible for O-antigen synthesis are located on the chromosome in most *Shigella* spp., some of these genetic determinants are encoded in plasmids or lysogenic bacteriophages in *S. sonnei*, *S. dysenteriae* 1 and *S. flexneri* 1–5.

*S. dysenteriae* 1, like *E. coli* O157:H7, produces a potent enterotoxin that has been implicated in producing some of the symptoms of HUS. The recruitment of the Shiga toxin genes, *stxA* and *stxB*, of *S. dysenteriae* has been postulated to be a consequence of a lysogenic bacteriophage. Over time, the phages became stably integrated due to loss of genetic function of essential phage genes, most likely as a result of rearrangement and transposition events. Shiga toxin (Stx) inhibits protein synthesis at the ribosomal 60S subunit of mammalian cells by inactivating ribosomal RNA (28S rRNA) and preventing elongation factor 2 from interacting with the ribosome (O'Brien and Holmes, 1996). Stxs are found infrequently in other *Shigella* spp. The Stx produced by *S. dysenteriae* and the Stx1 elaborated by *E. coli* O157:H7 are nearly identical, differing from each other by one amino acid.

Pathogenicity islands, clusters of genes that have a role in virulence, have been identified in several virulent bacteria including *Shigella*. These regions are typically large in size, 20–200 kb, and in addition to virulence genes, may have transpos-

**Table 17.4** Virulence loci of *Shigella*

| Locus | Plasmid-encoded genes | |
|---|---|---|
| | Protein | Role in virulence |
| *ipaA* | 70 kDa protein | Invasion; associates with vinculin |
| *ipaB* | 62 kDa protein | Invasion; lysis of vacuole; induction of apoptosis |
| *ipaC* | 43 kDa protein | Invasion; induces formation of filopodia and lamellipodial extensions |
| *ipaD* | 38 kDa protein | Invasion; forms anti-secretion complex with IpaB |
| *ipgC* | 17 kDa protein | Chaperone for IpaB and IpaC |
| *ipgD* | 59.8 kDa protein | Modulates invasion of bacteria into epithelial cells |
| *ipaH* | Family of proteins | Present in plasmid (five copies) and chromosome (seven copies) |
| *mxi/spa* | 20 proteins | Secretion of Ipa and other virulence proteins |
| *icsA (virG)* | 120 kDa protein | Actin polymerization for intracellular motility and intercellular spread |
| *sen* | Shet2 | Enterotoxin |
| *virB* | transcriptional activator | Temperature regulation of virulence genes |
| *virF* | Transcriptional activator | Temperature regulation of virulence genes |
| *osp* | Outer *Shigella* proteins | P roteins translocated by type III secretory system |
| *virR (hns)* | Histone-like protein | Repressor of virulence gene expression |
| *rfa*; *rfb* | Enzymes for LPS core and Unipolar localization of IcsA; O-antigen biosynthesis |
| *stx*[1] | Shiga toxin | Destruction of vascular tissue |
| *vacB (rnr)* | Exoribonuclease RnaseR | Posttranscriptional regulation of virulence gene expression |
| *cpxR* | Response regulator of *virF* | CpxA–CpxR two component system |
| *luc* | Aerobactin and receptor | Acquisition of iron in the host |
| *SodB* | Superoxide dismutase | Defense against oxygen-dependent killing in host |
| *set*[2] | ShET1 | Enterotoxin |
| *dsbA* | Disulfide bond catalyst | Facilitates secretion of IpaB and IpaC |
| *sigA* | 139.6 kDa exported | Intestinal fluid accumulation cytopathic protease |

[1]The *stx* locus and production of Shiga toxin is observed only in *S. dysenteriae* 1.
[2]The *set* locus and production of ShET1 is observed almost exclusively in *S. flexneri*.

able elements, e.g. IS elements, and other modes of mobility such as plasmid or bacteriophage genes, all indicating possible horizontal transfer of genetic information. Three chromosomal pathogenicity islands have been identified in *Shigella*: SHI-1 (Rajakumar *et al.*, 1997; Al-Hasani *et al.*, 2001), 46 kb in size, encodes for an enterotoxin gene (*set*); SHI-2, is 23.8 kb in size and contains the *iuc* locus, the genes for aerobactin synthesis and transport used in iron acquisition (Moss *et al.*, 1999.); SRL (*Shigella* resistance locus) is 42 kb in size and has the genetic determinants for several antibiotic resistance factors (Walker and Verma, 2002).

An interesting observation of early genetic studies by Formal and colleagues (1963), and from recent reports that select genetic loci that are not present in the *Shigella* chromosome but are found in the non-pathogenic *E. coli* chromosome (Maurelli *et al.*, 1998), suggest that the absence of particular genes in *Shigella* have had a direct consequence on its pathogenesis. The absence of these genetic determinants in *Shigella* spp. putatively occurred by either deletion or mutation and are presently referred to as "black holes" (Maurelli *et al.*, 1998). When a complementary gene to a genetic locus that is absent in *Shigella* is introduced into the pathogen, the presence of that newly acquired gene product in *Shigella* may inhibit the action of a particular virulence factor. For instance, the product of the *cadA* gene, lysine decarboxylase, yields cadaverine from lysine. Cadaverine inhibits the action of the *Shigella* enterotoxins that may be responsible for the diarrheal symptoms associated with shigellosis. Therefore, the presence of genetic determinants such as *cadA* and *ompT*, the latter encoding a protease that acts on IcsA preventing actin polymerization and subsequent movement of the bacterium through the epithelial cell (Nakata *et al*, 1993), may be considered "anti-virulence genes" (Maurelli *et al.*, 1998). Thus, in the evolution of *Shigella* as a human pathogen from a non-pathogenic *E. coli* lineage, the acquisition of the large virulence plasmid may have also involved the loss of specific incompatible genetic loci.

An important component of the regulatory cascade affecting the expression of the virulence genes is the *virR*/*hns* locus (Maurelli and Curtiss, 1988). A repressor gene product, VirR, interacts with two other regulatory genes, *virF* and *virB*, both are on the large virulence plasmid and mediates the expression of these two genes. Temperature regulation occurs at the level of transcription (Maurelli and Sansonetti, 1984); at 30°C, *virF* and *virB* are not expressed. At 37°C, the negative control of these two genes is abated and transcription of the virulence genes commences.

## Epidemiology

Diarrheal diseases are the leading worldwide cause of death among children. The World Health Organization estimates that 5 million deaths occur annually from diarrheal disease, and shigellae are responsible for 10% of these mortalities (Kotloff *et al.*, 1999). Epidemic dysentery, usually prolonged and large, caused by *S. dysenteriae* 1 (the only *Shigella* spp. to pro-

duce Shiga toxins), is a recurrent problem in many of the poorest areas of the world, notably in Africa, Central America, and parts of Asia. Many of these outbreaks are caused by multiple antibiotic resistant strains; the fatality rate of these infections can be as high as 20%. Furthermore, antimicrobial resistance appears to develop more quickly with *S. dysenteriae* 1 than in other *Shigella* spp (http://www.who.int/inf-fs/en/fact108. html).

Due to the low infectious dose of shigellae, the dissemination of the bacteria from person to person can be extremely swift and can be responsible for the high secondary attack rate when introduced within environs such as crowded and institutionalized populations. Children ages 1 to 6 are most susceptible to infections due to shigellae. This phenomenon is compounded in poorer developing nations because of the high numbers of malnourished children. These children face an increase in the attack and relapse rates and also greater mortality.

*S. sonnei* is predominantly found in industrialized countries and has become a significant problem in daycare centers and preschools. The reason for these incidences has been attributed to the poor and underdeveloped sanitary practices by toddlers and, in some cases, the daycare or preschool personnel, whose duties include attending to toddlers' sanitary needs and as food preparers, are not attentive about their own hygienic practices.

*S. dysenteriae* is frequently found in developing countries and causes the most severe and prolonged form of dysentery with the highest frequency of fatality than other shigellae. *S. flexneri* is found in more developed regions and also in significant numbers in countries such as the United States where 15% of foodborne outbreaks that are caused by *Shigella* are attributed to this pathogen; the majority of cases (85%) are due to *S. sonnei*. *S. boydii* is rarely isolated except in the Indian subcontinent and some cases have been reported recently from Europe and the United States.

In one study of pediatric and adult patients who presented bloody diarrhea in emergency rooms in the United States, *Shigella* was found as the leading cause of this particular symptom (15.3%) followed by *Campylobacter* (6.2%) and then *Salmonella* (5.8%; Talan *et al.*, 2001). Data from the 1998 FoodNet surveillance network found *Campylobacter*, *Salmonella* and *Shigella* as the most common isolates from patients with bacterial gastroenteritis (CDC, 1999). An increase in shigellosis amongst homosexual men has been reported in several countries with *S. sonnei* and *S. flexneri* associated with these illnesses (CDC, 2001; O'Sullivan *et al.*, 2002).

## Reservoirs

Unlike other bacterial human pathogens, which may have multiple animal or environmental niches, humans are the principal natural reservoir of *Shigella* infections. Infected individuals can shed $10^3$ to $10^9$ colony-forming units (CFU) per gram of stool during the acute phase and in convalescing patients, $10^2$ to $10^3$ CFU/g can be recovered. In many geographical areas, shigellosis outbreaks are seasonal with summer months showing the highest number of incidences. In warmer months, people's interactions increase; this augments the likelihood of symptomatic and asymptomatic carriers coming into contact with uninfected and susceptible population. Asymptomatic carriers of *Shigella* may exacerbate the spread and maintenance of this pathogen in developing countries. Two studies, one in Bangladesh (Hossain *et al.*, 1994) and the other in Mexico (Guerro *et al.*, 1994), show that *Shigella* was isolated from stool samples collected from asymptomatic children under the age of 5 years. *Shigella* are rarely found in infants under the age of 6 months.

Higher primates have also been known to harbor *Shigella* particularly at zoos and primate centers (Banish *et al.*, 1993, Line *et al.*, 1992). In one instance (Kennedy *et al.*,1993), three animal caretakers at a monkey house contracted diarrheal symptoms. Further examination showed that *S. flexneri* 1b was isolated from stool samples of these employees and that the four monkeys were shedding the identical serotype. The disease was spread by direct contact of the caretakers with excrement from the infected monkeys. In a recent study of free-ranging mountain gorillas in two national parks in Uganda, *S. flexneri*, *S. sonnei*, and *S. boydii* were isolated from preadult to adult gorillas for the first time during this particular study. Increase to human exposure may have contributed to anthropozoonotic transmission of these *Shigella* spp. (Nizeyi *et al.*, 2001).

## Outbreaks

### *Transmission and susceptible populations*
The primary means of human-to-human transmission of *Shigella* is by the fecal–oral route. Most cases of shigellosis are caused by the ingestion of fecal-contaminated food or water, and in the case of foods, the major factor for contamination is the poor personal hygiene of foodhandlers. From carriers, this pathogen can spread by several routes including food, fingers, feces, flies and a relatively new category, fomites, the latter considered to be inanimate objects, such as utensils and cutting surfaces. Flies usually transmit the bacteria from fecal matter to foods. Improper storage of contaminated foods is the second most common factor that accounted for foodborne outbreaks due to *Shigella* (Smith 1987). Other contributing factors are inadequate cooking, contaminated equipment and food obtained from unsafe sources (Bean *et al.*, 1990). To reduce the spread of shigellosis, infected patients, particularly in day-care centers, are monitored until stool samples are negative for *Shigella*.

*Shigella* is a frank pathogen capable of causing disease in otherwise healthy individuals. Certain populations, however, may be more predisposed to infection and disease due to the nature of transmission of the organism. The greatest frequency of illness occurs among children less than 6 years of age. In the United States, outbreaks of shigellosis and other diarrheal diseases in daycare centers are increasing as more single- and two-parent working families and working women turn to

these facilities to care for their children (Levine and Levine, 1994; Pickering *et al.*, 1986). Typical toddler behavior such as oral exploration of the environment and inadequate personal hygiene habits creates conditions ideally suited to bacterial, protozoal and viral pathogens that are spread by fecal contamination. Transmission of *Shigella* in this population is very efficient and the low infectious dose for causing disease increases the risk for shigellosis. Increased risk also extends to family contacts of daycare attendees (Weisman *et al.*, 1974).

Shigellosis can be endemic in other institutional settings, such as prisons, mental hospitals and nursing homes, where crowding and/or insufficient hygienic conditions create an environment for direct fecal–oral contamination. Crowded conditions and poor sanitation contribute to shigellosis being endemic in developing countries as well.

When natural or man-made disasters destroy the sanitary waste treatment and water purification infrastructure, developed countries assume the conditions of developing countries. These conditions place a population at risk for diarrheal diseases such as cholera and dysentery. Examples include the war in Bosnia-Herzegovina, famine and political upheaval in Somalia. Massive population displacement, e.g. refugees fleeing from Rwanda into Zaire in1994, can also lead to explosive epidemics of diarrheal disease caused by *Vibrio cholerae* and *S. dysenteriae* 1 (Goma Epid. Group, 1995).

## Foodborne outbreaks
As illustrated below, a wide range of foods have been contaminated with shigellae and the number of people affected in these examples of outbreaks of shigellosis varies from a few to many. In some cases, rapid dissemination of the pathogen occurred causing additional number of cases, not only within the immediate population but also there was a high secondary attack rate. Furthermore, with increasing globalization of commerce, the amount of foods being transported from one country to another has increased. In the context of foodborne illnesses, this raises the issue of how to ensure that imported foods are safe for the consumers.

### *Examples of foodborne outbreaks*
*Rainbow Family gathering – 1987*. At an annual gathering of the Rainbow Family, as many as half of the 12,700 people in attendance may have had shigellosis (Wharton *et al.*, 1990) with *S. sonnei* as the causative agent. Spread of the organism most likely occurred by the fecal–oral route in probably contaminated food or water with all the classic scenarios for rapid dissemination; a crowded environment and poor hygienic conditions. Cluster of outbreaks were reported in other states, probably due to attendees returning home and secondarily infecting other individuals. Overall, attendees from at least 26 states were positively identified with clinical symptoms.

*Cruise ships – 1989 and 1994*. In October 1989, 14% of the passengers and 3% of the crew members aboard a cruise ship reported having gastrointestinal symptoms (Lew *et al.*, 1991).

A multiple-antibiotic resistant strain of *S. flexneri* 2a was isolated from several ill passengers and crew. The source of this outbreak was identified as German potato salad. Contamination was introduced by infected food handlers, first in the country where the food was originally prepared and second by a member of the galley crew on the cruise ship. Another outbreak of shigellosis occurred in August 1994, on the cruise ship *SS Viking Serenade* (CDC, 1994). Thirty-seven percent of the passengers and 4% of the crew reported having diarrhea and one death did occur. In this outbreak, *S. flexneri* 2a was isolated from patients. The suspected source of contamination was spring onions.

*Military campaign Operation Desert Shield – 1990.* Diarrheal diseases during a military operation can obviously reduce the effectiveness of troops. In Operation Desert Shield, enteric pathogens were isolated from 214 US soldiers and, out of those, 113 cases were diagnosed with *Shigella*; *S. sonnei* was the prevalent of the four shigellae isolated (Hyams *et al.*, 1991). Shigellosis accounted for more time lost from military duties and was responsible for more severe morbidity than enterotoxigenic *E. coli*, the most common enteric pathogen isolated from USA troops in Saudia Arabia. The suspected source was contaminated fresh vegetables, notably, lettuce. Twelve heads of lettuce were examined and enteric pathogens were isolated from all.

*Local outbreak-moose soup in Alaska – 1991.* In September 1991, the Alaska Division of Public Health was contacted about a possible gastroenteritis outbreak (Gessner and Beller, 1994). In Galena, 25 people who had gathered at a local community event in which homemade foods were consumed contracted shigellosis. The implicated food was homemade moose soup. One of the five women who had prepared the soup reported that she had gastroenteritis before or at the time of preparing the soup. *S. sonnei* was isolated from one hospitalized patient.

*Multicountry outbreak-contaminated lettuce in Europe – 1994.* One hundred and ten culture confirmed cases of shigellosis caused by *S. sonnei* were reported in an outbreak in Norway in 1994 (Kapperud *et al.*, 1995) and possibly responsible for increases in the number of shigellosis cases in other European countries including the United Kingdom (Frost *et al.*, 1995) and Sweden. Iceberg lettuce from Spain, served in a salad bar, was suspected as the source of the multicountry outbreak based on epidemiological evidence. Furthermore, high numbers of coliform bacteria were evident on the lettuce indicating fecal contamination.

*Fresh parsley – 1998.* During the months of July and August, 1998, several public health departments in the United States and Canada were notified that there were outbreaks of shigellosis. Most of the affected people ate at restaurants that served chopped, uncooked parsley. The causative agent was identified as *S. sonnei* and one farm in Mexico was implicated as the source of contaminated parsley based on PFGE of isolated strains, epidemiologic, traceback and data from other investigations. The water supply was unchlorinated, prone to bacterial contamination (CDC, 1999) and was used for chilling the parsley in a hydrocooler and making ice.

*Bean dip – 2000.* An outbreak of shigellosis caused by the ingestion of contaminated five-layer (bean, salsa, guacamole, nacho cheese, and sour cream) party dip occurred in three west coast states in the United States (CDC, 2000) and affected at least 30 people. The causative agent, *S. sonnei*, was isolated from one layer of the dip.

One of the striking features about foodborne outbreaks caused by shigellae is that in many situations, contamination of foods may not have originated at the processing plant but rather the source can be traced to either a foodhandler or fecal containing water used for irrigation. As evident from the examples above, these incidents can occur by improper food handling from individuals to small town gatherings and picnics and larger scale outbreaks such as those on cruise ships and institutions.

## Susceptibility to physical and chemical agents and factors influencing survival

### Laboratory conditions

The ability of *Shigella* spp. to either grow or survive under a myriad of laboratory conditions in broth cultures are obviously influenced by temperature and pH, the latter also dependent on the type of acid supplemented in the medium. Various studies have addressed the impact of these environmental concerns on how storage conditions of foods can affect the growth or survival of the pathogen in this particular milieu. Although *Shigella* are considered more acid tolerant than some other enteric bacteria and this characteristic may enable these pathogens to transit the acidic conditions of human digestive system. Yet under some laboratory conditions and in fecal samples, shigellae are unable to grow and do not survive well in an acidic environment.

Growth of *Shigella* spp. in broth medium is observed over a range of temperatures. *S. sonnei* can grow at temperatures as low as 6°C and *S. flexneri* at 7°C and up to 48°C. Under laboratory conditions, *S. sonnei* and *S. flexneri* grow in culture media with nearly the same pH values, between 4.5 and 9.3. Under certain culture conditions with media supplemented with organic compounds, such as formic or acetic acid, salts (3.8–5.2% NaCl), or nitrite (300 to 700 ppm), no growth of shigellae is observed. Although no growth is observed, survival of shigellae can be extended depending on the temperature; lower temperature appears to support the survival time of *S. flexneri* in broth cultures at different pH values with different organic acids (Zaika, 2002a). Salt (NaCl) concentration and its affect on the growth and survival of *S. flexneri* and *S. sonnei*, is greatly influenced by temperature and pH. In media with a pH of 4, no growth is observed and survival of the bacteria declined. In brain heart infusion medium, growth of *S. sonnei* and *S. flexneri* was observed at a minimum pH of 4.50 and 4.75, respectively (Bagamboula *et al.*, 2002a). Overall, *S. flexneri* can grow and survive in broth medium supplemented with a range of salt concentrations under different pH > 5 and temperature, 12 to 37°C.

## Survival and growth in foods

Smith (Smith, 1987) has summarized which foods are commonly associated with foodborne outbreaks caused by *Shigella* spp. Recently, efforts have been directed to determine how well this pathogen grows and survives in foods with specific attention to produce/vegetables. Growth can most likely occur in foods under the right environmental conditions; temperatures above suggested proper storage settings, non-deleterious pH values, and lack of chemical and biological inhibitors. *Shigella* spp., as will be exemplified below, can survive in a broad milieu of food matrixes. For instance, *S. sonnei* and *S. flexneri* can survive at 4°C for 21 days in foods commonly implicated in outbreaks, such as cheese, potato salad and mayonnaise.

In foods, survival time is quite different at –20°C, 4°C, room temperature and a brief exposure to 80°C; *S. flexneri* and *S. sonnei* can survive for the longest period of time at room temperature. In acidic foods, such as citrus juices and carbonated soft drinks, the survival time for *S. flexneri* and *S. sonnei* is from 4 hrs to 10 days (Internation. Comm. Microbiol. Specif. Foods, 1996).

In a more detailed analysis, *S. sonnei* and *S. flexneri* did not grow at 22°C in either apple juice, pH 3.3 to 3.4 or tomato juice, pH 3.9 to 4.1. (Bagamboula *et al.*, 2002a) As for survival time at 7°C and 22°C, the number of strains that were detected depended on the strain and which juice was used in this particular study. For instance, in tomato juice, laboratory collection strains of *S. sonnei* were recovered at the end of the study period, 14 days, whereas clinical isolates were isolated after only 8 to 11 days. A clinical isolate of *S. sonnei* incubated in apple juice at 22°C were not recovered after 6 days and for the laboratory strains of *S. sonnei* and *S. flexneri*, they were not isolated after 6 or 8 days, respectively. At 7°C, results were different in both juices; survival time was extended to 10 to 14 days for both *Shigella* spp.(Bagamboula *et al.*, 2002a) For example, in citric juices (orange, grape, lemon), carbonated beverages and wine, variable recovery of shigellae was obtained after 1 to 6 days (see Internation. Comm. Microbiol. Specif. Foods, 1996). In neutral pH foods, such as butter or margarine, shigellae can be recovered after 100 days when stored frozen or at 6°C. *S. dysenteriae*, tested in orange juice at 4°C and grape juice at 20°C, survived up to 170 hours and 2 to 28 hours, respectively (Internation. Comm. Microbiol. Specif. Foods, 1996). Depending upon the *in vitro* conditions, *Shigella* sp. can survive in media with a pH range of 2 to 3 for several hours. Acid resistance seems to be modulated by a sigma factor encoded by the *rpoS* (*katF*) gene and at least two other genes, *hdeA* and *gadC*, are involved (Waterman and Small, 1996).

Few studies have been implemented to examine the survival and growth of *Shigella* in foods commonly associated with foodborne outbreaks. In order to address these concerns, a variety of produce was tested as to their effect on growth and survival of *Shigella* (Rafii *et al.*, 1995; Internation. Comm. Microbiol. Specif. Foods, 1996; Zaika and Scullen, 1996; Rafii and Lunsford, 1997; Bagamboula *et al.*, 2002b). Although different strains, inoculation numbers, incubation conditions and recovery/enumeration parameters were reported in these studies, some information has emerged that indicates how these pathogen may fare in foods. In most vegetables and other types of foods tested, e.g. carrot, cauliflower, radish, celery, broccoli, green pepper, onion, cabbage, strawberry, fresh fruit salad, pea, and broths (beef, chicken, vegetable), *Shigella* survived for more than 10 days at refrigerated temperatures (4–10°C). Composition of the food matrix and resident microbial flora had a pronounced effect on survival. For example, recovery of *S. flexneri* or *S. sonnei*, seeded onto with strawberries (pH 3.47) and fresh fruit salad (pH 3.73), 50–5000 CFU/g, and incubated at 4°C for 4–48 h was mixed. *S. sonnei* was recovered from both commodities whereas *S. flexneri* was only isolated from fruit salad and not strawberries. In some cases, lactic acid bacteria were identified as part of the microbial flora. Under conditions favoring growth, it was speculated that these bacteria lowered the pH of the environmental milieu and may have produce conditions unfavorable for both growth and survival of *Shigella*. As noted above, *Shigella* spp. do not grow in broth medium when the pH is less than 4.5. Growth of *Shigella* at elevated temperatures (12–37°C) was mixed. In most cases, growth was observed and in some foods, e.g. carrots, initial growth did occur followed by a significant decline in the number of recovered cells (Bagamboula *et al.*, 2002b).

Survival of shigellae in foods under acidic conditions depends on temperature and type of acid. *Shigella* can survive a temperature range of –20°C to room temperature. Survival of shigellae was longer in foods stored frozen or at refrigeration temperatures than at room temperature. In foods, such as salads with mayonnaise and some cheese products, *Shigella* survived for 13 to 92 days. These organisms can survive on dried surfaces for an extended period of time and in foods, e.g. shrimp, ice-cream, minced pork meat, stored frozen. Growth and survival rates of *Shigella* are impeded in the presence 3.8 to 5.2% NaCl at pH 4.8 to 5.0, in 300 to 700 mg/L NaNO$_2$, in 0.5 to 1.5 mg/L of sodium hypochlorite (NaClO) in water at 4°C. Data from a recent report (Zaika, 2002b) indicated that *S. flexneri* survived well in broth with pH values higher than 4 and with NaCl concentrations from 0.5% to 7% grown at temperatures of 12°C, 19°C, 28°C, and 37°C. However, growth was impeded in the presence of NaCl. They are sensitive to ionizing radiation and a reduction of 10$^7$ at 3 kGy is observed.

Factors that affect growth and survival of *Shigella* spp. in foods include not only pH, $a_w$, salt concentration and temperature, but chemical components of the food matrix and the indigenous microbial flora may also have a significant influence. Understanding the dynamics of the aforementioned factors, either as individual or collective components, may provide a more in-depth insight as to the affect that these environmental parameters have on the number of *Shigella* spp. present in the food during short or prolonged storage. Whether these conditions lead to either an increase in the total count of the pathogen or the maintenance of sufficient numbers of infectious agents, knowing the potential for contamination is important to prevent contaminated foods from reaching the consumer. For example, in some cases of shigellosis caused by the ingestion of contaminated produce, foods are rendered unsanitary at the site of production/harvest by fecal-laden irrigated water or poor sanitary hygienic practices by those responsible for harvest or by a food handler post harvest.

From several studies, its apparent that *Shigella* spp. survive well in most foods; the ability to increase in numbers may not be a significant point considering that the infectious dose is 10–200 organisms and the number of shigellae on the foods may greatly exceed this figure. From epidemiological data, it becomes obvious how one contaminated commodity can transverse the commercial food process maze and affect a great number of unsuspected consumers. Therefore, can methods be developed that would eradicate pathogens from foods yet retain the quality of that particular commodity? As simple an idea that this is, the concept to actually implement such a process becomes a complex issue. The food industry and regulatory agencies must grapple with whatever system is developed to reach a satisfactory approach that meets the demands of all concerned parties, including the health and safety of the consumer.

## Isolation and identification

### Bacteriological approach

Foods are not routinely examined for the presence of *Shigella* unless epidemiological data suggest that it may be the source of an outbreak. Vegetables and salads are common contaminated sources (Smith, 1987). United States government regulatory departments and agencies have implemented programs to test imported and domestic produce for the presence of select bacterial agents as a means to control the introduction of contaminated foods to the consumer. A survey of how extensive this problem is and how the US FDA is addressing this issue has been published (Beru and Salsbury, 2002; www.cfsan.fda.gov/~dms/prodact.html).

Foods can be contaminated in several ways, including fecal-contaminated irrigation water and, more likely, an infected foodhandler with poor personal hygiene. Unlike clinical samples of stool, with high numbers of *Shigella* and uniform composition, foods pose significant obstacles to successful isolation and subsequent identification of this bacterium. It is not unusual for a week to lapse before suspected foods may be analyzed for the presence of *Shigella*. In this time period, the food, if any still remains, may be further adulterated and the bacterial population may also be compromised to the point where recovery of viable organisms is virtually impossible.

The Bacteriological Analytical Manual (Andrews and Jacobsen, 2001; web site www.cfsan.fda.gov/~ebam/bam-6.html) details one method to isolate shigellae from foods. Twenty-five gram samples are added to 225 mL of *Shigella* broth supplemented with 3 μg/mL novobiocin (0.3 μg/mL is used for *S. sonnei*). After 10 min at room temperature, the mixture is incubated overnight under anaerobic conditions at 44°C for *S. sonnei* and at 42°C for the other *Shigella* spp. Alternatively, other enrichment broths can be used such as Gram-negative broth. Gram-negative broth is less inhibitory for *Shigella* because of the low amount of desoxycholate present in the media and the addition of mannitol that encourages the growth of *Shigella*.

As the analysis of suspected contaminated foods may not commence for a period of 7 to 10 days, the presence of injured, weak or damaged *Shigella* cells is probable (Smith and Palumbo, 1982). Selective media without bile salts and desoxycholate are recommended since the presence of these compounds may adversely affect the growth of impaired cells (Tollison and Johnson, 1985). Incubating foods seeded with *Shigella* spp. that may induce stress, e.g. reduced temperatures, recovery of *S. flexneri* is less than *S. sonnei*. In one report, recovery of *S. flexneri* was lower on Hektoen agar plates as compared to tryptone soya agar plates (Bagamboula *et al.*, 2002b).

A range of selective agar media is recommended to plate the overnight cultures. Two to three different selective media should be used to increase the chance of recovering *Shigella*. Growth of *Shigella* on MacConkey agar, a low selectivity medium, produces colonies that are 2 to 3 mm in diameter, translucent, slightly pink (*Shigella* are lactose negative), and may possess rough edges. Eosin methylene blue (EMB) and Tergitol-7 agar are alternative low selectivity agars containing lactose. Nonpigmented, semitranslucent colonies on EMB plates or bluish colonies on the yellowish-green Tergitol-7 agar are indicative of *Shigella*. Desoxycholate and xylose-lysine-desoxycholate (XLD) agars are intermediate selective media and are preferred media to isolate *Shigella* spp. *Shigella* colonies on XLD agar medium are typically 1 to 2 mm in diameter, translucent and red (alkaline). Although most *Shigella* do not ferment xylose, some species, e.g. *S. boydii* (variable), may be missed and therefore plating on XLD and MacConkey agar plates is recommended. *Shigella* spp. form reddish colonies on desoxycholate agar. Highly selective medium include *Salmonella-Shigella* and Hektoen agars. Some *Shigella* spp., such as *S. dysenteriae* type I, are unable to grow on highly selective *Salmonella-Shigella* medium. *Shigella* produce colorless, translucent colonies on this agar medium. Colonies on Hektoen agar appear to be green as do colonies from *Salmonella* spp; *E. coli* strains form yellow colonies. All presumptive colonies are inoculated into semisolid (motility) test agar. *Shigella* spp. are non-motile. Table 17.5 summarizes colony formation on selected agar medium for *Shigella*, *Salmonella* and *E. coli* strains.

Media containing chromogenic or fluorogenic indicators have been applied to isolating and detection regimens for a number of pathogens, notably *E. coli* O157:H7 and *Salmonella* spp. As for the former bacterium, the use of cefixime tellurite sorbitol MacConkey medium has greatly improved the rapid identification of this pathogen. Since most *E. coli* O157:H7 Shiga-toxin producing strains lack $\beta$-D-glucoronidase activity, the inclusion of the fluorogenic indicator for this enzyme aids as a marker to distinguish this particular strain of *E. coli* from other strains of the same genus. As for *Shigella*, several companies have attempted to develop a media containing a chromogenic indicator but presently, none of these have been made available commercially.

Biochemical tests are used to further identify *Shigella* spp. Suspected colonies that are Gram-negative, non-motile rods are inoculated onto lysine iron or Kliger iron agar and incubated 18 to 24 hours at 37°C. *Shigella* produce alkaline (red) slants, acid (yellow) butt, no $H_2S$, and no gas, with the exception of *S. flexneri* serotype 6, on these agars. Similar to other enteric bacteria, *Shigella* are oxidase-negative, ferment glucose and, except for *S. dysenteriae* type 1, are catalase positive. Further biochemical characterizations show that *Shigella* spp. are negative for $H_2S$ production, phenylalanine deaminase, sucrose and lactose (*S. sonnei* may after long incubation) fermentation, do not utilize citrate, acetate, KCN, malonate, inositol, adonitol, and salicin, and lack lysine decarboxylase. shigellae are negative for the Voges–Proskauer test (*S. sonnei* and *S. boydii* serotype 13 are positive); however, all shigellae are methyl red positive and are unable to produce acid from glucose and other carbohydrates (acid and gas production occur with *S. flexneri* serotype 6, *S. boydii* serotypes 13 and 14, and *S. dysenteriae* 3). One *Shigella* spp., *S. dysenteriae*, is catalase negative and has ornithine decarboxylase activity.

Tables 17.2 and 17.3 show key biochemical reactions to differentiate *Shigella* from *E. coli* and also to distinguish *Shigella* spp. from each other. Growth on Christensen citrate, sodium mucate, or acetate agar is one characteristic that discriminates between *E. coli* and *Shigella*; shigellae are unable to utilize citrate, acetate, or mucate as sole carbon source.

Other biochemical tests are used to identify the serotypes of *Shigella*. The ability to utilize mannitol, dulcitol, xylose, rhamnose, raffinose, glycerol, and indole and the presence of ornithine decarboxylase and o-nitrophenyl-$\beta$-D-galactopyranosidase have been used to physiologically discriminate between *Shigella* spp.

Serological testing using polyvalent antiserum is used to identify the *Shigella* groups A-D. A note of caution should be addressed. EIEC causes the same disease, bacillary dysentery, as do the shigellae. Some EIEC strains share homology with O-antigen structures of some *Shigella* serotypes (Tulloch *et al.*, 1973). Several serotypes of *S. dysenteriae*, *S. flexneri*, and *S. boydii* have reciprocal cross-reactivity with *E. coli* O antigens of the Alkalescens–Dispar bioserogroup or EIEC.

In some clinical laboratories, the application of PYR (L-pyrroglutamyl-aminopeptidase) test, usually associated with identifying gram positive cocci, e.g. *Streptococcus pyogenes*, can be applied in the identification of *Shigella*. Non-lactose fermenting colonies on MacConkey plates that are PYR negative are subsequently tested for urease production; colonies that yield negative results indicate either *Shigella* and *Salmonella* and can be further characterized serologically.

## DNA-based assays

Several types of DNA-based assays have been applied to detect *Shigella* spp. in foods. Initially, DNA probes and the polymerase chain reaction (PCR) were developed to detect *Shigella* from clinical samples. DNA probes, either DNA from virulence genes or short oligonucleotides that are specific for a particular genetic marker, can be used in a colony

**Table 17.5** Growth characteristics of *Shigella*, *E. coli* and *Salmonella* on selective media

| Organism | Medium | Colony Appearance |
|---|---|---|
| *Shigella* | MacConkey | Colorless (lactose non- fermentor); *S. sonnei* colonies are colorless to pink[1], translucent, flat with jagged edges |
| | Hektoen | Green and moist |
| | XLD | Colorless |
| *E. coli* | MacConkey | Lactose fermentor; flat, pink colonies surrounded by darker pink region (indicates sorbitol fermentors, non-sorbitol fermentors form colorless colonies). |
| | Hektoen | Yellow/yellow orange, salmon |
| | XLD | Yellow |
| *Salmonella* | MacConkey | Colorless (lactose non- fermenters) |
| | Hektoen | Green |
| | XLD | Red with black center |

[1]*S. sonnei* may ferment lactose slowly (> 40 h). Form I vs. Form II; smooth to rough look due to loss of 120 megadalton virulence plasmid.

hybridization format (Hill *et al.*, 2001; http://www.cfsan.fda.gov/~ebam/bam-24.html) or as confirmation step after PCR. The limitation of using DNA probes for colony blots is similar to conventional bacteriological for isolating shigellae from foods. There is a reduced likelihood of having isolated colonies of *Shigella* on agar plates from foods due to the physical state of the pathogen, the physical composition of the food matrix and the presence of competing microbial flora.

PCR, an *in vitro* amplification system, is a much more sensitive means of detecting the presence of microbial pathogens than conventional bacteriology. In some cases, cultivation of organisms at this time is not an option since there is not a practical culture protocol for isolating the pathogen, e.g. *Cyclospora*. As for using a PCR-based assay to detect *Shigella* in foods, primers are selected that specifically target one specific virulence gene, *ipaH* (Wang *et al.*, 1997). The advantage of these primers is that the *ipaH* gene is present in multiple copies in *Shigella*, found in the virulence plasmid and in the chromosome. Hence, the 5–12 target sequences increase the sensitivity of the assay as compared to primers targeting just one copy of a gene. Also, in instances when the large virulence plasmid is lost, these particular primers do target a copy of the *ipaH* gene in the chromosome, ensuring an opportunity for a successful reaction. Unlike other methods that may take several days to draw a conclusion regarding the presence of *Shigella* in foods, PCR assays can yield a result in less than 1 day.

One potential problem for PCR-based assays is the presence of inhibitors of the reaction deriving from the food matrix. To ensure that negative results are truly that, control reactions using washes or homogenates with target DNA added, are essential for accuracy. Therefore, template preparation is one of the key steps to assuring the correct result in analyzing foods by PCR. Two approaches are being entertained, whether or not to include an enrichment step prior to PCR. Several manufacturers provide kits and most, if not all, require enrichment in broth. This added step may be sufficient for some pathogens but shigellae may pose a more difficult task. Injured, weak, viable, but nonculturable cells may persist in the population and may not be resuscitated by enrichment. Therefore, false negative reactions may not be truly indicative of contamination. One alternative to enrichment uses a filter that lyses the bacterial cell and sequesters the DNA in the membrane and can be used directly as template in the reaction (Lampel *et al.*, 2000).

Real-time PCR (see Lampel and Levy for review of DNA-based assays) has the advantage of reducing the time for analysis. There are several different platforms that are available, each with their own strengths. In each system, the amplification of PCR products can be monitored in real time and in some cases, an extra step, such as using a probe within the reaction, can increase the specificity of the reaction in a very short period of time. Also, much of this type of analysis is automated, less chance of mistakes and reduced labor. As with conventional PCR, the same problems exist; template in sufficient quantity and free of PCR inhibitors prepared from all types of food matrices is critical. Most of these protocols require enrichment in broth for 6–18 h. An alternative to agarose gel electrophoresis is the application of an ELISA (enzyme-linked immunosorbent assay) test at the conclusion of a PCR run to detect the *ipaH* gene of *Shigella* spp. from stool specimens (Sethabutr *et al.*, 2000).

Another aspect of identifying causative agents of foodborne outbreaks is the ability to link the people affected in one region to other geographical locales and ultimately to the source of contamination. In this regard, pulse-field gel electrophoresis (PFGE) surpasses most other nucleic acid-based typing methods. Ribotyping, probing rRNA genes after digestion of total DNA with a specific endonuclease, has been used to establish a database to type and identify *Shigella* strains (Coimbra *et al.*, 2001). The Centers for Disease Control and Prevention (CDC) have established several surveillance programs to monitor foodborne outbreaks in the United States. These include PulseNet (http://www.cdc.gov/pulsenet/) and FoodNet (http://www.cdc.gov/foodnet/); others can be found at http://www.cdc.gov/and specifically for *Shigella* http://www.cdc.gov/ncidod/dbmd/phlisdata/shigella.htm. The World Health Organization (WHO; http://www.who.int/emc/) monitors outbreaks and publishes some of their findings in the *Weekly Epidemiological Record* (http://www.who.int/wer/index.html).

## Control

### Antibiotic treatment and prevention

Shigellosis is generally a self-limiting disease and patients can recover even in the absence of antimicrobial therapy. However, antimicrobial therapy shortens the duration of the illness and reduces the excretion of infectious organisms in feces (Bennish and Salam,1992) but it contributes only minimally to control of epidemic outbreaks. As with many bacterial pathogens, the incidence of multiple antibiotic-resistant strains is increasing throughout the world and has been noted in *Shigella*. Isolates resistant to sulfonamides, ampicillin, trimethoprim–sulfamethoxazole, tetracycline, chloramphenicol, and streptomycin have been reported in many geographical areas (Bennish and Salam, 1992). This phenomenon is particularly acute for *S. dysenteriae* 1, against which only fluoroquinolones are effective (Sack *et al.*,1997). Fluid replacement is administered for treatment of dehydration.

The single most effective means of preventing secondary transmission of shigellosis is handwashing. Foodhandling and preparation are important processes that deserve attention and persons with diarrhea should not handle food. Proper protocols for handwashing should be prominent and practiced. As noted above, the spread of shigellae within a population can be swift and large numbers of people can become ill from one foodhandler. Incidences of shigellosis in daycare centers have received much notice and in some jurisdiction, toddlers who are ill are not permitted to return until negative stool specimens span 24 hours.

## Vaccines

Over the past few years, several *Shigella* spp. have been constructed that have shown promise as vaccine strain (WHO, 1997). SC602 is an attenuated, oral vaccine strain of *S. flexneri* that was tested in a human study (Coster *et al.*, 1999). This strain has mutations that block iron uptake (*iuc* gene) and a deletion in the *icsA* gene that abolishes the intracellular and intercellular motility phenotypes. The *S. sonnei* vaccine strain, WRSS1, showed efficacy and was safe when a single dose of $10^3$ to $10^4$ CFU was administered. A 212-bp internal sequence of the *icsA* gene has been deleted from this strain. Recently, a *Shigella dysenteriae* 1 vaccine strain was constructed and tested in an animal model. This strain lacks the *stx* genes and a 10 kb fragment was deleted from the large virulence plasmid that removed the *icsA* gene (Venkatesan *et al.*, 2002).

While a limited challenge study showed protection among vaccines, one major drawback still to be resolved is the transient fever and mild diarrhea associated with administration of the vaccine (Coster *et al.*, 1999). One persistent problem faced by investigators attempting to develop a safe *Shigella* vaccine is to design a strain that is capable of inducing a protective immune response without producing unacceptable side-effects.

## References

Adler, B., Sasakawa, C., Tobe, T., Makino, S., Komatsu, K., and Yoshikawa, M. 1989. A dual transcriptional activation system for the 230 kb plasmid genes coding for virulence associated antigens of *Shigella flexneri*. Mol. Microbiol. 3: 627–635.

Al-Hasani, K., Rajakumar, K., Bulach, D., Robins-Browne, R., Adler, B., and Sakellaris, H. 2001. Genetic organisation of the *she* pathogenicity island in *Shigella flexneri* 2a. Microb. Pathog. 30: 1–8.

Andrews, W.H., and Jacobsen, A. 2001. Chapter 6. *Shigella*. In: Bacteriological Analytical Manual. 8th ed. International AOAC, Gaithersburg, MD, pp. 6.01–6.06.

Bagamboula, C.F., Uyttendaele, M., and Debevere, J. 2002a. Acid tolerance of *Shigella sonnei* and *Shigella flexneri*. J. Appl. Microbiol. Acid. 93: 479–486.

Bagamboula, C.F., Uyttendaele, M., and Debevere, J. 2002b. Growth and survival of *Shigella sonnei* and *S. flexneri* in minimal processed vegetables packed under equilibrium modified atmosphere and stored at 7°C and 12°C. Food Microbiol. 19: 529–536.

Banish, L.S., Sims, R., Bush, M., Sack, D., and Montali, R.J. 1993. Clearance of *Shigella flexneri* carriers in a zoologic collection of primates. J. Am. Vet. Med. Assoc. 203: 133–136.

Baudry, B., Maurelli, A.T., Clerc, P., Sadoff, J.C., and Sansonetti, P.J. 1987. Localization of plasmid loci necessary for the entry of *Shigella flexneri* into HeLa cells, and characterization of one locus encoding four immunogenic polypeptides. J. Gen. Microbiol. 133: 3403–3413.

Bean, N.H., and Griffin, P.M. 1990. Foodborne disease outbreaks in the United States, 1973-1987: Pathogens, vehicles, and trends. J. Food Protect. 53: 804–817.

Bensted, H.J. 1956. Dysentery Bacilli-Shigella. A brief historical review. Can. J. Microbiol. 2:163–174.

Bennish, M.L. 1991. Potentially lethal complications of shigellosis. Rev. Infect. Dis. 13(suppl 4):S319–324.

Bennish, M.L., and Salam, M.A. 1992. Rethinking options for the treatment of shigellosis. J Antimicrob. Chemother. 30: 243–247.

Bernardini, M.L., Mounier, J., d'Hauteville, H., Coquis-Rondon, M., and Sansonetti, P.J. 1989. Identification of *icsA*, a plasmid locus in *Shigella flexneri* that governs bacterial intra- and intercellular spread through interaction with F-actin. Proc. Natl. Acad. Sci. USA 86: 3867-3871.

Beru, N., and Salsbury, P. 2002. FDAs produce safety activities. Food Safety Magazine, 8: 14-19.

Boyd, J.S.K. 1931. Some investigations into the so-called non-agglutinable dysentery bacilli. J. Roy. Army Med. Corps, 66: 1–13.

Brenner, D.J., Fanning, G.E., Johnson, K.E., Citarella, R.V., and Falkow, S. 1969. Polynucleotide sequence relationships among members of the *Enterobacteriaceae*. J. Bacteriol. 98: 637–650.

Brewerton, D.A., Caffrey, M., Nicholls, A., Walters, D., Oates, J.K., and James, D.C.O. 1973. Reiter's disease and HLA-B27. Lancet i: 996–998.

Buchrieser, C., Glaser, P., Rusniok, C., Nedjari, H., d'Hauteville, F. Kunst, P. Sansonetti, and C. Parsot. 2000. The virulence plasmid pWR100 and the repertoire of proteins secreted by the type III secretion apparatus of *Shigella flexneri*. Mol. Microbiol. 38: 760–771.

Bunning, V.K., Raybourne, R.B., and Archer, D.L. 1988. Foodborne enterobacterial pathogens and rheumatoid disease. J. Appl. Bacteriol. Symp. 65(Suppl.):87S-107S.

Buysee, J.M., Stover, C.K., Oaks, E.V., Venkatesan, M., and Kopecko, D.J. 1987. Molecular cloning of invasion plasmid antigen (*ipa*) genes from *Shigella flexneri*: Analysis of *ipa* gene products and genetic mapping. J. Bacteriol. 169: 2561–2569.

Centers for Disease Control and Prevention. 1994. Outbreak of *Shigella flexneri* 2a infections on a cruise ship. Morb. Mortal. Wkly. Rep. 43: 657.

Centers for Disease Control and Prevention. 1999. Incidence of foodborne illnesses: preliminary data from the Foodborne Diseases Active Surveillance Network (FoodNet): United States, 1998. Morb. Mortal. Wkly. Rep. 48: 189–194.

Centers for Disease Control and Prevention. 2000. Public Health Dispatch: Outbreak of *Shigella sonnei* infections associated with eating a nationally distributed dip—California, Oregon, and Washington, January 2000. Morb. Mortal.Wkly. Rep. 49:60–61.

Centers for Disease Control and Prevention. 2001. *Shigella sonnei* outbreak among men who have sex with men-San Francisco, California, 2000–2001. Morb. Mortal. Wkly. Rep. 50: 922–926.

Cheasty, T., and Rowe, B. 1983. Antigenic relationships between the enteroinvasive *Escherichia coli* O antigens O28ac, O112ac, O124, O136, O143, O144, O152, and O164 and *Shigella* O antigens. J.Clin. Microbiol. 17: 681–684.

Coimbra, R.S., Nicastro, G., Grimont, P.A.D. and Grimont, F. 2001. Computer identification of *Shigella* species by rRNA gene restriction patterns. Res. Microbiol. 152: 47–55.

Coster, T. S., Hoge, C.W., Van DeVerg, L.L., Hartman, A.B., Oaks, E.V., Venkatesan, M.M., Cohen, D., Robin, G., Fontaine-Thompson, A., Sansonetti, P.J., and T.L. Hale, T.L. 1999. Vaccination against shigellosis with attenuated *Shigella flexneri* 2a strain SC602. Infect. Immun. 67: 3437–3443.

DuPont, H.L., Levine, M.M., Hornick, R.B., and Formal, S.B. 1989. Inoculum size in shigellosis and implications for expected mode of transmission. J. Infect. Dis. 159: 1126–1128.

Edwards, P.R. and Ewing W.H. 1972. Identification of *Enterobacteriaceae*. Minneapolis: Burgess Publishing Co.

*Enterobacteriaceae* sub-committee reports 1954. 1954. Intern. Bull. Bacteriol. Nomenclature and Taxonomy. 4: 1–94.

Ewing, W.H. 1949. Shigella nomenclature. J. Bacteriol. 57: 633–638.

Falkow, S., Schneider, H., Magnani, T.J., and Formal, S.B. 1963. Virulence of *Escherichia-Shigella* genetic hybrids for the guinea-pig. J. Bacteriol. 86: 1251–1258.

Flexner, S. 1900. On the aetiology of tropical dysentery. Middleton-Goldsmith Lecture, Philadelphia Med. J. 6: 414–424.

Formal, S., Schneider, H., Baron, H., and Formal, S.B. 1963. Virulence of *Escherichia-Shigella* genetic hybrids for the guinea-pig. J. Bacteriol. 86: 1251–1258.

Frost, J.A., McEvoy, M.B., Bentley, C.A., Andersson, Y., and Rowe, B. 1995. An outbreak of *Shigella sonnei* infection associated with consumption of iceberg lettuce. Emerg. Infect. Dis. 1: 26–29.

Galán, J.E. and Sansonetti, P.J. 1996. Molecular and cellular bases of *Salmonella* and *Shigella* interactions with host cells. In *Escherichia coli* and *Salmonella*: Cellular and Molecular Biology, Second Edition.

F.D. Neidhardt, R. Curtiss III, J.I. Ingraham, E.C.C. Lin, K.B. Low, Jr., B. Magasnik, W. Reznikoff, M. Riley, M. Schaechter, and H.E. Umbarger, eds. American Society for Microbiology, Washington, D.C., pp. 2757–2773.

Gao, L.-Y. and Kwaik, Y.A. 2000. The modulation of host cell apoptosis by intracellular bacterial pathogens. Trends Microbiol. 8: 306–313.

Gessner, B.D., and Beller, M. 1994. Moose soup shigellosis in Alaska. West. J. Med. 160: 430–433.

Goldberg, M.B. Barzu, O., Parsot, C., and Sansonetti, P.J. 1993. Unipolar localization and ATPase activity of IcsA, a Shigella flexneri protein involved in intracellular movement. J. Bacteriol. 175: 2189–2196, 1993.

Goma Epidemiology Group. 1995. Public health impact of Rwandan refugee crisis: what happened in Goma, Zaire, in July, 1994? Lancet 345: 339–344.

Guerrero, L., Calva, J.J., Morrow, A.L., Velazquez, F.R., Tuz-Dzib, F., Lopez-Vidal, Y., Ortega, H., Arroyo, H., Cleary, T.G., Pickering, L.K., and Ruiz-Palacios, G.M. 1994. Asymptomatic Shigella infections in a cohort of Mexican children younger than two years of age. Pediatr. Infect. Dis. J. 13: 597–602.

Hilbi, H. Chen, Y., Thirumalai, K., and Zychlinsky, A. 1997. The interleukin 1β-converting enzyme, caspase 1, is activated during Shigella flexneri-induced apoptosis in human monocyte-derived macrophages. Infect. Immun. 65: 5165–5170.

Hilbi, H. Moss, J.E., Hersh, D., Chen, Y., Arondel, J., Banerjee, S., Flavell, R.A., Yuan, J., Sansonetti, P.J. and Zychlinsky, A. 1998. Shigella-induced apoptosis is dependent on caspase-1 which binds to IpaB. J. Biol. Chem. 273: 32895–329000.

Hill, W.E., Datta, A.R., Feng, P., Lampel, K.A., and Payne, W.L. 2001. Identification of foodborne bacterial pathogens by gene probes. In Bacteriological Analytical Manual, 8th Edition. AOAC Internation. Chapter 24.

Hong, M. and Payne, S.M. 1997. Effect of mutations in Shigella flexneri chromosomal and plasmid-encoded lipopolysaccharide genes on invasion and serum resistance. Mol. Microbiol. 24: 779–791.

Hossain, M.A., Hasan, K.Z., and Albert, M.J. 1994. Shigella carriers among non-diarrhoeal children in an endemic area of shigellosis in Bangladesh. Trop. Geogr. Med. 46: 40–42.

Hueck, C.J. 1998. Type III protein secretion systems in bacterial pathogens of animals and plants. Microbiol. Mol. Biol. Rev. 62: 379–433.

Hyams, K.C., Bourgeois, A.L., Merrell, B.R., Rozmajzl, P., Escamilla, J., Thornton, S.A., Wasserman, G.M., Burke, A., Echeverria, P.,Green, K.Y., Kapikian, A.Z., and Woody, J.N. 1991. Diarrheal disease during Operation Desert Shield. N. Engl. J. Med. 325: 1423–1428.

International Commission on Microbiological Specifications for Foods. 1996. Microorganisms in Foods. 5. Shigella. Blackie Academic & Professional, London, United Kingdom, pp. 280-298.

Jin Q, Yuan, Z., Xu, J., Wang, Y. Shen, Y., Lu, W., Wang, J., Liu, H., Yang, J., Yang, F., Zhang, X., Zhang, J., Yang, G., Wu, H., Qu, D., Dong, J., Sun, L.,Xue, Y., Zhao, A., Gao, Y., Zhu, J., Kan, B., Ding, K., Chen, S., Cheng, H., Yao, Z., He, B., Chen, R., Ma, D., Qiang, B., Wen, Y. Hou, Y., and Yu, J. 2002. Genome sequence of Shigella flexneri 2a: insights into pathogenicity through comparison with genomes of Escherichia coli K12 and O157. Nucleic Acids Res. 30: 4432–4441.

Johnson, J.R. 1999. Shigella and E. coli. ASM News 65: 460–461.

Kapperud, G., Rorvik, L.M., Hasseltvedt, V., Hoiby, E.A., Iversen, B.G., Staveland, K., Johnsen, G., Leitao, J., Herikstad, H., Andersson, Y., Langeland, G., Gondrosen, B., and Lassen, J. 1995. Outbreak of Shigella sonnei infection traced to imported iceberg lettuce. J. Clin. Microbiol. 33: 609–614.

Karmali, M.A., Petric, M., Lim, C., Fleming, P.C., Arbus, G.S., and Lior, H. 1985. The association between idiopathic hemolytic syndrome and infection by Verotoxin-producing Escherichia coli. J. Infect. Dis.151: 775–782.

Kennedy, F.M., Astbury, J., Needham, J.R., and Cheasty, T. 1993. Shigellosis due to occupational contact with non-human primates. Epidemiol. Infect. 110: 247–257.

Kinsey, M.D., Formal, S.B., Dammin, G.J., and Giannella, R.A. 1976. Fluid and electrolyte transport in rhesus monkeys challenged intracecally with Shigella flexneri 2a. Infect. Immun. 14: 368–371.

Kotloff, K.L., Winickoff, J.P. Ivanoff, B., Clemens, J.D., Swerdlow, D.L., Sansonetti, P.J. Adak, G.K., and Levine, M.M. 1999. Global burden of Shigella infections: implications for vaccine development and implementation of control strategies. Bull. World Health Organ. 77: 651-666.

Kruse, W. 1900. Uber die Ruhr als Volkskrankheit und ihren Erreger. Deut. med Wochschr. 26: 637–639.

Lampel, K.A., Kornegay, L., and Orlandi, P.A. 2000. Improved template preparation for polymerase chain reaction based assays for the detection of foodborne bacterial pathogens. Appl. Environ. Microbiol. 66: 4539–4542.

Lampel, K.A., and Levy, D.D. 2002. DNA-based assays. In: Encyclopedia of Dairy Sciences. H. Roginske, J.W. Fuquay, and P.F. Fox, eds. Academic Press, London, UK, pp. 79–85.

Lan, R. and P.R. Reeves. 2002. Escherichia coli in disguise: molecular origins of Shigella. Microbes and Infect. 4: 1125–1132.

LaBrec, E.H., Schneider, H., Magnani, T.J., and Formal, S.B. 1964. Epithelial cell penetration as an essential step in the pathogenesis of bacillary dysentery. J. Bacteriol. 88: 1503–1518.

Lew, J.F., Swerdlow, D.L., Dance, M.E., Griffin, P.M., Bopp, C.A., Gillenwater, M.J., Mercatante, T., and Glass, R.I. 1991. An outbreak of shigellosis aboard a cruise ship caused by a multiple-antibiotic-resistant strain of Shigella flexneri. Am. J. Epidemiol. 134: 413–420.

Line, A.S., Paul-Murphy, J. Aucoin, D.P., and Hirsh, D.C. 1992. Enrofloxacin treatment of long-tailed macaques with acute bacillary dysentery due to multiresistant Shigella flexneri IV. Lab. Anim. Sci. 42: 240–244.

Levine, M.M., and Levine, O.S. 1994. Changes in human ecology and behavior in relation to the emergence of diarrheal diseases, including cholera. Proc. Natl. Acad. Sci. USA 91: 2390-2394.

Lopez, E.L., Diaz, M., Grinstein, S., Dovoto, S., Mendila-harzu, F., Murray, B.E., Ashkenazi, S., Rubeglio, E., Woloj, M., Vasquez, M. Turco, M., Pickering, L.K., and Cleary, T.G. 1989. Hemolytic uremic syndrome and diarrhea in Argentine children: the role of Shiga-like toxins. J. Infect. Dis. 160: 469–475.

Luria, S.E., and Burrous, J.W. 1957. Hybridization between Escherichia coli and Shigella. J. Bacteriol. 74: 461–476.

Maurelli, A.T., and Curtiss, R. 1984. Bacteriophage Mu dl (Apr lac) generates vir-lac operon fusions in Shigella flexneri 2a. Infect. Immun. 45: 642–648.

Maurelli, A.T., and Sansonetti, P.J. 1988. Identification of a chromosomal gene controlling temperature-regulated expression of Shigella virulence. Proc. Natl. Acad. Sci. USA 85: 2820–2824.

Maurelli, A.T., Baudry, B., d'Hauteville, H., Hale, T.L., and Sansonetti, P.J. 1985. Cloning of virulence plasmid DNA sequences involved in invasion of HeLa cells by Shigella flexneri. Infect. Immun. 49: 164–171.

Maurelli, A.T., Fernández, R.E., Bloch, C.A., Rode, C.K., and Fasano, A. 1998.'Black holes' and bacterial pathogenicity: A large genomic deletion that enhances the virulence of Shigella spp. and enteroinvasive Escherichia coli. Proc. Natl. Acad. Sci. USA 95: 3943-3948.

Ménard, R., Sansonetti, P.J., Parsot, C., and Vasselon, T. 1994. The IpaB and IpaC invasins of Shigella flexneri associate in the extracellular medium and are partitioned in the cytoplasm by a specific chaperon. Cell 76: 829–839.

Moss, J.E., Cardozo, T.J., Zychlinsky, A., and Groisman, E.A. 1999. The selC-associated SHI-2 pathogenicity island of Shigella flexneri. Mol. Microbiol. 33: 74–83.

Mounier, J., Vasselon, T., Hellio, R., Lesourd, M., and Sansonetti, P.J. 1992. Shigella flexneri enters human colonic Caco-2 epithelial cells through the basolateral pole. Infect. Immun. 60: 237–248.

Nakata, N., Tobe, T., Fukuda, I., Suzuki, T., Komatsu, K., Yoshikawa, M., and Sasakawa, C. 1993. The absence of a surface protease, OmpT, determines the intercellular spreading ability of Shigella: the relationship between the ompT and kcpA loci. Mol. Microbiol. 9: 459-468.

Niebuhr, K., Jouihri, N. Allaoui, A., Gounon, P., and Sansonetti, P.J. 2000. IpgD, a protein secreted by the type III secretion machinery of *Shigella flexneri*, is chaperoned by IpgE and implicated in entry focus formation. Mol. Microbiol. 38: 8–19.

Nizeyi, J.B., Innocent, R.B., Erume, J., Kalema, G.R., Cranfield, M.R., and Gracyzk, T.K. 2001. Camplyobacteriosis, salmonellosis, and shigellosis in free-ranging human-habituated mountain gorillas of Uganda. J. Wildl. Dis. 37: 239–244.

O'Brien, A.D., and Holmes, R.K. 1996. Protein toxins of *Escherichia coli* and *Salmonella*. In: *Escherichia coli* and *Salmonella*: Cellular and Molecular Biology, Second Edition. F.D. Neidhardt, R. Curtiss III, J.I. Ingraham, E.C.C. Lin, K.B. Low, Jr., B. Magasnik, W. Reznikoff, M. Riley, M. Schaechter, and H.E. Umbarger, eds. American Society for Microbiology, Washington, D.C., pp. 2788–2802.

O'Sullivan, B., Delpech, V., Pontivivo, G., Karagiannis, T., Marriott, D., Harkness, J., and McAnulty, J.M. 2002. Shigellosis linked to sex venues, Australia. Emerg. Infect. Dis. 8: 862-864.

Page, A.-L., and Parsot, C. 2002. Chaperones of the type III secretion pathway: jacks of all trades. Mol. Microbiol. 46: 1–11.

Perdomo, J.J., Cavaillon, J.M., Huere, M., Ohayon, H., Gounon, P., and Sansonetti, P.J. 1994. Acute inflammation causes epithelial invasion and mucosal destruction in experimental shigellosis. J Exp. Med. 180: 1307–1319.

Pickering, L.K., Bartlett, A.V., and Woodward, W.E. 1986. Acute infectious diarrhea among children in day-care: epidemiology and control. Rev. Infect. Dis. 8: 539–547.

Rafii, F., Holland, M.A., Hill, W.E., and Cerniglia, C.E.. 1995. Survival of *Shigella flexneri* on vegetables and detection by polymerase chain reaction. J. Food Prot. 58: 727–732.

Rafii, F., and Lunsford, P. 1997. Survival and detection of *Shigella flexneri* in vegetables and commercially prepared salads. J. AOAC. Int. 80: 1191–1197.

Raghupathy, P., Date, A., Shastry, J.C.M., Sudarsanam, A., Jadhav, M. 1978. Haemolytic-uraemic syndrome complicating shigella dysentery in south Indian children. Br. Med. J. 1: 1518–1521.

Rajakumar, K. C. Sasakawa, and B. Adler. 1997. Use of a novel approach, termed island probing, identifies the *Shigella flexneri she* pathogenicity island which encodes a homolog of the immunoglobulin A protease-like family of proteins. Infect. Immun. 65: 4606–4614.

Rout, W.R., Formal, S.B., Giannella, R.A., and Dammin, G.J. 1975. Pathophysiology of *Shigella* diarrhea in the rhesus monkey: intestinal transport, morphological, and bacteriological studies. Gastroenterology 68: 270–278.

Sack, R.B., Rahman, M., Yunus, M., and Khan, E.H. 1997. Antimicrobial resistance in organisms causing diarrheal disease. Clin. Infect. Dis. Suppl. 1:S102–5.

Sandlin, R.C., Lampel, K.A., Keasler, S.P., Goldberg, M.B., Stolzer, A.L. and Maurelli, A.T. 1995. A virulence of rough mutants of *Shigella flexneri*: Requirement of O-antigen for correct unipolar localization of IcsA in bacterial outer membrane. Infect. Immun. 63: 229-237.

Sandlin, R.C., Goldberg, M.B., and Maurelli, A.T. 1996. Effect of O side chain length and composition on the virulence of *Shigella flexneri*. Mol. Microbiol. 22: 63–73.

Sansonetti, P.J. 1999. *Shigella* plays dangerous games. ASM News 65: 611–617.

Sansonetti, P.J., Hale, T.L., Dammin, G.J., Kapfer, C., Collins, H.H., Jr., and Formal, S.B. 1983. Alterations in the pathogenicity of *Escherichia coli* K-12 after transfer of plasmid and chromosomal genes from *Shigella flexneri*. Infect. Immun. 39: 1392–1402.

Sansonetti, P.J., Kopecko, D.J., and Formal, S.B. 1981. *Shigella sonnei* plasmids: evidence that a large plasmid is necessary for virulence. Infect. Immun. 34: 75–83, 1981.

Sansonetti, P.J., Kopecko, D.J., and Formal, S.B. 1982. Involvement of a plasmid in the invasive ability of *Shigella flexneri*. Infect. Immun. 35: 852–860.

Sethabutr, O., Venkatesan, M., Yam, S., Pang, L.W., Smoak, B.L., Sang, W.K., Echeverria, P., Taylor, D.N., and Isenbarger, D.W. 2000. Detection of PCR products of the *ipaH* gene from *Shigella* and en-teroinvasive *Escherichia coli* by enzyme linked immunosorbent assay. Diagn. Microbiol. Infect. Dis. 37: 11–16.

Shiga, K. 1906. Observations on the epidemiology of dysentery in Japan. Phillpp. J. Sci. 1:485-500.

Simon, D.G., Kaslow, R.A., Rosenbaum, J., Kaye, R.L., and Calin, A. 1981. Reiter's syndrome following epidemic shigellosis. J. Rheumatol. 8: 969–973.

Smith, J.L. 1987. *Shigella* as a foodborne pathogen. J. Food. Prot. 50: 788–801.

Smith, J.L., and Palumbo, S.A. 1982. Microbial injury reviewed for the sanitarian. Dairy and Food Sanitation 2: 57–63.

Strong, R.P., and Musgrave, W.E. 1900. Preliminary note regarding the aetiology of the dysenteries of Manila. Rept. Surgeon-General of the Army, Washington.

Talan, D., Moran, G.J., Newdow, M., Ong, S., Mower, W.R., Nakase, J.Y., Pinner, R.W., and Slutsker, L. 2001. Etiology of bloody diarrhea among patients presenting to United States emergency departments: prevalence of *Escherichia coli* O157:H7 and other Eenteropathogens. Clin. Infect. Dis. 32: 573–580.

Tamano, K., Aizawa, S.-I., Katayama, E., Nonaka, T., Imajoh-Ohmi, S., Kuwae, A., Naga, S., and Sasakawa, C. 2000. Supramolecular structure of the *Shigella* type III secretion machinery: the needle part is changeable in length and essential for delivery of effectors. EMBO J. 19: 3876–3887.

Tobe, T. Yoshikawa, M., Mizuno, T., and Sasakawa, C. 1993. Transcriptional control of the invasion regulatory gene *virB* of *Shigella flexneri*: activation by *virF* and repression by H-NS. J. Bacteriol. 175: 6142–6149.

Tollison, S.B., and Johnson, M.G. 1985. Sensitivity to bile salts of *Shigella flexneri* sublethally heat stressed in buffer or broth. Appl. Environ. Microbiol. 50: 337.

Tran van Nhieu, G., Ben-Ze'ev, A., and Sansonetti, P.J. 1997. Modulation of bacterial entry into epithelial cells by association between vinculin and the *Shigella* IpaA invasin. EMBO J. 16: 2717–2729.

Tulloch, E.F., Ryan, K.J., Formal, S.B., and Franklin, F.A. 1973. Invasive enteropathic *Escherichia coli* dysentery. Ann Intern Med 79: 13–17.

Venkatesan, M.M., Goldberg, M.B., Rose, D.J. Grotbeck, E.J., Burland, V., and Blattner, F. 2001. Complete DNA Sequence and Analysis of the Large Virulence Plasmid of *Shigella flexneri*. Infect. Immun. 69: 3271–3285.

Venkatesan, M.M., Hartman, A.B., Newland, J.W., Ivanova, V. S., Hale, T.L., McDonough, M., and Butterton, J. 2002. Construction, characterization, and animal testing of WRSd1, a *Shigella dysenteriae* 1 vaccine. Infect. Immun. 70: 2950–2958.

Walker, J.C. and Verma, N.K. 2002. Identification of a putative pathogenicity island in *Shigella flexneri* using subtractive hybridisation of the *S. flexneri* and *Escherichia coli* genomes. FEMS Microbiol. Lett. 213: 257–264.

Wang, R.-F., Cao, W.-W., and Cerniglia, C.E. 1997. A universal protocol for PCR detection of 13 species of foodborne pathogens in foods. J Appl. Microbiol 83: 727–736, 1997.

Waterman, S.R. and Small, P.L. 1996. Identification of sigma S-dependent genes associated with the stationary-phase acid-resistance phenotype of *Shigella flexneri*. Mol. Microbiol. 21: 925-940.

Weissman, J.B., Schmerler, A., Weiler, P., Filice, G., Godby, N., and Hansen, I. 1974. The role of preschool children and day-care centers in the spread of shigellosis in urban communities. J. Pediatr. 84: 797–802.

Wharton, M., R.A. Spiegel, R.A., Horan, J.M., Tauxe, R.V., Wells, J.G., Barg, N., Herndon, J., Meriwether, R.A., MacCormack, J.N., and Levine, R.H. 1990. A large outbreak of antibiotic resistant shigellosis at a mass gathering. J. Infect. Dis. 162: 1324–1328.

WHO Weekly Epidemiological Record. 1997. Vaccine research and development 72: 73–80.

Zaika, L.L. 2002a. Effect of organic acids and temperature on survival of *Shigella flexneri* in broth at pH 4. J. Food Prot. 65: 1417–1421.

Zaika, L.L. 2002b. The effect of NaCl on survival of *Shigella flexneri* in broth as affected by temperature and pH. J. Food Prot. 65: 774–779.

Zaika, L.L. and Scullen, O.J. 1996. Growth of *Shigella flexneri* in foods: comparison of observed and predicted growth kinetics parameters. Int. J. Food Prot. 32: 91–102.

Zychlinsky, A., Prevost, M.C., and Sansonetti, P.J. 1992. *Shigella flexneri* induces apoptosis in infected macrophages. Nature 358: 167–169.

Zychlinsky, A., Kenny, B., Menard, R., Prevost, M.C., Holland, I.B., and Sansonetti, P.J. 1994. IpaB mediates macrophage apoptosis induced by *Shigella flexneri*. Mol. Microbiol. 11: 619–627.

# Diarrhea-inducing *Escherichia coli*

18

James L. Smith and Pina M. Fratamico

## Abstract

More information is available concerning *Escherichia coli* than any other organism, thus making *E. coli* the most thoroughly studied species in the microbial world. For many years, *E. coli* was considered a commensal of human and animal intestinal tracts with low virulence potential. Today, it is well known that many strains of *E. coli* act as pathogens inducing serious gastrointestinal diseases and even death in humans. There are six major categories of *E. coli* strains that cause enteric diseases in humans including the (1) enterohemorrhagic *E. coli*, which cause hemorrhagic colitis and hemolytic uremic syndrome, (2) enterotoxigenic *E. coli*, which induce traveler's diarrhea, (3) enteropathogenic *E. coli*, which cause a persistent diarrhea in children living in developing countries, (4) enteroaggregative *E. coli*, which provoke diarrhea in children, (5) enteroinvasive *E. coli* that are biochemically and genetically related to *Shigella* species and can induce diarrhea, and (6) diffusely adherent *E. coli*, which cause diarrhea and are distinguished by a characteristic type of adherence to mammalian cells. Genomic studies on the diarrhea-inducing *E. coli* have provided much information on the nature of the pathogenic mechanisms of these organisms and have provided information that can be used to design techniques for the detection of specific *E. coli* strains. In spite of extensive study, more information is needed concerning the interaction of the human host with pathogenic *E. coli*.

## Introduction

*Escherichia coli* is the most thoroughly studied organism in the microbial world and the information gained has been of benefit in many areas of the biological sciences. *E. coli* are gram-negative, non-spore forming, rod-shaped bacteria in the family *Enterobacteriaceae*. *E. coli* are motile by means of peritrichous flagella or may be nonmotile, and may have capsules or microcapsules. *Escherichia coli* is facultatively anaerobic, chemoorganotrophic, and grows optimally at 37°C (Brenner, 1984; Ørskov, 1984).

As an inhabitant of the gastrointestinal tract of humans and animals, *E. coli* is found in the environment, water, and food. It is not surprising that the organism is present in animal foods such as meat or milk, produce grown on land treated with animal manure or irrigated with fecally contaminated water, or in foods prepared by unsanitary food handlers.

The strains of *E. coli* that cause human disease can be broadly categorized as intestinal commensal, intestinal pathogenic, and extraintestinal pathogenic strains (Russo and Johnson 2003). Virulence factors have been acquired by intestinal pathogenic strains, which confer the ability to induce gastrointestinal diseases such as enteritis and colitis. Some *E. coli* strains acquired genes conferring the ability to cause extraintestinal diseases such as urinary tract, abdominal, and pelvic infections as well as septicemia, meningitis, and endocarditis. *E. coli* can be an opportunistic pathogen causing infection of surgical wounds, as well as infections in various organs and anatomical sites.

Many *E. coli* strains are known to act as intestinal pathogens but they vary in the mechanisms by which diarrhea is induced. Virulence mechanisms include the production of toxins and host cell attachment factors and/or intestinal cell invasion (Guerrant and Thielman, 1995). There are several different categories of diarrheic *E. coli*, and the most important or most studied are enterohemorrhagic (EHEC), enterotoxigenic (ETEC) and enteropathogenic (EHEC) *E. coli*. There are three other categories, including enteroinvasive, enteroaggregative, diffusely adherent *E. coli*, as well as some emerging minor categories (Guerrant and Thielman, 1995). The most important diarrheic strains (EHEC, ETEC, EPEC) will be discussed in detail and the other groups will be discussed briefly. Recently, new strains of diarrheic *E. coli* such as cell-detaching *E. coli* (CDEC) and necrotoxigenic *E. coli* (NTEC) have been described (see below). Due to the apparent ease of gene transfer in Gram-negative organisms (Clarke, 2001), food microbiologists and clinicians must be aware of newly emerging diarrheic *E. coli* strains.

## Enterohemorrhagic *Escherichia coli* (EHEC)

EHEC O157:H7 was first recognized as a cause of enteric disease following two outbreaks of hemorrhagic colitis in 1982 associated with undercooked hamburgers served at fast food restaurants. The term EHEC consists of serogroups of *E. coli* including O26, O111, O103, O104, O118, O145 (with various H antigen types) and others that share similar clinical, pathogenic, and epidemiologic features with *E. coli* O157:H7. *Escherichia coli* O157:H7 is the EHEC serotype accounting for the greatest proportion of disease cases. Numerous food- and water-borne outbreaks linked to EHEC have been documented (see Table 18.2). It is estimated that the O157:H7 serotype is responsible for greater than 73,000 cases of illness and 61 deaths each year in the United States, and non-O157 EHEC cause approximately 37,000 cases and 30 deaths each year (Mead *et al.*, 1999).

## Disease characteristics

*E. coli* O157:H7 infections lead to a mild non-bloody diarrhea or an acute grossly bloody diarrhea termed hemorrhagic colitis (HC); however, infection may also be asymptomatic (Ryan *et al.*, 1986; Griffin, 1995; Su and Brandt, 1995; Mead and Griffin, 1998; Stephan *et al.*, 2000). The incubation period can be as short as 1 to 2 days but generally ranges from 3 to 8 days. Less severe illness is seen in patients with non-bloody diarrhea, and they are less likely to develop systemic sequelae or to die. HC is marked by an acute onset of severe abdominal cramps followed by watery diarrhea that progresses to bloody diarrhea in approximately 1/4 to 3/4 of patients and lasts for 4 to 10 days. The areas in the gastrointestinal tract that are predominantly affected are the cecum and the ascending colon. Fever is usually absent or low-grade, stools are usually free of white blood cells, and vomiting occurs in approximately 50% of affected patients. Often associated with hemorrhagic colitis is an elevation of blood leukocytes, edema with "thumb printing", hemorrhage of the lamina propria, superficial ulceration, pseudomembrane formation, and necrosis of the superficial colonic mucosa.

In 2–7% of patients, primarily infants, children, and the elderly, *E. coli* O157:H7 infection can proceed to hemolytic uremic syndrome (HUS), a severe post-diarrheal systemic complication, which is the leading cause of acute renal failure in children in the United States (Mead and Griffin, 1998). Microangiopathic hemolytic anemia, thrombocytopenia, and renal insufficiency are the hallmarks of HUS (Su and Brandt, 1995). Complications of the central nervous system may occur in 30–50% of patients. The production of Shiga toxins (Stx) by EHEC strains plays an important role in the pathogenesis of hemorrhagic colitis and HUS. The binding of Shiga toxins to specific receptors on endothelial cells results in damage and death of the cells. There is deposition of platelets and fibrin leading to abnormal white blood cell adhesion, reduced blood flow in small vessels of the affected organs, increased coagulation, and thrombosis formation (O'Loughlin and Robins-Browne, 2001). Damage of red blood cells and platelets occurs with passage through narrowed blood vessels leading to hemolytic anemia and thrombocytopenia. Histological changes in the kidney include capillary wall thickening, endothelial cell swelling and thrombosis of capillaries in the glomeruli, resulting in necrosis of kidney tissue with complete occlusion of renal microvessels. Dialysis is required in roughly 50% of patients with HUS, approximately 3 to 5% die, and about 5% develop chronic renal failure, stroke, and other major sequelae (Mead and Griffin, 1998).

## Basis of pathogenicity

The production of one or more types of Shiga toxins, intestinal colonization, the production of enteropathogenic *E. coli* (EPEC)-like attaching and effacing (A/E) lesions mediated by genes located on the LEE locus (see below), and the presence of a plasmid of ca. 60 MDa (pO157) are major virulence factors associated with EHEC strains (LeBlanc, 2003). The most common cause of HC and HUS in the United States due to infection by EHEC serotype O157:H7; however, numerous other *E. coli* serotypes that produce Shiga toxins, referred to as Shiga toxin-producing *E. coli* (STEC), have caused HC and HUS (Meng *et al.*, 2001). The most important virulence factors in the pathogenesis of EHEC infection are the Shiga toxins (Stx), also known as verotoxins or verocytotoxins because of their cytopathic effect on Vero cells (African green monkey kidney cells). Cytotoxicity assays involving the addition of serial dilutions of the culture supernatants to Vero cell monolayers, followed by examination of the tissue culture by microscopy for cytotoxic effects has been be used to detect Shiga toxins (Konowalchuk *et al.*, 1977).

Shiga toxin is produced by *Shigella dysenteriae* type 1, and the gene is chromosomal. The Shiga toxin 1 (Stx1) produced by *E. coli* differs from Shiga toxin of *S. dysenteriae* by only one amino acid. In *E. coli*, Stx1 and Shiga toxin 2 (Stx2), previously called Shiga-like toxin 1 and 2 or referred to as Verotoxin 1 and 2, are encoded on the genomes of temperate bacteriophages. Variants of Stx1 and Stx2 have been identified (Table 18.1; Schmidt *et al.*, 2000; Schmidt, 2001; Zhang *et al.*, 2002; Bürk *et al.*, 2003; Leung *et al.*, 2003). An STEC strain may produce Stx1, Stx2, or a combination of one or both toxins and one of the variants. There is some evidence that Stx2-producing strains are potentially more virulent and important in the development of HUS than strains that produce Stx1 only, or that both produce Stx1 and Stx2 (Scotland *et al.*, 1987; Ostroff *et al.*, 1989). Stx1c was first associated with a sheep strain of *E. coli*, and the Stx2d variant was identified in a bovine strain (Brett *et al.*, 2003; Bürk *et al.*, 2003). Stx1c-producing strains represent of subset of STEC that have caused diarrheal illness in humans (Zhang *et al.*, 2002). These strains represent numerous serotypes and appear to lack *eae*, a gene in the LEE locus. A number of *stx*1c-positive strains possess the locus of proteolysis activity, a high pathogenicity island originally identified in pathogenic *Yersiniae*, and *saa*, a gene encoding an STEC autoagglutinating adhesin (Friedrich *et al.*, 2003). Edema disease in swine is caused by *E. coli* strains that produce Stx2e (Cornick, *et al.*, 1999). It was shown that *stx*2e is encoded on the genome of a Shiga toxin 2e-converting bac-

**Table 18.1** *E. coli* Shiga toxins

| Toxin type | Percent identity with Stx1 or Stx2 | | Receptor | Comments |
| | A subunit | B subunit | | |
|---|---|---|---|---|
| Stx1 | – | – | Gb$_3$ | A and B subunits of Stx1 are 55% and 57% identical to those of Stx2, respectively; Stx1 is 98% identical to Stx of *Shigella dysenteriae*, differing by only one amino acid in the A subunit |
| Stx1c | 97[a] | 97[a] | Gb$_3$ | Associated with human and ovine STEC; Stx1c-positive strains comprise a variety of serogroups, and isolates frequently lack *eae*, and possess the locus of proteolysis activity and *saa* (gene encoding an STEC agglutinating adhesin) (Friedrich *et al.*, 2003) |
| Stx1d | 93 | 92 | Gb$_3$ | Associated with human and bovine STEC |
| Stx2 | – | – | Gb$_3$ | Immunologically distinct from Stx1 |
| Stx2c | 100[b] | 97[b] | Gb$_3$ | Associated with STEC isolated from humans with diarrhea and HUS |
| Stx2d | 99 | 97 | Gb$_3$ | Cytotoxicity is activated 10- to 1000-fold by the elastase found in mouse or human intestinal mucus |
| Stx2e | 93 | 84 | Gb$_4$ | Associated with STEC that cause porcine edema disease and also in some human STEC |
| Stx2f | 63 | 57 | Not known | Associated with STEC isolated from feral pigeons |
| Stx2g (Vt2g) | | | Not known | The nucleic acid sequence of the *vt2g* A subunit showed 63 to 95% similarity to other *vt* gene A subunit sequences; the nucleic acid sequence of the *vt2g* B subunit showed 77 to 91% similarity to other *vt* gene B subunit sequences (Leung *et al.* 2003) |

[a]Percent identity with Stx1 A and B subunits.
[b]Percent identity with Stx2 A and B subunits.

teriophage in an ONT (non-typable):H⁻ *E. coli* strain isolated from a diarrheic patient (Muniesa *et al.*, 2000).

Shiga toxins are composed of a single A polypeptide and a B-pentamer that binds to the eukaryotic cell receptor, globotriaosylceramide (Gb$_3$), expressed on epithelial and endothelial cells. The receptor for Stx2e, however, is globotetraosylceramide (Gb$_4$). Binding of the B-pentamer to the glycolipid receptors leads to internalization of the holotoxin by endocytosis via clathrin-coated pits. The toxin is transported through the Golgi network to the endoplasmic reticulum and the nuclear membrane. In the Golgi, the 32-kDa A subunit is cleaved by a calcium-sensitive serine protease to an active 28-kDa peptide (A$_1$) and a 4-kDa peptide (A$_2$) joined by a disulfide bond. The disulfide bond linking the two peptides is reduced in the endoplasmic reticulum and the A1 fragment is translocated into the cytoplasm. The A1 subunit, an N-glycosidase, then acts on the 60S ribosomal subunit and removes a single adenine residue from the 28S rRNA of eukaryotic ribosomes. The aminoacyl-tRNA can then no longer bind to ribosomes resulting in an irreversible inhibition of protein synthesis. Shiga toxins contribute to the inflammatory responses and the pathology of EHEC diseases since the toxins can modulate the expression of chemokines and cytokines by human epithelial and endothelial cells (Cherla *et al.*, 2003; Matussek *et al.*, 2003). The development of bloody diarrhea and HUS may be due to induction of apoptosis (programmed cell death) by Shiga toxins in some cell types (Cherla *et al.*, 2003).

*E. coli* O157:H7 and other EHEC possess the LEE locus and produce A/E lesions in the intestinal mucosa similar to those caused by EPEC (LeBlanc, 2003). The LEE locus of *E. coli* O157:H7 strain EDL933 consists of 41 genes that are found in the same order and orientation as those in the EPEC O127:H6 strain. The average nucleotide identity between the LEE loci in the two strains is ca. 94%. The *esc* genes that encode the type III secretion system are highly conserved in EPEC and *E. coli* O157:H7; however, there is less similarity between the other LEE genes. The *eae* genes of EPEC and *E. coli* O157:H7 (strain 933) share 87% identity. The EHEC and EPEC intimin, the product of the *eae* gene that binds to enterocytes and the Tir (translocated intimin receptor) protein (see below in Enteropathogenic *E. coli*), is highly conserved in the N-terminal region but variable in the C-terminal 280 amino acid region (Frankel *et al.*, 1998). Both the EHEC and EPEC Tir bind intimin; however, the EHEC Tir is not tyrosine phosphorylated, indicating that tyrosine phosphorylation is not required for A/E lesion formation in EHEC (DeVinney *et al*, 1999c). At least 14 variants of intimin, including α1, α2, β1, β2, γ1, *eae*-ξ, and others have been identified using intimin type-specific PCR assays, and it is possible that different intimins in EPEC and EHEC may explain the different host tissue cell tropism (small bowel vs. large bowel, respectively) (Frankel *et al*, 1998; Blanco *et al.*, 2004).

Additional EHEC virulence factors are encoded on a ca. 60-MDa virulence plasmid. This plasmid is heterogenous and varies in size among EHEC strains, even within *E. coli* O157:H7 strains (LeBlanc, 2003). The complete DNA sequence of two pO157 virulence plasmids from strains EDL933 and RIMD 0509952 has been published (Burland *et al.*, 1998; Makino *et al.*, 1998). The plasmid from strain EDL 933 was a 92-kb F-like plasmid composed of 100 open reading frames. The plasmid is involved in the production of (a) an EHEC hemolysin (operon *ehxCABD*), that belongs to the RTX family of exoproteins; (b) KatP, a periplasmic catalase-peroxidase that functions to protect the bacterium against oxidative stress; (c) a serine protease (EspP), that may contribute to the cytotoxic activity and tissue destruction of EHEC; and (d) possesses a gene cluster related to the type II secretion pathway of gram-negative bacteria (Etp system), composed of a cluster of 13 genes, etpC through etpO. Another unusually large ORF of 3169 amino acids showed strong sequence similarity within the first 700 amino acids to the N-terminal activity-containing domain of the large clostridial toxin (LCT) gene family in Clostridium difficle that includes ToxA and ToxB. toxB is required for adherence of E. coli O157:H7 to epithelial cells (Tatsuno et al., 2001). The EHEC enterohemolysin induces the production of IL-1β from human monocytes and may mediate translocation of Stxs that further stimulate the production of IL-1β contributing to the development of HUS (Taneike et al., 2002).

Acid resistance is an additional virulence factor of E. coli O157:H7. The organism possesses at least three acid resistance systems that account for its ability to tolerate acid environments (Lin et al., 1996; see Growth and survival). The TAT (twin arginine translocation) system may be another EHEC virulence factor (Pradel et al., 2003). Deletion of the tatABC genes encoding the E. coli O157:H7 TAT system, involved in export of proteins across the cytoplasmic membrane, resulted in decreased secretion of Stx1 and abolition of the synthesis of flagella. A cytolethal distending toxin (cdt) gene cluster was identified in E. coli O157:H7 (6% of isolates examined) and in sorbitol-fermenting E. coli O157:H⁻ (87% of isolates) isolated from patients with diarrhea and HUS (Janka et al., 2003). Their studies suggested that cdt may have been acquired by phage transduction.

Membrane vesicles are released into the culture medium by E. coli O157:H7; the vesicles contain DNA encoding the eae, $stx_1$, $stx_2$, and uidA genes (Kolling and Matthews, 1999). Furthermore, these vesicles facilitated the transfer of virulence and antibiotic resistance genes to other enteric bacteria, and the recipient bacteria expressed those genes (Yaron et al., 2000). Thus, the formation of vesicles appears to be a mechanism for transport and transfer of genetic material and Shiga toxins. There is no evidence that E. coli O157:H7 is invasive in vivo; however, Matthews et al. (1997) showed that the pathogen invaded certain cell lines such as RPMI-4788 (human), MAC-T (bovine mammary secretory), and MDBK (bovine kidney, but not HeLa cells in vitro. The organism showed both localized and diffuse adherence to the culture cells, and microtubules were required for invasion. The ability to invade bovine mammary cells, for example, may be important in asymptomatic carriage of E. coli O157:H7 in cattle (Matthews et al., 1997). Oelschlaeger et al. (1994) showed that E. coli O157:H7 invaded human ileocecal (HCT-8) and bladder (T24) cell lines but not INT 407 intestinal cells. Microfilaments were involved in internalization of E. coli O157:H7 (Oelschlaeger et al., 1994).

Quorum sensing is a mechanism by which small signaling molecules termed autoinducers provide a means for cell-to-cell communication in response to cell population density and environmental conditions. By influencing the transcription of genes in the LEE operon, quorum sensing may control virulence in E. coli O157:H7 (Sperandio et al., 1999; Anand and Griffiths, 2003). Intestinal colonization of EHEC may be regulated by quorum sensing signals produced by nonpathogenic E. coli present in the intestine (Sperandio et al., 1999). Using E. coli K12 DNA arrays, hybridization patterns of cDNA from RNA extracted from E. coli O157:H7 and from its isogenic luxS (gene involved in synthesis of autoinducer 2, AI-2) mutant showed up-regulation of 235 genes and down-regulation of 169 genes comparing the wild-type and mutant strains (Sperandio et al., 2001). The up-regulated genes included those involved in chemotaxis, the SOS response, and the synthesis of flagella and Stx. Thus, quorum sensing in E. coli O157:H7 is a global regulatory system. In a subsequent publication, Sperandio et al. (2003) reported that a molecule termed AI-3, not AI-2, is the actual signal involved in quorum sensing in EHEC. The synthesis of AI-3 also depends on LuxS. Quorum sensing in E. coli O157:H7 may also involve bacterium-host communication since AI-3 cross-talks with the mammalian hormone, epinephrine (Sperandio et al., 2003).

## Treatment and vaccines

There is no established therapy for E. coli O157:H7 infection, but the development of several vaccines and other promising regimens are being evaluated. Treatment with antibiotics is controversial, since antimicrobial therapy may result in an increased risk for HUS (Mølbak et al., 2002). Antibiotics may increase the expression of the Shiga toxins, and/or bacterial injury caused by the antibiotic may result in increased release of preformed toxins. Treatments involving use of Synsorb Pk, a synthetic analog of the Shiga toxin receptor, Gb₃, bound to a calcinated diatomaceous material called Chromosorb are undergoing clinical trials (Takeda et al., 1999; Trachtman and Christen, 1999). The administration of Synsorb-Pk in combination with antibiotics may absorb sufficient amounts of toxin to prevent uptake into the circulatory system (Mulvey et al., 2002). Treatments under investigation include use of recombinant bacteria expressing a Shiga toxin receptor mimic, humanized Shiga toxin-neutralizing monoclonal antibodies, pooled bovine colostrum containing antibodies against Shiga toxins, intimin, and the EHEC-hemolysin, and bovine lactoferrin and its peptides (Shin et al., 1998; Huppertz et al., 1999; Paton et al., 2001; Yamagami et al., 2001). However, further studies using appropriate animal models are needed to elucidate the in vivo efficacy of the various proposed treatments.

A number of vaccine protocols for use in cattle and humans are being investigated (Horne et al., 2002). A plant cell-based intimin vaccine tested in mice gave an intimin-specific mucosal immune response and a reduced duration of shedding of E. coli O157:H7 (Judge et al., 2004). This plant-based vaccine system is being explored for oral administration to cattle in an effort to decrease shedding of the pathogen. Vaccination of cattle with type III secreted proteins reduced the duration of shedding and the level of *E. coli* O157:H7 in feces (Potter *et al.*, 2004). Furthermore, the prevalence of the organism in cattle was reduced in a clinical trial conducted under conditions of natural exposure in a feedlot setting. A vaccine consisting of liposomes incorporating monophosphoryl lipid A and antigens from an *E. coli* O157:H7 lysate induced IgG and IgA serum and mucosal antibody responses in immunized mice (Tana *et al.*, 2003). Other vaccine strategies including toxoid and O-specific polysaccharide-protein conjugate vaccines are under investigation for prevention of illness due to EHEC (Keusch *et al.*, 1998; Konadu *et al.*, 1998).

## Infectious dose

Investigation of foods implicated in disease outbreaks revealed that the infectious dose for EHEC is fewer than 50 organisms (Tilden *et al.*, 1996; Tuttle *et al.*, 1999). The calculated number of *E. coli* O157:H7 found in raw ground beef patties implicated in an outbreak that occurred in the western USA in November 1992 to February 1993 was 1.5 organisms per gram or 67.5 per patty (Tuttle *et al.*, 1999). In an outbreak that occurred in Australia associated with EHEC O111:H⁻, the contaminated fermented sausages contained fewer than one *E. coli* O111:H⁻ per 10 g (Paton *et al.*, 1996). That the infectious dose of EHEC is low is corroborated by the occurrence of water-borne outbreaks, outbreaks associated with visiting farms and petting zoos, and person-to-person transmission (Crump *et al.*, 2002; O'Donnell *et al.*, 2002; Olsen *et al.*, 2002).

## Antibiotic resistance

Antibiotic resistance has been increasing during the past 20 years in *E. coli* O157:H7 isolates (Schroeder *et al.*, 2002b; Wilkerson and van Kirk, 2004). The most common resistance found in bovine and human *E. coli* O157:H7 isolates is to tetracycline, followed by resistance to streptomycin and ampicillin (Wilkerson and van Kirk, 2004). Antibiotic resistance plasmids were readily transferred from a commensal *E. coli* strain to *E. coli* O157:H7 in bovine rumen fluid at a frequency exceeding that observed for mating in LB broth (Mizan *et al.*, 2002). In addition, antimicrobial resistance appears to be widespread in non-O157 STEC, also (Schroeder *et al.*, 2002a; DeCastro *et al.*, 2003). Selection pressure imposed by the use of antimicrobials including tetracycline derivatives, sulfa drugs, and penicillins in human and veterinary medicine is likely leading to the selection of antimicrobial-resistant strains of STEC (Schroeder *et al.*, 2002b).

## Animal models

While a number of animal species have been evaluated as models of EHEC infection, no system mirrors the entire spectrum of the disease as seen in humans. EHEC do not normally cause disease in cattle; however, colostrum-deprived, neonatal calves develop diarrhea 18 h following inoculation with $10^{10}$ CFU of bacteria. Calves can become colonized with the organism at

levels greater than or equal to $10^6$ CFU/g of intestinal tissue or feces. A/E lesions in the small or large intestine may develop (Dean-Nystrom, 2003). Colonization of EHEC O157:H7 in newborn calves required intimin (Dean-Nystrom *et al.*, 1998). The role of *stx₂*, *eae*, and *tir* in EHEC pathogenesis was studied using an infant rabbit model (Ritchie *et al.*, 2003). Colonization, A/E lesions, inflammation, and diarrhea did not occur in the infant rabbit when challenged with EHEC strains with deletions in the *tir* and *eae* genes. The presence of the *stx₂* gene increased the severity and duration of diarrhea but was not involved in attachment. Intragastric inoculation of Stx2 induced diarrhea and inflammation (Ritchie *et al.*, 2003). Oral infection of gnotobiotic piglets with Stx2-producing *E. coli* O157:H7 and O26:H11 led to gastrointestinal illness and thrombotic microangiopathy in the kidneys similar to that typically seen in humans with HUS (Gunzer *et al.*, 2002). Tzipori *et al.* (1995) compared an *E. coli* O157:H7 strain with a mutation in the *eaeA* gene with the wild-type strain and demonstrated that intimin (the product of *eaeA*) facilitated attachment to cells and affected the site of intestinal colonization in pigs. Vascular damage and necrosis in the intestines and brain developed in pigs injected intramuscularly with Stx1; these pathological effects were similar to those that develop in humans with EHEC disease (Dykstra *et al.*, 1993). An animal model based on the greyhound dog is being investigated due to similarities noted between EHEC disease and a condition in greyhounds called idiopathic cutaneous and renal glomerular vasculopathy (Fenwick and Cowan, 1998). Dogs with this illness exhibit renal changes similar to those seen in humans with HUS, and there is some evidence that the disease in greyhounds may be caused by Shiga toxin-producing *E. coli*. *E. coli* O157:H7 was isolated from diseased animals and the disease was observed in dogs administered Stx1 or Stx2 by intravascular inoculation. Baboons administered Stx1 by intravenous infusion develop renal failure and damage to the gastrointestinal mucosa (Melton-Celsa and O'Brien, 2003); the kidney lesions are similar to those seen in kidneys of humans with HUS. Diarrhea and A/E lesions developed in a macaque monkey model infected with *E. coli* O157:H7. Several mouse models have been developed to study EHEC pathogenesis (Melton-Celsa and O'Brien, 2003), and ferrets are being investigated as a model system to study EHEC-mediated HUS (Woods *et al.*, 2002). Colonization of the cecum and colon, and A/E lesions developed in 2 out of 7 chicks orally inoculated with *E. coli* O157:H7 (Beery *et al.*, 1985; Sueyoshi and Nakazawa, 1994). Use of an appropriate animal model will enhance the understanding of the pathophysiology of EHEC-induced HC and HUS and will assist in the development of strategies to prevent, control, and treat EHEC infection.

## Growth and survival

The minimum and maximum growth temperatures for *E. coli* O157:H7 in brain heart infusion broth were 10 and 45°C, respectively, and several strains grew slowly at 8°C (Palumbo *et al.*, 1995). *E. coli* O157:H7 has no unusual heat resistance;

however, the thermal resistance can be influenced by a number of factors, including pH, growth conditions and growth phase of the cells, and the method of heating. Juneja et al. (1999) found that exposure to increasing levels of NaCl increased the heat resistance of E. coli O157:H7, whereas increased levels of sodium pyrophosphate decreased heat resistance. The amount of fat in ground beef had an effect on thermal tolerance of E. coli O157:H7. The D values for beef containing 2% and 3% fat were 4.1 and 5.3 min, respectively at 57.2°C and 0.3 and 0.5 min, respectively, at 62.8°C (Line et al., 1991). Irradiation doses approved for red meats were sufficient to reduce the level of E. coli by several log values; however, the radiation source (electron beam versus gamma rays), the radiation temperature, and the oxygen permeability of the packaging material, influenced the $D_{10}$ values (López-González et al., 1999). A treatment consisting of dry heat in combination with an irradiation dose of 2.0 kGy eliminated E. coli O157:H7 from inoculated alfalfa and mung bean seeds. However, a dose of 2.5 kGy was necessary to eradicate the pathogen from radish seeds (Bari et al., 2003). Neither the germination percentage nor the length of alfalfa sprouts was affected by irradiation; however, irradiation resulted in radish and mung bean sprouts with decreased lengths compared to controls.

Although there is some strain variation, E. coli O157:H7 is relatively acid tolerant compared to other foodborne pathogens. pH levels supporting growth range from 4.4 to 9.0 and the organism can survive for extended periods in foods at pH levels of 3.5 to 5.5. E. coli O157:H7 survived in mayonnaise for 5 to 7 weeks at 5°C and for 10 to 31 days in apple cider at 8°C, and the organism survived for up to 2 months, with only a 100-fold reduction in cell numbers during fermentation, drying, and storage of fermented sausage (Glass et al., 1992; Zhao et al., 1993; Zhao et al., 1994). Warm (20°C) or hot (55°C) acetic, citric, and lactic acid sprays applied to raw beef did not noticeably reduce the levels of E. coli O157:H7 (Brackett et al., 1994). There are at least three systems for acid tolerance in E. coli O157:H7: an acid-induced oxidative system requiring the Rpos alternate sigma factor, an acid-induced arginine-dependent system, and a glutamate-dependent system (Lin et al., 1996). Therefore, several acid resistance mechanisms function in E. coli O157:H7 permitting survival under different acid stress conditions. Once induced, these systems remain active for prolonged periods of cold storage.

E. coli O157:H7 survived in inoculated tap and bottled spring and mineral water for 300 days or longer (Warburton et al., 1998). The organism survived for 14 days at less than 15°C in farm water stored outdoors, demonstrating E. coli O157:H7 may be transferred within a herd via farm water (McGee et al., 2002). E. coli O157:H7 persisted for 77, >226, and 231 days in manure-amended autoclaved soil stored at 5, 15, and 21°C, respectively (Jiang et al., 2002). In fallow soils, it survived for 25 to 41 days, for 47 to 96 days on rye roots, and for 92 days on alfalfa roots (Gagliardi and Karns, 2002). The presence of clay increased survival; whereas manure in the soil did not affect survival.

## Etiology of foodborne cases and outbreaks

The major reservoir for E. coli O157:H7 is cattle, and transmission from cattle to humans occurs via contaminated food or water. A survey to determine the distribution and prevalence of E. coli O157:H7 in cattle in four major feeder-cattle states in the United States indicated that 10.2% out of 10,662 fecal samples tested and 13.1% of water or water tank sediment samples were positive, with over 60% of feedlots having at least one positive water or water sediment sample (Sargeant et al., 2003). The overall prevalence of E. coli O157:H7 or O157: NM in feces and hides of fed cattle presented for slaughter at meat processing plants in the Midwestern United States was 28% (91/327) and 11% (38/355), respectively (Elder et al., 2000), which was a higher prevalence than reported in previous studies. Elder et al. (2000) suggested that the different isolation methods employed and time of year that samples were taken were the likely reasons for the differences in results.

Cattle harboring E. coli O157:H7 are generally free of disease; however, an infection in newborn calves may cause fatal ileocolitis. Tolerance to E. coli O157:H7 infection may be due to lack of Gb₃, the Shiga toxin receptor, in the bovine gastrointestinal tract (Pruimboom-Brees et al., 2000). The principal site of E. coli O157:H7 colonization in cattle is the terminal rectum (Grauke et al., 2002; Naylor et al., 2003). The mean duration of shedding in cattle inoculated with $5 \times 10^8$ CFU of E. coli O157:H7 was approximately 30 days after the first inoculation and 3 to 8 days after the second and third inoculations (Sanderson et al., 1999). E. coli O157:H7 has also been isolated from deer, pigs, horses, goats, sheep, cats, dogs, rabbits, poultry, and rats, and from birds such as ravens, doves, and seagulls (Meng et al., 2001; Feder et al., 2003). E. coli O157:H7 strains persisted longer in the gastrointestinal tracts of young adult sheep than enteropathogenic or enterotgoxigenic strains; the persistence of EHEC in ruminants probably explains why these animals are reservoirs of the pathogen (Cornick et al., 2000). The pathogen was isolated from houseflies collected from a school in Japan at which a disease outbreak occurred (Kobayashi et al., 1999). Feeding experiments showed that the bacteria were present in the fly's intestine. Flies shed E. coli O157:H7 for at least 3 days after feeding, which indicated that there was proliferation of the organism in the fly gut.

Foods of bovine origin, including undercooked ground beef, raw milk, and roast beef were associated with illness caused by E. coli O157:H7; however, goat cheese, venison jerky, and environmental contamination with sheep feces have also been linked with outbreaks (Meng et al., 2001; Ogden et al., 2002). An outbreak affecting 732 individuals, with 55 cases of HUS and 4 deaths occurred in late 1992 and early 1993 in the western USA and Canada (Bell et al., 1994). Undercooked ground beef, contaminated with E. coli O157:H7, served at multiple outlets of the same fast-food restaurant chain was implicated as the cause of outbreak. Outbreaks have been linked to other food vehicles such as apple cider, mayonnaise, pea salad, cantaloupe, lettuce, hard salami, and alfalfa and radish sprouts (Meng et al., 2001). Fruits and vegetables may become contaminated with cattle or other ruminant manure during harvesting and processing. Control measures include restricting the application of animal fecal wastes to crops used for direct human consumption. Since *E. coli* O157:H7 can tolerate acidic environments, the organism can survive in low pH foods, such as apple cider or fermented products. A large outbreak involving white radish sprouts occurred in Japan in 1996; 9578 individuals had symptoms, many of whom were school children. There were 90 cases of HUS, and 11 deaths (Bettelheim, 1997; Michino *et al.*, 1999). In addition, contaminated recreational water, well water, groundwater, and municipal water systems have also caused disease outbreaks. In Walkerton, Ontario, Canada, in May 2000, an estimated 2,300 individuals became seriously ill and 7 died due to exposure to drinking water contaminated with *E. coli* O157:H7 (Anonymous, 2000). The water became contaminated with cattle excrement that washed into the town's wells from a nearby farm during a flood weeks earlier. Outbreaks resulting from visits to agricultural fairs and petting zoos have also occurred, likely due to exposure to animals, in particular cattle or other ruminants, and the farm environment (CDC, 2001; Crump *et al.*, 2003). Person-to-person transmission of *E. coli* O157:H7 has led to HC or HUS in nursing homes, day-care centers, and within family members (Carter *et al.*, 1987; Al-Jader *et al.*, 1999). Table 18.2 lists a number of outbreaks caused by *E. coli* O157 and non-O157 strains.

## Diagnosis and methods for detection of EHEC

*E. coli* O157:H7 infections are diagnosed by isolating the pathogen from stools of patients presenting with bloody diarrhea or HUS onto sorbitol MacConkey agar (SMAC). However, since *E. coli* O157:H7 also causes non-bloody diarrhea, it has been recommended that stools from all patients with diarrhea should also be cultured. Demonstration of Shiga toxins in fecal filtrates is also useful for diagnosis. Since non-O157 EHEC may also cause HC or HUS, isolation and identification of the causative organism is necessary for epidemiological purposes. There are a number of strategies that have been described for the detection and isolation of *E. coli* O157:H7 in foods (Fratamico *et al.*, 2002; Deisingh and Thompson, 2004). *E. coli* O157:H7 does not ferment sorbitol within 24 hours; therefore, the colonies are colorless on SMAC. Sorbitol-negative colonies are isolated and characterized for responses to various biochemical parameters, for the presence of the O157 and H7 antigens, and for the presence of the Shiga toxin genes or Shiga toxin genes (Su and Brandt, 1995). Commercially available selective and differential agar media for isolation of *E. coli* O157:H7 include Rainbow Agar O157 (Biolog), CHROMagar O157 (Hardy Diagnostics), BCM O157:H7 Agar (Biosynth), and Fluorocult *E. coli* O157:H7 Agar (Merck). SMAC agar has been modified resulting in plating media with increased selectivity and ability to differentiate *E. coli* O157:H7 from colonies of other organisms. These include CT-SMAC, which contains potassium tellurite and cefixime

or CR-SMAC in which cefixime and rhamnose are added to SMAC, and SMA-BCIG containing the substrate for β-glucuronidase, 5-bromo-4-chloro-3-indoxyl-β-D-glucuronic acid cyclohexylammonium salt (BCIG). Presumptive positive colonies can be tested for the presence of the O157 and H7 antigens using commercially available latex agglutination kits or antisera including the RIM *E. coli* O157:H7 Latex Test (Remel) and the Immun°Card STAT *E. coli* O157 (Meridian Diagnostics) test.

Except for the production of Shiga toxins, there are generally no phenotypic markers that are shared by all non-O157 EHEC that can be utilized as a strategy to develop differential media to distinguish them from non-pathogenic *E. coli*. It is probable that the incidence of disease caused by non-O157 EHEC is underestimated due to the lack of reliable methods to detect them.

Conventional methods rely on culture and agar media to grow, isolate, and enumerate viable *E. coli* O157:H7. In recent years, however, a number of companies have developed methods for detection of *E. coli* O157:H7 and other pathogens that are specific, faster, and often more sensitive than conventional methods. Immunoassays and genetic-based assays such as the polymerase chain reaction (PCR) are examples of such rapid methods. Immunoassays rely on an antibody binding to a specific bacterial antigen. A number of immunoassay formats encompassing enzyme-linked immunosorbent or fluorescent assays, lateral flow immunoassays, and latex agglutination assays for detection of *E. coli* O157:H7 are commercially available (Feng, 2001; Fratamico *et al.*, 2002). Immunomagnetic separation (IMS) using magnetic particles coated with antibodies specific for *E. coli* O157 are available from Dynal, Inc. Alternatively, the particles can be purchased from a number of vendors, and the beads can be coated with the desired antibody. Use of IMS on food enrichments and other complex matrices results in concentration and sequestering of the target bacteria from non-target organisms and matrix components that interfere with subsequent detection systems (Fratamico and Crawford, 1999). IMS has been used in conjunction with plating onto selective agars and with PCR assays, and has been incorporated into commercially available kits including the PATH *IGEN E. coli* O157 test (Igen International) and the EHEC-Tek for *E. coli* O157:H7 (Organon Teknika) for detection of the pathogen in foods (Fratamico and Crawford, 1999). Shiga toxins can be detected in bacterial culture supernatants, food enrichments, and in stool samples using immunoassays, including colony immunoblot assays, latex agglutination, and antibody capture or toxin receptor-mediated enzyme linked immunosorbent assays. The VTEC RPLA toxin detection kit (Oxoid), the Premier EHEC (Meridian Diagnostics), and the RIDASCREEN Verotoxin kit (r-biopharm) are immunoassay-based kits for the detection of Stx1 and Stx2.

Nucleic acid-based detection systems, including the PCR and DNA hybridization techniques, rely on discrimination of *E. coli* O157:H7 DNA or RNA sequences from closely related organisms. Numerous PCR-based methods targeting genes

**Table 18.2** Outbreaks caused by O157 and non-O157 EHEC[a]

| Country | Serotype | Number affected | Vehicle of transmission | Reference |
|---|---|---|---|---|
| Japan | O157:H7 | 49 (13 asymptomatic) | Ikuro-sushi (rice with salmon roe) | Terajima *et al.* (1999) |
| Denmark, England, Finland, Sweden, Wales | O157 VTEC | 14 (one probable) | raw vegetables (holiday in Canary Islands) | Pebody *et al.* (1999) |
| Canada | O157:H7 | 7 | Hamburger | Health Canada (2000) |
| Canada | O157:H7 | 39 (HUS in 2) | Genoa salami | Williams *et al.* (2000) |
| England | O157 VTEC | 6 (HUS in 1) | Unpasteurized milk | PHLS Communicable Disease Surveillance Centre (2000a) |
| England | O157 VTEC | 7 | Person-to-person | PHLS Communicable Disease Surveillance Centre (2000b) |
| England | O157 VTEC | 8 (HUS in 1) | Delicatessen food items | PHLS Communicable Disease Surveillance Centre (2000c) |
| | O157 VTEC | 15 (HUS in 3) | ? | |
| United States | O157:H7 | 14 (HUS in 4) | Unpasteurized cider | Hilborn *et al.* (2000) |
| United States | O157:H7 | > 1000 (HUS in 2) | Unchlorinated well water at county fair | Yarze and Chase (2000) |
| Japan | O157 | 58 | "Whole roasted cow" | Yamamoto *et al.* (2001) |
| Scotland | O157 VTEC | 37 | Home-made cream cakes | O'Brien *et al.* (2001) |
| United States | O157:H7 | 82 (Hus in 4) | Alfalfa sprouts | Breuer *et al.* (2001) |
| United States | O157:H7 | 56 (HUS in 9) | Farm visit | CDC (2001) |
| Canada | O157:H7 | 5 (HUS in 2) | Unpasteurized goat milk | Health Canada (2002) |
| England | O157 VTEC | 114 (HUS in 3) | Pasteurized milk | Goh *et al.* (2002) |
| Ireland | O157:H7 | 11 (three asymptomatic) | Person-to-person in day care center | O'Donnell *et al* (2002) |
| Japan | O157 STEC | 28 | Grilled beef | Tsuji *et al.* (2002) |
| The Netherlands | O157 STEC | 1 (HUS) | Petting zoo | Heuvelink *et al.* (2002) |
| United States | O157:H7 | 26 | Ground beef | CDC (2002) |
| United States | O157:NM StEC | 3 (four probable) | Swallowing lake water | Feldman *et al* (2002) |
| United States | O157:H7 | 157 (HUS in 4) | Unchlorinated municipal drinking water | Olsen *et al.* (2002) |
| Canada | O157:H7 | 17 (HUS in 2) | Person-to-person in daycare center | Health Canada (2003) |
| Canada | O157:H7 (and *Campylobacter jejuni*) | >2000 (7 deaths) | Drinking water | Holme (2003) |
| England | O157 VTEC | 2 8 (probable) | Cucumber salad | Duffell *et al.* (2003) |
| Scotland | O157 | 20 (HUS in 1) | Environmental exposure at scout camp | Howie *et al.* (2003) |
| Sweden | O157 VTEC | 11 | Lettuce | Welinder-Olsson *et al.* (2003) |
| United States | O157:H7 | 37 (HUS in3) | Swimming in lake | Bruce *et al.* (2003) |
| United States | O157 STEC | 44 (HUS in 2) | Drinking water at agricultural fairs | Crump *et al.* (2003) |
| United States | O157 | 23 (HUS in 1) | Environmental exposure (dust, sawdust) at county fair | Varma *et al.* (2003) |
| Wales | O157 VTEC | 24 (person-to-person in 7) | Ice-cream and/or cotton candy | Payne *et al.* (2003) |
| Canada | O157:H7 | 143 (HUS in 6) | Hungarian salami | MacDonald *et al.* (2004) |
| United States | O157:H7 | 13 (HUS in 3) | Beef tacos | Jay *et al.* (2004) |
| United States | O157:H7 | 36 | Swallowing lake water while swimming | Samadpour *et al.* (2004) |
| France | O103:H2 | 6 | Not known | Mariani-Kurkdjian *et al.* (1993 |
| Italy | O111:NM | 9 | Not known | Caprioli *et al.* (1994) |
| United States | O104: H21 | 11 confirmed, 7 suspected | Milk | CDC (1995) |
| Australia | O111:H⁻ | 22 | Uncooked, semi-dry fermented sausage | Paton *et al.* (1999) |
| Japan | O118:H2 | 126 | Salad | Hashimoto *et al.* (1999) |

**Table 18.2** Continued

| Country | Serotype | Number affected | Vehicle of transmission | Reference |
|---------|----------|-----------------|-------------------------|-----------|
| Australia | O113:H21 | 3 | Not known | Paton *et al.* (1999) |
| Japan | O26:H11 | 1 (bloody diarrhea) 10 (asymptomatic) | Drinking water | Hoshina *et al.* (2001) |
| United States | O121:H19 | 11 | Swimming water | McCarthy *et al.* (2001) |
| Ireland | O26:H11 | 4 | Not known | McMaster *et al.* (2001) |
| Germany | O26:H11 | Six nonbloody diarrhea, five asymptomatic | "Seemerrolle" beef | Werber *et al.* (2002) |
| Germany | O26:H11 | 3 | Not known | Misselwitz *et al.* (2003) |

[a]Table 18.2 is an update of Table 10.4 presented by Meng *et al.* (2001).

such as $stx_1$, $stx_2$, *eae*, *uidA*, *hly*, *fliC*, *rfbE* and others have been described (Deisingh and Thompson, 2004). PCR assays have also been performed in a multiplex format in which more than one sequence is amplified simultaneously in a single reaction (Fratamico *et al.*, 2000; Campbell *et al.*, 2001). Real-time PCR assays employing fluorogenic probes to visualize amplification of target sequences during the reaction have been developed for detection of *E. coli* O157:H7 (Ibekwe and Grieve, 2003; Sharma and Dean-Nystrom, 2003). Commercially available PCR-based assays include the BAX *E. coli* O157:H7 (Dupont Qualicon), the Probelia PCR for *E. coli* O157:H7 (BioControl Systems), and the TaqMan *E. coli* O157:H7 detection kit (Applied Biosystems). More recently, seven specific genes of *E. coli* O157:H7 were detected using microarrays (Liu *et al.*, 2003). Biotin-labeled target DNA was obtained by incorporation of biotin-16-dUTP during multiplex PCR and hybridized to probes that were spotted onto glass slides, followed by staining with streptavidin-Cy3. Call *et al.* (2001) detected *E. coli* O157:H7 in chicken rinsate at a level of 55 CFU/mL by utilizing a combination of immunomagnetic capture, multiplex PCR, followed by detection of the PCR products using a microarray. Several other types of methods have also been described (Deisingh and Thompson, 2004).

## Evolution of *Escherichia coli* O157:H7

The *eae* and H7 flagellar gene (*fliC*) gene sequences are nearly identical in *E. coli* O157:H7 and O55:H7 (Reid *et al.*, 1999). Multilocus enzyme electrophoresis analyses indicate that *E. coli* O157:H7 evolved from a progenitor strain with serotype O55:H7 (Feng *et al.*, 1998). A model for the stepwise emergence of *E. coli* O157:H7 has been formulated through phylogenetic analyses based on enzyme allele profiles. An EPEC-like bacterium that ferments sorbitol and expresses β-glucuronidase was the immediate ancestor of *E. coli* O55:H7. The O55:H7 "clone" acquired a mutation at −10 in the β-glucuronidase gene, *uidA*, and also acquired the LEE pathogenicity island. Transduction by a toxin-converting bacteriophage with the $stx_2$ gene resulted in the Stx2-producing *E. coli* O55:H7 strain. A second mutation then occurred in the *uidA* gene at +92 resulting in a change of the O antigen from O55 to O157, possibly by lateral transfer and recombination of a region of the *rfb* (O antigen gene cluster) locus containing the

*rfbE* gene homologous to the perosamine synthetase gene of *Vibrio cholerae*. The EHEC virulence plasmid was acquired at this stage. Two distinct lines then evolved from this progenitor bacterium: a non-motile, sorbitol⁺, and β-glucuronidase⁺ Stx2-producing strain (O157:H⁻) and a sorbitol-negative, β-glucuronidase⁺ lineage that acquired the $stx_1$ gene (O157:H7, $stx_1^+$, $stx_2^+$). The latter lineage lost β-glucuronidase activity, which led to the immediate ancestor of the O157:H7 "clone" that has spread worldwide. Comparison of *gnd* gene sequences (located adjacent to the *E. coli* O antigen gene cluster, *rfb*) showed that *gnd* co-transferred with the adjacent *rfb* locus into *E. coli* O157 and O55 in distantly separated lineages. In addition, intragenic recombination may have contributed to allelic variation in this region of the O157 chromosome (Tarr *et al* 2000). In an attempt to understand the shift from O55 to O157, Wang, L. *et al.* (2002) sequenced the *E. coli* O55 O antigen genes and flanking sequences. Two recombination sites were identified, one within the *galF* gene and the other between the *hisG* and *amn* genes, thus providing evidence for the recombination event proposed for the evolution of the *E. coli* O157:H7 clone.

## Genomic analysis of *Escherichia coli* O157:H7

The genome of *E. coli* O157:H7 EDL933 has been sequenced, providing information on the evolution of this organism (Perna *et al.*, 2001). In addition, the sequence data allows identification of virulence genes and targets for the development of detection methods. When the O157:H7 genome is compared to that of *E. coli* K12, both share a similar backbone sequence of ca. 4.1 Mb; however, 1.34 Mb of DNA in *E. coli* O157:H7 is missing in K12, and 0.53 Mb of DNA in K12 is missing in O157:H7 (Perna *et al.*, 2001; Sperandio, 2001). The *E. coli* O157:H7 and K12 genomes consist of 5,416 and 4,405 genes, respectively. Both O157:H7 and K12 are punctuated by hundreds of islands or DNA segments of up to 88 kb in length, designated K-islands in K12 and O-islands in O157:H7. O-islands of diverse sizes, missing in K12, contain approximately 26% of the *E. coli* O157:H7 genes. Only 40% of the O-island genes can be assigned a function. The genes for the synthesis of fimbriae, iron uptake and utilization, and survival in different environments are found on smaller islands; putative virulence genes are encoded on nine large islands. Sequences related to known bacteriophages were identified in 18 multigenic regions. The sequence data indicate a high level of diversity between the O157:H7 and K12 genomes, and that the islands were probably acquired through horizontal gene transfer from other organisms. There is considerable variability in the presence, number, and location of O-islands encoding tellurite resistance within *E. coli* O157:H7 strains (Taylor *et al.*, 2002). Moreover, Shaikh and Tarr (2003) revealed that *E. coli* O157:H7 genomes possessed novel truncated bacteriophages and multiple *stx*$_2$ bacteriophage insertion sites. Several antibiotics promoted excision of bacteriophages, and bile salts attenuated excision.

## Genetic fingerprinting and outbreak investigation

Epidemiologic investigations of foodborne disease outbreaks are facilitated by the use of molecular typing methods to determine the genetic relatedness of bacterial isolates. Phenotypic methods such as serotyping, phage typing, or multilocus enzyme electrophoresis are gradually being replaced by genetic fingerprinting techniques such as ribotyping, random amplified polymorphic DNA, and pulsed field gel electrophoresis (PFGE).

The PFGE technique involves isolation of intact DNA followed by digestion with restriction enzymes, and analysis of the digestion products (typically 10 to 20 products ranging in size from 10 to 800 kb) that are separated by agarose gel electrophoresis with programmed variations in the direction and duration of the electric field. PFGE was used to determine the genetic relatedness of clinical and food isolates from an outbreak in 1993 caused by hamburgers contaminated with *E. coli* O157:H7. Identical phage types and PFGE patterns were demonstrated for all of the isolates associated with the multistate outbreak (Barrett *et al.*, 1994). It was concluded that the use of a standardized subtyping method would allow rapid comparison of isolates from different parts of the country and demonstrate a common source, leading to the prevention of further spread of infection.

Thus, in 1996, a national molecular subtyping network for foodborne disease surveillance, known as PulseNet, was established by the Centers for Disease Control and Prevention (CDC) in collaboration with the Association of Public Health Laboratories (Swaminathan *et al.*, 2001). Laboratories participating in PulseNet perform PFGE on outbreak isolates from humans and/or the suspected food and enter the data into an electronic database for rapid comparison of the fingerprint patterns. All 50 state public health laboratories, local public health laboratories, Food and Drug Administration and USDA Food Safety and Inspection laboratories participate in PulseNet. Such cooperation plays an integral role in the surveillance and investigation of foodborne outbreaks caused by *E. coli* O157:H7 (CDC, 2002). There has been an expansion of the PulseNet system, and it currently tracks non-typhoidal *Salmonella*, *Shigella*, *Listeria monocytogenes*, and *Campylobacter*. Noller *et al.* (2003) reported that a subtyping technique

known as multilocus variable-number tandem repeat analysis (MLVA), which targets short tandem repeats in the DNA at multiple loci, had a sensitivity equal to that of PFGE. The specificity of MLVA was greater than that of PFGE, since MLVA differentiated strains with unique PFGE fingerprints.

## Non-O157 STEC/EHEC

Over 200 STEC (also referred to as verocytotoxin-producing *E. coli* or VTEC) serotypes have been identified; 100 or more non-O157 O:H serotypes of STEC have been responsible for cases and outbreaks of HC and HUS (Nataro and Kaper, 1998; World Health Organization, 1998; Karmali, 2003; Table 18.2), and strains in this subset of STEC are termed EHEC. Non-O157 STEC serotypes are more prevalent than *E. coli* O157:H7 in Australia, Latin America, and many European countries (Nataro and Kaper, 1998; Elliott *et al.*, 2001; Werber *et al.*, 2002; Karmali, 2003). Twenty to 70 percent of STEC infections throughout the world are due to non-O157 STEC strains (World Health Organization, 1998). Examples of non-0157 STEC infections include (a) an *E. coli* O121:H19 outbreak of HUS at a lake in Connecticut (McCarthy *et al.*, 2001); (b) an *E. coli* O111:H$^-$ outbreak in Australia involving 21 cases with one fatality, linked to consumption of a locally produced semi-dry fermented sausage (Paton *et al.*, 1996): (c) an *E. coli* O111:H8 outbreak indistinguishable from that caused by *E. coli* O157:H7 in attendees at a youth camp (Brooks *et al.*, 2004); and (d) an *E. coli* O26:H11 multistate outbreak in Germany involving 11 case subjects and linked to a certain type of beef referred to as "Seemerrolle" (Werber *et al.*, 2002).

Although non-O157 STEC have caused HC and HUS, infection with some STEC strains may br asymptomatic or result in mild diarrhea. It is probable that these strains do not possess all of the virulence factors of *E. coli* O157:H7, and additional studies are needed to determine if that is true. Diarrhea was non-bloody, and 5 persons remained asymptomatic in an outbreak caused by STEC O26:H11 outbreak in Germany (Werber *et al.*, 2002). STEC possessing different *stx*$_2$ variants differed in their capacity to produce HUS. While all strains caused diarrhea, strains harboring *stx*$_{2c}$ induced HUS, whereas strains harboring *stx*$_{2d}$ or *stx*$_{2e}$ did not (Friedrich *et al.*, 2002). Using a subtractive genomic hybridization technique to identify virulence genes, Pradel *et al.* (2002) compared the DNA of the HUS-inducing *E. coli* O91:H21 strain with that of strains that were not associated with human illness. They found that the O91:H21 strain possessed fragments corresponding to previously identified unique sequences in *E. coli* O157:H7, in *Shigella flexneri*, and in enteropathogenic and STEC plasmids. Interestingly, the O91:H21 strain possessed three copies of the *stx*$_2$ gene. Pradel *et al.* (2002) suggested that highly pathogenic STEC strains acquired virulence genes through lateral gene transfer to a greater extent than strains with lesser virulence.

Determining the true incidence of disease caused by the non-O157 STEC is complicated, since it is necessary to detect the presence of the Shiga toxins or the *stx* genes in these strains. Unlike *E. coli* O157:H7, most non-O157 STEC cannot be detected by currently available selective and differential media. Park and coworkers (2002) plated stool specimens from patients onto SMAC, and in addition, they used a commercially available assay to detect Shiga toxins in the stool samples. *E. coli* O157:H7 was found in 45 out of 65 patient stool specimens and the remainder of the samples contained non-O157 STEC strains. The serotypes of the non-O157:H7 strains included O45:H2, O26:H12, O103:H2, O111:NM, O153:H2, O88:H25, O145:NM, and O96:H9. Park *et al.* (2002) recommended that all bloody stool specimens be tested for the presence of Shiga toxins. This recommendation is also emphasized by the CDC.

Several O antigen gene clusters involved in the synthesis of the O antigens of the different *E. coli* serogroups have recently been sequenced, and PCR assays targeting unique sequences within these regions have been developed to detect specific *E. coli* serogroups (Wang *et al.*, 2001a; Fratamico *et al.*, 2003). Suitable targets for serogroup specific PCR assays include the *E. coli wzx* (O antigen flippase) and *wzy* (O antigen polymerase) genes. Multiplex PCR assays targeting a serogroup specific region within the O antigen gene cluster as well as the *stx* or other virulence genes have also been reported (Wang, G. *et al.*, 2002). A plating medium consisting of washed sheep's blood agar and containing mitomycin C enhanced the ability to detect enterhemolysin-producing O157:H7 and non-O157 STEC strains (Sugiyama *et al.*, 2001). Magnetic particles linked with antibodies specific for *E. coli* O26, O103, O111, and O145 are commercially available (Dynal). Other methods for detection of STEC including immunological and DNA-based methods have been reported (Karch *et al.*, 1999; Bettelheim, 2003).

## Enterotoxigenic *Escherichia coli* (ETEC)

ETEC are a common cause of diarrhea with high morbidity and mortality in infants, young children, and the elderly in developing countries; however in industrialized countries, ETEC-induced diarrhea is uncommon. The primary cause of diarrhea in travelers to developing countries is ETEC infections (Cohen and Giannella, 1995; O'Brien and Holmes, 1996). Adult animals, unlike newborn and young domestic animals (calves, lambs, and pigs) are not susceptible to ETEC-induced diarrhea (Gyles, 1992).

The small intestine of both human adults and children is the site of ETEC colonization. The organisms attach to the enterocytic brush border via bacterial adherence factors, but there is no invasion or intestinal cell damage (Cohen and Gianella, 1995). Toxin(s) are secreted which induce the production of a non-inflammatory watery diarrhea; blood, mucus, and leukocytes are absent in stools. Symptoms include nausea and mild to moderate abdominal cramping without fever. Prolonged diarrhea in children leads to severe dehydration and malnutrition with high mortality. Traveler's diarrhea is usually a mild self-limited disease of 1 to 5 days' duration (Neil *et al.*, 1994; Cohen and Gianella, 1995).

ETEC infection can be transmitted by the ingestion of food and water contaminated by human feces or by an infected food handler with poor personal hygiene (Black *et al.*, 1981). The infective dose is ca. $10^8$ organisms (Levine *et al.*, 1977); therefore, person-to-person transmission of ETEC infection is rare. Humans are the predominant reservoir for ETEC (Doyle and Padhye, 1989).

Breastfeeding protects infants against ETEC infections (Black and Lanata, 1995; Pickering *et al.*, 1995; Clemens *et al.*, 1997). Oral rehydration is an effective therapy in ETEC-infected infants and children; rehydration prior to the occurrence of severe dehydration can be lifesaving. Antimotility agents and antibiotic therapy are not recommended since antibiotic resistance is common in developing countries and the offending organisms are not eliminated from the gut (Pickering *et al.*, 1995). Peristaltic removal of the pathogen is interfered with if antimotility agents are used to treat children. No ETEC vaccines are available (O'Brien and Holmes, 1996).

Since the use of antibiotics can lead to antibiotic resistance and changes in the intestinal flora, their use is generally not recommended for most cases of traveler's diarrhea. Prolonged or severe diarrhea can be treated with trimethoprim/sulfmethoxazole and in some cases, rehydration therapy may be required. Mild traveler's diarrhea can be treated with antidiarrheal drugs to restrict fluid accumulation and intestinal mobility (Berkow, 1992; Cohen and Giannella, 1995).

Water and food have been implicated in outbreaks of ETEC (Table 5.5 in Fratamico *et al.*, 2002). Foods implicated in outbreaks include salads, dipping sauces, or ready-to-eat items, including hog dogs, cold roast beef, cold turkey, brie cheese, i.e., foods that are served raw or foods that are cooked but served cold. ETEC is not considered to be a major cause of diarrheic foodborne outbreaks in the United States; however many cases may be missed since rapid and reliable tests for detection and identification of ETEC strains are not available. An 8-year study on the incidence of ETEC in the United States and on cruise ships indicated that from 1996 to 2003, 16 outbreaks occurred (Beatty *et al.*, 2004). In 11 of the outbreaks, the vehicle was identified and included drinking water, ice, various vegetables and salads, enchilada, tacos, tortilla chips, quesadillas, fajitas, chicken, lasagna, and catfish (Beatty *et al.*, 2004).

ETEC-induced diarrheal illness is caused by adherence and colonization of the intestinal mucosa by the bacteria and the production of the enterotoxins, heat-labile toxin (LT) and/or heat-stable toxin (ST) (Cohen and Giannella, 1995).

### The LT enterotoxins

There are two types of heat-labile toxins: LT-I and LT-II. Polyclonal antibodies raised against LT-I toxin neutralize cholera toxin (CT) but do not neutralize the activity of LT-II indicating that CT and LT-I are related toxins (O'Brien and Holmes,

1996). The CT-encoding genes are located on the genome of an integrated prophage, whereas the genes for LT-I are plasmid mediated (Davis and Waldor, 2003). The genes for LT-II are chromosomally encoded (O'Brien and Holmes, 1996). CT induces a more severe disease than ETEC that produce LT-I. Human disease is not mediated by LT-II-producing strains, and the toxin is found primarily in animal isolates (O'Brien and Holmes, 1996).

The LTs are oligomeric peptides similar to CT and consist of one A polypeptide subunit noncovalently bound to five B polypeptides. There is approximately 50% amino acid identity between the A subunits of LT-I and LT-II, but there is only about 10% identity between the LT-I and LT-II B subunits (O'Brien and Holmes, 1996). Like CT, the B subunits of LT-I are ring structures that bind to the host cell ganglioside GM1. Proteolytic nicking of the A subunit results in the A1 and A2 fragments. The A2 fragment is bound to the B subunits, and the A1 fragment is linked to A2 by a disulfide bond. After binding to the host cell, the toxin is endocytosed and translocated into the cell (Butterton and Calderwood, 1995; Nataro and Kaper, 1998). The adenylate cyclase situated on the membrane of intestinal epithelial cells is the cellular target of LT. The A1 fragment exhibits ADP-ribosyl transferase activity, and ADP ribosylation of a GTP-binding protein activates adenylate cyclase with an increase in cyclic AMP within the intestinal mucosa. Thus, there is a stimulation of chloride secretion and a decrease in sodium absorption. The increased ion content in the intestinal lumen leads to fluid and electrolyte loss with the production of a watery diarrhea (Butterton and Calderwood, 1995; Nataro and Kaper, 1998). Proteolytic nicking of the CT A subunit is necessary for activity (Butterton and Calderwood, 1995, while nicking is not necessary for an active LT-I A-subunit (Grant et al., 1994). However, nicking does enhance the biological and enzymatic activity of LT. An ETEC mutant lacking the nicking region of the LT A-subunit showed decreased induction of cyclic AMP and less diarrheal activity, indicating that nicking of the A-subunit may be necessary for optimum activity of the LT-I toxin (Tsuji et al. 1997). Other bacteria that produce LT-like toxins include *Klebsiella, Enterobacter, Aeromonas, Plesiomonas, Campylobacter,* and *Salmonella* (Cohen and Gianella, 1995).

## The ST enterotoxins

The monomeric ST toxins are subdivided into the STI and STII families. Two toxins are found in the STI family: an 18-amino-acid peptide of porcine origin, ST1a, and the other consists of 19 amino acids, ST1b, which is of human origin. There are three intramolecular disulfide bonds in the STI peptides. STI toxins are heat and acid stable, are water and methanol soluble, are not denatured by detergents, are resistant to proteases, and are poorly antigenic. Disruption of the disulfide bonds renders the toxins inactive (Cohen and Gianella, 1995; O'Brien and Holmes, 1996; Nataro and Kaper, 1998). Genes for both STI and STII are usually located on plasmids, but some are on transposable elements. Genes encoding LT, ST,

colonization factors, colicin, and antibiotic resistance may be present on the same plasmid (Cohen and Gianella, 1995; O'Brien and Holmes, 1996; Nataro and Kaper, 1998).

STI is synthesized intracellularly as a precursor protein consisting of a PRE-region (amino acid residues 1 to 19), PRO-region (amino acid residues 20 to 54) and a MATURE-region (amino acid residues 55 to 72). The precursor protein is translocated into the periplasmic space via Sec proteins of the type II secretion pathway. The PRE-region acts as a signal protein during translocation and is cleaved during or after translocation to the periplasmic space leaving the PRO- and MATURE-regions (Okamoto and Takahara, 1990). There, the PRO-region is cleaved, disulfide bonds are formed in the mature 18-amino acid toxin, and the toxin is translocated across the outer membrane (Yamanaka et al., 1994, 1997). The outer membrane protein, TolC, is involved in the translocation of STI from the periplasmic space, across the outer membrane and into the external environment (Yamanaka et al., 1998).

Guanylate cyclase C, located in the membrane of small intestine enterocytes, is the major receptor for STI. Binding of the enzyme by the toxin induces accumulation of cyclic GMP and secretion of chloride and water into the intestinal lumen (Cohen and Gianella, 1995; O'Brien and Holmes, 1996; Nataro and Kaper, 1998). The mammalian hormone, guanylin, aids in the regulation of fluid and electrolyte absorption in the gut and is homologous to ST1. They also bind to the same receptor on intestinal epithelial cells (Rabinowitz and Donnenberg, 1996). The STI toxins produce a reversible short-term effect mediated by guanylate cyclase, which is quick acting (within 5 minutes). However, the biological effect of CT and LT, mediated by activation of adenylate cyclase, is prolonged with a lag phase of about 1 hour and is reversible (Cohen and Gianella, 1995). LT-I and CT bind to adenylate cyclase from various tissues, whereas STI binds only to intestinal guanylate cyclase (Gyles, 1992).

STII is primarily found in ETEC strains isolated from pigs; it is methanol insoluble in comparison to STI, is a larger peptide than STI (5.1 kDa and 2 kDa, respectively), and the two toxins do not cross-react immunologically (O'Brien and Holmes, 1996). The action of STII is different from STI; STII induces secretion of bicarbonate ions and water into the intestinal lumen and causes an increase in intracellular $Ca^{2+}$ in intestinal cells (O'Brien and Holmes, 1996). STII is not an important contributor to human disease; however, a few cases of STII-induced human diarrhea have been reported (Lortie et al., 1991; Okamoto et al., 1993; Salyers and Whitt, 1994). Reviews by Dubreuil (1997) and Nair and Takeda (1998) discuss various aspects of the heat-stable enterotoxins.

*Citrobacter freundii, Yersinia enterocolitica,* and non-O1 *Vibrio cholerae* produce toxins similar to STI (Smith, 1988; Chaudhuri et al., 1998). An STIa-containing plasmid from ETEC could be transferred to species of *Shigella, Salmonella, Klebsiella, Enterobacter, Edwardsiella, Serratia,* and *Proteus* with stable maintenance of the plasmid and expression of toxin (Smith, 1988).

## EAST1 (EAEC heat-stable toxin 1)

EAST1 is a heat-stable enterotoxin of EAEC strains (Savarino et al., 1993). The gene for EAST1, *astA*, has been found in ETEC strains isolated from both humans and animals (Yamamoto and Echeverria, 1996; Yamamoto and Nakazawa, 1997). Savarino et al. (1996) detected the EAST1 gene in 100% of 75 *E. coli* O157:H7 strains, in 47% of 227 EAEC, in 41% of 149 ETEC, in 22% of 65 EPEC strains, and in 13% of 70 DAEC strains utilizing an *astA* DNA probe. In addition, *astA* was present in non-diarrhea-producing *E. coli* strains. Thus, the *astA* gene appears to be common in *E. coli*; however, it is uncertain if the EAST1 toxin plays a role in the pathogenesis of ETEC. EAST1 is a low molecular weight cysteine-rich, 38-amino acid polypeptide enterotoxin that is plasmid encoded, partially heat stable (63% of activity remained after 15 min at 65°C), and protease sensitive (Savarino et al., 1991; O'Brien and Holmes, 1996; Ménard and Dubreuil, 2002). While the EAST1 toxin does not cross-react serologically with STI, it shows homology with the receptor-binding domains of STI. It is probable that EAST1 shares the same receptor binding site of guanylate cyclase as STI, leading to cyclic GMP secretion (Savarino et al., 1991; O'Brien and Holmes, 1996; Ménard and Dubreuil, 2002). EAST1 appears to be a member of the STI family of heat-stable enterotoxins and contributes to the production of diarrhea through the secretion of cyclic GMP (Savarino et al., 1993). A review by Ménard and Dubreuil (2002) summarizes various aspects of EAST1.

## Colonization factors

The attachment of the ETEC strains to host cells is the first step in pathogenesis. The major adherence factors are the colonization factor antigens (CFAs), which include CFA I, II, and IV. The CFA genes are plasmid encoded, and CFAs, ST, and LT proteins may be encoded by the same plasmid (Cohen and Giannella, 1995; Mol and Oudega 1996). In addition to CFAI, CFAII, and CFAIV, a number of other CFAs have been described (Cassels and Wolf, 1995; Gaastra and Svennerholm, 1996; Grewal et al., 1997; Ricci et al., 1997). The CFA structures may be fimbrial rods, flexible fibrils, helical fibrils, or curly fibrils (Cassels and Wolf, 1995). The diarrheic effect in children infected with CFA-containing ETEC was similar to that in children infected with strains lacking CFAs (I, II, or IV) (López-Vidal et al., 1990). This result indicates that there are unidentified colonization factors involved in induction of diarrhea in those ETEC strains lacking CFAs. Norepinephrine stimulated growth and K99 pilus-mediated adhesion in ETEC strains (Lyte et al., 1997). K99 pili-expressing ETEC strains are pathogenic for lambs, calves, and pigs. The distal two-thirds region of the small intestine is highly innervated with adrenergic nerves, which produce norepinephrine at terminals present in the mucosal lining; this area of the small intestine is preferentially colonized by ETEC (Lyte et al., 1997). Since the K99 adhesion pili are virulence factors essential for ETEC colonization in animals (Parry and Rooke, 1985), it would be of interest to determine if intestinal norepinephrine also stimulates induction of CFAs in human strains of ETEC.

## Vaccines

Since both purified CT and LT (or the B subunit of the toxins) are potent mucosal immunogens, they have been used as oral adjuvants. LT is less toxic than CT and can be used at levels that does not induce diarrhea (Baqar et al., 1995). CT induces a $TH_2$ (T helper cells involved in the humoral immune response) response with production of the interleukin 4 (IL-4) and interleukin 5 (IL-5) cytokines as well as the immunoglobulins, IgA, $IgG_1$ and IgE. LT, however, induces a mixed $TH_1$ (T helper cells involved in the cellular immune response) and $TH_2$ response with production of interferon-gamma (IFN-γ) and the IL-4 and IL-5 cytokines. LT also induces an IgA, $IgG_1$, $IgG_2$, and $IgG_{2b}$ antibody response profile (Takahashi et al., 1996). The use of LT is preferable as a mucosal adjuvant since CT induces an IgE response, which can lead to an immediate-type allergic hypersensitivity reaction. Oral administration of LT and heat-killed *Campylobacter jejuni* stimulated both local and systemic *Campylobacter*-specific IgA and IgG in non-human primates (Baqar et al., 1995). Increased antiviral serum IgG and mucosal IgA was observed upon co-administration of LT with oral inactivated influenza vaccine to mice as compared to the vaccine alone (Katz et al., 1997). The control of human gastrointestinal and pulmonary diseases may be facilitated by using LT as an oral adjuvant to increase the secretion of secretory IgA on mucosal surfaces.

In developing countries, ETEC infections decrease as individuals become older, which suggests the development of protective immunity against ETEC infections. Therefore, it should be possible to develop vaccines against ETEC. Killed ETEC cells combined with recombinant cholera toxin B subunit, orally administered to children or adults, provided significant protection against ETEC infection (Savarino et al., 1999; Cohen et al., 2000; Quadri et al., 2000). However, it will be difficult to produce a vaccine with broad protective ability due to the large number of different adherence factors expressed by ETEC strains (Nataro and Kaper, 1998). Mason et al. (1998) constructed a synthetic gene coding for *E. coli* LT-B subunit for use in transgenic potatoes. Mice fed raw potatoes expressing the LT-B subunit protein produced high levels of serum and mucosal anti-LT-B immunoglobulins. It is conceivable that children and adults could be protected against diarrheic diseases through plant-derived vaccine antigens, particularly if present in raw edible fruits (Walmsley and Arntzen, 2000).

## Detection and identification

There are no distinctive serological or biochemical markers to differentiate toxin-producing strains from non-toxigenic strains of ETEC, therefore, it is necessary to identify the toxins produced by pathogenic ETEC strains. Nataro and Kaper (1998) describe a number of molecular diagnostic techniques that have been used to detect LT and ST. Tsen et al. (1998) developed a multiplex PCR, which allowed the simultaneous detection of the ETEC LTI and STII genes in skim milk and porcine stool. Monoclonal antibodies against various ETEC colonization factors were used to determine the prevalence of ETEC in children with diarrhea in Argentina (Viboud *et al.*, 1993). López-Saucedo *et al.* (2003) described a single multiplex polymerase chain reaction assay for detection of diarrheic *E. coli* including enterotoxigenic *E. coli*. DNA colony hybridization assays, including a pooled toxin (STIa, STIb, and LT) probe assay and individual probe assays were used to detect toxins and a number of different colonization factors in ETEC strains (Steinsland *et al.*, 2003).

## Enteropathogenic *E. coli* (EPEC)

In developing countries (not including China), an estimated 117 million infantile diarrheal episodes caused by EPEC occur each year with high mortality (Clarke *et al.*, 2002). Outbreaks of EPEC rarely occur in developed countries; however, outbreaks have been seen in daycare centers and pediatric wards (Vallance and Finlay, 2000). Transmission occurs primarily by the fecal–oral route; contaminated hands, food, and fomites serve as vehicles of infection. Children less than 2 years of age are at risk, but infants less than 6 months of age are particularly susceptible. In most cases, the diarrhea is self-limiting but severe cases can lead to prolonged illness with wasting and failure to thrive (Fagundes-Neto and Scaletsky. 2000). Acute EPEC infections are manifested by profuse watery, mucoid (but nonbloody) diarrhea, often accompanied with vomiting and fever, and severe cases can result in death (Vallance and Finlay, 2000; Willshaw *et al.*, 2000; Clarke *et al.*, 2002). Diarrhea is induced in adult volunteers by a dose of $10^8$ to $10^{10}$ CFU; however, the infectious dose in children is probably lower (Clarke *et al.*, 2002). The incubation period for EPEC infection in children is unknown. Infants are protected by breastfeeding, and they become infected following weaning due to weaning foods prepared with contaminated water.

The treatment of choice in mild cases is oral hydration, and parenteral rehydration is necessary in severe cases. Children may suffer several diarrheal episodes each year due to EPEC, and an effective vaccine is not available (Willshaw *et al.*, 2000; Clarke *et al.*, 2002). The reservoir for EPEC strains is the human gastrointestinal tract, and human serotypes are not found in animals.

The small bowel epithelium is the site of EPEC infection; the organisms bind loosely to the small bowel epithelial cells in a localized adherence pattern and inject virulence factors into the cells (Vallance *et al.*, 2002). Disease results when the translocated bacterial virulence factors interact with host cells components and alter the host cell signaling pathways (Vallance and Finlay, 2000). The EPEC adherence factor (EAF) plasmid is required for localized adherence. Densely packed three-dimensional clusters of bacterial cells adhering to the surface of tissue culture cells is the earmark of localized adherence. The EAF plasmid encodes the bundle-forming pilus (BFP), which is responsible for localized adherence and virulence. BFP mu-

tants have an impaired ability to induce diarrhea (Frankel *et al.*, 1998; Vallance and Finlay, 2000; Donnenberg and Whittam, 2001).

After initial attachment (localized adherence) of EPEC to the intestinal cell membrane, proteins are secreted, which result in intimate bacterial attachment and the formation of cuplike pedestals on the microvilli on which the bacteria rest with the accumulation of polymerized filamentous actin, α-actinin, talin, ezrin, and myosin light chains. The lesions are referred to as "attaching and effacing" (A/E) lesions and have been observed *in vitro* and *in vivo* (Donnenberg and Whittam, 2001). The genes mediating A/E pathology are situated on a 35-kb pathogenicity island, the locus of enterocyte effacement (LEE), which contains 41 open reading frames. The G+C content of the LEE region is 38.4% as compared to the *E. coli* chromosome which has a G+C content of 50.8% (Frankel *et al.*, 1998). The LEE genes are separated into three domains: (1) Tir (translocated intimin receptor) and the intimin outer membrane protein; (2) EspA-D, encoding secreted proteins and their chaperones; and (3) a region encoding a type III secretion system which translocates bacterial proteins directly into the host cell (Frankel *et al.*, 1998; Donnenberg and Whittam, 2001). The type III secretion system translocates the Tir protein which is inserted into the host cell plasma membrane. The Esps are involved in the translocation process. After being phosphorylated, the inserted Tir acts as the receptor for the intimin outer membrane protein. Intimin is the product of the *eae* gene located downstream of the *tir* gene in the LEE locus. Intimin is crucial for intimate adherence and A/E formation (DeVinney *et al.*, 1999a, 1999b; Donnenberg and Whittam, 2001). In this elegant manner, EPEC strains insert their own receptor (Tir) for the intimin adhesin protein, with resultant A/E lesion formation (DeVinney *et al.*, 1999a).

The EAST1 gene, *astA*, was present in 22% of 65 EPEC strains (Savarino *et al.*, 1996) but the significance of the toxin in EPEC is unknown. Some A/E producing EPEC strains have a gene (*lifA*) that encodes the toxin, lymphostatin. Lymphocyte proliferation and activation is inhibited by lymphostatin, and in addition, the toxin selectively inhibits the production of IL-2, IL-4, IL-5, and IFN-γ (Klapproth *et al.*, 2000). The expression of lymphostatin may suppress the immune response against the bacteria with prolongation of the infection. A prolonged infection would intensify the spread of the organism to other individuals. Nataro and Kaper (1998) and Vallance and Finlay (2000) have proposed a number of mechanisms to explain diarrhea causation by EPEC, but these proposed mechanisms have not been studied in enough detail to elucidate the diarrheic mechanism.

Since EPEC is a human pathogen, it does not infect most laboratory animals (Vallance and Finlay, 2000). However, A/E lesion-inducing *E. coli* strains have been isolated from rabbits (REPEC). The REPEC strains infect the small bowel of weanling rabbits and the rabbits suffer diarrhea and weight loss (DeVinney *et al.*, 1999a; Milon *et al.*, 1999). Unlike human EPEC, the pattern of adherence of REPEC is diffuse rather than localized; however, the REPEC LEE-encoded secreted proteins are similar to those of human EPEC (Tauschek *et al.*, 2002). *Citrobacter rodentium* produces A/E lesions in mice, but unlike human EPEC and REPEC, *C. rodentium* colonizes the large bowel (Higgins *et al.*, 1999). The organism induces a TH1 response with production of IL-ll and IL-12, IFN-γ, and α-tumor necrosis factor. However, *C. rodentium* does not induce diarrhea but induces intestinal epithelial cell hyperplasia (Higgins *et al.*, 1999).

Both the A/E and the localized adherence phenotypes of EPEC can be determined by the use of HEp-2 or HeLa cells (Nataro and Kaper, 1998). Genotypic assays based on the use of DNA probes and the PCR have been described to evaluate: (a) A/E (detection of the *eae* gene); (b) presence of the EAF plasmid (EAF or *bfpA* gene probes); and (c) the lack of Shiga toxin genes (use of gene probes or PCR primers targeting *stx* genes: discussed under Enterohemorrhagic *Escherichia coli*) (Nataro and Kaper, 1998). López-Saucedo *et al.* (2003) describe a single multiplex PCR reaction that could distinguish among EPEC, ETEC, EIEC, and Shiga toxin-producing *E. coli* based on amplification of specific virulence genes.

## Other diarrheic *Escherichia coli* strains

### Enterioinvasive *Escherichia coli* (EIEC)

Bacillary dysentery is induced by EIEC strains. These strains are genetically and biochemically similar to *Shigella* and cause *Shigella*-like disease symptoms. A hallmark of EIEC strains is cell invasion; they invade HeLa cells and induce keratoconjunctivitis in the guinea-pig eye (Sérény test) (Hale *et al.*, 1997; Nataro and Kaper, 1998).

An EIEC infection results in watery diarrhea in most patients; however, stools may contain blood and mucus. EIEC has an infective dose several log values higher than that observed with *Shigella* infections (Hale *et al.*, 1997), indicating that person-to-person spread is unusual, although it has been reported (Harris *et al.*, 1985).

EIEC attach to the epithelial cells of the colon and subsequently penetrate the enterocytes via endocytosis. There is intracellular multiplication of the bacterial cells in the endocytic vacuole followed by lysis of the vacuole (Nataro and Kaper, 1998). The attachment of cellular actin to one pole of the EIEC cell propels the organism through the cytoplasm into adjacent epithelial cells (Nataro and Kaper, 1998). Thus, EIEC is spread cell-to-cell without entering the extracellular milieu.

The genes necessary for invasion, multiplication, and survival within the colonic enterocytes are present on a 140-MDa plasmid (pINV). The Ipa proteins, IpaA-IpaD, necessary for the invasive phenotype, are encoded by the *ipa* (invasion plasmid antigen) genes present on the plasmid (Hale *et al.*, 1997; Nataro and Kaper, 1998). The virulence genes present on the EIEC plasmid are identical to those found on the *Shigella* 120- to 140-MDa invasion plasmid (Lan *et al.*, 2001). In addition,

12 of the 33 O-antigen forms in *Shigella* are identical to those found in EIEC strains (Wang *et al.*, 2001b).

Food- and water-borne outbreaks of EIEC are not common in industrialized nations. Foods that have caused outbreaks include French soft cheeses, potato salad, and guacamole (Gordillo *et al.*, 1992; Hale *et al.*, 1997; Willshaw *et al.*, 2000). Foods involved in outbreaks are generally uncooked or raw foods contaminated by infected workers or cooked foods that are not reheated after being handled by infected personnel. In developed countries, infections by EIEC do not appear to be important; however, EIEC infections are probably more important in developing countries, especially in young children. The reservoir for EIEC as well as *Shigella* is the human intestinal tract.

A PCR assay using enriched stool samples from children with acute diarrhea was more sensitive than stool culture or colony hybridization for detection of *Shigella* and EIEC (Dutta *et al.*, 2001). López-Saucedo *et al.* (2003) described a multiplex PCR that differentiated enterotoxigenic *E. coli*, enteropathogenic *E. coli*, Shiga toxin-producing *E. coli*, and EIEC in stool samples. While it is possible to differentiate between the various diarrheagenic *E. coli* strains via multiplex PCR, it is difficult to differentiate between *Shigella* and EIEC. A virulence antigen-specific, monoclonal antibody-based enzyme-linked immunoassay has been used to detect *Shigella* and EIEC (Pal *et al.*, 1997). In addition, DNA probes and primers have been developed and used to detect *Shigella* and EIEC by hybridization or by the polymerase chain reaction (PCR), respectively (Houng *et al.*, 1997).

### Cell-detaching *Escherichia coli* (CDEC)

The CDEC strains cause detachment of tissue culture cells from solid supports and produce an α-hemoylsin, pyelonephritis-associated pili, and cytotoxic necrotizing factor 1 (CNF-1) (Fábrega *et al.*, 2002; Okeke *et al.*, 2002). The CDEC may also possess virulence factors found in other categories of diarrheic *E. coli* (Clarke, 2001; Fábrega *et al.*, 2002; Okeke *et al.*, 2002). CDEC infections have been associated with pediatric diarrhea (Fábrega *et al.*, 2002; Okeke *et al.*, 2002).

### Necrotoxigenic *Escherichia coli* (NTEC)

The NTEC strains induce actin stress fiber formation and DNA synthesis in cell cultures, leading to the formation of multi-nucleated giant cells. Two toxins, CNF-1 in NTEC1 strains and cytotoxic necrotizing factor-2, CNF-2, in NTEC2 strains, are involved; in addition, many NTEC also produce cytolethal distending toxins (De Rycke *et al.*, 1999; Mainil *et al.*, 2003). Interestingly, in NTEC1 strains, genes that encode the CNF-1, α-hemolysin and P-fimbriae are located on a pathogenicity island (De Rycke *et al.*, 1999), suggesting that NTEC1 may be related to CDEC.

### Diffusely adherent *Escherichia coli* (DAEC)

The DAEC strains are poorly characterized, and the involvement of DAEC in diarrhea remains in question. The uniform attachment of bacteria to the surface of HeLa or HEp-2 cells is termed diffuse adherence, whereas if the bacteria adhere in groups at one or a few sites on the cell surface, the attachment is termed localized adherence (Scaletsky et al., 1984). A fimbrial adhesin, F1845, is believed to be responsible for diffuse HEp-2 cell adhesion by diarrheic E. coli isolates (Bilge et al., 1989). A DNA probe targeting the gene for F1845 fimbriae, daaC, has been developed to detect DAEC (Bilge et al., 1989). The gene is either chromosomal or plasmid-borne. While the probe is specific for this gene, it is not very sensitive in detecting DAEC strains suggesting that the diffuse-adherence pattern may be mediated by other adhesins (Willshaw et al., 2000). A second putative adhesin that mediates the diffuse adherence phenotype, AIDA-I (adhesin involved in diffuse adherence), is a 100 kDa cell surface protein (Benz and Schmidt, 1992).

Finger-like projections are found in some strains of DAEC. These projections jut from the surface of HEp-2 and Caco-2 cells and enclose the bacterial cells, protecting them from gentamicin. However, the DAEC are not located intracellularly (Cookson and Nataro, 1996). It is not known if the finger-like projections play a role in pathogenesis.

Infection by diarrheic strains of DAEC induces fever and vomiting, and stools are watery and mucoid. However, not all strains of DAEC produce diarrhea. Infections due to DAEC occur in underdeveloped countries, mainly in children between 48–60 months of age (Cookson and Nataro, 1996; Nataro and Kaper, 1998). Infants are rarely affected; nursing infants may be protected against DAEC since human milk proteins have been shown to inhibit the adherence of DAEC (Nascimento de Araújo and Giugliano, 2000). Most of the E. coli strains (100/262) isolated from stools of diarrheic infants, children, and adults in a French hospital were DAEC strains (Jallat et al., 1993). However, only one-third of the DAEC strains hybridized with the daaC probe, thereby, indicating that a great deal of heterogeneity exists in DAEC strains. Poitrineau et al. (1995) found that vomiting but not diarrhea was associated with the presence of DAEC in the stools of 24 diarrheic children. The hospital stay of children with F1845 DNA probe-positive DAEC was approximately three times longer as compared to children harboring other DAEC types. DAEC strains, therefore, vary widely in their level of pathogenicity.

### Enteroaggregative *Escherichia coli* (EAEC)

The EAEC strains do not secrete labile, stable, or Shiga toxins and adhere in an aggregative or "stacked brick" (AA phenotype) adhesion pattern to HEp-2 cells (Nataro et al., 1995; Law and Chart, 1998). EAEC strains may be either pathogenic or non-pathogenic. Diarrheic outbreaks associated with food or drinking water were shown to be due to EAEC; however, the organism was seldom isolated from the suspect vehicle (Cobeljic et al., 1996; Itoh et al., 1997; Smith et al., 1997; Nataro et al., 1998; Okeke and Nataro, 2001). The presence of EAEC has been demonstrated in baby formula (Morais et al., 1997) and tabletop sauces such as guacamole from Mexican-style restaurants (Adachi et al., 2002b).

In developing countries, EAEC strains are a common cause of persistent diarrhea in children. In industrialized countries, childhood EAEC infections are probably underreported and underdiagnosed (Okeke and Nataro, 2001). Nursing infants are likely protected against EAEC diarrhea because human milk protein components inhibit EAEC adhesion to HeLa cells (Nascimento de Araújo and Giugliano, 2000). EAEC can be a causative agent of diarrhea in adults who travel to developing countries (Adachi et al., 2001; Okeke and Nataro, 2001; Adachi et al., 2002a) and in HIV-infected individuals (Okeke and Nataro, 2001).

Infection with EAEC may be asymptomatic; however, in symptomatic patients, the diarrhea induced by EAEC is watery, may be protracted and is associated with abdominal pain. Borborygmus (rumbling due to gas), low-grade fever, vomiting, and dehydration may occur. Stools may contain gross mucus and blood; up to one-third of patients have grossly bloody stools (Nataro et al., 1998; Okeke and Nataro, 2001). Histologically, necrotic lesions are found in the ileal mucosa and a thick mucus gel covers the intestinal mucosa (Eslava et al., 1998). The inflammatory cytokines, IL-8 and IL-1β, induce mucosal inflammation (Okeke and Nataro, 2001; Greenberg et al., 2002). Pathogenicity of EAEC has been modeled using tissue culture (Nataro et al., 1996) and gnotobiotic piglets (Tzipori et al., 1992). Nataro et al. (1998) and Okeke and Nataro (2001) described a three-stage model for EAEC pathogenesis: adherence, mucus production, and elaboration of cytotoxins resulting in intestinal secretion and damage to the mucosa.

Malnutrition and growth retardation in infants and children may result from EAEC infections (Nataro et al., 1998; Steiner et al., 1998). The thick mucus covering the intestinal surface may promote tenacious colonization and lead to malnutrition. The persistent diarrhea in patients with EAEC infections may be due to the inability of malnourished individuals to repair damaged intestinal mucosa (Nataro et al., 1998; Okeke and Nataro, 2001). Oral hydration is an effective therapy (Law and Chart, 1998; Nataro et al., 1998; Smith and Cheasty, 1998).

Information concerning virulence factors in EAEC is limited and confusing. Several types of fimbriae are involved with aggregative attachment, and the pathology suggests that a toxin is involved in diarrhea induction. EAEC strains utilize heme or hemoglobin as the sole iron source and produce siderophores (Okeke et al., 2004). Genes associated with multiple iron utilization systems are present and may provide EAEC with a competitive advantage over other bacteria. A number of EAEC adherence factors have been demonstrated. Aggregative adherence fimbriae I (AAF/1) (Nataro et al., 1992) and AAF/II (Czeczulin et al., 1997) have been characterized. However, only a minority of EAEC strains possessed AAF/I and AAF/II when a DNA probe for these fimbriae was used (Czeczulin et al., 1997). Other aggregative fimbriae have been described by Knutton et al. (1992) and Collinson et al. (1992). In addition, some strains of EAEC express afimbrial adhesins (Okeke and Nataro, 2001). Two toxins that may be associated with virulence in the EAEC are the EAST1 toxin and the PET toxin. The plasmid-mediated low molecular weight heat-stable EAST1 toxin behaves similarly, *in vitro*, to the ETEC heat stable toxin (Savarino *et al.*, 1991). The *astA* gene encoding EAST1 is present in approximately 41% of EAEC strains (Okeke and Nataro, 2001). It is uncertain whether EAST1 induces diarrhea *in vivo* (Navarro-Garcia *et al.*, 1998). Rich *et al.* (1999) found no correlation between severity of diarrheic symptoms due to EAEC infection and the presence of the EAST1 toxin. A more detailed discussion of EAST1 is presented in the Enterotoxigenic *E. coli* (ETEC) section. A partially purified preparation of a high molecular weight (108 kDa), heat-labile protein toxin, isolated from EAEC, induced tissue damage, inflammation, and mucus secretion in isolated rat jejunum (Narvarro-García *et al.*, 1998). The gene (*pet*) for the 108-kDa toxin is located on the 65-MDa EAEC virulence plasmid, which also contains the genes for the aggregative phenotype, AA. The Pet toxin is an autotransporter highly homologous to other autotransporter proteins such as the EspP protease of EHEC and the cryptic protein EspC of EPEC (Eslava *et al.*, 1998). Pet is present in 18 to 44% of EAEC isolates (Okeke and Nataro, 2001). Morabito *et al.* (1998) isolated unusual EAEC strains that produced Stx2, had the AA phenotype, and possessed the *astA* gene for EAST1 but lacked the EHEC genes, *eaeA*, *hly*, and *katP*. These strains were involved in an outbreak of HUS.

EAEC strains constitute a very heterogeneous group comprising more than 50 "O" serogroups, thus serotyping is not useful for identification purposes (Chart *et al.*, 1997). Fimbriae vary from bundles of fine filaments (Knutton *et al.*, 1992), thin fimbriae (Collinson *et al.*, 1992), and bundle-forming fimbriae (Nataro *et al.*, 1992; Czeczulin *et al.*, 1997). In addition, not all EAEC express fimbriae (Chart *et al.*, 1997). The EAST1 and/or Pet toxins are not produced by all EAEC strains (Savarino *et al.*, 1996; Eslava *et al.*, 1998). EAEC strains were also heterogenous in their capability to induce diarrhea in adult volunteers. It is apparent that the heterogeneity of EAEC strains makes the diagnosis of EAEC-induced illnesses and strain identification difficult.

Demonstration of adhesion to HEp-2 cells is the most definitive assay for identification of EAEC (Law and Chart, 1998; Miqdady *et al.*, 2002). The technique is time consuming, cumbersome, subject to misinterpretation, and is only suitable for use in research laboratories. DNA probes (DebRoy *et al.*, 1994; Baudry *et al.*, 1990), PCR (Schmidt *et al.*, 1995) and multiplex PCR (Cerna *et al.*, 2003) techniques have been developed for the identification of EAEC strains. The probes and PCR tests were quite specific but failed to detect all EAEC with the typical AA phenotype. Therefore, the HEp-2 adherence assay is the only reliable method for verification of EAEC strains, but it does not distinguish between pathogenic and non-pathogenic strains.

Clearly, additional studies on EAEC are needed. Is the aggregative pattern a virulence factor or is it merely a diagnostic tool for detection of EAEC? What are the roles of the EAST1 and Pet toxins in pathogenesis of EAEC strains? Mutants unable to express these toxins or cloning of EAEC virulence factors in laboratory strains of *E. coli* should be generated for use in cell culture assays, animal models (described in section 7 of Law and Chart, 1998), and human volunteers. These types of studies should be useful in understanding the virulence of EAEC. It is clear that growth retardation and other growth deficits are associated with EAEC diarrhea (Nataro *et al.*, 1998; Steiner *et al.*, 1998). Does malnutrition predispose to EAEC-induced diarrhea? Is malabsorption of nutrients related to EAEC-induced mucosal damage? Does EAEC-induced stimulation of a mucus build-up on the intestinal mucosa impose a barrier to absorption of nutrients? Animal studies should clarify the putative role of EAEC in malnutrition.

## Control and prevention of *Escherichia coli*-induced disease

Diarrheagenic *E. coli* possess a wide range of specific virulence factors, and it is likely that others remain to be discovered. Both pathogenic and non-pathogenic *E. coli* strains are found in soil, water, and any other site contaminated with human or animal feces, thus it is difficult to prevent *E. coli* infections. Since *E. coli* infections are generally caused by the ingestion of contaminated food and water, steps must be taken by food manufacturing and water treatment personnel to prevent fecal contamination at all levels of food processing and water purification. If contamination is suspected, then the product must be subjected to a bactericidal step before it reaches the consumer.

Cattle are a major reservoir for *E. coli* O157:H7 and other EHEC. Currently, there are no validated strategies to reduce or control EHEC in cattle and other animals; however, certain farm management practices may provide practical means to reduce the prevalence of EHEC in cattle on the farm and on carcasses at slaughter. Possible interventions include the control of the presence of EHEC in cattle feed and in water troughs, competitive inhibition, in which animals are administered bacteria that compete with or are inhibitory to EHEC, vaccination, switching from a grain diet to an all hay diet prior to slaughter to reduce levels of acid-resistant *E. coli* O157:H7 in the colon of cattle, and reduction in the level of hide soiling in animals sent to slaughter (Hancock *et al.*, 2001). A major source of EHEC on carcasses is soiled hides. Chemical dehairing of hides resulted in a lower prevalence of *E. coli* O157:H7 on pre-evisceration carcasses as compared to conventional treatment of hides (1% and 50% prevalence, respectively) (Nou *et al.*, 2003). Furthermore, Bosilevac and coworkers (2004) have shown that treatment of cattle hides with a water wash followed by cetylpyridinium chloride treatment prior to stunning was capable of reducing hide prevalence of *E. coli* O157 from 50% to 80%, resulting in a 5% or less pre-evisceration carcass prevalence.

The consumer must also bear responsibility for keeping foods free from *E. coli* contamination. Raw and processed foods must be obtained from reputable suppliers and these products must be stored at appropriate temperatures until they are to be used. Prior to food preparation, hands, utensils, and countertops should be thoroughly cleaned. Foods must be kept separate to prevent cross-contamination of ready-to-eat foods by raw foods such as meat or produce. Dirty utensils may be a source of cross-contamination. Frozen foods should be thawed in the refrigerator and not at room temperature. Vegetables and fruits that are normally eaten raw should be washed before eating. Non-pasteurized milk or juices should be avoided. Red meats and poultry should be cooked to an internal temperature of 71.1°C and 82.2°C, respectively. Particular care must be taken with the cooking of ground beef patties. *E. coli* O157:H7 present in ground beef patties are inactivated if the patties are cooked to an internal temperature of 71.1°C. Only water treated with chlorine or other disinfectants should be used for drinking and cooking. Hot foods should be served promptly and cold foods should be removed from the refrigerator just prior to serving. Left-over perishable foods should be refrigerated within 2 hours.

Persons with diarrhea should not handle nor cook food. Infant caretakers should wash their hands with soap and water after changing soiled diapers to avoid spreading infections to others either directly or through food handling. All individuals should wash their hands after using the bathroom, and young children must be taught that handwashing is important.

Eating in restaurants is often a concern for the food safety conscious person. Restaurants should be selected with care. Dirty floors and windows, dirty table covers, dirty toilet facilities, and personnel with dirty aprons are indications of restaurant management that is not thoroughly committed to cleanliness and food safety. For a detailed discussion of safe food handling for the prevention of foodborne illness see Smith (1999).

## References

Adachi, J.A., Ericsson, C.D., Kiang, Z-D., DuPont, M.W., Pallegar, S.R., and DuPont, H.L. 2002a, Natural history of enteroaggregative and enterotoxigenic *Escherichia coli* infection among US travelers to Gluadalajara, Mexico. J. Infect. Dis. 185: 1681–1683.

Adachi, J.A., Jiang, Z-D., Mathewson, J.J., Verenkar, M.P., Thompson, S., Martinez-Sandoval, F., Steffen, R., Ericsson, C.D., and DuPont, H.L. 2001. Enteroaggregative *Escherichia coli* as a major etiologic agent in Traveler's diarrhea in 3 regions of the world. Clin. Infect. Dis. 32: 1706–1709.

Adachi, J.A., Mathewson, J.J., Jiang, Z-D., Eriesson, C.D., and Dupont, H.L. 2002b. Enteric pathogens in Mexican sauces of popular restaurants in Guadalajara, Mexico and Houston, Texas. Ann. Intern. Med. 136: 884–887.

Al-Jader, L., Salmon, R.L., Walker, A.M., Williams, H.M., Willshaw, G.A., and Cheasty, T. 1999. Outbreak of *Escherichia coli* O157 in a nursery: lessons for prevention. Arch. Dis.Child. 81: 60–63.

Anand, S.K. and Griffiths, M.W. 2003. Quorum sensing and expression of virulence in *Escherichia coli* O157:H7. Int. J. Food Microbiol. 85: 1–9.

Anonymous. 2000. Water-borne outbreak of gastroenteritis associated with a contaminated municipal water supply, Walkerton, Ontario, May-June 2000. Can. Commun. Dis. Rep. 26: 170–173.

Baqar, S., Bourgeois, A.L., Schultheiss, P.J., Walker, R.J., Rollins, D.M., Haberberger, R.L., and Pavlovskis, O.R. 1995. Safety and immunogenicity of a prototype oral whole-cell killed *Campylobacter* vaccine administered with a mucosal adjuvant in non-human primates. Vaccine 13: 22–28.

Bari, M.L., Nazuka, E., Sabina, Y., Todoriki, S., and Isshiki, K. 2003. Chemical and irradiation treatments for killing *Escherichia coli* O157:H7 on alfalfa, radish, and mung bean seeds. J. Food Prot. 66: 767–774.

Barrett, T.J., Lior, H., Green, J.H., Khakhria, R., Wells, J.G., Bell, B.P., Greene, K.D., Lewis, J., and Griffin, P.M. 1994. Laboratory investigation of a multistate foodborne outbreak of *Escherichia coli* O157:H7 by using pulsed-field gel electrophoresis and phage typing. J. Clin. Microbiol. 32: 3013–3017.

Baudry, B., Savarino, S.J., Vial, P., Kaper, J.B., and Levine, M.M. 1990. A sensitive and specific DNA probe to identify enteroaggregative *Escherichia coli*, a recently discovered diarrheal pathogen. J. Infect. Dis. 161: 1249–1251.

Beatty, M.E., Bopp, C.A., Wells, J.G., Greene, K.D., Puhr, N.A., and Mintz, E.D. 2004. Enterotoxin-producing *Escherichia coli* O169:H41, United States. Emerg. Infect. Dis. 10: 518–521.

Beery, J.T., Doyle, M.P., and Schoeni, J.L. 1985. Colonization of chicken cecae by *Escherichia coli* associated with hemorrhagic colitis. Appl. Environ. Microbiol. 49: 310–315.

Bell, B.P., Goldoft, M., Griffin, P.M., Davis, M.A., Gordon, D.C., Tarr, P.I., Bartleson, C.A., Lewis, J.H., Barrett, T.J., Wells, J.G., Baron, R., and Kobayashi, J. 1994. A multistate outbreak of *Escherichia coli* O157:H7-associated bloody diarrhea and hemolytic uremic syndrome from hamburgers – the Washington experience. J. Am. Med. Assoc. 272: 1349–1353.

Benz, I., and Schmidt, M.A. 1992. Isolation and serologic characterization of AIDA-I, the adhesin mediating the diffuse adherence phenotype of the diarrhea-associated *Escherichia coli* strain 2787 (O126:H27). Infect. Immun. 60: 13–18.

Berkow, R., ed. 1992. The Merck Manual of Diagnosis and Therapy, 16th ed., Merck Research Laboratories, Rahway, New Jersey.

Bettelheim, K.A. 1997. *Escherichia coli* O157 outbreak in Japan: lessons for Australia. Aust. Vet. J. 75: 108.

Bettelheim, K.A. 2003. Non-O157 verotoxin-producing *Escherichia coli*: a problem, paradox, and paradigm. Exp. Biol. Med. 228: 333–344.

Bilge, S.S., Clausen, C.R., Lau, W. and Moseley, S.L. 1989. Molecular characterization of a fimbrial adhesin, F1845, mediating diffuse adherence of diarrhea-associated *Escherichia coli* to HEp-2 cells. J. Bacteriol. 171: 4281–4289.

Black, R.E. and Lanata, C.F. 1995. Epidemiology of diarrheal diseases in developing countries. In: Infections of the Gastrointestinal Tract. M.J. Blaser, P.D. Smith, J.I. Ravdin, H.B Greenberg and R.L. Guerrant, eds. Raven Press, New York. p. 13–36.

Black, R.E., Merson, M.M., Rowe, B., Taylor, P.P., Alim, A.R.M.A., Gross, R.J. and Sack, D.A. 1981. Enterotoxigenic *Escherichia coli* diarrhoea: acquired immunity and transmission in an endemic area. Bull. WHO 59: 263–268.

Blanco, M., Blanco, J.E., Mora, A., Dahbi, G., Alonso, M.P., González, E A., Bernárdez, M.I. and Blanco, J. 2004. Serotypes, virulence genes, and intimin types of Shiga toxin (Verotoxin)-producing *Escherichia coli* isolates from cattle in Spain and identification of a new intimin variant gene (eaeξ). J. Clin. Microbiol. 42: 645–651.

Bosilevac, J.M., Arthur, T.M., Wheeler T.L., Shackelford, S.D., Rossman, M., Reagan, J.O. and Koohmaraie, M. 2004. Prevalence of *Escherichia coli* O157 and levels of aerobic bacteria and *Enterobacteriaceae* are reduced when hides are washed and treated with cetylpyridinium chloride at a commercial beef processing plant. J. Food Prot. 67: 646–650.

Brackett, R.E., Hao, Y.-Y., and Doyle, M.P. 1994. Ineffectiveness of hot acid sprays to decontaminate *Escherichia coli* O157:H7 on beef. J. Food Prot. 57: 198–203.

Brenner, D.J. 1984. Family 1. *Enterobacteriaceae*. In: Bergey's Manual of Systematic Bacteriology. Vol. 1. N.R. Krieg, ed. Williams & Wilkins, Baltimore, Maryland. p. 408–420.

Brett, K.N., Ramachandran, V., Hornitzky, M.A., Bettelheim, K.A., Walker, W.M. and Djordjevic, S.P. 2003. $stx_{1c}$ is the most common Shiga toxin 1 subtype among Shiga toxin-producing *Escherichia coli* isolates from sheep but not among isolates from cattle. J. Clin. Microbiol. 41: 926–936.

Breuer, T., Benkel, D.H., Shapiro, R.L., Hall, W.N., Winnett, M.M., Linnk, M.J., Neimann, J., Barrett, T.J., Dietrich, S., Downes, F.P., Toney, D.M., Pearson, J.L., Rolka, H., Slutsker, L., Griffin, P.M. and the Investigation Team. 2001. A multistate outbreak of *Escherichia coli* O157:H7 infections linked to alfalfa sprouts grown from contaminated seeds. Emerg. Infect. Dis. 7: 977–982.

Brooks, J.T., Bergmire-Sweat, D., Kennedy, M., Hendricks, K., Garcia, M., Marengo, L., Wells, J., Ying, M., Bibb, W., Griffin, P.M., Hoekstra, R.M. and Friedman, C.R. 2004. Outbreak of Shiga toxin-producing *Escherichia coli* O111:H8 infections among attendees of a high school cheerleading camp. Clin. Infect. Dis. 38: 190–198.

Bruce, M.G., Curtis, M.B., Payne, M.M., Gautom, R.K., Thompson, E.C., Bennett, A.L. and Kobayashi, J.M. 2003. Lake-associated outbreak of *Escherichia coli* O157:H7 in Clark County, Washington, August 1999. Arch. Pediatr. Adolesc. Med. 157: 1016–1021.

Bürk, C., Dietrich, R., Acar, G., Moravek, M., Bulte, M. and Martlbauer, E. 2003. Identification and characterization of a new variant of Shiga toxin 1 in *Escherichia coli* ONT:H19 of bovine origin. J. Clin. Microbiol. 41: 2106–2112.

Burland, V., Shao, Y., Perna, N.T., Plunkett, G., Sofia, H.J. and Blattner, F.R. 1998. The complete DNA sequence and analysis of the large virulence plasmid of *Escherichia coli* O157:H7. Nucleic Acids Res. 26: 4196–4204.

Butterton, J.R. and Calderwood, S.B. 1995. *Vibrio cholerae* O1. In: Infections of the Gastrointestinal Tract. M.J. Blaser, P.D. Smith, J.I. Ravdin, H.B Greenberg and R.L. Guerrant, eds. Raven Press, New York. p. 649–670.

Call, D.R., Brockman, F.J. and Chandler, D.P. 2001. Detecting and genotyping *Escherichia coli* O157:H7 using multiplexed PCR and nucleic acid microarrays. Int. J. Food Microbiol. 67: 71–80.

Campbell, G.R., Prosser, J., Glover, A. and Killham, K. 2001. Detection of *Escherichia coli* O157:H7 in soil and water using multiplex PCR. J. Appl. Microbiol. 91: 1004–1010.

Caprioli, A., Luzzi, I., Rosmini, F., Resti, C., Edefonti, A., Perfumo, F., Farina, C., Goglio, A., Gianviti, A. and Rizzoni, G. 1994. Community-wide outbreak of hemolytic-uremic syndrome associated with non-O157 verocytotoxin-producing *Escherichia coli*. J. Infect. Dis. 169: 208–211.

Carter, A.O., Borczyk, A.A., Carlson, J.A., Harvey, B., Hockin, J.C., Karmali, M.A., Krishnan, C., Dorn, D.A. and Lior, H. 1987. A severe outbreak of *Escherichia coli* O157:H7-associated hemorrhagic colitis in a nursing home. N. Engl. J. Med. 317: 1496–1500.

Cassels, F.J. and Wolf, M.K. 1995. Colonization factors of diarrheagenic *E. coli* and their intestinal receptors. J. Indust. Microbiol. 15: 214–226.

CDC (Centers for Disease Control and Prevention). 1995. Outbreak of acute gastroenteritis attributable to *Escherichia coli* serotype O104:H21 – Helena, Montana, 1994. Morb. Mort. Weekly Rep. 44: 501–503.

CDC (Centers for Disease Control and Prevention). 2001. Outbreaks of *Escherichia coli* O157:H7 infections among children associated with farm visits – Pennsylvania and Washington, 2000. Morb. Mort. Weekly Rep. 50: 293–297.

CDC (Centers for Disease Control and Prevention). 2002. Multistate outbreak of *Escherichia coli* O157:H7 infections associated with eating ground beef – United States, June–July 2002. Morb. Mort. Weekly Rep. 51: 637–639.

Cerna, J.F., Nataro, J.P. and Estrada-Garcia, T. 2003. Multiplex PCR for detection of three plasmid-borne genes of enteroaggregative *Escherichia coli* strains. J. Clin. Microbiol. 41: 2138–2140.

Chart, H., Spenser, J., Smith, H.R. and Rowe, B. 1997. Identification of entero-aggregative *Escherichia coli* based on surface properties. J. Appl. Microbiol. 83: 712–717.

Chaudhuri, A.G., Bhattacharya, J., Nair, G.G., Takeda, T. and Chakrabarti, M.K. 1998. Rise of cytosolic $Ca^{2+}$ and activation of membrane-bound guanylyl cyclase activity in rat enterocytes by heat-stable enterotoxin of *Vibrio cholerae* non-O1. FEMS Microbiol. Lett. 160: 125–129.

Cherla, R.P., Lee, S.-Y. and Tesh, V.L. 2003. Shiga toxins and apoptosis. FEMS Microbiol. Lett. 228: 159–166.

Clarke, S.C. 2001. Diarrhoeagenic *Escherichia coli* – an emerging problem? Diag. Microbiol. Infect. Dis. 41: 93–98.

Clarke, S.C., Haigh, R.D., Freestone, P.P.E. and Williams, P.H. 2002. Enteropathogenic *Escherichia coli* infection: history and clinical aspects. Br. J. Biomed. Sci. 59: 123–127.

Clemens, J.D., Rao, M.R., Chakraborty, J., Yunus, M., Ali, M., Kay, B., van Loon, F.P.L., Naficy, A. and Sack, D.A. 1997. Breastfeeding and risk of life-threatening enterotoxigenic *Escherichia coli* diarrhea in Bangladeshi infants and children. Pediatrics. 100 (6): e2.

Cobeljic, M., Miljkovic-Selimovic, B., Pasunovic-Todosijevic, D., Velickovic, Z., Lepsanovic, Z., Zec, N., Savic, D., Ilic, R., Konstantinovic, S., Javanovic, B. and Kostic, V. 1996. Enteroaggregative *Escherichia coli* associated with an outbreak of diarrhoea in a neonatal ward. Epidemiol. Infect. 117: 11–16.

Cohen, D., Orr, N., Haim, M., Ashkenazi, S., Robin, G., Green, M.S., Epphros, M., Sela, T., Slepon, R., Ashkenazi, I., Taylor, D.N., Svennerholm, A-M., Eldad, A. and Shemer, J. 2000. Safety and immunogenicity of two different lots of the oral, killed enterotoxigenic *Escherichia coli*-cholera toxin B subunit vaccine in Israeli young adults. Infect. Immun. 68: 4492–4497.

Cohen, M.B. and Gianella, R.A. 1995. Enterotoxigenic *Escherichia coli*. In: Infections of the Gastrointestinal Tract. M.J. Blaser, P.D. Smith, J.I. Ravdin, H.B. Greenberg and R.L. Guerrant, eds. Raven Press, New York, pp. 691–707.

Collinson, S.K., Emödy, L., Trust, T.J. and Kay, W.W. 1992. Thin aggregative fimbriae from diarrheagenic *Escherichia coli*. J. Bacteriol. 174: 4490–4495.

Cookson, S.T. and Nataro, J.P. 1996. Characterization of HEp-2 cell projection formation induced by diffusely adherent *Escherichia coli*. Microb. Pathog. 21: 421–434.

Cornick, N.A., Booher, S.L., Casey, T.A. and Moon, H.W. 2000. Persistent colonization of sheep by *Escherichia coli* O157:H7 and other *E. coli* pathotypes. Appl. Envrion. Microbiol. 66: 4926–4934.

Cornick, N.A., Matisse, I., Samuel, E., Bosworth, T. and Moon, H.W. 1999. Edema disease as a model for systemic disease induced by Shiga toxin-producing *E. coli*. Adv. Exp. Med. Biol. 473: 155–161.

Crump, J.A., Braden, C.R., Dey, M.E., Hoekstra, R.M., Rickelman-Apisa, J.M., Baldwin, D.A., De Fijter, S.J., Nowicki, S.F., Koch, E.M., Bannerman, T.L., Smith, F.W., Sarisky, J.P., Hochberg, N. and Mead, P.S. 2003. Outbreaks of *Escherichia coli* O157 infections at multiple county agricultural fairs: a hazard of mixing cattle, concession stands and children. Epidemiol. Infect. 131: 1055–1062.

Crump, J.A., Sulka, A.C., Langer, A.J., Schaben, C., Crielly, A.S., Gage, R., Baysinger, M., Moll, M., Withers, G., Toney, D.M., Hunter, S.B., Hoekstra, R.M., Wong, S.K., Griffin, P.M. and Van Gilder, T.J. 2002. An outbreak of *Escherichia coli* O157:H7 infections among visitors to a dairy farm. New Engl. J. Med. 347: 555–560.

Czeczulin, J.R., Balepur, S., Hicks, S., Phillips, A., Hall, R., Kothary, M.H., Navarro-García, F. and Nataro, J. P. 1997. Aggregative adherence fimbriae II, a second fimbrial antigen mediating aggregative adherence in enteroaggregative *Escherichia coli*. Infect. Immun. 65: 4135–4145.

Davis, B.M. and Waldor, M.K. 2003. Filamentous phages linked to virulence of *Vibrio cholerae*. Curr. Opin. Microbiol. 6: 35–42.

Dean-Nystrom, E.A. 2003. Bovine *Escherichia coli* O157:H7 infection model. Methods Mol. Med. 73: 329–338.

Dean-Nystrom, E.A., Bosworth, B.T., Moon, H.W. and O'Brien, A.D. 1998. *Escherichia coli* O157:H7 requires intimin for enteropathogenicity in calves. Infect. Immun. 66: 4560–4563.

DebRoy, C., Bright, B.D., Wilson, R.A., Yealy, J., Kumar, K. and Bhan, M.K. 1994. Plasmid-coded DNA fragment developed as a specific gene probe for the identification of enteroaggregative *Escherichia coli*. J. Med. Microbiol. 41: 393–398.

DeCastro, A.F.P., Guerra, B., Leomil, L., Aidar-Ugrinovitch, L. and Beutin, L. 2003. Multidrug-resistant Shiga toxin-producing *Escherichia coli* O118:H16 in Latin America. Emerg. Infect. Dis. 9:1027–1028.

De Rycke, J., Milon, A. and Oswald, E. 1999. Necrotoxic *Escherichia coli* (NTEC): two emerging categories of human and animal pathogens. Vet. Res. 30:221–233.

Deisingh, A.K. and Thompson, M. 2004 Strategies for the detection of *Escherichia coli* O157:H7 in foods. J. Appl. Microbiol. 96: 419–429.

DeVinney, R., Gauthier, A., Abe, A. and Finlay, B.B. 1999a. Enteropathogenic *Escherichia coli*: A pathogen that inserts its own receptor into host cells. Cell. Mol. Life Sci. 55: 961–976.

DeVinney, R., Knoechel, D.G. and Finlay, B.B. 1999b. Enteropathogenic *Escherichia coli*: cellular harassment. Curr. Opin. Microbiol. 2: 83–88.

DeVinney, R., Stein, M., Reinscheid, D., Abe, A., Ruschkowski, S. and Finlay, B.B. 1999c. Enterohemorrhagic *Escherichia coli* O157:H7 produces Tir, which is translocated to the host cell membrane but is not tyrosine phosphorylated. Infect. Immun. 67: 2389–2398.

Donnenberg, M.S. and Whittam, T.S. 2001. Pathogenesis and evolution of virulence in enteropathogenic and enterohemorrhagic *Escherichia coli*. J. Clin. Invest. 107: 539–548.

Doyle, M.P. and Padhye, V.V. 1989. *Escherichia coli*. In: Foodborne Bacterial Pathogens. M.P. Doyle, ed. Marcel Dekker, Inc., New York, pp. 235–281.

Dubreuil, J.D. 1997. *Escherichia coli* STb enterotoxin. Microbiology. 143: 1783–1785.

Duffell, E., Espié, E., Nichols, T., Adak, G. K., de Valk, H., Anderson, K. and Stuart, J.M. 2003. Investigation of an outbreak of E. coli O157 infections associated with a trip to France of schoolchildren from Somerset, England. Euro. Surveill. 8 (4): 81–86.

Dutta, S., Chatterjee, A., Dutta, P., Rajendran, K., Roy, S., Pramanik, K.C. and Bhattacharya, S.K. 2001. Sensitivity and performance characteristics of a direct PCR with stool samples in comparison to conventional techniques for diagnosis of *Shigella* and enteroinvasive *Escherichia coli* infection in children with acute diarrhoea in Calcutta, India. J. Med. Microbiol. 50: 667–674.

Dykstra, S.A., Moxley, R.A., Janke, B.H., Nelson, E.A. and Francis, D.H. 1993. Clinical signs and lesions in gnotobiotic pigs inoculated with Shiga-like toxin I from *Escherichia coli*. Vet. Pathol. 30: 410–417.

Elder, R.O., Keen, J.E., Siragusa, G.R., Barkocy-Gallagher, G.A., Koohmaraie, M. and Laegreid, W.W. 2000. Correlation of enterohemorrhagic *Escherichia coli* O157 prevalence in feces, hides, and carcasses of beef cattle during processing. Proc. Natl. Acad. Sci. USA 97: 2999–3003.

Elliott, E.J., Robins-Browne, R.M., O'Loughlin, E.V., Bennett-Wood, V., Bourke, J., Henning, P., Hogg, G. G., Knight, J., Powell, H., Redmond, D. and Contibutors to the Australian Paedriatic Surveillance Unit. 2001. Nationwide study of haemolytic uraemic syndrome: clinical, microbiological, and epidemiological features. Arch. Dis. Child. 85: 125–131.

Eslava, C., Navarro-García, F., Czeczulin, J.R., Henderson, I.R., Cravioto, A. and Nataro, J.P. 1998. Pet, an autotransporter enterotoxin from enteroaggregative *Escherichia coli*. Infect. Immun. 66: 3155–3163.

Fábrega, V.L.A., Ferreira, A.J.P., Da Silva Patrício, F.R., Brinkley, C. and Scaletsky, I.C.A. 2002. Cell-detaching *Escherichia coli* (CDEC) strains from children with diarrhea: Identification of a protein with toxigenic activity. FEMS Microbiol. Lett. 217: 191–197.

Fagundes-Neto, U. and Affonso Scaletsky, I.C. 2000. The gut at war: The consequences of enteropathogenic *Escherichia coli* infection as a factor of diarrhea and malnutrition. São Paulo Med. J. 118: 21–29.

Feder, I., Wallace, F.M., Gray, J.T., Fratamico. P., Fedorka-Cray, P.J., Pearce, R.A., Call, J.E., Perrine, R. and Luchansky, J.B. 2003. Isolation of *Escherichia coli* O157:H7 from intact colon fecal samples of swine. Emerg. Infect. Dis. 9: 380–383.

Feldman, K.A., Mohle-Boetani, J.C., Ward, J., Furst, K., Abbott, S.L., Ferrero, D.V., Olsen, A. and Werner, S.B. 2002. A cluster of *Escherichia coli* O157:Nonmotile infections associated with recreational exposure to lake water. Public Health Rep. 117: 380–385.

Feng, P. 2001. Development and impact of rapid methods for detection of foodborne pathogens. In: Food Microbiology: Fundamentals and Frontiers, 2nd ed. M.P. Doyle, L.R. Beuchat, and T.J. Montville, eds. ASM Press, Washington, D.C. p. 775–796.

Feng, P., Lampel, K. A., Karch, H. and Whittam, T.S. 1998. Genotypic and phenotypic changes in the emergence of *Escherichia coli* O157: H7. J. Infect. Dis. 177: 1750–1753.

Fenwick, B.W. and Cowan, L.A. 1998. Canine model of the hemolytic uremic syndrome. In: *Escherichia coli* O157:H7 and Other Shiga Toxin-Producing E. coli Strains. J.B. Kaper, and A.D. O'Brien, eds. American Society for Microbiology, Washington, D.C, pp. 268–277.

Frankel, G., Phillips, A.D., Rosenshine, I., Dougan, G., Kaper, J.B. and Knutton, S. 1998. Enteropathogenic and enterohemorrhagic *Escherichia coli*: More subversive elements. Mol. Microbiol. 30: 911–921.

Fratamico, P.M., Bagi, L.K. and Pepe, T. 2000. A multiplex polymerase chain reaction assay for rapid detection and identification of *Escherichia coli* O157:H7 in foods and bovine feces. J. Food Prot. 63: 1032–1037.

Fratamico, P.M., Briggs, C.E., Needle, D., Chen, C.-Y. and DebRoy, C. 2003. Sequence of the *Escherichia coli* O121 O-antigen gene cluster and detection of enterohemorrhagic E. coli O121 by PCR amplification of the *wzx* and *wzy* genes. J. Clin. Microbiol. 41: 3379–3383.

Fratamico, P.M. and Crawford, C.G. 1999. Detection by commercial immunomagnetic particle-based assays. In: Encyclopedia of Food Microbiology, vol. 1. R. Robison, C. Batt, and P. Patel, eds. Academic Press, London, pp. 654–663.

Fratamico, P.M., Smith, J.L. and Buchanan, R.L. 2002. *Escherichia coli*. In: Foodborne Diseases, 2nd. ed. D.O. Cliver and H.P. Riemann, eds. Academic Press, New York. p. 79–101.

Friedrich, A.W., Borell, J., Bielaszewska, M., Fruth, A., Tschäpe, H. and Karch, H. 2003. Shiga toxin 1c-producing Escherichia coli strains: phenotypic and genetic characterization and association with human disease. J. Clin. Microbiol. 41:2448–2453.

Friedrich, A.W., Bielaszewska, M., Zhang, W.L., Pulz, M., Kuczius, T., Ammon, A. and Karch, H. 2002. *Escherichia coli* harboring Shiga toxin 2 gene variants: frequency and association with clinical symptoms. J. Infect. Dis. 185: 74–84.

Friedrich, A.W., Borrell, J., Bielaszewska, M., Fruth, A., Tschäpe, H., and Karch, H. 2003. Shiga toxin 1c-producing *Escherichia coli* strains: phenotypic and genetic characterization and association with human disease. J. Clin. Microbiol. 41:2448–2453.

Gaastra, W. and Svennerholm, A.-M. 1996. Colonization factors of human enterotoxigenic *Escherichia coli* (ETEC). Trends Microbiol. 4: 444–452.

Gagliardi, J.V. and Karns, J.S. 2002. Persistence of *Escherichia coli* O157: H7 in soil and on plant roots. Environ. Microbiol. 4: 89–96.

Glass, K.A., Loeffelholz, J.M., Ford, J.P. and Doyle, M.P. 1992. Fate of *Escherichia coli* O157:H7 as affected by pH or sodium chloride and in fermented, dry sausage. Appl. Environ. Microbiol. 58: 2513–2516.

Goh, S., Newman, C., Knowles, M., Bolton, F.J., Hollyoak, V., Richards, S., Daley, P., Counter, D., Smith, H.R. and Keppie, N. 2002. E. coli phage type 21/28 outbreak in North Cumbria associated with pasteurized milk. Epidemiol. Infect. 129: 451–457.

Gordillo, M.E., Reeve, G.R., Pappas, J., Mathewson, J.J., DuPont, H.L. and Murray, B.E. 1992. Molecular characterization of strains of enteroinvasive *Escherichia coli* O143, including isolates from a large outbreak in Houston, Texas. J. Clin. Microbiol. 30: 889–893.

Grant, C.C.R., Messer, R.J. and Cieplak, W. 1994. Role of trypsin-like cleavage at arginine 192 in the enzymatic and cytotonic activities of *Escherichia coli* heat-labile enterotoxin. Infect. Immun. 62: 4270–4278.

Grauke, L.J., Kudva, I.T., Yoon, J.W., Hunt, C.W., Williams, C.J. and Hovde, C.J. 2002. Gastrointestinal tract location of *Escherichia coli* O157:H7 in ruminants. Appl. Environ. Microbiol. 68: 2269–2277.

Greenberg, D.E., Jiang, Z-D., Steffen, R., Verenker, M.P. and DuPont, H.L. 2002. Markers of inflammation in bacterial diarrhea among travelers, with a focus on enteroaggregative *Escherichia coli* pathogenicity. J. Infect. Dis. 185: 944–949.

Grewal, H.S., Valvatne, H., Bhan, M. K., van Duk, L., Gaastra, W. and Sommerfelt, H. 1997. A new putative fimbrial colonization factor, CS19, of human enterotoxigenic *Escherichia coli*. Infect. Immun. 65: 507–513.

Griffin, P.M. 1995. *Escherichia coli* O157:H7 and other enterohemorrhagic *Escherichia coli*. In: Infections of the gastrointestinal tract. M.J. Blaser, P.D. Smith, J.I. Ravdin, H.B. Greenberg and R.L. Guerrant, eds. Raven Press, Ltd., New York, pp. 739–761.

Guerrant, R.L. and Thielman, N.M. 1995. Types of *Escherichia coli* enteropathogens. In: Infections of the Gastrointestinal Tract. M.J. Blaser, P.D. Smith, J.I. Ravdin, H.B Greenberg and R.L. Guerrant, eds. Raven Press, New York. p. 687–690.

Gunzer, F., Hennig-Pauka, I., Waldmann, K. H., Sandhoff, R., Grone, H. J., Kreipe, H.H., Matussek, A. and Mengel, M. 2002. Gnotobiotic piglets develop thrombotic microangiopathy after oral infection with enterohemorrhagic *Escherichia coli*. Am. J. Clin. Pathol. 118: 364–375.

Gyles, C. L. 1992. *Escherichia coli* cytotoxins and enterotoxins. Can. J. Microbiol. 38: 734–746.

Hale, T.L., Echeverria, P. and Nataro, J.P. 1997. Enteroinvasive *Escherichia coli*. In: *Escherichia coli*: Mechanisms of Virulence. M. Sussman, ed. Cambridge University Press, Cambridge, UK, pp. 449–468.

Hancock, D., Besser, T., Lejeune, J., Davis, M. and Rice, D. 2001. The control of VTEC in the animal reservoir. Int. J. Food Microbiol. 66: 41–78.

Harris, J.R., Mariano, J., Wells, J.G., Payne, B.J., Donnell, H.D. and Cohen, M.L. 1985. Person-to-person transmission in an outbreak of enteroinvasive *Escherichia coli*. Am. J. Epidemiol. 122: 245–252.

Hashimoto, H., Mizukoshi, K., Nishi, M., Kawakita, T., Hasui, S., Kato, Y., Ueno, Y., Takeya, R., Okuda, N. and Takeda, T. 1999. Epidemic of gastrointestinal tract infection including hemorrhagic colitis attributable to Shiga toxin-producing *Escherichia coli* O118:H2 at a junior high school in Japan. Pediatrics 103(1): E2.

Health Canada. 2002. *Escherichia coli* O157 outbreak associated with the ingestion of unpasteurized goat milk in British Columbia, 2001. Can. Commun. Dis. Rep. 28 (1): 6–8.

Health Canada. 2003. Investigation of an *E. coli* O157:H7 outbreak in Brooks, Alberta, June-July 2002: The role of occult cases in the spread of infection within a daycare setting. Can. Commun. Dis. Rep. 29 (3): 21–28.

Health Canada. 2000. Outbreak of *Escherichia coli* O157:H7 leading to the recall of retail ground beef – Winnipeg, Manitoba, May 1999. Can. Commun. Dis. Rep. 26 (13):109–111.

Heuvelink, A.E., van Heerwaarden, C., Zwartkruis-Nahuis, J.T.M., van Oosterom, R., Edink, K., van Duynhoven, Y.T.H.P. and de Boer, E. 2002. *Escherichia coli* O157 infection associated with a petting zoo. Epidemiol. Infect. 129: 295–302.

Higgins, L.M., Frankel, G., Douce, G., Doughlas, G. and T.T. MacDonald 1999. *Citrobacter rodentium* infection in mice elicits a mucosal Th1 cytokine response and lesions similar to those in murine inflammatory bowel disease. Infect. Immun. 67: 3031–3039.

Hilborn, E.D., Mshar, P.A., Fiorentino, T.R., Dembek, Z.F., Barrett, T.J., Howard, R.T. and Cartter, M.L. 2000. An outbreak of *Escherichia coli* O157:H7 infections and haemolytic uraemic syndrome associated with consumption of unpasteurized apple cider. Epidemiol. Infect. 124: 31–36.

Holme, R. 2003. Drinking water contamination in Walkerton, Ontario: Positive resolutions from a tragic event. Water Sci. Technol. 47: 1–6.

Horne, C., Vallance, B.A., Deng, W. and Finlay, B.B. 2002. Current progress in enteropathogenic and enterohemorrhagic *Escherichia coli* vaccines. Expert Rev. Vaccines 1: 483–493.

Hoshina, K., Itagaki, A., Seki, R., Yamamoto, K., Masuda, S-I., Muku, T. and Okada, N. 2001. Enterohemorrhagic *Escherichia coli* O26 outbreak caused by contaminated natural water supplied by facility owned by local community. Jpn. J. Infect. Dis. 54: 247–248.

Houng, H-S., Sethabutr, O. and Echeverria, P. 1997. A simple polymerase chain reaction technique to detect and differentiate *Shigella* and enteroinvasive *Escherichia coli* in human feces. Diagn. Microbiol. Infect. Dis. 28: 19–25.

Howie, H., Mukerjee, A., Cowden, J., Leith, J. and Reid, T. 2003. Investigation of an outbreak of *Escherichia coli* O157 infection caused by environmental exposure at a scout camp. Epidemiol. Infect. 131: 1063–1069.

Huppertz, H.-I., Rutkowski, S., Busch, D.H., Eisebit, R., Lissner, R. and Karch, H. 1999. Bovine colostrum ameliorates diarrhea in infection with diarrheagenic *Escherichia coli*, Shiga toxin-producing *E. coli*, and *E. coli* expressing intimin and hemolysin. J. Pediatr. Gastroenterol. Nutr. 29: 452–456.

Ibekwe, A.M. and Grieve, C.M. 2003. Detection and quantification of *Escherichia coli* O157:H7 in environmental samples by real-time PCR. J. Appl. Microbiol. 94: 421–431.

Itoh, Y., Nagano, I., Kunishima, M. and Ezaki, T. 1997. Laboratory investigation of enteroaggregative *Escherichia coli* O untypeable:H10 associated with a massive outbreak of gastrointestinal illness. J. Clin. Microbiol. 35: 2546–2550.

Jallat, C., Livrelli, V., Darfeuille-Michaud, A., Rich, C. and Joly, B. 1993. *Escherichia coli* strains involved in diarrhea in France: high prevalence and heterogeneity of diffusely adhering strains. J. Clin. Microbiol. 31: 2031–2037.

Janka, A., Bielaszewska, M., Dobrindt, U., Greune, L., Schmidt, M.A. and Karch, H. 2003. Cytolethal distending toxin gene cluster in enterohemorrhagic *Escherichia coli* O157:H⁻ and O157:H7: characterization and evolutionary considerations. Infect. Immun. 71: 3634–3638.

Jay, M.T., Garrett, V., Mohle-Boetani, J.C., Barros, M., Farrar, J.A., Rios, R., Abbott, S., Sowadsky, R., Komatsu, K., Mandrell, R., Sobel, J. and Werner. S.B. 2004. A multistate outbreak of *Escherichia coli* O157: H7 infection linked to consumption of beef tacos at a fast-food restaurant chain. Clin. Infect. Dis. 39: 1–7.

Jiang, X., Morgan, J., and Doyle, M.. P. 2002. Fate of *Escherichia coli* O157:H7 in manure-amended soil. Appl. Environ. Microbiol. 68: 2605–2609.

Judge, N.A., Mason, H.S. and O'Brien, A.D. 2004. Plant cell-based intimin vaccine given orally to mice primed with intimin reduces time of *Escherichia coli* shedding in feces. Infect. Immun. 72: 168–175.

Juneja, V., Marmer, B., and Eblen, B.S. 1999. Predictive model for the combined effect of temperature, pH, sodium chloride, and sodium pyrophosphate on the heat resistance of *Escherichia coli* O157:H7. J. Food Safety 19: 147–160.

Karch, H., Bielaszewska, M., Bitzan, M. and Schmidt, H. 1999. Epidemiology and diagnosis of Shiga toxin-producing *Escherichia coli* infections. Diagn. Microbiol. Infect. Dis. 34: 229–243.

Karmali, M.A. 2003. The medical significance of Shiga toxin-producing *Escherichia coli* infections. Meth. Mol. Med. 73: 1–7.

Katz, J.M., Lu, X., Young, S.A. and Galphin, J.C. 1997. Adjuvant activity of heat-labile enterotoxin from enterotoxigenic *Escherichia coli* for oral administration of inactivated influenza virus vaccine. J. Infect. Dis. 175: 352–363.

Keusch, G.T., Acheson, D.W.K., Marchant, C. and McIver, J. 1998. Toxoid-based active and passive immunization to prevent and/or modulate hemolytic-uremic syndrome due to Shiga toxin-producing *Escherichia coli*. In: *Escherichia coli* O157:H7 and other Shiga toxin-producing *E. coli* strains. J.B. Kaper and A.D. O'Brien, eds. ASM Press, Washington, D.C., pp. 409–418.

Klapproth, J-M.A., Scaletsky, I.C.A., McNamara, B.P., Lai, L-C., Malstrom, C., James, S.P. and Donnenberg, M.S. 2000. A large toxin from pathogenic *Escherichia coli* strains that inhibits lymphocyte action. Infect. Immun. 68: 2148–2155.

Knutton, S., Shaw, R.K., Bhan, M.K., Smith, H.R., McConnell, M.M., Cheasty, T., Williams, P.H. and Baldwin, T.J. 1992. Ability of enteroaggregative *Escherichia coli* strains to adhere *in vitro* to human intestinal mucosa. Infect. Immun. 60: 2083–2091.

Kobayashi, M., Sasaki, T., Saito, N., Tamura, K., Suzuki, K., Watanabe, H. and Agui, N. 1999. Houseflies: not simple mechanical vectors of enterohemorrhagic *Escherichia coli* O157:H7. Am. J. Trop. Med. Hyg. 61: 625–629.

Kolling, G.L. and Matthews, K.R. 1999. Export of virulence genes and Shiga toxin by membrane vesicles of *Escherichia coli* O157:H7. Appl. Environ. Microbiol. 65: 1843–1848.

Konadu, E., Parke, J.C., Donohue-Rolfe, A., Calderwood, S.B., Robbins, J.B. and Szu, S.C. 1998. Synthesis and immunologic properties of O-specific polysaccharide-protein conjugate vaccines for prevention and treatment of infections with *Escherichia coli* O157 and other causes of hemolytic-uremic syndrome. In: *Escherichia coli* O157:H7 and other Shiga toxin-producing *E. coli* strains. J.B. Kaper and A.D. O'Brien, eds. ASM Press, Washington, D.C., pp. 419–424.

Konowalchuk, J., Spiers, J.I., and Stavric, S. 1977. Vero response to a cytotoxin of *Escherichia coli* cytotoxin. Infect. Immun. 20: 575–577.

Lan, R., Lumb, B., Ryan, D. and Reeves, P.R. 2001. Molecular evolution of large virulence plasmid in *Shigella* clones and enteroinvasive *Escherichia coli*. Infect. Immun. 69: 6303–6309.

Law, D. and Chart, H. 1998. Enteroaggregative *Escherichia coli*. J. Appl. Microbiol. 84: 685–697.

LeBlanc, J.J. 2003. Implication of virulence factors in *Escherichia coli* O157:H7 pathogenesis. Crit. Rev. Microbiol. 29: 277–296.

Leung, P. H., Peiris, J.S., Ng, W.W., Robins-Browne, R.M., Bettelheim, K.A. and Yan, W.C. 2003. A newly discovered verotoxin variant, VT2g, produced by bovine verocytotoxigenic *Escherichia coli*. Appl. Environ. Microbiol. 69: 7549–7553.

Levine, M.M., Caplan, E.S., Waterman, D., Cash, R.A., Hornick, R.B. and Snyder, M.J. 1977. Diarrhea caused by *Escherichia coli* that produce only heat-stable enterotoxin. Infect. Immun. 17: 78–82.

Lin, J., Smith, M.P., Chapin, K.C., Baik, H.S., Bennett, G.N. and Foster, J.W. 1996. Mechanisms of acid resistance in enterohemorrhagic *Escherichia coli*. Appl. Environ.Microbiol. 62: 3094–3100.

Line, J.E., Fain, A.R., Moran, A.B., Martin, L.M., Lechowich, R.V., Carosella, J.M. and Brown, W.L. 1991. Lethality of heat to *Escherichia coli* O157:H7: D-value and z-value determinations in ground beef. J. Food Prot. 54: 762–766.

Liu, H., Wang, H., Shi, Z., Liu, Q., Zhu, J., He, N., Wang, H. and Lu, Z. 2003. Identification of *Escherichia coli* O157:H7 with oligonucleotide arrays. Bull. Environ. Contam. Toxicol. 71: 826–832.

López-González, V., Murano, P.S., Brennan, R.E. and Murano, E.A. 1999. Influence of various packaging conditions on survival of *Escherichia coli* O157:H7 to irradiation by electron beam versus gamma rays. J. Food Prot. 62: 10–15.

López-Saucedo, C., Cerna, J.F., Villegas-Sepulveda, N., Thompson, R., Velazquez, F.R., Torres, J., Tarr, P.I. and Estrada-Garcia, T. 2003. Single multiplex polymerase chain reaction to detect diverse loci associated with diarrheagenic *Escherichia coli*. Emerg. Infect. Dis. 9: 127–131.

López-Vidal, Y., Calva, J.J., Trujillo, A., de Léon, A.P., Ramos, A., Svennerholm, A-M. and Ruiz-Palacios, G.M. 1990. Enterotoxin and adhesions of enterotoxigenic *Escherichia coli*: are they risk factors for acute diarrhea in the community? J. Infect. Dis. 162: 442–447.

Lortie, L-A., Dubreuil, J.D. and Harel, J. 1991. Characterization of *Escherichia coli* strains producing heat-stable enterotoxin b (STb) isolated from humans with diarrhea. J. Clin. Microbiol. 29: 656–659.

Lyte, M., Erickson, A.K., Arulanandam, B.P., Frank, C.D., Crawford, M.A. and Francis, D.H. 1997. Norepinephrine-induced expression of the K99 pilus adhesin of enterotoxigenic *Escherichia coli*. Biochem. Biophys. Res. Commun. 232: 682–686.

Mainil, J.G., Jacquemin, E. and Oswald, E. 2003. Prevalence and identity of *cdt*-related sequences in necrotoxigenic *Escherichia coli*. Vet. Microbiol. 94: 159–165.

MacDonald, D.M., Fyfe, M., Paccagnella, A., Trinidad, A., Louie, K. and Patrick, D. 2004. *Escherichia coli* linked to salami, British Columbia, Canada, 1999. Epidemiol. Infect. 132: 283–289.

Makino, K., Ishii, K., Yasunaga, T., Hattori, M., Yokoyama, K., Yutsudu, C.H., Kubota, Y., Yamaichi, Y., Iida, T., Yamamoto, K., Honda, T., Han, C-G., Ohtsubo, E., Kasamatsu, M., Hayashi, T., Kuhara, S. and Shinagawa, H. 1998. Complete nucleotide sequences of 93-kb and 3.3-kb plasmids of an enterohemorrhagic *Escherichia coli* O157:H7 derived from Sakai outbreak. DNA Res. 5: 1–9.

Mariani-Kurkdjian, P., Denamur, E., Milon, A., Picard, B., Cave, H., Lambert-Zechovsky, N., Loirat, C., Goullet, P., Sansonsetti, P.J. and Elion, J. 1993. Identification of a clone of *Escherichia coli* O103:H2 as a potential agent of hemolytic-uremic syndrome in France. J. Clin. Microbiol. 31: 296–301.

Mason, H. S., Haq, T.A., Clements, J.D. and Arntzen, C.J. 1998. Edible vaccine protects mice against *Escherichia coli* heat-labile enterotoxin (LT): Potatoes expressing a synthetic LT-B gene. Vaccine 16: 1336–1343.

Matthews, K. R., Murdough, P.A. and Bramley, A.J. 1997. Invasion of bovine epithelial cells by verocytotoxin-producing *Escherichia coli* O157:H7. J. Appl. Microbiol. 82: 197–203.

Matussek, A., Lauber, L., Bergau, A., Hansen, W., Rohde, M., Dittmar, K.E., Gunzer, M., Mengel, M., Glatzlaff, P., Hartmann, M., Buer, J. and Gunzer, F. 2003. Molecular and functional analysis of Shiga toxin-induced response patterns in human vascular endothelial cells. Blood 102: 1323–1332.

McCarthy, T.A., Barrett, N.L., Hadler, J.L., Salsbury, B., Howard, R.T., Dingman, D.W., Brinkman, C.D., Bibb, W.F. and Carter, M.L. 2001. Hemolytic-uremic syndrome and *Escherichia coli* O121 at a lake in Connecticut, 1999. Pediatrics. 108(4):E59.

McGee, P., Bolton, D.J., Sheridan, J.J., Earley, B., Kelly, G. and Leonard, N. 2002. Survival of *Escherichia coli* O157:H7 in farm water: its role as a vector in the transmission of the organism within herds. J. Appl. Microbiol. 93: 706–713.

McMaster, C., Roche, E.A., Willshaw, G.A., Doherty, A., Kinnear, W. and Cheasty, T. 2001. Verocytotoxin-producing *Escherichia coli* serotype O26:H11 outbreak in an Irish crèche. Eur.J. Clin. Microbiol. Infect. Dis. 20: 430–432.

Mead, P.S. and Griffin, P.M. 1998. *Escherichia coli* O157:H7. Lancet. 352: 1207–1212.

Mead, P.S., Slutsker, L., Dietz, V., McCaig, L.F., Bresee, J.S., Shapiro, C., Griffin, P.M. and Tauxe, R.V. 1999. Food-related illness and death in the United States. Emerg. Infect. Dis. 5: 607–625.

Melton-Celsa, A.R. and O'Brien, A.D. 2003. Animal models for STEC-mediated disease. Methods Mol. Med. 73: 291–305.

Ménard, L-P. and Dubreuil, J.D. 2002. Enteroaggregative *Escherichia coli* heat-stable enterotoxin 1 (EAST1): A new toxin with an old twist. Crit. Rev. Microbiol. 28: 43–60.

Meng, J., Doyle, M.P., Zhao, T. and Zhao, S. 2001. Enterohemorrhagic *Escherichia coli*. In: Food microbiology: fundamentals and frontiers, 2nd ed. M.P. Doyle, L.R. Beuchat, and T.J. Montville, eds. ASM Press, Washington, D.C. p. 193–213.

Michino, H., Araki, K., Minami, S., Takaya, S., Sakai, N., Miyazaki, M., Ono, A. and Yanagawa, H. 1999. Massive outbreak of *Escherichia coli* O157:H7 infection in school children in Sakai City, Japan, associated with consumption of white radish sprouts. Am. J. Epidemiol. 150: 787–796.

Milon, A., Oswald, E. and De Rycke, J. 1999. Rabbit EPEC: A model for the study of enteropathogenic *Escherichia coli*. Vet. Res. 30: 203–219.

Miqdady, M.S., Jiang, Z-D., Nataro, J.P. and DuPont, H.L. 2002. Detection of enteroaggregative *Escherichia coli* with formalin-preserved HEp-2 cells. J. Clin. Microbiol. 40: 3066–3067.

Misselwitz, J., Karch, H., Bielazewska, M., John, U., Ringelmann, F., Ronnefarth, G. and Patzer, L. 2003. Cluster of hemolytic-uremic syndrome caused by Shiga toxin-producing *Escherichia coli* O26: H11. Pediatr. Infect. Dis. J. 22: 349–354.

Mizan, S., Lee, M.D., Harmon, B.G., Tkalcic, S. and Maurer, J.J. 2002. Acquisition of antibiotic resistance plasmids by enterohemorrhagic *Escherichia coli* O157:H7 within rumen fluid. J. Food Prot. 65: 1038–1040.

Mol, O. and Oudega, B. 1996. Molecular and structural aspects of fimbriae biosynthesis and assembly in *Escherichia coli*. FEMS Microbiol. Rev. 19: 25–52.

Mølbak, K., Mead, P.S. and Griffin, P.M. 2002. Antimicrobial therapy in patients with *Escherichia coli* O157:H7 infection. J. Am. Med. Assoc. 288: 1014–1016.

Morais, T.B., Gomes, T.A.T. and Sigulem, D.M. 1997. Enteroaggregative *Escherichia coli* in infant feeding bottles. Lancet. 349: 1448–1449.

Morabito, S., Karch, H., Mariani-Kurkdjian, P., Schmidt, H., Minelli, F., Bingen, E. and Caprioli, A. 1998. Enteroaggregative, Shiga toxin-producing *Escherichia coli* O111:H2 associated with an outbreak of hemolytic-uremic syndrome. J. Clin. Microbiol. 36: 840–842.

Mulvey, G., Rafter, D.J. and Armstrong, G.D. 2002. Potential for using antibiotics combined with a Shiga toxin-absorbing agent for treating O157:H7 *Escherichia coli* infections. Can. J. Chem. 80: 871–874.

Muniesa, M., Recktenwald, J., Bielaszewska, M., Karch, H. and Schmidt, H. 2000. Characterization of a Shiga toxn 2e-converting bacteriophage from an *Escherichia coli* strain of human origin. Infect. Immun. 68: 4850–4855.

Nair, G.B. and Takeda, Y. 1998. The heat-stable enterotoxins. Microbial Pathogen. 24: 123–131.

Nascimento de Araújo, A. and Giugliano, L.G. 2000. Human milk fractions inhibit the adherence of diffusely adherent *Escherichia coli* (DAEC) and enteroaggregative *E. coli* (EAEC) to HeLa cells. FEMS Microbiol. Lett. 184: 91–94.

Nataro, J.P., Deng, Y., Cookson, S., Cravioto, A., S. Savarino, J., Guers, L.D., Levine, M.M. and Tacket, C.O. 1995. Heterogeneity of enteroaggregative *Escherichia coli* virulence demonstrated in volunteers. J. Infect. Dis. 171: 465–468.

Nataro, J.P., Deng, Y., Maneval, D.R., German, A.L., Martin, W.C. and Levine, M.M. 1992. Aggregative adherence fimbriae I of enteroaggregative *Escherichia coli* mediate adherence to HEp-2 cells and hemagglutination of human erythrocytes. Infect. Immun. 60: 2297–2304.

Nataro, J.P., Hicks, S., Philips, A.D., Vial, P.A. and Sears, C.L. 1996. T84 cells in culture as a model for enteroaggregative *Escherichia coli* pathogenesis. Infect. Immun. 64: 4761–4768.

Nataro, J.P. and Kaper, J.B. 1998. Diarrheagenic *Escherichia coli*. Clin. Microbiol. Revs. 11: 142–201.

Nataro, J.P., Steiner, T. and Guerrant, R.L. 1998. Enteroaggregative *Escherichia coli*. Emerg. Infect. Dis. 4: 251–261.

Navarro-García, F., Eslava, C., Villaseca, J.M., López-Revilla, R., Czeczulin, J.R., Srinivas, S., Nataro, J.P. and Cravioto, A. 1998. *In vitro* effects of a high-molecular weight heat-labile enterotoxin from enteroaggregative *Escherichia coli*. Infect. Immun. 66: 3149–3154.

Naylor, S.W., Low, J.C., Besser, T.E., Mahajan, A., Gunn, G.J., Pearce, M.C., McKendrick, I.J., Smith, D.G.E. and Gally, D.L. 2003. Lymphoid follicle-dense mucosa at the terminal rectum is the principal site of colonization of enterohemorrhagic *Escherichia coli* O157:H7 in the bovine host. Infect. Immun. 71: 1505–1512.

Neill, M.A., Tarr, P.I., Taylor, D.N. and Trofa, A.F. 1994. *Escherichia coli*. In: Foodborne Disease Handbook, Vol. 1. Y.H. Hui, J.R. Gorham, K.D. Murrell and D.O. Cliver, eds. Marcel Dekker, Inc., New York. p. 169–213.

Noller, A.C., McEllistrem, M.C., Pacheco, A.G.F., Boxrud, D.J. and Harrison, L.H. 2003. Multilocus variable-number tandem repeat analysis distinguishes outbreak and sporadic *Escherichia coli* O157:H7 isolates. J. Clin. Microbiol. 41: 5389–5397.

Nou, X., Rivera-Betancourt, M., Bosilevac, J.M., Wheeler, S.D., Shackelford, T.L., Gwartney, B.L., Reagan, J.O. and Koohmaraie, M. 2003. Effect of chemical dehairing on the prevalence of *Escherichia coli* O157:H7 and levels of aerobic bacteria and *Enterobacteriaceae* on carcasses in a commercial beef processing plant. J. Food Prot. 66: 2005–2009.

O'Brien, A.D. and Holmes, R.K. 1996. Protein toxins of *Escherichia coli* and *Salmonella*. In: *Escherichia coli* and *Salmonella*: Cellular and Molecular Biology, 2nd ed. F.C. Neidhardt, ed. ASM Press, Washington, D.C. p. 2788–2802.

O'Brien, S.J., Murdoch, P.S., Riley, A.H., King, I., Barr, M., Murdoch, S., Greig, A., Main, R., Reilly, W.J. and Thomson-Carter, F.M. 2001. A foodborne outbreak of vero cytotoxin-producing *Escherichia coli* O157:H-phage type 8 in hospital. J. Hosp. Infect. 49: 167–172.

O'Donnell, J.M., Thornton, L., McNamara, E.B., Predergast, T., Igoe, D. and Cosgrove, C. 2002. Outbreak of Vero cytotoxin-producing *Escherichia coli* O157 in child day care facility. Commun. Dis. Public Health. 5: 54–58.

Oelschlaeger, T.A., Barrett, T.J. and Kopecko, D.J. 1994. Some structures and processes of human epithelial cells involved in uptake of enterohemorrhagic *Escherichia coli* O157:H7 strains. Infect. Immun. 62: 5142–5150.

Ogden, I.D., Hepburn, N.F., MacRae, M., Strachan, N.J.C., Fenlon, D.R., Rusbridge, S.M. and Pennington, T.H. 2002. Long-term survival of *Escherichia coli* O157 on pasture following an outbreak associated with sheep at a scout camp. Lett. Appl. Microbiol. 34: 100–104.

Okamoto, K., Fujii, Y., Akashi, N., Hitotsubashi S., Kurazono, H., Karasawa, T. and Takeda, Y. 1993. Identification and characterization of heat-stable enterotoxin II-producing *Escherichia coli* from patients with diarrhea. Microbiol. Immunol. 37: 411–414.

Okamoto, K. and Takahara, M. 1990. Synthesis of *Escherichia coli* heat-stable enterotoxin STp as a pre-pro form and role of the pro sequence in secretion. J. Bacteriol. 172: 5260–5265.

Okeke, I N. and Nataro, J.P. 2001. Enteroaggregative *Escherichia coli*. Lancet Infect. Dis. 1: 304–313.

Okeke, I. N., Steinrück, H., Kanack, K.J., Elliott, S.J., Stunström, L., Kaper, J.B. and Lamikanra, A. 2002. Antibiotic-resistant cell-detaching *Escherichia coli* strains from Nigerian children. J. Clin. Microbiol. 40: 301–305.

Okeke, I.N., Scaletsky, I.C.A., Soars, E.H., Macfarlane, L.R. and Torres, A.G. 2004. Molecular epidemiology of the iron utilization genes of enteroaggregative *Escherichia coli*. J. Clin. Microbiol. 42: 36–44.

O'Loughlin, E.V. and Robins-Browne, R.M. 2001. Effect of Shiga toxin and Shiga-like toxins on eukaryotic cells. Microbes Infect. 3: 493–507.

Olsen, S.J., Miller, G., Breuer, T., Kenned, M., Higgins, C., Walford, J., McKee, G., Fox, K., Bibb, W. and Mead, P. 2002. A water-borne outbreak of *Escherichia coli* O157:H7 infections and hemolytic uremic syndrome: implications for rural water systems. Emerg. Infect. Dis. 8: 370–375.

Ørskov, F. 1984. Genus 1. *Escherichia*. In: Bergey's Manual of Systematic Bacteriology, Vol. 1. N.R. Krieg ed. Williams & Wilkins, Baltimore, Maryland, pp. 420–423.

Ostroff, S.M., Tarr, P.L., Neill, M.A., Lewis, J.H., Hargrett-Bean, N., and Kobayashi, J.M. 1989. Toxin genotypes and plasmid profiles as determinants of systemic sequelae in *Escherichia coli* O157:H7 infections. J. Infect. Dis. 160: 994–998.

Pal, T., Al-Sweih, N.A., Herpay, M. and Chugh, T.D. 1997. Identification of enteroinvasive *Escherichia coli* and *Shigella* strains in pediatric patients by an IpaC-specific enzyme-linked immunosorbent assay. J. Clin. Microbiol. 35: 1757–1760.

Palumbo, S.A., Call, J.E., Schultz, F.J. and Williams, A.C. 1995. Minimum and maximum temperatures for growth and verotoxin production by hemorrhagic strains of *Escherichia coli*. J. Food Prot. 58: 352–356.

Park, C.H., Kim, H.J. and Hixon, D.L. 2002. Importance of testing stool specimens for Shiga toxins. J. Clin. Microbiol. 40: 3542–3543.

Parry, S.H. and Rooke, D.M. 1985. Adhesins and colonization factors of *Escherichia coli*. In: The Virulence of *Escherichia coli*: reviews and methods. M. Sussman, ed. Academic Press, New York, pp. 79–155.

Paton, A.W., Ratcliff, R.M., Doyle, R.M., Seymour-Murray, J., Davos, D., Lanser, J.A. and Paton, J.C. 1996. Molecular microbiological investigation of an outbreak of hemolytic-uremic syndrome caused by dry fermented sausage contaminated with Shiga-like toxin-producing *Escherichia coli*. J. Clin. Microbiol. 34: 1622–1627.

Paton, J.C., Rogers, T.J., Morona, R. and Paton, A.W. 2001. Oral administration of formaldehyde-killed recombinant bacteria expressing a mimic of the Shiga toxin receptor protects mice from fatal challenge with Shiga-toxigenic *Escherichia coli*. Infect. Immun. 69: 1389–1393.

Paton, A.W., Woodrow, M.C., Doyle, R.M., Lanser, J.A. and Paton, J.C. 1999. Molecular characterization of a Shiga toxigenic *Escherichia coli* O113:H21 strain lacking *eae* responsible for a cluster of cases of hemolytic-uremic syndrome. J. Clin. Microbiol. 37: 3357–3361.

Payne, C.J.I., Petrovic, M., Roberts, R.J., Paul, A., Linnane, E., Walker, M., Kirby, D., Burgess, A., Smith, R.M.M., Cheasty, T., Willshaw, G. and Salmon, R.L. 2003. Vero cytotoxin-producing *Escherichia coli* O157 in farm visitors, North Wales. Emerg. Infect. Dis. 9: 526–530.

Pebody, R.G., Furtado, C., Rojas, A., McCarthy, N., Nylen, G., Ruutu, P., Leino, T., Chalmers, R., de Jong, B., Donnelly, M., Fisher, I., Gilham, C., Graverson, L., Cheasty, T., Willshaw, G., Navarro, M., Salmon, R., Leinikki, P., Wall, P. and Bartlett, C. 1999. An international outbreak of Vero cytotoxin-producing *Escherichia coli* O157 infection amongst tourists; a challenge for the European infectious disease surveillance network. Epidem. Infect. 123: 217–223.

Perna, N.T., Plunkett, G., Burland, V., Mau, B., Glasner, J.D., Rose, D.J., Mayhew, G.F., Evans, P.S., Gregor, J., Kirkpatrick, H.A., Posfai, G., Hackett, J., Klink, S., Boutin, A., Shao, Y., Miller, L., Grotbeck, E.J., Davis, N. W., Lim, A., Dimalanta, E.T., Potamousis, K.D., Apodaca, J., Anantharaman, T.S., Lin, J., Yen, G., Schwartz, D.C., Welch, R.A. and Blattner, F.R. 2001. Genome sequence of enterohemorrhagic *Escherichia coli* O157:H7. Nature. 409: 529–533.

PHLS Communicable Disease Surveillance Centre. 2000a. Outbreak of VTEC O157 infection linked to consumption of unpasteurized milk. Commun. Dis. Rep. CDR Wkly. 10 (23): 203, 206.

PHLS Communicable Disease Surveillance Centre. 2000b. Outbreak of VTEC O157 in South Yorkshire. Commun. Dis. Rep. CDR Wkly. 10 (40): 359.

PHLS Communicable Disease Surveillance Centre. 2000c. Two outbreaks of VTEC O157 infection in northern England. Commun. Dis. Rep. CDR Wkly. 10 (26): 229.

Pickering, L.K., Guerrant, R.L. and Cleary, T.G. 1995. Microorganisms responsible for neonatal diarrhea. In: Infectious Diseases of the Fetus and Newborn Infant, Fourth ed. J.S. Remington and J.O. Klein, ed. W. B. Saunders Co., Philadelphia, Pennsylvania. p. 1142–1222.

Poitrineau, P., Forestier, C., Meyer, M., Jallat, C., Rich, C., Malpuech, G. and de Champs, C. 1995. Retrospective case–control study of diffusely adhering *Escherichia coli* and clinical features in children with diarrhea. J. Clin. Microbiol. 33: 1961–1962.

Potter, A.A., Klashinsky, S., Li, Y., Frey, E., Townsend, H., Rogan, D., Erickson, G., Hinkley, S., Klopfenstein, T., Moxley, R.A., Smith, D.R. and Finlay, B.B. 2004. Decreased shedding of *Escherichia coli* O157:H7 by cattle following vaccination with type III secreted proteins. Vaccine. 22: 362–369.

Pradel, N., Leroy-Setrin, S., Joly, B. and Liverelli, V. 2002. Genomic subtraction to identify and characterize sequences of Shiga toxin-producing *Escherichia coli* O91:H21. Appl. Environ. Microbiol. 68: 2316–2325.

Pradel, N., Ye, C., Livrelli, V., Xu, J., Joly, B. and Wu, L.-F. 2003. Contribution of the twin arginine translocation system to the virulence of enterohemorrhagic *Escherichia coli* O157:H7. Infect. Immun. 71: 4908–4916.

Pruimboom-Brees, I.M., Morgan, T.W., Ackermann, M.R., Nystrom, E.D., Samuel, J.E., Cornick, N.A. and Moon, H. W. 2000. Cattle lack vascular receptors for *Escherichia coli* O157:H7 Shiga toxins. Proc. Natl. Acad. Sci. USA 97: 10325–10329.

Qadri, F., Wenneras, C., Ahmed, F., Asaduzzaman, M., Saha, D., Albert, J.J., Sack, R.B. and Svennerholm, A-M. 2000. Safety and immunogenicity of an oral, inactivated enterotoxigenic *Escherichia coli* plus cholera toxin B subunit vaccine in Bangladeshi adults and children. Vaccine. 18: 2704–2712.

Rabinowitz, R.P. and Donnenberg, M.S. 1996. *Escherichia coli*. In: Enteric Infections and Immunity. L.J. Paradise, M. Bendinelli and H. Friedman, eds. Plenum Press, New York. p. 101–131.

Reid, S.D., Selander, R.K. and Whittam, T.S. 1999. Sequence diversity of flagellin (*fliC*) alleles in pathogenic *Escherichia coli*. J. Bacteriol. 181: 153–160.

Ricci, L.C., de Faria, F.P., de S. Porto, P.S., de Oliveira, E.M.G. and de Castro, A.F.P. 1997. A new fimbrial putative colonization factor (PCF02) in human enterotoxigenic *Escherichia coli* isolated in Brazil. Res. Microbiol. 148: 65–69.

Rich, C., Favre-Bronte, S., Sapena, F., Joly, B. and Forestier, C. 1999. Characterization of enteroaggregative *Escherichia coli* isolates. FEMS Microbiol. Lett. 173: 55–61.

Ritchie, J.M., Thorpe, C.M., Rogers, A.B., and Waldor, M. K. 2003. Critical roles for *stx₂*, *eae*, and *tir* in enterohemorrhagic *Escherichia coli*-induced diarrhea and intestinal inflammation in infant rabbits. Infect. Immun. 71: 7129–7139.

Russo, T.A. and Johnson, J.R. 2003. Medical and economic impact of extraintestinal infections due to *Escherichia coli*: focus on an increasingly important endemic problem. Microbes Infect. 5: 449–456.

Ryan, C.A., Tauxe, R.V., Hosek, G.W., Wells, J.G., Stoesz, P.A., McFadden, H.W., Smith, P.W., Wright, G.F. and Blake, P.A. 1986. *Escherichia coli* O157:H7 diarrhea in a nursing home: clinical, epidemiological, and pathological findings. J. Infect. Dis. 154: 631–638.

Salyers, A.A. and Whitt, D.D. 1994. Bacterial Pathogenesis: A molecular approach. ASM Press, Washington, D.C.

Samadpour, M., Stewart, J., Steingart, K., Addy, C., Louderback, J., McGinn, M., Ellington, J. and Newman, T. 2004. Laboratory investigation of an *E. coli* O157:H7 outbreak associated with swimming in Battle Ground Lake, Vancouver, Washington. J. Environ. Health. 64: 16–20.

Sanderson, M.W., Besser, T.E., Gay, J.M., Gay, C.C. and Hancock, D.D. 1999. Fecal *Escherichia coli* O157:H7 shedding patterns of orally inoculated calves. Vet. Microbiol. 69: 199–205.

Sargeant, J.M., Sanderson, M.W., Smith, R.A. and Griffin, D.D. 2003. *Escherichia coli* O157 in feedlot cattle feces and water in four major feeder-cattle states in the USA. Prev. Vet. Med. 61: 127–135.

Savarino, S.J., Fasano, A., Robertson, D.C. and Levine, M.M. 1991. Enteroaggregative *Escherichia coli* elaborate a heat-stable enterotoxin demonstrable in an *in vitro* rabbit intestinal model. J. Clin. Invest. 87: 1450–1455.

Savarino, S.J., Fasano, A., Watson, J., Martin, B.M., Levine, M.M., Guandalini, S. and Guerry, P. 1993. Enteroaggregative *Escherichia coli* heat-stable enterotoxin 1 represents another subfamily of *E. coli* heat-stable toxin. Proc. Natl. Acad. Sci. USA 90: 3093–3097.

Savarino, S.J., Hall, E.R., Bassily, S., Brown, F.M., Youssef, F., Wierzaba, T.F., Peruski, L., El-Masry, N. A., Safwat, M., Rao, M., El Mohamady, H., Abu-Elyazeled, R., Naficy, A., Svennerholm, A.M., Jertborn, M., Lee, Y.J. and Clemens, J.D. 1999. Oral, inactivated whole cell enterotoxigenic *Escherichia coli* plus cholera toxin B subunit vaccine: Results of the initial evaluation in children. J. Infect. Dis. 179: 107–114.

Savarino, S.J., McVeigh, A., Watson, J., Cravioto, A., Molina, J., Echeverria, P., Bhan, M.K., Levine, M.M. and Fasano, A. 1996. Enteraggregative *Escherichia coli* heat-stable enterotoxin is not restricted to enteroaggregative *E. coli*. J. Infect. Dis. 173: 1019–1022.

Scaletsky, I.C.A., Silva, M.L.M. and Trabulsi, L.R. 1984. Distinctive patterns of adherence of enteropathogenic *Escherichia coli* to HeLa cells. Infect. Immun. 45: 534–536.

Schmidt, H. 2001. Shiga-toxin-converting bacteriophages. Res. Microbiol. 152: 687–695.

Schmidt, H., Knop, C., Franke, S., Aleksic, S., Heesemann, J. and Karch, H. 1995. Development of PCR for screening of enteroaggregative *Escherichia coli*. J. Clin. Microbiol. 33: 701–705.

Schmidt, H., Scheef, J., Morabito, S., Caprioli, A., Wieler, L.H. and Karch, H. 2000. A new Shiga toxin 2 variant (Stx2f) from *Escherichia coli* isolated from pigeons. Appl. Environ. Microbiol. 66: 1205–1208.

Schroeder, C.M., Meng, J., Zhao, S., DebRoy, C., Torcolini, J., Zhao, C., McDermott P.F., Wagner, D.D., Walker, R.D. and White, D.G. 2002a. Antimicrobial resistance of *Escherichia coli* O26, O103, O111, O128, and O145 from animals and humans. Emerg. Infect. Dis. 8: 1409–1414.

Schroeder, C.M., Zhao, C., DebRoy, C., Torcolini, J., Zhao, S., White, D.G., Wagner, D.D., McDermott, P.F., Walker, R.D. and Meng, J. 2002b. Antimicrobial resistance of *Escherichia coli* O157 isolated from humans, cattle, swine, and food. Appl. Environ. Microbiol. 68: 576–581.

Scotland, S.M., Willshaw, G.A., Smith, H.R., and Rowe, B. 1987. Properties of strains of *Escherichia coli* belonging to serogroup O157:

H7 with special reference to production of vero cytotoxins VT1 and VT2. Epidemiol. Infect. 99: 613–624.

Shaikh, N. and Tarr, P.I. 2003. *Escherichia coli* O157:H7 Shiga toxin-encoding bacteriophages: integrations, excisions, truncations, and evolutionary implications. J. Bacteriol. 185: 3596–3605.

Sharma, V.K. and Dean-Nystrom, E.A. 2003. Detection of enterohemorrhagic *Escherichia coli* O157:H7 by using a multiplex real-time PCR assay for genes encoding intimin and Shiga toxins. Vet. Microbiol. 93: 247–260.

Shin, K., Yamauchi, Y., Teraguchi, S., Hayasawa, H., Tomita, M., Otsuka, Y. and Yamazaki, S. 1998. Antibacterial activity of bovine lactoferrin and its peptides against enterohemorrhagic *Escherichia coli* O157:H7. Lett. Appl. Microbiol. 26: 407–411.

Smith, H.R. and Cheasty, T. 1998. Diarrhoeal diseases due to *Escherichia coli* and *Aeromonas*. In: Topley and Wilson's Microbiology and Microbial Infections. L. Collier, A. Balows and M. Sussman, eds. Arnold, London. p. 513–537.

Smith, H.R., Cheasty, T. and Rowe, B. 1997. Enteraggregative *Escherichia coli* and outbreaks of gastroenteritis in UK. Lancet. 350: 814–815.

Smith, J.L. 1999. Foodborne infections during pregnancy. J. Food Prot. 62: 818–829.

Smith, J.L. 1988. Heat-stable enterotoxins: properties, detection and assay. Devel. Indust. Microbiol. 29: 275–286.

Sperandio, V. 2001. Genome sequence of *E. coli* O157:H7. Trends Microbiol. 9: 159.

Sperandio, V., Mellies, J.L., Nguyen, W., Shin, S. and Kaper, J.B. 1999. Quorum sensing controls expression of the type III secretion gene transcription and protein secretion in enterohemorrhagic and enteropathogenic *Escherichia coli*. Proc. Natl. Acad. Sci. USA 96: 15196–15201.

Sperandio, V., Torres, A.G., Girón, J.A. and Kaper, J.B. 2001. Quorum sensing is a global regulatory mechanism in enterohemorrhagic *Escherichia coli* O157:H7. J. Bacteriol. 183: 5187–5197.

Sperandio, V., Torres, A.G., Jarvis, B., Nataro, J.P. and Kaper, J.B. 2003. Bacteria-host communication: the language of hormones. Proc. Natl. Acad. Sci. USA 100: 8951–8956.

Steiner, T.S., Lima A.A.M., Nataro, J.P. and Guerrant, R.L. 1998. Enteroaggregative *Escherichia coli* produce intestinal inflammation and growth impairment and cause interleukin-8 release from intestinal epithelial cells. J. Infect. Dis. 177: 88–96.

Steinsland, H., Valentiner-Branth, P., Grewal, H.M.S., Gaastra, W., Mølbak, K. and Sommerfelt, H. 2003. Development and evaluation of genotypic assays for the detection and characterization of enterotoxigenic *Escherichia coli*. Diagn. Microbiol. Infect. Dis. 45: 97–105.

Stephan, R., Ragettli, S. and Untermann, F. 2000. Prevalence and characteristics of verotoxin-producing *Escherichia coli* (VTEC) in stool samples from asymptomatic human carriers working in the meat processing industry in Switzerland. J. Appl. Microbiol. 88: 335–341.

Su, C., and Brandt, L.J. 1995. *Escherichia coli* O157:H7 infection in humans. Ann. Intern. Med. 123: 698–714.

Sueyoshi, M. and Nakazawa, M. 1994. Experimental infection of young chicks with attaching and effacing *Escherichia coli*. Infect. Immun. 62: 4066–4071.

Sugiyama, K., Inoue, K. and Sakazaki, R. 2001. Mitomycin-supplemented washed blood agar for the isolation of Shiga toxin-producing *Escherichia coli* other than O157:H7. Lett. Appl. Microbiol. 33: 193–195.

Swaminathan, B., Barrett, T.J., Hunter, S.B., Tauxe, R.V. and the CDC PulseNet Task Force. 2001. PulseNet: the molecular subtyping network for foodborne bacterial disease surveillance, United States. Emerg. Infect. Dis. 7: 382–389.

Takahashi, I., Marinaro, M., Kiyono, H., Jackson, R.J., Nakagawa, I., Fujihashi, K., Hamada, S., Clements, J.D., Bost, K.L. and McGhee, J.R. 1996. Mechanisms for mucosal immunogenicity and adjuvancy of *Escherichia coli* labile enterotoxin. J. Infect. Dis. 173: 627–635.

Takeda, T., Yoshino, K., Adachi, E., Sato, Y. and Yamagata, K. 1999. *In vitro* assessment of a chemically synthesized Shiga toxin receptor analog attached to Chromosorb P (Synsorb Pk) as a specific absorbing agent of Shiga toxin 1 and 2. Microbiol. Immunol. 43: 331–337.

Tana, W.S., Isoga, E. and Oguma, K. 2003. Induction of intestinal IgA and IgG antibodies preventing adhesion of verotoxin-producing *Escherichia coli* to Caco-2 cells by oral immunization with liposomes. Lett. Appl. Microbiol. 36: 135–139.

Taneike, I., Zhang, H-M., Wakisaka-Saito, N. and Yamamoto, T. 2002. Enterohemolysin operon of Shiga toxin-producing *Escherichia coli*: a virulence function of inflammatory cytokine production from human monocytes. FEBS Lett. 524: 219–224.

Tarr, P.I., Schoening, L.M., Yea Y-L., Ward, T.R., Jelacic, S. and Whittam, T.S. 2000. Acquisition of the *rfb-gnd* cluster in evolution of *Escherichia coli* O55 and O157. J. Bacteriol. 182: 6183–6191.

Tatsuno, I., Horie, M., Abe, H., Miki, T., Makino, K., Shinagawa, H., Taguchi, H., Kamiya, S., Hayashi, T. and Sasakawa, C. 2001. *toxB* gene on pO157 of enterohemorrhagic *Escherichia coli* O157:H7 is required for full epithelial cell adherence phenotype. Infect. Immun. 69: 6660–6669.

Tauschek, M., Strugnell, R.A. and Robins-Browne, R.M. 2002. Characterization and evidence of mobilization of the LEE pathogenicity island of rabbit-specific strains of enteropathogenic *Escherichia coli*. Mol. Microbiol. 44: 1533–1550.

Taylor, D.E., Rooker, M., Keelan, M., Ng, L-K., Martin, I., Perna, N.T., Burland, N.T.V. and Blattner, F.R. 2002. Genomic variability of O islands encoding tellurite resistance in enterohemorrhagic *Escherichia coli* O157:H7 isolates. J. Bacteriol. 184: 4690–4698.

Terajima, J., Izumiya, H., Iyoda, S., Tamura, K. and Watanabe, H. 1999. Detection of a multi-prefectural *E. coli* O157:H7 outbreak caused by contaminated ikura-sushi ingestion. Jpn. J. Infect. Dis. 52: 52–53.

Tilden, J., Young, W., McNamara, A.M., Custer, C., Bossel, B., Lambert-Fair, M.A., Majkowski, J., Vugia, D., Werner, S.B., Hollingsworth, J. and Morris, J. 1996. A new route of transmission for *Escherichia coli*: infection from dry fermented salami. Am. J. Public Health. 86: 1142–1145.

Trachtman, H. and Christen, E. 1999. Pathogenesis, treatment, and therapeutic trials in hemolytic uremic syndrome. Curr. Opin. Pediatr. 11: 162–168.

Tsen, H.Y., Jian, L.Z. and Chi, W.R. 1998. Use of a multiplex PCR system for the simultaneous detection of heat labile toxin I and heat stable toxin II genes of enterotoxigenic *Escherichia coli* in skim milk and porcine stool. J. Food Prot. 61: 141–145.

Tsuji, H., Oshibe, T., Hamada, K., Kawanish, S., Nakayama, A. and Nakajima, H. 2002. An outbreak of enterohemorrhagic *Escherichia coli* O157 caused by ingestion of contaminated beef at grilled meat-restaurant chain stores in the Kinki District in Japan: Epidemiological analysis by pulsed-field gel electrophoresis. Jpn. J. Infect. Dis. 55: 91–92.

Tsuji, T., Kamiya, Y.H., Kawamoto, Y. and Asano, Y. 1997. Relationship between a low toxicity of the mutant A subunit of enterotoxigenic *Escherichia coli* enterotoxin and its strong adjuvant action. Immunology. 90: 176–182.

Tuttle, J., Gomez, T., Doyle, M.P., Wells, J.G., Zhao, T., Tauxe, R.V. and Griffin, P.M. 1999. Lessons from a large outbreak of *Escherichia coli* O157:H7 infections: insights into the infectious dose and method of widespread contamination of hamburger patties. Epidemiol. Infect. 122: 185–192.

Tzipori, S., Gunzer, F., Donnenberg, M. S., DeMontigny, L., Kaper, J.B. and Donohue-Rolfe, A. 1995. The role of the *eaeA* gene in diarrhea and neurological complications in a gnotobiotic piglet model of enterohemorrhagic *Escherichia coli* infection. Infect. Immun. 63: 3621–3627.

Tzipori, S., Montanaro, J., Robins-Browne, R.M., Vial, P., Gibson, R. and Levine, M.M. 1992. Studies with enteroaggregative *Escherichia coli* in the gnotobiotic piglet gastroenteritis model. Infect. Immun. 60: 5302–5306.

Vallance, B.A., Chan, C., Robertson, M.L. and Finlay, B.B. 2002. Enteropathogenic and enterohemorrhagic *Escherichia coli* infections: Emerging themes in pathogenesis and prevention. Can. J. Gastroenterol. 16: 771–778.

Vallance, B.A. and Finlay, B.B. 2000. Exploitation of host cells by enteropathogenic *Escherichia coli*. Proc. Natl. Acad. Sci. USA 97: 8799–8806.

Varma, J.K., Greene, K.D., Reller, M.E., DeLong, S.M., Trottier, J., Nowicki S.F., Diorio, M., Koch, E.M., Bannerman, T.L., York, S.T., Lambert-Fair, M.A., Wells, J.G. and Mead, P.S. 2003. An outbreak of *Escherichia coli* O157 infection following exposure to a contaminated building. J. Am. Med. Assoc. 290: 2709–2712.

Viboud, G.I., Binsztein, N. and Svennerholm, A-M. 1993. Characterization of monoclonal antibodies against putative colonization factors of enterotoxigenic *Escherichia coli* and their use in an epidemiological study. J. Clin. Microbiol. 31: 558–564.

Walmsley, A.M. and Arntzen, C.J. 2000. Plants for delivery of edible vaccines. Curr. Opin. Biotechnol. 11: 126–129.

Wang, G., Clark, C.G. and Rodgers, F.G. 2002. Detection in *Escherichia coli* of the genes encoding the major virulence factors, the genes defining the O157:H7 serotype, and the components of the type 2 Shiga toxin family by multiplex PCR. J. Clin. Microbiol. 40: 3613–3619.

Wang, L., Briggs, C.E., Rothemund, D., Fratamico, P., Luchansky, J.B. and Reeves, P.R. 2001a. Sequence of the *E. coli* O104 antigen gene cluster and identification of O104 specific genes. Gene. 270: 231–236.

Wang, L., Huskic, S., Cisterne, A., Rothemund, D. and Reeves, P.R. 2002. The O-antigen gene cluster of *Escherichia coli* O55:H7 and identification of a new UDP-GlcNAc C4 epimerase gene. J. Bacteriol. 184: 2620–2625.

Wang, L., Qu, W. and Reeves, P.R. 2001b. Sequence analysis of four *Shigella boydii* O-antigen loci: implication for *Escherichia coli* and *Shigella* relationships. Infect. Immun. 69: 6923–6930.

Warburton, D.W., Austin, J.W., Harrison, B.H. and Sanders, G. 1998. Survival and recovery of *Escherichia coli* O157:H7 in inoculated bottled water. J. Food Prot. 61: 948–952.

Welinder-Olsson, C., Stenqvist, K., Badenfors, M., Brandberg, Å., Florén, K., Holmberg, M., Kjellin, E., Mårild, S., Studahl, A. and Kaijser, B. 2003. EHEC outbreak among staff at a children's hospital – use of PCR for verocytotoxin detection and PFGE for epidemiological investigation. Epidemiol. Infect. 132: 43–49.

Werber, D., Fruth, A., Liesegang, A., Littmann, M., Buchholz, U., Prager, R., Karch, H., Breuer, T., Tschäpe, H. and Ammon, A. 2002. A multistate outbreak of Shiga toxin-producing *Escherichia coli* O26:H11 infections in Germany, detected by molecular subtyping surveillance. J. Infect. Dis. 186: 419–422.

Williams, R.C., Isaacs, S., Decou, M.L., Richardson, E.A., Buffett, M.C., Slinger, R.W., Brodsky, M.H., Ciebin, B.W., Ellis, A., Hockin, J. and the *E. coli* O157:H7 Working Group. 2000. Illness outbreak associated with *Escherichia coli* O157:H7 in Genoa salami. Can. Med. Assoc. J. 162: 1409–1413.

Wilkerson, C. and van Kirk, N. 2004. Antibiotic resistance and distribution of tetracycline resistance genes in *Escherichia coli* O157:H7 isolates from humans and bovines. Antimicrob. Agents Chemother. 48: 1066–1067.

Willshaw, G.A., Cheasty, T. and Smith, H.R. 2000. *Escherichia coli*. In: The Microbiological Safety and Quality of Food, Vol II. B.M. Lund, T.C. Baird-Parker and G.W. Gould, eds. Aspen Publishers, Inc., Gaithersburg, Maryland. p. 1136–1177.

Woods, J.B., Schmitt, C.K., Darnell, S.C., Meysick, K.C. and O'Brien, A.D. 2002. Ferrets as a model system for renal disease secondary to intestinal infection with *Escherichia coli* O157:H7 and other Shiga toxin-producing *E. coli*. J. Infect. Dis. 185: 550–554.

World Health Organization. 1998. Zoonotic non-O157 Shiga toxin-producing *Escherichia coli* (STEC). Report of a WHO Scientific Working Group Meeting, 23 to 26 June 1998, Berlin, Germany [Online.] http://www.who.int/emc-documents/zoonoses/docs/whocsraph988.html/shigaindex.html

Yamagami, S., Motoki, M., Kimura, T., Isumi, H., Takeda, T., Katsuura, Y. and Matsumoto, Y. 2001. Efficacy of postinfection treatment with anti-Shiga toxin (Stx) 2 humanized monoclonal antibody TMA-15 in mice lethally challenged with Stx-producing *Escherichia coli*. J. Infect. Dis. 184: 738–742.

Yamamoto, J., Ishikawa, A., Miyamoto, M., Nomura, T., Uchimura, M. and Koiwai, K. 2001. Outbreak of enterohemorrhagic *Escherichia coli* O157 mass infection caused by "whole roasted cow". Jpn. J. Infect. Dis. 54: 88–89.

Yamamoto, T. and Echeverria, P. 1996. Detection of the enteroaggregative *Escherichia coli* heat-stable enterotoxin 1 gene sequence in enterotoxigenic *E. coli* strains pathogenic to humans. Infect. Immun. 64: 1441–1445.

Yamamoto, T. and Nakazawa, M. 1997. Detection and sequences of enteroaggregative *Escherichia coli* heat-stable enterotoxin 1 gene in enterotoxigenic *E. coli* strains isolated from piglets and calves with diarrhea. J. Clin. Microbiol. 35: 223–227.

Yamanaka, H., Kameyama, M., Baba, T., Fujii, Y. and Okamoto, K. 1994. Maturation pathway of *Escherichia coli* heat-stable enterotoxin I: requirements of DsbA for disulfide bond formation. J. Bacteriol. 176: 2906–2913.

Yamanaka, H., Nomura, T., Fujii, Y. and Okamoto, K. 1997. Extracellular secretion of *Escherichia coli* heat-stable enterotoxin I across the outer membrane. J. Bacteriol. 179: 3383–3390.

Yamanaka, H., Nomura, T., Fujii, Y. and Okamoto, K. 1998. Need for TolC, an *Escherichia coli* outer membrane protein, in the secretion of heat-stable enterotoxin I across the outer membrane. Microbial Pathogen. 25: 111–120.

Yaron, S., Kolling G.I., Simon, L. and Matthews, K.R. 2000. Vesicle-mediated transfer of virulence genes from *Escherichia coli* O157:H7 to other enteric bacteria. Appl. Environ. Microbiol. 66: 4414–4420.

Yarze, J.C. and Chase, M.P. 2000. *E. coli* O157:H7 – another waterborne outbreak! Am. J. Gastroenterol. 95: 1096.

Zhang, W., Bielaszewska, M., Kuczius, T. and Karch, H. 2002. Identification, characterization, and distribution of a Shiga toxin gene variant ($stx_{1c}$) in *Escherichia coli* strains isolated from humans. J. Clin. Microbiol. 40: 1441–1446.

Zhao, T. and Doyle, M.P. 1994. Fate of enterohemorrhagic *Escherichia coli* O157:H7 in commercial mayonnaise. J. Food Prot. 57: 780–783.

Zhao, T., Doyle, M.P. and Besser, R. 1993. Fate of enterohemorrhagic *Escherichia coli* O157:H7 in apple cider with and without preservatives. Appl. Environ. Microbiol. 59: 2526–2530.

# Clostridium botulinum and Clostridium perfringens

19

John S. Novak, Michael W. Peck, Vijay K. Juneja, and Eric A. Johnson

## Abstract

*Clostridium botulinum* produces extremely potent neurotoxins that result in the severe neuroparalytic disease, botulism. Although of lower lethality, the enterotoxin produced by *C. perfringens*, during sporulation of vegetative cells in the host intestine, still results in debilitating acute diarrhea and abdominal pain. Sales of refrigerated, processed foods of extended durability including *sous-vide* foods, chilled ready-to-eat meals, and cook-chill foods have increased over recent years. As a result of conditions accommodating growth, anaerobic spore-formers have been identified as the primary microbiological concern in these foods. There is also heightened awareness over possible intentional food source tampering with botulinum neurotoxin and the potential for genes encoding the toxins to be transferred to nontoxigenic clostridia. Similarly, enterotoxin produced by *C. perfringens* and the genomic location of the *cpe* gene has epidemiologic significance for understanding the capability to cause foodborne disease in humans. This chapter focuses on the unique characteristics and virulence factors of *C. botulinum* and *C. perfringens* that make them foodborne hazards in the food supply. The susceptibility of these bacterial spore-formers to physical and chemical agents is examined as well as recommended control measures. This information is useful in developing molecular strategies to study virulence genes and their regulation as a means to safer foods.

Mention of trade names or commercial products in this publication is solely for the purpose of providing specific information and does not imply recommendation or endorsement by the US Department of Agriculture.

## Introduction: *Clostridium botulinum*

Foodborne botulism is a rare but often severe neuroparalytic disease caused by the extremely potent botulinum neurotoxins (BoNTs) produced by *C. botulinum* and certain other species of neurotoxigenic clostridia (Dolman, 1964; Hatheway, 1993, 1995; Johnson and Goodnough, 1998; Johnson, 1999a). It is an intoxication resulting from consumption of pre-formed botulinum neurotoxin, with as little as 30 ng of neurotoxin sufficient to cause illness and even death. The consumption of as little as 0.1g of food in which *C. botulinum* or other neurotoxin-producing clostridia have grown can result in botulism (Lund and Peck, 2000). Currently the fatality rate is approximately 10% of cases; this is very high for a foodborne illness. Additionally, full recovery may take several weeks, several months or even longer.

Six physiologically and phylogenetically distinct Gram-positive spore-forming obligately anaerobic bacteria produce the botulinum neurotoxin (Table 19.1). The species *C. botulinum* is subdivided into four groups, while some strains of *C. baratii* and *C. butyricum* also form neurotoxin. The distinction between *C. botulinum* Groups I to IV is strong enough to justify the creation of four distinct species; however, the practice has been to retain the name of *C. botulinum* to emphasize the importance of neurotoxin production (Lund and Peck, 2000). For each of the six organisms, a non-neurotoxigenic phylogenetically equivalent organism is known (Hatheway, 1992). The neurotoxin genes are located on the chromosome, a bacteriophage, or a plasmid, and this is organism dependent (Table 19.1). The different physiology of the six organisms is reflected in the circumstances in which they present a hazard. For example, proteolytic *C. botulinum* (*C. botulinum* Group I) and non-proteolytic *C. botulinum* (*C. botulinum* Group II) are responsible for most cases of foodborne botulism (Lund and Peck, 2000). Proteolytic *C. botulinum* is a mesophile that produces heat resistant spores. It derives energy by degradation of proteins. Non-proteolytic *C. botulinum* is a psychrotroph and is saccharolytic. Spores of non-proteolytic *C. botulinum* are of moderate heat resistance. A recent finding has been the association of neurotoxigenic *C. baratii* and *C. butyricum* with foodborne botulism. Outbreaks of foodborne botulism involving neurotoxigenic *C. butyricum* type E have been reported in China, India and Italy (Anniballi *et al.*, 2002). In the USA in 2001, a case of foodborne botulism involving neurotoxigenic *C. baratii* type F was associated with consumption of spaghetti noodles and meat sauce by a 41-year-old woman. The woman eventually recovered, although she spent 12 weeks on a life support machine (Harvey *et al.*, 2002). Very rarely, cases of foodborne botulism have been reported involving *C. botulinum* Group III (Lund and Peck, 2000).

**Table 19.1** Characteristics of the six clostridia that produce the botulinum neurotoxin

| Neurotoxigenic organism | Neurotoxins formed | Location of neurotoxin gene | Non-neurotoxigenic equivalent organism |
|---|---|---|---|
| *C. botulinum* group I (proteolytic) | A, B, F | Chromosome | *C. sporogenes* |
| *C. botulinum* group II (non-proteolytic) | B, E, F | Chromosome | No name given |
| *C. botulinum* group III | C, D | Bacteriophage | *C. novyi* |
| *C. botulinum* group IV (*C. argentinense*) | G | Plasmid | *C. subterminale* |
| *C. baratii* | F | Chromosome | All typical strains |
| *C. butyricum* | E | Chromosome | All typical strains |

Seven botulinum neurotoxins (A to G) are produced, with the toxin type dependent on the producing organism (Table 19.1). The seven botulinum neurotoxins were originally distinguished on the basis of their antigenic response. More recently, the amino acid sequences and their mode of action have been established. All botulinum neurotoxins comprise a heavy chain and a light chain and are often associated with other proteins (e.g. haemagglutinin and non-toxin non-haemagglutinin). The light chains possess zinc endopeptidase activity, and specifically cleave proteins involved in neurotransmitter release, leading to flaccid paralysis of the muscle. If not treated, flaccid paralysis of the respiratory muscles can result in death.

## Epidemiology of foodborne botulism

Besides classical foodborne botulism, wound botulism was first reported in the United States in 1943, and intestinal botulism caused by infection and colonization of the gastrointestinal tracts of susceptible infants and adults was recognized in the 1970s and 1980s (Picket and Berg, 1976; Midura and Arnon, 1976; Lund and Peck, 2000). Botulism has also been noted to occur in humans that have an underlying illness or that have recently undergone intestinal surgery (Chia and Clark, 1986; Freedman and Armstrong, 1986; McCroskey and Hatheway, 1988; McCroskey and Hatheway, 1991; Griffin and Hatheway, 1997). Inhalational botulism has been reported to occur in laboratory workers, in nonhuman primates, and in rodents (van Holzer, 1962; Park and Simpson, 2003). Botulism is a rare disease, and consequently may be misdiagnosed. Infant botulism is the most prevalent form of botulism in the United States, but foodborne botulism is the most common form of the disease in most other geographic regions of the world (Dodds, 1993b; Hauschild, 1989, 1993). *C. botulinum* is not as adept as *C. tetani* in colonizing wounds, and wound botulism is rare compared to tetanus. However, wound botulism has been reported with some regularity (Merson and Dowell, 1973; Cherington and Ginsburg, 1975), and since the 1990s the numbers of cases of wound botulism have increased markedly, primarily in intravenous drug users in the USA and in certain European countries (MacDonald and Rutherford, 1985; CDC, 1995b; Passaro and Werner, 2000).

Although there is little historical record prior to the nineteenth century, anecdotal evidence indicates that botulism occurred in ancient cultures, and certain dietary laws and food processing methods probably evolved as a result of the disease (Hutt, 1984; Glass and Johnson, 2001). The word "botulism" is derived from the Latin word "*botulus*" meaning sausage, and the name "botulism" was given to a disease reported in central Europe in the 18th and 19th century that was frequently associated with consumption of blood sausage. It was recognized that botulism caused muscle paralysis, breathing difficulties, and had a fatality rate in excess of 50% (Johnson and Goodnough, 1998). In the early part of the nineteenth century, Justinus Kerner conducted research on the "sausage poison" and published monographs on the subject. From his clinical and experimental observations, Kerner concluded that: (i) the poison (toxin) develops in the sausage under anaerobic conditions; (ii) the poison (toxin) is a biological substance; (iii) the poison (toxin) is lethal even in small doses; and (iv) the poison (toxin) acts on the motor and autonomic nervous system (Erbguth, 2004). Kerner also made suggestions for the prevention and treatment of botulism. At the end of the nineteenth century, Emile van Ermengem first isolated a causative organism from home made raw salted ham and the spleen of a man who later died of botulism (van Ermengem, 1979). This outbreak, in Belgium, affected 23 musicians (three fatally). The isolated organism was initially called "*Bacillus botulinus*", and although these strains are now lost, they were probably *C. botulinum* Group II (Table 19.1). Over the next few decades botulism outbreaks were identified across the world. Many of these outbreaks were associated with the wider use (commercially and especially at home) of canning processes to extend shelf-life of food. For example, in the 7-year period from 1918 to 1924, there were 367 cases of botulism in the USA, of which 230 were fatal (Peck, 2004).

Through the understanding and implementation of effective control measures, the incidence of botulism is today generally much lower than in the early part of the twentieth century. Today most outbreaks of foodborne botulism are associated with home-prepared foods, where known control measures were not implemented (Hauschild, 1989, 1993; CDC, 1979, 1998; Johnson and Goodnough, 1998; Glass and Johnson, 2001; Sobel *et al.*, 2004). Epidemiologic and investigational studies have indicated that the home-prepared foods causing the most incidences of botulism included home-canned vegetables and meats, garlic, mushrooms and other vegetables packed in oil or in an anaerobic wrapping such as aluminum foil or plastic wraps; improperly processed or fermented meats, and fermented fish products and marine animal parts such as seal flippers in marine regions in the Northern hemisphere (Hauschild, 1989, 1993; Shapiro *et al.*, 1998; Lund and Peck, 2000; Franciosa *et al.*, 2003; Sobel, 2004). Examples of recent outbreaks of foodborne botulism associated with home-prepared foods are shown in Table 19.2.

Only occasionally has foodborne botulism involved commercial products, but the medical and economic consequences have been high. Recent outbreaks associated with commercial products in various regions of the world have been reviewed elsewhere (Hauschild, 1989, 1993; Rhodehamel, 1992; Hatheway, 1995; CDC, 1998; Lund and Peck, 2000; Glass and Johnson, 2001; Franciosa *et al.*, 2003; Sobel *et al.*, 2004). These commercial botulism incidences can be broadly divided into various classes: (a) restaurant/food service foods that were temperature abused (foil-wrapped baked potatoes, sautéed onions, garlic-in-oil, recontaminated process cheese sauce); (b) inadequate thermal processing or other deficiencies in commercial canning (mushrooms, canned tuna fish, beef stew, hazelnut yogurt); and (c) temperature abuse of commercial products which did not contain a secondary barrier to *C. botulinum* growth (garlic-in-oil, mascarpone cheese, clam chowder, black bean dip). The largest recorded outbreak in the United Kingdom was caused by type B toxin with 27 cases and one death (O'Mahony *et al.*, 1990). The outbreak was caused by yogurt prepared with toxin-contaminated hazelnut conserve. Heat-processing of the hazelnut preserve that was subsequently added to the prepared yogurt was inadequate to destroy *C. botulinum* spores. The largest restaurant-associated botulism outbreak in the past 25 years in the United States caused 30 cases including four with severe symptoms that required hospitalization and mechanical ventilation (Angulo *et al.*, 1998). The implicated food was a potato-based dip that was prepared using potatoes that were baked in aluminum foil and subsequently temperature abused prior to preparation of the dish. Potatoes wrapped in foil, baked, and then stored at ambient temperature have been a relatively common cause of restaurant-associated botulism.

Other recent outbreaks from commercial and restaurant prepared foods include paté and a processed meat called "casher," (1,400 cases of botulism including 19 deaths), canned chili in the USA (15 cases; Kalluri *et al.*, 2003); vacuum-packaged hot-smoked whitefish consumed in Germany (two cases; Korkeala *et al.*, 1998), canned macrobiotic food in Italy (one case; Franciosa *et al.*, 1997), Argentine meat roll "matambre" (nine cases; Villar *et al.*, 1999), and locally made cheese in Iran (27 cases, one death; Pourshafie *et al.*, 1998). Practices that have been documented to contribute to botulism include the incidence of *C. botulinum* spores in the food, poor hygienic conditions during processing, lack of competition by spoilage organisms in the food; inadequate heat processing or acidification, inadequate refrigeration during distribution and storage, and temperature abuse of refrigerated foods (Hauschild, 1989; Glass and Johnson, 2001; Setlow and Johnson, 2001). Further details of recent outbreaks of foodborne botulism associated with commercial products are included in Table 19.2. Botulism also occurs in many animals that are used for food consumption, and recent outbreaks in dairy cows have been postulated to affect the botulinal safety of foods (Cobb and Hogg, 2002).

While most reported outbreaks of foodborne botulism involving non-proteolytic *C. botulinum* are associated with strains forming type E toxin (Table 19.2), strains producing type B or F toxin are also important. For example, a high proportion of outbreaks in Europe have been associated with type B toxin, and it appears that many of these outbreaks are due to strains of non-proteolytic *C. botulinum* forming type B toxin (Hauschild, 1993; Lucke, 1984). In many of the botulism outbreaks in Europe, the toxin type has been readily identified, but isolation of the organism has either not been attempted or has not been possible.

Although reducing existing causes of foodborne botulism remains an important aim, it is also vital that *C. botulinum* and other neurotoxin-producing clostridia do not become emerging pathogens. It is therefore essential that as new approaches to food processing and new technologies are introduced, the foodborne botulism hazard is appropriately identified and controlled. Currently, concern exists about ensuring the continued safe production of refrigerated processed foods of extended durability (REPFEDs). Sales of these foods (also known as *sous-vide* foods, or chilled ready meals and cook-chill foods) have increased greatly in Europe over the last two decades (Peck, 1997; Peck, 1999). These foods address consumer demand in being of high quality, containing few preservatives and requiring minimal preparation time. Microbiological safety relies on a minimal heat treatment and appropriate refrigerated storage, and foodborne botulism has been identified as the principal microbiological safety hazard in these foods (Graham *et al.*, 1996a; Peck, 1997; Carlin *et al.*, 2000; Lindström *et al.*, 2003).

## Virulence factors in *C. botulinum*

Botulism is a true toxemia caused solely by the neurotoxins with characteristic neuroparalytic symptoms (van Ermengem, 1979; Hatheway, 1995; Johnson and Goodnough, 1998). Emile van Ermengem carried out a remarkable series of experiments (van Ermengem, 1979) of the disease and properties of "*B. botulinus*" and botulinum neurotoxin type B. This work has been reviewed (Johnson and Goodnough, 1998; Devries, 1999), and the principles that van Ermengem established still form the foundation of knowledge for prevention of foodborne botulism today. Briefly, he established that: (i) foodborne botulism is an intoxication and not an infection; (ii) the toxin is produced in food by a specific organism "*B. botulinus*"; (iii) the toxin is not inactivated by digestive enzymes; (iv) the toxin is labile to heat and alkali but is stable in acidic conditions; (v) the toxin is not produced in food with sufficient salt or acid; (vi) the organism produces heat-resistant endospores; (vii) animals vary in their susceptibility to botulinal toxin; and

**Table 19.2** Examples of recent outbreaks of foodborne botulism

| Year | Location | Product | Toxin type | Cases (deaths) | Factors | Reference |
|------|----------|---------|------------|----------------|---------|-----------|
| **Outbreaks involving proteolytic *C. botulinum*** | | | | | | |
| 1977 | Michigan, USA | Restaurant home-canned peppers | B | 59 | Underprocessed | Terranova *et al.* (1978) |
| 1978 | Colorado, USA | Potato salad | A | 7 | Baked potatoes held at room temperature for up to 5 days | Seals *et al.* (1981) |
| 1978 | New Mexico, USA | Restaurant potato salad | | 34 | Leftover baked potatoes; temperature abuse | MacDonald *et al.* (1986) |
| 1982 | California, USA | Commercial pot pie | A | 1 | Heated then temperature abuse | CDC (1983) |
| 1983 | Illinois, USA | Restaurant sautéed onions | A | 28(1) | Covered with oil; temperature abuse | MacDonald *et al.* (1985), Solomon and Kautter (1986) |
| 1985 | Canada | Commercial garlic-in-oil | B | 36 | Bottled; no preservatives; temperature abuse | St Louis *et al.* (1988) |
| 1986 | Taiwan | Commercial jars of heat processed- unsalted peanuts in water | A | 9(2) | Inadequate heat treatment | Chou *et al.* (1988) |
| 1987 | Canada | Bottled mushrooms | A | 11 | Underprocessing and/or inadequate acidification | CDC (1987), McLean *et al.* (1987) |
| 1988 | Florida, USA | Coleslaw | A | 4 | MAP shredded cabbage | Solomon *et al.* (1990) |
| 1989 | New York, USA | Commercial garlic-in-oil | A | 3 | Bottled; no preservatives; temperature abuse | Morse *et al.* (1990) |
| 1989 | United Kingdom | Commercial hazelnut yoghurt | B | 27(1) | Hazelnut conserve underprocessed | O'Mahoney *et al.* (1990) |
| 1989 | Italy | Black olives | B | 5 | Faulty preparation, lack of refrigeration | Fenicia *et al.* (1992) |
| 1993 | Georgia, USA | Restaurant commercial process cheese sauce | A | 8(1) | Recontamination; temperature abuse | Townes and Cieslak (1996) |
| 1993 | Italy | Commercial canned roasted eggplant in oil | B | 7 | Insufficient heat treatment; improper acidification 4.6 < pH < 5.1 | CDC (1995a) |
| 1994 | Texas, USA | Restaurant; potato dip ("skordalia") and aubergine dip ("meligianoslata") | A | 30 | Potatoes held at room temperature | Angulo *et al.* (1998) |
| 1994 | California, USA | Commercial clam chowder | A | 2 | No secondary barrier; temperature abuse | California Morbidity (1995) |
| 1994 | California, USA | Commercial black bean dip | A | 1 | No secondary barrier; temperature abuse | California Morbidity (1995) |
| 1996 | Italy | Commercial mascarpone cheese | A | 8(1) | No competitive microflora; pH 6–6.25; temperature abuse | Anon (1998), Franciosa and Pourshaban 1999) |
| 1997 | Italy | Homemade pesto/oil | B | 3 | pH 5.8, $a_w$ 0.97 | Chiorboli *et al.* (1997 |
| 1997 | Iran | Traditionally made cheese preserved in oil | A | 27(1) | Unsafe process | Pourshafie *et al.* (1998) |
| 1998 | Argentina | Meat roll (matambre) | A | 9 | Cooked and heat-shrink plastic wrap; temperature abuse | Villar *et al.* (1999) |
| 1998 | UK | Home-prepared bottled mushrooms in oil (imported from Italy) | B | 2(1) | Unsafe process | CDSC (1998); Roberts *et al.* (1998) |
| 2001 | Texas, USA | Commercially produced chili sauce | A | 16 | Temperature abuse at salvage store | Kalluri *et al.* (2003) |
| 2002 | South Africa | Commercially produced tinned pilchards | A | 2(2) | Corrosion of tin, permitted secondary contamination | Frean *et al.* (2004) |
| **Outbreaks involving non-proteolytic *C. botulinum*** | | | | | | |
| 1987 | New York, USA and Israel | Commercially produced, uneviscerated salted, air-dried fish ("kapchunka") | E | 8(1) | Lack of refrigeration | Slater *et al.* (1989) |

**Table 19.2** Continued

| Year | Location | Product | Toxin type | Cases (deaths) | Factors | Reference |
|------|----------|---------|-----------|---------------|---------|-----------|
| 1991 | Egypt | Commercially produced uneviscerated salted fish ("faseikh") | E | >91(18) | Putrefaction of fish before salting | Weber and Hibbs (1993) |
| 1992 | Alaska, USA | Commercially produced uneviscerated salted fish ("moloha") | E | 8 | Insufficient salt | CDC (1992) |
| 1995 | Canada | "Fermented" seal or walrus (four outbreaks) | E | 9 | Not reported | Proulx *et al.* (1997) |
| 1997 | Germany | Commercial hot-smoked vacuum-packed fish ("Raucherfisch") | E | 2 | Suspected temperature abuse | Jahkola and Korkeala (1997); Korkeala *et al.* (1998) |
| 1997 | Argentina | Home prepared cured ham | E | 6 | ? | Rosetti *et al.* (1999 |
| 1997 | Germany | Home-smoked vacuum-packed fish ("Lachsforellen") | E | 4 | Temperature abuse | Anon (1998) |
| 1998 | France | Frozen vacuum-packed scallops | E | 1 | Temperature abuse (?) | Boyer *et al.* (2001) |
| 1999 | Finland | Whitefish eggs | E | 1 | Temperature abuse | Lindström *et al.* (2004) |
| 1999 | France | Grey mullet | E | 1 | Temperature abuse (?) | Boyer *et al.* (2001) |
| 2001 | Australia | Reheated chicken | E | 1 | Poor temperature control | Mackle *et al.* (2001) |
| 2001 | Alaska, USA | Home prepared fermented beaver tail and paw | E | 3 | Temperature abuse | CDC (2001) |
| 2001 | Canada | Home-prepared fermented salmon roe (two outbreaks) | E | 4 | Unsafe process | Anon (2002) |
| 2002 | Alaska, USA | Home-prepared muktuk (from Beluga whale) | E | 8 | Unsafe process | CDC (2003) |
| 2003 | Germany | Home prepared salted air-dried fish | E | 3 | Temperature abuse (?) | Eriksen *et al.* (2004) |

(viii) animals can develop immunity to the botulinal toxins by exposure to inactive toxin or toxoid.

The dose of BoNT to cause botulism in humans is extremely low, and probably depends on many factors such as the serotype of toxin, the type of toxin complex, the route of exposure, protection by food during oral ingestion, and by many other factors (MacDonald *et al.*, 1986; Rhodehamel *et al.*, 1992; Johnson and Goodnough, 1998). An important consideration is the potential for botulism in infants and toddlers, as their relatively low body weight and size of the airway and diaphragm would make them more susceptible to botulism poisoning.

Classical studies demonstrated that BoNTs inhibited neurotransmitter (acetylcholine) release from motor neurons (Burgen *et al.*, 1949; Kao *et al.*, 1976). Irrespective of the source of BoNT (food, wound, infant, adult intestinal infection, inhalation), the neuroparalytic symptoms are similar. The hallmark clinical symptomatology of botulism is a bilateral and descending weakening and paralysis of skeletal muscles (Dickson, 1918; Cherington, 1998). The incubation time for onset of symptoms varies with the route of BoNT exposure, the serotype of BoNT, and the quantity of BoNT that enters the circulation. Wound botulism usually has a relatively long incubation period of 4 to 14 days reflecting the need for *C. botulinum* to grow in a wound and produce toxin (Birmingham *et* *al.*, 1994). Similarly, infant botulism has an incubation time of several days since the neurotoxigenic clostridia must colonize the intestinal tract, proliferate and produce toxin, which then is absorbed through the intestinal wall (Arnon, 1995, 2004). Classical foodborne botulism occurs through consumption of a food contaminated with preformed BoNT, and the majority of cases are caused by types A, B, and E (Table 19.2), and rarely by type F (Gangarosa and Danadio, 1971; Hauschild, 1989, 1993; Johnson and Goodnough, 1998; Lund and Peck, 2000).

The most severe and long-lasting foodborne botulism occurs with type A toxin. The incubation time of foodborne botulism varies with the BoNT serotype and quantity of toxin ingested, but the onset time is usually 12 to 36 hours following consumption of the toxic food. The incubation period can be as short as 2 hours when high quantities of toxin are ingested, or as long as several days to weeks, particularly with type B or E toxins and ingestion of low quantities of toxin (Gangarosa and Donadio, 1971; Cherington, 1998; Johnson and Goodnough, 1998; Shapiro *et al.*, 1998).

The site of action of neurotoxin is the presynapitc terminals of motoneurons, where after internalization into the nerve cytosol, the neurotoxin causes a blockade of the release of acetylcholine and regional flaccid paralysis of muscles innervated by the intoxicated nerves (reviewed in Montecucco and Schiavo, 1995; Humeau and Daussau, 2000; Schiavo and Montecucco, 1995; Turton *et al.*, 2002; Lalli *et al.*, 2003; Simpson, 2004). Generally, cranial nerves innervating the eyes are initially affected, and the first symptoms are blurred and double vision, dilated pupils, drooping eyelids, and a slow response to a light source (Cherington, 1998; CDC, 1998). The effects on the eyes are followed by a descending and progressive paralysis characterized by difficulty in swallowing, weakness of the neck, dry mouth and problems in speaking. The paralysis may further progress causing weakness of the upper and lower limbs. In severe cases respiratory muscle weakness occurs which requires mechanical ventilation in order to prevent death by suffocation (Cherington, 1998; Arnon *et al.*, 2001).

The major treatment of botulism is supportive care with careful attention being given to respiratory status. The fatality rate from foodborne botulism is in the range of 10% of cases, provided that adequate supportive care is provided (Cherington, 1998; CDC, 1998). Recovery from botulism is prolonged requiring weeks to months, but it is often complete and individuals regain normal muscle function (Arnon, 1995, 2004). Symptoms of fatigue, dry mouth, and blurred vision can persist for months, and recovery of autonomic abilities may take longer than neuromuscular functions. Due to the long recovery time and need for prolonged supportive care in an ICU, a large outbreak of botulism could compromise hospital capacity (Arnon *et al.*, 2001).

There is currently no countermeasure for treatment of botulism other than passive administration of equine antibodies, but this treatment is mostly effective during the period that the toxin is in the blood and before it binds to nerves, generally ≤18 hours after ingestion (CDC, 1998). For treatment of adults, currently the only antitoxin available is of equine origin, and hypersensitivity reactions occur in approximately 10 percent of adults on exposure. Antitoxin of human origin has been developed for treatment of infants with botulism and would be useful for treatment of adults (Arnon, 2004). Botulism Immune Globulin (BabyBIG®) has been demonstrated to reduce the severity of botulism in infants and reduce the hospital stay (Frankovich and Arnon, 1991; Arnon, 2004). There is an important need for countermeasures that could reverse intoxication once the toxin is internalized but this is a difficult target for a drug treatment. Due to the severity and long duration of the disease, botulism can have a profound emotional impact on the victim and friends and family (Cohen and Anderson, 1986).

While foodborne botulism is caused solely by the ingestion of preformed neurotoxins in foods, in intestinal or wound botulism, the organism must colonize the intestinal tract or a wound and it is likely that other virulence factors are involved in the disease process. Currently the genome of *C. botulinum* strain Hall A (ATCC 3502) is being completed at the Sanger Centre in the UK (Bennik *et al.*, 2002; 2003a,b). It is anticipated that this project should reveal putative virulence factors such as those found in the genome analyses of the pathogens *Clostridium tetani* and *Clostridium perfringens* (Shimizu *et al.*,

2002a; Brüggeman *et al.*, 2003; Brüggeman and Gottschalk, 2004). Genome analysis of *C. tetani* and *C. perfringens*, which are common species that colonize wounds or the intestinal tract, has indicated that they have genes for many putative virulence factors such as hemolysins, phospholipases, hyaluronidases, collagenases, proteinases, adhesins, fibronectin-binding protein genes, and several protein toxins. Identification of analogous putative virulence genes will facilitate an understanding of the traits involved in *C. botulinum* colonization of wounds and the intestine.

Several excellent reviews have recently been published on the biochemistry, genetics, structure, and cell biology of the botulinum neurotoxins and the reader is referred to these for fundamental information (Sugiyama, 1980; Habermann and Dreyer, 1986; DasGupta, 1989; Niemann, 1991; Montecucco and Schiavo, 1995; Popoff and Marvaud, 1999; Humeau and Doussau, 2000, Schiavo, 2000; Quinn and Minton, 2001; Johnson and Bradshaw, 2001; Turton *et al.*, 2002; Lalli *et al.*, 2003; Simpson, 2004). From a food safety perspective, particularly important aspects of the BoNTs include their production in foods, formation of stable complexes that facilitate oral passage in the intestinal tract, and stability characteristics of the toxins.

The BoNT-producing clostridia synthesize seven serotypes of BoNTs, designated A–G (Sugiyama, 1980; Sakaguchi, 1983). The toxins occur in complexes with nontoxic proteins (Sugiyama, 1980; Sakaguchi, 1983; Johnson and Bradshaw, 2001; Dineen *et al.*, 2003; Raffestin *et al.*, 2004). These nontoxic proteins in the toxin complexes impart stability to the labile BoNTs and facilitate passage through the gastric tract and oral toxicity (Ohishi *et al.*, 1977; Sakaguchi *et al.*, 1981; Sakaguchi, 1983). The complex forms of the toxins also increase their stability in foods and during manipulations such as purification and assays (Schantz and Johnson, 1992).

Considerable knowledge on the genetics and physiology of *C. botulinum* has been achieved during the past decade, and certain of these properties impact virulence and food safety. In particular, the genes encoding BoNT and nontoxic proteins of the toxin complexes are associated with mobile genetic elements and can be laterally transferred to clostridia previously considered nontoxigenic. Six evolutionary lineages of BoNT-producing clostridia have been proposed (Collins and Lawson, 1994; Collins and East, 1998). These results suggest that lineages have evolved independently and have gained the ability to produce BoNTs, emphasizing the ability of the toxin genes to be laterally transferred by phage or other mobile elements. The discovery of strains of *C. butyricum* and *C. baratii* that produce types E and F toxins, respectively (Hall and McCroskey, 1985; Aureli *et al.*, 1986; McCroskey and Hatheway, 1986; McCroskey and Hatheway, 1991; Meng and Yamakawa, 1997; Chaudry and Dhawan, 1998; Fenicia and Franciosa, 1999) emphasizes the ability for the genes encoding toxins to be transferred to clostridia previously thought to be nontoxigenic. The isolation of neurotoxigenic *C. butyricum* and *C. baratii* strains has seen an increased geographic spread during the past 10 years, suggesting that transfer of type E and F toxin genes may be a relatively common occurrence. The isolation and characterization of neurotoxigenic *C. butyricum* and *C. baratii* can be more tedious than for most serotypes of *C. botulinum*, and thus their detection may have been missed in certain cases of botulism. The genetic mechanism of toxin gene transfer is not known, but may involve transmissible plasmids or phages (Hauser and Gilbert, 1995). The transfer of genes for BoNT complexes is also supported by the diversity of toxin gene complexes within the various serotypes of *C. botulinum* (Johnson and Bradshaw, 2001; Dineen *et al.*, 2003), by the isolation of strains that produce more than one serotype of toxin (Gimenez and Gimenez, 1993; Barash and Arnon, 2004), and by the relatively common occurrence of "silent" toxin gene complexes in which the BoNT gene remnant is present but is not expressed due to mutations (Franciosa *et al.*, 1994; Hutson and Zhou, 1996). The lateral transfer and recombination of toxin cluster genes is an important pathogenic property of neurotoxigenic clostridia, but the mechanisms of these events have not been elucidated.

## Incidence of *C. botulinum*

The natural habitat of *C. botulinum* spores is soil and sediments (Popoff, 1995; Hatheway and Johnson, 1998). Foods cultivated in soils in geographic regions of high spore incidence including many vegetables, spices and related commodities may have high loadings. Other types of foods not directly harvested from soil may contain lower numbers of *C. botulinum* spores, including many meats such as raw poultry, beef, pork, and dairy products contain very low levels (Hauschild, 1989, 1993; Dodds, 1993a,b; ICMSF, 1998; Glass and Johnson, 2001). The prevalence of spores and their level in the environment and foods is related to the incidence of botulism.

Type A botulism is the predominant toxin type in the Western continental United States, China and Argentina and is associated mainly with vegetables. Outbreaks in Europe are generally associated with strains of *C. botulinum* forming type B toxin, and meat, yogurt, vegetables, and other foods have been involved as vehicles (Hauschild, 1993). Strains of *C. botulinum* forming type B toxin are also prevalent in Eastern USA soils (Hauschild, 1989, 1993; Dodds, 1993a,b). Strains forming type E toxin are associated with marine and freshwater foods, and are the predominant cause of botulism in cooler aquatic regions, including coastal regions of Canada, Alaska, and northern Japan (Hauschild, 1989; 1993; Dodds, 1993a,b).

Twenty-four percent of the soils tested in the United States were found to harbor botulinal spores (Dodds, 1993a,b). The incidence of spores was significantly greater in sediments and soils in and adjacent to Lake Michigan, and along the North American Pacific coast. Surveys revealed that 30–95% of these samples were positive, predominantly for strains forming type E toxin (Dodds, 1993a,b). The relative incidence of botulinal spores in soils and sediments are similar in analogous geographic and climatic regions of Canada, Central and South America, Europe, and Asia. Surveys in Australia and New Zealand reported a low incidence of spores for inland soil samples although sediments in southern Australia were heavily contaminated with spores of type strains where outbreaks of avian botulism have been reported.

The incidence of spores in some foods, however, may not correlate with the levels in the environment. Some foods such as honey often consist of blends from a variety of geographic regions, and it is difficult to define a relation between food origin and spore incidence. Spores can be inadvertently added to foods, particularly from dried vegetables or ingredients such as spices. Processing conditions and plant hygiene may also affect the contamination of foods by *C. botulinum* spores. Processing techniques must be carefully evaluated to verify that they will not increase the risk of botulinal growth in foods. In addition to processing considerations, the presence of competitive microflora and the formulation of foods is critical for controlling *C. botulinum* growth and toxin production in raw and minimally processed low acid foods (ICMSF, 1980; Hauschild, 1989, 1993; Peck *et al.*, 1995; Peck, 1997; Lund and Peck, 2000; Glass and Johnson, 2001).

## Detection and identity of BoNTs and isolation of neurotoxigenic clostridia

Since botulism is a true toxemia, the definitive diagnosis of botulism depends on the detection of BoNT in the patient's serum, feces, and/or in food that was consumed prior to onset of the disease (CDC, 1979; Hatheway, 1988; CDC, 1998; Johnson and Goodnough, 1998; Solomon and Johnson, 1995, 2001). Currently, the only reliable assay for BoNT is the mouse bioassay (AOAC Official Method 977.26; Schantz and Kautter, 1978; Schantz and Johnson, 1992) together with neutralization of mouse toxicity with standardized type-specific antitoxins (Bowmer, 1963) (current source is the Centers for Disease Control and Prevention, Atlanta, GA, USA). Botulinal toxin detection and identification have been described in detail in several excellent laboratory manuals and publications (Schantz and Kautter, 1978; CDC, 1979, 1998; Hatheway, 1988; Gimenez and Gimenez, 1995; Solomon and Johnson, 1995, 2001). Generally the quantal intraperitoneal assay is used (Schantz and Kautter, 1978; CDC, 1979, 1998; Hatheway, 1988; Solomon and Johnson, 1995, 2001), but a rapid tail vein injection assay has also been developed. The latter requires considerable skill to obtain consistent titers, and is limited to relatively high titers of toxin ($\geq 10^4$ mouse $LD_{50}$ per mL). Botulinal neurotoxin is considered to be the most poisonous substance known (Lamanna *et al.*, 1969; Sugiyama, 1980; Schantz and Johnson, 1992), and adequate measures are required to assure worker safety and toxin security during handling and manipulation of *C. botulinum* and its toxins (CDC, 1998; Malizio *et al.*, 2000).

Although the mouse bioassay is the "gold standard" for detection of BoNTs, it's drawbacks are well known including the need for trained researchers, for mice and approved animal facilities, the time required for a definitive endpoint and titer, and non-specific deaths in certain food and clinical samples. During the past 20 years, several laboratories have developed alternative assays based on the immunologic or catalytic properties of BoNTs. Among the first assays developed were various immunologic assays (Notermans and Nagel, 1989), in particular amplified enzyme-linked immunosorbent assays (Doellgast *et al.*, 1993; Goodnough and Hammer, 1993; Hatheway and Ferreira, 1996; Stringer *et al.*, 1999; Ferreira, 2001; Ferreira *et al.*, 2003; Szilagyi *et al.*, 2000), and certain platforms have been subjected to multilaboratory studies and have received approval for use by the AOAC (Ferreira *et al.*, 2003). Immunoassays have several drawbacks including reagent and equipment requirements, relatively low sensitivity, potential for detecting inactivated toxin, and requirement for several manipulations. The BoNTs and TeNTs were discovered during the past 20 years to be zinc metalloproteases that have catalytic activity on specific neuronal substrates (Montecucco and Schiavo, 1995). It has been possible to take advantage of their catalytic activity to develop enzyme assays including very sensitive fluorescence (Wictome *et al.*, 1999) and fluorescence resonance energy transfer detection (Anne *et al.*, 2001; Schmidt and Stafford, 2003). In all of the assays described above, BoNT standards should also be run for validation of the assay. Several research groups have developed biosensors for detection of BoNT *in vivo* or in clinical samples (Kalandakanond and Coffield, 2001; Liu *et al.*, 2003; Peruski and Johnson, 2002; Ligler *et al.*, 2003). Many of these biosensor platforms to date have low sensitivity or their efficacy has not been demonstrated with clinical (fecal or serum) or food samples.

Although botulism can only be confirmed by detection of BoNT, it is often desired to also isolate neurotoxigenic clostridia from food and clinical samples. Standard isolation methods for neurotoxigenic clostridia have been described (Dowell and Hawkins, 1974; Holdeman *et al.*, 1977; Willis, 1977; Dowell and Dezfulian, 1981; Cato *et al.*, 1986; Hatheway, 1988; CDC, 1979, 1998; Solomon and Johnson, 2001). Since neurotoxigenic clostridia generally form endospores in these samples, isolation can be facilitated by heating or treatment with chemicals (particularly ethanol) to reduce the populations of vegetative cells (Dowell *et al.*, 1964; Koransky *et al.*, 1978) and/or by inclusion of selective agents in the media (Dezfulian and McCroskey, 1981; Mills *et al.*, 1985). Neurotoxigenic clostridia are considered to be strict anaerobes, and exposure to an aerobic environment should be minimized. Occasionally growth of *C. botulinum* is inhibited by food or clinical components, and by bacteriocins or other toxic agents formed by competitor bacteria (Hall and Peterson, 1922; Kautter and Harmon, 1966; Lau *et al.*, 1974; Smith 1975, 1977; Graham, 1978; Dineen *et al.*, 2000). In these cases, specialized media containing trypsin can be used (Johnston *et al.*, 1964; Hatheway and Johnson, 1998), and enrichment is usually required. Anaerobic chambers are preferred for streaking of plates and single colony isolation. The cultures can be identified to species by appropriate taxo-

nomic methods, and BoNT should also be demonstrated to confirm the neurotoxic property of the isolate. PCR methods for detection of the BoNT gene in neurotoxigenic clostridia have also been developed (Fach and Hauser, 1993; Szabo and Pemberton, 1994; Ferreira and Hamdy, 1995; Franciosa and Fenicia, 1996; Lindström *et al.*, 2001; Wu and Huang, 2001), which are useful for detection of the gene in environmental, clinical, and food samples. However, PCR detection can be unreliable because of inhibitory compounds to the PCR assay, and have the potential drawback of false positive detection. Clinical samples may contain human viruses (HIV, hepatitis) or other pathogens, and appropriate worker precautions need to be followed and manipulations performed in a safe environment (CDC, 1998).

Epidemiologic analysis of botulism outbreaks can be assisted by phenotypic characterization using standard methods (Dowell and Hawkins, 1974; Holdeman *et al.*, 1977; Willis, 1977; Dowell and Dezfulian, 1981; Cato *et al.*, 1986; Hatheway, 1988; CDC, 1979, 1998; Solomon and Johnson, 2001), or by using molecular methods for genotypic analysis of the isolated neurotoxigenic clostridia from the food and from clinical samples. Several methods have been used for genotyping including restriction analysis of isolated genomic DNA, such as RFLP, RAPD, and pulsed field gel electrophoresis (PFGE) (Lin and Johnson, 1995; Hielm and Björkroth, 1998; Olive and Bean, 1999; Wang *et al.*, 2000). Of the various methods, most laboratories are using PFGE as it is most discriminating and providing consistent analyses. Determining the nucleotide sequence of regions of the gene encoding 16sRNA can also be useful in taxonomic identification and strain discrimination within the neurotoxigenic clostridia (Pourshaban *et al.*, 2002).

Biological safety and security is extremely important for conducting research with the neurotoxigenic clostridia. The laboratory should be designed according to recommended guidelines (CDC, 1998) A biosafety manual should be posted in the laboratory and should contain the proper emergency phone numbers and procedures for emergency response, spill control and decontamination. All personnel should be trained in these procedures. When performing experiments where aerosols may be created (e.g. centrifugation) special precautions need to be taken. A Class II or III biological safety cabinet or respiratory protection should be used during these procedures. The use of needles and syringes in bioassays also requires an extreme degree of caution.

Botulinum toxin has been considered as a potential bioterrorist agent that could be administered in foods, water, or aerosols (Arnon *et al.*, 2001). Botulinum toxin is absorbed through mucous membranes, and three cases of botulism were documented in three laboratory workers who apparently accidentally inhaled the toxin (van Holzer, 1962). Immunization is not feasible for protection of human populations from botulism due to the rarity of the disease, but pentavalent (A–E) toxoid is used for immunization of researchers and for certain military personnel. Despite being immunized, researchers must still follow careful laboratory practices in working with botulinum toxin, including avoiding aerosols through the use of biological safety cabinets and closed containers (CDC, 1998; Malizio *et al.*, 2000). The design and construction of safe and secure facilities and following good laboratory practices is important to prevent accidental exposure and inadvertent release of *C. botulinum* and its neurotoxins.

## Control measures for *Clostridium botulinum*

Proteolytic *C. botulinum* and non-proteolytic *C. botulinum* are responsible for most cases of foodborne botulism (Table 19.2). Since these two organisms differ physiologically (Table 19.3), they present a hazard in different types of foods. Proteolytic *C. botulinum* produces neurotoxins of type A, B, and F, or sometimes two different neurotoxins. The spores are of high heat resistance, and the organism has a minimum growth temperature of 10°-12°C. The canning process for low-acid foods is designed to inactivate spores of this organism. Many of these outbreaks have occurred when the full heat treatment has not been appropriately delivered to foods (Table 19.2). For example, the outbreak in the United Kingdom in 1989 was due to a failure to adequately heat the hazelnut conserve added to the yogurt. A strain of proteolytic *C. botulinum* type B then grew and produced neurotoxin in the hazelnut conserve. While, the restaurant associated outbreak in the USA in 1994 was attributed to inadequately baked potatoes that were subsequently added to potato and aubergine dips. In this outbreak, a strain of proteolytic *C. botulinum* type A formed toxin in the baked potato during cooling and subsequent storage at room temperature (Table 19.2).

Strains of non-proteolytic *C. botulinum* produce neurotoxins of type B, E or F. Spores of non-proteolytic *C. botulinum* are of moderate heat resistance, although measured heat resistance is considerably increased by the presence of lysozyme during recovery (Table 19.3). Non-proteolytic *C. botulinum* is able to multiply and form neurotoxin at temperatures as low as 3.0°C. Botulism outbreaks associated with non-proteolytic *C.*

botulinum have most frequently been associated with processed fish and fermented marine products (Table 19.2). Although most reported outbreaks of foodborne botulism involving non-proteolytic *C. botulinum* are associated with strains producing type E toxin, strains producing type B or F toxin are also important. For example, a high proportion of outbreaks in Europe have been due to type B toxin, and it many of these are due to strains of non-proteolytic *C. botulinum* producing type B toxin (Hauschild, 1993; Lucke, 1984). Proteolytic *C. botulinum* is able to grow and form neurotoxin at a lower pH, a lower water activity, and a higher NaCl concentration than is non-proteolytic *C. botulinum* (Table 19.3).

Preservative factors at a level that singly do not provide the required protection factor with respect to proteolytic *C. botulinum* and/or non-proteolytic *C. botulinum* can be combined in order that the required protection factor is achieved. Indeed, the safety of many foods relies on such combinations of preservative factors. Figure 19.1 illustrates how combinations of preservative factors (storage temperature, pH, NaCl concentration) restrict the growth domain of non-proteolytic *C. botulinum*. A number of predictive models have also been developed that describe the effect of different combinations of preservative factors on growth of proteolytic *C. botulinum* and non-proteolytic *C. botulinum*.

The severity of foodborne botulism requires that all steps are taken to minimize the incidence of this disease. At the present time, most cases are associated with home-prepared foods or occasionally with commercial products, where known control measures have not been implemented. However, it is essential that as new technologies and approaches to food processing are introduced, measures are in place to ensure that the foodborne botulism hazard is appropriately controlled, and that neurotoxin-producing clostridia do not become emerging pathogens. One group of foods of concern, in this respect, are REPFEDs. In recent years, sales of these foods have increased substantially in many European countries (ECFF, 1998; Chilled Food Association, 2004). These foods receive a moderate heat process (typical maximum of 75–95°C), and are

**Table 19.3** Effect of environmental factors on the growth and survival of the two clostridia most commonly responsible for foodborne botulism (from data in Lund and Peck, 2000)

|  | Proteolytic *C. botulinum* | Nonproteolytic *C. botulinum* |
|---|---|---|
| Neurotoxins formed | A, B, F | B, E, F |
| Minimum temperature for growth | 10–12°C | 3.0°C |
| Minimum pH for growth | 4.6 | 5.0 |
| Minimum water activity for growth |  |  |
| NaCl as humectant | 0.96 | 0.97 |
| Glycerol as humectant | 0.93 | 0.94 |
| NaCl concentration preventing growth | 10% | 5% |
| Spore heat resistance | $D_{121°C} = 0.21$ min | $D_{82.2°C} = 2.4/231$ min[a] |
| Spore radiation resistance | $D = 2.0–4.5$ kGy | $D = 1.0–2.0$ kGy |

[a]heat resistance data without/with lysozyme during recovery.

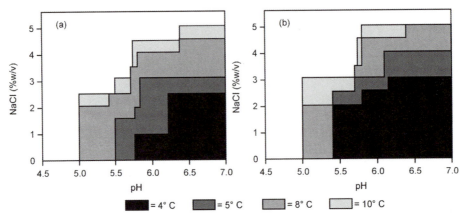

**Figure 19.1** Effect of incubation temperature on the growth domain of non-proteolytic *C. botulinum* (from data in Graham *et al.*, 1997). The shading indicates combinations of pH (adjusted with HCl) and NaCl that enabled growth and toxin formation from spores of non-proteolytic *C. botulinum* types B, E, and F in PYGS medium at 4°C, 5°C, 8°C, and 10°C within 3 weeks (a) and 6 weeks (b).

then cooled rapidly, and stored at refrigeration temperatures. Furthermore, these foods are often packed under vacuum or an anaerobic atmosphere, restricting growth of aerobic, but not anaerobic, microorganisms. This minimal process favors spore-forming microorganisms that grow in the absence of oxygen at refrigeration temperatures. In particular, concern exists about the potential for growth and neurotoxin production by non-proteolytic *C. botulinum* in the absence of a competing microflora, and the associated foodborne botulism hazard (Rhodehamel, 1992; Carlin and Peck, 1996a; Stringer and Peck, 1997; Peck, 1997; Carlin *et al.*, 2000).

In the last twenty years, developments in computing power and a widespread accessibility in the workplace to powerful computers and the internet have led to significant advances in the field of predictive microbiology. These advances have led to the development and availability of two types of tools, easily browsable databases and predictive microbiology applications that contribute to improved microbiological safety. One such database is ComBase (www.combase.cc), which is freely available on the Internet. It contains some twenty thousand original growth and survival curves and some 10,000 growth/survival parameters of foodborne pathogens and spoilage microorganisms. Among others, there are data on the combined effect of environmental factors on the growth, survival and death of both proteolytic *C. botulinum* and non-proteolytic *C. botulinum*.

Several predictive microbiology packages are also freely available on the Internet. These packages enable prediction of the likely response to combinations of important environmental factors (e.g. pH, temperature, NaCl). Two packages, Growth Predictor (www.ifr.ac.uk/Safety/GrowthPredictor/default.html) and the Pathogen Modeling Program (www.arserrc.gov/mfs/PMP6_Start.htm), contain predictive models for proteolytic *C. botulinum* and for non-proteolytic *C. botulinum*. These models, and the research on which they are based, and other predictive models have also been described in original publications. For example, for non-proteolytic *C. botulinum* the combined effect of incubation temperature, pH and NaCl

concentration on growth domain has been described (Graham *et al.*, 1997; Lund and Peck, 2000), and predictive models of the combined effect of different environmental factors on growth have been reported (Baker and Genigeorgis, 1992; Dodds, 1993a; Graham *et al.*, 1996b; Whiting and Oriente, 1997; Fernandez *et al.*, 2001).

Predictive models have also been developed for the combined effect of heating temperature and other environmental factors on the thermal death of spores of non-proteolytic *C. botulinum* (Peck *et al.*, 1993; Lund and Peck, 1994; Juneja and Eblen 1995; Juneja *et al.*, 1995a, 1995b). Additionally, other models have described the combined effect of heat treatment and subsequent incubation temperature on growth (Fernandez and Peck, 1997; Cawley *et al.*, 1998; Fernandez and Peck, 1999). The effect of other combinations of preservative factors on growth and survival of proteolytic *C. botulinum* and non-proteolytic *C. botulinum*, including other predictive models, has been summarized (Lund and Peck, 2000). A further development has been that of process risk models for REPFEDs (Barker *et al.*, 1999; Carlin *et al.*, 2000; Barker *et al.*, 2002, 2005). These process risk models use a probabilistic modeling approach. Techniques such as Markov Chain Monte Carlo methods and Bayesian Belief Networks are then used to combine probability distributions. This has enabled the risk of foodborne botulism presented by non-proteolytic *C. botulinum* in REPFEDs to be evaluated by considering the entire food chain.

There has been considerable interest in the use of protective cultures to ensure the safety of different foods (e.g. Rodgers, 2001). Growth and neurotoxin production by both proteolytic *C. botulinum* and non-proteolytic *C. botulinum* is inhibited by a range of lactic acid bacteria (Crandall and Montville, 1993; Skinner *et al.*, 1999; Rodgers *et al.*, 2003a). When tested under different conditions, however, species of *Bacillus* have been reported to promote growth of non-proteolytic *C. botulinum* (Carlin and Peck, 1996b), and also to inhibit growth of non-proteolytic *C. botulinum* (Lyver *et al.*, 1998). There has been much interest in the potential application of protective cultures to ensure the safety of REPFEDs (Gorris and Peck, 1998; Rodgers, 2003). It has been demonstrated that strains of non-proteolytic *C. botulinum* types B, E, and F were inhibited by a number of different lactic acid bacteria (Rodgers *et al.*, 2003a). For example, in a challenge test, *Pediococcus pentosaceus* was found to inhibit growth of non-proteolytic *C. botulinum* in a REPFED, seafood chowder (Rodgers *et al.*, 2003b).

## Introduction: *Clostridium perfringens*

*Clostridium perfringens* is commonly found in soil and dust, in the intestinal tract of humans and animals, in spices, on vegetable products, and in other raw and processed foods. The organism is frequently found in meats, generally through fecal contamination of carcasses, contamination from other ingredients, such as spices, or post-processing contamination. *C. perfringens* was detected in 36%, 80%, and 2% of fecal samples from cattle, poultry, and swine, respectively (Tschirdewahn *et al.*, 1991). *C. perfringens* was isolated from 43.1% of processed and unprocessed meat samples tested in one study, including beef, veal, lamb, pork, and chicken products (Hall and Angelotti, 1965). Many areas within broiler chicken processing plants are contaminated with the organism (Craven, 2001), and the incidence of *C. perfringens* on raw poultry ranges from 10–80% (Waldroup, 1996). *C. perfringens* was detected in 47.4% of raw ground beef samples (Ladiges *et al.*, 1974), and a mean level of 45.1 *C. perfringens* per $cm^2$ was detected on raw beef carcass surface samples (Sheridan *et al.*, 1996). *C. perfringens* was detected on 38.9% of commercial pork sausage samples (Bauer *et al.*, 1981), and on raw beef, equipment, and cooked beef in food service establishments (Bryan and McKinley, 1979).

Fifty-seven outbreaks and 2,772 cases of *C. perfringens* food poisoning were reported to the CDC between 1993–1997 (CDC, 2000). However, because most cases of *C. perfringens* food poisoning are mild and are not reported, the actual number of cases in the USA is estimated to be 248,520 per year (Mead *et al.*, 1999), with an estimated cost of $200 per case (Todd, 1989). The most commonly reported causes of such outbreaks are improper holding temperature of foods and inadequate cooking (CDC, 2000; Bean and Griffin, 1990; Bean *et al.*, 1997). Foodborne outbreaks of *C. perfringens* can be confirmed if $>10^5$ CFU/g of the organism are detected in the implicated food (Labbe and Juneja, 2002).

## Select environment for opportunism

Meat and poultry products were associated with the vast majority of *C. perfringens* outbreaks in the USA (Bryan, 1988; Taormina *et al.*, 2003), probably due to the fastidious requirement for more than a dozen amino acids and several vitamins for growth (Labbe and Juneja, 2002; Brynestad and Granum, 2002). Beef products were the vehicles for 33.9% of the *C. perfringens* outbreaks during this time period, although poultry products were also commonly implicated (Bryan, 1988). *C. perfringens* outbreaks linked to roast beef usually result from improper handling and mistreatment in food processing plant or food service environments (Bryan, 1988). *C. perfringens* outbreaks in general usually result from improper handling and preparation of foods at the home, retail, or food service level, and rarely involve commercial meat processors (Taormina *et al.*, 2003).

*Clostridium perfringens* food poisoning is one of the most common types of foodborne illness (Labbe, 1989), and typically occurs from the ingestion of $>10^8$ viable vegetative cells of the organism in temperature-abused foods (Labbe and Juneja, 2002). Acidic conditions encountered in the stomach may actually trigger initial stages of sporulation. Once in the small intestine, the vegetative cells sporulate, releasing an enterotoxin that causes diarrhea and abdominal pain. The typical incubation period before onset of symptoms is 8–24 h, and symptoms usually last less than 24 h. Fatalities are rare in healthy individuals.

Most outbreaks of *C. perfringens* food poisoning can be avoided by adequate cooking of meat products followed by holding at hot temperatures or rapid cooling. Although the American Public Health Association recommends that hot food be held at $\geq 60°C$, vegetative cells of *C. perfringens* can survive in many foods at 60–71°C (Tuomi *et al.*, 1974), and potentially growing during subsequent cooling.

## Growth requirements for *C. perfringens*

Although technically an anaerobe, *C. perfringens* is relatively aerotolerant. The optimal growth temperature range for the organism is 43–45°C, although growth can occur between 15-50°C (Labbe and Juneja, 2002). *C. perfringens* does not grow at proper refrigeration temperatures, and generally does not grow at temperatures below 15°C. The pH range for growth of *C. perfringens* is between pH 5.0–9.0.

*C. perfringens* is capable of extremely rapid growth in meat systems, which makes the organism a particular concern to meat processors as well as the food service industry. *C. perfringens* grew faster at 45°C in autoclaved ground beef than in broth media at the same temperature (Willardsen *et al.*, 1978; Willardsen *et al.*, 1979). One strain of *C. perfringens* had a generation time of 7.1 min in autoclaved ground beef held at 41°C, although the mean generation time for an 8-strain mixture ranged from 19.5 min at 33°C to 8.8 min at 45°C (Willardsen *et al.*, 1978). Because of its rapid growth, numbers of *C. perfringens* sufficient to cause illness can rapidly be reached under optimal storage conditions for meats and meat products.

Sodium nitrite levels used in cured meat products can inhibit growth of *C. perfringens* under certain conditions (Robach *et al.*, 1978). However, *C. perfringens* has been shown to survive and grow in the presence of curing salts (300 ppm sodium nitrite and 4–6% NaCl) at levels higher than those usually used in most curing operations (Gough and Alford, 1965). Most processed meat products contain between 2.75 and 3.25% NaCl (Maurer, 1983), and sodium nitrite is allowed in cured meats to a final level of 156 ppm. However, heating and acidic pH may increase the effectiveness of sodium nitrite against *C. perfringens* (Labbe, 1989; Riha and Solberg, 1975). Anaerobic conditions can also increase the antimicrobial effectiveness of sodium nitrite. Heating at 70–100°C increased the sensitivity of one strain of *C. perfringens* to both NaCl and sodium nitrite (Chumney and Adams, 1980).

The ability of *C. perfringens* to grow in meat systems, including many cured meats is well documented. For instance, growth from spores was rapid in frankfurters at 37°C and 23°C. Although growth was slower at lower temperatures, populations increased 2-log CFU/g after 2 days at 15°C and 3 days at 12°C (Solberg and Elkind, 1970). Huang (2002) described growth of *C. perfringens* in uncured cooked ground beef using multiple linear models. Juneja and Marmer (1996) investigated growth of *C. perfringens* in *sous-vide* turkey products containing sodium pyrophosphate and different NaCl levels. In that study, 3% NaCl delayed growth of the organism at 15°C and 28°C. Juneja et al. (1994a, 1994b) examined growth of vegetative *C. perfringens* cells in cooked ground beef and cooked turkey at various temperatures, and determined that *C. perfringens* was able to grow in both products under both anaerobic and aerobic conditions, although growth was faster in both products held under anaerobic conditions. In a related study, Juneja et al. (1996) showed that *C. perfringens* in cooked turkey grew slower when kept in a modified atmosphere containing certain combinations of $CO_2$, $O_2$, and $N_2$. Juneja and Majka (1995) detected growth of *C. perfringens* in cook-in-bag beef products with different pH and salt levels.

Slow cooking associated with low-temperature, long-time cooking can also result in growth of *C. perfringens* in foods. For instance, mean generation times in autoclaved ground beef during slow heating from 35 to 52°C ranged from 13 to 30 min with temperature increases of 6–12.5°C/h (Willardsen et al., 1978). Another study also demonstrated growth of the organism in autoclaved ground beef during linear temperature increases (4.1°C to 7.5°C/h) from 25°C to 50°C (Roy et al., 1981).

Studies have also described growth of *C. perfringens* during cooling of cooked uncured meat products. Shigehisa et al. (1985) documented *C. perfringens* spore germination, growth, and survival rates during heating and cooling of ground beef as well as in roast beef made with various muscle types. Steele and Wright (2001) evaluated growth of *C. perfringens* spores in turkey roasts cooked to an internal temperature of 72°C, followed by cooling in a walk-in cooler from 48.9°C to 12.8°C in 6, 8, or 10 h. Results of that study indicated that an 8.9-h cooling period was adequate to prevent growth of *C. perfringens* with a 95% tolerance interval. Juneja et al. (1994c) cooled inoculated cooked ground beef samples from 54.4 to 7.2°C at rates between 6 and 18 h, and concluded that pasteurized cooked beef should be cooled to 7.2C in 15 h or less to prevent growth of *C. perfringens*.

Recent studies have shown the efficacy of certain antimicrobial agents against the growth of *C. perfringens* during the cooling of meat products. For instance, Sabah et al. (2003a) found that 0.5–4.8% sodium citrate inhibited growth of *C. perfringens* in cooked vacuum-packaged restructured beef cooled from 54.4°C to 7.2°C within 18 h. The same laboratory demonstrated growth inhibition of the organism by oregano in combination with organic acids during cooling of *sous-vide* cooked ground beef products (Sabah et al., 2003b). Organic acid salts such as 1% sodium lactate, 1% sodium acetate, or 1% buffered sodium citrate (with or without sodium diacetate) inhibited germination and outgrowth of *C. perfringens* spores during the chilling of marinated ground turkey breast (Juneja and Thippareddi, 2003).

A limited amount of published research is available regarding growth of the pathogen in cooked cured meats during cooling. Taormina et al. (2003) inoculated bologna and ham batter with *C. perfringens* spores, followed by cooking and either cooling procedures typically used in industry, or extended chilling. In that study, growth of the organism was not detected in any of the products tested during chilling from 54.4 to 7.2°C. Another recent study demonstrated complete inhibition of *C. perfringens* growth in ground cooked ham formulations containing ≥ 3.1% NaCl during cooling from 54.4°C to ≤ 8.5°C in 15–21 h (Zaika, 2003).

Predictive bacterial growth models during cooling of food systems have been generated by other researchers using constant temperature data. Juneja et al. (2001) developed a predictive cooling model for cooked cured beef based on growth rates of the organism at different temperatures, which estimated that exponential cooling from 51°C to 11°C in 6, 8, or 10 h would result in an increase of 1.43, 3.17, and 11.8 log CFU/g, respectively. A similar model was later developed for cooked cured chicken (Juneja and Marks, 2002). Earlier studies in broth media led to a predictive model for the growth of *C. perfringens* at temperatures associated with the cooling of cooked meats (Juneja et al., 1999). Huang used different methods to estimate the growth kinetics of *C. perfringens* in ground beef during isothermal, square-waved, linear, exponential, and fluctuating cooling temperature profiles (Huang, 2003a; 2003b;, 2003c).

Blankenship et al. (1988) evaluated growth of *C. perfringens* in cooked chilli at five temperatures. *C. perfringens* grew in the chili held at 32°C for 6 h, or at 37–48°C for longer than 2 h. A simple model was developed to predict the growth of bacteria with time under exponential cooling conditions, and actual test data in cooling chili correlated closely with predicted results. Exponential cooling is more representative of actual cooling practices than linear cooling.

Numerous studies have examined the heat resistance of *C. perfringens* spores and/or vegetative cells. Heredia et al. (1997) found that a sublethal heat shock at 55°C for 30 min increased tolerance of both spores and vegetative cells to a subsequent heat treatment. Similar results were obtained by Juneja using both *C. perfringens* spores (Juneja et al., 2003) and vegetative cells (Juneja et al., 2001). Miwa et al. (2002) found that spores of enterotoxin-positive *C. perfringens* strains were more heat-resistant than enterotoxin-negative strains. Similarly, food poisoning isolates of the organism are generally more heat-resistant than *C. perfringens* isolates from other sources (Labbe, 1989).

Differentially expressed proteins in *C. perfringens* have recently been identified in response to heat shock in addition to the traditional molecular chaperones believed to play a role in protein folding and removal of damaged proteins (Novak *et al.*, 2001). One such protein was found to be glyceraldehyde-3-phosphate dehydrogenase, an enzyme important in the glycolytic pathway of anaerobic metabolism (Meijer, 1994). Another over-expressed product as a result of heat stress included rubrerythrin, believed to be a scavenger of oxygen radicals under stressful conditions (Lehman *et al.*, 1996). Several others require further elucidations.

## Dormant resilience

The heat resistance of bacterial endospores is well documented. *C. perfringens* spores frequently survive cooking processes used for most meat products. Heat resistance varies among strains of *C. perfringens*, although both heat-resistant and heat-sensitive strains can cause food poisoning (Labbe, 2001). Spores of *C. perfringens* strains associated with food poisoning are generally more heat-resistant than other strains of the organism (Roberts, 1968). This may be due to chromosomal, rather than plasmid *cpe* genes encoding for *C. perfringens* enterotoxin (CPE) (Sarker *et al.*, 2000). Cooking usually increases the anaerobic environment in food and reduces numbers of competing spoilage microorganisms, which is ecologically important as *C. perfringens* competes poorly with the spoilage flora of many foods. Cooking of foods can also heat-shock *C. perfringens* spores, since germination activation of *C. perfringens* spores can occur at temperatures between 60–80°C (Walker, 1975). Similar to spores of other bacterial species, spores of *C. perfringens* germinate at a higher rate after heat-shock. For instance, while only 3% of inoculated *C. perfringens* spores germinated in raw beef without prior heat shock, almost all spores germinated after the beef received a heat treatment (Barnes *et al.*, 1963).

The resilience of *C. perfringens*, documented to survive stressful conditions and temperatures as high as 100°C for more than 1 hour, is attributed to the formation of heat resistant spores (Rhodehamel and Harmon, 1998). That heat resistance is attained coincident with the formation of the spore coat layers (Labbe and Duncan, 1977). Extensive spore research over the last 30 years supported the establishment of different theories describing the enhanced survival of spore-forming bacteria despite the detrimental effects of heat (Setlow, 2000). A unifying theme appears to be the low water content of spores and associated means of reaching dormancy and resistance.

Spores are specialized structures, triggered by and very capable of surviving detrimental growth conditions. Structurally, they contain numerous outer protective layers, proteinaceous coats, and membranes. The centrally located core region of the spore contains ribosomes and nucleic acids complexed with calcium dipicolinic acid (DPA) in a dehydrated metabolically dormant state that provides protection from cooking (Murrell and Warth, 1965; Murrell, 1988). The surrounding cortex is composed of peptidoglycan that is similar, but not identical in structure to that in growing cells (Murrell and Warth, 1965; Murrell, 1988). Once conditions become favorable for growth, as occurs in temperature-abused foods, the spores germinate and the microorganism(s) proliferates. The times and temperatures necessary to completely destroy all spores would also negatively impact the food products, making them less natural and wholesome in the process.

Using differentially permeable radiolabeled solutes, the water content of spore components was analyzed and it was concluded that dehydration of the spore core was necessary for heat resistance (Beaman *et al.*, 1984). Buoyant density sedimentation in metrizamide gradients was applied to determine the water content of spore protoplasts (Lindsay *et al.*, 1985). It is generally accepted that the heat resistance in the spore is a consequence of low water activity ($a_w$) in the spore core, but it is acknowledged that other mechanisms must be in effect to protect germination enzymes present in peripheral locations outside the core (Miyata *et al.*, 1997).

Limiting the water activity of foods might impact the water available for chemical reactions within cellular structures of spores and vegetative cells. The levels of $a_w$ of foods or media could be decreased with the addition of NaCl, KCl, or glucose. A trend was established for *C. perfringens* that as the $a_w$ level was lowered, the rate and amount of growth were lessened (Strong *et al.*, 1970). The use of glucose to lower $a_w$ resulted in growth at lower $a_w$ levels (0.96) as compared to NaCl (0.975), whereas KCl was most effective comparatively in producing long lag times and the least amount of measurable *C. perfringens* growth (Strong *et al.*, 1970). The method of handling rehydrated dried foods is considered important with respect to *C. perfringens* as the pathogen has been isolated from a variety of food products of limited water content such as dried soup and sauce mixes due to the resilient spores (Strong *et al.*, 1970).

It has been shown that spores produced at higher temperatures (thermophiles) have higher heat resistance than those spores produced at moderate temperatures (mesophiles) and the resistance may be attributed to a drier spore (Gerhardt and Marquis, 1989). A direct relationship has been shown to exist between *C. perfringens* spore heat resistance and the temperature at which the spores are produced (Garcia-Alvarado *et al.*, 1992). The implication here is that spores, like vegetative cells, can become adapted to survival following sublethal high temperature exposures. A sublethal heat shock at 55°C for 30 min, applied as *C. perfringens* spores were in the process of being formed, resulted in spores with increased heat resistance (Heredia *et al.*, 1997; Heredia *et al.*, 1998). The acquired thermotolerance or adaptation was maintained transiently for 2 hours in metabolically active vegetative cells which may indicate metabolic turnover events such as the production and degradation of protein products (Heredia *et al.*, 1997). The duration to which dormant spores maintained the heat adaptation was not evaluated in these studies, and the possibility exists that spores remain heat-adapted until germination.

*C. perfringens* spores preheated to 100°C for 60 min were heat-resistant, but upon germination, the resultant vegetative cells sporulated poorly, whereas spores heated to 70°C for 10 min were far less heat resistant, but had subsequently increased sporulation potential of the vegetative cells (Nishida *et al.*, 1969). Commercial cooking frequently uses long time-low temperature cooking with temperatures increased gradually from 40° to 60°C in 4 hours (Smith *et al.*, 1980). A population of an 8 strain composite of *C. perfringens* vegetative cells inoculated into ground beef increased at 42°C, reaching maximum numbers at 55°C following 9.5 hours of increasing temperature (Smith *et al.*, 1980). Increased heating for an hour at 56°C resulted in a dramatic decrease in viable cells (Smith *et al.*, 1980). This does not imply that cooking of foods to 60°C will ensure safety from *C. perfringens* outbreaks as the cooking temperatures may kill vegetative cells, but the same temperatures can also serve to activate dormant spores which will germinate and multiply if the rate and extent of cooking is not sufficient (Adams, 1973).

DNA repair mechanisms have been shown to have a significant influence on heat resistance (Hanlin *et al.*, 1985). Heat resistance of spores is believed to be attributed in part to restricting the mobility of heat labile components of the spore core such as proteins and DNA (Gombas, 1983). Additional work has indicated that spore resistance to moist heat is affected by core water content, but the binding of small acid soluble proteins (SASPs) to DNA might be a mechanism as or more important in protecting spores from dry heat (Popham *et al.*, 1995; Setlow and Setlow 1995).

The involvement of dipicolinic acid (DPA) in heat resistance has long been thought important when correlated with the notable presence of DPA in spores and the measured acquisition of heat resistance following DPA synthesis (Balassa *et al.*, 1979). Localized predominantly in the spore core (Kozuka *et al.*, 1985), DPA complexed with $Ca^{2+}$, more recently might not be expected to play a direct role in spore heat resistance, but may stabilize the acquired resistance and be necessary for attaining full heat resistance of the spores (Paidhungat *et al.*, 2000). It has been suggested that mineralization of spores might also stabilize heat-labile biopolymers, whereas water or hydration counter balance the process (Beaman and Gerhardt, 1986). As a result, mineralization has been shown to increase heat resistance at lower killing temperatures of *Bacillus* spores, but much less at higher temperatures (Marquis and Bender, 1985).

## Regulations and recommendations

Due to its rapid growth in meat products and its ubiquity in meat products, growth of *C. perfringens* is used as a standard to assess the safety of cooling processes. The USDA/FSIS draft compliance guidelines for RTE meat and poultry products (USDA/FSIS 2001a) state that such products should be cooled at a rate sufficient to prevent more than a 1-$\log_{10}$ increase of *C. perfringens* cells. Guidelines also state that cooling from 54.4°C to 26.6°C should take no longer than 1.5 h, and cooling from 26.6°C to 4.4°C should take no longer than 5 h (USDA/FSIS, 2001a). Additional guidelines (USDA/FSIS, 2001b) allow for the cooling of certain cured cooked meats from 54.4°C to 26.7°C in 5 h, and from 26.7°C to 7.2°C in 10 h (USDA/FSIS, 2001b). Processors can use customized chilling processes as long as a < 1 $\log_{10}$ increase of *C. perfringens* in the finished products can be documented (Danler *et al.*, 2002).

The 2001 FDA Food Code dictates that cooked potentially hazardous foods such as meats should be cooled from 60°C to 21°C within 2 h, and from 60°C to 5°C within 6 h. In the UK, it is recommended that uncured cooked meats be cooled from 50°C to 12°C within 6 h, and from 12°C to 5°C within 1 h (Gaze *et al.*, 1998). Safe cooling times for cured meats may be up to 25% longer (Gaze *et al.*, 1998).

Rapid cooling after cooking and subsequent refrigeration are the most effective means for limiting growth of *C. perfringens* vegetative cells. Regardless of the precautions and regulatory guidelines imposed on the food industry, the last line of defense in ensuring food safety remains the responsibility of the retail food industry, institutional food service professionals, and cautious, informed consumers.

## Molecular approaches to the study of *C. perfringens*

As mentioned in an earlier section of this chapter, isolates of *Clostridium perfringens* can be categorized into five types (A, B, C, D, and E) based on the production of four extracellular toxins (alpha, beta, epsilon, and tau) (Petit *et al.*, 1999). Toxinotyping methods are based on mouse lethality and corrective seroprotection with neutralizing antibodies raised against a specific respective toxin (Petit *et al.*, 1999). *C. perfringens* type A infections cause gas gangrene and food poisoning, type B produce lamb dysentery and enterotoxemia, type C result in necrotic enteritis in animals and humans (pig bel disease), type D result in pulpy kidney disease in sheep and type E strains cause enteritis of rabbits (Rood and Cole, 1991). The major cause of human food poisoning is due to the sporulation-specific *C. perfringens* enterotoxin (CPE; Rood and Cole, 1991). Within the past 10 years, DNA-based techniques for genotyping using PCR and hybridizations have replaced serotyping as the more acceptable and accurate characterization scheme for categorizing *C. perfringens* isolates (Petit *et al.*, 1999).

## Plasmids as tools for gene transfer

The study of virulence gene expression in *C. perfringens* was first dependent upon the development of genetic systems and vectors for enabling gene manipulations or transfers within the spore-forming bacterium. Initially L forms of *C. perfringens*, described as polyethylene glycol-induced cell wall defective variants or "autoplasts", were used to enhance the transformation efficiency ($3.0 \times 10^{-5}$ to $4.4 \times 10^{-4}$ transformants per total viable cells) of the Gram-positive cell (Mahony *et al.*, 1988). Electroporation methods involving the application of high-intensity electric fields to permeabilize the lipid bilayer membrane of cells enabled the entry of nucleic acid macromolecules and a 100-fold increase in transformation frequencies (Allen and Blaschek, 1988, 1990; Phillips-Jones, 1990).

There have been significant advances in the construction of suitable vectors to shuttle DNA among strains of *C. perfringens* as well as other recipient species. The plasmid, pHR106, was constructed with origins of replication for both *C. perfringens* and *Escherichia coli*, a multiple cloning site, and ampicillin and chloramphenicol resistance markers (Roberts *et al.*, 1988). The transformation frequency using this construct was 1 transformant per $10^4$ viable *E. coli* cells and one transformant per $10^6$ viable *C. perfringens* cells (Roberts *et al.*, 1988). A similar plasmid shuttle vector, pAK201, was constructed and combined with electroporation to produce $10^6$ transformants per μg of DNA in *E. coli* and $10^4$ transformants per μg of DNA in *C. perfringens*, respectively (Kim and Blascheck, 1989). Subsequent improvements were made with the construct, pJIR418, that in addition to the chloramphenicol and erythromycin resistance genes contained the *lacZ* gene for blue-white colony, X-gal screening for inserted DNA and the complete multiple cloning restriction enzyme region from commercially available pUC18 (Sloan *et al.*, 1992). This construct was championed as a useful reporter based the reporter *catP* gene using a chloramphenicol acetyltransferase assay for analyses of promoters for other *C. perfringens* genes such as phospholipase C (*plc*) that results in the production of alpha toxin (Matsushita *et al.*, 1994). Two similar shuttle vectors to pJIR418 were also developed by Bannam and Rood (1993) and Bullifent *et al.* (1995). A bioluminescent reporter system was also constructed with the plasmid, pPS14, that contained the *luxA* and *luxB* genes of *Vibrio fischeri*, and enabled emitted light to be measured as an indication of promoter gene upregulation in *C. perfringens* (Phillips-Jones, 1993). Another shuttle vector, pJIR1457, originally developed for *C. perfringens*, was successfully transferred from *E. coli* to various strains of *C. botulinum* enabling expression studies of botulinum neurotoxins as well (Bradshaw *et al.*, 1998).

## Plasmids as a means of characterization

Most virulence genes in *C. perfringens* can be located on extrachromosomal replicons or plasmids (Katayama *et al.*, 1996). Plasmid profiling has been suggested for the differentiation and epidemiological characterization of different strains of *C. perfringens* (Phillips-Jones *et al.*, 1989; Mahony *et al.*, 1987; Eisgruber *et al.*, 1995). The plasmids from *C. perfringens* can be grouped into 3 classes according to molecular mass: 2–6 MDa,

9–10 MDa, and 30–40 MDa or higher (Solberg *et al.*, 1981). Advantages for plasmid profiling include that the method is convenient, fast, and does not require expensive equipment or restricted reagents (Eisgruber *et al.*, 1995). Disadvantages include loss of plasmids during culturing, 30% of *C. perfringens* do not contain plasmids and are untypable, and plasmid digestion due to endogenous nucleases (Blaschek and Klacik, 1984; Mahony *et al.*, 1986; Eisengruber *et al.*, 1995). Alternative methods of *C. perfringens* strain differentiations that have been suggested include toxin typing, biochemical characterizations, protein gel patterns, and fatty acid analyses using chromatographic techniques (Harpold *et al.*, 1985; Mulligan *et al.*, 1986; Krausse and Ullmann, 1991). However, none of these replacement methods offer the speed and convenience of DNA-based profiles using PFGE or ribotyping (Schalch *et al.*, 1998).

## Genome characterization

The physical and genetic maps of *C. perfringens* strain CPN50, shown to be associated with human disease, have been extensively characterized (Canard and Cole, 1989; Katayama *et al.*, 1995). *C. perfringens* was actually the first Gram-positive bacterium for which this was elucidated. Likewise, PFGE has been used to assess the genomes of several strains with respect to virulence gene regions compared to serotyping (Canard *et al.*, 1992). Results showed that toxinotypic classification of *C. perfringens* into groups based on serology did not correspond with differences in genomic organization (Canard *et al.*, 1992). Genomic differences were found between type A strains as well as type A strains compared to the type D or E strains (Canard *et al.*, 1992). Recent sequencing of the *C. perfringens* strain 13 genome has been completed and provides an essential resource for further DNA comparisons among different strains that could be used for molecular epidemiological analyses (Shimizu *et al.*, 2002a).

## Virulence gene isolation and regulation

Several studies have examined the expression of virulence characteristics in *C. perfringens* at the molecular level. The phospholipase C (alpha-toxin) gene of *C. perfringens* was isolated by cloning 3.2 to 12.8 kb DNA fragments from a partial *Hind*III digestion of *C. perfringens* type A strain NCTC 8237 genomic DNA into the pUC18 plasmid vector followed by transformation and expression in *E. coli* screening ampicillin resistant clones (Titball *et al.*, 1989). Alpha-toxin activity of *plc* transformants was detected by determining its effect on egg yolk lipoproteins (increase in tubidity) or sheep erythrocytes (hemolysis) (Titball *et al.*, 1989). Mutants of the *pfoA* for theta-toxin, and *plc* gene for alpha-toxin were constructed in *C. perfringens* by allelic exchange using a suicide plasmid that contained the respective gene region inactivated with an erythromycin resistance gene marker insert (Awad *et al.*, 1995). The mutants were virulence tested in mice to verify the disease-causing respective activities of the wild-type genes (Awad *et al.*, 1995).

The two-component *virR/virS* locus was identified that affected the expression of a number of extracellular toxins and enzymes (phospholipase C, protease, sialidase, and perfringolysin O) involved in *C. perfringens* virulence (Lyristis *et al.*, 1994). This was accomplished using transposon mutagenesis with Tn*916* to produce and isolate a pleiotropic mutant of *C. perfringens* that produced reduced levels of several toxins and enzymes (Lyristis *et al.*, 1994). A 4.3 kb *Pst*I fragment was isolated that contained the Tn*916* insertion into the *virR/virS* locus and was confirmed by complementation with a 4.3 kb *Pst*I DNA fragment cloned from the wild-type strain of *C. perfringens* (Lyristis *et al.*, 1994). The *virR* and *virS* genes were shown to comprise a single operon producing a 2.1-kb transcript identified using comparative Northern blot hybridizations of *virS* and *virR* DNA probes with total RNA preparations from wild-type strain 13 and isogenic *virR* and *virS* mutants of *C. perfringens* (Ba-Thein *et al.*, 1996). The *virR/virS* system was shown to positively regulate the production of alpha-toxin, theta-toxin, and kappa-toxin at the transcriptional level (Ba-Thein *et al.*, 1996). Genes other than toxin genes were shown to be globally regulated by *virR/virS* using a differential display technique where a plasmid library of *C. perfringens* chromosomal DNA was hybridized to cDNA probes prepared from total RNA of wild-type *C. perfringens* and a *virR* mutant (Banu *et al.*, 2000). Using gel mobility shift assays, the virR protein has been shown to bind DNA upstream of the *pfoA* gene that encodes perfringolysin O (Cheung and Rood, 2000). The C-terminal domain of the virR protein was found to have a highly conserved region with over 40 other proteins that were response regulators and transcriptional activators (McGowan *et al.*, 2002). It was this C-terminal domain region of the virR protein that was considered essential for DNA binding (McGowan *et al.*, 2002). A regulatory RNA molecule from the *virR/virS* operon was found to regulate expression of the alpha- and kappa-toxin genes in *C. perfringens* (Shimizu *et al.*, 2002b). Another unique gene, *virX*, was recently identified that regulates the expression of alpha-, kappa-, and theta-toxin genes in *C. perfringens* (Ohtani *et al.*, 2002). Overall, the *virR/virS* system controls many virulence and housekeeping genes both positively and negatively.

## *CPE* gene location determines foodborne type A illness

The location of the *cpe* gene, encoding *C. perfringens* enterotoxin, has been studied extensively with respect to the epidemiologic significance for disease production in humans. In strains of *C. perfringens* associated with human food poisoning, the *cpe* gene was found in the same chromosomal location next to a repetitive sequence, a *Hind*III repeat, and an open reading frame that may be part of an insert (Cornillot *et al.*, 1995). Strains of *C. perfringens* from domesticated livestock, the *cpe* gene was located on a large episome in close proximity to transposable element IS*1151* (Cornillot *et al.*, 1995). The first transposable elements in clostridia (Tn*4451* and Tn*4452*) were discovered several years earlier (Abraham and Rood, 1987). Enterotoxin production is considered a relatively rare characteristic present in approximately only 6% of tested *C. perfringens* strains (van Dammme-Jongsten *et al.*, 1989). It was suggested that insertion sequences or transposable elements might be involved in the transfer of virulence genes in *C. perfringens* (Cornillot *et al.*, 1995).

The initial findings were confirmed by others that the *cpe* gene in type A human food poisoning strains was on a 6.3-kb transposon flanked by copies of inverted repeats IS*1470* and IS*1469* (Brynestad *et al.*, 1997). It had been known for some time that strains of *C. perfringens* could suddenly lose or gain the ability to produce enterotoxin and there were certain regions of host DNA where insertions were more likely to occur (van Damme-Jongsten *et al.*, 1989). A question was raised as to whether heat used to treat foods could affect the movement and expression of *cpe* (Brynestad *et al.*, 1997). Additional isolates from various sources were compared genotypically using restriction fragment length polymorphism and PFGE (Collie and McClane, 1998). All of the food-poisoning isolates carried a chromosomal *cpe* gene, whereas all non-foodborne human gastrointestinal illness isolates carried an episomal *cpe* gene (Collie and McClane, 1998). The vegetative cells of chromosomal *cpe* isolates were found to exhibit two-fold higher heat resistance at 55°C than vegetative cells harboring plasmid *cpe* genes (Sarker *et al.*, 2000). The spores of chromosomal *cpe* isolates exhibited 60-fold higher heat resistance at 100°C than spores of plasmid *cpe* isolates (Sarker *et al.*, 2000). Heating was not found to cause plasmid loss, nor decreased expression of enterotoxin (Sarker *et al.*, 2000). Clearly, the chromosomal *cpe* location in strains of *C. perfringens* could be expected to enhance survival in inadequately heated foods.

Further investigations have shown the potential exists for *cpe*-positive isolates of *C. perfringens* to convert *cpe*-negative isolates of *C. perfringens* by conjugative plasmid transfer and increase the potential for causing gastrointestinal disease in normally avirulent strains (Brynestad *et al.*, 2001). Plasmid *cpe* isolates were found to cause most antibiotic-associated diarrhea, whereas chromosomal *cpe* isolates were found to cause most *C. perfringens* type A food poisoning cases (Sparks *et al.*, 2001). The presence of a similar plasmid *cpe* locus in *C. perfringens* strain F4969 and a non-related type A isolate, 452, from different origins suggested the horizontal transfer of a common plasmid to other isolates conferring virulence traits (Miyamoto *et al.*, 2002). Recently, a multiplex PCR assay was developed to distinguish type A isolates carrying a chromosomal or plasmid *cpe* gene (Miyamoto *et al.*, 2004). It was determined that the IS*1470* sequence rather than IS*1151* is more commonly present downstream of the plasmid *cpe* gene and the distinction could be used in molecular epidemiologic investigations of *cpe*-associated gastrointestinal illnesses (Miyamoto *et al.*, 2004).

The tools of molecular genetics have facilitated the characterization of virulence traits harbored by strains of *C. perfringens* and enabled the epidemiologic tracing of strains from foodborne outbreaks in order to plan preventative strategies. Future *C. perfringens* research should concentrate on toxin inactivation schemes, control of illness prevention conditions in food preparations, and efficient monitoring and traceability of strains of concern.

## Acknowledgments

Research in E.A.J.'s laboratory on *Clostridium botulinum* was supported by NIH, the College of Agriculture and Life Sciences, and by sponsors to the Food Research Institute, University of Wisconsin, Madison. Research in M.W.P.'s laboratory on foodborne clostridia is supported by the Competitive Strategic Grant of the BBSRC, the European Union (projects RASP, BACANOVA, and Genus Clostridium), and the Food Standards Agency.

## References

Abraham, L.J., and Rood, J.I. 1987. Identification of Tn4451 and Tn4452, chloramphenicol resistance transposons from *Clostridium perfringens*. J. Bacteriol. 169: 1579–1584.

Adams, D.M. 1973. Inactivation of *Clostridium perfringens* type A spores at ultrahigh temperatures. Appl. Microbiol. 26: 282–287.

Allen, S.P., and Blaschek, H.P. 1988. Electroporation-induced transformation of intact cells of *Clostridium perfringens*. Appl. Environ. Microbiol. 54: 2322–2324.

Allen, S.P., and Blaschek, H.P. 1990. Factors involved in the electroporation-induced transformation of *Clostridium perfringens*. FEMS Microbiol. Lett. 70: 217–220.

Angulo, F.J., Getz J., and Taylor, J.P. 1998. A large outbreak of botulism: The hazardous baked potato. J. Infect Dis. 178: 172–77.

Anne, C., Cornille, F., Lenoir, C., and Roques, B.P. 2001. High-throughput fluorigenic assay for determination of botulinum neurotoxin protease activity. Anal. Biochem. 291: 253–261.

Anniballi, F., Fenicia, L. Franciosa, G. and Aureli, P. 2002. Influence of pH and temperature on the growth and of toxin production by neurotoxinigenic strains of *Clostridium butyricum* type E. J. Food Prot. 65: 1267–1270.

Anon. 1998. Fallbericht: Botulismus nach dem Verzehr von geräucherten Lachsforellen. Epidemiol. Bull. 4/98:20.

Anon. 2002. Two outbreaks of botulism associated with fermented salmon roe – British Columbia – August 2001. Can. Commun. Dis. Rep. 28–06: 1–4.

Arnon, S.S. 1995. Botulism as an intestinal toxemia. In: Infections of the Gastrointestinal Tract. M.J. Blaser, P.D. Smith, J.I. Ravdin, Raven Press, New York, p. 257–271.

Arnon, S.S. 2004. Infant botulism. In: Textbook of Pediatric Infectious Diseases. R.D. Feigen, J.D. Cherry, eds. 5th ed., W. B. Saunders, Philadelphia.

Arnon S.S., Schechter, R., Inglesby, T.V., Henderson, D.A., Bartlett, J.G., Ascher, M.S., Eitzen, E., Fine, A.D., Hauer, H., Layton, M., Lillibridge, S., Osterholm, M.T., O'Toole, T., Parker, G., Perl, T.M., Russell, P.K., Swerdlow, D.L., and Tonat, K. 2001. Botulinum toxin as a biological weapon. Medical and public health management. JAMA 285: 1059–70.

Aureli, P., Fenicia, L., and Pasolini, B. 1986. Two cases of type E infant botulism caused by neurotoxigenic *Clostridium butyricum* in Italy. J Infect Dis. 154: 207–211.

Awad, M.M., Bryant, A.E., Stevens, D.L., and Rood, J.I. 1995. Virulence studies on chromosomal α-toxin and θ-toxin mutants constructed by allelic exchange provide genetic evidence for the essential role of α-toxin in *Clostridium perfringens*-mediated gas gangrene. Mol. Microbiol. 15: 191–202.

Baker, D.A. and Genigeorgis, C. 1992. Predictive modelling. In: *Clostridium botulinum*. Ecology and Control in Foods. A.H.W. Hauschild and K.L. Dodds, eds. Marcel Dekker, New York, pp. 343–406.

Balassa, G., Milhaud, P., and Raulet, E. 1979. A *Bacillus subtilis* mutant requiring dipicolinic acid for the development of heat-resistant spores. J. Gen. Microbiol. 110: 365–379.

Bannam, T.L., and Rood, J.I. 1993. *Clostridium perfringens-Escherichia coli* shuttle vectors that carry single antibiotic resistance determinants. Plasmid 229: 233–235.

Banu, S., Ohtani, K., Yaguchi, H., Swe, T., Cole, S.T., Hayashi, H., and Shimizu, T. 2000. Identification of novel virR/virS-regulated genes in *Clostridium perfringens*. Mol. Microbiol. 35: 854–864.

Barash, J.R., and Arnon, S.S. 2004. Dual-toxin producing strain of *Clostridium botulinum* type Bf isolated from a California patient with infant botulism. J. Clin. Microbiol. 42: 1713–1715.

Barker, G.C., Malakar, P.K., Del Torre, M., Stecchini, M.L. and Peck, M.W. 2005. Probabilistic representation of the exposure of consumers to *Clostridium botulinum* neurotoxin in a minimally processed potato product. Int. J. Food Microbiol. 100: 345–357.

Barker, G.C., Talbot, N.L.C. and Peck, M.W. 1999. Microbial risk assessment for sous-vide foods. In: Proceedings of Third European Symposium on Sous Vide. Alma Sous Vide Centre Belgium, pp. 37–46.

Barker, G.C., Talbot, N.L.C. and Peck, M.W. 2002. Risk assessment for *Clostridium botulinum*: A network approach. Int. J. Biodet. 50: 167–175.

Barnes, E.M., Despaul, J.E., and M. Ingram. 1963. The behavior of a food poisoning strain of *Clostridium welchii* in beef. J. Appl. Bacteriol. 26: 415–427.

Ba-Thein, W., Lyristis, M., Ohtani, K., Nisbet, I.T., Hayashi, H., Rood, J.I., and Shimizu, T. 1996. The virR/virS locus regulates the transcription of genes encoding extracellular toxin production in *Clostridium perfringens*. J. Bacteriol. 178: 2514–2520.

Bauer, F.T., Carpenter, J.A. and Reagan, J.O. 1981. Prevalence of *Clostridium perfringens* in pork during processing. J. Food Prot. 44: 279–283.

Beaman, T.C., and Gerhardt, P. 1986. Heat resistance of bacterial spores correlated with protoplast dehydration, mineralization, and thermal adaptation. Appl. Environ. Microbiol. 52: 1242–1246.

Beaman, T.C., Koshikawa, T., Pankratz, H.S., and Gerhardt, P. 1984. Dehydration partitioned within core protoplast accounts for heat resistance of bacterial spores. FEMS Microbiol. Lett. 24:47–51.

Bean, N.H. and Griffin, P.M. 1990. Foodborne disease outbreaks in the United States, 1973-1987: pathogens, vehicles, and trends. J. Food Prot. 9: 804–817.

Bean, N.H., Goulding, J.S., Daniels, M.T. and Angulo, F.J. 1997. Surveillance for foodborne disease outbreaks—United States, 1988–1992. J. Food Prot. 60: 1265–1286.

Bennik, M.H.J., Minton, N.P., Elmore, M., Barrell, B., Hutson, R.A., Parkhill, J. and Peck, M.W. 2002. Sequencing the genome of proteolytic *Clostridium botulinum* ATCC 3502 (Hall A). Oral presentation and abstract at "Inter-agency botulism research co-ordinating committee (IBRCC) meeting". Madison, USA.

Bennik, M.H.J., Mauchline, M., Bosveld, F., Elmore, M., Minton, N.P., Parkhill, J. and Peck, M.W. 2003a. Preliminary analysis of the genome sequence of proteolytic *Clostridium botulinum* ATCC 3502 (Hall A). Poster presentation and abstract at 152nd meeting of Society for General Microbiology. Edinburgh, UK.

Bennik, M.H.J., Mauchline, M., Bosveld, F., Elmore, M., Minton N.P., Parkhill, J. and Peck, M.W. 2003b. The genome sequence of proteolytic *Clostridium botulinum* ATCC 3502 (Hall A): some highlights. Invited oral presentation and abstract at Clostpath 2003. Woods Hole, MA, USA.

Birmingham, M.D., Walter, F.G., Haber, J., and Ekins, B.R. 1994. Wound botulism. Ann. Emerg. Med. 24: 1184–1187.

Blankenship, L.C., Craven, S.E., Leffler, R.G., and Custer, C. 1988. Growth of *Clostridium perfringens* in cooked chili during cooling. Appl. Environ. Microbiol. 53:1104–1108.

Blaschek, H.P., and Klacik, M.A. 1984. Role of DNase in recovery of plasmid DNA from *Clostridium perfringens*. Appl. Environ. Microbiol. 48: 178–181.

Bowmer, E.J. 1963. Preparation and assay of the international standards for *Clostridium botulinum* types A, B, C, D and E antitoxins. Bull WHO. 29: 701–709.

Boyer, A., Girault, C., Bauer, F., Korach, J.M., Salomon, J., Moirot, E., Leroy, J. and Bonmarchand, G. 2001. Two cases of foodborne botulism type E and review of epidemiology in France. Eur. J. Clin. Microbiol. Infect. Dis. 20: 192–195

Bradshaw, M., Goodnough, M.C., and Johnson, E.A. 1998. Conjugative transfer of the *Escherischia coli–Clostridium perfringens* shuttle vector pJIR1457 to *Clostridium botulinum* type A strains. Plasmid. 40: 233–237.

Brüggemann, H., Bäumer, S., Fricke, W.F., Wiezer, A., Liesegang, H., Decker, I., Herzberg, C., Marinez-Arias, R., Merkl, R., Henne, A., and Gottschalk, G. 2003. The genome sequence of *Clostridium tetani*, the causative agent of tetanus disease. Proc. Natl. Acad. Sci. USA 100: 1316–1321.

Brüggeman, H., and Gottschalk, G. 2004. Insights in metabolism and toxin production from the complete genome sequence of *Clostridium tetani*. Anaerobe 10: 53–68.

Bryan, F.L. 1988. Risks associated with vehicles of foodborne pathogens and toxins. J. Food Prot. 51: 498–508.

Bryan, F.L. and T.W. McKinley. 1979. Hazard analysis and control of roast beef preparation in foodservice establishments. J. Food Prot. 42: 4–18.

Brynestad, S. and P.E. Granum. 2002. *Clostridium perfringens* and foodborne infections. Int. J. Food Microbiol. 74: 195–202.

Brynestad, S., Synstad, B., and Granum, P.E. 1997. The *Clostridium perfringens* enterotoxin gene is on a transposable element in type A human food poisoning strains. Microbiology. 143: 2109–2115.

Brynestad, S., Sarker, M.R., McClane, B.A., Granum, P.E., and Rood, J.I. 2001. Enterotoxin plasmid from *Clostridium perfringens* is conjugative. Infect. Immun. 69: 3483–3487.

Bullifent, H.L., Moir, A., and Titball, R.W. 1995. The construction of a reporter system and use for the investigation of *Clostridium perfringens* gene expression. FEMS Microbiol. Lett. 131: 99–105.

Burgen, A.S.V., Dickens, F., and Zatman, L.F. 1949. The action of botulinum toxin on the neuro-muscular junction. J. Physiol. 109: 10–24.

California Morbidity, Division of Communicable Disease Control. 1995. Foodborne outbreaks in California, 1993–1994, p. 1–4, May 19, 1995. http://www.dhs.ca.gov/ps/dcdc/cm/950519CM.htm

Canard, B., and Cole, S.T. 1989. Genome organization of the anaerobic pathogen *Clostridium perfringens*. Proc. Natl. Acad. Sci. USA 86: 6676–6680.

Canard, B., Saint-Joanis, B., and Cole, S.T. 1992. Genomic diversity and organization of virulence genes in the pathogenic anaerobe *Clostridium perfringens*. Mol. Microbiol. 6: 1421–1429.

Carlin, F., Girardin, H., Peck, M.W., Stringer, S.C., Barker, G.C., Martinez, A. Fernandez, A., Fernandez, P., Waites, W.M., Movahedi, S., van Leusden, F., Nauta, M., Moezelaar, R., Del Torre, M. and Litman, S. 2000. Research on factors allowing a risk assessment of spore forming pathogenic bacteria in cooked chilled foods containing vegetables: a FAIR collaborative project. Int. J. Food Microbiol. 60: 117–135.

Carlin, F. and Peck, M.W. 1996a. Growth of, and toxin production by, non-proteolytic *Clostridium botulinum* in cooked vegetables at refrigeration temperatures. Appl. Environ. Microbiol. 62: 3069–3072.

Carlin, F. and Peck, M.W. 1996b. Metabiotic association between non-proteolytic *Clostridium botulinum* type B and foodborne *Bacillus* species. Sciences des Aliments 16: 545–551.

Cato, E.P., George, W.L., and Finegold, S.M. 1986. Genus *Clostridium*. Bergey's Manual of Systematic Bacteriology, P. Sneath, ed., Williams and Wilkins, Baltimore, pp. 1141–200.

Cawley, G.C., Peck, M.W. and Fernandez, P.S. 1998. A neural model of time to toxin production by non-proteolytic *Clostridium botulinum*. In: Proceedings of International Joint Conference on Neural Networks (IJCNN-98) [Sub-section of 1998 IEEE World Congress on Computational Intelligence], p. 101–105.

CDC. 1979. Botulism in the United States, 1899–1977. Handbook for epidemiologists, clinicians, and laboratory workers, U. S. Dept. Health, Education, Welfare, Public Health Service, Centers for Disease Control and Prevention, Atlanta, GA.

CDC. 1983. Botulism and commercial pot pie-California. MMWR 32: 39–40.

CDC. 1987. Restaurant-associated botulism from mushrooms bottled in-house – Vancouver, British Columbia, Canada. Morbid. Mortal. Wkly Rep. 36, 103.

CDC. 1992. Outbreak of type B botulism associated with an uneviscerated, salt-cured fish product – New Jersey, 1992. Morbid. Mortal. Wkly. Rep. 41: (29) 521–522.

CDC. 1995a. Type B botulism associated with roasted eggplant in oil – Italy, 1993. Morbid. Mortal. Weekly Rep. 44: 33–36.

CDC. 1995b. Wound botulism – California, 1995. Morbid. Mortal. Weekly Rep. 44: 890–2.

CDC. 1998. Botulism in the United States, 1899–1996. Handbook for Epidemiologists, Clinicians, and Laboratory Workers, Centers for Disease Control and Prevention, Atlanta, GA.

CDC. 2000. Surveillance for foodborne-disease outbreaks-United States, 1993–1997. MMWR 49: 1–58.

CDC. 2001. Botulism outbreak associated with eating fermented food – Alaska 2001. Morbid. Mortal. Wkly. Rep. 50(32): 680–682.

CDC. 2003. Outbreak of botulism type E associated with eating a beached whale – Western Alaska, July 2002. Morbid. Mortal. Wkly. Rep. 52(02): 24–26.

Chaudry, R., and Dhawan, B. 1998. Outbreak of suspected *Clostridium butyricum* botulism in India, Emerg. Infect. Dis. 4: 506–507.

Cherington, C. 1998. Clinical spectrum of botulism. Muscle and Nerve 21: 701–10.

Cherington, M., and Ginsburg, S. 1975. Wound botulism. *Arch Surg*. 110: 436–438.

Cheung, J.K., and Rood, J.I. 2000. The virR response regulator from *Clostridium perfringens* binds independently to two imperfect direct repeats located upstream of the pfoA promoter. J. Bacteriol. 182: 57–66.

Chia, J.K., and Clark, J.B. 1986. Botulism in an adult associated with foodborne intestinal infection with *Clostridium botulinum*. N. Engl. J. Med. 315: 239–41.

Chilled Food Association. 2004. www.chilledfood.org/Content/Market_Data.asp

Chiorboli, E., Fortina, G., and Bona, G. 1997. Flaccid paralysis caused by botulinum toxin type B after pesto ingestion. Pediatr. Infect. Dis. J. 16(7):725–726.

Chou, J.H., Hwang, P.H. and Malison, M.D. 1988. An outbreak of Type A foodborne botulism in Taiwan due to commercially preserved peanuts. Int. J. Epidemiol. 17: 899–902.

Chumney, R.K. and Adams, D.M. 1980. Relationship between the increased sensitivity of heat injured *Clostridium perfringens* spores to surface-active antibiotics and to sodium chloride and sodium nitrite. J. Appl. Bacteriol. 49: 55–63.

Cobb, S.P., and Hogg, R.A.. 2002. Suspected botulism in dairy cows and implications for the safety of human food. Vet. Rec. 150: 5–8.

Cohen, R.E., and Anderson, D.L. 1986. Botulism: emotional impact on patient and family. J. Psychomat. Res. 30: 321–326.

Collie, R.E., and McClane, B.A. 1998. Evidence that the enterotoxin gene can be episomal in *Clostridium perfringens* isolates associated with non-foodborne human gastrointestinal diseases. J. Clin. Microbiol. 36: 30–36.

Collins, M.D., and East, A.K. 1998. Phylogeny and taxonomy of the foodborne pathogen *Clostridium botulinum* and its neurotoxins. J. Appl. Microbiol. 84: 5–17.

Collins, M.D., and Lawson, P.A. 1994. The phylogeny of the genus *Clostridium*: proposal of five new genera and eleven new species combinations. Int. J. Syst Bacteriol. 44: 812–26.

Communication Disease Surveillance Centre (CDSC). 1998. Botulism associated with home-preserved mushrooms. Commun. Dis. Rep. CDR Wkly 8(18): 159 and 162.

Cornillot, E., Saint-Joanis, B., Daube, G., Katayama, S., Granum, P.E., Canard, B., and Cole, S.T. 1995. The enterotoxin gene (cpe) of *Clostridium perfringens* can be chromosomal or plasmid-borne. Mol. Microbiol. 15: 639–647.

Crandall, A.D. and Montville, T.J. 1993. Inhibition of *Clostridium botulinum* growth and toxigenesis in a model gravy system by co-inoculation with bacteriocin producing lactic acid bacteria. J. Food Prot. 56: 485–490.

Craven, S.E. 2001. Occurrence of *Clostridium perfringens* in the broiler chicken processing plant as determined by recovery in iron milk medium. J. Food Prot. 64: 1956–1960.

Danler, R.J., Boyle, E.A.E., Kastner, C.L., Thippareddi, H., Fung, D.Y.C. and Phebus, R.K. 2002. Effects of chilling rate on outgrowth of *Clostridium perfringens* spores in vacuum packaged cooked beef and pork. Presented at IFT Annual Meeting, Anaheim, CA.

DasGupta, B.R. 1989. The structure of botulinum neurotoxin. In: Botulinum Neurotoxin and Tetanus Toxin. L.L. Simpson, ed. Academic Press, San Diego, pp. 53–67.

Devries, P. P. 1999. On the discovery of *Clostridium botulinum*. J. Hist. Neurosci. 8: 43–50.

Dezfulian, M., and McCroskey, L.M. 1981. Selective medium for isolation of *Clostridium botulinum* from human feces. J. Clin. Microbiol. 13: 526–31.

Dickson, E.C. 1918. Botulism. A clinical and experimental study. Monograph No. 8 of the Rockefeller Inst. Med Res. Waverly Press, The William and Wilkins Co., Baltimore, USA.

Dineen, S.S., Bradshaw, M., and Johnson, E.A. 2000. Cloning, nucleotide sequence, and expression of the gene encoding bacteriocin boticin B from *Clostridium botulinum* strain 213B. Appl. Environ. Microbiol. 66: 5480–5483.

Dineen, S.S., Bradshaw, M. and Johnson, E.A. 2003. Neurotoxin gene clusters in *Clostridium botulinum* type A strains: sequence comparison and evolutionary implications. Curr. Microbiol. 46: 345–352.

Dodds, K.L. 1993a. An introduction to predictive microbiology and the development of probability models with *Clostridium botulinum*. J. Indust. Microbiol. 12, 139–143.

Dodds, K.L. 1993b. Worldwide incidence and ecology of infant botulism. In: *Clostridium botulinum*:Ecology and Control in Foods. A.H.W. Hauschild, and K.L. Dodds, eds. Marcel Dekker, New York. p. 105–17.

Doellgast, G., Triscott, G., and Beard, G. 1993. Sensitive enzyme-linked immunosorbent assay for detection of *Clostridium botulinum* neurotoxins A, B, and E using signal amplification via enzyme-linked coagulation assay. J. Clin. Microbiol. 31: 2402–2409.

Dolman, C.E. 1964. Botulism as a world health problem. H.H. Lewis, K. Casse, Jr., eds. Botulism. Proceedings of a symposium. U. S. Dept. Health, Education, and Welfare. Cincinnati, Ohio, USA, p. 5–30.

Dowell, V.R. Jr., and Dezfulian, M. 1981. Physiological characterization of *Clostridium botulinum* and development of practical isolation and identification procedures. Biomedical Aspects of Botulism. G.E.J. Lewis, eds. Academic Press, New York. p. 205-16.

Dowell, V.R. Jr, Hill, E.O., and Altemeier, W.A. 1964. Use of phenylethyl alcohol in media for isolation of anaerobic bacteria. J. Bacteriol. 88: 1811–13.

Dowell, V.R. Jr., and Hawkins, T.M. 1974. Laboratory Methods in Anaerobic Bacteriology. Centers for Disease Control, Atlanta.

Eisgruber, H., Wiedmann, M., and Stolle, A. 1995. Use of plasmid profiling as a typing method for epidemiologically related *Clostridium perfringens* isolates from food poisoning cases and outbreaks. Lett. Appl. Microbiol. 20: 290–294.

Erbguth, F.J. 2004. Historical notes of botulism, *Clostridium botulinum*, botulinum toxin, and the idea of the therapeutic use of the toxin. Movement Disorders 19(8): S2–S6.

Eriksen, T., Brantsaeter, A.B., Kiehl, W., and Steffens, I. 2004. Botulism infection after eating fish in Norway and Germany: two outbreak report. Eurosurveillance Weekly 8(3), 1–2.

ECFF European Chilled Food Federation. 1998. *Best practice in the chilled food industry*. Helsinki: The European Chilled Food Federation.

Fach, P., and Hauser, D. 1993. Polymerase chain reaction for the rapid identification of *Clostridium botulinum* type A strains and detection in food samples. J. Appl Bacteriol. 75: 234–9.

Fenicia, L., Ferrini, A.M., Aureli, P. and Padovan, M.T. 1992. Botulism outbreak from black olives. Ind. Aliment. 31: 307–308.

Fenicia, L., Franciosa, G. 1999. Intestinal toxemia botulism in two young people caused by *Clostridium butyricum* type E. Clin Infect Dis. 154: 207–11.

Fernandez, P.S. and Peck, M.W. 1997. A predictive model that describes the effect of prolonged heating at 70–80°C and incubation at refrigeration temperatures on growth and toxigenesis by non-proteolytic *Clostridium botulinum*. J. Food Prot. 60: 1064–1071.

Fernandez, P.S. and Peck, M.W. 1999. Predictive model that describes the effect of prolonged heating at 70–90°C and subsequent incubation at refrigeration temperatures on growth and toxigenesis by non-proteolytic *Clostridium botulinum* in the presence of lysozyme. Appl. Environ. Microbiol. 65: 3449–3457.

Fernandez, P.S., Baranyi, J. and Peck, M.W. 2001. A predictive model of growth from spores of nonproteolytic *Clostridium botulinum* in the presence of different $CO_2$ concentrations as influenced by chill temperature, pH and NaCl. Food Microbiol. 18: 453–462.

Ferreira, J.L. 2001. Comparison of amplified ELISA and mouse bioassay procedures for determination of botulinal toxins A, B, E, and F. J. AOAC Int. 84: 85–88.

Ferreira, J.L., and Hamdy, M.K. 1995. Detection of botulinal toxin genes: types A and E or B and F using the multiplex polymerase chain reaction. J Rapid Meth Automat Microbiol. 3: 177–83.

Ferreira, J.L., Maslanka, S., Johnson, E.A., and Goodnough, M.C. 2003. Detection of botulinal neurotoxins A, B, E, and F by amplified enzyme-linked immunosorbent assay: collaborative study. J AOAC Int. 86: 314–31.

Franciosa, G., Aureli, P., and Schechter, R. 2003. *Clostridium botulinum*. M.D. Miliotis, and J.W. Bier. International Handbook of Foodborne Pathogens. Marcel Dekker, Inc., New York, pp. 61–89.

Franciosa, G., Fenicia, L., Pourshaban, M., and Aureli, P. 1997. Recovery of a strain of *Clostridium botulinum* producing both neurotoxin A and neurotoxin B from canned macrobiotic food. Appl. Environ. Microbiol. 63: 1148–1150.

Franciosa, G., and Fenicia, L. 1996. PCR for detection of *Clostridium botulinum* type C in avian and environmental samples. J. Clin. Microbiol. 34: 882–5.

Franciosa, G., Ferreira, J.L., and Hatheway, C.L. 1994. Detection of type A, B, and E botulism neurotoxin genes in *Clostridium botulinum* and other *Clostridium* species by PCR – evidence of unexpressed type B toxin genes in type A toxigenic organisms. J. Clin. Microbiol. 32: 1911–17.

Franciosa, G., and Pourshaban, M. 1999. *Clostridium botulinum* spores and toxin in marscapone cheese and other milk products. J Food Protect. 62: 867–871.

Frankovich, T.L., and Arnon, S.S. 1991. Clinical trial of botulism immune globulin for infant botulism. West J. Med. 154: 103.

Frean, J., Arntzen, L., and van den Heever, J. 2004. Fatal type A botulism in South Africa, 2002. Trans. Royal Soc. Trop. Med. Hyg. 98: 290–295.

Freedman, M., and Armstrong, R.M. 1986. Botulism in a patient with a jejunoileal bypass. Ann Neurol. 20: 641–643.

Gangarosa, E.J., and Donadio, J.A. 1971. Botulism in the USA, 1899–1969. Am J Epidemiol. 93: 93–101.

Garcia-Alvarado, J.S., Labbe, R.G., and Rodriguez, M.A. 1992. Sporulation and enterotoxin production by *Clostridium perfringens* type A at 37 and 43°C. Appl. Environ. Microbiol. 58: 1411–1414.

Gaze, J.E., Shaw, R. and Archer, J. 1998. Identification and prevention of hazards associated with slow cooling of hams and other large cooked meats and meat products. Review No. 8. Campden and Chorleywood Food Research Association, Gloucestershire, UK

Gerhardt, P., and Marquis, R.E. 1989. Spore thermoresistance mechanisms. In: I. Smith, R.A. Slepecky, and P. Setlow, ed. Regulation of Prokaryotic Development. American Society for Microbiology, Washington, D.C., pp. 43–63.

Gill, D.M. 1982. Bacterial toxins: a table of lethal amounts. Microbiol Rev. 46: 86–94.

Giménez, D.F., and Giménez, J.A. 1993. Serological subtypes of botulinal neurotoxins. In: Botulism and Tetanus Neurotoxins: Neurotransmission and Biomedical Aspects. B.R. DasGupta, ed. Plenum Press, New York, pp. 421–431.

Giménez, D.F., and Giménez, J.A. 1995. The typing of botulinal neurotoxins. Int. J. Food Microbiol. 27: 1–9.

Glass, K.A., and Johnson, E.A. 2001. Formulating low-acid foods for botulinal safety. V.K. Juneja, and J.N. Sofos, eds. Control of Foodborne Organisms. Marcel-Dekker, New Yor, pp. 323–350.

Goodnough, M.C., and Hammer, B.A. 1993. Colony immunoblot assay of botulinal toxin. Appl. Environ. Microbiol. 59: 2339–2342.

Gombas, D.E. 1983. Bacterial spore resistance to heat. Food Technol. 37(11): 105–110.

Gorris, L.G.M. and Peck, M.W. 1998. Microbiological safety considerations when using hurdle technology with refrigerated processed foods of extended durability. In: *Sous-vide* and cook chill processing for the food industry. S. Ghazala, ed. p. 206–233. Aspen, Gaithersburg, USA.

Gough, B.J. and J.A. Alford. 1965. Effect of curing agents on the growth and survival of food-poisoning strains of *Clostridium perfringens*. J. Food Sci. 30: 1025–1028.

Graham, J.M. 1978. Inhibition of *Clostridium botulinum* type C by bacteria isolated from mud. J. Appl Bacteriol. 45: 205–211.

Graham, A.F., Mason, D.R., and Peck, M.W. 1996a. Inhibitory effect of combinations of heat treatment, pH and sodium chloride on growth from spores of non-proteolytic *Clostridium botulinum* at refrigeration temperatures. Appl. Environ. Microbiol. 62: 2664–2668.

Graham, A.F., Mason, D.R., and Peck, M.W. 1996b. A predictive model of the effect of temperature, pH, and sodium chloride on growth from spores of non-proteolytic *Clostridium botulinum*. Int. J. Food Microbiol. 31: 69–85.

Graham, A.F., Mason, D.R., Maxwell, F.J., and Peck, M.W. 1997. Effect of pH and NaCl on growth from spores of non-proteolytic *Clostridium botulinum* at chill temperatures. Lett. Appl. Microbiol. 24: 95–100.

Griffin, P.M., and Hatheway, C.L. 1997. Endogenous antibody production to botulinum toxin in an adult with intestinal colonization botulism and underlying Crohn's disease. J. Infect. Dis. 175: 633–7.

Habermann, E., and Dreyer, F. 1986. Clostridial neurotoxins: handling and the action at the molecular and cellular level. Curr. Top. Microbiol. Immunol. 129: 93–179.

Hall, H.E. and R. Angelotti. 1965. *Clostridium perfringens* in meat and meat products. Appl. Microbiol. 13: 352–357.

Hall, I.C., and Peterson, E. 1922. The effect of certain bacteria upon the toxin production of *Bacillus botulinum in vitro*. J Bacteriol. 8: 319–41.

Hall, J.D., and McCroskey, L.M. 1985. Isolation of an organism resembling *Clostridium barati* which produces type F botulinal toxin from an infant with botulism. J. Clin. Microbiol. 21: 654–5.

Hanlin, J.H., Lombardi, S.J., and Slepecky, R.A. 1985. Heat and UV light resistance of vegetative cells and spores of *Bacillus subtilis* rec⁻ mutants. J. Bacteriol. 163: 774–777.

Harpold, D.J., Wasilauskas, B.L., and O'Connor, M.L. 1985. Rapid identification of *Clostridium* species by high pressure liquid chromatography. J. Clin. Microbiol. 22: 962–967.

Harvey, S.M., Sturgeon, J., and Dassey, D.E. 2002. Botulism due to *Clostridium baratii* type F toxin. J. Clin. Microbiol. 40: 2260–2262.

Hatheway, C.L. 1988. Botulism. Laboratory Diagnosis of Infectious Diseases: Principles and Practice. A. Balows, W.H. Hausler, Jr. eds. Springer-Verlag, New York. p. 111–33.

Hatheway, C.L. 1992. *Clostridium botulinum* and other clostridia that produce botulinum neurotoxin. In: *Clostridium botulinum*. Ecology

and control in foods. A.H.W. Hauschild and K.L. Dodds, eds. Marcel Dekker, New York, pp 3–20.

Hatheway, C.L. 1993. *Clostridium botulinum* and other clostridia that produce botulinum neurotoxin. Food Sci Technol. New York 54: 3–20.

Hatheway, C.L. 1995. Botulism: the present status of the disease, In: Clostridial Neurotoxins. C. Montecucco, ed. Springer, Berlin. p. 55–75.

Hatheway, C.L., and Ferreira, J.L. 1996. Detection and identification of *Clostridium botulinum* neurotoxins. Adv. Exp. Med. Biol. 391: 481–498.

Hatheway, C.L., and Johnson, E.A. 1998. *Clostridium*: the spore-bearing anaerobes. In: Topley and Wilson's Microbiology and Microbial Infections. L. Collier, A. Balows, M. Sussman, eds. 9th Edition. Volume 2: Systematic Bacteriology. Arnold, London, pp. 731–782.

Hauschild, AHW. 1989. *Clostridium botulinum*. In: Foodborne Bacterial Pathogens. M. P. Doyle, ed. Marcel Dekker, New York, NY., pp. 111–189.

Hauschild, A.H.W. 1989. *Clostridium botulinum*. In: Foodborne bacterial pathogens. M.P. Doyle, ed. Marcel Dekker, New York. p. 112–189.

Hauschild, A.H.W. 1993. Epidemiology of human foodborne botulism, *Clostridium botulinum*: In: Ecology and Control in Foods. A.H.W. Hauschild, K.L. Dodds, eds. Marcel Dekker, New York. p. 69–104.

Hauser, D., and Gibert, M. 1995. Botulinal toxin genes, clostridial neurotoxin homology and genetic transfer in *Clostridium botulinum*. Toxicon. 33: 515–26.

Heredia, N.L., Garcia, G.A., Luevanos, R., Labbe, R.G., and Garcia-Alvarado, J.S. 1997. Elevation of the heat resistance of vegetative cells and spores of *Clostridium perfringens* type A by sublethal heat shock. J. Food Prot. 60: 998–1000.

Heredia, N.L., Labbe, R.G., and Garcia-Alvarado, J.S. 1998. Alteration in sporulation, enterotoxin production, and protein synthesis by *Clostridium perfringens* type A following heat shock. J. Food Prot. 61: 1143–1147.

Hielm, S., and Björkroth, J. 1998. Genomic analysis of *Clostridium botulinum* group II by pulsed-field gel electrophoresis. Appl. Environ. Microbiol. 64: 703–8.

Holdeman, L.V., Cato, E.P., and Moore, W.E.C. 1977. Anaerobe Laboratory Manual. 4th Edition. Department of Anaerobic Microbiology, Virginia Polytechnic Institute and State University, Blacksburg.

Huang, L. 2003a. Growth kinetics of *Clostridium perfringens* in cooked beef. J. Food Safety 23: 91–105.

Huang, L. 2003b. Estimation of growth of *Clostridium perfringens* in cooked beef under fluctuating temperature conditions. Food Microbiol. 20: 549–559.

Huang, L. 2003c. Dynamic computer simulation of *Clostridium perfringens* growth in cooked ground beef. Int. J. Food Microbiol. 87: 217–227.

Huang, L. 2002. Description of growth of *Clostridium perfringens* in cooked beef with multiple linear models. Food Microbiol. 19: 577–587.

Humeau, Y., and Doussau, F. 2000. How botulinum and tetanus neurotoxins block neurotransmitter release. Biochimie 82: 427–46.

Hutson, R.A., and Zhou, Y. 1996. Genetic characterization of *Clostridium botulinum* type A containing silent B neurotoxin gene sequences. J Biol Chem. 271:10786–92.

Hutt, P.B. 1984. Government regulation of the integrity of the food supply. Annu. Rev. Nutr. 4: 1–20, 1984.

International Commission for Microbiological Safety of Foods (ICMSF). 1980. Microbial ecology of foods: factors affecting life and death of microorganisms. Vol. 1, Academic Press, New York.

International Commission for Microbiological Safety of Foods (ICMSF). 1998. Microorganisms in Foods 6, Microbial Ecology of Food Commodities. Blackie Academic & Professional, London, pp. 66–111.

Jahkola, M. and Korkeala, H. 1997. Botulismi saksassa suomessa pakatusta savusiiasta. *Kansanterveys* 3/1997, 8–9.

Jiménez, D.F., and Giménez, J.A. 1995. The typing of botulinal neurotoxins. Int J Food Microbiol. 27: 1–9.

Johnson, E.A. 1999. Anaerobic Fermentations. In: Manual of Methods for Industrial Microbiology. A.L. Demain, and J. Davies, eds. 2nd Edition. ASM Press, Washington, DC.

Johnson, E.A., and Bradshaw, M. 2001. *Clostridium botulinum*: A metabolic and cellular perspective. Toxicon. 39: 1703–1722.

Johnson, E.A., and Goodnough, M.C. 1998. Botulism. In: Topley and Wilson's Microbiology and Microbial Infections, L. Collier, A. Balows, M. Sussman, eds. 9th Edition, Volume 3. Bacterial Infections. Arnold, London. p. 723–41.

Johnston, R., Harmon, S.M., and Kautter, D.A. 1964. Method to facilitate the isolation of *Clostridium botulinum* type E. J. Bacteriol. 88: 1521–2.

Juneja, V.K. and Eblen, B.S. 1995. Influence of sodium chloride on thermal inactivation and recovery of nonproteolytic *Clostridium botulinum* type B strain KAP B5 spores. J. Food Prot. 58: 813–816.

Juneja, V.K., Eblen, B.S., Marmer, B.S., Williams, A.C., Palumbo, S.A. and Miller, A.J. 1995a. Thermal resistance of nonproteolytic type B and type E *Clostridium botulinum* spores in phosphate buffer and turkey slurry. J. Food Prot. 58: 758–763.

Juneja, V.K., Marmer, B.S., Phillips, J.G. and Miller, A.J. 1995b. Influence of the intrinsic properties of food on thermal inactivation of spores of nonproteolytic *Clostridium botulinum*: development of a predictive model. J. Food Safety 15: 349–364.

Juneja, V.K. and Thippareddi, H. 2003. Inhibitory effects of organic acid salts on growth of *Clostridium perfringens* from spore inocula during chilling of marinated ground turkey breast. Poster presented at International Association for Food Protection annual meeting, New Orleans, LA.

Juneja, V.K., Novak, J.S., Huang, L., and Eblen, B.S. 2003. Increased thermotolerance of *Clostridium perfringens* spores following sublethal heat shock. Food Control 14: 163–168.

Juneja, V.K. and Marks, H.M. 2002. Predictive model for growth of *Clostridium perfringens* during cooling of cooked cured chicken. Food Microbiol. 19: 313–327.

Juneja, V.K., Novak, J.S., Marks, H.M., and Gombas, D.E. 2001. Growth of *Clostridium perfringens* from spore inocula in cooked cured beef: development of a predictive model. Innov. Food Sci. Emer. Technol. 2: 289–301.

Juneja, V.K., Whiting, R.C., Marks, H.M., and Snyder, O.P. 1999. Predictive model for growth of *Clostridium perfringens* at temperatures applicable to cooling of cooked meat. Food Microbiol. 16: 335–349.

Juneja, V.K. and Marmer, B.S. 1996. Growth of *Clostridium perfringens* from spore inocula in *sous-vide* turkey products. Int. J. Food Microbiol. 21: 115–123.

Juneja, V.K., Marmer, B.S., and Call, J.E. 1996. Influence of modified atmosphere packaging on growth of *Clostridium perfringens* in cooked turkey. J. Food Safety 16: 141–150.

Juneja, V.K. and Majka, W.M. 1995. Outgrowth of *Clostridium perfringens* spores in cook-in-bag beef products. J. Food Safety 15: 21–34.

Juneja, V.K., Call, J.E., Marmer, B.S., and Miller, A.J. 1994a. The effect of temperature abuse on *Clostridium perfringens* in cooked turkey stored under air and vacuum. Food Microbiol. 11: 187–193.

Juneja, V.K., Marmer, B.S., and Miller, A.J. 1994b. Growth and sporulation potential of *Clostridium perfringens* in aerobic and vacuum-packaged cooked beef. J. Food Prot. 57: 393–398.

Juneja, V.K., Snyder, O.P., and Cygnarowicz-Provost, M. 1994c. Influence of cooling rate on outgrowth of *Clostridium perfringens* spores and cooked ground beef. J. Food Prot. 57: 1063–1067.

Kalandakanond, S., and Coffield, J.A. 2001. Cleavage of intracellular substrates of botulinum toxins A, C, and D in a mammalian target tissue. J. Pharmacol. Exp.Therapeut. 296: 749–55.

Kalluri, P., Crowe, C., Reller, M., Gaul, L., Hayslett, J., Barth, S., Eliasberg, S., Ferreira, J., Holt, K., Bengston, S., Hendricks, K., and Sobel, J. 2003. An outbreak of foodborne botulism associated with food sold at a salvage store in Texas. Clin. Infect. Dis. 37: 1490-1495.

Kao, I., Drachman, D.B., and Price, D.L. 1976. Botulinum toxin: mechanism of presynaptic blockade. Science 193: 1256–8.

Katayama, S.-I., Dupuy, B., Garnier, T., and Cole, S. 1995. Rapid expansion of the physical and genetic map of the chromosome of *Clostridium perfringens* CPN50. J. Bacteriol. 177: 5680–5685.

Katayama, S., Dupuy, B., Daube, G., China, B., and Cole, S.T. 1996. Genome mapping of *Clostridium perfringens* strains with I-Ceu I shows many virulence genes to be plasmid-borne. Mol. Gen. Genet. 251: 720–726.

Kautter, D.A., and Harmon, S.M. 1966. Antagonistic effect on *Clostridium botulinum* type E by organisms resembling it. Appl Microbiol. 14: 616–22.

Kim, A.Y., and Blaschek, H.P. 1989. Construction of an *Escherichia coli-Clostridium perfringens* shuttle vector and plasmid transformation of *Clostridium perfringens*. Appl. Environ. Microbiol. 55: 360–365.

Koransky, J.R., Allen, S.D., and Dowell, V.R. Jr. 1978. Use of ethanol for selective isolation of sporeforming microorganisms. Appl Environ Microbiol. 35: 762–5.

Korkeala, H., Stengel, G., Hyytiä, Vogelsang, B., Bohl, A., Wihlman, H., Pakkala, P., and Hielm, S. 1998. Type E botulism associated with vacuum-packaged hot-smoked whitefish. Int.J. Food Microbiol. 43: 1–5.

Kozuka, S., Yasuda, Y., and Tochikubo, K. 1985. Ultrastructural localization of dipicolinic acid in dormant spores of *Bacillus subtilis* by immunoelectron microscopy with colloidal gold particles. J. Bacteriol. 162: 1250–1254.

Krausse, R., and Ullmann, U. 1991. A modified procedure for the identification of anaerobic bacteria by high performance liquid chromatography – quantitative analysis of short-chain fatty acids. Zbl. Bakt. 276: 1–8.

Labbe, R.G. 2001. *Clostridium perfringens*. In: Compendium of Methods for the Microbiological Examination of Foods. F.P. Downes and K. Ito, eds., 4th Edition. Am. Publ. Heath Ass.

Labbe, R.G. 1989. *Clostridium perfringens*. In: M.P. Doyle, ed., Foodborne Bacterial Pathogens. Marcel Dekker, New York.

Labbe, R.G., and Duncan, C.L. 1977. Spore coat protein and enterotoxin synthesis in *Clostridium perfringens*. J. Bacteriol. 131: 713–715.

Labbe, R.G. and Juneja, V.K. 2002. *Clostridium perfringens*. In: Foodborne Diseases, D.O. Cliver and H.P. Riemann, eds. 2nd Edition. Academic Press, Amsterdam.

Ladiges, W.C., Foster, J.F. and Ganz, W.M. 1974. Incidence and viability of *Clostridium perfringens* in ground beef. J. Milk Food Technol. 37: 622–623.

Lalli, G., Bohnert, S., Deinhardt, K., Verastegui, C., and Schiavo, G. 2003. The journey of tetanus and botulinum neurotoxins in neurons. Trends Microbiol. 11: 431–437.

Lamanna, C. 1969. The most poisonous poison. Science. 130: 763–72.

Lau, A.H.S., Wawirko, R.Z., and Chow, C.T. 1974. Purification and properties of boticin P produced by *Clostridium botulinum*. Can J Microbiol. 20: 385–390.

Lehmann, Y., Meile, L., and Teuber. 1996. Rubrerythrin from *Clostridium perfringens*: Cloning of the gene, purification of the protein, and characterization of its superoxide dismutase function. J. Bacteriol. 178: 7152–7158.

Ligler, F.S., Taitt, C.R., Shriver-Lake, L.C., Sapsford, K.E., Shubin, Y., and Golden, J.P. 2003. Array biosensor for detection of toxins. Anal. Bioanalyt. Chem. 377: 469–477.

Lin, W.J., and Johnson, E.A. 1995. Genome analysis of *Clostridium botulinum* type A by pulsed_field gel electrophoresis. Appl Environ Microbiol. 61: 4441–4447.

Lindsay, J. A., Beaman, T.C., and Gerhardt, P. 1985. Protoplast water content of bacterial spores determined by buoyant density sedimentation. J. Bacteriol. 163: 735–737.

Lindström, M., Heilm, S., Nevas, M., Tuisku, S. and Korkeala, H. 2004. Proteolytic *Clostridium botulinum* type B in the gastric content of a patient with type E botulism due to whitefish eggs. Foodborne Pathogens Dis. 1: 53–58.

Lindström, M., Nevas, M., Hielm, S., Lähteenmäki, L., Peck, M.W. and Korkeala, H. 2003. Thermal inactivation of nonproteolytic

*Clostridium botulinum* type E spores in model fish media and in vacuum-packaged hot-smoked fish products. Appl. Environ. Microbiol. 69: 4029–4036.

Lindström, M., Keto, R., Markkula, A., Nevas, M., Hielm, S., and Korkeala, H. 2001. Multiplex PCR assay for detection and identification of *Clostridium botulinum* types A, B, E, and F in food and fecal material. Appl. Environ. Microbiol. 67: 5694–5699.

Liu, W., Montana, V., Chapman, E.R., Mohideen, U., and Parpura. 2003. Botulinum toxin type B micromechanosensor. Proc. Natl. Acad. Sci. USA 100: 13621–13625.

Lucke, F.K. 1984. Psychrotrophic *Clostridium botulinum* strains from raw hams. Syst. Appl. Microbiol. 5: 274–279.

Lund, B.M. and Peck, M.W. 1994. Heat-resistance and recovery of non-proteolytic *Clostridium botulinum* in relation to refrigerated, processed foods with an extended shelf-life. J. Appl. Bacteriol. 76: 115s–128s.

Lund, B.M., and Peck, M.W. 2000. *Clostridium botulinum*. In: The Microbiological Safety and Quality of Food. B.M. Lund, T.C. Baird-Parker, G.W. Gould, eds. vol II, Aspen Publishers, Inc, Gaithersburg, Maryland, pp. 1057–1109.

Lyristis, M., Bryant, A.E., Sloan, J., Awad, M.M., Nisbet, I.T., Stevens, D.L., and Rood, J.I. 1994. Identification and molecular analysis of a locus that regulates extracellular toxin production in *Clostridium perfringens*. Mol. Microbiol. 12: 761–777.

Lyver, A., Smith, J.P., Nattress, F.M., Austin, J.W. and Blanchfield, B. 1998. Challenge studies with *Clostridium botulinum* type E in a value-added surimi product stored under modified atmosphere. J. Food Safety 18 1–23.

MacDonald, K.L., Cohen, M.L., and Blake, P.A. 1986. The changing epidemiology of adult botulism in the United States. Am. J. Epidemiol. 108: 150–156.

MacDonald, K.L., and Rutherford, G.W. 1985. Botulism and botulism-like illness in chronic drug abusers. Ann Intern Med. 102: 616–18.

MacDonald, K.L., Spengler, R.F., Hatheway, C.L., Hargrett, N.T., and Cohen, L.M. 1985. Type A botulism from sautéed onions: Clinical and epidemiological observations. JAMA 253: 1275–1278.

Mackle, I.J., Halcomb, E., and Parr, M.J.A. 2001. Severe adult botulism. *Anaesth. Intensive Care* 29: 297–300.

Mahony, D.E., Clark, G.A., Stringer, M.F., MacDonald, M.C., Duchesne, D.R., and Mader, J.A. 1986. Rapid extraction of plasmids from *Clostridium perfringens*. Appl. Environ. Microbiol. 51: 521–523.

Mahony, D.E., Stringer, M.F., Borriello, S.P., and Mader, J.A. 1987. Plasmid analysis as a means of strain differentiation in *Clostridium perfringens*. J. Clin. Microbiol. 25: 1333–1335.

Mahony, D.E., Mader, J.A., and Dubel, J.R. 1988. Transformation of *Clostridium perfringens* L forms with shuttle plasmid DNA. Appl. Environ. Microbiol. 54: 264–267.

Malizio, C.J., Goodnough, M.C., and Johnson, E.A. 2000. Purification of *Clostridium botulinum* type A neurotoxin. In: Bacterial Toxins: Methods and Protocols. O. Holst, ed. Meth Molec Biol, vol 145, Humana Press Inc, Totowa, New Jersey, pp. 27–39.

Marquis, R.E., and Bender, G.R. 1985. Mineralization and heat resistance of bacterial spores. J. Bacteriol. 161: 789–791.

Maurer, A.J. 1983. Reduced sodium usage in poultry muscle foods. Food Technol. 37: 60–65.

Masselli, R.A., and Bakshi, N. 2000. Botulism. Muscle and Nerve 23: 1137–44.

Matsushita, C., Matsushita, O., Koyama, M., and Okabe, A. 1994. A *Clostridium perfringens* vector for the selection of promoters. Plasmid 31: 317–319.

McCroskey, L.M., and Hatheway, C.L. 1988. Laboratory findings in four cases of adult botulism suggest colonization of the intestinal tract. J. Clin. Microbiol. 26: 1052–4.

McCroskey, L., and Hatheway, C.L. 1986. Characterization of an organism that produces type E botulinal toxin but which resembles *Clostridium butyricum* from the feces of an infant with type E botulism. J. Clin. Microbiol. 23: 201–2.

McCroskey, L.M., and Hatheway, C.L. 1991. Type F botulism due to neurotoxigenic *Clostridium baratii* from an unknown source in an adult. J. Clin. Microbiol. 29: 2618–20.

McGowan, S., Lucet, I.S., Cheung, J.K., Awad, M.M., Whisstock, J.C., and Rood, J.I. 2002. The FxRxHrS motif: a conserved region essential for DNA binding of the virR response regulator from *Clostridium perfringens*. J. Mol. Biol. 322: 997–1011.

McLean, H.E., Peck, S., and Blatherwick, F.J. 1987. Restaurant-associated botulism from in-house bottled mushrooms – British Columbia. Can. Commun. Dis. Rep. 13–8: 35–36.

Mead, P.S., Slutsker, L., Dietz, V., McCraig, L.F., Bresee, J.S., Shapiro, C., Griffin, P.M., and Tauxe, R.V. 1999. Food-related illness and death in the United States. Emerg. Infect. Dis. 5: 607–625.

Meijer, W.G. 1994. The Calvin cycle enzyme phosphoglycerate kinase of *Xanthobacter flavus* required for autotrophic $CO_2$ fixation is not encoded by the *cbb* operon. J. Bact. 176: 6120-6126.

Meng, X., and Yamakawa, K. 1997. Characterization of a neurotoxigenic *Clostridium butyricum* strain isolated from the food implicated in an outbreak of foodborne type E botulism. J. Clin Microbiol. 37: 1661–1669.

Merson, M.H., and Dowell, V.R. Jr. 1973. Epidemiologic, clinical and laboratory aspects of wound botulism. N. Engl. J Med. 289: 1105–10.

Midura, T.F. and Arnon, S.S. 1976. Infant botulism. Identification of *Clostridium botulinum* and its toxins in faeces. Lancet 2(7992):934–936.

Mills, D.C., Midura, T.F., and Arnon, S.S. 1985. Improved selective medium for the isolation of lipase-positive *Clostridium botulinum* from feces of human infants. J. Clin. Microbiol. 21: 947–50.

Miwa, N., Masuda, T., Kwamura, A., Terai, K. and Akiyama, M. 2002. Survival and growth of enterotoxin-positive and enterotoxin-negative *Clostridium perfringens* in laboratory media. Int. J. Food Microbiol. 72: 233–238.

Miyamoto, K., Chakrabarti, G., Morino, Y., and McClane, B.A. 2002. Organization of the plasmid cpe locus in *Clostridium perfringens* type A isolates. Infect. Immun. 70: 4261–4272.

Miyamoto, K., Wen, Q., and McClane, B.A. 2004. Multiplex PCR genotyping assay that distinguishes between isolates of *Clostridium perfringens* type A carrying a chromosomal enterotoxin gene (*cpe*) locus, a plasmid cpe locus with an IS1470-like sequence, or a plasmid *cpe* locus with an IS1151 sequence. J. Clin. Microbiol. 42: 1552–1558.

Miyata, S., Kozuka, S., Yasuda, Y., Chen, Y., Moriyama, R., Tochikubo, and K., Makino, S. 1997. Localization of germination-specific spore-lytic enzymes in *Clostridium perfringens* S40 spores detected by immunoelectron microscopy. FEMS Microbiol. Lett. 152:243–247.

Montecucco, C., and Schiavo, G. 1995. Structure and function of tetanus and botulinum neurotoxins.Quart Rev Biophys. 28: 423–472.

Morse, E.L., Pickard, L.K., Guzewich, J.J., Devine, B.D., and Shayegani, M. 1990. Garlic-in-oil associated botulism: episode leads to product modification. Am. J. Public Health 80: 1372-1373.

Mulligan, M.E., Halebian, S., Kwok, R.Y.Y., Cheng, W.C., Finegold, S.M., Anselmo, C.R., Gerding, D.N., and Peterson, L.R. 1986. Bacterial agglutination and polyacrylamide gel electrophoresis for typing *Clostridium difficile*. J. Infect. Dis. 153: 267–271.

Murrell, W.G. 1988. Bacterial spores-nature's ultimate survival package. In: Microbiology in Action. W.G. Murrell and I.R. Kennedy, eds. John Wiley & Sons, Inc., New York, pp. 311–346.

Murrell, W.G., and Warth, A.D. 1965. Composition and heat resistance of bacterial spores. In: Spores III. L.L. Campbell and H.O. Halvorson, eds. American Society for Microbiology, Washington, D.C., pp. 1–24.

Niemann, H. 1991. Molecular biology of clostridial neurotoxins. J.E. Alouf and J.H. Freer, eds., A sourcebook of bacterial protein toxins. Academic Press Ltd., London. p. 303–48.

Nishida, S., Seo, N., and Nakagawa, M. 1969. Sporulation, heat resistance, and biological properties of *Clostridium perfringens*. Appl. Microbiol. 17: 303–309.

Notermans, S., and Nagel, J. 1989. Assays for botulinum and tetanus toxins. Botulinum Neurotoxin and Tetanus Toxin. Ed. Simpson, L.L., Academic Press, San Diego 319–331.

Novak, J.S., Tunick, M.H., and Juneja, V.K. 2001. Heat treatment adaptations in *Clostridium perfringens* vegetative cells. J. Food Prot. 64: 1527–1534.

Ohishi, I., Sugii, S., and Sakaguchi, G. 1977. Oral toxicities of *Clostridium botulinum* toxins in response to molecular size. Infect Immun. 16: 107–9.

Ohtani, K., Bhowmik, S.K., Hayashi, H., and Shimizu, T. 2002. Identification of a novel locus that regulates expression of toxin genes in *Clostridium perfringens*. FEMS Microbiol. Lett. 209: 113–118.

Olive, D.M., and Bean, P. 1999. Principles and applications of methods for DNA-based typing of microbial organisms. J. Clin. Microbiol. 37: 1661–1669.

O'Mahony, M., Mitchell, E., and Gilbert, R.J. 1990. An outbreak of foodborne botulism with contaminated hazelnut yogurt. Epidemiol. Infect. 104: 389–395.

Paidhungat, M., Setlow, B., and Setlow, P. 2000. Characterization of spores of *Bacillus subtilis* which lack dipicolinic acid. J. Bacteriol. 182: 5505–5512.

Park, J.-B., and Simpson, L.L. 2003. Inhalational poisoning by botulinum toxin and inhalation vaccination with its heavy-chain component. Infect. Immun. 71: 1147–54.

Passaro, D.J., and Werner, S.B. 2000. Wound botulism associated with black tar heroin among injecting drug users. JAMA. 279: 859–863.

Peck, M.W. 1997. *Clostridium botulinum* and the safety of refrigerated processed foods of extended durability. Trends Food Sci. Technol. 8: 186–192.

Peck. M.W. 1999. Safety of *sous-vide* foods with respect to *Clostridium botulinum*. In: Proceedings of Third European Symposium on Sous Vide. Alma Sous Vide Centre Belgium. p. 171–194.

Peck, M.W. 2004. The good, the bad and the ugly – *Clostridium botulinum* neurotoxins. Microbiologist 5: 26–30.

Peck, M.W., Fairbairn, D.A. and Lund, B.M. 1993. Heat-resistance of spores of non-proteolytic *Clostridium botulinum* estimated on medium containing lysozyme. Lett. Appl. Microbiol. 16: 126–131.

Peck, M.W., Lund, B.M., Fairbairn, D.A., Kaspersson, A.S and Undeland, P. 1995. The effect of heat treatment on survival of, and growth from spores of non-proteolytic *Clostridium botulinum* at chill temperatures. Appl. Environ. Microbiol. 61: 1780–1785.

Peruski, A.H., Johnson, L.H., 3rd., and Peruski, L.F., Jr. 2002. Rapid and sensitive detection of biological warfare agents using time-resolved fluorescence assays. J. Immunol.Meth. 263.

Petit, L., Gilbert, M., and Popoff, M.R. 1999. *Clostridium perfringens*: toxinotype and genotype. Trends Microbiol. 7(3): 104–110.

Phillips-Jones, M.K., Iwanejko, L.A., and Longden, M.S. 1989. Analysis of plasmid profiling as a method for rapid differentiation of food-associated *Clostridium perfringens* strains. J. Appl. Bacteriol. 67: 243–254.

Phillips-Jones, M.K. 1990. Plasmid transformation of *Clostridium perfringens* by electroporation methods. FEMS Microbiol. Lett. 66: 221–226.

Phillips-Jones, M.K. 1993. Bioluminescence (lux) expression in the anaerobe *Clostridium perfringens*. FEMS Microbiol. Lett. 106: 265–270.

Pickett, J., and Berg, B. 1976. Syndrome of botulism in infancy: clinical and electrophysiologic study. N. Engl. J. Med. 295: 770–772.

Popham, D.L., Sengupta, S, and Setlow, P. 1995. Heat, hydrogen peroxide, and UV resistance of *Bacillus subtilis* spores with increased core water content and with or without major DNA-binding proteins. Appl. Environ. Microbiol. 61: 3633–3638.

Popoff, M.R. 1995. Ecology of neurotoxigenic strains of clostridia. Clostridial neurotoxins. C. Montecucco, ed. Springer, Berlin, pp. 1–29.

Popoff, M.R., and Marvaud, J.-C. 1999. Structural and genomic features of clostridial neurotoxins. J.E. Alouf, and J.H. Freer, eds. The Comprehensive Sourcebook of Bacterial Protein Toxins. Academic Press, London, pp. 174–201.

Pourshaban, M., Franciosa, G., Fenicia, L., and Aureli, P. 2002. Taxonomic identity of type E botulinum toxin-producing *Clostridium butyricum* strains by sequencing of a short 16S rDNA region. FEMS Microbiol Lett. 214: 119–125.

Pourshafie, M.R., Saifie, M., Shafiee, A., Vahdani, P., Aslani, M., and Salemian, J. 1998. An outbreak of foodborne botulism associated with contaminated locally made cheese in Iran. Scand. J. Infect. Dis. 30(1):92–94.

Proulx, J.F., Milot-Roy, V. and Austin, J. 1997. Four outbreaks of botulism in Ungava Bay, Nunavik, Quebec. Can. Commun. Dis. Rep. 23–4: 30–32.

Quinn, C.P., and Minton, N.P. 2001. Clostridial neurotoxins. In: Clostridia. H. Bahl, and P. Dürre, eds. Biotechnology and Medical Applications. Wiley-VCH, Weinheim, pp. 211–50.

Raffestin, S., Marvaud, J.C., Cerrato, R., Dupuy, B., and Popoff, M.R. 2004. Organization and regulation of the neurotoxin genes in *Clostridium botulinum* and *Clostridium tetani*. Anaerobe 10: 93–100.

Rhodehamel, E.J. 1992. FDA's concerns with sous vide processing. Food Technol. 46 (12): 73-76.

Rhodehamel, J., and Harmon, S. 1998. *Clostridium perfringens*. In: FDA Bacteriological Manual R.W. Bennett, ed., 8th Edition. AOAC International, Gaithersburg, MD, pp. 16.01–16.06.

Rhodehamel, E.J., Reddy, N.R., and Pierson, M.D. 1992. Botulism: the causative agent and its control in foods. Food Control. 3: 125–143.

Riha, W.E. and Solberg, M. 1975. *Clostridium perfringens* inhibition by sodium nitrite as a function of pH, inoculum size, and heat. J. Food Sci. 40:439–442.

Robach, M.C., F.J. Ivey, and C.S. Hickey. 1978. System for evacuating clostridial inhibition in cured meat products. Appl. Environ. Microbiol. 36: 210–211.

Roberts, E., Wales, S.M., Brett, M.M. and Bradding, P. 1998. Case report. Cranial-nerve palsies and vomiting. Lancet 352, 1674.

Roberts, I., Holmes, W.M., and Hylemon, P.B. 1988. Development of a new shuttle plasmid system for *Escherichia coli* and *Clostridium perfringens*. Appl. Environ. Microbiol. 54: 268-270.

Roberts, T.A. 1968. Heat and radiation resistance and activation of spores of *Clostridium welchii*. J. Appl Bact. 31: 133–144.

Rodgers, S. 2001. Preserving non-fermented refrigerated foods with microbial cultures – a review. Trends Food Sci. Technol. 12: 276–284.

Rodgers, S. 2003. Potential applications of protective cultures in cook-chill catering. Food Control 14: 35–42.

Rodgers, S., Peiris, P. and Casadei G. 2003a. Inhibition of non-proteolytic *Clostridium botulinum* with lactic acid bacteria and their bacteriocins at refrigeration temperatures. J. Food Prot. 66: 674–678.

Rodgers, S., Peiris, P., Kailasapathy, K. and Cox, J. 2003b. Inhibition of non-proteolytic *Clostridium botulinum* with lactic acid bacteria in extended shelf-life cook-chill soups. Food Biotechnol. 17: 39–52.

Rood, J.I., and Cole, S.T. 1991. Molecular genetics and pathogenesis of *Clostridium perfringens*. Microbiol. Rev. 55: 621–648.

Rosetti, F., Castelli, E., Labbe, J. and Funes, R. 1999. Outbreak of type E botulism associated with home-cured ham consumption. Anaerobe 5, 171–172.

Roy, R.J., Busta, F.F. and Thompson, D.R. 1981. Thermal inactivation of *Clostridium perfringens* after growth at several constant and linearly rising temperatures. J. Food Sci. 46: 1586–1591.

Sabah, J.R., Thippareddi, H., Marsden, J.L. and Fung, D.Y.C. 2003a. Use of organic acids for the control of *Clostridium* perfringens in cooked vacuum-packaged restructured roast beef during an alternative cooling procedure. J. Food Prot. 66: 1408–1412.

Sabah, J.R., V.K. Juneja, and D.Y.C. Fung. 2003b. Effect of spices and organic acids on the growth of *Clostridium perfringens* from spore inocula during cooling of *sous-vide* cooked ground beef products. Presented at IFT Annual Meeting, Chicago, IL.

Saint Louis, M.E., and Shaun, H.S. 1988. Botulism from chopped garlic: delayed recognition of a major outbreak. Ann Int Med. 108: 363–368.

Sakaguchi, G. 1983. *Clostridium botulinum* toxins. Pharmacol Ther. 19: 165–94.

Sakaguchi, G., Ohishi, I., and Kozaki, S. 1981. Purification and oral toxicities of *Clostridium botulinum* progenitor toxins. Biomedical aspects of botulism. Academic Press, Inc., New York, pp. 21–34.

Sarker, M.R., Shivers, R.P., Sparks, S.G., Juneja, V.K., and McClane, B.A. 2000. Compartive experiments to examine the effects of heating on vegetative cells and spores of *Clostridium perfringens* isolates carrying plasmid genes versus chromosomal enterotoxin genes. Appl. Environ. Microbiol. 66: 3234–3240.

Schalch, B., Eisgruber, H., Schau, H.P., Wiedmann, M., and Stolle, A. 1998. Strain differentiation of *Clostridium perfringens* by bacteriocin typing, plasmid profiling and ribotyping. J. Vet. Med. 45: 595–602.

Schantz, E.J., and Johnson, E.A. 1992. Properties and use of botulinum toxin and other microbial neurotoxins in medicine. Microbiol Rev. 56: 80–92.

Schantz, E.J., and Kautter, D.A. 1978. Standardized assay for *Clostridium botulinum* toxins. J. Assoc. Off. Anal. Chem. 61: 96–99.

Schiavo, G., and Montecucco, C. 1995. Tetanus and botulinum neurtoxins: isolation and assay. Meth Enzymol. 248: 643–52.

Schiavo, G. 2000. Neurotoxins affecting neuroexocytosis. Physiol. Rev. 80: 717–766.

Schmidt, J.J., and Stafford, R.G. 2003. Fluorigenic substrates for the protease activities of botulinum neurotoxins, serotypes A, B, and F. Appl. Environ. Microbiol. 69: 297–303.

Seals, J.E., Snyder, J.D., Edell, T.A., Hatheway, C.L., Johnson, C.J., Swanson, R.C., and Hughes, J.M. 1981. Restaurant-associated type A botulism: transmission by potato salad. Am. J. Epidemiol. 113(4): 436–444.

Setlow, P., and Johnson, E.A. 2001. Spores and their significance. In: Food Microbiology: Fundamentals and Frontiers. M.P. Doyle, L.R. Beuchat, and T. Montville, eds. ASM Press, Washington, DC.

Setlow, B., and Setlow, P. 1995. Small, acid-soluble proteins bound to DNA protect *Bacillus subtilis* spores from killing by dry heat. Appl. Environ. Microbiol. 61: 2787–2790.

Setlow, P. 2000. Resistance of bacterial spores. In: Bacterial Stress Responses. G. Storz and R. Hengge-Aronis, eds., American Society for Microbiology, Washington, D.C., pp. 217–230.

Shapiro, R.L., Hatheway, C.L., and Swerdlow, D.L. 1998. Botulism in the United States: a clinical and epidemiologic review. Ann. Intern. Med. 129: 221–8.

Sheridan, J.J., Buchanan, R.L. and Montville, T.J. eds. 1996. HACCP: An Integrated Approach to Assuring the Microbiological Safety of Meat and Poultry. Food & Nutrition Press, Trumbull, Conn.

Shigehisa, T., Nakagami, T. and Taji, S. 1985. Influence of heating and cooling rates on spore germination and growth of *Clostridium perfringens* in media and in roast beef. Jpn. J. Vet. Sci. 47: 259–267.

Shimizu, T., Ohtani, K., Hirakawa, H., Ohshima, K., Yamashita, A., Shiba, T., Ogasawara, N., Hattori, M., Kuhara, S., and Hayashi, H. 2002a. Complete genome sequence of *Clostridium perfringens*, an anaerobic flesh-eater. Proc. Natl. Acad. Sci. USA 99: 996–1001.

Shimizu, T., Yaguchi, H., Ohtani, K., Banu, S., and Hayashi, H. 2002b. Clostridial VirR/VirS regulon involves a regulatory RNA molecule for expression of toxins. Mol. Microbiol. 43: 257–265.

Simpson, L.L. 2004. Identification of the major steps in botulinum toxin action. Annu. Rev. Pharacol. Toxicol. 44: 167–193.

Skinner, G.E., Solomon, H.M. and Fingerhut, G.A. 1999. Prevention of *Clostridium botulinum* type A, proteolytic B and E toxin formation in refrigerated pea soup by *Lactobacillus plantarum*, ATCC 8014. J. Food Sci. 64: 724–727.

Slater, P.E., Addiss, D.G., and Cohen, A. 1989. Foodborne botulism: an international outbreak. Int. J. Epidemiol. 18: 693–696.

Sloan, J., Warner, T.A., Scott, P.T., Bannam, T.L., Berryman, D.I., and Rood, J.I. 1992. Construction of a sequenced *Clostridium perfringens-Escherichia coli* shuttle plasmid. Plasmid 27: 207–219.

Smith, L.B., Busta, F.F., and Allen, C.E. 1980. Effect of rising temperatures on growth and survival of *Clostridium perfringens* indigenous to raw beef. J. Food Prot. 43: 520–524.

Smith, L.D.S. 1975. Inhibition of *Clostridium botulinum* by strains of *Clostridium perfringens* isolated from soil. Appl Microbiol. 30: 319–323.

Smith, L.D.S. 1977. Botulism. In: The Organism, its Toxins, the Disease. 1st Edition. Charles C Thomas, Springfield, Illinois, USA.

Sobel, J., Tucker, N., Sulka, A., McLaughlin, J., and Maslanka, S. 2004. Foodborne botulism in the United States. 1990–2000. Personal communication.

Solberg, M., Blaschek, H.P., and Kahn, P. 1981. Plasmids in *Clostridium perfringens*. J. Food Safety 3: 267–289.

Solberg, M. and Elkind, B. 1970. Effect of processing and storage conditions on the microflora of *Clostridium perfringens*-inoculated frankfurters. J. Food Sci. 35: 126–129.

Solomon, H.M., and Johnson, E.A. 2001. *Clostridium botulinum* and its toxins. In: Compendium for the Microbiological Examination of Foods, F.P. Downes, and K. Ito, eds. 4th Edition. American Public Health Association, Washington, DC, pp. 317–324.

Solomon, H.M., and Kautter, D.A. 1986. Growth and toxin production by *Clostridium botulinum* in shredded cabbage at room temperature under modified atmosphere. J. Food Prot. 53: 831–833.

Solomon, H.M., Kautter, D.A., Lilly, T., and Rhodehamel, E.J. 1990. Outgrowth of *Clostridium botulinum* in shredded cabbage at room temperature under modified atmosphere. J. Food Prot. 53: 831–833.

Solomon, H.M., Rhodehamel, E.J., and Kautter, D.A. 1995. *Clostridium botulinum*. FDA Bacteriological Analytical Manual. 8th Edition. USA Food and Drug Administration, Washington, DC, pp. 17.01–10.

Sparks, S.G., Carman, R.J., Sarker, M.R., and McClane, B.A. 2001. Genotyping of enterotoxigenic *Clostridium perfringens* fecal isolates associated with antibiotic-associated diarrhea and food poisoning in North America. J. Clin. Microbiol. 39: 883–888.

Steele, F.M. and K.H. Wright. 2001. Cooling rate effect of outgrowth of *Clostridium perfringens* in cooked, ready-to-eat turkey breast roasts. Poultry Sci. 80: 813–816.

St. Louis, M.E., and Shuan, H.S. 1988. Botulism from chopped garlic: delayed recognition of a major outbreak. Ann. Intern. Med. 108: 363–368.

Stringer, S.C. and Peck, M.W. 1997. Combinations of heat treatment and sodium chloride that prevent growth from spores of non-proteolytic *Clostridium botulinum*. J. Food Prot. 60: 1553–1559.

Stringer, S.C., Haque, N. and Peck, M.W. 1999. Growth from spores of non-proteolytic *Clostridium botulinum* in heat treated vegetable juice. Appl. Environ. Microbiol. 65: 2136–2142.

Strong, D.H., Foster, E.F., and Duncan, C.L. 1970. Influence of water activity on the growth of *Clostridium perfringens*. Appl. Microbiol. 19: 980–987.

Sugiyama, H. 1980. *Clostridium botulinum* neurotoxin. Microbiol Rev. 44: 419–448.

Szabo, E.A., and Pemberton, J.M. 1994. Polymerase chain reaction for detection of *Clostridium botulinum* types A, B, and E in food, soil and infant feces. J Appl Bacteriol. 76: 539–545.

Szilagyi, M., Rivera, V.R., Neal, D., Merrill, G.A., and Poli, M.A. 2000. Development of sensitive colorimetric capture elisas for *Clostridium botulinum* neurotoxin serotypes A and B. Toxicon. 38: 381–389.

Taormina, P.J., G.W. Bartholomew, and W.J. Dorsa. 2003. Incidence of *Clostridium perfringens* in commercially produced cured raw meat product mixtures and behavior in cooked products during chilling and refrigerated storage. J. Food Prot. 66: 72–81.

Terranova, W., Breman, J.G., Locey, R.P., and Speck, S. 1978. Botulism type B: epidemiologic aspects of an extensive outbreak. Am. J. Epidemiol. 108(2): 150–156.

Titball, R.W., Hunter, S.E.C., Martin, K.L., Morris, B.C., Shuttleworth, A.D., Rubidge, T., Anderson, D.W., and Kelly, D.C. 1989. Molecular cloning and nucleotide sequence of the alpha-toxin (phospholipase C) of *Clostridium perfringens*. Infect. Immun. 57: 367–376.

Todd, E.C.D. 1989. Costs of acute bacterial foodborne disease in Canada and the United States. Int. J. Food Microbiol. 9: 313–326.

Townes, J.M., and Cieslak, P.R. 1996. An outbreak of type A botulism associated with a commercial cheese sauce. Ann. Intern. Med. 125: 558–563.

Tschirdewahn, B., Notermans, S., Wernars, K. and Untermann, F. 1991. The presence of enterotoxigenic *Clostridium perfringens* strains in faeces of various animals. Int. J. Food Microbiol. 14: 175–178.

Tuomi, S., Matthews, M.E. and Marth, E.H. 1974. Behavior of *Clostridium perfringens* in precooked chilled ground beef gravy during cooling, holding, and reheating. J. Milk Food Technol. 37: 494–498.

Turton, K., Chaddock, J.A., and Acharya, K.R. 2002. Botulinum and tetanus neurotoxins: structure, function, and therapeutic utility. Trends Biochem Sci. 27: 552–558.

USDA/FSIS. 2001a. Draft compliance guidelines for ready-to-eat meat and poultry products. Fed. Regist. 10 May 2001.

USDA/FSIS. 2001b. Performance standards for the production of certain meat and poultry products; final rule. Fed. Regist. 64: 732–749.

van Damme-Jongsten, M., Wernars, K., and Notermans, S. 1989. Cloning and sequencing of the enterotoxin gene. Antonie Leewenhoek J. Microbiol. 56: 181–190.

van Ermengem, E. 1979. Classics in infectious diseases. A new anaerobic *Bacillus* and its relation to botulism. Rev. Infect. Dis. 1: 701–719.

van Holzer, E. 1962. Botulisms durch inhalation. Med Klinik (Munchen). 57: 1735–1738.

Villar, R.G., Shapiro, R.L., Busto, S., Riva-Posse, C., Verdejo, G., Farace, M.I., Rosetti, F., San Juan, J.A., Julia, C.M., Becher, J., Maslanka, S.E., and Swerdlow, D.L. 1999. Outbreak of type A botulism and development of a botulism surveillance and antitoxin release system in Argentina. JAMA 281: 1334–1338.

Waldroup, A.L. 1996. Contamination of raw poultry with pathogens. World Poultry Sci J. 52: 7–25.

Walker, H.W. 1975. Foodborne illness from *Clostridium perfringens*. CRC Crit. Rev. Food Sci. Nutr. 7: 71–104.

Wang, X., Maegawa, T., Karasawa, T., Kozaki, S., Tsukamoto, K., Gyobu, Y., Yamakawa, K., Oguma, K., Sakaguchi, Y., and Nakamura, S. 2000. Genetic analysis of type E botulinum toxin-producing *Clostridium butyricum* strains. Appl Environ Microbiol. 66: 4992–4997.

Weber, J.T., and Hibbs, R.G. 1993. A massive outbreak of type E botulism associated with traditional salted fish in Cairo, Egypt. J. Infect. Dis. 167: 451–454.

Whiting, R.C. and Oriente, J.C. 1997. Time-to-turbidity model for non-proteolytic type B *Clostridium botulinum*. Int. J. Food Microbiol. 35: 49–60.

Wictome, M., Newton, K., Jameson, K., Hallis, B., Dunnigan, P., Mackay, E., Clarke, S., Taylor, R., Gaze, J., Foster, K., and Shone, C. 1999. Development of an *in vitro* bioassay for *Clostridium botulinum* type B neurotoxin in foods that is more sensitive than the mouse bioassay. Appl. Environ. Microbiol. 65: 3787–3792.

Willardsen, R.R., Busta, F.F., and Allen, C.E. 1979. Growth of *Clostridium perfringens* in three different beef media and fluid thioglycollate medium at static and constantly rising temperatures. J. Food Prot. 42: 144–148.

Willardsen, R.R., Busta, F.F., Allen, C.E., and Smith, L.B. 1978. Growth and survival of *Clostridium perfringens* during constantly rising temperatures. J. Food Sci. 43: 470–475.

Willis, A.T. 1977. Anaerobic Bacteriology: Clinical and Laboratory Practice. 3rd Edition. Butterworths, London.

Wu, H.C., and Huang, Y.L. 2001. Detection of *Clostridium botulinum* neurotoxin type A using immuno-PCR. Lett Appl Microbiol. 32: 321–5.

Zaika, L.L. 2003. Influence of NaCl content and cooling rate on outgrowth of *Clostridium perfringens* spores in cooked ham and beef. J. Food Prot. 66: 1599–1603.

# Bacillus cereus

## 20

Per Einar Granum

## Abstract

The *Bacillus cereus* group comprises six members: *B. anthracis*, *B. cereus*, *B. mycoides*, *B. pseudomycoides*, *B. thuringiensis*, and *B. weihenstephanensis*. These species are closely related and should be placed within one species, except for *B. anthracis*, which possesses specific large virulence plasmids. *B. cereus* is a normal soil inhabitant and is frequently isolated from a variety of foods, including vegetables, dairy products, and meat. It causes an emetic or a diarrheal type of food-associated illness that is becoming increasingly important in the industrialized world. Some patients may experience both types of illness simultaneously. The diarrheal type is most prevalent in the western hemisphere, whereas the emetic type is most prevalent in Japan. Desserts, meat dishes, and dairy products are the foods most frequently associated with diarrheal illness, whereas rice and pasta are the most common vehicles of emetic illness. The emetic toxin (cereulide) has been isolated and characterized; it is a small ring peptide synthesized non-ribosomally by a peptide synthetase. Three types of *B. cereus* enterotoxins involved in foodborne outbreaks have been identified. Two of these enterotoxins are three-component proteins and are related, while the last is a one-component protein (CytK). Deaths have been recorded both by strains that produce the emetic toxin and by a strain producing only CytK. Some strains of the *B. cereus* group are able to grow at refrigeration temperatures. These variants raise concern about the safety of cooked, refrigerated foods with an extended shelf life. *B. cereus* spores adhere to many surfaces and survive normal washing and disinfection (except for hypochlorite and UVC) procedures. *B. cereus* foodborne illness is likely underreported because of its relatively mild symptoms, which are of short duration. However, consumer interest in precooked chilled food products with a long shelf life may lead to products well suited for *B. cereus* survival and growth. The availability of such foods could increase the prominence of *B. cereus* as a foodborne pathogen.

## Characteristics and taxonomy

The aerobic endospore forming bacteria have traditionally been placed in the genus *Bacillus*. Over the past three decades, this genus has expanded to accommodate more than 100 species (Ash *et al.* 1991), and the family *Bacillaceae* has been divided into nine different genera (Bergey's Manual 2001; Taxonomic Outline: http://www.cme.msu.edu/Bergeys/). The first genus is *Bacillus*, where the members of the *B. cereus* group are found. There are six species that belong to this group including *B. anthracis* (Table 20.1). All of these species can probably cause food poisoning and will in most cases not be distinguished in routine food laboratories, apart from *B. anthracis* that is usually not hemolytic and sensitive to penicillin. It has also been suggested that these species are so closely related that they should be considered as one species (Carlson *et al.*, 1994, 1996). In this chapter, the *B. cereus* group (apart from *B. anthracis*) is dealt with as one species, although *B. thuringiensis* and *B. weihenstephanensis* will be dealt with separately. The members of the *B. cereus* group are Gram-positive, spore-forming, motile, aerobic rods, which grow well anerobically.

The six species of the *B. cereus* group (Table 20.1) are large (cell width > 0.9 μm) and produce central to terminal ellipsoid or cylindrical spores that do not distend the sporangia (Claus and Berkeley, 1986; Lechner *at al.*, 1998; Nakamura, 1998). These *Bacillus* species sporulate easily after 2 to 3 days on most media, and *B. cereus* and *B. thuringiensis* lose their motility during the early stages of sporulation.

## Growth

Food products that have not been autoclaved cannot be guaranteed free of *Bacillus* spp. spores; thus, the conditions for treatment and storage of the products will determine the possibility for germination and growth. In pasteurized foods, the conditions may allow growth of the members of the *B. cereus* group because competition from other vegetative bacteria is excluded. The vegetative cells all grow between 10°C and about 50°C (Claus and Berkeley, 1986). However, some *B. cereus* strains are able to grow at temperatures as low as 4 to 6°C (Christiansson *at al.*, 1989; van Netten *et al.*, 1990; Granum *at al.*, 1993a; Borge *et al.*, 2001; Stenfors and Granum, 2001). These psychrotolerant strains are found mainly in milk and dairy products, but are also isolated from other products that are heat treated below 100°C. Relatively low water activity

**Table 20.1** Criteria to differentiate between the members of the *Bacillus cereus* group

| Species | Colony morphology | Hemolysis | Mobility | Susceptible to penicillin | Parasporal crystal inclusion |
|---|---|---|---|---|---|
| *B. cereus* | White | + | + | – | -- |
| *B. anthracis* | White | – | - | + | – |
| *B. thuringiensis* | White/grey | + | + | – | + |
| *B. mycoides* | Rhizoid | (+) | - | – | – |
| *B. weihenstephanensis* | Separated from *B. cereus* by growth at < 7°C and not at 43°C and can be identified rapidly using rDNA or cspA (cold shock protein A) targeted PCR (Lechner *et al.*, 1998) | | | | |
| *B. pseudomycoides* | Not distinguishable from *B. mycoides* by physiological and morphological characteristics. Clearly separable based on fatty acid composition, and 16S RNA sequences (Nakamura, 1998) | | | | |

($a_w$ = 0.92) does not prevent *B. cereus* from growing, neither does salt in concentrations up to 7%. There are not many foods with a high enough pH to prevent growth of *B. cereus* (> 9.3), but growth is prevented below pH 4.5. Thus, foods such as yoghurt and similar products are not generally associated with *B. cereus* food poisoning (Granum and Baird-Parker, 2000). In summary, growth of *B. cereus* is supported in a large variety of foods if not maintained at temperatures below 4°C or above 60°C.

## Characteristics of disease, mechanism of pathogenesis, and virulence factors

### Diseases caused by *Bacillus cereus*

There are two different types of *B. cereus* food poisoning. The first type caused by an emetic toxin results in vomiting, while the second type, caused by one of three different enterotoxins, induces diarrhoea (Granum, 2001). In a small number of cases both types of symptoms are recorded (Kramer and Gilbert, 1989; Granum: Norwegian Reference Laboratory for *B. cereus*), due to production of both types of toxins. Although the enterotoxins that cause diarrhoea may be preformed in the foods, model experiments have shown that the enterotoxins are degraded on their way to the target cells in the ileum (Granum *at al.*, 1993a, 1993b). The characteristics of the two types of *B. cereus* food poisoning are given in Table 20.2.

In recent years, some reports on *B. cereus* food poisoning of specific concern have appeared, including two outbreaks with fatal outcomes. From Norway, there is one report of an outbreak where several people were affected after eating a stew. The infective dose was approximately $10^4$ to $10^5$ (only spores were found in the food), and 17 out of 24 individuals were affected. Three of the patients were hospitalized, one for 3 weeks (Granum, 1994). For these three patients the onset was late, after more than 24 h. Experiments using spores from the strain isolated after this outbreak showed that the spores were able to adhere to Caco-2 cells in culture, and that the adherence properties were linked to hydrophobicity and possibly to surface appendages. Other spores tested did not have this ability (Andersson *et al.*, 1998). The longer incubation period observed in this case was as expected, since the spores would first have to germinate. The emetic toxin, cereulide, was responsible

for the death of a 17-year-old Swiss boy due to fulminant liver failure (Mahler *et al.*, 1997). A large amount of *B. cereus* emetic toxin was found in the residue from the pan used to reheat the food (pasta) and in the boy's liver and bile. The newly discovered enterotoxin, cytotoxin K (CytK), is similar to the β-toxin of *Clostridium perfringens* (and other related toxins) and was the cause of the symptoms in a severe outbreak of *B. cereus* food poisoning in France in 1998 (Lund *et al.*, 2000). In this outbreak several people developed bloody diarrhea, and three died. This is the first described outbreak of *B. cereus* necrotic enteritis, although it is not nearly as severe as *C. perfringens* type C food poisoning (Brynestad and Granum 2002).

The infective dose of the members of the *B. cereus* group varies considerably due to large differences in levels of enterotoxin production among the strains (Granum and Baird-Parker, 2000). Spores contained in foods survive the stomach acid barrier, whereas most vegetative cells are destroyed. Some foods may protect the pathogens from the stomach acid barrier, while other foods give limited protection. Furthermore, infection with *B. cereus* is more likely if the dish is eaten late in a meal than if eaten early, since levels of vegetative cells may increase if the food is temperature abused. After the first recognised diarrheal outbreak of *B. cereus* food poisoning in Oslo (due to vanilla sauce), Professor Hauge isolated the strain, grew it to $4 \times 10^6$ CFU/mL, and drank 200 mL (Hauge, 1950; Hauge, 1955). After about 13 hours he developed abdominal pain and watery diarrhea that lasted for about 8 hours. This corresponded to an infective dose of about $8 \times 10^8$ *B. cereus*. After outbreaks, counts ranging from 200 to $10^9$ *B. cereus*/g (or mL) (Kramer and Gilbert, 1989; Granum and Baird-Parker, 2000) have been reported in the incriminated foods, giving total infective doses ranging from about $5 \times 10^4$ to $10^{11}$. Partly due to the big differences in the amount of enterotoxin produced by different strains and partly due to other factors mentioned above, the total infective dose seems to vary between about $10^5$ and $10^8$ viable cells or spores. Thus, any food containing greater than $10^3$ *B. cereus*/g cannot be considered completely safe for consumption.

The emetic toxin is produced in the foods before consumption, and counts of *B. cereus* ranging from about $10^5$ to up to $10^9$ have been reported from the dishes involved (Kramer and Gilbert, 1989). Since it is possible that the numbers of *B. cereus* were higher prior to analysis, these numbers must be evaluated

**Table 20.2** Characteristics of the two types of disease caused by *Bacillus cereus*[1]

|  | Diarrheal syndrome | Emetic syndrome |
|---|---|---|
| Infective dose | $10^5$–$10^7$ (total) | $10^5$–$10^8$ (cells per gram) |
| Toxin produced | In the small intestine of the host | Preformed in foods |
| Type of toxin | Protein | Cyclic peptide |
| Incubation period | 8–16 hours (occasionally > 24 hours) | 0.5–5 hours |
| Duration of illness | 12–24 hours (occasionally several days) | 6–24 hours |
| Symptoms | Abdominal pain, watery diarrhea and occasionally nausea. Bloody diarrhea has been observed | Nausea, vomiting and malaise (sometimes followed by diarrhea, due to additional enterotoxin production) |
| Foods most frequently implicated | Meat products, soups, vegetables, puddings/sauces and milk products | Fried and cooked rice, pasta, pastry and noodles |

[1]Based on Kramer and Gilbert (1986), Turnbull (1986), Shinagawa (1993), Granum (2001).

with care. The total amount of cereulide needed for emesis is about 10 μg/kg body weight of the patient (Paananen *et al.*, 2002).

## Mechanism of pathogenesis and virulence factors

The emetic type of *B. cereus* food poisoning is caused by a small ring peptide made of three repeats of four modified amino acids (and/or oxy-acids): [D-O-Leu-D-Ala-L-O-Val-L-Val]₃. The emetic toxin has been named cereulide and has a molecular mass of 1.2 kDa (Agata *et al.*, 1994, 1995b). The toxin is produced in foods, at temperatures ranging from about 10 to 40°C, under vegetative growth (Kramer and Gilbert, 1989). However, maximal production of emetic toxin seems to occur at temperatures between 12 and 22°C (Finlay *et al.*, 2000; Agata *et al.*, 2002; Haggblom *et al.*, 2002). Since these experiments were carried out under a relatively limited type of conditions and with only few strains, we still have to use this information with care. After the toxin is produced no treatment will destroy this stable molecule, including the stomach acid and the proteolytic enzymes of the intestinal tract (Kramer and Gilbert, 1989). The *B. cereus* cells may no longer be present in the foods when eaten, as occurs in reheated rice or pasta that has been stored at room temperature after a first heating (Mortimer and McCann, 1974). After released from the stomach into the duodenum, cereulide is bound to the 5-HT₃ receptor (Agata *et al.*, 1995b), and stimulation of the vagus afferent causes emesis (vomiting).

Since cereulide also acts as a $K^+$ ionophore, like valinomycin, it is able to cause inhibition of mitochondrial activity (inhibition of fatty acid oxidation). This effect of cereulide was the reason for the liver failure described after the death of the 17-year-old Swiss boy (Mahler *et al.*, 1997). Mice were injected intraperitoneally with synthetic cereulide, and the development of histopathological changes was examined (Yokoyama *et al.*, 1999). At high cereulide doses, massive degeneration of hepatocytes occurred. The serum values of hepatic enzymes were highest 2 to 3 days after the inoculation of cereulide, and rapidly decreased thereafter. General recovery from the pathological changes and regeneration of hepatocytes was observed

after 4 weeks. The properties of cereulide are summarized in Table 20.3.

The diarrheal syndrome of *B. cereus* is cased by three different enterotoxins. Two of these are three-component proteins. The three-component hemolysin (Hbl; consisting of three proteins: B, $L_1$ and $L_2$) with enterotoxin activity was the first to be fully characterised (Beecher and Wong, 1994a, 1997). This toxin also has dermonecrotic and vascular permeability activities and causes fluid accumulation in ligated rabbit ileal loops. Hbl has been suggested to be a primary virulence factor in *B. cereus* diarrhea (Beecher *et al.*, 1995) although this is probably not true, since only about half of the food poisoning strains produce this enterotoxin (Guinebretiere *et al.*, 2002; Stenfors *et al.*, 2002). It has been shown that all three components are necessary for maximal enterotoxin activity (Beecher *et al.*, 1995). It was first suggested that the B protein is the component that binds Hbl to the target cells, and that $L_1$ and $L_2$ have lytic functions (Beecher and Macmillan 1991). However, recently another model for the action of Hbl has been proposed, suggesting that the components of Hbl bind to target cells independently, and then constitute a membrane attacking complex resulting in a colloid osmotic lysis mechanism (Beecher and Wong, 1997). A 1:1:1 ratio of the three components seems to give the highest biological activity (Beecher *et al.*, 1995). Substantial heterogeneity has been observed in the components of Hbl, and individual strains produced various combinations of single or multiple bands of each component (Schoeni and Wong, 1999). This is probably due to DNA sequence variation in *hbl* genes, but this has to be established genetically.

A non-hemolytic three-component enterotoxin (Nhe) was characterized (Lund and Granum, 1996) after an outbreak caused by an *hbl*-negative strain involving 152 people. These three components were different from those of Hbl, although there were similarities. Binary combination of the components of this enterotoxin results in some biological activity (likely due to contamination with the third component), but not nearly as high as when all the components are present (Lund and Granum 1997). The ratio between these components is 10:10:1 (Lindbäck *et al.*, 2004), but more research is needed to provide a model of the mode of action of Nhe.

**Table 20.3** Properties of the emetic toxin: cereulide[1]

|  | Property/activity |
| --- | --- |
| Molecular mass | 1.2 kDa |
| Structure | Ring formed peptide |
| Isoelectric point | Uncharged |
| Antigenic | No |
| Biological activity on living primates | Vomiting, inhibits fatty acid oxidation in mitochondria |
| Receptor | 5-HT$_3$ (stimulation of the vagus afferent) |
| Ileal loop tests (rabbit, mouse) | None |
| Cytotoxic | Indirectly by inhibition of mitochondrial activity |
| HEp-2 cells | Vacuolation activity |
| Sperm cells | Inhibit movement in minutes |
| Stability to heat | 90 min at 121°C |
| Stability to pH | Stable at pH 2–11 |
| Effect of proteolysis (trypsin, pepsin) | None |
| Toxin produced | In food: rice and milk at 15–32°C. Best at room temperature or below |
| Production | By a peptide synthetase |

[1]Based on Kramer and Gilbert (1986), Shinagawa (1993), Agata *et al.* (1994), Agata *et al.* (1995), Mikkola *et al.* (1999; Haggblom *et al.* 2002).

Although there are similarities in the structure and identity in the sequences of Hbl and Nhe, components from one of the toxins cannot substitute for components of the other (Lund and Granum, 1997). The identity is highest in the N-terminal third of the proteins. The most pronounced similarities are found between *nheA* and *hblC*, *nheB* and *hblD*, and *nheC* and *hblA*. This is not only by direct comparison of the sequences, but also evident in the predicted transmembrane helices for the six proteins. NheA and HblC have no predicted transmembrane helices, while NheB and HblD have two each. Finally, NheC and HblA each have one predicted transmembrane helix, in the same position of the two proteins (Granum *et al.*, 1999).

Almost all tested *B. cereus/B thuringiensis* strains produce Nhe, and about 50% produce Hbl (Rivera *et al.*, 2000; Guinebretiere *et al.*, 2002). Presently we do not know how important each of toxins are in relation to food poisoning when both are present; however, Nhe seems to be somewhat more potent than Hbl against human epithelial cells (Choma and Granum, unpublished results).

The newly discovered enterotoxin, cytotoxin K (CytK), is similar to the β-toxin of *Clostridium perfringens* (and other related toxins), and a strain producing this toxin was the cause of the symptoms in a severe outbreak of *B. cereus* food poisoning in France in 1998 (Lund *et al.*, 2000). In this outbreak several people developed bloody diarrhea, and three died. It would be fair to call this an outbreak of *B. cereus* necrotic enteritis, although it was not nearly as severe as *C. perfringens* type C food poisoning (Brynestad and Granum 2002).

All three enterotoxins from the *B. cereus* group are cytotoxic cell membrane-active toxins that form channels or holes in the membranes (Granum and Brynestad, 1999; Granum, 2001). At high concentration the membrane system will disintegrate.

It has been shown in planar lipid bilayers that CytK is able to form pores, which are weakly anion selective and exhibit an open channel probability close to one (Hardy *et al.* 2001). The predicted minimum pore diameter is approximately 7 Å. CytK, like other β-barrel pore-forming toxins, spontaneously forms oligomers (hexa- or heptamers), which are resistant to SDS, but not to boiling.

Two other enterotoxins have been suggested: enterotoxin T and enterotoxin FM (Agata *et al.*, 1995a; Asano *et al.*, 1997). It has just been shown that enterotoxin T has no, or very low, activity compared to the three above described enterotoxins (Choma and Granum, 2002). Nothing is known about the role of enterotoxin FM in food poisoning, a "toxin" that has been cloned but not characterized biologically (Table 20.2). It has been shown that enterotoxin FM has sequence homology to a cell wall hydrolase from *B. subtilis* (Margot *et al.*, 1998), and is probably not an enterotoxin.

## Genetics of the virulence factors

It has been believed for some time that the emetic toxin is not synthesized ribosomally, and indeed it has just been shown to be made by a peptide synthetase (Agata and Ohta, 2002; Ehling-Schulz *et al.*, 2005), as expected.

All three proteins of the Hbl are transcribed from one operon (*hbl*) (Ryan *et al.*, 1997), and Northern blot analysis has shown an RNA transcript of 5.5 kb. *hblC* (transcribing L$_2$) and *hblD* (transcribing L$_1$) are separated by only 37 bp and encode proteins of 447 amino acids (aa) and 384 aa, respectively (Figure 20.1). L$_2$ has a signal peptide of 32 aa and L$_1$ of 30 aa. The B-protein, transcribed from *hblA* (although it is the third gene in the operon it was published first and thus called *hblA*), consists of 375 aa, with a signal peptide of 31 aa (Heinrichs *et al.*, 1993). The spacing between *hblD* and *hblA* is approximately 115 bp (overlapping sequence not published) (Ryan *et al.*, 1997). The spacing between *hblA* and *hblB*

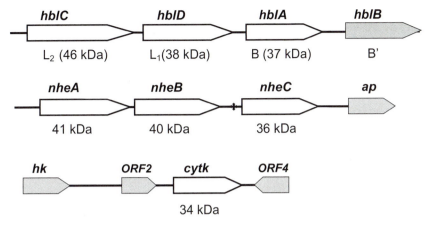

**Figure 20.1** Genetic organization of the three enterotoxins: *hbl*, *nhe* and *cytK* (Heinrichs *et al*., 1993; Ryan *et al*., 1997; Granum *et al*., 1999; Lund *et al*., 2000). There is an inverted repeat of 13 bp between *nheB* and *nheC*. An aminopeptidase gene (*ap*) is found about 600 bp downstream of *nheC*. *cytK* has an open reading frame (*ORF2*) upstream, encoding a protein of unknown function, and a histidine kinase (*hk*) further upstream. *ORF4* encodes for a long-chain-fatty-acid-CoA ligase (orientated in opposite direction to that of *cytK*). The arrowheads indicate the orientation of the genes. The name of the genes, proteins and size of the toxin components are indicated.

is 381 bp, and the length of *hblB* is not known (Heinrichs *et al*., 1993). However, based on the length of the RNA in the Northern blot, a size similar to that of *hblA* is suggested for *hblB*. The - and the putative B′ proteins are very similar in the first 158 aa (the known sequence of B′-protein based on the DNA-sequence). The function of the putative B′ protein is not yet known, but it is possible that it may substitute for the B-protein. The *hbl* operon was mapped to the unstable part of the *B. cereus* chromosome (Carlson *et al*., 1996).

The *nhe* operon contains three open reading frames (Figure 20.1): *nheA*, *nheB* and *nheC* (Granum *et al*., 1999). The two first gene products have been addressed earlier as the 45- and 39-kDa proteins, respectively. The last possible transcript from the *nhe* operon (*nheC*) has just been purified and shown to be the third component of Nhe as expected (Lindbäck *et al*., 2004). The three proteins transcribed from the *nhe* operon have properties as shown in Table 20.4. The correct size of the 45-kDa and 39-kDa proteins are 41 kDa and 39.8 kDa respectively. The size of NheC is 36.5 kDa. There is a gap of 40 bp between *nheA* and *nheB*, and a gap of 109 bp between *nheB* and *nheC*. In the 109 bp between *nheB* and *nheC*, there is an inverted repeat of 13 bp (Granum *et al*., 1999). This structure may result in little production of NheC, compared to production levels of NheA and NheB, due to a loop in the *nhe* mRNA that stabilizes the upstream region. This may explain why the NheC protein is not seen on 2-D SDS gels (Gohar *et al*., 2002), and indeed we have shown that much less NheC is required for full biological activity than required for NheA and NheB (Lindbäck *et al*., 2004).

The original CytK-producing *B. cereus* strain, which killed three people, did not contain genes for other known *B. cereus* enterotoxins. CytK is a protein of 34 kDa and is also haemolytic. The *cytK* is genetically organized as shown in Figure 20.1. We have also shown that CytK-like genes, with about 88% identity to the original sequence, are produced by about 30 to 40% of *B. cereus* strains. However, these CytK-like pro-

teins are less toxic than CytK of the original strain (Fagerlund *et al*., 2004).

For both Hbl and Nhe, maximal enterotoxin activity is found during late exponential or early stationary phase, and indeed it has been shown that both these enterotoxin operons are under regulation of *plcR* (Agaisse *et al*., 1999), a gene first described to regulate *plcA* (phospholipase C) (Lereclus *et al*., 1996). The binding sequence for this regulatory protein is TATGNAN₄TNCATG (Agaisse *et al*. 1999). This regulatory system seems to be common for many virulence factors in the *B. cereus* group as listed in Figure 20.2. In the original CytK-producing strain, the *cytK* gene was flanked by a histidine kinase that might be involved in regulation of CytK production (Lund *et al*., 2000), although a PlcR binding site is also found upstream of *cytK* (Figure 20.2).

## Reservoir, epidemiology, vehicles of infection, and outbreaks

Members of the *B. cereus* group are isolated from soil worldwide and are difficult to completely avoid in food products. It is, however, relatively simple to keep the numbers of *B. cereus* (and relatives) at an acceptable level in most foods. Spraying of *B. thuringiensis* is commonly used to protect crops against insect attacks in many parts of the world, which may cause contamination problems. And indeed, we have isolated up to $5 \times 10^3$ spores/g of a highly enterotoxic *B. thuringiensis* strain from broccoli imported to Sweden from an area where spraying is commonly used. It is certain that *B. thuringiensis* can cause food poisoning just like *B. cereus* (Ray, 1991; Drobniewski, 1993; Damgaard *et al*., 1996; Rivera *et al*., 2000), since the organism has caused food poisoning when administered to human volunteers (Ray, 1991). It has also been shown directly to be involved in food poisoning (Jackson *et al*., 1995).

*B. cereus* is easily spread to foods from its natural environment, especially to foods of plant origin (spices, rice and pasta). Through cross-contamination, it may then be spread to

**Table 20.4** Properties of the Nhe proteins[1]

| Proteins | Signal peptide | Active protein | Molecular weight of active protein | pI |
|---|---|---|---|---|
| NheA | 26 aa | 360 aa | 41.019 | 5.13 |
| NheB | 30 aa | 372 aa | 39.820 | 5.61 |
| NheC | 30 aa | 329 aa | 36.481 | 5.28 |

[1]Based on the reference (Granum et al., 1999).

other foods, such as meat products. We have also shown that when fed to cows in moderate numbers, much higher numbers were isolated from the feces (Torp et al., 2001). This can result in contamination of meat through slaughtering. The occurrence of B. cereus in milk and milk products is a result of contamination from soil and grass to the udder of the cows and into the raw milk. Through sporulation, B. cereus spores survive pasteurization, and after germination, the cells are free from competition from other vegetative cells. According to the classic literature (Claus and Berkeley, 1986), B. cereus is unable to grow at temperatures below 10°C, and cannot grow in milk and milk products stored at temperatures between 4 and 8°C. However, the psychrotolerant strains (mainly B. weihenstephanensis) that have evolved can grow at temperatures as low as 4 to 6°C (Lechner, et al., 1998; Stenfors and Granum, 2001).

If food becomes contaminated with B. cereus, improper food storage and handling may easily result in growth to levels that can result in food poisoning. This is frequently the case for heat-treated foods where only spores will survive and vegetative growth can occur without competition from other bacteria.

## Outbreaks

The number of outbreaks of B. cereus food poisoning is highly underestimated in the literature and official statistics. The main reason for this is probably the usually short duration of both types of diseases (usually less than 24 hours). The dominating type of illness caused by B. cereus differs from county to country. In Japan, the emetic type of illness is reported about 10 times more frequently than the diarrhoeal type (Shinagawa, 1993), while in Europe and North America the

diarrhoeal type is the most frequently reported (Kramer and Gilbert, 1989; Schmidt, 1995). This is probably due to eating habits, although milk has been reported to cause at least one large outbreak of the emetic type in Japan (Shinagawa, 1993). From several reports it is likely that some patients have suffered from both types of B. cereus food poisoning at the same time (Kramer and Gilbert, 1986). It is also clear that many B. cereus strains have the ability to produce both types of toxins (Kramer and Gilbert, 1989; Granum et al., 1993a; 1993b).

Foodborne diseases are reported differently from one country to another. This makes it difficult to compare the number of outbreaks from different countries. The percentage of outbreaks and cases associated with B. cereus in Japan, North America, and Europe varies from about 1% to 47% of the outbreaks and from about 0.7% to 33% of the cases (reports from different periods between 1960 and 1995) (Kramer and Gilbert, 1989; Schmidt, 1995). The highest number of outbreaks and cases are from the Iceland, Netherlands, and Norway. In Norway and Iceland it also reflects the fact that there are relatively few outbreaks of Salmonella spp. and Campylobacter spp., the two most frequently reported causes of foodborne disease in the UK (Schmidt, 1995). In the Netherlands in 1991, B. cereus was responsible for 27% of outbreaks in which the causative agent was identified. However, the incidence was only 2.8% of the total, since the majority of cases of food poisoning were of unknown aetiology (Schmidt, 1995). In addition, when the number of food poisoning cases is expressed as the number of cases per unit of population, many of the large regional differences in incidence would disappear.

Although milk is occasionally involved in B. cereus food poisoning of both types, it is not involved as frequently as

```
plcR    AATTTTATATATATTATGCATT--ATTTCATATCAAAAATTGTCGAATTCACATTATTGTAGTGGTA                        46

plcA    TTTAACTTTACTTCTATGCAAT--ATTTCATATTGATGAAATGTTATTTATTTTTTCTGAACTAGCTTATATTATATTCAAT           37

hbl     ATATATCTACATTTTATGCAAT--TATACATAACTAAATAAAGGTAA[N231]AAAAATTTAATGTTTTAATGAACAACAT            604

nhe     GTCGATAGAAAATTTATGCAATGTTATTCATACTAAGCAC[N100]TTTAACCATCTTAAATTATAT                          61

PlcR box            TATGNANN--NNTNCATA
```

**Figure 20.2** Alignment of the promoter regions of PlcR-regulated toxin genes. plcR is the pleiotropic regulator fist described to regulate plcA (phospholipase C) (Agaisse et al., 1999). hbl is the haemolytic enterotoxin, and nhe is the non haemolytic enterotoxin. The promoters from plcR, plcA, and hbl are from Økstad et al. (1999) and promoter from nhe is from Lindbäck et al. (2004). The PlcR binding sequence is indicated in bold.

might be expected. Since milk almost always contains *B. cereus*, sometimes at relatively high levels before consumed, we have to look for alternative explanations. There is a possibility that food poisoning involving *B. cereus*-contaminated milk is limited to one or two cases within a family, and therefore would not recognized as an outbreak. Are some milk drinkers protected against this type of food poisoning because the milk that they drink every day contains only small numbers of *B. cereus*? Although these explanations might contribute to our understanding, the main reason seems to be that milk contains predominantly psychrotolerant strains from the *B. cereus* group, mainly *B. weihenstephanensis*. Following examination of 50 *B. weihenstephanensis* strains, we found that only six of them produced enough enterotoxin to cause food poisoning (Stenfors *et al.*, 2002). It is also known that *B. cereus* can produce a protease that results in off-flavors that might limit the consumption of the product containing high numbers of the organism.

As pointed out earlier, members of the *B. cereus* group are isolated form many types of foods and this is reflected in the number of different dishes that have been involved in food poisoning throughout the world. Some examples of outbreaks are listed in Table 20.5.

In recent years food poisoning due to ingestion of *B. anthracis* infected meat has been reported (from CDC: JAMA 2000 Oct 4; 284 (13):1644–1646). Patients who did not develop anthrax symptoms instead developed diarrhoea, sometimes bloody, with fever and rashes. It is not yet known if the enterotoxins that are found in *B. cereus* are at least partly involved in the symptoms. It is; however, known that enterotoxin genes (*nhe*) are present in *B. anthracis* (from the total genome sequence), although the positive regulator for enterotoxin production (Figure 20.2) is non-functional.

## Susceptibility to physical and chemical agents and factors influencing survival

Strains from the members of the *B. cereus* group grow both aerobically and anaerobically at temperatures ranging from 4 to 50°C. Although all strains belonging to *B. weihenstephanensis* grow at 4 to 6°C, there are also psychrotolerant *B. cereus* stains (Stenfors and Granum, 2001). Some strains of the group (mainly *B. weihenstephanensis*) do not grow rapidly at 37°C and will not cause the diarrheal syndrome, since they would not grow fast enough in the small intestine of the host. Low pH values (< 4.5) will prevent growth of *B. cereus*, but preventing growth using high pH values is difficult as some strains can grow at pH levels up to 9.3. *B. cereus* is also able to grow at $a_w$ down to 0.92 and in 7% salt (Granum and Baird-Parker, 2000). The spores of *B. cereus* are not particularly heat resistant, but they can survive boiling. For *B. cereus*, the $D_{100°C}$ has mostly been reported to be between 1.2 to 8 min under various conditions. However, in oil the $D_{121°C}$ may be as high as 30 minutes (Kramer and Gilbert, 1989). The members of the *B. cereus* group are not competitive, but in heat-treated foods the spores will survive, and after cooling followed by spore germination, there is frequently no competition. Since these organisms can grow in most foods above pH 4.5 and at temperatures above 4°C, foods with a long shelf life (after mild heat treatment) must be treated with care.

A reservoir of *B. cereus* spores may build up in environments in the food industry and in all places where food is prepared. Specifically, adhesion of spores, which are very hydrophobic and have surface appendages, occurs on surfaces such as stainless steel (Figure 20.3) (Hachishuka *et al.*, 1984; Husmark and Rönner, 1992; Stalheim and Granum, 2001). The only way to control such a problem is by using suitable agents for disinfection at least once ever other month.

**Table 20.5** Examples of the variety of foods involved in *Bacillus cereus* food poisoning

| Type of food | Country | Number of people involved | Type of syndrome* |
| --- | --- | --- | --- |
| Barbecued chicken | Many countries | – | E, D |
| Cooked noodles | Spain | 13 | D |
| Cream cake | Norway | 5 | D |
| Fish soup | Norway | 20 | D |
| Lobster pâté | UK | – | D |
| Meat loaf | USA | – | D |
| Meat with rice | Denmark | > 200 | D |
| Milk | Many countries | – | E, D |
| Pea soup | The Netherlands | – | D |
| Sausages | Ireland, China | – | D |
| School lunch | Japan | 1877 | E |
| Scrambled egg | Norway | 12 | D |
| Several rice dishes | Many countries | – | E, D |
| Stew | Norway | 152 | D |
| Turkey | UK, USA | – | D |
| Vanilla sauce | Norway | > 200 | D |
| Wheat flour dessert | Bulgaria | – | D |

*E, emetic syndrome; D, diarrheal syndrome.

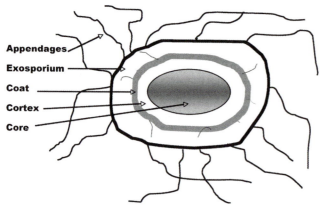

**Figure 20.3** The *B. cereus* spore with the different layers and appendages.

## Detection and analysis

Isolation of *B. cereus* from foods is quite simple, and involves plating of known amounts of food samples on selective agar and blood agar medium. The selective agar contains mannitol, and egg yolk medium, in addition to a dye that changes colour due to lack of acid production from mannitol (commercially available). Typical colonies of *B. cereus* will have a specific colour (blue or red, depending on the dye used) and are surrounded by an egg yolk reaction due to lecithinase. On blood agar plates large white colonies are observed, surrounded by a well defined haemolytic zone. Motility tests could also be carried out for *B. cereus*, which is motile until the first stage of sporulation. Occasionally we have observed that *B. cereus* only grows on blood agar plates when seeded directly from pasteurized milk. It is therefore safest to surface spread on both types of agar initially. If the number of colonies is about the same on both agar types, counting should be done from the blood agar plates. If *B. cereus* does not grow on selective medium initially, confirmation should be performed by picking 4 to 10 typical colonies from blood agar and streaking them onto selective agar. For routine testing, these methods should be sufficient for *B. cereus* verification. Toxin tests should be carried out either using serological methods or cytotoxicity tests on cells. Enumeration of spores is carried out by heat shocking food samples at 75 to 80°C for 10 to 20 minutes before plating on agar plates. By comparing results of direct plating and plating after heat shocking, the ratio between vegetative cells and spores can be determined.

Toxin detection is difficult, although immunological tests are available for one of the components in each of the triplet enterotoxins, Hbl and Nhe. For both tests the antibodies react with the protein transcribed by the first gene in each of the operons (HblC or $L_2$ and NheA). The antibodies in the kit from Oxoid react with HblC (Granum *et al.*, 1993a, 1993b; Beecher and Wong, 1994b), and NheA is detected by the Tecra kit (Granum and Lund, 1997, Lund and Granum, 1997). The main problem with these two kits is that both of these three component enterotoxins need all three components to be active, and we have seen strains that are positive by the two tests

that are not cytotoxic on epithelial cells (Granum, unpublished results). There is yet not a kit available for detection of CytK. The best way to detect the *B. cereus* enterotoxins is by using epithelial cells in culture, together with the commercial kits described above. PCR techniques are not always suitable for detection, since strains that show little or no production of the enterotoxins, have the genes in their chromosomes (Stenfors and Granum, 2002).

Unfortunately there is no commercial method for detection of the emetic toxin. However, a specific, sensitive, semiautomated, and quantitative Hep-2 cell culture-based 3-(4,5-dimethylthiazol-2-yl)-2,5-diphenyltetrazolium bromide assay for *B. cereus* emetic toxin has been developed (Finlay *et al.*, 1999). An even easier test to use for screening for emetic strains is the sperm test (Haggblom *et al.*, 2002). One colony of *B. cereus* is removed from an overnight blood agar plate incubated at room temperature. The *B. cereus* cells are then resuspended in 200 μL methanol, boiled (in a water bath) for about 10 minutes (until the methanol has evaporated) and then resuspended in 50 μL methanol. Twenty microliters of this extract is then added to viable sperm at 37°C for 5 to 10 min. The movement of sperm is then examined microscopically (at 37°C). Emetic strains inhibit sperm movement by inhibition of mitochondrial fatty acid oxidation (Mikkola *et al.*, 1999).

## Control

The control of *B. cereus* food poisoning is not difficult if present knowledge is applied, apart from in the dairy industry, where *B. cereus* causes major problems. Normal cleaning procedures are not sufficient for controlling the problem on surfaces, since build-up of spore "biofilms", specifically on stainless steel, will gradually increase the risk of contamination of food products, and therefore also food poisoning. The strong adhesion of *B. cereus* spores is mainly due to three characteristics: the high relative hydrophobicity, the low spore surface charge, and the spore morphology (appendages). At present the only way to overcome this problem, when first introduced, is through the use of hypochlorite (0.2% at pH 7 to 8) or UVC light, since neither low nor high pH cleaning is sufficient to control the problem.

There is a growing number of precooked long shelf life products on the market that are difficult to produce completely free from *Bacillus* spp. spores. If foods are not kept at temperatures between 6 and 60°C for too long, growth of the members of the *B. cereus* group species will not occur. Rapid cooling and proper reheating of cooked food is essential if the food is not consumed immediately. Long-term storage must be at temperatures below 8°C (or preferably between 4 and 6°C to prevent growth of *B. cereus*). Low-pH foods (pH < 4.5) can be considered safe from growth of *B. cereus*. All types of *Bacillus* spp. spores are commonly isolated from spices, cereals, and dried foods such as rice and pasta.

Presently the main problem with *B. cereus* seems to be in the dairy industry, where the keeping quality of milk is determined by the number of *B. cereus* cells/spores in the product (Andersson *et al.*, 1995). The bacterium may cause aggregation of the creamy layer of pasteurized milk, known as bitty cream, which is due to the lecithinase activity of *B. cereus*. Further, *B. cereus* is responsible for sweet curdling (without pH reduction), both in homogenized and non-homogenized low-pasteurized milk. It seems impossible to completely avoid the presence of *B. cereus* in milk where it contaminates the raw milk at the farm (Andersson *et al.*, 1995). Soiling of the udders of cows is the principal source of contamination of milk with *B. cereus*. Soil has been shown to contain $10^5$ to $10^6$ spores/g. It is therefore very important that the udder and the teats are cleaned to reduce the contamination of raw milk. Transport and further storage in the dairy may result in further contamination of the raw milk from *B. cereus* spores already present (adherent) in the tanks or pipelines (Anderson *et al.*, 1995). The vegetative bacteria are killed in the pasteurization process, but the spores survive. Pasteurization might activate at least some of the spores (heat activation), and they then might start germinating.

To control *B. cereus* in milk and milk products it is very important to trace the presence of spores from farmer to package. The storage temperature is the most important factor required to keeping the number of *B. cereus* spores low. An increase of just 2°C during storage, from 6°C to 8°C, increases the growth rate of *B. cereus* in milk tremendously. At the dairy the milk is, in general, kept at 4°C, and this ensures a good keeping quality. However, during distribution, where the energy costs often are valued higher than food quality, temperatures up to 8°C and above are common. Further, the consumer often exposes milk to higher temperatures for longer periods, for example at the breakfast table. The majority of the strains from the *B. cereus* group that grow at low temperatures (including *B. weihenstephanensis*) are usually low in enterotoxin production, although there are exceptions.

The emetic syndrome of *B. cereus* food poisoning is often connected with the consumption of rice in Chinese restaurants (Mortimer and McCann, 1974; Kramer and Gilbert, 1986). The predominance of episodes involving Chinese restaurants is linked to the common practice of saving portions of boiled rice from bulk cooking. The boiled rice is then usually stored at room temperature overnight, and *B. cereus* is then able to multiply. The same problem may occur when foods such as pasta and pizza are stored for long periods of time at room temperature.

## References

Agaisse, H., Gominet, M., Okstad, O.A., Kolsto, A.B, and Lereclus D. 1999. PlcR is a pleiotropic regulator of extracellular virulence factor gene expression in *Bacillus thuringiensis*. Mol. Microbiol. 32: 1043–1053.

Agata, N., and Ohta, M. 2002. Identification and molecular characterization of the genetic locus for biosynthesis of the emetic toxin, cereulide, of *Bacillus cereus*. Presentation Number: P-66, Poster Board Number: 428, 102nd General ASM Meeting, Salt Lake City, Utah May 19–23.

Agata, N., Mori, M. Ohta, M., Suwan, S., Ohtani, I., and Isobe, M. 1994. A novel dodecadepsipeptide, cereulide, isolated from *Bacillus cereus* causes vacuole formation in HEp-2 cells. FEMS Microbiol. Lett. 121: 31–34.

Agata, N., Ohta, M., Arakawa, Y., and Mori, M. 1995. The *bceT* gene of *Bacillus cereus* encodes an enterotoxic protein. Microbiology 141: 983–988.

Agata, N., M. Ohta, M. Mori, and M. Isobe. 1995. A novel dodecadepsipeptide, cereulide, is an emetic toxin of *Bacillus cereus*. FEMS Microbiol. Lett. 129: 17–20.

Agata, N., Ohta, M., and Yokoyama, K. 2002. Production of *Bacillus cereus* emetic toxin (cereulide) in various foods. Int. J. Food Microbiol. 73: 23–27.

Andersson, A., Rönner, U., and. Granum, P.E. 1995. What problems does the food industry have with the sporeforming pathogens *Bacillus cereus* and *Clostridium perfringens*? Int. J. Food Microbiol. 28: 145–156.

Andersson, A., Granum, P.E, and Rönner, U. 1998. The adhesion of *Bacillus cereus* spores to epithelial cells might be an additional virulence mechanism. Int. J. Food Microbiol. 39: 93-99.

Asano, S.I., Nukumizu, Y., Bando, H., Iizuka, T., and Yamamoto T. 1997. Cloning of novel enterotoxin genes from *Bacillus cereus* and *Bacillus thuringiensis*. Appl. Environ. Microbiol. 63: 1054–1057.

Ash, C., Farrow, J.A., Dorsch, M., Steckebrandt, E., and Collins M.D. 1991. Comparative analysis of *Bacillus anthracis*, *Bacillus cereus*, and related species on the basis of reverse transcriptase sequencing of 16S rRNA. Int. J. Syst. Bacteriol. 41: 343–346.

Beecher, D.J., and Macmillan, J.D. 1991. Characterization of the components of hemolysin BL from *Bacillus cereus*. Infect. Immun. 59: 1778–1784.

Beecher, D.J., Schoeni, J.L., and Wong, A.C.L. 1995. Enterotoxin activity of hemolysion BL from *Bacillus cereus*. Infect. Immun. 63: 4423–4428.

Beecher, D. J., and Wong, A.C.L. 1994. Improved purification and characterization of hemolysin BL, a hemolytic dermonecrotic vascular permeability factor from *Bacillus cereus*. Infect. Immun. 62: 980–986.

Beecher, D.J., and A.C.L. Wong. 1994. Identification and analysis of the antigens detected by two commercial *Bacillus cereus* diarrheal enterotoxin immunoassay kits. Appl. Environ. Microbiol. 60: 4614–4616.

Beecher, D.J., and Wong, A.C.L. 1997. Tripartite hemolysin BL from *Bacillus cereus*. Hemolytic analysis of component interaction and model for its characteristic paradoxical zone phenomenon. J. Biol. Chem. 272: 233–239.

Borge, G.A., Skeie, M., Langsrud T., and Granum, P.E. 2001. Growth and toxin profiles of *Bacillus cereus* isolated from different food sources. Int. J. Food Microbiol. 69: 237–246.

Brynestad, S., and Granum, P.E. 2002. *Clostridium perfringens* and foodborne infections. Int. J. Food Microbiol. 74: 195–202.

Carlson, C.R., Caugant, D.A., and Kolstø, A.-B. 1994. Genotypic diversity among *Bacillus cereus* and *Bacillus thuringiensis* strains. Appl. Environ. Microbiol. 60: 1719–1725.

Carlson, C.R., Johansen, T., and Kolstø, A.-B. 1996. The chromosome map of *Bacillus thuringiensis* subsp. *canadensis* HD224 is highly similar to that of *Bacillus cereus* type strain ATCC 14579. FEMS Microbiol. Lett. 141: 163–167.

Choma, C., and Granum, P.E. 2002. The Enterotoxin T (BcET) from *Bacillus cereus* can probably not contribute to food poisoning. FEMS Microbiol. Letters. 217: 215–219.

Christiansson, A., Naidu, A.S., Nilsson, I., Wadström, T., and Pettersson, H.-E. 1989. Toxin production by *Bacillus cereus* dairy isolates in milk at low temperatures. Appl. Environ. Microbiol. 55: 2595–2600.

Claus, D., and Berkeley, R.C.W. 1986. Genus *Bacillus*. In: The Bergey's Manual of Systematic Bacteriology, Vol. 2. P.H.A. Seneath, ed. The Williams and Wilkins Co., Baltimore, MD, pp. 1105–1139.

Damgaard, P.H., Larsen, H.D., Hansen, B.M., Bresciani, J., and Jørgensen, K. 1996. Enterotoxin-producing strains of *Bacillus thuringiensis* isolated from food. Lett. App. Microbiol. 23: 146–150.

Drobniewski, F.A. 1993. *Bacillus cereus* and related species. Clin. Microbiol. Rev. 6: 324–338.

Ehling-Schulz, N., Vukov, N., Schulz, A., Shaheen, R., Anderson, M., Martlbauer, E., and Scherer, S. 2005. Identification and partial characterization of the nonribosomal peptide synthetase genes responsible for cereulide production in emetic *Bacillus cereus*. Appl. Environ. Microbiol. 71: 105–113.

Fagerlund, A., Ween, O., Lund, T., Hardy, S.P., and Granum, P.E. 2004. Different cytotoxicity of CytK and CytK-like proteins from *B. cereus*. Microbiology 150: 2689–2697.

Finlay, W.M., Logan, N.A., and Sutherland, A.D. 1999. Semiautomated metabolic staining assay for *Bacillus cereus* emetic toxin. Appl. Environ. Microbiol. 65: 1811–1812.

Finlay, W.J., Logan, N.A., and Sutherland, A.D. 2000. *Bacillus cereus* produces most emetic toxin at lower temperatures. Lett. Appl. Microbiol. 31:385–389.

Gohar, M., Okstad, O.A., Gilois, N., Sanchis, V., Kolst Solidus In, Circle, A.B., Lereclus, D. 2002. Two-dimensional electrophoresis analysis of the extracellular proteome of *Bacillus cereus* reveals the importance of the PlcR regulon. Proteomics 2: 784–791.

Granum, P.E. 2001. *Bacillus cereus*. In: Food Microbiology. Fundamentals and Frontiers. M. Doyle, L. Beuchat, and T. Montville, eds. ASM Press, Washington DC, pp. 373–381.

Granum, P.E. 1994. *Bacillus cereus* and its toxins. Soc. Appl. Bacteriol. Symp. Ser. 23: 61S–66S.

Granum, P.E., and Baird-Parker, T.C. 2000. *Bacillus* spp. In: The Microbiological Safety and Quality of Food. B. Lund, T. Baird-Parker, and G. Gould, eds. Aspen Publishers MD, USA, pp. 1029–1039.

Granum, P.E., and Brynestad, S. 1999. Bacterial Toxins as Food Poisons. In: The Comprehensive Sourcebook of Bacterial Protein Toxins. J.E. Alouf, and J.H. Freer, eds. Academic Press, London, pp. 669–681.

Granum, P.E., Brynestad, S., and Kramer, J.M. 1993. Analysis of enterotoxin production by *Bacillus cereus* from dairy products, food poisoning incidents and non-gastrointestinal infections. Int. J. Food Microbiol. 17: 269–279.

Granum, P.E., Brynestad, S., O´Sullivan, K., and Nissen, H. 1993. The enterotoxin from *Bacillus cereus*: production and biochemical characterization. Neth. Milk and Dairy J. 47: 63–70.

Granum, P.E., and Lund, T. 1997. *Bacillus cereus* enterotoxins. FEMS Microbiol. Lett. 157:223–228.

Granum, P.E., O´Sullivan, K., and Lund, T. 1999. The sequence of the non-haemolytic enterotoxin operon from *Bacillus cereus*. FEMS Microbiol. Lett. 177: 225–229.

Guinebretiere, M.H., Broussoll,e V., and Nguyen-The, C. 2002. Enterotoxigenic profiles of food-poisoning and foodborne *Bacillus cereus* strains. J. Clin. Microbiol. 40: 3053–3056.

Hachishuka, Y., Kozuka, S., and Tsujikawa, M. 1984. Exosporia and appendages of spores of *Bacillus* species. Microbiol. Immunol. 28: 619–24

Haggblom, M.M., Apetroaie, C., Andersson, M.A., and Salkinoja-Salonen, M.S. 2002. Quantitative analysis of cereulide, the emetic toxin of *Bacillus cereus*, produced under various conditions. Appl. Environ. Microbiol. 68: 2479–2483.

Hardy, S.P., Lund, T., and Granum, P.E. 2001. CytK toxin of *Bacillus cereus* forms pores in planar lipid bilayers and is cytotoxic to intestinal epithelia. FEMS Microbiol. Lett. 197: 47–51.

Hauge, S. 1950. Matforgiftninger framkalt av *Bacillus cereus* (in Norwegian). Nordisk Hygienisk Tidskrift. 31: 189–206.

Hauge, S. 1955. Food poisoning caused by aerobic spore forming bacilli. J. Appl. Bacteriol. 18: 591–595.

Heinrichs, J.H., Beecher, D.J., MacMillan, J.M., and Zilinskas, B.A. 1993. Molecular cloning and characterization of the hblA gene encoding the B component of hemolysin BL from *Bacillus cereus*. J. Bacteriol. 175: 6760–6766.

Husmark, U., and Rönner, U. 1992. The influence of hydrophobic, electrostatic and morphologic properties on the adhesion of *Bacillus* spores. Biofouling 5: 335–344.

Jackson, S.G., Goodbrand, R.B., Ahmed, R., and Kasatiya, S. 1995. *Bacillus cereus* and *Bacillus thuringiensis* isolated in a gastroenteritis outbreak investigation. Lett. Appl. Microbiol. 21: 103–105.

Kramer, J.M., and Gilbert, R.J. 1989. *Bacillus cereus* and other *Bacillus* species. In: Foodborne Bacterial Pathogens. M.P. Doyle, ed. Marcel Dekker, New York, pp. 21–70.

Lechner, S., Mayr, R., Francic, K.P., Prub, B.M., Kaplan, T., Wieber-Gunkel, E., Stewart, G.A.S.B., and Scherer S. 1998. *Bacillus weihenstephanensis* sp. nov. is a new psychrotolerant species of the *Bacillus cereus* group. Int. J. Syst. Bacteriol. 48: 1373–1382.

Lereclus, D., Agaisse, H., Gominet, M., Salamitou, S., and Sanchis, V. 1996. Identification of a *Bacillus thuringiensis* gene that positively regulates transcription of the phosphatidylinositol-specific phospholipase C gene at the onset of the stationary phase. J. Bacteriol. 178: 2749–2756.

Lindbäck, T., Fagerlund, A., Rødland, M.A., and Granum, P.E. 2004. Characterization of the *Bacillus cereus* Nhe enterotoxin. Microbiology 150: 3959–3967.

Lund, T., and Granum, P.E. 1996. Characterisation of a non-haemolytic enterotoxin complex from *Bacillus cereus* isolated after a foodborne outbreak. FEMS Microbiol. Lett. 141: 151–156.

Lund, T., and. Granum, P.E. 1997. Comparison of biological effect of the two different enterotoxin complexes isolated from three different strains of *Bacillus cereus*. Microbiology 143:3329–3336.

Lund, T., De Buyser, M.L., and Granum, P.E. 2000. A new cytotoxin from *Bacillus cereus* that may cause necrotic enteritis. Mol. Microbiol. 38: 254–261.

Mahler, H., Pasi, A., Kramer. J.M., Schulte, P., Scoging, A.C., Bar, W., and Krahenbuhl. S. 1997. Fulminant liver failure in association with the emetic toxin of *Bacillus cereus*. N. Engl. J. Med. 336: 1142–1148.

Margot, P., Wahlen, M., Gholamhuseinian, A., Piggot, P., and Karamata, D. 1998. The lytE Gene of *Bacillus subtilis* 168 encodes a cell wall hydrolase. J. Bacteriol. 180: 749–752.

Mikkola, R., Saris, N.E., Grigoriev, P.A., Andersson, M.A., and Salkinoja-Salonen, M.S. 1999. Ionophoretic properties and mitochondrial effects of cereulide: the emetic toxin of *B. cereus*. Eur. J. Biochem. 263: 112–117.

Mortimer, P.R., and McCann, G. 1974. Food poisoning episodes associated with *Bacillus cereus* in fried rice. Lancet 1: 1043–1045.

Økstad, O.A., Gominet, M., Purnelle, B., Rose, M., Lereclus, D., and Kolstø. 1999. Sequence analysis of three *Bacillus cereus* loci carrying PlcR-regulated genes encoding degradative enzymes and enterotoxin. Microbiology. 145: 3129–3138.

Paananen, A., Mikkola, R., Sareneva, T., Matikainen, S., Hess, M., Andersson, M., Julkunen, I., Salkinoja-Salonen, M.S., and Timonen, T. 2002. Inhibition of human natural killer cell activity by cereulide, an emetic toxin from *Bacillus cereus*. Clin. Exp. Immunol. 129: 420-428.

Nakamura, L.K. 1998. *Bacillus pseudomycoides* sp. nov. Int. J. Syst. Bacteriol. 48: 1031–1035.

Ray, D.E. 1991. Pestisides derived from plants and other organisms. In: Handbook of Pesticide Toxology. W.J. Hayes, and E.R. Laws, Jr., eds. Academic Press Inc., New York. pp. 585-636.

Rivera, A.M.G., Granum, P.E., and Priest, F.G. 2000. Common occurrence of enterotoxin genes and enterotoxicity in *Bacillus thuringiensis*. FEMS Microbiol. Lett. 190: 151–155.

Ryan, P.A, Macmillan, J.M., and Zilinskas, B.A. 1997. Molecular cloning and characterization of the genes encoding the $L_1$ and $L_2$ components of hemolysin BL from *Bacillus cereus*. J. Bacteriol. 179: 2551–2556.

Schmidt, K. ed. 1995. WHO surveillance programme for control of foodborne infections and intoxications in Europe. Sixth Report – FAO/WHO Collaborating Centre for Research and Training in Food Hygiene and Zoonoses, Berlin.

Schoeni, J.L., and Wong, A.C.L. 1999. Heterogeneity observed in the components of hemolysin BL, an enterotoxin produced by *Bacillus cereus*. Int. J. Food Microbiol. 53: 159–167

Shinagawa, K. 1993. Serology and characterization of *Bacillus cereus* in relation to toxin production. Bull. Int. Dairy Fed. 287: 42–49.

Stalheim, T., and Granum, P.E. 2001. Characterisation of spore appendages from *Bacillus cereus* strains. J. Appl. Microbiol. 91: 839–845.

Stenfors, L.P., and Granum P.E. 2001. Psychrotolerant species from the *Bacillus cereus* group are not necessarily *B. weihenstephanensis*. FEMS Microbiol. Letters. 197: 223–228.

Stenfors, L.P., Mayr, R., Scherer, S., and Granum, P.E. 2002. Pathogenic potential of fifty *Bacillus weihenstephanensis* strains. FEMS Microbiol. Letters 215:47–51.

Torp, M., Holstad, G., and Granum, P.E. 2001. *Bacillus cereus* – feed and faeces as source to high numbers of spores in milk. Norsk veterinær-tidskrift. 113: 462–466. (In Norwegian)

Turnbull, P.C.B. 1986. *Bacillus cereus* Toxins. In: Pharmacology of Bacterial Toxins. International Encyclopedia of Pharmacology and Therapeutics, section 119. F. Dorner and J. Drews. eds. Pergamon Press, Oxford. p. 397–448.

van Netten, P., van de Moosdijk, A., van Hoensel, P., Mossel, D.A.A., and Perales, I. 1990 Psychrotrophic strains of *Bacillus cereus* producing enterotoxin. J. Appl. Bact. 69: 73–79.

Yokoyama, K., Ito, M., Agata, N., Isobe, M., Shibayama, K., Horii, T., and Ohta, M. 1999. Pathological effect of synthetic cereulide, an emetic toxin of *Bacillus cereus*, is reversible in mice. FEMS Immunol. Med. Microbiol. 24: 115–120.

# Terrorism and the Food Supply

Jeremy Sobel

21

## Abstract

Sabotage of foods by terrorists and criminals has occurred in the USA, though rarely. A multiplicity of suitable biological and chemical agents exists and the vast contemporary food supply is vulnerable. Prevention requires enhancement of food security. Since an outbreak caused by food sabotage would most likely be detected and handled by the existing public health system, minimization of casualties requires a robust standing public health infrastructure capable of detecting, investigation and controlling all foodborne disease outbreaks, intentional and unintentional, and providing appropriate medical resources.

## Introduction

The sabotage of food or beverage by contamination with the intention to assassinate individuals, incapacitate armies or demoralize populations has been practiced since antiquity. The vast and complex food supplies of nations are vulnerable to deliberate contamination. (Sobel *et al.*, 2002a, Khan *et al.*, 2001). Contamination of food or beverage with biological or chemical agents may serve the objectives of terrorists who seek to create panic, threaten civil order, or cause economic losses. Sabotage of crops or livestock may result in similar consequences.

Over 76 million foodborne illnesses are estimated to occur in the United States yearly (Mead *et al.*, 1999). Over 1,500 outbreaks of foodborne disease are reported to the Centers for Disease Control and Prevention (CDC) each year, and these represent but a fraction of actual events. The public health system and food safety regulatory apparatus have evolved over the past century to address such outbreaks; specific epidemiological, laboratory and legal approaches have been developed to detect, investigate and control these events (Sobel *et al.*, 2002b). The same personnel that handle naturally occurring foodborne disease in the course of their routine duties would almost certainly be the first to respond to an act of bioterrorism involving food (Sobel *et al.*, 2002a).

This chapter will focus on the public health and human illness aspects of sabotage of foods and beverages, not including water. The reader is referred to other sources on the topic of food security, which entails protecting the food supply from deliberate contamination in the first place (WHO, 2002, Lee *et al.*, 2003).

## Vulnerability of food supply

International and governmental authorities have recognized the threat of terrorism to the food supply (WHO, 2002; WHO 1970; US General Accounting Office, 1999; US Food and Drug Administration, 2003), and indeed biological contamination of foods by terrorists and criminals has occurred in the US in recent decades (Torok *et al.*, 1997; Kolavic *et al.*, 1997; Phills *et al.*, 1972). The modern food supply comprises thousands of classes of foods, domestically produced or imported. Ever-more centralized production and processing and wide distribution of products has resulted in unintentional foodborne disease outbreaks that increasingly occur over large, dispersed geographic areas, a situation that may delay recognition of an outbreak and complicate identification of the contaminated food (Hedberg *et al.*, 1994; Sobel *et al.*, 2001). Deliberate contamination of foods could produce a similar situation.

The potential consequences of an attack on the food supply can be inferred from examples of unintentional foodborne disease outbreaks. In 1994 about 224,000 persons in the United States were infected with *Salmonella enteritidis* from contaminated ice-cream (Hennesy *et al.*, 1996). In 1996, over 7,000 children in Sakai City, Japan, were infected with *Escherishia coli* O157:H7 from contaminated radish sprouts served in school lunches. The outbreak resulted in broad-reaching psychological trauma, including suicide (Mermin *et al.*, 1999). In 1985, over 170,000 persons were infected with *Salmonella Typhimurium* resistant to nine antimicrobial agents from contaminated pasteurized milk from a dairy plant in Illinois (Ryan *et al.*, 1987). However, as the mailings of *Bacillus anthracis* containing envelopes in the United States have demonstrated, even limited dissemination of biological agents using simple means, causing relatively few illnesses, can produce considerable public anxiety and challenge the public health system (CDC, 2001), and contamination with no cases of illness can produce severe economic loss (Grigg and Modeland, 1989).

Fool-proof protection of the food supply is impossible. Prevention falls under the rubric of food security and entails physical protection of the food supply along the "farm to table" continuum, including all stages of production, processing, transport, storage and retail (WHO, 2002; Lee et al., 2003). This challenge rests principally with food safety regulatory agencies, industry, and law enforcement. Approaches include identification of high risk foods and critical control points at which contamination could be carried out in the complex web of production and commerce and executing appropriate control measures. Should an attack occur, preparedness entails maximizing capacity to detect and investigate the consequent outbreak with the objective of identifying the contaminated food and removing it from circulation, advising the public and apprehending the perpetrators (Sobel et al., 2002a, Khan et al., 2001).

## Potential threat agents

The list of pathogens, chemicals and toxins that could cause disease ingestion is extensive. It is important to keep in mind that laboratory-based diagnosis and surveillance systems are geared to those that cause disease in natural settings.

The CDC strategic plan for Bioterrorism Preparedness and Response includes a list of critical biological agents for public health preparedness (CDC, 2000). The highest priority category of agents includes one naturally occurring foodborne toxin, *Clostridium botulinum* neurotoxin, which produces a flaccid paralysis that can result in death from respiratory arrest if untreated (Shapiro et al., 1998; CDC, 2000; Hatheway 1990), and *Bacillus anthracis*, which produces a high-mortality gastrointestinal illness in the developing world (Sirisanthana and Brown, 2002).

The category of second most critical biological agents for public health preparedness consists of organisms that are moderately easy to disseminate, cause moderate morbidity and low mortality, and require specific enhancement of diagnostic and surveillance capacities. This category includes several foodborne pathogens (Table 21.1). With proper therapy, these organisms are rarely lethal. Beyond this list is a variety of foodborne pathogens that could potentially be used, including viral and parasitic agents such as hepatitis A and *Cryptosporidium*. Additionally, various biological agents that have been weaponized may rarely cause unintentional foodborne disease, and their full potential for malicious contamination of food is not fully known. These include *Bacillus anthracis* (Erickson and Kornacki, 2003), *Yersinia pestis* (Butler et al., 1982), *Francisella tularensis* (Reintjes et al., 2002; Tarnvik and Berglund, 2003) and others.

Assorted chemical agents could be used to contaminate foods or beverages. Many are available in the form of pesticides, cleaning compounds or industrial solvents. The CDC list includes blood agents such as cyanide; heavy metals including arsenic, lead and mercury; and corrosive industrial chemicals and toxins (CDC, 2000). Naturally occurring biological toxins and synthetic chemicals have been weaponized for aerosolized dissemination but could produce illness by ingestion; these include aflotoxins and T-2 mycotoxins, saxitoxin, tetrodotoxin (Lee et al., 2003), and ricin (Franz and Jaax, 1997).

## Detection of an attack

Unless announced by the perpetrator, an attack would most likely be recognized by epidemiologic investigation of an outbreak. The potential for hoaxes is well recognized, and outlandish claims might accompany a small-scale contamination. As with any foodborne outbreak, early recognition and investigation is vital if the food vehicle has wide distribution, and prevention of additional cases may depend on identifying and recalling the yet-unconsumed food product. Additionally, prompt suspicion of the terroristic nature of the event will help direct the criminal investigation and bring into play the full array of federal resources available to counter bioterroristic attacks (US General Accounting Office, 1999). In a suspected bioterrorism event in the United States, the Federal Bureau of Investigations will assume overall leadership of the response (Institute of Medicine, 1999).

Outbreaks may be reported by astute clinicians cognizant of a cluster of patients with similar symptoms. Reporting such clusters immediately to public health authorities remains the fastest mode of detection, and training clinicians to rapidly report suspicious syndromes and disease clusters is a cornerstone of preparedness for biological terrorism and epidemics. Where cases are geographically dispersed, laboratory-based surveillance systems may detect increases in illnesses. For foodborne diseases, the Public Health Laboratory Information System (PHLS) electronically collects data on foodborne enteric pathogens, many of them on CDC's biological agents list (Bean et al., 1992). Computerized algorithms such as the *Salmonella* Outbreak Detection Algorithm (SODA) analyze disease trends for increases in the incidence of specific serotypes compared to historical baselines (Hutwagner et al., 1997; Mahon et al., 1997). A national molecular subtyping network, PulseNet, performs pulsed-field gel electrophoresis "fingerprinting" on isolates of select foodborne bacterial pathogens from patients, foods and farm animals and has detected many common-source outbreaks that occurred over widespread geographic areas without the focal increase in case counts required by less sensitive systems (Stephenson, 1997, Swaminathan et al., 2001; Sivapalasingam et al., 2000). In recent years, syndromic surveillance systems have been developed in several metropolitan areas (Greenko et al., 2003; Pavlin, 2003). These systems electronically monitor in near-real-time the rates of specific syndromes such as diarrhea, flu-like illnesses, pneumonia or neurological symptoms from emergency medical services calls, emergency room admission or discharge diagnoses, etc.. A unique surveillance system exists for botulism. A clinician suspecting a case must contact the state public health department in order to obtain the specific therapy, botulinum antitoxin, which is available in the USA only from CDC (Shapiro et al., 1998; CDC 1998).

**Table 21.1** Some potential foodborne biological terrorist agents and select characteristics

| Agent | Availability | Minimum infectious dose, secondary transmission | Clinical syndrome | Case-fatality | Other characteristics of microbe or illness |
|---|---|---|---|---|---|
| Botulinum toxin | Organism ubiquitous in environment; cultures require anaerobic conditions | $LD_{50} = 0.001$ µg/kg (Hatheway, 1990) | Descending paralysis, respiratory compromise | 5% (treated) (Shapiro et al., 1998) | 95% of patients require hospitalization; 25–60% of patients require incubation |
| Salmonella serotypes (excluding S. Typhi) | Clinical and research labs, culture collections, poultry, environmental sources | $10^3$ organisms (Blaser and Newman, 1982); limited secondary transmission | Acute diarrheal illness, 1–3% chronic sequelae | >1% (Benenson et al. (1995) | Organism hardy, prolonged survival in the environment |
| Salmonella Typhi | Clinical and research labs | $10^5$ organisms (Blaser and Newman, 1982); secondary transmission possible | Acute febrile illness, protracted recovery, 10% relapse, 1% intestinal rupture (WHO 1970). | 10% untreated, 1% treated (Benenson et al., 1995) | Clinical syndrome unfamiliar in USA. Long incubation period (1–3 weeks). Produces asymptomatic carrier state in 3% of cases |
| Shigella spp. | Clinical and research labs | $10^2$ organisms (Dupont et al., 1989); secondary transmission possible | Acute diarrhea, often bloody | For most common species in USA, < 1% (Benenson et al., 1995) | |
| Shigella dysenteriae Type 1 | Clinical and research labs | 10–100 organisms (Dupont et al., 1989); secondary transmission possible | Dysentery; seizures | Up to 20% (treated) (Benenson et al., 1995) | Causes dysentery, toxic megacolon, hemolytic–uremic syndrome, convulsions in children |
| E. coli O157:H7 | Clinical and research labs, bovine sources, farms | > 50 organisms (Mead and Griffin, 1998; Tilden et al., 1996); secondary transmission possible | Acute bloody diarrhea, 5% HUS, longer-term complications | 1% (Mead and Griffin, 1998) | Long-term sequelae: hypertension, stroke, renal insufficiency/failure, neurologic complications (Mead and Griffin, 1998; Griffin et al., 1994) |
| Vibrio cholerae | Clinical and research labs | $10^8$ organisms (Tauxe, 1992); secondary transmission possible | Acute life-threatening dehydrating diarrhea | Up to 50% untreated; 1% treated (Bennish, 1994) | Historically, causes massive water-borne epidemics in areas with poor sanitation |

## Recognition of a foodborne disease event as a terrorist attack or a criminal act

Epidemiologic clues to a deliberate, covert act of contamination are unusual relationships between person, time and place of the outbreak, or unusual or implausible combinations of pathogens and food vehicles (Treadwell et al., 2003). However, such features may be absent in an event of deliberate contamination, or they may occur in unintentional outbreaks. Epidemiologic features alone cannot prove a terroristic act; rather, they inform the investigators and may prompt consultation with law-enforcement agencies that may confirm or refute the possibility of malicious contamination.

The adequacy of response will depend on public health officials' capacity to respond to all foodborne disease outbreaks. Hence a cornerstone of preparedness is improving the public health infrastructure for detecting and responding to unintentional outbreaks: ensuring robust surveillance, improving laboratory diagnostic capacity for patient and food product samples, increasing trained staff for rapid epidemiologic investigations, and enhancing effective communications. Preparedness for such a situation additionally requires the capacity to respond to extraordinary demands on emergency services and medical resources.

### Diagnosis

A key factor in rapid diagnosis of the etiologic agent during the investigation of unexplained foodborne disease is ordering the appropriate diagnostic laboratory test. This requires that clinicians be familiar with the likely agents and their clinical presentations, that they not be hampered in ordering tests by cost concerns, and that they know how to contact public health sector consultants when needed.

Most foodborne pathogens on CDC's Strategic Plan for Bioterrorism Preparedness and Response are detectable by routine culture practices in state public health laboratories. Botulism is diagnosed in some state and municipal laboratories and at CDC. Identification of toxins requires testing of appropriate samples in specialized laboratories. CDC has developed the national Laboratory Response Network (LRN) for bioterrorism specializing in diagnosis of biological agents that includes public health, military, veterinary and commercial laboratories. This network provides standardized protocols for diagnosis and reagents, makes initial, rapid diagnoses and then refers specimens to appropriate specialty laboratories at CDC and elsewhere. The LRN provides surge capacity to handle increased numbers of samples anticipated in a bioterrorism event: over 120,000 samples were collected in the course of the anthrax mailings investigations of 2001.

### Response

In the United States, county, municipal, and in some cases state health departments are typically the first to be informed of outbreaks and to respond to investigate them. The state public health laboratory and a few municipal laboratories play a primary role in diagnosing the etiology of an outbreak.

Outbreaks with cases distributed over a wide geographic area, without clustering, may be recognized first by CDC through national laboratory-based surveillance systems, in which case CDC may play a coordinating role for a multistate investigation.

The objectives of the epidemiologic investigation of an outbreak of foodborne disease would not greatly change if intentional contamination is suspected. Identification of the etiologic agent, vehicle of transmission, and manner of contamination remain the most important aspects of an investigation, followed by timely implementation of control measures, including removal of the contaminated food from circulation and properly treating exposed persons (Sobel et al., 2002b). The familiar components of the investigation include formulation of case definitions, case finding, pooling and evaluation of data on potential exposures in different geographic locations, rapid development of standardized instruments and execution of case–control studies to identify specific food vehicles, collection of laboratory samples, transport and processing, collating information from tracebacks, coordination with law enforcement, food safety regulatory agencies, agencies involved in emergency medical response, and standardization of treatment and prophylaxis recommendations. CDC and federal food regulatory agencies, FDA and USDA, routinely collaborate on tracebacks of contaminated foods implicated in many of the approximately 1000 foodborne disease outbreaks reported annually in the US, and this norm would be followed in a bioterrorism event.

A sophisticated bioterrorist attack on the food supply could produce many casualties. In the USA, the medical components of the response to such an event is part of overall bioterrorism response preparedness and have been described elsewhere (Khan et al., 2000). Adequate stocks of antimicrobial drugs, antitoxins, other medications, and ventilators and other medical equipment are maintained in stockpiles and can be delivered rapidly. A biological terror attack targeting a food distributed over a wide geographic area could pose the challenge of assuring adequate medical supplies and personnel in far-flung locations. The effectiveness of the medical response will depend on timely epidemiologic surveillance data collected public health investigators to direct the medical resources to the casualties and their caretakers.

### Communications

Swift communication between health care providers, public health officials at various levels, and government agencies is an absolute requirement for a rapid, effective response to a bioterrorist attack on the food supply. Communication patterns similar to those used in coordination of multistate outbreak investigations will likely be effective for incidents of intentional contamination of food (Sobel et al., 2002b).

Clinicians, clinical laboratory staff, and coroners who identify suspected cases or clusters of illness must have lists of appropriate local contacts in order to notify the public health sector of their findings. Local health departments should notify state public health departments even as they begin their investigation locally. There are standing modalities used routinely to inform public health officials at the state and federal level of ongoing outbreaks and to coordinate multistate investigations. In the case of an intentional contamination of food, these communication systems would function as they do in regular outbreaks. Depending on the food affected the FDA or USDA's regulatory authorities would be engaged rapidly during a bioterrorist event linked to food. Communication between public health officials and food industries would be coordinated with the appropriate regulatory agency, that can request a recall of contaminated food from the market.

Intense media coverage of a bioterrorist event is to be expected. Skill and experience are required to transmit accurate information through the media about the nature and extent of the event, the suspected or implicated foods, and measures to take to prevent exposure or consequences of exposure. The accuracy, timeliness and consistency of the information provided may in part determine the success of control measures.

## Conclusions

Sabotage of food by terrorists and criminals has occurred in the USA, though rarely. A multiplicity of suitable biological and chemical agents exists and the vast contemporary food supply is vulnerable. Prevention requires enhancement of food security. Since an outbreak caused by food sabotage would most likely be detected and handled by the existing public health system, minimization of casualties requires a robust standing public health infrastructure capable of detecting, investigation and controlling all foodborne disease outbreaks, intentional and unintentional.

## References

Bean, M.H., Martin, S.M., and Bradford, H. 1992. PHLIS: An electronic system for reporting public health data from remote sites. Am. J. Public Health 82: 1273–1276.

Benenson A.S. (ed). 1995. Control of Communicable Diseases Manual, sixteenth edition. American Public Health Association, Washington, D.C.

Bennish, M.L. 1994. Cholera: Pathophysiology, clinical features, and treatment. In: Vibrio Cholerae and Cholera, Molecular to Global Perspectives. Wachsmuth, I.K., Blake, P.A., Olsvik, O., eds. ASM Press, Washington, D.C.: 229–256.

Blaser, M.J., and Newman, L.S. 1982. A review of human salmonellosis. I. Infective dose. Rev Infect Dis. 4:1096.

Butler, T., Fu, Y.S., Furman, L, Almeida, C, and Almeida, A. 1982. Experimental Yersinia pestis infection in rodents after intragastric inoculation and ingestion of bacteria. Infect. Immun. 36: 1160–7.

Centers for Disease Control and Prevention. 1998. Botulism in the United States, 1899–1996. Handbook for epidemiologists, clinicians, and laboratory workers. Centers for Disease Control and Prevention, Atlanta, GA.

Centers for Disease Control and Prevention. 2000. Biological and Chemical Terrorism: Strategic Plan for Preparedness and Response. Recommendations of the CDC Strategic Planning Workgroup. MMWR 49: 1–14.

Centers for Disease Control and Prevention. 2001. Update: Investigation of bioterrorism-related anthrax and interim guidelines for clinical evaluation of persons with possible anthrax. MMWR 50: 941–948.

Dupont, H.L., Levine, M.M., Hornick, R.B., et al. 1989. Inoculum size in shigellosis and implications for expected mode of transmission. J. Infect Dis. 159: 1126.

Erickson, M.C., and Kornacki, J.L. 2003. Bacillus anthracis: current knowledge in relation to contamination in food. J. Food Prot. 66: 691–699.

Franz, D.R., and Jaax, N.K. 1997. Ricin toxin. In: Medical Aspects of Chemical and Biological Warfare. F.R. Sidell, E.T. Takafuji, D.R. Franz, eds. Borden Institute, Walter Reed Army Medical Center, Washington, D.C., pp. 631–642.

Greenko, J., Mostashari, F., Fine, A., and Layton, M. 2003. Clinical evaluation of the Emergency Medical Services (EMS) ambulance dispatch-based syndromic surveillance system, New York City. J. Urban Health 80(2 suppl 1): i50–56.

Griffin, P.M., Bell, B.P., Cieslak, P.R., et al. 1994. Large outbreak of Escherichia coli O157:H7 infections in the western United States: the big picture. In: Recent Advances in Verocytotoxin-Producing Escherichia coli Infections. Karmali MA, Golglio AG, eds. Elsevier Science B.V., New York.

Grigg, B., and Modeland, V. The cyanide scare. In: A tale of two grapes. FDA Consumer July–August 1989;7–11.

Hatheway, C. L. 1990. Toxigenic clostridia. Clin. Microbiol. Rev. 3: 66–98.

Hedberg, C.W., MacDonald, K.L., and Osterholm, M.T. 1994. Changing epidemiology of foodborne disease: a Minnesota perspective. Clin. Infect. Dis. 18: 671–682.

Hennesy, T.W., Hedberg, C.W., Slutsker, L. et al. 1996. A national outbreak of Salmonella Enteritidis infections from ice-cream. N. Engl. J. Med. 334: 1281–6.

Hutwagner, L.C., Maloney, E.K., Bean, N.H., Slutsker, L., and Martin, S.M. 1997. Using laboratory-based surveillance data for prevention: an algorithm for detecting Salmonella outbreaks. Emerg. Infect Dis. 3: 395–400.

Institute of Medicine. 1999. Committee on R&D needs for improving civilian medical response to chemical and biological terrorism incidents. Chemical and Biological Terrorism. Research and development to improve civilian medical response. National Academy Press, Washington, D.C.

Khan, A.S. Swerdlow, D.L. and Juranek, D.D. 2001. Precautions against biological and chemical terrorism directed at food and water supplies. Public Health Reports 116: 3–14.

Khan, A.S., Morse, S., and Lillibridge, S. 2000. Public-health preparedness for biological terrorism in the USA. Lancet 356: 1179–1182.

Kolavic, S.A., Kimura, A., Simons, S.L., Slutsker, L., Barth, S., and Haley, C. 1997. An outbreak of Shigella dysenteriae type 2 among laboratory workers due to intentional food contamination. JAMA 278: 396–398.

Lee, R.V., Harbison, R.D., and Draughon, F.A. 2003. Food as a weapon. Food Prot. Trends 23: 664–674.

Mahon, B., Ponka, A., Hall, W., et al. An international outbreak of Salmonella infections caused by alfalfa sprouts grown from contaminated seed. J. Infect. Dis. 1997;175: 876–82.

Mead, P.S., Slutsker, L., Dietz, V., McCaig, L.F., Bresee, J.S., Shapiro C, et al. 1999. Food-Related Illness and Death in the United States. Emerg. Infect. Dis. 5: 607–625.

Mead, P.S., and Griffin, P. M. Escherichia coli O157:H7. Lancet 1998;352:1207–1212.

Mermin, J.H., and Griffin, P.M. 1999. Invited commentary: Public health crisis in crisis: outbreaks of Escherichia coli O157:H7 in Japan. Am. J. Epidemiol. 150: 797–803.

Pavlin, J.A. Investigation of disease outbreaks detected by 'syndromic' surveillance systems. J. Urban Health 80(2 suppl 1): i107–14.

Phills, J.A., Harrold, A.J., Whiteman, G.V., and Perelmutter, L. 1972. Pulmonary infiltrates, asthma, and eosinophilia due to Ascaris suum infestation in man. New Engl. J. Med. 286: 965–970.

Reintjes, R., Dedushaj, I., Gjini, A., Jorgensen, T.R., Cotter, B., Kieftucht, A., D'Ancona, F., Dennis, D.T., Kosoy, M.A., Mulliqui-Osmani, G., Grunow, R., Kalaveshi, A., Gashi, L., and Humolli, I. 2002. Tularemia outbreak investigation in Kosovo: a case control and environmental studies. Emerg. Infect. Dis. 8: 69–73.

Ryan, C.A., Nickels, M.K., Hargrett-Bean, N.T. et al. 1987. Massive outbreak of antimicrobial-resistant salmonellosis traced to pasteurized milk. JAMA, 258: 3269–74.

Shapiro, R., Hatheway, C., and Swerldlow, D. 1998. Botulism in the United States: a clinical and epidemiologic review. Ann. Intern. Med. 129: 221–228.

Sirisanthana, T., and Brown, A.E. 2002. Anthrax of the gastrointestinal tract. Emerg. Infect. Dis. 8: 649–651.

Sivapalasingam, S., Kimura, A., Ying, M., et al. 1999. A multistate outbreak of Salmonella Newport infections linked to mango consumption, November-December 1999. Latebreaker Abstract. 49th Annual Epidemic Intelligence Service (EIS) Conference, Centers for Disease Control and Prevention. Atlanta, GA 2000. Schedule Addendum, Latebreaker Abstracts.

Sobel, J., Swerdlow, D.L., Parsonnet, J. 2001. Is there anything safe to eat? In: Current Clinical Topics in Infectious Diseases. J.S. Remington, and MN Schwartz, eds. vol. 21. Blackwell Scientific Publications, Boston.

Sobel, J., Khan, A.S., and Swerdlow, D.S. 2002a. The threat of a biological terrorist attack on the United States food supply: The CDC Perspective. Lancet 359: 874–880.

Sobel, J., Griffin, P.M., Slutsker, L., Swerdlow, D.L., and Tauxe, R.V. 2002b. Investigation of multistate foodborne disease outbreaks. Public Health Reports 117: 8–19.

Stephenson, J. 1997. New Approaches for detecting and curtailing foodborne microbial infections. JAMA 277: 1337–1340.

Swaminathan, B., Barrett, T.J., Hunter, S.B., and Tauxe, R.V. 2001. PulseNet: The Molecular subtyping network for foodborne bacterial disease surveillance. United States. Emerg. Infect. Dis. 7: 382–389.

Tarnvik, A., and Berglund, L. 2003. Tularemia. Eur. Respir. J. 21: 361–373.

Tauxe, R.V. 1992. Letter from Peru: Epidemic cholera in Latin America [Letter]. JAMA 267: 1388–90.

Tilden, J., Young, W., McNamara, A.M., et al. 1996. A new route of transmission for Escherichia coli: infection from dry fermented salami. Am. J. Public Health 86: 1142–1145.

Torok, T., Tauxe, R.V., Wise, R.P., et al. 1997. A large community outbreak of Salmonella caused by intentional contamination of restaurant salad bars. JAMA 278: 389–395.

Treadwell, T.A., Koo, D. Kuker, K. Khan, A.S. 2003. Epidemiologic clues to bioterrorism. Public Health Reports. 118: 92–98.

USA Food and Drug Administration. 2003. Risk Assessment for Food Terrorism and Other Food Safety Concerns. October 13, 2003. www.cfsan.fda.gov/~dms/rabtact.html

USA General Accounting Office. Food safety: Agencies should further test plans for responding to deliberate contamination. GAO/RCED-00-3, October 27, 1999.

WHO. 2002. Terrorist Threats to Food, Guidelines for Establishing and Strengthening Prevention and Response Systems. World Health Organization, Geneva.

WHO. 1970. Health aspects of chemical and biological weapons. Report of a WHO group of consultants. Annex 5, Sabotage of Water Supplies. World Health Organization, Geneva, 113–120.

# Look What's Coming Down the Road: Potential Foodborne Pathogens

22

James L. Smith and Pina M. Fratamico

## Abstract

There are a number of factors involved in the emergence or re-emergence of pathogens associated with foodborne illness in the United States and other developed countries. These include environmentally related factors, such climate changes and deforestation, food-related factors, such as changes in food production and distribution practices, consumer-related factors, such as increased international travel and changes in eating habits, and pathogen-related factors, such as genetic changes in microorganisms as a result of exposure to environmental stresses. One major factor is the increased globalization of the food supply, resulting in transfer of pathogenic agents between countries. The use of antimicrobials for prophylaxis in animals has contributed to the emergence of bacterial strains resistant to multiple antibiotics. Potential emerging food-related diseases include hepatitis caused by the hepatitis E virus, intestinal spirochetosis caused by *Brachyspira pilosicoli*, gnathostomiasis caused by nematodes belonging to the genus *Gnathostoma*, and anisakidosis caused by fish nematodes. Other potential emerging pathogens include non-gastric *Helicobacter* spp., *Enterobacter sakazakii*, *non-jejuni/coli* species of *Campylobacter*, and non-O157 Shiga toxin-producing *Escherichia coli*. An increased awareness of emerging pathogens, consumer education, changes in food production and handling practices from farm to table, and improvements in microbiological detection methods will be needed to prevent the spread of emerging foodborne diseases.

## Introduction

Despite great improvements in sanitation, food storage, food manufacturing, slaughtering, regulatory oversight, consumer education and other areas of food safety and technology, illnesses caused by foodborne pathogens still represent a major public health concern in the United States and other developed countries. Compounding the problem is the ever-increasing globalization of the food supply. Consumer food tastes have become more sophisticated and cosmopolitan. The ease of international travel and the introduction of ethnic cuisines into developed countries by immigrant populations account for some changes in consumer eating habits; thus, there is a demand for food products not available in the United States and other developed countries (Swerdlow and Altekruse, 1998; Osterholm, 2000; Orlandi *et al.*, 2002). In addition, food habits have changed such that consumers want fresh fruit and vegetables all-year-round, and much of this demand must come from less-developed countries. Modern-day refrigerated transport has made it possible to ship foods rapidly from any part of the world to meet the demands of consumers. However, raw foods from the global market place, particularly from developing countries, may not always be grown, harvested, cleaned, processed, and shipped in a sanitary manner and thus may lead to the introduction of both new and old pathogens (Swerdlow and Altekruse, 1998; Osterholm, 2000; Orlandi *et al.*, 2002). For example, a large number of attendees at wedding receptions in Pennsylvania (Ho *et al.*, 2002) and Georgia (Murrow *et al.*, 2002) contacted cyclosporiasis from ingestion of imported raspberries from Central America. Imported fresh basil caused *Cyclospora cayetanensis* outbreaks in Missouri (Lopez *et al.*, 2001). In addition, imported fresh basil caused outbreaks due to *Shigella sonnei* in California, Florida, Massachusetts and Minnesota in the United States and in Alberta and Ontario in Canada, as well as enterotoxigenic *Escherichia coli* outbreaks in Minnesota (Naimi *et al.*, 2003). The source of these outbreaks was salads containing fresh basil; the basil was believed to have been imported from Mexico. An outbreak of typhoid fever occurred in Florida in individuals who consumed frozen fruit which had been processed in Central American plants (Katz *et al.*, 2002).

Other factors influence the ability to decrease the extent of foodborne illness. Today, food is grown and processed through large centralized facilities. For example, large farms where hundreds of beef cattle are fed have an ongoing problem of proper animal effluent disposal. These facilities also comprise large centralized meat slaughtering operations serving a large geographic area. A breakdown in the safe operation of such large facilities can result in massive foodborne outbreaks (Miller *et al.*, 1998; Swerdlow and Altekruse, 1998). In addition, many consumers are demanding raw or minimally processed foods because they believe that processing of foods diminishes the nutritive value; thus, processing conditions which would normally eliminate or reduce pathogens are circumvented (Orlandi *et al.*, 2002).

Microbial resistance to both antibiotics and biocides is an increasing problem, and there is a continual need for new antimicrobial compounds (Threfall *et al.*, 2000; Lipsitch, 2001; Russell, 2002). However, budget constraints in the pharmaceutical industry, due to the high costs incurred in testing new compounds, have led to a diminished introduction of new antibiotics and biocides (Drews and Ryser, 1997). Thus, the lack of effective antibiotics makes it difficult to treat patients with foodborne illnesses and other infections. In addition, the lack of new and more effective biocides may render hygienic control in food plant operations difficult. Also, due to modern medical advances, there is an increasing population of individuals whose immune status is deficient, and as a consequence, these individuals are more susceptible to foodborne illness (Smith, 1997). Particularly at risk is the ever expanding number of individuals over 65 years of age (Smith, 1998). Factors that facilitate the emergence of pathogens are presented in Table 22.1.

Based on current trends, it is likely that the acquisition of foods from the global market place will continue to be a source of fruits, vegetables and other foods contaminated with both familiar and "new" foodborne pathogens. Obtaining foods from other countries, particularly developing countries, potentially risky food behaviors of consumers, and the presence of a large immunocompromised population will, unfortunately, guarantee that foodborne illnesses will remain an ongoing problem.

## Hepatitis E virus: a potential foodborne pathogen from the global market place

Hepatitis E is the primary cause of acute viral hepatitis disease in developing countries located in Asia, Africa, and Latin America (Smith, 2001). The virus is transmitted by the fecal–oral route. Water borne epidemics and outbreaks of hepatitis E are common in tropical and subtropical countries; however, approximately half of the cases are sporadic (Smith, 2001; Hyams, 2002). The disease is rarely seen in developed countries but individuals from those countries may contact hepatitis E virus (HEV) when traveling in HEV-endemic areas. More recently, sporadic cases, not related to travel, have occurred in developed countries such as Australia, France, Germany, Greece, Italy, Japan, the Netherlands, New Zealand, the United Kingdom, and the United States (Smith, 2001;

**Table 22.1** Factors that may facilitate the emergence of pathogens[a]

*Environmentally related factors*

| | |
|---|---|
| 1 | Climate changes (increased temperatures, drought) |
| 2 | Deforestation (flooding) |
| 3 | Dam construction |

*Food-related factors*

| | |
|---|---|
| 1 | Changes in food production and distribution (the use of large facilities for animal production, slaughtering, meat portioning, product processing, shipping) |
| 2 | Processing modifications (decreased salt levels, decreased fat levels, decreased temperature treatment) |
| 3 | Alteration of packaging (type of package film or atmosphere) |
| 4 | Decontamination intervention (reduction of microbial contamination by intervention treatments of produce or carcasses) |
| 5 | Use of antibiotics in animal production |
| 6 | Global marketing (produce and other foods from developing countries may not be grown, harvested, processed and shipped under sanitary conditions) |

*Consumer-related factors*

| | |
|---|---|
| 1 | Rapid urbanization (overcrowding puts a strain on the public health and sanitation infrastructures) |
| 2 | Increased international travel |
| 3 | Increase in the number of immunocompromised individuals (increased life expectancy, advances in medical treatment, HIV and AIDS) |
| 4 | Eating habits (fewer meals at home, increased use of commercial ready-to-eat meals, increased interest in raw foods or minimally processed foods, increased interest in ethnic foods) |
| 5 | Basic lack of knowledge about proper food handling |

*Pathogen-related factors*

| | |
|---|---|
| 1 | Pathogen adaptation to various environments (adaptation to stress, antibiotics and biocides) |
| 2 | The acquisition of virulence traits by horizontal gene transfer |
| 3 | Increased recognition of pathogens (advances in molecular microbiology, immunology and pathogen detection methodology) |

[a]Samelis and Sofos (2003), Sharma *et al.* (2003).

Takahashi *et al.*, 2002; Worm *et al.*, 2002a; Okamoto *et al.*, 2003; Teich *et al.*, 2003; Widdowson *et al.*, 2003; Mansuy *et al.*, 2004). It is uncertain if these cases were food borne or water borne. While it is known that HEV is water borne, outbreaks of authentic foodborne hepatitis E infection have not been reported until recently. An outbreak of hepatitis E was reported from Japan in which uncooked Sika deer meat was implicated (Tei *et al.*, 2003). The nucleotide sequence of HEV RNA from frozen deer meat was identical to the HEV RNA isolated from the patients. The ingestion of undercooked wild boar meat was believed to be responsible for HEV-induced illness in 12 elderly Japanese men (Tamada *et al.*, 2004). Raw or undercooked swine liver has been implicated in hepatitis E outbreaks in Japan (Matsuda *et al.*, 2003; Okamoto *et al.*, 2003; Yazaki *et al.*, 2003). Detectable HEV RNA was present in 7/363 (1.9%) commercial packages of swine liver. Partial sequence analysis indicated that the HEV isolated from swine liver belonged to genotypes III or IV. Three of the swine liver HEV isolates were identical to patients' isolates; the remainder of the swine liver isolates were similar to Japanese human HEV isolates (Yazaki *et al.*, 2003).

In waterborne epidemics occurring in the developing world, the greatest number of clinical cases of hepatitis E is seen in the 15- to 40-year old population. The fatality rate ranges from 0.5% to 3%; however, the rate in pregnant women ranges from 15 to 25% (Smith, 2001; Worm *et al.*, 2002b). Sporadic acute viral hepatitis during pregnancy caused by hepatitis virus other than HEV was not as severe and did not show special predilection for pregnant women, in contrast to the disease induced by HEV (Khuroo and Kamili, 2003). The prevalence of anti-HEV IgG in blood donors from developed, industrialized countries is low, ranging from 0.0 to 3.3% (mean = 1.2%), whereas the prevalence among blood donors in developing countries is much higher, ranging from 7.2 to 24.5% (mean = 15.2%) (Smith, 2001). A high prevalence in blood donors would suggest that the virus is endemic in that country. Most HEV seropositive blood donors, particularly those from developed countries, do not recall being ill with hepatitis; therefore, the source of the infecting virus is unknown.

Hepatitis E virus consists of non-enveloped 27- to 30-nm spherical particles. The genome is a single strand of positive-sense RNA with three open reading frames (ORF1, ORF2, ORF3) and a length of approximately 7.2 kilobases (Smith, 2001; Worm *et al.*, 2002b). The virus is classified in the genus, *Hepatitis E-like virus* (Pringle, 1999). Hepatitis E virus exists as a single serotype but there are at least four genotypes. A HEV genotype was defined by Worm *et al.* (2002b) as those virus strains having nucleotide divergence of not more than 20% of the nucleotides in ORF2. Genotype I consists of strains isolated from Asia and Africa; genotype II includes strains from Mexico; strains from the United States, Europe (Greece, Italy and Spain) and Argentina comprise genotype III; and genotype IV consists of recently isolated Chinese strains (Pei and Yoo, 2002; Worm *et al.*, 2002b).

In most patients, hepatitis E presents as a mild jaundice with a self-limited course with an incubation period ranging from 15 to 60 days (mean = 40 days). Cholestasis (i.e., stoppage of bile flow) leads to elevated serum bilirubin and jaundice. Symptoms of hepatitis E infection include malaise, anorexia, abdominal pain, liver enlargement, vomiting and fever. Recovery takes 2–8 weeks after symptoms appear. Viremia and fecal excretion of the virus occur but generally do not exceed 2 weeks (Smith, 2001; Hyams, 2002; Worm *et al.*, 2002b). Chronic sequelae have not been observed. Jaundice-free subclinical disease with flu-like symptoms is more common than overt disease. The anti-HEV IgG titer increases during the acute phase of hepatitis E and decreases during convalescence; however, anti-HEV IgG is detectable for several years (Smith, 2001; Hyams, 2002; Worm *et al.*, 2002b). There is no effective treatment for hepatitis E but studies are under way which may lead to the development of a safe and effective vaccine (Koff, 2002; Purcell *et al.*, 2003). At the present time, hepatitis E can only be controlled by improving the hygienic conditions in HEV-endemic areas. This includes proper sewage disposal, proper water treatment, and a thorough education of individuals in proper personal and food/water hygiene (Smith, 2001; Hyams 2002; Worm *et al.*, 2002b).

Seropositivity to hepatitis E virus has been demonstrated in a number of both wild animals (old world and new world monkeys, rats and pigs) and domestic animals (cattle, swine, pet cats, and chickens) (Smith, 2001; He *et al.*, 2002; Usui *et al.*, 2004). Nucleotide sequencing data indicate that strains of swine HEV and human HEV are closely related (Garkavenko *et al.*, 2001; Okamoto *et al.*, 2001; van der Poel *et al.*, 2001; Huang *et al.*, 2002; Wang *et al.*, 2002; Wu *et al.*, 2002; Takahashi *et al.*, 2003). Swine HEV strains from Canada, Japan, Korea, the Netherlands, New Zealand, Spain, and the United States have been placed in HEV genotype III (Pei and Yoo, 2002; Worm *et al.*, 2002b; Choi *et al.*, 2003).

Hepatitis E viral RNA has been isolated from the sera and/or feces of infected swine (Smith, 2001). Individuals who work with swine have a higher prevalence of HEV antibody than non-swine workers (Hsieh *et al.*, 1999; Drobeniuc *et al.*, 2001; Withers *et al.*, 2002); thus, it is probable that swine are a source of infection to humans. Viral RNA was present in the feces, sera. liver, and bile of pigs inoculated with either a United States swine strain or a United States human strain (genotype III isolated from a sporadic case) of HEV (Halbur *et al.*, 2001). There were no signs of clinical illness nor elevation of liver enzymes or bilirubin in the infected pigs inoculated with either strain. Even though United States pigs could be infected with human genotype III HEV (Halbur *et al.*, 2001), they were not infected when inoculated with human HEV genotypes I or II strains (Meng *et al.*, 1998a), indicating that the United States swine HEV strain is closely related to genotype III human HEV strains. Hepatitis E virus seropositivity is common in swine and suggest that they are a major reservoir of HEV in both developed and developing countries (Table 22.2).

**Table 22.2** Anti-HEV IgG incidence in swine

| Country | Incidence of anti-HEV (no. positive/total no.) | IgG in swine (% positive) | Reference |
|---|---|---|---|
| Australia | 27/99 | 27.3 | Chandler et al. (1999) |
| Canada | 723/1710 | 42.3 | Meng et al. (1999), Yoo et al. (2001) |
| China | 352/491 | 71.7 | Meng et al. (1999), Wang et al. (2002) |
| Germany | 37/50 | 74.0 | Hartmann et al. (1998) |
| India | 122/284 | 43.0 | Arankalle et al. (2002) |
| Japan | 1448/2500 | 57.9 | Takahashi et al. (2003) |
| Korea | 96/404 | 23.8 | Meng et al. (1999), Choi et al. (2003) |
| Nepal | 48/159 | 30.2 | Clayson et al. (1995; 1996) |
| New Zealand | 54/72 | 75.0 | Garkavenko et al. (2001) |
| Taiwan | 102/275 | 37.0 | Hsieh et al. (1999) |
| Thailand | 23/75 | 30.7 | Meng et al. (1999) |
| United States | 221/327 | 67.6 | Meng et al. (1999), Withers et al. (2002) |

Inoculation of rhesus monkeys with a United States swine HEV strain led to seroconversion, fecal shedding of virus, and viremia. Clinically, the monkeys had a mild acute hepatitis with a slight elevation of liver enzymes (Meng et al., 1998b). Injection of a chimpanzee with the swine virus led to seroconversion and fecal shedding of virus, but viremia and hepatitis were not observed. The virus isolated from the feces of the primates was confirmed as the United States swine HEV strain (Meng et al., 1998b). Since non-human primates could be infected by the United States swine HEV, it is probable that humans can be infected also. The data obtained by Halbur et al. (2001) and Meng et al. (1998b) strongly suggest that swine are a reservoir of HEV and probably can transmit the virus to humans.

Rodents, particularly rats, can be seropositive for HEV (Smith, 2001; Hirano et al., 2003). Recently, He et al. (2002) detected HEV RNA in sera from viremic rats in Nepal. The HEV RNA isolated from the rats was closely related to genotype I HEV RNA isolated from Nepalese patients. He et al. (2002) hypothesized that rodents may also serve as a reservoir for HEV. Recently, Smith et al. (2002) demonstrated that the seroprevalence to HEV in homeless patients utilizing a free clinic in downtown Los Angeles was 13.6% (n = 200 patients). They also demonstrated that 98/134 (73.1%) of Norway rats in downtown Los Angeles were seropositive to HEV. Therefore, the life style of homeless people may allow close contact between them and rats, with the subsequent possibility of human acquisition of hepatitis E virus.

Hepatitis E virus has been demonstrated in raw sewage in both developing and industrialized countries (Smith, 2001; Vaidya et al., 2002; Clemente-Casares et al., 2003). The presence of HEV in sewage and in the feces of pigs indicate that untreated raw sewage and run-off water from swine farms could allow the entry of HEV into irrigation or coastal waters with possible viral contamination of produce and shellfish. Fecal contamination of pork carcasses may be another source of hepatitis E infection (Smith, 2001). Untreated waste water used for irrigation may be a source of hepatitis E in-

fection to agricultural workers. An anti-HEV seropositivity of 34.8% was demonstrated in Turkish farmers who worked in fields irrigated with untreated waste water as compared to 4.4% seropositivity in matched controls who did not do field work (Ceylan et al., 2003). Produce grown in fields irrigated with waste water could be contaminated with HEV and other pathogens and represent a source of infection to humans.

## *Taenia solium*: neurocysticercosis as a disease of immigrants

*Taenia solium*, the pork tapeworm, is endemic throughout Latin America, sub-Saharan Africa, India, and other Asian countries. The parasite is the cause of neurocysticercosis, the most common parasitic disease of the central nervous system, world-wide (Smith, 1994). The parasite has a two-host life cycle. The worm, 2 to 8 m in length, resides in the intestine of the definitive human host. The head (scolex) of *Taenia solium* attaches to the mucosa of the small bowel via hooklets and suckers and as the worm elongates, the terminal egg-bearing segments (proglottids) are released into the environment during defecation (Smith, 1994; Garg, 1998; Sciutto et al., 2000). Swine are the intermediate host for *T. solium*. Upon ingestion of the eggs by the pig, the egg membranes are digested by gastric enzymes with the liberation of infectious oncospheres. The oncospheres penetrate the wall of the small bowel and enter the circulatory system with eventual spread throughout the pig's body, particularly in skeletal muscle. The oncospheres develop into cysticerci (encysted larvae with an evaginated scolex) within 60 to 70 days. When humans eat raw or undercooked *T. solium*-infected pork, digestive juices dissolve the cyst walls of the ingested cysticerci releasing the scolex with eventual attachment to the small bowel and continuation of the life cycle of the tapeworm in the human (Smith, 1994; Garg, 1998; Sciutto et al., 2000).

Taeniasis, i.e., the tapeworm growing in the intestinal tract, is the disease resulting when humans ingest raw or poorly cooked *T. solium*-infected pork. The adult tapeworm is present only in humans. The patient is usually asymptomatic and most *T. solium* tapeworm carriers are unaware of their infection. Occasionally, however, patients may experience abdominal pain, hunger pains or chronic indigestion (Smith, 1994). If humans ingest the worm eggs, they develop cysticercosis just as pigs do, with deposition of cysticerci in various parts of the body. Deposition of the cysticerci in the brain can lead to neurocysticercosis with resulting neurological problems. Invasion of the central nervous system is seen in approximately 60% of the cases of cysticerci infestation, and neurological symptoms appear in about half of those patients within 2 months to 30 years (mean of 5 years). The most common clinical manifestation of neurocysticercosis is epilepsy (Smith, 1994; Garg, 1998; Sciutto *et al.*, 2000). Taeniasis can be treated with antihelmintics such as niclosamide, quinacrine or praziquantel (Smith, 1994). Neurocysticercosis can be treated with anticysticercal drugs such as praziquantel or albendazole; however, the use of these drugs for the treatment of neurocysticercosis is controversial (Garg, 1998; Sciutto *et al.*, 2000).

There is no information on the effect of cooking of pork products on the destruction of *T. solium* cysticerci; however, the USDA Food Safety and Inspection Service (FSIS) requires that a "lightly" infected pork carcass be cooked at an internal temperature of ≥76.7°C for 30 min to destroy the cysticerci (Smith, 1994). Freezing pork at −10°C for 14 days or radiation with 0.6 kGy also renders *T. solium* cysticerci non-infective (Smith, 1994). There is no information on temperature inactivation of *T. solium* eggs in contaminated food products (Smith, 1994).

Cysticercosis is a "pig-driven" disease. Pigs are inexpensive to raise since the animals are scavengers, are easily domesticated, are resistant to many adverse environmental conditions, and grow to a large size rapidly. Moreover, pork is a cheap source of protein for economically deprived countries (Soleto, 2003). The scavenging behavior of pigs combined with open-air defecation practiced by humans in developing countries results in pigs coming into contact with human feces, which may contain *T. solium* eggs. Therefore, preventing the infection of pigs with *T. solium* will break the parasitic cycle. *Taenia solium* is seldom seen in pigs in developed countries due to well-controlled swine husbandry practices, proper sewage disposal of human waste, and rigid meat inspection standards (Sciutto *et al.*, 2000). Cysticercosis and neurocysticercosis in humans is common in swine raising environments where there is poor sanitation, poor personal hygiene, lack of potable water, lack of sewage facilities and lack of adequate meat inspection (Smith, 1994; Garg, 1998; Sciutto *et al.*, 2000). The key factor for the elimination of *T. solium* in endemic countries lies in preventing pigs from ingesting human fecal waste; however, major changes will be required in terms of human behavior, sanitation, swine husbandry, and swine slaughtering inspection in endemic areas for this to occur.

In developed countries, most cases of neurocysticercosis are found in immigrants from *T. solium*-endemic countries or in individuals who have contact with food handled by an immigrant who is a tapeworm carrier. With the increased movement of immigrants and refugees from developing countries into developed countries, there has been a reintroduction of *T. solium*-associated diseases into developed countries (Schantz *et al.*, 1998). For example, Ong *et al.* (2002) studied neurocysticercosis prevalence in seizure patients admitted to 11 geographically diverse, university-associated, urban hospital emergency units in the United States. Neurocysticercosis was diagnosed in 2.1% (38/1801) of seizure patients admitted to the emergency departments. Neurocysticercosis was detected at 9/11 sites, and the disease was associated with Hispanic ethnicity (29/37 patients), being an immigrant, and prior exposure of the patient to a *T. solium*-endemic geographic area (Ong *et al.*, 2002). Most of the cases (25/38) of neurocysticercosis occurred in the southwestern United States (California, New Mexico and Arizona). The incidence of neurocysticercosis was 9.1% (29/320) among the Hispanic patients who demonstrated seizures (Ong *et al.*, 2002). Recently, Townes *et al.* (2004) reported on the cases of neurocysticercosis that occurred in Oregon, 1995–2000. Of 61 confirmed cases, 66% were male and 85% were of Hispanic origin, mostly from Mexico. The median age was 24 years at their first hospitalization. The mean annual incidence of neurocysticercosis was 2.0 per $10^6$ of the general population as compared to 31.0 per $10^6$ of the Hispanic population. Neurocysticercosis-related deaths in California, between 1989 and 2000, were studied by Sorvillo *et al.* (2004). During that period, there were 124 deaths attributed to neurocysticercosis. Sixty-six per cent of the deaths occurred in males and 72.6% occurred in people born in Mexico. The median age at death was 34.5 years. The death rate due to neurocysticercosis among the Hispanic population of California was 13.0 per $10^6$ of the general population as compared to a rate of 0.4 per $10^6$ for other racial/ethnic groups (Sorvillo *et al.*, 2004). Morbidity and mortality due to neurocysticercosis is more common in young Hispanic males than in other segments of the general population. There is also some indication that the acquisition of neurocysticercosis in the United States is increasing, particularly in pediatric patients with immigrant parents (Ong *et al.*, 2002). The presence of a family member with *T. solium* taeniasis increases the likelihood of other family members, particularly children, contacting cysticercosis.

A taeniasis-infected food worker, with poor personal hygiene, poses a threat to individuals who eat food prepared by that individual. Foods that do not require cooking before serving such as salads or sandwiches can become contaminated by *T. solium* eggs from an infected food preparer. It would be difficult to detect a neurocysticercosis outbreak, since there is usually a long lag period (up to 25 years) before symptoms appear (Smith, 1994).

Most of the pork consumed in the United States is domestic but about 4% of the pork consumed in the United States is imported (USA Census Bureau, 2000, Table No. 1121). Assuming that meat inspection at the port of entry is adequate, it would be virtually impossible to introduce cysticercotic pork from the global market place into the United States or other developed countries. Therefore, in the United States and other developed countries, neurocysticercosis is generally found in immigrants from *T. solium*-endemic countries or in individuals who have come in contact with a pork tapeworm carrier who is careless in personal hygiene.

## Chronic wasting disease: a threat to hunters?

Chronic wasting disease (CWD) is a prion-based disease (transmissible spongiform encephalopathy; TSE) that affects wild-ranging and captive deer and elk (Cervidae) in North America, particularly in the western United States. Recently, CWD was reported in free-ranging white-tailed deer in Wisconsin, representing the first report of CWD in free-ranging cervids east of the Mississippi (Joly *et al.*, 2003). The disease has been found in mule deer, black-tailed deer, white-tailed deer and Rocky Mountain elk (Hamir *et al.*, 2001). An estimated 15% of western free-ranging deer and elk are infected (Raymond *et al* 2000), whereas approximately 3% of Wisconsin wild deer are CWD-positive (Joly *et al.*, 2003). There is no evidence that farmed cervids transmitted CWD to the Wisconsin wild deer population. More recently, CWD has emerged among elk in game farms in Canada and in several western states in the United States (Williams and Miller, 2002). Commerce of farmed game cervids may allow the spread of CWD to other geographic areas.

What is the origin of CWD? Hamir *et al.* (2003) inoculated elk calves intracerebrally with brain suspensions from sheep naturally affected with scrapie. The authors demonstrated that sheep scrapie can be transmitted to elk by intracerebral inoculation and that the infection can result in severe, spongiform changes and accumulation of protease-resistant prions in the central nervous system (CNS). A limited survey of CNS sections from necropsied-scrapie inoculated elk indicated that the disease could not be distinguished from CWD. Hamir *et al.* (2003) speculated that the agents of scrapie and CWD have a common lineage, and that CWD may have originated as a cross-species transmission of sheep scrapie. There is a strong probability of a genetic basis for susceptibility of cervids to CWD; however, additional studies are needed to confirm this (Johnson *et al.*, 2003).

How CWD is transmitted remains unclear. The infective agent may be transmitted animal-to-animal (both intra- and interspecies) and environmentally via contaminated pastures. Miller *et al.* (2004) reported that CWD can be transmitted to mule deer under conditions where the grazing environment was contaminated by feces and decomposed carcasses of infected cervids. Maternal transmission is considered uncommon (Williams and Miller, 2002). A study of two captive mule deer populations indicated that horizontal (animal-to-animal) transmission of CWD is the major route of infection; maternal transmission made little or no contribution to infection

(Miller and Williams, 2003). Providing forage to captive deer and the concentration of deer during captivity may encourage transmission of CWD.

"Natural" transmission of CWD to domestic or laboratory animals has not been demonstrated; however, if experimental animals are inoculated intracerebrally with infected brain tissue, infection does occur (Hamir *et al.*, 2001). For example, cattle have been infected by intracerebral inoculation of brain tissue from CWD infected mule deer. However, only 3/13 cattle showed the presence of CWD prions three years after inoculation (Hamir *et al* 2001). This experiment is still ongoing. Gould *et al.* (2003) initiated a survey of adult cattle raised in CWD-endemic range areas of Colorado in order to assess the spread of CWD from free-ranging deer to cattle. Culled cattle, 4123 from 22 herds, that had spent at least four years in the endemic area, were autopsied and their brains examined for TSE lesions. There were no indications of TSE infection in the culled cattle, suggesting that direct or indirect transmission of CWD from free-ranging deer was unlikely (Gould *et al.*, 2003).

Chronic wasting disease, similar to other TSEs, has a long incubation period of approximately two years or more. Clinical CWD disease, in deer and elk, is characterized by emaciation, weight loss, excessive thirst, excessive urination, excessive salivation, changes in behavior (lack of coordination, separation from the herd, depression), paralysis, difficulties in swallowing and pneumonia. Clinical symptoms last from several weeks to eight months before death occurs (Williams and Miller, 2002).

In naturally infected deer with advanced CWD, the abnormal, protease-resistant infectious prions are present in lymph nodes, spleen, tonsils, Peyer's patches, and various regions of the brain (Sigurdson *et al.*, 2002; Williams and Miller, 2002). Since the clinical signs of CWD are not specific, post-mortem diagnosis is based on examination of cervid brain for spongiform lesions and for the presence of protease-resistant prions. Pre-clinical diagnosis is based on detection of protease-resistant prions in the tonsils of deer (Wild *et al.*, 2002; O'Rourke *et al.*, 2003). There is no antibody response to the CWD agent and, therefore, diagnosis by testing for serum anti-CWD antibody cannot be done (Williams and Miller, 2002). Recently, Safar *et al.* (2002) have described a sensitive conformation-dependent immunoassay which can differentiate between bovine spongiform encephalopathy (BSE) and CWD prions. They recommend the use of the assay on animals as a means to reduce human exposure to the prions.

There is no treatment for CWD and control measures rely on eradication of farmed herds that show the presence of CWD. Control of CWD in wild free-ranging animals is not possible at present (Williams and Miller, 2002). Creutzfeldt–Jakob disease (CJD) present in three young persons (≤30 years of age) who had histories of eating venison was recently investigated to determine if the CJD was related to CWD (Belay *et al.*, 2001). The authors concluded that there was no strong evidence of a causal link between the young patients' CJD and CWD. Several epidemiological studies indicated that there is no real evidence that CWD has caused prion disease in humans (Belay *et al.*, 2004). Using an *in vitro* system to detect conversion of normal prions to CWD prions, Raymond *et al.* (2000) were able to show that CWD prions converted normal cervid prions to the abnormal CWD form, but conversion of human normal prions to CWD prions by prions from infected cervids was >14-fold less efficient than inter-cervid conversion. Conversion of bovine normal prions to CWD prions was 5- to 12-fold less efficient than inter-cervid conversion (Raymond *et al.*, 2000). The *in vitro* results published by Raymond *et al.* (2000) indicate that conversion of normal human or bovine prions to CWD prions is very inefficient and has a low probability of occurrence. Nonetheless, since there is strong suspicion that the BSE prion can infect cattle and humans (Smolinski *et al.* (2003), prudence would indicate that steps to limit the exposure of humans and domestic animals to CWD should be in place.

Commerce, particularly international commerce, in cervids and cervid products should probably be discouraged at the present time. It is known that elks with CWD have been transported into Canada from the United States and into Korea from Canada (Sohn *et al.*, 2002). The institution of a global surveillance system for CWD among captive cervids in zoos and on game farms would provide an indication of the extent of the disease world-wide and also would give an indication of when control measures should be initiated.

It has not been established that infectious prions accumulate in the edible muscles of cattle infected with the BSE prion, sheep infected with the scrapie prion, or cervids infected with the CWD prion. However, recent studies suggest that infectious prions might be present in muscle foods from animals infected with TSE agents. Scrapie prions were found in the skeletal muscles of mice (Bosque *et al.*, 2002) and hamsters (Thomzig *et al.*, 2003) infected with the scrapie prion. These results suggest that muscle foods from TSE-infected animals may be a source of infectious prions to humans and that further studies should be done to address this issue. Hunters, meat processors, and taxidermists in the United States and Canada who handle cervid carcasses should take measures to decrease their exposure to the CWD agent. These measures include avoidance of handling and eating of the meat from obviously diseased cervids, wearing surgical gloves when dressing cervids, and thorough washing of butchering implements. The brain, spinal cord, lymph nodes, spleen, tonsils and eyes from cervids should be discarded, since these organs contain high levels of the CWD prions (Sigurdson *et al.*, 2002; Wild *et al.*, 2002; Williams and Miller, 2002). Surveillance, control, and eradication of CWD should be prime goals of wild-life managers and animal health agencies. Salman (2003) and Belay *et al.* (2004) haves written detailed reviews of CWD, which should be consulted for additional details of the disease.

## Gnathostomiasis: travelers, beware of raw meat and raw freshwater fish

Gnathostomiasis is a foodborne parasitic zoonosis induced by ingesting the larvae of different species of the nematode belonging to the genus *Gnathostoma*. The nematode is endemic in Asia and Latin America (Rusnak and Lucey, 1993; Ogata *et al.*, 1998; McCarthy and Moore, 2000). In developed countries, the disease is seen in returning travelers who had visited areas where *Gnathostoma* species are endemic and in immigrants from endemic countries. Recently, Moore *et al.* (2003) described several cases of gnathostomiasis diagnosed in a London hospital. These cases represented British citizens who had traveled to areas where *Gnathostoma* species were endemic as well as immigrants from endemic areas who had contacted the parasite before arriving to the United Kingdom.

Adult worms are present in the stomach wall of the animal definitive hosts, which include dogs, members of the feline family, mink, opossum, raccoon and otter. The parasite discharges eggs into the lumen of the stomach and eventually the eggs enter into the environment via feces. In the presence of water, first-stage larvae develop within the eggs (Rusnak and Lucey, 1993; McCarthy and Moore, 2000). These larvae are ingested by the first intermediate host, freshwater crustaceans (Cyclops). In the crustaceans, the larvae develop into early third-stage larvae. When crustaceans containing immature third-stage larvae are ingested by second intermediate hosts such as fresh water fish, reptiles, birds including domestic fowl, or mammals including squirrels, rabbits and pigs, the larvae mature into infectious third-stage larvae (Rusnak and Lucey, 1993; McCarthy and Moore, 2000). The larvae migrate to the muscles of the second intermediate host and encyst. Ingestion of an infected second intermediate host by a definitive host completes the parasitic cycle.

The third-stage larvae cannot complete their developmental cycle (become an adult worm) in humans; however, several days after ingestion, the larvae penetrate the intestinal wall and migrate through various tissues. The patient may exhibit eosinophilia, epigastric pain, vomiting, and anorexia. Cutaneous gnathostomiasis is the most frequent manifestation of the disease in humans (Rusnak and Lucey, 1993). Migration in the subcutaneous tissues causes intermittent, inflammatory, migratory, painful and pruritic (itching) swellings, and edema. The swellings can appear on any area of the body (Rusnak and Lucey, 1993; Diaz Camacho *et al.*, 1998; Ogata *et al.*, 1998; Vargas-Ocampo *et al.*, 1998; Rojas-Molina *et al.*, 1999; Grosbusch *et al.*, 2000; del Giudice *et al.*, 2001; Puente *et al.*, 2002). Mechanical damage to the tissue by the migrating larvae and the immune response against the parasite account for the pathology of gnathostomiasis (Rusnak and Lucey, 1993).

Visceral gnathostomiasis can involve deeper tissues and organs. The respiratory, gastrointestinal, genitourinary, ocular. auditory, and the CNS may also be affected (Punyagupta *et al.*, 1990; Rusnak and Lucey, 1993; Chandenier *et al.*, 2001; Montero *et al.*, 2001). When the larvae are present in the brain and spinal cord, serious neurological symptoms such as coma, paralysis, and seizures, as well as death, may result. The recommended drug for treatment of gnathostomiasis is albendazole (Grobusch *et al.*, 2000; McCarthy and Moore, 2000).

The most common source of gnathostomiasis to humans is consumption of raw or undercooked fish (Vargas-Ocampo *et al.*, 1998; Rojas-Molina *et al.*, 1999; León-Règagnon *et al.*, 2000; del Giudice *et al.*, 2001; Puente *et al.*, 2002); however, raw or undercooked pork, chicken, or other meats from a second intermediate host may also be a source of the parasite (Rusnak and Lucey, 1993). Gnathostomiasis can be prevented by avoiding raw or uncooked fish and meats, by drinking boiled water, and by wearing gloves when handling suspect foods (to prevent third-stage larval penetration of skin). Freezing foods at −20°C for 3–5 days also destroys the larvae (Rusnak and Lucey, 1993; Yoshimura, 1998).

While gnathostomiasis is considered to be a disease of Asian and Latin American countries, travelers who visit areas endemic for *Gnathostoma* species are at risk if they sample the traditional native cuisines, particularly if these foods include raw or under-cooked freshwater fish, pork, or poultry. Eradication of *Gnathostoma* species is not practical, and control of gnathostomiasis can only be achieved by raising public awareness to the dangers of eating raw or undercooked meats and fish when traveling overseas.

## Anisakidosis (anisakiasis): a disease associated with sashimi, sushi and other raw seafood dishes

The increasing popularity of "sushi bars" may eventually lead to an increase in the number of cases of anisakidosis in the United States and Europe. Fish nematode larvae of *Anisakis, Pseudoterranova*, or *Contracaecum* are responsible for anisakidosis in humans when infected, raw saltwater fish are eaten (Yoshimura, 1998; McCarthy and Moore, 2000). Sashimi (thinly sliced raw fish), sushi (cold rice dressed with vinegar, shaped in small cakes and topped or wrapped with raw fish slices), ceviche (raw fish marinated in lime or lemon juice), lomi lomi (raw salmon), sunomono (vinegar dressed salads, which may be garnished with raw seafood), Dutch green herring (raw herring), marinated fish, and cold-smoked fish may be sources of anisakidosis for the unwary consumer (Yoshimura, 1998; McCarthy and Moore, 2000).

Adult worms live in the stomachs of the definitive hosts which include whales, dolphins, seals and sea-lions. The eggs produced by the worms are excreted into the ocean via feces. The eggs hatch and stage 1 free-living larvae are ingested by krill (the first intermediate host) where the larvae moult from stage 2 to stage 3. Stage 3 larvae in infected krill are ingested by the second intermediate host, fish or squid. Ingestion of the infected fish or squid by marine mammals leads to the development of the adult worm. Humans are not suitable hosts for completion of the parasitic cycle (Bouree *et al.*, 1995; Buendia, 1997; Yoshimura, 1998).

*Anisakis simplex* is the most common cause of anisakidosis. There are two types of clinical presentation: parasitic infestation of the gastrointestinal tract and allergy. The former is characterized by symptoms that mimic other gastrointestinal distresses such as gastric tumors, ulcers, or acute appendicitis. Acute anisakidosis is seen in more than 97% of cases (McCarthy and Moore, 2000). Two to five hours after consuming infected fish, the patient has severe epigastric pain; nausea, vomiting, and fever are common. The worm larvae are either adherent to or are found within the gastric mucosa. Removal of the nematode larvae is curative. Rarely, the larvae may be found in the pharynx, esophagus, large intestine, or peritoneal cavity (Buendia, 1997; Yoshimura, 1998; McCarthy and Moore, 2000).

The second type of response to anisakidosis is allergy. Ingestion of the live parasites or food containing dead parasites (killed by cooking or pasteurization) may lead to sensitization followed by systemic IgE-mediated reactions in the infected individual. The allergic response may result in the appearance of urticaria (hives), angioedema (acute, intermittent edema swellings in skin and mucosa) or anaphylaxis (Alonso *et al.*, 1997; Buendia, 1997; McCarthy and Moore, 2000). Local allergic responses may include intestinal obstruction and mesenteric inflammation (mimicking appendicitis) (Takabe *et al.*, 1998; McCarthy and Moore, 2000). Other allergic responses to infestation by *Anisakis* can result in eosinophilic gastritis (Esteve *et al.*, 2000; McCarthy and Moore, 2000) and gastric hemorrhage (Gutiérrez-Ramos *et al.*, 2000). In addition, hypersensitivity to *A. simplex* is responsible for rheumatologic symptoms in patients (Cuende *et al.*, 1998). Foods infested with *A. simplex* and cooked to a temperature which kills the parasite can still invoke an allergic response (Alonso *et al.*, 1997).

A large number of different fish species are infested with anisakidosis-causing nematodes including anchovy, cod, flounder, herring, mackerel, redfish, rockfish, salmon, sardine, sea eel, sole, snapper, squid, and whiting (Bouree *et al.*, 1995; Buendia, 1997; Yoshimura, 1998; Griffiths, 1999). Herring and cod are particularly highly parasitized (Bouree *et al.*, 1995). The larvae in infested seafood are killed by freezing at −20°C for three days or by heating to 70°C internal temperature (Bouree *et al.*, 1995). At room temperature, *Anisakis* larvae can survive for several days in vinegar and at least one day in soy sauce or worcester sauce (Yoshimura, 1998). The larvae are not killed by gastric juice (Yoshimura, 1998).

The observed increase in anisakidosis is due to governmental protection of marine mammals as endangered species and to currently used methods of fish gutting (Bouree *et al.*, 1993). Fish are not gutted immediately after they are caught, but are kept in cold storage for a period of time before evisceration. The resultant delay allows the parasites to migrate from the fish gut into muscles (Bouree *et al.*, 1993). Perhaps the most important reason for the increase in anisakidosis is that eating of raw fish and other seafoods is "trendy". As long as it is the "in" thing, the incidence of anisakidosis will steadily increase.

## Intestinal spirochetosis: are *Brachyspira* foodborne?

Intestinal (or colonic) spirochetosis in humans, non-human primates, pigs, dogs, and fowl is caused by *Brachyspira pilosicoli* (*Serpulina pilosicoli*, *Anguillina coli*) (Zhang *et al.*, 2000). In humans, intestinal spirochetosis is diagnosed histologically. The colonic epithelium is heavily colonized by the organism, which is attached by one end to the epithelial surface to form a "false brush border". While most patients with intestinal spirochetosis are asymptomatic, the colonization of the colon may be associated with a variety of gastrointestinal symptoms including abdominal discomfort, chronic diarrhea, and rectal bleeding (Barrett, 1997). The massive colonization may interfere with intestinal reabsorption of essential nutrients (Josephson, 1998). Septicemia induced by *B. pilosicoli* has been noted, particularly in immunocompromised or critically ill patients (Fournié-Amazouz *et al.*, 1995; Trott *et al.*, 1997b; Kanavaki *et al.*, 2002). A volunteer who drank water containing *B. pilosicoli* became colonized and excreted the organism in his feces. In addition, the volunteer developed nausea, abdominal discomfort, and severe headaches (Oxberry *et al.*, 1998). Treatment of the volunteer with metronidazole led to clearing of the spirochetes with cessation of symptoms. Carriage of *B. pilosicoli* is uncommon in developed areas, but is common in developing countries (Trott, 1997a), Australian Aboriginals (Brooke *et al.*, 2001), immigrants from developing countries (Brooke *et al.*, 2001), and homosexuals (Trivett-Moore *et al.*, 1998). Immunocompromised populations may be more susceptible to spirochetal infection than the population at large.

*Brachyspira aalborgi* also induces intestinal spirochetosis in humans but is rarely isolated due to its fastidious nature and slow growth. The use of polymerase chain reaction (PCR) assays and fluorescent *in situ* hybridization has facilitated detection of *B. aalborgi* (Jensen *et al.*, 2001; Mikosza *et al.*, 2001b). *Brachyspira aalborgi* does not have the broad host range of *B. pilosicoli* and is found only in humans and non-human primates (Duhamel *et al.*, 1997). Patients with histologically identified intestinal spirochetosis due to infection with *B. aalborgi* may present with a variety of gastrointestinal symptoms (Padmanabhan *et al.*, 1996; Mikosza *et al.*, 1999; Kraaz *et al.*, 2000; Heine *et al.*, 2001; Mikosza *et al.*, 2001a). These symptoms can include persistent diarrhea, rectal bleeding, blood and mucus in stools, and abdominal pain.

*Brachyspira alborgi* or *B. pilosicoli* grow anaerobically but are oxygen tolerant. The cells are unicellular, motile by flagella, helicoidal, gram negative, and weakly β-hemolytic (Hovind-Hougen *et al.*, 1982; Trott *et al.*, 1996b). The spirochetes can be cultivated utilizing trypticase soy agar supplemented with 5% (v/v) defibrinated sheep blood incubated anaerobically (94% $N_2$:6% $CO_2$) for ≥15 days (Mikosza *et al.*, 2001b). The two species can be differentiated by using PCR assays specific for the 16S RNA gene. *Brachyspira pilosicoli* can survive for 119 days in soil and 210 days in soil mixed with 10% pig feces or in pig feces alone at 10°C (Boye *et al.*, 2001). It is probable that *B. pilosicoli* is present in soil frequented by animals suffering from intestinal spirochetosis. *Brachyspira pilosicoli* has been found in lake water, and subsequent studies indicated that the organism can survive in lake water for at least 66 days at 4°C and 4 days at 25°C (Oxberry *et al.*, 1998).

Porcine intestinal spirochetosis is a diarrheic disease induced by *B. pilosicoli*. Infected pigs present with a mucus-containing, non-bloody diarrhea; the pigs show poor feed conversion and reduced growth rates (Trott *et al.*, 1996b). Thus, porcine intestinal spirochetosis can represent a severe economic loss to the swine grower. *Brachyspira pilosicoli* infections have been found in chickens and have been associated with wet litter (watery feces), decreased egg production, and economic losses (Stephens and Hampson, 2001). Experimental infection of broiler breeder hens with *B. pilosicoli* resulted in a transient increase in the water content of feces and in delayed onset of egg laying with a significant reduction in egg production (Stephens and Hampson, 2002). *Brachyspira pilosicoli* has been isolated from farmed game birds such as partridges, pheasants, and mallards and various wild waterbirds (Oxberry *et al.*, 1998; Jansson *et al.*, 2001). Diarrhea in dogs has also been associated with *B. pilosicoli* infection (Duhamel *et al.*, 1998; Fellström *et al.*, 2001).

Intestinal spirochetosis was induced in chicks inoculated with either a human or canine strain of *B. pilosicoli* (Muniappa *et al.*, 1996). Intestinal spirochetosis with a watery, mucoid diarrhea developed in newly weaned pigs inoculated with either a human or porcine strain of *B. pilosicoli* (Trott *et al.*, 1996a). Mice inoculated with an avian or porcine strain of *B. pilosicoli* developed typical intestinal spirochetosis (end-on attachment of the bacteria to the mucosal surface of enterocytes); however, inoculation with a human strain of *B. pilosicoli* led to colonization of the mice but not to typical intestinal spirochetosis (Sacco *et al.*, 1997). These studies indicate that animals can be used to model human intestinal spirochetosis, indicating that strains of *B. pilosicoli* are closely related, not animal species specific, and that cross-infections from animals to humans may occur.

It is probable that both *B. aalborgi* and *B. pilosicoli* can be transmitted person-to-person via the fecal–oral route. Therefore, an infected food handler with poor personal hygiene could be the cause of an outbreak of intestinal spirochetosis. Eating meat from swine or fowl infected with *B. pilosicoli* may be a source of infection to consumers, as well as a source of cross-contamination to other food products. Abattoir workers may also be as risk for infection by *B. pilosicoli*. Run-off water from farms where intestinal spirochetosis is present in chickens or swine could contaminate streams leading to infection of domestic and wild animals who ingest the water. In addition, the pollution of irrigation water with farm run-off water could allow the contamination of vegetables and other crops.

It is uncertain whether *Brachyspira* infections are food borne. However, as fecal organisms, *B. pilosicoli* and *B. aalborgi* may be present in both water and food. As both a human and animal pathogen, *B. pilosicoli* could readily contaminate irrigation water used for the growth of produce and could be present on meat from infected animals. Spirochete-infected humans could be a source of infection via person-to-person transmission or by the handling of foods.

## *Helicobacter pylori* and other *Helicobacter* species as foodborne pathogens

Infection with *H. pylori* is common throughout the world; approximately 40% of adults in industrialized countries and >90% of adults in developing countries are infected (Kusters, 2001). Approximately 80% of infected individuals do not realize that they harbor the organism, and carriage can be lifelong without overt symptoms. The organism induces a strong inflammatory Th1-type immune response, which is non-protective (Kusters, 2001; Prinz *et al.*, 2003). Inflammation of the lining of the stomach results in gastritis. Symptoms of gastritis include loss of appetite, nausea, vomiting, and discomfort after eating (Kusters, 2001). Ten to 20% of infected patients will develop peptic ulcer disease which can be cured with antibiotic treatment (Prinz *et al.*, 2003). *Helicobacter pylori* is a significant risk factor for the development of gastric cancers in 0.1 to 4% of infected patients (Kusters, 2001; Prinz *et al.*, 2003).

*Helicobacter pylori* is known to enter into a viable but nonculturable (VBNC) state when subjected to low temperature and oxygen, antibiotics, nutrient deprivation and other environmental stresses. In the VBNC state, the cells change in morphology from rods to cocci (Velázquez and Feirtag, 1999; Engstrand, 2001; Gomes and de Martinis, 2004). Aleljung *et al.* (1996) and Wang *et al.* (1997) have demonstrated that the coccoid VBNC form can colonize the stomach and induce inflammation of the stomach mucosa in mice. It is difficult to culture *H. pylori* from feces and environmental sources, and it is probable that the VNC state may be the form that is excreted and the form which can survive in the environment (Weaver *et al.*, 1999). Winiecka-Krusnell *et al.* (2002) have shown that co-cultivation of *H. pylori* with the free-living ameba, *Acanthamoeba castellanii*, at 36.5°C, led to a 100-fold increase in bacterial numbers after 7 days. Similarly, Marciano-Cabral and Cabral (2003) have demonstrated that *H. pylori* multiplies within *Acanthamoeba* and is released into the environment in vesicles. Intracellular growth of *H. pylori* in *Acanthamoeba* may allow enhanced survival of the organism in the environment.

Since the organism is seldom cultured from environmental sources, PCR is useful for detection. *Helicobacter pylori* has been detected in human feces (Nilsson *et al.*, 1996; Parsonnet *et al.*, 1999; Weaver *et al.*, 1999; Kabir, 2001; Allaker *et al.*, 2002), vomitus (Parsonnet *et al.*, 1999), saliva (Parsonnet *et al.*, 1999; Allaker *et al.*, 2002), dental plaque (Allaker *et al.*, 2002; Gürbüz *et al.*, 2003), untreated municipal wastewater (Lu *et al.*, 2002), untreated well water (Sasaki *et al.*, 1999; Baker and Hegarty, 2001), river and pond water (Sasaki *et al.*, 1999), drinking water stored in earthenware pots (Bunn *et al.*, 2002), soil (Sasaki *et al.*, 1999), and flies (Sasaki *et al.*, 1999). The presence of *H. pylori* in water and soil may be due to con-

tamination with human feces. In addition, *H. pylori* has been detected in raw sheep (Dore *et al.*, 1999, 2001) and cow milk (Fujimura *et al.*, 2002).

*Helicobacter pylori* in spiked foods stored at 4°C do not grow, but the organisms can survive for periods ranging from two to five days albeit with a steady decline in numbers. *Helicobacter pylori*-spiked foods included leaf lettuce, raw chicken, pasteurized milk, sterile skim milk, tofu, and ground beef (Stevenson *et al.*, 2000; Poms and Tatini, 2001; Jiang and Doyle, 2002). Thus, *H. pylori* can survive for a few days in low-acid–high-moisture foods stored at refrigerated temperatures. However, *H. pylori* was rapidly killed in yoghurt stored at 4°C within a few hours (Poms and Tatini, 2001).

The reservoir for *H. pylori* is the human stomach (Cave, 1997); however, Gürbüz *et al.* (2003) suggest that dental plaque may be an additional reservoir site. While animals can be used as models of *H. pylori* infection (Lee, 1999), Brown *et al.* (2001) indicated that zoonotic transmission is not a major route for infection. Since the organism is confined to humans, the presence of *H. pylori* in animals may be due to human-to-animal transmission. Moreover, meat is not a likely source of *H. pylori*. Hopkins *et al.* (1990) and Webberley *et al.* (1992) found that the seroprevalence against *H. pylori* was similar in both meat-eating and non-meat-eating populations, suggesting that source of infection is not due to the eating of animal meats.

The oral–oral and fecal–oral routes are probably important means of transmission of *H. pylori* in institutionalized populations and in children (Brown, 2000). However, water contaminated with human feces, especially in third-world countries, is probably an important means of transmission. Uncooked vegetables and produce, irrigated or washed with water contaminated with human feces, could be a source of *H. pylori* infection (Velázquez and Feirtag, 1999; Vaira *et al.*, 2001). Houseflies could also transmit *H. pylori* to foods after contact with human feces (Brown, 2000). In addition, food could be contaminated by a *H. pylori*-infected foodhandler with poor personal hygiene. The route of transmission of *H. pylori* has never been determined with certainty and is an open question. The most likely route of transmission is contaminated water and indirectly, food in which contaminated water was used to grow, wash, or prepare the food.

*Helicobacter* species other than *H. pylori* have also been isolated from human cases of diarrhea. Isolates from diarrheic patients include *H. canadensis* (Fox *et al.*, 2000), *H. winghamensis* (Melito *et al.*, 2001), and *H. pullorum* (Stanley *et al.*, 1994; Steinbrueckner *et al.*, 1997). *Helicobacter pullorum* has been isolated from poultry (Stanley *et al.*, 1994), *H. canadensis* from wild geese (Waldenström *et al.*, 2003) and *H. bilis* and *H. hepaticus* from mice (Nilsson *et al.*, 2003); therefore, infection with these organisms may represent zoonotic infective agents. *H. winghamensis*, however, has only been found in humans (Melito *et al.*, 2001).

In a recent clinical report, Nilsson *et al.* (2003) noted that patients with autoimmune chronic liver diseases demonstrated a high prevalence of serum antibodies toward non-gastric *Helicobacter* species such as *H. pullorum*, *H. bilis*, and *H. hepaticus*. Antibodies against non-gastric helicobacters were observed in 59/76 (77.6%) of patients compared to 4/80 (5.0%) of healthy blood donors. How important are these other *Helicobacter* species clinically? The number of diarrheic cases associated with these pathogens is small but may be underreported. It is probable that as diagnostic tools improve for the identification and detection of *Helicobacter* species, there will be an increase in the detection of diarrheic cases induced by animal *Helicobacter* species (Solnick, 2003).

### *Enterobacter sakazakii*: babies at risk

*Enterobacter sakazakii* is a rod-shaped, motile, non-spore-forming, gram negative, facultative anaerobe that can cause life-threatening meningitis, bacteremia, and necrotizing enterocolitis or meningoencephalitis in newborns. It was first reported as a cause of neonatal meningitis in two infants in the United Kingdom in 1958 (Urmenyi and Franklin, 1961), and although infrequent, a number of outbreaks have occurred subsequently involving infants and neonates (Iversen and Forsythe, 2003). Mortality rates can be as high as 80% (Nazarowec-White and Farber, 1997). The source of the pathogen in the environment is unknown; however, it is found as an occasional contaminant of infant formula, which has been implicated as the vehicle of infection in a number of outbreaks. Iversen *et al.* (2004) showed that *E. sakazakii* grew in infant formula milk at 6°C with a doubling time of about 13 h, and the $D_{62}$-value of the organism in the formula was 0.3 s, thus the organism should be destroyed at standard pasteurization temperatures. The $D_{58}$-value of 12 strains of *E. sakazakii* added to rehydrated infant formula ranged from 30.5 to 591.9 s (Edelson-Mammel and Buchanan, 2004). Half of the strains had mean $D_{58}$-values ranging from 30.5 to 47.9 s whereas the mean values for the remaining strains ranged from 307.8 to 591.9 s. The z-value of the most heat resistant strain was 5.6°C. Dried infant formula containing the most heat resistant *E. sakazakii* strain rehydrated with water at ≥ 70°C showed a >4-log decrease in the bacterial level (Edelson-Mammel and Buchanan, 2004).

Enterobacter sakazakii in infant formula milk formed a biofilm on surfaces of materials commonly used for infant-feeding equipment and work surfaces, including latex, polycarbonate, and silicon (Iversen and Forsythe, 2003). *Enterobacter sakazakii* has also been isolated from different environmental sources and from foods such as rice seed, cheese and cured meat (Iversen and Forsythe, 2003). The organism was isolated from the guts of larvae of the stable fly, *Stomoxys calcitrans* (Hamilton *et al.*, 2003), and from Mexican fruit flies (Kuzina *et al.*, 2001) suggesting that insects may be a reservoir and may spread the organism in the environment. A small cluster of cases of neonatal infections was caused by a biochemical variant of *E. sakazakii* in a university hospital in Jerusalem (Block *et al.*, 2002). The isolates recovered from prepared formula and

from a kitchen blender were identical by pulsed-field gel electrophoresis. Isolation of the affected patients and replacement of powdered formulas with ready-to-use factory-prepared infant formula prevented additional cases of illness. Infections in immunocompromised adults have been reported, and one case involved a wound infected with *E. sakazakii* resistant to multiple antibiotics (Lai, 2001; Dennison and Morris, 2002). Detection of the organism can be performed using traditional culture techniques. The BAX® System PCR Assay for Screening *Enterobacter sakazakii* is also available from Dupont Qualicon. To determine the prevalence and clinical importance of *E. sakazakii*, however, additional research to identify environmental and food sources, virulence factors, and the mechanism of pathogenesis is needed.

### *Aeromonas* species: a foodborne pathogen in waiting?

For over two decades, *Aeromonas* species have been recognized as potential emerging foodborne pathogens. *Aeromonas* spp. are facultatively anaerobic, glucose-fermenting, oxidase-positive gram-negative rods within the family Aeromonadaceae, which currently consists of 14 different species (Isonhood and Drake, 2002). Most aeromonads grow optimally at 28°C and may grow at refrigeration temperatures; however, they are unlikely to grow in foods with pH values below pH 6 and that contain > 3 to 3.5% NaCl. Aeromonads have no unusual resistance to disinfectants, heat, or other common food processing procedures (Kirov, 2001). Aeromonads are aquatic organisms, found in fresh, stagnant, estuarine, and brackish water, and are also commonly found in drinking water. They are found associated with water-dwelling plants and animals, including fish and frogs. They have caused disease in amphibians, reptiles, shellfish, and snails, and are associated with hemorrhagic septicemia in fish. Aeromonads are also found in the intestinal tracts of many animals, including pigs, cows, sheep, and poultry; however, they are not commonly found in the gastrointestinal tract of humans (Kirov, 2001).

Since volunteer studies and suitable animal models of infection are lacking, and healthy carriers of *Aeromonas* spp. have been described, the role of *Aeromonas* as food- and water-borne pathogens is unsettled. The virulence factors and mechanism of pathogenesis of *Aeromonas* spp. are not well understood. It is likely that the presence of certain combinations of virulence factors renders some strains pathogenic in humans. Virulence factors include the production of S-layers, which are macromolecular arrays of protein subunits on the bacterial surface, the production of extracellular enzymes, including proteases, DNase, RNase, elastase, lecithinase, amylase, lipases, gelatinase, and chitinase, in addition to the production of siderophores, exotoxins, a cytotoxic enterotoxin, hemolysins, cytotonic enterotoxins, and adhesins (Kirov, 2001). A number of studies have shown that some strains of certain *Aeromonas* species can cause gastrointestinal illness. *Aeromonas hydrophila* caused gastroenteritis associated with oysters, and *Aeromonas* spp., including *A. hydrophila*, *A. caviae*, *A. veronii*, and *A. jandaei*, were identified as a cause of traveler's diarrhea (Abeyta *et al.*, 1986; Vila *et al.*, 2003). In addition to causing enteric disease, aeromonads have caused a variety of extraintestinal infections, including wound infections, septicemia, and endocarditis in immunocompetent, as well as immunocompromised individuals.

## Multiple antibiotic-resistant foodborne pathogens

The problem of antimicrobial resistance in enteric pathogens has become a serious clinical and public health concern. The emergence of multiple drug-resistant enteric pathogens, including *Salmonella* and *Shigella*, has resulted in an increase in morbidity and mortality, in particular, in developing countries. Although, some resistance traits are intrinsic, i.e., part of the normal bacterial chromosome, most resistance genes are acquired through gene transfer from other bacteria via conjugation, transduction, transformation, and transposition (Bennish and Levy, 1995). Many resistance genes are transferred on discrete segments of DNA called transposons, which move between various genetic structures intracellularly or between bacterial cells. Integrons, which are genetic structures that consist of a gene encoding a site-specific integrase and a recombination site for insertion of genes cassettes, are also involved in mobilization of resistance genes. *Salmonella* Typhimurium definitive type 104 (DT104) resistant to ampicillin, chloramphenicol, streptomycin, sulfonamides, and tetracycline (ACSSuT resistance phenotype), emerged in recent years as a major cause of illness in humans and animals in Europe, in particular in the United Kingdom. Glynn *et al.* (1998) reported that the prevalence of multiply-resistant *S.* Typhimurium DT104 in the United States increased from 0.6% in 1979–1980 to 34% of human isolates in 1996. The genes conferring the ACSSuT phenotype in *S.* Typhimurium and other serovars are located on a 14-kb region found within a 43-kb chromosomal genomic island designated *Salmonella* genomic island I (SGI1) (Briggs and Fratamico, 1999; Carattoli *et al.*, 2002; Doublet *et al.*, 2003). On either end of SGI are 18-bp direct repeats, indicating that the island was likely acquired through site specific recombination events. Furthermore, bacteria can acquire whole plasmids (R plasmids) carrying drug resistance genes from other bacteria. Small pT181-like plasmids carrying tetracycline resistance genes occur in *Staphylococcus aureus*. These small plasmids may be incorporated into larger plasmids or into the bacterial chromosome via the IS257 insertion element (Sørum and L'Abée-Lund, 2002).

Bacterial resistance to an antibiotic develops in response to the selective pressure exerted by exposure to the drug, and the extensive use of antibiotics in humans, animals, and plants has resulted in the establishment of a pool of resistance genes in the environment. Although estimates vary, the total amount of antibiotic usage in the United States for medical and agricultural purposes has been estimated at 35 to 50 million pounds per year, and 70% of all antibiotics are for non-therapeutic use in livestock, mainly for the purposes of growth promotion and disease prevention in cattle, swine, and poultry (Mellon *et al.*, 2001). This incidence of resistance to antibiotics, including tetracycline, erythromycin, and fluoroquinolones, in foodborne pathogens has been increasing in recent years. For example, one study conducted in Canada showed that resistance to tetracycline in *C. jejuni* isolates increased from 19.1 to 55.7% between 1985 and 1995 (Gaudreau and Gilbert, 1998), and resistance to quinolones also increased markedly during the same time period. In the Netherlands, increased resistance of poultry and human isolates of *C. jejuni* to fluoroquinolones coincided with the approval of the use of enrofloxacin, a fluoroquinolone antibiotic, in poultry in 1987 (Endtz *et al.*, 1991). Similarly, in the United States, resistance to fluoroquinolones increased in poultry and human *Campylobacter* isolates following the approval by the Food and Drug Administration for the use of this group of antibiotics in chickens and turkeys (Falkow and Kennedy, 2001). In response to public health concerns over the emergence of antibiotic resistant pathogens, in 2002, the European Union member states agreed that by January 2006 the use of antibiotics as growth promoters in food animals would no longer be permitted. As of April 1998, antimicrobial growth promoters were no longer used in Danish poultry, cattle, and pigs weighing more than 35 kg, resulting in dramatic decreases in resistance to antibiotics, including vancomycin, tylosin, virginiamycin, avoparcin, and avilamycin, particularly in *Enterococcus faecium* isolates from poultry (Wegener, 2003).

Using PCR assays targeting specific genes involved in tetracycline resistance to assess the mobilization of the genes among bacteria in waste lagoons and groundwater underlying swine farms, Chee-Sanford *et al.* (2001) showed that groundwater is a potential reservoir for the dissemination of resistance genes in the environment. Results of another study indicated that the use of oxytetracycline to control bacterial diseases in fruit orchards may lead to an increase in the number of commensal orchard bacteria possessing resistance plasmids to tetracycline, streptomycin, and sulfonamides with potential transfer of the plasmids to other organisms (Schnabel and Jones, 1999). It is evident that overuse of antibiotics in human medicine and in agriculture has led to a dramatic increase in antibiotic resistance in pathogenic, as well as non-pathogenic bacteria. Research to develop alternatives to antibiotics that may be utilized as animal growth promoters, such as use of probiotic cultures or enzymes that break down feed components to improve nutrient utilization, is needed. Although the reversal of resistance is a challenge, the development of new antibiotics will provide more options for treatment of human and animal diseases.

## Other potential emerging foodborne pathogens

*Campylobacter* is the most common cause of diarrheal illness worldwide, and *C. jejuni* is the predominant species isolated from humans, followed by *C. coli*. However, a number of other species of *Campylobacter* including *C. lari*, *C. fetus*, and *C. upsaliensis* are recognized as human pathogens, and new species of *Campylobacter* are being identified on a regular basis (Lastovica and Skirrow, 2000). When looked for, these species have been isolated and recognized as disease agents; however, procedures used for isolation *C. jejuni* or *C. coli* may not support the growth of the non-*jejuni/coli* species. Thus, the true prevalence in the food supply and the incidence of illness due to the non-*jejuni/coli* species are currently unknown.

The genus *Arcobacter* consists of organisms previously identified as aerotolerant campylobacters or campylobacter-like organisms. *Arcobacter* spp. have been isolated from humans, animals, food, and water. *Arcobacter cryoaerophilus* and *A. butzleri* have been associated with disease in animals and gastrointestinal illness in humans; however, the clinical significance and mechanism of pathogenesis of *Arcobacter* spp. requires further investigation (Meng and Doyle, 1997).

In addition to the six known categories of diarrheagenic *E. coli* (diffusely adherent, enteroinvasive, enteropathogenic, enterotoxigenic, enteroaggregative, and enterohemorrhagic *E. coli*), new categories are emerging, likely through horizontal transfer of pathogenicity islands and different sets of virulence genes. Recently described diarrheagenic *E. coli* include the non-O157:H7 Shiga toxin-producing *E. coli*, atypical enteropathogenic *E. coli*, necrotoxic *E. coli*, and cytolethal distending toxin-producing *E. coli* (Meng and Doyle, 1997; De Rycke *et al.*, 1999; Clarke, 2001; Trabulsi *et al.*, 2002).

## Conclusions

Disease-causing microorganisms will continue to emerge, re-emerge and persist. Factors that may be involved in the emergence of new pathogens or the re-emergence of old pathogens are listed in Table 22.1. Some of these factors, singly or collectively, play a role in the emergence of new foodborne pathogens. Globalization with widespread and rapid movement of agricultural and biological products or livestock and other animals, and foreign travel for business or leisure elevates the exposure of individuals to previously unknown disease agents and the spread of these agents. Globalization subjects individuals to agricultural products from developing countries where raw foods may not be grown, harvested, and shipped in a sanitary manner. Blood, vaccines, and other biological products obtained from the global marketplace may be a source of infection. Many foodborne pathogens are zoonotic organisms, and animals shipped from global markets may expose individuals to new pathogens. Travelers may eat foods that are not thoroughly cooked or drink untreated water, and as a consequence, they may become ill and take a new pathogen home with them.

The changing eating habits of people in developed countries may expose individuals to foodborne pathogens, both old and new. There is a demand for fresh produce and fruit all-year-round, necessitating the shipment of these foods from the markets of developing countries. There is also an increased demand for raw foods, minimally processed foods and convenience foods, and these items can expose people to foodborne pathogens. People are eating away from home more, and these consumers have less control over the food they eat and how it is prepared.

Genetic (chromosomal defects of immunological function), biological (age, chronic disease, HIV), and technological (blood transfusions, cancer chemotherapy, transplantation, kidney dialysis) factors can induce immunosuppression in humans. Immunocompromised individuals are more susceptible to infections by both new and old pathogens.

The genetic promiscuity of microorganisms allows them to acquire virulence and drug resistance factors from other microorganisms. These new traits result in an increased pathogenicity, an increased abililty to infect new hosts or resistance to antimicrobial drugs. The selective pressure of antimicrobial drugs allows rare resistant mutants to grow and survive in the host. Antibiotic abuse by the food animal industries has likely led to increased resistance in the microbial population of animal gastrointestinal tracts, and this resistance can be readily transferred to other microorganisms. Exposure of microorganisms to sublethal physical and chemical stresses common to food processing technologies induces the adaptation of microorganisms to more stringent stress events with resultant failure to eliminate the organisms from the food environment. The ability of microorganisms to easily adapt and change plays an extremely important role in the emergence and re-emergence of foodborne pathogens.

The response to a new foodborne pathogen depends on a multidisciplinary approach, which must involve various areas of government, academia and industry for effective control and prevention. It will be necessary to improve and enlarge both global and local responses to foodborne infections, as well as expand and improve global and local disease surveillance. To meet new foodborne microbial threats, there must be a re-building and sustaining of public health facilities. Rapid, sensitive, and specific diagnostic tests must be developed to detect and monitor the emergence of new pathogens. The training of additional health professionals in areas of epidemiology and field-based research and the continued development of new antimicrobial compounds and vaccines are necessary to protect the public against microbially induced infections. Since many foodborne pathogens have their origin in animals, the study of zoonotic diseases and their control must be supported. To effectively fight new foodborne pathogens, key personnel in government agencies, food production and food service industries, and microbiology and food science departments of academia throughout the world must develop collaborative infectious disease research programs to facilitate monitoring for emerging foodborne threats and to determine strategies for their control and/or prevention.

## References

Abeyta, C., Kaysner, C.A., Wekell, M.M., Sullivan, J.J., and Stelma, G.N. 1986. Recovery of *Aeromonas hydrophila* from oysters implicated in an outbreak of foodborne illness. J. Food Prot. 49: 643–646.

Aleljung, P., Nilsson, H-O., Wang, X., Nyberg, P., Mörner, T., Warsame, I. and Wadström, T. 1996. Gastrointestinal colonisation of BALB/cA mice by *Helicobacter pylori* monitored by heparin magnetic separation. FEMS Immunol. Med. Microbiol. 13: 303–309.

Allaker, R.P., Young, K.A., Hardie, J.M., Domizio, P. and Meadows, N.J. 2002. Prevalence of *Helicobacter pylori* at oral and gastrointestinal site in children: Evidence for possible oral-to-oral transmission. J. Med. Microbiol. 51: 312–317.

Alonso, A., Daschner, A. and Moreno-Ancillo, A. 1997. Anaphylaxis with *Anisakis simplex* in the gastric mucosa. N. Engl. J. Med. 337: 350–351.

Arankalle, V.A.,Chobe, L.P., Joshi, M.V., Chadha, M.S., Kundu, B. and Walimbe, A.M. 2002. Human and swine hepatitis E viruses from Western India belong to different genotypes. J. Hepatol. 36: 417–425.

Baker, K.H. and Hegarty, J.P. 2001. Presence of *Helicobacter pylori* in drinking water is associated with clinical infection. Scand. J. Infect. Dis. 33: 744–746.

Barrett, S.P. 1997. Human intestinal spirochaetosis. In: Intestinal Spirochaetes in Domestic Animals and Humans. D. J. Hampson and T. B. Stanton, eds. CAB International, Wallingford, UK, pp. 243–266.

Belay, E.D., Gambetti, P., Schonberger, L.B., Parchi, P., Lyon, D.R., Capellari, S., McQuiston, J. H., Bradley, K., Dowdle, G., Crutcher, J.M. and Nichois, C.R. 2001. Creutzfeld-Jakob disease in unusually young patients who consumed venison. Arch. Neurol. 58: 1673–1678.

Belay, E.D., Maddox, R.A., Williams, E.S., Miller, M.W., Gambetti, P. and Schonberger, L.B. 2004. Chronic wasting disease and potential transmission to humans. Emerg. Infect. Dis. 10: 977–984.

Bennish, M.L. and Levy, S.B. 1995. Antimicrobial resistance in enteric pathogens. In: Infections of the gastrointestinal tract. M. J. Blaser, P. D. Smith, J. I. Ravdin, H. B. Greenberg, and R. L. Guerrant, eds. Raven Press, Ltd., New York. p. 1499–1523.

Block, C., Peleg, O., Minster, N., Bar-Oz, B., Simhon, A., Arad, I. and, Shapiro, M. 2002 Cluster of neonatal infections in Jerusalem due to unusual biochemical variant of *Enterobacter sakazakii*. Eur. J. Clin. Microbiol. Infect. Dis. 21: 613–616.

Bosque, P.J., Ryou, C., Telling, G., Peretz, D., Legname, G., DeArmond, S.J. and Prusiner, S.B. 2002. Prions in skeletal muscle. Proc. Natl. Acad. Sci. USA 99: 3812–3817.

Bouree, P., Paugam, A. and Petithory, J-C. 1995. Anisakidosis: Report of 25 cases and review of the literature. Comp. Immun. Microbiol. Infect. Dis. 18: 75–84.

Boye, M., Baloda, S.B., Lester, T.D. and Møller, K. 2001. Survival of *Brachyspira hyodysenteriae* and *B. pilosicoli* in terrestrial microcosms. Vet. Microbiol. 81: 33–40.

Briggs, C.E. and Fratamico, P.M. 1999. Molecular characterization of an antibiotic resistance gene cluster of *Salmonella typhimurium* DT104. Antimicrob. Agents Chemother. 43: 846–849.

Brooke, C.J., Clair, A.N., Mikosza, A.S.J., Riley, T.V. and Hampson, D.J. 2001. Carriage of intestinal spirochaetes by humans: epidemiological data from Western Australia. Epidem. Infect. 127: 369–174.

Brown, L.M. 2000. *Helicobacter pylori*: Epidemiology and routes of transmission. Epidemiol. Rev. 22: 283–297.

Brown, L.M., Thomas, T.L., Ma, J.L., Chang, Y.S., You, W.C., Liu, W.D., Zhang, L. and Gail, M.H. 2001. *Helicobacter pylori* infection in rural China: Exposure to domestic animals during childhood and adulthood. Scand. J. Infect. Dis. 33: 686–691.

Buendia, E. 1997. *Anisakis*, anisakidosis, and allergy to *Anisakis*. Allergy 52: 481–482.

Bunn, J.E.G., MacKay, W.G., Thomas, J.E., Reid, D.C. and Weaver, L. T. 2002. Detection of *Helicobacter pylori* DNA in drinking water biofilms: Implications for transmission in early life. Lett. Appl. Microbiol. 34: 450–454.

Carattoli, A., Filetici, E., Villa, L., Dionisi, A.M., Ricci, A., and Luzzi, I. 2002. Antibiotic resistance genes and *Salmonella* genomic island 1 in *Salmonella enterica* serovar *Typhimurium* isolated in Italy. Antimicrob. Agents Chemother. 46: 2821–2828.

Cave, D.R. 1997. How is *Helicobacter pylori* transmitted? Gastroenterology 113(Suppl. 6): S9–S14.

Ceylan, A., Ertem, M., Ilcin, E. and Ozekinci, T. 2003. A special risk group for hepatitis E infection: Turkish agricultural workers who use untreated waste water for irrigation. Epidemiol. Infect. 131: 753–756.

Chandenier, J., Husson, J., Canaple, S., Gondry-Jouet, C., Dekumyoy, P., Danis, M., Riveau, G., Hennequin, C., Rosa, A. and Raccurt, C.P. 2001. Medullary gnathostomiasis in a white patient: use of immunodiagnosis and magnetic resonance imaging. Clin. Infect. Dis. 32: e154-e157.

Chandler, J.D., Riddell, M.A., Li, F., Love, R. J. and Anderson, D.A. 1999. Serological evidence for swine hepatitis E virus infection in Australian pig herds. Vet. Microbiol. 68: 95–105.

Chee-Sanford, J.C., Aminov, R.I., Krapac, I.J., Garrigues-Jeanjean, N., and Mackie, R.I. 2001 Occurrence and diversity of tetracycline resistance genes in lagoons and groundwater underlying two swine production facilities. Appl. Environ. Microbiol. 67: 1494–1502.

Choi, I-S., Kwon, H-J., Shin, N-R. and Yoo, H.S. 2003. Identification of swine hepatitis E virus (HEV) and prevalence of anti-HEV antibodies in swine and human populations in Korea. J. Clin. Microbiol. 41: 3602–3608.

Clarke, S.C. 2001. Diarrhoeagenic *Escherichia coli* – an emerging problem. Diagn. Microbiol. Infect. Dis. 41: 93–98.

Clayson, E.T., Innis, B.L., Myint, K.W.A., Narupiti, S., Vaughn, D.W., Giri, S., Ranabhat, P. and Shrestha, M.P. 1995. Detection of hepatitis E virus infections among domestic swine in the Kathmandu Valley of Nepal. Am. J. Trop. Med. Hyg. 53: 228–232.

Clayson, E.T., Snitbhan, R., Ngarmpochana, M., Vaughn, D.E. and Shrestha, M.P. 1996. Evidence that the hepatitis E virus (HEV) is a zoonotic virus: detection of natural infections among swine, rats, and chickens in an area endemic for human disease. In: Enterically transmitted hepatitis E viruses. Y. Buisson, P. Coursaget and M. Kane (eds.), La Simarre, Joué-lès-Tours, France. p. 329–335.

Clemente-Casares, P., Pina, S., Buti, M., Jardi, R., Martin, M., Bofil-Mas, S. and Girones, R. 2003. Hepatitis E virus epidemiology in industrialized countries. Emerg. Infect. Dis. 9: 448–454.

Cuende, E., Audicana, M.T., García, M., Anda, M., Fernández de Corres, L., Jímenez, C. and Vesga, J.C. 1998. Rheumatic manifestations in the course of anaphylaxis caused by *Anisakis simplex*. Clin. Exp. Rheumatol. 16: 303–304

del Giudice, P., Dellamonica, P., Durant, J., Rahelinrina, V., Grobusch, M.P., Janitschke, K., Hahan-Guedj, A. and Fichoux, Y.E. 2001. A case of gnathostomiasis in a European traveller returning from Mexico. Brit. J. Dermatol. 145: 487–489.

Dennison, S.K. and Morris, J. 2002. Multiresistant Enterobacter sakazakii wound infection in an adult. Infect. Med. 19: 533–535.

De Rycke, J., Milon, A. and Oswald, E. 1999. Necrotoxic *Escherichia coli* (NTEC): two emerging categories of human and animal pathogens. Vet. Res. 30: 221–233.

Diaz Camacho, S.P., Ramos, M.Z., Torrecillas, E.P., Ramirez, I.O., Velazquez, R.C., Gaxiola, A.F., Heredia, J.B., Willms, K., Akahane, H., Ogata, K. and Nawa, Y. 1998. Clinical manifestations and immunodiagnosis of gnathostomiasis. Am. J. Trop. Med. Hyg. 59: 908–915.

Dore, M.P., Sepulveda, A.R., El-Zimaity, H., Yamaoka, Y., Osato, M.S., Mototsugu, K., Nieddu, A.M., Realdi, G. and Graham, D.Y. 2001. Isolation of *Helicobacter pylori* from sheep – implications for transmission to humans. Appl. J. Gastroenterol. 96: 1396–1401.

Dore, M.P., Sepulveda, A.R., Osato, M.S., Realdi, G. and Graham, D.Y. 1999. *Helicobacter pylori* in sheep milk. Lancet 354: 132.

Doublet, B., Lailler, R., Meunier, D., Brisabois, A., Boyd, D., Mulvey, M.R., Chaslus-Dancla, E., and Cloeckaert, A. 2003. Variant *Salmonella ge-*

nomic island 1 antibiotic resistance gene cluster in *Salmonella enterica* serovar Albany. Emerg. Infect. Dis. 9: 585–591.

Drews, J. and Ryser, S. 1997. The role of innovation in drug development. Nature Biotechnol. 15: 1318–1319.

Drobeniuc, J., Favorov, M.O., Shapiro, C.N., Bell, B.P., Mast, E.E., Dadu, A., Culver, D., Iarovoi, P., Robertson, B.H. and Margolis, H.S. 2001. Hepatitis E virus antibody prevalence among persons who work with swine. J. Infect. Dis. 184: 1594–1597.

Duhamel, G.E., Elder, R.O., Muniappa, N., Mathiesen, M.R., Wong, V.J. and Tarara, R.P. 1997. Colonic spirochetal infection in nonhuman primates that were associated with *Brachyspira aalborgi*, *Serpulina pilosicoli* and unclassified flagellated bacteria. Clin. Infect. Dis. 25(Suppl. 2): S186–S188.

Duhamel, G.E., Trott, D.J., Muniappa, N., Mathiesen, M.R., Tarasiuk, K., Lee, J.I. and Hampson, D.J. 1998. Canine intestinal spirochetes consist of *Serpulina pilosicoli* and a newly identified group provisionally designated "*Serpulina canis*" sp. nov. J. Clin. Microbiol. 36: 2264–2270.

Edelson-Mammel, S.G. and Buchanan, R.L. 2004. Thermal inactivation of *Enterobacter sakazakii* in rehydrated infant formula. J. Food Prot. 67:L 60–63.

Endtz, H.P., Ruijs, G.J., van Klingeren, B., Jansen, W.H., van der Reyden. T. and Mouton. R.P. 1991. Quinolone resistance in *Campylobacter* isolated from man and poultry following introduction of fluoroquinolone in veterinary medicine. J. Antimicrob. Chemother. 27: 199–208.

Engstrand, L. 2001 *Helicobacter* in water and water-borne routes of transmission. J. Appl. Microbiol. 90 (Symposium Supplement): 80S-84S.

Esteve, C., Resano, A., Diaz-Tejeiro, P. and Fernández-Benitez, M. 2000. Eosinophilic gastritis due to *Anisakis*: A case report. Allergol. Immunopathol. (Madrid) 28: 21–23.

Falkow, S. and Kennedy, D. 2001. Antibiotics, animals, and people – again! Science. 291: 397.

Fellström, C., Pettersson, B., Zimmerman, U., Gunnarsson, A. and Feinstein, R. 2001. Classification of *Brachyspira* spp. isolated from Swedish dogs. Anim. Health Res. Rev. 2: 75–82.

Fournié-Amazouz, E., Baranton, G., Carlier, J.P., Chambreluil, G., Cohadon, F., Collin, P., Golugleon Jolivet, A., Hermès, I., Lemarie, C. and Saint Girons, I. 1995. Isolations of intestinal spirochaetes from the blood of human patients. J. Hosp. Infect. 30: 160–162.

Fox, J.G., Chien, C.C., Dewhirst, F.E., Paster, B.J., Shen, Z., Melito, P.L., Woodward, D.L. and Rodgers, F.G. 2000. *Helicobacter canadensis* sp. nov. isolated from humans with diarrhea as an example of an emerging pathogen. J. Clin. Microbiol. 38: 2546–2549.

Fujimura, S., Kawamura, T., Kato, S., Tateno, H. and Watanabe, A. 2002. Detection of *Helicobacter pylori* in cow's milk. Lett. Appl. Microbiol. 35: 504–507.

Garg, R. K. 1998 Neurocysticercosis. Postgrad. Med. J. 74: 321–326.

Garkavenko, O., Obriadina, A., Meng, J., Anderson, D.A., Benard, H.J., Schroeder, B.A., Khudyakov, Y.E., Fields, H.A. and Croxson, M.C. 2001. Detection and characterization of swine hepatitis E virus in New Zealand. J. Med. Virol. 65: 525–529.

Gaudreau, C. and Gilbert, H. 1998 Antimicrobial resistance of clinical strains of *Campylobacter jejuni* subsp. *jejuni* isolated from 1985 to 1997 in Quebec, Canada. Antimicrob. Agents Chemother. 42: 2106–2108.

Glynn, M.K., Bopp, C., Dewitt, W., Dabney, P., Molktar, M. and Angulo, F.J. 1998. Emergence of multidrug-resistant *Salmonella enterica* serotype *Typhimurium* DT104 infections in the United States. N. Engl. J. Med. 338: 1333–1338.

Gomes, B. C. and de Martinis, E.C.P. 2004. The significance of *Helicobacter pylori* in water, food and environmental samples. Food Control. 15: 397–403.

Gould, D.H., Voss, J.L., Miller, M.W., Bachand, A.M., Cummings, B.A. and Frank. A.A. 2003. Survey of cattle in northeast Colorado for evidence of chronic wasting disease: Geographical and high-risk targeted sample. J. Vet. Diagn. Invest. 15: 274–277.

Griffiths, J. K. 1999. Exotic and trendy cuisine. In: Infections of Leisure, 2nd ed. D. Schlossberg, ed. ASM Press, Washington, DC, pp. 307–333.

Grobusch, M.P., Bergmann, F., Teichmann, D. and Klein, E. 2000. Cutaneous gnathostomiasis in a woman from Bangladesh. Int. J. Infect. Dis. 4: 51–54.

Gürbüz, A.K., Özel, A.M., Yazgan, Y., Çelik, M. and Yildirim, Ş. 2003. Oral colonization of *Helicobacter pylori*: Risk factors and response to eradication therapy. Southern Med. J. 96: 244–247.

Gutiérrez-Ramos, R., Guillén-Bueno, R., Madero-Jarabo. R. and Cuéllar del Hoyo, C. 2000. Digestive haemorrhage in patients with anti-*Anisakis* antibodies. Eur. J. Gastroenterol. Hepatol. 12: 337–343.

Halbur, P.G., Kasorndorkbua, C., Gilbert, C., Guenette, D., Potters, M.B., Purcell, R.H., Emerson, S.U., Toth, T.E. and Meng, X.J. 2001. Comparative pathogenesis of infection of pigs with hepatitis E viruses recovered from a pig and a human. J. Clin. Microbiol. 39: 918–923.

Hamilton, J.V., Lehane, M.J. and Braig, H.R. 2003. Isolation of *Enterobacter sakazakii* from midgut of *Stomxys clacitrans*. Emerg. Infect. Dis. 9: 1355–1356.

Hamir, A.N., Cutlip, R.C., Miller, J.M., Williams, E.S., Stack, M.J., Miller, M.W., O'Rourke, K.I. and Chaplin, M.J. 2001. Preliminary findings on the experimental transmission of chronic wasting disease agent of mule deer to cattle. J. Vet. Diagn. Invest. 13: 91–96.

Hamir, A.N., Miller, J.M., Cutlip, R.C., Stack, M.J., Chaplin, M.J. and Jenny, A.L. 2003. Preliminary observations on the experimental transmission of scrapie to elk (*Cervus elaphus nelsoni*) by intracerebral inoculation. Vet. Pathol. 40: 81–85.

Hartmann, W.J., Frösner, C.G. and Eichenlaub, D. 1998. Transmission of hepatitis E in Germany. Infection 26: 409.

He, J., Innis, B.L., Shrestha, M.P., Clayson, E.T., Scott, R.M., Linthicum, K.J., Musser, G.G., Gigliotti, S.C., Binn, L.N., Juschner, R.A. and Vaughn, D.W. 2002. Evidence that rodents are a reservoir of hepatitis E virus for humans in Nepal. J. Clin. Microbiol. 40: 4493–4498.

Heine, R.G., Ward, P.B., Mikosza, A.S.J., Bennett-Wood, B., Robins-Browne, R.M. and Hampson, D.J. 2001. *Brachyspira aalborgi* infection in four Australian children. J. Gastroenterol. Hepatol. 16: 872–875.

Hirano, M., Ding, X., Li, T-C., Takeda, N., Kawabata, H., Koizumi, N., Kadosaka, T., Goto, I., Masuzawa, T., Nakamura, M., Taira, K., Kuroki, T., Tanikawa, T., Watanabe, H. and Abe, K. 2003. Evidence for widespread infection of hepatitis E virus among wild rats in Japan. Hepatol. Res. 27: 1–5.

Ho, A.Y., Lopez, A.S., Eberhart, M.G., Levenson, R., Finkel, B.S., Da Silva, A.J., Roberts, J.M., Orlandi, P.A., Johnson, C.C. and Herwaldt, B.L. 2002. Outbreaks of cyclosporiasis associated with imported raspberries, Philadelphia, Pennsylvania, 2000. Emerg. Infect. Dis. 8: 783–788.

Hopkins, R.J., Russell, R.G., O'Donnaghue, J.M., Wasserman, S.S., Lefkowitz, A. and Morris, J.G. 1990. Seroprevalence of *Helicobacter pylori* in Seventh-Day Adventists and other groups in Maryland. Lack of association with diet. Arch. Intern. Med. 150: 2347–2348.

Hovind-Hougen, K., Birch-Andersen, A., Henrik-Nielsen, R., Orholm, M., Pedersen, J.O., Teglbjærg, P.S. and Thaysen, E.H. 1982. Intestinal spirochetosis: Morphological characterization and cultivation of the spirochete *Brachyspira aalborgi* gen. nov., sp. nov. J. Clin. Microbiol. 16: 1127–1136.

Hsieh, S.Y., Meng, J.J., Wu, Y.H., Liu, S.T., Tam, A.W., Lin, D.Y. and Liaw, Y.F. 1999. Identity of a novel swine hepatitis E virus in Taiwan forming a monophyletic group with Taiwan isolates of human hepatitis E virus. J. Clin. Microbiol. 37: 3828–3834.

Huang, F.F., Haqsheanas, G., Guenette, D.K., Halbur, P.G., Schommer, S.K., Pierson, F.W., Toth, T.E. and Meng, X.J. 2002. Detection by reverse transcription-PCR and genetic characterization of field isolates of seine hepatitis E virus from pigs in different geographic regions of the United States. J. Clin. Microbiol. 40: 1326–1332.

Hyams, K.C. 2002. New Perspectives on hepatitis E. Curr. Gastroenterol. Repts. 4: 302–307.

Isonhood, J.H. and Drake, M. 2002. *Aeromonas* species in foods. J. Food Prot. 65: 575–582.

Iversen, C. and Forsythe, S. 2003. Risk profile of *Enterobacter sakazakii*, and emergent pathogen associated with infant milk formula. Trends Food Sci. Technol. 14: 443–454.

Iversen, C., Lane, M. and Forsythe, S.J. 2004. The growth profile, thermotolerance and biofilm formation of *Enterobacter sakazakii* grown in infant formula milk. Lett. Appl. Microbiol. 38: 378–382.

Jansson, D.S., Bröjer, C., Gravier-Widén, D., Gunnarsson, A. and Fellström, C. 2001. *Brachyspira* spp. (*Sperpulina* spp.): A review and results from a study of Swedish game birds. Anim. Health Res. Rev. 2: 93–100.

Jensen, T.K., Boye, M., Ahrens, P., Korsager, B., Teglbjærg, P.S., Lindboe, C. F. and Møller, K. 2001. Diagnostic examination of human intestinal spirochetosis by fluorescent *in situ* hybridization for *Brachyspira aalborgi*, *Brachyspira pilosicoli*, and other species of the genus *Brachyspira* (*Serpulina*). J. Clin. Microbiol. 39: 4111–4118.

Jiang, X. and Doyle, M.P. 2002. Optimizing enrichment culture conditions for detecting *Helicobacter pylori* in foods. J. Food Prot. 65: 1949–1954.

Johnson, C., Johnson, J., Clayton, M., McKenzie, D. and Aiken, J. 2003. Prion protein gene heterogeneity in free-ranging white-tailed deer within the chronic wasting disease affected region of Wisconsin. J. Wildlife Dis. 39: 576–581.

Joly, D.O., Ribic, J.A., Langenberg, J.A., Beheler, K., Batha, C.A., Dhuey, B.J., Rolley, R.E., Bartelt, G., van Deelen, T.R. and Samuel, M.D. 2003. Chronic wasting disease in free-ranging Wisconsin white-tailed deer. Emerg. Infect. Dis. 9: 599–601.

Josephson, S.L. 1998. Spirochaetosis. In: Topley & Wilson's Microbiology and Microbial Infections, 9th ed., Vol. 3. L. Collier, A. Balows and M. Sussman, eds. Arnold, London, UK, pp. 871–884.

Kabir, S. 2001. Detection of *Helicobacter pylori* in faeces by culture, PCR and enzyme immunoassay. J. Med. Microbiol. 50: 1021–1029.

Kanavaki, S. Mantadakis,E., Thomakos,N., Perfanis, A., Matsiota-Bernard, P., Karabela, S. and Samonis, G. 2002. *Brachyspira* (*Serplina*) *pilosicoli* spirochetemia in an immunocompromised patient. Infection 30: 175–177.

Katz, D. J., Cruz, M.A., Trepka, M.J., Suarez, J.A., Fiorella, P.D. and Hammond, R.M. 2002. An outbreak of typhoid fever in Florida associated with an imported frozen fruit. J. Infect. Dis. 186: 234–239.

Khuroo, M. S. and Kamili, S. 2003. Aetiology, clinical course and outcome of sporadic acute viral hepatitis in pregnancy. J. Viral Hepatitis 10: 61–69.

Kirov, S. M. 2001. *Aeromonas* and *Plesiomonas* species. In: Food microbiology: fundamentals and frontiers, 2nd ed. M. P. Doyle, L. R. Beuchat, and T. J. Montville, eds., ASM Press, Washington, D.C, pp. 301–327.

Koff, R. S. 2003. Hepatitis vaccines: Recent advances. Int. J. Parasitol. 33: 517–523.

Kraaz, W., Pettelrsson, B., Thunberg, U., Engstrand, L. and Fellsström, C. 2000. *Brachyspira aalborgi* infection diagnosed by culture and 16S ribosomal DNA sequencing using human colonic biopsy specimens. J. Clin. Microbiol. 38: 3555–3560.

Kusters, J.G. 2001. Recent developments in *Helicobacter pylori* vaccination. Scand. J. Gastroenterol. 234 (Supplement): 15–21.

Kuzina, L.V., Peloquin, J.J., Vacek, D.C. and Miller, T.A. 2001. Isolation and identification of bacteria associated with adult laboratory Mexican fruit flies, *Anastrepha ludens* (Diptera: Tephritidae). Curr. Microbiol. 42:290–294.

Lai, K.K. 2001. *Enterobacter sakazakii* infections among neonates, infants, children, and adults. Medicine. 80: 113–122.

Lastovica, A.J. and Skirrow, M.B. 2000. Clinical significance of *Campylobacter* and related species other than *Campylobacter jejuni* and *C. coli*. In: *Campylobacter*, 2nd ed. I. Nachamkin and M. J. Blaser, eds. ASM Press, Washington, D.C, pp. 89–120.

Lee, A. 1999. Animal models of *Helicobacter* infection. Mol. Med. Today 5: 500–501.

León-Règagnonj, García-Prieto, L., Osorio-Sarabia, D. and Jiménez-Ruiz, A. 2000. Gnathostomiasis in fish from Tres Palos Lagoonk, Guerrero, Mexico. Emerg. Infect. Dis. 6: 429–430

Lipsitch, M. 2001. The rise and fall of antimicrobial resistance. Trends Microbiol. 9: 438–444.

Lopez, A.S., Dodson, D.R., Arrowood, M.J., Orlandi, P.A., de Silva, A.J., Bier, J.W., Hanauer, S.D., Kuster, R.L., Oltman, S., Baldwin, M.S., Won, K.Y., Nace, E.M., Eberhard, M.L. and Herwaldt, B.L. 2001. Outbreak of cyclosporiasis associated with basil in Missouri in 1999. Clin. Infect. Dis. 32: 1010–1017.

Lu, Y., Redlinger, T.E., Avitia, R., Galindo, A. and Goodman, K. 2002. Isolation and genotyping of *Helicobacter pylori* from untreated municipal wastewater. Appl. Environ. Microbiol. 68: 1436–1439.

Mansuy, J.M., Peron, J.M., Bureau, C., Alric, L., Vinel, J.P. and Izopot, J. 2004. Immunologically silent autochthonous acute hepatitis E virus infection in France. J. Clin. Microbiol. 42: 912–913.

Marciano-Cabral, F. and Cabral, G. 2003. *Acanthamoeba* spp. as agents of disease to humans. Clin. Microbiol. Rev. 16: 273–307.

Matsuda, H., Okada, K., Takahashi, K. and Mishiro, S. 2003. Severe hepatitis E virus infection after ingestion of uncooked liver from a wild boar. J. Infect. Dis. 188: 944.

McCarthy, J. and Moore, T.A. 2000. Emerging helminth zoonoses. Internat. J. Parasitol. 30: 1351–1360.

Melito, P.L., Munro, C., Chipman, P.R., Woodward, D.L., Booth, T.F. and Rodgers, F.G. 2001. *Helicobacter winghamensis* sp. nov., a novel *Helicobacter* sp. isolated from patients with gastroenteritis. J. Clin. Microbiol. 39: 2412–2417.

Mellon, M., Benbrook, C. and Benbrook, K. 2001. Hogging it: estimates of antimicrobial abuse in livestock. Cambridge: Union of Concerned Scientists Publications.

Meng, J. and Doyle, M.P. 1997. Emerging issues in microbiological food safety. Annu. Rev. Nutr. 17: 255–275.

Meng, X.-J., Dea, S., Engle, R.E., Friendship, R., Lyoo, Y.S., Sirinarumitr, T., Urairong, K., Wang, D., Wong, D., Yoo, D., Zhang, Y., Purcell, R.H. and Emerson, S.U. 1999. Prevalence of antibodies to the hepatitis E virus in pigs from countries where hepatitis E is common or is rare in the human population. J. Med. Virol. 59: 297–202.

Meng, X.-J., Halbur, P.G., Haynes, J.S., Tsarera, T.S., Bruna, J.D., Royer, R.L., Purcell, R.H. and Emerson, S.U. 1998a. Experimental infection of pigs with the newly identified swine hepatitis E virus (swine HEV), but not with human strains of HEV. Arch. Virol. 143: 1405–1415.

Meng, X.-J., Halbur, P.B., Shapiro, M.S., Govindarajan, S., Bruna, J.D., Mushahwar, I.K., Purcell, R.H. and Emerson, S.U. 1998b. Genetic and experimental evidence for cross-species infection by swine hepatitis E virus. J. Virol. 72: 9714–9721.

Mikosza. A.S.J. La, T., Brooke, C.J., Lindboe, C.F., Ward, P.B., Heine, R.G., Guccion, J.G., de Boer, W.B. and Hampson, D.J. 1999. PCR amplification from fixed tissue indicates frequent involvement of *Brachyspira aalborgi* in human intestinal spirochetosis. J. Clin. Med. 37: 2093–2098.

Mikosza, A.S.J., La, T., de Boer, W.B. and Hampson, D.J. 2001a. Comparative prevalence of *Brachyspira* and *Brachyspira* (*Serpulina*) *pilosicoli* as etiologic agents of histologically identified intestinal spirochetosis in Australia. J. Clin. Microbiol. 39: 347–350.

Mikosza, A.S.J., La, T., Margawani, K.R., Brooke, C.J. and Hampson, D.J. 2001b. PCR detection of *Brachyspira aalborgi* and *Brachyspira pilosicoli* in human faeces. FEMS Microbiol. Lett. 197: 167–170.

Miller, A.J., Smith, J.L. and Buchanan, R.L. 1998. Factors affecting the emergence of new pathogens and research strategies leading to their control. J. Food Safety 18: 243–263.

Miller, M.W. and Williams, E.D. 2003 Horizontal prion transmission in mule deer. Nature 425: 35–36.

Miller, M.W., Williams, E.S., Hobbs, N.T. and Wolfe, L.L. 2004. Environmental sources of prion transmission in mule deer. Emerg. Infect. Dis. 10: 1003–1006.

Montero, E., Montero, J., Rosales, M.J. and Mascaró, C. 2001. Human gnathostomiasis in Spain: First Report in humans. Acta Tropica 78: 59–62.

Moore, D.A.J., McCroddan, J., Dekumyoy, P. and Chiodini, P.L. 2003. Gnathostomiasis: An emerging imported disease. Emerg. Infect. Dis. 9: 647–650.

Muniappa, N., Duhamel, G.E., Mathiesen, M.R. and Bargar, T.W. 1996. Light microscopic and ultrastructural changes in the ceca of chicks inoculated with human and canine *Serpulina pilosicoli*. Vet. Pathol. 33: 542–550.

Murrow, L.B., Blake, P. and Krekman, L. 2002. Outbreak of cyclosporiasis in Fulton County, Georgia. Georgia Epidemiol. Rept. 18 (01): 1–2.

Naimi, T.S., Wicklund, J.H., Olsen, S.J., Krause, G., Wells, J.G., Bartkus, J.M., Boxrud, D.J., Sullivan, M., Kassenborg, H., Besser, J.M., Mintz, E.D., Osterholm, M.T. and Hedberg, C.W. 2003. Concurrent outbreaks of *Shigella sonnei* and enterotoxigenic *Escherichia coli* infections associated with parsley: Implications for surveillance and control of foodborne illness. J. Food Prot. 66: 535–541.

Nazarowec-White, M. and Farber, J.M. 1997. *Enterobacter sakazakii*: a review. Int. J. Food Microbiol. 34: 1–3-113.

Nilsson, H-O., Aleljung, P., Nilsson, I., Tyszkiewicz, T. and Wadström, T. 1996. Immunomagnetic bead enrichment and PCR for detection of *Helicobacter pylori* in human stools. J. Microbiol. Meth. 27: 73–79.

Nilsson, I., Kornilovska, I., Lindgren, S., Ljungh, Å. and Waldström, T. 2003. Increased prevalence of seropositivity for non-gastric *Helicobacter* species in patients with autoimmune liver disease. J. Med. Microbiol. 59: 949–953.

Ogata, K., Nawa, Y., Akahane, H., Diaz Camacho, S.P., Lamothe-Argumedo, R. and Cruz-Reyes, A. 1998. Short report: Gnathostomiasis in Mexico. Am. J. Trop. Med. Hyg. 58: 316–318.

Okamoto, H., Takahashi, M. and Nishizawa, T. 2003. Features of hepatitis E virus infection in Japan. Intern. Med. 42: 1065–1071.

Okamoto, H., Takahashi, M., Nishizawa, T., Fukai, K., Muramatsu, U. and Yoshikawa, A. 2001. Analysis of the complete genome of indigenous swine hepatitis E virus isolated in Japan. Biochem. Biophys. Res. Comm. 289: 929–936.

Ong, S., Talan, D.A., Moran, G.J., Mower, W., Newdow, M., Tsang, V.C.W., Pinner, R.W. and the Emergency ID Net Study Group. 2002. Neurocysticercosis in radiographically imaged seizure patients in USA emergency departments. Emerg. Infect. Dis. 8: 608–613.

Orlandi, P.A., Chu, D-M.T., Bier, J.W. and Jackson, G.J. 2002. Parasites and the food supply. Food Technol. 56(4): 72–81.

O'Rourke, K.I., Zhuang, D., Lyda, A., Gomez, G., Willians, E.S., Tuo, W. and Miller, M.W. 2003. Abundant PrP^{CWD} om tonsil from mule deer with preclinical chronic wasting disease. J. Vet. Diagn. Invest. 15: 320–323.

Osterholm, M. T. 2000. Emerging infections – another warning. N. Engl. J. Med. 342: 1280–1281.

Oxberry, S.L., Trott, D.J. and Hampson, D.J. 1998. *Serpulina pilosicoli*, waterbirds and water: Potential sources of infection for humans and other animals. Epidemiol. Infect. 121: 219–225.

Padmanabhan, V., Dahlstrom, J., Maxwell, L., Kaye, G., Clarke, A. and Barratt, P.J. 1996. Invasive intestinal spirochetosis: a report of three cases. Pathology 28: 283–286.

Parsonnet, J., Shmuely, H. and Haggerty, T. 1999. Fecal and oral shedding of *Helicobacter pylori* from healthy infected adults. J. Am. Med. Assoc. 282: 2240–2245.

Pei, Y. and Yoo, D. 2002. Genetic characterization and sequence heterogeneity of a Canadian isolate of swine hepatitis E virus. J. Clin. Microbiol. 40: 4021–4029.

Poms, R.E. and Tatini, S.R. 2001. Survival of *Helicobacter pylori* in ready-to-eat foods at 4°C. Int. J. Food Microbiol. 63: 281–286.

Pringle, C.R. 1999. Virus taxonomy at the XIth International Congress of Virology, Sydney, Australia, 1999. Arch. Virol. 144: 2065–2070.

Prinz, C., Hafsi, N. and Voland, P. 2003. *Helicobacter pylori* virulence factors and the host immune response: Implications for therapeutic vaccination. Trends Microbiol. 11: 134–138.

Puente, S., Gárate, T., Grobusch, M.P., Janitschke, K., Bru, F., Rodriguez, M. and González-Lahoz, J.M. 2002. Two cases of imported gnathostomiasis in Spanish women. Eur. J. Clin. Microbiol. Infect. Dis. 21: 617–620.

Punyagupta, S., Bunnag, T. and Juttijudata, P. 1990. Eosinophilic meningitis in Thailand. Clinical and epidemiological characteristics of 162 patients with myeloencephalitis probably caused by *Gnathostoma spinigerum*. J. Neurol. Sci. 96: 242–256.

Purcell, R.H., Nguyen, H., Shapiro, M., Engle, R.E., Govindarajan, S., Blackwelder, W.C., Wong, D.C., Prieels, J-P. and Emerson, S.U. 2003. Pre-clinical immunogenicity and efficacy trial of a recombinant hepatitis E vaccine. Vaccine. 21: 2607–2615.

Raymond, G.J., Bossere, A., Raymond, L.D., O'Rourke, K.I., McHolland, L.E., Bryant, P.K., Miller, M.W., Williams, E.S., Smits, M. and Caughey, B. 2000. Evidence of a molecular barrier limiting susceptibility of humans, cattle and sheep to chronic wasting disease. EMBO J. 19: 4425–4430.

Rojas-Molina, N., Pedraza-Sanchez, S., Torres-Bibiano, B., Meza-Martinez, H. and Escobar-Gutierrez, A. 1999. Gnathostomiasis, an emerging foodborne zoonotic disease in Acapulco, Mexico. Emerg. Infect. Dis. 5: 264–266.

Rusnak, J.M. and Lucey, D.R. 1993. Clinical gnathostomiasis: Case report and review of the English-language literature. Clin. Infect. Dis. 16: 33–50.

Russell, A.D. 2002. Antibiotic and biocide resistance in bacteria: Introduction. J. Appl. Microbiol. 92(Suppl., Symposium Series #92): 1S–3S.

Sacco, R.E., Trampel, D.W. and Wannemuehler, M.J. 1997. Experimental infection of C3H mice with avian, porcine, or human isolates of *Serpulina pilosicoli*. Infect. Immun. 65: 5349–5353.

Safar, J.G., Scott, M., Monaghan, J., Deering, C., Didorenko, S., Vergara, J., Ball, H., Legname, G., Leclerc, E., Solforosi, L., Serban, H., Groth, D., Burton, D.R., Prusiner, S.B. and Williamson, R.A. 2002. Measuring prions causing bovine spongiform encephalopathy or chronic wasting disease by immunoassays and transgenic mice. Nature Biotechnol. 20: 1147–1150.

Salman, M.D. 2003. Chronic wasting disease in deer and elk: Scientific facts and findings. J. Vet. Med. Sci. 65: 761–768.

Samelis, J. and Sofos, J.N. 2003. Strategies to control stress-adapted pathogens. In: Microbial stress adaptation and food safety., A.E. Yousef and V.K. Juneja, eds. CRC Press, Boca Raton, FL, USA, pp. 303–351.

Sasaki, K., Tajiri, Y., Sata, M., Fujii, Y., Matsubara, F., Zhao, M., Shimizu, S., Toyonaga, A. and Tanikawa, K. 1999. *Helicobacter pylori* in the natural environment. Scand. J. Infect. Dis. 31: 275–279.

Schantz, P.M., Wilkins, P.P. and Tsang, V.V.W. 1998. Immigrants, imaging, and immunoblots: The emergence of neurocysticercosis as a significant public health problem. In: Emerging Infections 2. W.M. Scheid, W.A. Craig and J.M. Hughes, eds. ASM Press, Washington DC, pp. 213–242.

Schnabel, E.L. and Jones, A.L. 1999. Distribution of tetracycline resistance genes and transposons among phylloplane bacteria in Michigan apple orchards. Appl. Environ. Microbiol. 65: 4898–4907.

Sciutto, E., Fragoso, G., Fleury, A., Laclette, J.P., Sotelo, J., Aluja, A., Vargas, L. and Larralde, C. 2000. *Taenia solium* disease in humans and pigs: An ancient parasitosis disease rooted in developing countries and emerging as a major health problem of global dimensions. Microbes Infect. 2: 1875–1890.

Sharma, S., Sachdeva, P. and Virdi, J.S. 2003. Emerging water-borne pathogens. Appl. Microbiol. Biotechnol. 61: 424–428.

Sigurdson, C.J., Barillas-Mury, C., Miller, M.W., Oesch, B., van Keulen, L.J.M., Langeveid, J.P.M. and Hoover, E.A. 2002. PrP^{CWD} lymphoid cell targets in early and advanced chronic wasting disease in mule deer. J. Gen. Virol. 83: 2617–2628.

Smith, H.M., Reporter, R., Rood, M.P., Linscott, A.J., Mascola, L.M., Hogrefs, W. and Purcell, R.H. 2002. Prevalence study of antibody to ratborne pathogens and other agents among patients using a free clinic in downtown Los Angeles. J. Infect. Dis. 186: 1673–1676.

Smith, J.L. 2001. A review of hepatitis E virus. J. Food Prot. 64: 572–586.

Smith, J.L. 1998. Foodborne illness in the elderly. J. Food Prot. 61: 1229–1239.

Smith, J.L. 1997. Long-term consequences of foodborne toxoplasmosis: Effect on the unborn, the immunocompromised, the elderly and the immunocompetent. J. Food Prot. 60: 1595–1611.

Smith, J.L. 1994. *Taenia solium* neurocysticercosis. J. Food Prot. 57: 831–844.

Smolinski, M.S., Hamburg, M.A. and Lederberg, J., eds. 2003. Microbial threats to health: Emergence, detection, and response. The National Academies Press, Washington, DC.

Sohn, H.-J., Kim, J.-H., Choi, K.-S., Naii, J.-J., Joo, Y.-S., Jean, Y.-H., Ahn, S.-W., Kim, O.-K., Kim, D.-Y. and Balachandran, A. 2002. A case of chronic wasting disease in an elk imported to Korea from Canada. J. Vet. Med. Sci. 64: 855–858.

Soleto, J. 2003. Neurocysticercosis. Eradication of cysticercosis is an attainable goal. Br. Med. J. 326: 511–512.

Solnick, J.V. 2003. Clinical significance of *Helicobacter* species other than *Helicobacter pylori*. Clin. Pract. 36: 349–354.

Sørum, H. and L'Abée-Lund, T.M. 2002. Antibiotic resistance in food-related bacteria – a result of interfering with the global web of bacterial genetics. Int. J. Food Microbiol. 78: 43–56.

Sorvillo, F.J., Portigal, L., DeGiorgio, C., Smith, L., Waterman, S.H., Berlin, G.W., and Ash, L.R. 2004. Cysticercosis-related deaths, California. Emerg. Infect. Dis. 10: 465–469.

Stanley, J., Linton, D., Burnens, A.P., Dewhirst, F.E., On, S.L.W., Porter, A., Owen, R.J. and Costas, M. 1994. *Helicobacter pullorum* sp. nov. – genotype and phenotype of a new species isolated from poultry and from human patients with gastroenteritis. Microbiology. 140: 3441–3449.

Steinbrueckner, B., Haerter, G., Pelz, K., Weiner, S., Rump, J.A., Deissler, W., Bereswell, S. and Kist, M. 1997. Isolation of *Helicobacter pullorum* from patients with enteritis. Scand. J. Infect. Dis. 29: 315–318.

Stephens, C.P. and Hampson, D.J. 2002. Experimental infection of broiler breeder hens with the intestinal spirochaete *Brachyspira (Serpulina) pilosicoli* causes reduced egg production. Avian Pathol. 31: 169–175.

Stephens, C.P. and Hampson, D.J. 2001. Intestinal spirochete infections of chickens: A review of disease associations, epidemiology and control. Anim. Health Res. Rev. 2: 83–91.

Stevenson, T.H., Bauer, N., Lucia, L.M. and Acuff, G.R. 2000. Attempts to isolate *Helicobacter* from cattle and survival of *Helicobacter pylori* in beef products. J. Food Prot. 63: 174–178.

Swerdlow, D.L. and Altekruse, S.F. 1998. Foodborne diseases in the global village: What's on the plate for the 21st century. In: Emerging Infections 2. W. M. Scheld, W. A. Craig and J. M. Hughes, eds. ASM Press, Washington, D.C., pp. 273–294.

Takabe, K., Ohki, S., Kunihiro, O., Sakashita, T., Endo, I., Ichikawa, Y., Sekido, H., Amano, T., Nakatan, Y., Suzuki, K. and Shimada, H. 1998. Anisakidosis: A cause of intestinal obstruction from eating sushi. Am. J. Gastroenterol. 93: 1172–1173.

Takahashi, M., Nishizawa, T., Miyajima, H., Gotanda, Y., Iita, T., Tsuda, F. and Okamoto, H. 2003. Swine hepatitis E virus strains in Japan form four phylogenetic clusters comparable with those of Japanese isolates of human hepatitis E virus. J. Gen. Virol. 84: 851–862.

Takahashi, M., Nishizawa, T., Yoshikawa, A., Sato, S., Isoda, N., Ido, K., Sugano, K. and Okamoto, H. 2002. Identification of two distinct genotypes of hepatitis E virus in a Japanese patient with acute hepatitis who had not traveled abroad. J. Gen. Virol. 83: 1931–1940.

Tamada, Y., Yano, K., Yatsuhashi, H., Inoue, O., Mawatari, F. and Ishibashi, H. 2004. Consumption of wild boar linked to cases of hepatitis E. J. Hepatol. 40: 869–870.

Tei, S., Kitajima, N., Takahashi, K. and Mishiro, S. 2003. Zoonotic transmission of hepatitis E virus from deer to human beings. Lancet. 362: 371–373.

Teich, N., Tannapfel, A., Ammon, A., Ruf, B.R., van der Poel, W.H.M., Mössner, J. and Liebert, U.G. 2003. Sporadische akute Hepatitis E in Deutschland: eine su selten errante Erkrankung? Z. Gastroenterol. 41: 419–423.

Thomzig, A., Knatzel, C., Lenz, G., Krüger, D. and Beekes, M. 2003. Widespread PrP$^{Sc}$ accumulation in muscles of hamsters orally infected with scrapie. EMBO Reports 4: 530–533.

Threfall, E.J., Ward, L.R., Frost, J.A. and Willshaw, G.A. 2000. The emergence and spread of antibiotic resistance in foodborne bacteria. Int. J. Food Microbiol. 62: 1–5.

Townes, J.M., Hoffman, C.J. and Kohn, M.A. 2004. Neurocysticercosis in Oregon, 1995–2000. Emerg. Infect. Dis. 10: 508–510.

Trabulsi, L.R., Keller, R. and Gomes, T.A.T. 2002. Typical and atypical enteropathogenic *Escherichia coli*. Emerg. Infect. Dis. 8: 508–513.

Trivett-Moore, N.L., Gilbert, G.L., Law, C.L.H., Trott, D.J. and Hampson, D.J. 1998. Isolation of *Serpulina pilosicoli* from rectal biopsy specimens showing evidence of intestinal spirochetosis. J. Clin. Microbiol. 36: 261–265.

Trott, D.J., Combs, B.G., Mikosza, A.S.J., Oxberry, S.L., Robertson, J.D., Passey, M., Taime, J., Sehuko, R., Alpers, M.P. and Hampson, D.J. 1997a. The prevalence of *Serpulina pilosicoli* in humans and domestic animals in the Eastern Highlands of Papua New Guinea. Epidem. Infect. 119: 369–379.

Trott, D.J., Huxtable, C.R. and Hampson, D.J. 1996a. Experimental infection of newly weaned pigs with human and porcine strains of *Serpulina pilosicoli*. Infect. Immun. 64: 4648–4654.

Trott, D.J., Jensen, N.S., Saint Girons, I., Oxberry, S.L., Stanton, T.B., Lindquist, D. and Hampson, D.J. 1997b. Identification and characterization of *Serpulina pilosicoli* isolates recovered from the blood of critically ill patients. J. Clin. Microbiol. 35: 482–485.

Trott, D.J., Stanton, T.B., Jensen, N.S., Duhamel, G.E., Johnson, J.L. and Hampson, D.J. 1996b. *Serpulina pilosicoli* sp. nov., the agent of porcine intestinal spirochetosis. Int. J. Syst. Bacteriol. 46: 206–215.

USA Census Bureau, Statistical Abstracts of the United States: 2000. (120th ed.), Washington, D.C.

Urmenyi, A.M.C. and Franklin, A.W. 1961. Neonatal death from pigmented coliform infection. Lancet 1: 313–315.

Usui, R., Kobayashi, E., Takahashi, M., Nishizawa, T. and Okamoto, H. 2004. Presence of antibodies to hepatitis E virus in Japanese pet cats. Infection 32: 57–58.

Vaidya, S.R., Chitambar, S.D. and Arankalle, V.A. 2002. Polymerase chain reaction-based prevalence of hepatitis A, hepatitis E and TT viruses in sewage from an endemic area. J. Hepatol. 37: 131–136.

Vaira, D., Holton, J., Ricci, C., Menegatti, M., Gatta, L., Berardi, S., Tampieri, A. and Miglioli, M. 2001. Review article: The transmission of *Helicobacter pylori* from stomach to stomach. Aliment. Pharmacol. Ther. 15 (Suppl. 1): 33–42.

van der Poel, W.H.M., Verschoor, F., van der Heide, R., Herrera, M-I., Vivo, A., Kooreman, M. and de Roda Husman, A. M. 2001. Hepatitis E virus sequences in swine related to sequences in humans, the Netherlands. Emerg. Infect. Dis. 7: 970–976.

Vargas-Ocampo, F., Alarcón-Rivera, E. and Alvarado-Alemán, F.J. 1998. Human gnathostomiasis in Mexico. Int. J. Dermatol. 37: 441–444.

Velázquez, M. and Feirtag, J.M. 1999. *Helicobacter pylori*: Characteristics, pathogenicity, detection methods and mode of transmission implicating foods and water. Int. J. Food Microbiol. 53: 95–104.

Vila, J., Ruiz, J., Gallardo, F., Vargas, M., Soler, L., Figueras, J.J. and Gascon, J. 2003. *Aeromonas* spp. and traveler's diarrhea: clinical features and antimicrobial resistance. Emerg. Infect. Dis. 9: 552–555.

Waldenström, J., On, S.L.W., Ottvall, R., Hasselquest, D., Harrington, C.S. and Olsen, B. 2003. Avian reservoirs and zoonotic potential of the emerging human pathogen, *Helicobacter canadensis*. Appl. Environ. Microbiol. 69: 7523–7526.

Wang, X., Sturegård, E., Rupar, R., Nilsson, H-O., Aleljung, P.A., Carlén, B., Willen, R. and Wadström, T. 1997. Infection of BALB/c A mice by spiral and coccoid forms of *Helicobacter pylori*. J. Med. Microbiol. 46: 657–663.

Wang, Y-C., Zhang, H-Y., Xia, N-S., Peng, G., Lan, H-Y., Zhuang, H., Zhu, Y-H., Li, S-W., Tian, K-G., Gu, W-J., Lin, J-X., Wu, X., Li, H-M. and Harrison, T.J. 2002. Prevalence, isolation, and partial sequence analysis of hepatitis E virus from domestic animals in China. J. Med. Virol. 67: 516–521.

Weaver, L.T., Shepherd, A.J., Doherty, C.P., McColl, K.E.L. and Williams, C.L. 1999. *Helicobacter pylori* in the faeces? Quart. J. Med. 92: 361–364.

Webberley, M.J., Webberley, J.M., Newell, D.G., Lowe, P. and Melikian, V. 1992. Seroepidemiology of *Helicobacter pylori* infection in vegans and meat-eaters. Epidem. Infect. 108: 457–462.

Wegener, H.C. 2003. Ending the use of antimicrobial growth promoters is making a difference. ASM News 69: 443–448.

Widdowson, M-A., Jaspers, W.J.M., van der Poel, W.H.M., Verschoor, F., de Roda Husman, A.M., Winter, H.L.J., Zaaijer, H.L. and Koopmans, M. 2003. Cluster of cases of acute hepatitis associated with hepatitis E virus infection acquired in the Netherlands. Clin. Infect. Dis. 36: 29–33.

Wild, M.A., Spraker, T.R., Sigurdson, C.J., O'Rouke, K.I. and Miller, M.W. 2002. Preclinical diagnosis of chronic wasting disease in captive mule deer (*Odocoileus hemionus*) and white-tailed deer (*Odocoileus virginianus*) using tonsillar biopsy. J. Gen. Virol. 83: 2629–2634.

Williams, E.S. and Miller, M.W. 2002. Chronic wasting disease in deer and elk in North America. Rev. Sci. Tech. Off. Int. Epiz. 21: 305–316.

Winiecka-Krusnell, J., Sreiber, K., von Euler, A., Engstrand, L. and Linder, E. 2002. Free-living amoebae promote growth and survival of *Helicobacter pylori*. Scand. J. Infect. Dis. 34: 253–256.

Withers, M.R., Correa, M.T., Morrow, M., Stebbins, M.E., Seriwatana, J., Webster, W.D., Boak, M.B. and Vaughn, D.W. 2002. Antibody level of hepatitis E virus in North Carolina swine workers, non-swine workers, swine and murids. Am. J. Trop. Med. Hyg. 66: 384–388.

Worm, H.C., Schlauder, G.G. and Brandstätter, G. 2002a. Hepatitis E and its emergence in non-endemic areas. Wien. Klin. Wochenschr. 114: 663–670.

Worm, H.C., van der Poel, W.H.M. and Brandstätter, G. 2002b. Hepatitis E: An overview. Microbes Infect. 4: 657–666.

Wu, J-C., Chen, C-M., Chiang, T-Y., Tsai, W-H., Jeng, W-J., Sheen, I-J., Lin, C-C. and Meng, X-J. 2002. Spread of hepatitis E virus among different-aged pigs: Two-year survey in Taiwan. J. Med. Virol. 66: 488–492.

Yazaki, Y., Mizuo, H., Takahashi, M., Nishizawa, T., Sasaki, N., Gotanda, Y. and Okamoto, H. 2003. Sporadic acute or fulminant hepatitis E in Hokkaido, Japan, may be foodborne, as suggested by the presence of hepatitis E virus in pig liver as food. J. Gen. Virol. 84: 2351–2357.

Yoo, D., Willson, P., Pei, Y., Hayes, M.A., Deckert, A., Dewey, C.E., Friendship, R.M., Yoon, Y., Gottschalk, M., Yason, C. and Giulivi, A. 2001. Prevalence of hepatitis E virus antibodies in Canadian swine herds and identification of a novel variant of swine hepatitis E virus. Clin. Diag. Lab. Immunol. 8: 1213–1219.

Yoshimura, K. 1998. *Angiostrongylus (Parastrongylus)* and less common nematodes. In: Topley & Wilson's Microbiology and Microbial Infections, 9th ed., vol. 5, Arnold, London, UK. p. 635–659.

Zhang, P., Cheng, X. and Duhamel, G.E. 2000. Cloning and DNA sequence analysis of an immunogenic glucose-galactose Mg1B lipoprotein homologue from *Brachyspira pilosicoli*, the agent of colonic spirochetosis. Infect. Immun. 68: 4559–4565.

# Index

# Books of Related Interest

**Full details of all these books at: www.horizonbioscience.com**